Term	Meaning	Example
cysto-	Bladder or sac	Cystocele (hernia of a bladder)
-cyte-	Cell	Erythrocyte (red blood cell)
cyto-	Cell	Cytoskeleton (supportive fibers inside a cell)
de-	Away from	Dehydrate (remove water)
derm-	Skin	Dermatology (study of the skin)
di-	Two	Diploid (two sets of chromosomes)
dia-	Through, apart, across	Diapedesis (ooze through)
dis-	Reversal, apart from	Dissect (cut apart)
-duct-	Draw	Abduct (lead away from)
-dynia	Pain	Mastodynia (breast pain)
dys-	Difficult, bad	Dysmentia (bad mind)
e-	Out, away from	Eviscerate (take out viscera)
ec-	Out from	Ectopic (out of place)
ecto-	On outer side	Ectoderm (outer skin)
-ectomy	Cut out	Appendectomy (cut out the appendix)
-edem-	Swell	Myoedema (swelling of a muscle)
em-	In	Empyema (pus in)
-emia	Blood	Anemia (deficiency of blood)
en-	In	Encephalon (in the brain)
endo-	Within	Endometrium (within the uterus)
entero-	Intestine	Enteritis (inflammation of the intestine)
epi-	Upon, on	Epidermis (on the skin)
erythro-	Red	Erythrocyte (red blood cell)
eu-	Well, good	Euphoria (well-being)
ex-	Out, away from	Exhalation (breathe out)
exo-	Outside, on outer side	Exogenous (originating outside)
extra-	Outside	Extracellular (outside the cell)
-ferent	Carry	Afferent (carrying to the central nervous system)
-form	Expressing resemblance	Fusiform (resembling a fusion)
gastro-	Stomach	Gastrodynia (stomach ache)
-genesis	Produce, origin	Pathogenesis (origin of disease)
gloss-	Tongue	Hypoglossal (under the tongue)
glyco-	Sugar, sweet	Glycolysis (breakdown of sugar)
-gram	A drawing	Myogram (drawing of a muscle contraction)
-graph	Instrument that records	Myograph (instrument for measuring muscle contraction)
hem-	Blood	Hemolysis (breakdown of blood)
hemi-	Half	Hemiplegia (paralysis of half of the body)
hepato-	Liver	Hepatitis (inflammation of the liver)
hetero-	Different, other	Heterozygous (different genes for a trait)
hist-	Tissue	Histology (study of tissues)
homeo-	Same	Homeostasis (state of staying the same)
hydro-	Wet, water	Hydrocephalus (fluid within the head)
hyper-	Over, above, excessive	Hypertrophy (overgrowth)
hypo-	Under, below, deficient	Hypotension (low blood pressure)
-ia	Expressing condition	Neuralgia (pain in nerve)
-iatr-	Treat, cure	Pediatrics (treatment of childern)
-id	Expressing condition	Flaccid (state of being weak)
im-	Not	Impermeable (not permeable)
in-	In, into	Injection (forcing fluid into)
infra-	Below	Infraorbital (below the eye)
inter-	Between	Intercostal (between the ribs)

ANATOM

ESSENTIALS OF
ANATOMY AND PHYSIOLOGY

ROD R. SEELEY, Ph.D.
Professor of Physiology
Idaho State University

TRENT D. STEPHENS, Ph.D.
Professor of Anatomy and Embryology
Idaho State University

PHILIP TATE, D.A. (Biological Education)
Instructor of Anatomy and Physiology
Phoenix College
Maricopa Community College District

St. Louis Baltimore Boston Chicago London Philadelphia Sydney

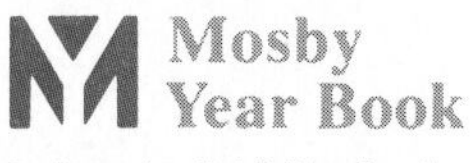

Dedicated to Publishing Excellence

Publisher: Edward F. Murphy
Editor: Deborah Allen
Developmental Editor: Robert J. Callanan
Project Manager: Patricia Tannian
Production Editor: Teresa Breckwoldt
Book Designer: Susan E. Lane

Cover Photograph: © Mehau Kulyk/SPL/Photo Researchers, Inc.

New Illustrations by: Christine Oleksyk
John Daugherty
Scott Bodell
Michael Schenk
Eileen Draper
John Hagen

Credits for all materials used by permission appear after the glossary.

Printed in the United States of America

Mosby–Year Book, Inc.
11830 Westline Industrial Drive
St. Louis, MO 63146

Library of Congress Cataloging in Publication Data

Seeley, Rod R.
Essentials of anatomy and physiology / Rod R. Seeley, Trent D. Stephens, Philip Tate.
p. cm.
A study guide, laboratory manual, instructor's manual/testbank, a set of full-color transparencies, and a computerized test bank software package are available to accompany this text.
Includes bibliographical references.
Includes index.
0-8016-0227-0
1. Human Physiology. 2. Human anatomy. I. Stephens, Trent D. II. Tate, Philip. III. Title.
[DNLM: 1. Anatomy. 2. Physiology. QS 4 S452e]
OP34.5.S418 1991
612--dc20
DNLM/DLC
for Library of Congress 90-13437
CIP

C/VH/VH 9 8 7 6 5 4 3 2 1

■ THIS BOOK IS DEDICATED TO OUR FAMILIES FOR THEIR SUPPORT AND ENCOURAGEMENT ■

Rod R. Seeley, Professor of Physiology, Idaho State University

With a bachelor's degree in zoology from Idaho State University and a master's degree and doctorate in zoology from Utah State University, Rod Seeley has built a solid reputation as a widely published author of journal and feature articles, a popular public lecturer, and an award-winning instructor. Very much involved in the methods and mechanisms that help students learn, he contributes to this text his teaching expertise and proven ability to communicate effectively in any medium.

Trent D. Stephens, Professor of Anatomy and Embryology, Idaho State University

A versatile educator, Trent Stephens teaches such courses as basic human anatomy, neuro-anatomy, and physiology. His skill as a biological illustrator has greatly influenced every illustration in this text. With bachelor's and master's degrees in zoology from Brigham Young University and a doctorate in anatomy from the University of Pennsylvania, Trent Stephens uses his background to reach diverse audiences. His students continually rate him highly on their evaluations—you will too!

Philip Tate, Instructor of Anatomy and Physiology, Phoenix College

From the community college to the private 4-year college, Phil Tate has taught anatomy and physiology to all levels of students: nursing and allied health, physical education, and biology majors. At San Diego State University, Phil earned bachelor's degrees in mathematics and zoology and a master's degree in ecology. He earned his doctorate in biological education from Idaho State University.

PREFACE

This text is based on the notion that the facts and relationships that represent the essence of anatomy and physiology can be presented in an understandable and inviting manner. Information is presented in an explanatory way, with the relationship between different elements highlighted to produce a conceptual framework that ties all the individual bits of information together. Students begin with simple, easy-to-understand facts and build systematically upon them until a complete understanding is achieved.

Two conceptual themes are emphasized throughout the text: ***The Relationship between Structure and Function*** and ***Homeostasis.*** Just as the structure of a hammer makes it well-suited for the function of pounding nails, the forms of specific cells, tissues, and organs within the body allow them to perform specific functions effectively. For example, muscle cells contain proteins that make contraction possible, and bone cells surround themselves with a mineralized matrix that provides strength and support. Knowledge of structure and function relationships makes it easier to understand anatomy and physiology and greatly enhances appreciation for the subject.

Homeostasis, the maintenance of an internal environment within an acceptably narrow range of values, is necessary for the survival of the human body. For instance, if the blood delivers inadequate amounts of oxygen to body cells, heart and respiration rates increase until oxygen delivery becomes adequate. The emphasis in this book is on how the mechanisms that normally maintain homeostasis operate. However, because failure of these mechanisms also illustrates how they work, pathological conditions that result in dysfunction, disease, and possibly death are also presented. A consideration of pathology adds relevance and interest that makes the material more meaningful.

Teaching human anatomy and physiology in the 1990s is a greater challenge than before. There is an ever-increasing need for students to learn relevant, up-to-date information. But, there is also an increasing recognition that knowledge of content alone is only a beginning. It is essential that anatomy and physiology students develop their ability to solve practical, real-life anatomy and physiology problems. Though today's knowledge will change and expand, problem-solving skills will always be an asset for students.

The themes—the relationship between structure and function, and homeostasis—combined with a problem-based learning approach, make this text quite unique among anatomy and physiology texts.

A Systematic Presentation of Content

This text provides a solid foundation in anatomy and a thorough coverage of physiology. Content is explained with an emphasis on a conceptual framework that ties individual bits of information together. Explanations start with simple, easy-to-understand facts and are developed in a logical sequence. Students can see the "big picture" without being overwhelmed with details.

Understanding the material has a number of benefits. It makes the information meaningful and thus reduces the amount of memorization necessary to master it. Understanding also makes it possible to use basic knowledge to solve problems. Reinforcing content through application improves long-term retention, because once the information is understood and used, it "belongs" to the student.

Relevant Clinical Examples

Clinical information should never be an end in itself. In some texts, mere clinical description or medical terminology represents a significant portion of the material. This text provides clinical examples to both promote interest and demonstrate relevance, but clinical information is used primarily to illustrate the application of basic knowledge, a skill that goes beyond mere acquaintance with either clinical or basic anatomy and physiology content. As a result, students taking the course are encouraged in their professional and private lives to effectively use the knowledge they have gained through comprehending and solving basic clinical problems.

Analysis of Practical Problems

At best, some anatomy and physiology texts include a few "thought" questions that, for the most part, involve a restatement or a summary of content. Yet once students understand the material well enough to state it in their own words, it only seems logical for them to proceed to the next step—that is, to apply the knowledge to hypothetical situations. This text features two sets of problem-solving questions in every chapter, ***Predict Questions*** and ***Concept Questions*** (to be highlighted in more detail later in this preface). Answers and explanations for these questions are provided. The explanations illustrate the methods used to solve problems and provide a model for the development of problem-solving skills. The acquisition of such skills is necessary for a complete understanding of anatomy and physiology; it is fun; and it makes it possible for the student to deal with the many problems that occur as a part of professional and everyday life.

Illustration Program

The statement, "A single picture is worth ten thousand words," is especially true in anatomy and physiology. Structure/function relationships become immediately apparent in the well-designed, accurate illustrations in this text. To maximize the effectiveness of the illustrations, they have been placed as close as possible to the narrative where they are cited, and special attention has been devoted to the figure legends, which summarize or emphasize the important features of each illustration. Although the anatomical drawings are accurate and the physiological flow diagrams and graphs are conceptually clear, the illustrations accomplish more than just presentation of important information. They have been designed to be nonintimidating and aesthetically pleasing, features that encourage the student to spend time with the illustrations for maximum learning and pleasure. All the art work in this textbook is in full color, making the illustrations attractive and emphasizing the important structures. Another unique feature is the use of Three Dimensional Drawings, making the figures more realistic and dynamic.

In addition to the illustrations, numerous photographs bring a dimension of realism to the text. In most cases, photographs are accompanied by line drawings that emphasize important features of the photograph.

Developmental Story

No matter how innovative our original vision for this text may have been, there is no doubt that, without the help of numerous instructors who were willing to help us implement our ideas and hone the results to near-perfection, we would have not been able to produce this text. It was our goal to produce a text embodying our unique ideas, which would also be judged suitable for widespread classroom use. Fortunately, many of our reviewers were in agreement with our goals, for they too had often experienced frustrations with their existing texts.

After *Anatomy and Physiology* was published, it became apparent that instructors like the problem-based learning approach, the example-rich narrative, and the three-dimensional illustrations. It also became apparent that there are a lot of courses that need a less comprehensive text—either due to time restrictions or a less clinically oriented student population. With this in mind, we began writing *Essentials of Anatomy and Physiology*, a book that combines the "essence" of the problem-solving approach and the outstanding illustrations of *Anatomy and Physiology* with the essential concepts necessary to understand this subject.

Our first-draft manuscript was carefully reviewed by a panel of nine instructors, individuals who thoughtfully helped us identify just what the "truly essential concepts" of anatomy and physiology are. Special attention was paid to ensure that concepts were honed down to an appropriate level of understanding for their students without becoming erroneous or ambiguous. The second draft was reviewed by yet another panel of six reviewers, who also were indispensable in our effort to produce a useful, accurate anatomy and physiology text. Finally, two skilled instructors, Linda M. Peck from the University of Findlay (Ohio) and Darrell Wiens from the University of Northern Iowa, spent considerable time and effort searching the manuscript for errors and ambiguities. Dr. Peck and Dr. Wiens had perhaps the most overpowering reviewer responsibility, for theirs were the final sets of eyes to examine the material before it was published in final form.

The illustration program enjoyed a similar developmental process. All illustrations were carefully reviewed to ensure that the labeling and legends were written at a similar level to the text. Inasmuch as the illustrations can be as important as the narrative in an anatomy and physiology text, it was essential that they be accorded such treatment.

Obviously, it has been our goal to produce a text that is useful, as well as free of errors, ambiguities, and typographical errors.

Learning Aids

The text must be an effective teaching tool. Because each student may learn best in a different way, a variety of teaching and learning aids are provided. This enables students to organize the material in their minds, determine the main points, and evaluate the progress of their learning.

Objectives. Each chapter begins with a series of learning objectives. The objectives are not a detailed cataloguing of everything to be learned in the chapter. Rather, they emphasize the important facts, topics, and concepts to be covered. The chapter objectives are a conceptual framework to which additional material will be added as the chapter is read in detail.

Vocabulary aids. Learning anatomy and physiology is, in many ways, like learning a new language. To communicate effectively, a basic terminology, dealing with important or commonly used facts and concepts, must be mastered. At the beginning of each chapter are the ***Key Terms,*** a list of some of the more important new words to be learned along with their definitions. Throughout the text, these and additional terms are presented in ***Boldface Print.*** In cases where it is instructionally valuable, the ***Derivation*** or ***Origin*** of the word is given. In their original language words are often descriptive, and knowing the original meaning can enhance understanding and make it easier to remember the definition of the word. Common prefixes, suffixes, and combining forms of many biological terms appear on the inside covers of the text and provide additional information on the derivation of words. When pronunciation of the word is complex, a ***Pronunciation Key*** is presented. Often simply being able to pronounce a word correctly is the key to remembering it. The ***Glossary,*** which collects the most important terms into one location for easy reference, also has a pronunciation guide.

Asides. The aside is a brief statement following the discussion of an important concept. It clarifies the concept by presenting an example of the concept in action. For example, the aside may illustrate the normal response of a system to exercise, or it might describe a pathological condition that shows how a system responds to an abnormal situation. The advantage of the aside is that it appears right after the presentation of the concept. In this way the relevance of the concept is immediately apparent, helping the student to better appreciate and understand it.

Boxed essays. The boxed essays are expanded versions of the asides that permit a more detailed or complete coverage of a topic. Subjects covered include pathologies, current research, sports medicine, exercise physiology, pharmcology, and clinical applications. They are designed not only to illustrate the chapter content but to stimulate interest as well.

Predict questions. While the aside or boxed essay can illustrate how a concept works, the predict question requires the application of the concept. When reading a text it is very easy to become a "passive" learner; everything seems very clear to passive learners until they attempt to use the information. The predict question converts the "passive" learner into an "active" learner who must use new information to solve a problem. The answer to this kind of question is not a mere restatement of fact, but rather a prediction, an analysis of the data, the synthesis of an experiment, or the evaluation and weighing of the important variables of the problem. For example, "Given a stimulus, predict how a system will respond." Or, "Given a clinical condition, explain why the observed symptoms occurred." Predict questions are practice problems that help to develop the skills necessary to answer the concept questions at the end of the chapter. In this regard, not only are possible answers given for the predict questions, but explanations that demonstrate the process of problem-solving are provided.

Tables. The book contains many tables that have several useful functions. They provide more specific information than that included in the text discussion, allowing the text to concentrate on the general or main points of a topic. The tables also summarize some aspect of the chapter's content, providing a convenient way to find information quickly. Often, a table is designed to accompany an illustration so that a written description and a visual presentation are combined to communicate information effectively.

Chapter summary. As the student reads the chapters, details can often obscure the overall picture. The chapter summary is an outline that briefly states the important facts and concepts and provides a perspective of the "big picture."

Content review. The content questions are another method used in this text to turn the "passive" learner into an "active" learner. The questions systematically cover the content and require students to summarize and restate the content in their own words.

Concept review. Following the mastery of the content review and therefore chapter content, the concept review requires the application of that content to new situations. These are not essay questions that involve the restatement or summarization of chapter content. Instead, they provide additional practice in problem-solving and promote the development and acquisition of problem-solving skills.

Appendices. Appendix A is a table of measurements. Reference to this table will help the student to relate the metric system to the more familiar English system when determining the size or weight of a structure.

Appendix B is a table of routine clinical texts along with their normal values and clinical significance. Reference to this table will provide students with the homeostatic values of many common substances in the blood and urine. Also, the importance of laboratory testing in the diagnosis and/or treatment of illnesses becomes readily apparent to the students.

Supplements

Any textbook can be used alone, but thoughtfully developed supplements increase its effectiveness for both student and instructor because they are designed to support the pedagogical model developed in the text. This text is accompanied by a wide range of supplements designed to complement the text.

Study guide. The study guide by James Kennedy and Philip Tate of Phoenix College supports the problem-based learning approach of the text. It introduces the student to the content of anatomy and physiology using word parts, matching, labeling, and completion exercises. A Mastery Learning Activity consisting of multiple-choice questions emphasizes comprehension of the material, evaluates progress, and prepares the student for classroom testing. In addition, a Final Challenges section consisting of essay questions provides practice with questions similar to the predict and concept questions of the textbook. Answers are given for all exercises, and explanations are furnished for the Mastery Learning Activity and the Final Challenges. Carefully reviewed by experienced instructors who currently teach anatomy and physiology, the student guide provides the reinforcement and practice so essential for the student's success in the course.

Laboratory manual. The laboratory manual, developed by Kevin Patton of St. Charles County Community College, divides the material typically covered in an anatomy and physiology lab into 42 subunits. Selection of the subunits and the sequence of their use permits the design of a laboratory course that is well integrated with the emphasis and sequence of the lecture material. As with the textbook and the study guide, basic content is introduced first, and gradually more complex activities are developed. This laboratory manual also contains boxed hints, safety boxes, separate lab reports, and coloring exercises. An Instructor's Manual to accompany the lab manual is available. Once again, the suggestions and corrections of reviewers with wide experience in teaching anatomy and physiology have been incorporated in this supplement.

Instructor's manual/testbank. The instructor's manual, written by James Kennedy and Robert R. Smith, has many features to assist in the development of a well-integrated course. It suggests ways to organize the material and is keyed to relevant transparencies, boxed essays, and illustrations, and to the laboratory exercises. Major points that deserve emphasis are included, hints on how to reinforce concepts are given, typical problem areas are noted along with ways to deal with the problems, and possible topics for discussion are considered. Answers for the concept questions at the end of each chapter in the text are included as are 60 transparency masters, including key diagrams from the text, as well as additional useful material for handouts. For the novice instructor or teaching assistant there is a "Teaching Survival Guide." Each chapter includes a list of relevant audiovisual resources. The manual also contains an extensive listing of laboratory supply houses. Perhaps the most unique feature of this manual is the conversion notes that detail the differences in terms of organization and coverage between our text and several of the leading texts now on the market. This is an ideal tool to assist you in converting your lecture notes from your current text to the Seeley/Stephens/Tate text. The Testbank, written by James Kennedy of Phoenix College, has been carefully designed to complement the textbook and the study guide. The test bank contains over 1000 test items, including multiple choice, completion, matching, and essay questions. Each question is classified according to level of difficulty, and answers are provided. All the questions have been carefully reviewed and painstakingly polished to offer the best possible evaluative tool.

Transparencies. A set of full-color transparencies, which emphasize the major anatomical structures and physiological processes covered in the textbook, are available to qualified adopters. These transparencies have been selected from illustrations in the text and provide a common vehicle for communication between the lecturer and the student. Larger, bolder labels have been added to enhance classroom efficacy.

Test-generating system. Qualified adopters of this text may request Diploma II, a computerized test bank package, compatible for use on the IBM PC, Apple IIc, or Apple IIe microcomputers. This software package is a unique combination of user-friendly computerized aids for the instructor:

- **EXAM** allows the user to select test-bank items either manually or randomly; to add, edit, or delete test items that include multiple-choice, short-answer, or matching options; to scramble answers; and to print exams with or without saving them for future use.
- **GRADEBOOK** is a computerized record keeper that saves student names (up to 250), assignments (up to 50), and related grades in a format similar to that used in manual grade books. Statistics on individual or class performance, test weighing, and push-button grade curving are features of this software.
- **PROCTOR** uses items from the test bank or instructor-generated questions for student review. Student scores can be merged into GRADEBOOK.
- **CALENDAR** makes class planning and schedule management quick and convenient.

Human body systems software. Developed by Kevin Patton and Kathryn Baalman of St. Charles County Community College, this interactive software program allows the student to experience applications relevant to anatomy and physiology. The software is divided into 11 body system modules. Each module contains an introduction, a tutorial with practice review questions, a practical applications section, and a final quiz. It is available on IBM to qualified adopters.

Acknowledgments

The efforts of many people are required to produce a modern textbook. It is difficult to adequately acknowledge the contributions of all people who have played a role. The encouragement and emotional support of our families were essential for the completion of this project. Their tolerance of the many evenings and weekends dedicated to this project and the many hours they enthusiastically endured discussions of the text were essential for its completion.

We wish to express our gratitude to the staff of Mosby–Year Book, Inc., for their steadfast help and encouragement. It has clearly been more than a vocation to them. Deborah Allen and Robert Callanan have worked with us in an untiring fashion to bring this work to completion. Their effort and contributions, as well as the many others who have influenced the design and production of this text, are greatly appreciated.

We thank the team of artists who have contributed to the text. Their attention to detail and their artistic contributions have made the text attractive as well as an effective teaching tool.

We sincerely thank the reviewers. This book was conscientiously reviewed by people who are exceptional teachers and who are dedicated to excellence. Their constructive comments and suggestions have added substantially to the quality of the text.

ROD SEELEY
TRENT STEPHENS
PHILIP TATE

Reviewers

Sue Anderson, *Lansing Community College*
Michael Autori, *University of Bridgeport*
Lawrence Baum, *Northeast Louisiana University*
Minyon Bond, *Phoenix College*
Armando De La Cruz, *Mississippi State University*
Marlene Donovan, *Hocking Technical College*
Ralph Ferges, *Palomar College*
Gloria Hillert, *Triton College*
Joel Ostroff, *Brevard Community College*
Mark Page, *Santa Barbara City College*
Linda M. Peck, *University of Findlay*
Frank Peek, *State University of New York—Morrisville*
Robert R. Smith, *St. Louis Community College—Forest Park*
James L. Smothers, *University of Louisville*
Linda Strause, *University of California—San Diego*
Rita Schwieterman, *Dayton School of Practical Nursing*
Darrell Wiens, *University of Northern Iowa*

CONTENTS IN BRIEF

CONTENTS

ESSENTIALS OF
ANATOMY AND PHYSIOLOGY

1

INTRODUCTION TO THE HUMAN BODY

After reading this chapter you should be able to:

1 Explain the importance of understanding the relationship between structure and function.

2 Define anatomy and physiology.

3 Describe the seven levels of organization of the body and give the major characteristics of each level.

4 List the 11 organ systems and give the major functions of each.

5 Define homeostasis and explain why it is important.

6 Diagram a negative-feedback system and a positive-feedback system and describe their relationship to homeostasis.

7 Describe the anatomical position.

8 Define the directional terms for the human body and use them to locate specific body structures.

9 Name and describe the three major planes of the body or an organ.

10 Define the axial and appendicular regions of the body and describe the subdivisions of each region.

11 Describe the major trunk cavities.

12 Describe the serous membranes and give their functions.

anatomical position Position in which a person is standing erect with the feet forward, arms hanging to the sides, and the palms of the hands facing forward.

anatomy [Gr. *ana*, up + *tome*, a cutting] Scientific discipline that investigates the structure of the body.

appendicular (ap′pen-dik′u-lar) [L. *appendo*, to hang something on] Relating to an appendage, as the limbs and their associated girdles.

axial (ak′se-al) [L. *axle*, axis] Relating to the head, neck, and trunk as distinguished from the extremities.

coronal plane (ko-ro′nal) [Gr. *korone*, crown] Plane separating the body into anterior and posterior portions; frontal section.

homeostasis (ho′me-o-sta′sis) [Gr. *homoio*, like + *stasis*, a standing] State or maintenance of equilibrium in the body with respect to functions and the composition of fluids.

mesentery (mes′en-tĕr′e) [Gr. *mesos*, middle + *enteron*, intestine] Double layer of peritoneum extending from the abdominal wall to the abdominopelvic organs; conveys blood vessels and nerves to abdominopelvic organs; holds and supports abdominopelvic organs.

negative feedback Mechanism by which any deviation from an ideal normal value is resisted or negated; returns a parameter to its normal range and thereby maintains homeostasis.

physiology [Gr. *physis*, nature + *logos*, study] Scientific discipline that deals with the processes or functions of living things.

positive feedback Mechanism by which any deviation from an ideal normal value is made greater.

sagittal plane (saj′ĭ-tal) [L. *sagitta*, the flight of an arrow] Plane running vertically through the body and dividing it into right and left parts.

serous membrane Thin sheet consisting of epithelium and connective tissue that lines cavities not opening to the outside of the body; does not contain glands but does secrete serous fluid.

tissue [L., to weave] Collection of similar cells and the substances between them.

transverse plane Plane separating the body into superior and inferior parts; a cross section.

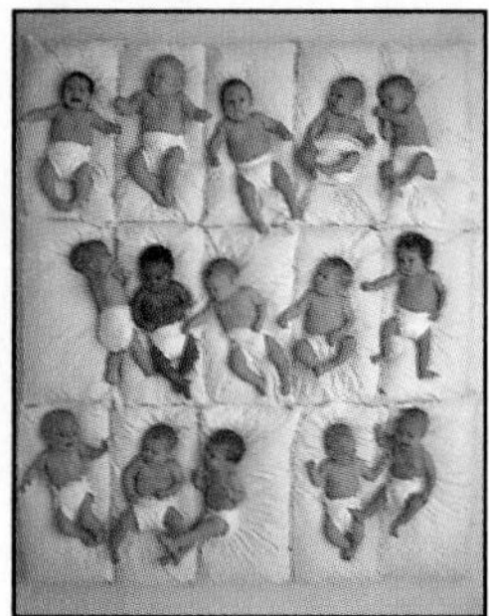

Human anatomy and physiology is the study of the structure and function of the human body. Knowledge of the structure and function of the body makes it possible to predict how the body will respond to a stimulus. For example, knowledge of the circulatory and nervous systems will help you to understand why it is important to control high blood pressure to prevent a stroke. This knowledge also will help you understand the symptoms seen in a person who has suffered a stroke.

The study of anatomy and physiology is essential for those who plan a career in the health sciences, because a sound knowledge of structure and function is necessary for health professionals to perform their duties adequately. Knowledge of anatomy and physiology is also beneficial to the nonprofessional. This background improves your ability to evaluate your physiological activities, understand recommended treatments, critically evaluate advertisements and reports in the popular literature, and interact with health professionals.

ANATOMY

Anatomy is the scientific discipline that investigates the body's structure. It is a dynamic discipline that covers a wide range of studies, including the form of structures, their microscopic organization, and the processes by which structures develop. In addition, anatomy examines the relationship between the structure of a body part and its function. Just as a hammer's structure makes it well-suited for pounding nails, the structure of body parts allows them to perform specific functions effectively. For example, bones can provide strength and support because bone cells surround themselves with a hard, mineralized matrix. Understanding the relationship between structure and function makes it easier to understand and appreciate anatomy.

Systemic anatomy is the study of the body by systems and is the approach taken in this and most other undergraduate textbooks. Examples of systems are the circulatory, nervous, skeletal, and muscular systems. **Regional anatomy** is the study of the body's organization by areas and is the approach taken in most medical and dental schools. Within each region, such as the head, abdomen, or arm, all systems are studied simultaneously.

Surface anatomy is the study of external features, such as bony projections, which serve as landmarks to locate deeper structures (for examples of external landmarks see Chapters 6 and 7). **Anatomic imaging** involves the use of x-rays, ultrasound, magnetic resonance imaging (MRI), and other technologies to create pictures of internal structures. Both surface anatomy and anatomic imaging provide important information useful in diagnosing disease.

For much of the history of anatomy, public sentiment made it difficult for anatomists to obtain human bodies for dissection. In the early 1800s, the benefits of human dissection for training physicians became apparent, and the need for cadavers increased beyond the ability to acquire them legally. Thus arose the resurrectionists, or body snatchers. For a fee, and no questions asked, they would remove bodies from graves and provide them to medical schools. Because the bodies were not always in the best condition, an enterprising man named William Burke went one step farther. He murdered 17 people and sold their bodies to a medical school. When discovered and convicted, Burke was hanged and then publicly dissected. Discovery of Burke's activities so outraged the public that sensible laws regulating the acquisition of cadavers were soon passed, and this dark chapter in the history of anatomy was closed.

PHYSIOLOGY

Physiology is the scientific discipline that deals with the processes or functions of living things. It is important in physiology to recognize structures as dynamic rather than unchanging. The major goals of physiology are understanding and predicting the body's responses to stimuli and understanding how the body maintains conditions within a narrow range of values in the presence of a continually changing environment.

Physiology is divided according to the organisms involved, the levels of organization within a given organism, or the specific system studied. **Cellular** and **systemic physiology** are examples of physiology that emphasize specific organizational levels, whereas **human physiology** refers to the study of a specific organism, the human.

STRUCTURAL AND FUNCTIONAL ORGANIZATION

The body can be studied at seven structural levels: chemical, organelle, cellular, tissue, organ, organ system, and organism (Figure 1-1; Table 1-1).

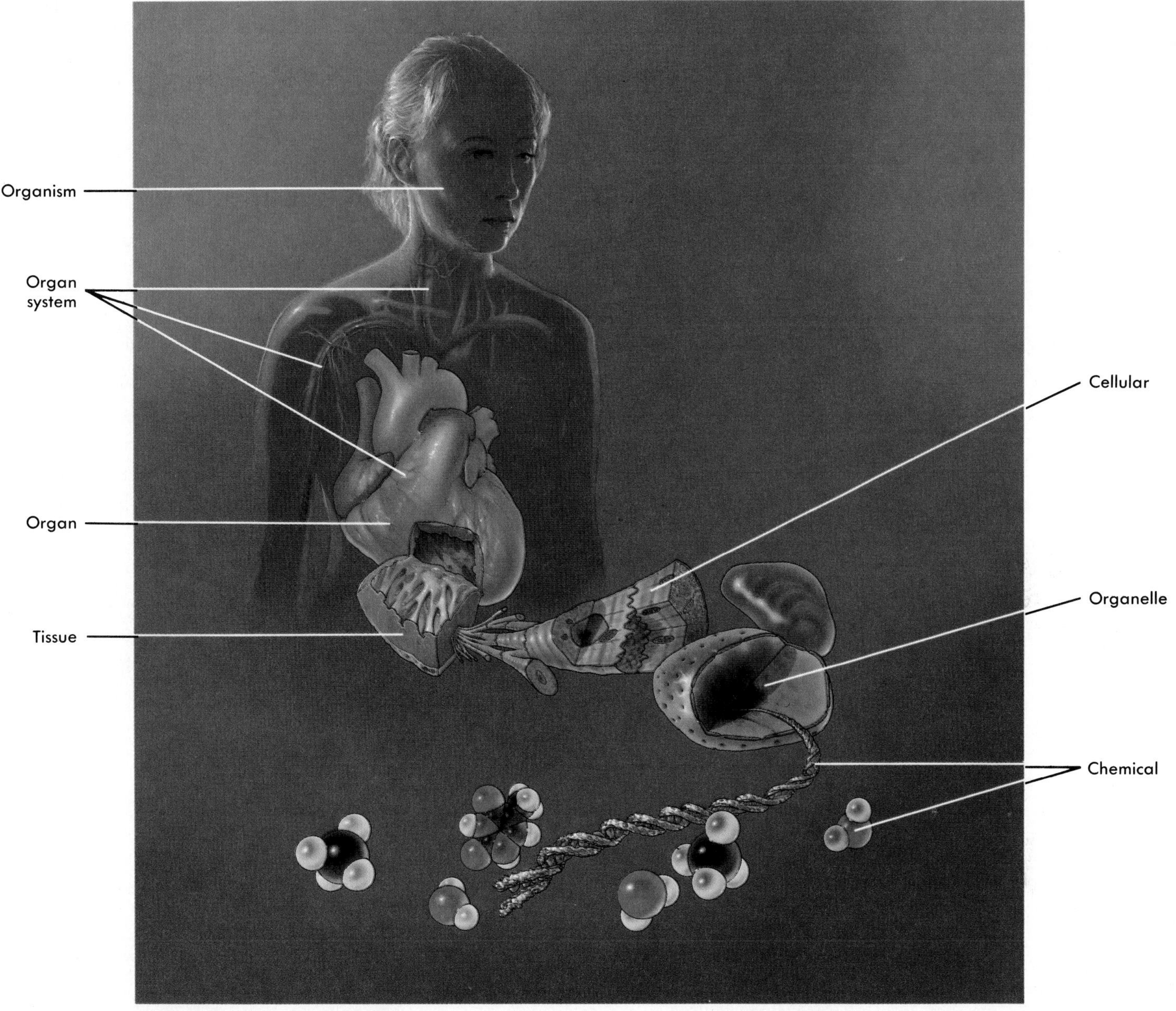

FIGURE 1-1 Levels of organization.
The seven levels of organization are the chemical, organelle, cellular, tissue, organ, organ system, and organism.

Chemical

The structural and functional characteristics of all organisms are determined by their chemical makeup (a brief overview of chemistry is presented in Chapter 2). The **chemical** level of organization involves interactions between atoms and their combinations into molecules. The function of a molecule is related intimately to its structure. For example, collagen molecules are strong, ropelike fibers that give skin structural strength. If collagen loses this ropelike structure, the skin becomes fragile and is torn easily.

Organelle

An **organelle** is a structure contained within a cell that performs one or more specific functions. For example, the nucleus is an organelle containing the cell's hereditary information. Organelles are discussed in Chapter 3.

TABLE 1-1

Organ Systems of the Body

SYSTEM	MAJOR COMPONENTS	FUNCTIONS
Integumentary	Skin, hair, nails, and sweat glands	Protects, regulates temperature, prevents water loss, and produces vitamin D precursors
Skeletal	Bones, associated cartilage, and joints	Protects, supports, and allows body movement, produces blood cells, and stores minerals
Muscular	Muscles attached to the skeleton	Produces body movement, maintains posture, and produces body heat
Nervous	Brain, spinal cord, nerves, and sensory receptors	A major regulatory system: detects sensation, controls movements, controls physiological and intellectual functions
Endocrine	Endocrine glands such as the pituitary, thyroid, and adrenal glands	A major regulatory system: participates in the regulation of metabolism, reproduction, and many other functions
Cardiovascular	Heart, blood vessels, and blood	Transports nutrients, waste products, gases, and hormones throughout the body; plays a role in the immune response and the regulation of body temperature
Lymphatic	Lymph vessels, lymph nodes, and other lymph organs	Removes foreign substances from the blood and lymph, combats disease, maintains tissue fluid balance, and absorbs fats

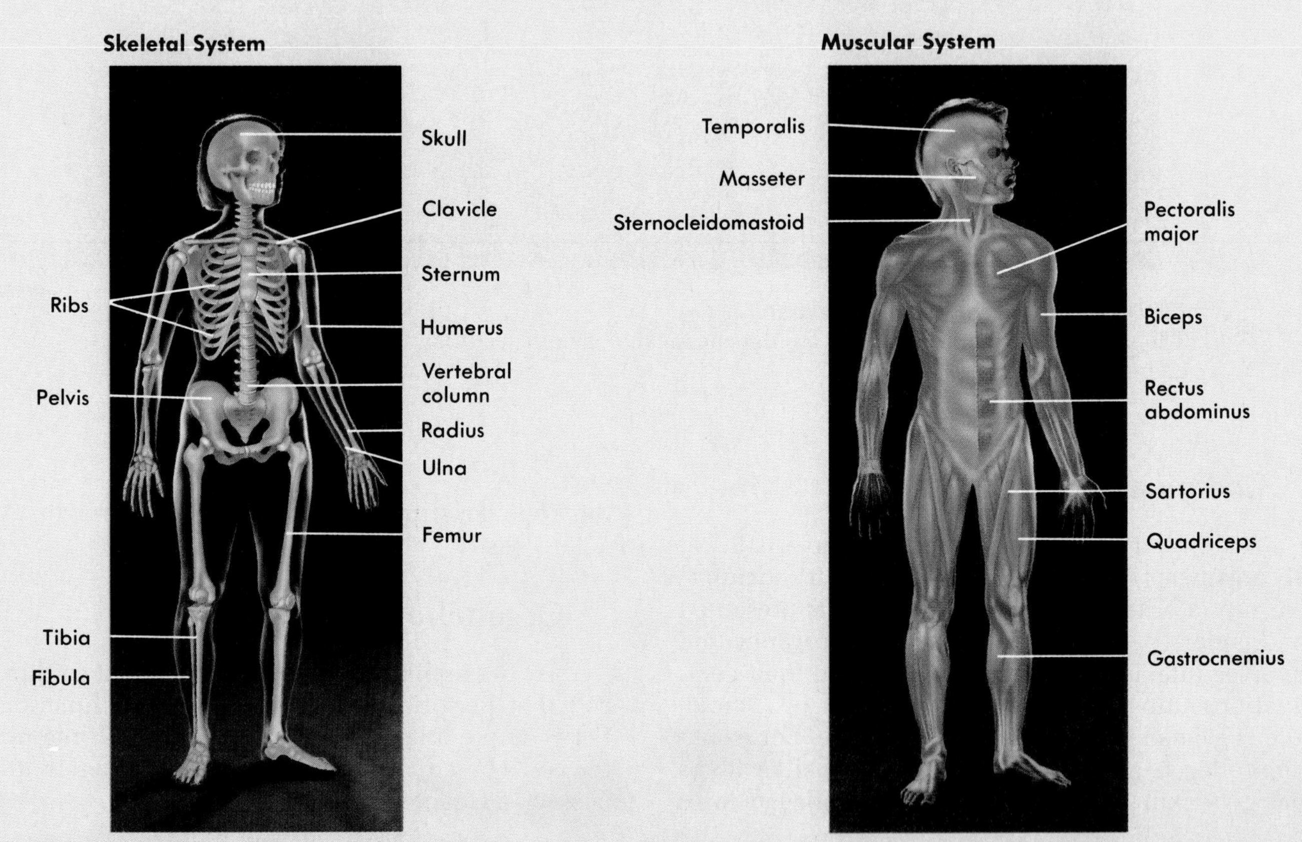

Nervous System

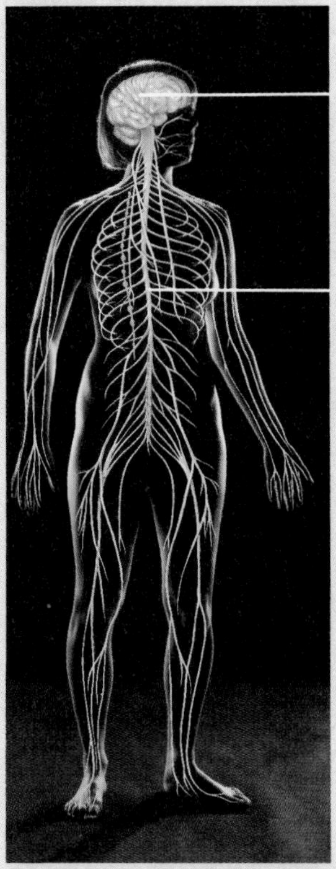

Endocrine System

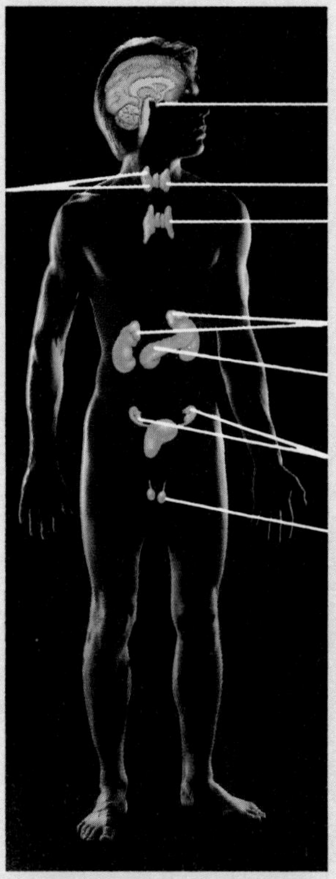

Cardiovascular System

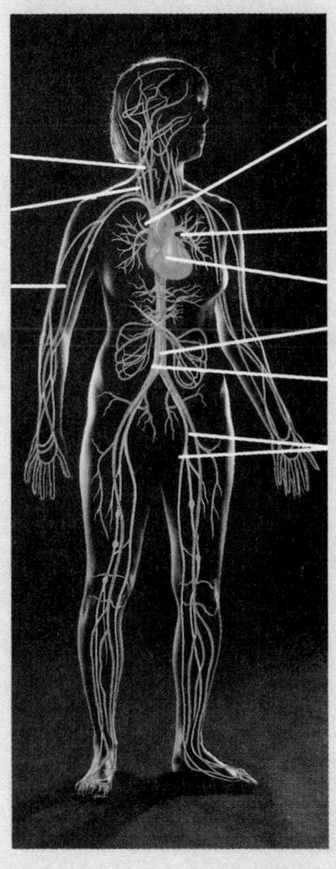

Lymphatic System

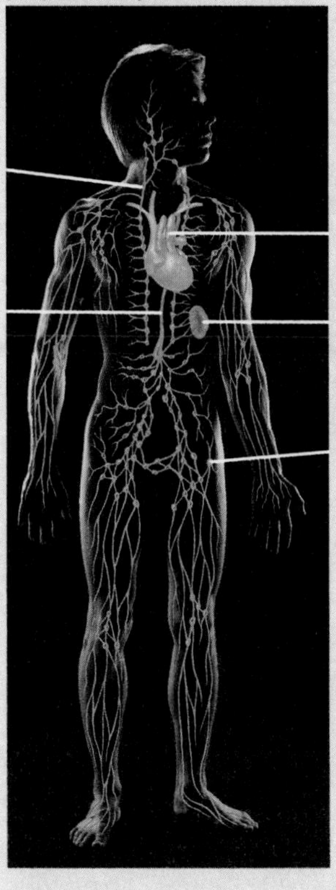

Continued.

TABLE 1-1—cont'd

Organ Systems of the Body

SYSTEM	MAJOR COMPONENTS	FUNCTIONS
Respiratory	Lungs and respiratory passages	Exchanges gases (oxygen and carbon dioxide) between the blood and the air and regulates blood pH
Digestive	Mouth, esophagus, stomach, intestines, and accessory structures	Performs the mechanical and chemical processes of digestion, absorption of nutrients, and elimination of wastes
Urinary	Kidneys, urinary bladder, and the ducts that carry urine	Removes waste products from the circulatory system; regulates blood pH, ion balance, and water balance
Reproductive	Gonads, accessory structures, and genitals of males and females	Performs the processes of reproduction and controls sexual functions and behaviors

Respiratory System

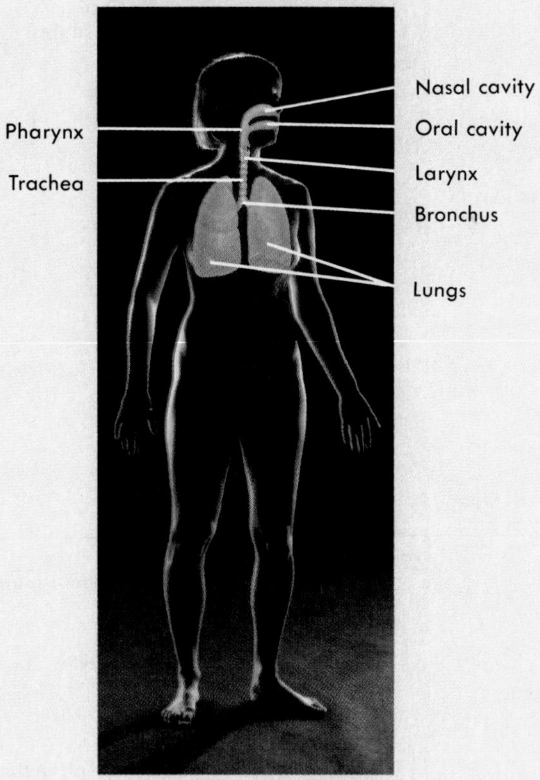

Digestive System

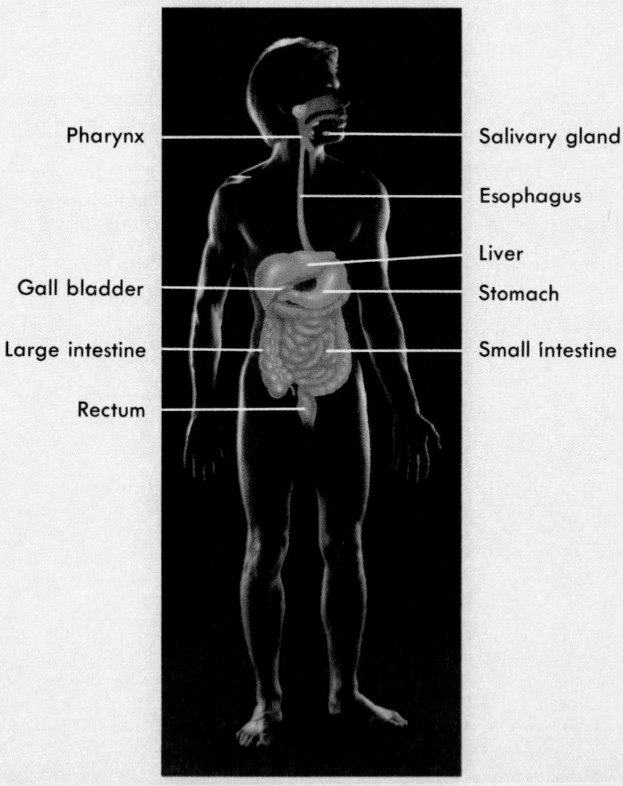

Urinary System

Kidney
Ureter
Bladder
Urethra

Reproductive System—Male

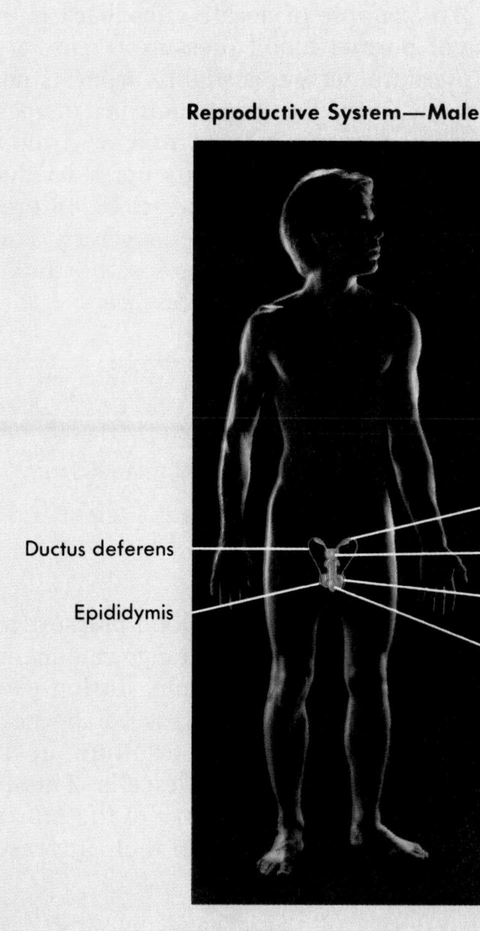

Reproductive System—Female

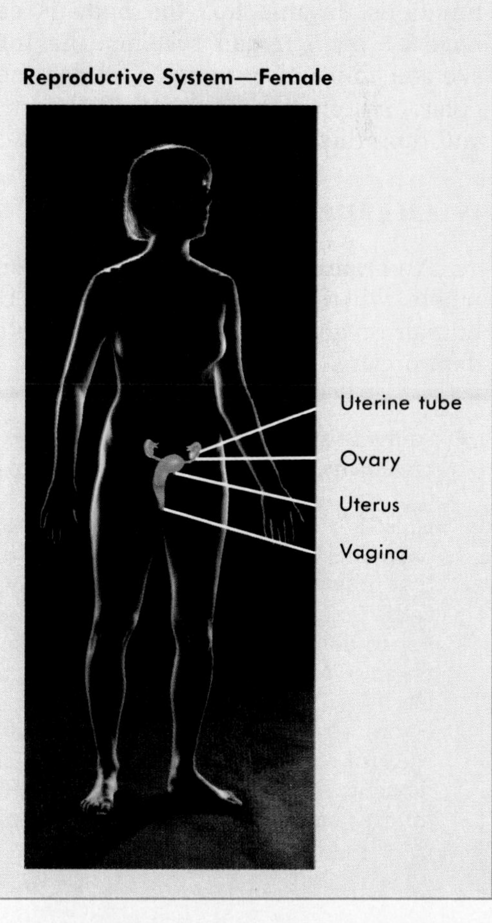

Cell

Cells are the basic living units of all plants and animals. Although cell types differ in their structure and function, they have many characteristics in common. Knowledge of these characteristics and their variations is essential to a basic understanding of anatomy and physiology. The cell is discussed in Chapter 3.

Tissue

A group of cells with similar structure and function plus the extracellular substances located between them is a **tissue.** The many tissues that make up the body are classified into four primary tissue types: epithelial, connective, muscle, and nervous. Tissues are discussed in Chapter 4.

Organ

Organs are composed of two or more tissue types that together perform one or more common functions. The skin, stomach, eye, and heart are examples of organs.

Organ System

An **organ system** is a group of organs classified as a unit because of a common function or set of functions. In this text the body is considered to have 11 major organ systems: the integumentary, skeletal, muscular, nervous, endocrine, cardiovascular, lymphatic, respiratory, digestive, urinary, and reproductive systems (see Table 1-1).

Organism

An **organism** is any living thing considered as a whole, whether composed of one cell or many. The human organism is a complex of mutually dependent organ systems.

> Humans share many characteristics with other organisms, and much of the knowledge about humans has come from studying other organisms. For example, the study of molecules and organelles in bacteria (single-celled organisms) has provided much information about human cells. Some biomedical research, however, cannot be done on isolated cells. For example, great progress in open heart surgery was made possible by perfecting techniques on other mammals before attempting them on humans. However, because other organisms are also different from humans, the ultimate answers to questions about humans can be obtained only from humans.

HOMEOSTASIS

Homeostasis (ho′me-o-sta′sis) is the existence and maintenance of a relatively constant environment within the body. Each cell of the body is surrounded by a small amount of fluid, and the normal function of that cell depends on the maintenance of its fluid environment within a narrow range of conditions, including volume, temperature, and chemical content. If the fluid surrounding cells deviates from homeostasis, the cells and possibly the individual may die.

All the organ systems contribute to the cellular environment and are controlled so that this environment remains relatively constant. For example, the digestive, respiratory, circulatory, and urinary systems are regulated so that each cell in the body receives adequate oxygen and nutrients and so that waste products do not accumulate to a toxic level.

Negative Feedback

The systems of the body are regulated by **negative-feedback** mechanisms that function to maintain homeostasis. "Negative" means that any deviation from a normal value is resisted or negated. Negative feedback does not prevent variation but maintains that variation within a normal range (Figure 1-2, *A*). An example of negative feedback is the maintenance of normal blood pressure (Figure 1-2, *B*). If blood pressure decreases slightly from its normal value, negative-feedback mechanisms increase blood pressure and return it to normal; or if blood pressure increases slightly above its normal value, negative-feedback mechanisms decrease blood pressure. As a result, blood pressure constantly rises and falls around the normal value, establishing a normal range of values for blood pressure.

> **1 Describe the consequences when a negative-feedback mechanism cannot return the value of some parameter such as blood pressure to its normal range.** ?

Although parameters such as blood pressure are maintained within a set range, the range can change in a beneficial manner. For example, during exercise the normal ranges for blood pressure and heart rate are increased significantly, resulting in increased delivery of blood to muscle cells. This increases oxygen and nutrient delivery to the muscle cells, which is required to maintain their increased rate of activity.

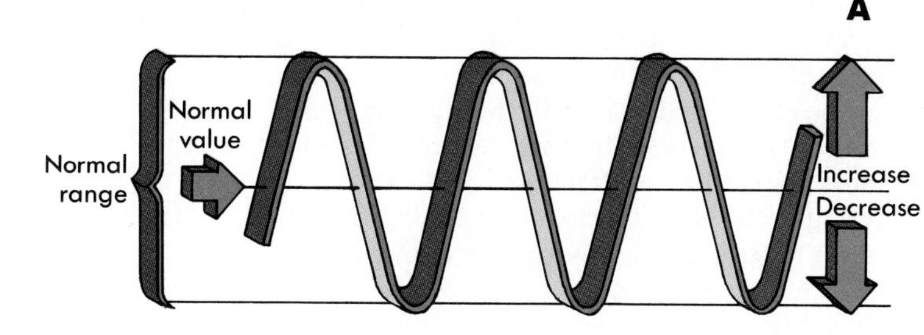

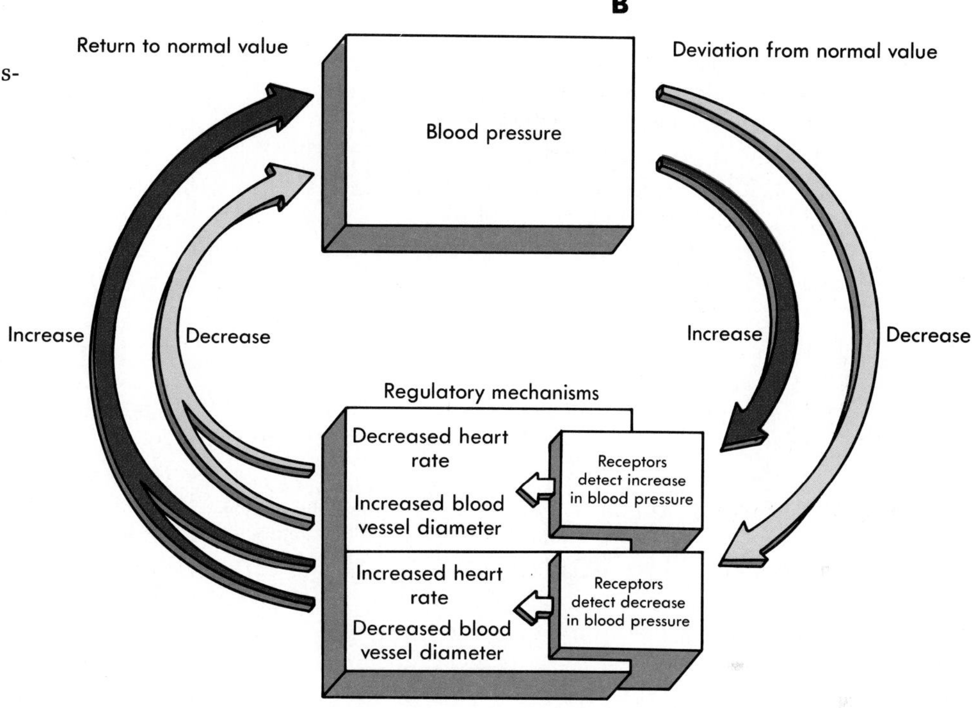

FIGURE 1-2 ■ Negative feedback.
A Values for the condition fluctuate above and below a normal value within a normal range. As long as the values remain within the normal range, homeostasis is maintained.
B Example of negative feedback: maintenance of blood pressure. An increase in blood pressure is detected by receptors, initiating regulatory changes that cause heart rate to decrease and blood vessels to increase in diameter. These events decrease blood pressure. Conversely, a decrease in blood pressure causes heart rate to increase and blood vessel diameter to decrease, resulting in increased blood pressure.

■ The idea that the body maintains a balance (homeostasis) can be traced back to ancient Greece. It was believed that the body supported four juices: the red juice of blood, the yellow juice of bile, the white juice secreted from the nose and lungs, and the black juice in the pancreas. Disease was thought to be caused by an excess of one of the juices, and normally the body would attempt to heal itself by expelling the excess juice. For example, mucus would be expelled from the nose of a person with a cold. This led to the practice of bloodletting to restore the body's normal balance of juices. Tragically, in the eighteenth and nineteenth centuries bloodletting went to extremes. When bloodletting did not result in improvement of the patient, it was taken as evidence that not enough blood had been removed to restore a healthy balance of the body's juices. The obvious solution was to let more blood, undoubtedly causing many deaths. Eventually the failure of this approach became obvious and the practice was abandoned. Fortunately we now understand better how the body maintains homeostasis. ■

Because good health requires homeostasis, illness results when negative-feedback mechanisms are disrupted. Medical therapy seeks to overcome illness by aiding the negative-feedback process. One example is a blood transfusion that reverses a constantly decreasing blood pressure and thus restores homeostasis.

Positive Feedback

■ **Positive-feedback** responses are not homeostatic and are rare in healthy individuals. "Positive" implies that when a deviation from a normal value occurs, the response of the system is to make the deviation larger (Figure 1-3, *A*). Therefore positive feedback usually creates a "vicious cycle" leading away from homeostasis and, in some cases, resulting in death. For example, if blood pressure declines to a sufficiently low value, such as may occur following extreme blood loss, too little blood flows

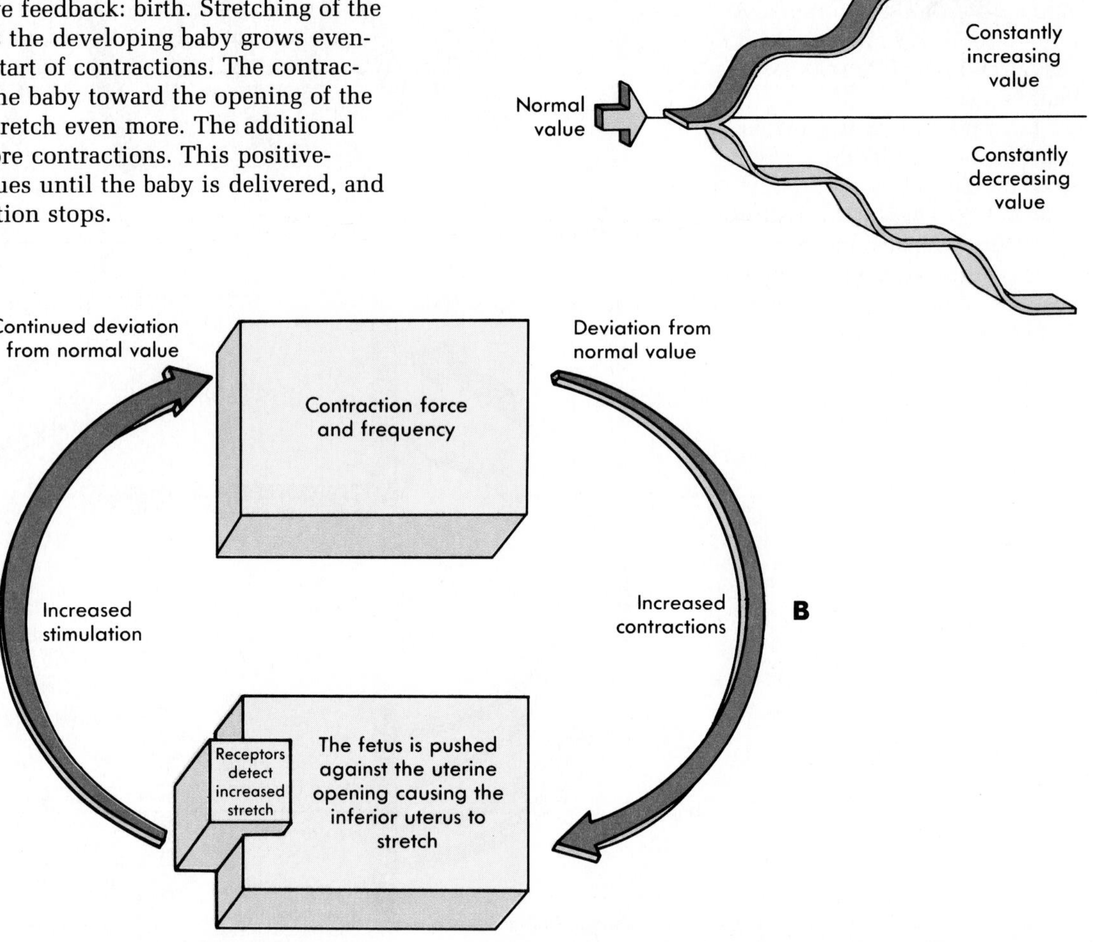

FIGURE 1-3 Positive feedback.
A Values deviate further and further away from the normal value (in either a positive or negative direction).
B Example of positive feedback: birth. Stretching of the uterus *(bottom box)* as the developing baby grows eventually stimulates the start of contractions. The contractions *(top box)* push the baby toward the opening of the uterus, causing it to stretch even more. The additional stretching initiates more contractions. This positive-feedback cycle continues until the baby is delivered, and the stretching stimulation stops.

back to the heart. Consequently, the heart pumps less blood, and blood pressure drops further. The additional decrease in blood pressure causes the heart to pump even less blood, which again decreases the blood pressure. The process continues until the blood pressure is too low to sustain life or until the positive-feedback cycle is interrupted by medical intervention, such as a blood transfusion.

A few positive-feedback mechanisms do operate in the body under normal conditions, but in all cases they are eventually limited in some way. Birth is an example of a normally occurring positive-feedback mechanism (Figure 1-3, *B*). Near the end of pregnancy, the uterus is stretched by the baby's large size and weight. This stretching, especially around the opening of the uterus, stimulates contractions of the uterine muscles. The uterine contractions push the baby against the opening of the uterus, stretching it further. This stimulates additional contractions that result in additional stretching. This positive-feedback sequence ends only when the baby is delivered from the uterus and the stretching stimulus is eliminated.

2 Is the sensation of thirst associated with a negative- or a positive-feedback mechanism? Explain. (Hint: What parameter is being regulated in this case?) ?

TERMINOLOGY AND THE BODY PLAN

When you begin studying anatomy and physiology, the number of new words can seem overwhelming. Learning is easier and more interesting if you pay attention to the origin, or etymology, of new words. Most of the terms are derived from Latin or Greek, and the terms are descriptive in the original languages. For example, anterior in Latin means to go before. Therefore the anterior surface of the body is the belly surface or that which goes before when we are walking.

Words are often modified by adding a prefix or suffix. The suffix "-itis" means an inflammation, so appendicitis is an inflammation of the appendix. As new terms are introduced in this text, their meanings are often explained. The glossary and the list of word roots, prefixes, and suffixes on the inside cover of the textbook also provide additional information.

Directional Terms

Directional terms refer to the body in the anatomical position regardless of its actual position. The term **anatomical position** refers to a person standing erect with the feet forward, arms hanging to the sides, and the palms of the hands facing forward. In human anatomy, up is replaced by **superior,** down by **inferior,** front by **anterior,** and back by **posterior.** Directional terms are used to describe the position of structures in relation to other structures or body parts. For example, the neck is superior to the chest, but it is inferior to the head. Important directional terms are presented in Table 1-2 and are illustrated in Figure 1-4.

3 Provide the correct directional term for the following statement. When a man is standing on his head, his nose is __________ to his mouth. **?**

Planes

At times it is conceptually useful to discuss the body in reference to a series of planes (imaginary flat surfaces) passing through it (Figure 1-5, *A*). A **sagittal** (saj′ĭ-tal) plane runs vertically through the body and separates it into right and left portions. The word sagittal literally means "the flight of an arrow" and refers to the way the body would be split by an arrow passing anteriorly to posteriorly. If the plane divides the body into equal right and left halves, it is a **midsagittal** (mid′saj′ĭ-tal) plane. A **transverse** or **horizontal** plane runs parallel to the surface of the ground and divides the body into superior and inferior parts. A **frontal** or **coronal** (ko-ro′nal) plane runs vertically from right to left and

TABLE 1-2

Directional Terms for Humans

TERMS	ETYMOLOGY*	DEFINITION	EXAMPLE
Right		Toward the body's right side	The right ear
Left		Toward the body's left side	The left eye
Inferior	L., lower	A structure lower than another	The nose is inferior to the forehead
Superior	L., higher	A structure higher than another	The mouth is superior to the chin
Anterior	L., to go before	Toward the front of the body	The teeth are anterior to the throat
Posterior	L. *posterus*, following	Toward the back of the body	The throat is posterior to the teeth
Dorsal	L. *dorsum*, back	Toward the back (synonymous with posterior)	The spine is dorsal to the breastbone
Ventral	L. *ventr*, belly	Toward the belly (synonymous with anterior)	The navel is ventral to the spine
Proximal	L. *proximus*, nearest	Closer to the point of attachment to the body than another structure	The elbow is proximal to the wrist
Distal	L. *di* + *sto*, to be distant	Farther from the point of attachment to the body than another structure	The wrist is distal to the elbow
Lateral	L. *latus*, side	Away from the midline of the body	The nipple is lateral to the breastbone
Medial	L. *medialis*, middle	Toward the middle or midline of the body	The bridge of the nose is medial to the eye
Superficial	L. *superficialis*, surface	Toward or on the surface	The skin is superficial to muscle
Deep	O.E. *deop*, deep	Away from the surface, internal	The lungs are deep to the ribs

*Origin and meaning of the word: *L.*, Latin; *O.E.*, Old English.

A

B

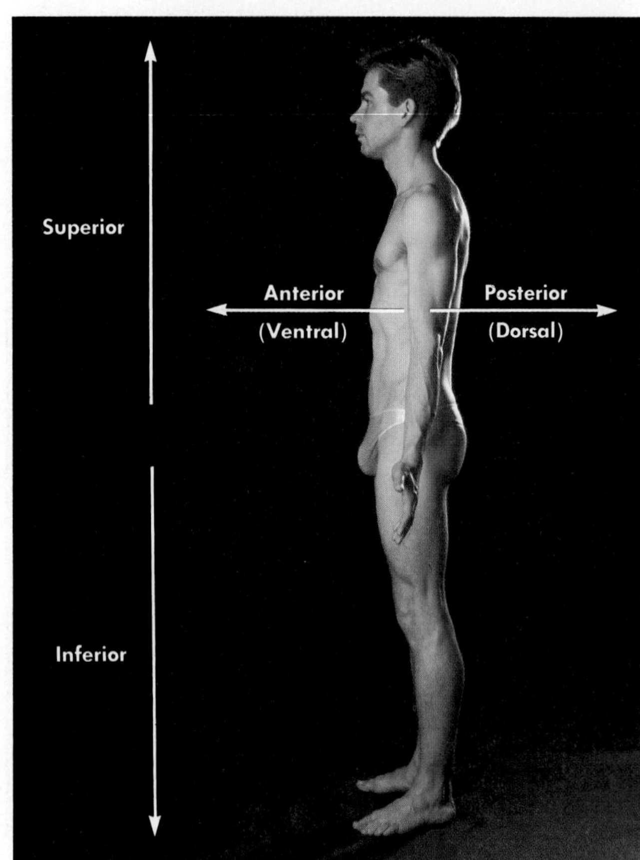

divides the body into anterior and posterior portions.

Organs often are sectioned to reveal their internal structure (Figure 1-5, *B*). A cut through the long axis of the organ is a **longitudinal** section, and a cut at a right angle to the long axis is a **transverse** or **cross** section. If a cut is made across the long axis at other than a right angle, it is called an **oblique** section.

Body Regions

The body is commonly divided into several regions (Figure 1-6). The first major division is between the **appendicular** (ap′pen-dik′u-lar) and **axial** (ak′se-al) regions. The appendicular region includes the limbs and their girdles (the bony structures that attach the limbs to the body). The upper limb is divided into the arm, forearm, wrist, and hand. The **arm** extends from the shoulder to the elbow, and the **forearm** extends from the elbow to the wrist. The upper limb is attached to the body by the **shoulder,** or **pectoral** (pek′to-ral) **girdle.** The lower limb is divided into the thigh, leg, ankle, and foot. The **thigh** extends from the hip to the knee, and the **leg** extends from the knee to the ankle. The lower limb is attached to the body by the **hip,** or **pelvic** (pel′vik) **girdle.** Note that the terms arm and leg, contrary to popular usage, refer to only a portion of the respective limb.

The axial region of the body consists of the **head, neck,** and **trunk.** The trunk can be divided into the **thorax** (chest), **abdomen** (region between the thorax and pelvis), and **pelvis** (the inferior end of the trunk associated with the hips).

The abdomen often is subdivided superficially into four **quadrants** by two imaginary lines—one horizontal and one vertical—that intersect at the navel (Figure 1-7, *A*). The quadrants formed are the upper-right, upper-left, lower-right, and lower-left quadrants. In addition to these quadrants, the abdomen sometimes is subdivided into nine **regions** by four imaginary lines—two horizontal and two vertical. These four lines create an imaginary tic-tac-toe figure on the abdomen (Figure 1-7, *B*). The quadrants or regions are commonly used by clinicians as reference points for locating the underlying organs. For example, the pain of an acute appendicitis is usually located in the lower-right quadrant.

FIGURE 1-4 ■ Directional terms.
A Directional terms from the front. Arrows point in the indicated direction.
B Directional terms from the side. Arrows point in the indicated direction.

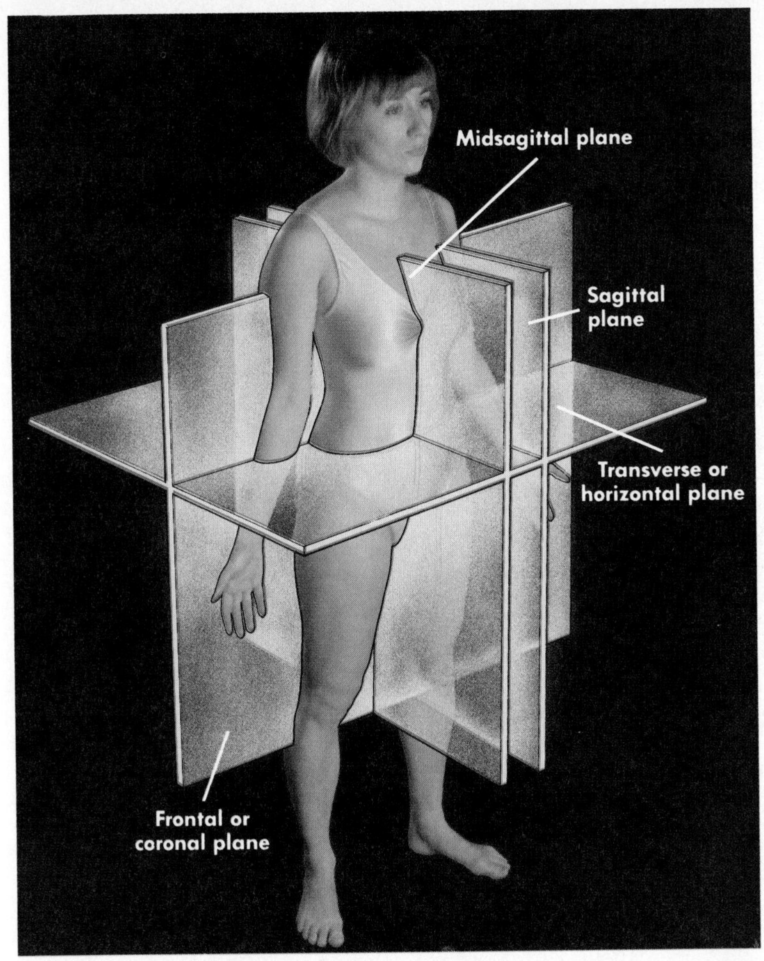

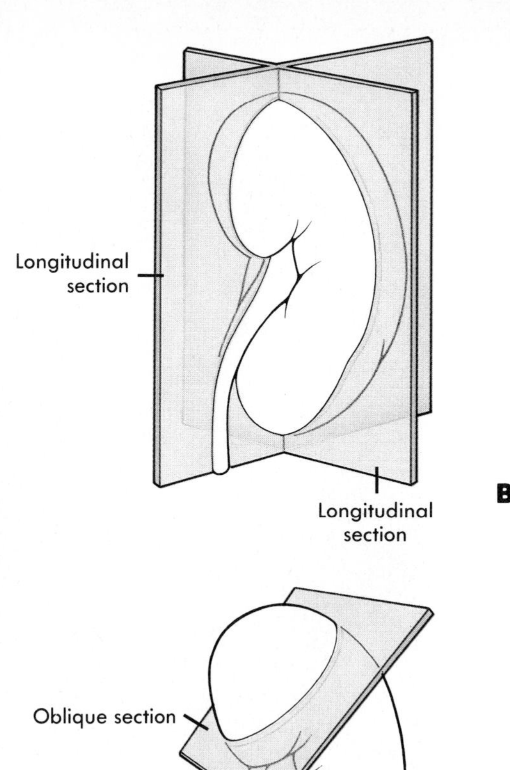

FIGURE 1-5 ■ Planes of section.
Planes are indicated by "glass" sheets.
A The whole body.
B A single organ (the kidney).

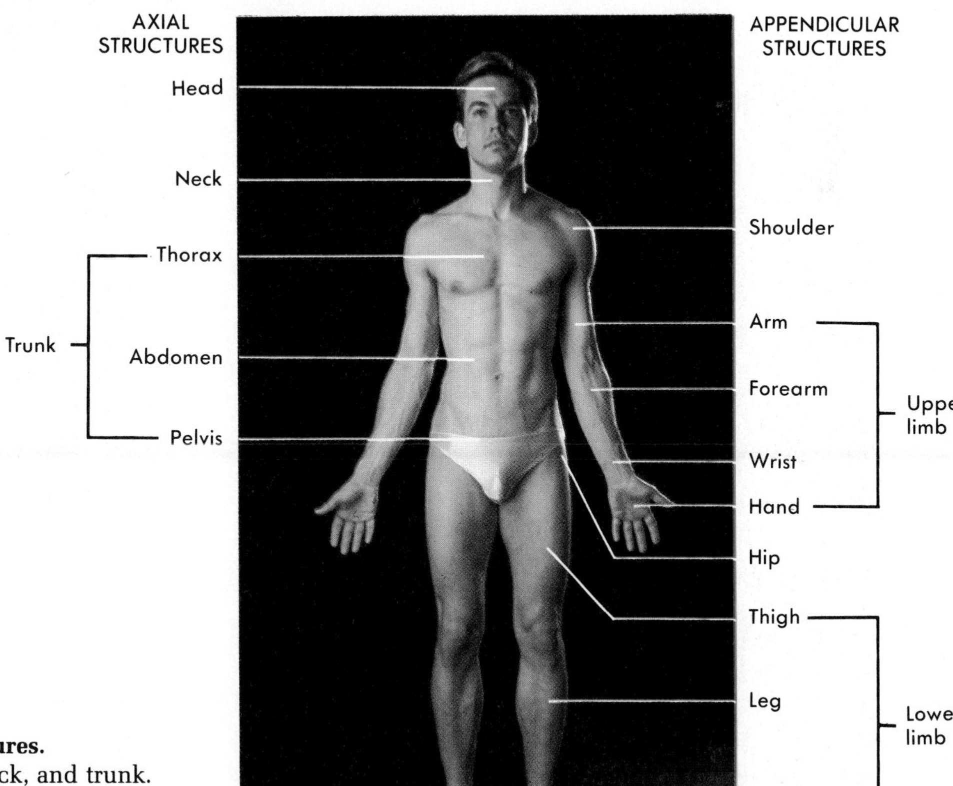

FIGURE 1-6 ■ Body regions and structures.
The axial region consists of the head, neck, and trunk. The appendicular region consists of the limbs and their girdles.

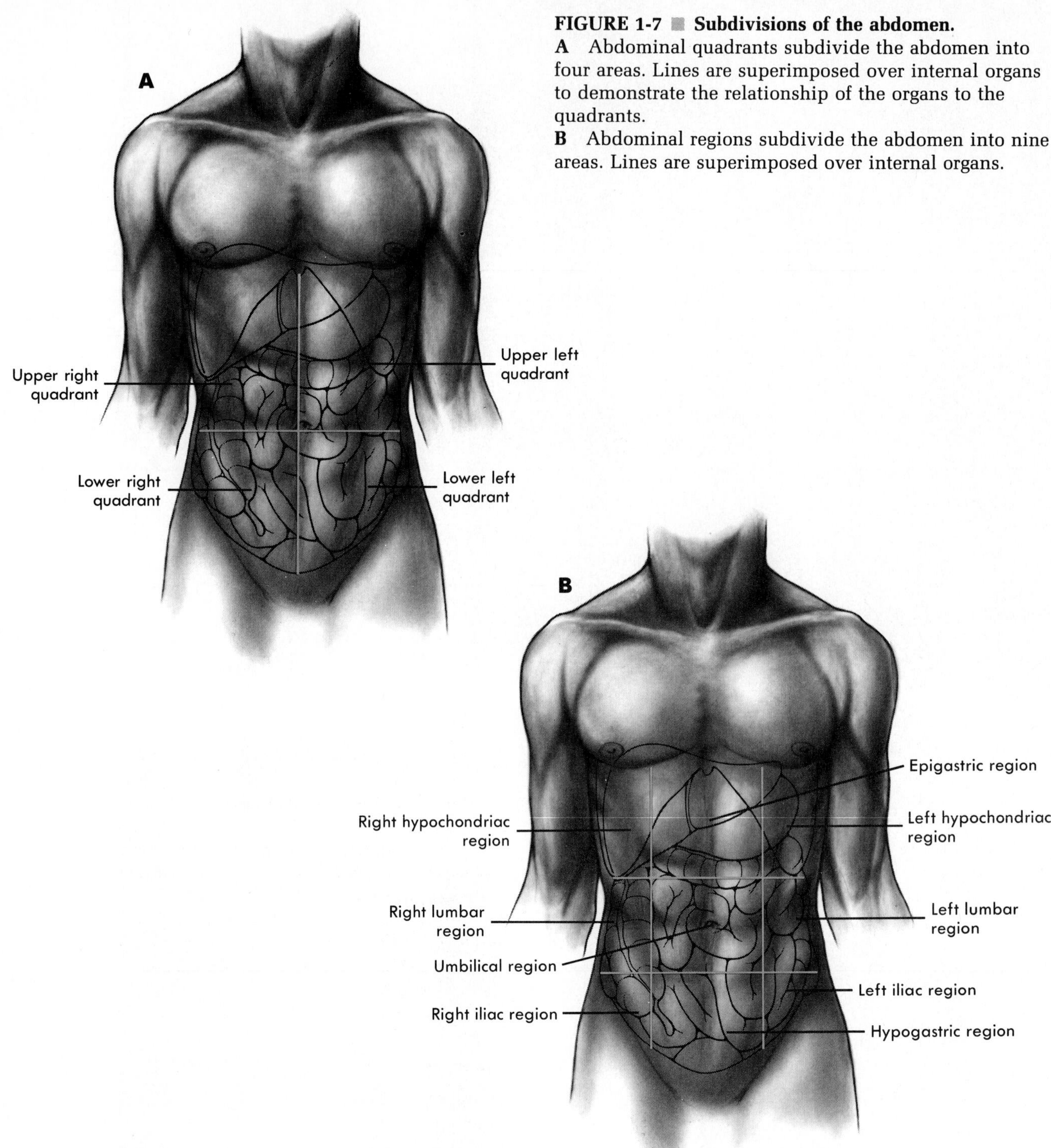

FIGURE 1-7 Subdivisions of the abdomen.
A Abdominal quadrants subdivide the abdomen into four areas. Lines are superimposed over internal organs to demonstrate the relationship of the organs to the quadrants.
B Abdominal regions subdivide the abdomen into nine areas. Lines are superimposed over internal organs.

Body Cavities

The body contains many cavities, such as the nasal, cranial, abdominal, and bone marrow cavities. Some of these cavities open to the outside of the body, and some do not. Undergraduate anatomy and physiology textbooks often describe a dorsal cavity, in which the brain and spinal cord are found, and a ventral body cavity that contains all the trunk cavities. This concept is not described in standard works on anatomy, however, and therefore is not emphasized here. Discussion in this chapter is limited to the major trunk cavities that do not open to the outside.

The trunk contains three large cavities: the **thoracic cavity,** the **abdominal cavity,** and the **pelvic cavity** (Figure 1-8). The thoracic cavity is surrounded by the rib cage and is separated from the

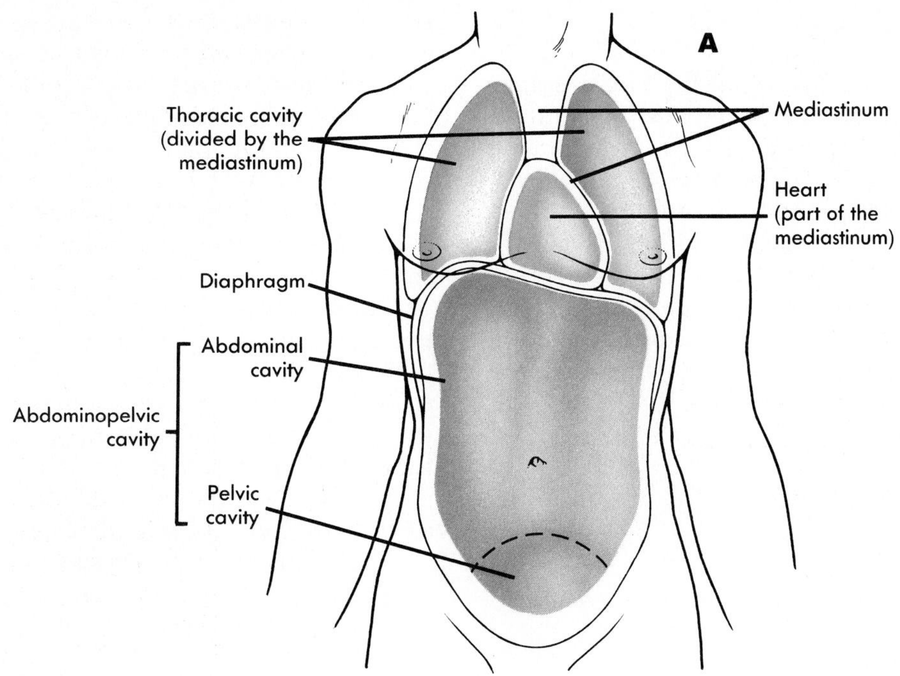

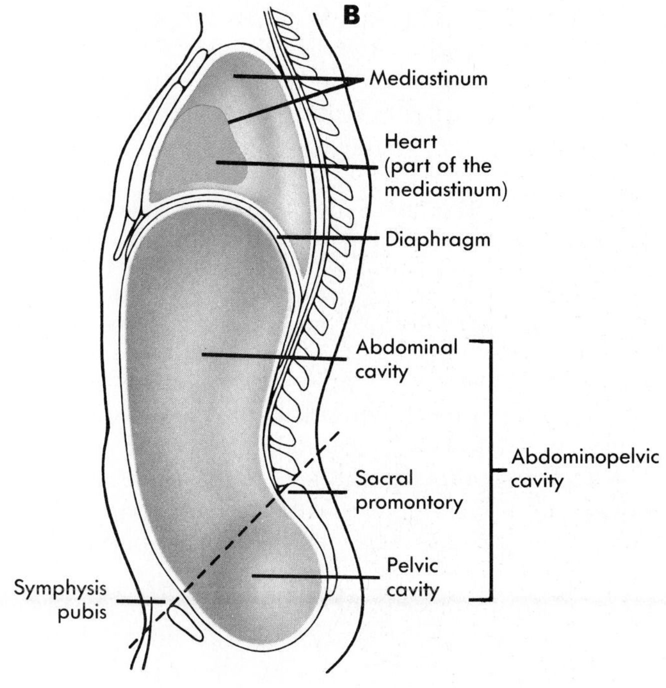

FIGURE 1-8 Trunk cavities.
A Anterior view showing the major trunk cavities. The diaphragm separates the thoracic cavity from the abdominal cavity. The mediastinum, which includes the heart, divides the thoracic cavity.
B Sagittal view of trunk cavities. Imaginary plane shows the division between the abdominal and pelvic cavities.

abdominal cavity by the muscular diaphragm. The thoracic cavity is divided into two parts by a median structure called the **mediastinum** (me′de-as-ti′-num; a wall). The mediastinum is a partition containing the heart, trachea, esophagus, and other structures. The lungs are located on either side of the mediastinum.

The abdominal and pelvic cavities are not physically separated and sometimes are called collectively the abdominopelvic cavity. The abdominal cavity is bounded primarily by the abdominal muscles and contains the stomach, intestines, liver, spleen, pancreas, and kidneys. The pelvic cavity is a small space enclosed by the bones of the pelvis and contains the urinary bladder, part of the large intestine, and the internal reproductive organs.

Serous Membranes

Serous membranes line the trunk cavities and cover the organs of the trunk cavities. To understand the relationship between a serous membrane and its organ, imagine an inflated balloon into which a fist has been pushed (Figure 1-9, *A*). The walls of the balloon represent the serous membranes and the fist represents the organ (Figure 1-9, *B*). The portion of the serous membrane in contact with the organ is referred to as **visceral** (vis′er-al; organ), and the outer part of the membrane is referred to as **parietal** (pă-ri′ĕ-tal; wall). There is a cavity or space between the visceral and parietal membranes that normally is filled with a thin, lubricating film of serous fluid produced by the membranes. If an organ rubs against another organ or against the body wall, the serous fluid and smooth serous membranes function to reduce friction.

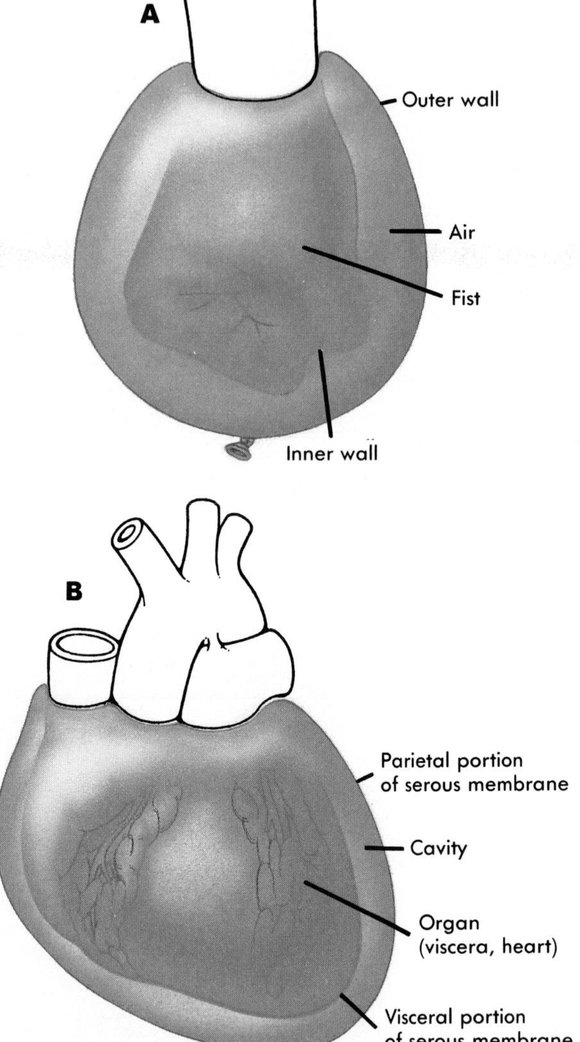

FIGURE 1-9 Serous membranes.
A Fist pushing into a balloon illustrates the relationship of the visceral and parietal serous membranes to organs.
B An organ (the heart) and its parietal and visceral membranes (pericardium).

The thoracic cavity contains three serous membrane-lined cavities: two pleural cavities and a pericardial cavity. The **pericardial** (pĕr-ĭ-kar′de-al) **cavity** surrounds the heart (Figure 1-10, *A*). The heart is covered by the visceral pericardium and is contained within a connective tissue sac that is lined with the parietal pericardium. The pericardial cavity, which contains pericardial fluid, is located between the visceral and parietal pericardium. Each lung is surrounded by a **pleural** (ploor′al) **cavity** and is covered by visceral pleura (Figure 1-10, *B*). The inner surface of the thorax is covered by the parietal pleura. The pleural cavity is located between the visceral and parietal pleura and contains pleural fluid.

The **peritoneal** (pĕr′ĭ-to-ne′al) **cavity** is located within the abdominopelvic cavity (Figure 1-10, *C*). Many of the organs of the abdominopelvic cavity are covered by visceral peritoneum and the wall of the abdominopelvic cavity is lined with parietal peritoneum. The space between the two membranes contains peritoneal fluid and is called the peritoneal cavity.

> The serous membranes can become inflamed—often as a result of an infection. Pleurisy is inflammation of the pleura, pericarditis is inflammation of the pericardium, and peritonitis is inflammation of the peritoneum.

The parietal peritoneum is connected to the visceral peritoneum of many abdominopelvic organs by a double-layered membrane called a **mesentery** (mes′en-tĕr′e) (see Figure 1-10, *C*). The mesenteries anchor the organs to the body wall and provide a pathway for nerves and blood vessels to reach the organs. Other abdominopelvic organs are more closely attached to the body wall and do not have mesenteries. They are covered by the parietal peritoneum and are said to be **retroperitoneal** (rĕ′tro-pĕr′ĭ-to-ne′al; behind the peritoneum). The retroperitoneal organs include the kidneys, adrenal glands, pancreas, portions of the intestines, and the urinary bladder (see Figure 1-10, *C*).

4 Explain how an organ can be located within the abdominopelvic cavity but not be within the peritoneal cavity.

FIGURE 1-10 Serous cavities.
A Frontal section showing the pericardial cavity surrounding the heart.
B Frontal section showing the pleural cavities surrounding the lungs.
C Sagittal section through the abdominopelvic cavity showing the peritoneal cavity, peritoneum, mesenteries, and retroperitoneal organs.

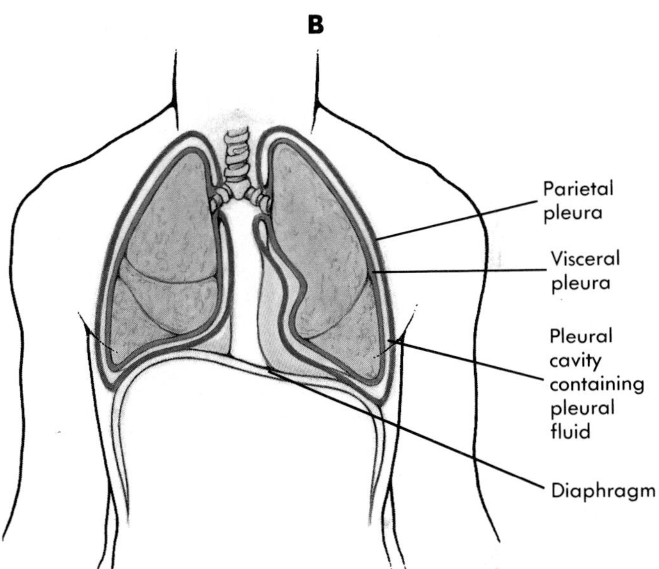

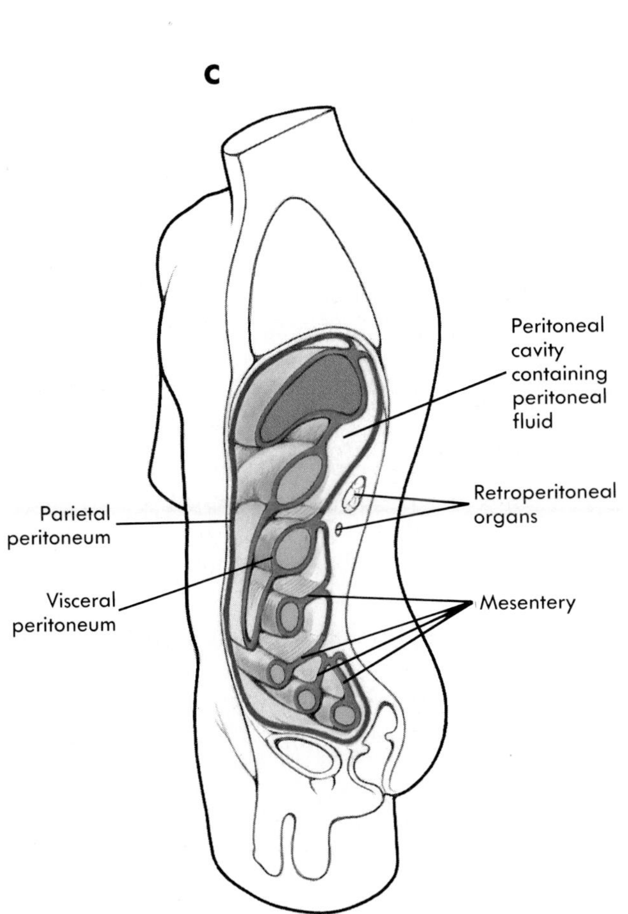

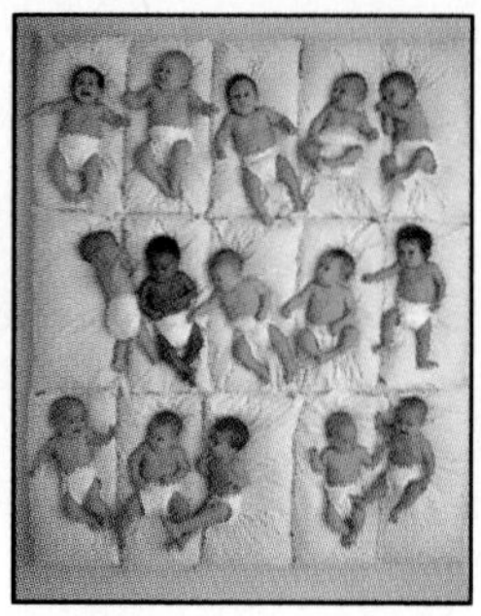

SUMMARY

A knowledge of anatomy and physiology can be used to predict the body's responses to stimuli when healthy or diseased.

ANATOMY

Anatomy is the study of the body's structures.

Systemic anatomy is the study of the body by systems.

Surface anatomy uses superficial structures to locate deeper structures, and anatomic imaging is a noninvasive method for examining deep structures.

PHYSIOLOGY

Physiology is the study of the body's processes and functions.

STRUCTURAL AND FUNCTIONAL ORGANIZATION

The human body can be organized into seven levels: chemical (atoms and molecules), organelle (small structures within cells), cell, tissue (groups of similar cells), organ (two or more tissues), organ system (groups of organs), and organism.

The 11 organ systems are the integumentary, skeletal, muscular, nervous, endocrine, cardiovascular, lymphatic, respiratory, digestive, urinary, and reproductive systems (see Table 1-1).

HOMEOSTASIS

Homeostasis is a state of equilibrium in which body functions, fluids, and other factors of the internal environment are maintained within a range of values suitable to support life.

Negative-feedback mechanisms operate to restore homeostasis.

Positive-feedback mechanisms increase deviations from normal.

Although a few positive-feedback mechanisms normally exist in the body, most positive-feedback mechanisms are harmful.

TERMINOLOGY AND THE BODY PLAN

Directional Terms

A human standing erect with the feet forward, arms hanging to the sides, and palms facing forward is in the anatomical position.

Directional terms always refer to the anatomical position regardless of the body's actual position (see Table 1-2).

Planes

A sagittal plane divides the body into left and right parts, a transverse plane divides the body into superior and inferior parts, and a frontal plane divides the body into anterior and posterior parts.

A longitudinal section divides an organ along its long axis, a transverse section cuts an organ at a right angle to the long axis, and an oblique section cuts across the long axis at an angle other than a right angle.

Body Regions

The body can be divided into appendicular (limbs and girdles) and axial (head, neck, and trunk) regions.

The abdomen can be divided superficially into four quadrants or nine regions that are useful for locating internal organs or describing the location of a pain.

Body Cavities

The thoracic cavity is bounded by the ribs and the diaphragm. The mediastinum divides the thoracic cavity into two parts.

The abdominal cavity is bounded by the diaphragm and the abdominal muscles.

The pelvic cavity is surrounded by the pelvic bones.

Serous Membranes

The trunk cavities are lined by serous membranes.

- The parietal portion of a serous membrane lines the wall of the cavity, and the visceral portion covers the internal organs.
- The serous membranes secrete fluid that fills the space between the parietal and visceral membranes. The serous membranes protect organs from friction.
- The pericardial membrane surrounds the heart, the pleural membranes surround the lungs, and the peritoneal membranes line the abdominopelvic cavity and cover its organs.

Mesenteries are parts of the peritoneum that hold the abdominal organs in place and provide a passageway for blood vessels and nerves to organs.

Retroperitoneal organs are found "behind" the parietal peritoneum. The kidneys, urinary bladder, and pancreas are examples of retroperitoneal organs.

CONTENT REVIEW

1. Define anatomy, surface anatomy, anatomic imaging, and physiology.
2. List seven structural levels at which the body can be studied conceptually.
3. Define tissue. What are the four primary tissue types?
4. Define organ and organ system. What are the 11 organ systems of the body and their functions?
5. What does the term homeostasis mean? If a deviation from homeostasis occurs, what kind of mechanism restores homeostasis?
6. Define positive feedback. Why are positive-feedback mechanisms generally harmful?
7. Why is knowledge of the etymology of anatomical and physiological terms useful?
8. What is the anatomical position? Why is it important to remember the anatomical position when using directional terms?
9. Define the following directional terms: inferior, superior, anterior, posterior, dorsal, ventral, proximal, distal, lateral, medial, superficial, and deep.
10. Define sagittal, midsagittal, transverse, and frontal planes.
11. Define longitudinal, transverse, and oblique sections of an organ.
12. List the subdivisions of the appendicular and axial regions.
13. Describe the four-quadrant method of subdividing the abdominal region. What is the purpose of this method?
14. Define the following cavities: thoracic, abdominal, pelvic, and abdominopelvic. What is the mediastinum?
15. What is the difference between the visceral and parietal layers of a serous membrane? What function do serous membranes perform?
16. Name the serous membranes associated with the lungs, heart, and abdominopelvic organs.
17. Define a mesentery. What does the term retroperitoneal mean?

CONCEPT REVIEW

1. The following observations were made on a patient who had suffered a bullet wound:

   Heart rate elevated and rising.
   Blood pressure very low and dropping.
   After bleeding was stopped and a blood transfusion was given, blood pressure increased.

   Which of the following statements are consistent with these observations?
   A. Negative-feedback mechanisms are occasionally inadequate without medical intervention.
   B. The transfusion interrupted a positive-feedback mechanism.
   C. The transfusion interrupted a negative-feedback mechanism.
   D. The transfusion was not necessary.
   E. A and B
2. During physical exercise, respiration rate increases. Two students are discussing the mechanisms involved: Student A claims they are positive feedback, and Student B claims they are negative feedback. Do you agree with Student A or Student B, and why?
3. Complete the following statements, using the correct directional terms for a human being.
   A. The navel is __________ to the nose.
   B. The heart is __________ to the breastbone (sternum).
   C. The ankle is __________ to the knee.
   D. The ear is __________ to the eye.
4. During pregnancy, which would increase more in size, the mother's abdominal or pelvic cavity? Explain.
5. A bullet enters the left side of a man, passes through the left lung, and lodges in the heart. Name in order the serous membranes through which the bullet passes.

ANSWERS TO  PREDICT QUESTIONS

1 *p. 8* When a negative-feedback mechanism fails to return a value to its normal level, the value will continue to deviate from its normal range. Homeostasis is not maintained in this situation and the person's health can be threatened.

2 *p. 10* The thirst sensation is associated with a decrease in body fluid levels. The thirst mechanism causes the person to drink fluids, which returns the fluid level to normal. Thirst therefore is a sensation involved in negative-feedback control of body fluids.

3 *p. 11* When a man is standing on his head, his nose is superior to his mouth. Remember that directional terms refer to a person in the anatomical position and do not refer to the body's current position.

4 *p. 16* There are two ways. First, the visceral peritoneum wraps around organs. Thus the peritoneal cavity surrounds the organ, but the organ is not inside the peritoneal cavity. The peritoneal cavity contains only peritoneal fluid. Second, retroperitoneal organs are in the abdominopelvic cavity, but they are between the wall of the abdominopelvic cavity and the peritoneal membrane.

2

THE CHEMISTRY OF LIFE

After reading this chapter you should be able to:

1 Define atom and element.
2 Name the subatomic particles of an atom and describe how they are organized.
3 Interpret a chemical formula.
4 Describe three types of chemical bonds.
5 Using symbols, explain synthesis, decomposition, and exchange reactions.
6 Distinguish between exergonic and endergonic reactions.
7 List the factors that affect the rate of chemical reactions.
8 Describe how enzymes work.
9 Explain how equilibrium is achieved by reversible reactions.
10 Describe the pH scale and its relationship to acidity and alkalinity.
11 Explain why buffers are important.
12 List the properties of water that make it important for living organisms.
13 Describe four important types of organic molecules and their functions.

acid Any substance that is a proton donor; any substance that releases hydrogen ions.

atom [Gr. *atomos*, indivisible, uncut] Smallest particle into which an element can be divided using conventional methods; composed of neutrons, protons, and electrons.

base Any substance that is a proton acceptor; any substance that binds to hydrogen ions.

buffer A chemical that resists changes in pH when either an acid or a base is added to a solution containing the buffer.

covalent bond Chemical bond that is formed when two atoms share a pair of electrons.

electron Negatively charged particle found in the orbitals of atoms.

endergonic reaction (en-der-gon′ik) [L. *endo-*, inside + Gr. *ergon*, work] A chemical reaction resulting in the absorption of energy.

enzyme (en′zīm) [Gr. *en*, in + *zyme*, leaven] A protein molecule that increases the rate of a chemical reaction without being permanently altered.

exergonic reaction (ek′ser-gon′ik) [L. *exo-*, outside + Gr. *ergon*, work] A chemical reaction resulting in the release of energy.

ion (i′on) Atom or group of atoms carrying an electrical charge due to loss or gain of one or more electrons.

ionic bond Chemical bond that is formed when one atom loses an electron and another atom accepts that electron.

molecule Two or more atoms of the same or different type joined by a chemical bond.

neutron [L. *neuter*, neither] Electrically neutral particle found in the nucleus of atoms.

proton [Gr. *protos*, first] Positively charged particle found in the nucleus of atoms.

Chemistry is the scientific discipline that deals with the composition and the structure of substances and with the many reactions they undergo. A basic knowledge of chemical principles is essential for understanding anatomy and physiology. For example, the physiological processes of digestion, muscle contraction, and generation of nerve impulses can be described in chemical terms.

BASIC CHEMISTRY

Matter is anything that occupies space. An **element** is matter composed of atoms of only one kind. **Atoms** are the smallest particles into which an element can be divided using conventional chemical means. For example, oxygen is an element composed of only oxygen atoms.

The Structure of Atoms

The three major types of subatomic particles that make up atoms are **neutrons, protons,** and **electrons.** Neutrons have no electrical charge, protons have a positive electrical charge, and electrons have a negative electrical charge. The number of protons in an atom is equal to the number of electrons. Consequently atoms are electrically neutral, with neither a positive nor negative charge.

Protons and neutrons are organized in atoms to form a central **nucleus,** and the electrons move around the nucleus (Figure 2-1). Although it is impossible to know precisely where any given electron is located at any particular moment, the region where each electron is most likely to be found is that electron's **orbital.** These three-dimensional orbitals are often represented as a series of concentric circles around the nucleus.

The number of protons in an atom is called its **atomic number.** Different elements have different numbers of protons, and therefore different atomic numbers. Elements commonly found in the human body are listed in Table 2-1. Note that the symbols used in the table can be used to refer to elements or to individual atoms.

Electrons and Chemical Bonds

An atom's chemical behavior largely is determined by the electrons in the outermost orbitals. **Chemical bonds** are formed when the outermost electrons are transferred or shared between atoms. The resulting combination of atoms is called a **molecule.** If a molecule has two or more different kinds of atoms, it may be referred to as a **compound.** A molecule can be represented by a **chemical formula,** which consists of the symbols of the atoms in the molecule plus a subscript denoting the number of each type of atom. For example, the chemical formula for glucose (a sugar) is $C_6H_{12}O_6$. Thus glucose has 6 carbon, 12 hydrogen, and 6 oxygen atoms.

Ionic bonds

An **ionic bond** results when one atom loses an electron and another atom accepts that electron. For example, sodium (Na) can lose an electron that can be accepted by chlorine (Cl). The molecule that is formed is sodium chloride (NaCl), or table salt (Figure 2-2).

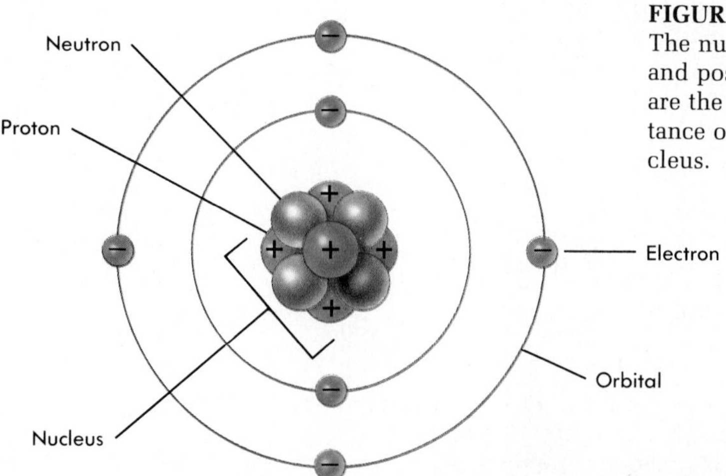

FIGURE 2-1 **Model of an atom.**
The nucleus contains neutrons, which have no charge, and positively charged protons. The concentric circles are the orbitals, which represent the approximate distance of the negatively charged electrons from the nucleus.

TABLE 2-1

Some Common Elements

ELEMENT	SYMBOL	ATOMIC NUMBER	AMOUNT IN HUMAN BODY BY WEIGHT (%)	AMOUNT IN HUMAN BODY BY NUMBER OF ATOMS (%)
Hydrogen	H	1	9.5	63.0
Carbon	C	6	18.5	9.5
Nitrogen	N	7	3.3	1.4
Oxygen	O	8	65.0	25.5
Sodium	Na	11	0.2	0.3
Phosphorus	P	15	1.0	0.22
Sulfur	S	16	0.3	0.05
Chlorine	Cl	17	0.2	0.03
Potassium	K	19	0.4	0.06
Calcium	Ca	20	1.5	0.31
Iron	Fe	26	Trace	Trace
Iodine	I	53	Trace	Trace

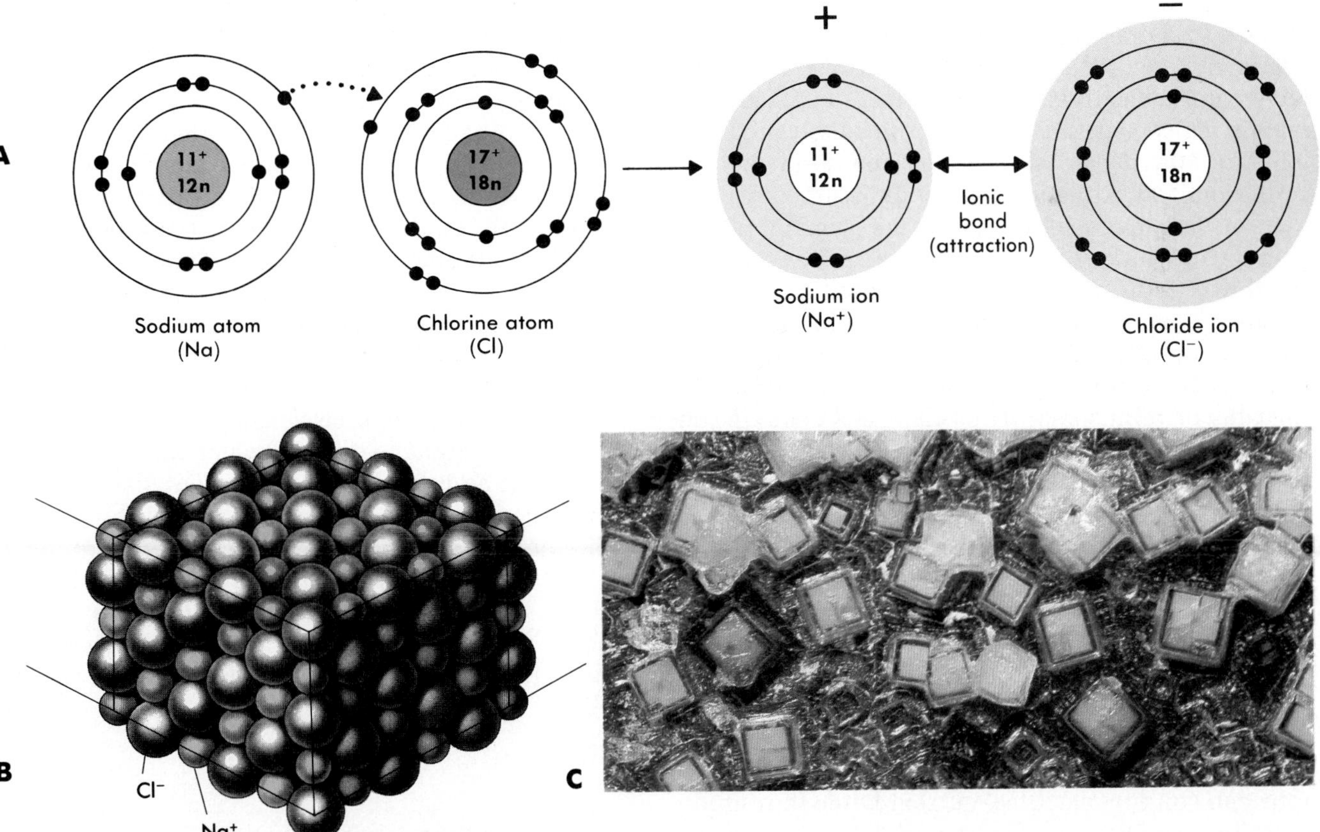

FIGURE 2-2 ■ Ionic bond.
A Sodium loses an electron, chlorine gains an electron, and they combine through an ionic bond to form sodium chloride (table salt).
B and C The sodium chloride is organized into crystals with a cube shape.

TABLE 2-2

Important Ions

COMMON IONS	SYMBOL	FUNCTION
Calcium	Ca^{2+}	Component of bones and teeth, necessary for blood clotting and muscle contraction
Sodium	Na^{+}	Helps maintain membrane potentials (electrical charge differences across a membrane) and water balance
Potassium	K^{+}	Helps maintain membrane potentials
Hydrogen	H^{+}	Helps maintain acid-base balance
Hydroxide	OH^{-}	Helps maintain acid-base balance
Chloride	Cl^{-}	Helps maintain acid-base balance
Bicarbonate	HCO_3^{-}	Helps maintain acid-base balance
Ammonium	NH_4^{+}	Helps maintain acid-base balance
Phosphate	PO_4^{3-}	Component of bone and teeth, involved in energy exchange and acid-base balance
Iron	Fe^{2+}	Necessary for red blood cell formation and function
Magnesium	Mg^{2+}	Necessary for enzymes

Atoms are electrically neutral because they have an equal number of protons and electrons. After an atom donates an electron, it has one more proton than it has electrons and is positively charged. After an atom accepts a donated electron, it has one more electron than it has protons and is negatively charged. These charged atoms are called **ions** (i'ons). Positively and negatively charged ions remain close together because oppositely charged ions are attracted to each other. The bond that results from this attraction is an ionic bond.

Ions are denoted by using the symbol of the atom from which the ion was formed. The charge of the ion is indicated by a superscripted plus ($^{+}$) or minus ($^{-}$) sign. For example, the sodium ion is Na^{+} and the chloride ion is Cl^{-}. If more than one electron has been lost or gained, a number is used with the plus or minus sign. Thus Ca^{2+} is a calcium ion formed by the loss of two electrons. Table 2-2 lists some of the important ions found in the human body.

1 If an iron (Fe) atom lost three electrons, what would be the charge of the resulting ion? Write the symbol for this ion. ?

Substances that produce ions when dissolved are sometimes referred to as **electrolytes** because ions can conduct electrical current when they are in solution. An electrocardiogram (ECG) is a recording of the electrical currents produced by the heart. These currents can be detected by electrodes on the surface of the body because the ions in the body's fluids conduct the electrical currents.

Covalent bonds

A **covalent bond** results when two atoms share a pair of electrons. Two hydrogen atoms, for example, can share their electrons to form a hydrogen molecule (Figure 2-3, *A*). A carbon atom can share four of its electrons with other atoms, forming four covalent bonds (Figure 2-3, *B*). Most commonly, only one pair of electrons is shared by two atoms, and this is referred to as a **single covalent bond.** Occasionally an atom shares two or three pairs of electrons with another atom to form a **double covalent bond** or a **triple covalent bond.** The large variety of molecules in the human body results from the covalent bonds formed between carbon atoms, and between carbon atoms and hydrogen, oxygen, and nitrogen atoms.

Hydrogen atoms also can share electrons with an oxygen atom to form a water molecule (Figure 2-3, *C*). The hydrogen atoms do not share the electrons equally with the oxygen atom, however, and the electrons tend to spend more time around the oxygen atom than the hydrogen atoms. This unequal sharing of electrons is called a **polar covalent bond,** because the unequal sharing of electrons results in one end (pole) of the molecule having a charge opposite to the other end.

2 In a water molecule the two hydrogen atoms form one end of the molecule and the oxygen atom forms the other end. Which end of the water molecule is negatively charged? ?

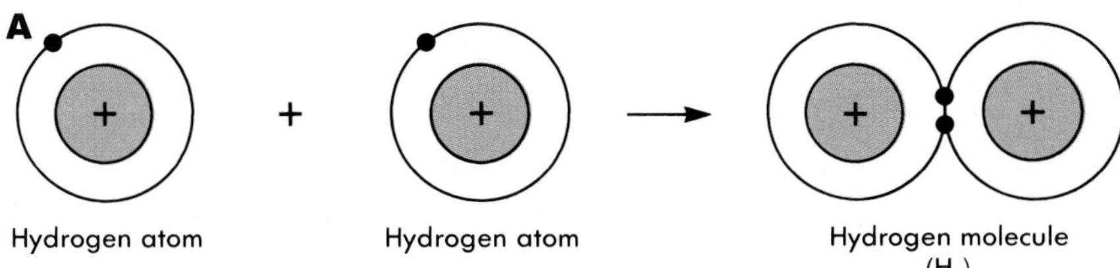

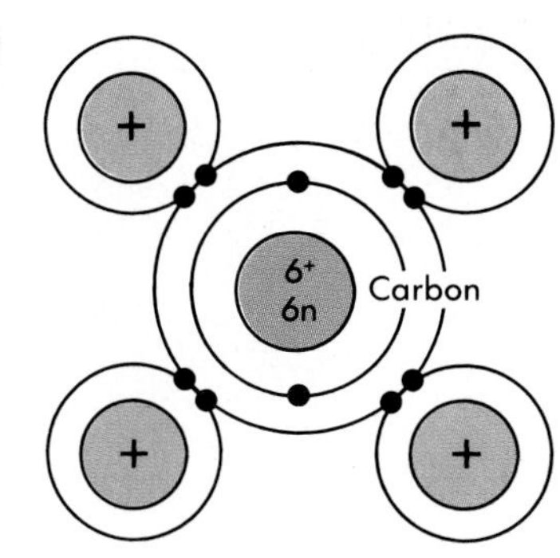

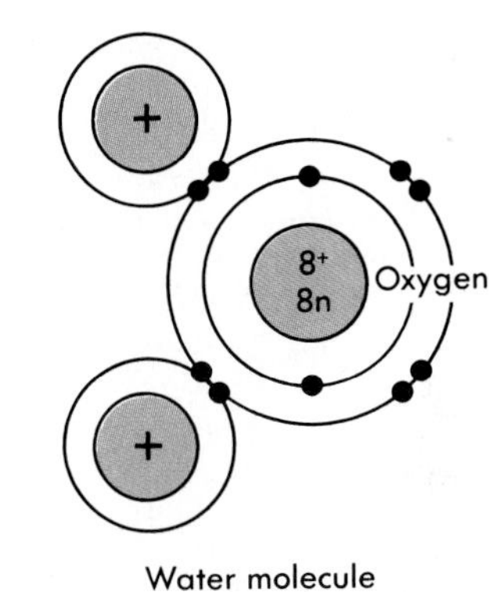

FIGURE 2-3 Covalent bond.
A Each hydrogen atom has a single electron. The hydrogen atoms form a covalent bond and become a hydrogen molecule when the two electrons are shared between the two hydrogen atoms.
B A methane molecule consists of four hydrogen atoms and a carbon atom. Each hydrogen atom shares its electron with the carbon atom and the carbon atom shares four electrons with the hydrogen atoms.
C A water molecule consists of two hydrogen atoms and an oxygen atom. Each hydrogen atom shares its electron with the oxygen atom, and the oxygen atom shares two of its electrons with the hydrogen atoms.

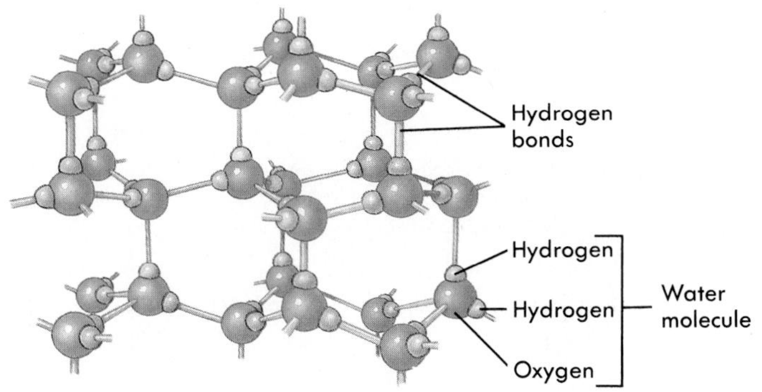

FIGURE 2-4 Hydrogen bonds.
The hydrogen and oxygen atoms of water form polar covalent bonds. The positively charged hydrogen end of one water molecule forms a hydrogen bond with the negatively charged oxygen end of another water molecule.

Hydrogen bonds

Molecules with polar covalent bonds are weakly attracted to ions or other polar covalent molecules. This weak attraction is called a **hydrogen bond.** For example, the positively charged end of a water molecule could be attracted to a negatively charged ion or to the negatively charged end of another polar covalent molecule (Figure 2-4). This weak attraction to the water molecules causes substances to separate, or **dissociate**, in water (see Figure 2-9). Hydrogen bonds also play an important role in determining the shape of complex molecules, because the hydrogen bonds between different polar parts of the molecule hold the molecule together.

CHEMICAL REACTIONS

A **chemical reaction** is the process by which atoms or molecules interact to form or break chemical bonds. The atoms or molecules present before the chemical reaction occurs are the **reactants** and those produced by the chemical reaction are the **products.** For example, the reactants sodium and chlorine combine to form the product sodium chloride.

When two or more atoms, ions, or molecules combine to form a new and larger molecule, the process is called a **synthesis reaction.** This can be represented symbolically as follows:

$$A + B \rightarrow AB$$

Clinical Applications of Atomic Particles

Protons, neutrons, and electrons are responsible for the chemical properties of atoms. They also are involved with other properties that can be useful in a clinical setting. Understanding these properties has led to the development of methods for examining the inside of the body.

Isotopes (i′so-tōpz) are two or more elements that have the same number of protons and electrons, but a different number of neutrons. For example, hydrogen has no neutrons, and its isotope deuterium has one. Water made with deuterium is called heavy water because of the weight of the "extra" neutron. Because isotopes have the same number of electrons, they are very similar in their chemical behavior. The nuclei of some isotopes are stable and do not change. Radioactive isotopes, however, have unstable nuclei that lose neutrons or protons. Several different kinds of radiation can be produced when neutrons and protons, or the products formed by their breakdown, are ejected from the nucleus of the isotope.

The radiation given off by some radioactive isotopes can penetrate and destroy tissues. Rapidly dividing cells are more sensitive to radiation than slowly dividing cells. Radiation is used to treat cancerous (malignant) tumors because cancer cells divide rapidly. If the treatment is effective, the cancerous cells are killed with tolerable destruction of healthy tissue. Radioactive isotopes also are used in diagnosis. The radiation can be detected, and the movement of the atoms throughout the body can be traced. For example, the thyroid gland normally transports iodine into the gland and uses the iodine in the formation of thyroid hormones. Radioactive iodine can be used to determine if iodine transport is normal.

Radiation can be produced in ways other than changing the nucleus of atoms. X-rays are radiation formed when electrons lose energy by moving from a higher energy orbital to a lower energy orbital. X-rays are used in examination of bones to determine if they are broken, and of teeth to see if they have caries (cavities). Mammograms are low-energy x-rays of the breast that can be used to detect tumors because the tumors are slightly more dense than normal tissue.

Computers are used to analyze a series of x-rays, each made at a slightly different body location. The picture of each x-ray "slice" through the body is assembled by the computer to form a three-dimensional image. A computerized axial tomography (CAT) scan is an example of this technique (Figure 2-A). CAT scans are used extensively to detect tumors and other abnormalities in the body.

Magnetic resonance imaging (MRI) is another method for looking into the body (Figure 2-B). The patient is placed in a very powerful magnetic field that aligns the hydrogen nuclei. Radiowaves given off by the hydrogen nuclei then are used by a computer to make an image of the body. Because MRI affects hydrogen, it is very effective for visualizing soft tissues that contain a lot of water. MRI technology is used to detect tumors and other abnormalities in the body.

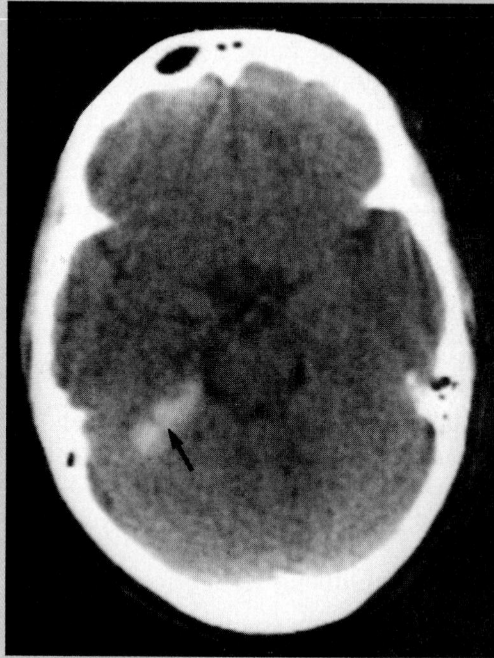

FIGURE 2-A ■ CAT Scan.
CAT scan of patient with a cerebral hemorrhage *(arrow)*.

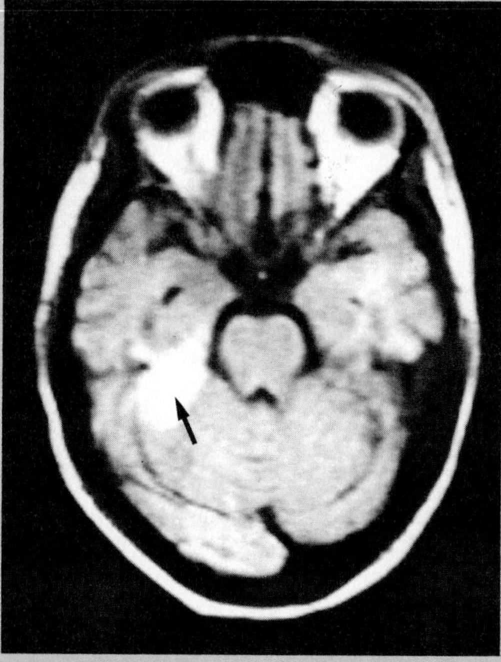

FIGURE 2-B ■ MRI.
MRI of the same patient with a cerebral hemorrhage *(arrow)*.

An example of a synthesis reaction in the body is the formation of carbon dioxide (CO_2) from from carbon (C) and oxygen (O_2).

$$C + O_2 \rightarrow CO_2$$

In a **decomposition reaction,** larger molecules are broken down into smaller molecules, ions, or atoms. A decomposition reaction is the reverse of a synthesis reaction, and can be represented in this way:

$$AB \rightarrow A + B$$

The breakdown of water (H_2O) to hydrogen (H_2) and oxygen (O_2) molecules.

$$2H_2O \rightarrow 2H_2 + O_2$$

An **exchange reaction** is partly a decomposition and partly a synthesis reaction. In the decomposition part of an exchange reaction, larger molecules are broken down. In the synthesis part of the reaction, the products of the decomposition reaction are formed into new molecules. The symbolic representation of an exchange reaction is:

$$AB + CD \rightarrow AC + BD$$

The reaction of hydrochloric acid (HCl) with sodium hydroxide (NaOH) to form table salt (NaCl) and water (H_2O) is an exchange reaction.

$$HCl + NaOH \rightarrow NaCl + H_2O$$

Chemical reactions are important because of the products they form and the energy changes they produce. Energy exists in chemical bonds as stored energy. As a result of **exergonic** (ek'ser-gon'ik) **reactions,** the products contain less stored energy than the reactants, and energy is released (Figure 2-5, A). Some of the energy is released as heat. The body temperature of humans is maintained by heat produced in this fashion. The rest of the energy is used to drive processes such as muscle contraction or to produce new molecules.

3 Why does body temperature increase during exercise? ?

An example of an exergonic reaction is the breakdown of the molecule adenosine triphosphate (ATP) to adenosine diphosphate (ADP) and a phosphate group (P).

$$ATP \rightarrow ADP + P + Energy$$

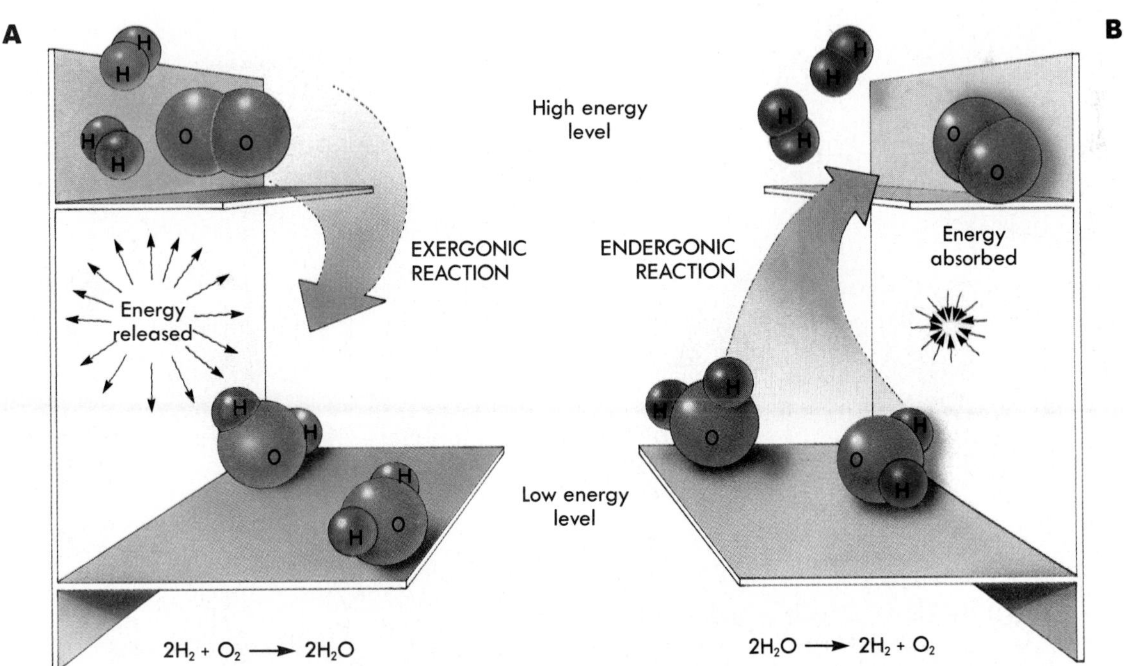

FIGURE 2-5 Energy and chemical reactions.
A Exergonic reaction in which energy is released.
B Endergonic reaction in which energy is required for the reaction to proceed.
In each figure the upper shelf represents a higher energy state and the lower shelf represents a lower energy state.

The energy that makes almost all life on earth possible ultimately comes from the sun. Plants, in the process of photosynthesis, capture the energy in sunlight. Through endergonic reactions, plants convert the sun's energy into chemical bonds in glucose. Plants, and organisms that eat plants, break down glucose through exergonic reactions. The energy released fuels the chemical reactions of life.

The ATP molecules store energy that is released when ATP is broken down. Because ATP molecules can store and provide energy, they are called the energy currency of the cell.

In **endergonic** (en-der'gon'ik) **reactions** the products contain more energy than the reactants (Figure 2-5, *B*). Therefore these reactions require the input of energy from another source. The energy released by exergonic reactions during the breakdown of food molecules is the energy source for endergonic reactions in the body. The synthesis of the molecules found in living organisms (for example, fats and proteins) is the result of endergonic reactions.

Rate of Chemical Reactions

The rate at which a chemical reaction proceeds is influenced by the nature of the reacting substances, their concentration, the temperature, and enzymes.

Substances differ in their ability to react with other substances. Iron, for example, reacts slowly with oxygen to form rust. The components of dyna-

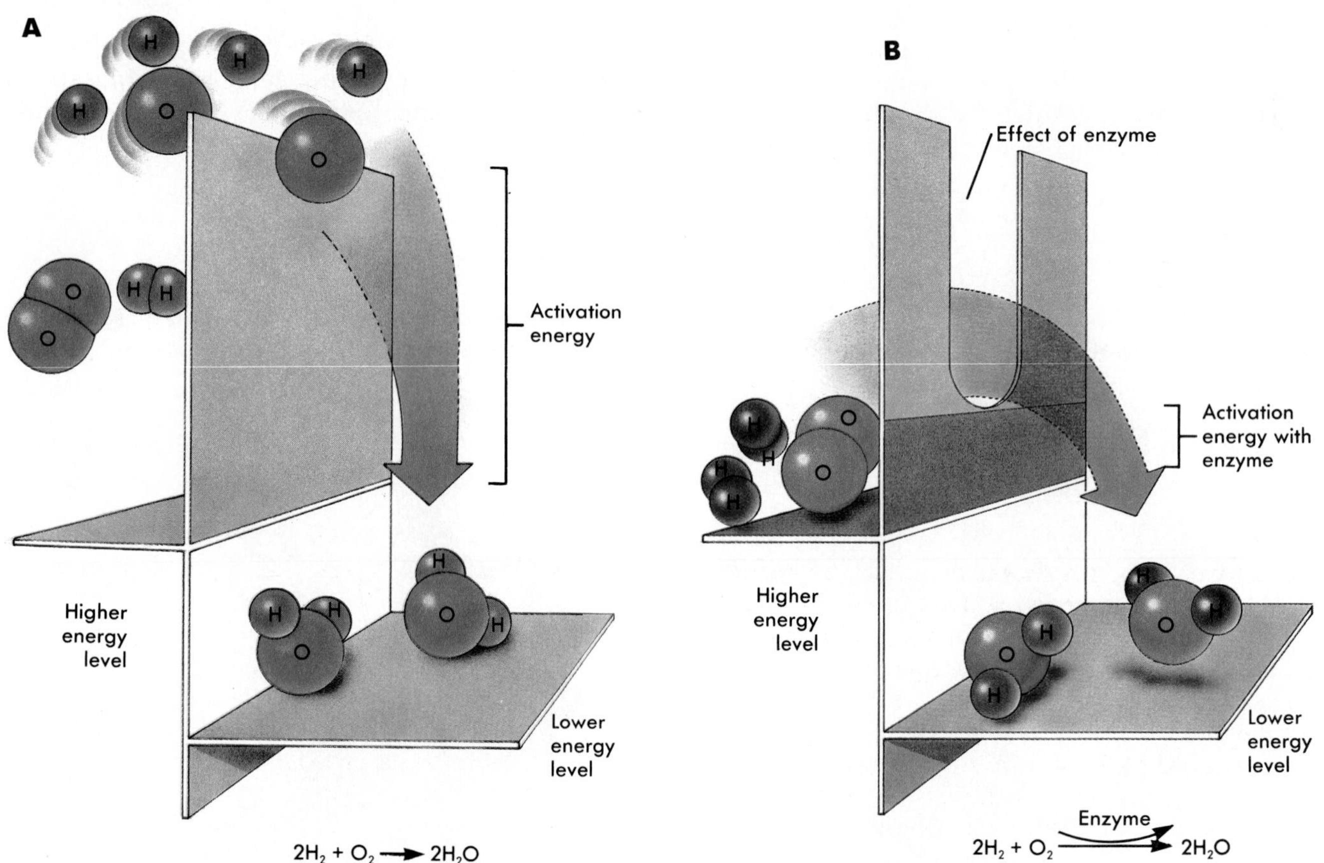

FIGURE 2-6 Activation energy and enzymes.
A Activation energy is needed to change hydrogen and oxygen to water. The upper shelf represents a higher energy state and the lower shelf represents a lower energy state. The "wall" extending above the upper shelf represents the activation energy. Even though energy is given up moving from the upper to the lower shelf, the activation energy "wall" must be overcome before the reaction can proceed.
B The enzyme lowers the activation energy, making it easier for the reaction to proceed.

mite, on the other hand, react violently with each other in a fraction of a second.

Within limits, the greater the concentration of reactants, the greater the rate at which a chemical reaction will occur. This occurs because as the concentration increases, the reacting molecules are more likely to come into contact with one another. For example, the normal concentration of oxygen inside cells enables oxygen to come into contact with other molecules, producing the chemical reactions necessary for life. If the oxygen concentration decreases, then the rate of chemical reactions decreases. This can impair cell function and even result in cell death.

The speed of chemical reactions also increases when the temperature is increased. When a person has a fever of only a few degrees, reactions occur throughout the body at a faster rate. This results in increased activity in the organ systems such as increased heart and respiratory rates. When body temperature drops, reactions slow. The sluggish movement of very cold fingers results largely from the reduced rate of chemical reactions in cold muscle tissue.

At normal body temperatures, most chemical reactions would take place to slowly to sustain life if it were not for the body's enzymes. **Enzymes** (en′zīmz) are protein molecules that act as catalysts. A **catalyst** is a substance that increases the rate at which a chemical reaction proceeds. The chemical reaction does not permanently change the catalyst or decrease the amount of the catalyst. Catalysts increase the rate of chemical reactions by lowering the **activation energy,** which is the energy necessary to start a chemical reaction. For example, heat in the form of a spark is required to start the reaction between oxygen and gasoline. Most of the chemical reactions that occur in the body have a high activation energy that is decreased by enzymes (Figure 2-6). The lowered activation energy enables reactions to proceed at a rate that sustains life.

The three-dimensional shape of enzymes is critical for their normal function. According to the **lock and key model** of enzyme action, the shapes of the enzyme and the reactants allow the enzyme to bind easily to the reactants. Bringing the reactants very close to each other reduces the activation energy for the reaction. Because the enzyme and the reactants must fit together, enzymes are very specific for the reactions they control, and each enzyme controls only one chemical reaction. After the reaction takes place, the enzyme is released and can be used again (Figure 2-7).

The chemical events of the body are regulated primarily by mechanisms that control either the concentration or activity of enzymes. The rate at which enzymes are produced in cells, or whether the enzymes are in an active or inactive form, determines the rate of each chemical reaction.

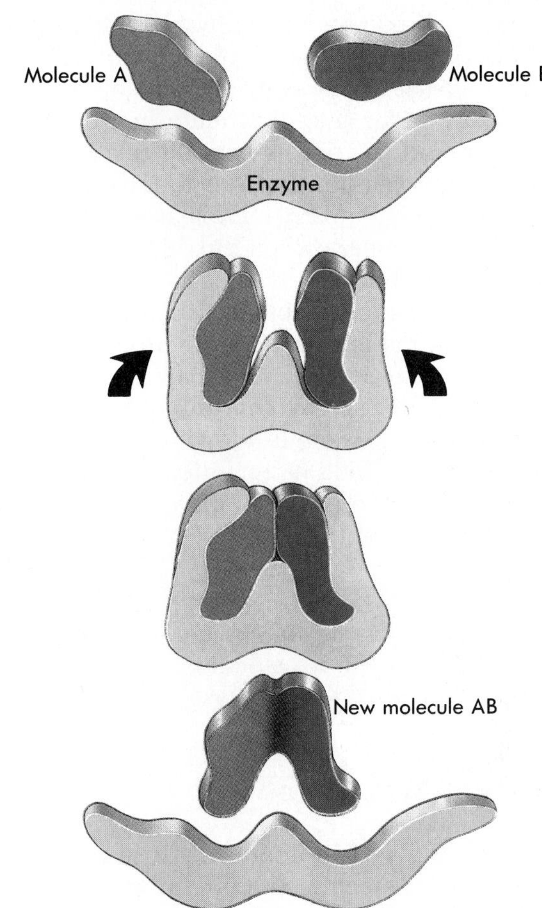

FIGURE 2-7 **Enzyme action.**
The enzyme brings the two reacting molecules together. This is possible because the reacting molecules "fit" the shape of the enzyme (lock and key model). After the reaction the unaltered enzyme can be used again.

Reversible Reactions

In a **reversible reaction** the reaction can proceed from reactants to products and from products to reactants. When the rate of product formation is equal to the rate of reactant formation, the reaction is said to be at equilibrium. At **equilibrium** the amount of the reactants and products relative to each other tends to remain constant. For example, the reaction between carbon dioxide (CO_2) and water (H_2O) to form hydrogen ions (H^+) and bicarbonate ions (HCO_3^-) is reversible (indicated by a double-ended arrow):

$$CO_2 + H_2O \rightleftarrows H^+ + HCO_3^-$$

If carbon dioxide is added to water, the amount of carbon dioxide relative to the amount of hydrogen ions increases. However, the reaction of carbon dioxide with water produces more hydrogen ions, and the amount of carbon dioxide relative to the amount of hydrogen ions returns to the equilibrium value. Conversely, adding hydrogen ions results in the formation of more carbon dioxide, and the equilibrium is maintained.

Maintaining a constant level of hydrogen ions in the body fluids is necessary for proper functioning of the nervous system. This maintenance can be accomplished, in part, by controlling blood carbon dioxide levels. For example, slowing the respiration rate would cause blood carbon dioxide levels to increase, which would cause an increase in hydrogen ion concentration in the blood.

4 If the respiration rate increases, carbon dioxide is removed from the blood. What effect would this have on blood hydrogen ion levels?

ACIDS AND BASES

An **acid** is a proton donor. Because a hydrogen atom without its electron is a proton, any substance that releases hydrogen ions in water is an acid. For example, hydrochloric acid (HCl) in the stomach forms hydrogen (H^+) ions and chloride (Cl^-) ions.

$$HCl \rightarrow H^+ + Cl^-$$

A **base** is a proton acceptor. For example, sodium hydroxide (NaOH) forms sodium (Na^+) ions and hydroxide (OH^-) ions. It is a base because the hydroxide ion is a proton acceptor that binds with a hydrogen ion to form water.

$$NaOH \rightarrow Na^+ + OH^-$$

The pH Scale

The **pH scale** (Figure 2-8), which ranges from 0 to 14, indicates the hydrogen ion concentration of a solution. Pure water is defined as a neutral solution. A **neutral solution** has an equal number of hydrogen ions and hydroxide ions and has a pH of 7. Solutions with a pH less than 7 are **acidic,** and they have a greater concentration of hydrogen ions than hydroxide ions. **Alkaline** or **basic solutions** have a pH greater than 7, and they have fewer hydrogen ions than hydroxide ions. As the pH value becomes smaller, the solution is more acidic, and as the pH value becomes larger, the solution is more basic.

The symbol *pH* stands for the power (p) of hydrogen ion (H^+) concentration. The power is a factor of 10, which means that a change in the pH of a

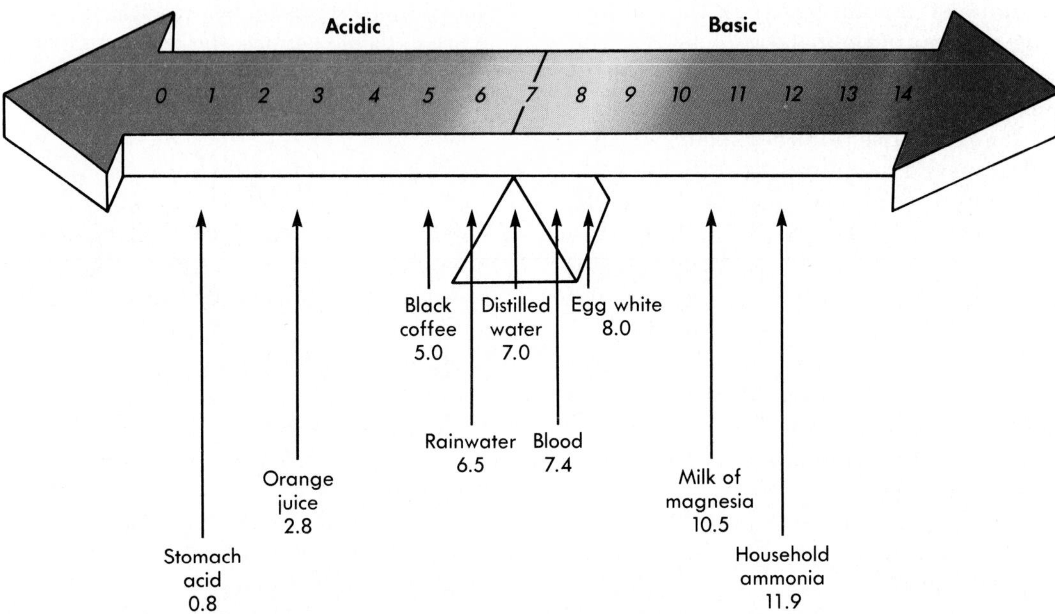

FIGURE 2-8 The pH scale.
A pH of 7 is considered to be neutral, so the scale is depicted as balancing at that point. Values to the left (below 7) are acidic. The lower the value, the more acidic the solution. Values to the right (above 7) are basic or alkaline. The higher the value, the more basic the solution. Representative fluids and their approximate pH values are listed below the figure.

solution by one pH unit represents a tenfold change in the hydrogen ion concentration. For example, a solution of pH 6 has 10 times the number of hydrogen ions as a solution with a pH of 7. Thus small changes in pH represent large changes in hydrogen ion concentration.

The normal pH range for human blood is 7.35 to 7.45. The condition of acidosis results if blood pH drops below 7.35. The nervous system becomes depressed and the individual becomes disoriented and possibly comatose. Alkalosis results if blood pH rises above 7.45. The nervous system becomes overexcitable, and the individual may be extremely nervous or have convulsions. Both acidosis and alkalosis can result in death.

Salts

A **salt** is a molecule consisting of a positive ion other than hydrogen and a negative ion other than hydroxide. Salts are formed by the reaction of an acid and a base. For example, hydrochloric acid combines with sodium hydroxide to form the salt sodium chloride.

$$HCl + NaOH \rightarrow NaCl + H_2O$$

Salts dissociate to form positively and negatively charged ions when dissolved in water (Figure 2-9).

Buffers

The chemical behavior of many molecules changes as the pH of the solution in which they are dissolved changes. An organism's survival depends on its ability to regulate body fluid pH within a narrow range. One way normal body fluid pH is maintained is through the use of buffers. A **buffer** is a chemical that resists changes in pH when either an acid or a base is added to a solution containing the buffer. When an acid is added to a buffered solution, the buffer either removes hydrogen ions or adds a base to the solution. This prevents a decrease in the pH of the solution.

5 If a base is added to a solution, would the pH of the solution increase or decrease? If the solution were buffered, what responses from the buffer would prevent the change in pH? ?

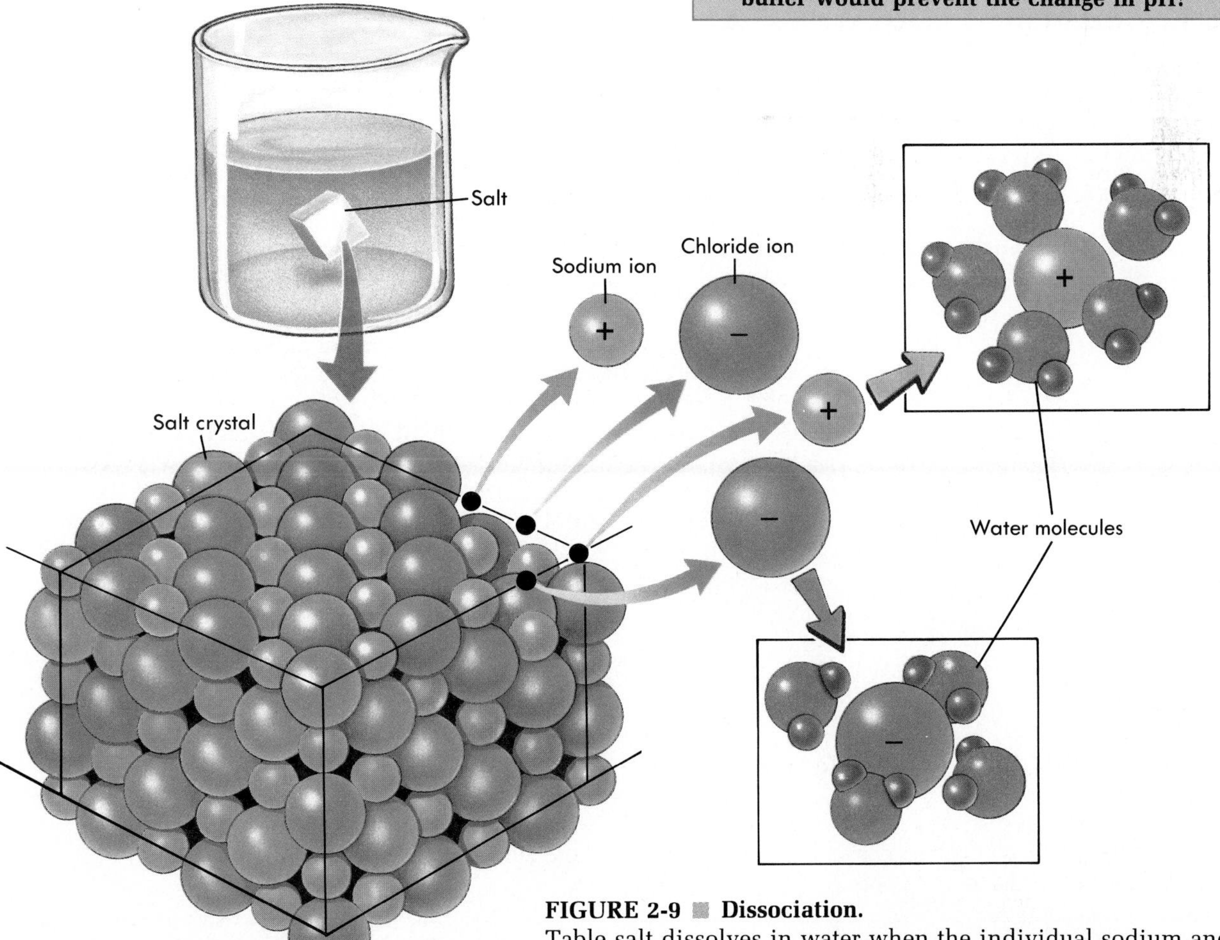

FIGURE 2-9 Dissociation.
Table salt dissolves in water when the individual sodium and chloride ions of salt each are surrounded by water molecules.

WATER

A molecule of water consists of one atom of oxygen joined by covalent bonds to two atoms of hydrogen. Water has many useful properties for living organisms.

1. Water can absorb large amounts of heat and remain a liquid. Blood, which is mostly water, can transfer heat effectively from deep within the body to the body's surface. Blood is warmed deep in the body and then flows to the surface where the heat is released. In addition, water evaporation in the form of sweat carries large amounts of heat away from the body.
2. Water is an effective lubricant. For example, tears protect the eye's surface from the rubbing of the eyelids.
3. Water is necessary in many chemical reactions. For example, it is used during the digestion of food.
4. Many different substances dissolve in water. Blood transports nutrients, gases, and waste products within the body. When ionic substances dissolve in water, the positive and negative ions separate or dissociate (see Figure 2-9). The polar water molecules surround the positive and negative ions, keeping them in solution. When the ions are in solution they can react with other molecules, so this property of water makes possible many of the body's chemical reactions.

ORGANIC MOLECULES

Organic molecules are those that contain carbon. **Inorganic molecules** are all other molecules. An exception is carbon dioxide, which is traditionally considered an inorganic molecule. Important large organic molecules in humans are carbohydrates, lipids, proteins, and nucleic acids (Table 2-3).

TABLE 2-3

Important Organic Molecules and Their Functions

MOLECULE	ELEMENTS	BUILDING BLOCKS	FUNCTION	EXAMPLES
Carbohydrate	C, H, O	Monosaccharides	Energy	Monosaccharides can be used as energy sources. Glycogen (polysaccharide) is an energy storage molecule.
Lipid	C, H, O (P, N in some)	Glycerol and fatty acids (for fats)	Energy	Fats can be stored and broken down later for energy; per unit of weight fats yield twice as much energy as carbohydrates.
			Structure	Phospholipids and cholesterol are important components of cell membranes.
			Regulation	Steroid hormones regulate many physiological processes, e.g., estrogen and testosterone are responsible for many of the differences between males and females.
Protein	C, H, O, N (S in most)	Amino acids	Regulation	Enzymes control the rate of chemical reactions. Hormones regulate many physiological processes, e.g., thyroid hormone affects metabolic rate.
			Structure	Collagen fibers form a structural framework in many parts of the body.
			Energy	Proteins can be broken down for energy; per unit of weight they yield the same energy as carbohydrates.
			Contraction	Actin and myosin in muscle are responsible for muscle contraction.
			Transport	Hemoglobin transports oxygen in the blood.
Nucleic acid	C, H, O, N, P	Nucleotides	Regulation	DNA directs the activities of the cell.
			Heredity	Genes are pieces of DNA that can be passed from one generation to the next generation.
			Protein synthesis	RNA is involved in protein synthesis.

Carbohydrates

Carbohydrates are small to very large molecules that are composed of carbon, hydrogen, and oxygen atoms. The amount of hydrogen relative to oxygen is the same in most carbohydrates as it is in water: two hydrogen atoms for each oxygen atom. The number of oxygen atoms in most carbohydrates is equal to the number of carbon atoms. For example, the chemical formula for glucose is:

$$C_6H_{12}O_6$$

The smallest carbohydrates are **monosaccharides** (Gr., one sugar) or **simple sugars.** Glucose (blood sugar) and fructose (fruit sugar) are important carbohydrate energy sources for many of the body's cells (Figure 2-10, *A*). Larger carbohydrates are formed by chemically binding monosaccharides. For this reason, monosaccharides are considered the building blocks of carbohydrates. **Disaccharides** (Gr., two sugars) are formed when two monosaccharides join. For example, glucose and fructose combine to form sucrose (table sugar) (see Figure 2-10, *A*). **Polysaccharides** (Gr., many sugars) consist of many monosaccharides bound in long chains. Glycogen, or animal starch, is a polysaccharide of glucose (Figure 2-10, *B*). When cells containing glycogen need energy, the glycogen is broken

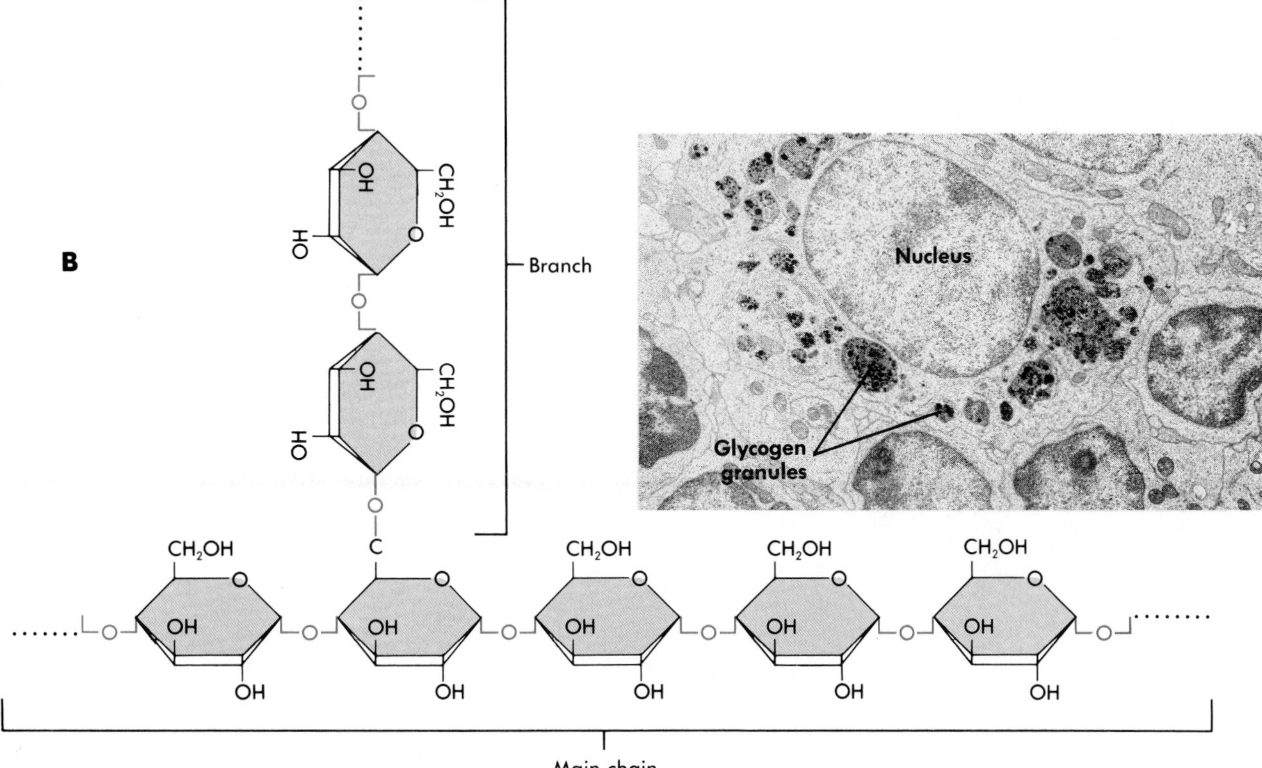

FIGURE 2-10 Carbohydrates.
A Glucose and fructose are monosaccharides that combine to form the disaccharide sucrose.
B Glycogen is a polysaccharide formed by combining many glucose molecules. The photograph shows glycogen granules in a liver cell.

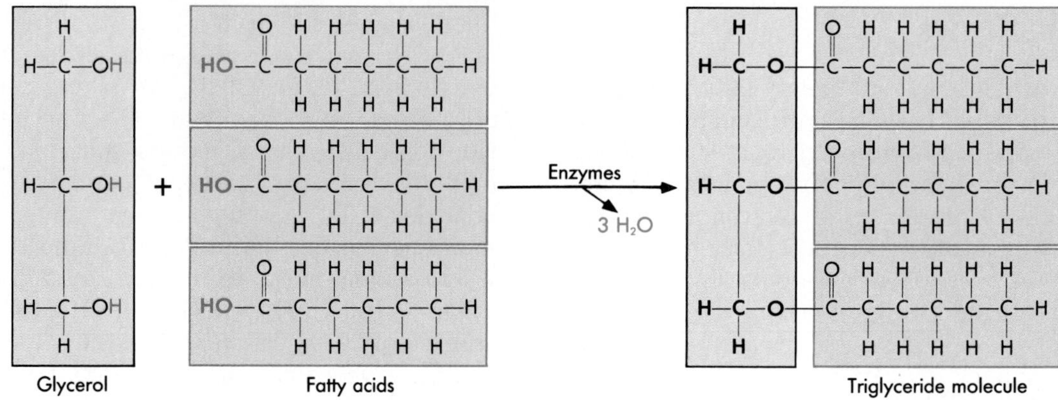

FIGURE 2-11 Fats.
Glycerol and fatty acids combine to form a triglyceride, the most common type of fat.

down into individual glucose molecules that can be used as energy sources. Plant starch, also a polysaccharide of glucose, can be ingested and broken down into glucose.

Lipids

Lipids are defined as substances that dissolve in nonpolar solvents such as alcohol or acetone, but not in water (a polar solvent; see p. 26). Lipids are composed mainly of carbon, hydrogen, and oxygen, but other elements such as phosphorus and nitrogen are minor components of some lipids. Lipids contain a lower proportion of oxygen to carbon than do carbohydrates.

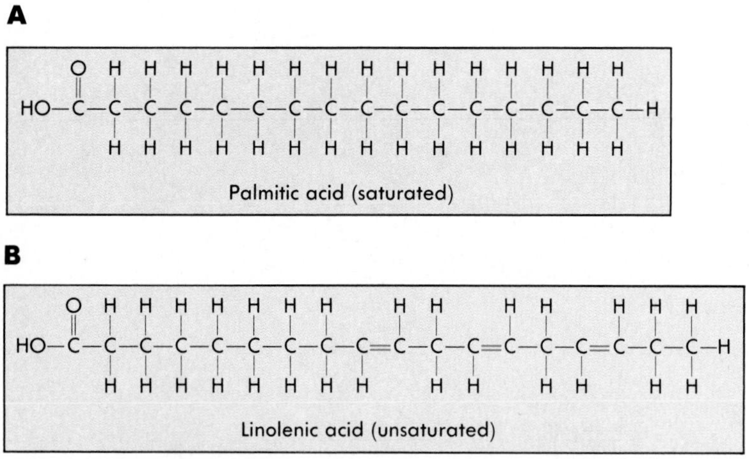

A

Palmitic acid (saturated)

B

Linolenic acid (unsaturated)

FIGURE 2-12 Saturated and unsaturated fats.
A Palmitic acid is saturated, having only single covalent bonds between carbon atoms (indicated by a single line).
B Linolenic acid is unsaturated, having some double covalent bonds between carbon atoms (indicated by double lines).

Fats, phospholipids, and steroids are examples of lipids. **Fats** are important energy storage molecules that also give the body padding and insulation. The building blocks of fats are glycerol (glis′er-ol) and fatty acids (Figure 2-11). **Glycerol** is a three-carbon molecule with a hydroxyl group (—OH) attached to each carbon atom, and **fatty acids** consist of a carbon chain with a carboxyl group attached at one end. A carboxyl group consists of both an oxygen atom and a hydroxyl group attached to a carbon atom (—COOH). The carboxyl group is responsible for the acidic nature of the molecule because it releases hydrogen ions into solution. **Triglycerides** are the most common type of fat molecule, and they have three fatty acids bound to a glycerol molecule.

Fatty acids differ from each other according to the length and the degree of saturation of their carbon chains. Most naturally occurring fatty acids contain 14 to 18 carbon atoms. A fatty acid is **saturated** if it contains only single covalent bonds between the carbon atoms. The carbon chain is **unsaturated** if it has one or more double covalent bonds (Figure 2-12). Unsaturated fats are believed to be the best type of fats in the diet, because saturated fats may contribute more to the development of atherosclerosis, a disease of blood vessels.

Proteins

All **proteins** contain carbon, hydrogen, oxygen, and nitrogen, and most have some sulfur. The building blocks of proteins are **amino acids,** which are organic acids containing an amino group ($—NH_2$) and a carboxyl group (Figure 2-13, *A*). There are 20 basic types of amino acids. Humans can synthesize 12 of these from simple organic molecules, but the remaining eight "essential amino acids" must be ob-

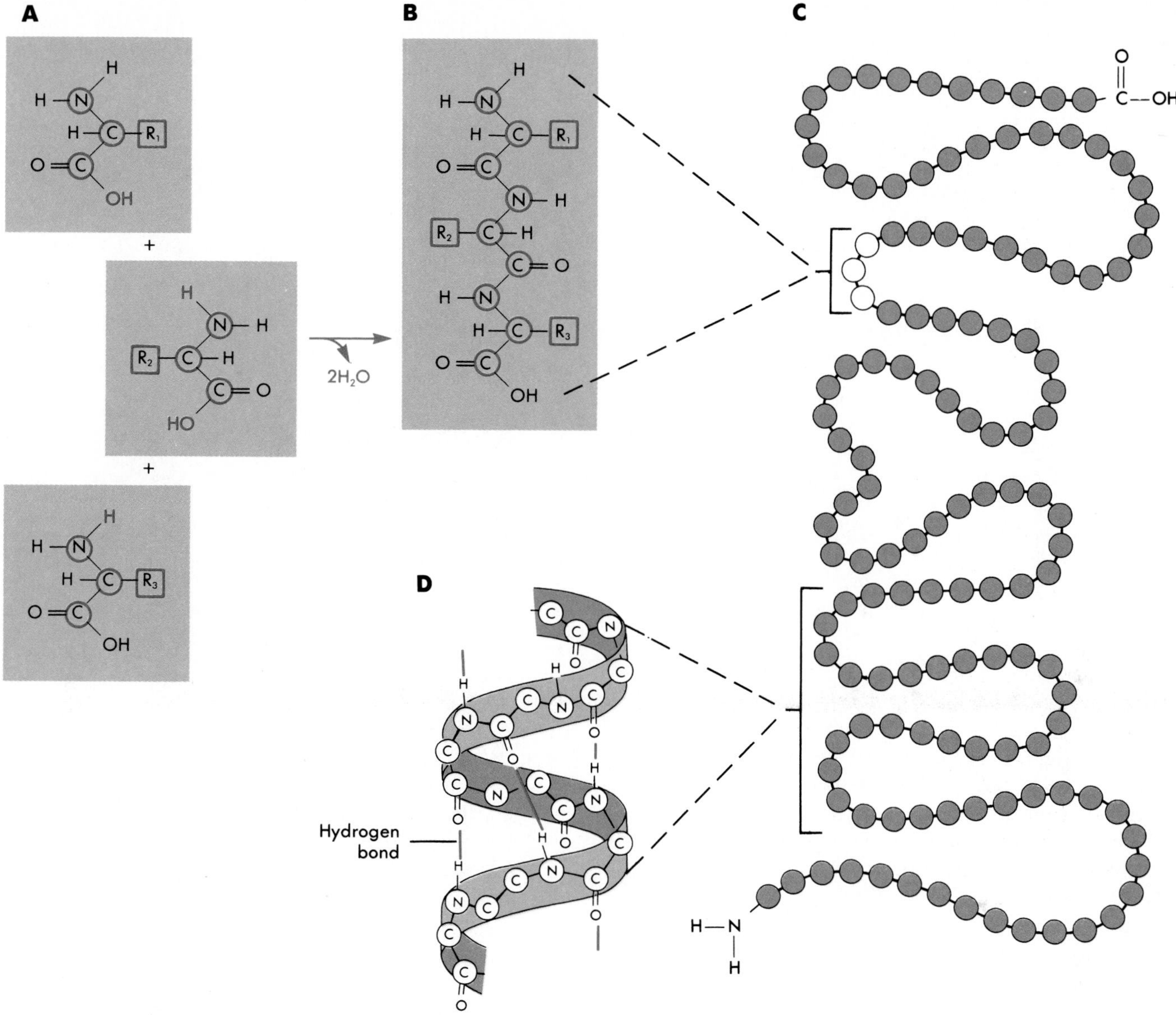

FIGURE 2-13 **Proteins.**

A Individual amino acids with amino groups (—NH_2) and carboxyl groups (—COOH).
B The individual amino acids are joined.
C A protein consisting of a chain of amino acids (represented by the circles).
D A three-dimensional representation of the amino acid chain. Hydrogen bonds between different amino acids maintain the shape of the protein. Denaturation breaks the hydrogen bonds, resulting in a change of shape that makes the protein nonfunctional.

tained in the diet. Although there are only 20 amino acids, they can combine in many different ways to form proteins with unique structural and functional features (Figure 2-13, *B* and *C*).

Proteins perform many important functions. For example, enzymes are proteins that regulate the rate of chemical reactions, structural proteins provide the framework for many of the body's tissues, and muscles contain proteins that are responsible for muscle contraction. The ability of proteins to perform their functions depends on their shape. If the hydrogen bonds that maintain the protein's shape are broken, the protein becomes nonfunctional (Figure 2-13, *D*). This change in shape is called **denaturation,** and it can be caused by abnormally high temperatures or changes in pH of the body's fluids.

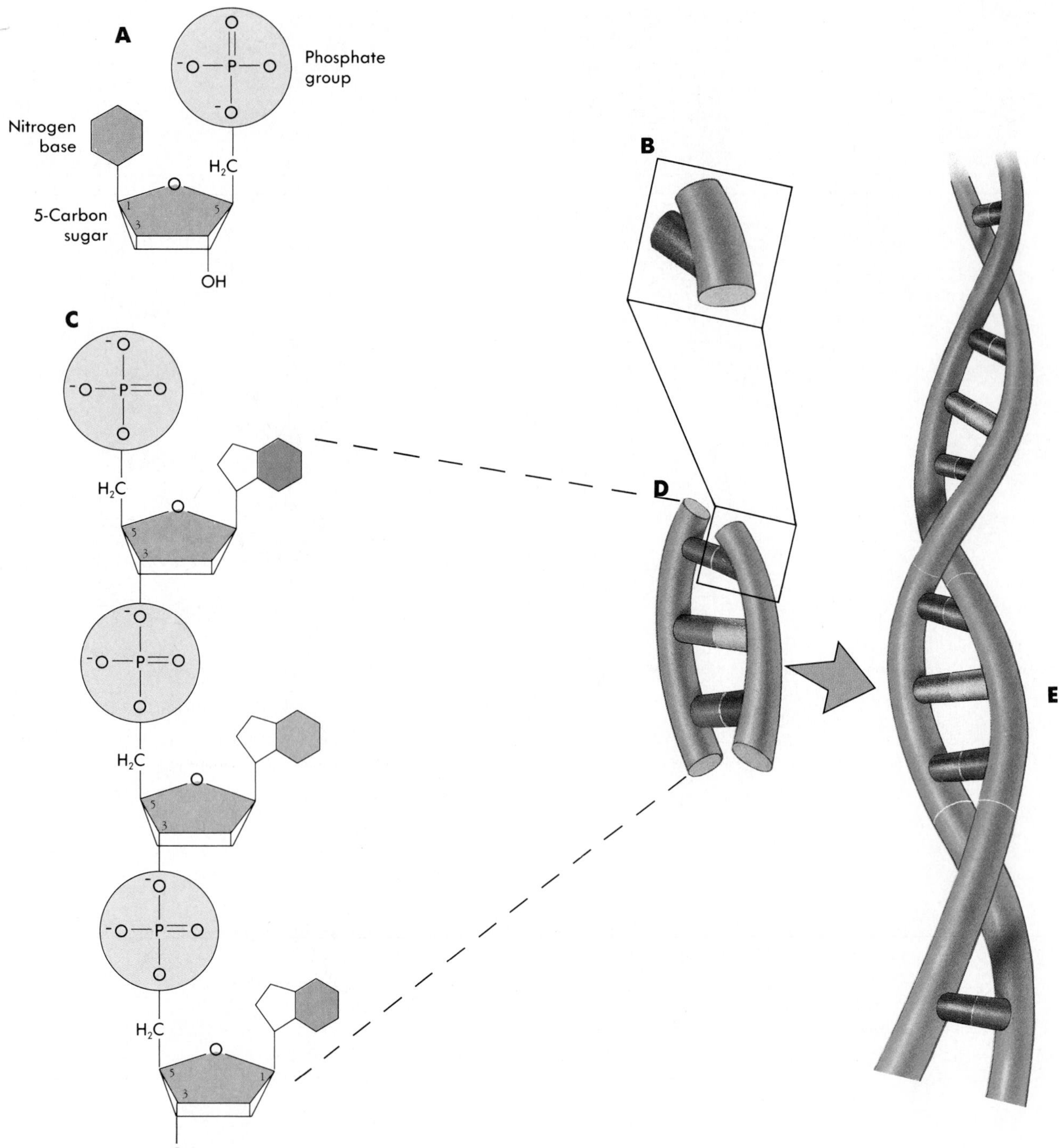

FIGURE 2-14 Nucleic acids.
A A nucleotide, the basic building block of nucleic acids, consists of a five-carbon sugar with an attached nitrogen base and phosphate group.
B A pictorial representation of a single nucleotide.
C Nucleotides join to form a chain.
D The nitrogen bases of two chains of nucleotides are joined together by hydrogen bonds.
E Deoxyribonucleic acid (DNA) is composed of two parallel strands of nucleotides that coil together to form a helix.

Nucleic Acids

Nucleic acids are large molecules composed of carbon, hydrogen, oxygen, nitrogen, and phosphorus. The building blocks of nucleic acids are nucleotides, which are organic molecules containing a five-carbon sugar, a nitrogen base, and a phosphate group (Figure 2-14). In **deoxyribonucleic** (de-oks'ĭ-ri'bo-nu-kle'ik) **acid (DNA)** the nucleotides contain the sugar deoxyribose. The nucleotides form two chains that coil around each other to form a double helix. DNA is the genetic material of the cell, and it is responsible for controlling cell activities. **Ribonucleic** (ri'bo-nu-kle'ik) **acid (RNA)** is a single strand of nucleotides that contains the sugar ribose. Three different types of RNA are involved in protein synthesis (see Chapter 3).

SUMMARY

Chemistry is the study of the composition and structure of substances and the reactions they undergo.

BASIC CHEMISTRY

The atom is the smallest unit of matter that is unable to be altered by chemical means.

An element is matter composed of one kind of atom.

The Structure of Atoms

Atoms consist of neutrons, positively charged protons, and negatively charged electrons.

Electrons are found in orbitals around the nucleus. The nucleus consists of protons and neutrons.

The atomic number is the number of protons in an element.

Electrons and Chemical Bonds

A molecule is two or more atoms joined by chemical bonds. A compound is a molecule with two or more different kinds of atoms.

An ionic bond results when an electron is transferred from one atom to another atom.

A covalent bond results when a pair of electrons are shared between atoms. A polar covalent bond is an unequal sharing of electron pairs.

A hydrogen bond is the weak attraction between oppositely charged ends of two polar covalent molecules.

CHEMICAL REACTIONS

Types of chemical reactions.

- A synthesis reaction is the combination of atoms, ions, or molecules to form a new, larger molecule.
- A decomposition reaction is the breakdown of larger molecules into smaller molecules, ions, or atoms.
- An exchange reaction is a decomposition reaction, in which molecules are broken down, and a synthesis reaction, in which the products of the decomposition reaction combine to form new molecules.

Energy is released in exergonic reactions; energy is absorbed in endergonic reactions.

Rate of Chemical Reactions

The rate of chemical reactions increases when the concentration of the reactants increases, temperature increases, or an enzyme is present.

Enzymes speed up chemical reactions without being altered permanently. Enzymes are specific, bind to reactants according to the lock and key model, and function by lowering activation energy.

Reversible Reactions

- In a reversible reaction, the reactants can form products, or the products can form reactants.
- The amount of reactants relative to products remains constant at equilibrium.

ACIDS AND BASES

- Acids are proton (hydrogen ion) donors and bases are proton acceptors.

The pH Scale

- A neutral solution has an equal number of hydrogen ions and hydroxide ions and has a pH of 7.
- An acidic solution has more hydrogen ions than hydroxide ions and has a pH of less than 7.
- A basic solution has less hydrogen ions than hydroxide ions and has a pH greater than 7.

Salts

- A salt is formed when an acid reacts with a base.

Buffers

- Chemicals that resist changes in pH are buffers.

WATER

- Water effectively transfers heat within the body, acts as a lubricant, is necessary in many chemical reactions, and dissolves many substances.
- Substances held together by ionic bonds dissociate in water to form ions.

ORGANIC MOLECULES

- Organic molecules contain carbon, inorganic molecules do not.
- Carbohydrates give the body energy. Monosaccharides are the building blocks that form more complex carbohydrates such as disaccharides and polysaccharides.
- Lipids provide energy (fats), are structural components (phospholipids), and regulate physiological processes (steroids). The building blocks of fats are glycerol and fatty acids.
- Proteins regulate chemical reactions (enzymes), are structural components, and cause muscle contraction. The building blocks of proteins are amino acids. Denaturation of proteins changes their shape and makes them nonfunctional.
- Nucleic acids include DNA, the genetic material, and RNA, which is involved in protein synthesis. The building blocks of nucleic acids are nucleotides, which consist of a sugar (deoxyribose or ribose), a nitrogen base, and a phosphate group.

CONTENT REVIEW

1. Define chemistry. Why is an understanding of chemistry important?
2. Diagram the structure of an atom and label the parts. Compare the charges of the subatomic particles.
3. What is a chemical formula?
4. Distinguish between ionic, covalent, polar covalent, and hydrogen bonds. Define an ion.
5. Define a chemical reaction. Describe synthesis, decomposition, and exchange reactions, giving an example of each.
6. Contrast exergonic and endergonic reactions.
7. Name three ways that the rate of chemical reactions can be increased.
8. Describe the action of enzymes in terms of activation energy and the lock and key model.
9. What is meant by the equilibrium condition in a reversible reaction?
10. What is an acid and what is a base? Describe the pH scale.
11. Define a salt. What is a buffer, and why are buffers important?
12. List four functions that water performs in the human body.
13. Define an organic and an inorganic molecule.
14. Name the four major types of organic molecules. Give a function for each.

CONCEPT REVIEW

1. If an atom of iodine (I) gained an electron, what would be the charge of the resulting ion? Write the symbol for this ion.
2. In which of the following ways would two water molecules align with each other: the hydrogen atoms of one water molecule would be close to the hydrogen atoms of the other water molecule; or, the hydrogen atoms of one water molecule would be close to the oxygen atom of the other water molecule? Explain.
3. For each of the following chemical equations, determine if a synthesis reaction, a decomposition reaction, an exchange reaction, or dissociation has taken place.
   A. $NaCl \rightarrow Na^+ + Cl^-$
   B. $ADP + P \rightarrow ATP$
   C. $NaHCO_3 + HCL \rightarrow H_2CO_3 + NaCl$
   D. $H_2O \rightarrow H^+ + OH^-$
4. A mixture of chemicals was warmed slightly. As a consequence, although no more heat was added, the solution became very hot. Explain what happened to make the solution so hot.
5. In terms of exergonic and endergonic reactions, explain why eating food is necessary for increasing muscle mass.
6. Given that the hydrogen ion concentration of a solution is based on the following reversible reaction:

   $$CO_2 + H_2O \rightleftarrows H^+ + HCO_3^-$$

   What will happen to the pH of the solution if $NaHCO_3$ (sodium bicarbonate) is added to the solution? (Hint: the sodium bicarbonate will dissociate to form Na^+ and HCO_3^- ions).

ANSWERS TO  PREDICT QUESTIONS

1 *p. 27* Because atoms are electrically neutral, the iron (Fe) atom has the same number of protons and electrons. The loss of three electrons would mean the iron atom has three more protons than electrons and therefore a charge of plus three. The correct symbol would be Fe^{3+}.

2 *p. 27* Because the electrons spend more time around the oxygen atom, the oxygen end of the molecule is slightly negative compared to the more positive hydrogen end of the molecule.

3 *p. 29* During exercise muscle contractions increase. This requires the release of stored energy in exergonic reactions. Some of the energy is used to drive muscle contractions and some of it is released as heat. Because the rate of these reactions increases during exercise, more heat is produced than when at rest, and body temperature increases.

4 *p. 32* Carbon dioxide and water are in equilibrium with hydrogen ions and bicarbonate ions. A decrease in carbon dioxide causes some hydrogen ions to react with bicarbonate ions to form carbon dioxide and water. Consequently the hydrogen ion concentration decreases.

5 *p. 33* Adding a base to a solution would make the solution more basic, and the pH would increase. If the buffer removed the base from the solution or added hydrogen ions to the solution, the increase in pH would be resisted.

3

CELL STRUCTURES AND THEIR FUNCTIONS

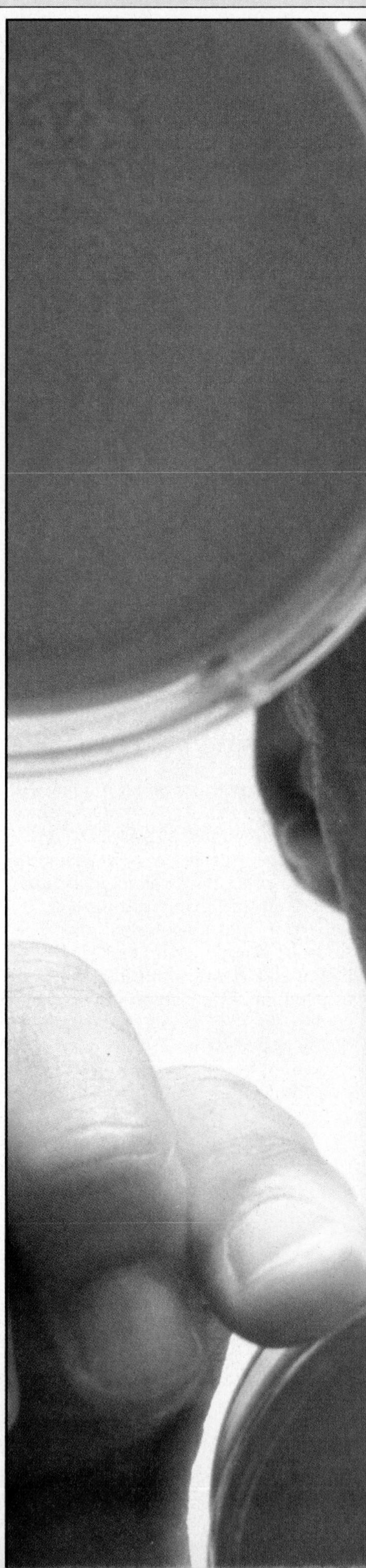

After reading this chapter you should be able to:

1 Describe the structure of the cell membrane. Explain why the cell membrane is more permeable to lipid-soluble substances and small molecules than to large water-soluble substances.
2 Explain the role of osmosis and of osmotic pressure in controlling the movement of water across the cell membrane. Compare isotonic, hypertonic, and hypotonic solutions.
3 Define mediated transport and compare the processes of facilitated diffusion and active transport.
4 Describe the structure and function of the nucleus and nucleoli.
5 Compare the structure and function of rough and smooth endoplasmic reticulum.
6 Describe the roles of the Golgi apparatuses and secretory vesicles in secretion.
7 Explain the role of lysosomes in digesting material taken into cells by phagocytosis.
8 Describe the structure and function of mitochondria.
9 Compare the structure and function of cilia, flagella, and microvilli.
10 Describe the process of protein synthesis.
11 Explain what is accomplished during mitosis and differentiation.

active transport Carrier-mediated process that requires ATP and can move substances against a concentration gradient.

chromosome (kro'mo-sōm) [Gr. *chroma*, color + Gr. *soma*, body] One of the bodies (normally 46 in humans) in the cell nucleus that carry the cell's genetic information.

cytoplasm (si'to-plazm) Cellular material surrounding the nucleus.

diffusion [L. *diffundo*, to pour in different directions] Tendency for solute molecules to move from an area of high concentration to an area of low concentration in solution; the product of the constant random motion of all atoms, molecules, or ions in a solution.

endoplasmic reticulum (en'do-plaz'mik re-tik'u-lum) Membranous network inside the cytoplasm; rough has ribosomes attached to the surface; smooth does not have ribosomes attached.

facilitated diffusion Carrier-mediated process that does not require ATP and moves substances into or out of cells from a high to a low concentration.

Golgi apparatus (gōl'je) Stacks of flattened sacks, formed by membranes, that concentrate and package materials for secretion from the cell.

mitochondria (mi'to-kon'dre-ah) [Gr. *mitos*, thread + *chandros*, granule] Small, spherical, rod-shaped or thin filamentous structures in the cytoplasm that are sites of ATP production.

mitosis (mi-to'sis) [Gr., thread] Division of the nucleus. Process of cell division that results in two daughter cells with exactly the same number and type of chromosomes as the parent cell.

nucleus [L., inside of a thing] Cell organelle containing most of the cell's genetic material.

osmosis (os-mo'sis) [Gr. *osmos*, thrusting or an impulsion] Diffusion of solvent (water) through a selectively permeable membrane from a less concentrated solution to a more concentrated solution.

plasma membrane (plaz'mah) Cell membrane; outermost component of the cell, surrounding and binding the rest of the cell contents.

ribosome Small, spherical, cytoplasmic organelle where protein synthesis occurs.

The trillions of cells that make up the human body are responsible for its structure and function. All cells in each human originate from a single fertilized egg, and during development they become specialized to form nerve, muscle, bone, fat, blood, and other cell types. Each cell is a highly organized unit. Within cells there are specialized structures called **organelles** that perform specific functions (Figure 3-1). At some time in their existence each cell contains a **nucleus,** an organelle containing the cell's genetic material, which controls the cell. The living material surrounding the nucleus is called **cytoplasm** (si'to-plazm), which contains many other types of organelles. The cytoplasm is enclosed by the **plasma** (plaz'mah) membrane.

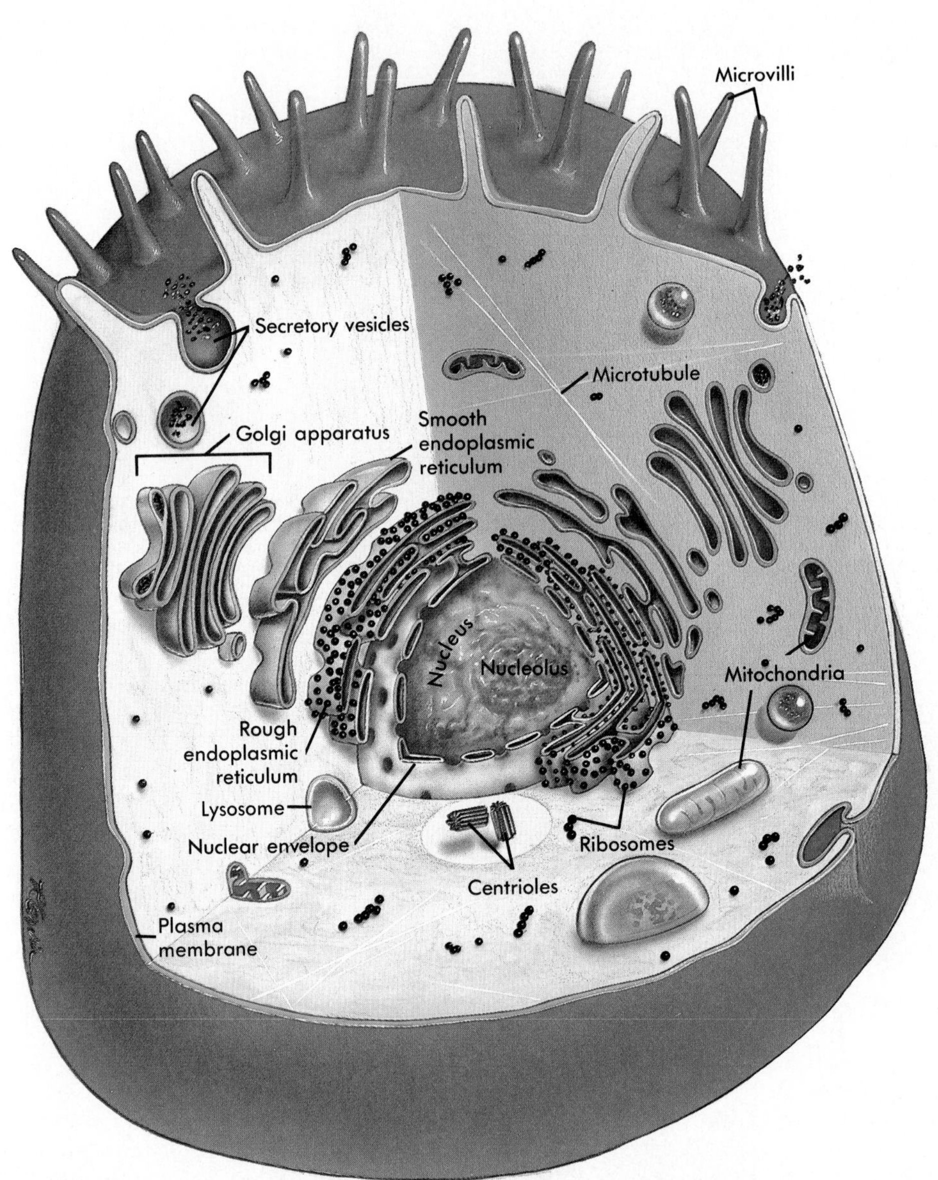

FIGURE 3-1 ■ Generalized cell showing the major organelles.
No single cell contains all organelle types. In addition, some cells may contain many organelles of one type and another cell may contain very few.

STRUCTURE OF THE PLASMA MEMBRANE

The plasma membrane, or **cell membrane,** is the outermost component of a cell. The cell membrane encloses the cytoplasm and forms a boundary between material inside the cell and material outside the cell. Substances outside the cell are called **extracellular** substances, and substances inside the cell are called **intracellular** substances. In addition to enclosing and supporting the cell contents, the cell membrane is a selective barrier that determines what moves into and out of the cell. Many of the organelles within cells have membranes similar in structure and function to the cell membrane.

The major molecules that make up the cell membrane are phospholipids and proteins. In addition, the membrane contains other molecules such as cholesterol, carbohydrates, water, and ions. Studies of the arrangement of molecules in the cell membrane have given rise to a model of its structure called the **fluid mosaic model.** The cell membrane consists of a double layer of phospholipid molecules with other lipids such as cholesterol interspersed among them. The phospholipid molecules are arranged so that the polar, phosphate-containing ends face outward on either side of the membrane and the nonpolar, fatty acid ends face each other to form a double layer (Figure 3-2).

The polar ends of the phospholipid molecules are exposed to water inside and outside the cell, and the nonpolar ends form a lipid barrier between the inside and outside of the cell.

The double layer of phospholipid molecules has a liquid quality, and protein molecules "float" on both the inner and the outer surfaces. Carbohydrate molecules are bound to some protein molecules. Some of the protein molecules penetrate the double layer of lipid from one surface to the other. The proteins function as membrane channels, carrier molecules, receptor molecules, enzymes, or structural supports in the membrane.

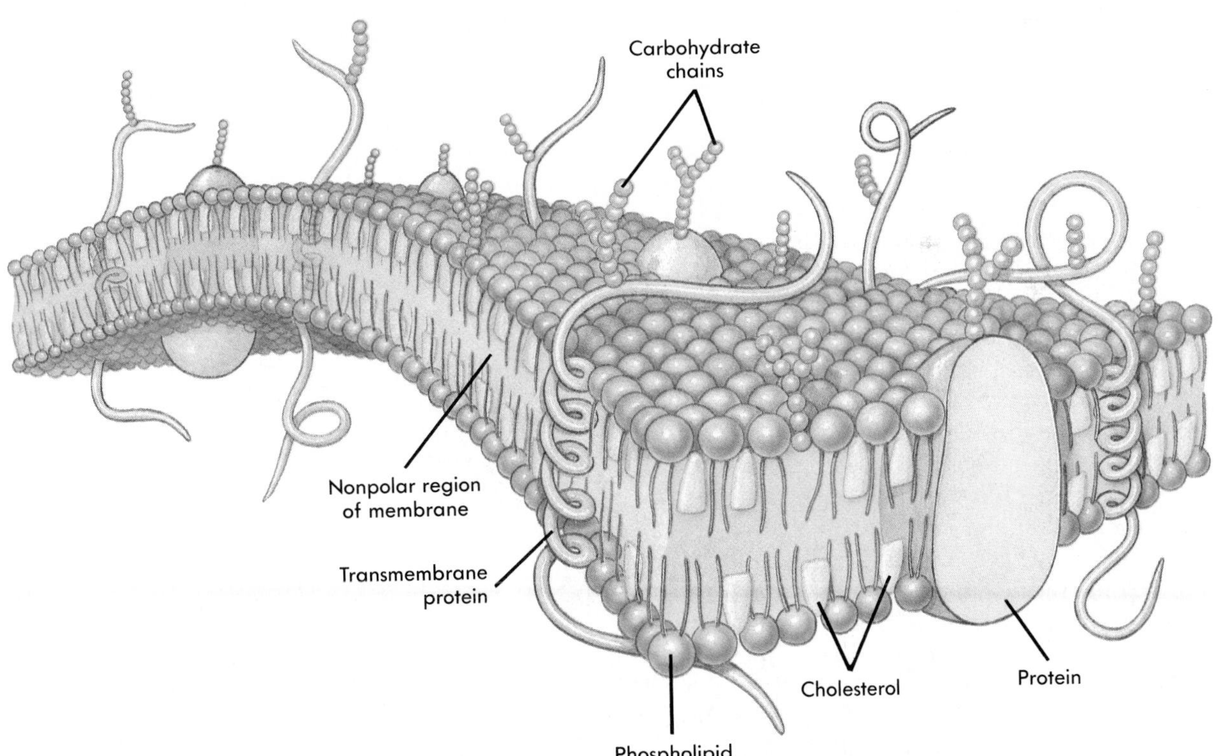

FIGURE 3-2 ■ The cell membrane.
A model of the cell membrane composed of a double layer of phospholipid molecules with proteins "floating" in the membrane. The nonpolar end of each phospholipid molecule is directed toward the center of the membrane, and the polar end of each phospholipid molecule is directed toward the water environment either outside or inside the cell. Cholesterol molecules are interspersed among the phospholipid molecules. Groups of proteins may form membrane channels, carrier molecules, or receptor molecules.

MOVEMENT THROUGH THE PLASMA MEMBRANE

The plasma membrane is **selectively permeable,** allowing some substances to pass through it but not others. Intracellular material has a different composition from extracellular material, and the survival of cells depends on maintaining the difference. Substances such as enzymes, glycogen, and potassium ions are found at higher concentrations intracellularly, and sodium, calcium, and chloride ions are found in greater concentrations extracellularly. In addition, nutrients must enter cells continually, and waste products must exit. Because of the plasma membrane's permeability characteristics and its ability to transport certain molecules, cells are able to maintain proper intracellular concentrations of molecules. Rupture of the membrane, alteration of its permeability characteristics, or inhibition of transport processes can disrupt the normal intracellular concentration of molecules and lead to cell death.

Molecules that are soluble in lipids pass through the plasma membrane by dissolving in the double layer of phospholipid. The phospholipid layers act as a barrier to most substances that are not soluble in lipids. Certain small molecules that are not lipid soluble, however, can pass through membrane channels, and certain large molecules that are not lipid soluble are transported by carrier molecules through the plasma membrane.

Plasma **membrane channels,** consisting of large protein molecules, extend from one surface of the plasma membrane to the other (see Figure 3-2). There are several channel types, and each type allows certain molecules to pass through it. The size, shape, and charge of molecules determines whether they can pass through each kind of channel. For example, sodium ions pass through sodium channels, and potassium and chloride ions pass through potassium and chloride channels, respectively.

Larger molecules that are not lipid soluble, such as glucose and amino acids, cannot pass through the plasma membrane in significant amounts unless they are transported by special carrier molecules. Substances that are transported across the plasma membrane by carrier molecules are said to be transported by **carrier-mediated processes.** The carrier molecules are proteins that extend from one side of the cell membrane to the other. They bind to molecules to be transported and move them across the plasma membrane. Each carrier molecule transports a specific type of molecule. For example, carrier molecules that transport glucose across the plasma membrane will not transport amino acids, and carrier molecules that transport amino acids will not transport glucose.

Diffusion

A **solution** is either a liquid or a gas and consists of one or more substances called **solutes** dissolved in the predominant liquid or gas, which is called the **solvent. Diffusion** can be viewed as the tendency for solute molecules to move from an area of high concentration to an area of low concentration in solution (Figure 3-3, *A* and *B*). Examples of diffusion are the movement and distribution of smoke or perfume throughout a room in which there are no air currents, or of a dye throughout a beaker of still water.

Diffusion is a product of the constant random motion of all atoms, molecules, or ions in a solution. Because there are more solute particles in an area of high concentration than in an area of low concentration, and because particles move randomly, the chances are greater that solute particles will move from the higher toward the lower concentration than in the opposite direction. At equi-

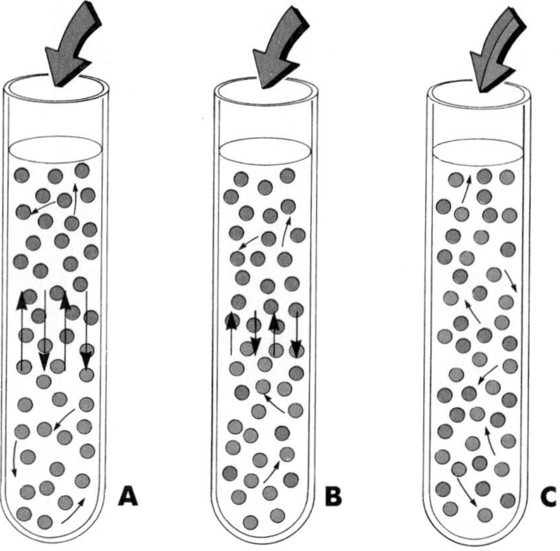

FIGURE 3-3 Diffusion.
A One solution (*red,* representing one type of molecule) is layered onto a second solution (*blue,* representing a second type of molecule). The molecules in each solution move at random; however, because there are no red molecules in the blue solution and no blue molecules in the red solution, the molecules of one solution tend to move into the other solution. Therefore the net movement is substantial, and the concentration gradient is high.
B One solution (*red*) is added to another solution (*red* and *blue*) already containing some red molecules. The net movement is less than in **A** because the concentration difference is not as great.
C One solution (*red* and *blue*) is added to an identical solution (*red* and *blue*). An equilibrium with no net movement occurs because there is no concentration gradient.

librium, the net movement of solutes stops, although the random molecular motion continues, and the movement of solutes in any one direction is balanced by an equal movement in the opposite direction (Figure 3-3, C).

A **concentration gradient** is a measure of the difference in the concentration of a solute in a solvent between two points. For a given distance, the concentration gradient is equal to the higher concentration minus the lower concentration of a solute. Movement down a concentration gradient, or with a concentration gradient, describes the movement of solute molecules from a higher toward a lower concentration of solute molecules, and movement up a concentration gradient, or against a concentration gradient, describes the movement of solute molecules from a lower toward a higher concentration of solute molecules. The concentration gradient is said to be steeper when the concentration gradient is large.

Diffusion is an important means by which substances move through the extracellular and intracellular fluids in the body. In addition, substances that can pass either through the lipid layers of the cell membrane or through membrane channels diffuse through the cell membrane. Some nutrients enter and some waste products leave the cell by diffusion. The normal intracellular concentrations of many substances depends on diffusion. For example, if the extracellular concentration of oxygen is reduced, not enough oxygen will diffuse into the cell, and normal cell function cannot occur.

1 Urea is a toxic waste produced inside cells. It diffuses from cells into the blood and is eliminated from the body by the kidneys. What would happen to the intracellular and extracellular concentration of urea if the kidneys stopped functioning? ?

Osmosis

Osmosis (os-mo′sis) is the diffusion of water (a solvent) across a **selectively permeable membrane,** such as the cell membrane, from a region of high water concentration to a region of lower water concentration. Osmosis is important to cells because large volume changes caused by water movement can disrupt normal cell functions. Diffusion of water occurs when the cell membrane is either less permeable to or not permeable to solutes, and when a concentration difference exists across the cell membrane. Water diffuses from a solution with a high concentration of water across the cell membrane into a solution with a lower concentration of water. The ability to predict the direction of water movement depends upon knowing which solution on either side of a membrane has the highest concentration of water.

The concentration of a solution is not expressed in terms of water, however, but in terms of solute. For example, if sugar solution A is twice as concentrated as sugar solution B, then solution A has twice as much sugar (solute) as solution B. As the concentration of a solution increases, the amount of water (solvent) proportionately decreases. Thus, water diffuses from the less concentrated solution, with fewer solute molecules and more water molecules, into the more concentrated solution, with more solute molecules and fewer water molecules.

Osmotic pressure is the force required to prevent the movement of water across a selectively permeable membrane. Thus osmotic pressure is a measure of the tendency for water to move by osmosis across a selectively permeable membrane. It can be measured by placing a solution in a bag made of a selectively permeable membrane, connecting a vertical tube to the bag, and immersing the bag in distilled water (Figure 3-4, A). Water

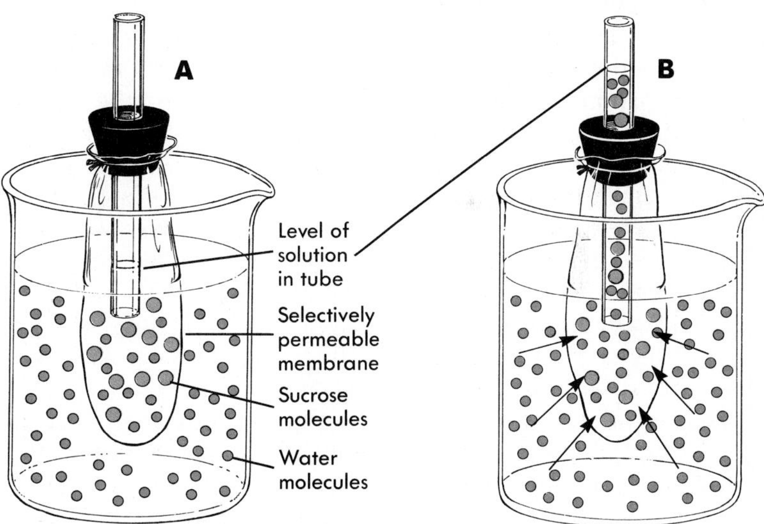

FIGURE 3-4 Osmosis.
A A solution containing molecules of a certain size such as sucrose (table sugar) is placed into a selectively permeable bag such as cellophane that allows water molecules to pass in and out but retains the sucrose molecules within the bag. The top of the bag is closed with a cork, which has a glass tube extending through it, and the bag is placed into distilled water.
B Because the bag contains sucrose molecules as well as water molecules, the concentration of water is less in the bag than in the beaker. The water molecules diffuse into the bag. Because the sucrose molecules cannot leave the bag, the total fluid volume inside the bag increases, and fluid moves up the glass tube as a result of osmosis. The height of the water in the tube is proportional to the osmotic pressure of the fluid in the bag.

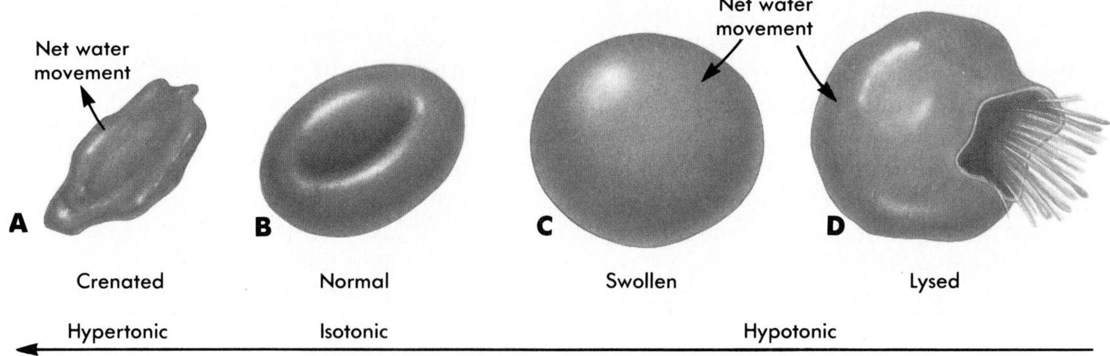

FIGURE 3-5 Effects of hypertonic, isotonic, and hypotonic solutions on red blood cells.
A Hypertonic solutions with high ion concentrations result in crenation (shrinkage) of the cell.
B Isotonic solutions with equal ion concentrations inside and outside the cell result in normal-shaped cells.
C and **D** Hypotonic solutions with low ion concentrations result in swelling and finally, lysis of cells.

molecules move by osmosis through the membrane into the bag, forcing the solution to rise from the bag into the tube (Figure 3-4, *B*).

The movement of water into the bag increases the hydrostatic pressure in the bag, which is proportional to the height of the water in the tube. Net movement of water stops when the hydrostatic pressure in the bag causes water to move out of the bag at the same rate water diffuses into the bag. The higher the concentration (more solutes, less water) of a solution in the bag, the greater the movement of water across the membrane into the solution and the higher the solution rises in the tube. The height of the column of water in the tube is proportional to the solution's osmotic pressure.

Three terms describe solutions in which cells may have a tendency to shrink or swell. When a cell is immersed in an **isotonic** (i'so-ton'ik) solution, the concentration of various solutes and water are the same on both sides of the plasma membrane. Therefore, the cell neither shrinks nor swells (Figure 3-5). When a cell is immersed in a **hypertonic** (hi'per-ton'ik) solution, the solution has a higher concentration of solutes and a lower concentration of water than the cytoplasm of the cell. Water moves by osmosis from the cell into the hypertonic solution, resulting in cell shrinkage or **crenation** (kre-na'shun) (see Figure 3-5, *A*). When a cell is placed in a **hypotonic** (hi'po-ton'ik) solution, the solution usually has a lower concentration of solutes and a higher concentration of water than the cytoplasm of the cell. Water moves by osmosis into the cell, causing it to swell. If the cell swells enough it can rupture, a process called **lysis** (see Figure 3-5, *C* and *D*). Solutions injected into the circulatory system or tissues must be isotonic because crenation or swelling disrupts the normal function of cells and can lead to cell death.

Filtration

Filtration is the passage of a solution through a membrane in response to a mechanical pressure difference, such as blood pressure, across a membrane. Normally larger molecules such as proteins cannot pass through membranes, while smaller molecules and water can. The initial step in urine formation in the kidney involves filtration. Blood cells and large particles such as proteins remain in the kidney blood vessels while water and smaller molecules are forced to pass through the filtration membrane into the kidney tubules.

Mediated Transport Mechanisms

Many nutrient molecules, such as amino acids and glucose, cannot enter the cell by the process of diffusion, and many substances produced in cells, such as proteins, cannot leave the cell by diffusion. **Mediated transport mechanisms** involve carrier molecules within the plasma membrane that function to move large, water-soluble molecules or electrically charged molecules across the plasma membrane (Figure 3-6, *A*). After a molecule to be transported binds to a carrier molecule on one side of the membrane, the three-dimensional shape of

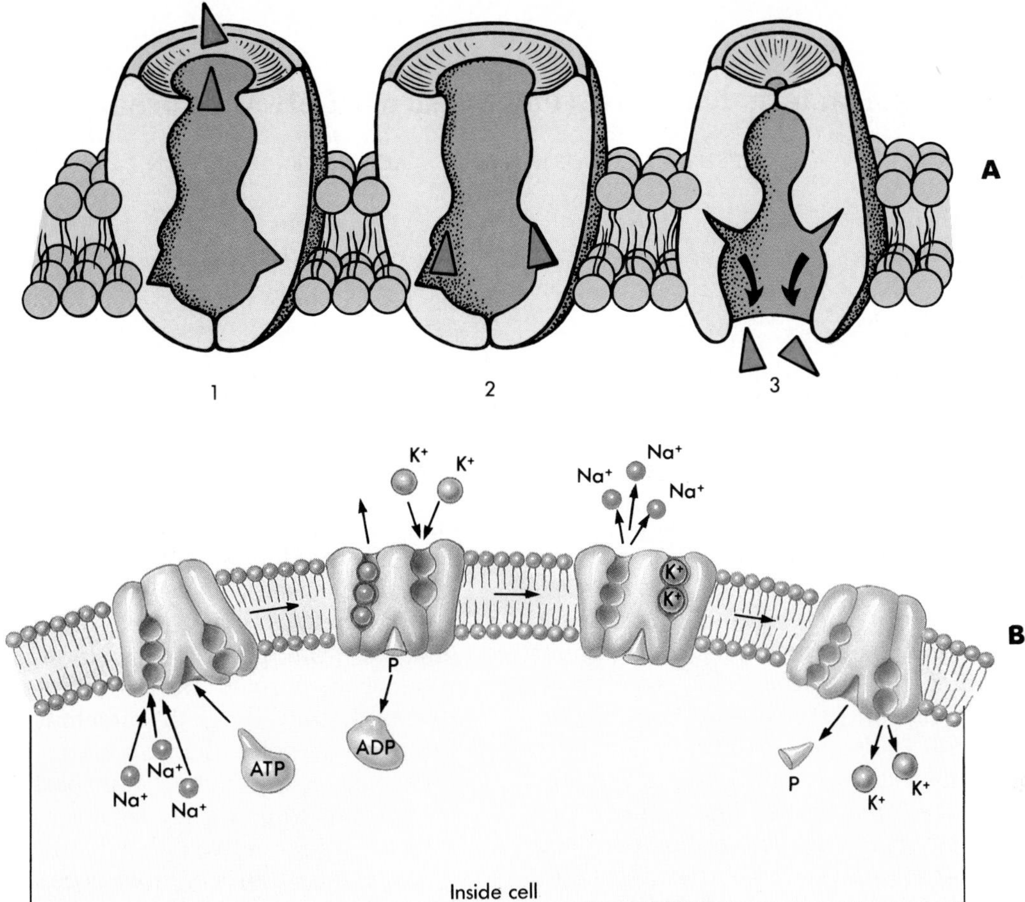

FIGURE 3-6 Mediated transport mechanisms.
A A molecule binds to a transport protein in the cell membrane, a conformational (shape) change in the transport protein brings the molecule inside the cell, and the molecule is released.
B An exchange pump. The system is similar to that shown in **A** except that two molecules (in this case sodium [Na^+]and potassium [K^+] ions) are exchanged. The K^+ ions outside the cell and the Na^+ ions inside the cell bind to a transport protein in the cell membrane. A conformational change occurs requiring energy. The K^+ ions are brought inside the cell, the Na^+ ions are carried outside, and the molecules are released. Energy in the form of ATP is required in this process because Na^+ and K^+ ions are transported from a low to a high concentration. Note that Na^+ and K^+ ions are not exchanged in equal numbers across the cell membrane.

the carrier molecule changes, and the transported molecule is moved to the opposite side of the plasma membrane. The transported molecule then is released by the carrier molecule, which then resumes its original shape and is available to transport another molecule. There are two kinds of mediated transport: facilitated diffusion and active transport.

Facilitated diffusion

Facilitated diffusion is a mediated transport process that moves substances into or out of cells from a high to a low concentration (Table 3-1). Because movement is with the concentration gradient, it does not require metabolic energy in the form of adenosine triphosphate (ATP) (see discussion of mitochondria).

2 The transport of glucose into and out of most cells occurs by facilitated diffusion. Because transport occurs from a high to a low concentration, glucose cannot accumulate within these cells at a higher concentration than is found outside the cell. Once glucose enters a cell, it is rapidly converted to other molecules such as glucose phosphate or glycogen. What effect does this conversion have on the cell's ability to transport glucose? ?

TABLE 3-1

Types and Characteristics of Movement across Cell Membranes

CHARACTERISTIC	DIFFUSION AND OSMOSIS	FACILITATED DIFFUSION	ACTIVE TRANSPORT
Requires ATP	No	No	Yes
Movement occurs with concentration gradient	Yes	Yes	No*
Movement occurs against concentration gradient	No	No	Yes

*Active transport normally moves substances against a concentration gradient, but it also can move molecules with a concentration gradient.

Active transport

Active transport is a carrier-mediated process that moves substances from regions of lower concentration to regions of higher concentration against the concentration gradient (see Table 3-1). Consequently, active transport processes accumulate substances on one side of the cell membrane at concentrations many times greater than those on the other side. Active transport requires energy in the form of ATP. If ATP is not available, active transport stops immediately. Examples include the transport of glucose and amino acids from the small intestine into the blood. In other cases, the active transport mechanism may exchange one substance for another. For example, the sodium-potassium exchange pump moves sodium ions out of cells and potassium ions into cells (Figure 3-6, *B*). The result is a higher concentration of sodium ions outside the cell and a higher concentration of potassium ions inside the cell.

Endocytosis and Exocytosis

Endocytosis (en′do-si′to-sis) includes both phagocytosis (fag′o-si-to′sis) and pinocytosis (pin′o-si-to′sis) and refers to the bulk uptake of material through the plasma membrane by the formation of a **vesicle** (ves′ĭ-kl). A vesicle is a membrane-bound droplet found within the cytoplasm of a cell. A portion of the cell membrane wraps around a particle or droplet and fuses so that the particle or droplet is surrounded by the membrane. That portion of the membrane then "pinches off" so the droplet, surrounded by a membrane, is within the cytoplasm of the cell, and the plasma membrane is left intact (Figure 3-7).

Phagocytosis means cell eating and applies to endocytosis when solid particles are ingested. The cytoplasm extends around and engulfs the particle. White blood cells and some other cell types phagocytize bacteria, cell debris, and foreign particles. Phagocytosis is an important means by which white blood cells take up and destroy harmful substances that have entered the body.

Pinocytosis means cell drinking. It is distinguished from phagocytosis in that much smaller vesicles are formed, they contain liquid rather than particles, and the cell membrane invaginates to form the vesicles that are taken into the cell. Pinocytosis is a common transport mechanism and occurs in certain kidney cells, epithelial cells of the intestine, liver cells, and cells that line capillaries.

In some cells, secretions accumulate within vesicles. These secretory vesicles then move to the plasma membrane where the vesicle membrane fuses with the cell membrane, and the content of the vesicle is eliminated from the cell. This process is called **exocytosis** (eks-o-si-to′sis) (Figure 3-8). Secretion of digestive enzymes by the pancreas, mucus by the salivary glands, and milk from the mammary glands are examples of exocytosis. In many respects the process is similar to endocytosis but occurs in the opposite direction. Endocytosis results in the uptake of materials by cells and exocytosis in the expulsion of material from cells.

ORGANELLES AND CELL FUNCTIONS

Cells contain several specialized structures called organelles, each of which performs functions important to the cell's survival (Table 3-2). The number and type of organelles within each cell determine the cell's specific structure and functions. Cells secreting large amounts of protein contain well-developed organelles that synthesize and secrete protein. Cells actively transporting substances such as sodium ions across their cell membranes contain highly developed organelles that produce ATP. The following sections describe the structure and main functions of the major organelles found in cells.

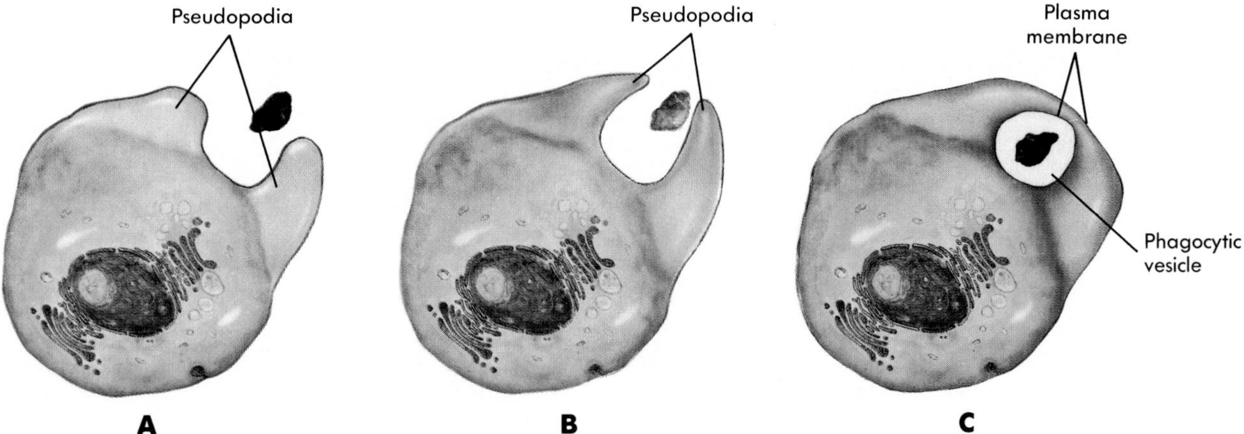

FIGURE 3-7 ■ Phagocytosis.
A and **B** Cell processes (pseudopodia) extend from the cell and surround the particle to be taken into the cell by phagocytosis,
C Once the pseudopodia have surrounded the particle, they fuse to form a vesicle, which contains the particle. The vesicle then is internalized in the cell.

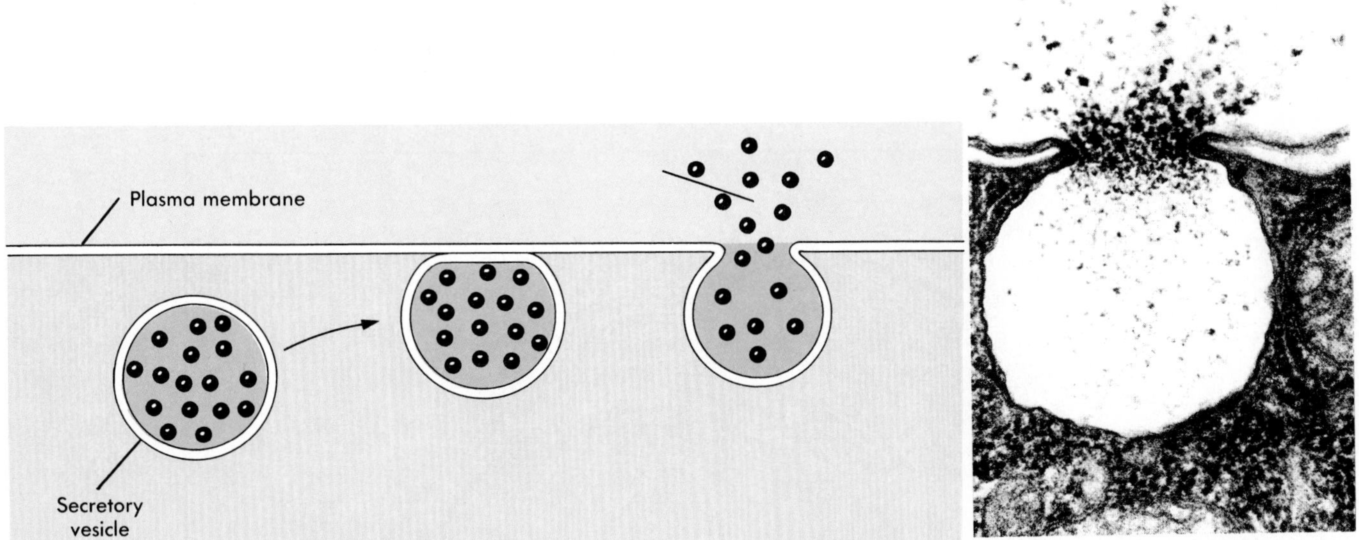

FIGURE 3-8 ■ Exocytosis.
Secretory products accumulate within vesicles whose membranes fuse with the plasma membrane, releasing their contents to the cell surface.

Nucleus

The **nucleus** is a large organelle usually located near the center of the cell (see Figure 3-1). All cells of the body have a nucleus at some point in their life cycle, although some cells, such as red blood cells, lose their nuclei as they mature. Other cells contain more than one nucleus, such as osteoclasts (a type of bone cell) and skeletal muscle cells.

The nucleus is surrounded by two membranes with a narrow space between them. At many points on the surface of the nucleus, the inner and outer membranes come together to form **nuclear pores** (Figure 3-9). The nucleus contains the cell's genetic material in the form of 23 pairs of **chromosomes** (kro′mo-sōmz) characteristic of human cells. DNA

TABLE 3-2

Organelles and Their Locations and Functions

ORGANELLES	LOCATION AND FUNCTION(S)
Nucleus	Usually near center of the cell; contains genetic material of cell (DNA) and nucleoli; the site of ribosome and messenger RNA synthesis
Nucleolus	In the nucleus; site of ribosomal RNA and ribosomal proteins synthesis
Rough endoplasmic reticulum (ER)	In cytoplasm; many ribosomes attached to ER; site of protein synthesis
Smooth endoplasmic reticulum (ER)	In cytoplasm; site of lipid synthesis
Golgi apparatus	In cytoplasm; modifies protein structure and packages proteins in secretory vesicles
Secretory vesicle	In cytoplasm; contains materials produced in the cell; formed by the Golgi apparatus; secreted by exocytosis
Lysosome	In cytoplasm; contains enzymes that digest material taken into the cell
Mitochondria	In cytoplasm; site of oxidative metabolism and the major site of ATP synthesis
Microtubule	In cytoplasm; supports cytoplasm; assists in cell division and forms components of cilia and flagella
Cilia	On cell surface with many on each cell; cilia move substances over the surface of cells
Flagella	On cell surface with one per cell; propels sperm cells
Microvilli	Extensions of cell surface with many on each cell; increase the surface area of cells

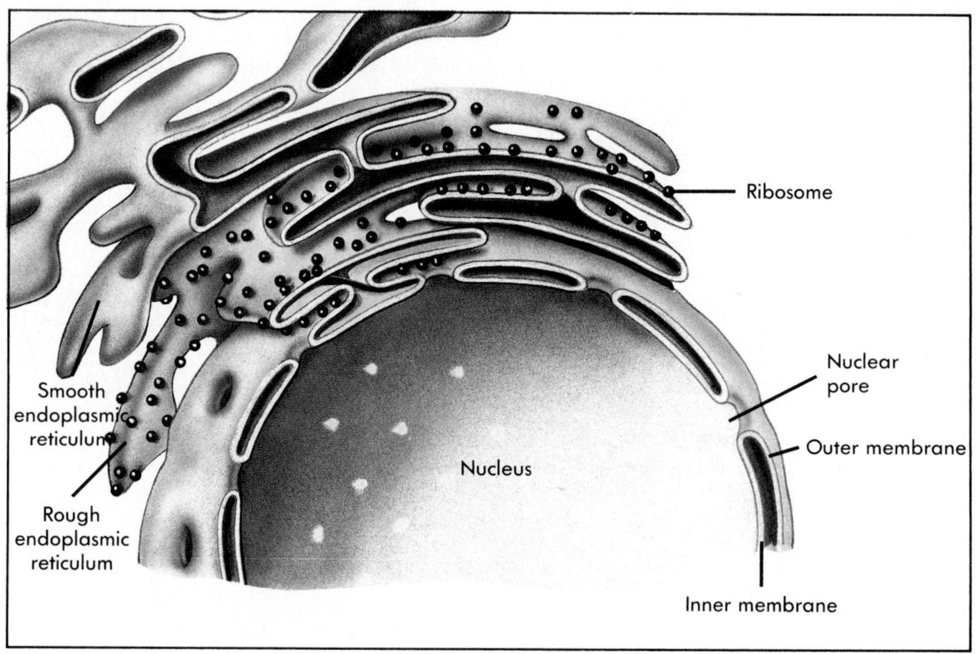

FIGURE 3-9 The nuclear membrane and endoplasmic reticulum.
An inner and an outer membrane surround the nucleus. There are several nuclear pores where the inner and outer nuclear membranes become fused. The outer membrane is continuous with the endoplasmic reticulum (ER). Rough ER has ribosomes attached to its membrane and smooth ER has no ribosomes attached to it. Some cells contain predominantly smooth ER, and others contain predominantly rough ER.

molecules within the chromosomes form the genes that determine the structure and function of each cell and of the individual.

Nucleoli and Ribosomes

Nucleoli (nu-kle'o-li) number from one to four per nucleus. They are rounded, dense, well-defined nuclear bodies with no surrounding membrane. Ribosomal proteins produced in the cytoplasm move through the nuclear pores into the nucleus and to the nucleoli. The proteins are joined to ribosomal RNA (rRNA), produced within the nucleoli, to form large and small ribosomal subunits (Figure 3-10). The ribosomal subunits then move from the nucleus through the nuclear pores into the cytoplasm, where one large and one small subunit join to form individual **ribosomes.** The ribosomes function as sites of protein synthesis. Some ribosomes become associated with endoplasmic reticulum to form rough endoplasmic reticulum, whereas other ribosomes remain free in the cytoplasm.

Rough and Smooth Endoplasmic Reticulum

The **endoplasmic reticulum** (en'do-plaz'mik re-tik'u-lum) **(ER)** is a series of membranes that extend from the outer nuclear membrane into the cytoplasm. **Rough ER** is endoplasmic reticulum with ribosomes attached to it (see Figure 3-9). Because rough ER has numerous ribosomes, a large amount of rough ER in a cell indicates that it synthesizes protein. On the other hand, ER without ribosomes is called **smooth ER.** Smooth ER is a site for lipid synthesis in cells.

The Golgi Apparatus

The **Golgi** (gōl'je) **apparatus** (named for Camillo Golgi, an Italian histologist) concentrates and packages materials for secretion from the cell. It consists of closely packed stacks of curved, membrane-bound sacs (Figure 3-11). Proteins produced at the ribosomes enter the Golgi apparatuses from the en-

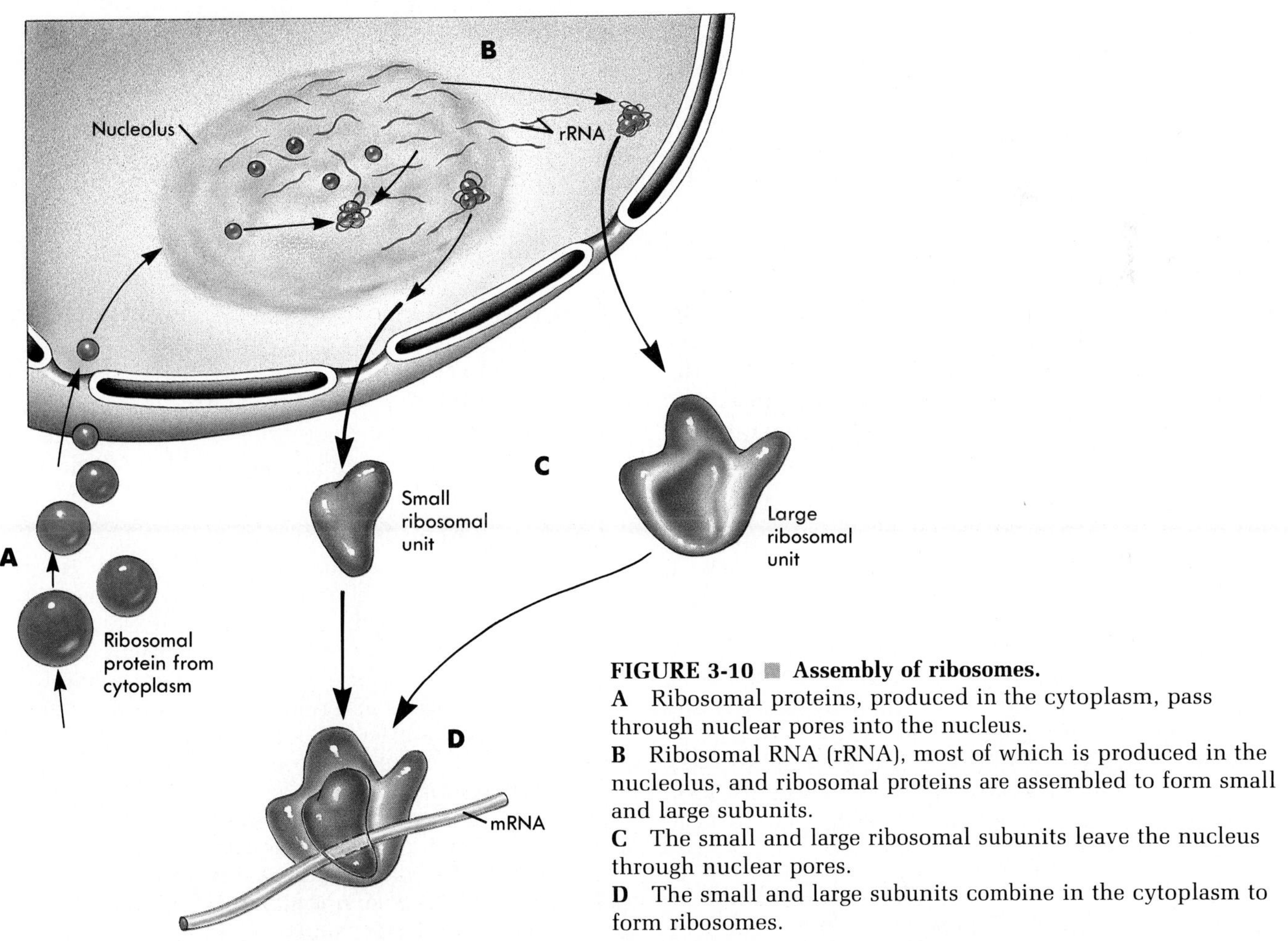

FIGURE 3-10 ■ **Assembly of ribosomes.**
A Ribosomal proteins, produced in the cytoplasm, pass through nuclear pores into the nucleus.
B Ribosomal RNA (rRNA), most of which is produced in the nucleolus, and ribosomal proteins are assembled to form small and large subunits.
C The small and large ribosomal subunits leave the nucleus through nuclear pores.
D The small and large subunits combine in the cytoplasm to form ribosomes.

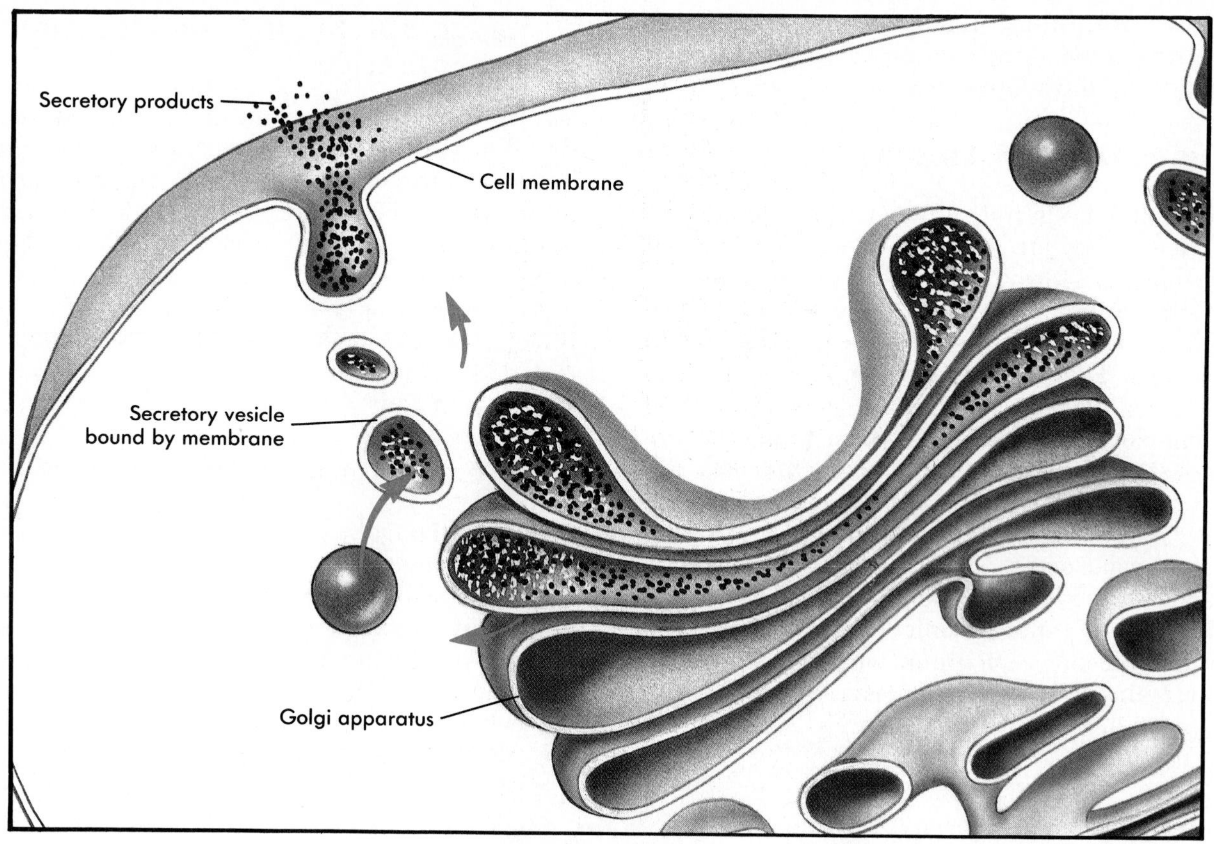

FIGURE 3-11 ■ The Golgi apparatus.
The Golgi apparatus concentrates and, in some cases, modifies protein molecules produced by the rough ER and then packages them in secretory vesicles. In exocytosis, a secretory vesicle is formed inside the cell. The vesicle moves to the cell membrane, fuses with the membrane, opens to the outside, and dumps its contents into the extracellular space.

doplasmic reticulum. The Golgi apparatus concentrates and, in some cases, chemically modifies the proteins by attaching carbohydrate or lipid molecules to them. The proteins then are packaged into secretory vesicles, which pinch off from the margins of the Golgi apparatus. The Golgi apparatus is present in larger numbers and is most highly developed in cells that secrete protein, such as the cells of the salivary glands or the pancreas.

Secretory Vesicles

■ The **secretory vesicles** pinch off from the Golgi apparatus and move to the surface of the cell (see Figure 3-11). Their membranes then fuse with the cell membrane, and the contents of the vesicles are released to the exterior by exocytosis. In many cells secretory vesicles accumulate in the cytoplasm and quickly are released to the exterior when the cell receives a signal. For example, secretory vesicles containing the hormone insulin remain in the cytoplasm of pancreas cells until rising blood levels of glucose act as a stimulus for their release.

Lysosomes

■ **Lysosomes** (li′so-sōmz) are membrane-bound vesicles containing a variety of enzymes that function as intracellular digestive systems (see Figure 3-1). Phagocytic vesicles fuse with lysosomes, and enzymes within the lysosomes digest the contents of the phagocytic vesicles. White blood cells take

■ Some diseases result from nonfunctional lysosomal enzymes. For example, *Pompe's disease* results from the inability of lysosomal enzymes to break down the carbohydrate glycogen. Glycogen accumulates in large amounts in the heart, liver, and skeletal muscles. Glycogen accumulation in the heart muscle cells often leads to heart failure. Lipid storage disorders are often hereditary and are characterized by the accumulation of large amounts of lipid in phagocytic cells. These cells take up the lipid by phagocytosis, but they lack the enzymes required to break down the lipid droplets. Symptoms include enlargement of the spleen and liver and replacement of bone marrow by the affected cells. ■

up bacteria in phagocytic vesicles, and the enzymes within lysosomes destroy the bacteria. Also, when tissues are damaged, ruptured lysosomes within the damaged cells release their enzymes and digest both healthy and damaged cells. The released enzymes are responsible for part of the resulting inflammation.

Mitochondria

Mitochondria (mi′to-kon′dre-ah) are small bean-shaped or rod-shaped organelles with inner and outer membranes separated by a space. The outer membranes have a smooth contour, but the inner membranes have numerous infoldings that project like shelves into the interior of the mitochondria. Mitochondria are the major sites of ATP production within cells (Figure 3-12). ATP is the major energy source for most endergonic chemical reactions within the cell, and cells with a large energy requirement have more mitochondria than cells with a lower energy requirement. Mitochondria carry out **oxidative metabolism** in which oxygen is required to allow the reactions that produce ATP to proceed. Cells that carry out extensive active transport contain many mitochondria, and when muscles enlarge as a result of exercise, the number of mitochondria within the muscle cells increases and provides the additional ATP required for muscle contraction.

Microtubules

Microtubules are hollow tubules composed primarily of protein units called **tubulin** (Figure 3-13). Microtubules perform a variety of roles, such as helping to provide support to the cytoplasm of cells, assisting in the process of cell division, and forming essential components of certain organelles such as cilia and flagella.

Cilia, Flagella, and Microvilli

Cilia (sil′e-ah) project from the surface of cells and are capable of moving. They vary in number from one to thousands per cell. Cilia have a cylindrical shape, contain specialized microtubules, and are enclosed by the cell membrane. Cilia are numerous on surface cells that line the respiratory tract. Their coordinated movement moves mucus, in which dust particles are embedded, upward and away from the lungs. The action helps keep the lungs clear of debris.

Flagella (flă-jel′ah) have a structure similar to cilia but are much longer, and there is usually only one per cell. For example, each spermatozoon (sperm cell) has one flagellum, which functions to propel the spermatozoon.

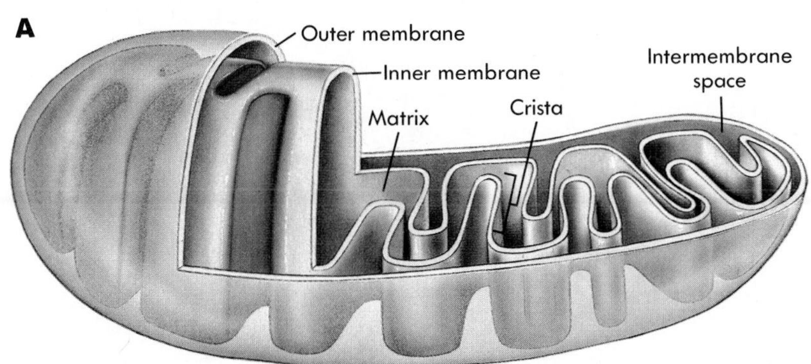

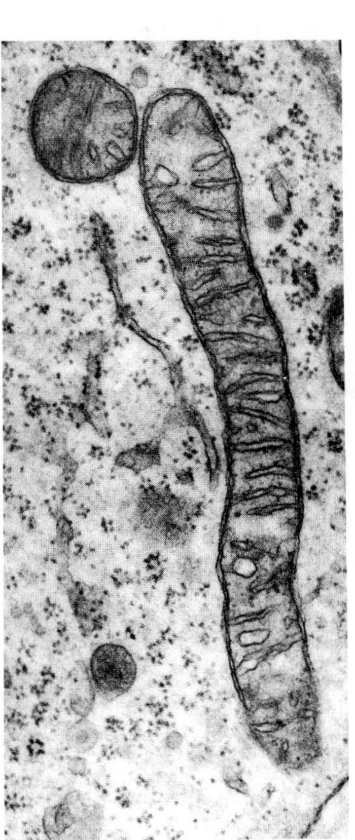

FIGURE 3-12 ■ **Mitochondria.**
A Typical mitochondrion structure.
B Electron micrograph of a mitochondrion in longitudinal and cross sections.

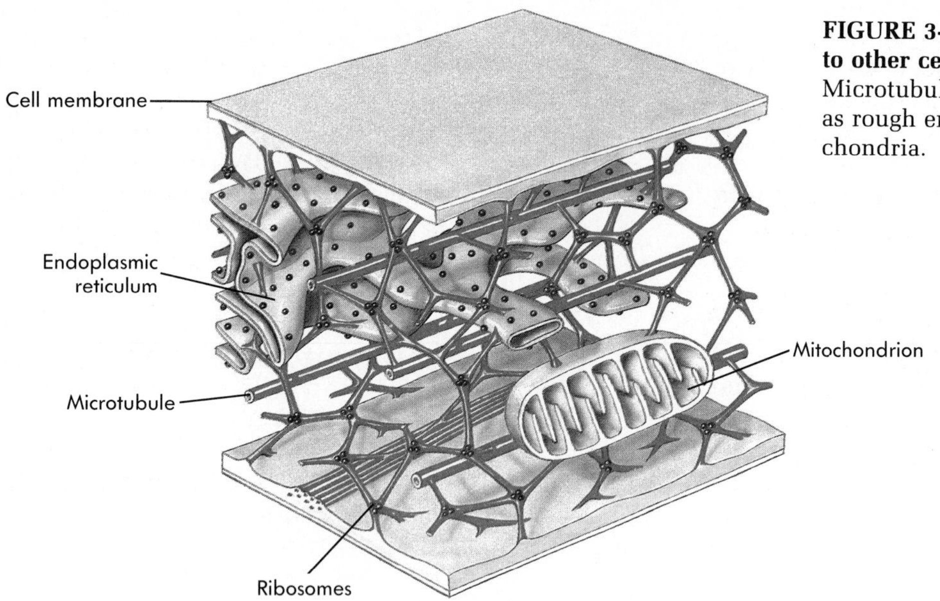

FIGURE 3-13 ■ Relationship of microtubules to other cell organelles.
Microtubules among other cell organelles such as rough endoplasmic reticulum and mitochondria.

Microvilli (mi′kro-vil′e) are specialized extensions of the cell membrane (see Figure 3-1). They are presented here with cilia and flagella because of their similar appearance. Microvilli are cylindrical extensions of the cell membrane, but they do not move, and they do not contain the specialized microtubules found in cilia. Microvilli are numerous on cells that have them, and they function to increase the surface area of cells. Microvilli are abundant on the surface of cells that line the intestine, kidney, and other areas where absorption is an important function.

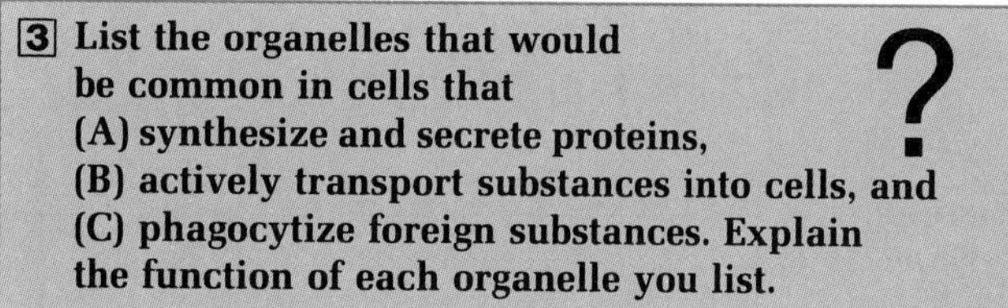
3 List the organelles that would be common in cells that (A) synthesize and secrete proteins, (B) actively transport substances into cells, and (C) phagocytize foreign substances. Explain the function of each organelle you list. ?

PROTEIN SYNTHESIS

■ DNA directs the production of proteins within cells. The proteins produced in a cell function as enzymes regulating chemical reactions or as structural components inside and outside of cells. DNA's ability to direct protein synthesis allows it to control the activities and the structure of cells and, therefore, the structural and functional characteristics of the entire organism. Whether an individual has blue eyes, brown hair, or sickle cell anemia is determined ultimately by the sequence of nucleotide bases in DNA molecules. For example, in sickle cell anemia alteration of the DNA that con-

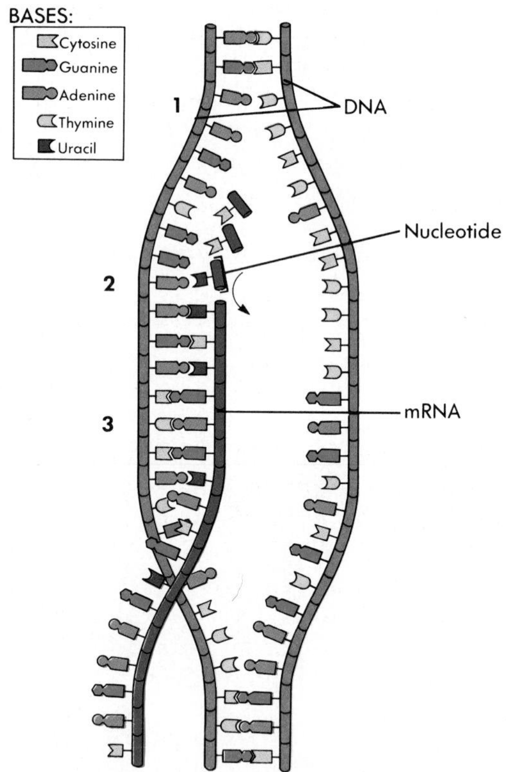

FIGURE 3-14 ■ mRNA synthesis.
Transcription of mRNA in the cell nucleus. *(1)* A segment of the DNA chain is opened, and *(2)* RNA nucleotides pair with the appropriate DNA nucleotides. *(3)* As a result, a long strand of mRNA is formed, and the sequence of nucleotides that make it up is determined by the sequence of nucleotides in the DNA molecule.

trols the structure of one protein chain of hemoglobin results in a replacement of one amino acid with a different amino acid. As a result, the structure of the entire protein is changed enough to cause it to function abnormally.

The events leading to protein synthesis begin in the nucleus and are completed in the cytoplasm. In the nucleus, DNA determines the structure of **messenger RNA (mRNA).** The mRNA molecules move out of the nucleus through the nuclear pores and into the cytoplasm where mRNA molecules determine the structure of proteins. The process starts with the **transcription** of mRNA (Figure 3-14). Transcription occurs when the double strands of a DNA segment separate, and RNA nucleotides pair with DNA nucleotides. DNA nucleotides containing adenine, thymine, cytosine, and guanine pair with RNA nucleotides containing uracil, adenine, guanine, and cytosine, respectively. An enzyme then catalyzes reactions that form chemical bonds between the RNA nucleotides to form a long mRNA segment. The number and sequence of nucleotides in the mRNA are determined by the number and sequence of nucleotides in the segments of DNA that are transcribed. After an mRNA segment has been transcribed, portions of the mRNA molecule may be removed, or two or more mRNA molecules may be combined.

Once produced, the mRNA molecules, which contain information required to determine the sequence of amino acids in proteins, pass through the nuclear pores out of the nucleus and into the cytoplasm. The information is carried in groups of three nucleotides called **codons,** and each codon codes for a specific amino acid. For example, the nucleotide sequence of uracil, cytosine, and adenine (UCA) of mRNA codes for the amino acid serine. There are 64 possible mRNA codons, but only 20 amino acids are in proteins. As a result, more than one codon may code for the same amino acid. For example, CGA, CGG, CGT, and CGC code for the amino acid alanine, and UUU and UAC code for phenylalanine. Some codons do not code for amino acids but perform other functions. UAA acts as a signal for stopping the production of a protein.

Protein synthesis requires two types of RNA in addition to mRNA: **transfer RNA (tRNA)** and **ribosomal RNA (rRNA).** There is one type of tRNA for each mRNA codon. A series of three nucleotides of each tRNA molecule, the **anticodon,** pairs with the codon of the mRNA. Another portion of each tRNA molecule binds to a specific amino acid. For example, tRNA that pairs with the UUU codon of mRNA has the anticodon AAA and binds only to the amino acid phenylalanine.

The anticodons of tRNAs pair with the codons of mRNA while the mRNA is attached to a ribosome. The amino acids bound to the tRNAs then are joined to each other by an enzyme associated with the ribosome. The enzyme causes the formation of a chemical bond, called a **peptide bond,** between the adjacent amino acids to form a **polypeptide chain,** consisting of many amino acids bound together by peptide bonds. The polypeptide chain then becomes folded to form the three-dimensional structure of the protein molecule. The synthesis of polypeptide chains at the ribosome, in response to information contained in the mRNA molecules, is called **translation** (Figure 3-15). A protein may consist of a single polypeptide chain or two or more polypeptide chains that are joined after each chain is produced on separate ribosomes.

4 Explain how changing one nucleotide within a DNA molecule of a cell could change the structure of a protein produced by the cell. ?

CELL DIVISION AND DIFFERENTIATION

Mitosis

Each cell of the human body contains 46 chromosomes. The 46 chromosomes make up 23 pairs called a **diploid** (dip'loyd) number. Of the 23 pairs, one pair is the sex chromosomes, which consist of two **X chromosomes** if the person is a female, or an X chromosome and a **Y chromosome** if the person is a male. The remaining 22 pairs of chromosomes are called **autosomes.** The sex chromosomes determine the individual's sex, and the autosomes determine most other characteristics.

All cells of the body, except those that give rise to reproductive cells, divide by **mitosis** (mi-to'sis) (Figure 3-16). First, the genetic material within a cell is **replicated,** or duplicated, and second, the cell divides to form two cells with the same amount and type of DNA as the "parent" cell. Growth and repair depend on mitosis. New cells added to the body as it grows and new cells that replace old or injured cells are produced by mitosis.

The period between active cell division is called **interphase** (see Figure 3-15, *A*). During interphase, DNA is dispersed throughout the nucleus and is not organized in the form of observable chromosomes. DNA is replicated during interphase. At the end of interphase each cell has two complete sets of genetic material.

Mitosis follows interphase. For convenience, mitosis is divided into four stages. Although each

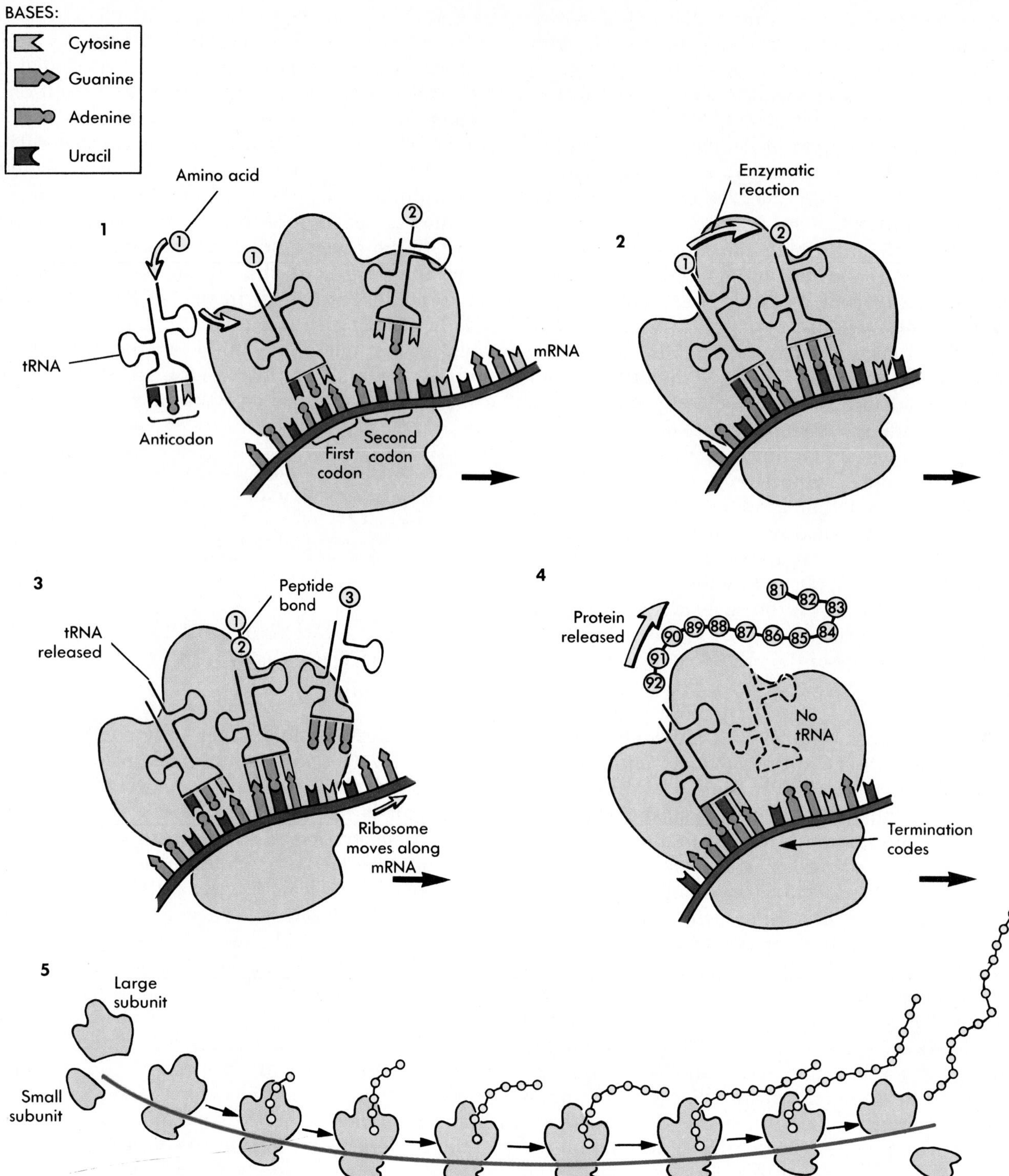

FIGURE 3-15 ■ Protein synthesis.
Translation. *(1)* In the cytoplasm the mRNA strand becomes associated with the ribosomes. Amino acids attach to specific tRNAs, and the tRNAs associate with the mRNA on the ribosome so that the three bases of tRNA (anticodon) pair with three bases of the mRNA (codon). *(2)*, An enzyme within the ribosome joins the two amino acids attached to the tRNAs on the ribosome to form a peptide bond. *(3)*, Amino acid separates from one tRNA, which is released from the ribosome as another tRNA enters the ribosome and pairs with the next codon of the mRNA. *(4)*, The peptide chain is formed as amino acids from tRNAs are joined by peptide bonds until a special codon stops chain elongation, and the protein chain separates from the ribosome. The ribosome then separates into its two subunits and separates from the mRNA. *(5)*, A schematic view of the entire process.

stage represents major events, the process of mitosis is continuous. Learning each of the stages is helpful, but the most important concept is how each cell produced obtains the same number and type of chromosomes as the parent cell.

The first stage of mitosis is **prophase** (see Figure 3-16, *B*). In this stage the genetic material condenses to form visible chromosomes. After interphase, each chromosome is made up of two separate strands called **chromatids** (kro'mah-tidz), which are linked at one point by a specialized region called the **centromere** (sen'tro-mēr). Replication of the genetic material during interphase results in the two identical chromatids of each chromosome. Also during prophase, microtubules called spindle fibers extend from the centromeres to the centrioles (see Figure 3-1). **Centrioles** are small organelles that divide and migrate to each pole of the nucleus. In late prophase, the nucleolus and nuclear envelope disappear.

The second stage of mitosis is **metaphase** (see Figure 3-16, *C*). During metaphase, the chromosomes align along the center of the cell.

The third stage of mitosis is **anaphase** (see Figure 3-16, *D*). At the beginning of anaphase the centromeres separate. When the centromeres separate each chromatid then is referred to as a chromosome. So, when the centromeres divide, the chromosome number doubles to form two identical sets of 46 chromosomes. The two sets of 46 chromosomes are pulled toward the poles of the cell by the spindle fibers. At the end of anaphase the chromosomes have reached the poles of the cell, and the cytoplasm begins to divide.

The fourth stage of mitosis is **telophase** (see Figure 3-16, *E*). The chromosomes in each of the daughter cells become organized to form two separate nuclei. The chromosomes begin to unravel and resemble the genetic material during interphase. Following telophase the cytoplasm of the two cells complete division and two separate daughter cells are produced (see Figure 3-16, *F*).

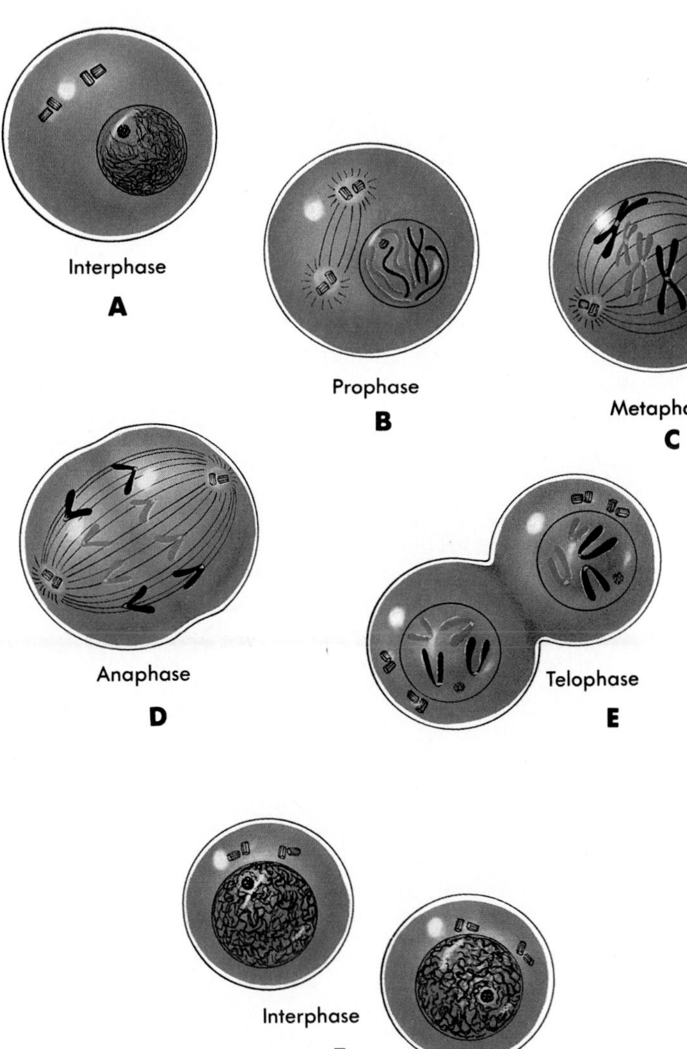

FIGURE 3-16 ■ Mitosis.
A Interphase. DNA replication occurs at this stage.
B Prophase. The centrioles move to opposite ends of the cell, the DNA and associated proteins condense into chromosomes, and the nucleolus and the nuclear envelope disappear.
C Metaphase. The chromosomes align in the center of the cell in association with the spindle fibers.
D Anaphase. The chromosomes move toward the centrioles at each end of the cell.
E Telophase. The chromosomes disperse, the nuclear envelope forms, and following telophase the cytoplasm divides into two cells.
F Interphase.

Relationships between Cell Structure and Cell Function

Each cell is well adapted for the functions it performs, and the abundance of organelles in each cell reflects the cell's function. For example, epithelial cells that line the larger-diameter respiratory passages secrete mucus and transport the mucus toward the throat, where it is either swallowed or expelled from the body by coughing. Particles of dust and other debris suspended in the air become trapped in the mucus. Thus the production and transport of mucus from the respiratory passages function to keep the respiratory passages clean. Cells of the respiratory passages have abundant rough endoplasmic reticulum, Golgi apparatuses, and secretory vesicles. Cilia are on the surface of the respiratory epithelial cells. The ribosomes on the rough endoplasmic reticulum are the sites where the proteins, which are a major component of mucus, are produced. The Golgi apparatuses package the proteins and other components of mucus into secretory vesicles, which move to the surface of the epithelial cells. Their contents are released onto the surface of the epithelial cells. Cilia then propel the mucus toward the throat.

In people who smoke, the prolonged exposure of the respiratory epithelium to the irritation of tobacco smoke causes the respiratory epithelial cells to change in structure and function. The cells may flatten and form several layers of epithelial cells. The flat epithelial cells no longer contain abundant rough endoplasmic reticulum, Golgi apparatuses, secretory vesicles, or cilia. The cells are adapted to protect the underlying cells from irritation, but they no longer function to clean the respiratory passages. Extensive replacement of normal epithelial cells in respiratory passages is correlated with chronic inflammation of the respiratory passages (bronchitis), which often exists in people who smoke heavily.

Differentiation

Each individual begins as a single fertilized egg. The trillions of cells that ultimately make up the body of an adult stem from that single fertilized egg. All the cells in an individual's body therefore contain the same complement of DNA that encodes all of the genetic information for that individual. Not all cells look and function alike even though their genetic information is identical. Bone cells, for example, do not look like or function as fat cells or red blood cells.

The process by which cells develop specialized structures and functions is called **differentiation.** Early in development, cells begin to differentiate to give rise to the different cell types. Differentiation continues with each mitotic division until specific mature cells are formed.

During differentiation of a cell some portions of DNA are active while other portions of the DNA are inactive. The active and inactive sections of DNA differ with each cell type. The portion of DNA that is responsible for the structure and function of a bone cell is a different portion of DNA from that responsible for the structure and function of a fat cell. Differentiation, then, results from the selective activation and inactivation of DNA. We do not fully understand the mechanisms that determine which portions of DNA are active in any one cell type, but it is the resulting differentiation that produces the many cell types which function together to make a person.

5 Cancer cells divide continuously. The normal mechanisms that regulate whether cell division occurs or ceases do not function properly in cancer cells. Do cancer cells, such as breast cancer cells or skin cancer cells, look like mature breast cells or mature skin cells? Explain. ?

SUMMARY

Cells are highly organized units composed of living material.

The nucleus contains genetic material, and cytoplasm is living material outside the nucleus.

STRUCTURE OF THE PLASMA MEMBRANE

The plasma membrane forms the outer boundary of the cell. It determines what enters and leaves the cell.

The plasma membrane is composed of a double layer of lipid molecules in which proteins float. The proteins function as membrane channels, carrier molecules, receptor molecules, enzymes, and structural components of the membrane.

MOVEMENT THROUGH THE PLASMA MEMBRANE

Lipid-soluble molecules pass through the plasma membrane readily by dissolving in the lipid portion of the membrane.

Small molecules may pass through membrane channels.

Large molecules that are not lipid soluble may be transported through the membrane by carrier molecules.

Diffusion

Diffusion is the movement of a substance from an area of high concentration to an area of low concentration, until there is a uniform distribution of molecules.

For a given distance, a concentration gradient is equal to the higher concentration minus the lower concentration of a solute in a solution.

The rate of diffusion increases with an increase in the concentration gradient, an increase in temperature, or a decrease in molecule size.

Osmosis

Osmosis is the diffusion of a solvent (water) across a selectively permeable membrane.

Osmotic pressure is a measure of the tendency of water to move across the selectively permeable membrane.

Cells placed in an isotonic solution neither swell nor shrink. In a hypertonic solution they shrink (crenation), and in a hypotonic solution they swell (and may undergo lysis).

Filtration

Filtration is the passage of a solution through a membrane in response to a mechanical pressure difference.

Mediated Transport Mechanisms

Mediated transport is the movement of a substance across a membrane by means of a carrier molecule. The substances transported tend to be large, water-soluble molecules.

Facilitated diffusion moves substances from a high to a low concentration and does not require energy in the form of ATP.

Active transport can move substances from a low to a high concentration and requires ATP. An exchange pump is an active transport mechanism that simultaneously moves two substances in opposite directions across the cell membrane.

Endocytosis and Exocytosis

Endocytosis is the movement of materials into cells by the formation of a vesicle.

- Phagocytosis is the movement of solid material into cells by the formation of a vesicle.
- Pinocytosis is similar to phagocytosis, except that the material ingested is much smaller and is in solution.

Exocytosis is the secretion of materials from cells by vesicle formation.

ORGANELLES AND CELL FUNCTIONS

Nucleus

The nuclear envelope consists of two separate membranes with nuclear pores.

DNA and associated proteins are found inside the nucleus. DNA is the hereditary material of the cell and controls the activities of the cell. DNA determines the structure of messenger RNA, which, in turn, determines the structure of proteins.

Nucleoli and Ribosomes

Nucleoli consist of RNA and proteins and are the sites of ribosomal subunit assembly.

Ribosomes are the sites of protein synthesis.

Rough and Smooth Endoplasmic Reticulum

Rough endoplasmic reticulum is endoplasmic reticulum with ribosomes attached. It is a major site of protein synthesis.

Smooth endoplasmic reticulum does not have ribosomes attached and is a major site of lipid synthesis.

The Golgi Apparatus

The Golgi apparatus is a series of closely packed membrane sacs that function to concentrate and package into secretory vesicles the lipids and proteins produced by the endoplasmic reticulum.

Secretory Vesicles

Secretory vesicles are membrane-bound sacs that carry substances from the Golgi apparatus to the plasma membrane, where the vesicle contents are released by exocytosis.

Lysosomes

Membrane-bound sacs containing enzymes are called lysosomes. Within the cell the lysosomes break down phagocytized material.

Mitochondria

Mitochondria are the major sites of ATP production, which cells use as an energy source. Enzymes that carry out oxidative metabolism (requires oxygen) are contained in the mitochondria.

Microtubules

Microtubules are hollow tubes that support the cytoplasm and are components of cilia and flagella.

Cilia, Flagella, and Microvilli

Cilia move substances over the surface of cells.

Flagella are much longer than cilia and propel spermatozoa.

Microvilli increase the surface area of cells and aid in absorption.

PROTEIN SYNTHESIS

The sequence of nucleotides in DNA determines the sequence of nucleotides in mRNA; the mRNA moves through the nuclear pores to ribosomes.

Transfer RNAs, which carry amino acids, interact at the ribosome with mRNA. Nucleotides of tRNA bind to the codons of mRNA, and the amino acids are joined to form a protein.

Cell activity is regulated by enzymes (proteins), and DNA controls enzyme production.

CELL DIVISION AND DIFFERENTIATION

Mitosis

Cell division occurs by mitosis in all tissues except those that produce reproductive cells; mitosis produces new cells for growth and tissue repair.

DNA replicates during interphase, the time between cell division.

Mitosis is divided into four stages:

- Prophase. Each chromosome consists of two chromatids joined at the centromere.
- Metaphase. Chromosomes align at the center of the cell.
- Anaphase. Chromatids separate at the centromere and migrate to opposite poles.
- Telophase. The two new nuclei assume their normal structure and cell division is completed, producing two new daughter cells.

Differentiation

Differentiation, the process by which cells develop specialized structures and functions, results from the selective activation and inactivation of DNA.

CONTENT REVIEW

1. Define cytoplasm and cell organelle.
2. Describe the structure of the plasma membrane. What functions does it perform?
3. How do lipid-soluble molecules, small molecules that are not lipid-soluble, and large molecules that are not lipid-soluble cross the cell membrane?
4. Define diffusion. How do concentration differences, temperature, and molecule size affect the rate of diffusion?
5. Define osmosis and osmotic pressure.
6. What happens to cells that are placed in isotonic solutions? In hypertonic or hypotonic solutions? What are crenation and lysis?
7. Define filtration.
8. What is mediated transport? How are facilitated diffusion and active transport similar, and how are they different?
9. Describe phagocytosis, pinocytosis, and exocytosis. What do they accomplish?
10. Describe the structure of the nucleus and nuclear pores. Name the organelles found in the nucleus and give their functions.
11. Where are ribosomes assembled, and what kinds of molecules are found in them?
12. What is endoplasmic reticulum? Compare the functions of rough and smooth endoplasmic reticulum.
13. Describe the Golgi apparatus, and state its function.
14. Where are secretory vesicles produced? What are their contents, and how are they released?
15. What is the function of the lysosomes?
16. Describe the structure and function of mitochondria.
17. What are some functions of microtubules?
18. Describe the structure and function of cilia, flagella, and microvilli.
19. Describe how proteins are synthesized and how the structure of DNA determines the structure of proteins.
20. Define autosome, sex chromosome, and diploid number. How do the sex chromosomes of males and females differ?
21. Describe what happens during interphase and each phase of mitosis. What kind of tissues undergo mitosis?

CONCEPT REVIEW

1. A man's body was found floating in the salt water of San Francisco Bay, which has about the same osmotic concentration as body fluids. When seen during an autopsy, the cells in his lung tissues were clearly swollen. Choose the most logical conclusion.
   A. He probably drowned in the bay.
   B. He probably drowned, but in fresh water rather than in the bay.
   C. He did not drown.
   D. The data do not allow you to determine if the person drowned in the bay.
2. A dialysis membrane is selectively permeable, and substances smaller than proteins are able to pass through it. If you wanted to use a dialysis machine to remove only urea (a small molecule) from blood, what could you use for the dialysis fluid? The dialysis machine allows blood to flow past one side of the dialysis membrane and the dialysis fluid flows on the other side of the membrane.

A. A solution that is isotonic and contains only protein.
B. A solution that is isotonic and contains the same concentration of substances as blood, except for having no urea in it.
C. Distilled water.
D. Blood.

3. A researcher wanted to determine the nature of the transport mechanism that moved substance X into a cell. She could measure only the concentration of substance X in the extracellular fluid and within the cell, so she did a series of experiments and gathered the data presented here.

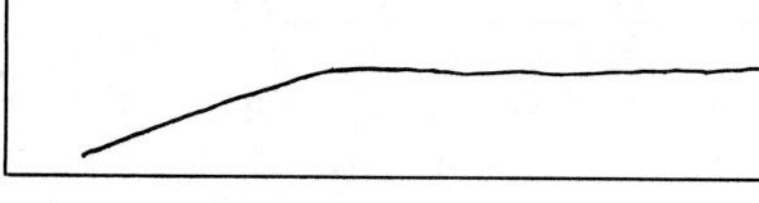

At point *A* the extracellular concentration of substance X was equal to the intracellular concentration of substance X. Choose the transport process that is consistent with the data.
A. Diffusion
B. Active transport
C. Facilitated diffusion
D. There is inadequate information to make a judgment.

4. Given the following data from electron micrographs of a cell, predict the major function of the cell:
Moderate number of mitochondria
Well-developed rough endoplasmic reticulum
Moderate number of lysosomes
Well-developed Golgi apparatuses
Numerous vesicles
No microvilli at the cell surface

5. Assume that you can measure the rate at which a protein is released from a cell in response to a stimulus. In addition, you have been given the problem of determining whether the proteins are stored in the cell some time prior to secretion or if they are synthesized and immediately secreted. You have a drug that inhibits messenger RNA synthesis. How could you solve the problem?

ANSWERS TO ? PREDICT QUESTIONS

1 *p. 47* Urea is produced continually by metabolizing cells and diffuses from the cells into the interstitial spaces and from the interstitial spaces into the blood. If the kidneys stop eliminating urea, it begins to accumulate in the blood. Because the concentrations increase in the blood, urea cannot diffuse from the interstitial spaces. Because urea accumulates in the interstitial spaces, it cannot diffuse from the cells. The urea finally reaches concentrations high enough to be toxic to cells, causing cell damage followed by cell death.

2 *p. 50* Glucose transported by facilitated diffusion across the cell membrane will move from a high to a low concentration. If glucose molecules are converted quickly to some other molecule as they enter the cell, a large concentration difference is maintained, and so, glucose transport into the cell continues proportional to the magnitude of the concentration difference.

3 *p. 56* (A) Cells specialized to synthesize and secrete proteins would have abundant rough ER, because this is an important site of protein synthesis. Well-developed Golgi apparatuses would exist to package proteins in secretory vesicles, and numerous secretory vesicles would be present.

(B) Cells highly specialized to actively transport substances into the cell would have a large surface area exposed to the fluid from which substances are actively transported. Numerous mitochondria would be present near the membrane across which active transport occurs.

(C) Cells highly specialized to phagocytize foreign substances would have numerous lysosomes in their cytoplasm and evidence of phagocytic vesicles.

4 *p. 58* By changing a single nucleotide within a DNA molecule, a change also would occur in the nucleotide sequence of messenger RNA produced from that segment of DNA. The change in mRNA would result in a different codon, and a different amino acid would be placed in the amino acid chain for which the messenger RNA codes. Because a change in the amino acid sequence of a protein could change its structure, one substitution of a nucleotide in a DNA chain could result in altered protein structure and function.

5 *p. 60* Because they are continuously undergoing mitosis, cancer cells generally appear to be undifferentiated. Instead of dividing and then undergoing differentiation, they continue to divide and do not differentiate. One measure of the severity of cancer is related to the degree of differentiation the cancer cells do undergo. Those that are more differentiated divide more slowly and are less dangerous than those that do not differentiate as much.

4

TISSUES, GLANDS, AND MEMBRANES

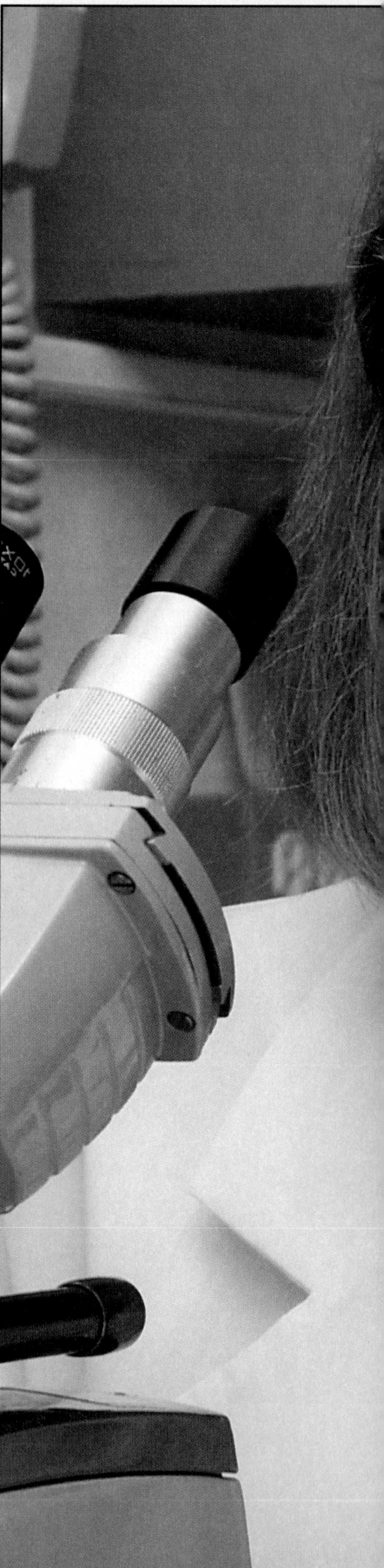

After reading this chapter you should be able to:

1 List the characteristics of epithelial tissue.
2 Classify and give an example of the major types of epithelium.
3 Explain the functional significance for epithelium of the following: cell layers, cell shapes, free cell surfaces, and connections between cells.
4 Define and categorize glands.
5 Describe the basis for classifying connective tissue and give examples of each major type.
6 Name the three types of muscle and list their functions.
7 State the functions of nervous tissue and describe a neuron.
8 List the structural and functional characteristics of mucous and serous membranes.
9 Describe the process of inflammation and explain why inflammation protects the body.
10 Describe the major events involved in tissue repair.

connective tissue One of the four major tissue types; consists of cells surrounded by large amounts of extracellular material; functions to hold other tissues together and provides a supporting framework for the body.

epithelial tissue One of the four major tissue types; consists of cells with a basement membrane (exceptions are lymph vessels and liver sinusoids), little extracellular material, and no blood vessels; covers the surfaces of the body and forms glands.

extracellular matrix Nonliving chemical substances located between cells; consists of protein fibers, ground substance, and fluid.

gland A single cell or a multicellular structure that secretes substances into the blood, into a cavity, or onto a surface.

inflammatory response Complex sequence of events involving chemicals and immune system cells that results in the isolation and destruction of foreign substances such as bacteria; symptoms include redness, heat, swelling, pain, and disturbance of function.

mucous membrane (mu′kus) Thin sheet consisting of epithelium and connective tissue that lines cavities that open to the outside of the body; many contain mucous glands, which secrete mucus.

muscle tissue One of the four major tissue types; consists of cells with the ability to contract; includes skeletal, cardiac, and smooth muscle.

nervous tissue One of the four major tissue types; consists of neurons, which have the ability to conduct action potentials, and neuroglia, which are support cells.

serous membrane (sēr′us) Thin sheet consisting of epithelium and connective tissue that lines cavities not opening to the outside of the body; does not contain glands but does secrete serous fluid.

tissue [L. *texo*, to weave] A collection of cells with similar structure and function, and the substances between the cells.

tissue repair Substitution of viable cells for damaged or dead cells by regeneration or replacement.

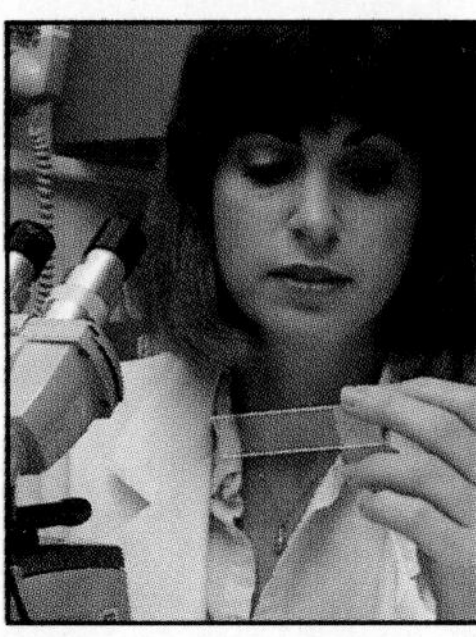

A **tissue** is a group of cells with similar structure and function plus the extracellular substances located between the cells. The microscopic study of the structure of tissues is called **histology** (his-tol'o-je). Knowledge of tissue structure and function is important in understanding how individual cells are organized to form tissues, and how tissues are organized to form organs, organ systems, and the complete organism. There are four basic tissue types: epithelial, connective, muscular, and nervous tissue. This chapter emphasizes epithelial and connective tissues. Muscular and nervous tissues are considered in more detail in later chapters.

EPITHELIAL TISSUE

Epithelium (plural: epithelia) covers surfaces of the body or forms glands. Surfaces of the body include the skin on the outside of the body and the lining of cavities inside the body such as the digestive tract and blood vessels. Epithelium consists almost entirely of cells that have very little extracellular material between them. Although there are some exceptions, most epithelia have a **free surface,** which is not in contact with other cells, and a **basement membrane,** which attaches the epithelial cells to underlying tissues (Figure 4-1). The basement membrane is a mixture of carbohydrates and proteins secreted by epithelial cells and cells of the underlying tissue. Epithelium does not have blood vessels, so gases and nutrients that reach the epithelium must diffuse across the basement membrane from the underlying tissues.

Classification

Epithelia are classified according to the number of cell layers and the shape of the cells. **Simple epithelium** consists of a single layer of cells, and **stratified epithelium** consists of more than one layer of epithelial cells (some cells sit on top of other cells). Categories of epithelium based on cell shape are **squamous** (skwa'mus; flat), **cuboidal** (cubelike), and **columnar** (tall and thin). In most cases, an epithelium is given two names such as simple squamous, simple columnar, or stratified squamous. When epithelium is stratified, it is named according to the shape of cells at the free surface.

Simple squamous epithelium is a single layer of thin, flat cells (Figure 4-2, *A*). Because substances easily pass through this thin layer of tissue, it is often found where diffusion takes place. For example, the respiratory passages end as small sacs called alveoli. The alveoli consist of simple squamous epithelium that allows oxygen from the air to diffuse into the body, and for carbon dioxide to diffuse out of the body into the air.

Simple squamous epithelium also functions to prevent abrasion between organs in the thoracic and abdominopelvic cavities. The outer surfaces of the organs are covered with simple squamous epithelium that secretes a slippery fluid. The fluid lu-

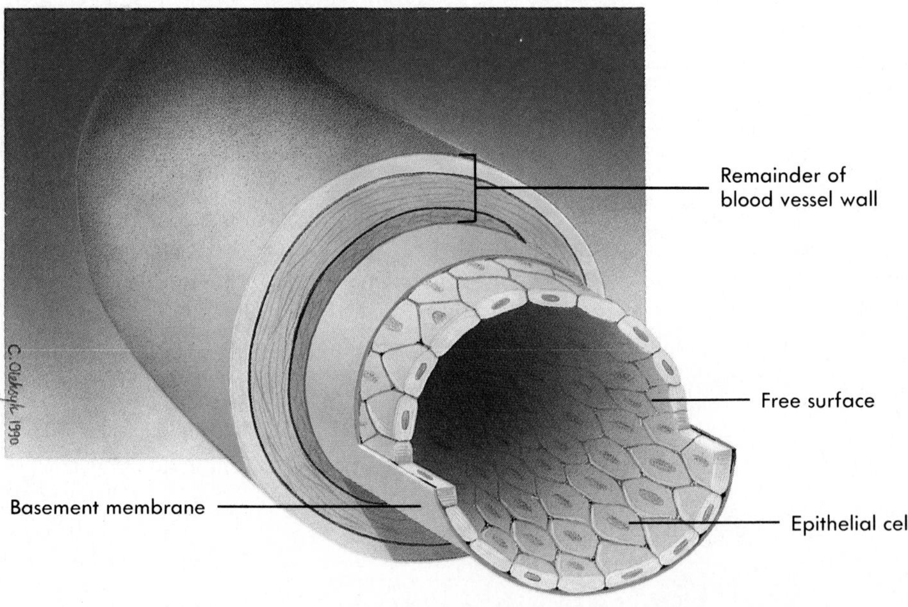

FIGURE 4-1 ■ Characteristics of epithelium.
Epithelium lining a blood vessel illustrates the following characteristics of epithelium: little extracellular material between cells, a free surface, a basement membrane, and no blood vessels directly supplying the epithelium.

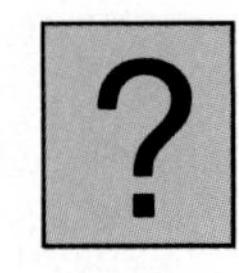

FIGURE 4-2, cont'd ■ **Epithelial Tissues**

	LOCATION	STRUCTURE/FUNCTION

D ■ **Stratified squamous**

Skin, mouth, throat, esophagus, anus, vagina, and cornea

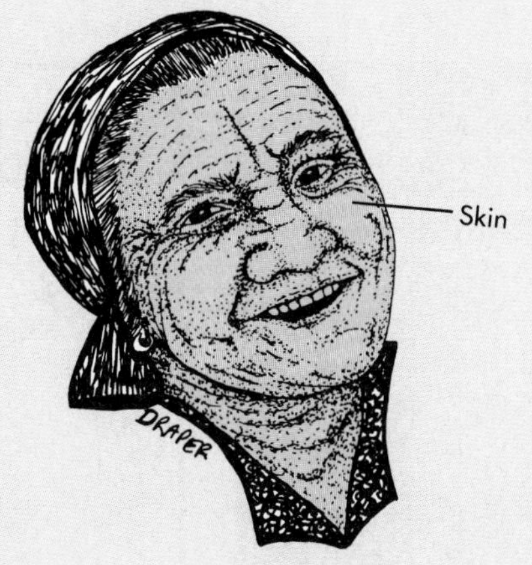

STRUCTURE
Many layers of cells in which the basal layer is cuboidal and becomes flattened at the free surface

FUNCTION
Protection against abrasion and infection

E ■ **Pseudostratified**

Nasal cavity, nasal sinuses, auditory tubes, pharynx, larynx, trachea, and bronchi of lungs

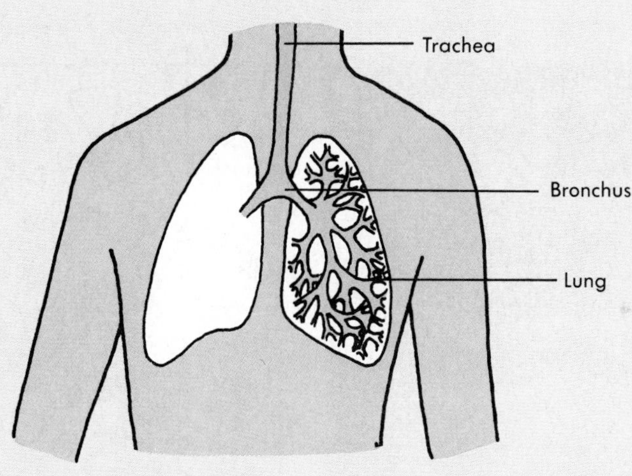

STRUCTURE
Single layer of cells. Some cells are tall and thin and reach the free surface, and others do not. The nuclei of these cells are at different levels and appear stratified. The cells are almost always ciliated and are associated with goblet cells which produce mucus.

FUNCTION
Movement of mucus (or fluid) that contains foreign particles

F ■ **Transitional**

Urinary bladder, ureters, and superior urethra

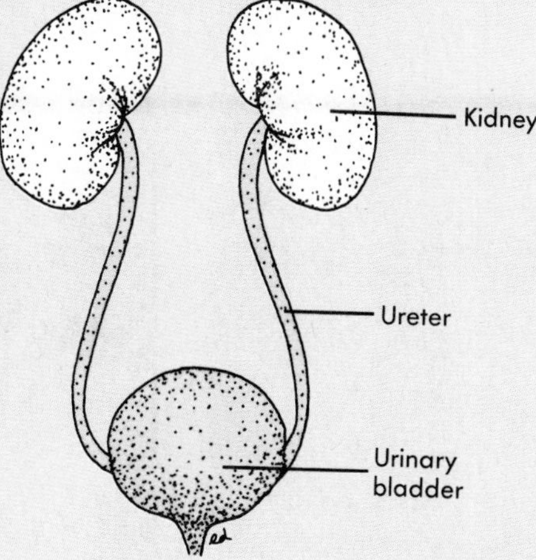

STRUCTURE
Stratified cells that appear cubelike when the organ or tube is relaxed and appear squamous when the organ or tube is distended by fluid

FUNCTION
Accommodation of fluid fluctuations in an organ or tube; protection against the caustic effects of urine

HISTOLOGY

Capillary endothelium (× 1000)

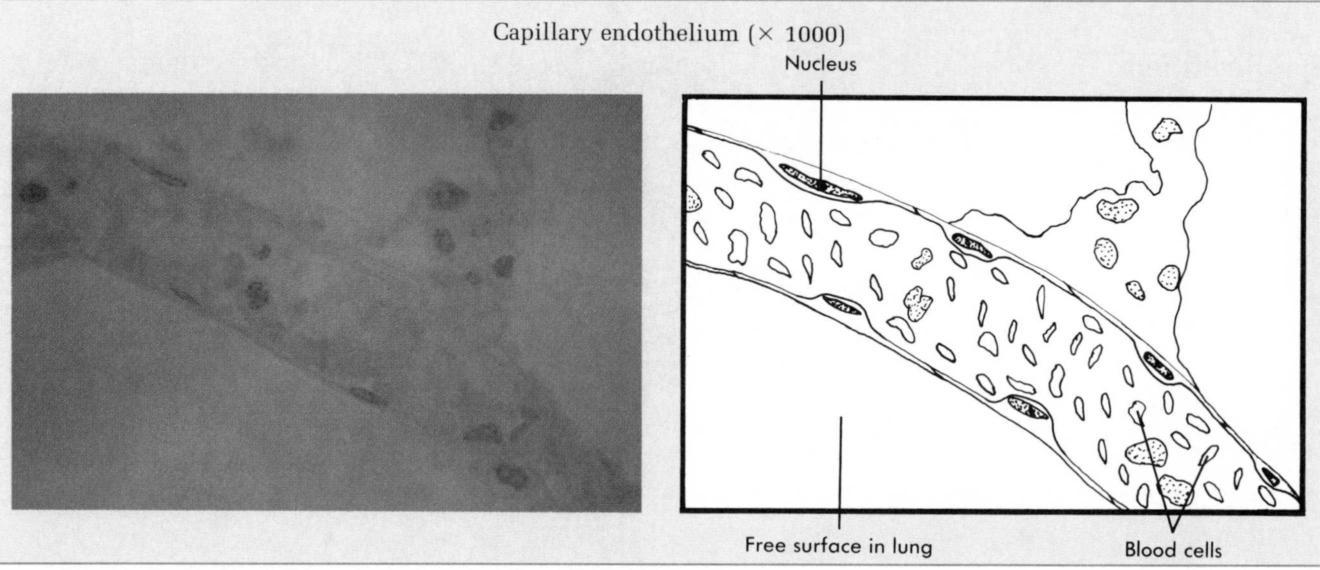

Kidney tubule (× 260)

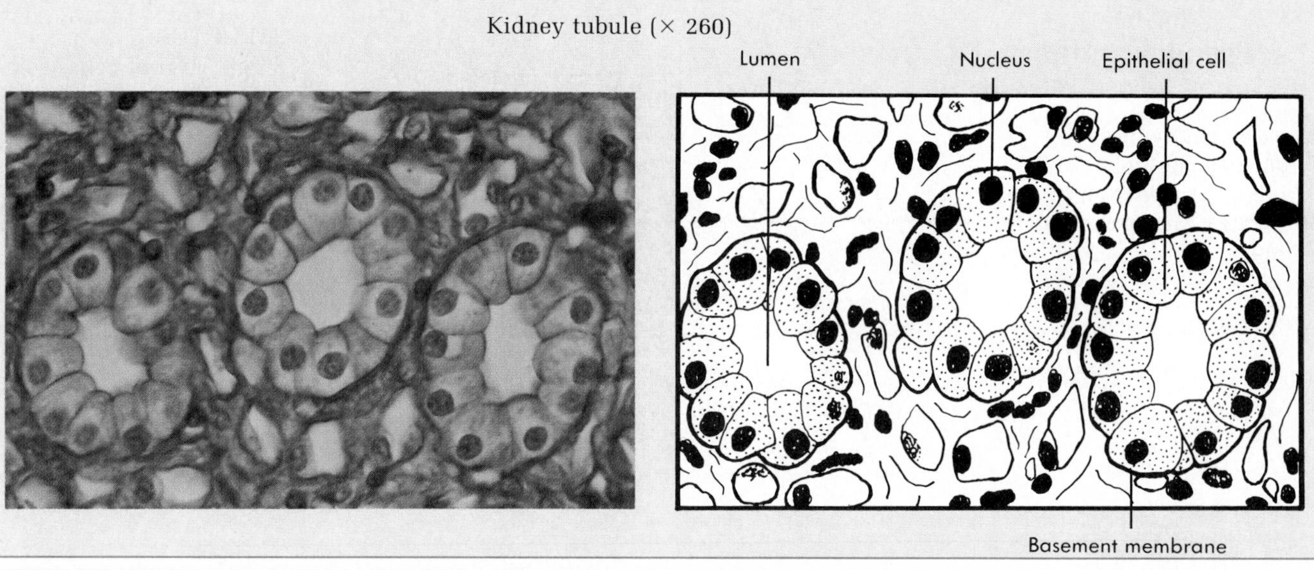

Small intestine (× 1000)

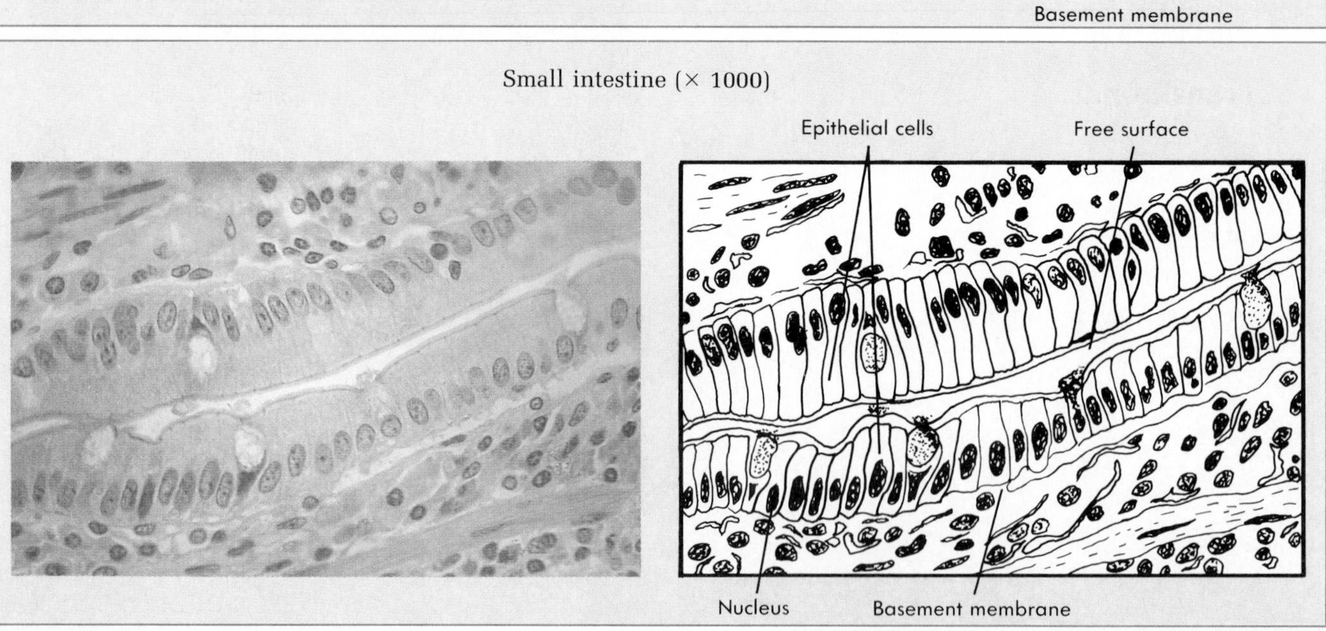

Continued.

FIGURE 4-2 ■ Epithelial Tissues

	LOCATION	STRUCTURE/FUNCTION

A ■ **Simple squamous**

Lining of blood vessels, heart, lymph vessels, and serous membranes, alveoli (air sacs) of lungs, and kidney tubules (Bowman's capsule, thin segment of the loop of Henle)

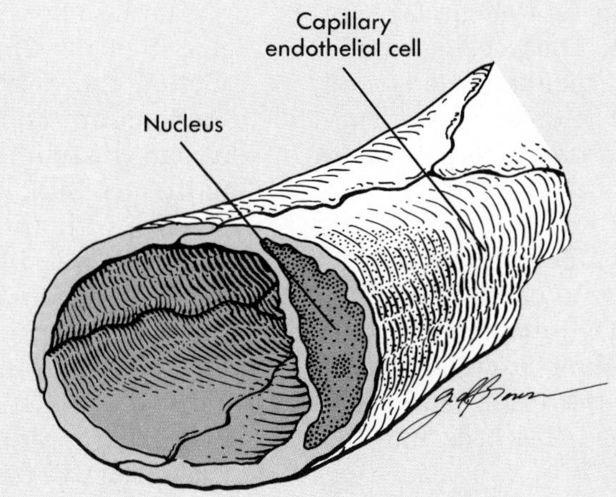

STRUCTURE
Single layer of thin, flat cells

FUNCTION
Diffusion, filtration, and protection against friction (secrete serous fluid)

B ■ **Simple cuboidal**

Kidney tubules, glands and their ducts, choroid plexus of the brain, terminal bronchioles of lungs, and surface of the ovaries and retina

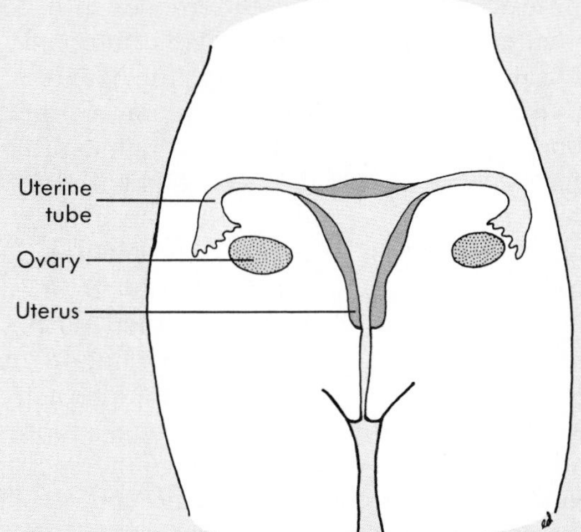

STRUCTURE
Single layer of cube-shaped cells; some cells have microvilli (kidney tubules) or cilia (terminal bronchioles of lungs)

FUNCTION
Secretion and absorption by cells of the kidney tubules; secretion by cells glands and the choroid plexus; movement of mucus-containing particles out of the terminal bronchioles by ciliated cells

C ■ **Simple columnar**

Stomach, intestines, glands, some ducts, bronchioles of lungs, auditory tubes, uterus, and uterine tubes

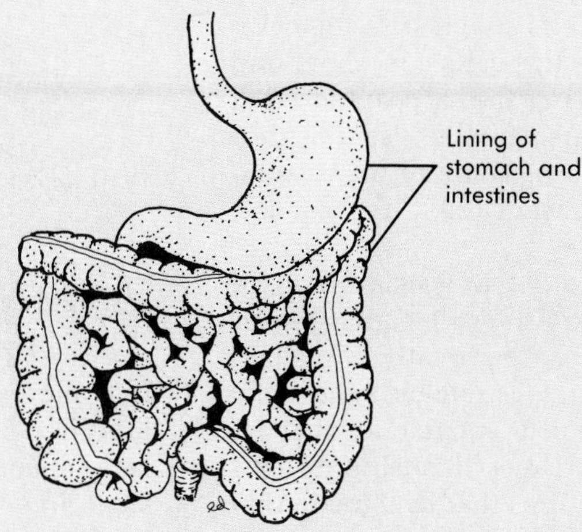

STRUCTURE
Single layer of tall, narrow cells; some have microvilli (stomach, intestines, and glands) or cilia (bronchioles of lungs, auditory tubes, uterus, and uterine tubes)

FUNCTION
Secretion by cells of the stomach, intestines, and glands; absorption by cells of the intestine; movement of mucus by ciliated cells clears the lungs and is partially responsible for the movement of the egg through the uterine tubes

bricates the surfaces between the organs, preventing damage from friction when the organs rub against each other.

Simple cuboidal epithelium is a single layer of cubelike cells (Figure 4-2, *B*). These cells have a greater volume than simple squamous cells and contain more cell organelles. The organelles of simple cuboidal cells produce carrier molecules and ATPs that control the movement of materials into and out of the cells by active transport and facilitated diffusion (see Chapter 3). Thus movement of materials across a layer of simple cuboidal epithelium can be regulated. In kidney tubules, for example, simple cuboidal epithelium excretes waste products from the body into the urine, and reabsorbs useful materials from the urine back into the body.

Simple columnar epithelium is a single layer of cells that are tall and thin (Figure 4-2, *C*). These large cells contain organelles that enable them to perform complex functions. For example, the simple columnar epithelium of the small intestine produces and releases digestive enzymes that complete the process of digesting food. The columnar cells then absorb the digested foods by active transport, facilitated diffusion, or simple diffusion.

Stratified squamous epithelium forms a thick epithelium because it consists of many layers of cells (Figure 4-2, *D*). The deepest cells are columnar and are capable of dividing and producing new cells. As these newly formed cells are pushed to the surface they become flat and thin. Stratified squamous epithelium forms the outer layer of the skin, providing protection against abrasion. As cells at the surface are damaged and rubbed away, they are replaced by cells formed in the deeper layers.

Pseudostratified epithelium is a special type of simple epithelium (Figure 4-2, *E*). The term "pseudo" means false, so this type of epithelium appears to be stratified but is not. It consists of one layer of cells with all the cells attached to the basement membrane. Some of the cells are short and some of them are tall, giving the appearance that there are two or more layers of cells. Pseudostratified epithelium is found lining some of the respiratory passages such as the nasal cavity, trachea, and bronchi.

Transitional epithelium is a special type of stratified epithelium (Figure 4-2, *F*) consisting of many layers of cells that can be greatly stretched. As transitional epithelium is stretched, the cells change shape from cuboidal to squamous, and the number of cell layers decreases. Transitional epithelium is found lining cavities that can greatly expand such as the urinary bladder.

Functional Characteristics

Cell layers and cell shapes

The number of cell layers and the shape of the cells in a specific type of epithelium reflect the function the epithelium performs. Two important functions are controlling the passage of materials through the epithelium and protecting the underlying tissues. Simple epithelium, with its single layer of cells, is found in organs where the principal functions are diffusion (for example, alveoli of lungs), filtration (for example, kidneys), secretion (for example, glands), or absorption (for example, intestines). The movement of materials through a stratified epithelium would be hindered by its many layers. Stratified epithelium is well adapted for a protective function. As the outer cell layers are damaged, they are replaced by cells from deeper layers. Stratified squamous epithelium is found in areas of the body where abrasion can occur such as in the skin, anus, and vagina.

Differences in function are also reflected in cell shape. Cells involved in diffusion are normally flat and thin, for example, the alveoli of the lungs are simple squamous epithelium. Materials from blood diffuse through capillary walls into the surrounding tissues. Cells with the major function of secretion or absorption are usually cuboidal or columnar. Their greater size enables them to contain the organelles responsible for their function. The stomach, for example, is lined with simple columnar epithelium. These cells contain many secretory vesicles filled with **mucus** (mu'kus), which is a clear, viscous material. The large amounts of mucus produced by the simple columnar epithelium protect against digestive enzymes and acid found in the stomach. An ulcer or irritation in the epithelium and underlying tissue can develop if this protective mechanism fails.

1 What type of epithelium would you expect to find in capillaries (small blood vessels that are the site of exchange of substances between the blood and tissues) and in the mouth? ?

Free cell surfaces

Most epithelia have a free surface, which is not in contact with other cells and faces away from underlying tissue. The characteristics of the free surface reflect the functions performed by the free surface. **Smooth** surfaces reduce friction. For example, the lining of blood vessels is simple squamous epi-

Text continued on p. 75.

HISTOLOGY

Skin (× 250)

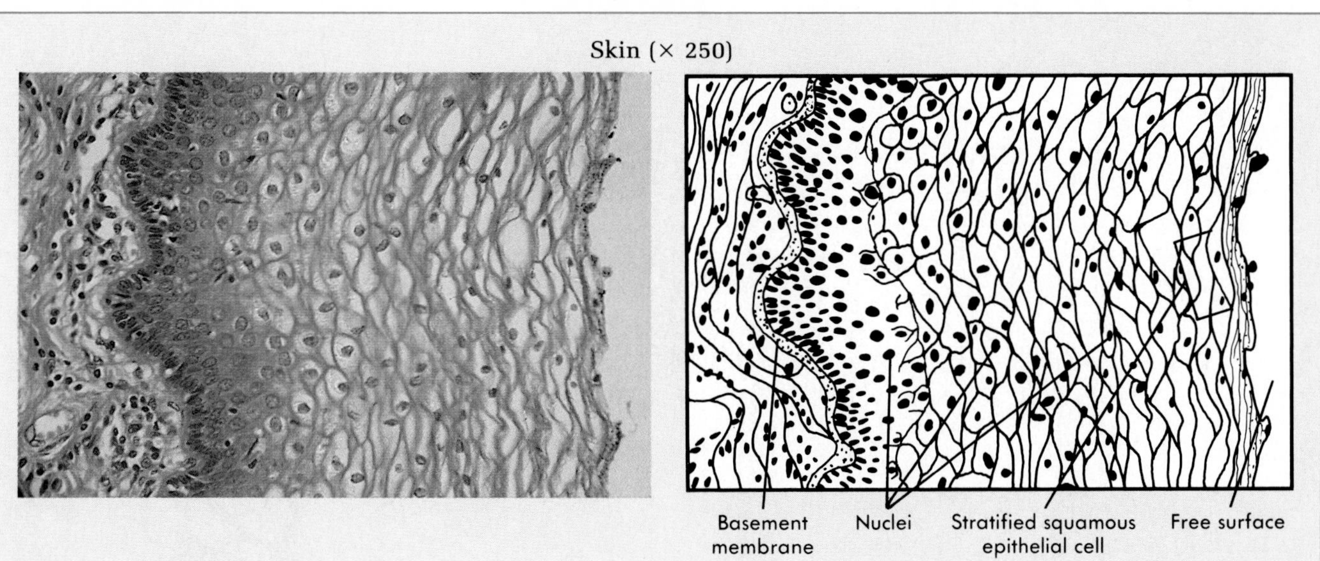

Trachea (× 500)

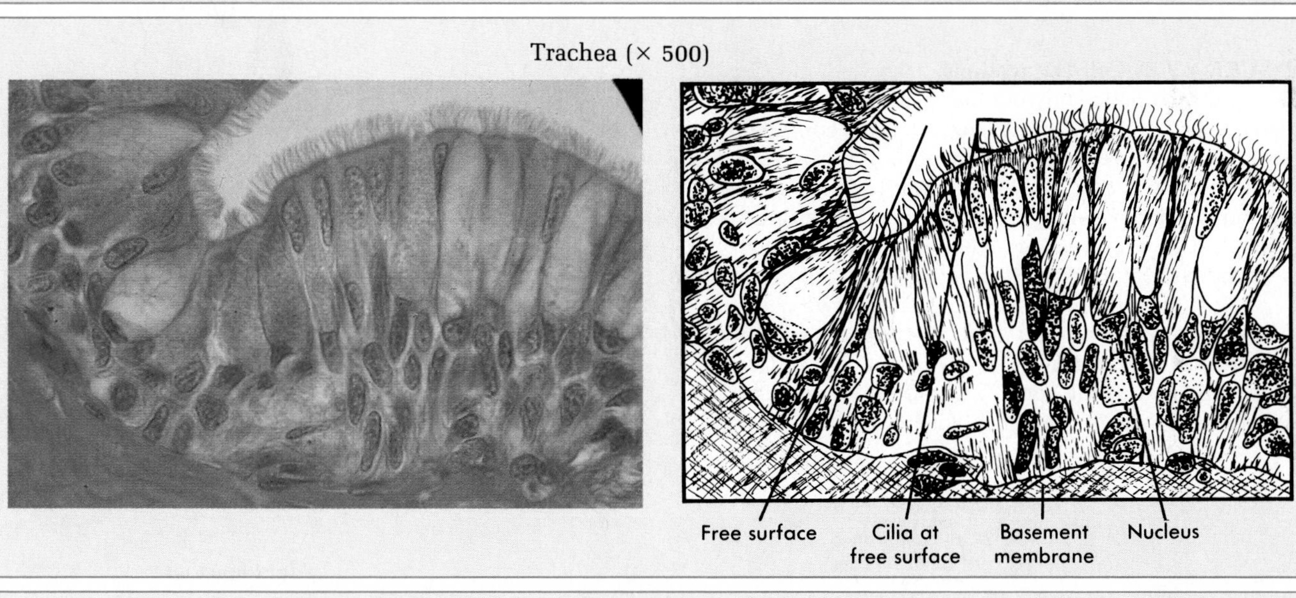

Urinary bladder (× 240)

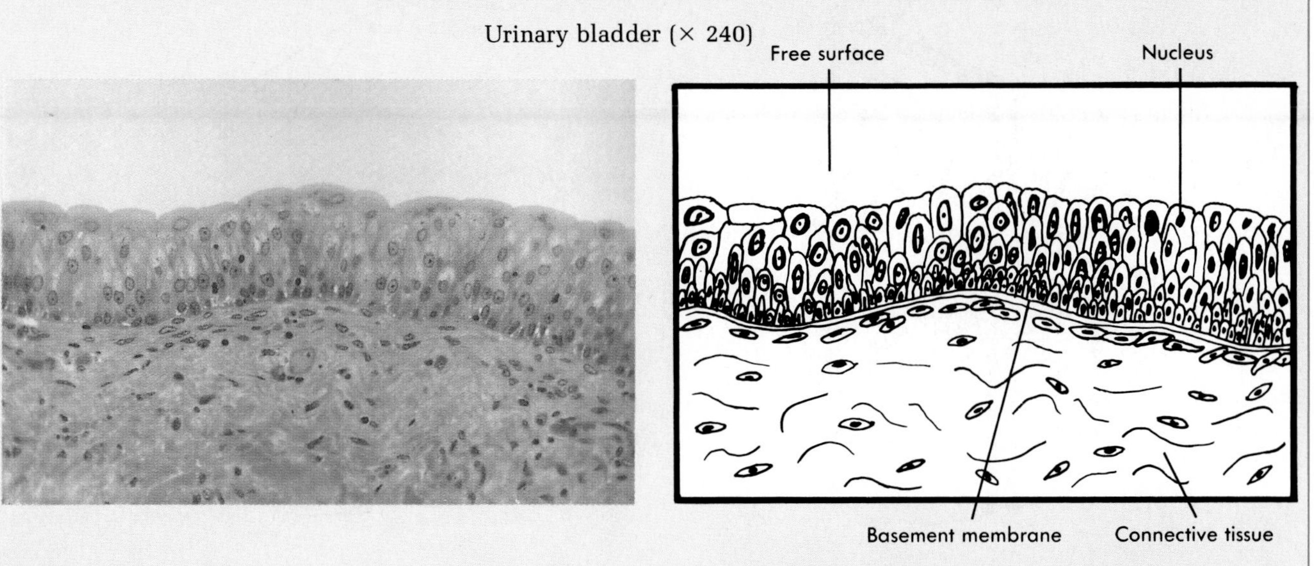

A

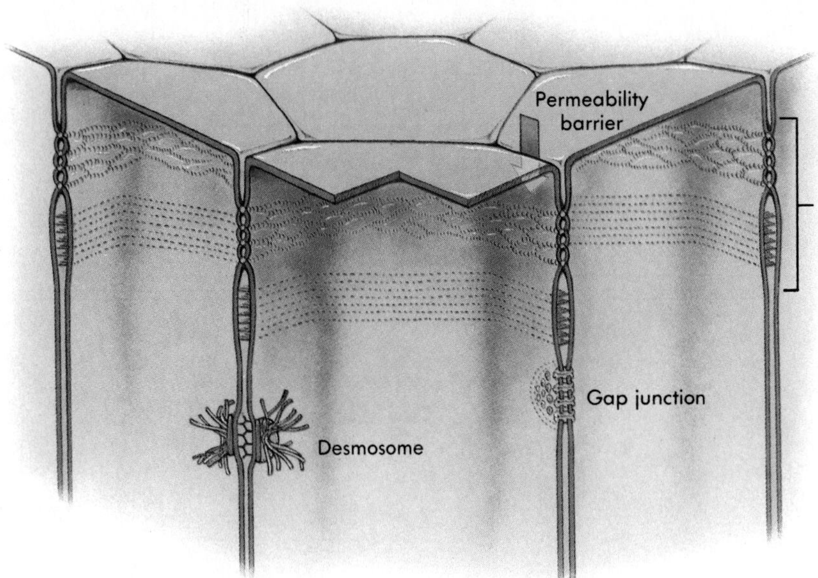

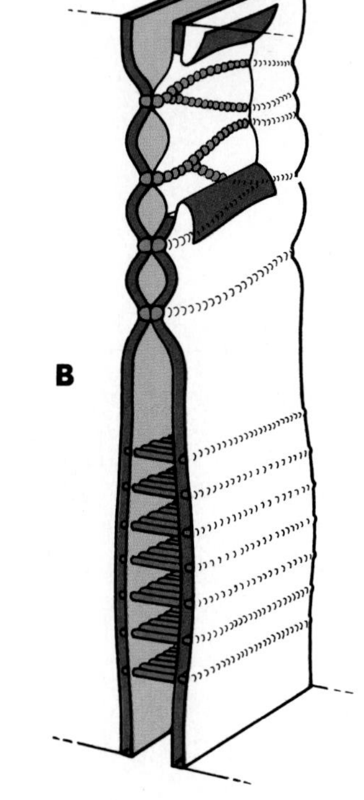

FIGURE 4-3 ■ Cell connections.
A Epithelial cells showing the location of different types of cell connections. Tight junctions completely encircle each cell, forming a permeability barrier (the space between cells is exaggerated in order to illustrate the tight junctions). Desmosomes bind cell membranes together. Gap junctions are small channels between cells that are involved with intercellular communication.
B A portion of a tight junction between two cell membranes. The tight junction effectively seals the two membranes together.
C Details of a desmosome. They are disk-shaped structures joined by protein filaments.
D Details of a gap junction.

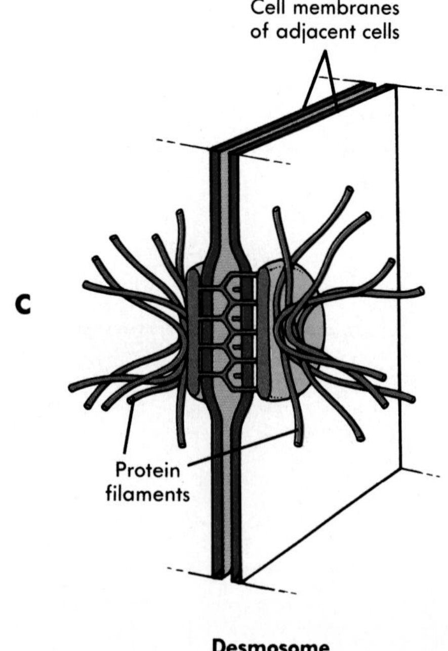

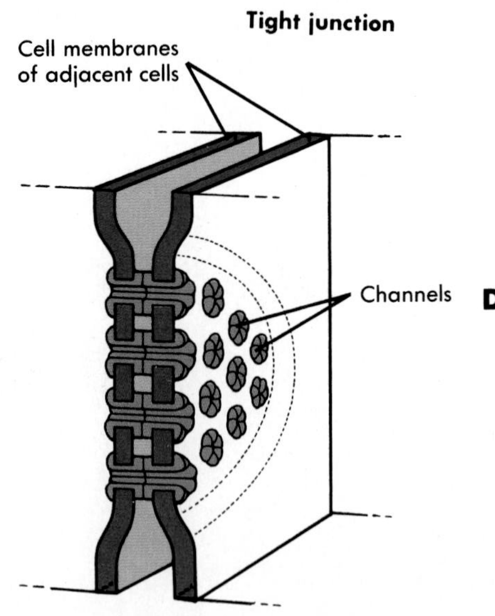

thelium with a smooth surface that reduces friction as blood flows through the vessels. Microvilli are cylindrical extensions of the cell membrane. Normally many microvilli are on each cell, and they function to increase the cell surface area. They are found on the free surface of cells involved in absorption or secretion such as the lining of the small intestine. **Cilia** (see Chapter 3) propel materials along the surface of cells. The nasal cavity is lined with pseudostratified ciliated epithelium. Intermixed with the ciliated cells are specialized mucus-producing cells called **goblet cells.** Dust and other materials are trapped in the mucus that covers the epithelium, and movement of the cilia moves the mucus with its entrapped particles to the back of the throat where it is swallowed. The constant movement of mucus helps to keep the nasal passages clean.

Cell connections

Epithelial cells are connected to each other in several ways. **Tight junctions** bind adjacent cells together and form permeability barriers (Figure 4-3, *A* and *B*). Because tight junctions completely surround each cell, they prevent the passage of materials between cells. Tight junctions force materials to pass through the cells, which can regulate what materials cross the epithelial layer. Tight junctions are found in the lining of the intestines and most other simple epithelia. **Desmosomes** (dez'mo-sōmz) are mechanical links that function to bind cells together (Figure 4-3, *C*). Many desmosomes are found in epithelia subjected to stress, such as the stratified squamous epithelium of the skin. **Gap junctions** are small channels that allow materials to pass from one epithelial cell to an adjacent epithelial cell. Most epithelial cells are connected to each other by gap junctions, and it is believed the gap junction provides a means of intercellular communication (Figure 4-3, *D*).

Glands

A **gland** is a single cell or a multicellular structure that secretes substances onto a surface, into a cavity, or into the blood. Most glands are composed primarily of epithelium. Glands with ducts are called **exocrine** (ek'so-krin) glands (Figure 4-4). They can be **simple,** with ducts that have no branches, or **compound,** with ducts that have many branches. The end of a duct can be **tubular** or expanded into a saclike structure called an **acinus** (as'ĭ-nus) or **alveolus** (al-ve'o-lus). Secretions from the glands pass through the ducts onto a surface or into an organ. For example, sweat (from sweat glands) and oil (from sebaceous glands) flow onto the skin surface.

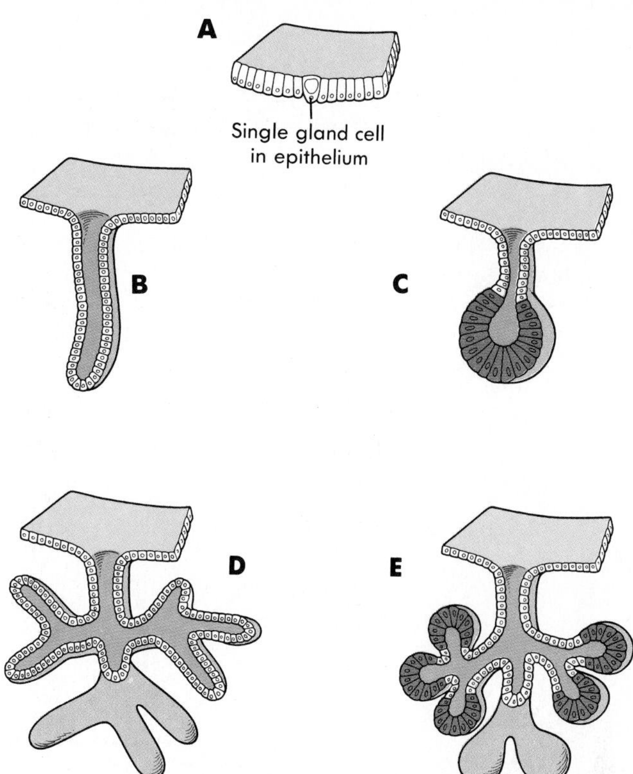

FIGURE 4-4 Types of exocrine glands.
A Unicellular.
B Simple straight tubular.
C Simple acinar or alveolar (ending in a saclike structure).
D Compound tubular.
E Compound acinar or alveolar.

Endocrine (en'do-krin) glands have no ducts and empty their secretions directly into the blood. These secretions, called hormones, are carried by the blood to other parts of the body. Endocrine glands are discussed more fully in Chapter 10.

CONNECTIVE TISSUE

Connective tissue functions to join together cells and other tissues (Figure 4-5). It also provides a supporting framework for the body (for example, bone) and transports substances (for example, blood). Connective tissue is characterized by large amounts of extracellular matrix that separates cells from each other. The **extracellular matrix** has three major components: (1) protein fibers, (2) ground substance consisting of nonfibrous protein and other molecules, and (3) fluid. Connective tissue

Text continued on p. 80.

FIGURE 4-5 ■ Connective Tissues

	LOCATION	STRUCTURE/FUNCTION

A

■ Dense connective

Tendons (attach muscle to bone), ligaments (attach bone to bones), dermis of the skin, and organ capsule

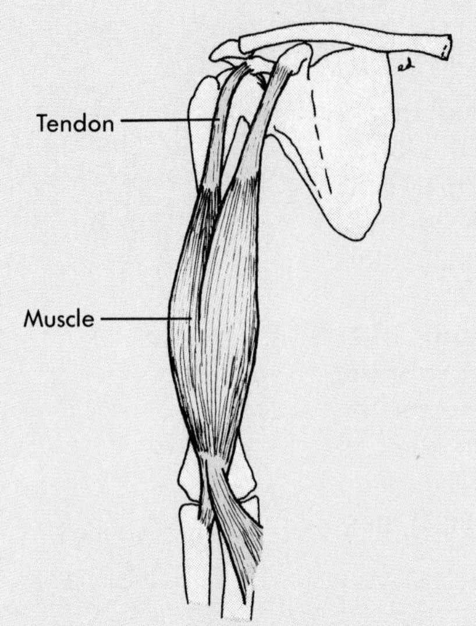

STRUCTURE
Matrix consists almost entirely of collagen fibers produced by fibroblasts. The fibers can all be oriented in the same direction (tendons and ligaments), or in many different directions (dermis and capsules).

FUNCTION
Able to withstand great pulling forces in the direction of fiber orientation

B

■ Loose, or areolar

Widely distributed throughout the body, it is the substance on which most epithelial tissue rests. It is the packing between glands, muscles, and nerves and attaches the skin (dermis) to underlying tissues.

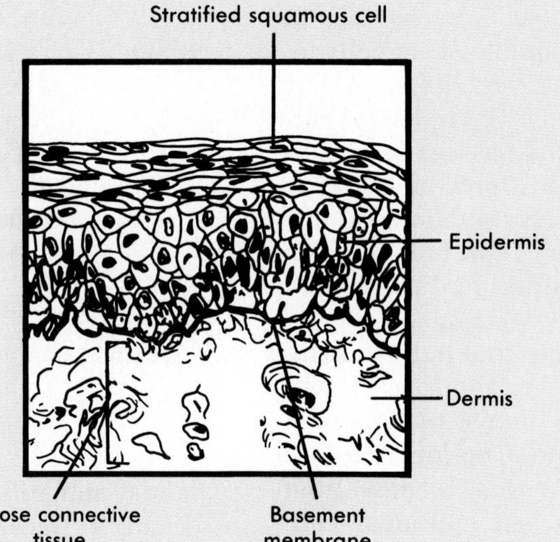

STRUCTURE
Cells (fibroblasts, macrophages, and lymphocytes) within a fine network of mostly collagen fibers; the cells and fibers are separated from each other by fluid-filled spaces

FUNCTION
Loose packing, support, and nourishment for the structures with which it is associated

C

■ Adipose tissue

Under skin, around organs such as the heart and kidneys, in the breasts, and in bones

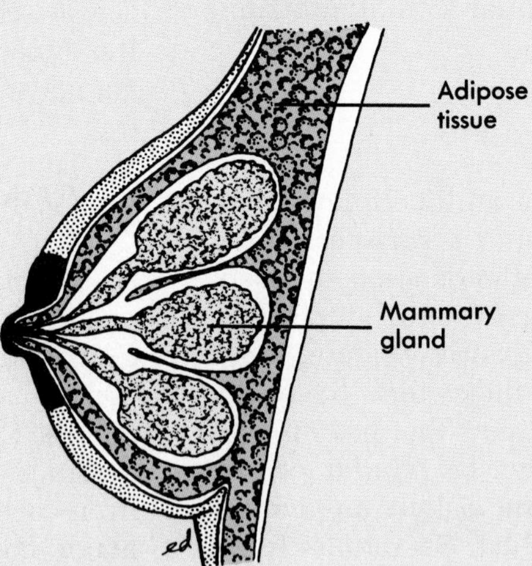

STRUCTURE
Little extracellular material between adipose cells; the cells are so full of lipids that the cytoplasm is pushed to the periphery of the cell

FUNCTION
Energy storage, packing material that provides protection, and heat insulator

HISTOLOGY

Tendon (× 400)

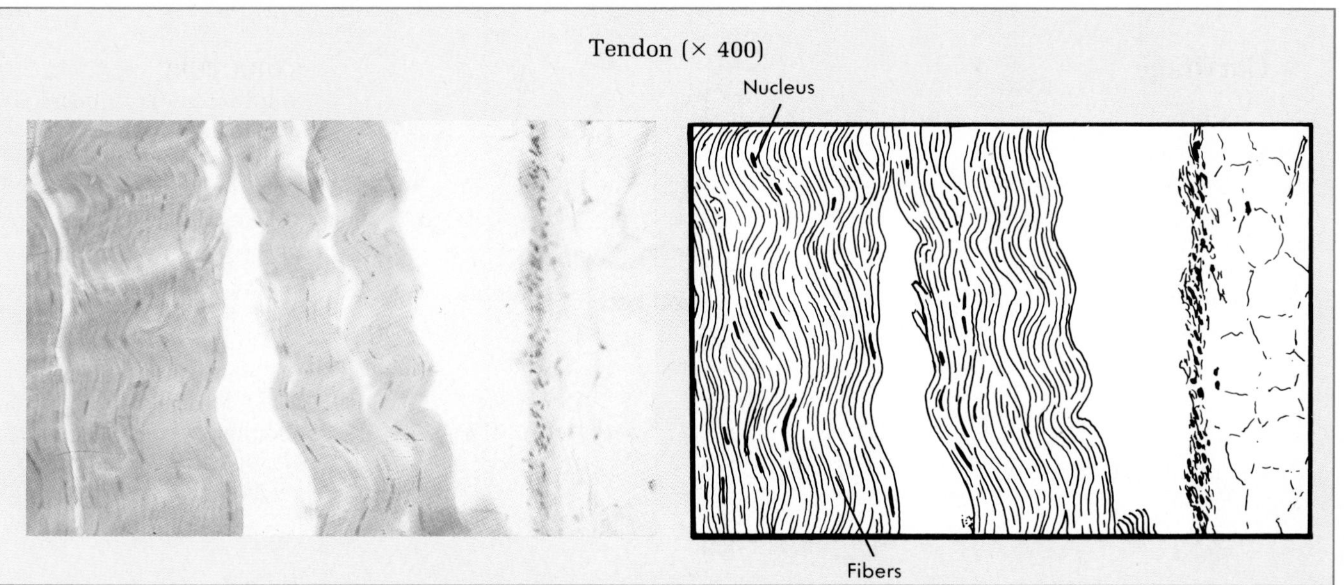

Loose connective tissue (× 300)

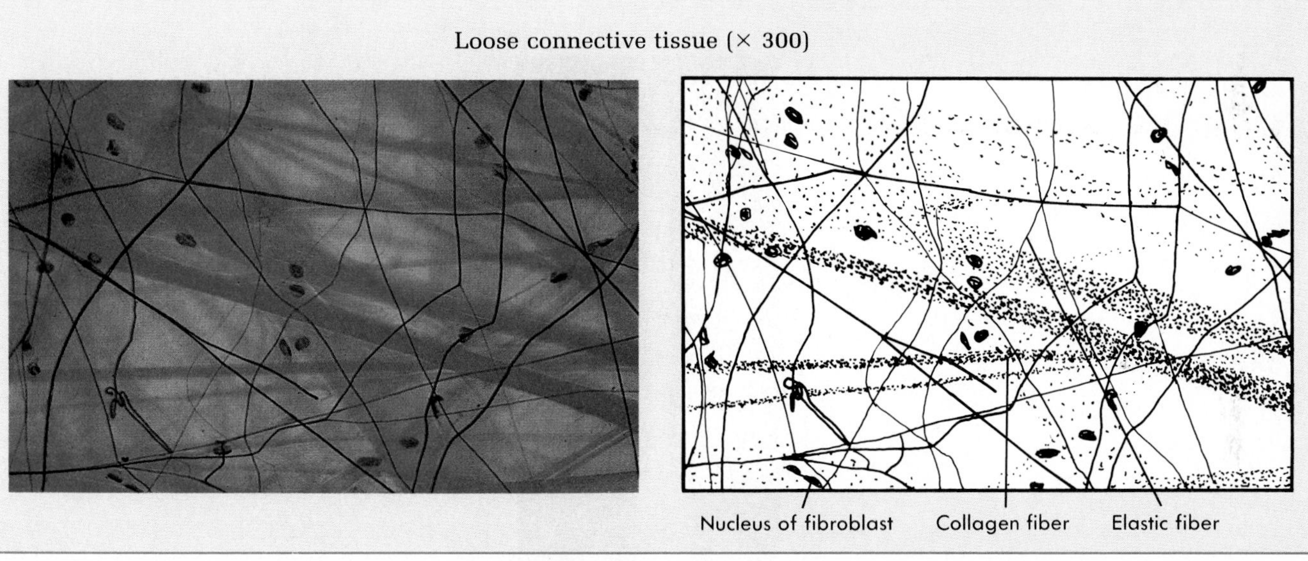

Adipose tissue (× 300)

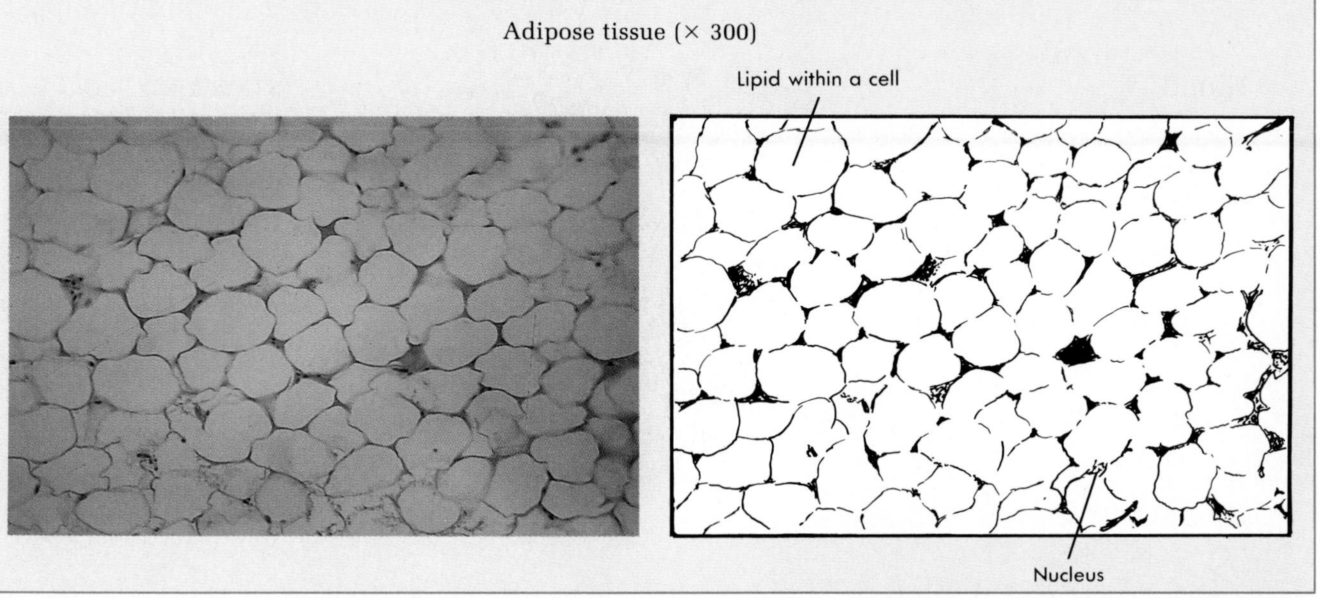

Continued.

LOCATION | **STRUCTURE/FUNCTION**

D

■ Cartilage

Hyaline cartilage is found in the costal cartilages of ribs, cartilage rings of the respiratory tract, nasal cartilages, covering the ends of bones, growth (epiphyseal) plates of bones, and the embryonic skeleton. Fibrocartilage is found in intervertebral disks, symphysis pubis, and articular disks (mensici of knees). Elastic cartilage is found in the external ear.

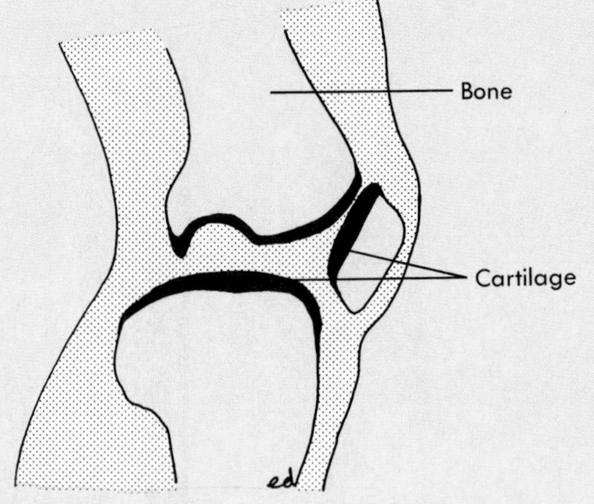

STRUCTURE
Solid matrix with fibers dispersed throughout the ground substance. Chondrocytes are found within lacunae.

FUNCTION
Hyaline cartilage forms a smooth surface in joints, a site of bone growth, and the embryonic skeleton. Fibrocartilage can withstand great pressure, and elastic cartilage returns to original shape when bent.

E

■ Bone tissue

Bones

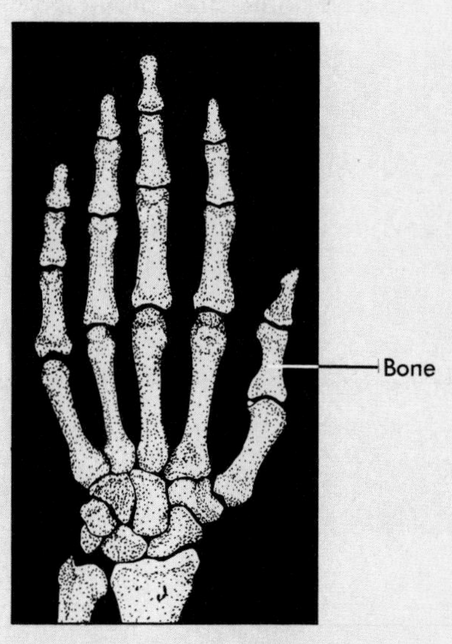

STRUCTURE
Hard, mineralized matrix with osteocytes located within lacunae

FUNCTION
Provides great strength and support and protects internal organs such as the brain

F

■ Blood

Within the blood vessels and heart

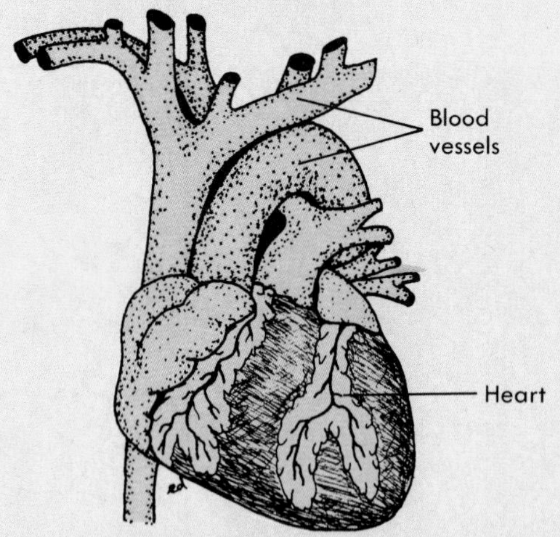

STRUCTURE
Blood cells within a fluid matrix called plasma

FUNCTION
Transports oxygen, carbon dioxide, hormones, nutrients, waste products, and other substances; protects the body from infection and is involved in temperature regulation

HISTOLOGY

Hyaline cartilage (× 400)

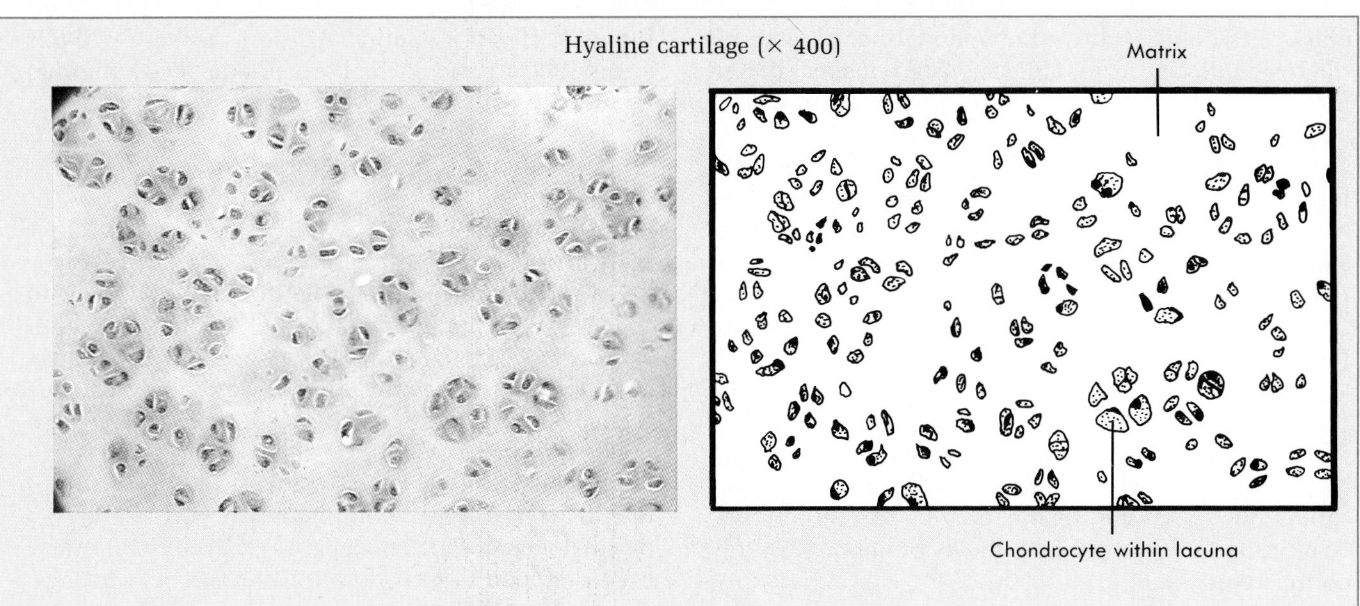

Compact bone (× 400)

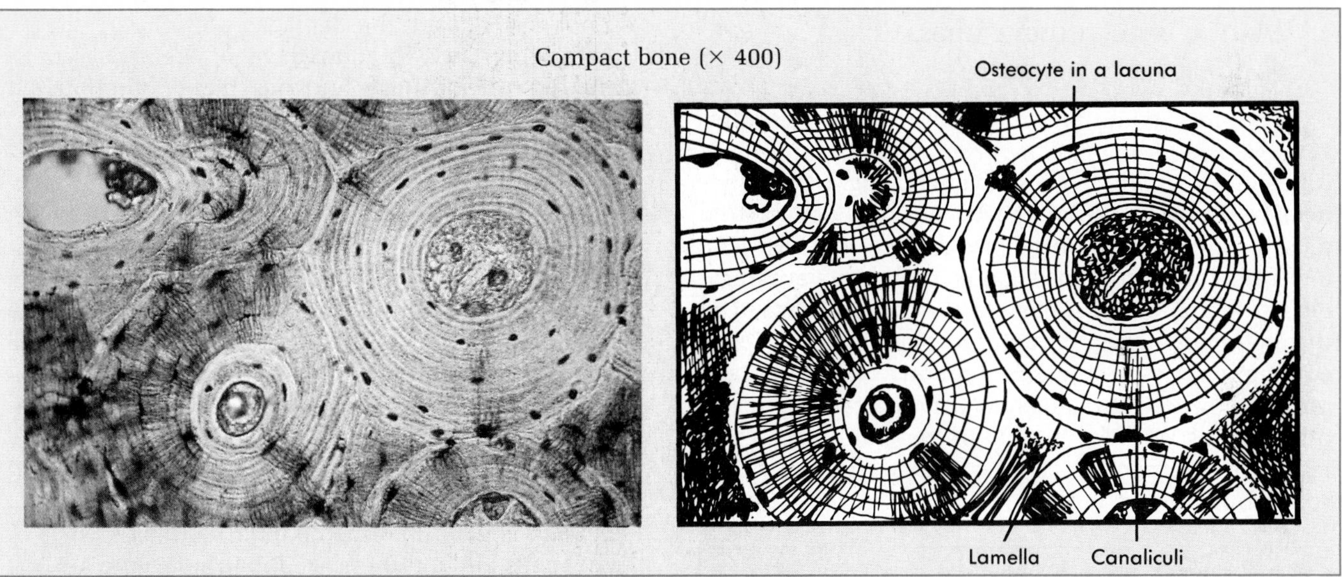

Blood (× 1000)

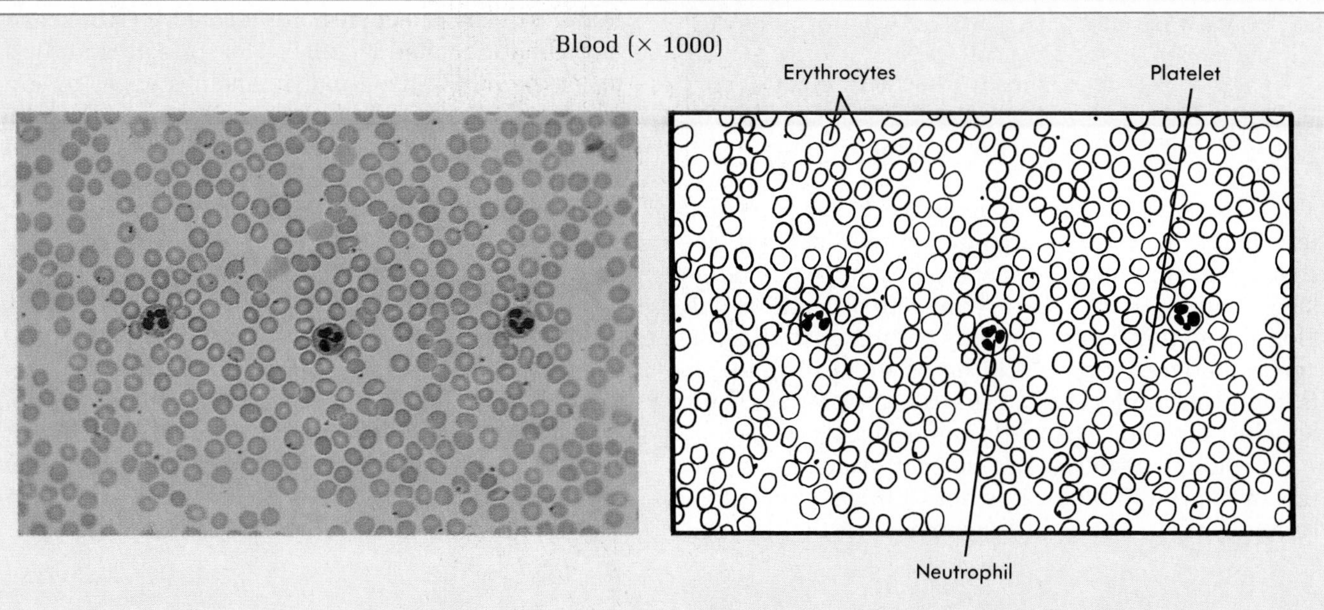

cells are named according to their functions. **Blast** cells produce the matrix, **cyte** cells maintain it, and **clast** cells break it down for remodeling. For example osteoblasts (os'te-o-blasts; osteo means bone) form bone, osteocytes (os'te-o-sīts) maintain bone, and osteoclasts (os'te-o-klasts) break down bone. Cells associated with the immune system are also found in connective tissue. **Macrophages** are large cells, capable of moving about and ingesting foreign substances, including microorganisms, that are found in the connective tissue. **Mast** cells are nonmotile cells that release chemicals promoting inflammation.

Classification

The nature of the extracellular matrix determines the functional characteristics of the connective tissue and is used as a means of classifying the connective tissue.

Matrix with protein fibers as the primary feature

The most common protein in the body is **collagen** (kol'lă-jen), which resembles microscopic ropes. It is flexible but resists stretching. **Dense connective tissue** has an extracellular matrix consisting mostly of collagen fibers (see Figure 4-5, *A*). The few cells found in dense connective tissue are fibroblasts, which are responsible for the production of the collagen fibers. Structures made up of dense connective tissue include tendons, which attach muscles to bones; ligaments, which attach bones to other bones; and the dermis of the skin, which is a layer of connective tissue under the stratified squamous epithelium.

2 In tendons, collagen fibers are oriented parallel to the length of the tendon. In the skin, collagen fibers are oriented in many directions. What are the functional advantages of the fiber arrangements in tendons and in the skin? ?

In contrast to dense connective tissue, the protein fibers in **loose** or **areolar** (ah-re'o-lar) **connective tissue** are widely separated from each other (see Figure 4-5, *B*). Loose connective tissue is the "loose packing" material of the body, which fills the spaces between organs and holds them in place. It is found around glands, muscles, and nerves, and attaches the skin to underlying tissues.

Although **adipose** (ad'ĭ-pōs; fat) **tissue** has a matrix with protein fibers, it is not a typical connective tissue. There is very little matrix, and the adipose cells are large and closely packed together (Figure 4-5, *C*). Adipose cells are filled with lipids and function to store energy. Adipose tissue also pads and protects parts of the body and acts as a thermal insulator.

Matrix with both protein fibers and ground substance

Cartilage (kar'ti-lij) is composed of cartilage cells, or **chondrocytes** (kon'dro-sits), located in spaces called **lacunae** (lă-ku'ne) within an extensive matrix (Figure 4-5, *D*). Collagen in the matrix gives cartilage strength, and the ground substance of the matrix traps water, which enables the cartilage to spring back after being compressed. Cartilage is relatively rigid and provides support, but if it is bent or slightly compressed it will resume its original shape. Cartilage heals slowly after an injury because blood vessels do not penetrate the cartilage. Thus cells and nutrients necessary for tissue repair do not easily reach the damaged area.

Hyaline (hi'ă-lin) **cartilage** is the most abundant type of cartilage and has many functions. It covers the ends of bones where bones come together to form joints, and provides a smooth, resilient surface that can withstand compression. Hyaline cartilage also forms the costal cartilages, which attach the ribs to the sternum (breast bone). **Fibrocartilage** has more collagen than hyaline cartilage. In addition to withstanding compression, it is able to resist pulling or tearing forces. It is found in the disks between vertebrae (bones of the back) for example. **Elastic cartilage** contains an elastic protein fiber in addition to collagen and ground substance, and is able to recoil to its original shape when bent. The external ear contains elastic cartilage.

Bone is a hard connective tissue that consists of living cells and a mineralized matrix (Figure 4-5, *E*). Bone cells, or **osteocytes,** are located within spaces in the matrix called lacunae. The strength and rigidity of the mineralized matrix enables bones to support and protect other tissues and organs of the body. The two types of bone, **compact** and **cancellous** (kan'sĕ-lus), are considered in greater detail in Chapter 6.

Fluid matrix

Blood is unique because the matrix is liquid, enabling blood cells to move about freely (Figure 4-5, *F*). Some blood cells even leave the blood and wander into other tissues. The liquid matrix enables blood to flow rapidly through the body carrying food, oxygen, waste products, and other materials. Blood is discussed more fully in Chapter 11.

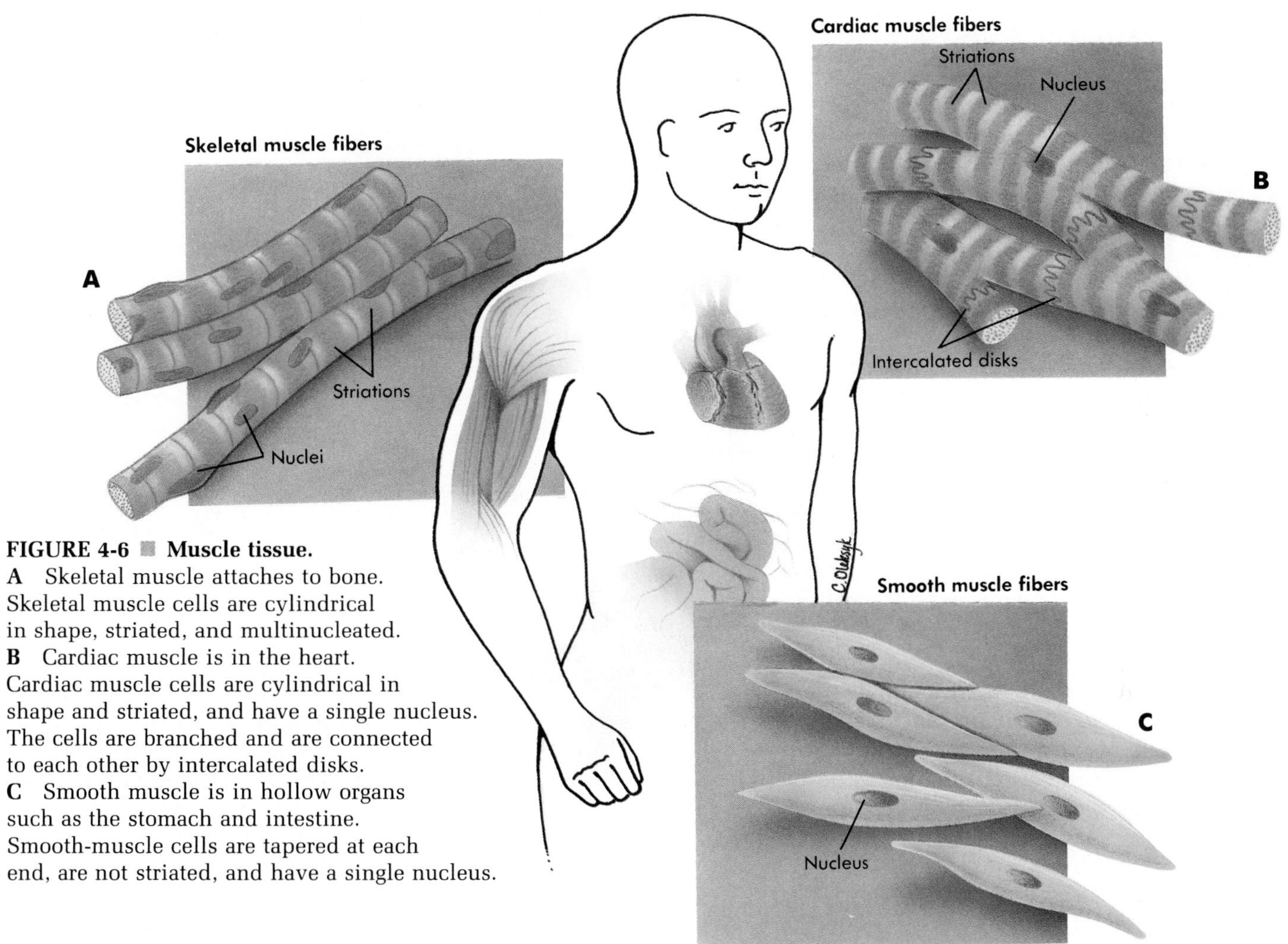

FIGURE 4-6 ■ **Muscle tissue.**
A Skeletal muscle attaches to bone. Skeletal muscle cells are cylindrical in shape, striated, and multinucleated.
B Cardiac muscle is in the heart. Cardiac muscle cells are cylindrical in shape and striated, and have a single nucleus. The cells are branched and are connected to each other by intercalated disks.
C Smooth muscle is in hollow organs such as the stomach and intestine. Smooth-muscle cells are tapered at each end, are not striated, and have a single nucleus.

MUSCLE TISSUE

The main characteristic of **muscle tissue** is its ability to contract or shorten, making movements possible. Muscle contraction is due to contractile proteins found inside muscle cells (see Chapter 7). The length of muscle cells is greater than their diameter. Because they often resemble tiny threads, muscle cells are called **muscle fibers.**

The three types of muscle are skeletal, cardiac, and smooth muscle. **Skeletal muscle** is what normally is thought of as "muscle" (Figure 4-6, *A*). It is the meat of animals and comprises about 40% of a person's body weight. Skeletal muscle, as the name implies, attaches to the skeleton and causes body movements. Skeletal muscle is normally under voluntary (conscious) control. Skeletal muscle cells tend to be long, cylindrical cells with several nuclei per cell. Some skeletal muscle cells extend the length of an entire muscle. Skeletal muscle cells are **striated** or banded because of the arrangement of contractile proteins within the cells (see Chapter 7).

Cardiac muscle is the muscle of the heart and is responsible for pumping blood (Figure 4-6, *B*). It is under involuntary (unconscious) control. Cardiac muscle cells are cylindrical in shape, but are much shorter in length than skeletal muscle cells. Cardiac muscle cells are striated with one nucleus per cell, and they often are branched and connected to each other by intercalated (in-ter′kă-la-ted) disks. The intercalated disks are important in coordinating the contractions of the cardiac muscle cells (see Chapter 12).

Smooth muscle forms the walls of hollow organs (except the heart), and also is found in the skin and the eyes (Figure 4-6, *C*). It is responsible for a number of functions such as movement of food through the digestive tract and emptying of the urinary bladder. Smooth muscle is under involuntary control. Smooth muscle cells are tapered at each end and have a single nucleus. They are not striated.

NERVOUS TISSUE

Nervous tissue forms the brain, spinal cord, and nerves. It is responsible for coordinating and controlling many of the body's activities, for example, the conscious control of skeletal muscles and the unconscious regulation of cardiac muscle. Awareness of ourselves and the external environment, emotions, reasoning skills, and memory are other functions performed by nervous tissue. Many of these functions depend upon the ability of nervous tissues to communicate with each other by electrical signals called action potentials.

Nervous tissue consists of neurons, which are responsible for the action potential conduction, and support cells. The **neuron** or **nerve cell** is composed of three parts (Figure 4-7). The **cell body** contains the nucleus and is the site of general cell functions. **Dendrites** (den'drītz) and **axons** (ak'sonz) are nerve cell processes (extensions). Dendrites usually receive action potentials and conduct them toward the cell body, whereas the axon (only one per neuron) usually conducts action potentials away from the cell body. **Neuroglia** (nu-rog'le-ah) are the support cells of the nervous system, and they function to nourish, protect, and insulate the neurons. Nervous tissue is considered in greater detail in Chapter 8.

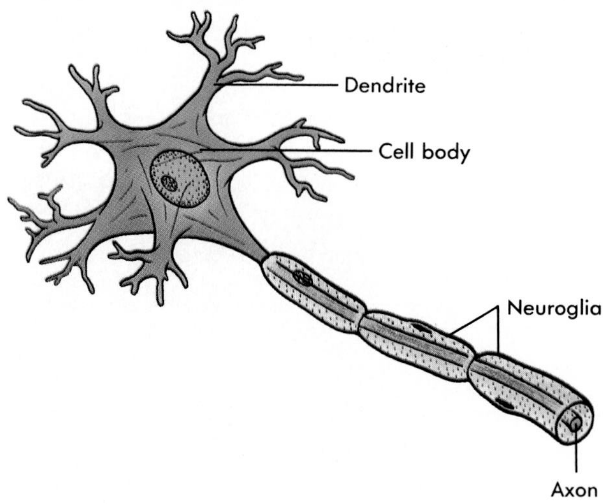

FIGURE 4-7 Nervous tissue.
The neuron consists of dendrites, a cell body, and a long axon. Neuroglia, or support cells, surround the axon.

MEMBRANES

A **membrane** is a thin sheet or layer of tissue that covers a structure or lines a cavity. Most membranes consist of epithelium and the connective tissue on which the epithelium rests. The two major categories of membranes are the mucous membranes and the serous membranes.

Mucous Membranes

Mucous (mu'kus) **membranes** consist of various kinds of epithelium resting on a thick layer of loose connective tissue. They line cavities that open to the outside of the body such as the digestive, respiratory, excretory, and reproductive tracts. Many, but not all, mucous membranes have mucous glands, which secrete mucus. The functions of mucous membranes vary depending on their location and include protection, absorption, and secretion. For example, the stratified squamous epithelium of the oral cavity (mouth) performs a protective function, whereas the simple columnar epithelium of the intestine functions to absorb nutrients and secrete digestive enzymes and mucus.

Serous Membranes

Serous (sēr'us) **membranes** consist of simple squamous epithelium resting on a delicate layer of loose connective tissue. Serous membranes line the trunk cavities and cover the organs located within the trunk cavities (see Chapter 1). The serous membranes secrete serous fluid, which covers the surface of the membranes. The smooth surface of the epithelial cells of the serous membranes, and the lubricating qualities of the serous fluid, combine to prevent damage from abrasion when organs in the thoracic or abdominopelvic cavities rub against each other. The serous membranes are named according to their location: the **pleural** (ploor'al) **membranes** are associated with the lungs, the **pericardial** (pĕr-ĭ-kar'de-al) **membranes** are associated with the heart, and the **peritoneal** (pĕr'ĭ-to-ne'al) **membranes** are located in the abdominopelvic cavity.

Other Membranes

In addition to mucous and serous membranes, there are several other membranes in the body. The skin or **cutaneous membrane** is stratified squamous epithelium and dense connective tissue (see Chapter 5). Other membranes are made up of only connective tissue. **Synovial** (sĭ-no've-al) **membranes** line the inside of joint cavities (the space where bones come together within a movable joint), and the **periosteum** (pĕr'e-os'te-um) surrounds bone. These connective tissue membranes are discussed in Chapter 6.

INFLAMMATION

The **inflammatory response,** or **inflammation,** occurs when tissues are damaged (Figure 4-8). It mobilizes the body's defenses, isolates and destroys microorganisms, and removes foreign materials and damaged cells so that tissue repair can proceed. Inflammation produces five major symptoms: redness, heat, swelling, pain, and disturbance of function. Although unpleasant, the processes producing the symptoms are usually beneficial.

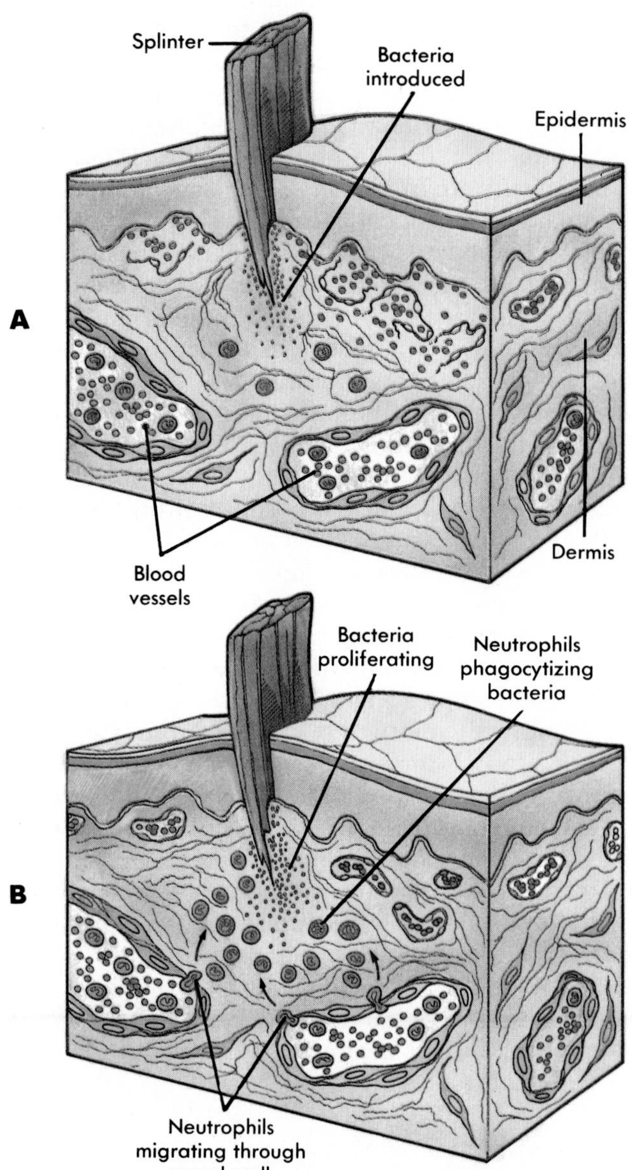

FIGURE 4-8 Inflammation.
A A splinter in the skin causes tissue damage and introduces bacteria. Dilated blood vessels (capillaries) cause skin to become red, and fluid from the vessels causes swelling.
B Neutrophils leave the vessels and arrive at the site of bacterial infection where they begin to phagocytize the bacteria.

Following an injury, chemical substances called **mediators of inflammation** are released or activated in the injured tissues and adjacent blood vessels. The mediators include histamine (his'tă-mēn), kinins (ki'ninz), prostaglandins (pros'tă-glan'dinz), leukotrienes (lu'ko-trēnz), and others. Some mediators cause vasodilation (expansion of blood vessels). This produces the symptoms of redness and heat, as occur when a person blushes. Vasodilation is beneficial because it increases the speed with which blood cells and other substances important for fighting infections and repairing the injury are brought to the injury site.

Mediators of inflammation also increase the permeability of blood vessels, allowing materials and blood cells to move out of the vessels and into the tissue where they can deal directly with the injury. **Edema** (e-de'mah), or swelling, of the tissues results when proteins and water from the blood move into the tissues. One of the proteins, fibrin, diffuses into the tissue and forms a fibrous network that "walls off" the site of injury from the rest of the body. This can help to prevent the spread of infectious agents. One type of blood cell that enters the tissues is **neutrophils,** phagocytic white blood cells that fight infections by ingesting bacteria. They also ingest tissue debris, clearing the area for tissue repair. Neutrophils are killed in this process and can accumulate as a mixture of dead cells and fluid, called **pus.**

Pain associated with inflammation is produced in several ways. Nerve cell endings are stimulated by direct damage and by some mediators of inflammation to produce pain sensations. In addition, the increased pressure in the tissue caused by edema and accumulation of pus can cause pain.

Pain, limitation of movement resulting from edema, and tissue destruction all contribute to the disturbance of function. This disturbance of function can be adaptive because it warns the person to protect the injured area from further damage.

> Sometimes the inflammatory response lasts longer or is more intense than is desirable, and drugs are used to suppress the symptoms by inhibiting the mediators of inflammation. For example, the effects of histamines released in hay fever are suppressed by antihistamines. Aspirin is an effective antiinflammatory agent and relieves pain by preventing the synthesis of prostaglandins.

TISSUE REPAIR

Tissue repair is the substitution of viable cells for dead cells, and it can occur by regeneration or replacement. In **regeneration,** the new cells are the same type as those that were destroyed, and normal function is usually restored. In **replacement,** a new type of tissue develops that eventually causes scar production and the loss of some tissue function. The tissues involved and the nature of the wound determine the type of tissue repair that dominates.

Cells can be classified into three groups based on their ability to divide and produce new cells. **Labile** cells (for example, the skin and mucous membranes) continue to divide throughout life. Damage to these cells can be repaired completely by regeneration. **Stable** cells (for example, connective tissue and glands, including the liver and pancreas) do not actively replicate after growth ceases. They do retain the ability to divide after an injury and are capable of regeneration. **Permanent** cells (for example, neurons and skeletal muscle cells) cannot undergo mitosis and divide. If they are killed, they usually are replaced by connective tissue. This does not mean permanent cells can't recover from a limited amount of damage. For example, if the axon of a neuron is damaged it is possible for the neuron to grow a new axon.

In addition to the type of cells involved, the severity of an injury can influence whether repair is by regeneration or replacement. Given that a tissue can repair by regeneration, if the injury is severe some of the tissue may be repaired by replacement. Generally, the more severe the injury, the greater the likelihood that repair involves replacement. If the edges of the wound are close together, the wound heals primarily by regeneration, and the process is called **primary union.** If the edges are not close together, or if there has been extensive tissue loss, the wound heals by regeneration and replacement, and the process is called **secondary union.**

Repair of the skin can be used to illustrate tissue repair (Figure 4-9). In primary union, the wound fills with blood and a clot forms (see Chapter 11). The **clot** contains a threadlike protein, fibrin, which binds the edges of the wound together and stops any bleeding. The surface of the clot dries to form a **scab,** which seals the wound and helps to prevent infection. An inflammatory response is activated to fight infectious agents in the wound and to help the repair process. Vasodilation (increased blood vessel diameter) brings blood cells and other substances to the injury area, and increased blood vessel permeability allows them to enter the tissue. The area is "walled off" by the fibrin, and neutrophils enter the tissue from the blood.

While the inflammatory response proceeds, the epithelium at the edge of the wound undergoes regeneration and migrates under the scab. After a few days, the epithelial cells from the edges meet, and eventually the epithelium is restored. After the epithelium is repaired, the scab is sloughed off (shed). Meanwhile, a second type of phagocytic cell, called a **macrophage,** removes the dead neutrophils, cellular debris, and the decomposing clot. Fibroblasts from the surrounding connective tissue migrate into the area, producing collagen and other extracellular matrix components. Capillaries grow from blood vessels at the edge of the wound and revascularize the area. The result is the replacement of the clot by a delicate connective tissue called **granulation tissue,** which consists of fibroblasts, collagen, and capillaries. Eventually normal connective tissue replaces the granulation tissue. Sometimes a large amount of granulation tissue persists as a **scar,** which at first is bright red because of the vascularization of the tissue. The scar turns from red to white as collagen accumulates and the blood vessels are compressed.

Repair by secondary union is similar to healing by primary union, but there are some differences. Because the wound edges are far apart, the clot may not completely close the gap, and the epithelial cells take much longer to regenerate and cover the wound. With increased tissue damage the degree of the inflammatory response is greater, there is more cell debris for the phagocytes to remove, and the risk of infection is greater. Much more granulation tissue forms, and **wound contracture,** a result of the contraction of fibroblasts in the granulation tissue, pulls the edges of the wound closer together. Although wound contracture reduces the size of the wound and speeds healing, it can lead to disfiguring and debilitating scars.

3 Explain why it is advisable to suture large wounds. ?

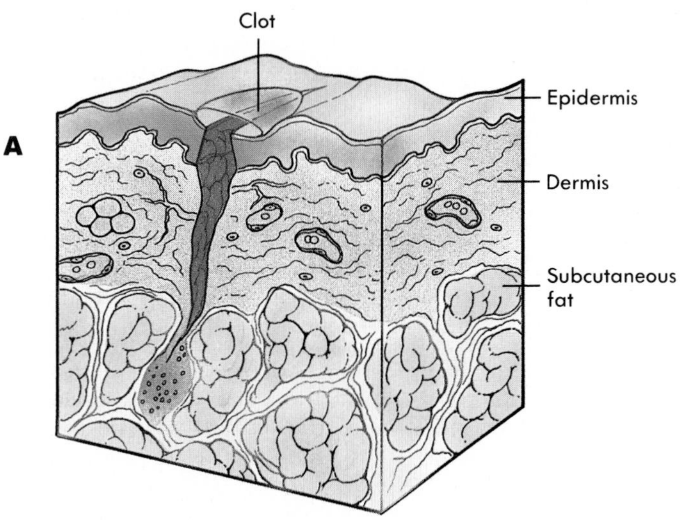

FIGURE 4-9 ■ **Tissue repair.**
A Fresh wound cuts through the epithelium (epidermis) and underlying connective tissue (dermis), and a clot forms.
B Approximately 1 week after the injury a scab is present, and epithelium is growing into the wound.
C Approximately 2 weeks after the injury the epithelium has grown completely into the wound, and granulation tissue has formed.
D Approximately 1 month after the injury the wound has closed completely, the scab has been sloughed, and the granulation tissue is being replaced.

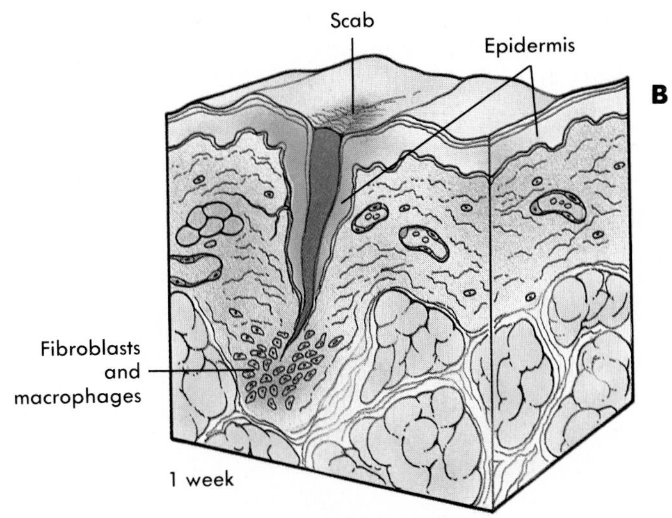

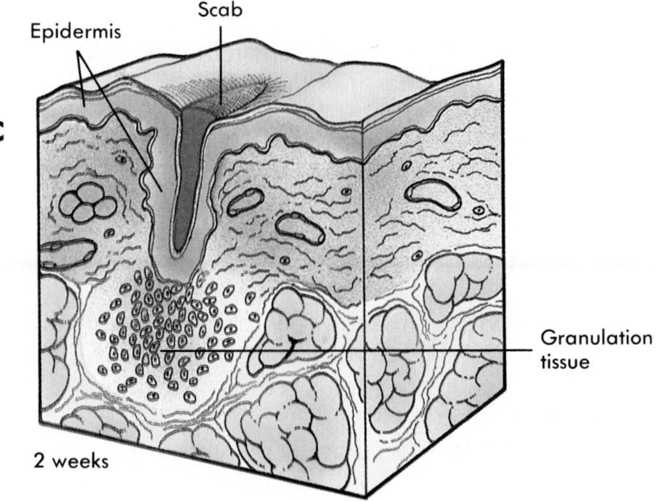

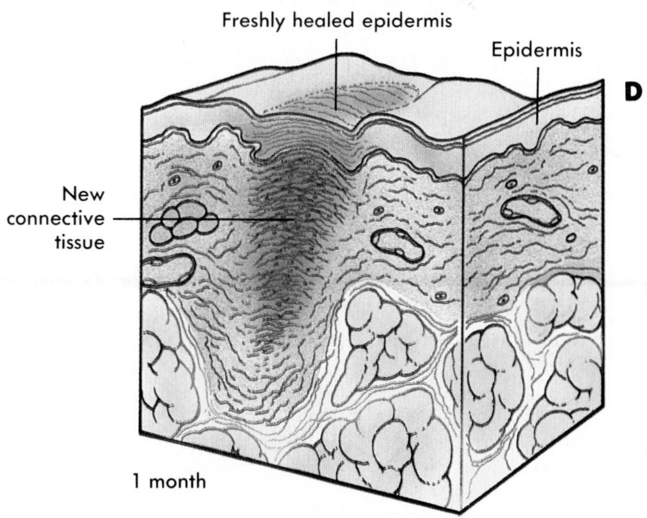

Cancer Tissue

Cancer refers to a malignant, spreading tumor and the illness that results from such a tumor. A **tumor** is any swelling, but modern usage has limited the term to swellings that involve neoplastic tissue. **Neoplasm** (ne'o-plazm) means "new growth" and refers to abnormal tissue growth resulting from cellular divisions that continue after normal cell division of the tissue has stopped or slowed considerably. A neoplasm can be **benign** (be-nin'; kind), with little likelihood of spreading, or, **malignant** (mă-lig'nant; with malice or intent to cause harm), with the ability to spread. Although benign tumors are usually less dangerous than malignant tumors, they can cause problems. As the tumor enlarges it can compress surrounding tissues and impair their functions. In some cases, for example brain tumors, the result can be death.

Malignant tumors can spread by local growth and expansion or by **metastasis** (mĕ-tas'tă-sis; moving to another place), which occurs when tumor cells separate from the main mass and are carried by the lymphatic or circulatory system to a new site where a second neoplasm is formed. The illness associated with cancer usually occurs as the tumor invades and destroys the healthy surrounding tissue, eliminating its function.

Malignant neoplasms lack the normal growth control that most other adult tissues have. This breaking loose from normal control involves the genetic machinery and can be induced by viruses, environmental toxins, and other causes. Cancer therapy concentrates on trying to confine and then kill the malignant cells. This is accomplished by killing the tissue with radiation or lasers, by removing the tumor surgically, by treating the patient with drugs that selectively kill rapidly dividing cells, or by stimulating the patient's immune system to destroy the tumor. **Oncology** (ong-kol'o-je; tumor study) is the study of cancer and its associated problems.

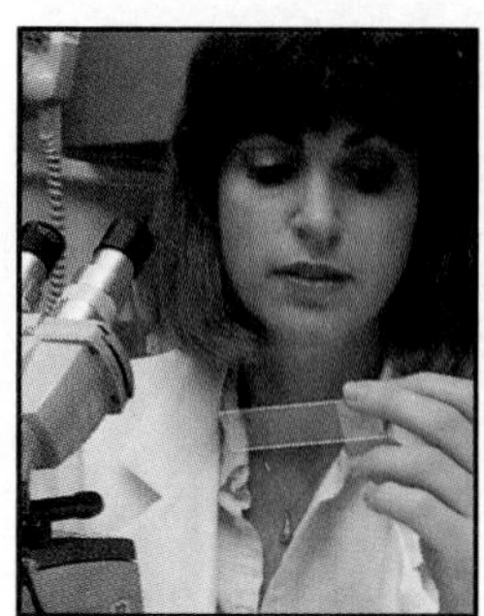

SUMMARY

A tissue is a group of cells with similar structure and function plus the extracellular substances located between the cells.

Histology is the study of tissues.

EPITHELIAL TISSUE

Epithelial tissue covers surfaces, usually has a basement membrane, has little extracellular material, and has no blood vessels.

Classification

Simple epithelium has one layer of cells, whereas stratified epithelium has more than one layer of cells.

Pseudostratified epithelium is simple epithelium that appears to have two or more cell layers.

Transitional epithelium is stratified epithelium that can be greatly stretched.

Functional Characteristics

Simple epithelium is involved with diffusion, secretion, or absorption. Stratified epithelium serves a protective role. Squamous cells function in diffusion. Cuboidal or columnar cells, which contain more organelles, secrete or absorb.

A smooth free surface reduces friction, microvilli increase surface area, and cilia move materials over the cell surface.

Desmosomes mechanically bind cells together, tight junctions form a permeability barrier, and gap junctions allow intercellular communication.

Glands

A gland is a single cell or a multicellular structure that secretes.

Exocrine glands have ducts and endocrine glands do not.

CONNECTIVE TISSUE

Connective tissue holds cells and tissues together.
Connective tissue has an extracellular matrix consisting of protein fibers, ground substance, and fluid.
Blast cells form the matrix, cyte cells maintain it, and clast cells break it down.

Classification

Connective tissue that has protein fibers as the primary feature of the extracellular matrix are dense connective tissue (tendons, ligaments, and dermis of skin), loose connective tissue ("loose packing" material of the body), and adipose tissue.
Connective tissue that has protein fibers and ground substance as important features of the extracellular matrix are cartilage and bone.
Connective tissue that has a fluid matrix is blood.

MUSCLE TISSUE

Muscle tissue is specialized to shorten or contract.
The three types of muscle tissue are skeletal, cardiac, and smooth muscle.

NERVOUS TISSUE

Nervous tissue is specialized to conduct action potentials (electrical signals).
Neurons conduct action potentials and neuroglia support the neurons.

MEMBRANES

Mucous membranes line cavities that open to the outside of the body (digestive, respiratory, excretory, and reproductive tracts). They contain glands and secrete mucus.
Serous membranes line trunk cavities that do not open to the outside of the body (pleural, pericardial, and peritoneal cavities). They do not contain glands, but do secrete serous fluid.
Other membranes include the cutaneous membrane (skin), synovial membranes (line joint cavities), and periosteum (around bone).

INFLAMMATION

The function of the inflammatory response is to isolate and destroy harmful agents.
The inflammatory response produces five symptoms: redness, heat, swelling, pain, and disturbance of function.

TISSUE REPAIR

Tissue repair is the substitution of viable cells for dead cells. Tissue repair occurs by regeneration or replacement.
Labile cells divide throughout life and can undergo regeneration.
Stable cells do not ordinarily divide but can regenerate if necessary.
Permanent cells do not divide, and if killed, repair is by replacement.
Tissue repair by primary union occurs when the edges of wound are close together. Secondary union occurs when the edges are far apart.

CONTENT REVIEW

1. Define tissue and histology.
2. Where is epithelium located? What are four characteristics of epithelial tissue?
3. How is epithelial tissue classified according to number of cell layers and cell shape? What is pseudostratified and transitional epithelium?
4. What kinds of functions would a single layer of epithelium be expected to perform? A stratified layer? Give an example of each.
5. Contrast the functions performed by squamous cells with cuboidal or columnar cells. Give an example of each.
6. What is the function of an epithelial free surface that is smooth, one that has microvilli, and one that has cilia?

7. Name the ways in which epithelial cells are connected to each other and give the function for each way.
8. Define gland. Distinguish between an exocrine and an endocrine gland.
9. What is the function of connective tissue? How does it differ from other types of tissue?
10. What are the three major components of the extracellular matrix of connective tissue? How are they used to classify connective tissue?
11. Explain the difference between connective tissue cells that are termed blast, cyte, and clast cells.
12. Describe dense connective tissue and give three examples.
13. How is adipose tissue different from other connective tissues? List the functions of adipose tissue.
14. Describe the components of cartilage. Give an example of hyaline cartilage, fibrocartilage, and elastic cartilage.
15. Describe the components of bone.
16. Functionally, what is unique about muscle? Which of the muscle types is under voluntary control? What tasks does each type perform?
17. Functionally, what is unique about nervous tissue? What do neurons and neuroglia accomplish? What is the difference between an axon and a dendrite?
18. Compare mucous and serous membranes according to the type of cavity they line and their secretions. Name the serous membranes associated with the lungs, heart, and abdominopelvic organs.
19. What is the function of the inflammatory response? Name the five symptoms of inflammation and explain how each is produced.
20. Define tissue repair. What is the difference between repair by regeneration and by replacement?
21. Differentiate between labile cells, stable cells, and permanent cells. Give an example of each type. What is the significance of these cell types to tissue repair?
22. Describe the process of tissue repair. Contrast healing by primary union and secondary union.

CONCEPT REVIEW

1. Normally the trachea (the "wind pipe"; a tube that conducts air between the lungs and throat) is lined with pseudostratified ciliated epithelium. Constant, heavy smoking can result in the destruction of the normal epithelium in the trachea and its replacement with another type of epithelium. What type of epithelium would be expected as a replacement tissue? If the replacement epithelium was not ciliated, how could this affect normal tracheal function?
2. The blood-brain barrier is a specialized epithelium in capillaries that prevents many materials from passing from the blood into the brain. What kind of cell connections would be expected in the blood-brain barrier?
3. One of the functions of the pancreas is to produce digestive enzymes that are secreted into the small intestine. How many cell layers, and what cell shape, cell surface, and type of cell-to-cell connections would be expected in the epithelium responsible for producing the digestive enzymes?
4. Some dense connective tissue has elastic fibers in addition to collagen fibers. This enables a structure to stretch and then recoil to its original shape. Examples are certain ligaments that hold together the vertebrae (bones of the back). When the back is bent (flexed) the ligaments are stretched. How does the elastic nature of these ligaments help the back to function? How would the fibers in the ligaments be organized?

5. The aorta is a large blood vessel that is attached to the heart. When the heart beats, blood is ejected into the aorta, which expands to accept the blood. The wall of the aorta is constructed with dense connective tissue that has elastic fibers. How would the fibers be arranged?
6. Antihistamines block the effect of a chemical mediator, histamine, that is released during the inflammatory response. Would administering antihistamines be beneficial?

ANSWERS TO  PREDICT QUESTIONS

1 *p. 69* Simple squamous epithelium is found in capillaries. It facilitates diffusion of gases and other substances between the blood and tissues surrounding the capillaries. Stratified squamous epithelium protects against abrasion in the mouth.

2 *p. 80* When a muscle contracts, the pull it exerts is transmitted along the length of its tendons. The tendons need to be very strong in that direction, but not strong in others. Therefore the collagen fibers, which are like microscopic ropes, all are arranged in the same direction to maximize their strength. In the skin, collagen fibers are oriented in many directions because the skin can be pulled in many directions.

3 *p. 84* Suturing large wounds brings the edges of the wounds close together so healing by primary union can occur. Thus healing will be more rapid, there is less danger of infections, less scar tissue is formed, and wound contracture is greatly reduced.

5

THE INTEGUMENTARY SYSTEM

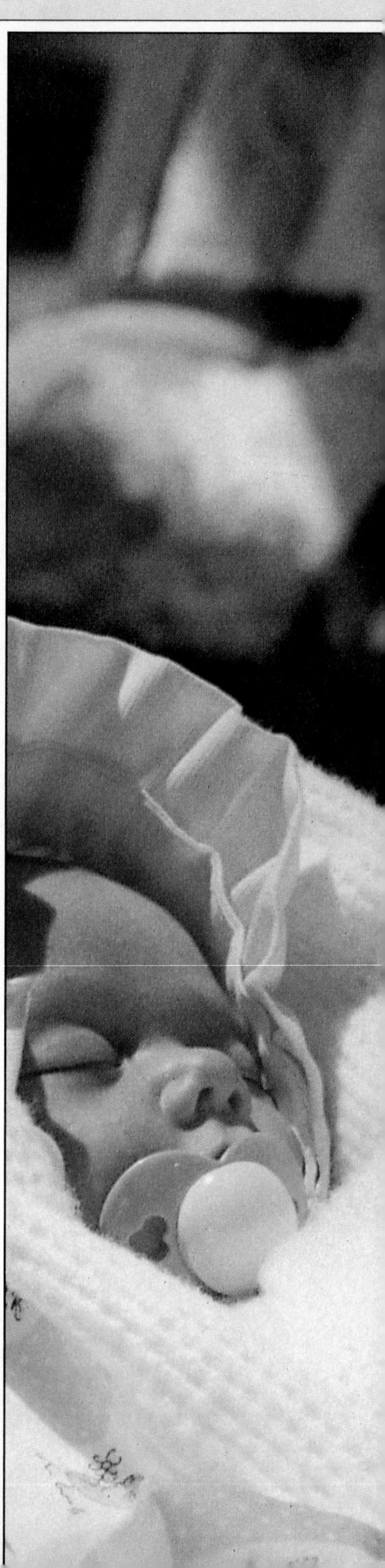

After reading this chapter you should be able to:

1 Describe the structure and function of the hypodermis, dermis, and epidermis.
2 Define epidermal strata and relate them to the process of keratinization.
3 Explain how melanin, carotene, blood, and collagen affect skin color.
4 Describe the structure of a hair and discuss the phases of hair growth.
5 Name the glands of the skin and describe the secretions they produce.
6 Describe the parts of a nail and explain how nails grow.
7 Discuss the functions of skin, hair, glands, and nails.
8 List the changes the integumentary system undergoes with age.
9 Explain how the integumentary system can be used as a diagnostic aid.
10 Classify burns based on the amount of skin damage they produce.
11 Name and define the types of skin cancer.

arrector pili (ah-rek'tor pī'le) — [L., that which raises hair] Smooth muscle attached to the hair follicle and dermis that raises the hair by contracting.

dermis (der'mis) — [Gr. *derma*, skin] Dense connective tissue that forms the deep layer of the skin; responsible for the structural strength of the skin.

epidermis (ep'ĭ-der'mis) — [Gr. *epi*, upon + *derma*, skin] Outer portion of the skin formed of epithelial tissue that rests on the dermis; resists abrasion and forms a permeability barrier.

full-thickness burn — Burn that destroys the epidermis and the dermis and sometimes the underlying tissue as well; sometimes called a third-degree burn.

hair — A threadlike outgrowth of the skin consisting of columns of dead keratinized epithelial cells.

hypodermis (hi'po-der'mis) — [Gr. *hypo*, under + *dermis*, skin] Loose connective tissue under the dermis that attaches the skin to muscle and bone.

keratinization (kĕr'ah-tin-ĭ-za'shun) — Production of keratin and changes in the structure and shape of epithelial cells as they move to the skin surface.

melanin (mel'ah-nin) — [Gr. *melas*, black] Brown to black pigment responsible for skin and hair color.

nail — A thin, horny plate at the ends of the fingers and toes, consisting of several layers of dead epithelial cells containing a hard keratin.

partial-thickness burn — Burn that damages only the epidermis (first-degree burn) or the epidermis and part of the dermis (second-degree burn).

sebaceous gland (sē-ba'shus) — [L. *sebum*, tallow] Gland of the skin that produces sebum; usually associated with a hair follicle.

stratum — [L., bed cover, layer] Layer of tissue.

sweat gland — Usually a secretory organ that produces a watery secretion called sweat that is released onto the surface of the skin; some sweat glands, however, produce an organic secretion.

vitamin D — Fat-soluble vitamin produced from precursor molecule in skin exposed to ultraviolet light; increases calcium and phosphate uptake in the intestine.

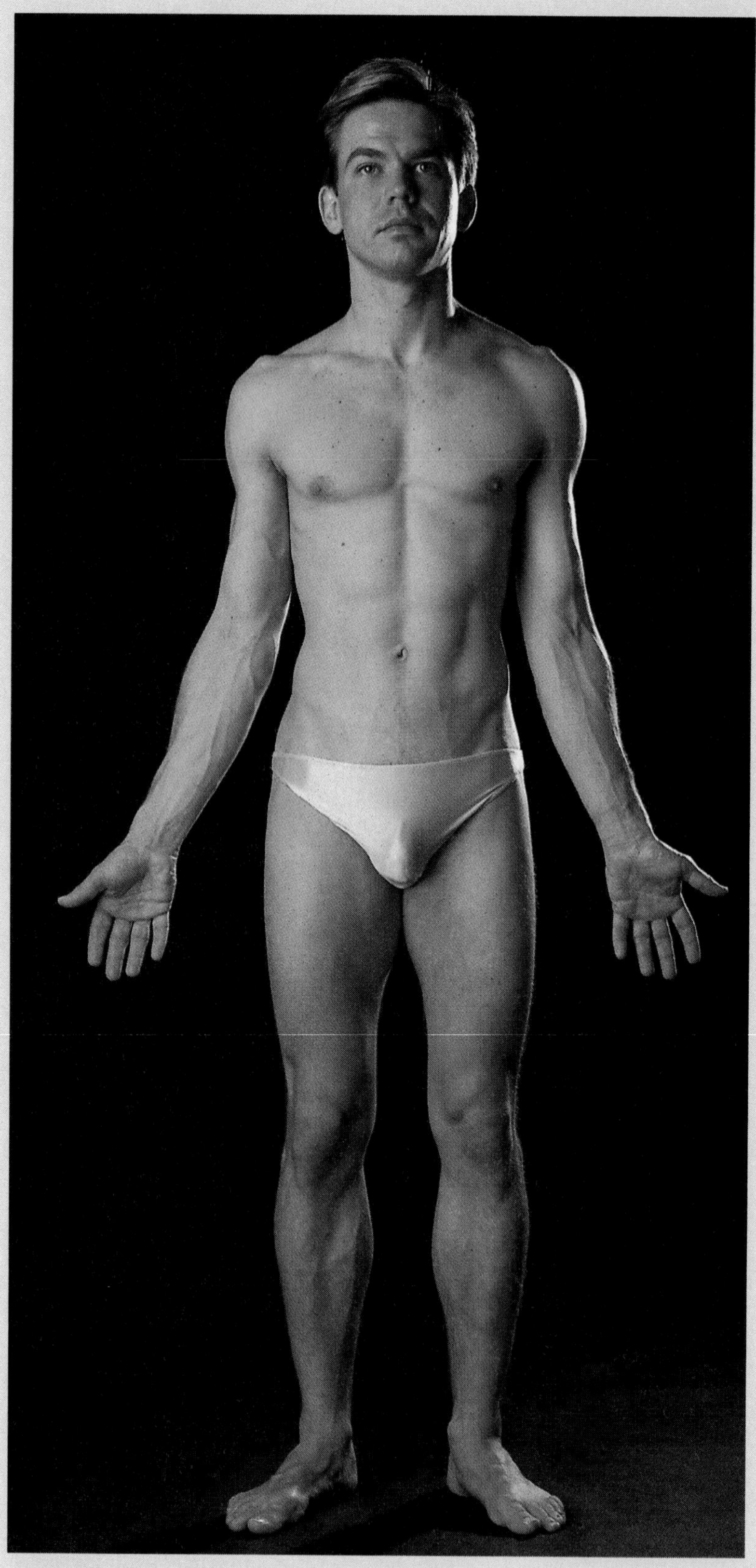

FIGURE 5-1 ■ The integumentary system.

The Integumentary System

MAJOR COMPONENTS

Skin, hair, nails, and sweat glands

MAJOR FUNCTIONS

Protects against abrasions and ultraviolet light

Prevents the entry of microorganisms and harmful substances

Reduces water loss

Regulates body temperature

Produces vitamin D precursors

Provides sensory information regarding heat, cold, pressure and pain

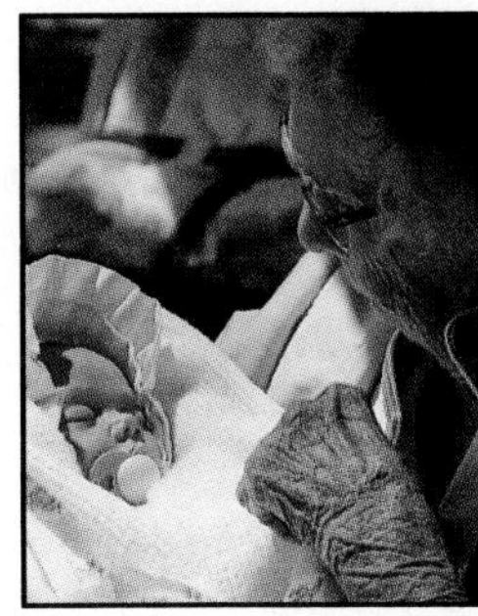

The integumentary system, consisting of the skin and accessory structures such as hair, nails, and glands, is the most extensive organ system in the body. Because skin is located on the outside of the body, it is an organ system with which most people are already familiar. The integumentary system has many functions, including the protection of internal structures, prevention of infectious agent entry, regulation of temperature and production of vitamin D (Figure 5-1).

HYPODERMIS

The **hypodermis** (hi'po-der'mis; under the dermis) or **subcutaneous** (sub'ku-ta'ne-us) **tissue** is not part of the integumentary system, but usually is considered with it because the hypodermis attaches the skin to underlying bone or muscle. The hypodermis is loose connective tissue that contains about half the body's stored fat, although the amount and location vary with age, sex, and diet. Fat in the hypodermis functions as padding and insulation, and it is responsible for some of the structural differences between men and women.

The hypodermis can be used to estimate total body fat. The skin and hypodermis are pinched at selected locations, and the thickness of the fold of skin and underlying hypodermis is measured. The thicker the fold, the greater is the amount of total body fat. Clinically, the hypodermis is the site of subcutaneous injections.

SKIN

The skin is made up of two major tissue layers. The **dermis** (der'mis; skin) is a layer of dense connective tissue, and the **epidermis** (ep'ĭ-der'mis; upon the dermis) is a layer of epithelial tissue that rests on the dermis (Figure 5-2).

Dermis

The dermis is dense connective tissue with fibroblasts, a few fat cells, and macrophages. The dermis is divided into two indistinct layers, the deeper **reticular** (rĕ-tik'u-lar) **layer** and the more superficial **papillary** (pap'ĭ-lĕr'e) **layer** (Figure 5-3). The reticular layer is the main fibrous layer of the dermis and consists of collagen and elastic fibers. It is responsible for most of the structural strength of the skin. The collagen fibers are oriented in many dif-

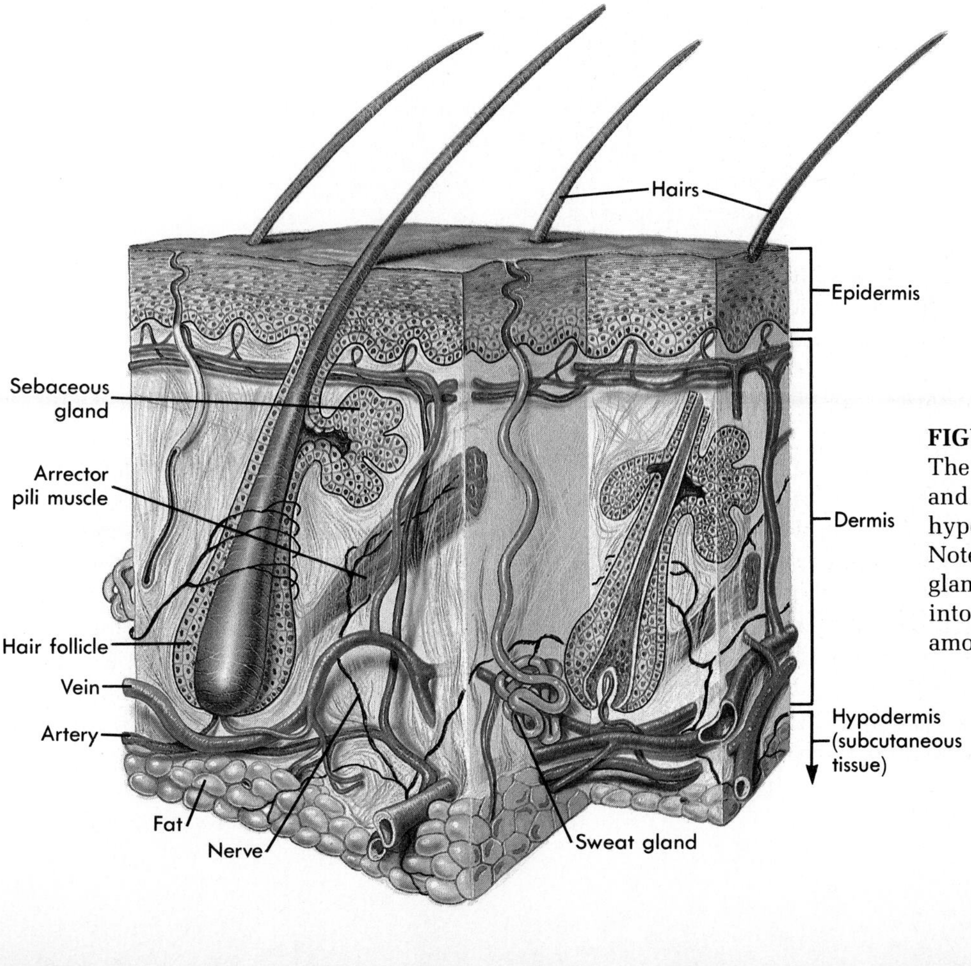

FIGURE 5-2 Skin and hypodermis. The skin, consisting of the epidermis and the dermis, is connected by the hypodermis to underlying structures. Note the accessory structures (hairs, glands, and arrector pili) that project into the hypodermis, and the large amount of fat in the hypodermis.

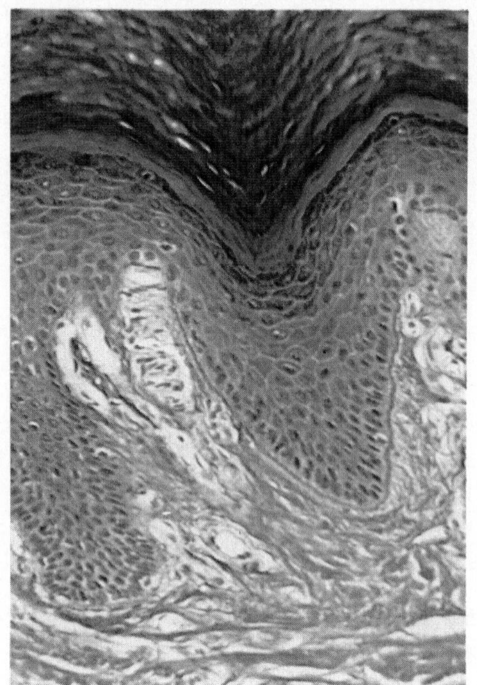

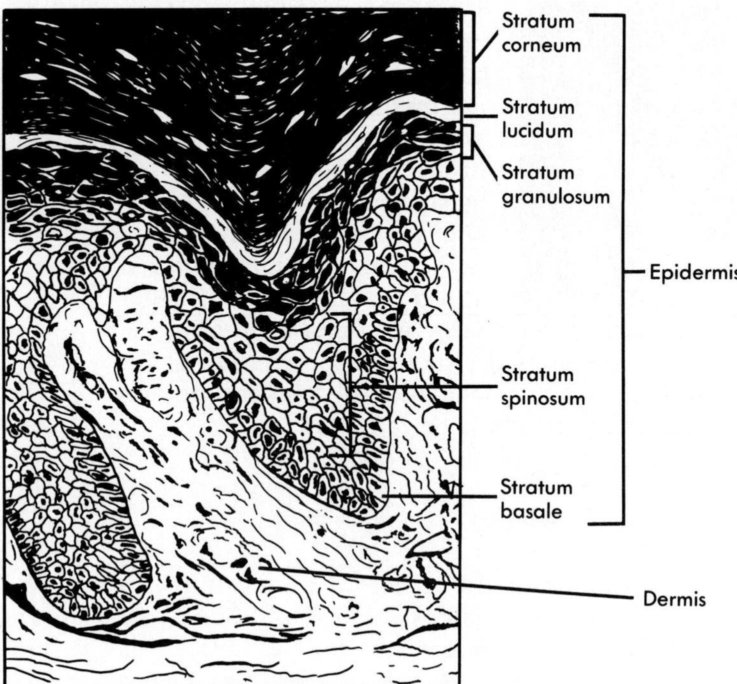

FIGURE 5-3 ■ Dermis and epidermis.
The dermis consists of two layers of connective tissue, the papillary and reticular layers. The epidermis is stratified squamous epithelium divided into strata. Cells are produced in the stratum basale and undergo keratinization as they move toward the surface of the skin to become stratum corneum cells.

ferent directions in order to resist stretch in many directions. There are more collagen fibers oriented in some directions than others, however. This produces **cleavage** or **tension lines** in the skin, and the skin is most resistant to stretch along the cleavage lines (Figure 5-4). It is important for surgeons to be aware of cleavage lines. An incision made across the cleavage lines is likely to gap, producing considerable scar tissue, but an incision made parallel with the lines tends to gap less and produce less scar tissue (see Chapter 4). If the skin is overstretched, the dermis can be damaged, leaving lines that are visible through the epidermis. These lines, called **striae** (stri′e) or **stretch marks,** can develop on the abdomen and breasts of a woman during pregnancy.

The papillary layer derives its name from projections called **papillae** (pă-pil′e) that extend into the epidermis (see Figure 5-3). The papillae of the hands, the soles of the feet, and the tips of the digits are in parallel, curving ridges that shape the overlying epidermis into fingerprints and footprints. The ridges increase friction and improve the grip of the hands and feet. The papillary layer contains many blood vessels that supply the overlying avascular epidermis with nutrients, remove waste products, and aid in regulating body temperature.

■ The dermis is that portion of an animal hide from which leather is made. The epidermis is removed, and the fibrous dermis is preserved by tanning. Clinically, the dermis in humans is sometimes the site of injections, such as the tuberculin skin test. ■

Epidermis

■ The epidermis is stratified squamous epithelium separated from the dermis by a basement membrane. Cells are produced in the deepest layers of the epidermis by mitosis. As new cells are formed, they push older cells to the surface where they slough or flake off. The outermost cells protect the cells underneath, and the deeper replicating cells replace cells lost from the surface. During their movement the cells undergo **keratinization** (kĕr′ah-tin-ĭ-za′shun), a process that changes the cells' shape and chemical composition. During keratinization these cells eventually die and produce an outer layer of cells that resists abrasion and forms a permeability barrier.

Although keratinization is a continuous process, distinct layers called **strata** are recognized (see

Figure 5-3). The deepest stratum, the **stratum basale** (bah-să'le), consists of columnar cells that undergo mitotic divisions approximately every 19 days. One daughter cell becomes a new stratum basale cell and can divide again. The other cell is pushed toward the surface, a journey that takes about 40 to 56 days. During this time the cell undergoes keratinization as it becomes a stratum corneum cell.

The **stratum corneum** (kor'ne-um) consists of dead, squamous **cornified cells** that have undergone keratinization. These cells are filled with the hard protein keratin, which gives the stratum corneum its structural strength. The cornified cells are also coated and surrounded by lipids that help prevent fluid loss through the skin.

1 What kinds of substances could easily pass through the skin by diffusion? What kinds would have difficulty?

The stratum corneum is composed of as many as 25 or more layers of dead squamous cells joined by desmosomes. Eventually the desmosomes break apart, and the cells are sloughed from the skin. Dandruff is an example of stratum corneum cells sloughed from the surface of the scalp. In skin subjected to friction, the number of layers in the stratum corneum greatly increases, producing a thickened area called a **callus** (kal'us). Over a bony prominence, the stratum corneum can thicken to form a cone-shaped structure called a **corn,** which can be quite painful.

Skin Color

Skin color is determined by pigments in the skin, by blood circulating through the skin, and by the thickness of the stratum corneum. **Melanin** (mel'ah-nin), a brown-to-black pigment, is responsible for most skin color. Melanin is produced by **melanocytes** (mel'ă-no-sīts) in the stratum basale. The melanocytes transfer melanin to other cells, so

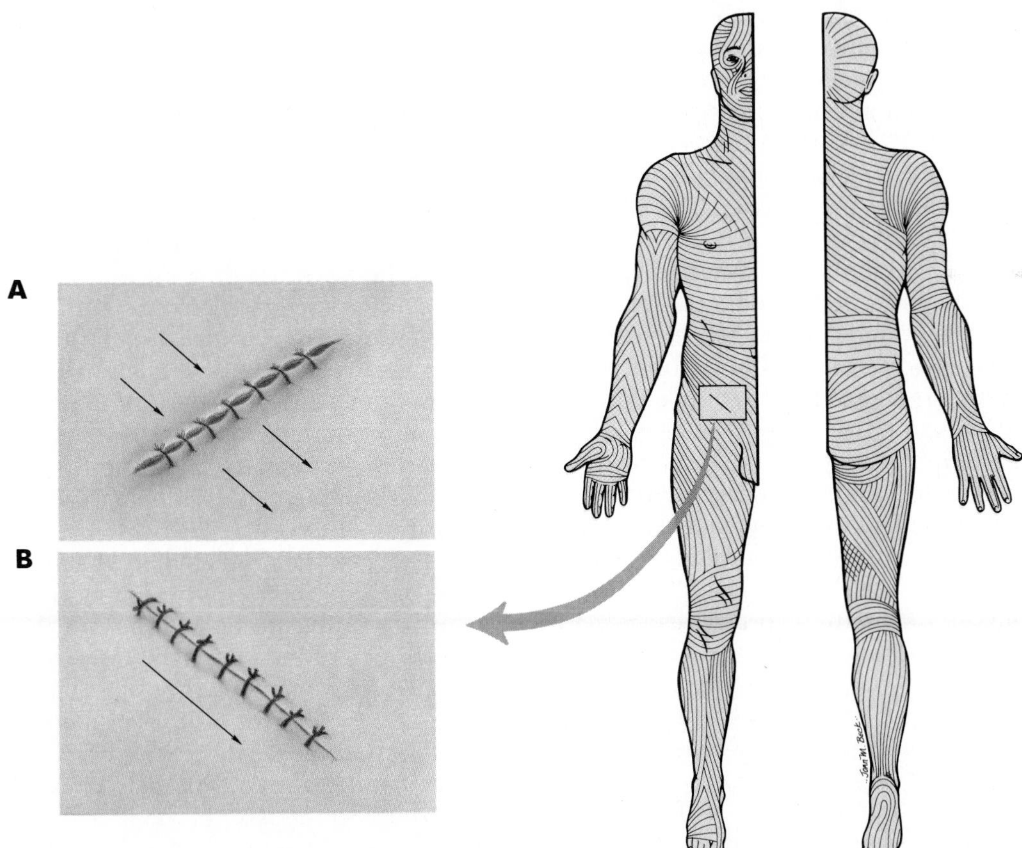

FIGURE 5-4 Cleavage lines.
The orientation of collagen fibers produces cleavage, or tension, lines in the skin.
A An incision made across cleavage lines can gap, increasing the time necessary for healing, and resulting in increased scar tissue formation.
B An incision made parallel to cleavage lines results in less gap, faster healing, and less scar tissue.

eventually all the cells in the deeper layers of the epidermis become pigmented. There are large amounts of melanin in some regions of the skin, such as freckles, moles, the genitalia, the nipples, and the areolae (ă-re'o-le; pigmented, circular areas around the nipples). Other areas, such as the lips, palms of the hands, and soles of the feet, have less melanin. Racial variations in skin color, such as black, brown, yellow, and white, are determined by the amount and distribution of melanin produced. All races have about the same number of melanocytes.

Melanin production is determined by genetic factors, hormones, and exposure to light. Genetic factors are responsible for the amounts of melanin produced in different races. Although many genes are responsible for skin color, a single mutation can prevent the manufacture of melanin. For example, **albinism** (al'bĭ-nizm) is a recessive genetic trait that causes a deficiency or absence of melanin. Albinos have fair skin, white hair, and unpigmented irises in the eyes. During pregnancy certain hormones cause an increase in melanin production in the mother, darkening the nipples, areolae, and genitalia. The cheekbones, forehead, and chest can also darken, resulting in "the mask of pregnancy," and a dark line of pigmentation can appear on the midline of the abdomen. Exposure to ultraviolet light, for example in sunlight, also stimulates melanocytes to increase melanin production. The result is a suntan.

The location of pigments and other substances in the skin affects the color produced. As light passes through a substance, some wavelengths of light are reflected or scattered to make a blue color. The color of the sky is an example of blue color produced by light reflecting from dust particles. In the skin, collagen fibers of the dermis also scatter light to produce a blue color. Thus the deeper within the dermis or hypodermis any dark pigment is located, the bluer the pigment appears because of the light-scattering effect of the overlying tissue. This effect causes the blue color of tattoos, bruises, and some superficial blood vessels.

Carotene (kar'o-tēn) is a yellow pigment found in plants such as squash and carrots. Humans normally ingest carotene and use it as a source of vitamin A. Carotene is lipid soluble, and when consumed it accumulates in the lipids of the stratum corneum and in the fat cells of the dermis and hypodermis. This gives the skin a slight yellowish tint. If large amounts of carotene are consumed, the skin can become quite yellowish.

Blood flowing through the skin imparts a reddish hue, and when blood flow increases (for example, during blushing, anger, and the inflammatory response), the red color intensifies. A decrease in blood flow such as occurs in shock can make the skin appear pale. A decrease in the blood oxygen content produces a bluish color called **cyanosis** (si-ă-no'sis). **Birthmarks** are congenital (present at birth) disorders of the blood vessels (capillaries) in the dermis.

2 Explain the differences in skin color between:
(A) the palms of the hands and the lips;
(B) the palms of the hands of a person who does heavy manual labor and one who does not; and
(C) the anterior and posterior surfaces of the forearm.

ACCESSORY SKIN STRUCTURES

Hair

The presence of **hair** is one of the characteristics common to all mammals. If the hair is thick and covers most of the body surface, it is called fur. In humans, hair is found everywhere in the skin except the palms, soles, lips, nipples, parts of the genitalia, and the distal segments of the fingers and toes. The **shaft** of the hair protrudes above the surface of the skin, whereas the **root** and **hair bulb** are below the surface (Figure 5-5). A hair has a hard **cortex** that surrounds a softer center, the **medulla.** The cortex is covered by the **cuticle,** a single layer of overlapping cells that holds the hair in the hair follicle. The **hair follicle** is an extension of the epidermis deep into the dermis. The hair follicle can play an important role in tissue repair. If the surface epidermis is damaged, the epithelial cells within the hair follicle can divide and serve as a source of new epithelial cells.

The hair is produced in the hair bulb, which rests on a dermal papilla. Blood vessels within the papilla supply the hair bulb with the nourishment needed to produce the hair. Hair is produced in cycles. During the growth stage, hair is formed by cells within the hair bulb. These cells, like the cells of the stratum basale in the skin, divide and undergo keratinization. The hair grows longer as cells are added to the base of the hair root. Thus the hair root and shaft consist of columns of dead keratinized epithelial cells. During the resting stage, growth stops, and the hair is held in the hair follicle. When the next growth stage begins, a new hair is formed, and the old hair falls out. The duration of each stage depends on the hair. Eyelashes grow for

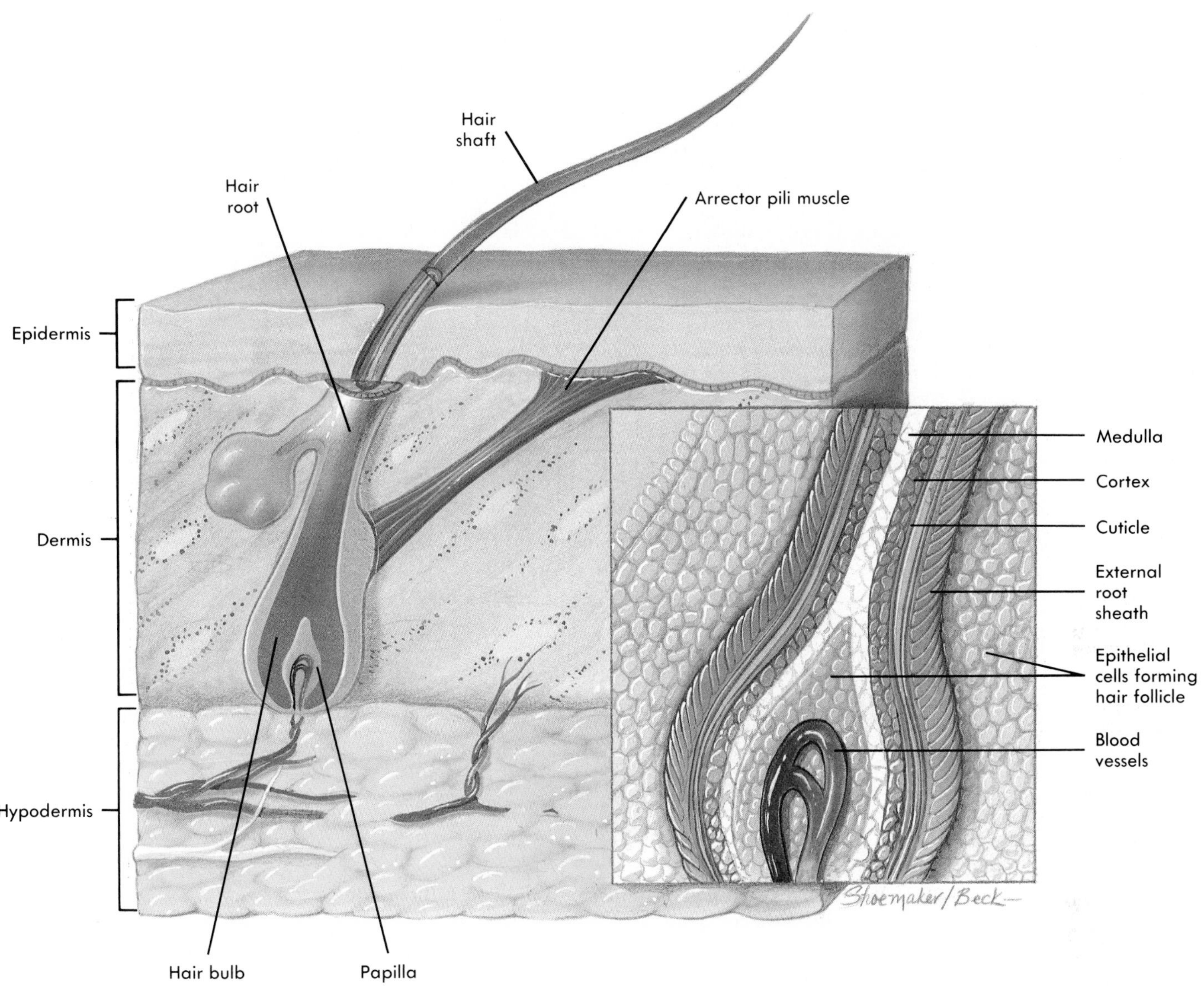

FIGURE 5-5 ■ Hair within a hair follicle.
The hair has been sectioned to reveal its internal components.

about 30 days and rest for 105 days, whereas scalp hairs grow for 3 years and rest for 1 to 2 years.

The loss of hair normally means that the hair is being replaced, because the old hair falls out of the hair follicle when the new hair begins to grow. In some men, however, a permanent loss of hair results in "pattern baldness." Actually the hair is not lost. It is replaced by a very short, transparent hair, which for practical purposes is invisible. This conversion occurs when male sex hormones act on the hair follicles of men who have the genetic predisposition of "pattern baldness."

Hair color is determined by varying amounts and types of melanin. The production and distribution of melanin by melanocytes occurs in the hair bulb by the same method as in the skin. With age the amount of melanin in hair can decrease, causing the hair to become faded or white (that is, having no melanin). Gray hair is usually a mixture of unfaded, faded, and white hairs.

3 Marie Antoinette's hair supposedly turned white overnight after she heard she was to be sent to the guillotine. Is it reasonable to believe this story? ?

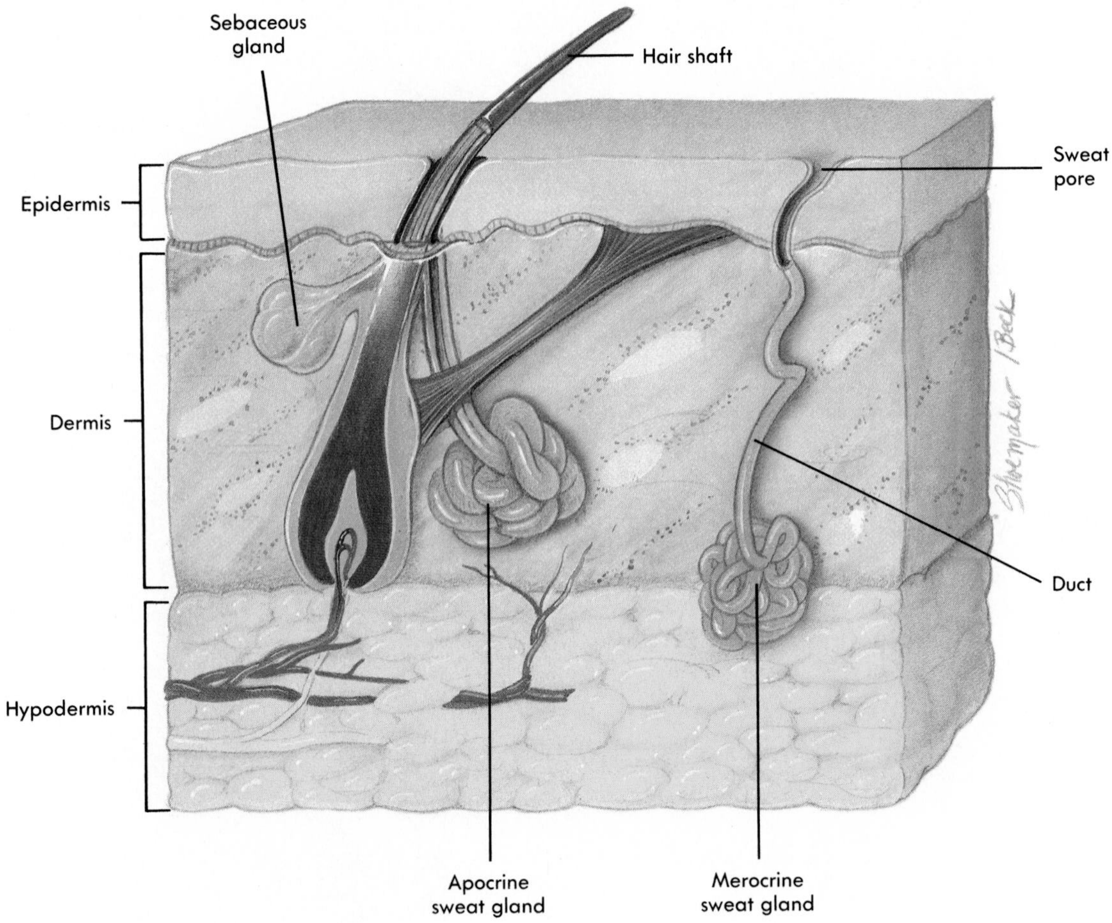

FIGURE 5-6 Glands of the skin.
Sebaceous and apocrine sweat glands empty into the hair follicle. Merocrine sweat glands empty onto the surface of the skin.

Muscles

Associated with each hair follicle are smooth muscle cells, the **arrector pili** (ah-rek'tor pĭ'le) (see Figure 5-5). Contraction of the arrector pili causes the hair to "stand on end" and also produces a raised area of skin called "goose flesh." In animals with fur, contraction of the arrector pili is beneficial because it functionally increases the thickness of the fur by raising the hairs. In the cold, the thicker layer of fur traps air and becomes a better insulator. Or, the thickened fur can make the animal appear larger and more ferocious, which might deter an attacker. It is unlikely that humans, with their sparse amount of hair, derive any important benefit from contraction of their arrector pili.

Glands

The major glands of the skin are the **sebaceous** (sĕ-ba'shus) **glands** and the **sweat glands** (Figure 5-6). Most sebaceous glands are connected by a duct to the upper part of a hair follicle. They produce **sebum,** an oily, white substance rich in lipids. The sebum lubricates the hair and the surface of the skin, which prevents drying and protects against some bacteria.

There are two kinds of sweat glands. **Merocrine** (mĕr'o-krin) **sweat glands** are located in almost every part of the skin, and are most numerous in the palms and soles. They produce a secretion that is mostly water with a few salts. Merocrine sweat glands have a duct that opens onto the surface of the skin through sweat pores. When the body temperature starts to rise above normal levels, the sweat glands produce sweat, which evaporates and cools the body. Sweat can also be released in the palms, soles, axillae (armpits), and other places due to emotional stress.

> Emotional sweating is used in lie detector (polygraph) tests because sweat gland activity usually increases when a person tells a lie. The sweat produced, even small amounts, can be detected because the salt solution conducts electricity and lowers the electrical resistance of the skin.

Apocrine (ap'o-krin) **sweat glands** produce a thick secretion rich in organic substances. They open into hair follicles, but only in the axillae and genitalia. Apocrine sweat glands become active at puberty because of the influence of sex hormones. Their secretion, which is essentially odorless when released, is quickly broken down by bacteria to cause what is commonly known as body odor.

Nails

The distal ends of the digits of humans (and other primates) have **nails,** whereas reptiles, birds, and most mammals have claws or hooves. The nail is a thin plate, consisting of layers of dead stratum corneum cells that contain a very hard type of keratin. The visible part of the nail is the **nail body,** and the part of the nail covered by skin is the **nail root** (Figure 5-7). The **eponychium** (ep-on-nik'i-um) or **cuticle** is stratum corneum that grows onto the nail body. The nail grows from the **nail matrix,** located under the proximal end of the nail. A small part of the nail matrix, the **lunula** (lu'nu-lah), can be seen through the nail body as a whitish, crescent-shaped area at the base of the nail. Unlike hair, nails grow continuously and do not have a resting stage.

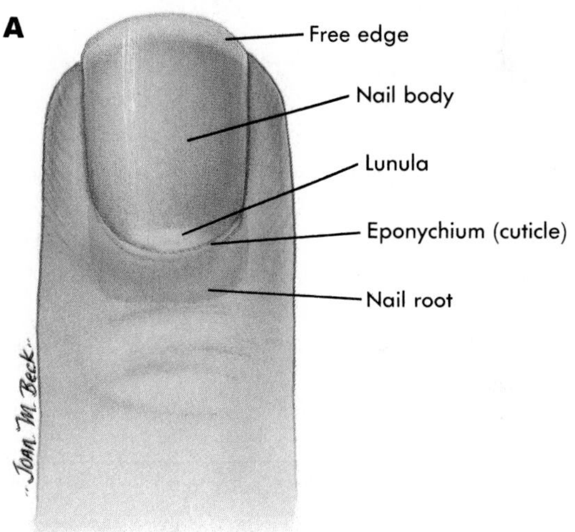

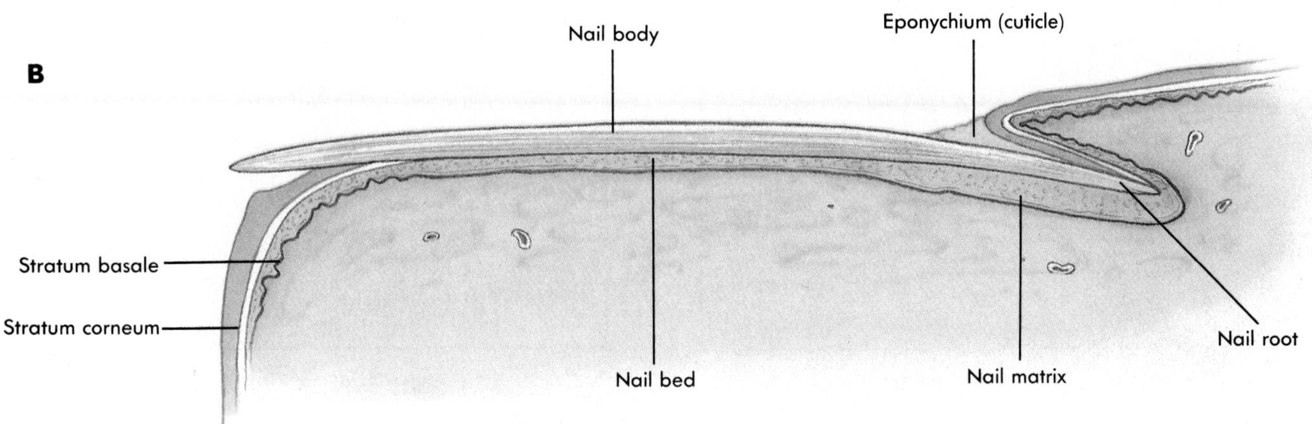

FIGURE 5-7 Nail.
A Dorsal view.
B Lateral view. The stratum basale and a few layers of cells are indicated by purple. Most of the epidermis is absent from the nail bed.

FUNCTIONS OF THE INTEGUMENTARY SYSTEM

Protection

The integumentary system performs many protective functions. The intact skin plays an important role in preventing water loss, and it prevents the entry of microorganisms and other foreign substances into the body. Secretions from skin glands also produce an environment unsuitable for some microorganisms. The stratified squamous epithelium of the skin protects underlying structures against abrasion. Melanin absorbs ultraviolet light and protects underlying structures. Hair provides protection in several ways: On the head it acts as a heat insulator, eyebrows keep sweat out of the eyes, eyelashes protect the eyes from foreign objects, and hair in the nose and ears prevents the entry of dust and other materials. The nails protect the ends of the digits from damage and can be used in defense.

Temperature Regulation

Body temperature normally is maintained at about 37° C (98.6° F). Regulation of body temperature is important because chemical reactions within the body can be speeded up or slowed down by changes in the body's temperature. Even slight changes in temperature can make enzymes operate less effeciently and disrupt the normal rates of chemical changes in the body.

Exercise, fever, or an increase in environmental temperature tend to raise body temperature. Homeostasis requires the loss of excess heat. Blood vessels (arterioles) in the dermis dilate and enable more blood to flow through the skin, thus transferring heat from deeper tissues to the skin, where the heat is lost by radiation (infrared energy), convection (air movement), or conduction (direct contact with an object) (Figure 5-8, *A*). Sweat that spreads over the surface of the skin and evaporates also carries away heat and reduces body temperature.

If body temperature begins to drop below normal, heat can be conserved by constriction of dermal blood vessels, which reduces blood flow to the skin. Thus less heat is transferred from deeper structures to the skin, and heat loss is reduced (Figure 5-8, *B*). However, with smaller amounts of warm blood flowing through the skin, the skin temperature decreases. If the skin temperature drops below about 15° C (59° F), blood vessels dilate as a protective mechanism to prevent tissue damage from the cold.

4 You may have noticed that on very cold winter days peoples' noses and ears turn red. Can you explain why this happens? ?

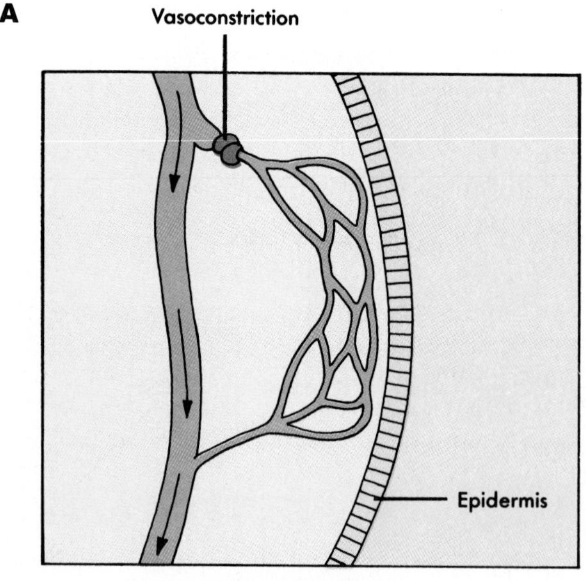

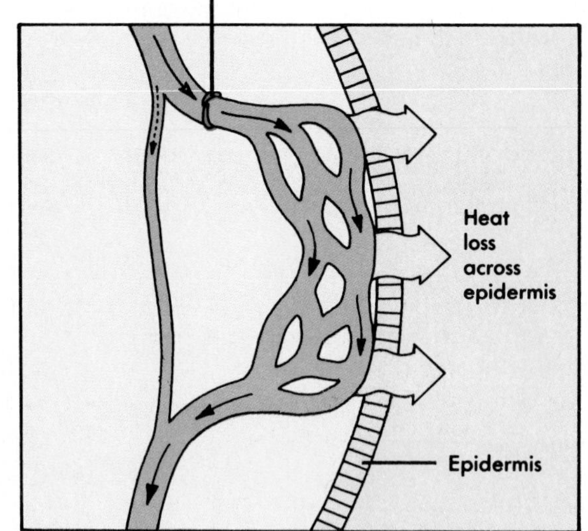

FIGURE 5-8 **Heat exchange in the skin.**

A Blood vessels in the dermis dilate (vasodilate), allowing more blood close to the surface where heat is lost from the body.

B Blood vessels in the dermis constrict (vasoconstrict), reducing blood flow and heat loss.

Vitamin D Production

When the skin is exposed to ultraviolet light, a precursor molecule of **vitamin D** is formed. The precursor is carried by the blood to the liver, where it is modified, and then to the kidneys where the precursor is modified further to form active vitamin D. If exposed to enough ultraviolet light, humans can produce all the vitamin D they need. Many people need to ingest vitamin D, however, because clothing and indoor living reduce exposure to ultraviolet light. Natural sources of vitamin D include liver (especially fish liver), egg yolks, and dairy products.

Adequate levels of vitamin D are necessary because vitamin D stimulates calcium and phosphate uptake in the intestines. These substances are necessary for normal bone metabolism (see Chapter 6) and normal muscle function (see Chapter 7).

Sensation

The skin has receptors in the epidermis and dermis that can detect pain, heat, cold, and pressure (see Chapter 8). Although hair does not have a nerve supply, movement of the hair can be detected by sensory receptors around the hair follicle.

Excretion

Excretion is the removal of waste products from the body. In addition to water and salts, sweat contains a small amount of waste products such as urea, uric acid, and ammonia. Even though large amounts of sweat can be lost from the body, the sweat glands do not play a significant role in excretion.

EFFECTS OF AGING ON THE INTEGUMENTARY SYSTEM

As the body ages, the blood flow to the skin is reduced, and the skin becomes thinner and appears more transparent. Because of decreased amounts of collagen in the dermis, skin is more easily damaged and repairs more slowly. Loss of elastic fibers in the skin and loss of fat from the hypodermis cause the skin to sag and wrinkle. A decrease in the activity of sebaceous and sweat glands results in dry skin and poor ability to regulate body temperature. The number of melanocytes generally decreases, but in some areas, the number of melanocytes increases to produce age spots. (Age spots are different from freckles, which are caused by increased melanin production.) White or gray hair also results because of a decrease in or a lack of melanin production. Skin that is exposed to sunlight ages more rapidly than nonexposed skin, so avoiding overexposure to sunlight and using sun blockers is advisable.

THE INTEGUMENTARY SYSTEM AS A DIAGNOSTIC AID

The integumentary system is useful in diagnosis because it is observed easily and often reflects events occurring in other parts of the body. For example, cyanosis, a bluish color caused by decreased blood oxygen content, is an indication of impaired circulatory or respiratory function. A yellowish skin color, **jaundice** (jawn′dĭs), can occur when the liver is damaged by a disease such as viral hepatitis. Normally the liver secretes bile pigments, which are products of the breakdown of worn out red blood cells, into the intestine. Bile pigments are yellow, and a buildup of bile pigments in the blood and tissues can indicate that the liver is not functioning properly.

Rashes and lesions in the skin can be symptoms of problems elsewhere in the body. For example, scarlet fever results from a bacterial infection in the throat. The bacteria release a toxin into the blood that causes a pink-red rash in the skin. The development of a rash can also indicate an allergic reaction to foods or drugs such as penicillin.

BURNS

Burns are classified according to the depth of the burn (Figure 5-9). In **partial-thickness burns** some portion of the stratum basale remains viable, and regeneration of the epidermis occurs from within the burn area as well as from the edges of the burn. Partial-thickness burns are divided into **first-** and **second-degree burns.**

First-degree burns involve only the epidermis and are red and painful, and slight edema can be present. They can be caused by sunburn or brief exposure to hot or cold objects and heal without scarring in about a week.

Second-degree burns damage the epidermis and the dermis. If there is minimal dermal damage, symptoms include redness, pain, edema, and blisters. Healing takes about 2 weeks, and there is no scarring. If the burn goes deep into the dermis, however, the wound appears red, tan, or white, can take several months to heal, and might scar. In all second-degree burns the epidermis regenerates from epithelial tissue in hair follicles and sweat glands, as well as from the edges of the wound.

In **full-thickness** or **third-degree burns** the epidermis and the dermis are completely destroyed, and recovery occurs from the edges. Third-degree burns often are surrounded by areas of first- and second-degree burns. Although the first- and second-degree burn areas are painful, the region of third-degree burn is usually painless because sensory receptors in the epidermis and dermis have been destroyed. Third-degree burns appear white, tan, brown, black, or deep cherry red.

The skin normally prevents loss of water from the body and prevents the entry of microorganisms into the body. Deep partial-thickness and full-thickness burns over large areas of the body result in excessive fluid loss. Proper fluid replacement is necessary to prevent dehydration, shock, and possibly death. Because microorganisms can infect the body more easily through these wounds, patients are also maintained under aseptic conditions and are given antibiotics. Burns of these types also take a long time to heal, and they form scar tissue with disfiguring and debilitating wound contracture. To prevent these complications and to speed healing, skin grafts are performed. For example, in a split skin graft the epidermis and part of the dermis are removed from another part of the body and placed over the burn. Interstitial fluid from the burn nourishes the graft until blood vessels grow into the graft and supply it with nourishment. Meanwhile, the donor tissue produces new epidermis from epithelial tissue in the hair follicles and sweat glands, such as occurs in superficial second-degree burns.

FIGURE 5-9 ■ Burns.
Parts of the skin damaged by different types of burns. Partial-thickness burns are subdivided into first-degree burns (damage to only the epidermis) and second-degree burns (damage to the epidermis and part of the dermis). Full-thickness or third-degree burns destroy the epidermis, dermis, and sometimes deeper tissues.

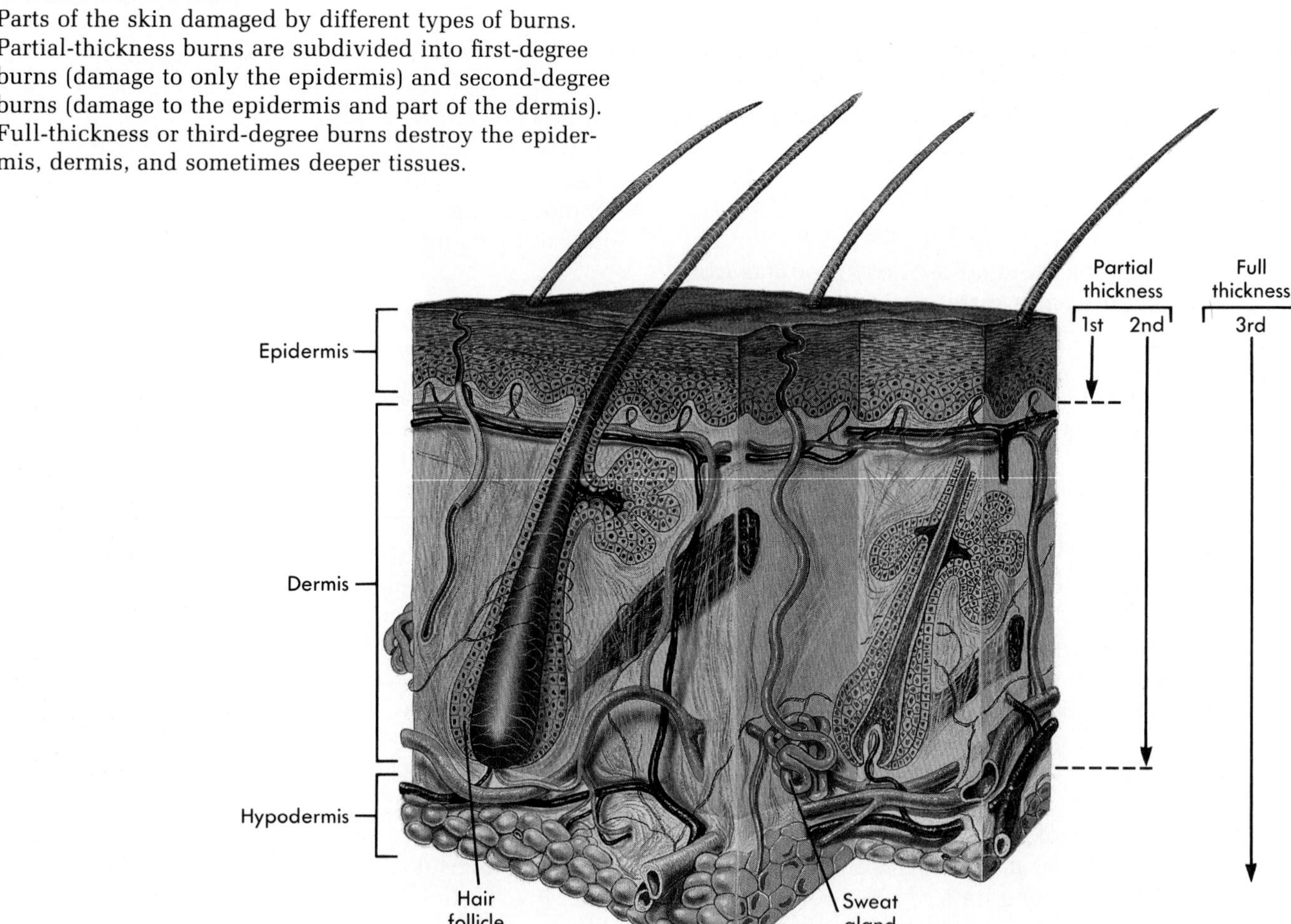

SKIN CANCER

Skin cancer is the most common type of cancer. Although chemicals and radiation (x-rays) are known to induce cancer, the development of skin cancer most often is associated with exposure to ultraviolet light from the sun. Consequently, most skin cancers develop on the face, neck, or hands.

Basal cell carcinoma, the most frequent skin cancer, begins with cells in the stratum basale and extends into the dermis to produce an open ulcer (Figure 5-10, *A*). Surgical removal or radiation therapy cures this type of cancer. Fortunately there is little danger that this type of cancer will spread, or metastasize, to other areas of the body. **Squamous cell carcinoma** develops from cells immediately superficial to the stratum basale. Normally these cells undergo little or no cell division. In squamous cell carcinoma, the cells continue to divide as they produce keratin. Typically, the result is a nodular, keratinized tumor confined to the epidermis (Figure 5-10, *B*). If untreated the tumor can invade the dermis, metastasize, and cause death. **Malignant melanoma** is a rare form of skin cancer that arises from melanocytes, usually in a preexisting mole. A mole is simply an aggregation or "nest" of melanocytes. The melanoma can appear as a large, flat spreading lesion or as a deeply pigmented nodule (Figure 5-10, *C*). Metastasis is common, and unless diagnosed and treated early in development, this cancer is often fatal.

A

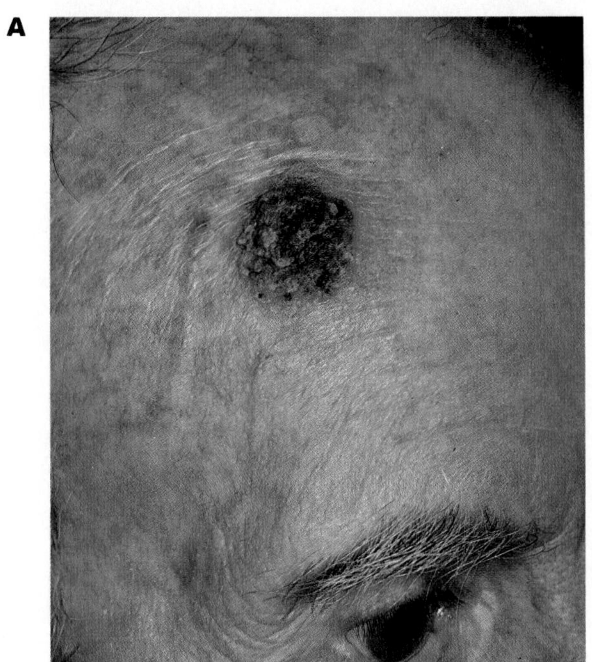

B

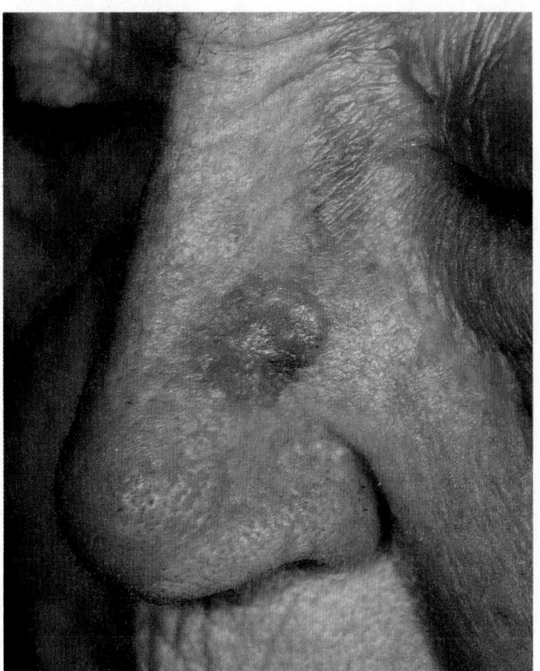

C

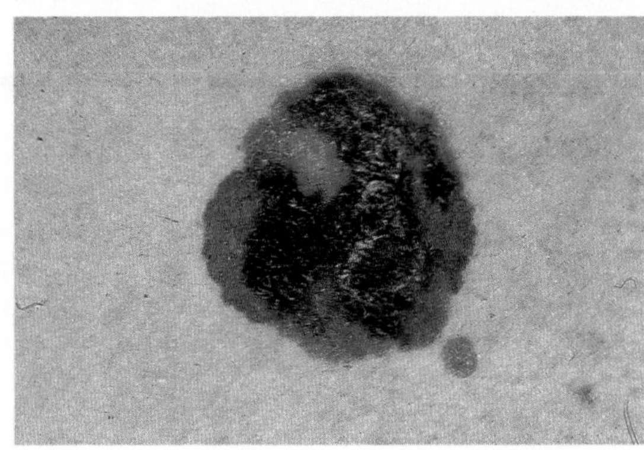

FIGURE 5-10 ■ Cancer of the skin.
A Basal cell carcinoma.
B Squamous cell carcinoma.
C Malignant melanoma.

Diseases of the Skin

BACTERIAL INFECTIONS

Impetigo is a skin disease caused by *Staphylococcus aureus*. It usually affects children, producing small blisters containing pus that easily rupture to form a thick, yellowish crust. The bacteria are transmitted by direct contact (touching) and enter the skin through abrasions or small breaks in the skin.

Acne is a disorder of the hair follicles and sebaceous glands. Four factors are believed to be involved: hormones, sebum, abnormal keratinization, and the bacterium *Propionibacterium acnes*. The lesions begin with the overproduction of epidermal cells in the hair follicle. These cells are shed from the wall of the hair follicle, and they stick to each other to form a mass of cells mixed with sebum that blocks the hair follicle. During puberty, hormones, especially testosterone, stimulate the sebaceous glands and sebum production increases. Because both the adrenal glands and the testes produce testosterone, the effect is seen in males and females. An accumulation of sebum behind the blockage produces a whitehead. A blackhead develops when the accumulating mass of cells and sebum pushes through the opening of the hair follicle. Although there is general agreement that dirt is not responsible for the black color, the exact cause of the black color in blackheads is disputed. A pimple results if the wall of the hair follicle ruptures, forming an entry way into the surrounding tissue. *P. acnes* and other bacteria stimulate an inflammatory response that results in the formation of a red pimple filled with pus. If tissue damage is extensive, scarring occurs.

Decubitis (de-ku'bi-tus) **ulcers,** also known as bedsores or pressure sores, develop in patients who are immobile (for example, bedridden or confined to a wheelchair). The weight of the body, especially in areas over bony projections such as the hip bones and heels, compresses tissue and causes **ischemia** (is-ke'me-ah) or reduced circulation. The lack of blood flow results in the destruction or **necrosis** (nĕ-kro'sis) of the hypodermis and the skin. After the skin dies, bacteria gain entry to produce an infected ulcer.

VIRAL INFECTIONS

Interestingly, many of the viral skin diseases do not enter the body through the skin. Instead the viruses enter through the respiratory system where they incubate for about 2 weeks. Then they are carried by the blood to the skin where they cause lesions. Examples are **rubeola** (measles), **rubella** (German measles), and **chickenpox.** Rubeola can be a dangerous disease because it can develop into pneumonia, or the virus can invade the brain and cause damage. Rubella is a mild disease but can prove dangerous if contracted during pregnancy. The virus can cross the placenta and damage the fetus. Possible outcomes are deafness, cataracts, heart defects, mental retardation, and death of the fetus. Chickenpox is a mild disease if contracted as a child. **Herpes zoster** or **shingles** is a disease caused by the chickenpox virus that occurs after the childhood infection. The virus remains dormant within nerve cells. Trauma, stress, or another illness somehow activates the virus, which moves through the nerve to the skin where it causes lesions along the nerve's pathway.

Cold sores or **fever blisters** are caused by a virus related to the chickenpox virus. The initial infection usually does not produce symptoms. Dormant viruses can become active, however, and produce lesions in the skin around the mouth and in the mucous membrane of the mouth. The virus is transmitted by oral or respiratory routes. A related virus is transmitted by sexual contact and produces genital lesions (genital herpes).

Warts are uncontrolled growths caused by the human papilloma virus. Usually the growths are benign and disappear spontaneously, or they can be removed by a variety of techniques. The viruses are transmitted by direct contact with contaminated objects or an infected person. They can also be spread by scratching. The virus enters the skin and incubates for several weeks before the wart appears.

FUNGAL INFECTIONS

Ringworm is a fungal infection that affects the keratinized portion of the skin, hair, and nails. It produces patch scaling and an inflammatory response. The lesions are often circular with a raised edge and in ancient times were thought to be caused by worms. Several species of fungus cause ringworm in humans, and they usually are described by their location on the body. In the scalp the condition is ringworm, in the groin it is jock itch, and in the feet it is athlete's foot.

ECZEMA AND DERMATITIS

Eczema (ek'zĕ-mah) and **dermatitis** (der'mă-ti'tis) describe inflammatory conditions of the skin. Causes of the inflammation can be allergy, infection, poor circulation, or exposure to physical factors such as chemicals, heat, cold, or sunlight.

PSORIASIS

The cause of **psoriasis** (so-ri'ă-sis) is unknown, but there may be a genetic component. In psoriasis there is increased cell division in the stratum basale, abnormal keratinization, and elongation of the dermal papillae toward the skin surface. The result is a thicker-than-normal stratum corneum that sloughs to produce large, silvery scales. If the scales are scraped away, bleeding occurs from the blood vessels at the top of the dermal papillae. Psoriasis is a chronic disease that can be controlled but as yet has no cure.

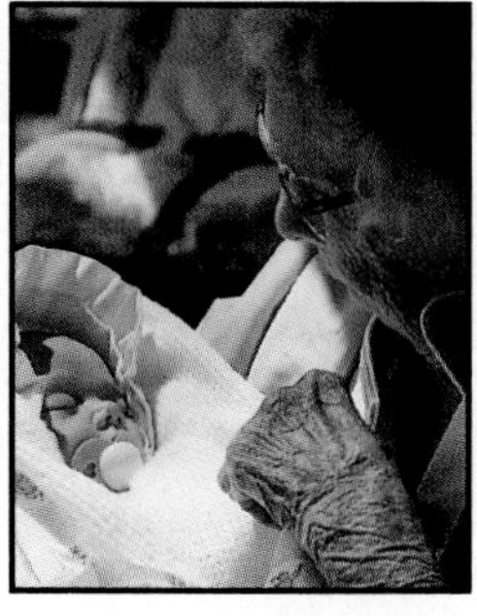

SUMMARY

The integumentary system consists of the skin, hair, glands, and nails.

HYPODERMIS

The hypodermis is loose connective tissue that attaches the skin to underlying tissues.

About half of the body's fat is stored in the hypodermis.

SKIN

Dermis

The dermis is dense connective tissue.

The reticular layer provides structural strength, and the blood vessels of the papillary layer supply the epidermis with nutrients.

Epidermis

The epidermis is stratified squamous epithelium divided into strata.

- Cells are produced in the stratum basale.
- The stratum corneum is many layers of dead, cornified, squamous cells. The most superficial layers are sloughed.

Keratinization is the transformation of stratum basale cells into stratum corneum cells.

- Structural strength results from keratin inside the cells, and from desmosomes, which hold the cells together.
- Permeability characteristics result from lipids surrounding the cells.

Skin Color

Melanocytes produce melanin, that is responsible for different racial skin colors. Melanin production is determined genetically but can be modified by hormones and ultraviolet light (tanning).

Carotene, a plant pigment ingested as a source of vitamin A, can cause the skin to appear yellowish.

Scattering of light by collagen produces a bluish color.

Increased blood flow produces a red skin color, whereas a decreased blood flow causes a pale skin color. Decreased blood oxygen results in the blue color of cyanosis.

ACCESSORY SKIN STRUCTURES

Hair

Hairs are columns of dead, keratinized epithelial cells. Each hair consists of a shaft (above the skin), root (below the skin), and hair bulb (site of hair cell formation).

Hairs have a growth phase and a resting phase.

Muscles

Contraction of the arrector pili, which are smooth muscles, causes hair to "stand on end" and produces "goose flesh."

Glands

Sebaceous glands produce sebum, which oils the hair and the surface of the skin.

Merocrine sweat glands produce sweat that cools the body.

Apocrine sweat glands produce an organic secretion that can be broken down by bacteria to cause body odor.

Nails

The nail consists of the nail body and nail root.

The nail matrix produces the nail, which is stratum corneum containing hard keratin.

FUNCTIONS OF THE INTEGUMENTARY SYSTEM

Protection

The skin prevents the entry of microorganisms, acts as a permeability barrier, and provides protection against abrasion and ultraviolet light.

Temperature Regulation

Through dilation and constriction of blood vessels, the skin controls heat loss from the body.

Evaporation of sweat cools the body.

Vitamin D Production

Ultraviolet light stimulates the production of a precursor molecule in the skin that is modified by the liver and kidneys into vitamin D.

Vitamin D increases calcium uptake in the intestines.

Sensation

The skin contains sensory receptors for pain, heat, cold, and pressure.

Excretion

Skin glands remove small amounts of waste products but are not important in excretion.

EFFECTS OF AGING ON THE INTEGUMENTARY SYSTEM

Blood flow to the skin is reduced, the skin becomes thinner, and elasticity is lost.

Sweat and sebaceous glands are less active, and the number of melanocytes decreases.

THE INTEGUMENTARY SYSTEM AS A DIAGNOSTIC AID

The integumentary system is easily observed and often reflects events occurring in other parts of the body (for example, cyanosis, jaundice, rashes).

BURNS

Partial-thickness burns damage only the epidermis (first-degree burn) or the epidermis and the dermis (second-degree burn).

Full-thickness burns (third-degree burns) destroy the epidermis, dermis, and usually underlying tissues.

CANCER

Basal cell carcinoma involves the cells of the stratum basale and is readily treatable.

Squamous cell carcinoma involves the cells immediately superficial to the stratum basale and can metastasize.

Malignant melanoma involves melanocytes and is often fatal.

CONTENT REVIEW

1. Name the components of the integumentary system.
2. What type of tissue is the hypodermis, and what are its functions?
3. What type of tissue is the dermis? Name the two layers of the dermis and give their functions.
4. What kind of tissue is the epidermis? In which stratum of the epidermis are new cells formed? In which stratum are they sloughed?
5. Define keratinization. What structural changes does keratinization produce to make the skin resistant to abrasion and water loss?
6. Name the cells that produce melanin. What happens to the melanin after it is produced?
7. Describe the factors that determine the amount of melanin produced in the skin.
8. How do melanin, carotene, collagen, and blood affect skin color?
9. Define the root, shaft, and hair bulb of a hair. What kind of cells are found in a hair?
10. What is a hair follicle? Why is it important in the repair of skin?
11. Where does hair grow? What are the stages of hair growth?
12. What happens when the arrector pili of the skin contract?
13. What secretion is produced by the sebaceous glands? What is the function of the secretion?

14. Which glands of the skin are responsible for cooling the body? Which glands are involved in producing body odor?
15. Name the parts of a nail. Where is the nail produced, and what kind of cells make up a nail? What is the lunula? Describe nail growth.
16. How does the integumentary system provide protection?
17. How does the integumentary system assist in the regulation of body temperature?
18. Describe the production of vitamin D by the body. What is the function of vitamin D?
19. List the types of sensations detected by receptors in the skin.
20. Name the substances excreted by skin glands. Is the skin an important site of excretion?
21. What changes occur in the skin as a result of age?
22. Why is the skin a useful diagnostic aid? Give three examples.
23. Define the different categories of burns. How is repair accomplished after each type?
24. What is the most common cause of skin cancer? Describe three types of skin cancer and the risks of each type.

CONCEPT REVIEW

1. A woman has stretch marks on her abdomen, yet she states that she has never been pregnant. Is this possible?
2. Harry Fastfeet, a Caucasian, jogs on a cold day. What color would you expect his skin to be (1) after going outside and just before starting to run, (2) during the run, and (3) 5 minutes after the run?
3. Given what you know about the cause of acne, propose some ways to prevent or treat the disorder.
4. Consider the following statement: Dark-skinned children are more susceptible to rickets (insufficient calcium in the bones) than fair-skinned children. Defend or refute this statement.
5. Pulling on hair can be quite painful, yet cutting hair is not painful. Explain.

ANSWERS TO ? PREDICT QUESTIONS

1 *p. 95* Because the permeability barrier is composed mainly of lipids surrounding the epidermal cells, substances that are lipid soluble could diffuse through the barrier easily. This fact is used as a basis for administering some medications through the skin. On the other hand, water-soluble substances would have difficulty diffusing through the skin. The skin's lipid barrier prevents water loss from the body.

2 *p. 96* (A) The lips are pinker or redder than the palms of the hand. Several explanations for this are possible. There could be more blood vessels in the lips, there could be increased blood flow in the lips, or the blood vessels could be easier to see through the epidermis of the lips. The last possibility actually explains most of the difference in color between the lips and palms. The epidermis of the lips is thinner and not as heavily keratinized as that of the palms. In addition, the papillae containing the blood vessels in the lips are "high" and closer to the surface.

(B) A person who does manual labor has a thicker stratum corneum (and possibly calluses) than a person who does not perform manual labor. The thicker epidermis masks the underlying blood vessels, and the palms do not appear as pink. Additionally, carotene accumulating in the lipids of the stratum corneum might give the palms a yellowish cast.

(C) The posterior surface of the forearm appears darker because of the tanning effect of ultraviolet light from the sun. The posterior surface of the forearm usually is exposed to more sunlight than the anterior surface of the forearm.

3 *p. 97* It is not reasonable to believe the story. Hair color is due to melanin that is added to the hair in the hair bulb as the hair grows. The hair itself is dead. To turn white, the hair must grow out without the addition of melanin. This, of course, takes more time than one night.

4 *p. 100* On cold days skin blood vessels of the ears and nose may dilate, bring warm blood to the ears and nose, and thus prevent tissue damage from the cold. The increased blood flow makes the ears and nose appear red.

6

THE SKELETAL SYSTEM—BONES AND JOINTS

After reading this chapter you should be able to:

1 List and describe the components of the skeletal system.
2 Describe the components of the connective tissue matrix and state the function of each.
3 Describe the structure of compact and cancellous bone.
4 Explain bone ossification, growth, remodeling, and repair.
5 Describe the main features of the skull as seen from the lateral, frontal, internal, and basal views.
6 Describe the vertebral column, the general features of each vertebra, and differences between vertebrae from each region of the vertebral column.
7 List the bones of the thoracic cage, including the three types of ribs.
8 Name and describe the bones of the pectoral girdle and upper limb.
9 Name and describe the bones of the pelvic girdle and lower limb.
10 List and describe the various types of joints.
11 Describe the major types of joint movement.

appositional growth [L. *ap* + *pono*, to put or place] To place one layer of bone, cartilage, or other connective tissue against an existing layer; increases the width or diameter of bones.

articulation A place where two bones come together; a joint.

cancellous bone (kan'sĕ-lus) Bone with a latticelike appearance; spongy bone.

compact bone Bone that is more dense and has fewer spaces than cancellous bone.

endochondral growth Growth of cartilage, which is then replaced by bone.

foramen (fo-ra'men) A hole; referring to a hole or opening in a bone.

girdle A bony ring or belt that attaches a limb to the body.

lamella (lă-mel'ah) A thin sheet or layer of bone.

matrix The substance between the cells of a tissue.

ossification (os'ĭ-fĭ-ka'shun) Bone formation.

paranasal sinus Air-filled cavity within certain skull bones that connects to the nasal cavity; the four sets of paranasal sinuses are the frontal, maxillary, sphenoid, and ethmoid.

synovial fluid (sĭ-no've-al) [G. *syn*, coming together + *ovia*, resembling egg albumin] A somewhat viscous substance serving as a lubricant in movable joints, tendon sheaths, and bursae.

trabecula (tră-bek'u-lah) [G., beam] A beam or plate of cancellous bone.

The skeletal system consists of bones and their associated connective tissues, including cartilage, tendons, and ligaments. Because **bone** is rigid, it helps maintain the shape of the body, protects organs, and provides a system of levers upon which muscles act to produce body movements. Bone also functions as a site for mineral storage and blood cell formation. **Cartilage** (kar'tĭ-lij), on the other hand, is somewhat rigid but more flexible than bone. Cartilage is abundant in the embryo and the fetus, where it provides a model for most of the adult bones and is a major site of skeletal growth in the embryo, fetus, and child. In the adult, the surfaces of bones within movable joints are covered with cartilage. Cartilage also provides a firm yet flexible support within certain structures such as the nose, external ear, ribs, and trachea. Tendons and ligaments are strong bands of fibrous connective tissue; **tendons** attach muscles to bones, and **ligaments** attach bones to bones.

CONNECTIVE TISSUE

Connective tissue consists of cells separated from each other by an extracellular **matrix.** The cells produce and maintain the extracellular matrix, which is largely responsible for the characteristics of the connective tissues (see Chapter 4). The characteristics of the matrix are based on the nature and relative quantities of the molecules it contains: collagen, proteoglycan, other organic molecules, water, and minerals. **Collagen** (kol'lă-jen) is a tough, ropelike protein. **Proteoglycans** (pro'te-o-gli'kanz) are large molecules consisting of polysaccharides attached to proteins and resembling bottle brushes. Proteoglycans can attract and retain large amounts of water.

The extracellular matrix of tendons and ligaments is made up primarily of collagen, making these structures like tough ropes. Cartilage matrix contains collagen and proteoglycan. Collagen makes cartilage very tough, while the water-filled proteoglycans make it smooth and resilient, so that cartilage is an excellent shock absorber.

Bone matrix contains collagen and minerals, including calcium and phosphate. The matrix resembles reinforced concrete. Collagen, like reinforcing steel bars, lends flexible strength to the matrix. The mineral component, like concrete, gives the matrix compression (weight-bearing) strength.

1 What would happen to a bone if all the mineral were removed? What would happen if all the collagen were removed? ?

GENERAL FEATURES OF BONE

There are four types of bone, described by their shape as long, short, flat, and irregular. They will be described in more detail later in this chapter. Each long bone consists of a shaft, called the **diaphysis** (di-af'ĭ-sis), and an **epiphysis** (e-pif'ĭ-sis) at each end of the bone (Figure 6-1, *A*). During growth of a long bone, a cartilaginous **epiphyseal,** or growth, **plate,** is the site of growth in bone length. When bone growth stops, the epiphyseal plate is replaced by bone and is called the **epiphyseal line** (Figure 6-1, *B*).

Bones contain cavities such as the large **medullary cavity** in the diaphysis, and smaller cavities in the epiphyses of long bones and in the interior of other bones. These spaces are filled with either yellow or red marrow. Marrow is the soft tissue in the center of the bone. Yellow marrow consists mostly of fat, and red marrow consists of blood-forming cells (see Chapter 11).

The outer surface of bone is covered by a connective tissue layer called the **periosteum** (per'e-os'te-um), which contains blood vessels and nerves (Figure 6-1, *C*). The surface of the medullary cavity is lined with a thinner connective tissue membrane, the **endosteum** (en-dos'te-um).

There are two major types of bone, based on their histological structure. **Compact bone** is mostly solid matrix and cells with few spaces, whereas **cancellous** (kan'sĕ-lus) **bone** has many spaces within a lacy network of bone.

Compact Bone

Compact bone (Figure 6-2) consists of cells called **osteocytes** (os'te-o-sītz) located within spaces called **lacunae** (lă-ku'ne), which are between thin sheets of matrix called **lamellae** (lă-mel'e). The osteocytes are connected to each other by tiny cell processes that extend across the lamellae through tiny canals called **canaliculi** (kan-ă-lik'u-le). Osteocytes receive nutrients from blood vessels that enter the bone from the periosteum and endosteum. Vessels that run parallel to the long axis of the bone are contained within **haversian** (hă-ver'shan) **canals** surrounded by concentric rings of lamellae. Each haversian canal, with the lamellae and osteocytes surrounding it, is called a **haversian system.**

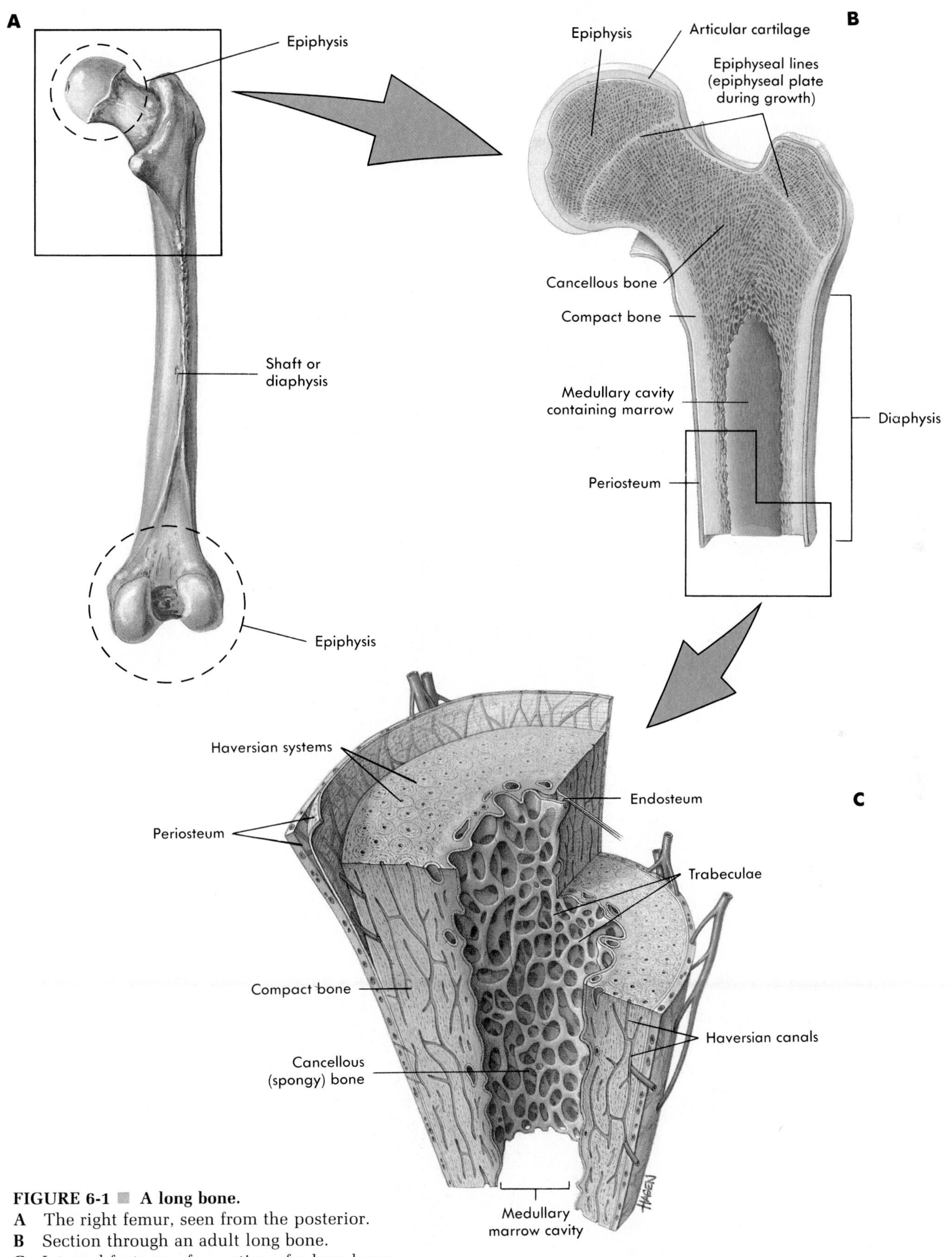

FIGURE 6-1 ■ A long bone.
A The right femur, seen from the posterior.
B Section through an adult long bone.
C Internal features of a portion of a long bone.

Skeletal Disorders

GROWTH AND DEVELOPMENTAL DISORDERS

Giantism is a condition of abnormally increased size that usually involves long bones at the epiphyseal plate where excessive endochondral growth occurs. **Dwarfism,** the condition in which a person is abnormally small, is often the result of improper growth in the epiphyseal plates (Figure 6-A). **Osteogenesis imperfecta,** a group of genetic disorders producing very brittle bones that are easily fractured, occurs because insufficient collagen is formed to properly strengthen the bones.

Rickets is a condition involving growth retardation resulting from nutritional deficiencies either in minerals (calcium and phosphate, necessary for normal ossification) or in vitamin D (necessary for calcium and phosphate absorption). The condition results in bones that are soft, weak, and easily broken. Rickets most often occurs in children who get inadequate amounts of sunlight (necessary for vitamin D production by the body) and whose diets are deficient in vitamin D.

FIGURE 6-A ■ Giant and dwarf. Both are adults.

BACTERIAL INFECTIONS

Osteomyelitis is bone inflammation that often results from bacterial infection, and it can lead to complete destruction of the bone. Tuberculosis was once contracted from contaminated milk and once inside the body the disease could spread from the lungs to the bones. Because of milk pasteurization, tuberculosis is now rare in the United States, and *Staphylococcus* (staph) infections, introduced into the body through wounds, are a more common cause of osteomyelitis.

TUMORS

There are many types of bone **tumors** with a wide range of resultant bone defects. Tumors may be benign or malignant. Malignant bone tumors may metastasize (spread) to other parts of the body or may result from metastasizing tumors elsewhere.

DECALCIFICATION

Osteomalacia (os′te-o-mă-la′shĭ-ah), or the softening of bones, results from calcium depletion from bones. If the body has an unusual need for calcium (for example, during pregnancy when fetal growth requires large amounts of calcium), it may be removed from the mother's bones, which consequently soften and weaken. Osteomalacia may be associated with adult rickets.

Osteoporosis (os′te-o-po-ro′sis), or porous bone, results from reduction in the overall quantity of bone tissue. In older people, the normal production of bone may lag behind the resorption rate, and bones may take on a mottled or porous appearance. Osteoporosis can be a severe problem in older people because it results in bones that are easily fractured. Postmenopausal osteoporosis may occur in women aged 50 to 65. It is thought that the decrease in estrogen levels as a result of menopause are responsible for this form of osteoporosis. Osteoporosis is not limited to older people, however. It can also become a problem in young athletes who "overtrain."

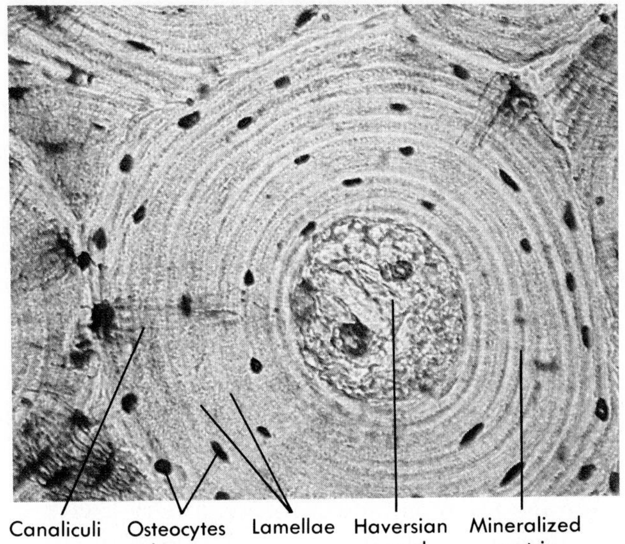

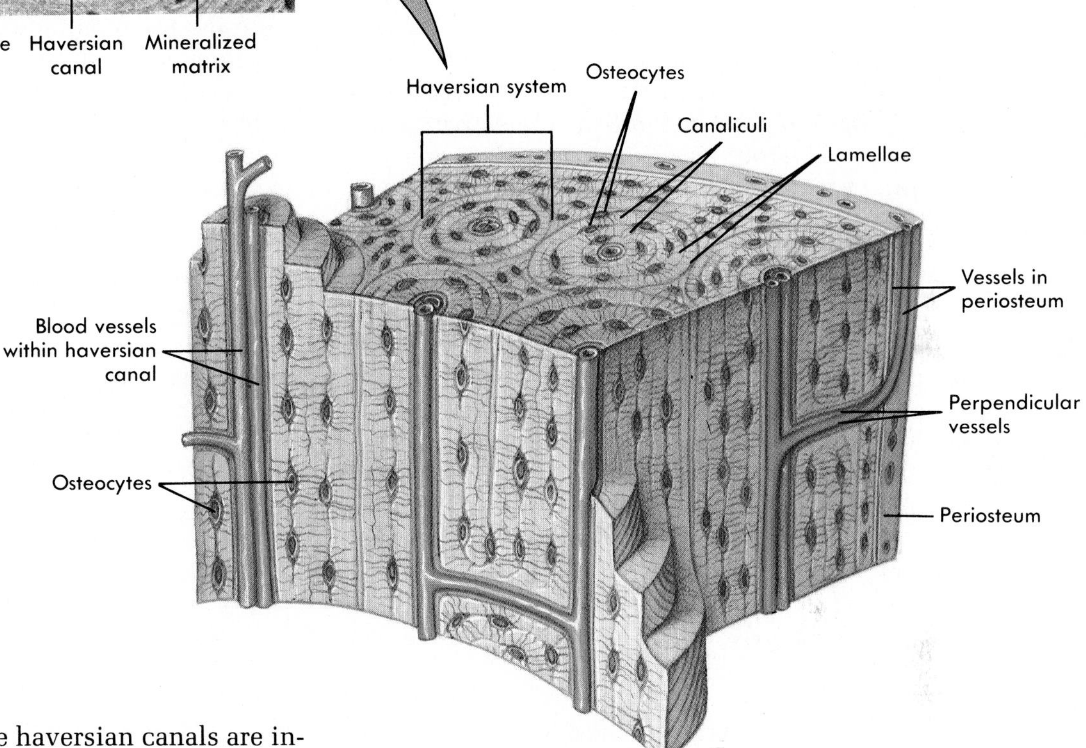

FIGURE 6-2 ■ Fine structure of compact bone. The upper illustration shows a photomicrograph of compact bone.

The blood vessels of the haversian canals are interconnected by a network of vessels running perpendicular to the long axis of the bone. Blood vessels in the periosteum and endosteum connect with these perpendicular vessels and thereby supply blood to the vessels in the haversian canals. Nutrients leave the vessels of the haversian canals and diffuse to the osteocytes via the canaliculi. Waste products diffuse in the opposite direction.

Cancellous Bone

■ Cancellous (spongy) bone (see Figure 6-1, *B*; Figure 6-3) consists of interconnecting rods or plates of bone called **trabeculae** (tră-bek′u-le; beam). The spaces between the trabeculae are filled with marrow. Each trabecula consists of several lamellae with osteocytes between the lamellae. Usually no blood vessels penetrate the trabeculae, and the trabeculae have no haversian canals. Nutrients exit vessels in the marrow and pass by diffusion to the cells of the trabeculae.

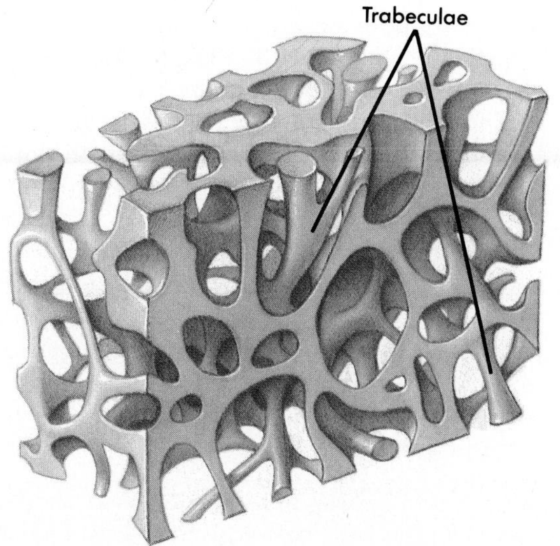

FIGURE 6-3 ■ Cancellous bone.
Spaces between the trabeculae are filled with marrow.

Bone Ossification

Ossification (os'ĭ-fĭ-ka'shun) is the formation of bone by **osteoblasts** (os'te-o-blastz; bone-forming cells). After an osteoblast becomes completely surrounded by bone matrix, it is called an osteocyte (a mature bone cell). Ossification occurs by two processes, each involving preexisting connective tissue. Bone formation that occurs within connective tissue membranes is **intramembranous** (within membranes) **ossification,** and bone formation that occurs in cartilage is **endochondral** (inside cartilage) **ossification.** Both types of bone formation result in the same type of bone, which is histologically indistinguishable.

Intramembranous ossification occurs when osteoblasts begin to produce bone in connective tissue membranes. This occurs primarily in the bones of the skull. Osteoblasts line up on the surface of connective tissue fibers and begin depositing bone matrix to form trabeculae. These trabeculae may remain as such or may enlarge and fuse to form compact bone.

The bones at the base of the skull and most of the remaining skeletal system develop from cartilage models (having the general shape of the mature bone) through the process of endochondral ossification. During this process, cartilage cells proliferate, enlarge, and die. Osteoblasts invade the spaces left by the dying cartilage cells and begin to form bone lamellae.

Bone Growth

Bone growth can occur by **appositional growth,** the formation of new bone on the surface of existing bone, or by **endochondral growth,** the growth of cartilage in the epiphyseal plate and its eventual replacement by bone. Appositional growth results as osteoblasts deposit bone matrix on the surface of bones between the periosteum and the existing bone. Endochondral growth occurs in the epiphyseal plate. Cartilage cells proliferate in the epiphyseal plate, line up in columns parallel to the long axis of the bone, causing elongation of the bone, and then enlarge and die. The dying cartilage is replaced by osteoblasts, which start forming bone. Appositional growth is responsible for the increase in diameter of long bones and most growth of other bones, but bone elongation, the major source of increased height in the individual, must occur by endochondral growth.

> **2 What would happen to a person if endochondral bone growth did not occur?** ?

Bone Remodeling

Bone remodeling involves the removal of old bone by cells called **osteoclasts** (os'te-o-klastz) and the deposition of new bone by osteoblasts. Bone remodeling occurs in all bone and is responsible for bone growth, changes in bone shape, the adjustment of bone to stress, bone repair, and calcium ion regulation in the body. As a long bone increases in length and diameter, the size of the marrow cavity also increases; otherwise, the bone would be very heavy. As a bone grows in diameter, new bone is added to the outer surface as older bone is removed from the inner surface of the medullary cavity.

Bone is the major storage site for calcium in the body. Blood calcium levels must be maintained within narrow limits for functions such as muscle contraction and the conduction of action potentials to occur normally. Calcium is removed from bones when blood calcium levels decrease, and it is replaced when dietary calcium is adequate. This removal and deposition occurs under hormonal control (see Chapter 10). If too much bone is deposited, the bones become thick or have abnormal spurs or lumps that may interfere with normal function. Too little bone formation or too much bone removal weakens the bones and makes them susceptible to fracture.

Bone Repair

When a bone is broken, blood vessels in the bone are also damaged. The vessels bleed, and a clot forms in the damaged area. Two to three days after the injury, blood vessels and cells from surrounding tissues invade the clot. Some of these cells produce a fibrous network between the broken bones, which holds the bone fragments together and fills the gap between the fragments. Other cells produce cartilage in the fibrous network. The zone of tissue repair between the two bone fragments is called a **callus.**

Osteoblasts enter the callus and begin forming cancellous bone. Cancellous bone formation in the callus is usually complete 4 to 6 weeks after the injury. Immobilization of the bone is critical up to this time, because movement can refracture the delicate new matrix. The cancellous bone is slowly remodeled to form compact and cancellous bone, and the repair is complete. Total healing of the fracture may require several weeks or even months, depending on several conditions. If bone healing occurs properly, the healed region is usually even stronger than the adjacent bone.

Bone Fractures

Bone fractures (Figure 6-B) can be classified as **open** if the bone protrudes through the skin, or **closed** if the skin is not perforated. If the fracture totally separates the two bone fragments, the fracture is called **complete,** and if not, it is called **incomplete.** A **comminuted** (kom′ĭ-nu-ted) fracture is one in which the bone breaks into more than two fragments. Fractures can also be classified according to the direction of the fracture line as **linear** (parallel to the long axis), **transverse** (at right angles to the long axis), or **oblique** (at an angle other than a right angle to the long axis).

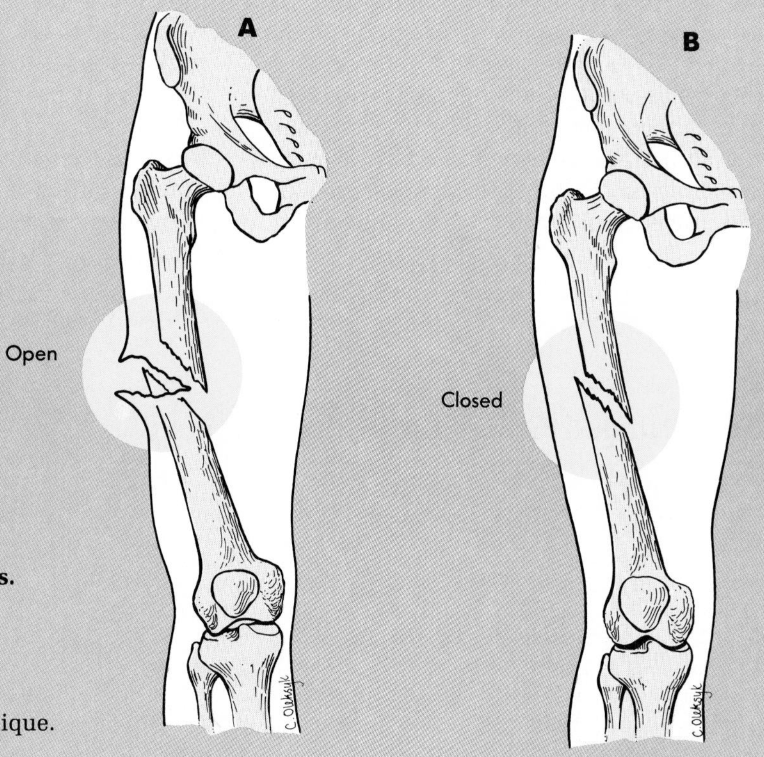

FIGURE 6-B ■ Bone fractures.
A Open.
B Closed.
C Complete and incomplete.
D Comminuted.
E Linear, transverse, and oblique.

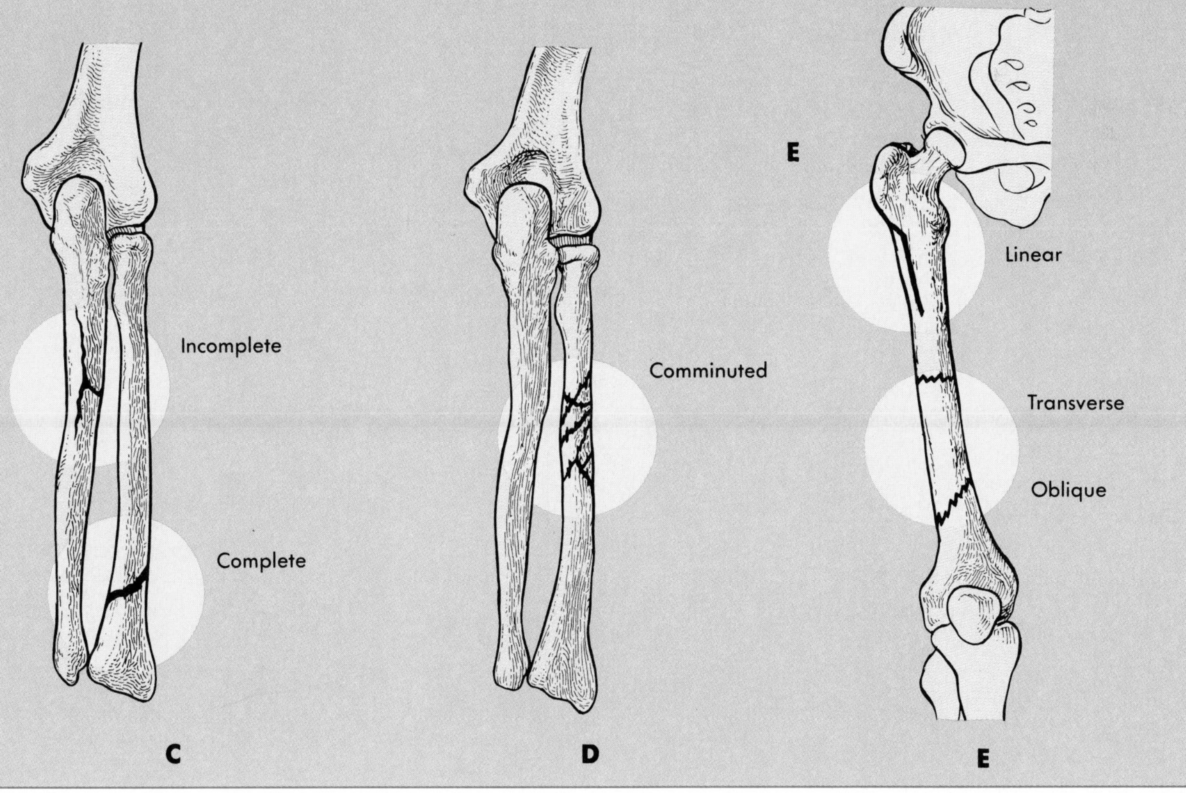

GENERAL CONSIDERATIONS OF BONE ANATOMY

It is traditional to list 206 bones in the average adult skeleton (Table 6-1 and Figure 6-4), although the actual number varies from person to person and decreases with age as some bones become fused. Individual bones can be classified according to their shape as long, short, flat, or irregular. Long bones are longer than they are wide. Most of the bones of the upper and lower limbs are long bones. Short bones are approximately as broad as they are long, such as the bones of the wrist and ankle. Flat bones have a relatively thin, flattened shape. Examples of flat bones are certain skull bones, ribs, scapulae (shoulder blades), and the sternum. Irregular bones include the vertebrae and facial bones with shapes that do not fit readily into the other three categories.

Several common terms are used to describe features of bones (Table 6-2). For example, a hole in a bone is called a **foramen** (fo-ra′men). If the hole is elongated into a tunnel-like passage though the

TABLE 6-1

Number of Named Bones Listed by Category

BONES		NUMBER
AXIAL SKELETON		
SKULL		
Cranial vault		
Paired	Parietal	2
	Temporal	2
Unpaired	Frontal	1
	Occipital	1
	Sphenoid	1
	Ethmoid	1
Face		
Paired	Maxilla	2
	Zygomatic	2
	Palatine	2
	Nasal	2
	Lacrimal	2
	Inferior nasal concha	2
Unpaired	Mandible	1
	Vomer	1
Auditory ossicles	Malleus	2
	Incus	2
	Stapes	2
	TOTAL SKULL	28
HYOID		1
VERTEBRAL COLUMN		
Cervical vertebrae		7
Thoracic vertebrae		12
Lumbar vertebrae		5
Sacrum		1
Coccyx		1
	TOTAL VERTEBRAL COLUMN	26
RIB CAGE		
Ribs		24
Sternum		1
	TOTAL RIB CAGE	25
	TOTAL AXIAL SKELETON	80

BONES		NUMBER
APPENDICULAR SKELETON		
PECTORAL GIRDLE		
Scapula		2
Clavicle		2
UPPER LIMB		
Humerus		2
Ulna		2
Radius		2
Carpals		16
Metacarpals		10
Phalanges		28
	TOTAL UPPER LIMB AND GIRDLE	64
PELVIC GIRDLE		
Coxa		2
LOWER LIMB		
Femur		2
Tibia		2
Fibula		2
Patella		2
Tarsals		14
Metatarsals		10
Phalanges		28
	TOTAL LOWER LIMB AND GIRDLE	62
	TOTAL APPENDICULAR SKELETON	126
	TOTAL BONES	206

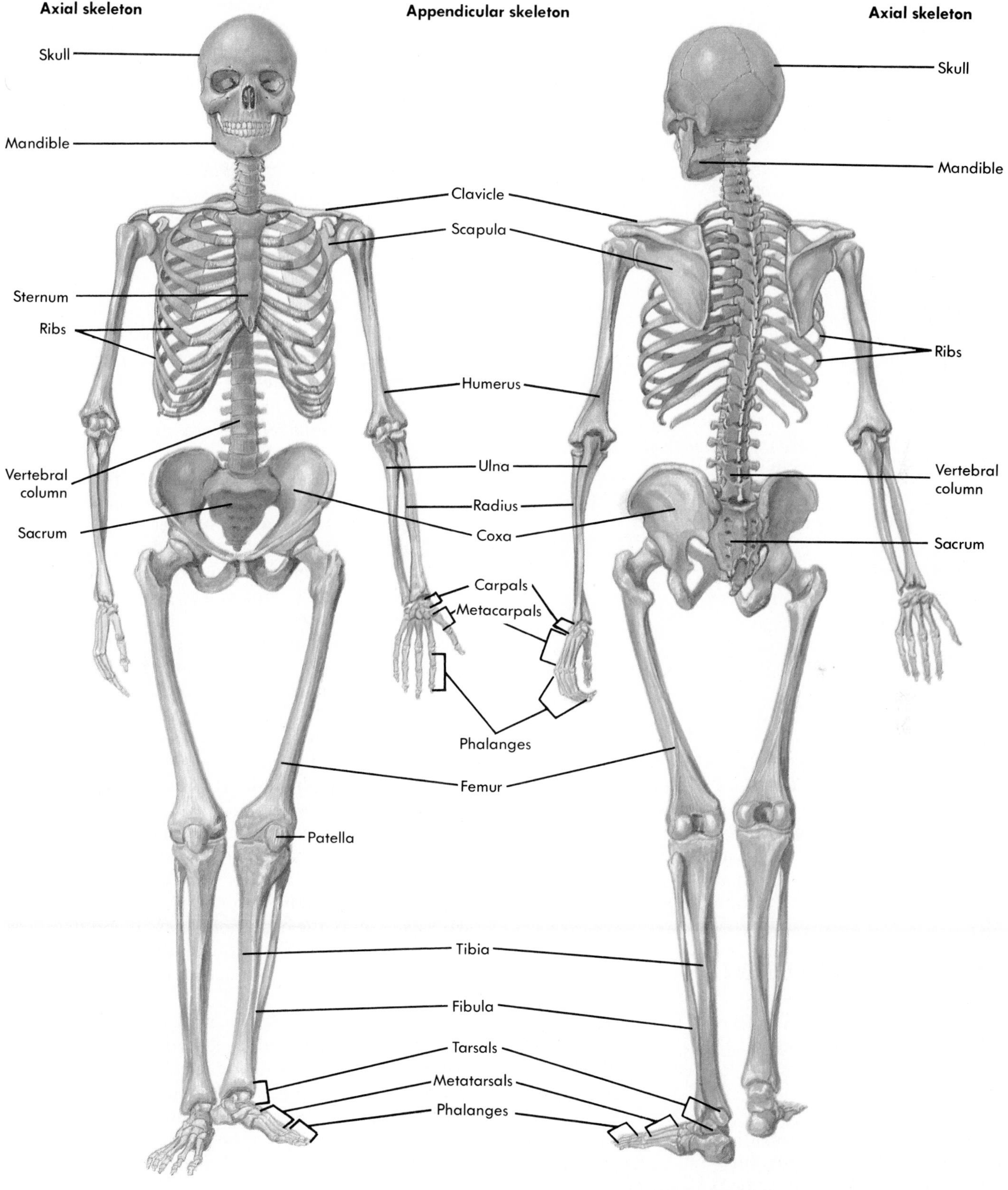

FIGURE 6-4 ■ The complete skeleton, not in anatomical position.
A Anterior view.
B Posterior view.

TABLE 6-2

General Anatomical Terms for Various Features of Bones

TERM	DESCRIPTION
MAJOR FEATURES	
Body, Shaft	Main portion
Head	Enlarged (often rounded) end
Neck	Constricted area between head and body
Condyle	Smooth, rounded articular surface
Facet	Small, flattened articular surface
Crest	Prominent ridge
Process	Prominent projection
Tubercle or tuberosity	Knob or enlargement
Trochanter	Large tuberosity found only on the proximal femur
Epicondyle	Enlargement near or above a condyle
OPENINGS OR DEPRESSIONS	
Foramen	Hole
Canal, Meatus	Tunnel
Fissure	Cleft
Sinus	Cavity
Fossa	Depression

bone, it is called a **canal** or a **meatus** (me-a'tus). A depression in a bone is called a **fossa.** A lump on a bone is called a **tubercle** or **tuberosity,** and a projection from a bone is called a **process.** The smooth, rounded end of a bone, where it forms a joint with another bone, is called a **condyle** (kon'dīl).

AXIAL SKELETON

The **axial skeleton** is divided into the skull, vertebral column, and thoracic cage.

Skull

The bones of the skull (see Table 6-1) are divided into two groups: those of cranial vault and those of the face. The **cranial vault,** or brain case, consists of 8 bones that immediately surround and protect the brain. The **facial bones** (14) form the structure of the face in the anterior skull but do not contribute to the cranial vault. There are also 3 auditory ossicles in each middle ear (6 total) to bring the total number of bones in the skull to 28.

The **hyoid bone** (hi'oyd) is not part of the skull but is attached to the skull and larynx by muscles and ligaments. It serves as the attachment point for several important neck and tongue muscles (see Figure 7-10).

We often find it convenient to think of the skull, excluding the mandible, as a single unit. The major features of the intact skull are described as follows.

Lateral view

The **parietal** (pă-ri'ĕ-tal; wall) bone and **temporal** (the term refers to time; the hairs of the temples turn white, indicating the passage of time) bone form a large portion of the side of the head (Figure 6-5). The **squamous suture** joins these bones (a suture is a joint uniting bones of the skull). Anteriorly, the parietal bone is joined to the **frontal** (forehead) bone by the **coronal suture,** and posteriorly it is joined to the **occipital** (ok-sĭ'pĭ-tal; back of the head) bone by the **lambdoid** (lam'doyd; shaped like the Greek letter *lambda*, λ) **suture.** A prominent feature of the temporal bone is a large opening, the **external auditory meatus,** a canal that enables sound waves to reach the eardrum. The **mastoid** (mas'toyd; resembling a breast) **process** of the temporal bone can be seen and felt as a prominent lump just posterior to the ear. Important neck muscles involved in rotation of the head attach to the mastoid process.

Part of the **sphenoid** (sfĕ'noyd; wedge-shaped) bone is immediately anterior to the temporal bone. The sphenoid bone, although appearing to be two paired bones (one on each side of the skull), is actually a single bone that extends completely across the skull. The sphenoid bone resembles a butterfly, with its body in the center of the skull and its wings extending to the sides of the skull. Anterior to the sphenoid bone is the **zygomatic** (zi-go-mat'ik; yoke) bone, or cheek bone, which can be easily felt. The **zygomatic arch,** which consists of joined processes from the temporal and zygomatic bones, forms a bridge across the side of the face.

The **maxilla** (upper jaw) is anterior to the zygomatic bone to which it is joined. The **mandible** (lower jaw) is inferior to the maxilla and articulates with the temporal bone. The maxilla contains the superior set of teeth and the mandible contains the inferior teeth.

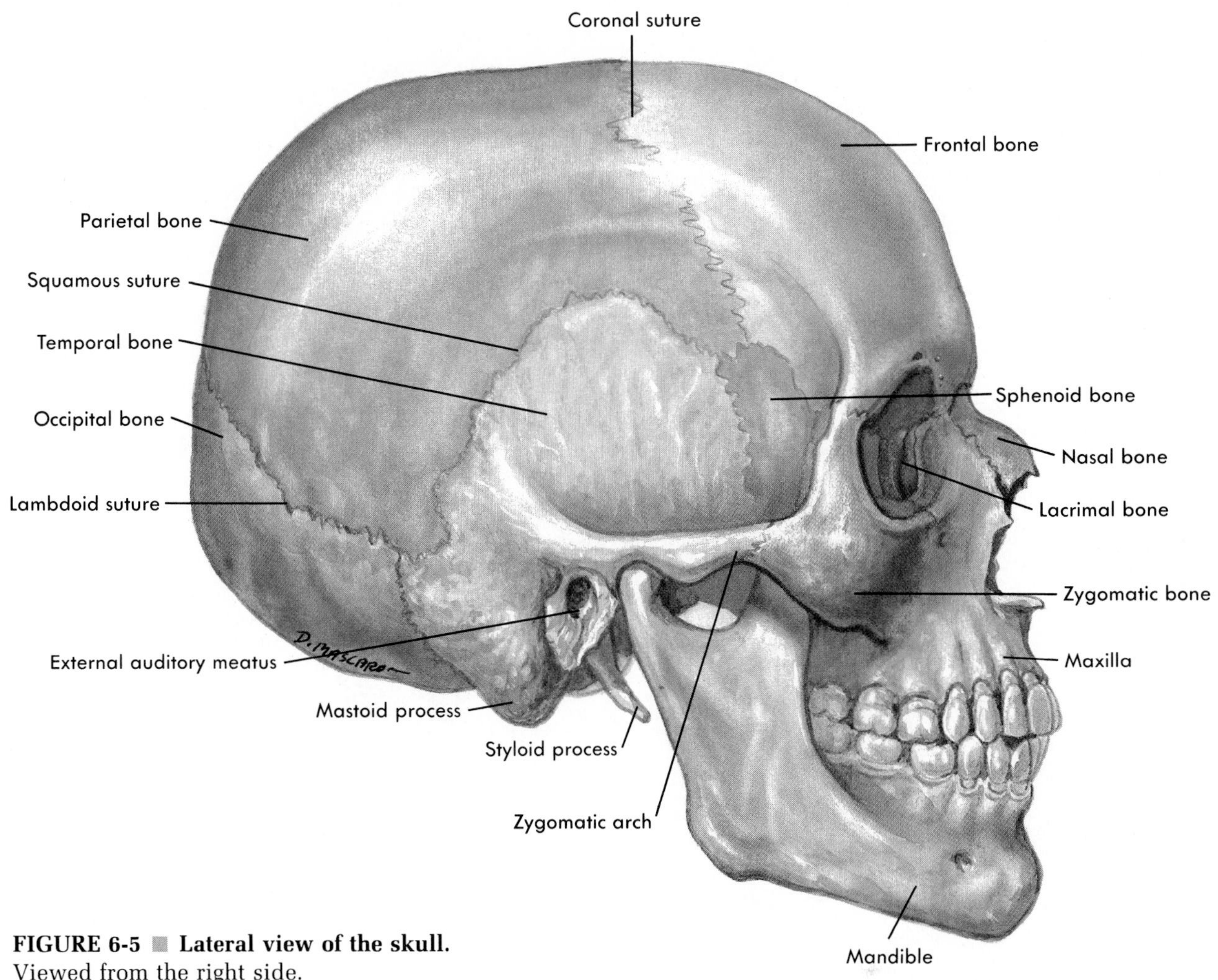

FIGURE 6-5 **Lateral view of the skull.** Viewed from the right side.

Frontal view

The major structures seen from the frontal view are the frontal bone, the zygomatic bones, the maxillae, and the mandible (Figure 6-6, *A*). The teeth are very prominent in this view. Many bones of the face may be easily felt through the skin of the face (Figure 6-6, *B*).

From this view, the most prominent openings into the skull are the **orbits** (eye sockets) and the **nasal cavity.** The orbits are cone-shaped fossae so named because of the rotation of the eyes within them. The bones of the orbits provide both protection for the eyes and attachment points for the muscles that move the eyes.

Each orbit (Figure 6-6, *C*) has several openings through which structures communicate with other cavities. The largest of these are the **superior** and **inferior orbital fissures.** They provide openings through which nerves and vessels communicate with the orbit or pass to the face. The optic nerve, for the sense of vision, passes from the eye through the **optic foramen** and enters the cranial vault. The **nasolacrimal canal** (cannot be seen in the illustration) passes from the orbit into the nasal cavity. It contains a duct that carries tears from the eyes to the nasal cavity. A small **lacrimal** (lak′rĭ-mal; tear) bone can be seen in the orbit just above the opening of this canal (see Figures 6-5 and 6-6, *C*).

> **3 Why does your nose run when you cry?** ?

The nasal cavity is divided into right and left halves by a **nasal septum** (sep′tum; wall) (see Figure 6-6, *A*). The bony part of the nasal septum consists primarily of the **vomer** (vo′mer; shaped like a

FIGURE 6-6 ■ The skull.
A Seen from the front.

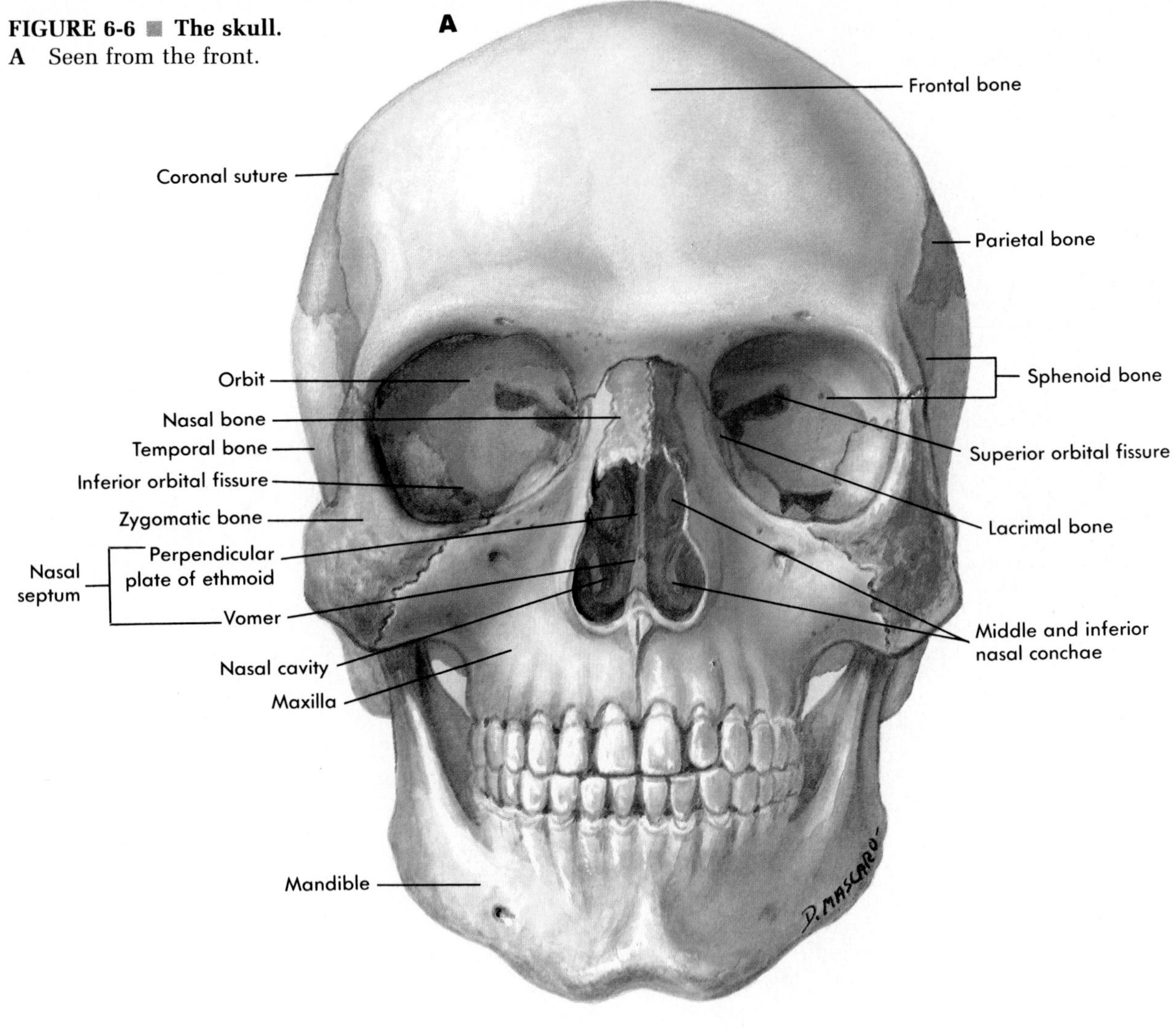

plowshare) inferiorly and the perpendicular plate of the **ethmoid** (eth′moyd; sieve-shaped) bone superiorly. The anterior portion of the nasal septum is formed by cartilage.

The external portion of the nose is formed mostly of cartilage. The bridge of the nose is formed by the **nasal** bones.

The lateral wall of the nasal cavity has three bony shelves, the **nasal conchae** (kon′ke; resembling a conch shell). The inferior nasal concha is a separate bone, and the middle and superior conchae are projections from the ethmoid bone. The conchae function to increase the surface area in the nasal cavity. The increased surface of the overlying epithelium facilitates moistening and warming of the air inhaled through the nose.

Several of the bones associated with the nasal cavity have large cavities within them, called the **paranasal sinuses** (Figure 6-7), which open into the nasal cavity. The sinuses decrease the weight of the skull and act as resonating chambers during voice production. Compare the normal voice to the voice of a person who has a cold and whose sinuses are "stopped up." The sinuses are named for the bones in which they are located and include the frontal, maxillary, ethmoid, and sphenoid sinuses.

The skull has additional sinuses, the mastoid air cells, which are located inside the mastoid processes of the temporal bone. These air cells open into the middle ear instead of into the nasal cavity. An auditory tube connects the middle ear to the nasopharynx.

B

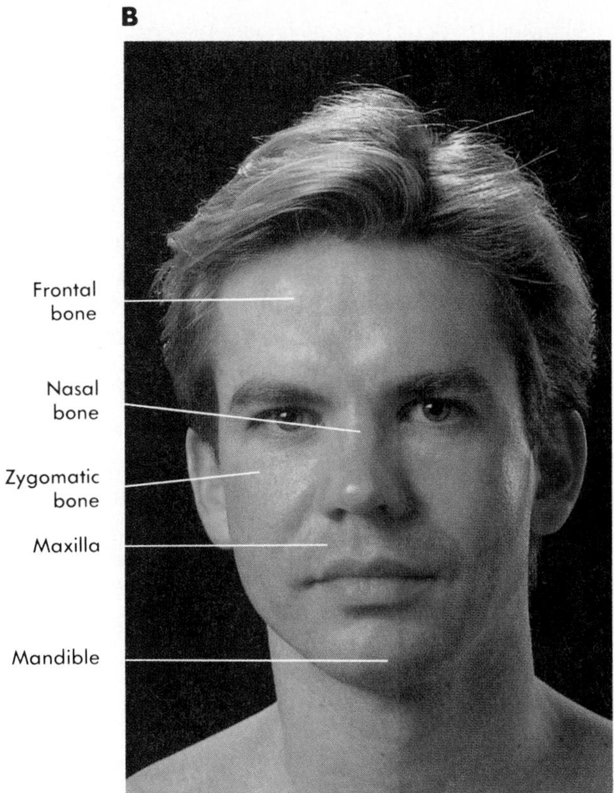

FIGURE 6-6, cont'd ■ The skull.
B Bony landmarks of the face.
C The right orbit.

C

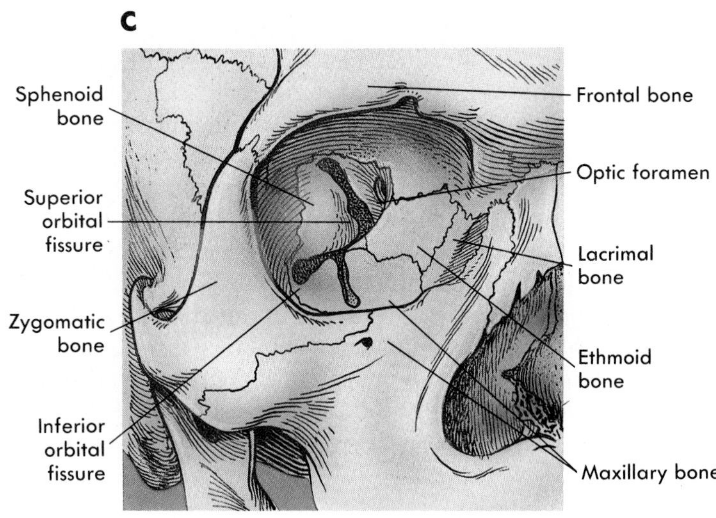

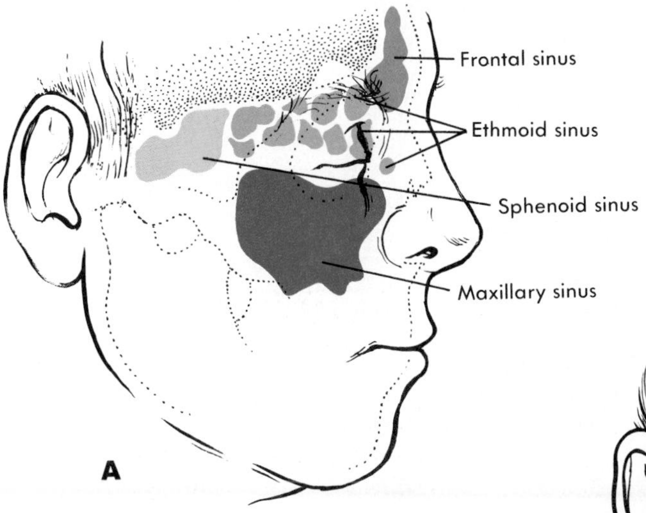

Frontal sinus
Ethmoid sinus
Sphenoid sinus
Maxillary sinus
A

FIGURE 6-7 ■ The paranasal sinuses.
A Lateral view.
B Anterior view.

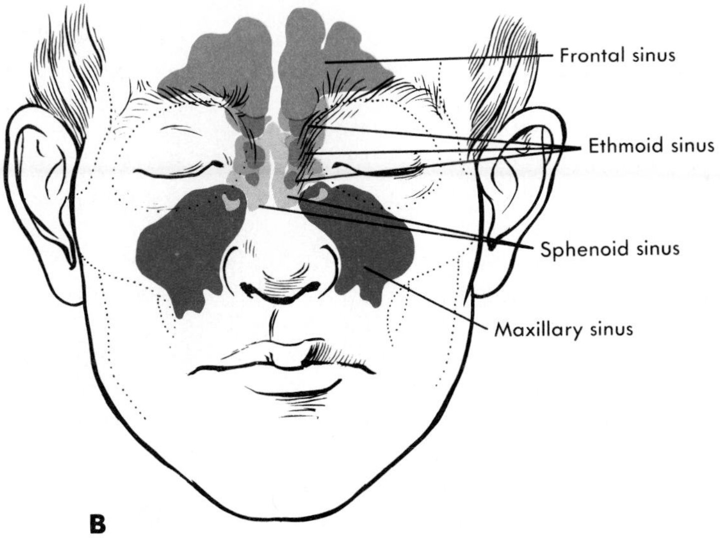

Frontal sinus
Ethmoid sinus
Sphenoid sinus
Maxillary sinus
B

Interior of the cranial vault

■ The floor of the cranial vault, when viewed from above with the roof cut away (Figure 6-8), can be divided roughly into three fossae (anterior, middle, and posterior), which are formed as the developing skull conforms to the shape of the brain. The bones forming the floor of the cranial vault are the frontal, ethmoid, sphenoid, temporal, and occipital. Several foramina can be seen in the floor of the middle fossa (named mostly for their shape; see Figure 6-8). These allow passage of nerves and blood vessels through the skull. One large foramen, the **foramen magnum,** through which the spinal cord communicates with the brain, is located in the posterior fossa. The central region of the sphenoid bone is modified into a structure resembling a saddle, the **sella turcica** (sel'ah tur'sĭ-kah; Turkish saddle), which is occupied by the pituitary gland.

Base of skull seen from below

■ Seen from below, with the mandible removed (Figure 6-9), many of the same foramina can be seen in the base of the skull that were seen in the interior of the skull. Other specialized structures, such as processes for muscle attachments, can also be seen. The foramen magnum passes through the occipital bone near the center of the skull base. **Occipital condyles,** the smooth points of articulation (joint) between the skull and the vertebral column, are located beside the foramen magnum.

Two long, pointed **styloid** (sti'loyd; stylus or pen-shaped) **processes** project from the floor of the temporal bone. Muscles involved in movement of the tongue, the hyoid bone, and the pharynx (throat) originate from this process. The **mandibular fossa,** where the mandible articulates with the temporal bone, is anterior to the mastoid process.

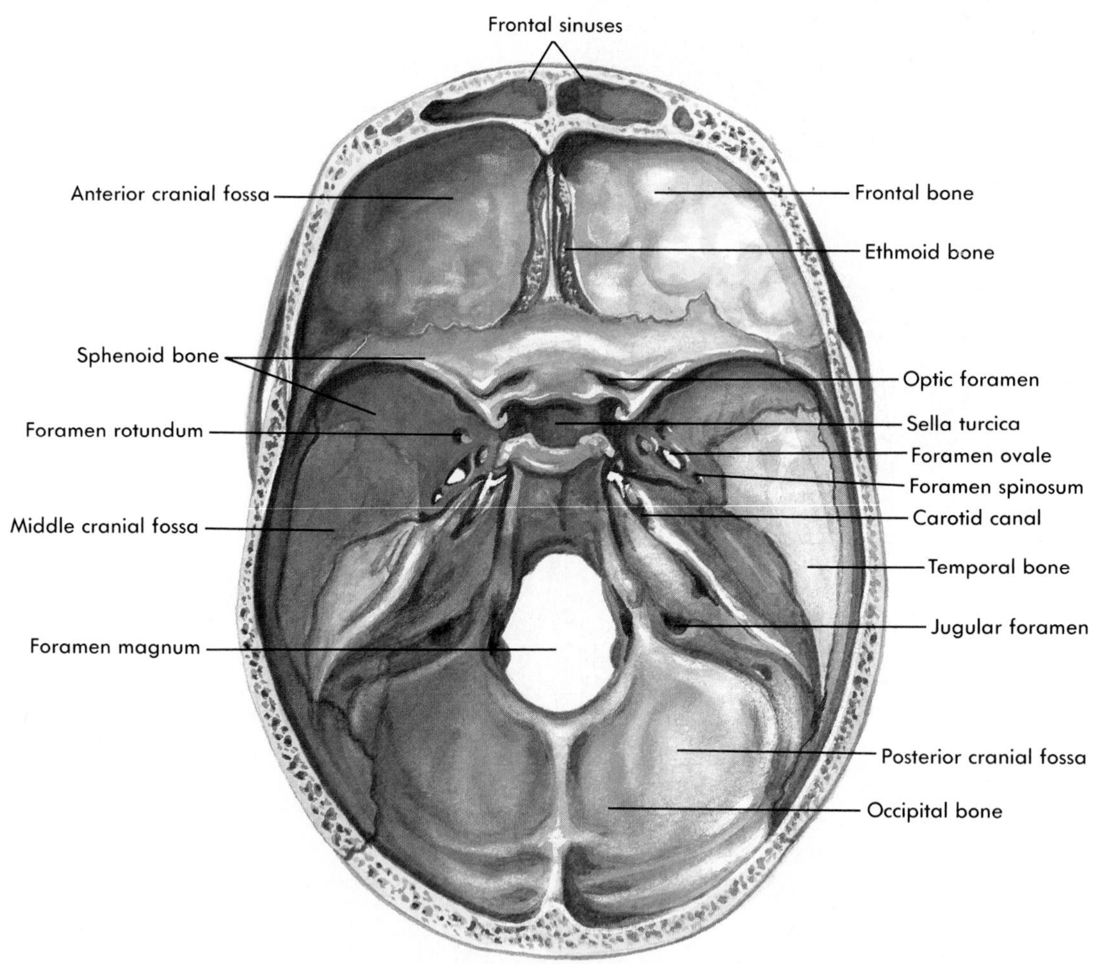

FIGURE 6-8 ■ Floor of the cranial vault.
The roof of the skull has been removed, and the floor is viewed from above.

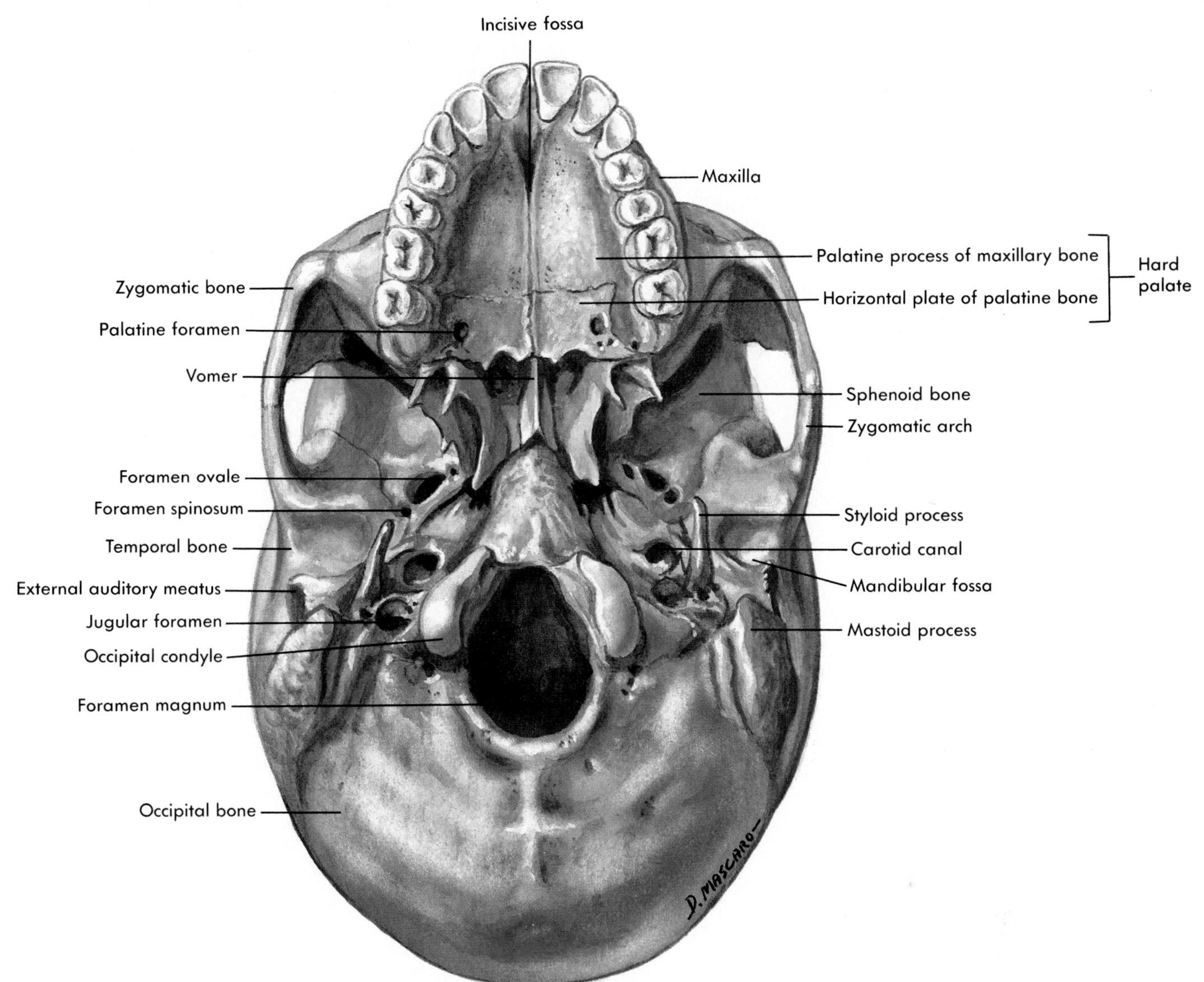

FIGURE 6-9 ■ Base of the skull.
Viewed from below, mandible removed.

The **hard palate** forms the floor of the nasal cavity. The anterior two thirds is formed by the maxillae and the posterior one third by the **palatine** bones. The connective tissue and muscles that make up the **soft palate** extend posteriorly from the hard or bony palate. The hard and soft palates function to separate the nasal cavity and nasopharynx from the mouth, enabling us to eat and breathe at the same time.

Vertebral Column

The **vertebral column** performs five major functions: (1) supports the weight of the head and trunk; (2) protects the spinal cord; (3) allows spinal nerves to exit the spinal cord; (4) provides a site for muscle attachment; and (5) permits movement of the head and trunk.

The vertebral column (Figure 6-10) consists of five regions, usually containing 26 bones (see Table 6-1): 7 **cervical vertebrae,** 12 **thoracic vertebrae,** 5 **lumbar vertebrae,** 1 **sacral bone,** and 1 **coccygeal** (kok-sij'e-al) **bone.** The adult vertebral column has four major curvatures (see Figure 6-10). The cervical region curves anteriorly, the thoracic region curves posteriorly, the lumbar region curves anteriorly, and the sacral and coccygeal regions, together, curve posteriorly.

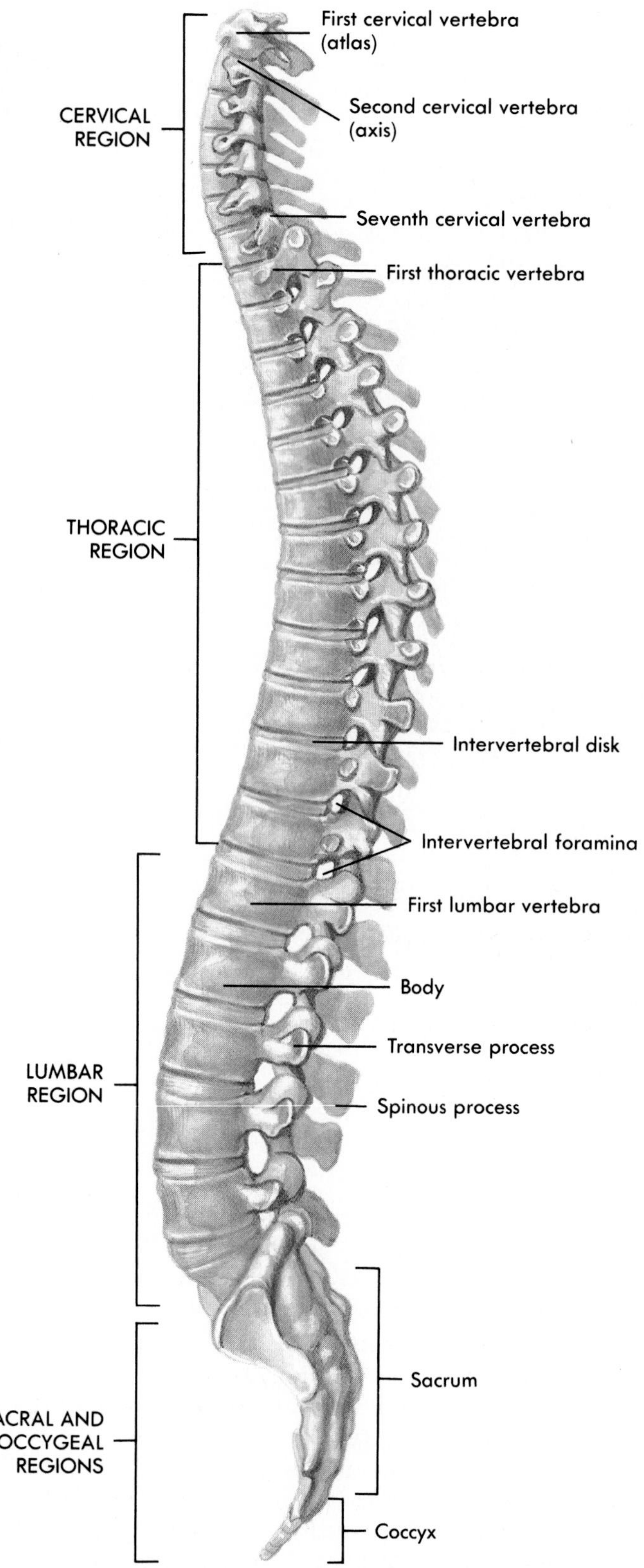

FIGURE 6-10 ■ Vertebral column. Viewed from the left side.

■ Abnormal vertebral curvatures are not uncommon. *Kyphosis* (ki-fo'sis; hunchback) is an abnormal posterior curvature of the spine, *lordosis* (lōr-do'sis; swayback) is an abnormal anterior curvature of the spine, and *scoliosis* (sko-le-o'sis) is an abnormal lateral curvature of the spine. ■

Each vertebra consists of a body, an arch, and various processes (Figure 6-11). The weight-bearing portion of each vertebra is the **body.** The **vertebral arch** surrounds a large opening called the **vertebral foramen.** The vertebral foramina, collectively, form the **vertebral canal** where the spinal cord is located. The vertebral canal protects the spinal cord from injury. Each vertebral arch has two parts: the **pedicle** (ped'ĭ-kl; foot), which is attached to the body, and the **lamina** (lam'ĭ-nah; thin plate). Spinal nerves exit the spinal cord through the **intervertebral foramina,** which are formed by notches in the pedicles of adjacent vertebrae (see Figure 6-10). Each vertebra has a superior and inferior **articular process** where the vertebrae articulate (join) each other. Each articular process has a smooth "little face" called an **articular facet** (fas'et). A **transverse process** extends laterally from each side of the arch, and a single **spinous process** projects dorsally from where the two laminae meet. The spinous processes can be seen and felt as a series of lumps down the midline of the back (see Figure 6-16). The transverse and spinous processes provide attachment sites for muscles that move the vertebral column.

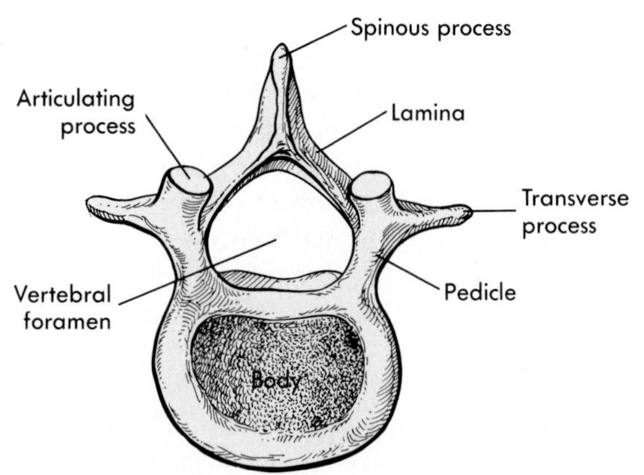

FIGURE 6-11 ■ A typical vertebra. Seen from a superior view.

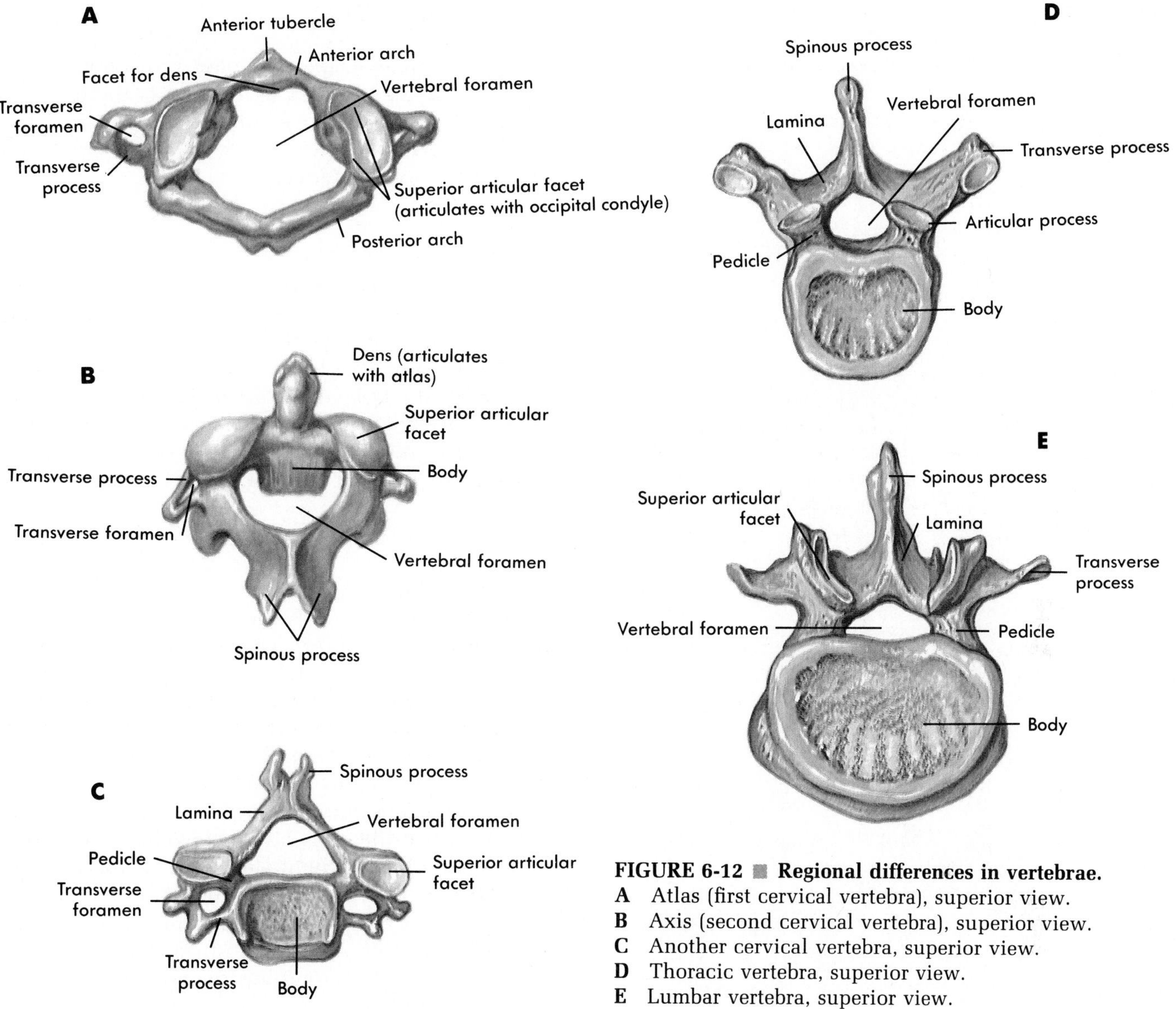

FIGURE 6-12 ■ Regional differences in vertebrae.
A Atlas (first cervical vertebra), superior view.
B Axis (second cervical vertebra), superior view.
C Another cervical vertebra, superior view.
D Thoracic vertebra, superior view.
E Lumbar vertebra, superior view.

Regional differences in vertebrae

The **cervical vertebrae** (Figure 6-12, *A-C*) have very small bodies, partly split spinous processes, and a transverse foramen in each transverse process, through which the vertebral arteries pass. The first cervical vertebra (Figure 6-12, *A*) is called the **atlas** because it holds up the head, as Atlas in classical mythology held up the world. Movement between the atlas and the occipital bone is responsible for a "yes" motion of the head. The second cervical vertebra (Figure 6-12, *B*) is called the **axis** because a considerable amount of rotation occurs at this vertebra, as in shaking the head "no."

The **thoracic vertebrae** (Figure 6-12, *D*) possess long, thin spinous processes that are directed inferiorly. The thoracic vertebrae also have extra articular facets where they articulate with the ribs.

The **lumbar vertebrae** (Figure 6-12, *E*) have large, thick bodies and heavy, rectangular transverse and spinous processes.

The five sacral vertebrae are fused into a single bone called the **sacrum** (Figure 6-13). The spinous processes of the first four sacral vertebrae form the median sacral crest. The spinous process of the fifth vertebra does not form, leaving a **sacral hiatus** at the inferior end of the sacrum, which is often the site of caudal anesthetic injections given just before childbirth. The anterior edge of the body of the first sacral vertebra bulges to form the **sacral promon-**

tory (see Figure 6-21), a landmark that can be felt during a vaginal examination, and it is used as a reference point during measurement of the pelvic outlet.

The **coccyx** (kok'siks; shaped like a cuckoo bill), or tailbone, usually consists of four more or less fused vertebrae.

■ Because the cervical vertebrae are relatively delicate and have small bodies, dislocations and fractures are more common in this area than in other regions of the vertebral column. Because the lumbar vertebrae have massive bodies and carry a large amount of weight, ruptured intervertebral disks are more common in this area than in other regions of the column. Each invertebral disk is made up of a ring of fibrous connective tissue with a softer center of semifluid tissue. The weight of the body can compress the disk causing the fibrous ring to bulge, or even break. This allows the vertebrae to come close together and compress the nerves exiting the intervertebral foramina. The coccyx is easily broken in falls during which a person sits down hard on a solid surface. Also, a mother's coccyx may be fractured during childbirth. ■

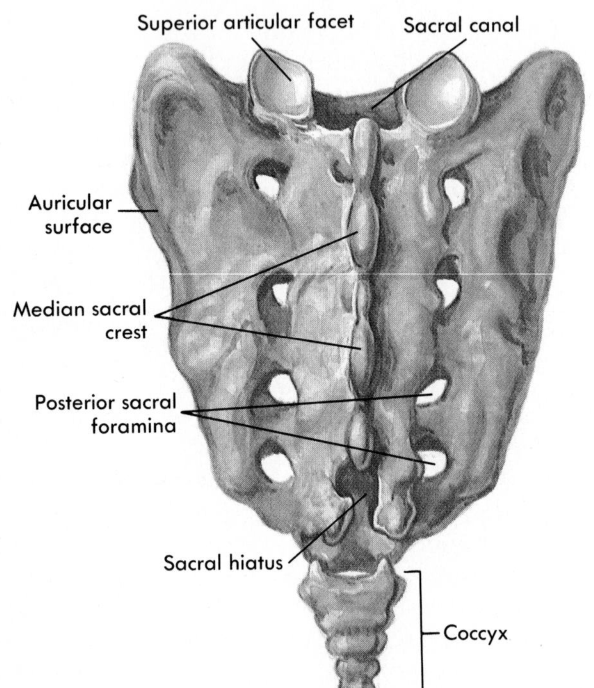

FIGURE 6-13 ■ Posterior view of the sacrum and coccyx.

Thoracic Cage

■ The **thoracic cage** or **rib cage** protects the vital organs within the thorax and prevents the collapse of the thorax during respiration. It consists of the thoracic vertebrae, the ribs with their associated cartilages, and the sternum.

Ribs and costal cartilages

■ The 12 pairs of ribs (Figure 6-14) can be divided into true and false ribs. The superior seven, called the **true ribs,** attach directly to the sternum by means of costal cartilages. The inferior five, called **false ribs,** do not attach directly to the sternum. Ribs 8 through 10 attach to the sternum by a common cartilage, and the eleventh and twelfth ribs, called the **floating ribs,** do not attach to the sternum.

Sternum

■ The **sternum** (see Figure 6-14), or the breastbone, is divided into three parts: the **manubrium** (ma-nu'bre-um; handle), the **body,** and the **xiphoid** (zī'foyd; sword) **process.** The sternum resembles a sword, with the manubrium forming the handle, the body forming the blade, and the xiphoid process forming the tip. At the superior end of the sternum there is a depression, called the **jugular notch,** located between the ends of the clavicles where they join the sternum. A slight elevation, called the **sternal angle,** can be felt at the junction of the manubrium and body of the sternum. This junction is an important landmark as it identifies the location of the second rib. This identification allows the ribs to be counted and, for example, allows location of the heart's apex, which is between the fifth and sixth ribs in a living person.

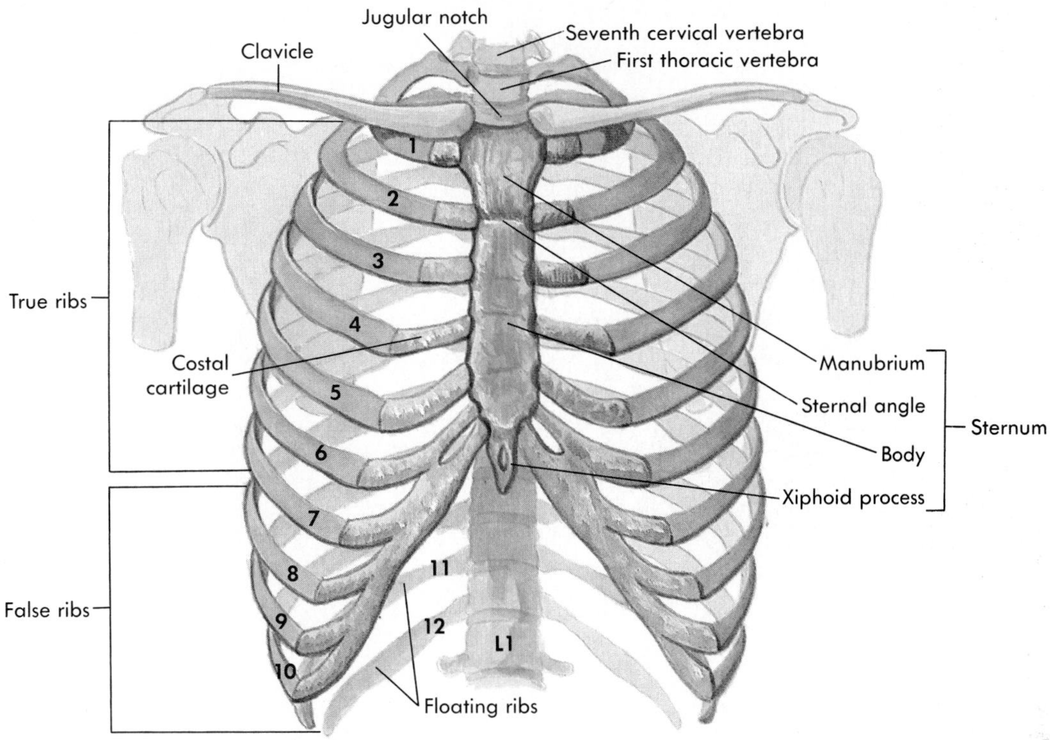

FIGURE 6-14 Anterior view of the thoracic cage.

APPENDICULAR SKELETON

The **appendicular** (ap'pen-dik'u-lar) **skeleton** consists of the bones of the upper and lower limbs as well as the girdles, which attach the limbs to the axial skeleton.

Pectoral Girdle

The **pectoral** (pek'to-ral), or **shoulder girdle** consists of two bones that attach the upper limb to the body: the **scapula** (skap'u-lah), or shoulder blade (see Figures 6-4, 6-15, *A* and *B*, and 6-16), and the **clavicle** (klav'ĭ-kl), or collar bone (Figure 6-15, *C*). The scapula is a flat, triangular bone with three borders, three angles, and three fossae, where muscles attaching to the arm originate. A ridge, called the **spine,** runs across the posterior surface of the scapula. A process, called the **acromion** (a-kro'me-un), extends from the scapular spine to form the point of the shoulder. The clavicle attaches to the scapula at the acromion. The proximal end of the clavicle is attached to the sternum, providing the only bony attachment of the upper limb to the body. The coracoid process curves below the clavicle and provides attachment for arm and chest muscles. Between these two processes is a depression called the **glenoid fossa** where the head of the humerus connects to the scapula.

Upper Limb

The upper limb consists of the bones of the arm, forearm, wrist, and hand.

Arm

The **arm** is the region between the shoulder and the elbow and contains the **humerus** (Figure 6-17). At the proximal end, the humerus has a smooth, rounded **head,** which attaches the humerus to the scapula. Just lateral to the head are two tubercles, a **greater** tubercle and a **lesser** tubercle, to which muscles attach to hold the humerus to the scapula. The distal end of the humerus is modified into specialized condyles which connect the humerus to the **forearm** bones. **Epicondyles** on the distal end of the humerus, just lateral to the condyles, provide attachment sites for forearm muscles.

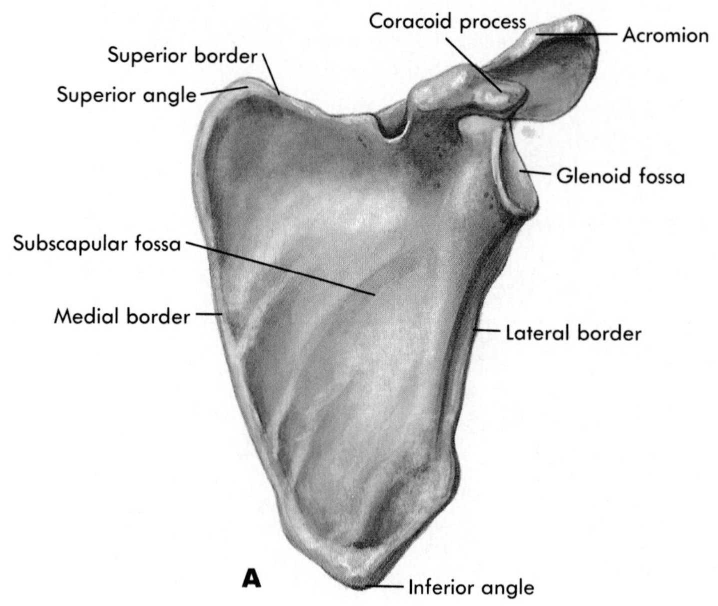

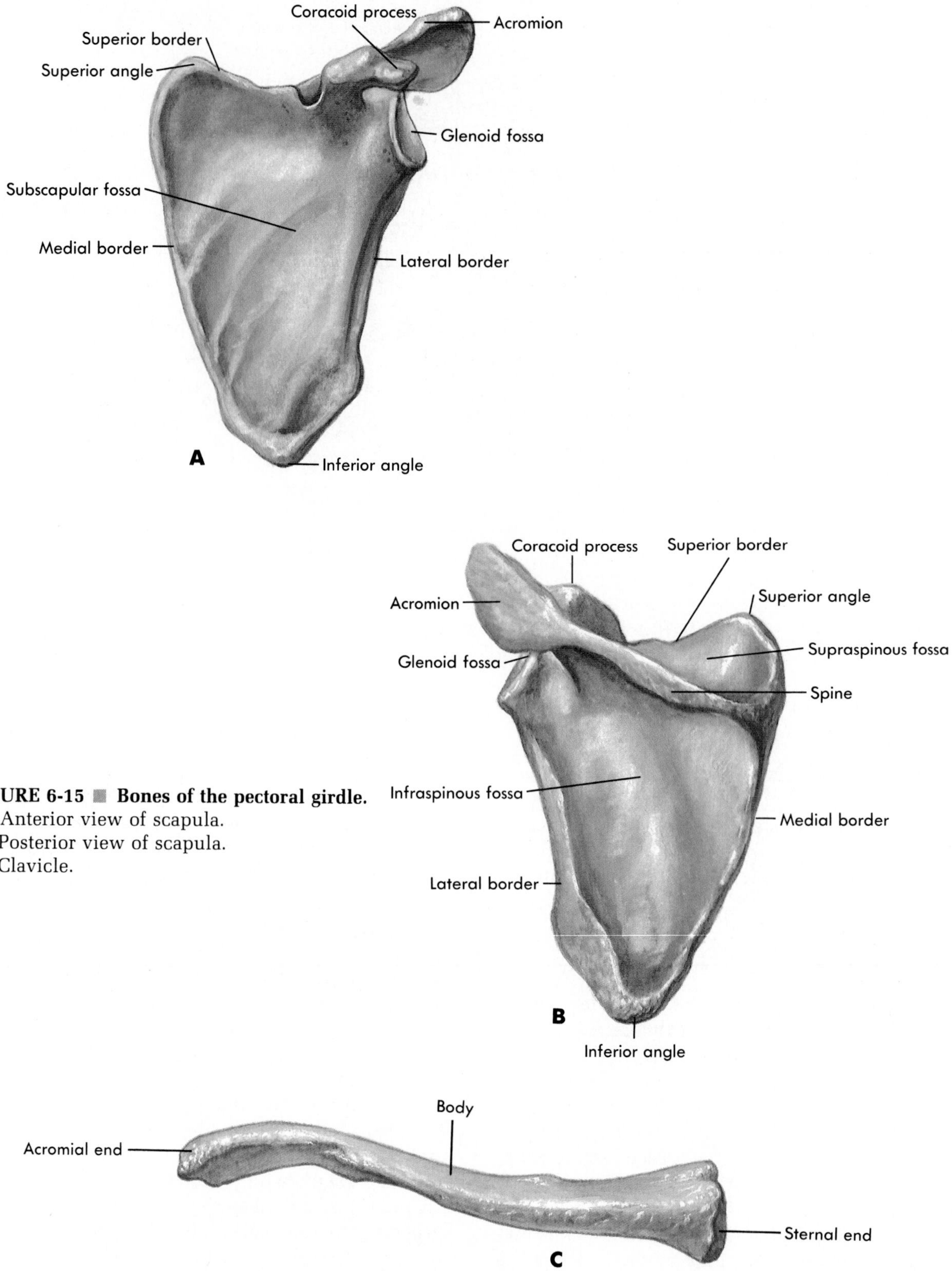

FIGURE 6-15 ■ Bones of the pectoral girdle.
A Anterior view of scapula.
B Posterior view of scapula.
C Clavicle.

Spinous process of
seventh cervical vertebra
Scapula
Medial border
of scapula
Inferior angle
of scapula
Lumbar spinous
processes

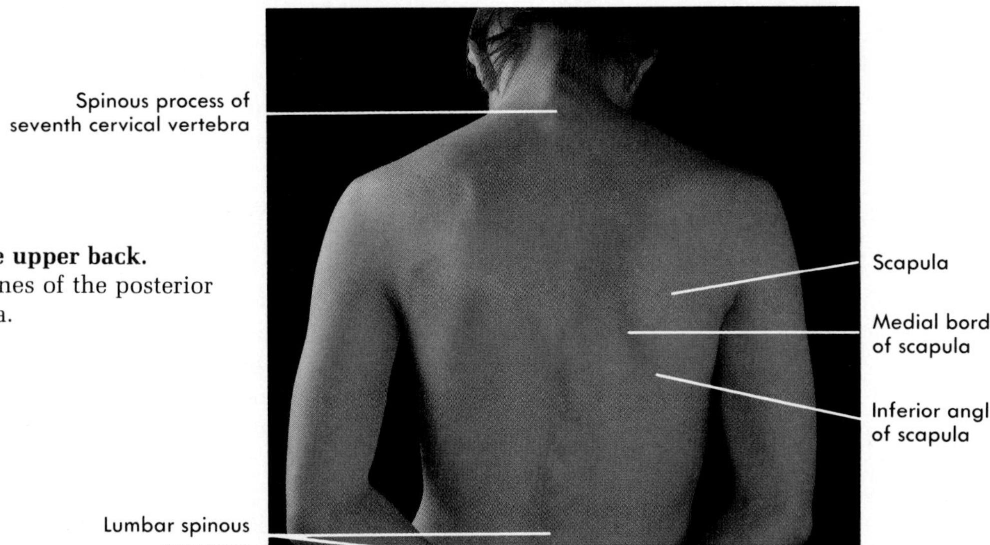

FIGURE 6-16 Bones of the upper back. Surface anatomy showing bones of the posterior vertebral column and scapula.

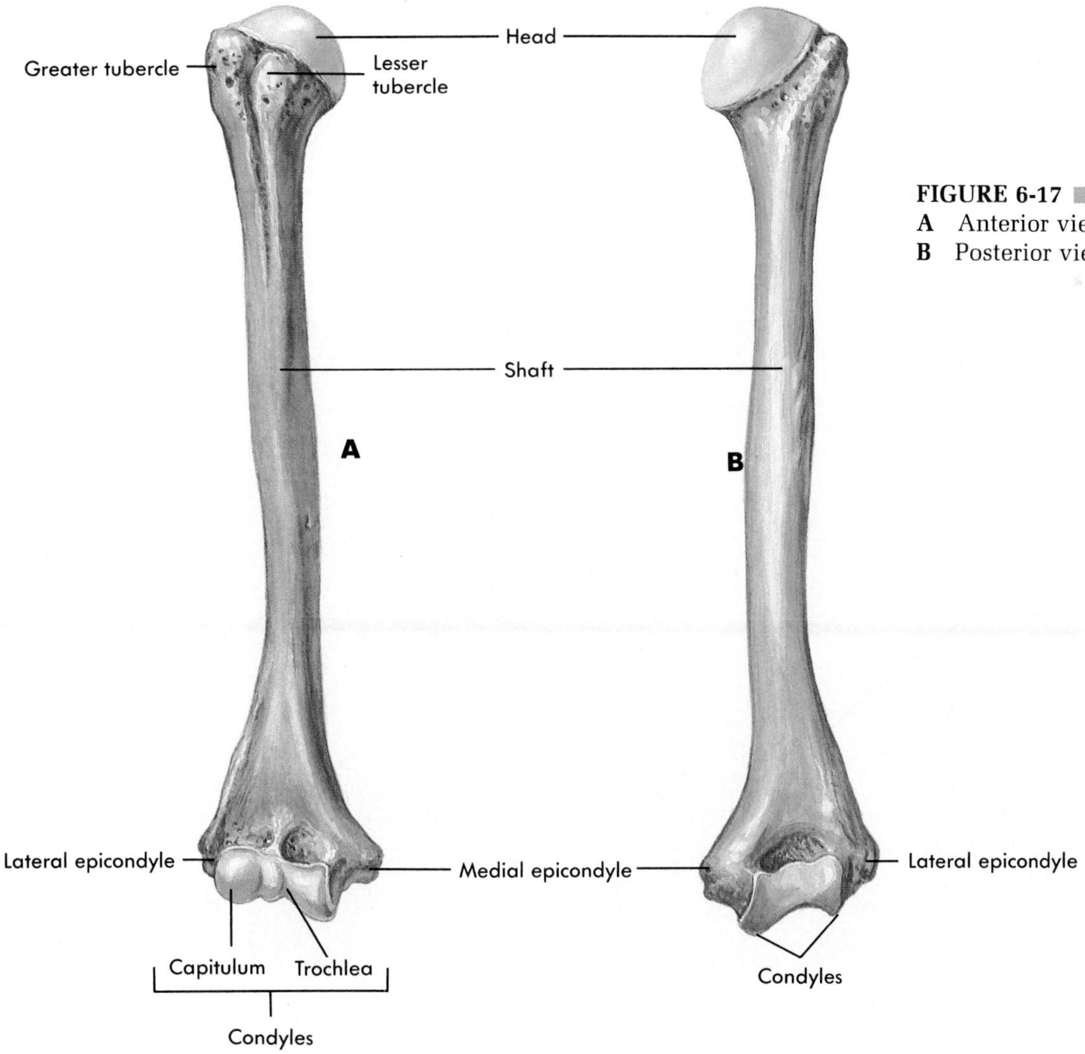

Head
Greater tubercle
Lesser
tubercle
Shaft
A
B
Lateral epicondyle
Medial epicondyle
Lateral epicondyle
Capitulum
Trochlea
Condyles
Condyles

FIGURE 6-17 Right humerus.
A Anterior view.
B Posterior view.

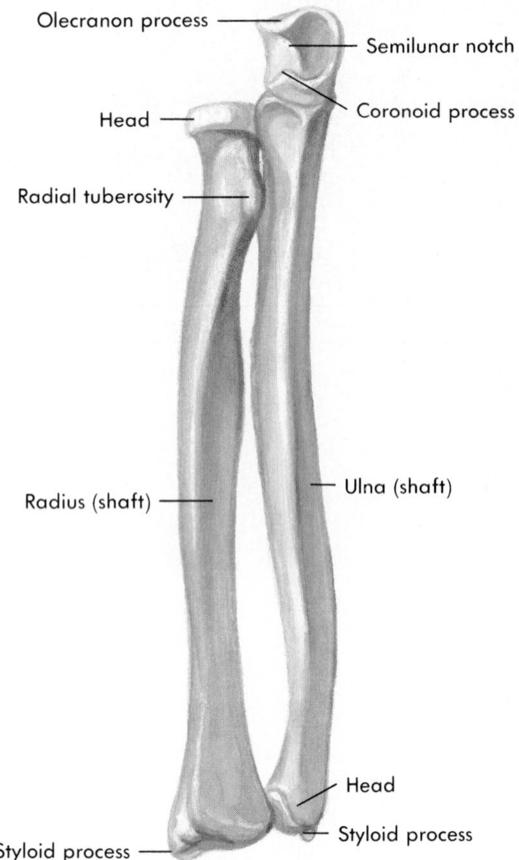

FIGURE 6-18 ■ Right ulna and radius. Anterior view.

Forearm

■ The **forearm** has two bones, the **ulna** on the medial side of the forearm (the side with the little finger) and the **radius** on the lateral (thumb) side of the forearm (Figure 6-18). The proximal end of the ulna forms a **semilunar notch** that fits tightly over the end of the humerus, forming the elbow joint. Just proximal to the semilunar notch is an extension of the ulna, called the **olecranon** (o-lek′ră-non; elbow), which can be felt as the point of the elbow (Figure 6-19). Just distal to the semilunar notch is a **coronoid process,** which helps complete the "grip" of the ulna on the distal end of the humerus. The distal end of the ulna forms a head, which articulates with the bones of the wrist, and a **styloid process** (shaped like a stylus, or pen) is located on its medial side. The ulnar head can be seen as a prominent lump on the posterior ulnar side of the wrist. The radius attaches by a head at its proximal end to both the humerus and the ulna, but it does not attach as firmly to the humerus as does the ulna. Just distal to the radial head is a **radial tuberosity,** where one of the arm muscles, the biceps brachii, attaches. The distal end of the radius articulates with the wrist bones. A styloid process is located on the lateral side of the distal end. The styloid processes of the radius and ulna provide attachments for the ligaments of the wrist.

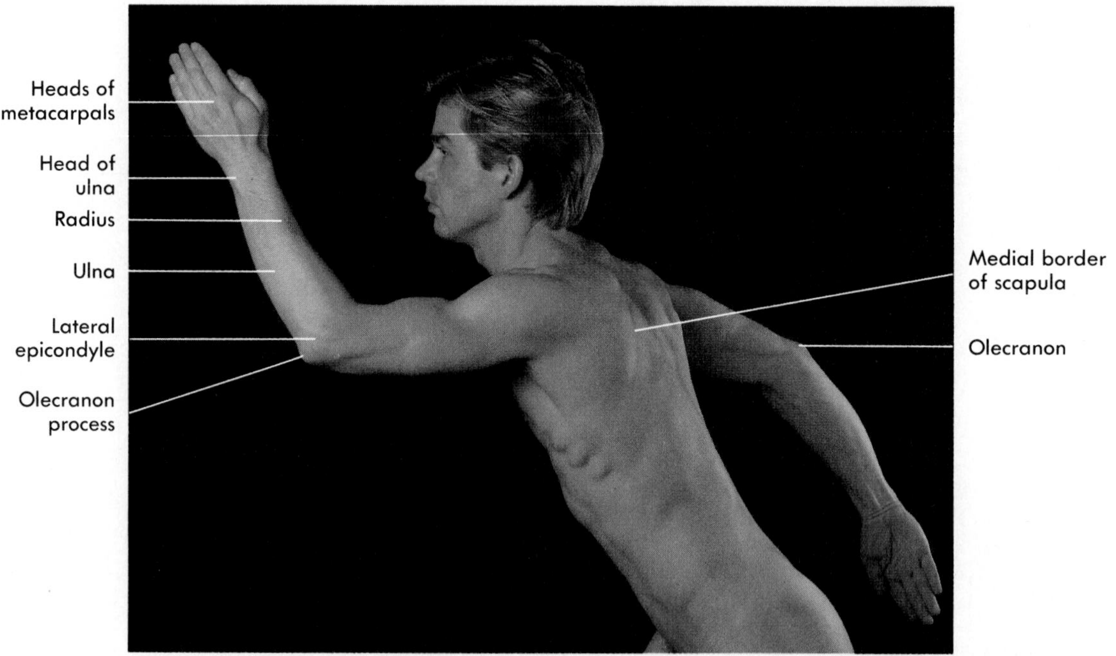

FIGURE 6-19 ■ Left upper limb. Surface anatomy.

Wrist

■ The **wrist** is a relatively short region between the forearm and hand and is composed of eight **carpal** (kar′pul) bones (Figure 6-20). The eight carpal bones are the scaphoid, lunate, triquetrum, pisiform, trapezium, trapezoid, capitate, and hamate.

> ■ The bones and ligaments on the anterior side of the wrist form a *carpal tunnel,* which does not have much "give." Tendons and nerves pass through the carpal tunnel to the hand. If fluid accumulates in the carpal tunnel as a result of overuse or trauma, it may apply pressure to a major nerve passing through the tunnel. The pressure on this nerve causes *carpal tunnel syndrome,* which consists of tingling, burning, and numbness in the hand. ■

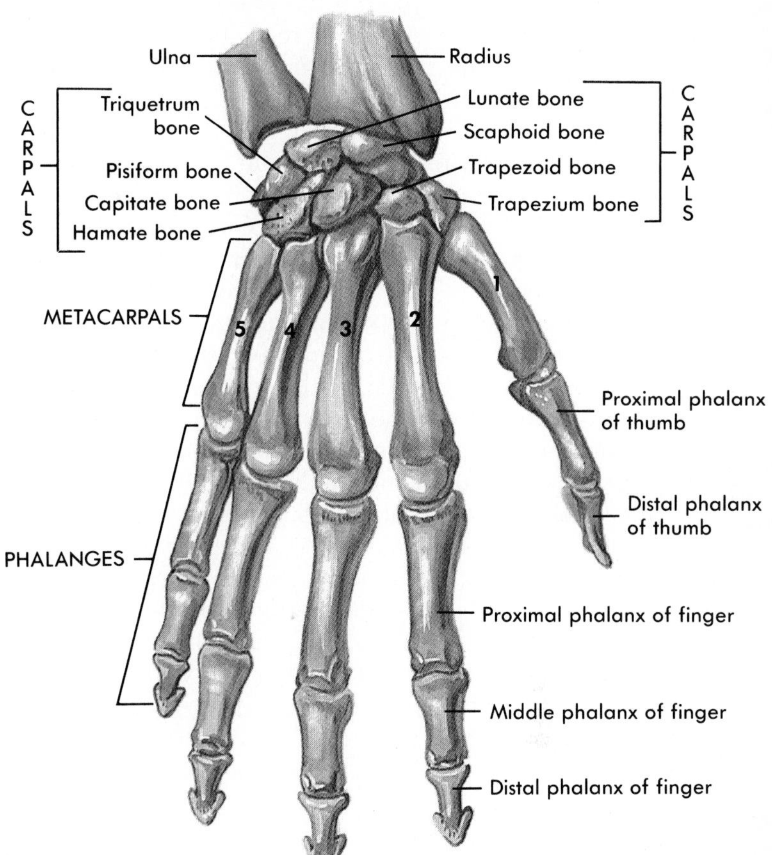

FIGURE 6-20 ■ Right wrist and hand. Posterior view.

Hand

■ Five **metacarpals** are attached to the carpal bones and form the bony framework of the hand (see Figure 6-20). Each finger, called a **digit,** consists of three small bones called **phalanges** (fă-lan′-jēz; the Greek "phalanx" is a wedge of soldiers holding their spears, tips outward, in front of them). The thumb has two phalanges.

Pelvic Girdle

■ The **pelvic girdle** (Figure 6-21) is a ring of bones formed by the sacrum and two **coxae** (kok′se). Each coxa is formed by three bones fused to each other to form a single bone; the **ilium** (il′e-um; groin), the **is-**

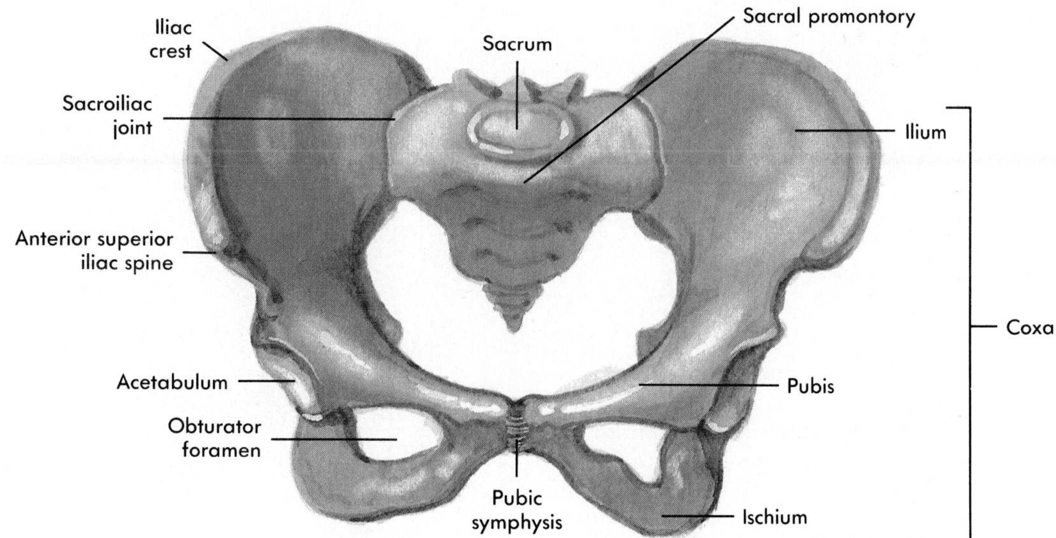

FIGURE 6-21 ■ Anterior view of pelvis.

chium (ish′e-um; hip), and the **pubis** (pu′bis; refers to the genital hair). An **iliac crest** can be seen along the superior margin of each ilium, and an **anterior superior iliac spine,** an important hip landmark, is located at each ilium's anterior end. The coxae join each other anteriorly at the pubic symphysis and join the sacrum at the sacroiliac joint. The **acetabulum** (a′sŭ-tab′u-lum, vinegar cup) is the socket of the hip joint, and the **obturator foramen** is the large hole in the coxa. The term obturator means closed or stopped up and refers to the fact that the foramen is closed off by muscles and other structures. The male pelvis can be distinguished from the female pelvis in that even though the male pelvis is usually larger and more massive, the female pelvis is broader (Figure 6-22 and Table 6-3). Both the pelvic inlet and the pelvic outlet (the inferior opening of the pelvis) of the female pelvis are larger than those of the male pelvis. The pelvic inlet is marked by the pelvic brim laterally and anteriorly and by the sacral promontary posteriorly. The pelvic outlet is bounded by the ischial spines laterally, by the pubic symphysis anteriorly, and by the coccyx posteriorly.

TABLE 6-3

Differences between Male and Female Pelvis

AREA	DESCRIPTION OF DIFFERENCE
General	Female pelvis somewhat lighter in weight and wider laterally, but shorter superiorly to inferiorly and less funnel-shaped; less obvious muscle attachment points in female than in male
Sacrum	Broader in female with the inferior portion directed more posteriorly; the promontory projects less anteriorly in female
Pelvic inlet	Heart-shaped in male; oval in female
Pelvic outlet	Broader and more shallow in female
Subpubic angle	Less than 90° in male; 90° or more in female
Ilium	More shallow and flared laterally in female
Ischial spines	Farther apart in female
Ischial tuberosities	Turned laterally in female and medially in male (not shown in Figure 6-22)

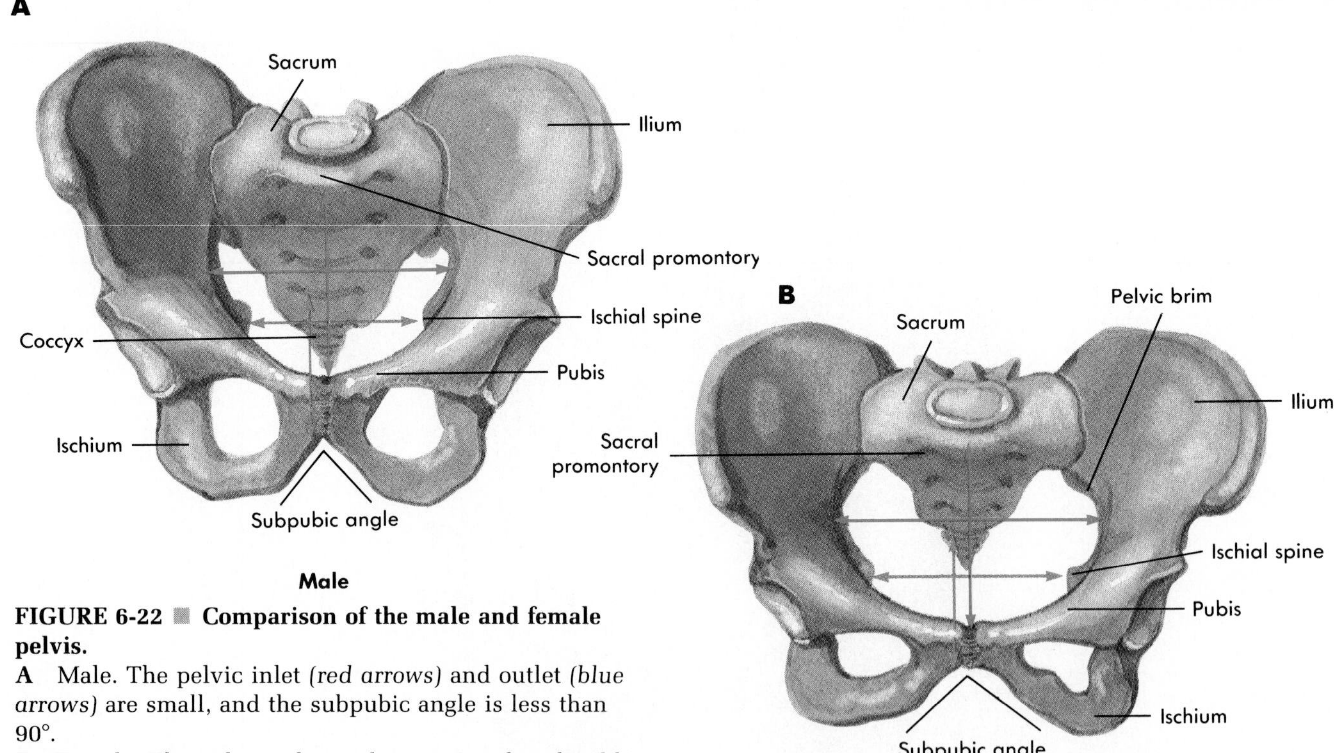

FIGURE 6-22 ■ Comparison of the male and female pelvis.
A Male. The pelvic inlet *(red arrows)* and outlet *(blue arrows)* are small, and the subpubic angle is less than 90°.
B Female. The pelvic inlet *(red arrows)* and outlet *(blue arrows)* are larger, and the subpubic angle is 90° or greater (see Table 6-3).

Lower Limb

The lower limb consists of the bones of the thigh, leg, ankle, and foot.

Thigh

The **thigh** (Figure 6-23, *A* and *B*), the region between the hip and the knee, contains a single bone, the **femur,** attached at one end to the pelvis and at the other end to the tibia. The head of the femur articulates with the acetabulum of the coxa, and the condyles articulate with the tibia. Epicondyles, located medial and lateral to the condyles, provide points of muscle attachment. The femur can be distinguished from the humerus by its long neck located between the head and the **trochanters** (tro'-kan-terz; runners). The trochanters are points of muscle attachment. The **patella,** or kneecap, is located within the major anterior tendon of the thigh muscles and enables the tendon to turn the corner over the knee. Epicondyles on the distal end of the femur serve as muscle attachment sites.

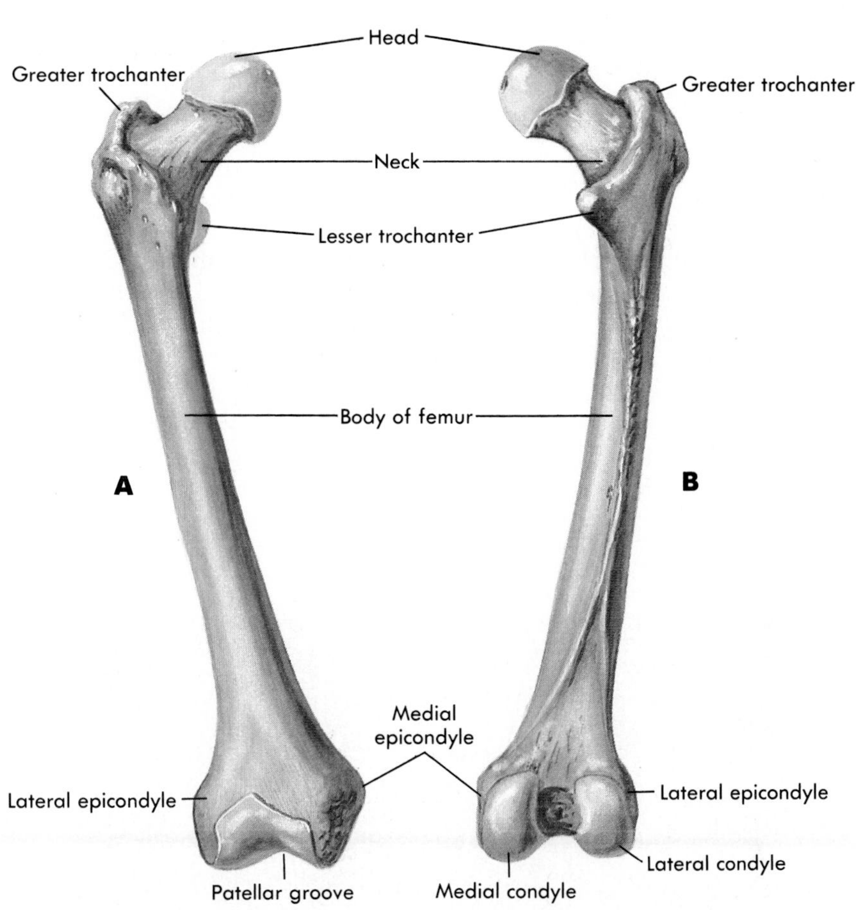

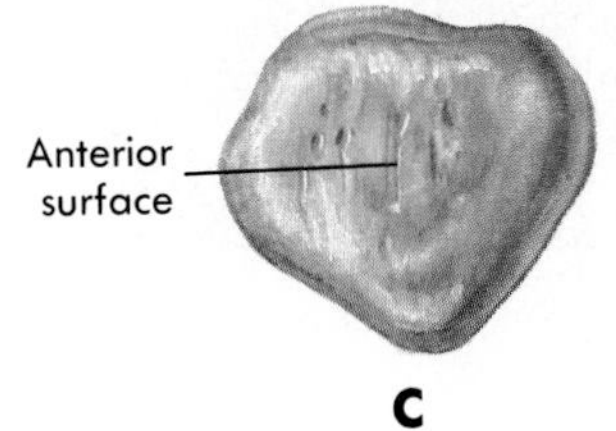

FIGURE 6-23 The right femur and patella.
A Anterior view.
B Posterior view.
C The patella.

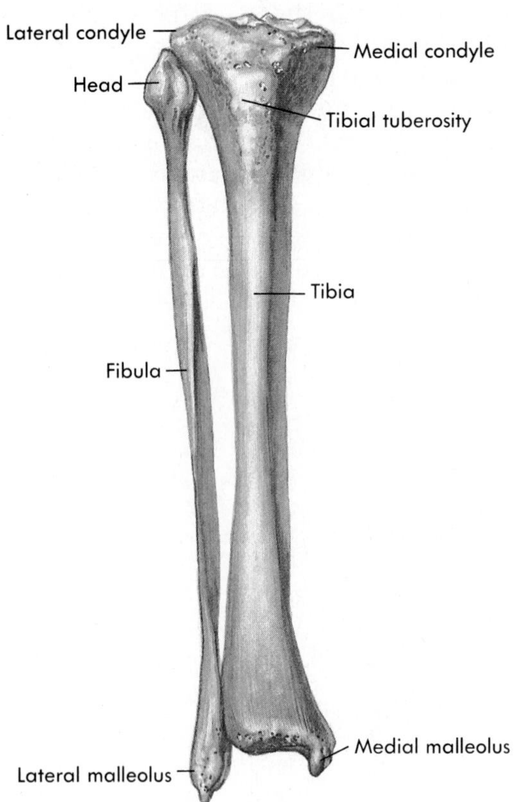

FIGURE 6-24 ■ Right tibia and fibula. Anterior view.

Leg

■ The **leg** (Figure 6-24), the region between the knee and the ankle, contains two bones, the **tibia** (tib'ĭ-ah; shin bone) and the **fibula** (fib'u-lah; resembling a clasp or buckle). The tibia is the larger of the two and supports most of the weight of the leg. The condyles of the femur rest on the flat proximal, condylar end of the tibia. Just distal to the condylar end of the tibia, on its anterior surface, is the **tibial tuberosity,** where the muscles of the anterior thigh ultimately insert. The fibula does not articulate with the femur but attaches by its head to the proximal end of the tibia. The distal ends of the tibia and fibula unite to form a partial socket that articulates with a bone of the ankle (the talus). A prominence can be seen on each side of the ankle (Figure 6-25). These are the medial **malleolus** (mă-le'o-lus; a hammer or mallet) of the tibia and the lateral malleolus of the fibula.

Ankle

■ The **ankle** consists of seven **tarsal** (tar'sal; the sole of the foot) bones (Figure 6-26). The tarsal bones are the talus, calcaneus, cuboid, and navicular; and the medial, intermediate, and lateral cuneiforms. The **talus** (tal'us; ankle bone) articulates with the tibia and fibula to form the ankle joint. The **calcaneus** (kal-ka'ne-us; heel) is located inferior to the talus and protrudes posteriorly to form the heel.

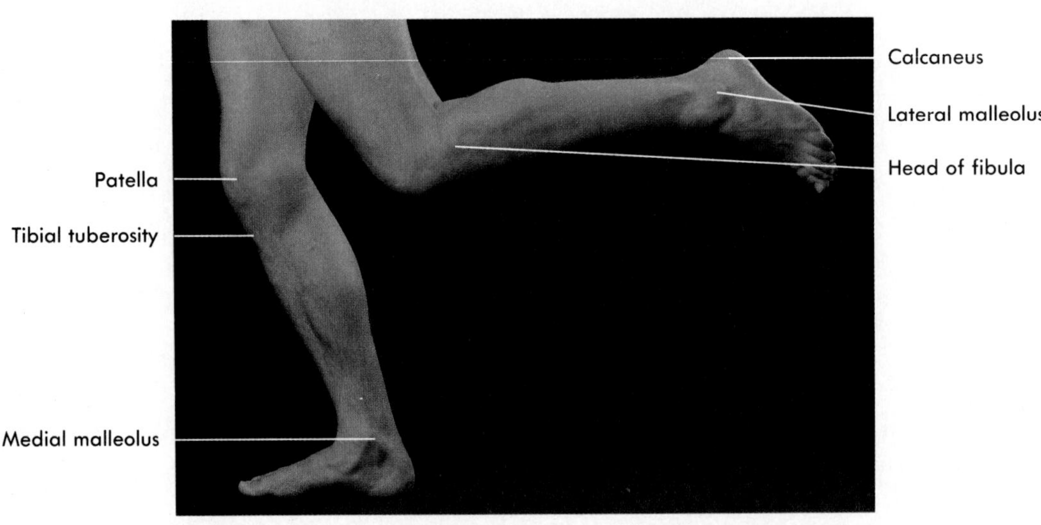

FIGURE 6-25 ■ The lower limbs. Surface anatomy.

Foot

The **metatarsals** and phalanges of the foot are arranged in a manner very similar to the metacarpals and phalanges of the hand (see Figure 6-26). The metatarsals are somewhat longer than the metacarpals, while the phalanges are considerably shorter than those of the hand. The heads formed at the distal ends of the metatarsals make up the ball of the foot. There are two primary **arches** in the foot, formed by the positions of the tarsals and the metatarsals, and held in place by ligaments. A longitudinal arch extends from the heel to the ball of the foot, and a transverse arch extends across the foot. The arches function similar to the springs of a car, allowing the foot to give and spring back.

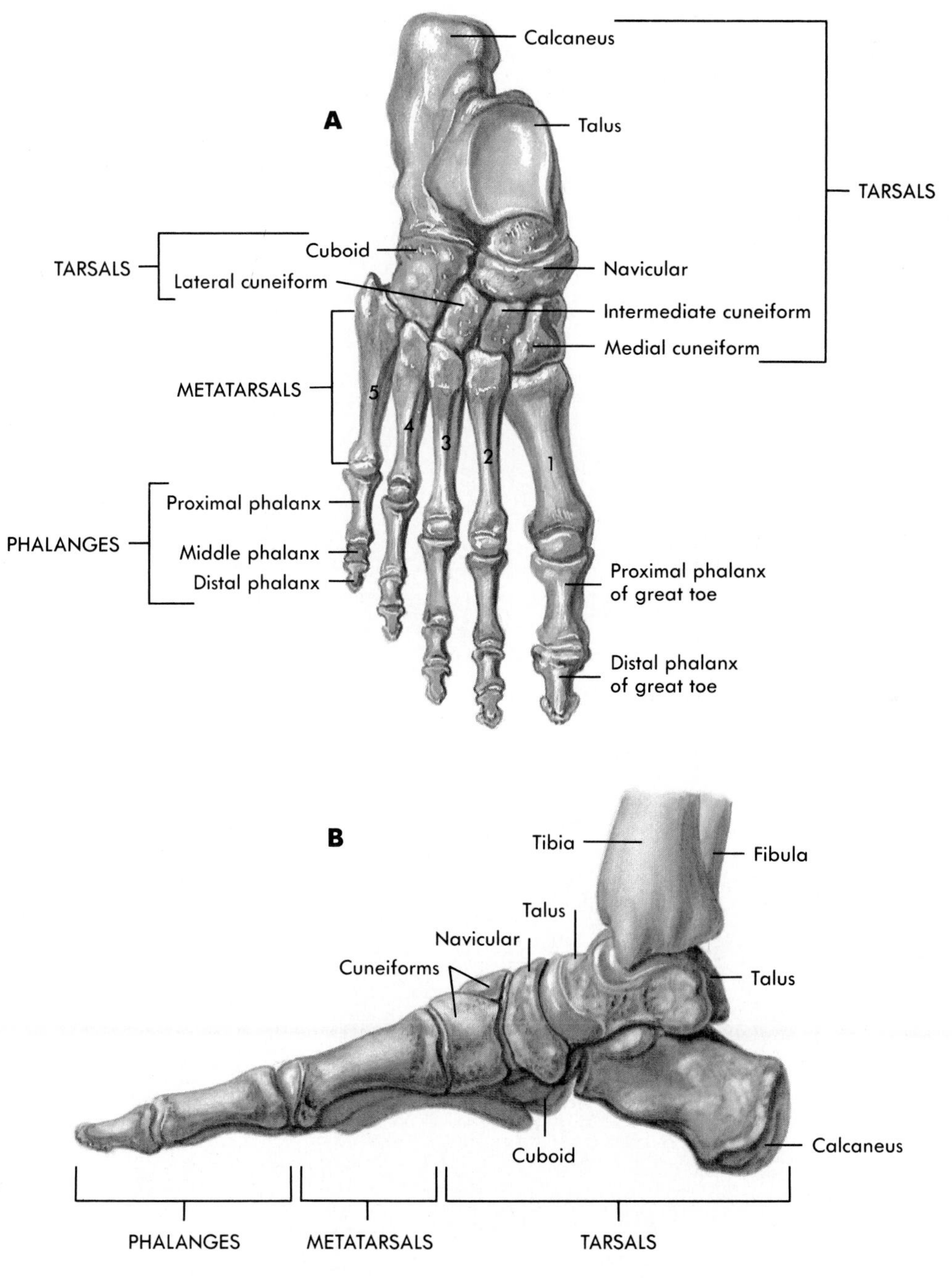

FIGURE 6-26 ■ Bones of the right foot.
A Dorsal view.
B Medial view.

ARTICULATIONS

An **articulation,** or joint, is a place where two bones come together. A joint is usually considered movable, but that is not always the case. Many joints exhibit limited movement, and others are completely, or almost completely, immovable.

Joints are classified according to the major connective tissue type that binds the bones together and whether there is a fluid-filled joint capsule. The three major classes of joints are fibrous, cartilaginous, and synovial.

Fibrous Joints

Fibrous joints consist of two bones that are united by fibrous tissue and that exhibit little or no movement. Joints in this group are further classified on the basis of structure as sutures, syndesmoses, or gomphoses. **Sutures** (su′churz) are fibrous joints between the bones of the skull. In a newborn, some portions of the sutures are quite wide and are called **fontanels** (fon′tă-nelz; soft spots) (Figure 6-27). They allow flexibility in the skull during the birth process, as well as growth of the head after birth.

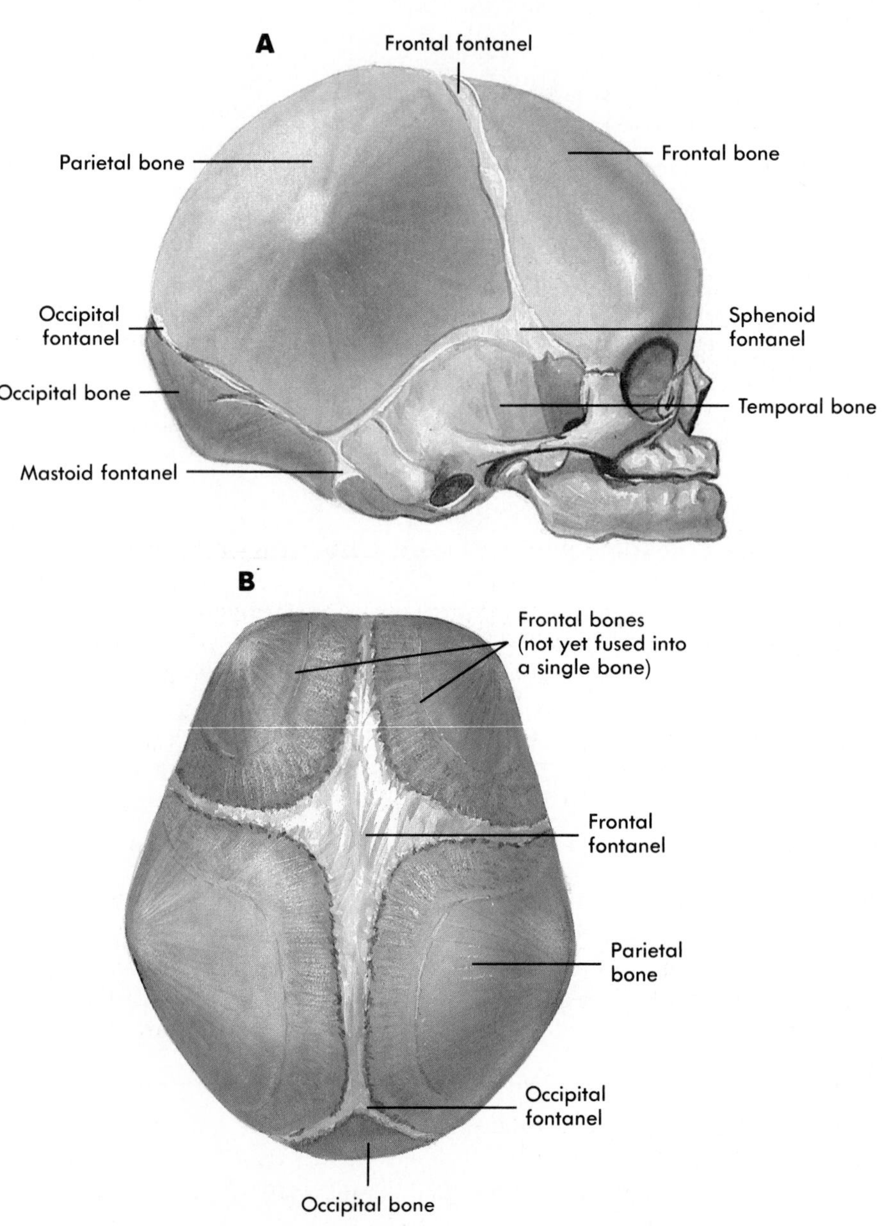

FIGURE 6-27 Fetal skull showing the fontanels.
A Lateral view.
B Superior view.

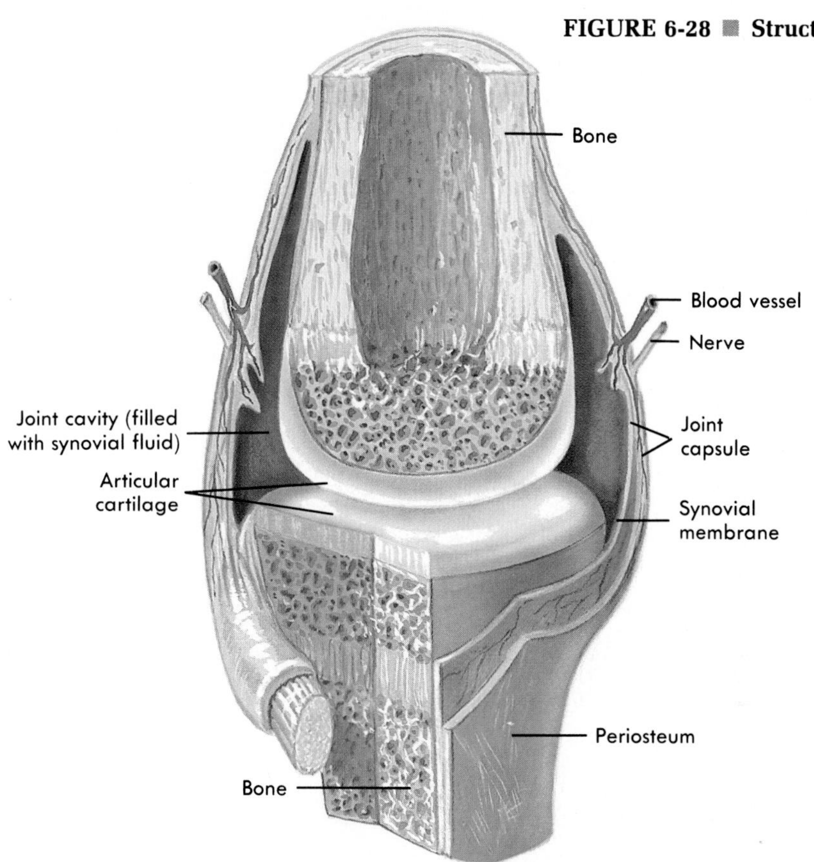

FIGURE 6-28 ■ **Structure of a synovial joint.**

Syndesmoses (sin′dez-mo′sēz) are fibrous joints where the bones are separated by some distance and are held together by ligaments. An example is the fibrous membrane connecting the radius and ulna. **Gomphoses** (gom-fo′sēz) consist of pegs fitted into sockets and held in place by ligaments. The joint between a tooth and its socket is an example of a gomphosis.

Cartilaginous Joints

Cartilaginous joints unite two bones by means of cartilage. Only slight movement can occur at these joints. The cartilages between the ribs and sternum are examples.

Synovial Joints

Synovial (sĭ-no′ve-al; G. *syn*, coming together + *ovia*, resembling egg albumin) **joints** are freely movable joints that contain **synovial fluid** in a cavity surrounding the ends of articulating bones. Most joints that unite the bones of the appendicular skeleton are synovial, whereas many of the joints that unite the bones of the axial skeleton are not. This pattern reflects the greater mobility of the appendicular skeleton compared to the axial skeleton.

Several features of synovial joints are important to their function (Figure 6-28). The articular surfaces of bones within synovial joints are covered with a thin layer of **articular cartilage,** which provides a smooth surface where the bones meet. The **joint cavity** is filled with synovial fluid. The cavity is enclosed by a **joint capsule,** which helps to hold the bones together and, at the same time, allows for movement. Portions of the joint capsule may be thickened to form ligaments. In addition, ligaments and tendons outside the joint capsule contribute to the strength of the joint.

A **synovial membrane** lines the joint capsule everywhere except over the articular cartilage. The membrane produces synovial fluid, which is a complex mixture of polysaccharides, proteins, fat, and cells. Synovial fluid forms a thin lubricating film covering the surfaces of the joint. In certain synovial joints, the synovial membrane may extend as a pocket or sac, called a **bursa** (bur′sah; pocket). Bursae function to reduce friction in areas of the body where structures would rub together, such as where a tendon crosses a bone.

Joint Disorders

ARTHRITIS

Arthritis (Figure 6-C), the inflammation of a joint, is the most common and best known of the joint disorders, affecting 10% of the world's population. There are at least 20 different types of arthritis, which differ in the cause and progress. Causes include infectious agents, metabolic disorders, trauma, and immune disorders.

Rheumatoid arthritis affects about 3% of all women and about 1% of all men in the United States. It is a general connective tissue disorder that affects the skin, vessels, lungs, and other organs, but it is most pronounced in the joints. It is severely disabling and most commonly destroys small joints such as those in the hands and feet. The initial cause is unknown but may involve a transient infection or an autoimmune disease (an immune reaction to one's own tissues). There may also be a genetic predisposition. In rheumatoid arthritis the synovial membrane and associated connective tissue cells proliferate, forming a pannus (clothlike layer) in the joint capsule, which can grow into the articulating surfaces of the bones, destroying the articular cartilage. In advanced stages, the bones forming the joint can become fused.

DEGENERATIVE JOINT DISEASE

Degenerative joint disease (DJD), also called osteoarthritis, results from the gradual "wear and tear" of a joint that occurs with advancing age. Slowed metabolic rates with increased age also seem to contribute to DJD. It is very common in older individuals and affects 85% of all people in the United States over the age of 70. It tends to occur in the weight-bearing joints such as the knees and is more common in overweight individuals. Mild exercise retards joint degeneration and enhances mobility.

A

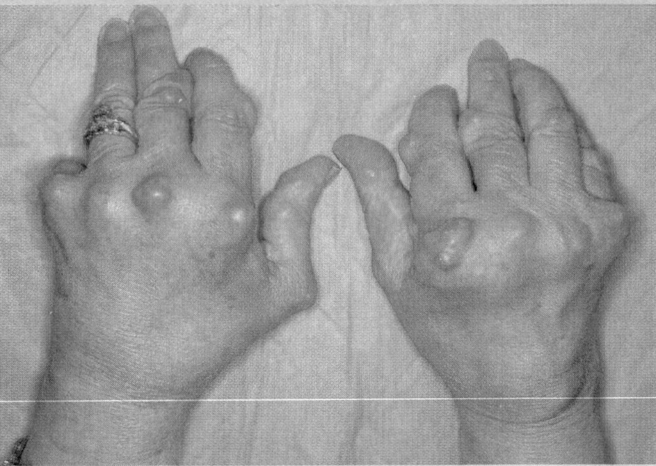

B

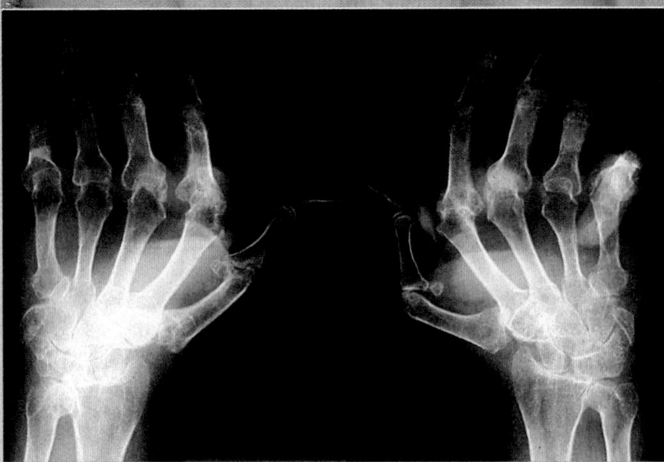

FIGURE 6-C ■ Rheumatoid arthritis.
A Photograph of hands with rheumatoid arthritis.
B X-ray of the same hands.

GOUT

Gout is caused by an increase in uric acid in the body. Uric acid is a waste product, which can accumulate as crystals in various tissues, including the kidneys and joint capsules. Gout is more common in males than in females.

Frequently, only one or two joints are affected. The most commonly affected joints (85% of the cases) are the base of the great toe and other foot and leg joints. Any joint may ultimately be involved, and damage to the kidney from crystal formation occurs in almost all advanced cases.

BURSITIS AND BUNIONS

Bursitis is the inflammation of a bursa. The bursae around the shoulders and elbows are common sites of bursitis. A **bunion** is a bursitis that develops over the joint at the base of the great toe. Bunions are frequently irritated by shoes that rub on them.

JOINT REPLACEMENT

As a result of recent advancements in biomedical technology, many joints of the body can now be replaced by artificial joints. Joint replacement, or arthroplasty, was developed in the late 1950s. It is used in patients with joint disorders to eliminate unbearable pain and to increase joint mobility. Degenerative joint disease is the leading disease requiring joint replacement, accounting for two thirds of the patients. Rheumatoid arthritis accounts for more than half the remaining cases.

Artificial joints usually are composed of metal (for example, stainless steel, titanium alloys, or cobalt-chrome alloys) in combination with modern plastics (for example, high-density polyethylene, silicone rubber, or elastomer). The bone of the articular area is removed on one side (hemireplacement) or both sides (total replacement) of the joint, and the artificial articular structures are attached to the bone. The smooth metal surface rubbing against the smooth plastic surface provides a low-friction contact with a range of movement that depends on the design.

Types of synovial joints

Synovial joints are classified according to the shape of the adjoining articular surfaces (Figure 6-29). **Plane,** or **gliding joints,** consist of two opposed flat surfaces that glide over each other. Examples of these joints are the articular processes between vertebrae. **Saddle joints** consist of two saddle-shaped articulating surfaces oriented at right angles to one another. Movement in these joints can occur in two planes. The joint at the base of the thumb is an example. **Hinge joints** permit movement in one plane only. They consist of a convex cylinder of one bone applied to a corresponding concavity of the other bone. Examples are the elbow and knee joints (Figure 6-30, *A*). The knee joint contains shock-absorbing fibrocartilage pads, called **menisci. Pivot joints** restrict movement to rotation around a single axis. Each pivot joint consists of a cylindrical bony process that rotates within a ring

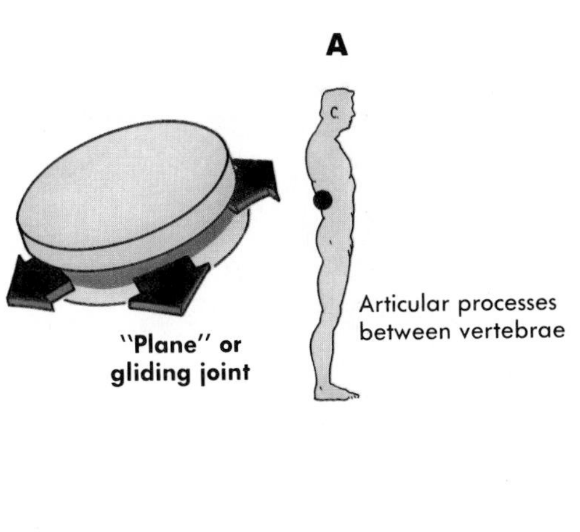

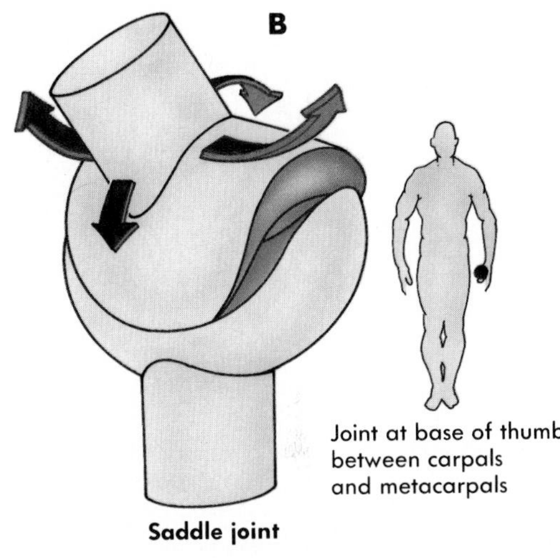

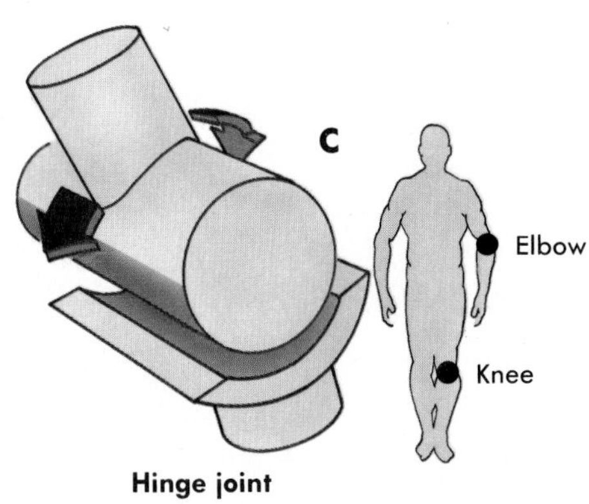

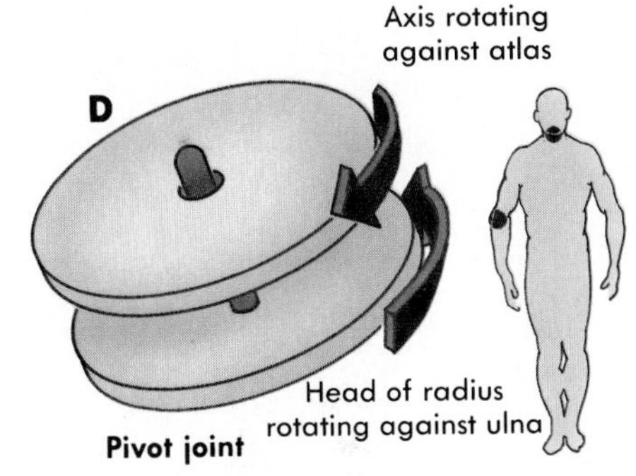

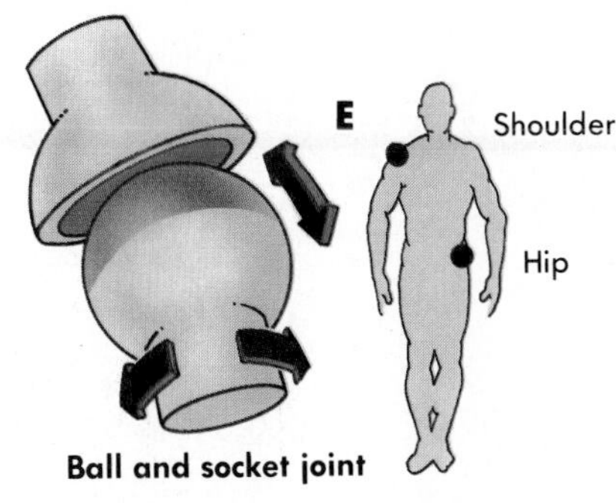

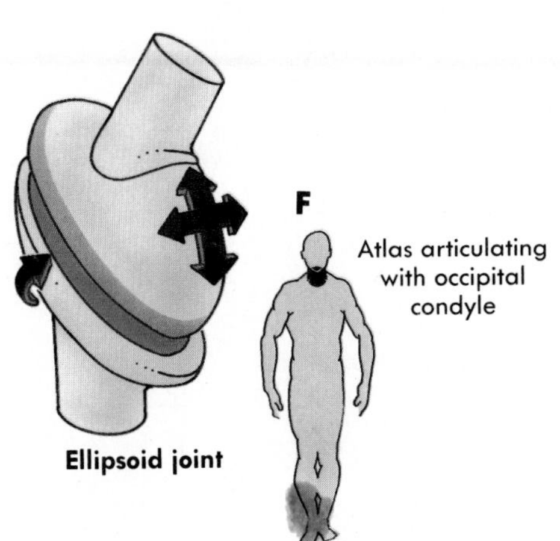

FIGURE 6-29 ■ **Types of synovial joints.**

FIGURE 6-30 ■ Three skeletal joints.
A Sagittal section through the right knee joint.
B Frontal section through the right shoulder.
C Frontal section through the right hip joint.

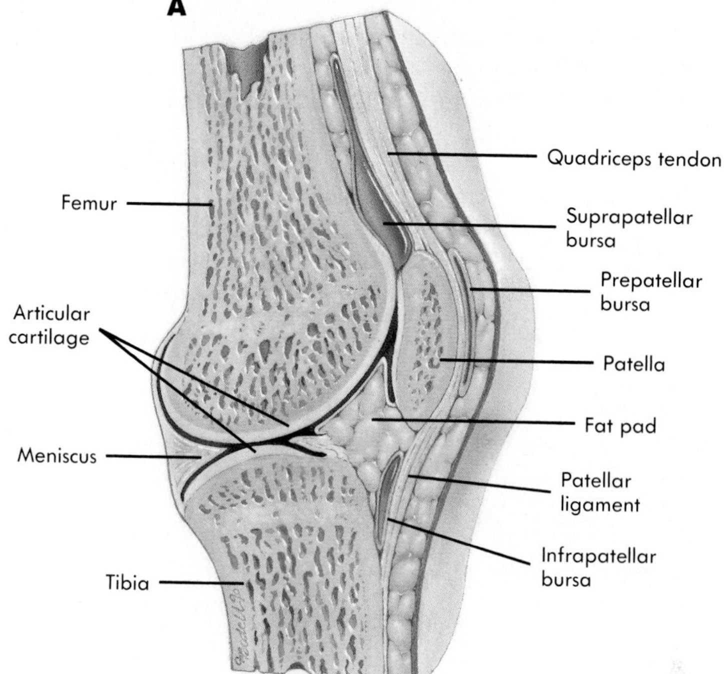

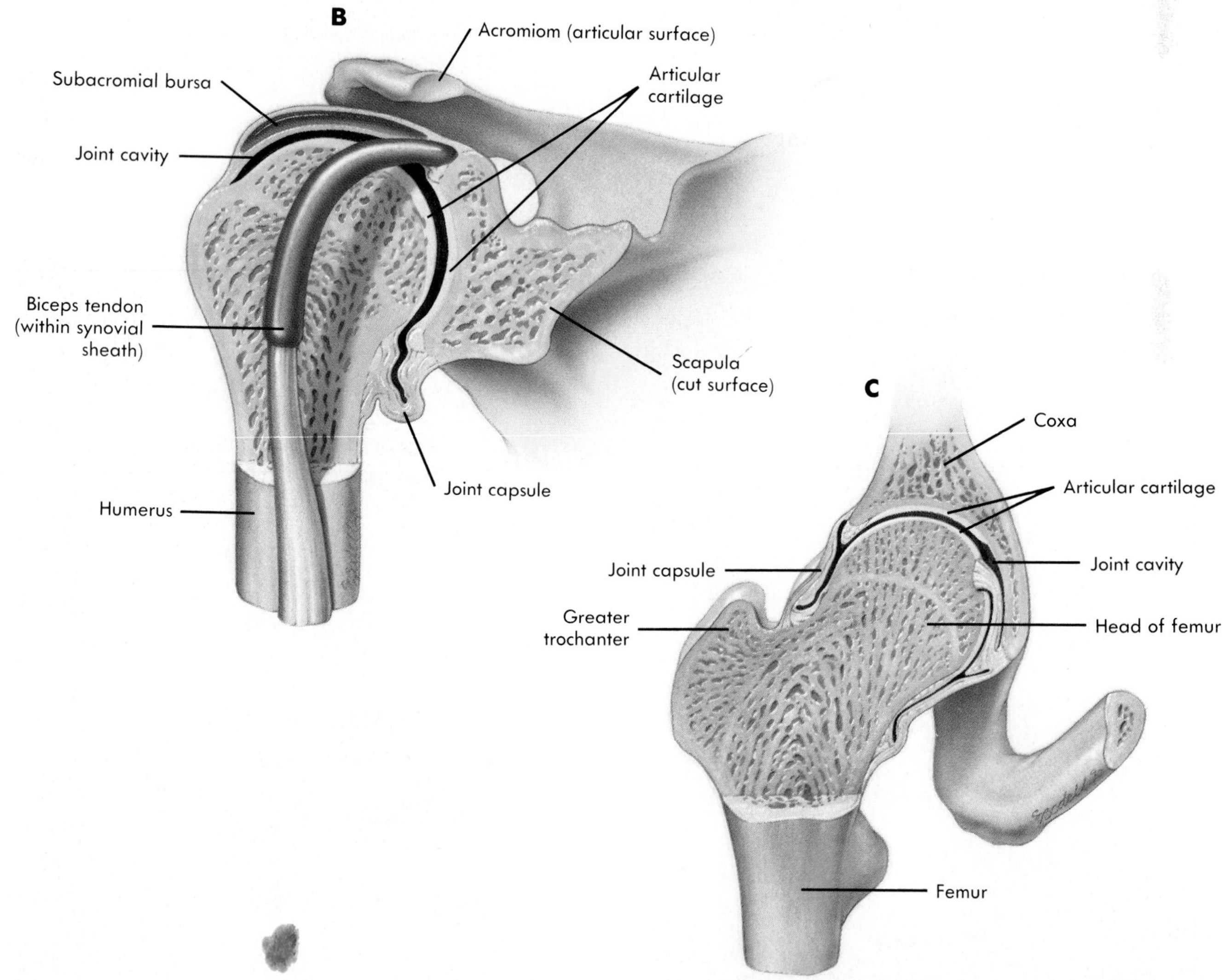

FIGURE 6-31 ■ Movements.
A Flexion and extension of the elbow.
B Flexion and extension of the neck.
C Abduction and adduction of the fingers.
D Pronation and supination of the forearm and hand.
E Medial and lateral rotation of the humerus.
F Circumduction of the shoulder.

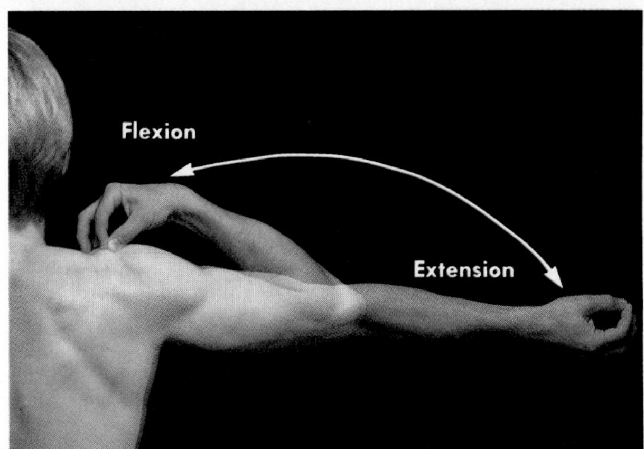

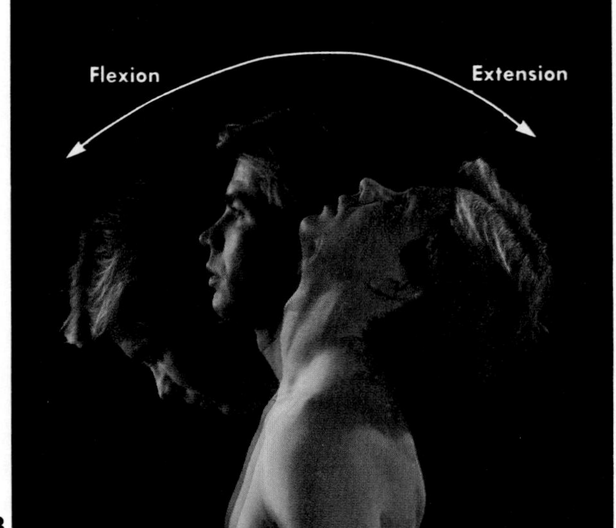

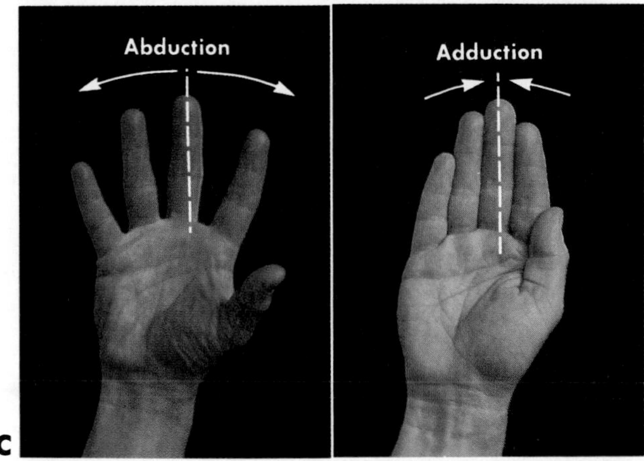

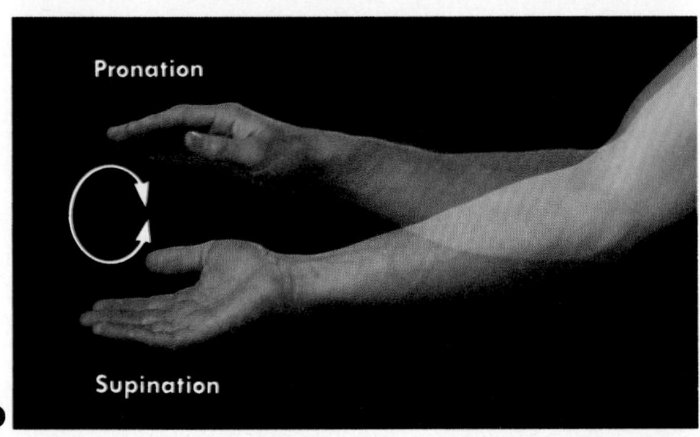

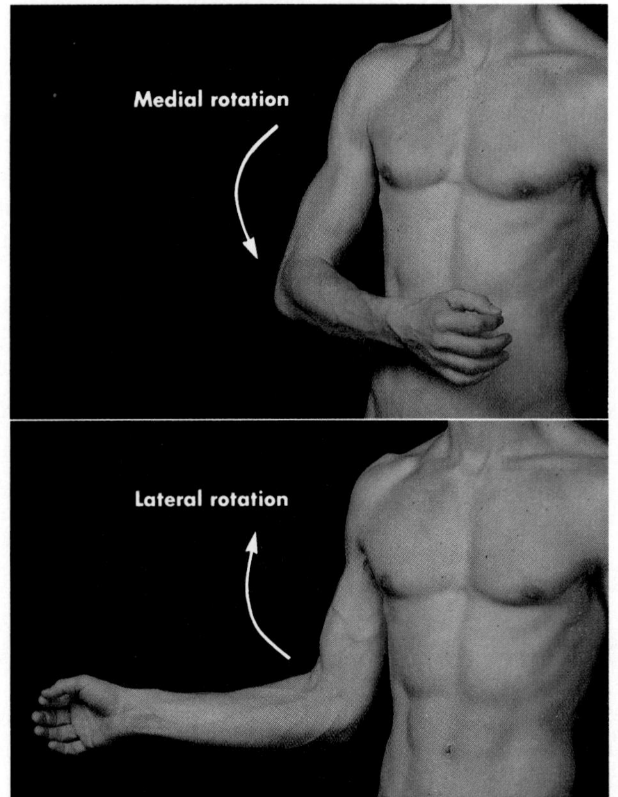

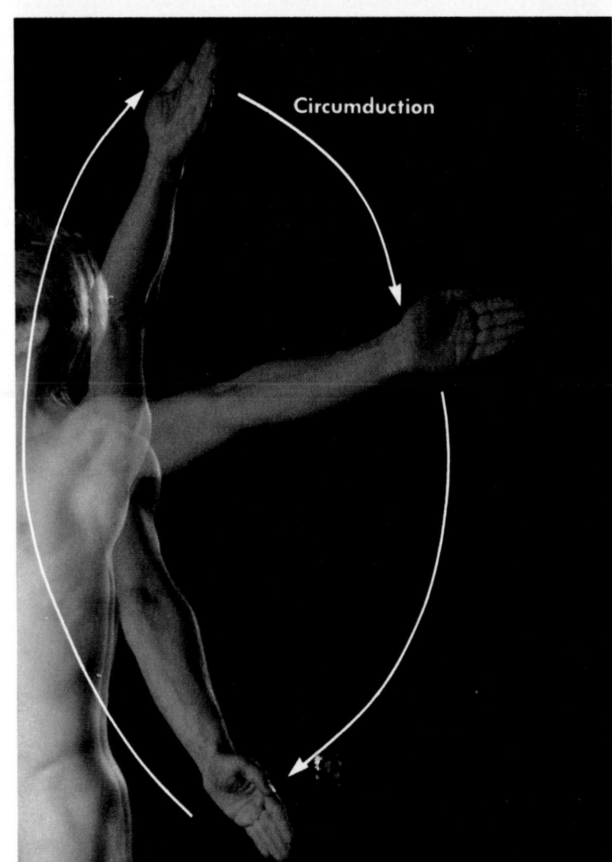

composed partly of bone and partly of ligament. The rotation that occurs between the axis and atlas when shaking the head "no" is an example.

Ball-and-socket joints consist of a ball (head) at the end of one bone and a socket in an adjacent bone into which a portion of the ball fits. This type of joint allows a wide range of movement in almost any direction. Examples are the shoulder and hip joints (Figure 6-30, *B* and *C*). **Ellipsoid** or **condyloid joints** are elongated ball-and-socket joints. The shape of the joint limits its range of movement nearly to a hinge motion, but in two planes. The joint between the occipital condyles of the skull and the atlas of the vertebral column is an example.

■ A *sprain* results when the bones of a joint are forcefully pulled apart and the ligaments around the joint are pulled or torn. A *separation* exists when the bones remain apart after an injury to a joint. A *dislocation* is when the head of one bone is pulled out of the socket in a ball-and-socket or ellipsoid joint. ■

Types of Movement

■ The types of movement occurring at a given joint are related to the structure of that joint. Some joints are limited to only one type of movement, whereas others permit movement in several directions. Most movements are accompanied by movements in the opposite direction and are therefore listed in pairs (Figure 6-31).

Flexion literally means to bend; **extension** means to straighten. As an example, flexion of the elbow is bending the elbow; extension straightens it. For many joints, however, flexion and extension are not quite that simple. For example, how does one define flexion and extension of the neck or shoulder using the above description? The most accurate way to describe flexion and extension is to imagine a frontal plane through the center of the body. Flexion moves a structure in the anterior direction, anterior to the frontal plane, and extension moves a structure in the posterior direction, posterior to the imaginary plane. For the knee and structures inferior to the knee, the rule is reversed: flexion moves a structure in the posterior direction and extension moves a structure in the anterior direction. It may also be helpful to think of the fetal position, because in such a position, nearly all the synovial joints are flexed.

Abduction (to take away) of most joints is movement away from the midline; **adduction** (to bring together) is movement toward the midline. Moving the legs away from the midline of the body, as in the outward movement of "jumping jacks," is abduction, and bringing the legs back together is adduction.

Pronation (pro-na'shun) is rotation of the forearm so that the palm faces posteriorly, or down, if the elbow is flexed; **supination** (su'pin-a'shun) is rotation of the forearm so that the palm faces anteriorly, or up, if the elbow is flexed.

Rotation is the turning of a structure around its long axis, as in shaking the head "no."

Circumduction occurs at freely movable joints such as the shoulder. In circumduction, the arm moves so that it describes a cone with the shoulder joint at the apex.

Most movements that occur in the course of normal activities are combinations of movements. A complex movement can be described by naming the individual movements involved.

4 What combination of movements is required at the shoulder and elbow joints for a person to move his right arm from the anatomical position and touch the side of his head? ?

SUMMARY

The skeletal system consists of bone, cartilage, tendons, and ligaments.

CONNECTIVE TISSUE

Connective tissue consists of matrix and the cells that produce matrix, which contains collagen and proteoglycans.

Varying amounts of collagen, proteoglycan, and mineral in the matrix determine the characteristics of the connective tissue.

GENERAL FEATURES OF BONE

Long bones consist of a diaphysis (shaft), epiphyses (ends), and epiphyseal (growth) plates.

Compact Bone

Compact bone tissue consists of haversian systems.

Haversian systems consist of osteocytes organized into lamellae surrounding haversian canals.

Cancellous Bone

Cancellous bone tissue consists of trabeculae without haversian canals.

Bone Ossification

Bone ossification is either intramembranous or endochondral.

Intramembranous ossification occurs within connective tissue membranes.

Endochondral ossification occurs within cartilage.

Bone Growth

Bone growth is either appositional (on the surface of existing bone) or endochondral (within cartilage).

Bone Remodeling

Bone remodeling consists of removal of old bone and deposition of new bone.

Bone Repair

During bone repair, cells move in and form a callus, which is replaced by bone.

GENERAL CONSIDERATIONS OF BONE ANATOMY

There are 206 bones, which can be classified by shape.

AXIAL SKELETON

The axial skeleton includes the skull, vertebral column, and rib cage.

Skull

The skull consists of 22 bones, divided between the cranial vault and face.

From a lateral view, the parietal, temporal, and sphenoid bones can be seen.

From a frontal view, the orbits and nasal cavity can be seen, as well as associated bones and structures, such as the frontal bone, zygomatic bone, maxilla, and mandible.

The interior of the cranial vault contains three fossae with several foramina.

Seen from below, the base of the skull reveals numerous foramina and other structures, such as processes for muscle attachment.

Vertebral Column

The vertebral column contains 7 cervical, 12 thoracic, and 5 lumbar vertebrae, plus 1 sacral and 1 coccygeal bone.

Each vertebra consists of a body, an arch, and processes.

Regional differences in vertebrae are as follows: Cervical vertebrae have transverse foramina; thoracic vertebrae have long spinous processes and attachment sites for the ribs; lumbar vertebrae have square transverse and spinous processes; the sacrum is a single, fused bone.

Thoracic Cage

The thoracic cage consists of thoracic vertebrae, ribs, and sternum.

There are 12 ribs; 7 true and 5 false (two of the false ribs are also called floating ribs).

The sternum consists of the manubrium, body, and xiphoid process.

APPENDICULAR SKELETON

The appendicular skeleton consists of the bones of the upper and lower limbs and their girdles.

Pectoral Girdle

The pectoral girdle includes the scapula and clavicle.

Upper Limb

The upper limb consists of the arm (humerus), forearm (ulna and radius), wrist (eight carpal bones), and hand (five metacarpals, three phalanges in each finger, and two phalanges in the thumb).

Pelvic Girdle

The pelvic girdle is made up of the sacrum and two coxae. Each coxa consists of an ilium, ischium, and pubis.

Lower Limb

The lower limb includes the thigh (femur), leg (tibia and fibula), ankle (seven tarsals), and foot (metatarsals and phalanges, similar to the bones in the hand).

ARTICULATIONS

An articulation is a place where bones come together.

Fibrous Joints

Fibrous joints consist of bones united by fibrous connective tissue. They allow little or no movement.

Cartilaginous Joints

Cartilaginous joints consist of bones united by cartilage, and they exhibit slight movement.

Synovial Joints

Synovial joints consist of articular cartilage over the uniting bones, a joint cavity lined by a synovial membrane and containing synovial fluid, and a joint capsule. They are highly movable joints.

Synovial joints can be classified as plane, saddle, hinge, pivot, ball-and-socket, or ellipsoid.

Types of Movement

The major types of movement include flexion/extension, abduction/adduction, pronation/supination, rotation, and circumduction.

CONTENT REVIEW

1. What are the components of the skeletal system? List their functions.
2. Name the major types of fibers and molecules found in the extracellular matrix of the skeletal system. How do they contribute to the functions of tendons, ligaments, cartilage, and bones?
3. Define the terms diaphysis, epiphysis, epiphyseal plate, medullary cavity, periosteum, and endosteum.
4. Describe the structure of compact bone. How do nutrients reach the osteocyte in compact bone?
5. Describe the structure of cancellous bone. What are trabeculae?
6. Define intramembranous and endochondral ossification.
7. How does bone grow in width? How do long bones grow in length?
8. What is accomplished by bone remodeling? How does bone repair occur?
9. Define the axial skeleton and the appendicular skeleton.
10. Name the bones of the cranial vault and the face.
11. Give the locations of the paranasal sinuses. What are their functions?
12. What is the function of the hard palate?
13. Through what foramen does the brain connect to the spinal cord?
14. How do the vertebrae protect the spinal cord? Where do spinal nerves exit the vertebral column?
15. Name and give the number of each type of vertebra. Describe the characteristics that distinguish the different types from each other.
16. What is the function of the thoracic cage? Name the parts of the sternum. Distinguish true, false, and floating ribs.

17. Name the bones that make up the pectoral girdle, arm, forearm, wrist, and hand. How many phalanges are in each finger and in the thumb?
18. Define the pelvis. What bones fuse to form each coxa? Where and with what bones do the coxae articulate?
19. Name the bones of the thigh, leg, ankle, and foot.
20. Define the term articulation or joint. Name the three major classes of joints.
21. Describe the structure of a synovial joint. How do the different parts of the joint function to permit joint movement?
22. On what basis are synovial joints classified? Describe the different types of synovial joints and give examples of each. What movements do each type of joint allow?
23. Describe and give examples of flexion/extension, abduction/adduction, and supination/pronation.

CONCEPT REVIEW

1. A 12-year-old boy fell while playing basketball. The physician explained that the head (epiphysis) of the femur was separated from the shaft (diaphysis). Although the bone was set properly, by the time the boy was 16 it was apparent that the injured lower limb was shorter than the normal one. Explain why this difference occurred.
2. Justin Time leaped from his hotel room to avoid burning to death in a fire. If he landed on his heels, what bone was likely to fracture? Unfortunately for Justin, a 240-pound fireman, Hefty Stomper, ran by and stepped heavily on the distal part of Justin's foot (not the toes). What bones now could be broken?
3. One day while shopping, Ms. A. Bargin picked up her 3-year-old son, Some, by his right wrist and lifted him into a shopping cart. She heard a clicking sound and Some immediately began to cry and hold his elbow. Given that lifting the child caused a dislocation at the elbow, which is more likely: dislocation of the radius or dislocation of the ulna?

ANSWERS TO  PREDICT QUESTIONS

1 *p. 110* If all the mineral were removed, the bone would become so flexible that it could be tied into a knot (this can be accomplished by soaking a bone in vinegar for an extended time). The bone would be so weak that it could not support weight.

If all the collagen were removed, the bone would become very brittle and easily broken. This is a problem with many older people whose bones are fragile and easily broken.

2 *p. 114* If endochondral bone growth failed to occur, the bone would be normal in diameter (or even greater in diameter than normal) but would be much shorter than normal. This is the condition seen in one type of dwarfism, where the head and trunk are more normal in size, but the long bones of the limbs are very short.

3 *p. 119* Tears are produced in lacrimal glands in the superior lateral corner of the orbit. The tears run across the surface of the eye and enter the duct that passes through the nasolacrimal canal to the nasal cavity. The extra moisture in the nasal cavity causes the "runny nose."

4 *p. 140* Shoulder flexion and abduction and elbow flexion. This would allow you to touch the medial side of the little finger to the side of the head. Partial pronation of the forearm is required to touch the side of the head with the palmar surface of the head or fingers.

7

THE MUSCULAR SYSTEM

After reading this chapter you should be able to:

1 Illustrate the microscopic structure of a muscle. Produce diagrams that illustrate the arrangement of myofilaments, myofibrils, and sarcomeres.

2 Describe the events that result in muscle contraction and relaxation in response to an action potential in a motor neuron.

3 Distinguish between aerobic and anaerobic muscle contraction.

4 Distinguish between fast-twitch and slow-twitch muscles and explain the function for which each type is best adapted.

5 Distinguish between skeletal, smooth, and cardiac muscle.

6 Define the following terms and give an example of each: origin, insertion, synergist, antagonist, and prime mover.

7 Describe various facial expressions and list the major muscles causing each.

8 Describe mastication, tongue movement, and swallowing, and list the muscles or groups of muscles involved in each.

9 Describe the muscles of the trunk and the actions they accomplish.

10 Describe the movements of the arm, forearm, and hand, and list the muscle groups involved in each movement.

11 Describe the movements of the thigh, leg, and foot, and list the muscle groups involved in each movement.

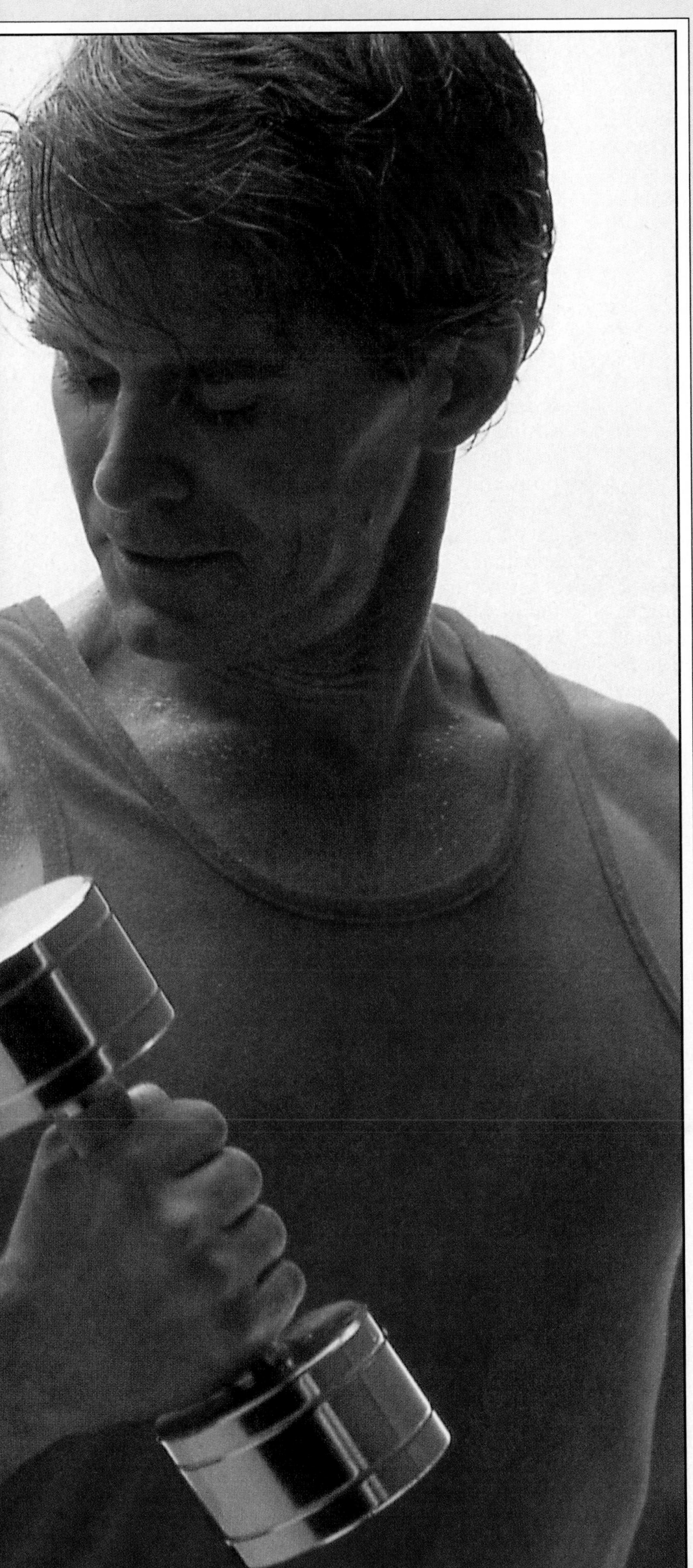

aerobic (a-ro′bik) Metabolism in the presence of oxygen.

anaerobic (an′ă-ro′bik) Metabolism in the absence of oxygen.

antagonist A muscle that acts in opposition to another muscle.

insertion The more movable attachment point of a muscle.

isometric contraction Muscle contraction where the length of the muscle does not change, but the amount of tension increases.

isotonic contraction Muscle contraction where the amount of tension is constant, and the muscle shortens.

motor unit A neuron and all the skeletal muscle fibers it innervates.

muscle twitch Contraction of an entire muscle in response to a stimulus that causes an action potential in one or more muscle fibers.

myofibril A fine longitudinal fibril of skeletal muscle consisting of sarcomeres, composed of thick (myosin) and thin (actin) myofilaments, placed end to end.

myofilament An ultramicroscopic protein thread that helps form myofibrils in skeletal muscle. Thin myofilaments are composed of actin and thick myofilaments are composed of myosin.

neuromuscular junction The synaptic junction between a nerve axon and a muscle fiber.

origin The less movable attachment point of a muscle.

oxygen debt The amount of oxygen required to convert the lactic acid produced during anaerobic respiration to glucose and to replenish creatine phosphate stores.

sarcomere (sar′ko-mēr) [Gr. *sarco*, flesh, means muscle + *meros*, part] The part of a myofibril formed of actin and myosin myofilaments, extending from Z line to Z line; the structural and functional unit of a muscle.

sliding filament mechanism The mechanism by which muscle contraction occurs wherein actin and myosin myofilaments slide past each other.

synergist (sin′er-jist) A muscle that works with another muscle to cause a movement.

As described in Chapter 4, there are three types of muscle tissue: skeletal, cardiac, and smooth. This chapter will deal primarily with the structure and function of skeletal muscle. The skeletal muscles are attached to the bones and are typically under concious control.

CHARACTERISTICS OF SKELETAL MUSCLE

Skeletal muscle, with its associated connective tissue, comprises approximately 40% of the body's weight. It is responsible for locomotion, facial expressions, posture, and other body movements. Skeletal muscle has four major functional characteristics: contractility, excitability, extensibility, and elasticity. **Contractility** is the ability of skeletal muscle to shorten with force. When they contract, skeletal muscles cause movement of the structures to which they are attached. Although skeletal muscles shorten forcefully during contraction, they lengthen passively. Therefore, movement in the opposite direction requires an antagonistic force (in the opposite direction) such as that produced by another muscle or gravity. **Excitability** is the capacity of skeletal muscle to respond to a stimulus. Normally skeletal muscle contracts as a result of stimulation by nerves. **Extensibility** means that skeletal muscles can be stretched. After a contraction, skeletal muscles can be stretched to their normal resting length and beyond to a limited degree. **Elasticity** is the ability of skeletal muscles to recoil to their original resting length after they have been stretched. In addition to these features, the metabolism that occurs in the body's large mass of muscle tissue produces heat essential for the maintenance of normal body temperature.

Structure

Each skeletal muscle is surrounded by a connective tissue sheath called the **epimysium** (Figure 7-1). Another layer of connective tissue, called **fascia** (fash'e-ah), located outside the epimysium, also surrounds and separates muscles. A muscle is composed of numerous visible bundles called **muscle fasciculi** (fă-sik'u-li), which are surrounded by loose connective tissue called the **perimysium** pēr'ĭ-mis'ĭ-um). The fasciculi are composed of single muscle cells called fibers. Each **muscle fiber** is a single cylindrical cell containing several nuclei located at the periphery of the muscle fibers. Each fiber is surrounded by a connective tissue sheath called an **endomysium** (en'do-miz'ĭ-um).

The cytoplasm of each muscle fiber is filled with myofibrils (see Figure 7-1). Each **myofibril** is a thread-like structure that extends from one end of the muscle fiber to the other. Myofibrils consist of two major kinds of protein fibers: actin and myosin myofilaments. **Actin myofilaments,** or thin myofilaments, resemble two minute strands of pearls twisted together. **Myosin myofilaments,** or thick myofilaments, resemble bundles of minute golf clubs.

The actin and myosin myofilaments form highly ordered units called **sarcomeres** (sar'ko-mērz), which are joined end to end to form the myofibrils (see Figure 7-1). The sarcomere is the basic structural and functional unit of the muscle. Each sarcomere extends from one **Z line** to another Z line. Each Z line is a network of protein fibers forming an attachment site for actin myofilaments. The arrangement of the actin and myosin myofilaments gives the myofibril a banded appearance. A light **I band,** which consists only of actin myofilaments, spans each Z line and ends at the myosin myofilaments. A darker, central region in each sarcomere, called an **A band,** extends the length of the myosin myofilaments. The actin and myosin myofilaments overlap for some distance at both ends of the A band. In the center of each sarcomere is a second light zone, called the **H zone,** which consists only of myosin myofilaments. The myosin myofilaments are anchored in the center of the sarcomere at a dark-staining band, called the **M line.** The alternating I bands and A bands of the sarcomeres are responsible for the striations seen in skeletal muscle cells observed through the microscope (Figure 7-2).

Membrane Potentials

The outside of most cell membranes is positively charged compared to the inside of the cell membrane, which is negatively charged. This charge difference across the membrane of an unstimulated cell membrane is called the **resting membrane potential** (see Chapter 8 for a more detailed discussion). Thus a small voltage difference, or potential, can be measured across the resting cell membrane.

When a muscle cell or nerve cell is stimulated, the membrane characteristics are changed for a very brief time. Positively charged sodium ions rush into the cell, causing the inside of the cell membrane to become positive (depolarization). The normal permeability characteristics are quickly reestablished, and the cell membrane returns to its resting mem-

FIGURE 7-1 ■ Parts of a muscle.
A The hand showing some of the hand muscles.
B Enlargement of one hand muscle. A muscle is composed of muscle fasciculi. The fasciculi are composed of bundles of individual muscle fibers (muscle cells).
C Each muscle fiber contains myofibrils in which the banding patterns of the sarcomeres are seen.
D The myofibrils are composed of actin myofilaments and myosin myofilaments, which are formed from thousands of individual actin and myosin molecules.

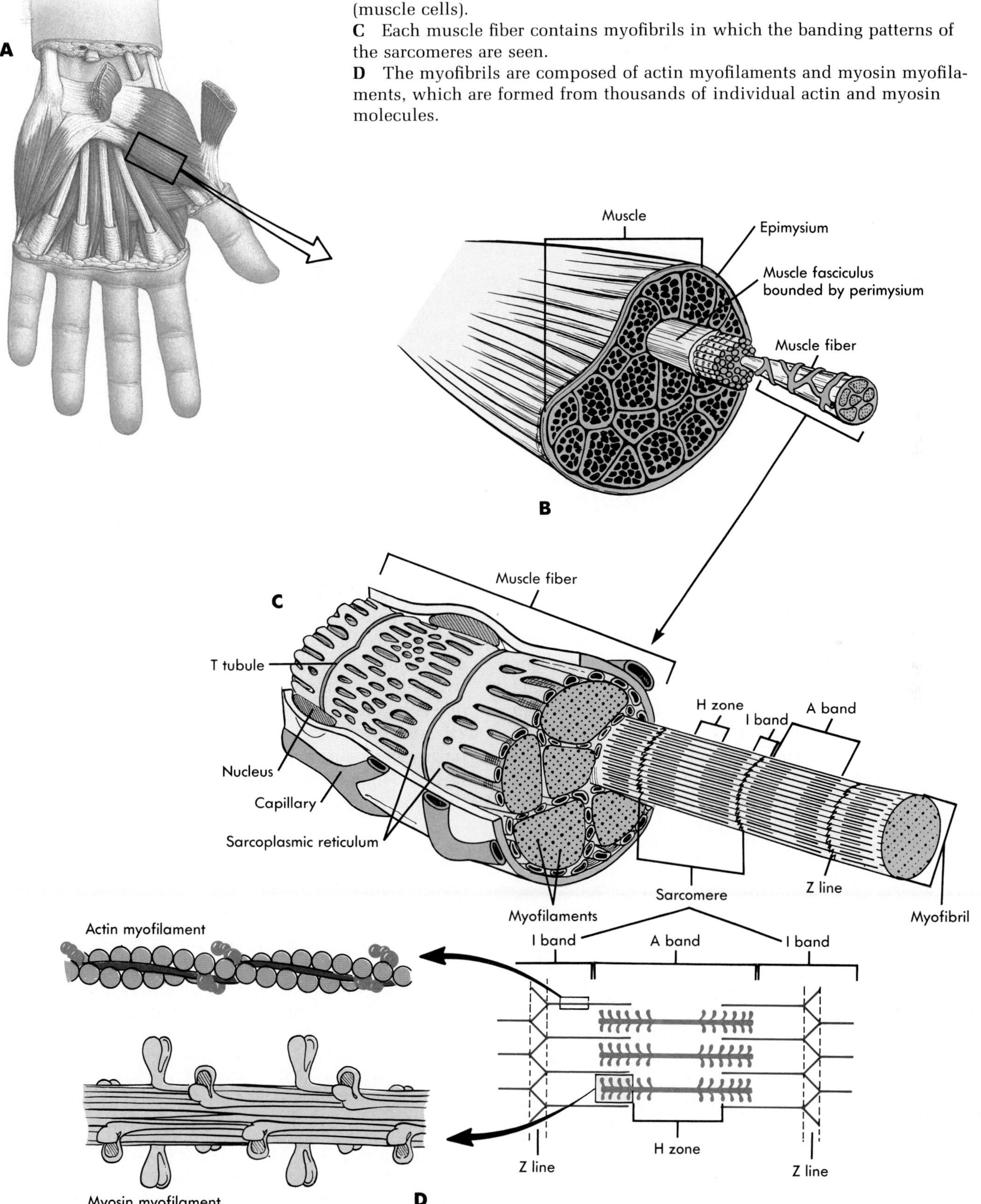

FIGURE 7-2 Electron micrograph of skeletal muscle. Several sarcomeres are shown in several adjacent muscle fibers.

brane potential (repolarization). This brief reversal of the charge across the cell membrane is called an **action potential.**

Nerve Supply

Motor neurons are nerve cells that carry action potentials to skeletal muscle fibers. Axons from these neurons enter the muscles and branch repeatedly. Each branch projects toward one muscle fiber and forms a **neuromuscular junction** or **synapse** (sin'aps) near the center of the muscle fiber (Figure 7-3). A single motor neuron and all the skeletal muscle fibers it innervates are called a **motor unit.** Many motor units form a single muscle.

A neuromuscular junction is formed by an enlarged nerve terminal resting in an indentation of the muscle cell membrane. The enlarged nerve terminal is the **presynaptic terminal,** the space between the presynaptic terminal and the muscle cell is the **synaptic cleft,** and the muscle cell membrane is the **postsynaptic terminal.** Each presynaptic terminal contains many small vesicles, called **synaptic vesicles.** These vesicles contain **acetylcholine** (as'ĕ-til-ko'lēn), which functions as a **neurotransmitter,** a substance released from a presynaptic terminal that diffuses across the synaptic cleft and binds to the postsynaptic terminal, causing a change in the postsynaptic cell (in this case, the muscle cell).

When an action potential reaches the nerve terminal, it causes the synaptic vesicles to release acetylcholine into the synaptic cleft by exocytosis. The acetylcholine diffuses across the synaptic cleft and binds to receptor molecules in the muscle cell membrane. The combination of acetylcholine with its receptor causes an influx of sodium ions into the muscle cell. This influx initiates an action potential in the muscle cell which causes it to contract. The acetylcholine released into the synaptic cleft between the neuron and muscle cell is rapidly broken down by an enzyme, acetylcholinesterase. This enzymatic breakdown ensures that one action potential in the neuron yields only one action potential in the skeletal muscle, and only one contraction of the muscle cell.

Anything that affects the production, release, or degradation of acetylcholine or its ability to bind to receptors on the muscle cell membrane will also affect the transmission of action potentials across the neuromuscular junction. Some insecticides inhibit the enzymes that break down acetylcholine. Consequently, acetylcholine accumulates in the space between the neuron and muscle cell, and acts as a constant stimulus to the muscle fiber. The insects die, partly because their muscles contract and cannot relax. Other poisons such as *curare* (ku-rah're) bind to the acetylcholine receptors on the muscle cell membrane preventing acetylcholine from binding to them. Therefore, the muscle cannot be stimulated by acetylcholine and does not contract.

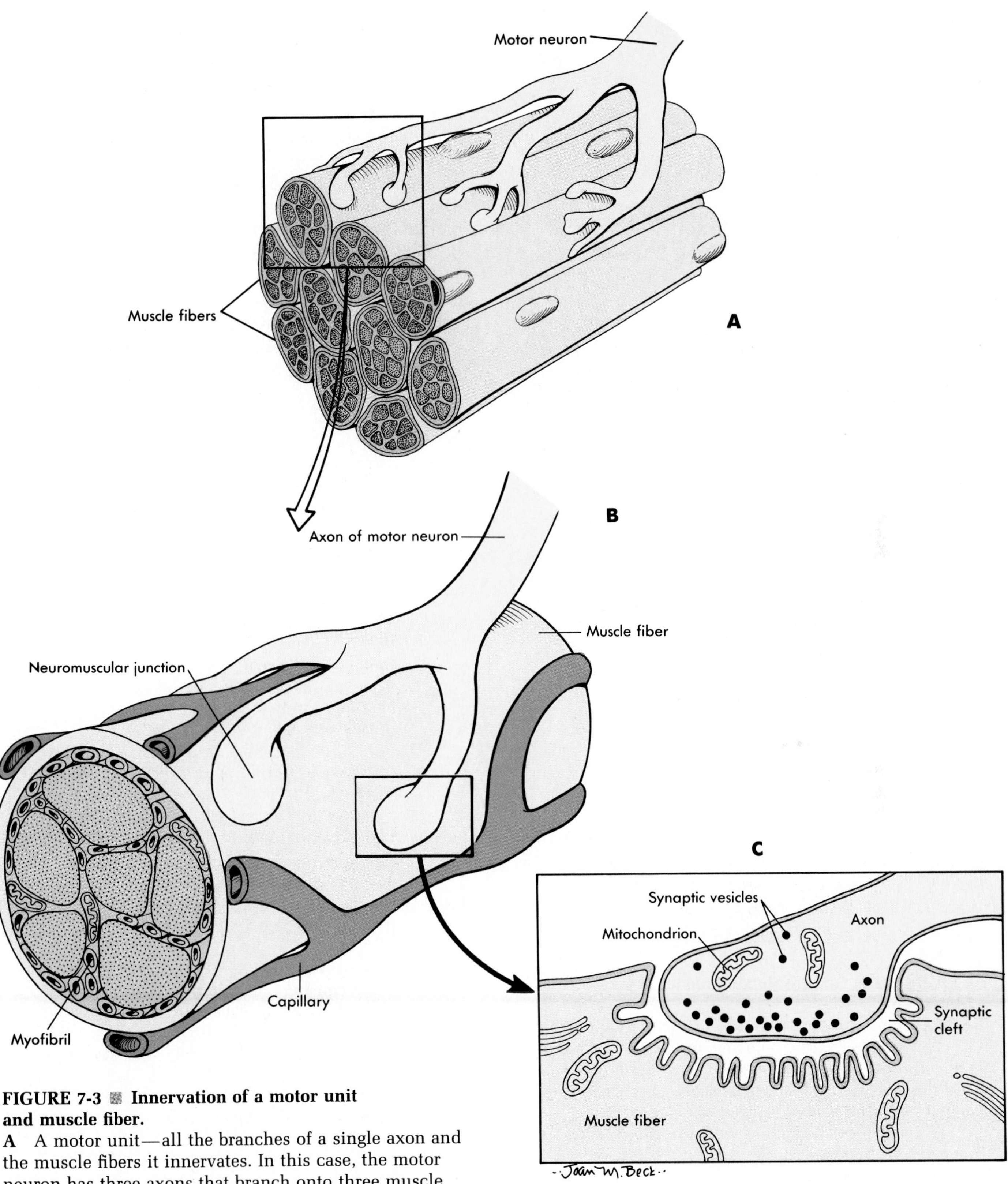

FIGURE 7-3 ■ Innervation of a motor unit and muscle fiber.
A A motor unit—all the branches of a single axon and the muscle fibers it innervates. In this case, the motor neuron has three axons that branch onto three muscle fibers to constitute one motor unit. The other fibers are included to show context.
B An enlarged view of the termination of axon branches on a single muscle fiber.
C An enlarged cross section of one neuromuscular junction.

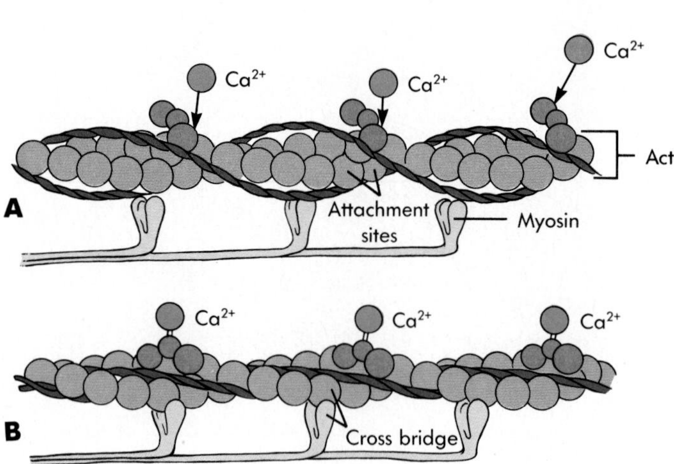

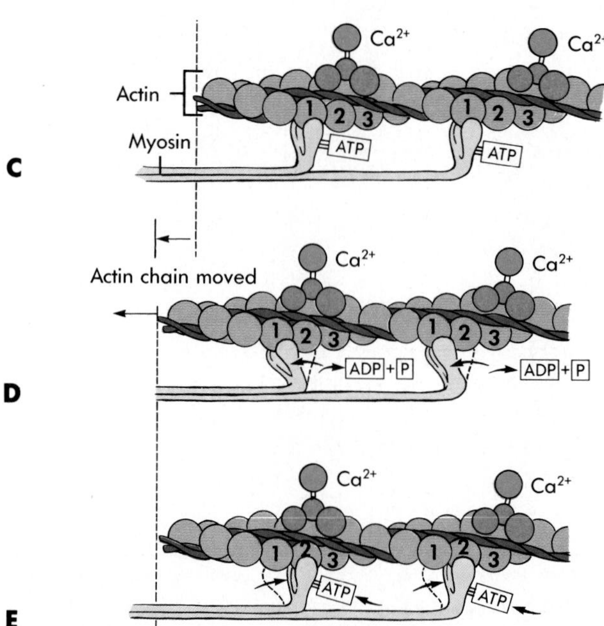

FIGURE 7-4 Sliding filament model of muscle contraction.

A Calcium ions *(Ca^{2+})* bind to the actin myofilament.

B This binding causes a change in the configuration of the actin myofilament, which exposes attachment sites, and the myosin myofilament attaches to the actin myofilament by forming cross bridges.

C A molecule of ATP attaches to the globular head of a myosin myofilament. Myosin then binds to exposed active sites on the actin molecules.

D ATP degrades to ADP *(ADP + P)* and the released energy causes the myosin head to bend, sliding the actin myofilament past the myosin myofilament.

E Another ATP attaches to the myosin head; myosin loses its attachment to actin and recoils to its original shape. The cycle is then repeated. (See also Table 7-1).

Muscle Contraction

Muscle contraction occurs as actin and myosin myofilaments slide past one another causing the sarcomeres to shorten. Shortening of the sarcomeres causes the muscle to shorten. The sliding of actin myofilaments past myosin myofilaments during contraction is called the **sliding filament mechanism** of muscle contraction. The H zones and I bands shorten during contraction, but the A bands do not change in length.

Muscle cells have a network of minute tubules, called **T tubules,** that function as relays between the cell membrane and a specialized endoplasmic reticulum, called the **sarcoplasmic reticulum,** within the cell (see Figure 7-1). Action potentials initiated

TABLE 7-1

Summary of Skeletal Muscle Contraction (see Figure 7-4)

1. An action potential travels along a neuronal axon to a neuromuscular junction, or synapse.
2. Acetylcholine is released from the synaptic vesicles of the neuron.
3. Acetylcholine diffuses across the synaptic cleft and binds to receptor molecules in the muscle cell membrane.
4. Sodium ions diffuse into the muscle cell, initiating an action potential in the muscle cell.
5. Calcium ions are released from the sarcoplasmic reticulum.
6. Calcium ions bind to actin myofilaments and expose myosin attachment sites.
7. The heads of myosin myofilaments attach to the actin myofilaments, forming cross bridges.
8. The heads of the myosin myofilaments bend, causing the actin myofilaments to slide over the surface of the myosin myofilaments.
9. The bending action of the myosin heads requires energy. ATP, bound to the myosin heads, is broken down, releasing energy to produce the movement.
10. Another ATP binds to the myosin head, causing it to release the actin myofilament.
11. Calcium ions are actively transported (requiring ATP) back into the sarcoplasmic reticulum
12. If the muscle continues to contract, step 1 through 11 are repeated as long as action potentials are transmitted along the neurons

in the cell membrane travel along these T tubules and cause the sarcoplasmic reticulum to release calcium ions. The calcium ions bind to actin myofilaments and cause attachment sites on the actin myofilaments to be exposed (Figure 7-4, *A*). The exposed attachment sites bind to the heads of the myosin myofilaments to form cross bridges between the actin and myosin myofilaments (Figure 7-4, *B*). The heads of the myosin myofilaments bend, forcing the actin myofilaments to slide over the surface of the myosin myofilaments (Figure 7-4, *C* and *D*).

The bending of the myosin heads requires energy. Energy is supplied to the muscles in the form of adenosine triphosphate (ATP), a high-energy molecule produced from the energy that is released during the digestion of food (see Chapter 17). ATP is attached to the heads of the myosin myofilaments. During the movement of actin and myosin, the ATP bound to the myosin is broken down and energy is released. Part of the energy is required for the movement of the cross bridges and part is released as heat. The heat released during muscle contraction increases body temperature. This is why a person becomes warm during exercise and shivers when it is cold. Another ATP molecule binds to myosin and the myosin head is released from the active site, returns to its original shape, and attaches to the next active site on the actin myofilament (Figure 7-4, *E*). The cycle of cross bridge formation, movement, and cross bridge release is then repeated (Table 7-1). When ATP is not available after a person dies, the cross bridges that have formed are not released and a condition called **rigor mortis** occurs.

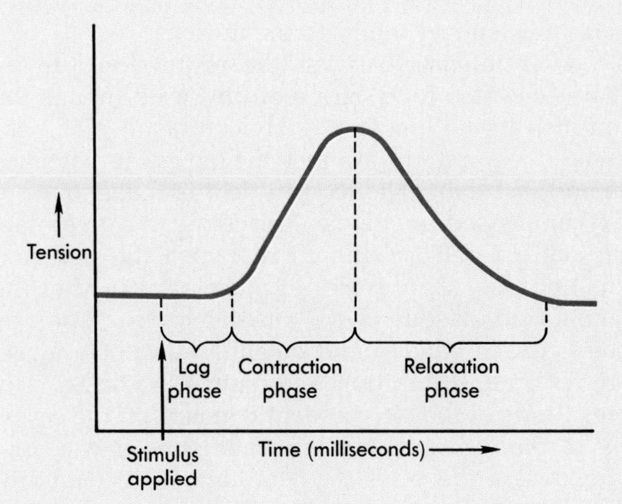

FIGURE 7-5 ■ Muscle twitch.
Muscle activity is recorded as the amount of tension per unit time.

Muscle relaxation occurs as the calcium ions are actively transported back into the sarcoplasmic reticulum (a process that requires ATP). As a consequence, the binding sites on the actin molecules are covered so that cross bridges cannot re-form.

1 Predict the consequences of having the following conditions develop in a muscle in response to a stimulus: (A) inadequate ATP is present in the muscle fiber before a stimulus is applied; (B) adequate ATP is present within the muscle fiber, but action potentials occur at a frequency so great that calcium ions are not transported back into the sarcoplasmic reticulum between individual action potentials. ?

Muscle twitch, tetany, and recruitment

A **muscle twitch** is a contraction of an entire muscle in response to a stimulus that causes an action potential in one or more muscle fibers (Figure 7-5). A muscle fiber will not respond to stimulus until that stimulus reaches a level called **threshold,** at which point the muscle fiber will contract maximally. This phenomenon is called the **all-or-none response.** The time between application of a stimulus to a motor neuron and the beginning of contraction is the **lag phase.** The time of contraction is the **contraction phase.** The time during which the muscle relaxes is the **relaxation phase.**

If successive stimuli are given to a muscle, successive twitches occur (Figure 7-6, *1*). As the stimulus frequency increases, muscle twitches may occur so frequently that the muscle does not have enough time to fully relax before another twitch is initiated (Figure 7-6, *2* and *3*). If the stimulus frequency increases even more, no relaxation occurs between muscle twitches (Figure 7-6, *4* and *5*). This condition, where the muscle remains contracted without relaxing, is called **tetany.**

A small contraction force in a muscle is produced when only a few of its motor units are stimulated, and each contracts in an all-or-none fashion. Simultaneous stimulation of more motor units produces a greater contraction force because more motor units are contracting. This increase in the number of motor units being activated is called **recruitment.** Maximum force is produced when all motor units of a muscle are stimulated. If all motor units in a muscle are stimulated simultaneously, a quick, jerking motion occurs. If motor units are recruited gradually so that some are stimulated and held in tetany while others are relaxing, a slower, much smoother, sustained contraction occurs.

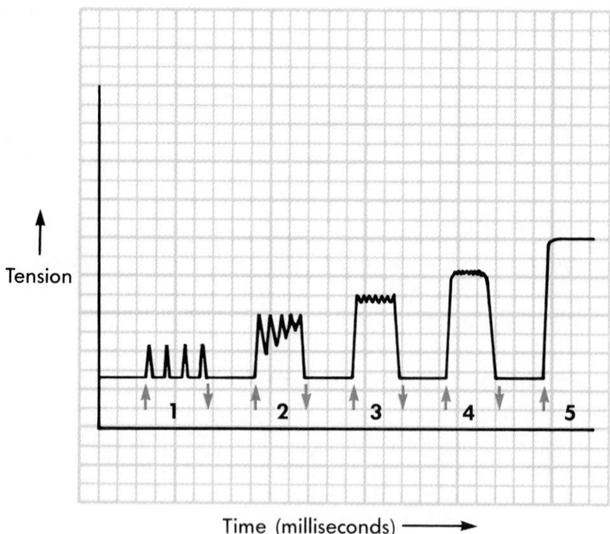

FIGURE 7-6 ■ Muscle twitch—multiple stimuli. Multiple stimuli result in multiple twitches *(1)*. More frequent stimuli do not allow complete relaxation between twitches so that the individual twitches seem to fuse together *(2* and *3)*. With even greater stimulus frequency *(4* and *5)*, no relaxation can be seen between twitches and the muscle remains continually contracted (tetany).

Energy requirements for muscle contraction

■ The ATP required to provide energy for muscle contraction is produced primarily in numerous mitochondria located within the muscle cells between the myofibrils. ATP is a very short-lived molecule and rapidly degenerates to the more stable ADP, plus phosphate. Therefore it is necessary for muscle cells to constantly produce ATP.

While muscle cells are at rest, they cannot stockpile ATP in preparation for periods of activity. The cells, however, can store another high-energy molecule, **creatine phosphate.** During periods of inactivity, as excess ATP is produced in the muscle cell, the energy contained in the ATP is used to synthesize creatine phosphate. During periods of activity, the energy stored in creatine phosphate can be accessed quickly and used to produce ATP, which can then be used in muscle contraction.

ATP is produced by anaerobic or aerobic cellular respiration. **Anaerobic respiration,** which occurs in the absence of oxygen, results in the breakdown of glucose to yield ATP and lactic acid. **Aerobic respiration** requires oxygen and breaks down glucose by glycolysis (see Chapter 17) to produce ATP, carbon dioxide, and water. Compared to anaerobic respiration, aerobic respiration is much more efficient. The metabolism of a glucose molecule by aerobic respiration theoretically can produce approximately 18 times as much ATP as produced by anaerobic respiration. In addition, aerobic respiration can utilize a greater variety of nutrient molecules to produce ATP than can anaerobic respiration. For example, aerobic respiration can use fatty acids to generate ATP. Although anaerobic respiration is less efficient than aerobic respiration, it occurs faster and is important when creatine phosphate supplies are depleted and when oxygen availability limits aerobic respiration. By utilizing many glucose molecules, anaerobic respiration can rapidly produce much ATP, but only for a short period of time.

Resting muscles or muscles undergoing long-term exercise (such as long-distance running) depend primarily on aerobic respiration for ATP synthesis. Although some glucose is used as an energy source, fatty acids are a more important energy source during sustained exercise as well as during resting conditions. On the other hand, during short periods of intense exercise (such as sprinting), the energy stored in creatine phosphate is used first to produce ATP. Once the creatine phosphate stores are depleted anaerobic respiration takes over rapidly metabolizing available glucose and providing enough ATP to support intense muscle contraction for a short period of time (approximately 15 to 20 seconds). Anaerobic metabolism is ultimately limited by depletion of glucose and a buildup of lactic acid within the muscle fiber. Lactic acid is an end product of anaerobic respiration. Lactic acid can diffuse out of the muscle cell into the blood, allowing anaerobic respiration to proceed somewhat longer than it would otherwise. Lactic acid can irritate muscle fibers, causing short-term pain. Muscle pain that lasts for a couple of days, however, indicates mechanical injury to the muscle.

After intense exercise, the respiration rate remains elevated for a period of time even though the muscles are no longer actively contracting. This increased respiration provides the oxygen to pay back the oxygen debt. The **oxygen debt** is the amount of oxygen needed in chemical reactions to convert lactic acid to glucose and to replenish the depleted creatine phosphate stores in muscle cells. After the lactic acid produced by anaerobic respiration is converted to glucose and creatine phosphate levels are restored, respiration rate returns to normal. The magnitude of the oxygen debt depends on the severity of the exercise, the length of time it was sustained, and the physical condition of the individual. The capacity of an individual in poor physical condition to perform metabolism is much lower than that of a well-trained athlete. With exercise and training, a person's ability to carry out both aerobic and anaerobic activities is enhanced.

2 After a 1-mile run with a sprint at the end, a runner continues to breathe heavily for a period of time. Indicate the type of metabolism functioning to produce energy during the run, during the sprint, and after the run. ?

Muscle **fatigue** results when ATP is used during muscle contraction faster than it can be produced in the muscle cells, and lactic acid builds up faster than it can be removed. As a consequence, ATP levels are too low for tetanic contractions to continue and muscle fibers cannot produce their maximum force of contraction. Under conditions of extreme muscular fatigue, muscles may become incapable of either contracting or relaxing. This condition, called **physiological contracture,** is caused by lack of ATP to bind to myosin myofilaments. The binding of ATP to the myosin heads is necessary for cross bridge release between the actin and myosin. When ATP levels are extremely low, the cross bridges between the actin and myosin myofilaments cannot be broken and the muscle cannot relax.

Types of muscle contractions

Muscle contractions are classified as either isometric or isotonic. In **isometric** (equal distance) **contractions,** the length of the muscle does not change, but the amount of tension increases during the contraction process. Isometric contractions are responsible for the constant length of the postural muscles of the body, such as the muscles of the back. On the other hand, in **isotonic** (equal tension) **contractions,** the amount of tension produced by the muscle is constant during contraction, but the length of the muscle changes. Movements of the arms or fingers are predominantly isotonic contractions. Most muscle contractions are a combination of isometric and isotonic contractions in which the muscles shorten some distance and the degree of tension increases.

Muscle tone

Muscle tone refers to the constant tension produced by muscles of the body for long periods of time. Muscle tone is responsible for keeping the back and legs straight, the head held in an upright position, and the abdomen from bulging. Muscle tone depends on a small percentage of all the muscle fibers contracting isometrically out of phase with each other at any point in time.

Slow and fast fibers

Muscles are sometimes classified as either fast-twitch or slow-twitch muscles. Fast-twitch muscle fibers contract quickly and fatigue quickly, whereas slow-twitch muscle fibers contract more slowly and are more resistant to fatigue. Fast-twitch muscles are well adapted to perform anaerobic metabolism,

Disorders of Muscle Tissue

Muscle disorders are caused by disruption of the normal nerve supply to a muscle, degeneration and replacement of muscle cells, injury, lack of use, or disease.

CRAMPS

Cramps are painful, spastic contractions of muscle that are usually the result of an irritation within a muscle. Local inflammation from buildup of lactic acid or connective tissue inflammation can cause contraction of muscle fibers surrounding the irritated region.

HYPERTROPHY AND ATROPHY

Exercise causes muscular hypertrophy, which is increased muscle size resulting from an increase in the number of myofibrils. Conversely, extreme disuse of muscle results in muscular atrophy (decreased muscle size). Severe atrophy involves the permanent loss of skeletal muscle fibers and the replacement of those fibers by connective tissue. Immobility resulting from damage to the nervous system may lead to muscular atrophy.

MUSCULAR DYSTROPHY

Muscular dystrophy refers to a group of muscle diseases that destroy skeletal muscle tissue. The diseases usually are inherited and are characterized by the progressive degeneration of muscle fibers leading to atrophy and eventual replacement by connective tissue.

MYASTHENIA GRAVIS

Myasthenia gravis, which usually begins in the face, is a muscular weakness not accompanied by atrophy. It is a chronic, progressive disease resulting from the destruction of acetylcholine receptors in the neuromuscular junction. Abnormal antibodies that bind to and destroy acetylcholine receptors can be identified in many people who have myasthenia gravis. Because of the decrease in the number of acetylcholine receptors, not as many action potentials are transmitted to the muscle cell, and the muscle is weaker as a result.

TENDINITIS

As the name implies, tendinitis is an inflammation of a tendon and/or its attachment point. It usually occurs in athletes who overtax the muscle to which the tendon is attached.

TABLE 7-2

Comparison of Muscle Types

FEATURE	SKELETAL MUSCLE	CARDIAC MUSCLE	SMOOTH MUSCLE
Location	Attached to bone	Heart	Walls of hollow organs, blood vessels, and glands
Cell shape	Long, cylindrical	Branched	Spindle-shaped
Nucleus	Multiple, peripheral	Single, central	Single, central
Special features		Intercalated disks	
Striations	Yes	Yes	No
Spontaneous contractions	No	Yes	Yes
Control	Voluntary	Involuntary	Involuntary
Function	Move the whole body	Heart contraction to propel blood through the body	Compression of ducts and tubes

whereas slow-twitch muscles are better suited for aerobic metabolism.

The white meat of a chicken's breast is comprised mainly of fast-twitch fibers. The muscles are adapted to contract rapidly for a short time but fatigue quickly. Chickens normally do not fly long distances. They spend most of their time walking. Ducks, on the other hand, fly for much longer periods of time and for greater distances. The red or dark meat of a chicken's leg or a duck's breast is composed of slow-twitch fibers. The darker appearance is due partly to the dark color of the enzyme system involved in aerobic metabolism, partly to a richer blood supply, and partly to the presence of **myoglobin,** which stores oxygen temporarily. Myoglobin may continue to release oxygen in a muscle even when a sustained contraction has interrupted the continuous flow of blood.

Humans exhibit no clear separation of slow-twitch and fast-twitch muscle fibers in individual muscles. Most muscles have both types of fibers, although the number of each type varies in a given muscle. The large postural muscles contain more slow-twitch fibers, whereas muscles of the upper limb contain more fast-twitch fibers. The distribution of the fibers in a given muscle is constant for each individual and is established before birth. People who are good sprinters have a greater percentage of fast-twitch muscle fibers in their lower limbs, whereas good long-distance runners have a higher percentage of slow-twitch fibers. Athletes who are able to perform a variety of anaerobic and aerobic exercises tend to have a more balanced mixture of fast-twitch and slow-twitch muscle fibers.

Neither fast-twitch nor slow-twitch muscle fibers can normally be converted to muscle fibers of the other type. Nevertheless, training can increase the capacity of both types of muscle fibers to perform more efficiently. Intense exercise resulting in anaerobic metabolism increases muscular strength and mass and has the greater effect on fast-twitch muscle fibers. Aerobic exercise increases the vascularity of muscles and causes enlargement of slow-twitch muscle fibers. Aerobic metabolism also can convert fast-twitch muscle fibers that fatigue readily to fast-twitch muscle fibers that resist fatigue. This conversion is accomplished by increasing the number of mitochondria in the muscle cells and increasing the blood supply to the fast-twitch muscle fibers.

The number of cells in a skeletal muscle remains relatively constant following birth. Therefore, enlargement of muscles after birth is the result of an increase in the size of the existing muscle fibers.

SMOOTH MUSCLE AND CARDIAC MUSCLE

Smooth muscle cells are small and spindle shaped, with one nucleus per cell (Table 7-2). They contain less actin and myosin than do skeletal muscle cells, and the myofilaments are not organized into sarcomeres. As a result, smooth muscle cells are not striated. Smooth muscle cells contract more slowly than skeletal muscle cells and do not develop an oxygen debt. The resting membrane potential of some smooth muscle cells fluctuates with slow depolarization and repolarization phases. These fluctuations may lead to spontaneously generated action potentials (a condition called autorhythmicity). Smooth muscle is under involuntary control, whereas skeletal muscle is under voluntary motor control. Some hormones can stimulate smooth muscle to contract.

Cardiac muscle shares some characteristics with both smooth and skeletal muscle (see Table 7-2). Cardiac muscle cells are long, striated, and branching, with only one nucleus per cell. The rate of cardiac muscle contraction is intermediate between smooth and skeletal muscle. Cardiac muscle contraction is autorhythmic. Cardiac muscle does not develop an oxygen debt and does not fatigue. As with smooth muscle, cardiac muscle is under involuntary control. Cardiac muscle cells are connected to each other by **intercalated disks,** which facilitate stimulus conduction between the cells. This cell-to-cell connection allows cardiac muscle cells to function as a unit.

MUSCLE ANATOMY
General Principles

Most muscles extend from one bone to another and cross at least one joint. Muscle contraction causes most body movements by pulling one of the bones toward the other across the movable joint. Some muscles are not attached to bone at both ends. For example, some facial muscles attach to the skin, which moves as the muscles contract.

The points of attachment of each muscle are its origin and insertion. At these attachment points, the muscle is connected to the bone by a **tendon.** The **origin,** also called the **head,** is the most stationary end of the muscle. The **insertion** is the end of the muscle attached to the bone undergoing the greatest movement. The portion of the muscle between the origin and the insertion is the **belly** (Figure 7-7). Some muscles have multiple origins, or heads, such as the biceps with two heads and the triceps with three heads.

Muscles tend to function together to accomplish specific movements. For example, the deltoid, biceps, and pectoralis major all help flex the arm. Further, many muscles are members of more than one group, depending on the type of movement being considered. For example, the anterior part of the deltoid muscle functions with the flexors of the arm, whereas the posterior part functions with the extensors of the arm. Muscles that work together to cause movement are **synergists** (sin′er-jistz), and a muscle working in opposition to another muscle is called an **antagonist.** The brachialis and biceps brachii are synergists in flexing the forearm; the triceps brachii is the antagonist and extends the forearm. Among a group of synergists, if one muscle plays the major role in accomplishing the desired movement, it is the **prime mover.**

Nomenclature

Most muscles have names that are descriptive. Some muscles are named according to their location (*pectoralis*, anterior chest), origin and insertion (*brachioradialis*, from the arm to the radius), number of heads (*biceps*, two heads), or function (flexor). Other muscles are named according to their size (*vastus*, large), shape (*deltoid*, triangular), or orientation of fasciculi (*rectus*, straight). Recognizing the descriptive nature of muscle names makes learning those names much easier. The most superficial muscles are shown in Figure 7-8.

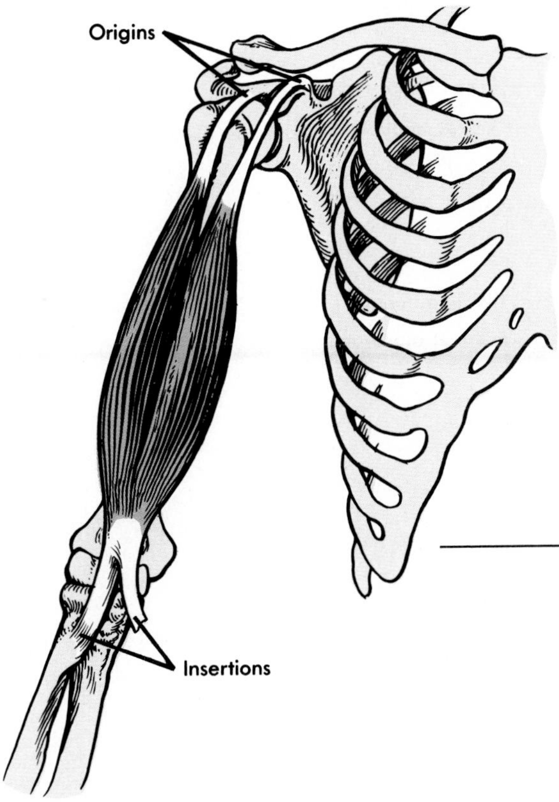

Origins

Insertions

FIGURE 7-7 ■ Muscle attachment.
Muscles are attached to bones by tendons. The biceps brachii has two heads that originate on the scapula. The biceps tendon inserts into the radial tuberosity.

FIGURE 7-8 ■ **The muscular system.**
A Anterior view.

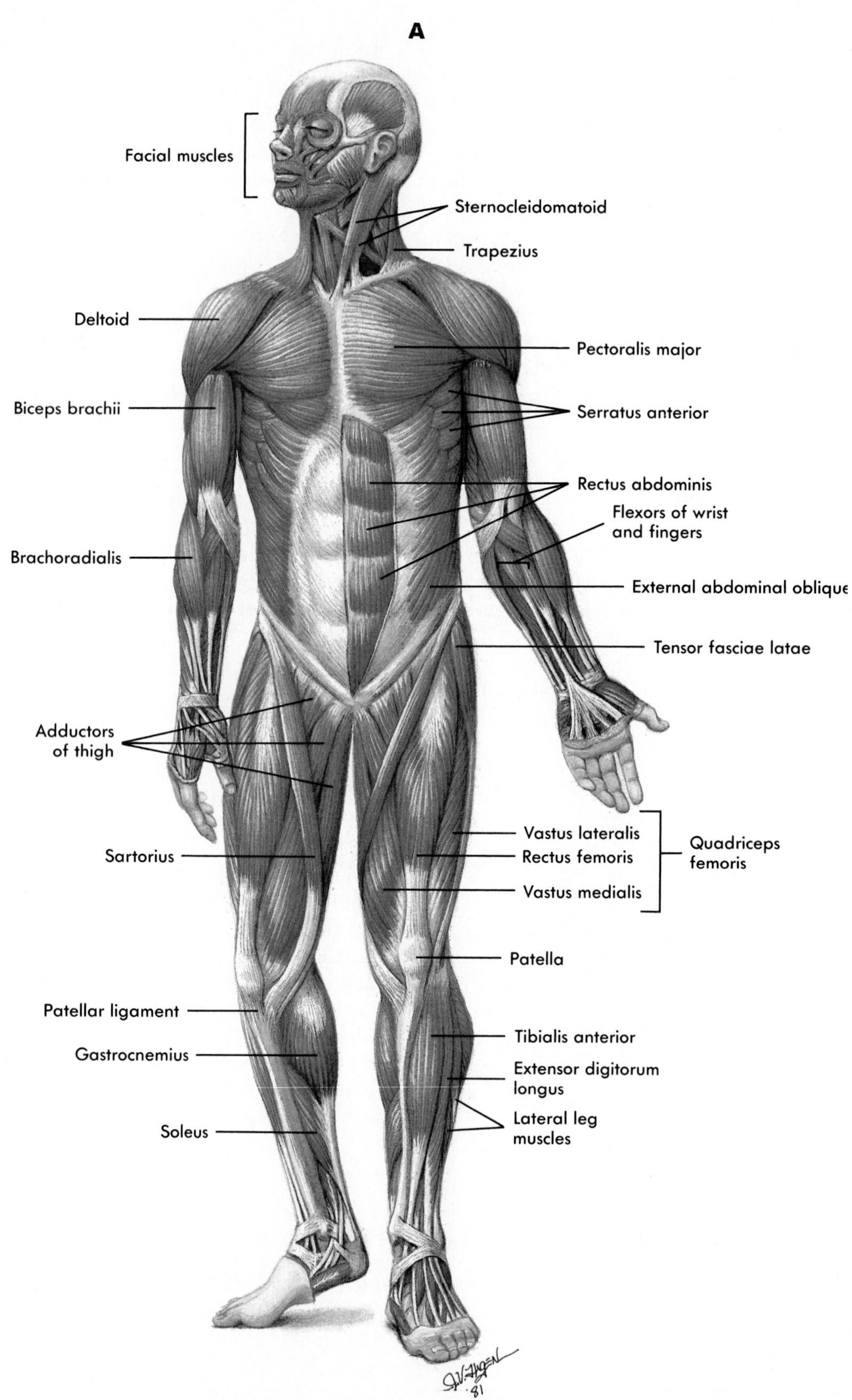

FIGURE 7-8, cont'd ■ The muscular system.
B Posterior view.

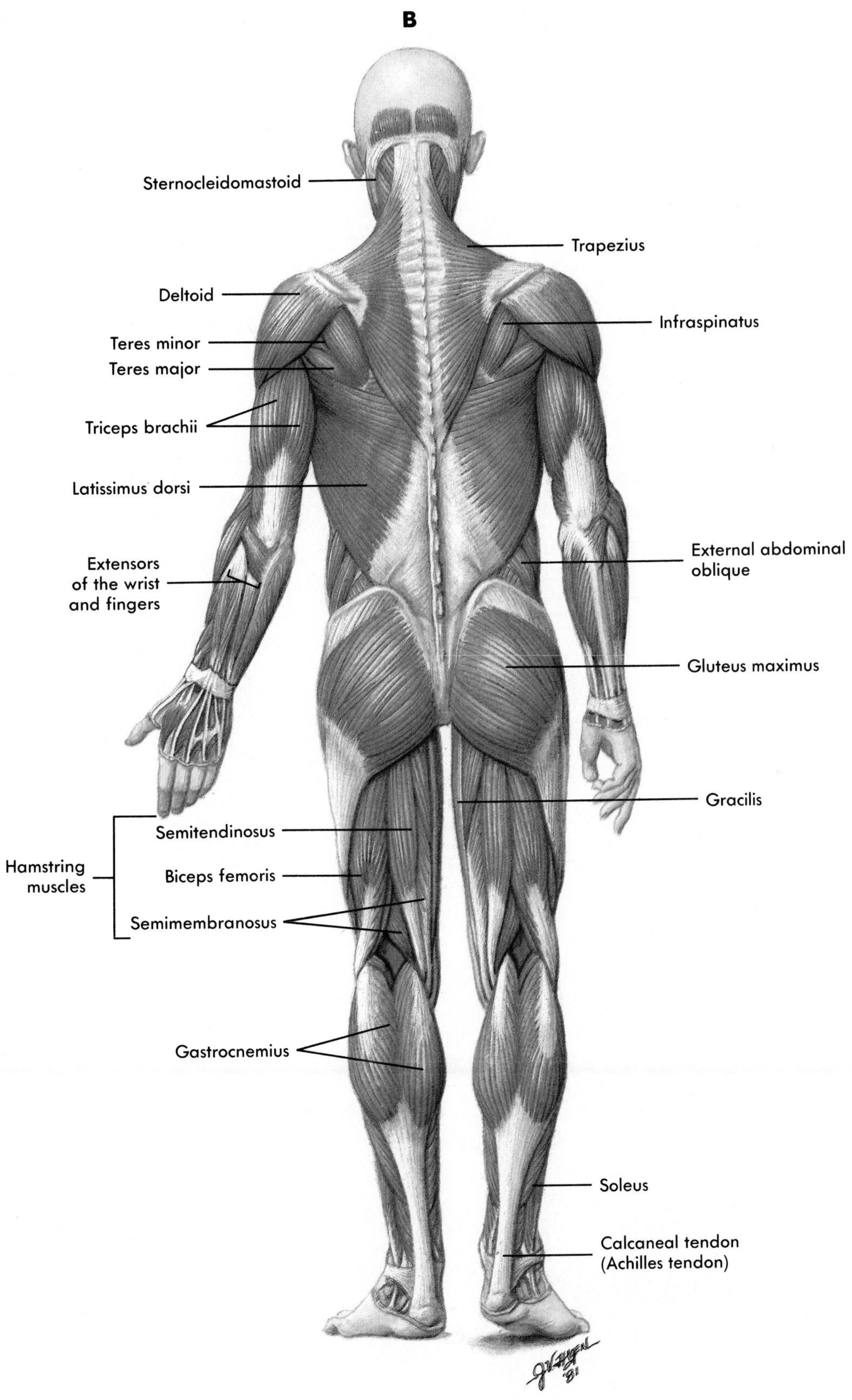

TABLE 7-3

Muscles of Facial Expression (see Figure 7-9)

MUSCLE	ORIGIN	INSERTION	ACTION
Occipitofrontalis	Occipital bone	Skin of eyebrow and nose	Moves scalp; elevates eyebrows
Orbicularis oculi	Maxilla and frontal bones	Encircles eye, and inserts near origin	Closes eye
Orbicularis oris	Maxilla and mandible	Skin around the lips	Closes lips
Buccinator	Maxilla and mandible	Corner of mouth	Flattens cheeks
Zygomaticus major and minor	Zygomatic bone	Corner of mouth	Elevate corner of mouth
Levator labii superioris	Maxilla	Upper lip	Elevates upper lip
Depressor anguli oris	Mandible	Lower lip near corner of mouth	Depresses corner of mouth

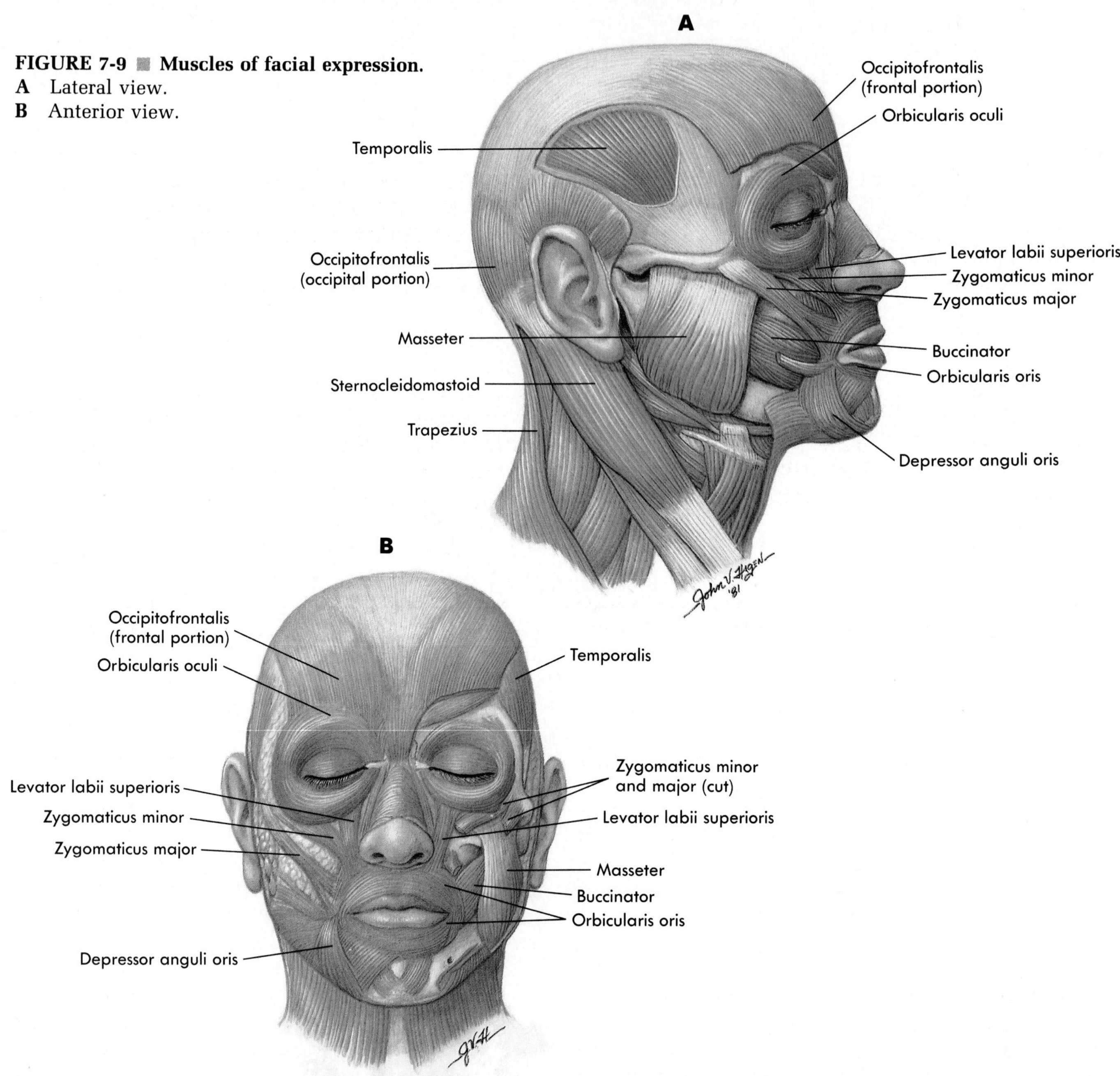

FIGURE 7-9 ■ **Muscles of facial expression.**
A Lateral view.
B Anterior view.

TABLE 7-4

Muscles of Mastication (see Figure 7-9, *A*)

MUSCLE	ORIGIN	INSERTION	ACTION
Temporalis	Temporal region on side of head	Mandible	Closes jaw
Masseter	Zygomatic arch	Mandible	Closes jaw
Pterygoids (Not shown in illustration)	Inferior side of skull	Mandible	One closes jaw; one opens jaw

Muscles of the Head and Neck

The muscles of the head and neck include those involved in facial expression and mastication (chewing), the tongue muscles, muscles of swallowing and those of the larynx, muscles moving the eye, and neck muscles.

Facial expression

Several muscles act on the skin around the eyes and eyebrows (Table 7-3 and Figure 7-9). The **occipitofrontalis** raises the eyebrows. The **orbicularis oculi** (ōr-bĭk′ū-lăr′ŭs ŏk′ū-lī) closes the eyelids and causes "crow's feet" wrinkles in the skin at the lateral corners of the eyes.

Several other muscles function in moving the lips and the skin surrounding the mouth (see Figure 7-9). The **orbicularis oris** (or′us) and **buccinator** (buk′sĭ-na′-tor), the kissing muscles, pucker the mouth. The buccinator also flattens the cheeks as in whistling or blowing a trumpet, and is, therefore, sometimes called the trumpeter's muscle. Smiling is accomplished primarily by the **zygomaticus** (zi′go-mat′ĭ-kus) muscles. Sneering is accomplished by the **levator labii** (la′be-i) **superioris,** and frowning or pouting largely by the **depressor anguli oris.**

3 Harry Wolf, a notorious flirt, on seeing Sally Gorgeous, raises his eyebrows, winks, whistles, and smiles. Name the facial muscles he used to carry out this communication. Sally, thoroughly displeased with this exhibition, frowns and sneers in disgust. What muscles did she use? ?

Mastication

The four pairs of muscles of chewing, or **mastication** (mas′tĭ-ka-shun), are some of the strongest muscles of the body (Table 7-4). The **temporalis** and **masseter** muscles (see Figure 7-9, *A*) can be easily seen and felt on the side of the head during mastication. The **pterygoid** (tĕr′ĭ-goyd) muscles, consisting of two pair, are deep to the mandible.

Tongue and swallowing muscles

The tongue is very important in mastication and speech. The tongue moves food around in the mouth; with the buccinator muscle, it holds the food in place while the teeth grind the food; and it pushes food up to the palate and back toward the pharynx to initiate swallowing. The tongue consists of a mass of **intrinsic muscles** (entirely within the tongue), which function to change its shape, and **extrinsic muscles** (outside of the tongue but attached to it), which move the tongue (Table 7-5 and Figure 7-10).

Swallowing involves a number of structures and their associated muscles, including the hyoid muscles, soft palate, pharynx, and larynx. The **hyoid muscles** are divided into a suprahyoid group (superior to the hyoid bone) and an infrahyoid group (inferior to the hyoid) (see Table 7-5 and Figure 7-10). When the suprahyoid muscles hold the hyoid bone in place from above, the infrahyoid muscles can elevate the larynx. To observe this effect, place your hand on your larynx (Adam's apple) and swallow.

The muscles of the **soft palate** close the posterior opening to the nasal cavity during swallowing, preventing food and liquid from entering the nasal cavity. Swallowing is accomplished by elevation of the pharynx and larynx and constriction of the pharynx. The **pharyngeal elevators** elevate the pharynx and the **pharyngeal constrictors** constrict from superior to inferior, forcing the food into the esophagus. Pharyngeal muscles also open the auditory tube, which connects the middle ear with the pharynx. Opening the auditory tube equalizes the pressure between the middle ear and the atmosphere. This is why it is sometimes helpful to chew gum or swallow when ascending or descending a mountain in a car or changing altitude in an airplane.

TABLE 7-5

Tongue and Swallowing Muscles (see Figure 7-10)

MUSCLE	ORIGIN	INSERTION	ACTION
Tongue muscles			
Intrinsic	Inside tongue	Inside tongue	Changes shape of tongue
Extrinsic	Bones around oral cavity	Onto tongue	Moves the tongue
Hyoid muscles			
Suprahyoid	Base of skull, mandible	Hyoid bone	Elevates or stabilizes hyoid
Infrahyoid	Sternum, larynx	Hyoid bone	Depresses or stabilizes hyoid
Soft palate (Not shown in illustration)	Skull or soft palate	Palate, tongue, or pharynx	Moves soft palate, tongue, or pharynx
Pharyngeal muscles			
Elevators (Not shown in illustration)	Soft palate and auditory tube	Pharynx	Elevate pharynx
Constrictors	Larynx and hyoid	Pharynx	Constrict pharynx

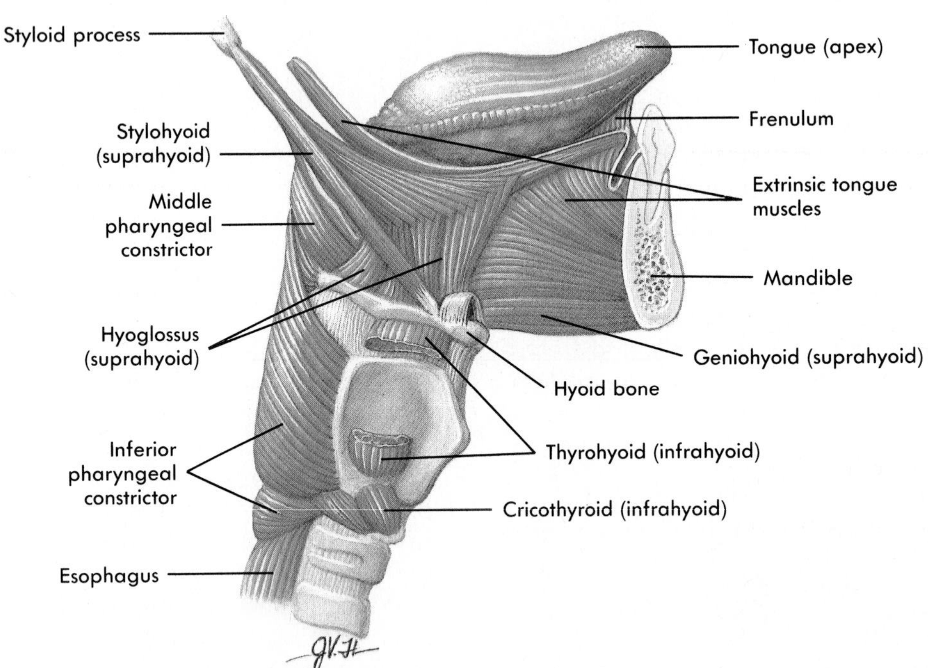

FIGURE 7-10 ■ Tongue and swallowing muscles.
Muscles of the tongue, hyoid, pharynx, and larynx as seen from the right.

TABLE 7-6

Neck Muscles (see Figures 7-8, *A* and 7-14, *A*)

MUSCLE	ORIGIN	INSERTION	ACTION
Deep neck muscles*			
Flexors	Anterior side of vertebrae	Base of skull	Flex head and neck
Extensors	Posterior side of vertebrae	Base of skull	Extend head and neck
Sternocleidomastoid	Sternum and clavicle	Mastoid process of skull	Rotates and extends head
Trapezius	Skull and upper vertebral column	Scapula	Extends head and neck

*Not shown in illustrations.

Neck muscles

The deep neck muscles (Table 7-6) include neck flexors, located along the anterior sides (bodies) of the vertebral bodies, and neck extensors, which are located posteriorly. Rotation and abduction of the head are accomplished by muscles of the lateral and posterior portions of the neck. The **sternocleidomastoid** (ster'no-kli'do-mas'toyd) muscle (see Figure 7-9, *A*), the prime mover of the lateral group, is easily seen on the anterior and lateral sides of the neck. Torticollis (meaning a twisted neck), or wry neck, may result from injury to one of the sternocleidomastoid muscles. It is sometimes caused by damage to a baby's neck muscles during a difficult birth and usually can be corrected by exercising the muscle.

4 Shortening of the right sternocleidomastoid muscle would rotate the head in which direction? ?

Trunk Muscles

Trunk muscles include those moving the vertebral column, those of the thorax and abdominal wall, and those of the pelvic floor.

Muscles moving the vertebral column

In humans, the back muscles are very strong to maintain erect posture. The **erector spinae** group of muscles on each side of the back are the muscles primarily responsible for keeping the back straight and the body erect (Table 7-7). **Deep back muscles,** located between the spinous and transverse processes of adjacent vertebrae, are responsible for several movements of the vertebral column such as flexion, extension, abduction, and rotation.

Thoracic muscles

The muscles of the thorax (Table 7-8 and Figure 7-11) are involved almost entirely in the process of breathing. The **external intercostals** (in'ter-kos'talz) elevate the ribs during inspiration. The **internal intercostals** contract during forced expiration.

The major movement produced in the thorax during quiet breathing, however, is accomplished by the dome-shaped **diaphragm** (di'ă-fram). When it contracts, the dome is flattened, causing the volume of the thoracic cavity to increase and resulting in inspiration.

Abdominal wall muscles

The muscles of the anterior abdominal wall (Table 7-9 and Figure 7-12) flex and rotate the vertebral column, compress the abdominal cavity, and hold

TABLE 7-7

Muscles Moving the Vertebral Column

MUSCLE	ORIGIN	INSERTION	ACTION
Erector spinae	Vertebrae, pelvis	Superior vertebrae and ribs	Extends vertebral column
Deep back muscles	Vertebrae	Vertebrae	Flex or extend vertebral column

TABLE 7-8

Thoracic Muscles (see Figure 7-11)

MUSCLE	ORIGIN	INSERTION	ACTION
External intercostals	Ribs	Next rib below origin	Inspiration
Internal intercostals	Ribs	Next rib above origin	Forced expiration
Diaphragm	Interior of body wall	Central tendon of diaphragm	Inspiration

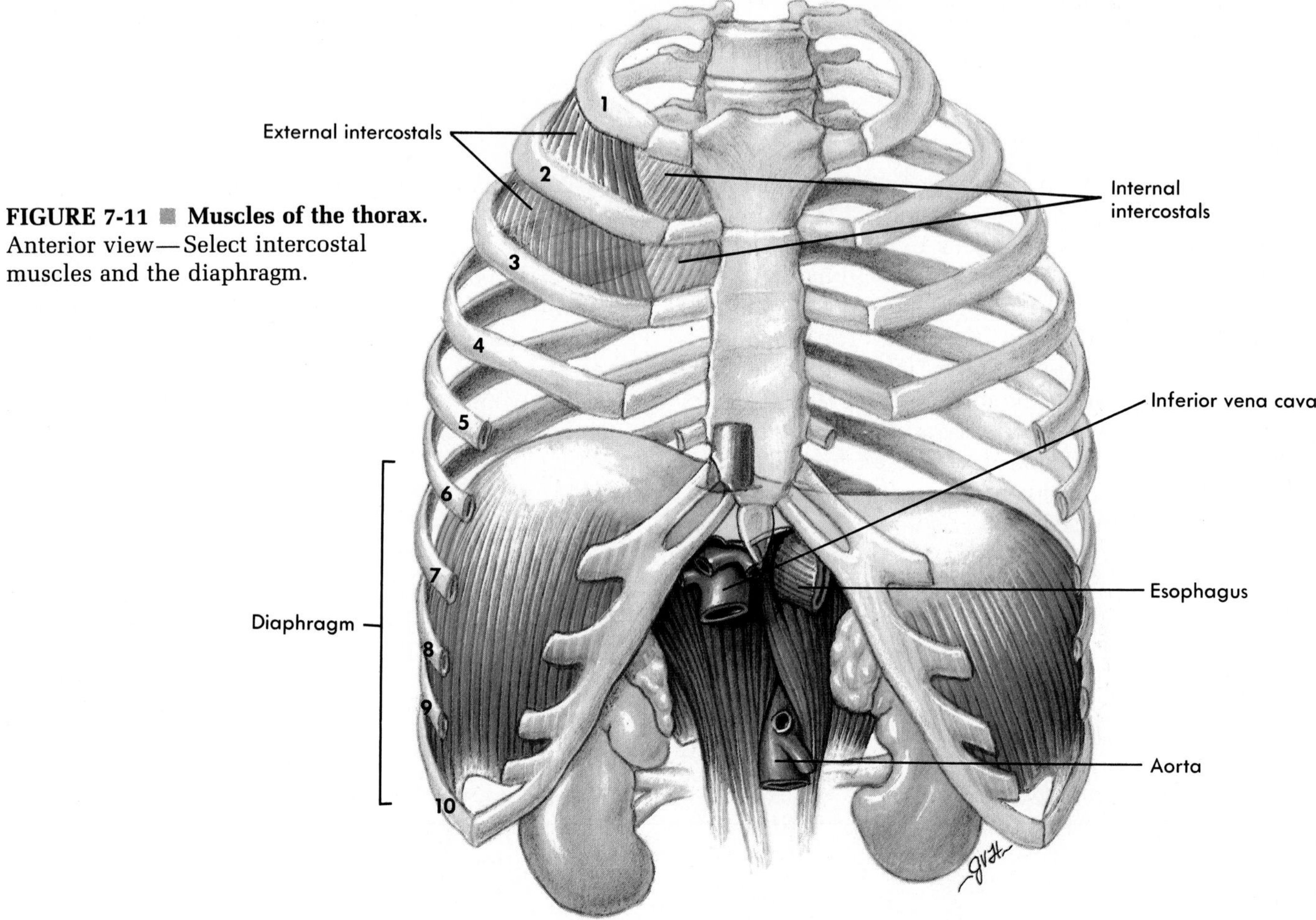

FIGURE 7-11 ■ **Muscles of the thorax.** Anterior view—Select intercostal muscles and the diaphragm.

TABLE 7-9

Muscles of the Abdominal Wall (see Figure 7-12)

MUSCLE	ORIGIN	INSERTION	ACTION
Rectus abdominis	Pubis	Rib cage and sternum	Flexes vertebral column
External abdominal oblique	Rib cage	Iliac crest and fascia of rectus abdominis	Compresses abdomen; flexes and rotates vertebral column
Internal abdominal oblique	Pelvis	Lower ribs and fascia of rectus abdominis	Compresses abdomen; flexes and rotates vertebral column
Transversus abdominis	Vertebrae, pelvis, ribs	Lower ribs, fascia of rectus abdominis, and pubis	Compresses abdomen

FIGURE 7-12 ■ Muscles of the anterior abdominal wall.
A Superficial view.
B Deeper view, with the external oblique removed on the viewer's left, and the rectus abdominis, external oblique, and internal oblique removed on the viewer's right.

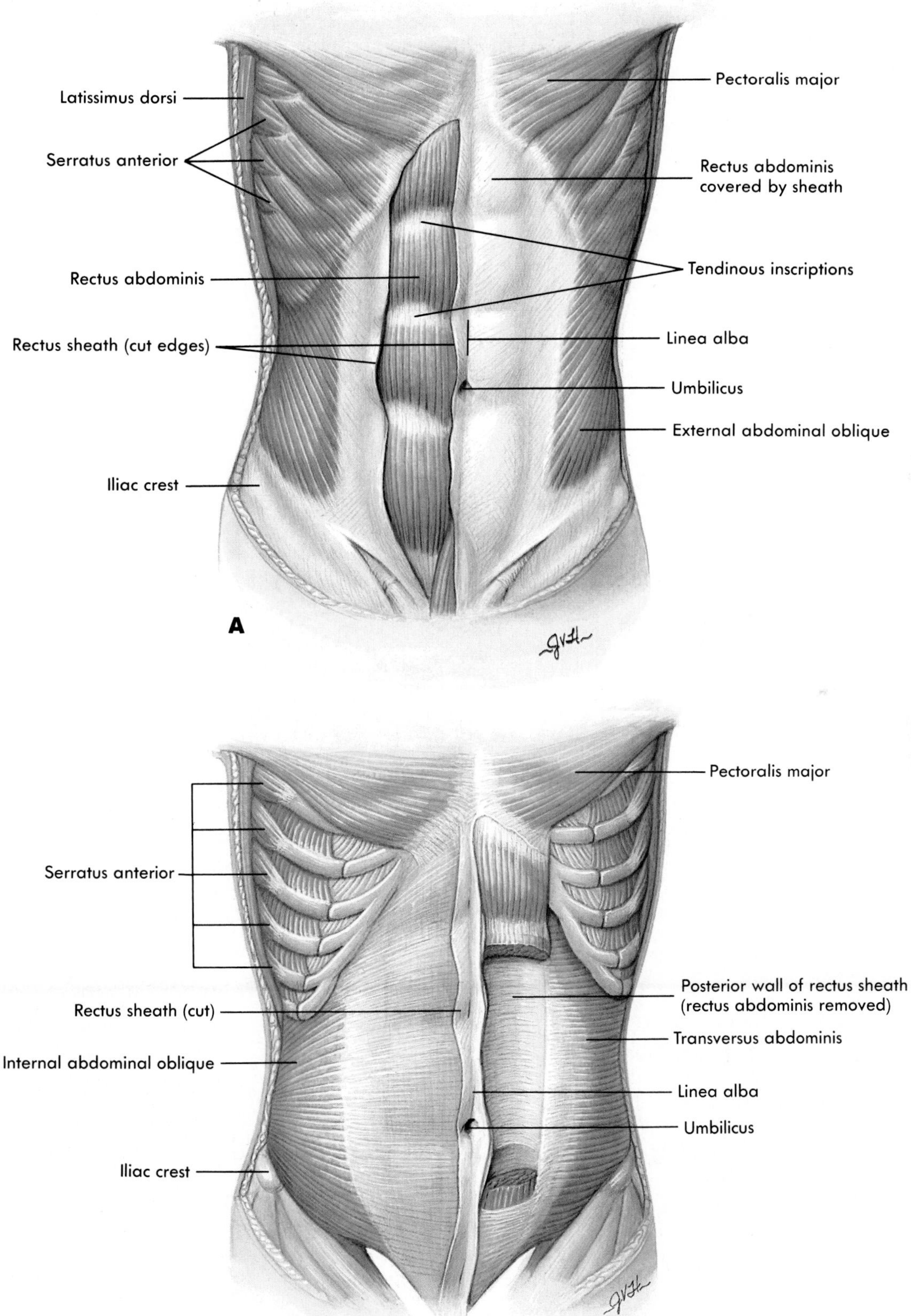

in and protect the abdominal viscera. In a relatively muscular person with little fat, a vertical linear indentation is visible, extending from the sternum, through the navel to the pubis. This tendinous area of the abdominal wall, called the **linea alba** (lin'e-ah al'bah; white line), consists of white connective tissue rather than muscle. On each side of the linea alba is the **rectus abdominis** muscle. **Tendinous inscriptions** (in-scrip'shunz) cross the rectus abdominis at three or more locations, causing the abdominal wall of a well-muscled person to appear segmented. Lateral to the rectus abdominis are three layers of muscle. From superficial to deep, these muscles are the **external abdominal oblique, internal abdominal oblique,** and **transversus abdominis** muscles. The fasciculi of these three muscle layers are oriented in opposite directions to one another and when these muscles contract, they compress the abdominal contents.

Pelvic floor muscles

The pelvis is a ring of bone with an inferior opening that is closed by a muscular floor through which the anus and the openings of the urinary tract and reproductive tract penetrate. Most of the pelvic floor, also referred to as the **pelvic diaphragm,** is formed by the **levator ani** (a'ne) muscle. The area inferior to the pelvic floor is the **perineum** (per'ĭ-ne'um), which is somewhat diamond shaped. The anterior half of the diamond is the urogenital triangle, and the posterior half is the anal triangle. The urogenital triangle contains a number of muscles associated with the male or female reproductive structures (Table 7-10 and Figure 7-13). Several of these muscles help regulate urination and defecation.

TABLE 7-10

Muscles of the Pelvic Floor (see Figure 7-13)

MUSCLE	ORIGIN	INSERTION	ACTION
Levator ani	Pubis and ischium	Sacrum and coccyx	Elevates anus
Ischiocavernosus	Ischium	Clitoris or penis	Compresses base of clitoris or penis
Bulbospongiosus			
Male	Bulb of penis	Central tendon of perineum	Constricts urethra; aids in erection of penis
Female	Central tendon of perineum	Base of clitoris	Aids in erection of clitoris

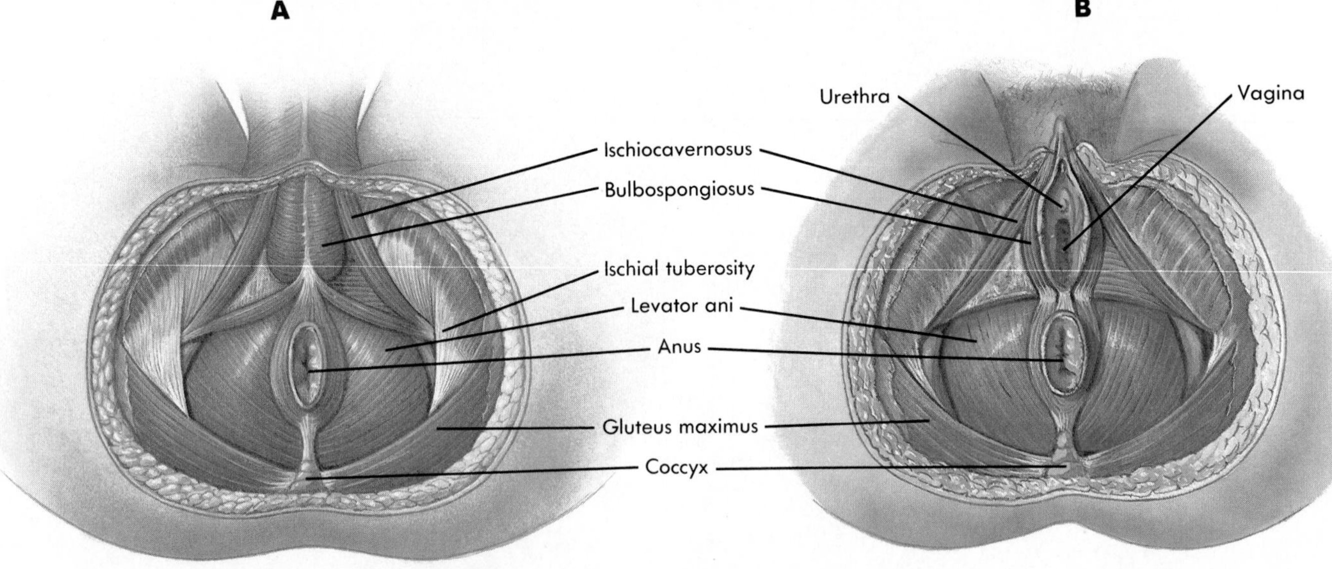

FIGURE 7-13 Muscles of the pelvic floor and reproductive structures.
A Male, inferior view.
B Female, inferior view.

Upper Limb Muscles

The muscles of the upper limb include those that attach the limb and girdle to the body and those that are in the arm, forearm, and hand.

Scapular movements

The connection of the upper limb to the body is accomplished primarily by muscles. The muscles that attach the scapula to the thorax and move the scapula include the **trapezius, levator scapulae** (skap'u-le), **rhomboids** (rom-boy'dz), **serratus** (ser-a'tus) **anterior,** and **pectoralis minor** (Table 7-11 and Figure 7-14). These muscles act as fixators to hold the scapula firmly in position when the muscles of the arm contract. The scapular muscles also move the scapula into different positions, thereby increasing the range of movement of the limb. The trapezius (Figure 7-14, *A*) forms the upper line from each shoulder to the neck, and the origin of the serratus anterior from the first eight or nine ribs can be seen along the lateral thorax.

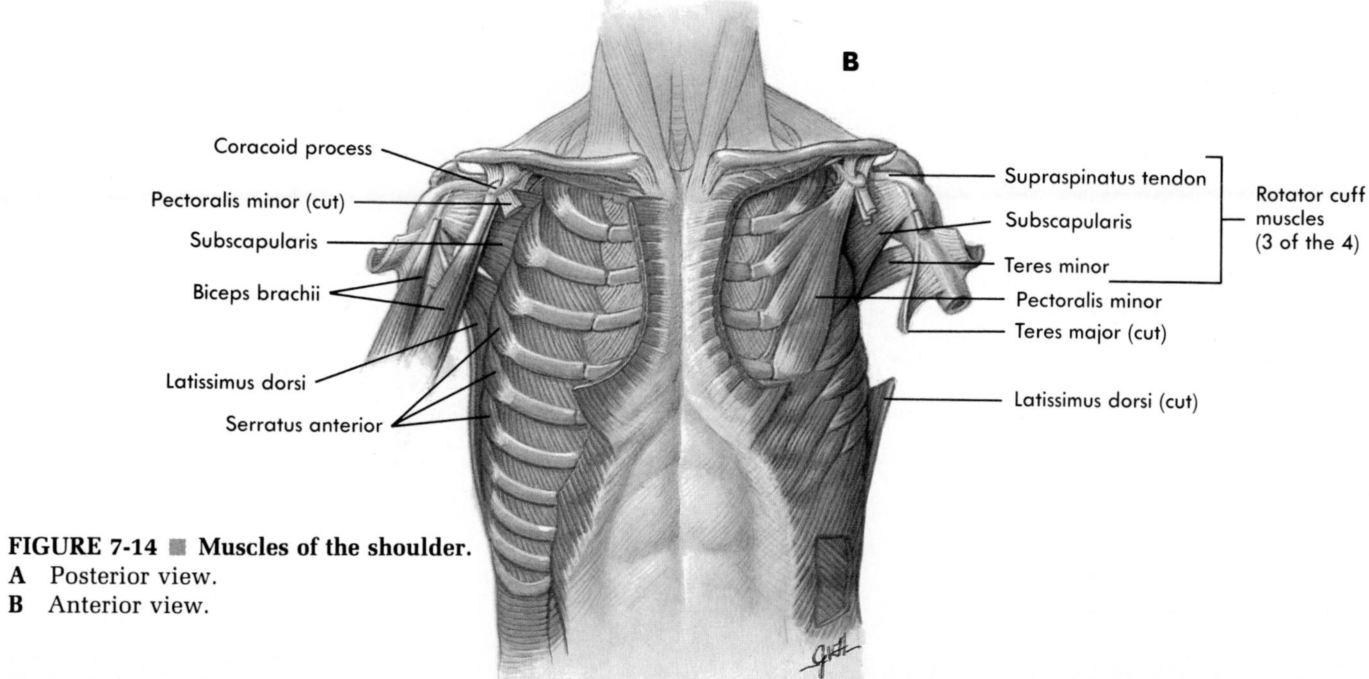

FIGURE 7-14 ■ Muscles of the shoulder.
A Posterior view.
B Anterior view.

TABLE 7-11

Scapular Movements (see Figures 7-12 and 7-14)

MUSCLE	ORIGIN	INSERTION	ACTION
Trapezius	Skull and vertebrae	Scapular spine	Rotates scapula
Levator scapulae	Vertebrae	Superior angle scapula	Pulls scapula superiorly
Rhomboids	Vertebrae	Medial border of scapula	Draws scapula toward vertebral column
Serratus anterior	Ribs	Medial border of scapula	Pulls scapula anteriorly
Pectoralis minor	Ribs	Coracoid process of scapula	Pulls scapula inferiorly

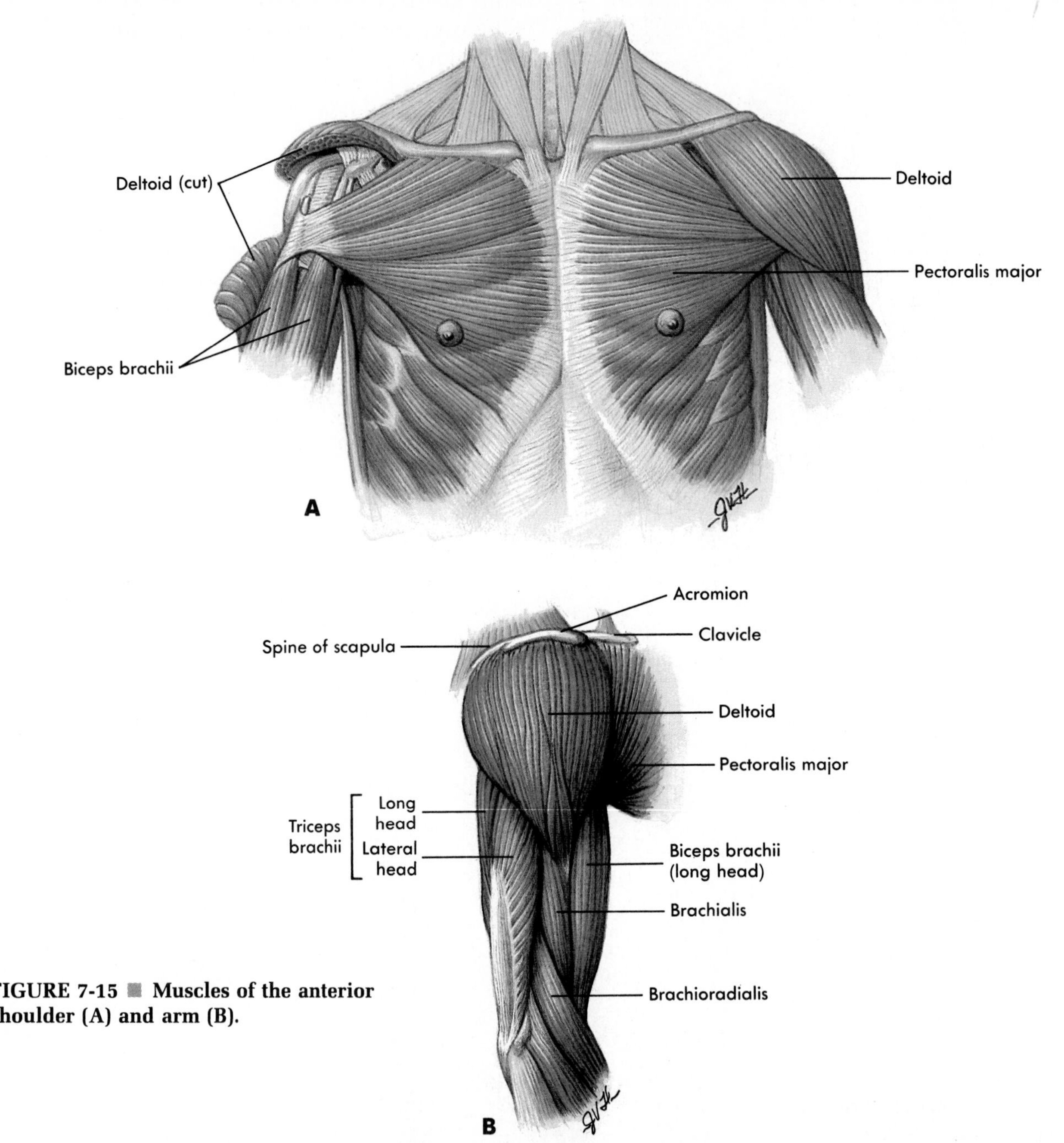

FIGURE 7-15 ■ Muscles of the anterior shoulder (A) and arm (B).

TABLE 7-12

Arm Movements (see Figures 7-8, *B*; 7-14, *B*; and 7-15)

MUSCLE	ORIGIN	INSERTION	ACTION
Pectoralis major	Ribs and clavicle	Shaft of humerus	Adducts and flexes arm
Teres major	Scapula	Humerus	Extends, adducts, and rotates arm
Latissimus dorsi	Vertebrae	Shaft of humerus	Extends, medially rotates, and adducts arm
Rotator cuff muscles			
Infraspinatus	Scapula	Greater tubercle of humerus	Extends and rotates arm
Supraspinatus	Scapula	Greater tubercle of humerus	Abducts arm
Subscapularis	Scapula	Lesser tubercle of humerus	Extends and rotates arm
Teres minor	Scapula	Greater tubercle of humerus	Adducts and rotates arm
Deltoid	Scapula and clavicle	Humerus	Flexes, extends, and abducts arm

Arm movements

The arm is attached to the thorax by the **pectoralis major** and **latissimus dorsi** (lah-tis'ĭ-mus dor'se) muscles (Table 7-12 and Figure 7-15, and see Figure 7-8, *B*). The pectoralis major adducts and flexes the arm. The latissimus dorsi medially rotates, adducts, and powerfully extends the arm. Because a swimmer uses these three motions during the power stroke of the crawl, the latissimus dorsi is often called the swimmer's muscle. Another group of four muscles, called the **rotator cuff muscles,** attaches the humerus to the scapula and forms a cuff or cap over the proximal humerus (see Table 7-12 and Figure 7-14, *B*). A rotator cuff injury involves damage to one or more of these muscles or their tendons. The **deltoid** muscle attaches the humerus to the scapula and clavicle, and is the major abductor of the upper limb. The pectoralis major forms the upper chest, and the deltoid forms the rounded mass of shoulder (see Figure 7-17). The deltoid is a common site for administering injections.

Forearm movements

The arm can be divided into anterior and posterior compartments. The **triceps brachii,** the primary extensor of the forearm, occupies the posterior compartment (Table 7-13 and Figure 7-15, *B*). The anterior compartment is occupied mostly by the **biceps brachii** and the **brachialis,** the primary flexors of the forearm. The **brachioradialis,** which is actually a posterior forearm muscle, helps flex and supinate the forearm.

Supination and pronation

Supination of the forearm, turning the forearm so that the palm is up, is accomplished by the **supinator** and the **biceps brachii** (Tables 7-13 and 7-14, and see Figure 7-16), which tends to supinate the forearm while flexing it. Pronation, turning the forearm so that the palm is down, is a function of two **pronator** muscles.

TABLE 7-13

Arm Muscles (see Figure 7-15)

MUSCLES	ORIGIN	INSERTION	ACTION
Triceps brachii	Humerus and scapula	Olecranon of ulna	Extends forearm
Biceps brachii	Scapula: superior to glenoid fossa, and coracoid process	Radial tuberosity	Flexes and supinates forearm
Brachialis	Humerus	Coronoid process of ulna	Flexes forearm

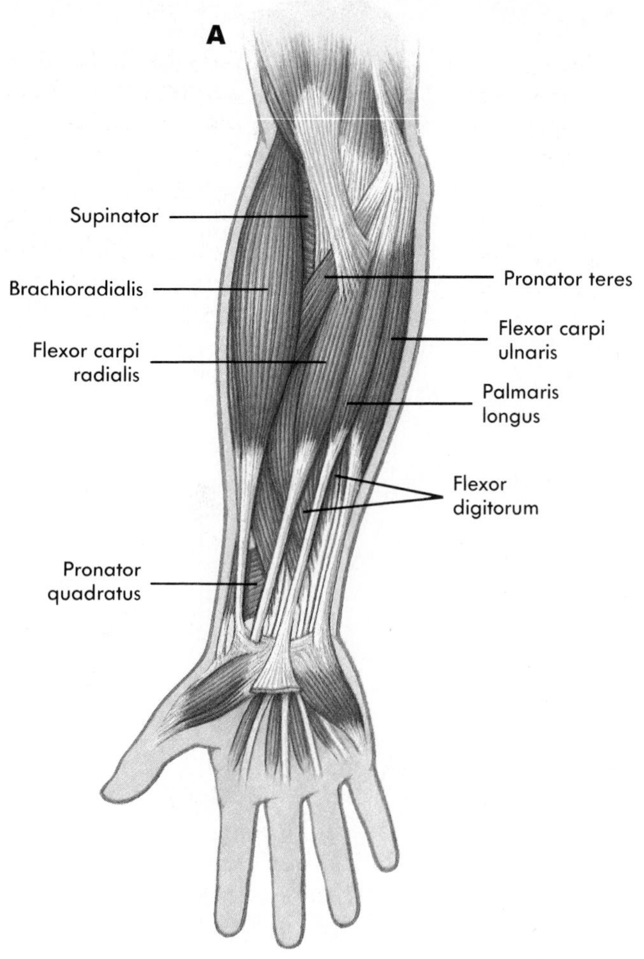

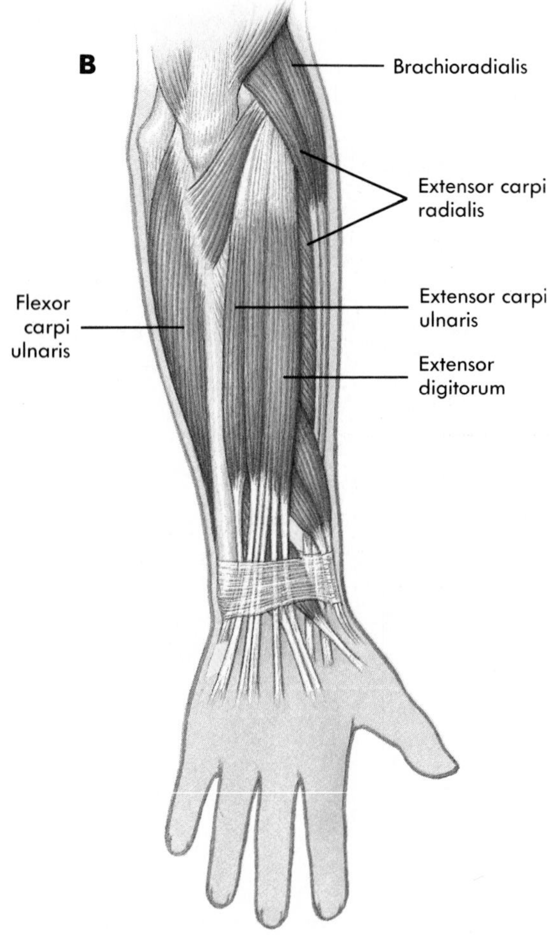

FIGURE 7-16 ■ Muscles of the forearm.
A Anterior view.
B Posterior view.

Wrist and finger movements

■ The 20 muscles of the forearm can also be divided into anterior and posterior groups. Only a few of these muscles, the most superficial, are listed in Table 7-14 and are illustrated in Figure 7-16. Most of the anterior forearm muscles are responsible for flexion of the wrist and fingers, whereas most of the posterior forearm muscles cause extension. A strong band of fibrous connective tissue, the **retinaculum** (ret′i-nak′u-lum; bracelet), covers the flexor and extensor tendons and holds them in place around the wrist so that they do not "bowstring" during muscle contraction.

The **flexor carpi** (kar′pi; wrist) muscles flex the wrist, and **extensor carpi** muscles extend the wrist. The wrist flexors and extensors are visible on the anterior and posterior surfaces of the forearm. The tendon of the carpal flexor on the radial side of the forearm is used as a landmark for locating the radial pulse. Flexion of the fingers is the function of the **flexor digitorum** (flexor of the digits, or fingers); extension is accomplished by the **extensor digitorum.** The tendons of this muscle are very visible on the dorsum of the hand (Figure 7-17). The thumb has

TABLE 7-14

Forearm Muscles (see Figure 7-16)

MUSCLE	ORIGIN	INSERTION	ACTION
ANTERIOR			
Pronators (two muscles)	Ulna	Radius	Pronate forearm
Flexor carpi ulnaris	Medial epicondyle of humerus	Carpal	Flexes and abducts (ulnar deviates) wrist
Flexor carpi radialis	Medial epicondyle of humerus	Metacarpals	Flexes and abducts (radial deviates) wrist
Flexors digitorum (two muscles)	Medial epicondyle of humerus, ulna, radius	Phalanges	Flex fingers
POSTERIOR			
Brachioradialis	Lateral epicondyle of humerus	Distal radius	Flexes and pronates forearm
Supinator	Ulna	Radius	Supinates forearm
Extensor carpi ulnaris	Lateral epicondyle of humerus	Metacarpal	Extends and abducts wrist
Extensors carpi radialis (two muscles)	Lateral epicondyle of humerus	Metacarpals	Extend and abduct wrist
Extensor digitorum	Lateral epicondyle of humerus	Phalanges	Extends fingers

its own set of flexors, extensors, adductors, and abductors. The little finger also has some of its own muscles.

> Forceful extension of the wrist repeated over a period of time such as occurs in a tennis backhand may result in inflammation and pain where the muscles attach to the lateral humeral epicondyle (the common point of origin for most extensors of the wrist and hand). This condition is sometimes referred to as "tennis elbow."

There are 19 muscles, called **intrinsic hand muscles,** located within the hand. **Interossi** muscles, located between the metacarpals, are responsible for abduction and adduction of the fingers. Other intrinsic hand muscles are responsible for many other movements of the thumb and fingers. These muscles account for the masses at the base of the thumb and little finger and the fleshy region between the metacarpals of the thumb and index finger.

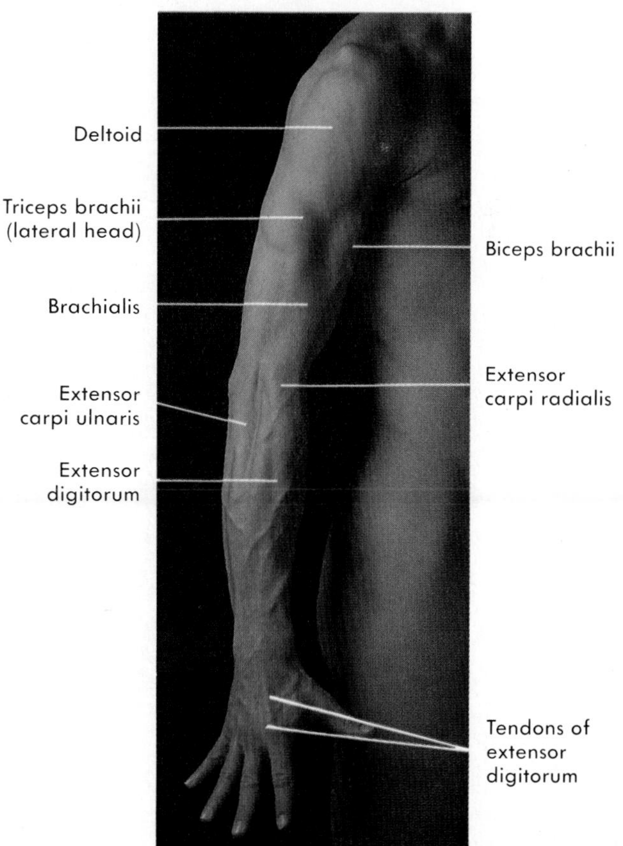

FIGURE 7-17 ■ Surface anatomy of the right upper limb.

TABLE 7-15

Muscles Moving the Thigh (see Figure 7-18)

MUSCLE	ORIGIN	INSERTION	ACTION
Iliopsoas	Iliac fossa and vertebrae	Lesser trochanter of femur	Flexes and medially rotates thigh
Tensor fascia latae	Anterior superior iliac spine	Lateral condyle of tibia	Abducts thigh
Gluteus maximus	Ilium, sacrum, and coccyx	Lateral side of femur	Extends, abducts, and laterally rotates thigh
Gluteus medius	Ilium	Greater trochanter of femur	Abducts and medially rotates thigh

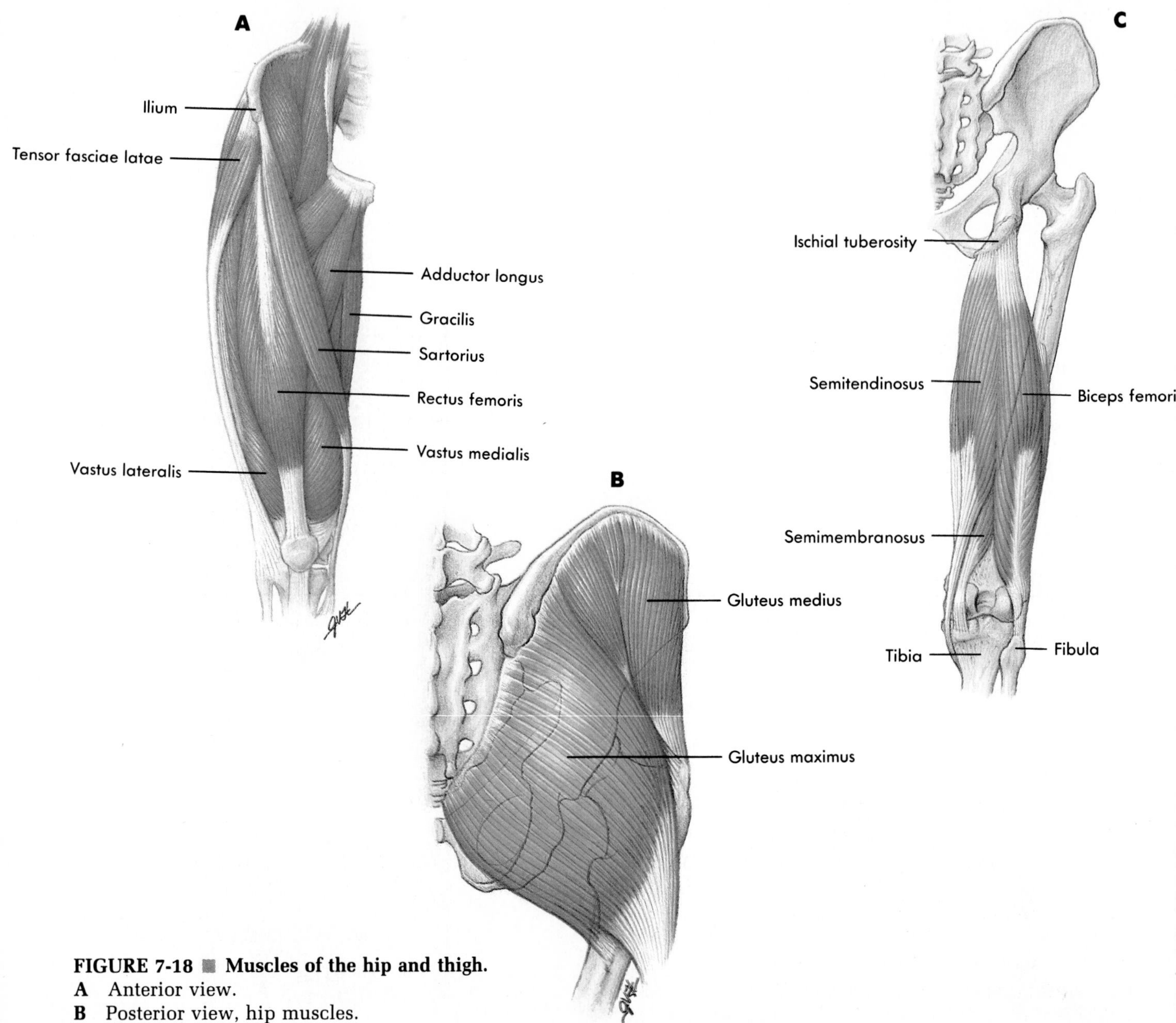

FIGURE 7-18 ■ Muscles of the hip and thigh.
A Anterior view.
B Posterior view, hip muscles.
C Posterior view, thigh muscles.

Lower Limb Muscles

The muscles of the lower limb include those located in the hip, thigh, leg, and foot.

Thigh movements

Several hip muscles originate on the coxa and insert onto the femur (Table 7-15 and Figure 7-18). The anterior muscle, the **iliopsoas,** flexes the thigh. The posterior and lateral hip muscles consist of the **gluteal** muscles and the **tensor fascia latae.** The tensor fascia latae is so named because it tenses the lateral fascia, a thick band of connective tissue on the lateral side of the thigh, and by so doing, it can abduct the thigh. The **gluteus maximus** contributes most of the mass that can be seen as the buttocks, and the **gluteus medius,** a common site for injections, creates a smaller mass just superior and lateral to the maximus (see Figure 7-20). The gluteus maximus functions maximally to extend the thigh when the thigh is flexed at a 45-degree angle.

5 Explain why in the sprinter's stance and the bicycle racing posture the thigh is flexed at a 45-degree angle.

In addition to the hip muscles, some of the muscles located in the thigh also attach to the coxa and can cause movement of the thigh. There are three groups of thigh muscles: the anterior thigh muscles, which flex the thigh; the posterior thigh muscles, which extend the thigh; and the medial thigh muscles, which adduct the thigh.

Leg movements

The anterior thigh muscles are the **quadriceps femoris** (four muscles) and the **sartorius** (Table 7-16 and see Figure 7-18, *A*). The quadriceps femoris muscles are the primary extensors of the leg. They have a common insertion, the patellar tendon, on and around the patella. The patellar ligament is an extension of the patellar tendon onto the tibial tuberosity. The patellar ligament is the point that is tapped with a rubber hammer when testing the knee-jerk reflex in a physical examination. The sartorius is called the "tailor's muscle" because it flexes the thigh and leg, and rotates the thigh laterally for sitting cross-legged, as tailors used to sit while sewing.

The posterior thigh muscles are called **hamstring** muscles, and they are responsible for flexing the leg (see Table 7-16 and Figures 7-8, *B* and 7-18,

TABLE 7-16

Leg Movements (see Figure 7-18)

MUSCLE	ORIGIN	INSERTION	ACTION
Quadriceps femoris			
Rectus femoris	Ilium	Tibial tuberosity via patellar tendon	Extends leg; flexes thigh
Vastus lateralis	Femur	Tibial tuberosity via patellar tendon	Extends leg
Vastus medialis	Femur	Tibial tuberosity via patellar tendon	Extends leg
Vastus intermedius (Not shown in illustration)	Femur	Tibial tuberosity via patellar tendon	Extends leg
Sartorius	Anterior superior iliac spine	Medial side of tibial tuberosity	Flexes thigh; flexes and medially rotates leg
Hamstring muscles			
Biceps femoris	Ischium and femur	Head of fibula	Flexes leg; extends thigh
Semimembranosus	Ischium	Medial condyle of tibia	Flexes leg; extends thigh
Semitendinosus	Ischium	Tibia	Flexes leg; extends thigh
Adductor muscles			
Adductor longus	Pubis	Femur	Adducts thigh
Gracilis	Pubis	Tibia	Adducts thigh

B). Their tendons are easily felt and seen on the medial and lateral posterior aspect of a slightly bent knee (see Figure 7-20). The hamstrings were so named because these tendons in hogs or pigs could be used to suspend hams during curing. Some animals such as wolves often bring down their prey by biting through the hamstrings. Therefore, "to hamstring" someone is to render them helpless. A "pulled hamstring" consists of tearing one or more of these muscles or their tendons, usually where the tendons attach to the coxa.

The medial thigh muscles, the **adductor** muscles, are, as the name implies, involved primarily in adduction of the thigh.

Ankle and toe movements

■ The 13 muscles in the leg, with tendons extending into the foot, can be divided into three groups: anterior, posterior, and lateral. As with the forearm, only the most superficial muscles are listed in Table 7-17 and are illustrated in Figure 7-19. The anterior muscles (Figure 7-19, *A*) are extensor muscles involved in dorsiflexion of the foot and extension of the toes.

The superficial muscles of the posterior compartment of the leg (Figure 7-19, *B*), the **gastrocnemius** and **soleus,** form the bulge of the calf (posterior leg; Figure 7-20). They join to form the common **calcaneal** (kal-ka′ne-al), or *Achilles tendon.* These muscles are flexors and are involved in plantar flexion of the foot. The deep muscles of the posterior compartment plantar flex and invert the foot and flex the toes.

■ The Achilles tendon derives its name from a hero of Greek mythology. As a baby, Achilles was dipped into magic water that made him invulnerable to harm everywhere it touched his skin. His mother, however, holding him by the back of his heel, overlooked submerging his heel under the water. Consequently, his heel was vulnerable and proved to be his undoing; he was shot in the heel with an arrow at the battle of Troy and died. Thus, saying that someone has an "Achilles heel" means he has a weak spot that can be attacked. ■

The lateral muscles of the leg, called the **peroneus** muscles (Figure 7-19, *C*), are primarily everters (turning the lateral side of the foot outward) of the foot, but they also aid in plantar flexion.

The 20 muscles located within the foot, called the **intrinsic foot** muscles, flex, extend, abduct, and adduct the toes. They are arranged in a manner similar to the intrinsic muscles of the hand.

TABLE 7-17

Muscles Moving the Ankle and Toes (see Figure 7-19)

MUSCLE	ORIGIN	INSERTION	ACTION
ANTERIOR LEG MUSCLES			
Tibialis anterior	Tibia	Medial cuneiform and first metatarsal	Extends foot (dorsiflexes)
Extensor digitorum longus	Tibia and fibula	Phalanges of four lateral toes	Extends toes
POSTERIOR LEG MUSCLES			
Gastrocnemius	Medial and lateral epicondyle of femur	Calcaneus	Flexes foot (plantar flexes)
Soleus	Tibia and fibula	Calcaneus	Flexes foot (plantar flexes)
LATERAL LEG MUSCLES			
Peroneus muscles (three muscles)	Tibia and fibula	Tarsals and metatarsals	Flex and evert foot

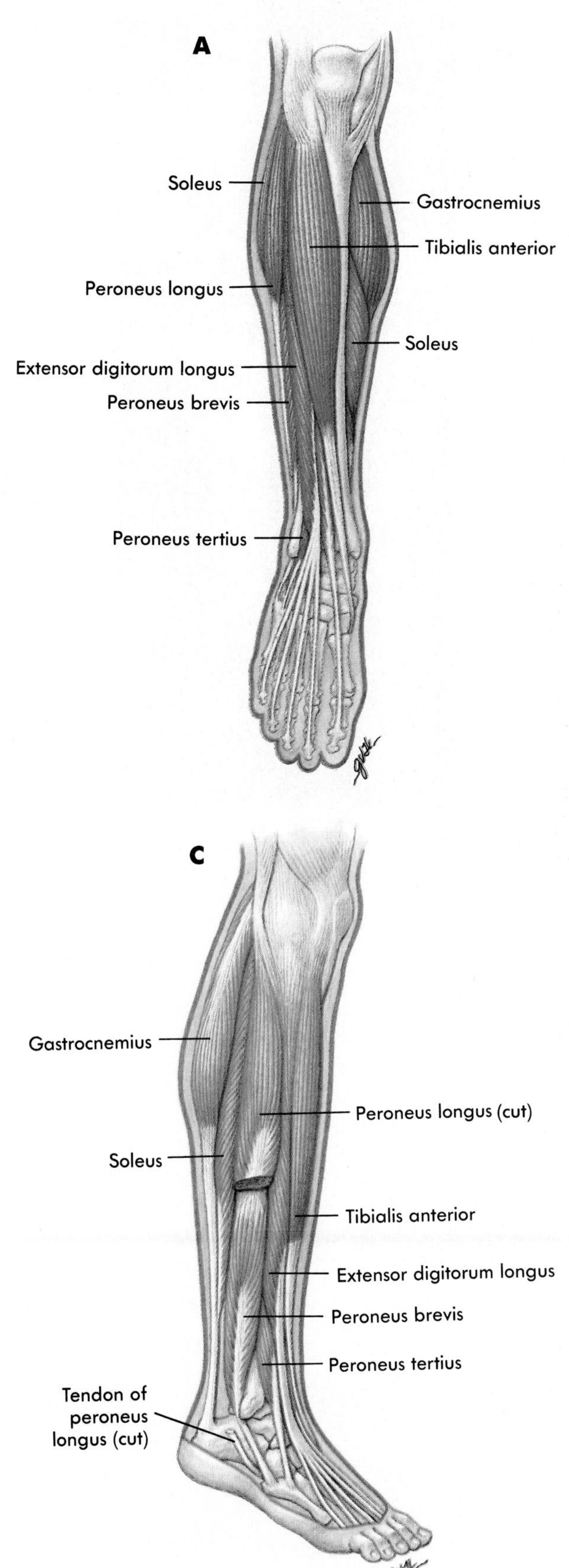

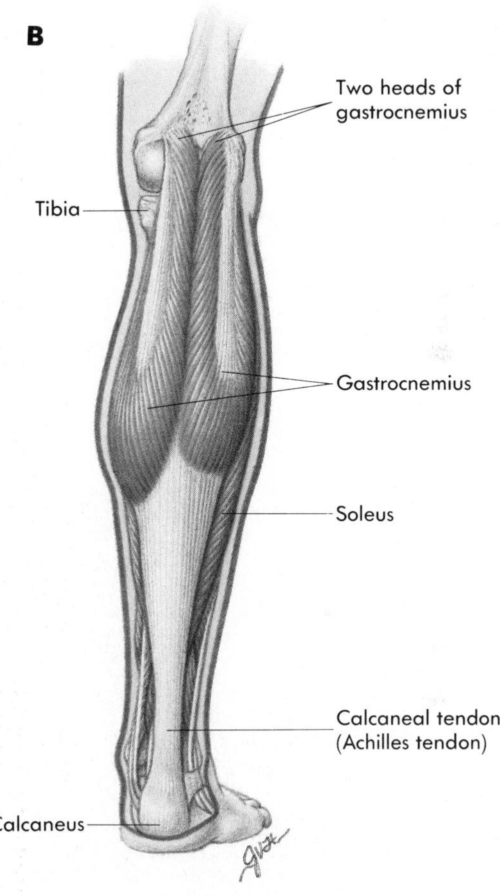

FIGURE 7-19 ■ Superficial muscles of the leg.
A Anterior view.
B Posterior view.
C Lateral view.

FIGURE 7-20 ■ Surface anatomy of the lower limb.
A Anterior view.
B Posterior view.

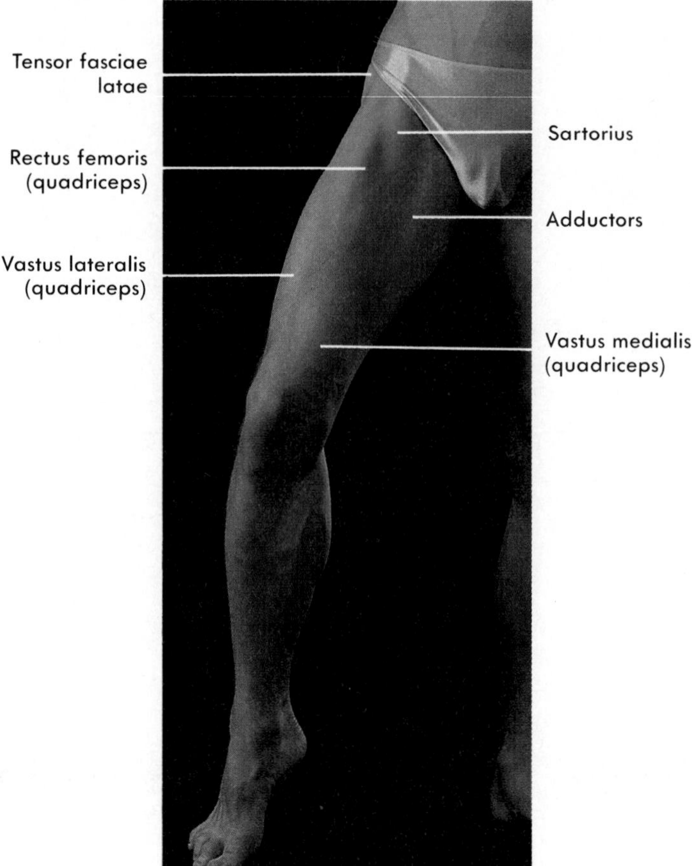

Gluteus medius
Gluteus maximus
Vastus lateralis (quadriceps)
Tendon of biceps femoris
Tendons of semimembranous and semitendinosus
Gastrocnemius
Soleus
Calcaneal (Achilles) tendon
B

SUMMARY

This chapter deals primarily with skeletal muscles.

CHARACTERISTICS OF SKELETAL MUSCLE

Skeletal muscle has contractility, excitability, extensibility, and elasticity.

Muscles shorten forcefully.

Structure

Each skeletal muscle fiber is a single cell containing numerous myofibrils.

Myofibrils are composed of actin and myosin myofilaments.

Sarcomeres are joined end to end to form myofibrils.

Muscle fibers are organized into fasciculi, and fasciculi are organized into muscles by associated connective tissue. Fascia separates muscles.

Membrane Potentials

Most cell membranes have a positive charge on the outside.

Action potentials are a brief reversal of the membrane charge.

Nerve Supply

Motor neurons carry action potentials to skeletal muscles where the neuron and muscle cells form neuromuscular junctions.

Neurons release acetylcholine, which binds to receptors on muscle cell membranes and stimulates muscles to contract.

Muscle Contraction

Action potentials travel along T tubules to the sarcoplasmic reticulum where they cause the release of calcium ions.

Calcium ions, released from the sarcoplasmic reticulum, bind to the actin myofilament, exposing attachment sites.

Myosin forms cross bridges with the exposed actin attachment sites.

The myosin molecules bend, causing the actin molecules to slide past; this is the sliding filament mechanism.

Muscle twitch, tetany, and recruitment

A muscle twitch consists of a lag phase, contraction phase, and relaxation phase.

Tetany occurs when stimuli occur so rapidly that a muscle does not relax between twitches.

Small contraction forces are generated when small numbers of motor units are recruited, and greater contraction forces are generated when large numbers of motor units are recruited.

Energy requirements for muscle contraction

Energy is produced by anaerobic (without oxygen) or aerobic (with oxygen) respiration.

After intense exercise, the rate of aerobic metabolism remains elevated to repay the oxygen debt.

Types of muscle contraction

Muscles contract either isometrically (muscle length stays the same, but tension increases) or isotonically (muscle length decreases, but tension remains the same).

Muscle tone

Muscle tone consists of a small percentage of muscle fibers contracting isometrically and is responsible for posture.

Slow and fast fibers

Muscles contain a combination of slow-twitch and fast-twitch fibers.

Slow-twitch fibers are better suited for aerobic metabolism and fast-twitch fibers are adapted for anaerobic metabolism.

Sprinters have more fast-twitch fibers, while distance runners have more slow-twitch fibers.

SMOOTH MUSCLE AND CARDIAC MUSCLE

Smooth muscle is not striated, contracts more slowly than skeletal muscle, is autorhythmic, and is under involuntary control.

Cardiac muscle is striated, is autorhythmic, and is under involuntary control.

MUSCLE ANATOMY

General Principles

Most muscles have an origin on one bone, an insertion onto another bone, and cross at least one joint.

Muscles working together are synergists, muscles working in opposition are antagonists.

A prime mover is the one muscle of a synergistic group that is primarily responsible for the movement.

Nomenclature

Muscles are named according to a number of different factors.

Muscles of the Head and Neck

Muscles of facial expression are associated primarily with the mouth and eyes.

Four pairs of muscles are involved in mastication.

Tongue movements involve intrinsic and extrinsic muscles.

Swallowing involves suprahyoid and infrahyoid muscles, plus muscles of the soft palate, pharynx, and larynx.

Neck muscles move the head.

Trunk Muscles

Erector spinae muscles hold the body erect.

Intercostal muscles and the diaphragm are involved in breathing.

Muscles of the abdominal wall flex and rotate the vertebral column, compress the abdominal cavity, and hold in the abdominal viscera.

Muscles form the inferior floor of the pelvis.

Upper Limb Muscles

The upper limb is attached to the body primarily by muscles.

Arm movements are accomplished by pectoral, rotator cuff, and deltoid muscles.

The forearm is flexed and extended by anterior and posterior arm muscles respectively.

Supination and pronation of the forearm are accomplished by supinators and pronators in the forearm.

Movements of the wrist and fingers are accomplished by most of the 20 forearm muscles and 19 intrinsic muscles in the hand.

Lower Limb Muscles

Hip muscles flex, extend, and abduct the thigh.

Thigh muscles flex, extend, and abduct the thigh. They also flex and extend the leg.

Muscles of the leg and foot can be considered as being somewhat similar to those of the forearm and hand.

CONTENT REVIEW

1. Define contractility, excitability, extensibility, and elasticity.
2. What is a muscle fiber?
3. Describe the composition of a myofibril.
4. What is a sarcomere?
5. What are fasciculi?
6. Describe the resting membrane potential and how it is produced.
7. Describe the production of an action potential.
8. What is a neuromuscular junction? What happens there?
9. Describe the sliding filament mechanism of muscle contraction.
10. Explain how an action potential results in a muscle contraction.
11. Define muscle twitch and tetany.

12. Describe the two ways energy is produced in skeletal muscle.
13. Compare isometric and isotonic contraction.
14. What is muscle tone?
15. Compare slow-twitch and fast-twitch muscle fibers.
16. How do smooth muscles and cardiac muscles differ from skeletal muscle?
17. Define origin, insertion, synergist, antagonist, and prime mover.
18. Describe the muscles of facial expression.
19. What is mastication? What muscles are involved?
20. What are intrinsic and extrinsic tongue muscles?
21. What muscles are involved in swallowing?
22. Describe the functions of the muscles of the anterior abdominal wall.
23. What muscles are involved in respiration?
24. What primarily is responsible for attaching the upper limb to the body?
25. Describe, by muscle groups, movements of the arm, forearm, and hand.
26. Describe, by muscle groups, movements of the thigh, leg, and foot.

CONCEPT REVIEW

1. Harvey Leche milked cows by hand each morning before school. One morning he overslept and had to hurry to get to school on time. As he was milking the cows as fast as he could, his hands became very tired, and then for a short time he could neither release his grip nor squeeze harder. Explain what happened.
2. A researcher was investigating the fast-twitch vs. slow-twitch composition of muscle tissue in the gastrocnemius muscle (in the calf of the leg) of athletes. Describe the general differences this researcher would see when comparing the muscles from athletes who were outstanding in the following events: 100-meter dash, weight lifting, the 10,000-meter run.
3. Describe an exercise routine that would build up each of the following groups of muscles: anterior arm, posterior arm, anterior forearm, anterior thigh, posterior leg, and abdomen.
4. Sherri Speedster started a 100-meter dash but fell to the ground in pain. Examination of her right lower limb revealed the following symptoms: the leg was held in a slightly flexed position, but she could not flex it voluntarily; she could extend the leg with difficulty, but this caused her considerable pain; and there was considerable pain and bulging of the muscles in the posterior thigh. Explain the nature of her injury.

ANSWERS TO  PREDICT QUESTIONS

1 *p. 153* (A) If ATP levels are low in a muscle fiber before stimulation, there will be a reduced force of contraction, because there would be insufficient ATP for all motor units to contract.

(B) If action potentials occur at a frequency so great that calcium is not released between individual action potentials, the muscle will not relax.

2 *p. 155* During a 1-mile run, aerobic metabolism is the primary source of ATP production for muscle contraction. Anaerobic metabolism provides the short (15 to 20 seconds) burst of energy for the sprint at the finish. After the race, aerobic metabolism is elevated for a time to repay the oxygen debt, causing the heavy breathing after the race.

3 *p. 161* Raising eyebrows—occipitofrontalis;
winking—orbicularis oculi;
whistling—orbicularis oris and buccinator;
smiling—zygomaticus;
frowning—depressor anguli oris;
sneering—levator labii superioris.

4 *p. 163* Shortening the right sternocleidomastoid muscle rotates the head to the left and also slightly elevates the chin.

5 *p. 173* In the sprinter's stance and the bicycle racing posture the thigh is flexed at a 45-degree angle because at that angle the gluteus maximus functions at its maximum in extending the thigh, thus providing maximum force.

8

THE NERVOUS SYSTEM

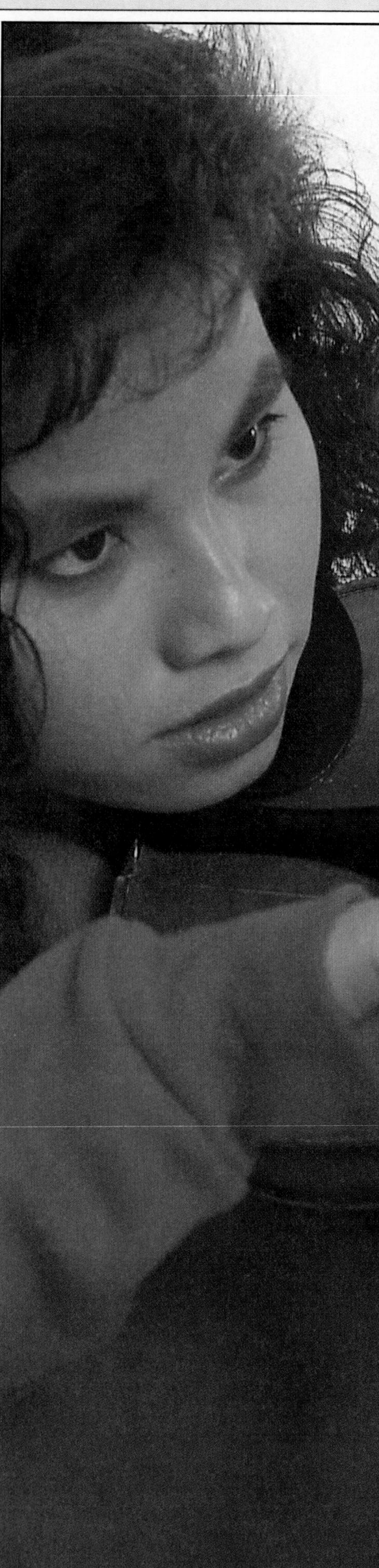

After reading this chapter you should be able to:

1 List the divisions of the nervous system and describe the characteristics of each.
2 Describe the structure of neurons and the function of their components. Describe the location, structure, and general function of neuroglia cells.
3 Define and describe the structure of a nerve, nerve tract, nucleus, and ganglion.
4 Explain a resting membrane potential and how an action potential is generated and propagated.
5 Describe the structure and function of a synapse.
6 List the parts of the brainstem and diencephalon, and give their functions.
7 Describe the major functional areas of the cerebral cortex and explain their interactions. Describe sensory, short-term, and long-term memory.
8 Describe the major functions of the basal ganglia, limbic system, and cerebellum.
9 List the parts of a reflex arc and describe its function.
10 List and describe the major ascending and descending pathways of the spinal cord.
11 Describe the three meningeal layers surrounding the central nervous system; the four ventricles of the brain; and the origin and circulation of the cerebrospinal fluid.
12 List the various types of cranial nerves and give a brief description of their function.
13 Define the term plexus and describe the three primary plexuses, including their branches.
14 Contrast the structure of the autonomic nervous system and the somatomotor nervous system. Name the two divisions of the autonomic nervous system and describe the differences between them.

action potential The all-or-none change in membrane potential in an excitable tissue that is propagated as an electrical signal.

cerebellum (sĕr′ĕ-bel′um) [L., little brain] A part of the brain attached to the brainstem at the pons and important in maintaining muscle tone, balance, and coordination of movements.

cerebrum [L., brain] The largest part of the brain, consisting of two hemispheres and including the cortex, nerve tracts, and basal ganglia.

ganglion (gan′gle-on) [Gr., knot] A group of nerve cell bodies in the peripheral nervous system.

meninges (mĕ-nin′jēz) [Gr., membrane] The connective tissue membranes that surround and protect the brain and spinal cord.

neuroglia (nu-rog′le-ah) [Gr. *neuro*, nerve + *glia*, glue] Cells that play a support role in the nervous system; also called glia.

neuron [Gr., nerve] A nerve cell.

neurotransmitter [Gr. *neuro*, nerve + L. *tramitto*, to send across] A chemical that is released by a presynaptic cell into the synaptic cleft and that acts upon the postsynaptic cell.

parasympathetic division Subdivision of the autonomic nervous system involved in involuntary (vegetative) functions such as digestion, defecation, and urination.

proprioception (pro′pre-o-sep′shun) [L. *proprius*, one's own + *capio*, to take] Sensation of the position of the body and its various parts.

somatomotor (so-mă′to-mo′tor) [Gr. *soma*, body] A type of motor (efferent) neuron of the peripheral nervous system that innervates skeletal muscle.

somesthetic cortex (so′mes-thet′ik) [Gr. *soma*, body + *aisthesis*, sensation] That portion of the cerebral cortex involved with the conscious perception and localization of body sensations.

sympathetic division Subdivision of the autonomic nervous system generally involved in preparing the body for immediate physical activity.

synapse [Gr. *syn*, together + *haptein*, to clasp] Junction between a nerve cell and some other cell.

thalamus [Gr., a bedroom] A large mass of gray matter making up the bulk of the diencephalon; involved in the relay of sensory input to the cerebrum.

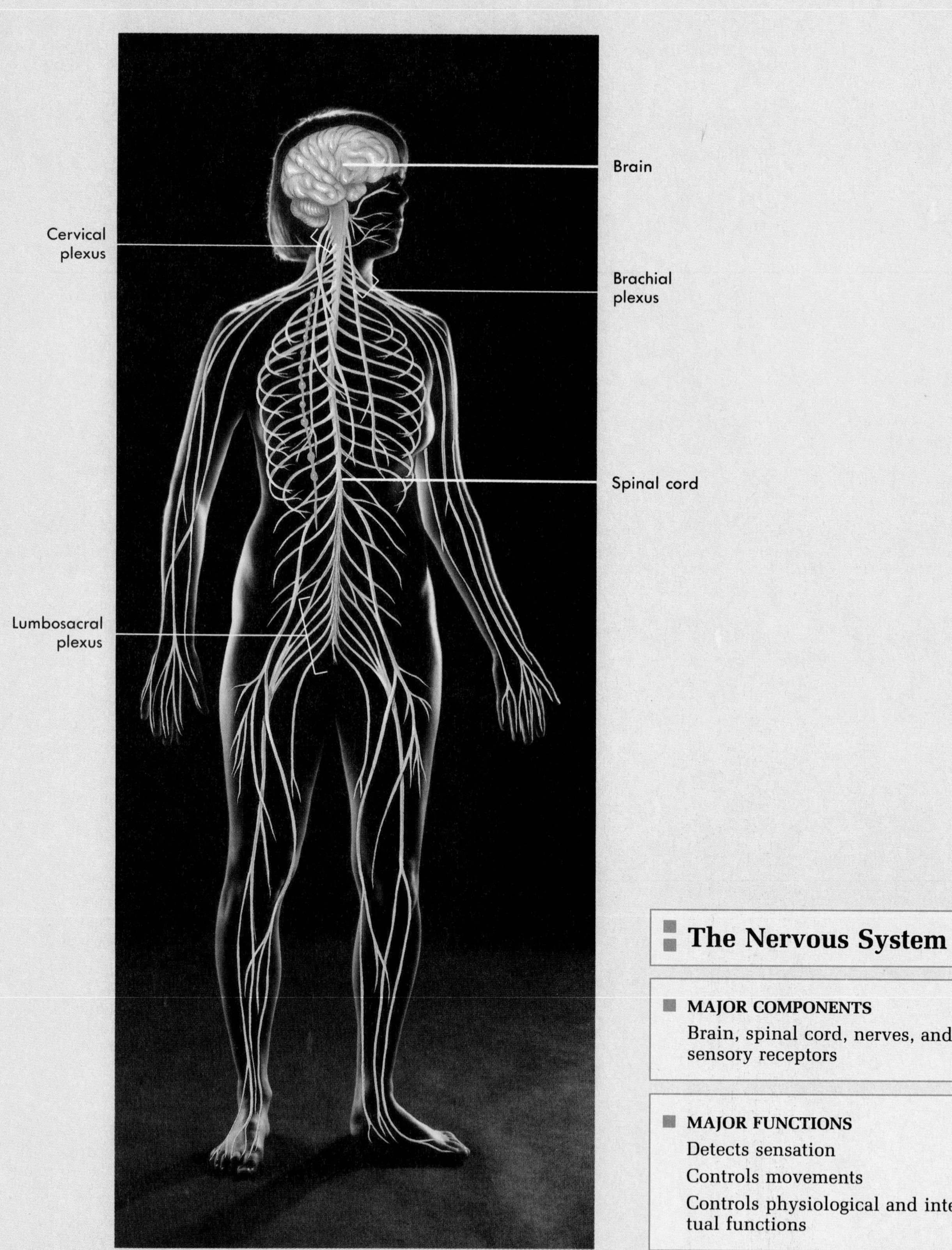

FIGURE 8-1 ■ The nervous system.
The brain, spinal cord, and peripheral nerves.

The Nervous System

MAJOR COMPONENTS

Brain, spinal cord, nerves, and sensory receptors

MAJOR FUNCTIONS

Detects sensation

Controls movements

Controls physiological and intellectual functions

The nervous system is one of the body's major regulatory and coordinating systems. The nervous system (Figure 8-1) is the seat of all mental activity, including consciousness, memory, and thinking. Homeostasis is maintained to a large degree by the nervous system's activities, which depend on the nervous system's ability to detect, interpret, and respond to changes in internal and external conditions.

DIVISIONS OF THE NERVOUS SYSTEM

The nervous system can be divided into the central and the peripheral nervous systems (Figure 8-2). The central nervous system (CNS) consists of the brain and spinal cord. The peripheral nervous system (PNS) consists of nerves, with their receptors, synapses, and ganglia, which lie outside the CNS.

The PNS has two subdivisions: The **afferent division** transmits action potentials from sensory organs to the CNS. The nerve fibers that transmit action potentials from the periphery to the CNS are **afferent fibers.** The **efferent division** transmits action potentials from the CNS to effector organs such as muscles and glands. The nerve fibers that transmit action potentials from the CNS toward the periphery are **efferent fibers.**

The efferent division can be further subdivided into the somatomotor nervous system, which transmits impulses from the CNS to skeletal muscle, and the autonomic nervous system (ANS), which transmits action potentials from the CNS to smooth muscle, cardiac muscle, and glands.

CELLS OF THE NERVOUS SYSTEM

Cells of the nervous system are neurons and neuroglia.

Neurons

Neurons, or **nerve cells** (Figure 8-3), receive stimuli and transmit action potentials to other neurons or to effector organs. Each neuron consists of a cell body and two types of processes: dendrites and axons.

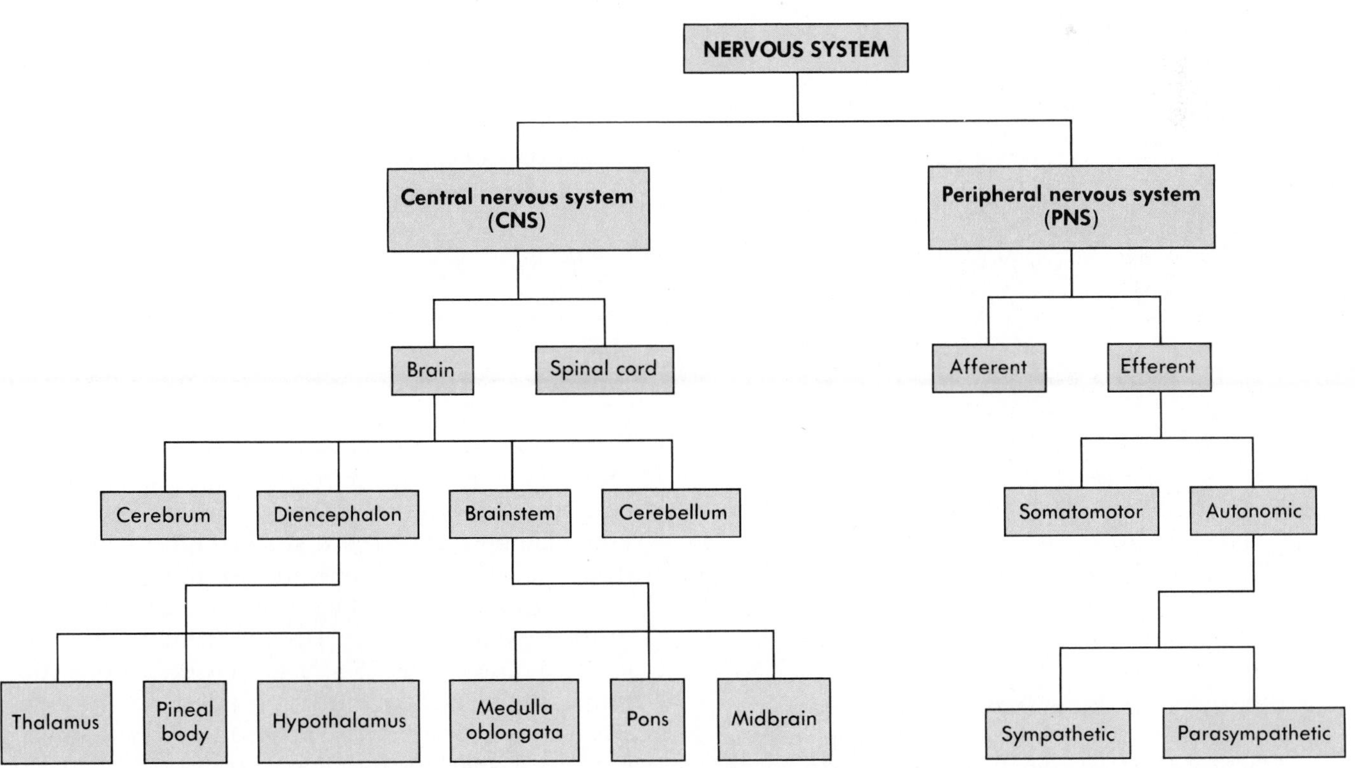

FIGURE 8-2 **Divisions of the nervous system.**

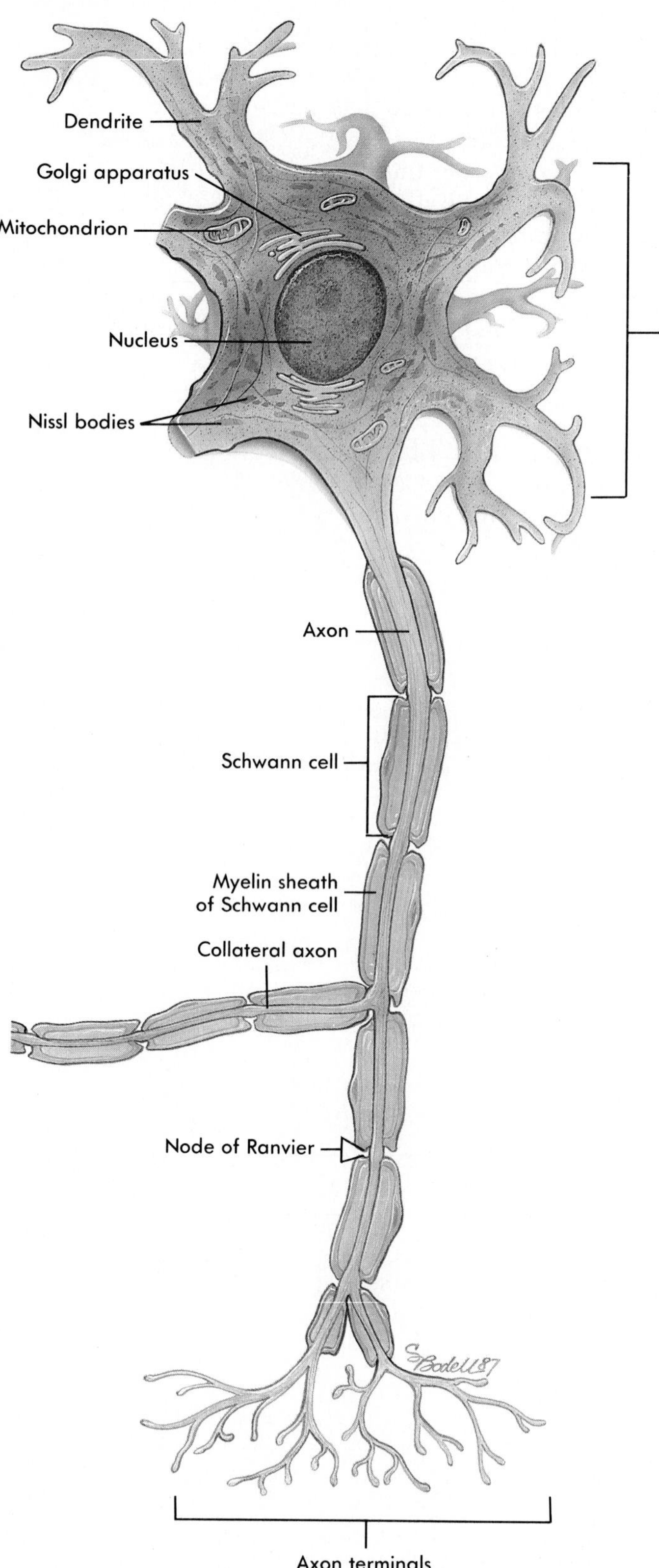

FIGURE 8-3 ■ Structural features of a neuron. Neurons consist of a dendrite, a cell body and an axon. Dendrites carry nervous impulses toward the cell body, and axons carry impulses away from the cell body.

Each neuron cell body contains a single nucleus. Because the nucleus is the source of information for protein synthesis, an axon will die if it is separated from the cell body because it then would have no connection to the nucleus, and no new protein synthesis could occur in the axon. Extensive rough endoplasmic reticulum (RER) and Golgi apparatuses surround the nucleus. Large numbers of neurofilaments and microtubules course through the cytoplasm in all directions and separate the rough ER into distinct areas in the cell body. These areas of rough ER are called **Nissl** (nis'l) **bodies.**

Dendrites (den'drītz; trees) are short, often highly branching cytoplasmic extensions that are tapered from their bases at the neuron cell body to their tips. Most dendrites are extensions of the nerve cell body, but dendritelike structures also project from the peripheral ends of some afferent axons as axon terminals. Dendrites usually function to receive information from other neurons or sensory receptors and carry the information, as local potentials, toward the neuron cell body. If these local potentials reach a certain level (threshold), an action potential results.

A single **axon** arises from each nerve cell body. An axon may remain as a single structure or may branch to form **collateral axons.** Each axon has a constant diameter and may vary in length from a few millimeters to more than a meter. Axons conduct action potentials that arise near the nerve cell body, in the case of motor neurons, or that arise in peripheral dendritic processes, in the case of most sensory neurons. Axons usually function to carry information away from the nerve cell body. Many axons are surrounded by an insulating layer of cells called the **myelin sheath** (described in more detail later in the chapter).

Types of neurons

■ Three categories of neurons exist based on their shape: **multipolar neurons, bipolar neurons,** and **unipolar neurons** (Figure 8-4 and Table 8-1). Most of the neurons within the CNS, including nearly all motor neurons, are multipolar. Bipolar neurons are often components of sensory organs, such as the eye and nose. Most other sensory neurons are unipolar. Unipolar neurons have a single process extending from the cell body. This process divides into two processes a short distance from the cell body. One process extends to the periphery, and the other process extends to the CNS. The two extensions function as a single axon with the small dendritic extensions at the periphery. The axon receives sensory information at the periphery and transmits that information to the CNS.

FIGURE 8-14 ■ Brainstem.
A Anterior view.
B Posterolateral view.

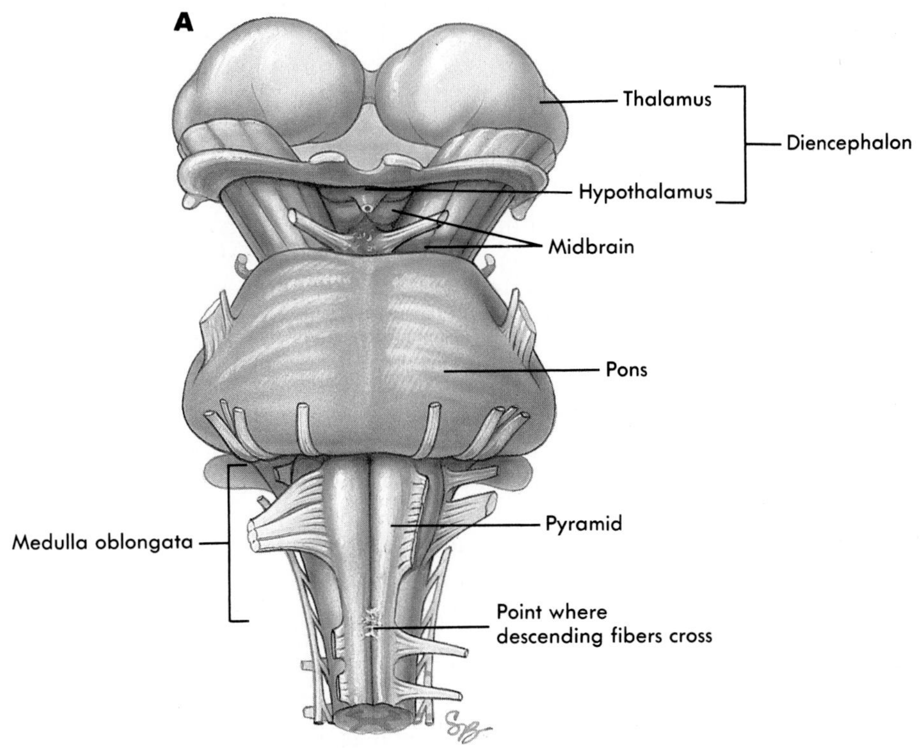

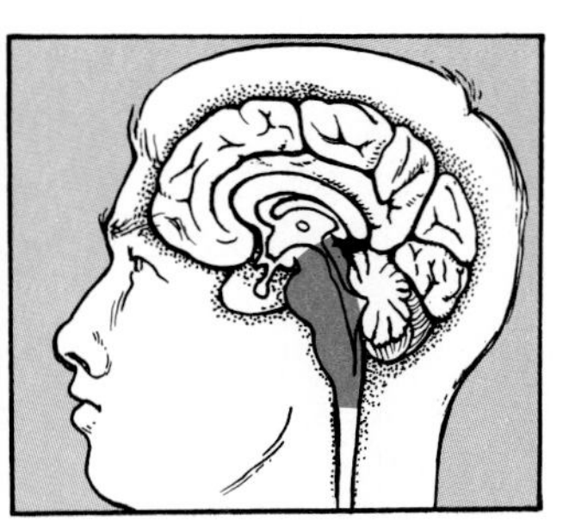

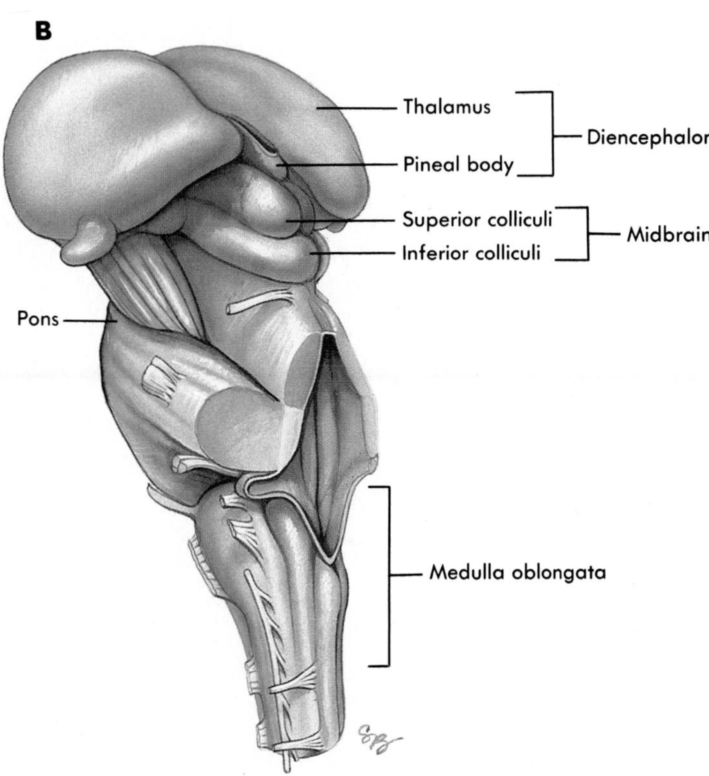

Midbrain

The **midbrain,** just superior to the pons, is the smallest region of the brainstem (see Figure 8-14). The roof of the midbrain consists of four mounds called the **colliculi** (kol-lik′u-le; hills). The two inferior colliculi are involved in hearing and are an integral portion of the auditory pathways in the CNS. The two superior colliculi are involved in visual reflexes. The rest of the midbrain consists largely of ascending tracts from the spinal cord to the cerebrum and descending motor tracts from the cerebrum to the spinal cord.

Reticular formation

Scattered throughout the brainstem is a group of nuclei collectively called the **reticular formation,** which plays an important role in arousing and maintaining consciousness through the sleep/wake cycle. Stimuli such as an alarm clock ringing, sudden bright lights, ammonia (smelling salts), or cold water being splashed on the face can arouse consciousness. Conversely, removal of visual or auditory stimuli may lead to drowsiness or sleep. General anesthetics function by suppressing the reticular activating system. Damage to cells of the reticular formation may result in coma.

Diencephalon

The **diencephalon** (Figure 8-15) is the part of the brain between the brainstem and the cerebrum. Its main components are the thalamus, pineal body, and hypothalamus.

Thalamus

The **thalamus** is by far the largest portion of the diencephalon. It consists of a cluster of nuclei and is shaped somewhat like a yo-yo, with two large, lateral portions connected in the center by a small intermediate mass (see Figure 8-14). Most sensory input that ascends through the cord and brainstem projects to the thalamus, where afferent neurons synapse with thalamic neurons, which in turn send their axons to the cerebral cortex. The thalamus also has other functions such as influencing mood. The thalamus also registers an unconscious, unlocalized, uncomfortable perception of pain.

Pineal body

The **pineal body** (pi′ne-al; pinecone-shaped) is located posterior to the thalamus (see Figure 8-15). It is an endocrine gland that may influence the onset of puberty. It also may play a role in controlling

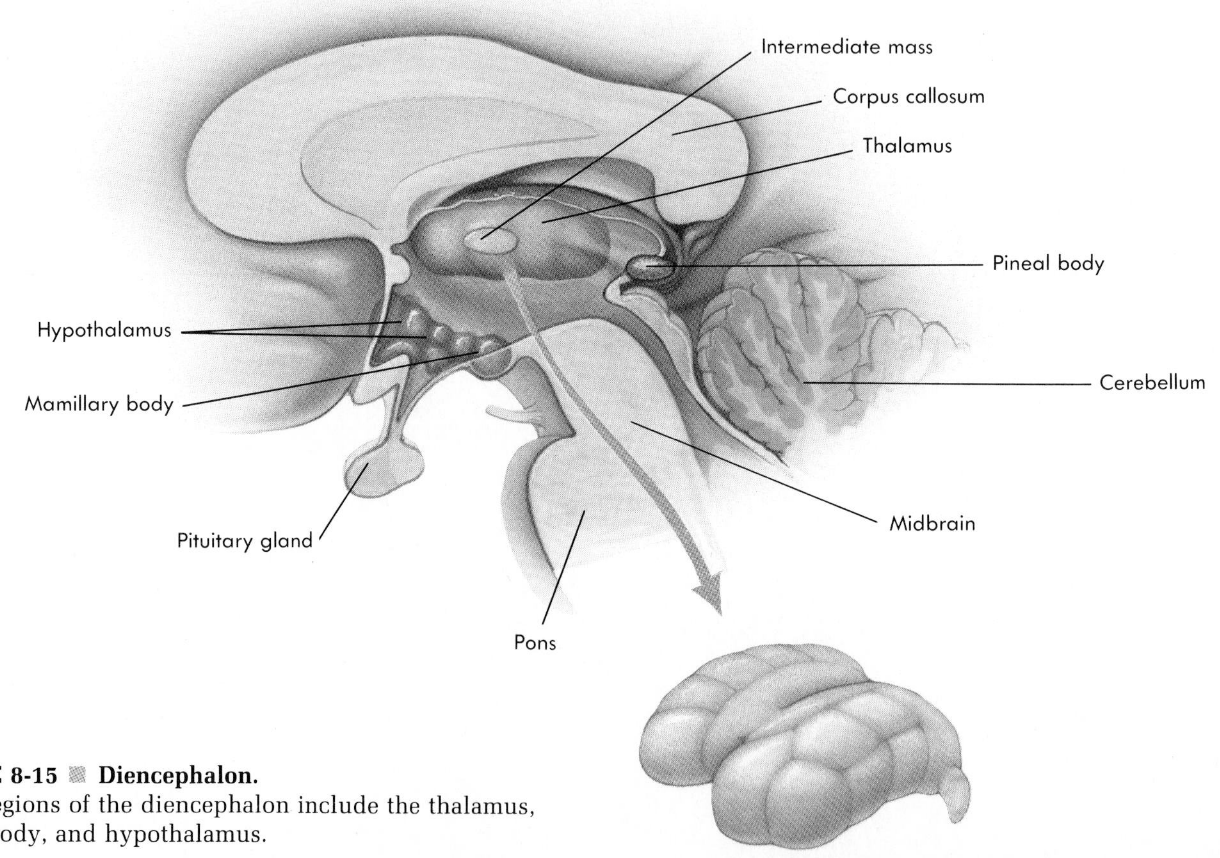

FIGURE 8-15 Diencephalon.
Major regions of the diencephalon include the thalamus, pineal body, and hypothalamus.

some long-term cycles that are influenced by light intensity. In other animals, it is known to influence annual behaviors such as migration in birds, as well as changes in fur color and density and reproductive changes in some mammals (see Chapter 10).

Hypothalamus

The **hypothalamus** is the most inferior portion of the diencephalon and contains several small nuclei (see Figure 8-15). These nuclei are very important in maintaining homeostasis. The hypothalamus plays a central role in the control of body temperature, hunger, and thirst. Sensations such as sexual pleasure, feeling relaxed and "good" after a meal, rage, and fear are related to hypothalamic functions. Emotional responses, such as "nervous perspiration," involve the hypothalamus. In this instance, the hypothalamus initiates perspiration in response to stress. An example of an apparent confused role in the hypothalamus is when some people feel hungry and eat when they are depressed. A funnel-shaped stalk, the **infundibulum,** extends from the floor of the hypothalamus to the pituitary gland. The hypothalamus plays an important role in controlling the endocrine system by regulating the pituitary gland's hormone secretion (see Chapter 10). The **mamillary bodies,** which are involved in olfactory reflexes, that is, emotional responses to odors and memory, form externally visible swellings on the posterior portion of the hypothalamus.

Cerebrum

The **cerebrum** is the largest portion of the brain (Figure 8-16). It is divided into left and right hemispheres by a **longitudinal fissure.** The most conspicuous features on the surface of each hemisphere are numerous folds called **gyri** (ji′ri; singular: gyrus), which greatly increase the surface area of the cortex, and intervening grooves called **sulci** (sul′si; singular: sulcus).

Each cerebral hemisphere is divided into lobes (see Figure 8-16), which are named for the skull bones overlying them. The **frontal lobe** is important in voluntary motor function, motivation, aggression, mood, and olfactory (smell) reception. The **parietal lobe** is the principal center for the reception and evaluation of most sensory information, such as touch, balance, and taste. The frontal and parietal lobes are separated by a prominent sulcus called the **central sulcus.** The **occipital lobe** functions in the reception and integration of visual input and is not distinctly separate from the other lobes. The

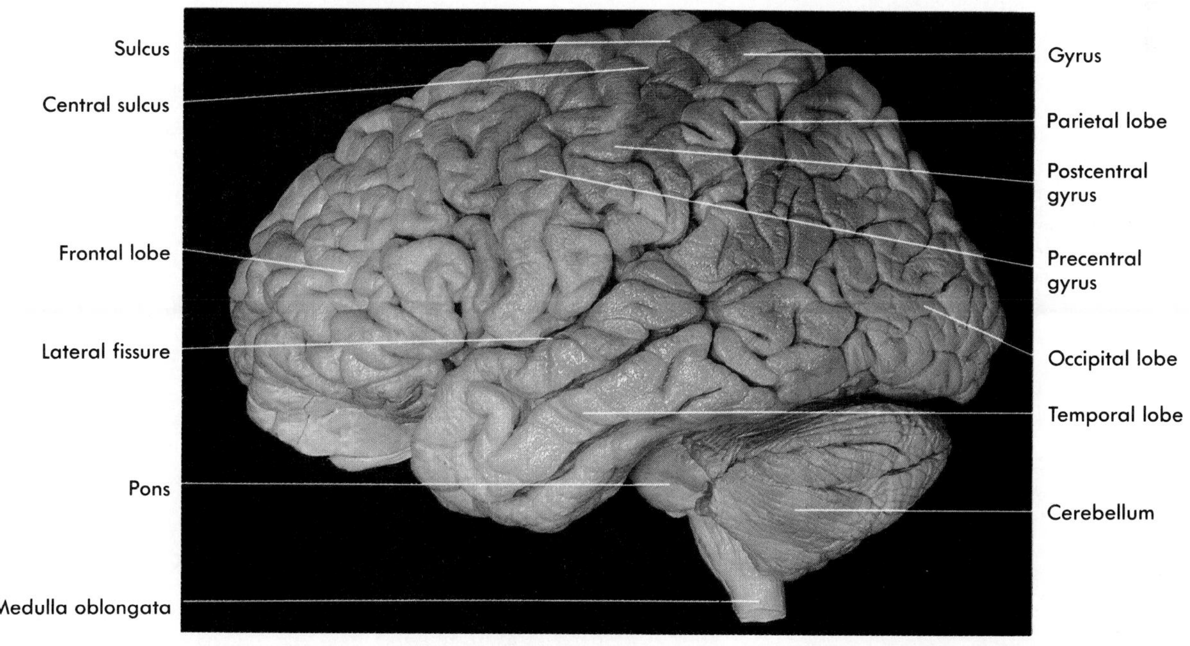

FIGURE 8-16 ■ **Cerebral cortex.**
Lateral view of the cerebrum showing the gyri and lobes.

temporal lobe evaluates olfactory and auditory (hearing) input and plays an important role in memory. Its anterior and inferior portions are referred to as the "psychic cortex," and they are associated with functions such as abstract thought and judgement. The temporal lobe is separated from the rest of the cerebrum by a **lateral fissure.**

Cerebral cortex

■ Sensory pathways project to specific regions of the cerebral cortex (Figure 8-17), called **primary sensory areas,** where those sensations are perceived. The **primary somesthetic** (so'mes-thet'ik) **cortex,** or **general sensory area** is located in the parietal lobe posterior to the central sulcus. Afferent fibers carrying general sensory input such as pain, pressure, and temperature synapse in the thalamus, and thalamic neurons relay the information to the primary somesthetic cortex. Other primary sensory areas include the visual cortex in the occipital lobe, the auditory area in the temporal lobe, and the olfactory cortex in the frontal lobe.

Cortical areas immediately adjacent to the primary sensory centers, called **association areas,** are involved in the process of recognition. For example, afferent action potentials originating in the retina of the eye reach the visual cortex, where the image is "seen." Action potentials then pass from the visual cortex to the visual association cortex, where the present visual information is compared to past visual experience ("Have I seen this before?"). Based on this comparison, the visual association cortex "decides" whether or not the visual input is recognized and passes judgement concerning the significance of the input. You pay less attention, for example, to a person you have never seen before than to someone you know.

The **primary motor area** is located in the posterior portion of the frontal lobe (see Figure 8-17). Efferent action potentials initiated in this region control voluntary movements of skeletal muscles. The **premotor area** of the frontal lobe is the staging area where motor functions are organized before they are actually initiated in the primary motor area. For example, if a person decides to take a step, the neurons of the premotor area are first stimulated, and the determination is made there as to which muscles must contract, in what order, and to what degree. Impulses are then passed to the primary motor area, which actually initiates each planned movement.

The motivation and the foresight to plan and initiate movements occur in the anterior portion of the frontal lobes, the **prefrontal area.** This is an area of association cortex that is well developed only in primates, especially in humans. It is involved in motivation and regulation of emotional

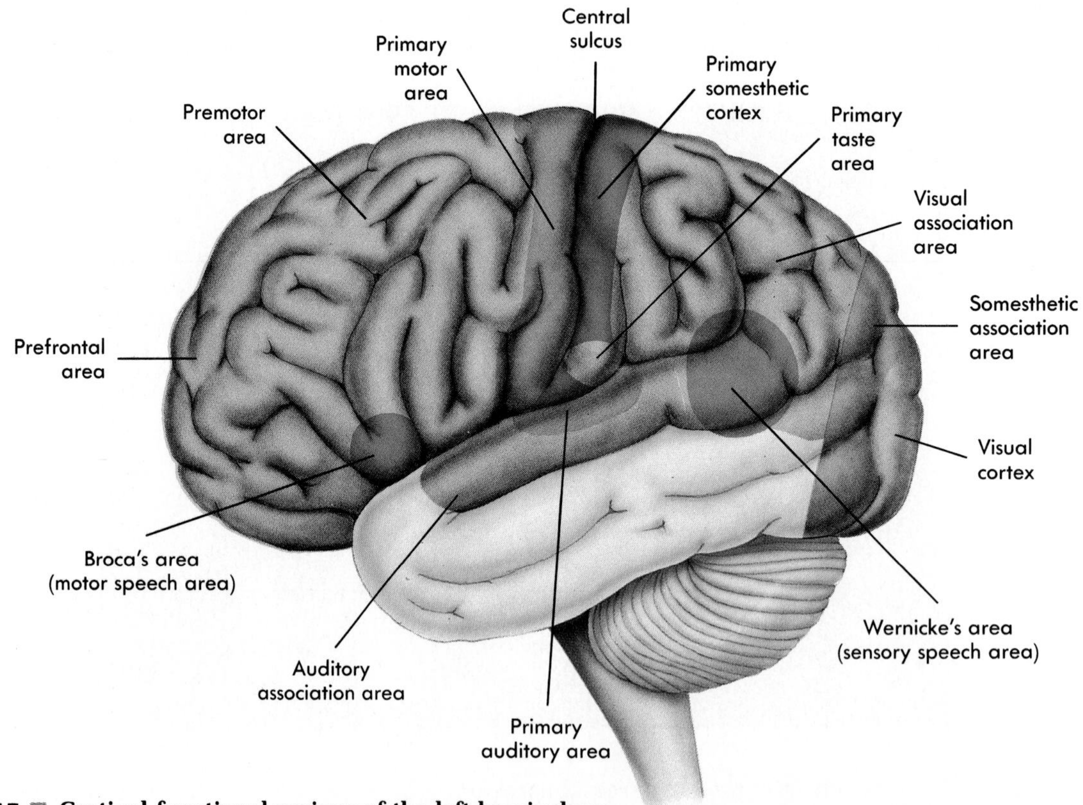

FIGURE 8-17 ■ Cortical functional regions of the left hemisphere.

behavior and mood. The large size of this area in humans may account for our relatively well-developed forethought and motivation, and for our emotional complexity.

In relation to its involvement in motivation, the prefrontal area is the functional center for aggression. In the past, one method used to eliminate uncontrollable aggression in mental hospital patients was to surgically remove or destroy the prefrontal regions of the brain (prefrontal or frontal lobotomy). This operation was successful in eliminating aggression, but it also eliminated the motivation to do much else and destroyed the personality.

Speech

In the vast majority of people, the speech area is in the left cortex. Two major cortical areas are involved in speech: **Wernicke's area** (sensory speech area), a portion of the parietal lobe, and **Broca's area** (motor speech area) in the inferior portion of the frontal lobe (see Figure 8-17). Damage to these parts of the brain, or to associated brain regions, may result in **aphasia** (a-fa'ze-ah), absent or defective speech or language comprehension.

To repeat a word that one hears requires the functional integrity of the following pathway. Action potentials from the ear reach the auditory cortex, where the word is heard; the word is recognized in the auditory association cortex and is comprehended in portions of Wernicke's area. Action potentials representing the word are then conducted through association fibers that connect Wernicke's and Broca's areas. In Broca's area the word is formulated as it is to be repeated; impulses then go to the premotor cortex where the movements are programmed and finally to the primary motor area where the proper movements are triggered.

Speaking a written word is somewhat similar. The information enters the visual cortex, passes to the visual association cortex where it is recognized, and continues to Wernicke's area, where it is understood and formulated as it is to be spoken. From Wernicke's area it follows the same route for repeating audibly received words: via association fibers to Broca's area, to the premotor area, and then to the primary motor area.

2 Propose the pathway needed for a blindfolded person to name an object placed in her right hand. ?

Brain waves

Electrodes placed on a person's scalp and attached to a recording device can record the brain's electrical activity, producing an **electroencephalogram (EEG)** (Figure 8-18). These electrodes are not sensitive enough to detect individual action potentials but can detect the simultaneous action potentials in large numbers of neurons. As a result, the

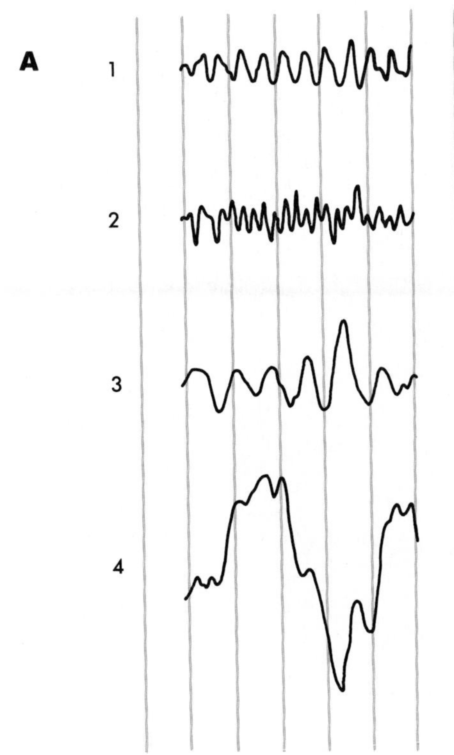

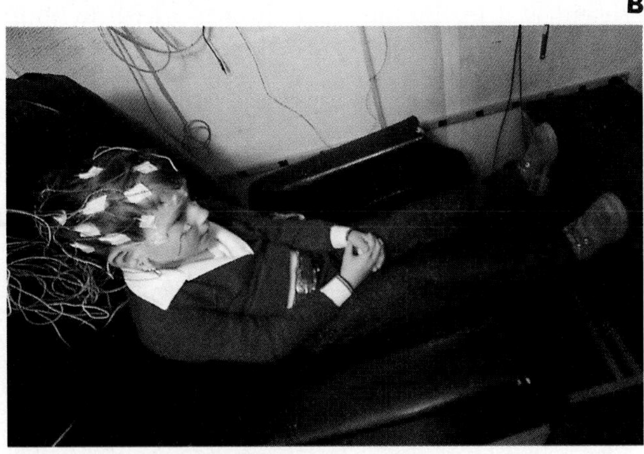

FIGURE 8-18 ■ Electroencephalogram (EEG).
A Four electroencephalogram tracings. *1*, Alpha waves, often seen in a relaxed individual with eyes closed. *2*, Beta waves, typical of an alert individual. *3*, Theta waves, seen in the first stage of sleep. *4*, Delta waves, characteristic of deep sleep.
B Patient with electrodes attached to his head.

EEG displays wavelike patterns known as **brain waves.** This electrical activity is constant, but the intensity and frequency of electrical discharge differs from time to time based on the state of brain activity.

Distinct EEG patterns occur with specific brain disorders such as epileptic seizures. Neurologists use these patterns to diagnose and determine the treatment for the disorders.

Memory

Memory can be divided into at least three types: sensory, short-term (or primary), and long-term. **Sensory memory** is the brief retention of sensory input received by the brain while something is scanned, evaluated, and acted upon. This type of memory lasts less than a second.

If a given piece of data held in sensory memory is considered valuable enough, it is moved within the temporal lobe from sensory memory into **short-term memory,** where information is retained for a few seconds to a few minutes. This memory is limited primarily by the number of bits of information that can be stored at any one time, which is usually about seven bits of information. Have you ever wondered why telephone numbers are seven digits long? More bits can be stored when the numbers are grouped and separated by spaces, such as when adding an area code. When new information is presented, old information, previously stored in short-term memory, is eliminated. What happens to a telephone number you just looked up if you are distracted? If the temporal lobe is damaged, the transition from sensory to short-term memory may not occur, and the person will always live only in the present and in the more remote past, with memory already stored before the injury. This person is unable to add new memory.

Certain pieces of information are transferred from short-term to **long-term memory,** some of which may become permanent. Long-term memory may involve a physical change in neuron shape. A whole series of neurons, called **memory engrams** or memory traces, are probably involved in the long-term retention of a given piece of information, thought, or idea. Rehearsal of information assists in the transfer of information from short-term to long-term memory.

Right and left hemispheres

The right cerebral hemisphere controls muscular activity in and receives sensory input from the left half of the body. The left cerebral hemisphere controls muscles and receives input from the right half of the body. Sensory information received by one hemisphere is shared with the other through connections between the two hemispheres called **commissures** (kom′i-shurs; a joining together). The largest of these commissures is the **corpus callosum** (kor′pus kah-lo′sum; callous body), a broad band of nerve tracts (white matter) at the base of the longitudinal fissure (see Figure 8-13).

Language and perhaps other functions, such as artistic activities, are not shared equally between the two hemispheres. The left hemisphere is thought to be the analytic hemisphere, emphasizing such skills as mathematics and speech. The right hemisphere is thought to be involved primarily in functions such as three-dimensional or spatial perception and musical ability. Some people believe that the "dominant" left hemisphere can repress the artistic abilities of the right hemisphere.

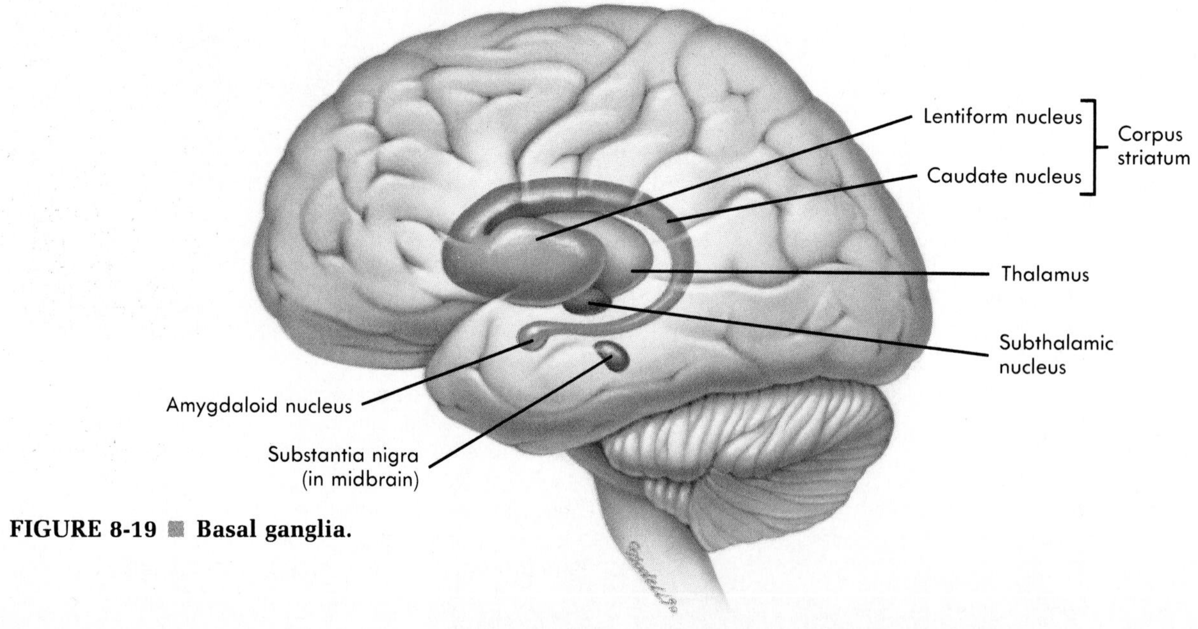

FIGURE 8-19 Basal ganglia.

Basal ganglia

■ The **basal ganglia** are a group of functionally related nuclei (Figure 8-19). Two primary nuclei are the **corpus striatum,** located deep within the cerebrum, and the **substantia nigra,** a group of darkly pigmented cells located in the midbrain. The basal ganglia play an important role in posture and in planning and coordinating motor movements. Complex neural connections link the basal ganglia with the cerebral cortex. The neurotransmitter dopamine is produced in the substantia nigra, and it exerts an inhibitory influence on the corpus striatum. The major effect of the basal ganglia is to decrease muscle tone and inhibit muscular activity. Disorders of the basal ganglia, such as Parkinson's disease and cerebral palsy, result in increased muscle tone and in exaggerated, uncontrolled movements.

Limbic system

■ The olfactory cortex and certain deep portions of the cerebrum and diencephalon are grouped together under the title **limbic system** (lim'bik; a boundary) (Figure 8-20). The limbic system responds to olfactory stimulation by initiating responses necessary for survival, such as hunger and thirst. It influences emotions, visceral responses to emotions, motivation, and mood. The limbic system is connected to and functionally associated with the hypothalamus. Lesions in the limbic system may result in voracious appetite, increased (often perverse) sexual activity, and docility (including loss of normal fear and anger responses).

Cerebellum

■ **Cerebellum** (sĕr'ĕ-bel'um) (see Figures 8-13 and 8-16) means little brain. The cerebellar cortex is composed of gray matter, and it has gyri and sulci, but the gyri are much smaller than those of the cerebrum. Internally, the cerebellum consists of nuclei and nerve tracts. The cerebellum is involved in balance, maintenance of muscle tone, and coordination of fine motor movement. If the cerebellum is damaged, muscle tone decreases, and fine motor movements become very clumsy.

A major function of the cerebellum is that of a **comparator** (an instrument that makes comparisons) (Figure 8-21). Impulses from the cerebral motor cortex descend into the spinal cord to initiate voluntary movements. Collateral branches are also sent from the motor cortex to the cerebellar cortex, giving information representing the intended movement. Simultaneously, impulses from proprioceptive neurons (providing information about the posi-

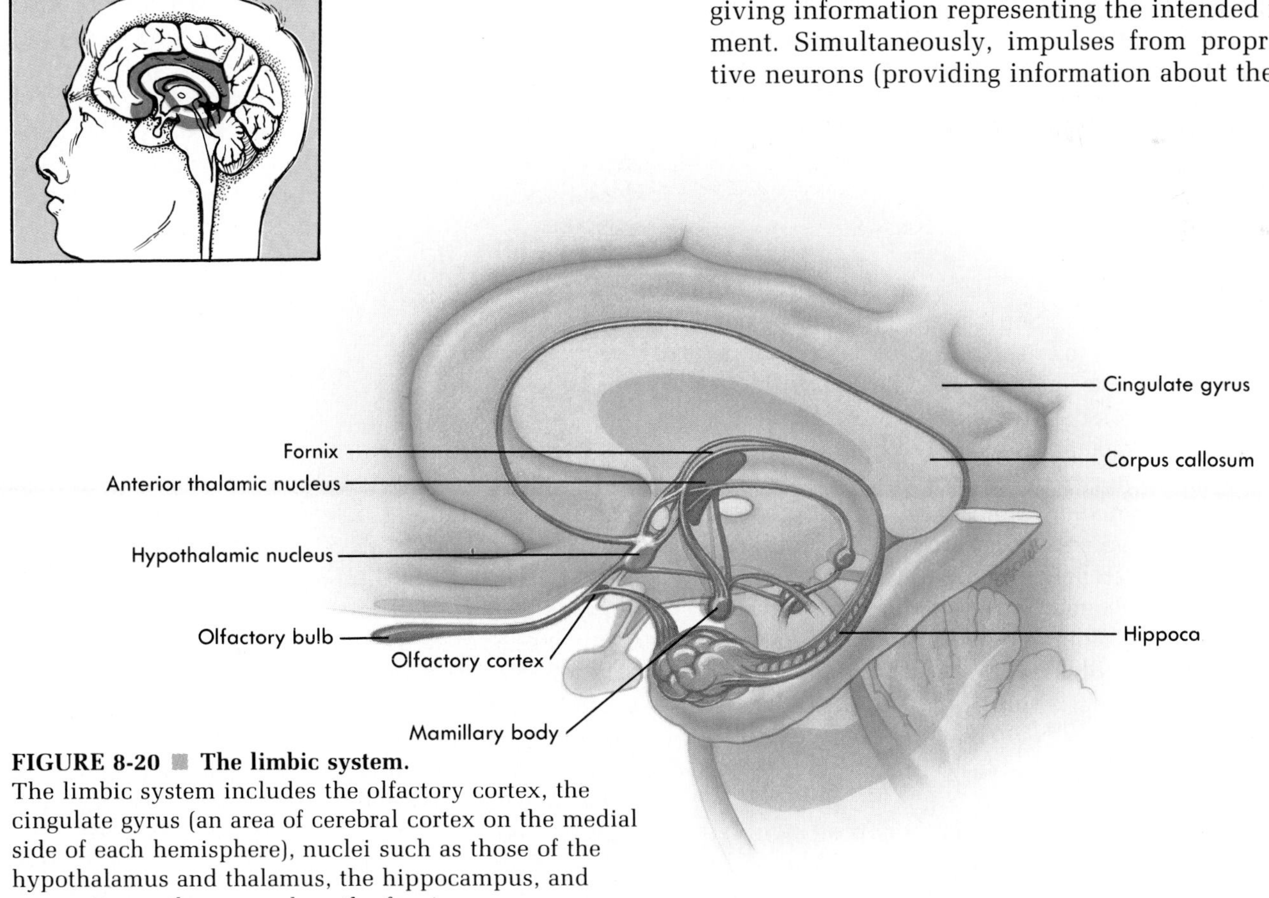

FIGURE 8-20 ■ The limbic system.
The limbic system includes the olfactory cortex, the cingulate gyrus (an area of cerebral cortex on the medial side of each hemisphere), nuclei such as those of the hypothalamus and thalamus, the hippocampus, and connecting pathways such as the fornix.

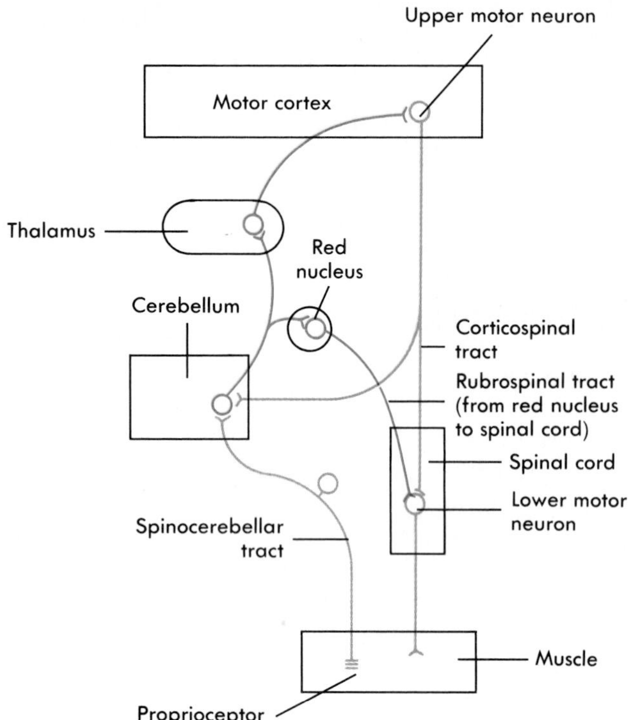

FIGURE 8-21 ■ Cerebellar comparator function. Impulses from the motor cortex of the cerebrum stimulate motor neurons in the spinal cord, which innervate skeletal muscle. At the same time, impulses are sent from the motor cortex to the cerebellum to tell the cerebellum what movement is intended. The cerebellum compares the intended movement with the actual movement and sends impulses to motor neurons in the spinal cord, which adjust the movement toward what is intended.

tion of the body parts) that innervate the joints and tendons of the structure being moved, reach the cerebellar cortex. The cerebellum compares the intended movement from the motor cortex with the actual movement from the moving structure. If a difference is detected, the cerebellum sends impulses to the motor neurons in the spinal cord to correct the discrepancy. The result is smooth and coordinated movements. For example, if you close your eyes, the cerebellar comparator function allows you to touch your nose smoothly and easily with your finger. If the cerebellum is not functioning, your finger will tend to overshoot the target. One effect of alcohol is that it directly inhibits the cerebellum. Dysfunction of the cerebellar comparator can be understood by observing the actions of someone who is drunk.

Another function of the cerebellum involves learning a motor skill such as playing the piano or riding a bicycle. When such a skill is being learned, the cerebrum is directly involved in initiating the various movements. Once the cerebellum "learns" these skills, much of the movement can be accomplished automatically by the cerebellum.

Spinal Cord

The **spinal cord** extends from the foramen magnum at the base of the skull to the second lumbar vertebra. The cord consists of a central gray portion and a peripheral white portion (Figure 8-22). The gray matter, seen in cross section, is shaped like the letter H with **posterior** (dorsal) **horns,** containing sensory neurons and **anterior** (ventral) **horns,** containing motor neurons. In the thoracic and upper lumbar regions, an additional gray horn, the **lateral horn,** contains sympathetic autonomic neurons. Axons carrying impulses to (ascending) or from (descending) the brain are grouped by function as nerve pathways (tracts) within the white portion of the spinal cord.

Dorsal (posterior) and ventral (anterior) roots exit the spinal cord. The **dorsal root** consists of afferent nerve processes that carry action potentials to the spinal cord, and the **ventral root** consists of efferent nerve processes that carry action potentials away from the spinal cord. The dorsal roots have **dorsal root ganglia** (also called spinal ganglia), which contain the cell bodies of the afferent neurons. The axons of these neurons project into the posterior horn, where they synapse with other neurons, or ascend or descend in the spinal cord. The dorsal and ventral roots unite to form **spinal nerves.**

> **3 Explain why the dorsal root ganglia are larger in diameter than the spinal nerves.** ?

Reflexes

Reflexes are responses to stimuli that do not involve conscious thought. Reflexes allow us to react to a stimulus more quickly than would be possible if conscious thought were involved. The **reflex arc,** (see Figure 8-22) is the basic functional unit of the nervous system and is the smallest, simplest pathway capable of receiving a stimulus and yielding a response. A reflex arc has several basic components: a sensory receptor, an afferent or sensory neuron, association neurons (neurons located between and communicating with the afferent and efferent neurons), an efferent or motor neuron, and an effector organ. Most reflexes involve the spinal cord or brainstem and don't involve higher brain centers.

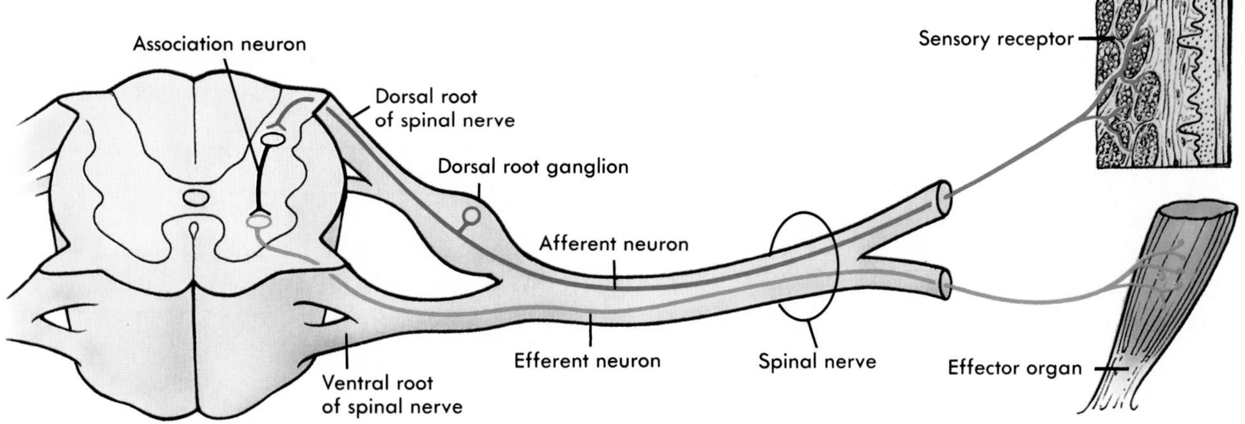

FIGURE 8-22 ■ Reflex arc.
Basic diagram of a reflex arc, including the sensory receptor, afferent neuron, association neuron, efferent neuron, and effector organ.

An example of a reflex can be seen when a person's finger touches a hot stove. Sensory receptors in the skin detect the heat. Afferent neurons carry this information as action potentials to the spinal cord, where they synapse with association neurons. The association neurons, in turn, synapse with efferent neurons in the spinal cord that send action potentials along their axons to flexor muscles in the upper limb. These muscles contract and pull the finger away from the stove. No conscious thought is required for this reflex, and withdrawal of the finger from the the painful stimulus begins before the person is consciously aware of any pain.

Pathways

■ The names of most ascending (afferent/sensory) and descending (efferent/motor) pathways in the central nervous system reflect their general function (Table 8-2 and Figure 8-23). Pathways are usually given composite names in which the first half of the name indicates their origin and the second half indicates their termination. The names of ascending pathways, therefore, usually begin with *spino-*, indicating that they originate in the spinal cord. For example, a spinothalamic tract is one that originates in the spinal cord and terminates in the thalamus. The names of descending pathways usually begin with *cortico-*, indicating that they begin in the cerebral cortex. The corticospinal tract is a descending tract that originates in the cerebral cortex and terminates in the spinal cord.

Most ascending pathways consist of two or three neurons in sequence from the periphery to the brain. The primary neuron in almost every case has its cell body located in a dorsal root ganglion or in a sensory cranial nerve ganglion, and its axon extends into the spinal cord, where it synapses with a secondary neuron. The secondary neuron ascends within one of the spinal cord tracts to a termination in the brainstem, cerebellum, or thalamus. There are often short association neurons located between the primary and secondary neurons. Almost all neurons relaying information to the cerebrum terminate in the thalamus. A tertiary neuron then relays the information from the thalamus to the cerebral cortex.

The major ascending pathways involved in the conscious perception of external stimuli are the spinothalamic tracts and the dorsal column (see Table 8-2). The **spinothalamic tracts** carry general sensory perceptions such as pain, temperature, and light touch, sensations from the skin. The **dorsal column** carries **proprioceptive** (pro′pre-o-sep′tiv) information from joints and tendons and two-point discrimination (the ability to distinguish between two points on the skin being touched simultaneously) from the skin (Figure 8-24). The dorsal column is not named according to the scheme described above but is named for the columns of white fibers in the dorsal portion of the spinal cord. **Proprioception** is the sensation of body position. The ascending pathways carrying information of which we are not consciously aware include the

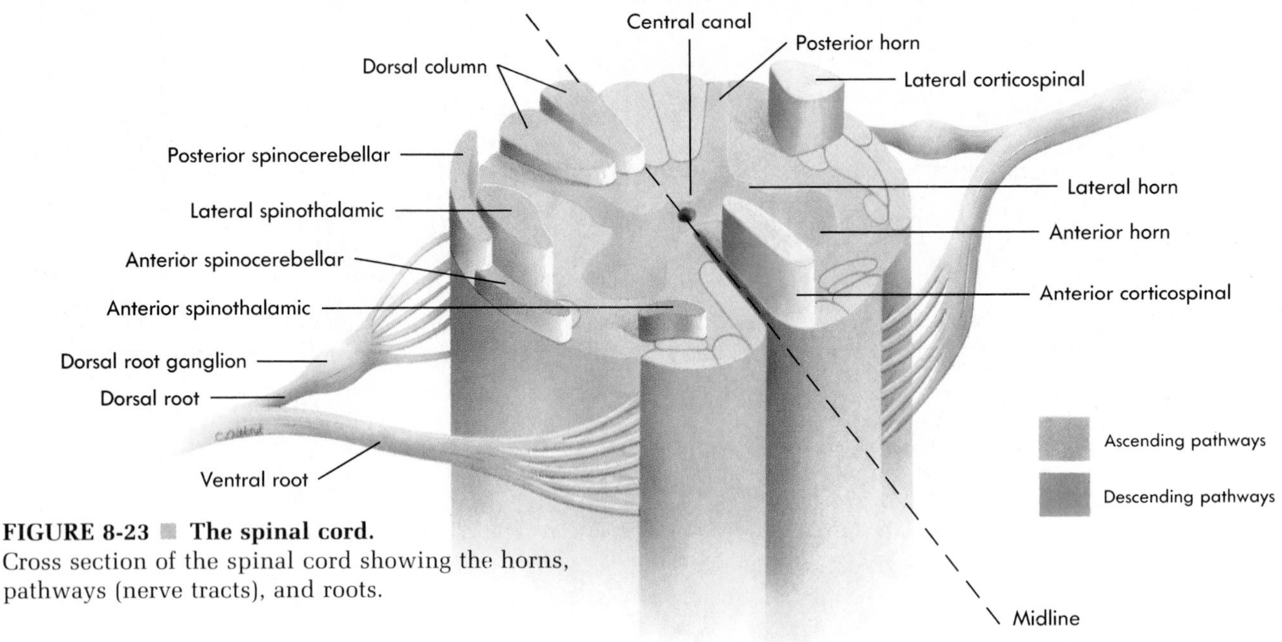

FIGURE 8-23 The spinal cord.
Cross section of the spinal cord showing the horns, pathways (nerve tracts), and roots.

TABLE 8-2

Examples of Spinal Cord Pathways (see Figures 8-23 to 8-25)

PATHWAY	FUNCTION
ASCENDING	
Spinothalamic	Pain, temperature, light touch, pressure, tickle, and itch sensations
Dorsal column	Proprioception, two-point discrimination, deep pressure, and vibration
Spinocerebellar	Proprioception to cerebellum
DESCENDING	
Corticospinal	Muscle tone and skilled movement, especially of the hands
Extrapyramidal	Unconscious control of general body movements

spinocerebellar tracts, carrying proprioceptive information to the cerebellum.

The voluntary motor system consists of two primary groups of neurons: lower motor neurons and upper motor neurons. **Lower motor neurons** are in either the anterior horn of the spinal cord central gray matter or the cranial nerve nuclei of the brainstem, and the axons from both groups extend through nerves to skeletal muscles. **Upper motor neurons** originate in the cerebral cortex, cerebellum, and brainstem, and their axons descend into the spinal cord, where they modulate the activity of the lower motor neurons. Their fibers compose the descending motor pathways.

The major descending pathway involved in conscious motor control is the **corticospinal tract** (Figure 8-25). This pathway contains axons of upper motor neurons that pass through the pyramids of the medulla oblongata, most of which cross to the opposite side, and synapse with lower motor neurons in the anterior horn of the spinal cord. As a result of the crossing over of axons, each half of the brain controls the opposite side of the body. The pathways involved in the unconscious control of motor function are collectively referred to as extrapyramidal because their fibers do not pass through the pyramids but take other routes through the medulla oblongata. An example is the pathway that an axon from the red nucleus takes to reach the spinal cord in the cerebellar comparator system. This pathway is called the rubrospinal tract (*rubro* means red). (see Figure 8-21).

Meninges

Three connective tissue layers, the **meninges** (mĕ-nin′jēz) (Figure 8-26), surround and protect the brain and spinal cord. The most superficial and thickest layer is the **dura mater** (du′rah ma′ter; tough mother). Folds of dura mater extend into the longitudinal fissure between the two cerebral hemispheres and between the cerebrum and cerebellum.

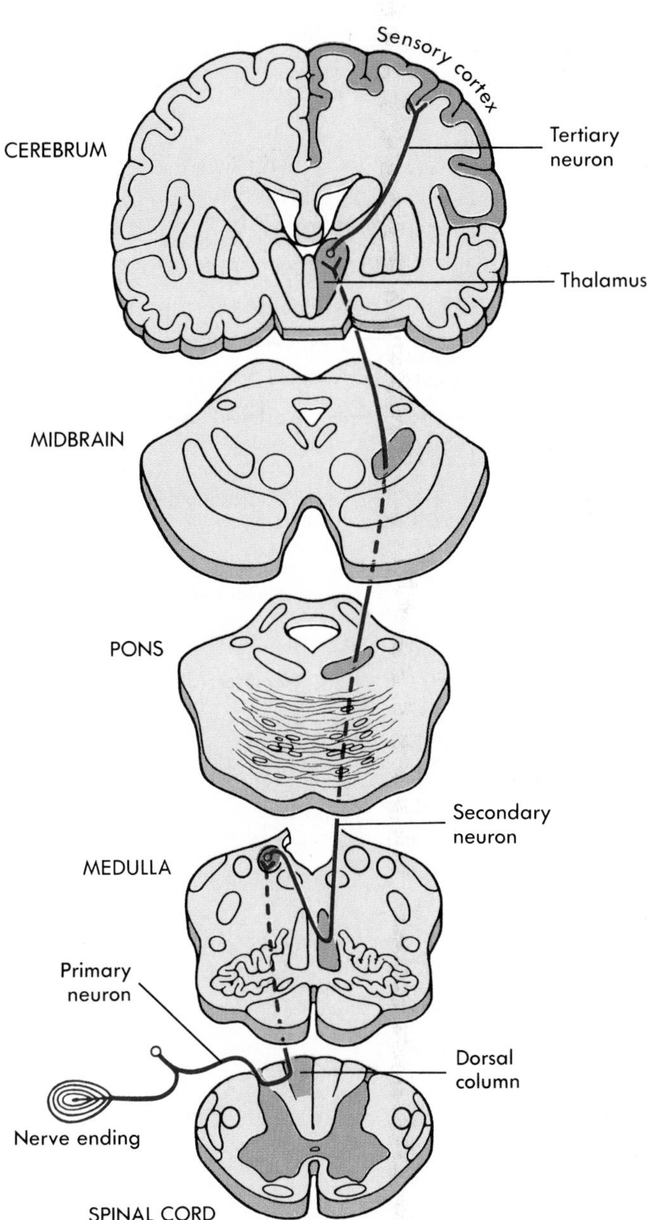

FIGURE 8-24 ■ Example of an ascending pathway. The dorsal column system. This example shows the sensory pathway for touch and pressure. Information from nerve endings in the skin travels to the medulla where it crosses to the other side of the body on its way to the primary somesthetic cortex.

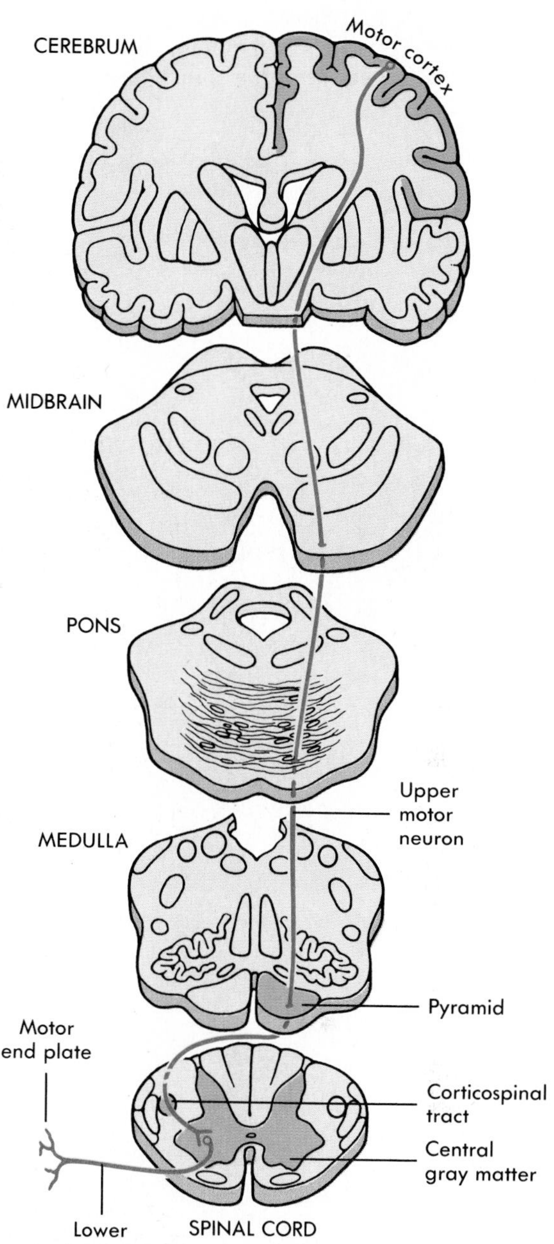

FIGURE 8-25 ■ Example of a descending pathway. The pyramidal system. Information controlling voluntary movement (such as turning a page) travels from the motor cortex to the medulla where it crosses to the opposite side of the body before traveling to the muscle.

The dura mater contains spaces called **dural sinuses,** which collect blood from the small veins of the brain. The dural sinuses empty into the jugular veins, which exit the skull.

The dura mater of the spinal cord is surrounded by an **epidural space** (see Figure 8-26). Epidural anesthesia of the spinal nerves is induced by injecting anesthetics into this space.

The second meningeal layer is the very thin, wispy **arachnoid** (ar-ak′noyd; spiderlike; that is, cobwebs) **layer.** The third meningeal layer, the **pia mater** (pe′ah; affectionate mother), is very tightly bound to the surface of the brain and spinal cord. Between the arachnoid layer and the pia mater is the **subarachnoid space,** which is filled with cerebrospinal fluid and contains blood vessels.

FIGURE 8-26 Meninges.
A Meningeal coverings of the brain.
B Meningeal coverings of the spinal cord.

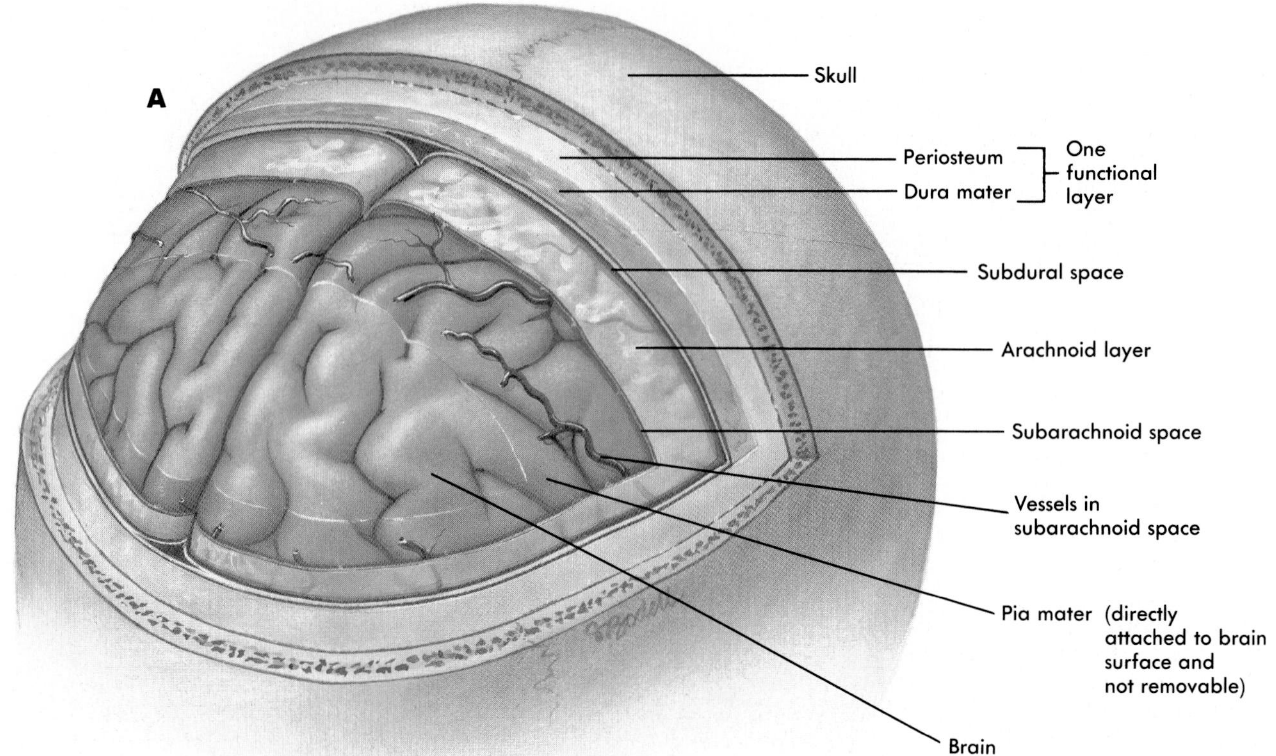

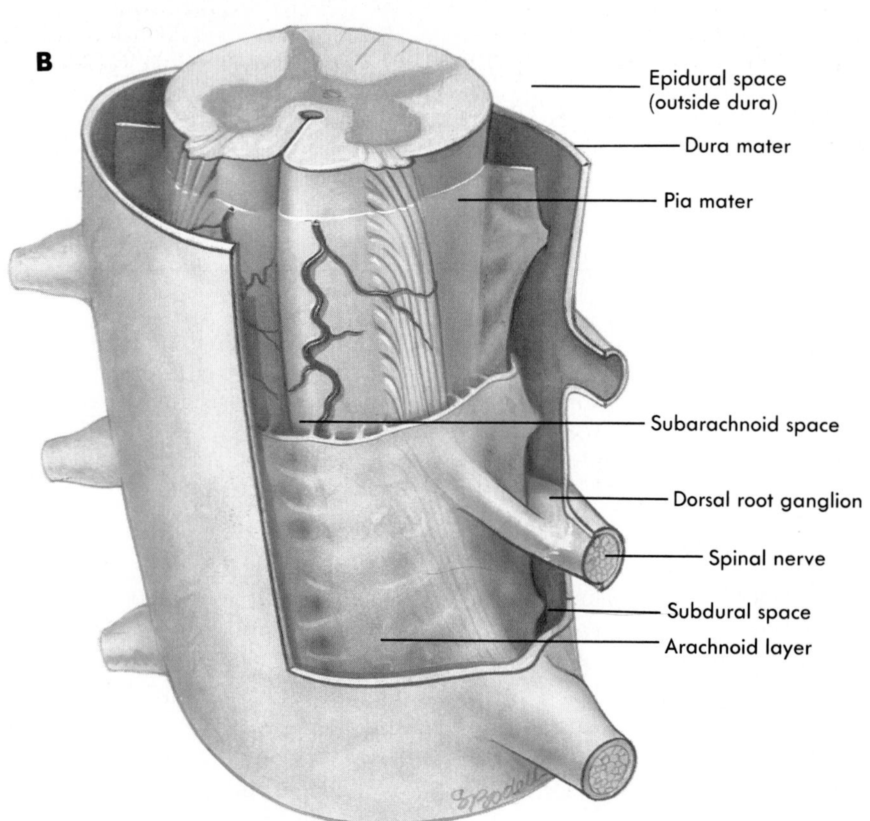

The spinal cord extends only to approximately the level of the second lumbar vertebra. Spinal nerves surrounded by meninges extend to the end of the vertebral column. Because there is no spinal cord in the inferior portion of the vertebral canal, a needle can be introduced into the subarachnoid space inferior to the end of the spinal cord to induce spinal anesthesia (spinal block) or to take a sample of cerebrospinal fluid (spinal tap) without damaging the cord. The cerebrospinal fluid may be examined for infectious agents (meningitis) or for blood (hemorrhage). A radiopaque substance may be injected into this area, and a myelograph (x-ray of the spinal cord) may be taken to visualize spinal cord defects or damage.

Ventricles

The CNS contains fluid-filled cavities, called ventricles, that may be quite small in some areas and large in others (Figure 8-27). Each cerebral hemisphere contains a relatively large cavity, the **lateral ventricle.** A smaller midline cavity, the **third ventricle,** is in the center of the diencephalon between the two halves of the thalamus and is connected by foramina (holes) to the lateral ventricles. The **fourth ventricle** is located at the base of the cerebellum and is connected to the third ventricle by a narrow canal, the **cerebral aqueduct.** The fourth ventricle is continuous with the **central canal** of the spinal cord.

Cerebrospinal Fluid

Cerebrospinal fluid (CSF) bathes the brain and spinal cord, providing a protective cushion around the CNS. It is produced as a blood filtrate by the **choroid** (ko′royd; lacy) **plexus**, a specialized structure in the ventricles. CSF fills the ventricles, the central canal, and the subarachnoid space. The cerebrospinal fluid flows from the lateral ventricles into the third ventricle and then through the cerebral aqueduct into the fourth ventricle. It exits from the interior of the brain through foramina in the walls of the fourth ventricle and enters the subarachnoid space. Masses of arachnoid tissue, **arachnoid granulations,** penetrate into the superior sagittal sinus (a dural sinus in the longitudinal fissure) (Figure 8-28), and cerebrospinal fluid passes from the subarachnoid space into the blood though these granulations.

Blockage of the foramina of the fourth ventricle or of the cerebral aqueduct may result in accumulation of cerebrospinal fluid in the ventricles, a condition known as **hydrocephalus.** The accumulation of fluid inside the brain causes pressure that compresses the nervous tissue and dilates the ventricles. Compression of the nervous tissue usually results in irreversible brain damage. If the skull bones are not completely ossified when the hydrocephalus occurs, such as in a fetus or newborn, the pressure may also cause severe enlargement of the head. Hydrocephalus is treated by placing a drainage tube (shunt) into the brain ventricles in an attempt to eliminate the high internal pressures.

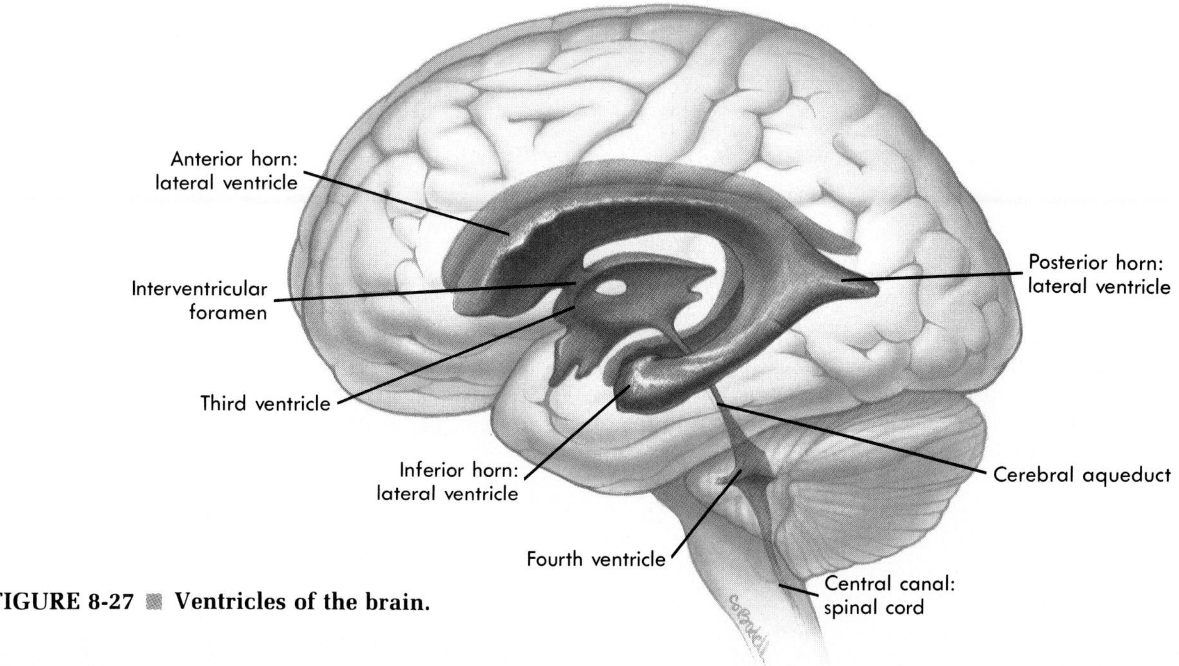

FIGURE 8-27 **Ventricles of the brain.**

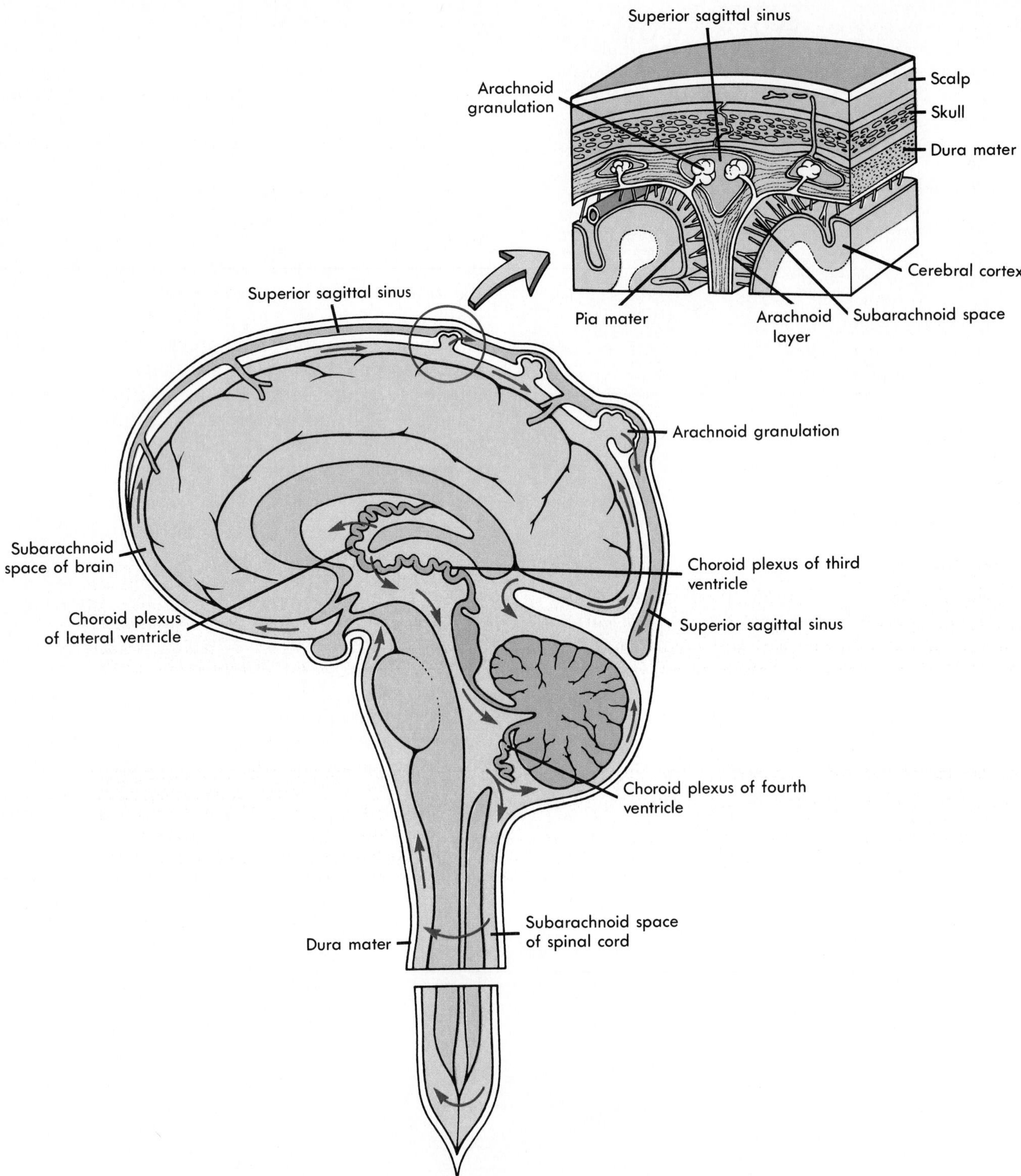

FIGURE 8-28 ■ Flow of cerebrospinal fluid.
CSF flows through the ventricles as shown. It exits the fourth ventricle and enters the subarachnoid space. CSF passes back into the blood through the arachnoid granulations, which penetrate the superior sagittal sinus.

Central Nervous System Disorders

INFECTIONS

Encephalitis is an inflammation of the brain, most often caused by a virus and less often by bacteria or other agents. A large variety of symptoms result, including fever, coma, and convulsions. Encephalitis may also result in death. **Meningitis** is an inflammation of the meninges. It may be caused by either a viral infection or a bacterial infection. Symptoms usually include stiffness in the neck, headache, and fever. In severe cases, meningitis may also cause paralysis, coma, or death.

Rabies is a viral disease transmitted by the bite of an infected animal such as a skunk, bat, or dog. The rabies virus infects the brain, salivary glands, muscles, and connective tissue. When the patient attempts to swallow, the effort can produce pharyngeal muscle spasms. Sometimes even the thought of swallowing water or the sight of water can induce the spasms. Thus the term hydrophobia, the fear of water, is applied to the disease. The brain infection results in abnormal excitability, aggression, and, in the later stages, paralysis and death.

Tabes dorsalis is a progressive disorder occurring as a result of a syphilis infection. It may occur many years after the original infection. *Tabes* means a wasting away, and *dorsalis* refers to the dorsal roots and dorsal regions of the spinal cord. The symptoms include ataxia, which is the inability to coordinate voluntary muscle activity, such as walking, because of a loss of proprioceptive function, carried in the dorsal column of the cord; anesthesia (loss of sensation) as a result of dorsal root damage; and eventually paralysis, as the infection spreads.

Multiple sclerosis (MS), although of unknown cause, is possibly viral in origin. The disease results in inflammation and an increased activity of the immune system (hyperimmunity). The inflammation and immune response result in localized brain lesions and demyelination of the brain and spinal cord. The myelin sheaths around axons become sclerotic, or hard, resulting in poor conduction of action potentials. Its symptomatic periods are separated by periods of apparent remission. With each recurrence, however, many neurons are permanently damaged. Progressive symptoms of the disease include exaggerated reflexes, tremor, nystagmus (tremorous movement of the eyes), and speech defects.

DYSKINESIAS

Dyskinesias are a group of disorders involving the basal ganglia that result in brisk, jerky, purposeless movements resembling fragments of voluntary movements. **Sydenham's chorea** (St. Vitus' dance) is a disease usually associated with a toxic or infectious disorder that apparently causes temporary dysfunction of the basal ganglia in children. **Huntington's chorea** is a dominant hereditary disorder that begins in middle life and causes progressive degeneration of the basal ganglia in affected persons.

Cerebral palsy is a general term referring to defects in motor functions or coordination resulting from several types of brain damage, which may be caused by abnormal brain development or birth-related injury. Some symptoms of cerebral palsy are related to basal ganglia dysfunction. One of the features of cerebral palsy is the presence of slow, writhing, aimless movements. When the face, neck, and tongue muscles are involved, characteristics are grimacing, protrusion and writhing of the tongue, and difficulty in speaking and swallowing.

Parkinson's disease, characterized by muscular rigidity, tremor, a slow, shuffling gait, and general lack of movement, is caused by a lesion in another part of the basal ganglia. A resting tremor called "pill-rolling" is characteristic of Parkinson's disease and consists of circular movement of the opposed thumb and index finger tip. The increased muscular rigidity in Parkinson's disease results from defective inhibition of muscle tone by some of the basal ganglia. In this disease, dopamine, an inhibitory neurotransmitter substance, is deficient. Parkinson's disease can be treated with L-dopa, a precursor to dopamine that will cross the blood-brain barrier from the capillaries of the brain into the brain tissue. Dopamine cannot cross the barrier.

Cerebellar lesions result in a spectrum of characteristic functional disorders. There is a tendency to overshoot, to point past a mark that one tries to touch with the finger. A cerebellar tremor is an "intention tremor," that is, the more carefully one tries to control a given movement, the greater the tremor becomes. For example, when a person with a cerebellar tremor tries to drink a glass of water, the closer the glass comes to the mouth, the shakier the movement becomes. This type of tremor is in direct contrast to basal ganglia tremors described previously, in which the resting tremor largely or completely disappears during purposeful movement.

OTHER DISORDERS

Tumors of the brain develop from glial cells and not from neurons. Symptoms vary widely, depending on the location of the tumor, but may include headaches, neuralgia (pain along the distribution of a peripheral nerve), paralysis, seizures, coma, and death.

Stroke is a term meaning a sudden blow, suggesting the speed with which this type of defect can occur. It is also referred to clinically as a **cerebrovascular accident** (CVA) and is caused by hemorrhage, thrombosis (a clot in a blood vessel), embolism (a piece of clot that has broken loose and floats through the circulation until it reaches a vessel too small for it to pass through, which it blocks), or vasospasm of the cerebral blood vessels. These causes result in an **infarct,** a local area of cell death caused by a lack of blood supply. Symptoms depend on the location, but may include anesthesia or paralysis on the side of the body opposite the cerebral infarct.

Continued.

Central Nervous System Disorders—cont'd

OTHER DISORDERS—cont'd

Senility once was thought to be a normal part of aging. The Latin term *senile* means old age. The severe symptoms of senility—general intellectual deficiency, mental deterioration (called dementia), memory loss, short attention span, moodiness, and irritability—result from several specific disease states. **Alzheimer's disease** is a severe type of senility, often affecting people under age 50. It results in general mental deterioration. The exact cause is unknown.

Tay-Sachs disease is a hereditary lipid-storage disorder of infants that primarily affects neurons of the CNS and results in severe brain disfunction. Symptoms include paralysis, blindness, and death, usually before age 5.

Epilepsy is actually a group of brain disorders that have seizure episodes in common. The seizure, a sudden massive neuronal discharge, can be either partial or complete, depending on the amount of brain involved and whether or not consciousness is impaired. The neuronal discharges may stimulate muscles innervated by the nerves involved, resulting in involuntary muscle contractions, that is, convulsions.

Headaches have a variety of causes that can be grouped into two basic classes: extracranial and intracranial. Extracranial headaches can be caused by inflammation of the paranasal sinuses, dental irritations, eye disorders, or tension in the muscles moving the head and neck. Intracranial headaches may result from inflammation of the brain or meninges, vascular problems, mechanical damage, or tumors.

PERIPHERAL NERVOUS SYSTEM

The **peripheral nervous system (PNS)** collects information and relays it by way of afferent fibers to the central nervous system, where it is evaluated. Efferent fibers in the peripheral nervous system relay information from the central nervous system to various parts of the body, primarily muscles and glands, regulating activity in those structures. The peripheral nervous system can be divided structurally into two parts: a cranial part, consisting of 12 pairs of nerves, and a spinal part, consisting of 31 pairs of nerves.

Cranial Nerves

The 12 **cranial nerves** (Figure 8-29) are listed in Table 8-3. There are three general categories of cranial nerve function: (1) sensory (afferent), (2) motor (efferent), and (3) parasympathetic (efferent). Sensory functions include the special senses such as vision and the more general senses such as touch and pain in the face. Motor functions refer to the control of skeletal muscles in the head and neck through motor neurons. Parasympathetic (part of the autonomic nervous system) function involves the regulation of glands, smooth muscle, and cardiac muscle. Some cranial nerves have only one of the three functions, whereas others have more than one. Cranial nerves with more than one category of function are called mixed nerves.

Three cranial nerves (I, II, and VIII) are sensory only, and four cranial nerves (IV, VI, XI, and XII) are motor only.

The trigeminal (V) nerve has sensory and motor functions. It has the greatest general sensory distribution of all the cranial nerves. It is the only cranial nerve supplying sensory innervation to the skin of the face. The skin over all the rest of the body is innervated by spinal nerves. Injections of anesthetic administered by a dentist are designed to block sensory transmission from the teeth by branches of the trigeminal nerve. The dental branches of the trigeminal nerve are probably anesthetized more often than any other nerve in the body.

The oculomotor nerve (III) is motor and parasympathetic. The facial (VII), glossopharyngeal (IX), and vagus (X) nerves have all three functions: sensory, motor, and parasympathetic. The vagus nerve is perhaps the most important parasympathetic nerve in the body. It is very important in regulating the functions of the thoracic and abdominal organs. Functions such as regulating heart rate, respiration rate, and digestion are all controlled by the vagus nerve.

Spinal Nerves

The **spinal nerves** arise along the spinal cord from the union of the dorsal root (containing afferent/sensory fibers) and ventral root (containing efferent/motor fibers) (see Figures 8-22 and 8-23). Most of the spinal nerves exit the vertebral column between adjacent vertebrae, and all the spinal nerves are mixed nerves, containing both afferent and efferent fibers. The efferent fibers are both so-

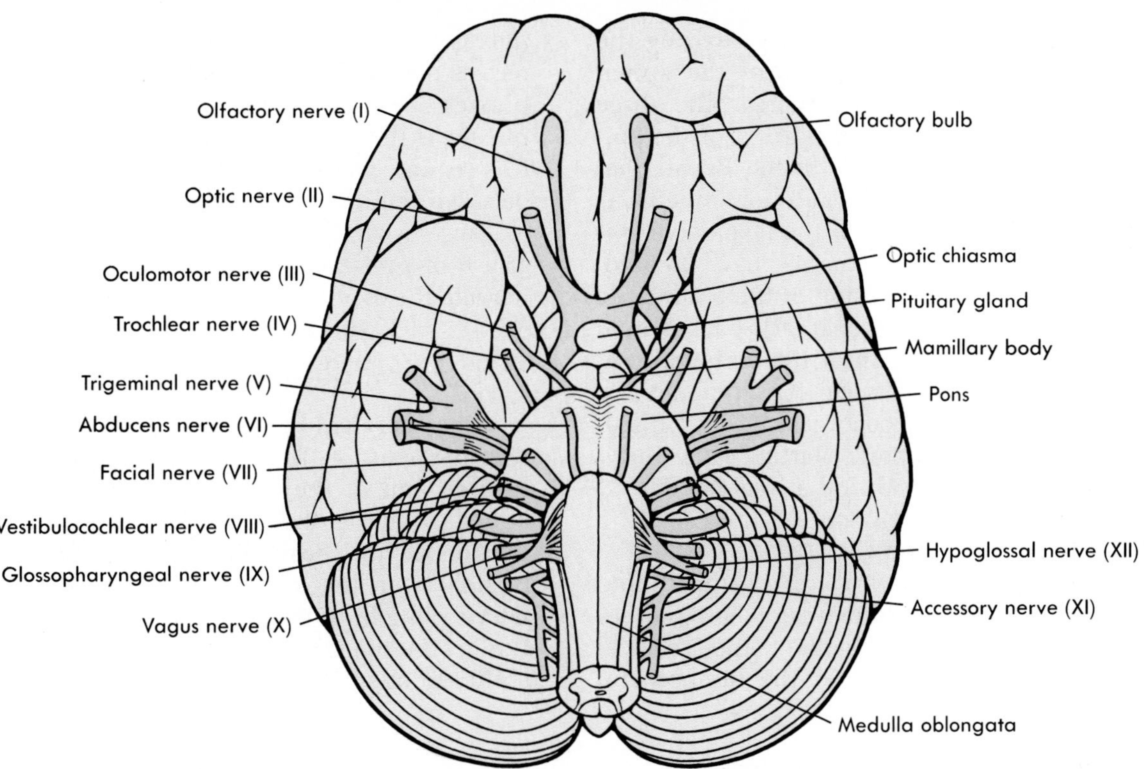

FIGURE 8-29 ■ Cranial nerves.
Inferior surface of the brain showing the origins of the cranial nerves.

TABLE 8-3

Cranial Nerves and Their Functions (see Figure 8-29)

NUMBER	NAME	GENERAL FUNCTION*	SPECIFIC FUNCTION
I	Olfactory	S	Smell
II	Optic	S	Vision
III	Oculomotor	M,P	Motor to four of six eye muscles and upper eyelid; parasympathetic: constricts pupil, thickens lens
IV	Trochlear	M	Motor to one eye muscle
V	Trigeminal	S,M	Sensory to face and teeth; motor to muscles of mastication (chewing)
VI	Abducens	M	Motor to one eye muscle
VII	Facial	S,M,P	Sensory: taste; motor to muscles of facial expression; parasympathetic to salivary and tear glands
VIII	Vestibulocochlear	S	Hearing and balance
IX	Glossopharyngeal	S,M,P	Sensory: taste and touch to back of tongue; motor to pharyngeal muscle; parasympathetic to salivary glands
X	Vagus	S,M,P	Sensory to pharynx, larynx and viscera; motor to palate, pharynx, and larynx; parasympathetic to viscera of thorax and abdomen
XI	Accessory	M	Motor to two neck and upper back muscles
XII	Hypoglossal	M	Motor to tongue muscles

*S, sensory; M, motor; P, parasympathetic.

matomotor and autonomic. The nerves exiting the thoracic and upper lumbar regions of the spinal cord carry sympathetic fibers, whereas the three middle nerves exiting the sacral portion of the spinal cord carry parasympathetic fibers. Recall that the primary cell bodies of these autonomic neurons are located in the lateral horn of the respective spinal cord level. All other spinal nerves have no autonomic component (see discussion of the autonomic nervous system later in this chapter). Each spinal nerve is named for the region of the vertebral column where it emerges—cervical (C), thoracic (T), lumbar (L), sacral (S), and coccygeal (Cx). The spinal nerves are also numbered (starting superiorly) according to their order within that region. The 31 pairs of spinal nerves are, therefore, C1 to C8, T1 to T12, L1 to L5, S1 to S5, and Cx.

Most of the spinal nerves are organized into three **plexuses** (plek'sus-ez; braids; where nerves come together and then separate): the cervical plexus, the brachial plexus, and the lumbosacral plexus (Table 8-4 and Figure 8-30). The major nerves of the neck and limbs are branches of these plexuses. Spinal nerves T2 to T11 do not join a plexus as there are no limbs attached to the body in this region. These nerves extend around the thorax between the ribs as simple arcs, giving off branches to muscles and skin. Nerves innervating a given group of skeletal muscles (efferents) often also innervate the overlying skin (afferents) (see Table 8-4).

Cervical plexus

The **cervical plexus** originates from spinal nerves C1 to C4. Branches from this plexus innervate several of the muscles attached to the hyoid bone, as well as the skin of the neck and posterior portion of the head. One of the most important branches of the cervical plexus is the **phrenic nerve,** which innervates the diaphragm. Contraction of the diaphragm is largely responsible for the ability to breathe (see Chapter 15).

TABLE 8-4

Plexuses of the Spinal Nerves (see Figure 8-30)

PLEXUS	ORIGIN	MAJOR NERVES	MUSCLES INNERVATED	SKIN INNERVATED
Cervical	C1-C4		Several neck muscles	Neck and posterior head
		Phrenic	Diaphragm	
Brachial	C5-T1	Axillary	Two shoulder muscles	Part of shoulder
		Radial	Posterior arm and forearm muscles	Posterior arm, forearm, and hand
		Musculocutaneous	Anterior arm muscles	Radial surface of forearm
		Ulnar	Two anterior forearm muscles, most of the intrinsic hand muscles	Ulnar side of hand
		Median	Most anterior forearm muscles, some intrinsic hand muscles	Radial side of hand
Lumbosacral	L1-S4	Obturator	Medial thigh muscles	Medial thigh
		Femoral	Anterior thigh muscles (extensors)	Anterior thigh, medial leg and foot
		Sciatic		
		Tibial	Posterior thigh muscles (flexors), anterior and posterior leg muscles, most foot muscles	Sole of foot
		Common peroneal	Lateral thigh and leg, some foot muscles	Anterior and lateral leg, and dorsal foot

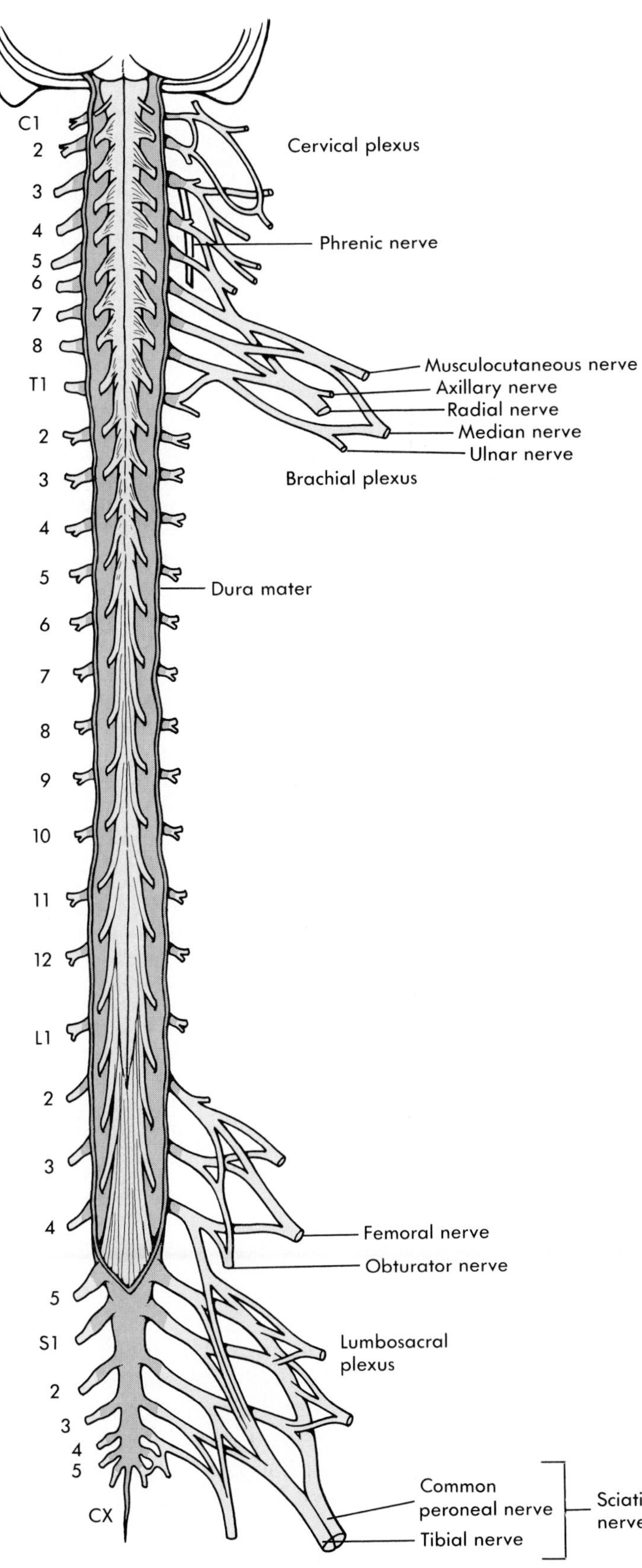

FIGURE 8-30 ■ Spinal nerves.
Spinal cord, the spinal nerves, their plexuses, and their branches.

> **4** **The phrenic nerve may be damaged where it descends along the neck, or during open thorax or open heart surgery. Explain how damage to the right phrenic nerve would affect the diaphragm. Describe the effect on breathing of completely severing the spinal cord in the thoracic region (below the exit point of the phrenic nerve) versus the upper cervical region (above the exit point of the phrenic nerve).** ?

Brachial plexus

■ The **brachial plexus** originates from spinal nerves C5 to T1. Five major nerves emerge from the brachial plexus to supply the upper limb and shoulder. The **axillary nerve** innervates two shoulder muscles and the skin over part of the shoulder. The **radial nerve** innervates all the muscles located in the posterior arm and forearm. It innervates the skin over the posterior surface of the arm, forearm, and hand. The **musculocutaneous nerve** innervates the anterior muscles of the arm and the skin over part of the forearm. The **ulnar nerve** innervates two anterior forearm muscles and most of the intrinsic hand muscles. It also innervates the skin over the ulnar side of the hand. The **median nerve** innervates most of the anterior forearm muscles and some of the intrinsic hand muscles. It also innervates the skin over the radial side of the hand.

Lumbosacral plexus

■ The **lumbosacral plexus** originates from spinal nerves L1 to S4. Four major nerves exit the plexus to supply the lower limb. The **obturator nerve** innervates the muscles of the medial thigh and the skin over the same region. The **femoral nerve** innervates the anterior thigh muscles and the skin over the anterior thigh and medial side of the leg. The **tibial nerve** innervates the posterior thigh muscles, the anterior and posterior leg muscles, and most of the intrinsic foot muscles. It also innervates the skin over the sole of the foot. The **common peroneal nerve** innervates the muscles of the lateral thigh and leg, and some intrinsic foot muscles. It innervates the skin over the anterior and lateral leg, and the dorsal surface of the foot. The tibial and common peroneal nerves are bound together within a connective tissue sheath and are, as such, called the **sciatic nerve.**

Peripheral Nervous System Disorders

ANESTHESIA AND NEURITIS

Anesthesia is the loss of sensation. It may be a pathological condition, or it may be induced to facilitate surgery or some other medical action. **Neuralgia** is a condition involving severe spasms of throbbing or stabbing pain along the pathway of a nerve. **Neuritis** is an inflammation of a nerve resulting from any one of a number of causes, including injury or infection. In sensory nerves, neuritis may result in anesthesia and/or neuralgia. In motor nerves, neuritis can result in the loss of motor function. **Trigeminal neuralgia** involves the trigeminal nerve and consists of sharp bursts of pain in the face. **Facial palsy** involves the facial nerve and results in unilateral paralysis of the facial muscles. The affected side of the face droops because of the absence of muscle tone. **Sciatica** is a neuralgia of the sciatic nerve, with pain radiating down the back of the thigh and leg. The most common cause of sciatica is a herniated lumbar disk putting pressure on the spinal nerves forming the lumbosacral plexus.

INFECTIONS

Tetanus is a bacterial infection found in soil contaminated with animal wastes, and it is often introduced into the body through an open wound. The bacteria produce a potent neurotoxin. The toxin prevents muscle relaxation so that the body becomes rigid. The jaw muscles are affected early in the disease, and the jaw cannot be opened (lockjaw). Death results from spasms in the diaphragm and other respiratory muscles.

Botulism is a disease caused by the toxin of an anaerobic bacterium, which grows in environments such as improperly processed cans of food. Botulism toxin is the most potent natural toxin known. The toxin blocks the release of acetylcholine in synapses. As a result, muscles cannot contract and the person may die from respiratory and/or cardiac failure.

Leprosy is a disease of the peripheral nervous system that can infect and kill skin and other cells. As a result, disfiguring nodules form on the body, and tissue necrosis may occur. Even though the disease can be treated with sulfone drugs, there are still millions of affected people in Asia and Africa. Leprosy is not highly contagious and is usually transmitted from lesions or clothing of an infected person through cuts or abrasions on the skin of another. The time from infection to the appearance of symptoms may be several years. The disease leprosy itself is usually not fatal, but patients may die from complications.

Herpes is a family of viral diseases characterized by skin lesions. The viruses apparently reside in the ganglia of sensory nerves and cause lesions along the course of the nerve. Herpes simplex I causes lesions of the lips and nose. The lesions are prone to occur during times of decreased resistance, such as during a cold. For this reason, they are called cold sores or fever blisters. Herpes simplex II (genital herpes) is responsible for a venereal (sexually transmitted) disease causing lesions on the external genitalia. Herpes zoster causes chicken pox in children and shingles in older adults.

Poliomyelitis (polio) is a viral infection of the central nervous system, but it damages the motor neurons, which extend into the peripheral nervous system. Without stimulation from the CNS, the muscles are paralyzed and they atrophy, or waste away.

OTHER DISORDERS

Myotonic dystrophy is a dominant hereditary disease characterized by muscle weakness, dysfunction, and atrophy, and by visual impairment as a result of nerve degeneration.

Neurofibromatosis is also a genetic disorder, in which neurofibromas (benign tumors along peripheral nerve tracts) occur in early childhood, and result in large skin growths, which may result in substantial disfigurement. The most famous victim of this disorder was the so-called "Elephant Man."

Myasthenia gravis is an autoimmune disorder in which the immune system attacks neuromuscular junctions, resulting in muscle weakness and abnormal fatigability.

AUTONOMIC NERVOUS SYSTEM

Efferent neurons can be divided into two systems—**somatomotor** (so-mă′to-mo′tor; motor to the body) **nervous system** and the **autonomic nervous system (ANS)**—which differ structurally and functionally. Single axons of somatomotor neurons (lower motor neurons) extend from the central nervous system to skeletal muscles. Autonomic pathways, on the other hand, consist of two neurons in series extending from the central nervous system to the innervated organs. The first neuron is called the **preganglionic neuron,** and the second neuron is called the **postganglionic neuron.** The neurons are so named because preganglionic neurons synapse with postganglionic neurons in ganglia outside of the central nervous system. Autonomic neurons innervate smooth muscle, cardiac muscle, and glands. Autonomic functions are usually unconsciously controlled.

The autonomic nervous system is composed of sympathetic and parasympathetic divisions (Table 8-5 and Figure 8-31). Increased activity in sympathetic neurons generally prepares the individual for

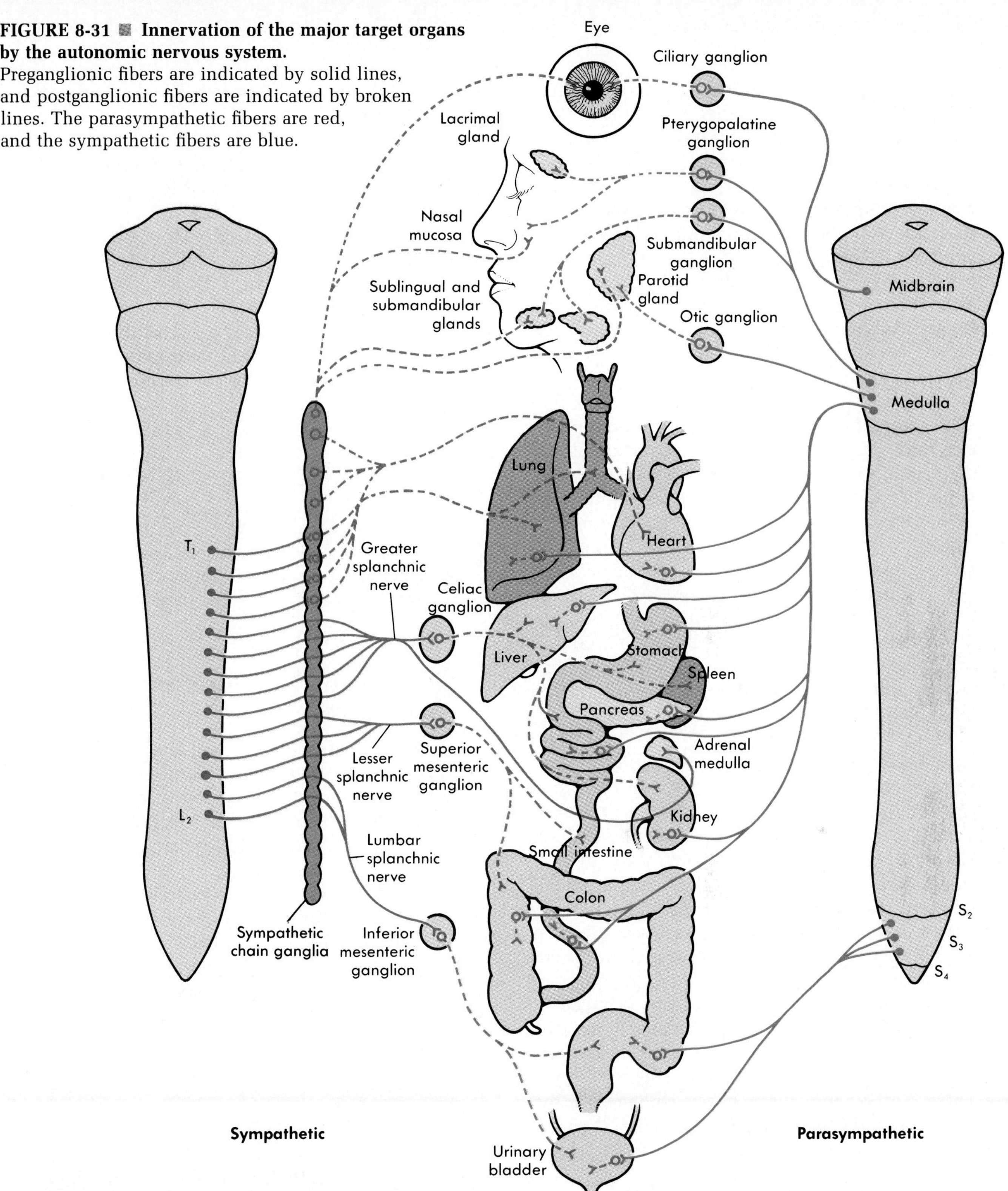

FIGURE 8-31 ■ Innervation of the major target organs by the autonomic nervous system. Preganglionic fibers are indicated by solid lines, and postganglionic fibers are indicated by broken lines. The parasympathetic fibers are red, and the sympathetic fibers are blue.

TABLE 8-5

Autonomic Nervous System (see Figure 8-31)

DIVISION	LOCATION OF PREGANGLIONIC CELL BODY	LOCATION OF POSTGANGLIONIC CELL BODY	FUNCTION
Sympathetic	T1-L2	Sympathetic chain ganglia or collateral ganglia	"Fight-or-flight"; prepares the body for physical activity
Parasympathetic	Cranial nerves III, VII, IX, X; spinal nerves S2-S4	Terminal ganglia near or embedded in the walls of target organs	Stimulates vegetative activities; slows heart and respiration rates; constricts pupil, thickens lens

physical activity, whereas parasympathetic stimulation generally activates vegetative functions.

Most organs that receive autonomic neurons are innervated by both the parasympathetic and sympathetic divisions. Sweat glands and blood vessels, however, are innervated by sympathetic neurons almost exclusively, whereas the smooth muscles associated with the lens of the eye are innervated only by parasympathetic neurons. In most cases, the influence of the two autonomic divisions is opposite on structures that receive dual innervation.

Sympathetic Division

The **sympathetic division** of the autonomic nervous system prepares a person for physical activity by increasing the heart rate, blood pressure (by decreasing the diameter of blood vessels supplying the viscera), and respiration, and by stimulating perspiration. The sympathetic division also stimulates the release of glucose from the liver for energy, and it stimulates mental activities. At the same time, it inhibits digestive activities. The sympathetic division is sometimes referred to as the "fight-or-flight" system because, as it increases the physical activity of the body, it prepares the person to either stand and face a threat, or leave as quickly as possible.

Cell bodies of sympathetic preganglionic neurons are in the lateral horn of the spinal cord gray matter (see Figure 8-23) between the first thoracic (T1) and the second lumbar (L2) segments. The axons of the preganglionic neurons exit through ventral roots and project to autonomic ganglia, called **sympathetic chain ganglia.** The ganglia are connected to each other and form a chain along both sides of the spinal cord. Some preganglionic neurons synapse in the sympathetic chain ganglia, but others pass through these ganglia and synapse in **collateral ganglia** located nearer target organs (the celiac, superior mesenteric, and inferior mesenteric ganglia shown in Figure 8-31). Postganglionic neurons arise in the sympathetic chain ganglia or in the collateral ganglia and project to target tissues: glands, smooth muscles, or cardiac muscle.

Parasympathetic Division

The **parasympathetic division** of the autonomic nervous system stimulates vegetative activities such as digestion, defecation, and urination. At the same time, it slows the heart rate and respiration. It also causes the pupil of the eye to constrict and the lens to thicken, both of which help in close vision, such as in reading a book.

Preganglionic cell bodies of the parasympathetic division are located either within brainstem nuclei of the oculomotor (III), facial (VII), glossopharyngeal (IX), or vagus (X) cranial nerves, or within the lateral horn in the S2 to S4 regions of the spinal cord.

Axons of the preganglionic neurons extend through spinal nerves or cranial nerves to **terminal ganglia** located either near their target organs (in the head) (see Figure 8-31) or embedded in the walls of target organs (in the rest of the body). The axons of the postganglionic neurons extend a relatively short distance from the terminal ganglia to the target organ.

5 List some of the responses stimulated by the autonomic nervous system in
(A) a person who is extremely angry and
(B) a person who has just finished eating and is now relaxing.

Autonomic Neurotransmitter Substances

The sympathetic and parasympathetic nerve endings secrete one of two neurotransmitters. All preganglionic neurons of the sympathetic and parasympathetic divisions and all postganglionic neurons of the parasympathetic division secrete **acetylcholine.** Most postganglionic neurons of the sympathetic division secrete **norepinephrine** (some secrete acetycholine). Many body functions can be stimulated or inhibited by drugs that either mimic these neurotransmitters or prevent the neurotransmitters from activating their target tissues.

Raynaud's disease involves the spasmodic contraction of blood vessels in the periphery, especially in the digits, and results in pale, cold hands that are prone to ulcerations and gangrene as a result of poor circulation. This condition may be caused by exaggerated sensitivity of the blood vessels to sympathetic innervation. Preganglionic denervation (cutting the preganglionic neurons) is occasionally performed to alleviate the condition. *Dysautonomia* is an inherited condition involving reduced tear secretion, poor vasomotor control, trouble in swallowing, and other symptoms. It is the result of poorly controlled autonomic reflexes.

Biofeedback and Meditation

Biofeedback takes advantage of electronic instruments or other techniques to monitor and change subconscious activities, many of which are regulated by the ANS. Skin temperature, heart rate, and nerve impulse patterns within the brain (brain waves) are monitored electronically. By watching the monitor and using biofeedback techniques, a person can learn to consciously reduce his heart rate and blood pressure or reduce the severity of migraine headaches.

Some people use biofeedback methods to relax by learning to reduce the heart rate or change the pattern of brain waves. The severity of stomach ulcers, high blood pressure, anxiety, or depression may be reduced by using biofeedback techniques.

Meditation is another technique that influences autonomic functions. Although it has not been documented, it has been claimed that meditation can improve one's spiritual well being, consciousness, and holistic view of the universe. It has been established, however, that meditation can influence autonomic functions. Meditation techniques are useful in some people in reducing heart rate, blood pressure, severity of ulcers, and other symptoms that are frequently associated with stress.

SUMMARY

The nervous system regulates and coordinates body functions.

DIVISIONS OF THE NERVOUS SYSTEM

The central nervous system (CNS) consists of the brain and spinal cord, whereas the peripheral nervous system (PNS) consists of nerves and ganglia.

The afferent division of the PNS transmits action potentials to the CNS; the efferent division carries action potentials away from the CNS.

The somatomotor nervous system innervates skeletal muscle and is mostly under voluntary control. The autonomic nervous system innervates cardiac muscle, smooth muscle, and glands, and it is mostly under involuntary control.

The autonomic nervous system has sympathetic and parasympathetic divisions.

CELLS OF THE NERVOUS SYSTEM

Neurons

Neurons receive stimuli and transmit action potentials. Neurons consist of a cell body, dendrites, and an axon.

Neurons are multipolar, bipolar, or unipolar.

Neuroglia

Neuroglia are the support cells of the nervous system. They include astrocytes, microglia, ependymal cells, oligodendrocytes, and Schwann cells.

Myelin Sheaths

Axons are either unmyelinated or myelinated.

Organization of Nervous Tissue

Nervous tissue consists of white matter and gray matter. White matter forms nerve tracts in the CNS and nerves in the PNS. Gray matter forms nuclei in the CNS and ganglia in the PNS.

PROPAGATION OF IMPULSES

Membrane Potentials and Action Potentials

A resting membrane potential results from the charge difference that exists across the membrane of nearly all cells.

An action potential occurs when the charge across the cell membrane is briefly reversed.

The Synapse

The synapse is the point of contact between two neurons. An action potential arriving at the synapse causes the release of a neurotransmitter, which diffuses across the synaptic cleft and binds to the receptors of the postsynaptic terminal.

Neuronal Circuits

Neuronal circuits are either diverging or converging.

CENTRAL NERVOUS SYSTEM

The central nervous system consists of the brain and spinal cord. The brain consists of the brainstem, diencephalon, cerebrum, and cerebellum.

Brainstem

The brainstem contains several nuclei as well as ascending and descending tracts.

The medulla oblongata contains nuclei that control such activities as heart rate, breathing, swallowing, and balance.

The pons contains relay nuclei between the cerebrum and cerebellum.

The midbrain is involved in hearing and in visual reflexes.

The reticular formation is scattered throughout the brainstem and is involved in maintaining consciousness, and in the sleep/wake cycle.

Diencephalon

The diencephalon consists of the thalamus (main sensory relay center), pineal body (may play a role in sexual maturation), and hypothalamus (the homeostatic control center of the body).

Cerebrum

The cerebrum has two hemispheres divided into lobes. The lobes are the frontal, parietal, occipital, and temporal.

Many CNS functions can be localized to specific areas of the cortex. Association cortex is involved in the recognition of information.

Speech involves Wernicke's area, Broca's area, and the interactions between them and other cortical areas.

An EEG monitors brain waves, which are a summation of the brain's electrical activity.

Memory consists of sensory (less than one second), short-term (lasting a few minutes), and long-term (permanent) memory.

Each hemisphere controls the opposite half of the body. Commissures connect the two hemispheres. The left hemisphere is thought to be the dominant, analytical hemisphere, and the right hemisphere is thought to be dominant for spatial perception and musical ability.

The basal ganglia are cerebral nuclei that inhibit extraneous muscular activity.

The limbic system is located deep within the cerebrum and is involved with the emotional and visceral response to smell.

Cerebellum

The cerebellum is involved in balance and muscle coordination. Its main function is as a comparator, comparing the intended action with what is occurring, and modifying to eliminate differences.

Spinal Cord

The spinal cord has a central gray portion organized into horns, and a peripheral white portion forming nerve tracts.

Roots of spinal nerves extend out of the cord.

Reflexes

Reflexes are the functional units of the nervous system.

A reflex arc consists of a sensory receptor, afferent neuron, association neurons, efferent neuron, and effector organ.

Pathways

The major ascending pathways are the spinothalamic (general sensory perception) and the dorsal column (proprioception).

The voluntary motor system consists of lower and upper motor neurons. The primary descending pathway for axons from upper motor neurons is the corticospinal pathway.

Meninges

Three connective tissue meninges cover the CNS: the dura mater, arachnoid layer, and pia mater.

Ventricles

The brain and spinal cord contain fluid-filled cavities: the lateral ventricles in the cerebral hemispheres, a third ventricle in the diencephalon, a cerebral aqueduct in the midbrain, a fourth ventricle at the base of the cerebellum, and a central canal in the spinal cord.

Cerebrospinal Fluid

Cerebrospinal fluid is formed in the choroid plexuses in the ventricles, it exits through the fourth ventricle, and it reenters the blood through arachnoid granulations in the superior sagittal sinus.

PERIPHERAL NERVOUS SYSTEM

The peripheral nervous system (PNS) consists of afferent and efferent fibers contained in spinal and cranial nerves.

Cranial Nerves

There are 12 cranial nerves, three with only sensory function (S), four with only motor function (M), and five with mixed functions. Four of the cranial nerves have parasympathetic function (P).

The cranial nerves are: olfactory (I; S), optic (II; S), oculomotor (III; M, P), trochlear (IV; M), trigeminal (V; S, M), abducens (VI; M), facial (VII; S, M, P), vestibulocochlear (VIII; S), glossopharyngeal (IX; S, M, P), vagus (X; S, M, P), accessory (XI; M), and hypoglossal (XII; M).

Spinal Nerves

The spinal nerves exit from the cervical, thoracic, lumbar, and sacral regions.

The nerves are grouped into plexuses.

- The phrenic nerve, which supplies the diaphragm, is the most important branch of the cervical plexus.
- The brachial plexus supplies nerves to the upper limb.
- The lumbosacral plexus supplies nerves to the lower limb.

AUTONOMIC NERVOUS SYSTEM

The autonomic nervous system contains preganglionic and postganglionic neurons.

The autonomic nervous system has sympathetic and parasympathetic divisions.

Sympathetic Division

The sympathetic division is involved in preparing the person for action, by increasing heart rate, blood pressure, and respiration rate.

Preganglionic cell bodies of the sympathetic division lie in the thoracic and upper lumbar regions of the spinal cord.

Postganglionic cell bodies are located in the sympathetic chain ganglia or in collateral ganglia.

Parasympathetic Division

The parasympathetic division is involved in vegetative activities such as the digestion of food, defecation, and urination.

Preganglionic cell bodies of the parasympathetic division are associated with some of the cranial and sacral nerves.

Postganglionic cell bodies are located in terminal ganglia, located either near or within target organs.

Autonomic Neurotransmitter Substances

All autonomic preganglionic and parasympathetic postganglionic neurons secrete acetylcholine.

Most sympathetic postganglionic neurons secrete norepinephrine.

CONTENT REVIEW

1. Describe the CNS and PNS anatomically and functionally.
2. Define the afferent and efferent divisions of the PNS, and the somatomotor and autonomic nervous systems.
3. What are the functions of neurons? Name the three parts of a neuron and describe their functions.
4. List the three types of neurons based upon their shapes.
5. Define neuroglia. Name and describe the functions of the different neuroglia.
6. What are the differences between unmyelinated and myelinated axons? Which propagates action potentials more rapidly? Why?
7. For nerve tracts, nerves, nuclei, and ganglia, name the cells or parts of cells found in each, state if they are white or gray matter, and name the part (CNS or PNS) of the nervous system in which they are found.
8. Describe the function of the synapse, starting with an action potential in the presynaptic neuron and ending with the generation of an action potential in the postsynaptic neuron.
9. Name the four parts of the brainstem and describe their functions.
10. Name the three main components of the diencephalon, describing their functions.
11. Name the five lobes of the cerebrum, describing their locations and functions.
12. Describe the cerebral cortex locations of the special and general senses and their association areas. How do the association areas interact with the primary areas?
13. Describe the process required to speak a word that is seen or heard.
14. Name the three types of memory and describe the processes that result in long-term memory.
15. Describe the function of the basal ganglia.
16. What is the function of the limbic system?
17. Describe the comparator activities of the cerebellum.
18. Describe the spinal cord gray matter. Where are sensory and motor neurons located in the gray matter?
19. Differentiate dorsal root, ventral root, and spinal nerve. Which contain sensory fibers, and which contain motor fibers?
20. Name the five components of a reflex arc and explain the operation of a reflex arc.
21. What are the functions of the spinothalamic tracts, dorsal column, spinocerebellar tracts, and corticospinal tracts?
22. Distinguish upper and lower motor neurons.
23. Describe the three meninges that surround the CNS.
24. Describe the production and circulation of the cerebrospinal fluid. Where does the cerebrospinal fluid return to the blood?
25. What are the three principal functional categories of the cranial nerves? List the nerves within each functional category.
26. Describe the spinal nerves by name and number.
27. Name the main plexuses and the major nerves derived from each.
28. Define preganglionic and postganglionic neuron.
29. Contrast the somatomotor nervous system and the autonomic nervous system in terms of the number of efferent neurons, conscious versus unconscious control, and types of effector organs.
30. Contrast the functions of the sympathetic and parasympathetic nervous systems.
31. What kinds of neurons (sympathetic or parasympathetic, preganglionic or postganglionic) are found in the following:
    A. Cranial nerve nuclei III, VII, IX, and X.
    B. Lateral horn of the thoracic spinal cord gray matter.
    C. Lateral horn of the sacral spinal cord gray matter.
    D. Collateral ganglia.
    E. Terminal ganglia.

CONCEPT REVIEW

1. Louis Ville was accidentally struck in the head with a baseball bat. He fell to the ground unconscious. Later, when he had regained consciousness, he was unable to remember any of the events that happened during the 10 minutes before the accident. Explain.
2. A patient suffered brain damage in an automobile accident. It was suspected that the cerebellum was the part of the brain affected. Based on what you know about cerebellar function, how could you determine that the cerebellum was involved?
3. The left lung of a cancer patient was removed. To reduce the empty space left in the thorax after the lung was removed, the diaphragm on the left side was paralyzed to allow the abdominal viscera to push the diaphragm upward into the space. What nerve would be cut to paralyze the left half of the diaphragm?
4. Name the nerve that, if damaged, would produce the following symptoms:
   A. The elbow and wrist on one side are held in a flexed position and cannot be extended.
   B. The patient is unable to extend the leg (as in kicking a ball) on one side.
5. Name the cranial nerve that, if damaged, would produce the following symptoms:
   A. The patient is unable to move the tongue.
   B. The patient is unable to see on one side.
   C. The patient is unable to feel one side of the face.
   D. The patient is unable to move the facial muscles on one side.
   E. The pupil of one eye is dilated and will not constrict.

ANSWERS TO  PREDICT QUESTIONS

1 *p. 192* Nuclei within the medulla oblongata regulate heart rate, blood vessel diameter, breathing, swallowing, vomiting, coughing, sneezing, balance, and coordination. While all these functions are important, loss of some of them may not necessarily result in death. Loss of cardiovascular regulation or loss of breathing regulation, however, could result in death. Because both blood flow and respiration are vital functions, interference with either of them may remove the person's functions from the normal homeostatic range. If not corrected, the loss of homeostasis could result in death.

2 *p. 197* If a person holds an object in her right hand, sensations from the skin of the hand are sent to the somesthetic cortex of the left hemisphere. The information is then passed to the somesthetic association cortex, where the object is recognized. Impulses then travel from there to Wernicke's area, where the object is given a name. From there, impulses travel to Broca's area, where the spoken word is initiated. Impulses from Broca's area travel to the premotor and then motor areas, where impulses are initiated that stimulate the muscles necessary to formulate the word.

3 *p. 200* Dorsal root ganglia are larger in diameter because they contain nerve cell bodies, which are larger than the axons of the spinal nerves.

4 *p. 211* Damage to the right phrenic nerve would result in absence of muscular contraction in the right half of the diaphragm. Because the phrenic nerves originate in the cervical region of the spinal cord, damage to the spinal cord in the thoracic region would not affect breathing. Damage to the upper cervical region, however, would cut the connection between the upper and lower motor neurons. This would eliminate phrenic nerve function and interfere with breathing.

5 *p. 214* (A) A person who is extremely angry has activated the sympathetic nervous system and the expected responses might include increased heart rate, blood pressure, respiration, and perspiration.

(B) In a person who has just finished eating and is now relaxing, the parasympathetic nervous system would be functioning. The responses might be decreased heart rate and respiration. There would be an increase in digestive activities, such as secretion and motility in the digestive tract. If the person was relaxing and reading a newspaper, the pupils of the eyes would be constricted.

9

THE GENERAL AND SPECIAL SENSES

After reading this chapter you should be able to:

1 Define general and special senses.
2 List the major types of general senses and briefly describe each.
3 Describe olfactory neurons and explain how airborne molecules can stimulate an action potential in the olfactory nerves.
4 Outline the structure and function of a taste bud.
5 List the accessory structures of the eye and explain their functions.
6 Name the tunics of the eye, list the parts of each tunic, and give the functions of each part.
7 Explain the differences in function of the rods and cones.
8 Describe the chambers of the eye and the fluids they contain.
9 Explain how images are focused on the retina.
10 Describe the structures of the outer and middle ear and state the function of each.
11 Describe the anatomy of the cochlea and explain how sounds are detected.
12 Explain how the structures of the vestibule and semicircular canals function in static and kinetic equilibrium.

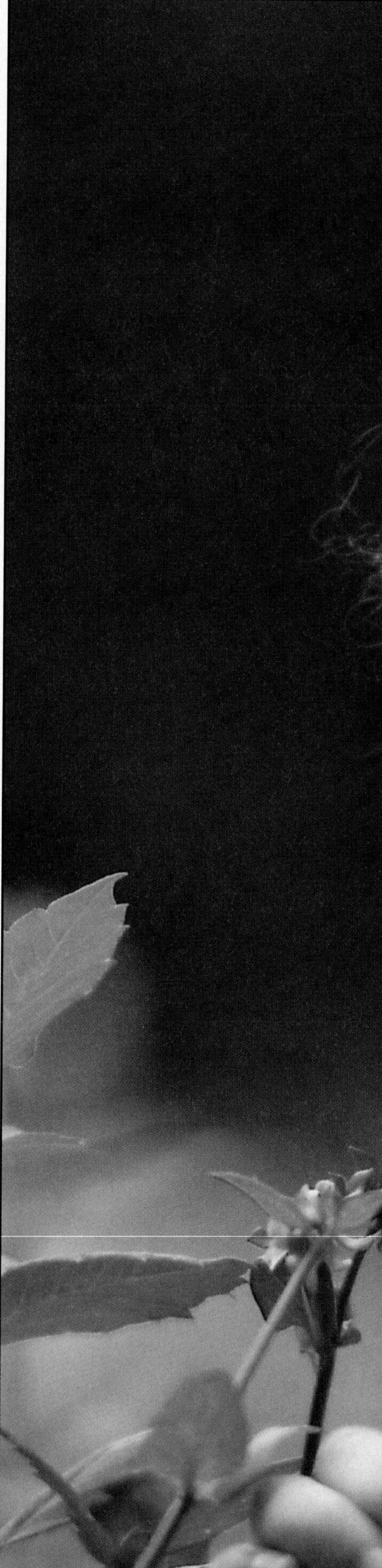

ciliary body (sil'e-ăr-e) Structure continuous with the choroid layer of the eye at its anterior margin that contains smooth muscle cells; attached to the lens by suspensory ligaments; regulates thickness of the lens.

cochlea (ko'kle-ah) The portion of the inner ear involved in hearing; shaped like a snail shell

cornea (kor'ne-ah) Transparent, anterior portion of the fibrous tunic of the eye through which light enters the eye.

cupula (ku'pu-lah) [L. *cupa*, a tub] Gelatinous mass that overlies the hair cells of the cristae ampullares of the semicircular canals.

endolymph [Gr. *endo-*, inside + L. *lympha*, clear fluid] The fluid inside the membranous labyrinth of the inner ear.

iris Specialized part of the vascular tunic of the eye; the "colored" part of the eye that can be seen through the cornea; consists of smooth muscles that regulate the amount of light entering the eye.

labyrinth (lab'ĭ-rinth) A series of membranous and bony tunnels in the temporal bone; part of the inner ear involved in hearing and balance.

macula (mak'u-lah) One of the sensory structures in the vestibule, consisting of hair cells and a gelatinous mass embedded with otoliths; responds to gravity.

perilymph [Gr. *peri-*, around + L. *lympha*, clear fluid] Fluid contained between the bony labyrinth and the membranous labyrinth of the inner ear.

pupil Opening in the iris of the eye through which light passes.

retina The inner, light-sensitive tunic of the eye; nervous tunic.

sclera (skler'ah) The dense, white, opaque posterior portion of the fibrous tunic of the eye; white of the eye.

taste bud Sensory structure found mostly on the tongue, that functions as a taste receptor.

tunic [L., coat] An enveloping layer of the wall of the eye; the three eye tunics are the fibrous, vascular, and nervous tunics.

The "senses" result from the effect of stimuli on sensory receptors, transmission of information in the form of action potentials from the receptors to the cerebral cortex, and integration of that information within the cerebral cortex. The senses can be characterized as general or as special. Sensations such as touch, temperature, pain, and proprioception (sense of position) are referred to as general sensations because their receptors are found throughout the body. Taste, smell (olfaction), sight (vision), hearing, and balance (equilibrium) are usually referred to as the special senses. These senses are dependent upon highly localized organs with very specialized sensory cells.

GENERAL SENSES

Receptors for the **general senses** are distributed throughout the body. Many of these receptors (nerve endings) are associated with the skin (Figure 9-1); others are associated with deeper structures such as tendons, ligaments, and muscles; and others can be found in both the skin and deeper structures. There are several types of nerve endings that respond to a wide range of stimuli, including pain, temperature changes, touch, pressure, and position. The neurons with which these endings are associated relay action potentials to the thalamus and on to the primary somesthetic cortex (see Chapter 8).

The simplest and most common sensory nerve endings are **free nerve endings** (see Figure 9-1), which are distributed throughout almost all parts of the body. These nerve endings respond to several stimuli, including pain, temperature, itch, and movement. Temperature sensations are actually detected by specialized free nerve endings called cold receptors and warm receptors. When cold or hot temperatures become excessive, below 12° C (54° F) or above 47° C (117° F), these cold or hot receptors no longer function, and pain receptors are stimulated. That is why it is sometimes difficult to distinguish very cold from very hot objects touching the skin.

The remaining nerve endings (see Figure 9-1) are more complex than free nerve endings, and many of them are enclosed by capsules. **Merkel's disks** are small, superficial nerve endings involved in detecting light touch and superficial pressure. **Hair follicle receptors,** associated with hairs, are also involved in detecting light touch. Light touch receptors are very sensitive but are not very discriminative (the point being touched cannot be specifically identified). Receptors for fine, discriminative touch, called **Meissner's corpuscles,** are located just deep to the epidermis. These receptors are very specific in localizing tactile sensations. Deeper tactile receptors, called **Ruffini's end-organs,** play an important role in detecting continuous touch or pressure in the skin. The deepest receptors, and the receptors associated with tendons and joints, are called **Pacinian corpuscles.** These receptors relay information concerning deep pressure, vibration, and position (proprioception).

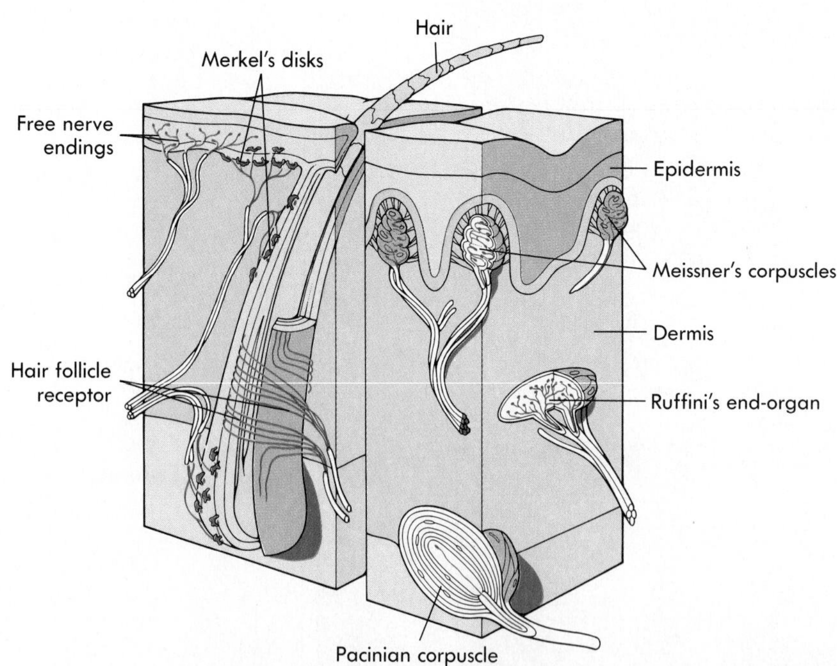

FIGURE 9-1 Sensory nerve endings in the skin.

Pain

Pain is a sensation characterized by a group of unpleasant perceptual and emotional experiences. There are two types of pain sensation: (1) rapidly conducted impulses, which give rise to sharp, well-localized, pricking, or cutting pain, followed by (2) more slowly propagated impulses, which give rise to diffuse, burning, or aching pain.

Superficial pain is highly localized as a result of the simultaneous stimulation of pain receptors and tactile receptors in the skin. Deep or visceral pain is not highly localized because of the absence of tactile receptors in the deeper structures, and it is normally perceived as a diffuse pain.

Pain impulses can be suppressed by chemical anesthetics injected near the nerve (local anesthesia). They can also be suppressed if consciousness is inhibited in the reticular formation (general anesthesia). Pain impulses can also be influenced by inhibitory impulses that are carried by descending neurons of the dorsal column system (see Chapter 8). These neurons are stimulated by tactile receptors and may act to "gate" pain impulses that travel in the lateral spinothalamic tracts. The descending neurons synapse with and inhibit neurons in the dorsal horn that give rise to the lateral spinothalamic tract. Increased activity in the dorsal column system tends to close the gate, reducing pain impulses transmitted in the lateral spinothalamic tract.

The gate-control theory may explain the physiological basis for several methods that have been used to reduce the intensity of pain. Dorsal column neurons may be stimulated when the skin is rubbed vigorously and when the limbs are moved. This may explain why vigorously rubbing a large area around a source of pain tends to reduce its intensity. Exercise normally decreases the sensation of pain, and exercise programs are important components in the clinical management of chronic pain. Acupuncture and acupressure procedures may also stimulate dorsal column neurons.

Referred pain

Referred pain is a painful sensation in a region of the body that is not the source of the pain stimulus. Most commonly, referred pain is sensed in the skin or other superficial structures when internal organs are damaged or inflamed (Figure 9-2). This occurs because afferent neurons from the superficial area to which the pain is referred and the neurons from the deep, visceral area in which the actual damage occurs converge onto the same spinal nerves, and these nerves project to the same area of the cerebral cortex. The brain cannot distinguish between the two sources, and the painful sensation is referred to the most superficial structures innervated (such as the skin).

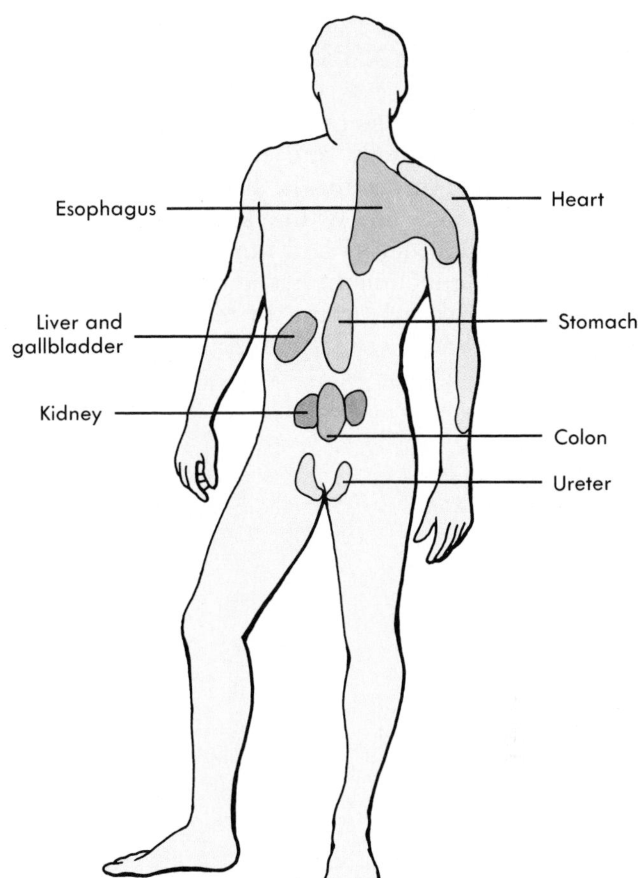

FIGURE 9-2 Areas of referred pain on the body surface.
Pain from the indicated internal organs is referred to the surface areas shown in the figure.

Referred pain is clinically useful in diagnosing the actual cause of the painful stimulus. For example, heart attack victims often feel cutaneous pain radiating from the left shoulder down the arm.

> Phantom pain occurs in people who have had appendages amputated. Frequently these people perceive intense pain in the amputated structure as if it were still there. If a sensory neuron is stimulated at any point along its pathway, action potentials are initiated and are propagated toward the central nervous system. Integration in the CNS results in the perception of pain that is projected to the original site of the sensory receptors for that pathway, even if those sensory receptors are no longer present. A similar phenomenon can be easily demonstrated by bumping the ulnar nerve as it crosses the elbow (the funny bone). A sensation of pain is felt in the fourth and fifth digits, even though the neurons were stimulated at the elbow.

SPECIAL SENSES

The sensations of smell and taste are closely related, both structurally and functionally, and are both stimulated by chemicals. The sense of vision is stimulated by light and is unique among the special senses in both structure and function. Hearing and balance both function in response to mechanical stimulation. Hearing occurs in response to sound waves, and balance occurs in response to gravity or motion.

OLFACTION

Olfaction (ol-fak'shun; the sense of smell) occurs in response to airborne molecules called odors that enter the nasal cavity (Figure 9-3). **Olfactory neurons** are bipolar neurons within the olfactory epithelium of the nasal cavity. The dendrites of the olfactory neurons extend to the epithelial surface of the nasal cavity, and their ends are modified into bulbous enlargements. These enlargements possess extremely long cilia, which lie in a thin mucous film on the epithelial surface.

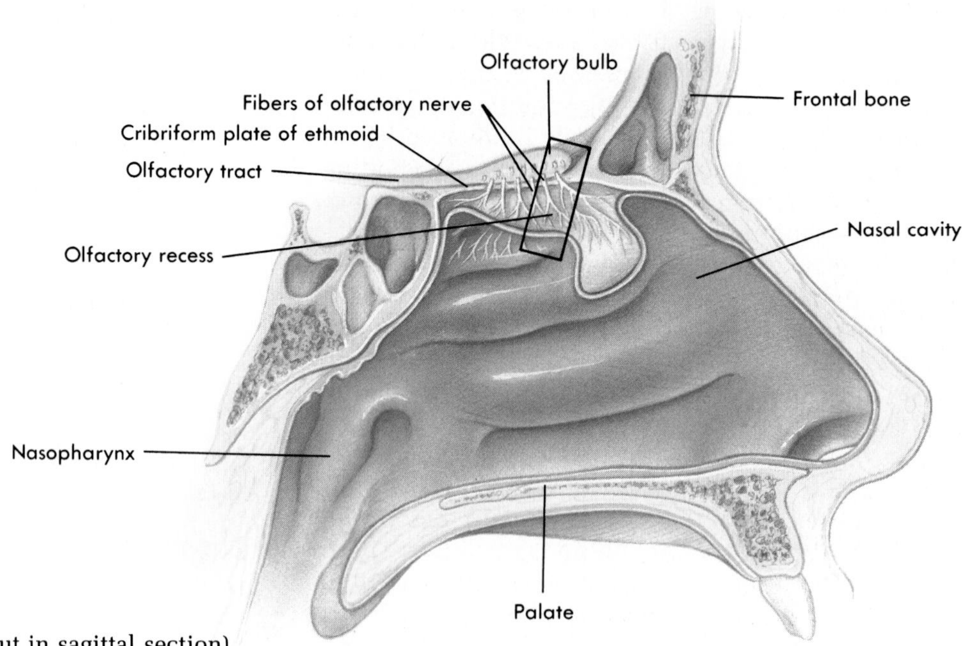

FIGURE 9-3 Nasal cavity.
Lateral wall of the nasal cavity (cut in sagittal section) showing the olfactory region. An olfactory neuron is highlighted above and shown enlarged below.

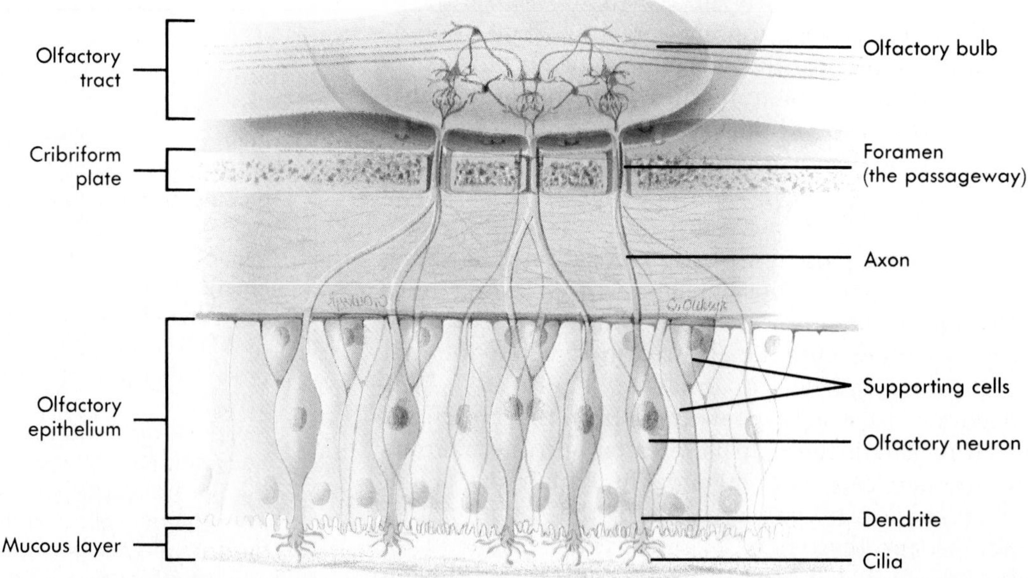

Airborne molecules become dissolved in the fluid on the surface of the epithelium and bind to receptor molecules of the ciliary membrane. The exact nature and site of this interaction is not fully understood. The olfactory neurons react by depolarizing. The threshold for the detection of odors is very low, so very few molecules are required to initiate an action potential. It has been proposed that a wide variety of detectable odors are actually combinations of a smaller number (perhaps as few as seven) of primary odors.

Axons from olfactory neurons (cranial nerve I) pass through foramina of the cribriform plate and enter the **olfactory bulb,** where they synapse with association neurons that relay action potentials to the brain through the **olfactory tracts.** Each olfactory tract terminates in an area of the brain called the **olfactory cortex,** located within the posterior, inferior portion of the frontal lobe.

TASTE

The sensory structures that detect **taste** stimuli are the **taste buds** (Figure 9-4). Taste buds are oval structures located on the surface of certain **papillae,** which are enlargements on the surface of the tongue. Each taste bud consists of two types of cells. Specialized epithelial cells form the exterior supporting capsule of the taste bud; the interior of each bud consists of about 40 **taste cells.** Each taste cell contains hair-like processes, called **taste hairs,** that extend into a tiny opening in the epithelium of the taste bud, called a **taste pore.** Receptors on the hairs are stimulated by dissolved substances to initiate action potentials that are carried by afferent neurons to the temporal lobe of the cerebral cortex.

Taste sensations from the anterior two thirds of the tongue are carried by the facial nerve (cranial nerve VII). Taste sensations from the posterior one third of the tongue are carried by the glossopharyngeal nerve (cranial nerve IX). In addition, the vagus nerve (cranial nerve X) carries some taste sensation from the root of the tongue.

Taste sensations can be divided into four basic types: sour, salty, bitter, and sweet. Even though there are only four primary taste sensations, a fairly large number of different tastes can be perceived, presumably by combining the four basic taste sensations. Many taste sensations, however, are strongly influenced by olfactory sensations. This influence can be demonstrated by comparing the taste of some food before and after pinching your nose. It is easy to detect that the sense of taste is reduced while the nose is pinched.

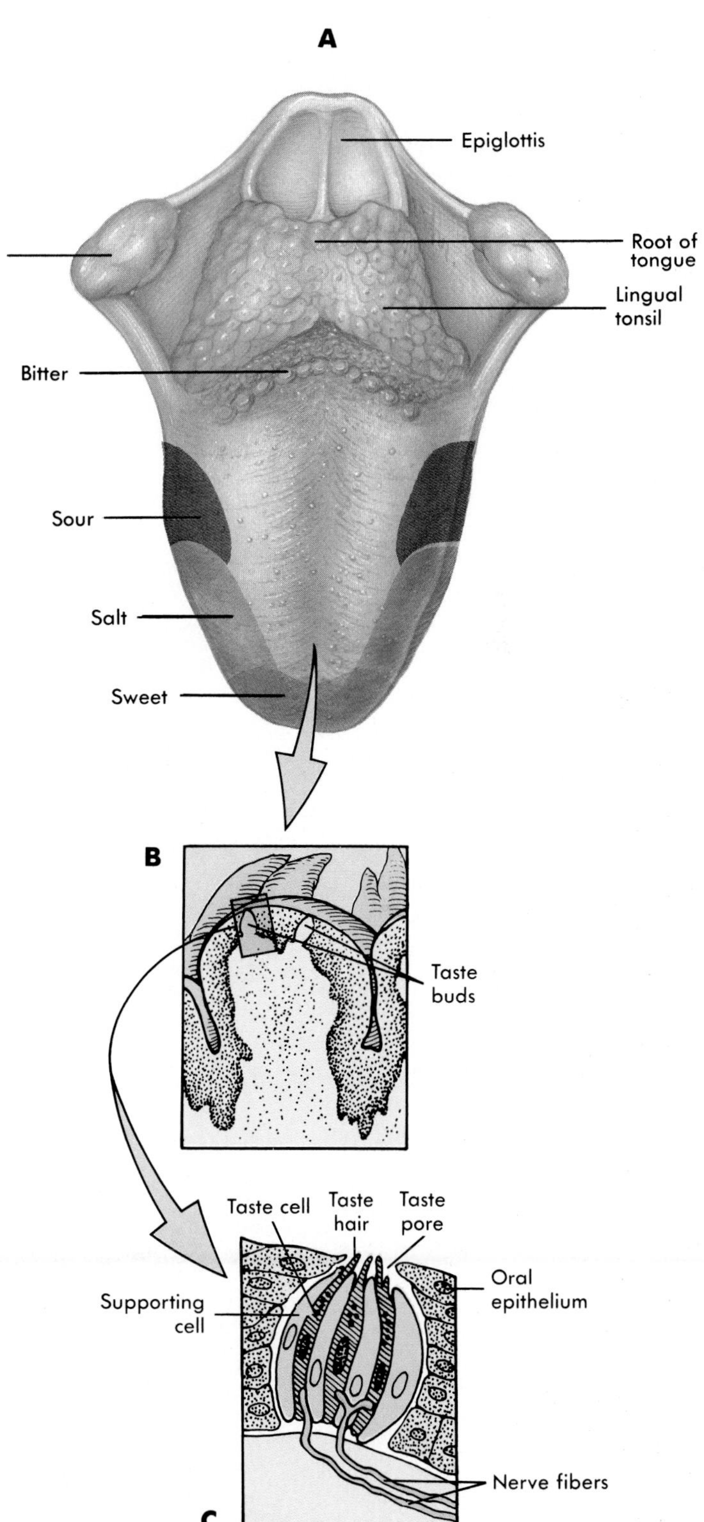

FIGURE 9-4 ■ The tongue.
A Dorsal surface of the tongue and regions of the tongue sensitive to various tastes.
B Section through a papilla with taste buds on the side.
C Enlarged view of a section through a taste bud.

> **1 Why does food not taste as good when you have a cold?** ?

Although all taste buds are able to detect all four of the basic taste sensations, each taste bud is usually most sensitive to one class of taste stimuli. The stimulus type to which each taste bud responds most strongly is related to its position on the tongue. Taste buds at the tip of the tongue react more strongly to sweet and salty taste stimuli, taste buds at the back of the tongue react more strongly to bitter taste stimuli, and taste buds on the side of the tongue react most strongly to sour taste stimuli (see Figure 9-4).

VISION

The visual system includes the eyes and the accessory structures, as well as the optic nerves, tracts, projections, and cortex where visual impulses are interpreted. The eyes respond to light and initiate afferent signals that are transmitted from the eyes to the brain by the optic (visual) nerves and tracts. The accessory structures, such as the eyebrows, eyelids, eyelashes, and tear glands, help protect the eyes from direct sunlight and damaging particles. Much of the information we obtain about the world around us is detected by the visual system. Our education is largely based on visual input and depends on our ability to read words and numbers. Visual input includes information about light and dark, movement, color, and hue.

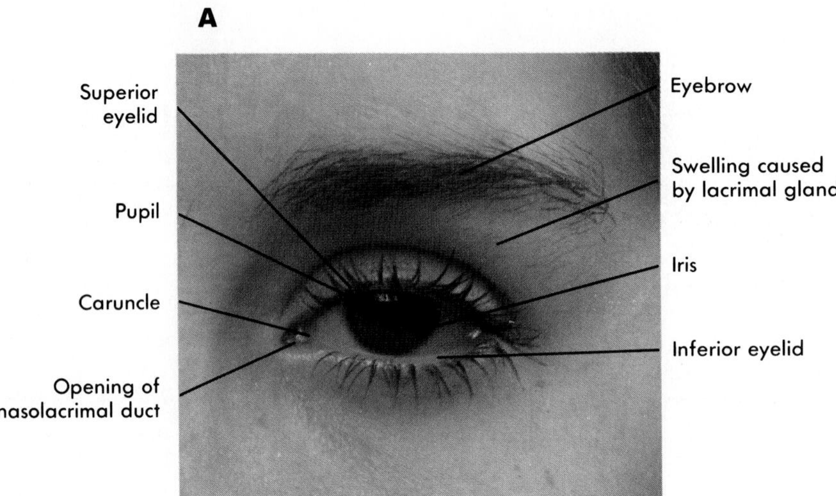

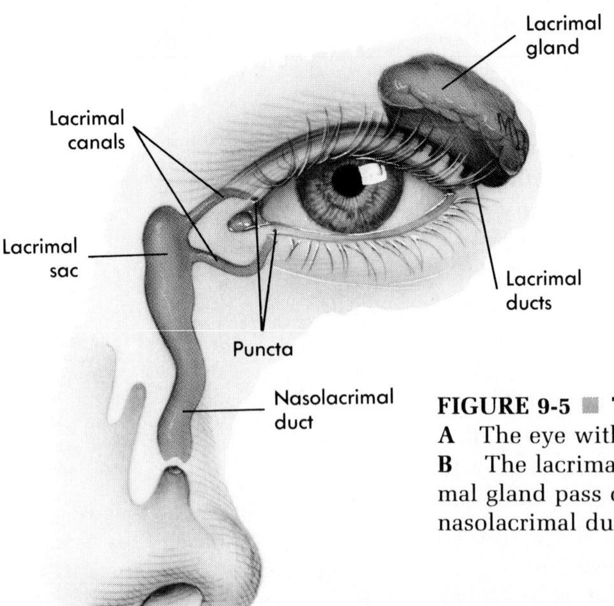

FIGURE 9-5 ■ The eye.
A The eye with its accessory structures.
B The lacrimal structures. Tears produced in the lacrimal gland pass over the surface of the eye and enter the nasolacrimal duct to the nasal cavity.

Accessory Structures

Accessory structures (Figure 9-5) protect, lubricate, and move the eye. They include the eyebrows, eyelids, conjunctiva, lacrimal apparatus, and extrinsic eye muscles.

Eyebrows

The **eyebrows** protect the eyes by preventing perspiration, which can irritate the eyes, from running down the forehead and into the eyes. They also help shade the eyes from direct sunlight.

Eyelids

The **eyelids** with their associated lashes protect the eyes from foreign objects. If an object suddenly approaches the eye, the eyelids protect the eye by closing and then opening quite rapidly (blink reflex). Blinking, which normally occurs about 20 times per minute, also helps to keep the eyes lubricated by spreading tears over the surface of the eye.

Conjunctiva

The **conjunctiva** (kon-junk-ti'vah) is a thin, transparent mucous membrane that covers the inner surface of the eyelids, and the anterior surface of the eye. Conjunctivitis is an inflammation or some other irritation of the conjunctiva (see the essay on p. 232).

Lacrimal apparatus

The **lacrimal** (lak'rĭ-mal) **apparatus** (see Figure 9-5) consists of a lacrimal gland situated in the superior lateral corner of the orbit and a nasolacrimal duct in the inferior medial corner of the orbit. The **lacrimal gland** produces tears, which pass over the anterior surface of the eye. Most of the fluid produced by the lacrimal glands evaporates from the surface of the eye, but excess tears are collected in the medial corner of the eye by **lacrimal canaliculi.** These canals open into the **nasolacrimal duct,** which opens into the nasal cavity. Tears serve to lubricate the eye and to cleanse it. In addition, tears contain an enzyme that helps combat eye infections.

2 Explain why it is often possible to "taste" a medication, such as eyedrops, that has been placed into the eyes. Why does a person's nose "run" when he cries? ?

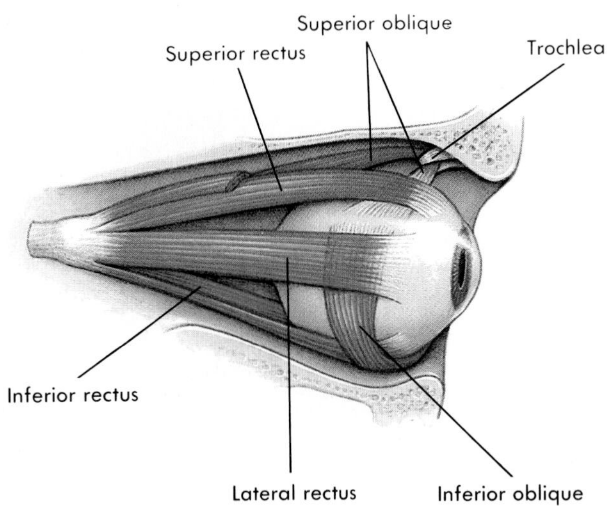

FIGURE 9-6 Eye muscles.
Extrinsic muscles of the right eye as seen from a lateral view with the lateral wall of the orbit removed.

Extrinsic eye muscles

Movement of each eyeball is accomplished by six muscles, the extrinsic eye muscles (Figure 9-6 and see Chapter 7). Four of these muscles run more or less straight from their origins posterior to the eye to attach to the four quadrants of the eyeball. They are the superior, inferior, medial, and lateral **rectus muscles.** Two muscles, the **superior** and **inferior oblique muscles,** are placed at an angle to the eyeball.

The superior oblique muscle is innervated by the trochlear nerve (cranial nerve IV). The nerve is so named because the superior oblique muscle goes around a little pulley, or trochlea, as it passes to the eye. The lateral rectus muscle is innervated by the abducens nerve (cranial nerve VI), so named because the lateral rectus muscle abducts the eye. The other four extrinsic eye muscles are innervated by the oculomotor nerve (cranial nerve III).

Anatomy of the Eye

The eyeball can be thought of as consisting of two hollow, fluid-filled spheres compressed against each other. The larger, posterior sphere accounts for about five sixths of the surface of the eye. The much smaller anterior sphere accounts for about one sixth of the total surface area. The chambers within these spheres and the fluid within each chamber will be discussed later in this chapter.

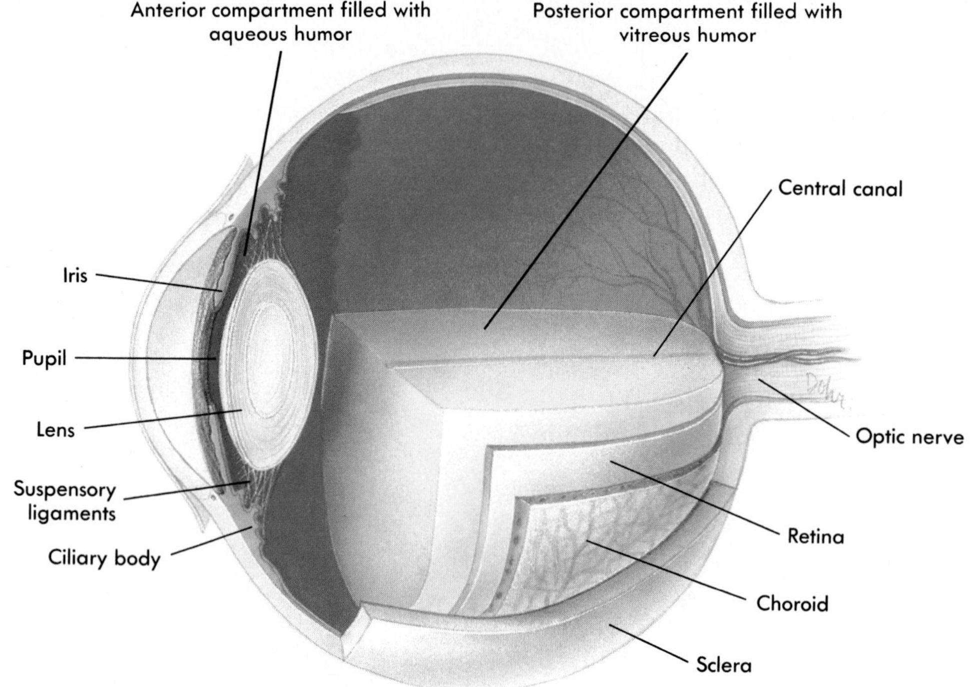

FIGURE 9-7 ■ Layers and compartments of the eye.
Sagittal section through the eye demonstrating its layers and compartments.

The wall of the eye consists of three layers or **tunics** (Figure 9-7). The outer or **fibrous tunic** consists of the sclera and cornea. The middle or **vascular tunic** consists of the choroid, ciliary body, and iris. The inner or **nervous tunic** consists of the retina.

Fibrous tunic

The **sclera** (skler'ah) is the firm, opaque, white, outer connective tissue layer of the posterior five sixths of the eye. The sclera helps maintain the shape of the eye, protects the internal structure, and provides attachment sites for the extrinsic eye muscles. A small portion of the sclera can be seen as the "white of the eye." The **cornea** (kor'ne-ah) is the avascular, transparent, anterior one sixth of the eye that permits light to enter the eye. As part of the focusing system of the eye, it also bends or refracts the entering light.

> The cornea was one of the first organs to be successfully transplanted. Several characteristics make the cornea relatively easy to transplant: It is easily accessible and relatively easily removed; it is avascular and therefore does not require as extensive a circulation as other tissues; and it is less immunologically active and is therefore less likely to be rejected than other tissues.

Vascular tunic

The middle tunic of the eye is called the **vascular tunic** because it is the layer containing most of the blood vessels of the eye. The posterior portion of the vascular tunic, associated with the scleral portion of the eye, is the **choroid** (ko'royd). This is a very thin structure consisting of a vascular network and large numbers of melanin-containing pigment cells, so that it appears black in color. The black color absorbs the entering light so that it is not reflected and does not interfere with vision.

Anteriorly the vascular tunic consists of the ciliary body and iris. The **ciliary** (sil'e-ăr-e) **body** is continuous with the anterior margin of the choroid. The ciliary body contains smooth muscles called **ciliary muscles** (the intrinsic eye muscles), which attach to the lens by **suspensory ligaments.** The **lens** is a flexible, biconvex, transparent disc.

The **iris** is the colored portion of the eye. It is attached to the anterior margin of the ciliary body. The iris is a contractile structure consisting mainly of smooth muscle that surrounds an opening called the **pupil.** Light passes through the pupil, and the iris regulates the amount of light entering the eye by controlling the diameter of the pupil. Parasympathetic stimulation (via cranial nerve III) causes the iris to contract, while sympathetic stimulation causes it to dilate. As light intensity increases, the pupil contracts; as light intensity decreases, the pupil dilates.

Nervous tunic

■ The **retina** or **nervous tunic** is the innermost tunic of the eye. It consists of an outer **pigmented retina,** which, with the choroid, keeps light from being reflected back into the eye, and an inner **sensory retina,** which responds to light (Figure 9-8). The sensory retina contains photoreceptor cells called **rods** and **cones,** and numerous association neurons (some of which are named in Figure 9-8). Over most of the retina, rods are 20 times more common than cones. Rods are very sensitive to light and can function in very dim light, but do not provide a very sharp image. Cones require much more light and

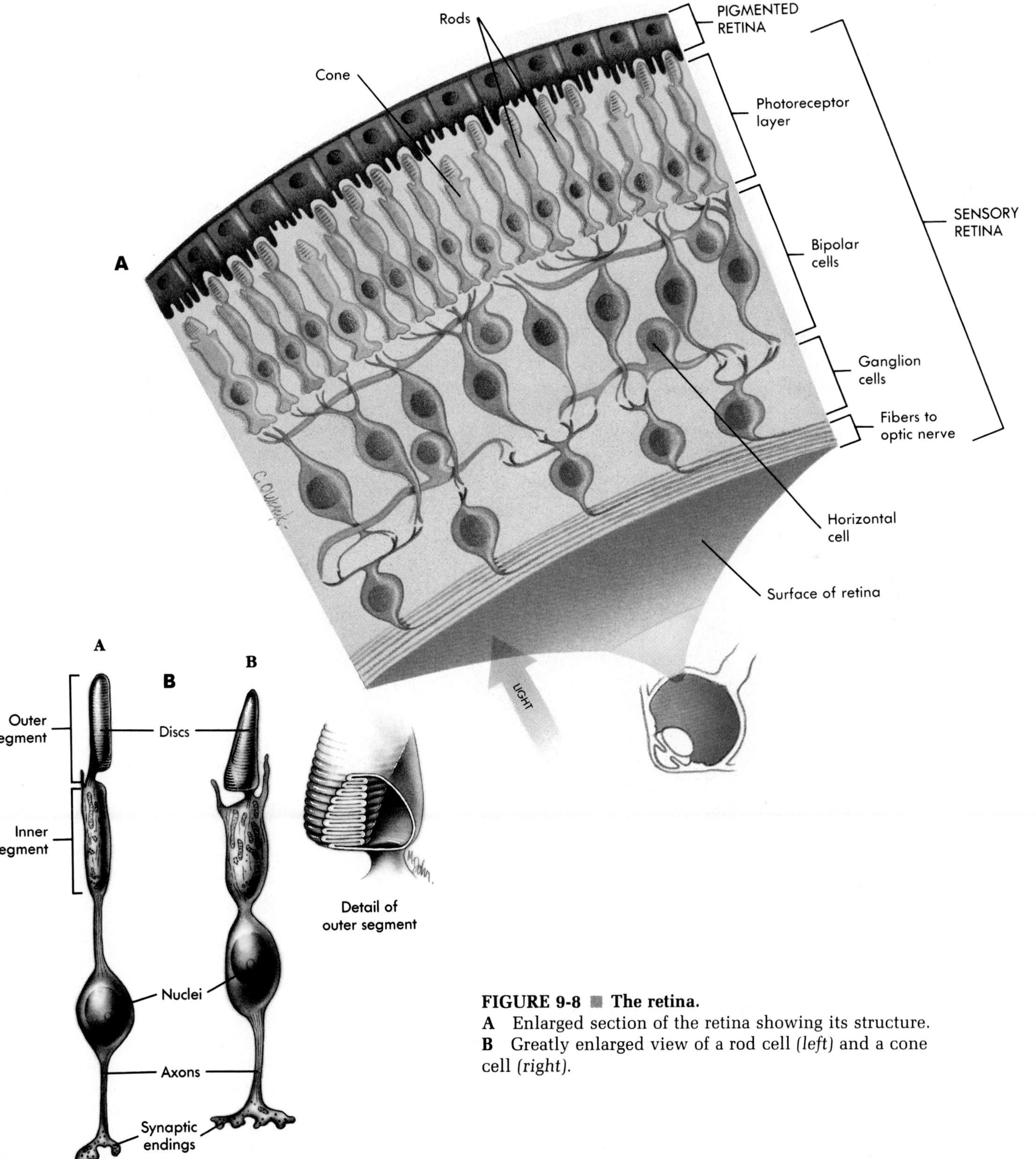

FIGURE 9-8 ■ The retina.
A Enlarged section of the retina showing its structure.
B Greatly enlarged view of a rod cell *(left)* and a cone cell *(right).*

provide us with a very clear image. Cones are also responsible for color vision. There are three types of cones, each sensitive to a different color: blue, green, or red. The many colors that we can see result from various functional combinations of these three types of cones.

> **3 In dim light, colors seem to fade and objects seem to become colored by shades of gray. Explain why this happens.** ?

Rod cells contain a photosensensitive pigment called **rhodopsin,** which is made up of the protein **opsin** in loose chemical combination with a pigment called **retinal** (Figure 9-9). Cone cells contain a slightly different photosensitive pigment. When light strikes a rod cell, retinal changes shape and loses its attachment to the opsin molecule. Because of the retinal detachment, opsin "opens up" with a release of energy. This reaction is somewhat like a spring (opsin) being held by a trigger (retinal). This change in rhodopsin stimulates a response in the rod cell. Retinal then completely detaches from rhodopsin. Energy (ATP) is required to reattach retinal to rhodopsin and, at the same time, to return rhodopsin to the shape that it had before being stimulated by light.

Manufacture of retinal in rod cells requires vitamin A. A person with a vitamin A deficiency may have difficulty seeing, especially in dim light, a condition called night blindness.

When the posterior region of the retina is examined with an ophthalmoscope, two interesting features can be observed (Figure 9-10). First, near the center of the posterior retina is a small yellow spot called the **macula lutea** (mak'u-lah lu'te-ah). In the center of the macula lutea is a small pit, the **fovea** (fo've-ah) **centralis.** The fovea centralis is the portion of the retina where light is normally focused when the eye is looking directly at an object. The fovea centralis is the portion of the retina with the highest concentration of cones, and with the greatest visual acuity, the ability to see images most clearly.

> **4 Given the function of the fovea centralis, explain why there are more cones than rods in this part of the retina.** ?

Just medial to the macula lutea is a white spot, the **optic disc,** through which a number of fairly large blood vessels enter the eye and spread over the surface of the retina. This is also the spot where axons from the sensory retina meet, pass through the outer two tunics, and exit the eye as the **optic nerve** (cranial nerve II). The optic nerves project to

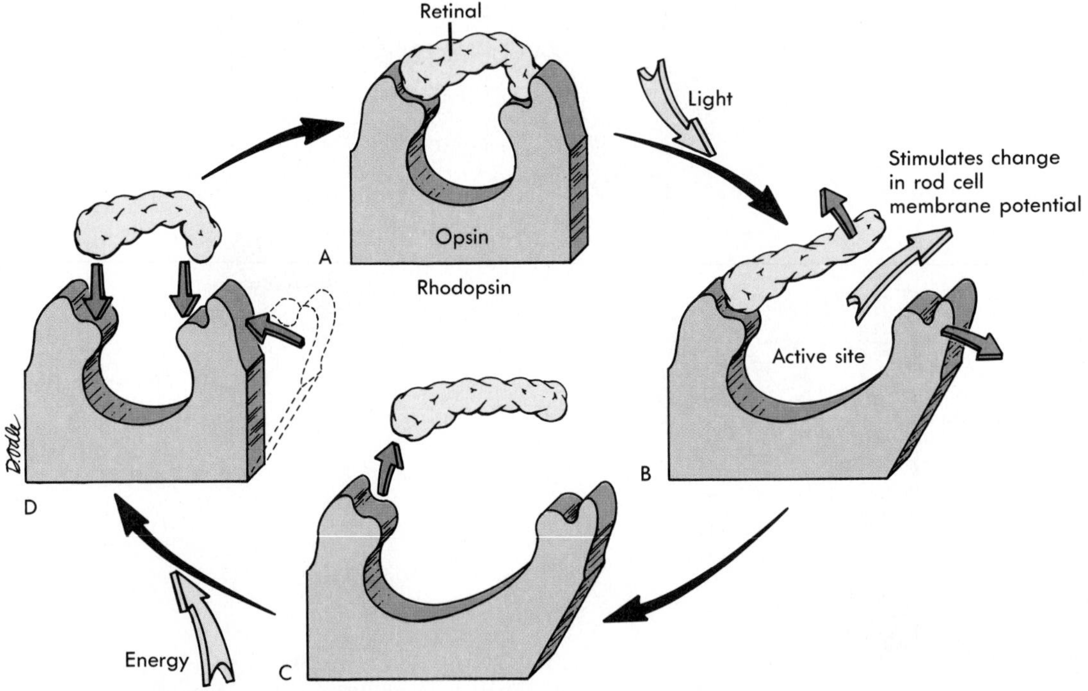

FIGURE 9-9 ■ Rhodopsin cycle.
Retinal is attached to opsin *(A)*. Light causes retinal to change shape *(B)*. Opsin molecule also changes shape (opens), exposing an active site, and stimulates a change in the rod cell membrane potential. Retinal separates from opsin *(C)*. Energy is required to bring opsin back to its original form and to attach retinal to it *(D)*.

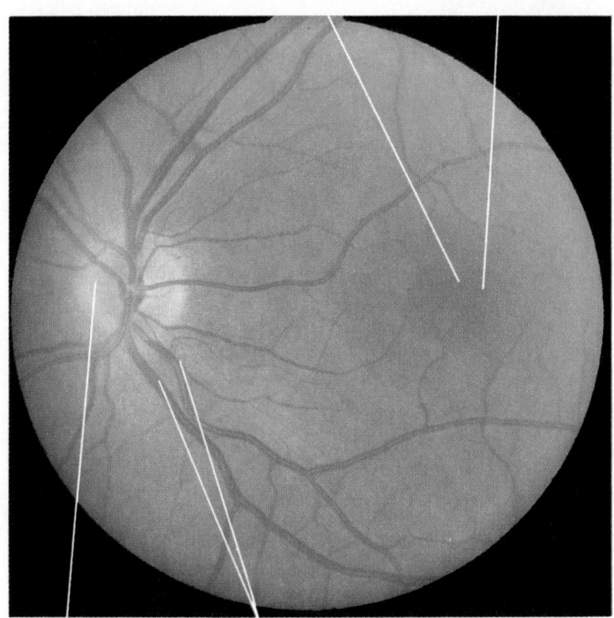

FIGURE 9-10 Ophthalmoscopic view of the retina. This view shows the posterior wall as seen through the pupil. Notice the vessels entering the eye through the optic disc, and notice the macula lutea with the fovea centralis.

the thalamus, and nerves in the thalamus project to the visual cortex in the occipital lobes of the brain. The optic disc contains no photoreceptor cells and does not respond to light; it is therefore called the **blind spot** of the eye.

Compartments of the eye

The two major compartments of the eye, one inside each of the two spheres that make up the eyeball, are separated by the lens (see Figure 9-7). The anterior compartment is filled with **aqueous humor** (watery fluid), which helps maintain pressure within the eye, refracts (bends) light, and provides nutrients to the inner surface of the eye. Aqueous humor is produced by the ciliary body as a blood filtrate and is returned to the circulation through a venous ring at the base of the cornea. Pressure within the eye resulting from the presence of aqueous humor keeps the eye inflated much like the air in a basketball. If circulation of the aqueous humor is blocked so that pressure in the eye increases, a defect called glaucoma can result (see the essay on p. 232).

The posterior compartment of the eye is filled with a transparent jellylike substance, the **vitreous** (vit're-us) **humor.** The vitreous humor also helps maintain pressure within the eye and holds the lens and the retina in place. It also functions to refract the light. Unlike the aqueous humor, the vitreous humor does not circulate.

Functions of the Complete Eye

The eye functions much like a camera. The iris allows light into the eye, and the light is focused by the cornea, lens, and humors onto the retina. The light striking the retina is converted into action potentials that are relayed to the brain.

Light refraction

An important characteristic of light is that it can be refracted (bent). As light passes from air to some other, more dense substance, the rays are bent. If the surface of a lens is concave, the light rays diverge; if the surface is convex, they converge. As the light rays converge, they finally reach a point where they cross. The crossing point is called the **focal point** (Figure 9-11), and causing light to converge is called **focusing.**

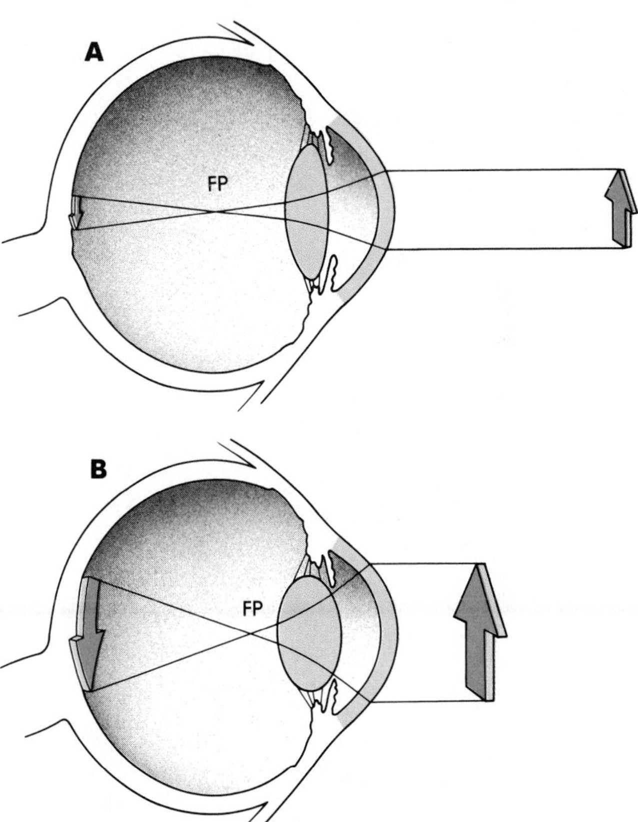

FIGURE 9-11 Ability of the lens to focus images on the retina of the left eye.
The focal point *(FP)* is where light rays cross.
A Distant image: The lens is flattened, and the image is focused on the retina.
B Near vision (accommodation): The lens is more rounded, and the image is focused on the retina.

Eye Disorders

INFECTIONS

Conjunctivitis is an inflammation of the conjunctiva, usually resulting from a bacterial infection. **Contagious conjunctivitis** (pinkeye) occurs primarily in children. It can be transmitted by hand contact, flies, or contaminated water (such as in swimming pools). **Neonatal gonorrheal ophthalmia** is a severe form of conjunctivitis that is contracted by an infant as it passes through the birth canal of a mother with gonorrhea. This infection carries a high risk of blindness. The treatment of newborn eyes with silver nitrate is effective in preventing the disease. **Chlamydial conjunctivitis** is contracted as an infant passes through the birth canal of a mother with a chlamydial infection. This infection is not affected by silver nitrate, so in many places, newborns are treated with antibiotics against both chlamydia and gonorrhea. **Trachoma** is the greatest single cause of blindness in the world today. The disease, also caused by chlamydia, is transmitted by hand contact, flies, or objects such as towels. It is a conjunctivitis that leads to scarring of the cornea and blindness. It is most common in arid parts of Africa and Asia.

A **chalazion** (ka-la′ze-on) is a cyst caused by infection of the glands along the edge of the eyelid. A **stye** is an infection of an eyelash hair follicle.

DEFECTS OF FOCUS

Myopia, or nearsightedness (ability to see close objects but not distant ones), is a defect of the eye in which the focal point is too near to the lens, and the image is focused in front of the retina (Figure 9-A, *1*). Myopia is corrected by a concave lens that spreads out the light rays coming to the eye so that when the light is focused by the eye, it is focused on the retina (Figure 9-A, *2*).

Hyperopia, or farsightedness, is a disorder in which the focal point is too far from the lens, and the image is focused "behind" the retina (Figure 9-A, *3*). Hyperopia is corrected by a convex lens that causes light rays to converge as they approach the eye, and to focus on the retina (Figure 9-A, *4*).

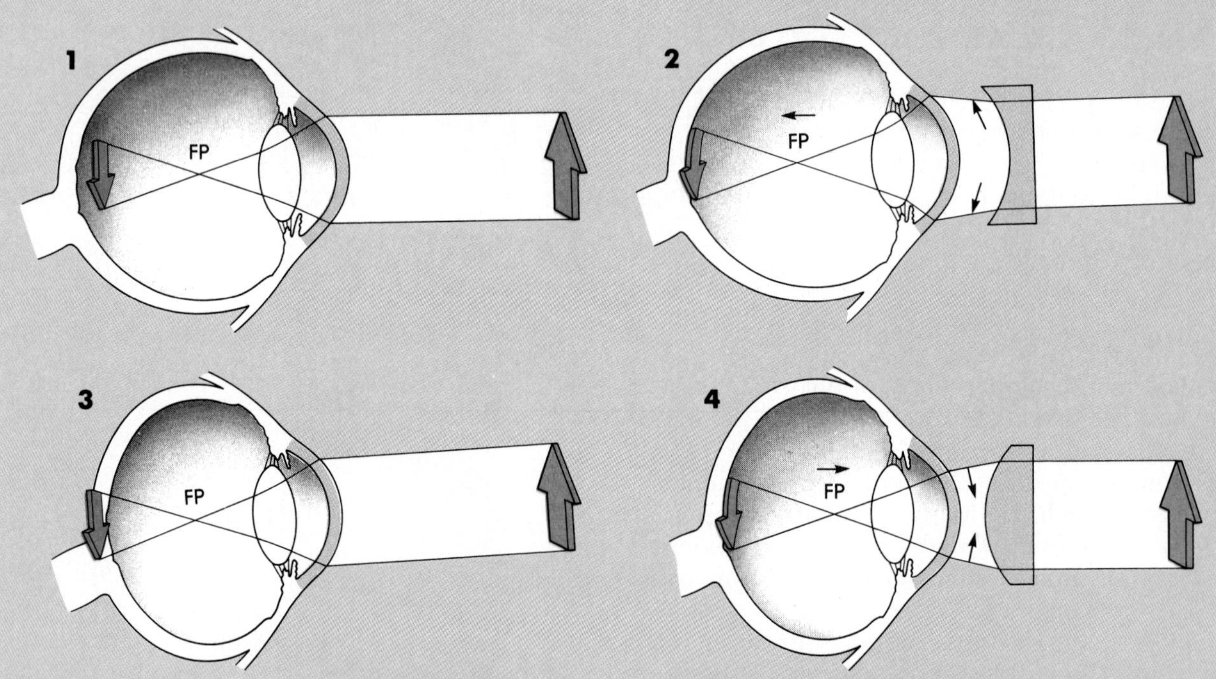

FIGURE 9-A Visual disorders and their correction by various lenses.
FP is the focal point.
1 Myopia (nearsightedness).
2 Correction of myopia with a concave lens.
3 Hyperopia (farsightedness).
4 Correction of hyperopia with a convex lens.

Presbyopia is the decrease in the ability of the eye to accommodate for near vision; this occurs as a normal part of aging. With age, the lens becomes less flexible. The average age of onset of presbyopia is the mid-forties. Presbyopia can be corrected by the use of "reading glasses" or by bifocals, which have different lenses in the top and bottom. The bottom half of a bifocal lens is more convex to accommodate for near vision when the person reads, and the top half is less convex for distant vision.

Astigmatism is a defect in which the cornea or lens is not uniformly curved, and the image is not sharply focused. Glasses may be made to adjust for the abnormal curvature as long as the curvature is not too irregular. If the curvature of the cornea or lens is too irregular, the condition is difficult to correct.

STRABISMUS

Strabismus is a condition in which one eye or both eyes are directed medially or laterally. The condition may result from abnormally weak eye muscles.

COLOR BLINDNESS

Color blindness is the absence of perception of one or more colors by the cone cells (Figure 9-B). There may be a complete loss of color perception or only a decrease in perceptive ability. The loss may involve perception of all three colors, or of one or two colors. Most forms of color blindness occur more frequently in males (an X-linked trait; see Chapter 20). In Western Europe, about 8% of all males have some form of color blindness, whereas only about 1% of the females are color blind.

1

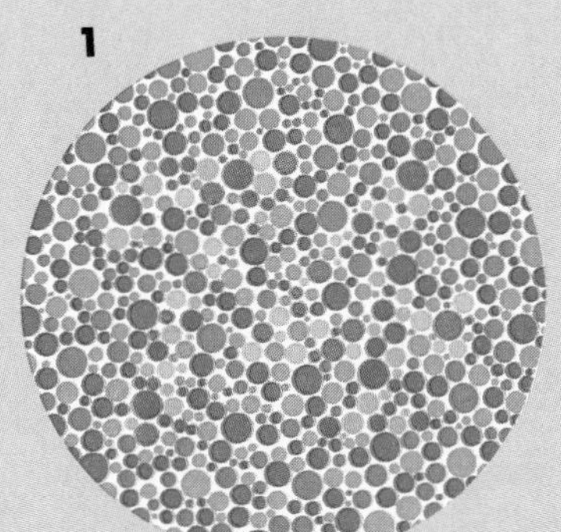

2

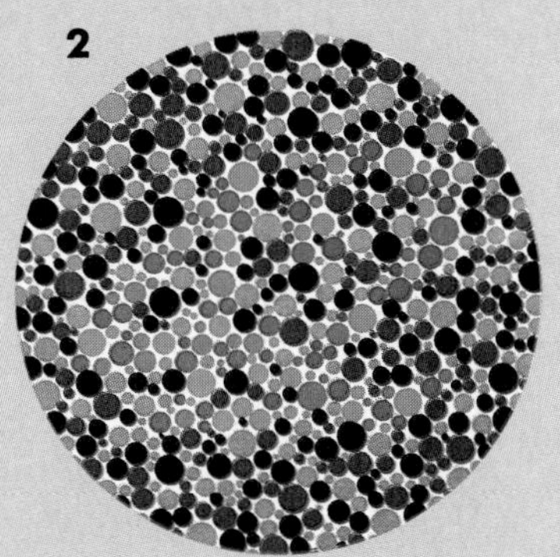

FIGURE 9-B ■ Color blindness charts.
1 A person with normal vision can see the number 74, whereas a person with red-green color blindness sees the number 21.
2 A person with normal vision can see the number 42. A person with red color blindness sees the number 2, and a person with green color blindness sees the number 4.

BLINDNESS

Cataract is the most common cause of blindness in the United States. It is a condition in which clouding of the lens occurs as the result of advancing age, infection, or trauma. Excess exposure to ultraviolet radiation may be a factor in causing cataracts, so the wearing of sunglasses in bright sunshine is recommended. A certain amount of clouding occurs in 95% of people over age 65. Surgery to remove a cataract is actually the removal of the lens. More than 400,000 cataracts are removed in the United States each year.

Glaucoma is a condition involving excessive pressure buildup in the aqueous humor. Glaucoma results from an interference with normal reentry of aqueous humor into the blood stream, or an overproduction of aqueous humor. The increased pressure within the eye can close off the blood vessels entering the eye and may destroy the retina or optic nerve, resulting in blindness.

Diabetes is a major cause of blindness in the United States. Diabetes can result in optic nerve degeneration, cataracts, and retinal detachment. These defects are often caused by blood vessel degeneration and hemorrhage, which are common in diabetic patients.

Retinal detachment, the separation of the sensory retina from the pigmented retina, is a relatively common problem that can result in complete blindness. If a hole or tear occurs in the retina, fluid may accumulate between the sensory and pigmented retina. As a result, the sensory retina may become detached from the pigmented retina and degenerate, resulting in loss of vision.

Focusing of images on the retina

The cornea is a convex structure, and as light rays pass from the air through the cornea, they converge (see Figure 9-11). Additional convergence occurs as light passes through the aqueous humor, lens, and vitreous humor. The greatest contrast in media density is between the air and the cornea. Therefore the greatest amount of convergence occurs at that point. The shape of the cornea and its distance from the retina are fixed, however, so that no adjustment in focus can be made by the cornea. Fine adjustments in focus are accomplished by changing the shape of the lens.

When the ciliary muscles are relaxed, the suspensory ligament of the choroid maintains elastic pressure on the lens, keeping it relatively flat and allowing for distant vision (see Figure 9-11, *A*). When an object is brought closer than 20 feet (about 6 ½ meters) to the eye, the ciliary muscles contract as a result of parasympathetic stimulation, pulling the choroid toward the lens. This reduces the tension on the suspensory ligaments of the lens and allows the lens to assume a more spherical form because of its own internal elastic nature (see Figure 9-11, *B*). The spherical lens then has a more convex surface, causing greater refraction of light. This process is called **accommodation,** and it enables the eye to focus on objects closer than 20 feet on the retina.

When a person's vision is tested, a chart is placed 20 feet from the eye, and the person is asked to read a line that has been standardized for normal vision. If the person can read the line, he has 20/20 vision, which means that he can see at 20 feet what people with normal vision see at 20 feet. If, on the other hand, the person only can read the letters at 20 feet that people with normal vision see at 40 feet, the person's eyesight is 20/40.

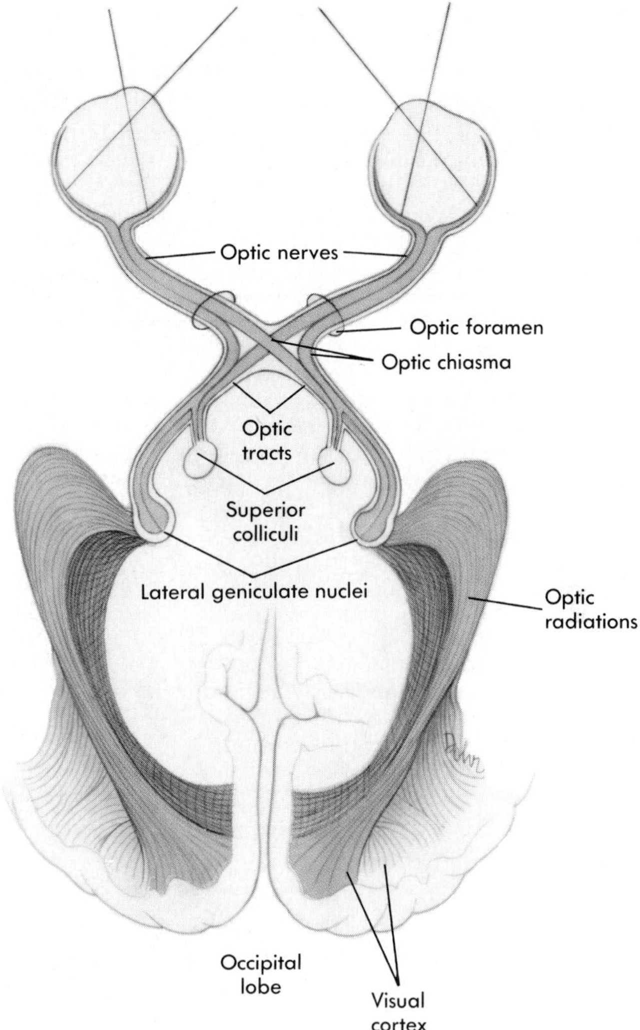

FIGURE 9-12 Visual central nervous system pathways (superior view).
Neuronal fibers from the eyes travel through the optic nerves to the optic chiasma, where some of the fibers cross. The neurons synapse in the lateral geniculate nuclei of the thalamus, and neurons from there form the optic radiations, which project to the visual cortex.

Neuronal Pathways

The **optic nerve** (Figure 9-12) leaves the eye and exits the orbit through the optic foramen to enter the cranial vault. Just inside the cranial vault the optic nerves are connected to each other at the **optic chiasma** (ki'az-mah). Axons from the medial portion of the retina cross through the optic chiasma and project to the opposite side of the brain. Axons from the lateral portion of the retina pass through the optic nerves and project to the brain on the same side of the body without crossing.

Beyond the optic chiasma the route of the ganglionic axons is called the **optic tract** (see Figure 9-12). Most of the optic tract axons terminate in the **lateral geniculate nucleus** of the thalamus. Some axons do not terminate in the thalamus but separate from the optic tract to terminate in the superior colliculi, the center for visual reflexes (see Chapter 8). Neurons of the lateral geniculate nucleus form the fibers of the **visual radiations,** which project to the **visual cortex** in the occipital lobe (see Figure 9-12).

HEARING AND BALANCE

The organs of hearing (auditory or acoustic organs) and balance or equilibrium can be divided into three portions: external, middle, and inner ear (Figure 9-13). The external and middle ears are involved in hearing only, whereas the inner ear functions in both hearing and balance.

Auditory Structures and Their Functions

External ear

The **auricle** (aw'rĭ-kl; ear) is the fleshy part of the external ear on the outside of the head. The auricle opens into the **external auditory meatus** (me-a'tus), a passageway that leads to the eardrum. The auricle directs sound waves toward the external auditory meatus. The meatus is lined with hairs and **ceruminous** (sĕ-roo'mĭ-nus) **glands,** which produce **cerumen,** a modified sebum commonly called earwax. The hairs and cerumen help prevent foreign objects from reaching the delicate eardrum.

The **tympanic** (tim-pan'ik) **membrane** or **eardrum** is a thin membrane that separates the external from the middle ear. Sound waves reaching the tympanic membrane through the external auditory meatus cause it to vibrate.

Middle ear

Medial to the tympanic membrane is the air-filled cavity of the middle ear. Two membrane-covered openings, the **oval window** and the **round window** on the medial side of the middle ear, connect the middle ear with the inner ear. There are two openings into the middle ear that are not covered by membranes. One opens into the mastoid air cells in the mastoid process of the temporal bone. The other, the **auditory,** or **Eustachian tube,** opens into the pharynx and enables air pressure to be equalized between the outside air and the middle ear cavity.

Unequal pressure between the middle ear and the outside environment can distort the eardrum, dampen its vibrations, and make hearing difficult. Distortion of the eardrum also stimulates pain fibers associated with that structure. That distortion is

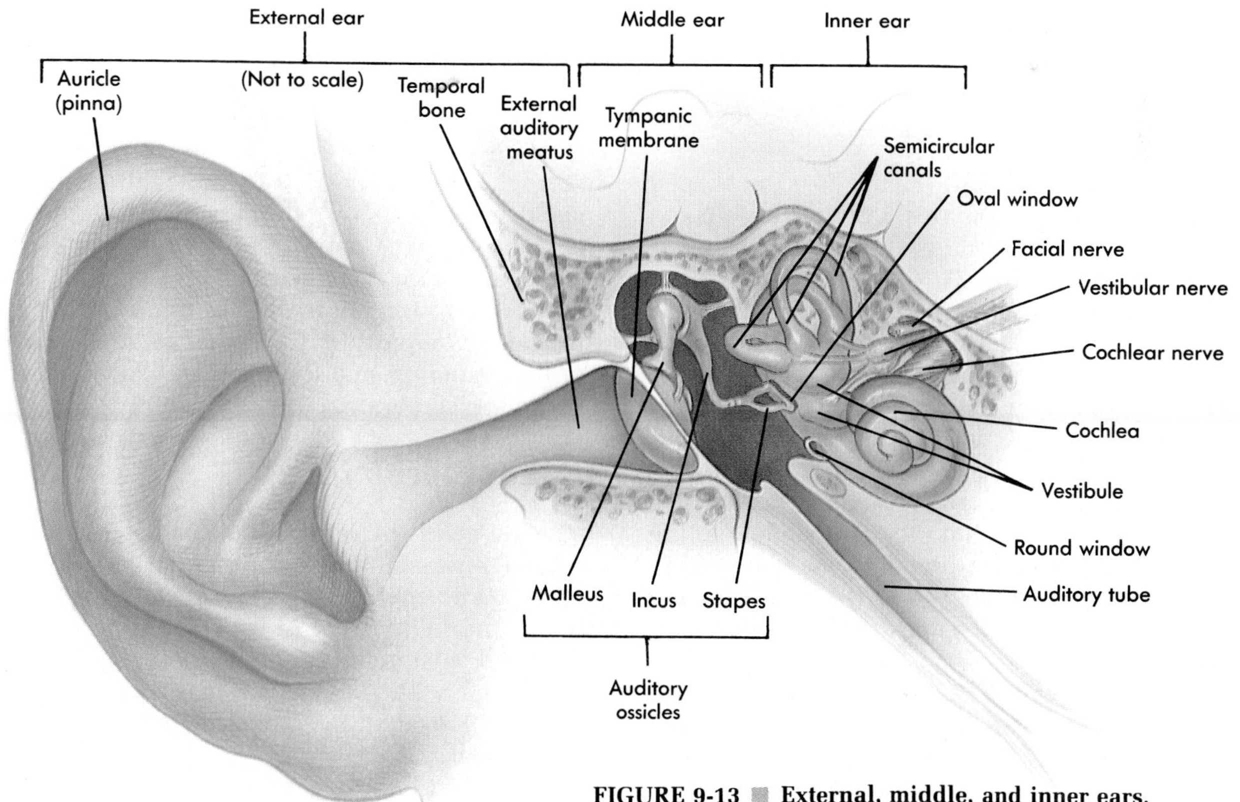

FIGURE 9-13 External, middle, and inner ears.

why, as a person changes altitude, sounds seem muffled and the eardrum may become painful. These symptoms can be relieved by opening the auditory tube, allowing air to enter the middle ear. Swallowing, yawning, chewing, and holding the nose and mouth shut while gently trying to force air out of the lungs are methods that can be used to open the auditory tube.

The middle ear contains three **auditory ossicles** (ear bones): the **malleus** (mal'e-us; hammer), **incus** (ing'kus; anvil), and **stapes** (sta'pēz; stirrup), which transmit vibrations from the eardrum to the membrane of the oval window. As the vibrations are transmitted from the malleus to the incus, they are amplified about twentyfold.

Inner ear

The inner ear consists of interconnecting tunnels and chambers within the temporal bone, called the **bony labyrinth** (lab'ĭ-rinth; maze). Inside the bony labyrinth is a similarly shaped but smaller set of membranous tunnels and chambers called the **membranous labyrinth.** The membranous labyrinth is filled with a clear fluid called **endolymph** (en'do-limf), and the space between the membranous and bony labyrinth is filled with a fluid called **perilymph.** The bony labyrinth can be divided into three regions: the cochlea, vestibule, and semicircular canals. The cochlea is involved in hearing, and the vestibule and semicircular canals are involved primarily in balance.

Hearing

The **cochlea** (ko'kle-ah) (Figure 9-14) is shaped like a snail shell and contains a bony core shaped like a screw. The thread of this screw shape is called the **spiral lamina.** A Y-shaped membranous complex divides the cochlea into three portions. The base of the Y is the **spiral lamina.** One branch of the Y is the **vestibular membrane,** and the other branch is the **basilar membrane.** The space between these latter two membranes is called the **cochlear duct.** The space above the Y is called the **scala vestibuli,** and the space below the **scala tympani.** The scala vestibuli extends from the oval window to the apex of the cochlear spiral, and the scala tympani extends from the apex to the round window. The two scalae are continuous at the apex of the cochlea.

Inside the cochlear duct is a specialized structure called the **organ of Corti.** The organ of Corti contains specialized sensory cells called **hair cells,** which have hairlike projections (microvilli). The hair tips are embedded within an acellular gelatinous shelf called the **tectorial** (tek-tor'e-al) **membrane,** which is attached to the spiral lamina.

Hair cells have no axons of their own, but each hair cell is covered by synaptic terminals of sensory neurons, the cell bodies of which are located within the **cochlear ganglion.** Afferent fibers of these neurons join to form the **cochlear nerve.** This nerve joins the vestibular nerve to become the **vestibulocochlear nerve** (cranial nerve VIII), which carries impulses to the brain.

Sound waves are collected by the auricle and are conducted through the external auditory meatus toward the tympanic membrane. Sound waves strike the tympanic membrane and cause it to vibrate. This vibration causes vibration of the three ossicles of the middle ear, and by this mechanical linkage vibration is amplified and is transferred to the membrane of the oval window (Figure 9-15).

Vibrations of the oval window membrane produce waves in the perilymph of the cochlea. The two scalae can be thought of as a continuous tube with the oval window at one end and the round window at the other end. The vibrations of the oval window cause movement of the perilymph, which pushes against the membrane of the round window. This phenomenon is similar to pushing against a rubber diaphragm on one end of a fluid-filled glass tube. If the tube has a rubber diaphragm on each end, the fluid can move. Waves in the perilymph cause the vestibular membrane to vibrate. This vibration creates waves in the endolymph and vibration of the basilar membrane. As the basilar membrane vibrates, the hairs embedded in the tectorial membrane bend, causing stimulation of the hair cells. The hair cells induce action potentials in the cochlear nerves.

The cochlear nerves, whose cell bodies are located in the cochlear ganglion, send axons to the **cochlear nucleus** in the brainstem. Neurons in the cochlear nucleus project to the superior olivary nucleus, and from there neurons project to the **inferior colliculus** in the midbrain. From the inferior colliculus, fibers project to the **medial geniculate nucleus** of the thalamus, and from there to the auditory cortex of the cerebrum.

Equilibrium

The sense of equilibrium, or balance, has two components: static equilibrium and kinetic equilibrium. **Static equilibrium** is associated with the vestibule and is involved in evaluating the position of the head relative to gravity. **Kinetic equilibrium** is associated with the semicircular canals and is involved in evaluating the change in rate of head movements.

The vestibule can be divided into two chambers, the **utricle** and the **saccule** (Figure 9-16, *A*). Each chamber contains specialized patches of epi-

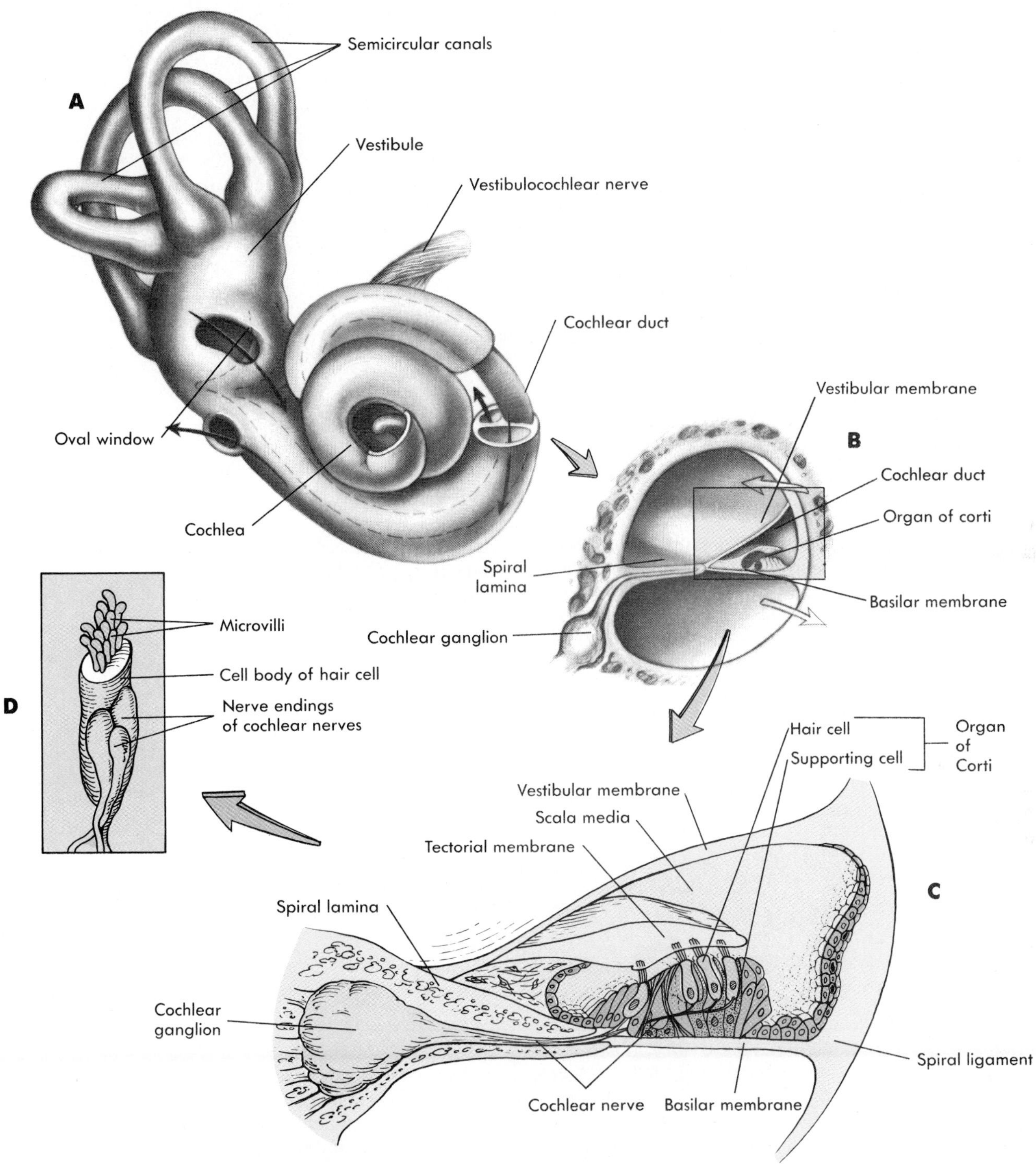

FIGURE 9-14 ■ The cochlea.
A Location of the cochlea, with a segment removed to show the cochlear duct.
B An enlarged cross section of the cochlea to show the location of the duct.
C Greatly enlarged view of the organ of Corti.
D A single hair cell with its hairs.

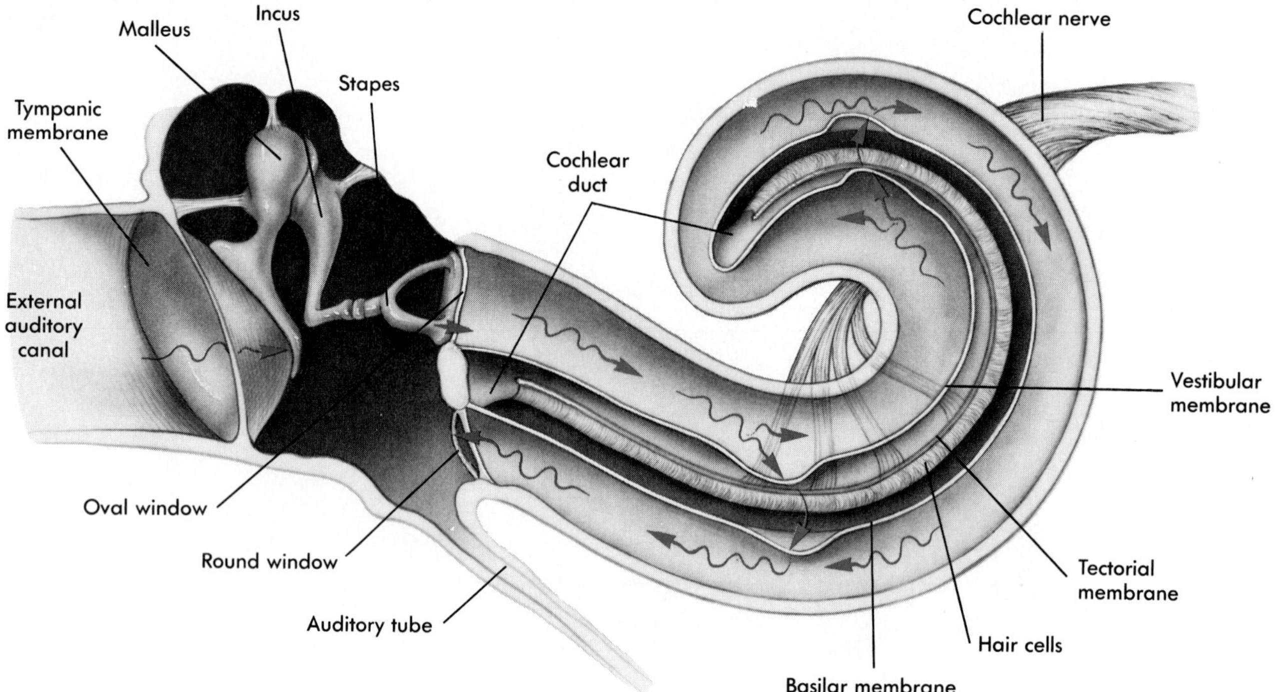

FIGURE 9-15 ■ Effect of sound waves on cochlear structures.
Sound waves strike the tympanic membrane and cause it to vibrate. This vibration causes the three bones of the middle ear to vibrate, which causes the membrane of the oval window to vibrate. This vibration causes the perilymph to vibrate, which causes the fluid in the cochlear duct and the basilar membrane to vibrate. Sound is detected by the hair cells of the organ of Corti, which is attached to the basilar membrane.

FIGURE 9-16 ■ Structure and function of the macula.
A The location of the utricular and saccular maculae within the vestibule.
B Enlarged view of a section through the macula showing the otoliths.
C and **D** The maculae respond to changes in position of the head relative to gravity, for example, as a person bends over from a vertical position **(C)**, the maculae are displaced by the pull of gravity **(D)**.

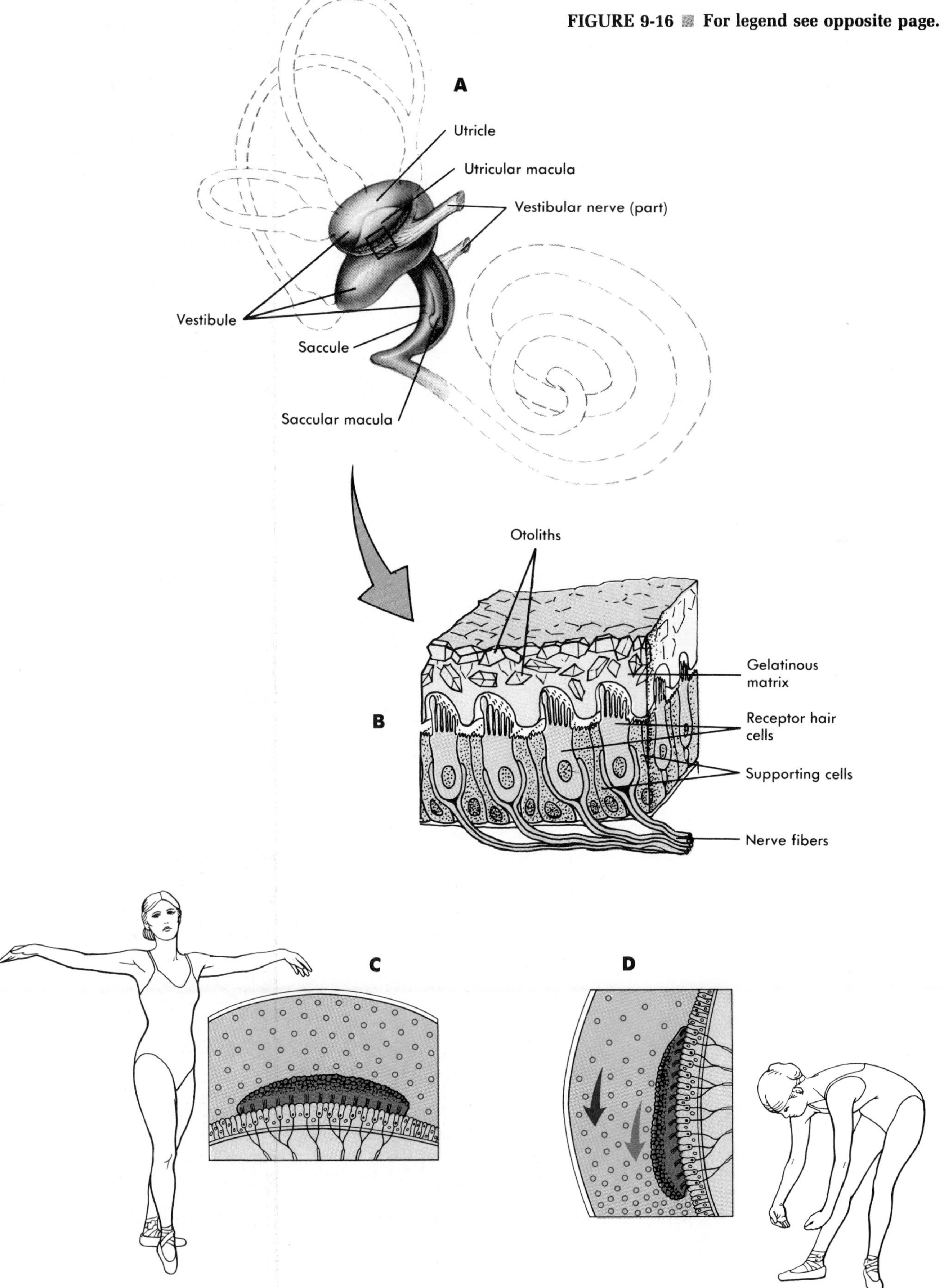

FIGURE 9-16 ■ For legend see opposite page.

thelium called the **maculae** (mak'u-le). The maculae resemble the organ of Corti and contain hair cells. The hairs (microvilli) of these cells are embedded into a gelatinous mass weighted by **otoliths** (weights) composed of protein and calcium carbonate (Figure 9-16, *B*). The mass moves in response to gravity, bending the hair cells and initiating action potentials in the associated neurons. The action potentials from these neurons are relayed by way of the vestibulocochlear nerve (cranial nerve VIII) to the brain where they are interpreted as movement of the head. For example, when a person bends over, the maculae are displaced by gravity and the resultant action potentials provide information to the brain concerning the position of the head relative to gravity (Figure 9-16, *C* and *D*).

There are three **semicircular canals,** which are involved in kinetic equilibrium and placed at nearly right angles to each other. The placement of the semicircular canals enables a person to detect movements in almost any direction. The base of each semicircular canal is expanded into an **ampulla.** Within each ampulla the epithelium is specialized to form a **crista ampullaris** (Figure 9-17, *A*

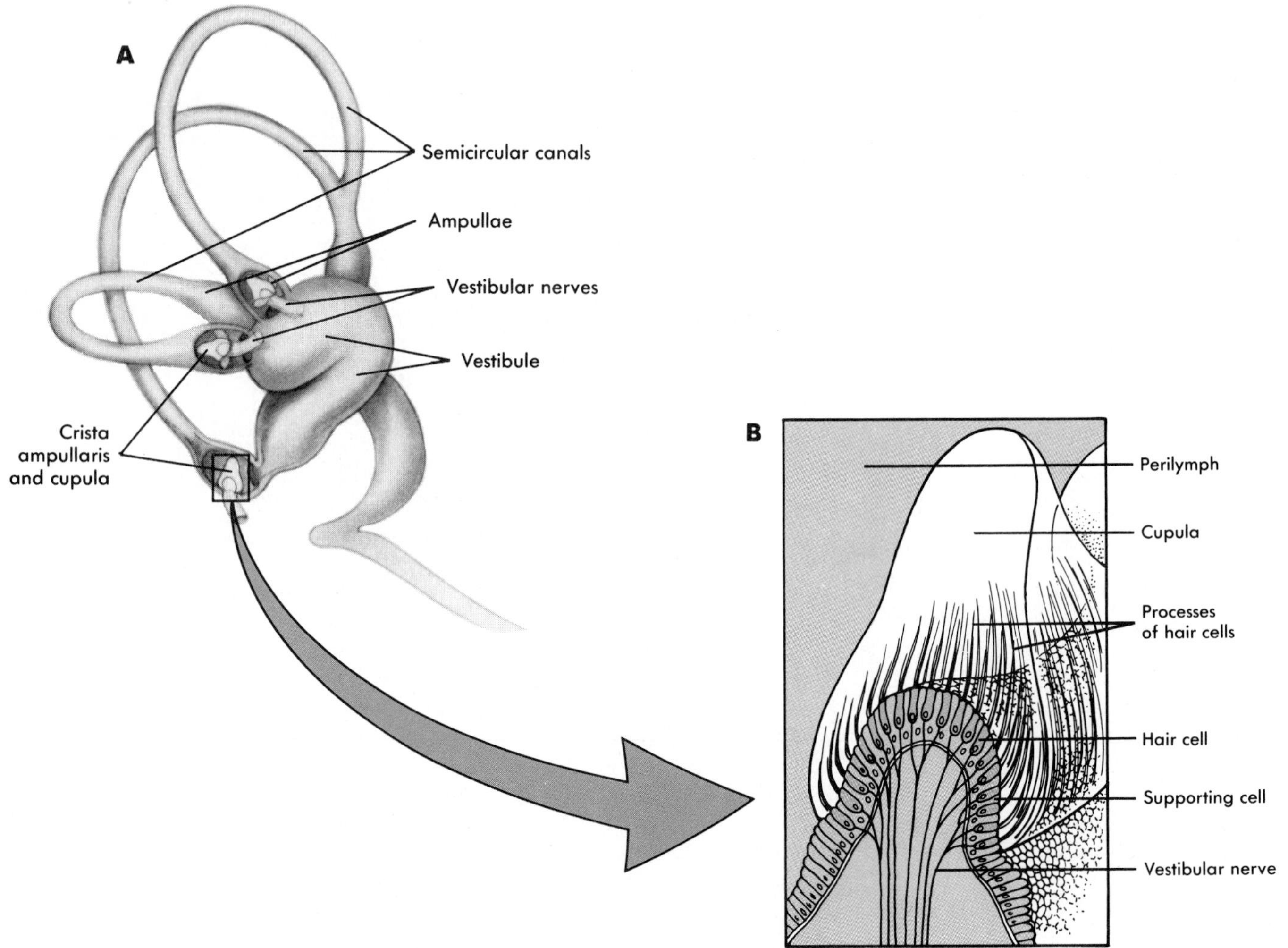

FIGURE 9-17 ■ **Structure and function of the crista ampularis.**
A Semicircular canals showing location of the crista ampularis.
B Enlargement of a section through the crista ampullaris showing the hair cells.
C and D As a person begins to spin from rest **(C)**, the endolymph in the semicircular canals tends to stay in place as the body and the crista ampularis begin to move. As a result, the crista ampularis is displaced in a direction opposite to the direction of spin **(D)**.
E When the person stops spinning, the endolymph continues to move and displaces the crista ampularis in the direction of the spin.

and *B*). Each crista consists of a ridge of epithelium with a curved gelatinous mass, the **cupula** (ku′pu-lah), suspended over the crest. The cupula is structurally and functionally very similar to the maculae, except that there are no otoliths in the cupula. The hairlike processes of the crista hair cells are embedded in the cupula. The cupula is a float that is displaced by fluid movement within the semicircular canals (Figure 9-17, *C* to *E*). As the head begins to move in a given direction (acceleration), the endolymph tends to remain stationary. This difference causes the cupula to be displaced in a direction opposite to that of the movement of the head. As movement continues the fluid "catches up." When movement of the head stops (deceleration), the fluid tends to continue to move. Movement of the head relative to the endolymph (acceleration) or movement of the endolymph relative to the head (deceleration) causes movement of the cupula and causes the hairs to bend. Bending the hairs initiates depolarization in the hair cells. This depolarization initiates action potentials in the vestibular nerves, which join the cochlear nerves and relay the information to the brain.

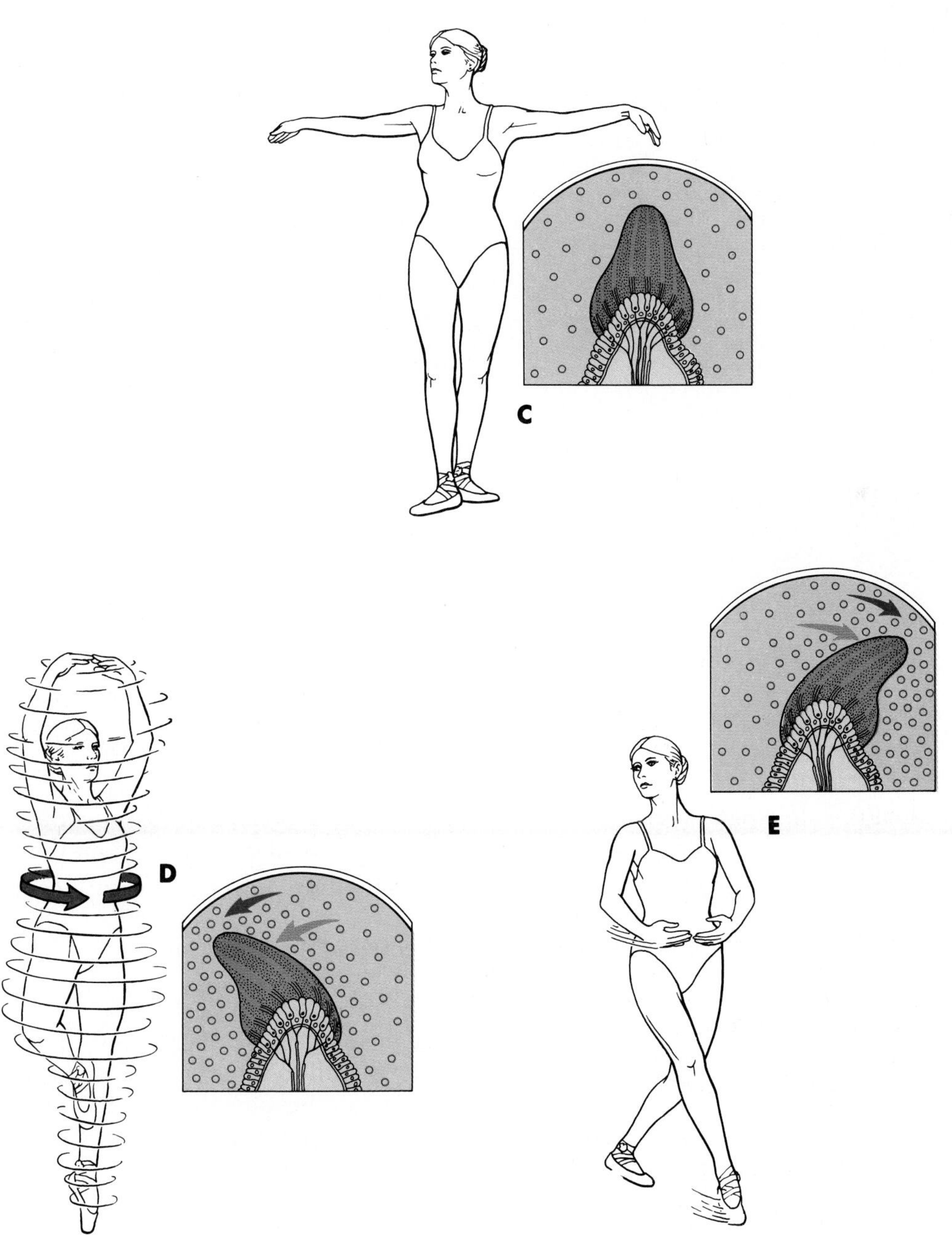

Ear Disorders

DEAFNESS

Deafness can have many causes. In general, there are two categories of deafness: conduction and sensorineural (or nerve) deafness. **Conduction deafness** involves a mechanical deficiency in transmission of sound waves from the outer ear to the spiral organ. Hearing aids may help people with such hearing deficiencies by boosting the sound volume reaching the ear. Some conduction deafness can be corrected surgically. For example, **otosclerosis** is an ear disorder in which bone grows over the oval window and immobilizes the stapes. This disorder can be surgically corrected by breaking away the bony growth and the stapes. The oval window membrane is replaced by a fat pad or synthetic membrane, and the stapes is replaced by a small metal rod connected to the oval window at one end and to the incus at the other end.

Sensorineural deafness involves the spiral organ or nerve pathways and is more difficult to correct. Research is currently being conducted on ways to replace the hearing pathways with electrical circuits. One approach involves the direct stimulation of the cochlear nerve by electrical impulses. The mechanism consists of a microphone for picking up sound waves, a microelectronic processor for converting the sound into electrical signals, a transmission system for relaying the signals to the inner ear, and a long, slender electrode that is threaded into the cochlea. This electrode delivers electrical signals directly to the cochlear nerve.

EAR INFECTIONS

Infections of the middle ear **(otitis media)** are quite common in young children. These infections usually result from the spread of infection from the mucous membrane of the pharynx through the auditory tube. The symptoms of low-grade fever, lethargy, and irritability are not often recognized by the parent as signs of middle ear infection. The infection also can cause a temporary decrease or loss of hearing because fluid buildup can dampen the eardrum. In extreme cases, the infection can damage or rupture the eardrum.

MOTION SICKNESS

Motion sickness consists of nausea and weakness caused by stimulation of the semicircular canals during motion (for example, in a boat, automobile, or airplane).

SPACE SICKNESS

Space sickness is a balance disorder occurring in zero gravity and resulting from unfamiliar sensory input to the brain. The brain must adjust to these unusual signals, or severe symptoms may result, such as headache and dizziness. Space sickness is unlike motion sickness in that motion sickness results from excessive stimulation to the brain, and space sickness results from too little stimulation as a result of weightlessness.

SUMMARY

The five special senses are olfaction, taste, vision, hearing, and balance.

GENERAL SENSES

Receptors for general senses such as pain, temperature, touch, pressure, and position are scattered throughout the body.

Pain

Pain is an unpleasant sensation with a fast component and a slow component.

Pain can be "gated," referred, or phantom.

SPECIAL SENSES

OLFACTION

Olfactory neurons have enlarged distal ends with long cilia. The cilia have receptors that respond to dissolved substances.

Axons of the olfactory neurons enter the olfactory bulb. Olfactory tracts carry impulses from the olfactory bulbs to the olfactory cortex of the brain.

The wide range of detectable odors results from combinations of a few primary odors.

TASTE

Taste buds contain gustatory cells with hairs that extend into pores. Receptors on the hairs detect dissolved substances.

There are four basic types of taste: sour, salty, bitter, and sweet.

The facial nerves carry taste from the anterior two thirds of the tongue, the glossopharyngeal from the posterior one third of the tongue, and the vagus from the root of the tongue.

VISION

Accessory Structures

The eyebrows prevent perspiration from entering the eyes.

The eyelids protect the eyes from foreign objects.

The conjunctiva covers the inner eyelid and the anterior surface of the eye.

Lacrimal glands produce tears that flow across the surface of the eye. Tears lubricate and protect the eye. Excess tears pass into the nasal cavity.

The extrinsic eye muscles move the eyeball.

Anatomy of the Eye

The fibrous tunic is the outer layer of the eye. It consists of the sclera and cornea.

The vascular tunic is the middle layer of the eye. It consists of the choroid, ciliary body, and iris.

The lens is held in place by the suspensory ligaments, which are attached to the ciliary muscles.

The retina (nervous tunic) is the inner layer of the eye and contains neurons sensitive to light.

- Rods are responsible for vision in low illumination (night vision).
- Cones are responsible for color vision and visual acuity.
- The fovea centralis is the area of greatest visual acuity.
- The optic disc or blind spot is the location where nerves exit the eye, and blood vessels enter.

The anterior compartment of the eye is filled with aqueous humor, whereas the posterior compartment is filled with vitreous humor.

Functions of the Complete Eye

Light passing through a concave surface diverges. Light passing through a convex surface converges.

Converging light rays cross at the focal point and are said to be focused.

The cornea, aqueous humor, lens, and vitreous humor all refract light. The cornea is responsible for most of the convergence, whereas the lens can adjust the focus by changing shape (accommodation).

Neuronal Pathways

Nerve fibers pass through the optic nerves to the optic chiasma, where some of the fibers cross.

Optic tracts, from the chiasma, lead to the lateral geniculate nucleus.

Visual radiations extend from the lateral geniculate nucleus to the visual cortex in the occipital lobe.

HEARING AND BALANCE

Auditory Structures and Their Functions

The external ear consists of the auricle and external auditory meatus.

The middle ear connects the external and inner ear.

- The eardrum is stretched across the external auditory meatus.
- The malleus, incus, and stapes connect the eardrum to the oval window of the inner ear.
- The auditory or Eustachian tube connects the middle ear to the pharynx and functions to equalize pressure. The middle ear is also connected to the mastoid air cells.

The inner ear has three parts: the semicircular canals, vestibule, and cochlea.

Hearing

The cochlea is a snail-shaped canal.

The cochlea is divided into three compartments by the vestibular and basilar membranes.

The organ of Corti consists of hair cells that attach to the tectorial membrane.

Sound waves are funneled by the auricle down the external auditory meatus, causing the eardrum to vibrate.

The eardrum vibrations are passed along the ossicles to the oval window of the inner ear.

Movement of the oval window membrane causes the perilymph to move the vestibular membrane, which causes the endolymph to move the basilar membrane. Movement of the basilar membrane causes movement of the hair cells in the spiral organ and generation of action potentials, which travel along the vestibulocochlear nerve.

Equilibrium

Static equilibrium evaluates the position of the head relative to gravity.

Maculae, located in the vestibule, consist of hair cells with the hairs embedded in a gelatinous mass that contains otoliths.

The gelatinous mass moves in response to gravity.

Kinetic equilibrium evaluates movements of the head.

There are three semicircular canals in the inner ear, arranged perpendicularly to each other. The ampulla of each semicircular canal contains a crista ampullaris, which has hair cells with hairs embedded in a gelatinous mass, the cupula.

CONTENT REVIEW

1. Describe the process by which an airborne molecule initiates an action potential in an olfactory neuron.
2. How is the sense of taste related to the sense of smell?
3. What are the four primary tastes? Where are they concentrated in the tongue? How do they produce many different kinds of taste sensations?
4. Describe the following structures and state their functions: eyebrows, eyelids, conjunctiva, lacrimal apparatus, and extrinsic eye muscles.
5. Name the three layers (tunics) of the eye. Describe the structures composing each layer, and explain the functions of these structures.
6. Name the two compartments of the eye, the substances that fill each compartment, and the function of the substances.
7. Describe the lens of the eye, how the lens is held in place, and how the shape of the lens is changed.
8. Describe the arrangement of cones and rods in the fovea centralis and in the periphery of the eye.
9. What is the blind spot of the eye, and what causes it?
10. What causes the pupil to constrict and dilate?
11. What causes light to refract? What is a focal point?
12. Define accommodation. What does accommodation accomplish?
13. Name the three regions of the ear, name the structures found in each region, and state the functions of the structures.
14. Describe the relationship between the tympanic membrane, the ear ossicles, and the oval window of the inner ear.
15. Describe the structure of the cochlea.
16. Starting with the auricle, trace sound into the inner ear to the point where action potentials are generated in the vestibulocochlear nerve.
17. Describe the macula and its function.
18. What is the function of the semicircular canals? Describe the crista ampullaris and its mode of operation.

CONCEPT REVIEW

1. An elderly man with normal vision developed cataracts. He was surgically treated by removing the lenses of his eyes. What kind of glasses would you recommend he wear to compensate for the removal of his lenses?
2. On a camping trip, Starr Gazer was admiring all the stars that could be seen in the night sky. She noticed an interesting little cluster of dim stars at the edge of her vision. When she looked directly at that part of the sky, however, she could not see the cluster. When she looked toward the stars but not directly at them, she could see them. Explain what was happening.
3. Skin divers are subject to increased pressure as they descend to the bottom of the ocean. Sometimes this pressure can lead to damage to the ear and loss of hearing. Describe the normal mechanisms that adjust for changes in pressure, suggest some conditions that might interfere with pressure adjustment, and explain how the increased pressure might cause loss of hearing.
4. If a vibrating tuning fork is placed against the mastoid process of the temporal bone, the vibrations will be perceived as sound, even if the external auditory meatus is plugged. Explain how this could happen.

ANSWERS TO 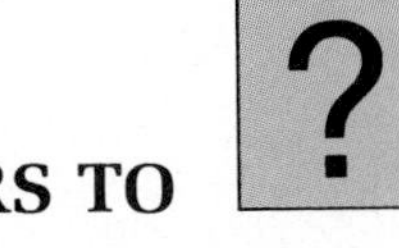 PREDICT QUESTIONS

1 *p. 226* Much of taste is based on olfactory function. A cold may include a stuffy nose, which may interfere with olfaction, and thus with taste.

2 *p. 227* Medication placed into the eyes may pass through the nasolacrimal duct into the nasal cavity, where its odor can be detected. Because much of our taste sensation is actually smell, the medication is perceived to have a taste that is detected. When a person cries, excess tears pass from the eyes through the nasolacrimal ducts into the nasal cavity.

3 *p. 230* In dim light, cones, which are involved in color perception but are less sensitive to dim light, tend to quit functioning, leaving only the rods, which can function in dim light but cannot perceive color.

4 *p. 230* The fovea is the point of greatest visual acuity. There are more cones than rods in the fovea centralis because cones are the cells with the greatest acuity.

10

THE ENDOCRINE SYSTEM

After reading this chapter you should be able to:

1 Compare the means by which the nervous and endocrine systems regulate body functions.
2 List the major categories of hormones based on their chemical structure and describe how they interact with tissues to produce a response.
3 Describe three methods for regulating the release of hormones.
4 State the location of each of the endocrine glands in the body.
5 List the hormones produced by each of the endocrine glands and describe their effects on the body.
6 Describe how the hypothalamus regulates hormone secretion from the pituitary.
7 Describe how the pituitary regulates the secretion of hormones from other endocrine glands.
8 Choose a hormone and use it to explain how negative feedback results in homeostasis.

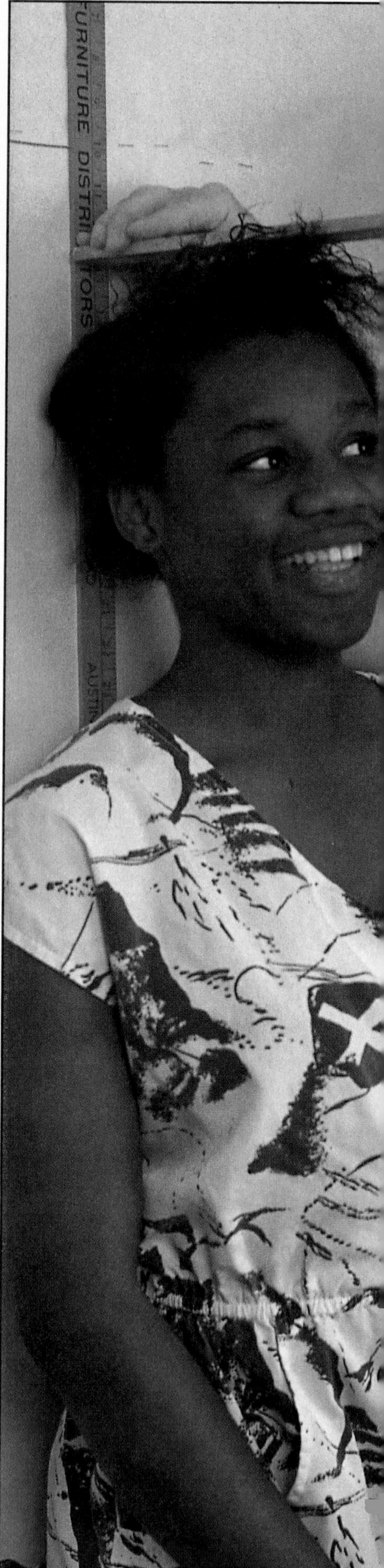

endocrine (en′do-krin) [Gr. *endon*, inside + *krino*, to separate] Ductless gland that secretes internally, usually into the circulation.

hormone [Gr. *hormon*, to set into motion] A substance secreted by endocrine tissues into the blood that acts on a target tissue to produce a specific response.

hypothalamic-pituitary portal system A series of blood vessels that carry blood from the area of the hypothalamus to the anterior pituitary; they originate from capillary beds in the hypothalamus and terminate as a capillary bed in the anterior pituitary.

hypothalamus (hi′po-thal′ă-mus) [Gr. *hypo*, under, below + *thalamus*, bedroom] Important autonomic and endocrine control center of the brain located beneath the thalamus.

pituitary, (pit-u′ĭ-tĕr-e) **or hypophysis** (hi-pof′ĭ-sis) [Fr., an undergrowth] An endocrine gland attached to the hypothalamus; secretes hormones that influence the function of several other endocrine glands and tissues.

receptor A molecule in the membrane or cytoplasm of cells of a target tissue to which a hormone binds. The combining of a hormone with its receptor initiates a response in the target tissue.

releasing hormone Hormone that is released from neurons of the hypothalamus and flows through the hypothalamic-pituitary portal system to the anterior pituitary; functions to regulate the secretion of hormones from the cells of the anterior pituitary gland.

target tissue Tissue upon which a hormone acts.

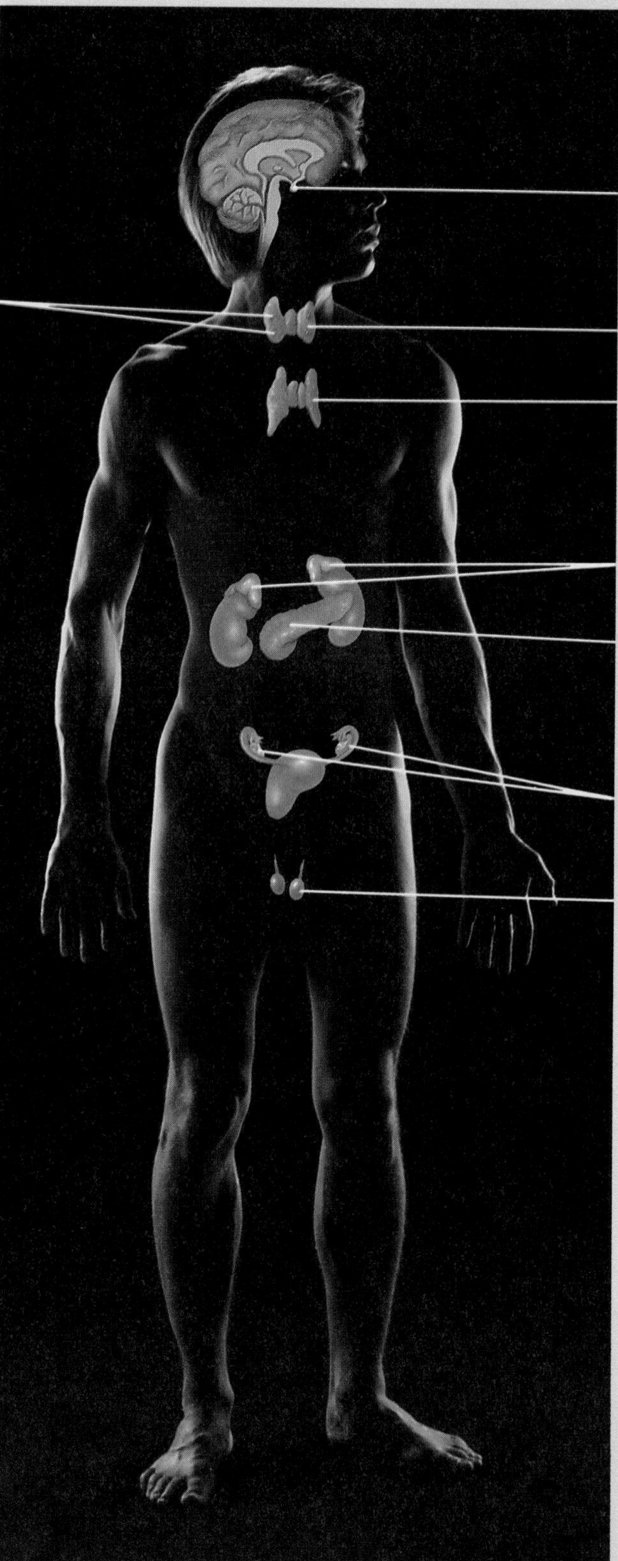

The Endocrine System

MAJOR COMPONENTS

Endocrine glands, such as the pituitary, thyroid, and adrenal glands

MAJOR FUNCTIONS

Regulates metabolism and growth

Regulates absorption of nutrients

Regulates fluid balance and ion concentration

Regulates the body's response to stress

Regulates sexual characteristics, reproduction, birth, and lactation

FIGURE 10-1 ■ Endocrine glands and their location in the human body.

The nervous and the endocrine systems are the two major regulatory systems in the body. Together they regulate and coordinate the activity of nearly all other body structures. Although the structures and functions of the nervous and endocrine systems are interrelated, they differ in several ways. The nervous system controls structures by sending action potentials along axons while the endocrine system acts by sending chemical signals through the circulatory system. The nervous system usually acts more quickly and has a short-term effect, whereas the endocrine system usually responds more slowly and has a longer lasting effect. In general, each nervous stimulus controls a specific tissue or organ while the endocrine system has a more general effect on the body.

HORMONES

The term **endocrine** (en′do-krin) is derived from the Greek words *endo* and *krino* meaning within and to separate. The word implies that endocrine glands secrete their chemical signals into the blood. The chemical signals are transported in the blood from the endocrine glands to tissues some distance from the endocrine glands. Examples of endocrine glands are the thyroid gland and the adrenal glands. In contrast, **exocrine glands** secrete their products into ducts, which then carry the secretory products to an external or internal surface, such as the skin or digestive tract. Examples of exocrine glands are the sweat glands and the salivary glands. The **endocrine system** consists of all the body's endocrine glands (Figure 10-1).

Chemical signals secreted by endocrine glands are called **hormones,** a term derived from the Greek word *hormon* meaning to set into motion. Traditionally a hormone is defined as a substance that is produced in minute amounts by a collection of cells, is secreted into the circulatory system to be transported some distance, and acts on tissues at another site in the body to influence their activity in a specific way. All hormones exhibit most of these characteristics.

Hormones are distributed in the blood to all parts of the body, but only certain **target tissues** respond to each type of hormone. A target tissue for a hormone is made up of cells that have **receptors** for the hormone. Each hormone can bind only to its receptors and cannot influence the function of cells that do not have receptors for the hormone (Figure 10-2).

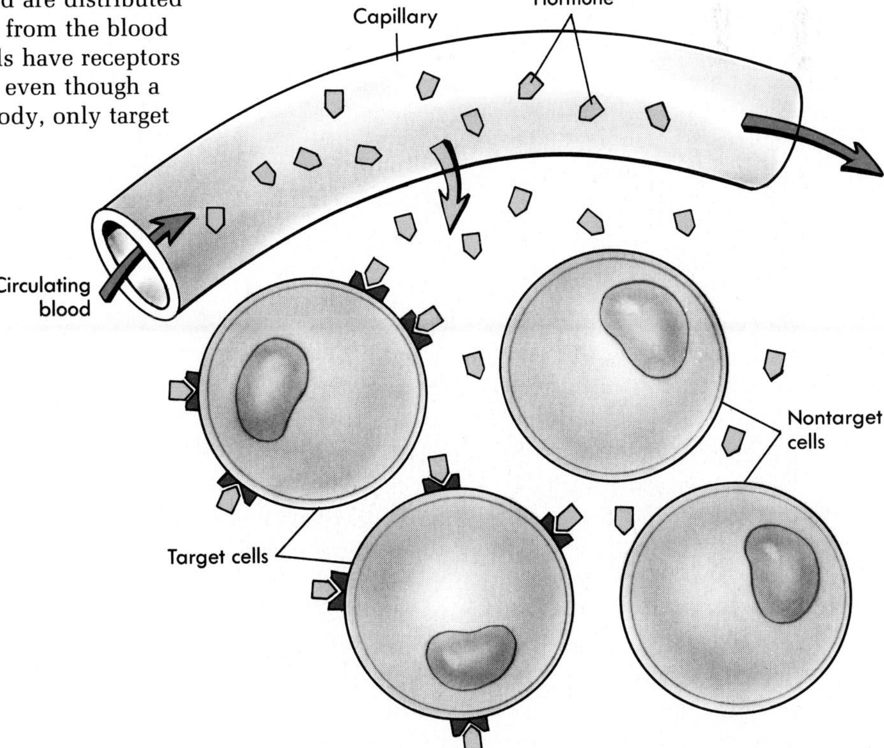

FIGURE 10-2 ■ Target cell response to hormones. Hormones are secreted into the blood and are distributed throughout the body, where they diffuse from the blood into the interstitial fluid. Only target cells have receptors to which hormones can bind. Therefore, even though a hormone is distributed throughout the body, only target cells for that hormone can respond to it.

Chemistry

Hormones fall into the following chemical categories:

1. Peptides and amino acid derivatives.
   A. Peptide hormones consist of short chains of amino acids. The hormones of the posterior pituitary gland that control functions such as milk "let-down" in the breast and urine volume are examples of short chains of amino acids.
   B. Some hormones are proteins consisting of many amino acids bound together by peptide bonds. Some of the protein hormones have carbohydrate molecules bound to them. Most hormones of the anterior pituitary gland, which control functions such as growth, metabolism, and reproductive functions, are examples of protein hormones.
   C. Some hormones consist of single amino acids that have been chemically modified.

> Specific hormones are given as a treatment for certain illnesses. Hormones that are soluble in lipids, such as steroids, can be taken orally. The lipid-soluble hormones diffuse across the wall of the intestine into the circulatory system. In contrast, protein hormones cannot diffuse across the wall of the intestine because they are not lipid soluble. Protein hormones are broken down to individual amino acids before they are transported across the wall of the digestive system. The normal structure of a protein hormone is destroyed when it is taken into the digestive system, and its physiological activity is lost. Consequently, protein hormones must be injected rather than taken orally. The most commonly administered protein hormone is insulin, which is prescribed for diabetes mellitus.

2. Lipid hormones
   A. Steroid hormones are lipids, all of which are derived from cholesterol. The steroids all have a structure that varies only slightly among the different types. The small differences, however, give each type of steroid unique functional characteristics. Steroid hormones are produced mostly by the adrenal cortex and the gonads (testis and ovary).
   B. Prostaglandin-like compounds are derived from the fatty acid arachidonic acid. The prostaglandin-like compounds are produced by many tissues and generally have local effects. They play important roles in regulating smooth muscle contractions and inflammation.

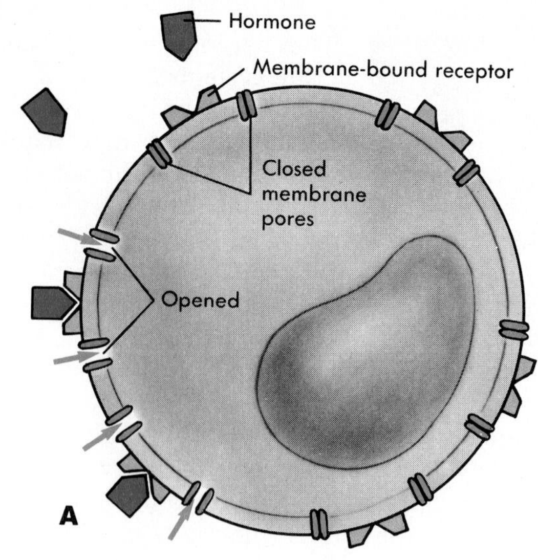

Hormone Action

Most peptide and protein hormones cannot diffuse through the cell membrane. Instead, they bind to receptors located on the surface of the cell membrane. The combination of a hormone with its receptor may alter the permeability of the cell's membrane, resulting in a response; or the combination may cause the production of a substance inside the cell. In this latter case, the hormone is called the **first messenger,** and the substance inside the cell is called the **second messenger.** The first messenger binds to a receptor on the cell membrane, which causes a second messenger to be produced inside the cell. Enzymes activated by the second messenger catalyze reactions that produce the characteristic response of the target cell to the hormone (Figure 10-3). For example, epinephrine binds to receptors for epinephrine within the cell membrane of liver cells. The occupied receptors activate an enzyme on the inner surface of the cell membrane called adenylate cyclase, which catalyzes the following reaction:

$$\text{ATP} \xrightarrow[\text{Cyclase}]{\text{Adenylate}} \text{Cyclic AMP} + \text{PPi (two phosphates)}$$

Cyclic AMP (cAMP) functions as a second messenger. Cyclic AMP diffuses throughout the cell's cytoplasm and binds to specific enzymes activating them. The activated enzymes increase the activity of yet other enzymes. The final result is the breakdown of glycogen by liver enzymes to individual glucose molecules, which are released into the circulatory system.

Hormones that stimulate the synthesis of a second-messenger molecule often produce rapid responses (within minutes) because the second messenger influences the activity of existing enzymes.

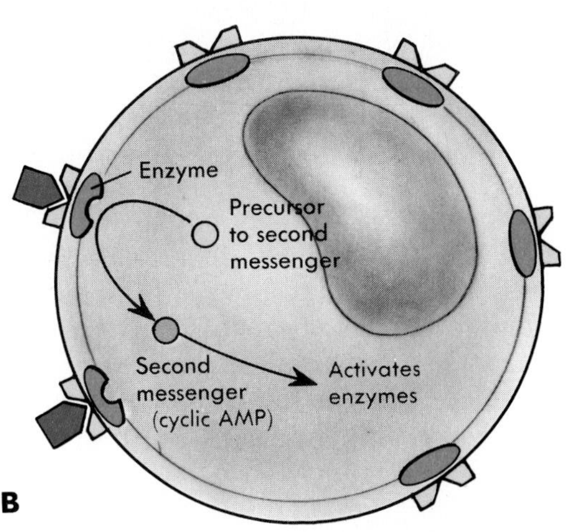

FIGURE 10-3 ■ Membrane-bound receptors.
Some hormone receptors are attached to the cell membrane and are called membrane-bound receptors. Hormones in the interstitial fluid bind to the hormone receptors, and the combination of the hormones and receptors initiates a response in the target cells.
A Some hormones bind to membrane-bound receptors, and the cell responds by altering the permeability of the cell membrane to ions such as sodium and calcium. The combination of the hormone with the receptor causes pores in the cell membrane to open *(arrows)*.
B Some hormones bind to membrane-bound receptors, and the cell responds by activating an enzyme that synthesizes "second messenger" molecules inside the cell. The second messengers, such as cyclic AMP, interact with enzymes within the cell to alter their activity. The altered activity of the enzymes produces the cell's response to the hormone.

In addition, because a few second-messenger molecules can activate many enzymes, an amplification system exists.

Lipid-soluble hormones such as steroids can pass through the cell membrane by diffusion. The receptor molecules float freely in the cytoplasm or the nucleus of target cells. By the process of diffusion, lipid-soluble hormones cross the cell membrane into the cytoplasm. The hormones may bind to receptors located in the cytoplasm or in the nucleus. If the hormone binds to a receptor in the cytoplasm, the receptor and hormone diffuse into the nucleus. Once within the nucleus the hormone, bound to the receptor, activates specific genes within the DNA. These genes produce messenger RNA (mRNA) (Figure 10-4). The mRNA moves to the cytoplasm and directs the synthesis of new proteins at the ribosomes, which produce the cell's response to the hormone. For example, the steroid aldosterone causes its target cells within the kidney to increase the rate of sodium chloride transport. Newly synthesized proteins produced in the target cells are responsible for the increased rate of sodium chloride transport.

In cells that synthesize new protein molecules as part of their response, there is a delay of several hours between the time a hormone binds to its receptor and the time the response is observed. During this time, mRNA and protein synthesis occurs.

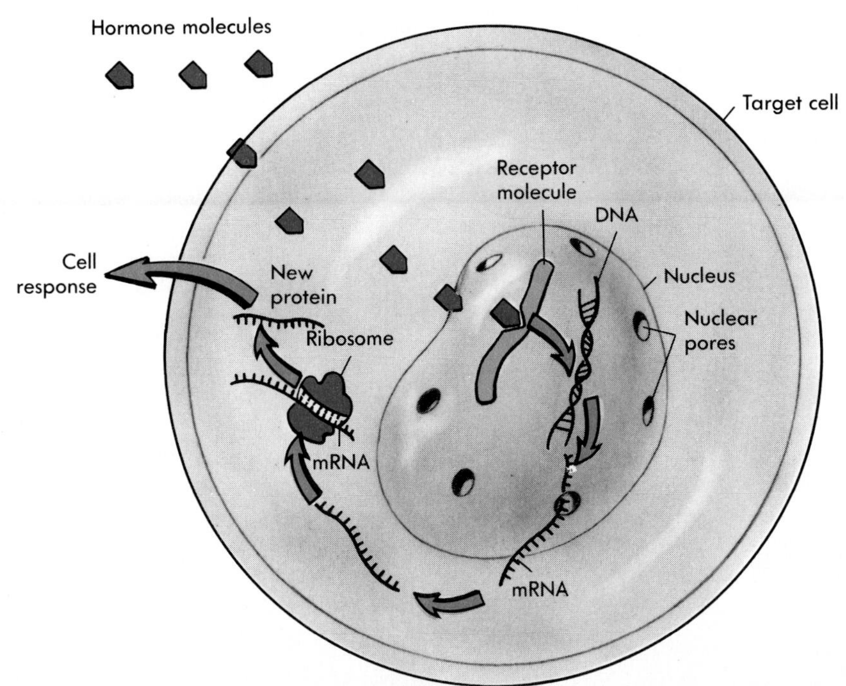

FIGURE 10-4 ■ Intracellular receptor model.
Some hormones are lipid soluble and can diffuse through the cell membrane and enter the cytoplasm of the cell. Once inside, the hormones bind to receptor molecules, usually within the nucleus of the cells. The combination of the hormones with their receptor molecules causes messenger RNA to be produced. The messenger RNA molecules leave the nuclei, pass into the cytoplasm, and bind to ribosomes, where they direct the synthesis of specific proteins. The proteins synthesized on the ribosomes carry out the cell's response to the hormone.

The effect of hormones on target tissues may be to change the rate of protein synthesis, the rate of secretion, or the permeability of the membrane. Thus, hormones have a dramatic effect on processes such as growth, metabolism, water balance, blood pressure, reproduction, and behavior.

Regulation of Hormone Secretion

The secretion of hormones is controlled by negative-feedback mechanisms (see Chapter 1). Negative-feedback mechanisms keep the body functioning within a narrow range of values consistent with life. For example, insulin is a hormone that regulates the concentration of blood glucose, or blood sugar. When blood glucose levels increase after a meal, insulin is secreted. Insulin acts on several target tissues and causes them to take up glucose, causing blood glucose levels to decline. As the blood glucose levels begin to fall, however, the rate at which insulin is secreted falls also. As insulin levels fall, the rate at which glucose is taken up by the tissues decreases, keeping the blood glucose levels from declining too much. This negative-feedback mechanism counteracts increases and decreases in blood glucose to maintain homeostasis.

TABLE 10-1

Endocrine Glands, Hormones, and Their Target Tissues

GLAND	HORMONE	TARGET TISSUE	RESPONSE
Pituitary gland			
Posterior	Antidiuretic hormone (ADH)	Kidney	Increases water reabsorption (less water is lost as urine)
	Oxytocin	Uterus	Increases uterine contractions
		Mammary gland	Increases milk "let-down" from mammary glands
Anterior	Growth hormone	Most tissues	Increase protein synthesis, breakdown of lipids, and release of fatty acids from cells; increase blood glucose levels.
	Thyroid-stimulating hormone (TSH)	Thyroid gland	Increases thyroid hormone secretion
	Adrenocorticotropic hormone (ACTH)	Adrenal cortex	Increases secretion of glucocorticoid hormones such as cortisol; increases skin pigmentation at high concentrations
	Melanocyte-stimulating hormone (MSH)	Melanocytes in skin	Increase melanin production in melanocytes to make the skin darker in color
	Luteinizing hormone (LH) or interstitial cell stimulating hormone (ICSH)	Ovary in females, testis in males	Ovulation and progesterone production in the ovary, testosterone synthesis and support for sperm production in testis
	Follicle-stimulating hormone (FSH)	Follicles in ovary in females, seminiferous tubules in males	Follicle maturation and estrogen secretion in ovary; spermatogenesis in testis
	Prolactin	Ovary and mammary gland in females, testis in males	Milk production in women; increases sensitivity to LH in males
Thyroid gland	Thyroid hormones	Most cells of the body	Increase metabolic rates, essential for normal process of growth and maturation
	Calcitonin	Primarily bone	Decreases rate of breakdown of bone; prevents large increase in blood calcium levels
Parathyroid glands	Parathyroid hormone	Bone, kidney	Increases rate of bone breakdown by osteoclasts; increases vitamin D synthesis, essential for maintenance of normal blood levels of calcium

The secretion rate for some hormones is directly controlled by the blood levels of chemicals. For example, insulin secretion is controlled by blood levels of glucose, and secretion of parathyroid hormone is controlled by blood levels of calcium. In addition, the secretion rate of some hormones is controlled by another hormone. For example, hormones from the pituitary glands act on the ovary and the testis, causing these structures to secrete sex hormones. Finally, some hormones, which are called **neurohormones**, are released directly from nerve cells. Control of the secretion of neurohormones is maintained by the nervous system. An example of a neurohormone is epinephrine, which is released from the adrenal medulla.

THE ENDOCRINE SYSTEM

The endocrine system consists of ductless glands that secrete hormones directly into the circulatory system (Table 10-1 and see Figure 10-1). The endocrine glands are supplied by an extensive network of blood vessels, and organs with the richest blood supply include endocrine glands such as the adrenal and thyroid glands.

TABLE 10-1—cont'd

Endocrine Glands, Hormones, and Their Target Tissues

GLAND	HORMONE	TARGET TISSUE	RESPONSE
Adrenal glands			
Adrenal medulla	Epinephrine mostly, some norepinephrine	Heart, blood vessels, liver, fat cells	Increase cardiac output; increase blood flow to skeletal muscles and heart; increase release of glucose and fatty acids into blood; in general, prepare for physical activity
Adrenal cortex	Mineralocorticoids (aldosterone)	Kidneys; to lesser degree, intestine and sweat glands	Increase rate of sodium transport into body; increase rate of potassium excretion; secondarily favor water retention
	Glucocorticoids (cortisol)	Most tissues (e.g., liver, fat, skeletal muscle, immune tissues)	Increase fat and protein breakdown; increase glucose synthesis from amino acids; increase blood nutrient levels; inhibit inflammation and immune response
	Adrenal androgens	Most tissues	Insignificant in males; increase female sexual drive, pubic hair and axillary hair growth in women
Pancreas	Insulin	Especially liver, skeletal muscle, fat tissue	Increases uptake and use of glucose and amino acids
	Glucagon	Primarily liver	Increases breakdown of glycogen, release of glucose into the circulatory system
Reproductive organs			
Testes	Testosterone	Most cells	Aid in spermatogenesis, maintenance of functional reproductive organs, secondary sexual characteristics, sexual behavior
Ovaries	Estrogens and progesterone	Most cells	Aid in uterine and mammary gland development and function, external genitalia structure, secondary sexual characteristics, sexual behavior, and menstrual cycle
Uterus, ovaries, inflamed tissues	Prostaglandins	Most tissues	Mediate inflammatory responses; increase uterine contractions, ovulation, and more
Thymus gland	Thymosin	Immune tissues	Immune system development and function
Pineal body	Melatonin	At least the hypothalamus	Inhibits secretion of gonadotropin-releasing hormone, thereby inhibiting reproduction

Some glands of the endocrine system perform functions in addition to hormone secretion. For example, the endocrine portion of the pancreas has cells that secrete hormones; and, in addition, a major exocrine portion of the pancreas secretes digestive enzymes. The ovaries and the testes have endocrine portions that secrete hormones, and they also produce eggs and sperm, respectively.

THE ENDOCRINE GLANDS AND THEIR HORMONES

The Pituitary and Hypothalamus

The **pituitary** (pit-u'ĭ-tĕr-e), or the **hypophysis** (hi-pof'ĭ-sis), is a small gland about the size of a pea (Figure 10-5). It rests in a depression of the sphenoid bone below the hypothalamus of the brain. The **hypothalamus** (hi'po-thal'ă-mus) is an important autonomic and endocrine control center of the brain located beneath the thalamus. The pituitary is located behind the optic chiasma and is connected to the hypothalamus by a stalk called the **infundibulum** (in-fun-dib'u-lum). The gland is divided into two parts: The **anterior pituitary** is made up of epithelial cells derived from the embryonic oral cavity, whereas the **posterior pituitary** is an extension of the brain and is made up of nerve cells. The hormones secreted from each lobe of the pituitary are listed in Table 10-1.

Hormones from the pituitary gland control the function of many other glands in the body such as the gonads, thyroid gland, and adrenal cortex. The pituitary gland also secretes hormones that influence growth, kidney function, delivery of infants, and milk production by the breast. The pituitary

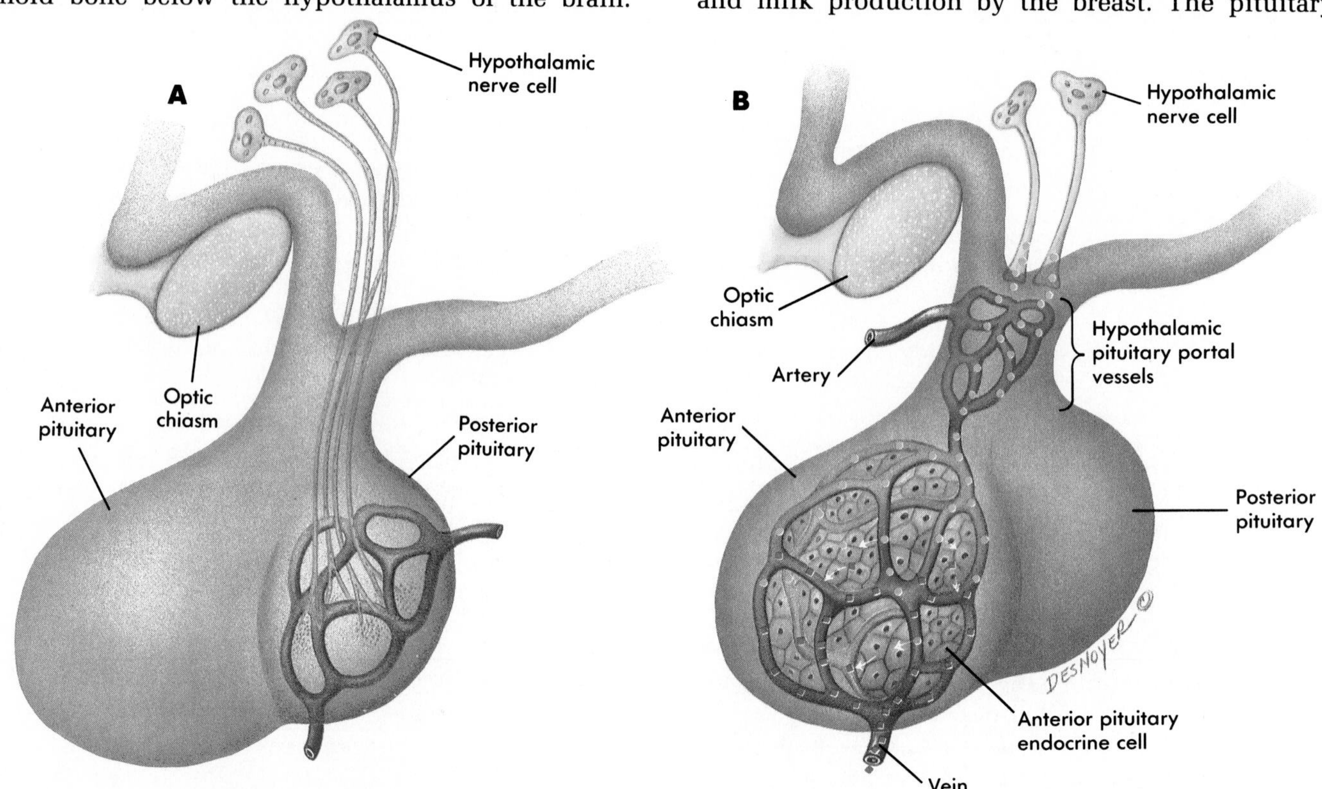

FIGURE 10-5 ■ The hypothalamus and pituitary.
The figure illustrates the relationship between the hypothalamus of the brain and the pituitary gland.
A Substances called releasing hormones or releasing factors are secreted from the hypothalamic neurons as a result of certain stimuli acting on the brain. They pass through the hypothalamic-pituitary portal system to the anterior pituitary. Within the anterior pituitary the releasing hormones influence the secretion of anterior pituitary hormones. Hormones secreted from the anterior pituitary pass through the blood and influence the activity of their target tissues.
B Some neurosecretory cells of the hypothalamus have axons that extend to the posterior pituitary and make up the secretory portion of the posterior pituitary. In response to stimulation, hormones are released from the neurosecretory cells of the posterior pituitary gland and pass through the blood to target tissues.

gland historically was referred to as the **master gland** of the body because it controls the function of so many other glands. It is now known, however, that the pituitary gland is controlled in two ways by the hypothalamus of the brain. The secretion of hormones from the anterior pituitary is controlled by **releasing hormones** produced by the hypothalamus (see Figure 10-5). The releasing hormones enter a capillary bed in the hypothalamus and are transported through veins to a second capillary bed in the anterior pituitary. There they leave the blood and bind to receptors involved with the regulation of anterior pituitary hormone secretion. The capillary beds and veins that transport the hormones are called the **hypothalamic-pituitary portal system.**

Secretion of hormones from the posterior pituitary is controlled by nervous system stimulation of nerve cells within the hypothalamus (see Figure 10-5, *B*). These nerve cells have their cell bodies in the hypothalamus, and they send their axons through the infundibulum to the posterior pituitary. Hormones are produced in the cell bodies and are transported within the axons to the posterior pituitary where they are stored in the axon endings. When these nerve cells are stimulated, action potentials from the hypothalamus travel along the axons to the axon endings and cause the release of hormones from the posterior pituitary.

Within the hypothalamus and pituitary, the nervous and endocrine systems are closely interrelated. Emotions such as joy and anger as well as chronic stress influence the endocrine system through the hypothalamus. Hormones of the endocrine system can influence the functions of the hypothalamus and other parts of the brain.

Hormones of the anterior pituitary

Growth hormone (GH) stimulates the growth of bones, muscles, and other organs by increasing protein synthesis. It also resists protein breakdown during periods of food deprivation and favors fat breakdown. A young person suffering from a deficiency of this hormone will remain a small, though normally proportioned, person called a **pituitary dwarf.** This condition can be treated by administering growth hormone. If excess growth hormone is present before bones complete their growth in length, exaggerated bone growth occurs. The result is **giantism,** and the person becomes abnormally tall. If excess hormone is secreted after growth in bone length is complete, growth in bone diameter continues. As a result, the facial features and hands become abnormally large, a condition called **acromegaly** (ak′ro-meg′al-e).

> Because growth hormone is a protein, it is difficult to artificially produce using conventional techniques. Two sources of growth hormone exist for people who suffer from a low rate of growth hormone secretion. Growth hormone can be extracted from the pituitary glands of people who have died. In addition, human genes for growth hormone have been successfully introduced into bacteria. The gene in the bacteria causes growth hormone synthesis, and the growth hormone can be extracted from the medium in which the bacteria are grown. Thus, modern genetic engineering has provided a source of human growth hormone.

The secretion of growth hormone is controlled by two releasing hormones from the hypothalamus. One releasing hormone stimulates growth hormone secretion and the other inhibits it. Most people have a rhythm of growth hormone secretion with daily peak levels occurring during deep sleep. The blood growth hormone level does not become greatly elevated during periods of rapid growth, although children tend to have somewhat higher blood levels of growth hormone than adults. In addition to growth hormone, genetics, nutrition, and sex hormones also influence growth.

Thyroid-stimulating hormone (TSH) binds to receptors on cells of the thyroid gland and causes the cells to secrete thyroid hormone. When too much TSH is secreted it causes the thyroid gland to enlarge and secrete too much thyroid hormone. When too little TSH is secreted, the thyroid gland decreases in size and too little thyroid hormone is secreted. The rate of TSH secretion is controlled by a releasing hormone from the hypothalamus.

Adrenocorticotropic (ă-dre′no-kor′tĭ-ko-tro′pik) **hormone (ACTH)** binds to receptors on cells in the cortex of the adrenal glands. ACTH increases the secretion of a hormone from the adrenal cortex called **cortisol** (kor′tĭ-sol), and ACTH is required to keep the adrenal cortex from degenerating. ACTH also binds to melanocytes in the skin and increases skin pigmentation. One symptom of too much ACTH secretion is a darkening of the skin. The rate of ACTH secretion is controlled by a releasing hormone from the hypothalamus.

Gonadotropins (gon′ă-do-tro′pinz) are hormones that bind to the gonads (ovaries and testes). They regulate the growth, development, and functions of the gonads. **Luteinizing** (lu′te-ĭ-nīz-ing) **hormone (LH)** causes ovulation and sex hormone secretion from the ovary in females and sex hormone secretion from the testis in males. LH is sometimes referred to as **interstitial cell–stimulat-**

ing hormone (ICSH) in males because it stimulates interstitial cells of the testis to secrete testosterone. **Follicle-stimulating hormone (FSH)** stimulates the development of eggs in the ovary and sperm cells in the testis. Without LH and FSH the ovary and testis decrease in size, no longer produce eggs or sperm, and no longer secrete hormones. A single releasing hormone from the hypothalamus controls the secretion of both LH and FSH.

Prolactin (pro-lak'tin) helps promote development of the breast during pregnancy and stimulates the production of milk in the breast following pregnancy. Prolactin may also make the gonads more sensitive to the effects of LH. The regulation of prolactin secretion is complex, and several substances released from the hypothalamus may regulate its secretion. There are two releasing hormones, one that increases prolactin secretion and one that decreases it.

Melanocyte-stimulating hormone (MSH) binds to receptors on melanocytes and causes them to synthesize melanin. Oversecretion of MSH causes the skin to darken. The structure of MSH is similar to ACTH, and both hormones cause the skin to darken. Regulation of MSH is not well understood, but there appear to be two releasing hormones from the hypothalamus, one that increases MSH secretion and one that decreases it.

Hormones of the posterior pituitary

Antidiuretic (an'tĭ-di-u-rĕ-tik) **hormone (ADH)** increases water reabsorption by kidney tubules, with the result that less water is lost as urine. ADH can also cause blood vessels to constrict when secreted in large amounts. Consequently, it is sometimes called **vasopressin.**

Oxytocin (ok-sĭ-to'sin) causes contraction of the muscle of the uterus and milk ejection or milk "let-down" from the breasts in lactating women. Commercial preparations of oxytocin are given under certain conditions to assist in childbirth and to constrict uterine blood vessels following childbirth.

The Thyroid Gland

The **thyroid** (thi'royd) **gland** is made up of two lobes of thyroid tissue connected by a narrow band. The lobes are located on either side of the trachea just below the larynx (Figure 10-6, *A*). The thyroid gland is one of the largest endocrine glands. It is highly vascular, appears more red than surrounding tissues, and is surrounded by a connective tissue capsule. The thyroid gland contains numerous **thyroid follicles,** which are small spheres with walls that consist of single layers of cuboidal epithelial cells (see Figure 10-6, *C*). The center of each thyroid follicle is filled with proteins to which thyroid hormones are attached. The cells of the thyroid follicles synthesize thyroid hormones, which are stored in the follicles. Between the follicles a network of loose connective tissue contains capillaries and scattered **parafollicular** (păr'ah-fo-lik'u-lar) **cells.**

The main function of the thyroid gland is to secrete **thyroid hormones,** such as thyroxine, which regulate the rate of metabolism in the body (see Table 10-1). Without a normal rate of thyroid hormone secretion, growth and development cannot proceed normally. A lack of thyroid hormones is called **hypothyroidism.** In infants, hypothyroidism can result in **cretinism,** a condition in which the person is mentally retarded and has a short stature with abnormally formed skeletal structures. In adults, the lack of thyroid hormones results in a reduced rate of metabolism, sluggishness, and a reduced ability to perform routine tasks. An elevated rate of thyroid hormone secretion is known as **hyperthyroidism.** It results in an elevated rate of metabolism, extreme nervousness, and chronic fatigue.

Thyroid hormone secretion is regulated by TSH from the anterior pituitary. Small fluctuations occur in blood TSH levels on a daily basis, with a small increase at night. Increasing blood levels of TSH increase the synthesis and secretion of thyroid hormones, and decreasing blood levels of TSH decrease the synthesis and secretion of thyroid hormones. The thyroid hormones, in turn, have a negative-feedback effect on the hypothalamus and pituitary so that increasing levels of thyroid hormones inhibit TSH secretion. Decreasing thyroid hormone levels allow additional TSH to be secreted. Because of the negative-feedback effect, the thyroid hormones fluctuate within a narrow concentration range in the blood.

The body requires iodine to synthesize the thyroid hormones. The iodine is taken up by the thyroid follicles where hormone synthesis occurs. One thyroid hormone, called **thyroxine** (thi-rok'sin), contains four iodine atoms and is abbreviated T_4. The other thyroid hormone contains three iodine atoms and is abbreviated T_3. If the quantity of iodine present is not sufficient, the production and secretion of the thyroid hormones decreases. This decrease stimulates the pituitary to secrete TSH in large amounts. The excess TSH stimulates the thyroid gland and causes it to enlarge. An enlarged thyroid gland is called a **goiter** (goy'ter).

1 What would happen if a person's immune system produced a large amount of an antibody, which was so much like TSH it could bind to thyroid cells and act like TSH? ?

A

Anterior

Thyroid cartilage

Thyroid gland

Trachea

B

Posterior

Thyroid gland

Parathyroid glands

Trachea

C

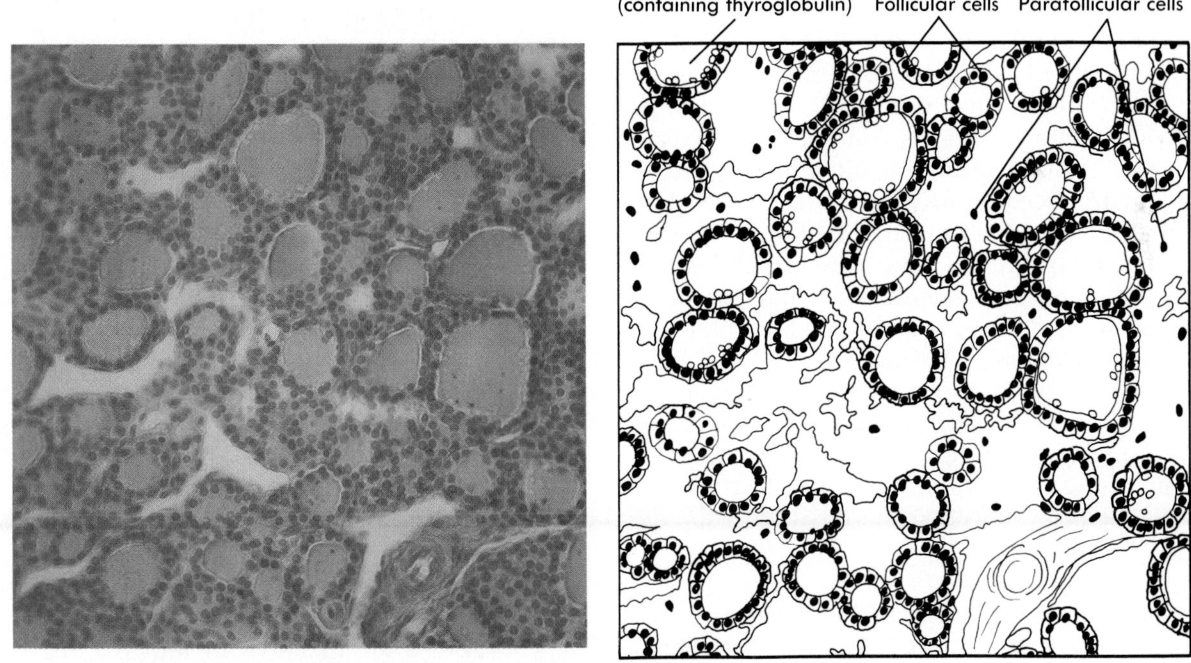

FIGURE 10-6 ■ The thyroid and parathyroid glands.
A An anterior view of the thyroid gland.
B A posterior view of the thyroid gland with the small parathyroid glands embedded in the thyroid tissue.
C Light micrograph and drawing of thyroid follicles and parafollicular cells.

The thyroid gland secretes a hormone called **calcitonin** (kal-sĭ-tōn′nin) in addition to the thyroid hormones. Calcitonin is synthesized by the parafollicular cells of the thyroid gland. If the blood concentration of calcium ions becomes too high, calcitonin is secreted, causing calcium ion levels to decrease to their normal range. Calcitonin reduces the rate at which calcium is absorbed by the intestine, decreases the rate of calcium released from bone, and increases the amount of calcium lost in the urine. Although calcitonin is important in reducing increased blood calcium levels, it does not elevate blood calcium levels when they are low.

The Parathyroid Glands

Four tiny **parathyroid** (păr-ă-thi′royd) **glands** are embedded in the posterior wall of the thyroid gland (see Figure 10-6, *B*). The parathyroid glands secrete a hormone called **parathyroid hormone (PTH),** which is essential for the regulation of blood calcium levels (see Table 10-1). PTH increases the absorption of calcium from the intestine by causing an increase in active vitamin D formation. It also increases the resorption (breakdown) of bone tissue to release calcium into the circulatory system and decreases the rate at which calcium is lost in the urine. When blood levels of calcium decline, PTH secretion increases. PTH acts on its target tissues to raise blood calcium levels to normal.

Active vitamin D increases absorption of calcium by the intestine and raises blood levels of calcium. It is produced from precursors in the skin that are modified by the liver and the kidney. Ultraviolet light acting on the skin is required for the first stage of vitamin D synthesis, and the final stage of synthesis in the kidney is stimulated by PTH. Vitamin D can also be supplied in the diet.

> **2 Explain why a lack of vitamin D results in bones that are softer than normal.** ?

PTH is more important than calcitonin in regulating blood levels of calcium. If too little calcium is consumed in the diet, blood levels of calcium begin to fall. The decreasing blood levels of calcium stimulate PTH secretion.

Increasing blood levels of calcium cause a decrease in PTH secretion. The decreased PTH secretion results in a reduction in blood levels of calcium. In addition, increasing blood levels of calcium stimulate calcitonin secretion. The calcitonin also causes blood levels of calcium to decline.

In summary, blood levels of calcium are regulated by PTH, which increases blood calcium levels when they are low, and calcitonin, which decreases blood calcium levels when they are high. Active vitamin D is also important in maintaining normal blood calcium levels (Figure 10-7).

The Adrenal Glands

The **adrenal** (ă-dre′nal) **glands,** or **suprarenals,** are two small glands, each of which is located on top of a kidney (see Figure 10-1 and Table 10-1). Each adrenal gland has an inner part, called the **adrenal medulla,** and an outer part, called the **adrenal cortex.** The adrenal medulla and the adrenal cortex function as separate endocrine glands.

The adrenal medulla

The principal hormone released from the adrenal medulla is **epinephrine** or **adrenalin,** but small amounts of **norepinephrine** are also released. The epinephrine and norepinephrine are released in response to stimulation by the sympathetic nervous system, which becomes most active when a person is physically excited (Figure 10-8). Epinephrine and norepinephrine are referred to as the **fight-or-flight** hormones because they prepare the body for vigorous physical activity. Some of the major effects of the hormones released from the adrenal medulla are:

1. Stimulate smooth muscle in the walls of arteries supplying the internal organs and the skin, but not those supplying skeletal muscle. Constriction of the blood vessels causes blood pressure to increase and blood flow to internal organs and the skin to decrease. Blood flow through skeletal muscles increases.
2. Increase the heart rate, which also causes the blood pressure to increase.
3. Increase the metabolic rate of tissues, especially skeletal muscle, cardiac muscle, and nervous tissue.
4. Dilate the bronchioles through relaxation of the smooth muscle cells in their walls. This allows air to move in and out of the lung with greater ease.
5. Cause the breakdown of glycogen to glucose in the liver, the release of the glucose into the blood, and the release of fatty acids from fat cells. The glucose and fatty acids are used to maintain the increased rate of metabolism in tissues.

Responses to hormones from the adrenal medulla reinforce the effect of the sympathetic ner-

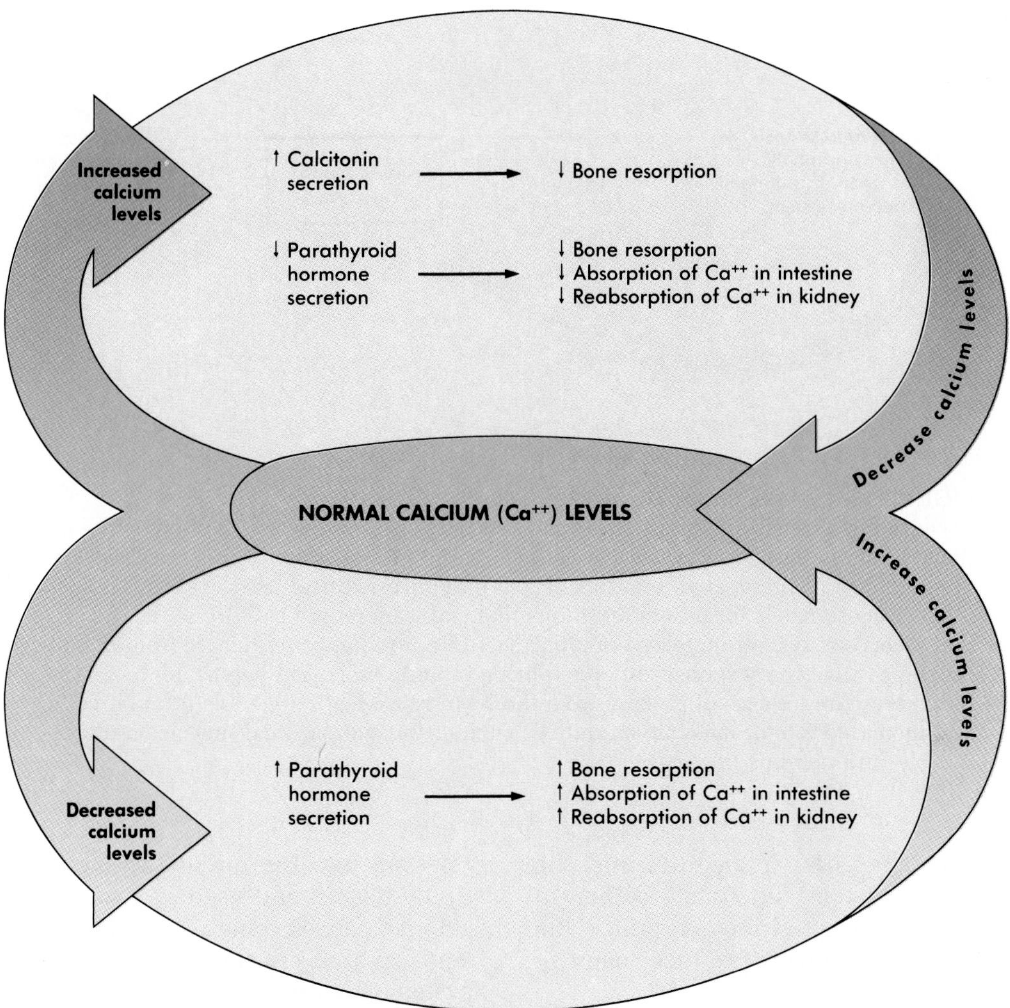

FIGURE 10-7 ■ Regulation of blood levels of calcium.
When blood calcium levels increase, the rate of parathyroid hormone secretion decreases, and the rate of calcitonin secretion increases. The response of target tissues to these changes is to cause the blood calcium levels to decline toward their normal range. When blood calcium levels decrease, the rate of parathyroid hormone secretion increases. The response of target tissues to the change is to increase blood calcium levels to their normal range.

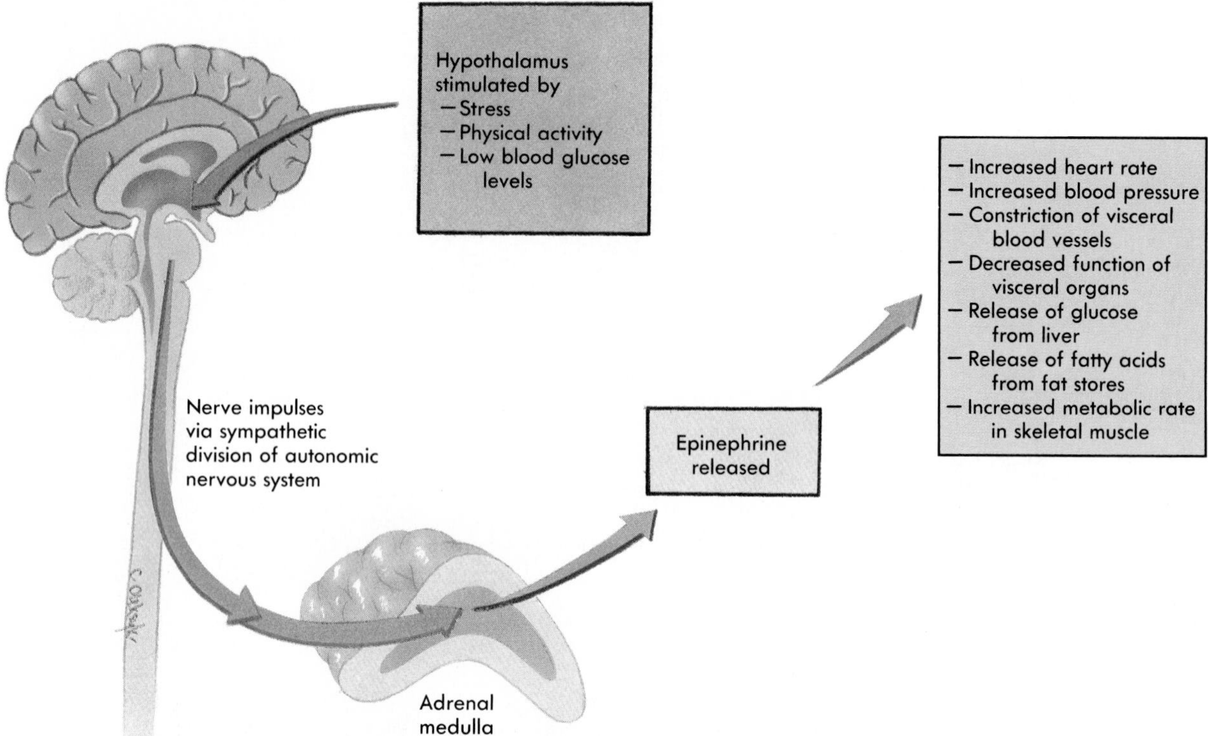

FIGURE 10-8 ■ Regulation of adrenal medullary secretions.
Because the adrenal medulla is a specialized portion of the sympathetic division of the autonomic nervous system, sympathetic stimulation results in increased secretion of epinephrine and smaller amounts of norepinephrine. Stress, physical activity, low blood glucose levels, or other conditions that cause increased activity of the sympathetic nervous system increase epinephrine and norepinephrine release from the adrenal medulla. The responses to epinephrine include increased heart rate, increased blood pressure, release of glucose from the liver, release of fatty acids from fat cells, and increased rate of metabolism in cells such as those of skeletal muscle, cardiac muscle, and nervous tissues.

vous system. Thus, the adrenal medulla and the sympathetic nervous system function together to prepare one for physical activity, to produce the "fight-or-flight" response, and to produce many of the responses to stress.

The adrenal cortex

■ Three classes of steroid hormones are secreted from the adrenal cortex. Each class has a unique set of structural and functional characteristics.

The **glucocorticoids** help regulate blood nutrient levels in the body. The major glucocorticoid hormone is **cortisol** (kor′tĭ-sol), which initiates the breakdown of protein and fat and initiates their conversion to forms that can be used as energy sources by the body. For example, when blood glucose levels decline, cortisol secretion increases. Cortisol acts on the liver, causing it to convert amino acids to glucose, and it acts on adipose tissue, causing fat stored in fat cells to be broken down to fatty acids. The glucose and fatty acids are released into the circulatory system and are taken up by tissues and used as a source of energy. Cortisol also causes protein to be broken down to amino acids, which are then released into the circulatory system (Figure 10-9).

In times of stress, cortisol is secreted in larger than normal amounts. It aids the body in responding to stressful conditions by providing energy sources for tissues. Cortisol also reduces the inflammatory response. **Cortisone,** a steroid closely related to cortisol, is often given as a medication to reduce inflammation such as occurs during certain allergic responses and injuries.

Adrenocorticotropic hormone (ACTH) from the anterior pituitary regulates the secretion of cortisol from the adrenal cortex. Without ACTH, the adrenal cortex degenerates and loses all of its secretory capability, but ACTH is more important in the minute-to-minute regulation of cortisol secretion than in the regulation of other hormones of the adrenal cortex.

The second class of hormones from the adrenal

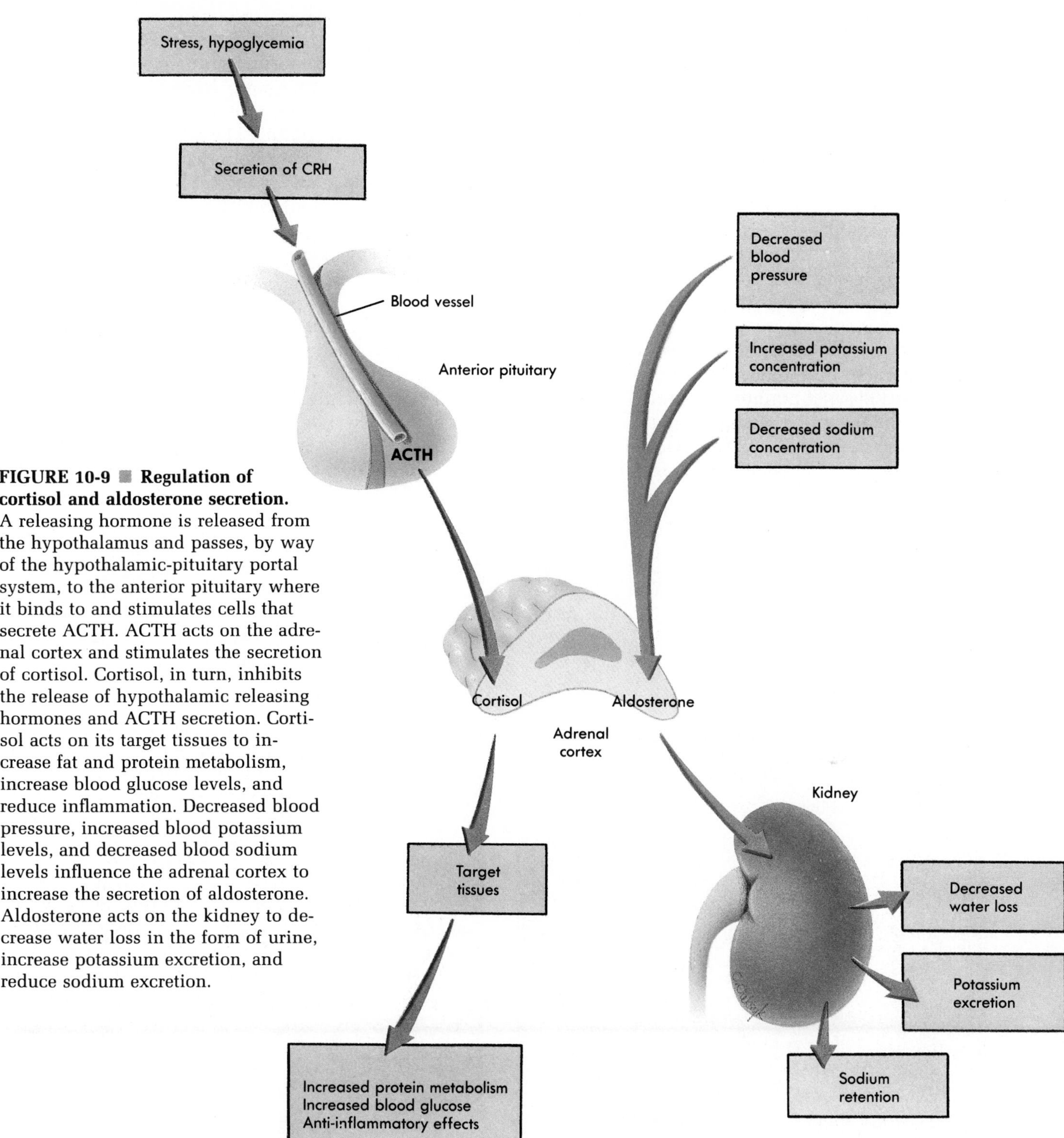

FIGURE 10-9 ■ Regulation of cortisol and aldosterone secretion. A releasing hormone is released from the hypothalamus and passes, by way of the hypothalamic-pituitary portal system, to the anterior pituitary where it binds to and stimulates cells that secrete ACTH. ACTH acts on the adrenal cortex and stimulates the secretion of cortisol. Cortisol, in turn, inhibits the release of hypothalamic releasing hormones and ACTH secretion. Cortisol acts on its target tissues to increase fat and protein metabolism, increase blood glucose levels, and reduce inflammation. Decreased blood pressure, increased blood potassium levels, and decreased blood sodium levels influence the adrenal cortex to increase the secretion of aldosterone. Aldosterone acts on the kidney to decrease water loss in the form of urine, increase potassium excretion, and reduce sodium excretion.

cortex, the **mineralocorticoids,** helps regulate blood volume and blood levels of potassium and sodium ions. **Aldosterone** (al-dos′-ter-ōn) is the major hormone of this class. It acts primarily on the kidney, but it also affects the intestine, sweat glands, and salivary glands. Aldosterone causes sodium ions and water to be retained in the body and increases the rate at which potassium is eliminated from the body.

Blood levels of potassium and sodium ions act directly on the adrenal cortex to influence aldosterone secretion, although the adrenal gland is much more sensitive to changes in blood potassium levels. The rate of aldosterone secretion increases when blood potassium levels increase, when blood sodium levels decrease, or when blood pressure decreases. Changes in blood pressure do not directly affect the rate of aldosterone secretion by the adrenal cortex. Low blood pressure causes the release of a protein called **renin** (re'nin) from the kidney. Renin causes a blood protein called **angiotensinogen** (an'je-o-tin-sin-o-jin) to be converted to **angiotensin I** (an'je-o-tin-sin). An active protein called **angiotensin II** is formed from angiotensin I. Angiotensin II acts on the adrenal cortex to increase aldosterone secretion (causing retention of sodium and water), and it constricts blood vessels. Together these effects help raise blood pressure.

The third class of hormones secreted by the adrenal cortex is the **androgens** [Gr. *aner*, male]. They are named for their ability to stimulate the development of male sexual characteristics. In adult males, most androgens are secreted by the testes. If the secretion of sex hormones from the adrenal cortex is abnormally high, exaggerated male characteristics develop in both males and females. This condition is most apparent in females and in males before puberty when the effects are not masked by the secretion of androgens by the testes.

The Pancreas, Insulin, and Diabetes

The endocrine portion of the pancreas (pan'kre-us) consists of **pancreatic islets** (islets of Langerhans) dispersed among the exocrine portion of the pancreas (Figure 10-10). The islets secrete two hormones that function to help regulate blood levels of nutrients, especially blood glucose. These hormones are insulin and glucagon (see Table 10-1).

It is very important to maintain blood glucose levels within a normal range of values. If blood glucose declines below its normal range, the nervous system malfunctions because glucose is the nutrient that is the nervous system's main source of energy. In addition, fats are broken down rapidly by other tissues, which provides an alternative energy source, and acidic substances are released into the circulatory system as byproducts. The acidic substances cause the pH of the body fluids to decrease below normal, a condition called acidosis. If blood glucose levels are too high, the kidneys produce large volumes of urine containing substantial amounts of glucose. Because of the rapid loss of water in the form of urine, dehydration may result.

Insulin (in'su-lin) is released from the pancreatic islets in response to elevated blood glucose levels, increased parasympathetic stimulation, and in-

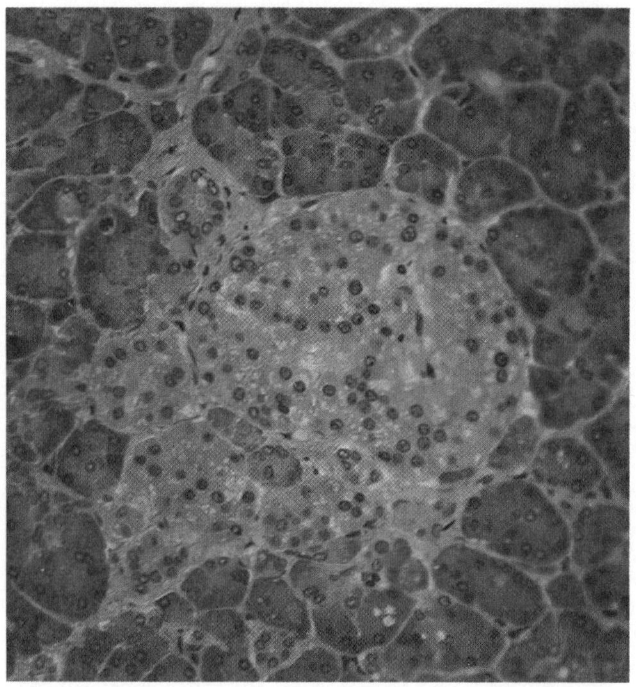

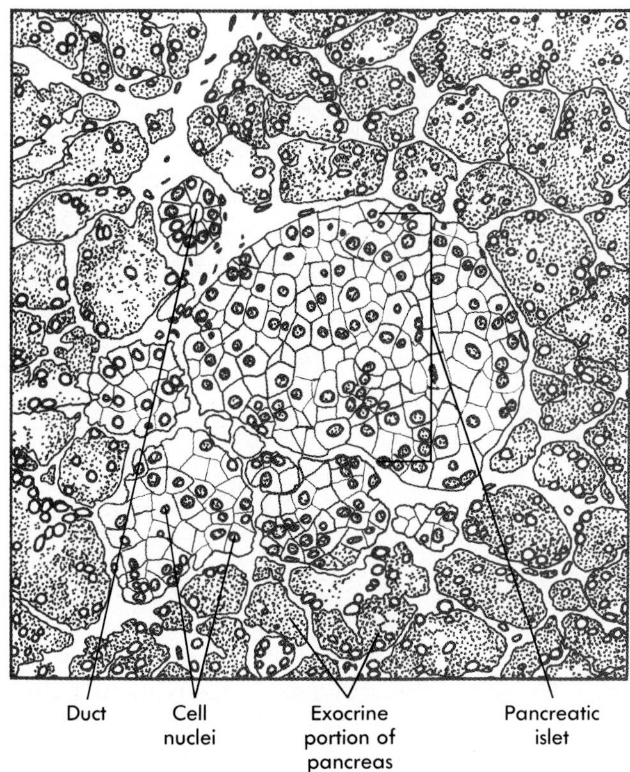

FIGURE 10-10 ■ Structure of the pancreas.
The endocrine portion of the pancreas is made up of scattered pancreatic islets, which secrete insulin and glucagon.

TABLE 10-2

Effects of Insulin and Glucagon on Target Tissues

TARGET TISSUE	INSULIN RESPONSES	GLUCAGON RESPONSES
Skeletal muscle, cardiac muscle, cartilage, bone fibroblasts, blood cells, and mammary glands	Increases glucose uptake and glycogen synthesis; increases uptake of amino acids	Little effect
Liver	Increases glycogen synthesis; increases use of glucose for energy	Causes rapid increase in the breakdown of glycogen to glucose and release of glucose into the blood; increases the formation of glucose from amino acids and, to some degree, from fats; increases metabolism of fatty acids
Adipose cells	Increases glucose uptake, glycogen synthesis, fat synthesis	High concentrations cause breakdown of fats; probably unimportant under most conditions
Nervous system	Little effect except to increase glucose uptake in the satiety center	No effect

creased blood levels of amino acids. The major target tissues for insulin are the liver, adipose tissue, muscles, and the satiety (sa'tī-ĕ-tē) center in the hypothalamus. Insulin increases the rate of glucose and amino acid uptake in these tissues. Glucose is converted to glycogen or fat, and the amino acids are used to synthesize protein. The effects of insulin on target tissues are summarized in Table 10-2.

Diabetes mellitus can result from any of the following: secretion of too little insulin from the pancreas, insufficient numbers of insulin receptors on target cells, or defective receptors that do not respond normally to insulin. In people who have diabetes mellitus, tissues cannot take up glucose effectively, causing blood glucose levels to become very high. The excess glucose is excreted in the urine. In addition, fats and proteins are broken down to provide an energy source for metabolism, which results in the wasting away of body tissues. Rapid fat breakdown produces acidic byproducts, which cause acidosis, a lowering of blood pH.

People who have untreated diabetes mellitus have an increased appetite, because glucose cannot enter cells of the satiety center of the brain without insulin. Thus, the satiety center responds as if there were very little blood glucose, even though blood glucose levels are very high. Symptoms of diabetes are high blood glucose levels, an exaggerated appetite, wasting away of body tissue, lack of energy, and the production of large amounts of urine containing glucose.

When too much insulin is present, such as occurs when a diabetic is injected with too much insulin or has not eaten after an insulin injection, blood glucose levels become very low. The brain, which depends primarily on glucose for an energy source, malfunctions. This condition is called *insulin shock*. Disorientation, convulsions, and loss of consciousness may result.

Glucagon (glu'kă-gon) is released from the pancreatic islets when blood glucose levels are low. Glucagon acts primarily on the liver to cause the conversion of glycogen stored in the liver to glucose. The glucose is then released into the blood to increase blood glucose levels. After a meal, when blood glucose levels are elevated, glucagon secretion is reduced.

3 Predict how the rate of insulin and glucagon secretion would be affected following a large meal rich in carbohydrates, and after 12 hours without eating. ?

Insulin and glucagon function together to regulate blood glucose levels (Figure 10-11). When blood glucose levels increase, insulin secretion increases and glucagon secretion decreases. When blood glucose levels decrease, the rate of insulin secretion declines and the rate of glucagon secretion

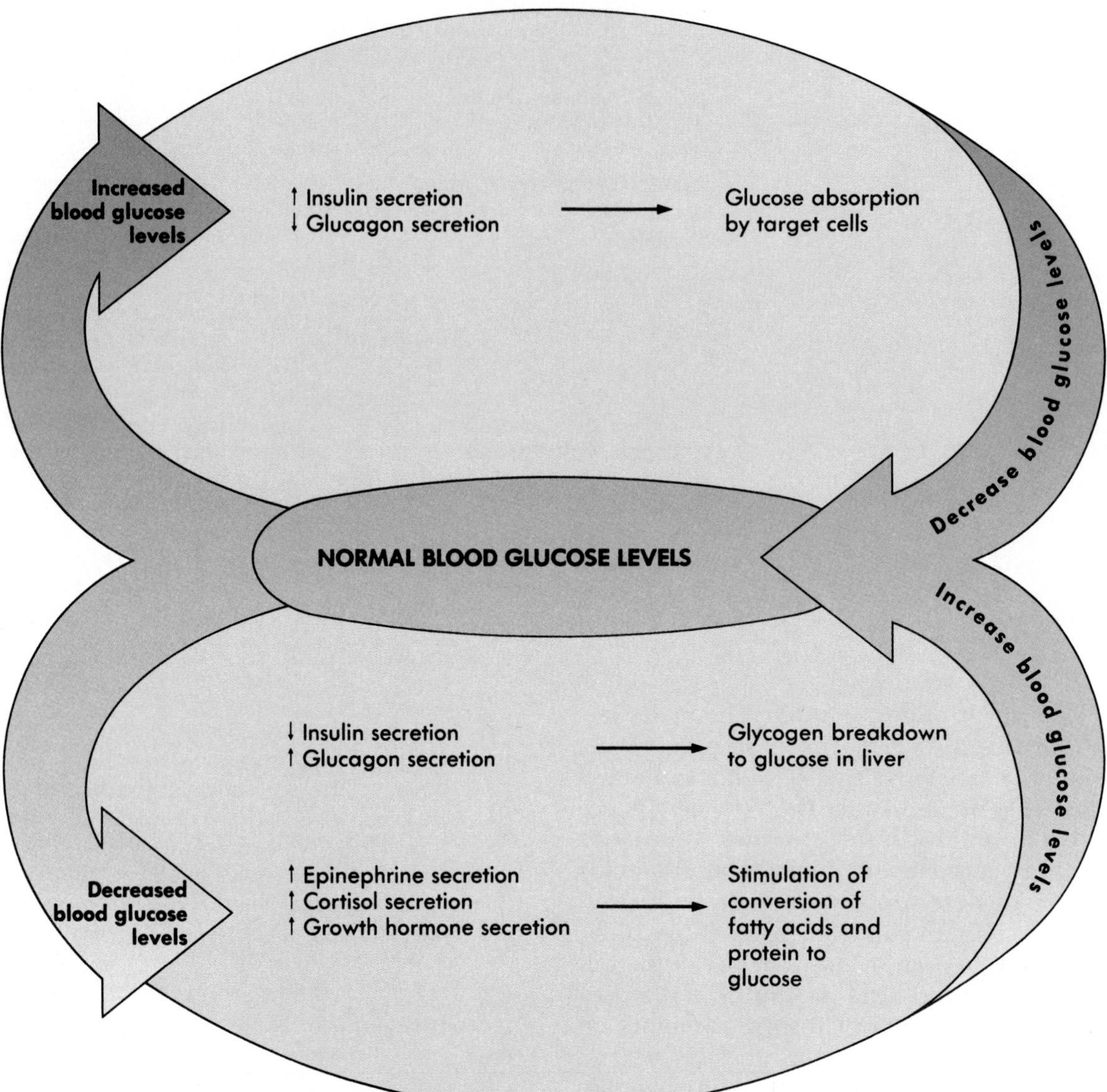

FIGURE 10-11 ■ Regulation of blood glucose levels.
Increased blood glucose levels stimulate insulin secretion from the pancreatic islets and decrease glucagon secretion. Insulin acts on target tissues causing them to take up glucose. Consequently, blood glucose levels decrease toward normal values. Decreased blood glucose levels cause a decrease in insulin secretion and an increase in glucagon secretion. Glucagon causes the liver to release glucose into the blood, raising blood glucose levels. Decreased blood glucose also results in increased epinephrine release from the adrenal medulla, cortisol release from the adrenal cortex, and growth hormone release from the anterior pituitary. These hormones increase fat and protein breakdown and cause fatty acids and amino acids to increase in the blood. They also cause amino acids and some components of fat to be converted to glucose. Growth hormone reduces protein breakdown and favors fat breakdown.

increases. Other hormones such as epinephrine, the glucocorticoids, and growth hormone also function to maintain blood levels of nutrients. When blood glucose levels decrease, these hormones are secreted at a greater rate. They cause the breakdown of protein and fat and the synthesis of glucose to help increase blood levels of nutrients. Growth hormone slows protein breakdown and favors fat breakdown.

The Testes and Ovaries

The testes of the male and the ovaries of the female secrete sex hormones in addition to producing sperm or eggs. The hormones produced by these organs play an important role in the development of sexual characteristics. Structural and functional differences between males and females and the ability to reproduce depend on the sex hormones. (See Figure 10-1 and Table 10-1).

The main hormone produced in the male is **testosterone.** It is responsible for the growth and development of the male reproductive structures, muscle enlargement, growth of body hair, voice changes, and increased male sexual drive.

In the female, there are two main classes of hormones that affect sexual characteristics, the **estrogens** and **progesterone.** Together these hormones contribute to the development and function of female reproductive structures and other female sexual characteristics. These characteristics include enlargement of the breasts and distribution of fat, which influences the shape of the hips, breasts, and legs. The female menstrual cycle is controlled by the cyclical release of estrogens and progesterone from the ovary, and the secretion of these hormones is controlled, in turn, by anterior pituitary hormones.

A releasing hormone from the hypothalamus increases luteinizing hormone (LH) and follicle-stimulating hormone (FSH) secretion from the anterior pituitary. The LH and FSH cause the secretion of hormones of the ovary and testis. The hormones from the ovaries and testes, in turn, have a negative-feedback effect on both the hypothalamus and anterior pituitary, which function to keep the blood levels of reproductive hormones within a normal range of values.

The Thymus Gland

The thymus gland lies in the upper part of the thoracic cavity above the heart (see Figure 10-1 and Table 10-1). It is important in the function of the immune system. Part of the thymus gland's function is to secrete a hormone called **thymosin** (thi′mo-sin), which helps in the development of certain white blood cells called T cells. T cells help protect the body against infection by foreign organisms. The thymus gland is most important early in life, and it becomes smaller in the adult. If an infant is born without a thymus gland, the immune system does not develop normally, and the body is less capable of fighting infections (see Chapter 14).

The Pineal Body

The **pineal** (pin′e-al) **body** is a small pine-cone shaped structure located superior and posterior to the thalamus of the brain (see Chapter 8). The pineal body produces a hormone called **melatonin,** which is thought to decrease the secretion of LH and FSH by decreasing the release of hypothalamic-releasing hormones (see Table 10-1). Thus, melatonin acts to inhibit the functions of the reproductive system by decreasing LH and FSH secretion. Animal studies have demonstrated that the amount of light controls the melatonin secretion rate. In many animals, short day length causes an increase in melatonin secretion, while longer day length causes a decrease in melatonin secretion. Some evidence suggests that melatonin may play an important role in the onset of puberty in humans. Tumors may develop in the pineal body. In some cases the tumors increase pineal secretions and in other cases they decrease pineal secretions.

> **4 Predict the effect on a young person's reproductive system of a tumor that destroys the ability of the pineal body to secrete melatonin.** ?

OTHER HORMONES

Cells in the lining of the stomach and small intestine secrete hormones that stimulate the production of digestive juices from the pancreas, stomach, and liver. These hormones aid the process of digestion by causing secretion of digestive juices when food is present in the digestive system but not at other times. Hormones secreted from the small intestine also help regulate the rate at which food passes from the stomach into the small intestine, so that food enters the small intestine at an optimal rate (see Chapter 16).

The **prostaglandins** (pros′tă-glan′dinz) are a group of lipid hormones that are widely distributed in tissues of the body. Unlike most hormones, they

Hormones and Stress

Stress in the form of disease, physical injury, or emotional anxiety initiates a specific response from the body that involves the nervous and endocrine systems. The stressful condition influences the hypothalamus, and through the hypothalamus the sympathetic division of the autonomic nervous system is activated. The sympathetic division prepares one for physical activity. It increases heart rate and blood pressure, shunts blood from the gut and other visceral structures to skeletal muscles, and increases the rate of metabolism in other tissues, especially in skeletal muscle. Part of the sympathetic nervous system's response is due to the release of epinephrine from the adrenal medulla.

In addition to sympathetic responses, stress causes the release of ACTH from the pituitary. ACTH acts on the adrenal cortex to cause the release of glucocorticoids. These hormones increase blood glucose levels and break down protein and fat, making nutrients readily available to tissues.

Although the ability to respond to stress is adaptive for short periods of time, responses triggered by stressful conditions are harmful if they occur for long periods of time. Prolonged stress can lead to hypertension (elevated blood pressure), heart disease, ulcers, inhibited immune system, and other conditions. Humans are frequently exposed to prolonged mental stress of high-pressure jobs or the inability to meet monetary obligations or social expectations. Although responses to stress prepare one for physical activity, frequently physical activity is not an appropriate response to the situation causing the stress. Long-term exposure to stress under conditions in which physical activity and emotions must be constrained may be harmful. Techniques that effectively reduce responses to stressful conditions are of substantial advantage to people who are exposed to chronic stress. Biofeedback, meditation, or other relaxation exercises are useful. Getting adequate rest, relaxation, and regular physical exercise are important in maintaining good health and reducing the response to stressful situations.

are usually not transported long distances in the circulatory system, but have their effects in the tissues where they are produced. Some prostaglandins cause relaxation of smooth muscle, such as dilation of blood vessels, and others cause contraction. Prostaglandins produced during the delivery of a baby cause uterine smooth muscle to contract. Because of their action on the uterus, prostaglandins have been used medically to initiate abortion. Prostaglandins also play a role in inflammation. They are released by damaged tissues and cause blood vessel dilation, swelling, and pain. The ability of aspirin and related substances to reduce pain and inflammation, help prevent painful cramping of uterine smooth muscle, and treat headache may be a result of their inhibitory effect on prostaglandin synthesis.

SUMMARY

The nervous and endocrine systems are the two major regulatory systems in the body.

The nervous system sends electrical signals through nerves; the endocrine system sends chemical signals through the circulatory system.

The endocrine system has a more general effect, acts more slowly, and has a longer-lasting effect than the nervous system.

HORMONES

Endocrine glands produce hormones that are released into the circulatory system and that travel some distance, where they act on target tissues to produce a response.

A target tissue for a given hormone has receptor molecules for that hormone.

Chemistry

Hormones are basically either peptides or lipids.

Hormone Action

Some hormones bind to receptors on the cell membrane and cause permeability changes or the production of a second messenger inside the cell. Others enter the cell and bind to receptors inside the cell.

The combining of hormones with their receptors results in a response.

Regulation of Hormone Secretion

The secretion of hormones is controlled by negative-feedback mechanisms.

Secretion of hormones from a specific gland is controlled by blood levels of some chemical, another hormone, or nerve impulses.

THE ENDOCRINE SYSTEM

The endocrine system consists of ductless glands.

Some glands of the endocrine system perform more than one function.

THE ENDOCRINE GLANDS AND THEIR HORMONES

The Pituitary and Hypothalamus

The pituitary is connected to the hypothalamus of the brain by the infundibulum. It is divided into the anterior and posterior pituitary.

Secretions from the anterior pituitary are controlled by releasing hormones that pass through the hypothalamic-pituitary portal system from the hypothalamus.

Neurohormones secreted from the posterior pituitary are controlled by nervous impulses that pass from the hypothalamus through the infundibulum.

The hormones released from the anterior pituitary are: growth hormone (GH), thyroid-stimulating hormone (TSH), adrenocorticotropic hormone (ACTH), luteinizing hormone (LH), follicle-stimulating hormone (FSH), prolactin, and melanocyte-stimulating hormone (MSH).

Hormones released from the posterior pituitary include antidiuretic hormone (ADH), and oxytocin.

The Thyroid Gland

The thyroid gland secretes thyroid hormones, which control the metabolic rate of tissues, and it secretes calcitonin, which helps regulate blood calcium levels.

The Parathyroid Glands

The parathyroid glands secrete parathyroid hormone, which helps regulate blood levels of calcium. Active vitamin D also helps regulate blood levels of calcium.

The Adrenal Glands

The adrenal medulla secretes primarily epinephrine and some norepinephrine. These hormones help prepare the body for physical activity.

The adrenal cortex secretes three classes of hormones.

Glucocorticoids reduce inflammation and break down fat and protein, making them available as energy sources to other tissues.

Mineralocorticoids help regulate sodium and potassium levels and water volume in the body. Renin helps regulate blood pressure by increasing angiotensin II and aldosterone production. These hormones cause blood vessels to constrict and enhance sodium and water retention by the kidney.

Adrenal androgens control sexual drive in females.

The Pancreas, Insulin, and Diabetes

The pancreas secretes insulin in response to elevated levels of blood glucose and amino acids. Insulin increases the rate at which many tissues, including adipose tissue, liver, and skeletal muscles, take up glucose and amino acids.

The pancreas secretes glucagon in response to reduced blood glucose and increases the rate at which the liver releases glucose into the blood.

The Testes and Ovaries

The testes secrete testosterone, and the ovaries secrete estrogens and progesterone. These hormones help control reproductive processes.

LH and FSH from the pituitary gland control ovarian and testicular functions.

The Thymus Gland

The thymus gland secretes thymosin, which enhances the immune system's ability to function.

The Pineal Gland

The pineal gland secretes melatonin, which may help regulate the onset of puberty by acting on the hypothalamus.

OTHER HORMONES

Hormones secreted by cells in the stomach and intestine help regulate stomach, pancreatic, and liver secretions.

The prostaglandins are hormones that have a local effect, produce numerous effects on the body, and play a role in inflammation.

CONTENT REVIEW

1. What are the major functional differences between the endocrine and the nervous systems?
2. Define endocrine gland and hormone.
3. What makes one tissue a target tissue and another not a target tissue for a hormone?
4. Into what chemical categories can hormones be classified?
5. Compare the means by which hormones that can and cannot cross the cell membrane produce a response.
6. Name three ways that hormone secretion is regulated.
7. Describe how secretions of the anterior and posterior pituitary hormones are controlled.
8. What are the functions of growth hormone? What happens when there is too little or too much growth hormone?
9. Describe the effect of gonadotropins on the ovary and testis.
10. What are the functions of the thyroid hormones, and how is their secretion controlled?
11. Explain how calcitonin, parathyroid hormone, and vitamin D are involved in maintaining blood calcium levels.
12. List the hormones secreted from the adrenal gland, give their functions, and compare the means by which the secretion rate of each is controlled.
13. What are the major functions of insulin and glucagon? How is their secretion regulated?
14. List the effects of testosterone, progesterone, and estrogen.

15. What hormones are produced by the thymus gland and pineal body? Name the effects of these hormones.
16. List the effects of prostaglandins. How is aspirin able to reduce the severity of the inflammatory response?
17. Describe the response of the nervous and endocrine systems to stressful situations.

CONCEPT REVIEW

1. Predict the long-term effects of a prolonged diet that contains too little calcium on (a) the secretion rates of the hormones that regulate blood levels of calcium and (b) on body structures.
2. What would be the consequences if the adrenal cortex degenerated and was no longer capable of secreting hormones?
3. Explain why a doctor would not want a pregnant woman to take aspirin late in pregnancy.
4. How would the levels of glucocorticoids, epinephrine, insulin, and glucagon be altered after 3 hours of vigorous exercise such as occurs in the average lay person who runs a marathon? What are the effects of these changes in hormone levels?
5. Explain how relaxation exercises such as meditation could reduce responses to stressful conditions.

ANSWERS TO  PREDICT QUESTIONS

1 *p. 256* An antibody that is similar to TSH is the cause of many cases of oversecretion of the thyroid gland (hyperthyroidism). Symptoms associated with hypersecretion of thyroid hormone become obvious. Because the antibody acts like TSH, the rate of synthesis and secretion of thyroid hormone is increased. In addition, antibody production is not sensitive to the inhibitory effect (negative feedback) of thyroid hormones on TSH secretion. Consequently, the antibody levels remain high, and so does the rate of thyroid hormone secretion.

2 *p. 258* Insufficient vitamin D results in insufficient calcium absorbed by the intestine. As a result, blood levels of calcium begin to fall. In response to the low blood calcium levels, parathyroid hormone is secreted from the parathyroid glands. Parathyroid hormone acts primarily on bone, causing bone to be broken down and calcium to be released into the blood to maintain blood levels of calcium within the normal range. Eventually so much calcium is removed from bones that they become soft, fragile, and easily broken. The condition is most obvious in children, where the bones are bent and deformed, and it is called rickets.

3 *p. 263* After a large meal, glucose enters the blood from the intestine. The increasing blood glucose stimulates insulin secretion and decreases glucagon secretion.

After 12 hours without eating, blood glucose levels would tend to decrease. Decreasing blood glucose levels would result in a decreased rate of insulin secretion and would stimulate glucagon secretion. The glucagon acts on the liver to cause it to release glucose into the blood.

4 *p. 265* The pineal body secretes melatonin, which inhibits the release of reproductive hormones by acting on the hypothalamus of the brain. If the pineal body secretes less melatonin, it no longer should have an inhibitory effect on the hypothalamus. As a result, reproductive hormones could be secreted in greater amounts, which would result in exaggerated development of the reproductive system in young people with this condition. The evidence for this mechanism is not as clear in humans as it is in other animals.

11

BLOOD

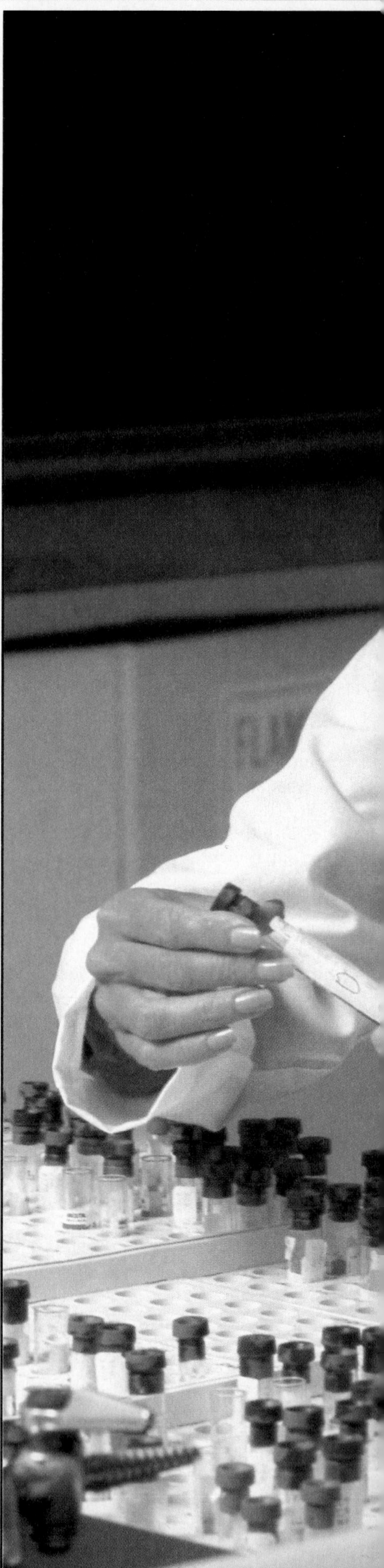

After reading this chapter you should be able to:

1 List the functions of blood.
2 Name the components of plasma and give their functions.
3 Describe the origin and production of the formed elements.
4 Describe the structure, function, and life history of erythrocytes.
5 Compare the structures and functions of the five different types of leukocytes.
6 Describe the origin and structure of platelets.
7 Explain the formation and function of platelet plugs and clots.
8 Describe the regulation of clot formation and how clots are removed.
9 Explain the basis of ABO and Rh incompatibilities.
10 Describe diagnostic blood tests and the normal values for the tests, and give examples of disorders that produce abnormal test values.

anticoagulant (an′tĭ-ko-ag′u-lant) — Chemical that prevents coagulation or blood clotting; antithrombin is an example.

clot retraction — Condensation of the clot into a denser, compact structure.

erythrocyte (ĕ-rith′ro-sīt) — [Gr. *erythro*, red + *kytos*, cell] Red blood cell; biconcave disk that contains hemoglobin, which transports oxygen and carbon dioxide; erythrocyte does not have a nucleus.

erythropoietin (ĕ-rith′ro-poy′ĕ-tin) — Protein hormone that stimulates erythrocyte formation in red bone marrow.

fibrin (fi′brin) — [L. *fibra*, fiber] A threadlike protein fiber derived from fibrinogen by the action of thrombin; forms a clot, that is, a network of fibers that traps blood cells, platelets, and fluid, which stops bleeding.

fibrinolysis (fi′brĭn-ol′ĭ-sis) — [L. *fibra*, fiber + Gr. *lysis*, dissolution] The breakdown of a clot by plasmin.

hematocrit (hem′ă-to-krit) — [Gr. *hemato*, blood + *krino*, to separate] The percentage of total blood volume composed of erythrocytes.

hemoglobin (he′mo-glo′bin) — A substance in erythrocytes consisting of four globin proteins, each with an iron-containing red pigment heme; transports oxygen and carbon dioxide.

leukocyte (lu′ko-sīt) — [Gr. *leukos*, white + *kytos*, cell] White blood cell; nucleated cell involved in immunity. The five types of leukocytes are neutrophils, eosinophils, basophils, lymphocytes, and monocytes.

plasma — Fluid portion of blood; blood minus the formed elements.

thrombocyte (throm′bo-sīt) — [Gr. *thrombos*, clot + *kytos*, cell] A cell fragment involved in platelet plug and clot formation; also called a platelet.

Cells are metabolically active and, as a result they require constant nutrition and continuous waste removal. Most cells, however, are located some distance from nutrient sources such as the digestive tract and sites of waste disposal such as the kidneys. The cardiovascular system provides the necessary connection between various tissues. The heart pumps blood through blood vessels that extend throughout the body. The blood is important in the maintenance of homeostasis in several ways: (1) it transports oxygen, nutrients, enzymes, and hormones to tissues, (2) it carries carbon dioxide and waste products away from tissues, (3) it plays a central role in temperature, fluid, electrolyte, and pH regulation, (4) it protects the body from disease-causing microorganisms, foreign substances, and tumors, and (5) it clots to prevent blood loss when blood vessels are damaged.

Blood is classified as a connective tissue, consisting of cells and cell fragments surrounded by a liquid matrix. The cells and cell fragments are the **formed elements,** and the fluid matrix is the **plasma.** The total blood volume in the average adult is about 4 to 5 liters in females and 5 to 6 liters in males. Blood makes up about 8% of the body's total weight.

PLASMA

Plasma is a pale yellow fluid that accounts for over half of the total blood volume. It consists of 92% water and 8% suspended or dissolved substances such as proteins, ions, nutrients, gases, waste products, and regulatory substances (Table 11-1). Plasma proteins include albumin, globulins, and fibrinogen. Albumin makes up 60% of the plasma proteins. The osmotic pressure of blood (see Chapter 3) is primarily the result of albumin and sodium ions. The water balance between blood and tissues is determined by the movement of water into and out of the blood by osmosis. Globulins account for 36% of the plasma proteins. These molecules include antibodies and other chemicals that function in immunity (see Chapter 14) and that transport molecules. Fibrinogen constitutes 4% of plasma proteins and is responsible for the formation of blood clots (see discussion later in the chapter). When the proteins that produce clots are removed from plasma, the remaining fluid is called **serum** (sēr′um).

Plasma volume remains relatively constant. Normally water intake through the digestive tract closely matches water loss through the kidneys, lungs, digestive tract, and skin. The suspended and dissolved substances come from the liver, kidneys, intestines, endocrine glands, and immune tissues such as the spleen. These substances are also regulated and maintained within narrow limits.

TABLE 11-1

Composition of Plasma

PLASMA COMPONENTS	PERCENT OF TOTAL PLASMA VOLUME	FUNCTIONS AND EXAMPLES
Water	91.5	Acts as a solvent and suspending medium for blood components
Proteins	7.0	Maintain osmotic pressure (albumin), destroy foreign substances (antibodies and complement), transport molecules (globulins), and form clots (fibrinogen)
Ions	0.9	Involved in osmotic pressure (sodium ions), membrane potentials (sodium and potassium ions), and acid-base balance (hydrogen, hydroxide, and bicarbonate ions)
Nutrients	0.3	Source of energy and "building blocks" of more complex molecules (glucose, amino acids, triglycerides)
Gases	0.1	Aerobic respiration (oxygen and carbon dioxide)
Waste products	0.1	Breakdown products of protein metabolism (urea and ammonia salts), red blood cells (bilirubin), and anaerobic respiration (lactic acid)
Regulatory substances	0.1	Catalyze chemical reactions (enzymes) and stimulate or inhibit many body functions (hormones)

FORMED ELEMENTS

About 95% of the volume of the formed elements consists of **red blood cells (RBCs)** or **erythrocytes** (ĕ-rith′ro-sītz). The remaining 5% consists of **white blood cells (WBCs)** or **leukocytes** (lu′ko-sītz) and cell fragments called **platelets** or **thrombocytes** (throm′bo-sītz). Erythrocytes are the most common of the formed elements in blood. They are 700 times more numerous than leukocytes and 17 times more numerous than platelets. The formed elements of the blood are outlined and illustrated in Table 11-2.

Production of Formed Elements

The process of blood cell production is called **hematopoiesis** (hem′ă-to-poy-e′sis). In the fetus, hematopoiesis occurs in many tissues such as the liver, thymus gland, spleen, lymph nodes, and red bone marrow. After birth, hematopoiesis is confined primarily to red bone marrow, but some leukocytes

TABLE 11-2

Formed Elements of the Blood

CELL TYPE	DESCRIPTION	FUNCTION
Erythrocyte	Biconcave disk; no nucleus; 7-8 micrometers in diameter	Transports oxygen and carbon dioxide
Leukocytes		
Neutrophil	Spherical cell; nucleus with two to four lobes connected by thin filaments; cytoplasmic granules stain a light pink or reddish purple; 12-15 micrometers in diameter	Phagocytizes microorganisms
Basophil	Spherical cell; nucleus with two indistinct lobes; cytoplasmic granules stain blue-purple; 10-12 micrometers in diameter	Releases histamine, which promotes inflammation, and heparin, which prevents clot formation
Eosinophil	Spherical cell; nucleus often bilobed; cytoplasmic granules stain orange-red or bright red; 10-12 micrometers in diameter	Releases chemicals that reduce inflammation; attacks certain worm parasites
Lymphocyte	Spherical cell with round nucleus; cytoplasm forms a thin ring around the nucleus; 6-8 micrometers in diameter	Produces antibodies and other chemicals responsible for destroying microorganisms; responsible for allergic reactions, graft rejection, tumor control, and regulation of the immune system
Monocyte	Spherical or irregular cell; nucleus round, or kidney or horse-shoe shaped; contains more cytoplasm than does lymphocyte; 10-15 micrometers in diameter	Phagocytic cell in the blood; leaves the circulatory system and becomes a macrophage, which phagocytizes bacteria, dead cells, cell fragments, and debris within tissues
Platelet	Cell fragments surrounded by a cell membrane and containing granules; 2-5 micrometers in diameter	Forms platelet plugs; release chemicals necessary for blood clotting

FIGURE 11-1 Hematopoiesis.
The stem cell or hemocytoblast gives rise to the cell lines that produce the formed elements.

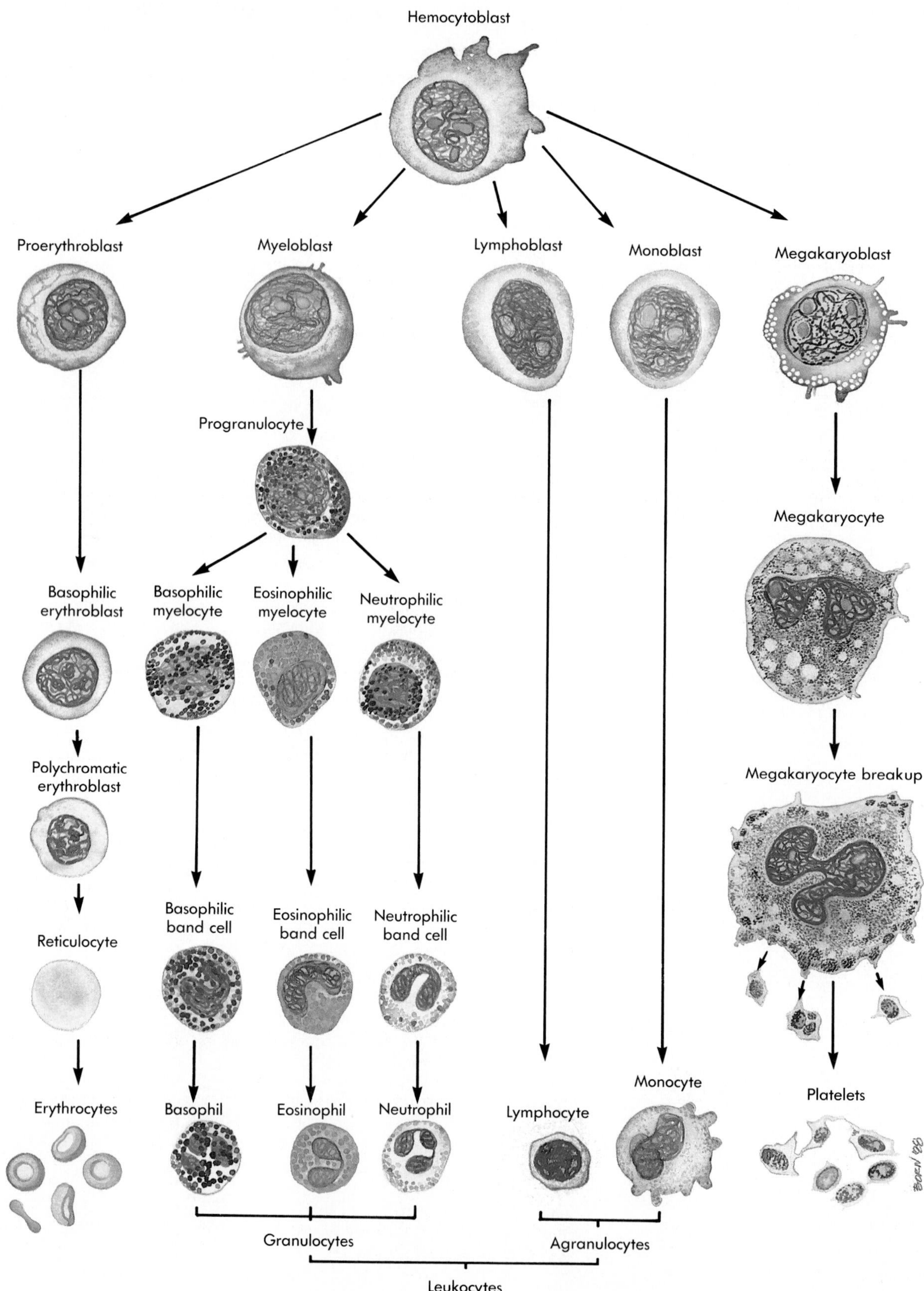

are produced in lymphatic tissue (see Chapter 15).

All the formed elements of blood are derived from a single population of cells called **stem cells** or **hemocytoblasts** (he′mo-cy′to-blastz). These stem cells differentiate to give rise to different cell lines, each of which ends with the formation of a particular type of formed element (Figure 11-1). The development of each cell line is regulated by a specific growth factor. That is, the type of formed element derived from the stem cells, and how many formed elements are produced, is determined by the growth factor.

■ An undesirable side effect of some cancer therapies is the destruction of red bone marrow. Following treatment for cancer, growth factors are used to stimulate the rapid regeneration of the red bone marrow. Although not a cure for cancer, the use of growth factors can speed recovery from the cancer therapy. ■

Erythrocytes

Normal erythrocytes are biconcave disks with edges that are thicker than the center area (Figure 11-2). During their development, erythrocytes lose their nuclei and most of their organelles. Consequently, they are unable to divide. Erythrocytes live for about 120 days in males and 110 days in females. The main component of an erythrocyte is the pigmented protein **hemoglobin** (he′mo-glo′bin), which accounts for about one third of the cell's volume and is responsible for its red color.

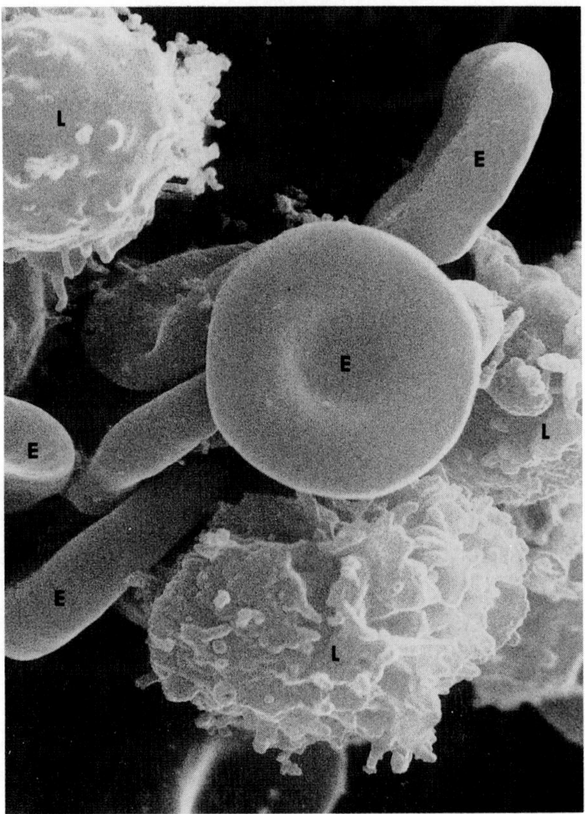

FIGURE 11-2 ■ Erythrocytes and leukocytes. The erythrocytes *(E)* are biconcave disks with a thin center. The leukocytes *(L)* are spherical and have cytoplasmic extensions.

Function

The primary functions of erythrocytes are to transport oxygen from the lungs to the various tissues of the body and to assist in the transport of carbon dioxide from the tissues to the lungs. Oxygen transport is accomplished by hemoglobin, which consists of four protein chains and four heme groups. Each protein, called a **globin,** is bound to one **heme** (hēm), a red-pigmented molecule. Each heme contains one iron atom that is necessary for the normal function of hemoglobin. When hemoglobin is exposed to oxygen, one oxygen molecule binds to the iron atom of each heme. Hemoglobin that is bound to oxygen is bright red in color, whereas hemoglobin without bound oxygen is a darker red color. Hemoglobin is responsible for 97% of the oxygen transport in blood. The remaining 3% is transported dissolved in plasma.

Because iron is necessary for oxygen transport, it is not surprising that two thirds of the body's iron is found in hemoglobin. Small amounts of iron are required in the diet to replace the small amounts lost in the urine and feces. Women need more dietary iron than men because women lose more iron as a result of blood loss during menstruation.

■ Carbon monoxide is a gas produced by the incomplete combustion of gasoline. It binds to the iron in hemoglobin approximately 210 times as fast as oxygen and does not tend to dissociate. As a result, the hemoglobin bound to carbon monoxide no longer transports oxygen. Nausea, headache, unconsciousness, and death are possible consequences of prolonged exposure to carbon monoxide. ■

Carbon dioxide transport involves hemoglobin, carbonic anhydrase, and plasma. Carbon dioxide can bind to the globin part of hemoglobin. About 20% of the carbon dioxide in blood is transported bound to hemoglobin or other blood proteins. The enzyme **carbonic anhydrase,** found inside erythrocytes, is involved in the transport of 72% of carbon dioxide. This enzyme catalyzes a reaction that converts carbon dioxide into a bicarbonate ion (see Chapter 15). The remaining 8% of carbon dioxide is transported dissolved in plasma.

Life history of erythrocytes

■ Under normal conditions, about 2.5 million erythrocytes are destroyed every second. Fortunately, new erythrocytes are produced as rapidly as old erythrocytes are destroyed. Stem cells called hemocytoblasts form **proerythroblasts,** which give rise to the erythrocyte cell line (see Figure 11-1). Erythrocytes are the final cells produced from a series of cell divisions. After each cell division the newly formed cells change or differentiate to become different from the original parent cell. The process of cell division requires the vitamins B_{12} and folic acid, which are necessary for the synthesis of DNA. Cell differentiation requires iron because hemoglobin is manufactured within the cells. Lack of these vitamins or iron can interfere with normal erythrocyte production.

Erythrocyte production is stimulated by low blood oxygen levels. Typical causes of low blood oxygen are decreased numbers of erythrocytes, decreased or defective hemoglobin, diseases of the lungs, high altitude, inability of the cardiovascular system to deliver blood to tissues, and increased tissue demands for oxygen (for example, during endurance exercises). Low blood oxygen levels increase erythrocyte production by increasing the formation of the glycoprotein **erythropoietin** (ĕ-rith′ro-poy′ĕ-tin). Erythropoietin is produced in an inactive form by the liver and released into the blood. When oxygen levels in the blood reaching the kidneys decrease, the kidneys release **renal erythropoietic factor** into the blood. This enzyme converts the inactive erythropoietin into active erythropoietin, which stimulates red bone marrow to produce more erythrocytes (Figure 11-3). Thus when oxygen levels in the blood decrease, the production of erythrocytes is increased. This increases the ability of the blood to transport oxygen. The mechanism returns blood oxygen levels to normal

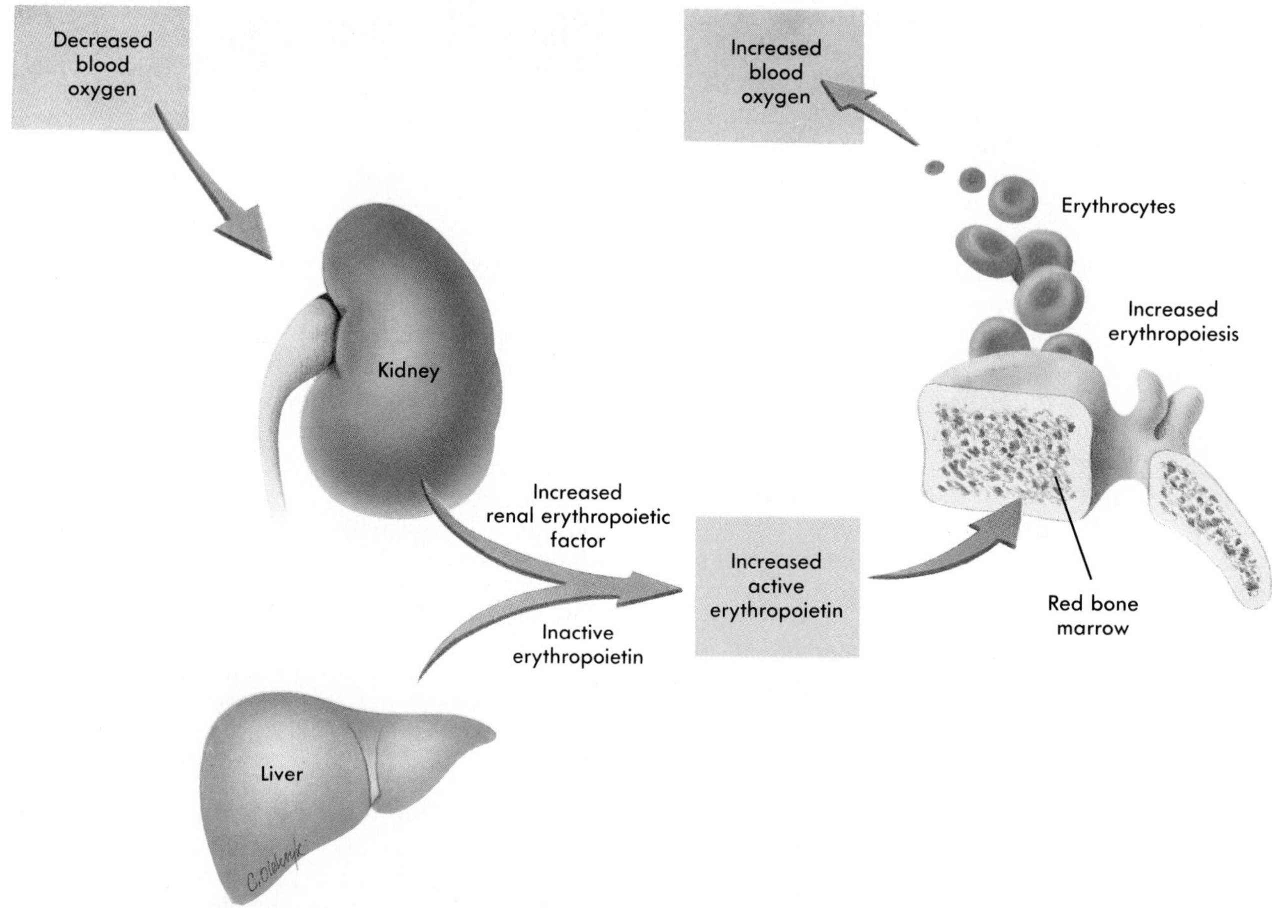

FIGURE 11-3 ■ Erythrocyte production.
In response to decreased blood oxygen, the kidneys release renal erythropoietic factor. The renal erythropoietic factor converts inactive erythropoietin from the liver into active erythropoietin, which stimulates erythrocyte production in the red bone marrow.

and maintains homeostasis by increasing the delivery of oxygen to tissues, including the kidneys. Conversely, if blood oxygen levels increase, less renal erythropoietic factor is produced, less erythropoietin is activated, and erythrocyte production decreases.

> **1 Cigarette smoke produces carbon monoxide. If a nonsmoker smoked a pack of cigarettes a day for a few weeks, what would happen to the number of erythrocytes in the person's blood? Explain.** ?

Old, damaged, or defective erythrocytes are removed from the blood by macrophages (large "eating" cells) located in the spleen and liver (Figure 11-4). Within the macrophage the globin part of the molecule is broken down into amino acids that are reused to produce other proteins. The iron is released and can be used to produce new hemoglobin in the red bone marrow. Only small amounts of iron are required in the daily diet because the iron is recycled. The heme groups are converted to bilirubin, which is taken up by the liver. The bilirubin normally is excreted into the small intestine as part of the bile (see Chapter 17). If the excretion of bile is hindered, bilirubin builds up in the circulation and produces jaundice, a yellowish color of the skin. After it is in the intestine, bilirubin is converted by bacteria into other pigments. Some of these pigments give feces their brown color, whereas others are absorbed from the intestine into the blood and are excreted by the kidneys in the urine, giving urine its characteristic yellow color.

Leukocytes

Leukocytes or white blood cells are spherical cells that are whitish in color because they lack hemoglobin (see Figure 11-2). They are larger than erythrocytes, and they have a nucleus (see Table

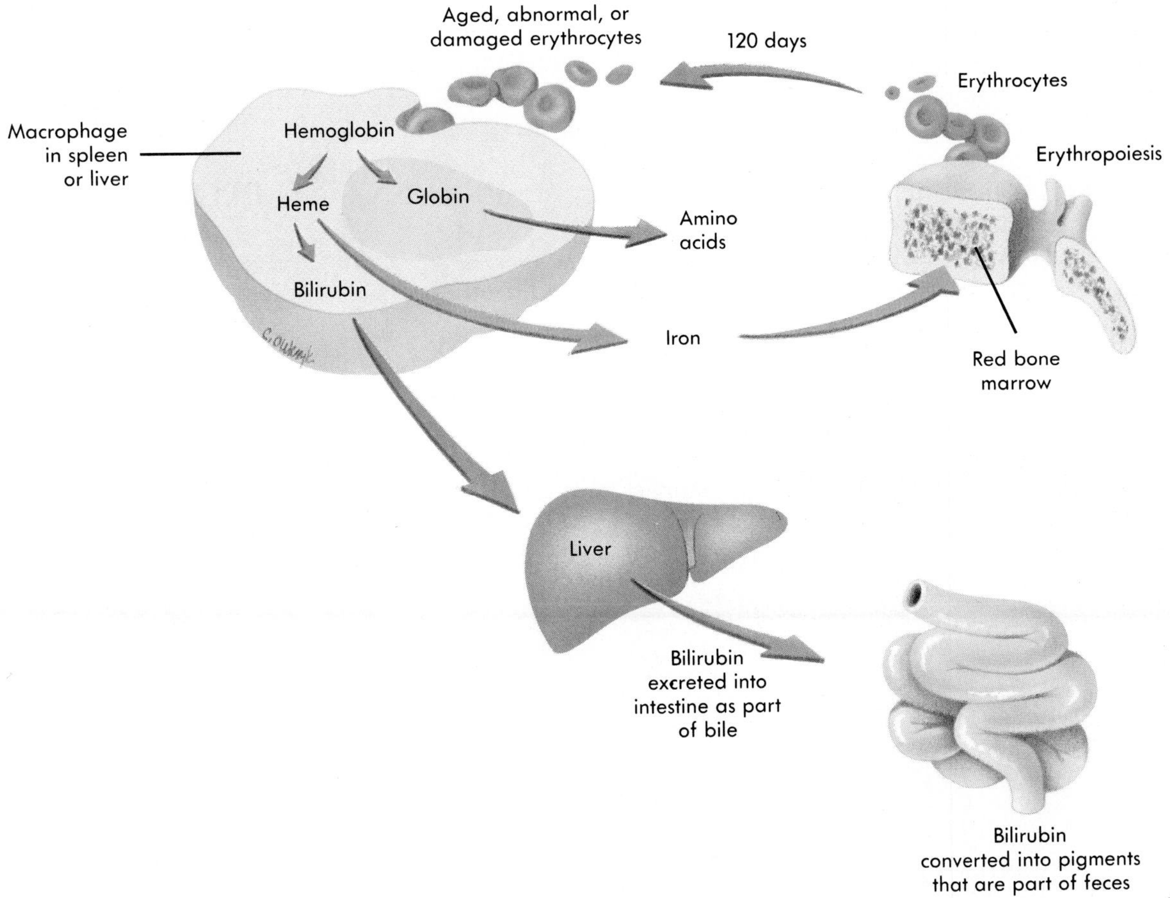

FIGURE 11-4 ■ Breakdown of hemoglobin.
Macrophages in the spleen and liver break down hemoglobin. Globin is converted to amino acids, and heme is converted to iron and bilirubin. Iron is used in the production of new hemoglobin, and the bilirubin is excreted as part of the bile.

11-2). Although leukocytes are components of the blood, the blood serves primarily as a means to transport these cells to other tissues of the body. Leukocytes can leave the blood and move by **ameboid** (ah-me'boid) **movement** through the tissues. In this process, the cell projects a cytoplasmic extension that attaches to an object. Then the rest of the cell flows into the extension. Leukocytes function to protect the body against invading microorganisms, and they remove dead cells and debris from the tissues by phagocytosis.

Leukocytes are named according to their appearance in stained preparations. Those containing large cytoplasmic granules are **granulocytes** (gran'u-lo-sītz), and those with very small granules that cannot be easily seen with the light microscope are **agranulocytes.**

There are three kinds of granulocytes. **Neutrophils** (nu'tro-filz) are the most common type of leukocyte (Figure 11-5). They usually remain in the blood for a short time (10 to 12 hours), move into other tissues, and phagocytize microorganisms and other foreign substances. Dead neutrophils, cell debris, and fluid can accumulate at sites of infections as pus. **Basophils** (ba'so-filz) and **eosinophils** (e-o-sin'o-filz) are involved in regulating the inflammatory response (see Chapters 4 and 14). Basophils release histamine and other chemicals that promote inflammation. Basophils also contain heparin, which prevents the formation of clots. Eosinophils release chemicals that reduce inflammation. In addition, chemicals from eosinophils are involved with the destruction of certain worm parasites.

There are two kinds of agranulocytes. **Lymphocytes** (lim'fo-sītz) are the smallest of the leukocytes (see Figure 11-5). There are several types of lymphocytes, and they play an important role in the body's immune response. Their diverse activities involve the production of antibodies and other chemicals that destroy pathogens, contribute to allergic reactions, reject grafts, control tumors, and regulate the immune system. Chapter 14 considers these cells in more detail. **Monocytes** (mon'o-sītz) are the largest of the leukocytes. They may leave the blood and transform to become **macrophages** (mak'ro-fāj'ez), which phagocytize bacteria, dead neutrophils, cell fragments, and any other debris within the tissues.

Platelets

Platelets or **thrombocytes** (throm'bo-sītz) are minute fragments of cells consisting of a small amount of cytoplasm surrounded by a cell membrane (see Figure 11-5). They are produced in the red bone marrow from **megakaryocytes** (meg'ă-kăr'e-o-sītz), which are large cells (see Figure 11-1). Small fragments of these cells break off and enter the blood as platelets. Platelets play an important role in hemostasis, which prevents blood loss.

PREVENTING BLOOD LOSS

When a blood vessel is damaged, blood can leak into other tissues and interfere with normal tissue

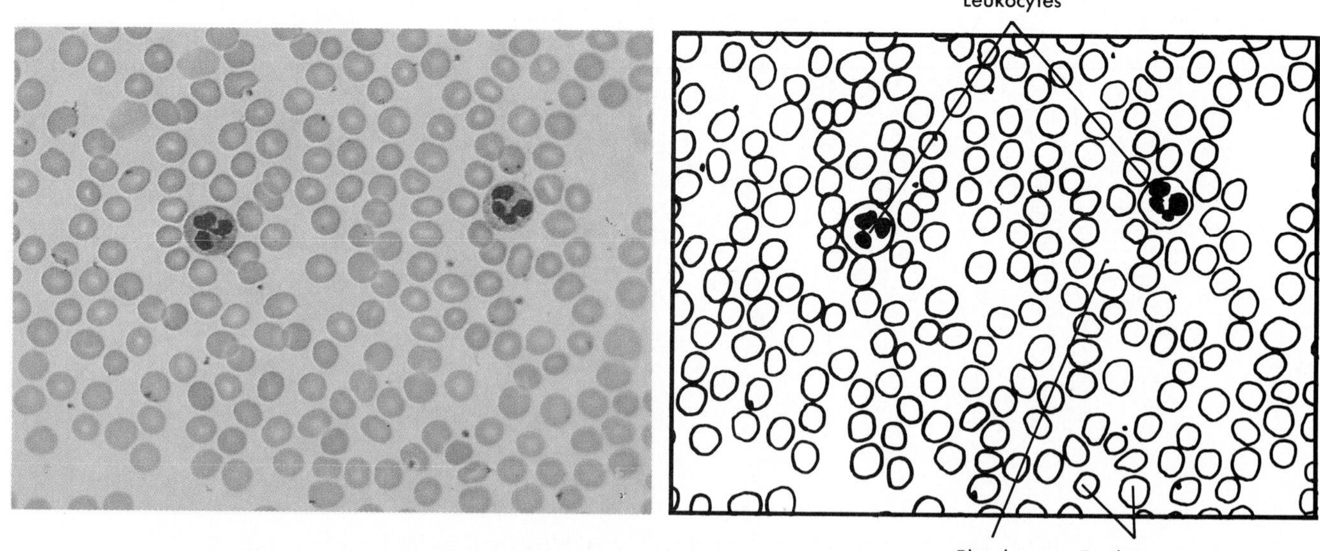

FIGURE 11-5 Standard blood smear.
A thin film of blood is spread on a microscope slide, and the blood cells are stained. The erythrocytes are pink, the leukocytes have a purple-stained nucleus, and the platelets are purple-stained cell fragments. (× 1000)

function, or blood can be lost from the body. Small amounts of blood loss from the body can be tolerated, but new blood must be produced to replace the lost blood. If large amounts of blood are lost, death can occur. Fortunately, when a blood vessel is damaged, several events help to prevent loss of blood.

Blood Vessel Constriction

When a blood vessel is damaged, smooth muscle in the vessel wall contracts. In small vessels, the resulting constriction of the vessel can close off the vessel completely and stop the flow of blood through the vessel. Contraction of the smooth muscle is mediated through neural responses to pain and through chemicals released by platelets.

Platelet Plugs

Small tears occur in the smaller blood vessels each day, and **platelet plug** formation quickly closes the tears. When a vessel is damaged, the epithelial lining is torn and the underlying connective tissue is exposed. Platelets are activated by the exposed connective tissue, especially the collagen fibers, and the platelets stick to the connective tissue and to each other. The accumulating mass of platelets forms a platelet plug that seals the vessel shut.

When the platelets are activated, they release a number of chemicals that act to decrease blood loss. For example, serotonin causes blood vessels to constrict and reduce blood flow. Other chemicals stimulate the synthesis of prostaglandins that function to activate additional platelets, causing them to attach to the platelet plug.

Blood Clotting

Platelet plugs alone are not sufficient to close large tears or cuts in blood vessels. When a blood vessel is severely damaged, **blood clotting** or **coagulation** (ko-ag'u-la-shun) results in the formation of a clot. A **clot** is a network of threadlike protein fibers, called **fibrin** (fi'brin), that traps blood cells, platelets, and fluid.

The formation of a blood clot depends on a number of proteins found within plasma called **clotting factors.** Normally the clotting factors are inactive and do not cause clotting. Following injury, however, the clotting factors are activated to produce a clot. This is a complex process involving many chemical reactions, but it can be summarized in three main stages (Figure 11-6).

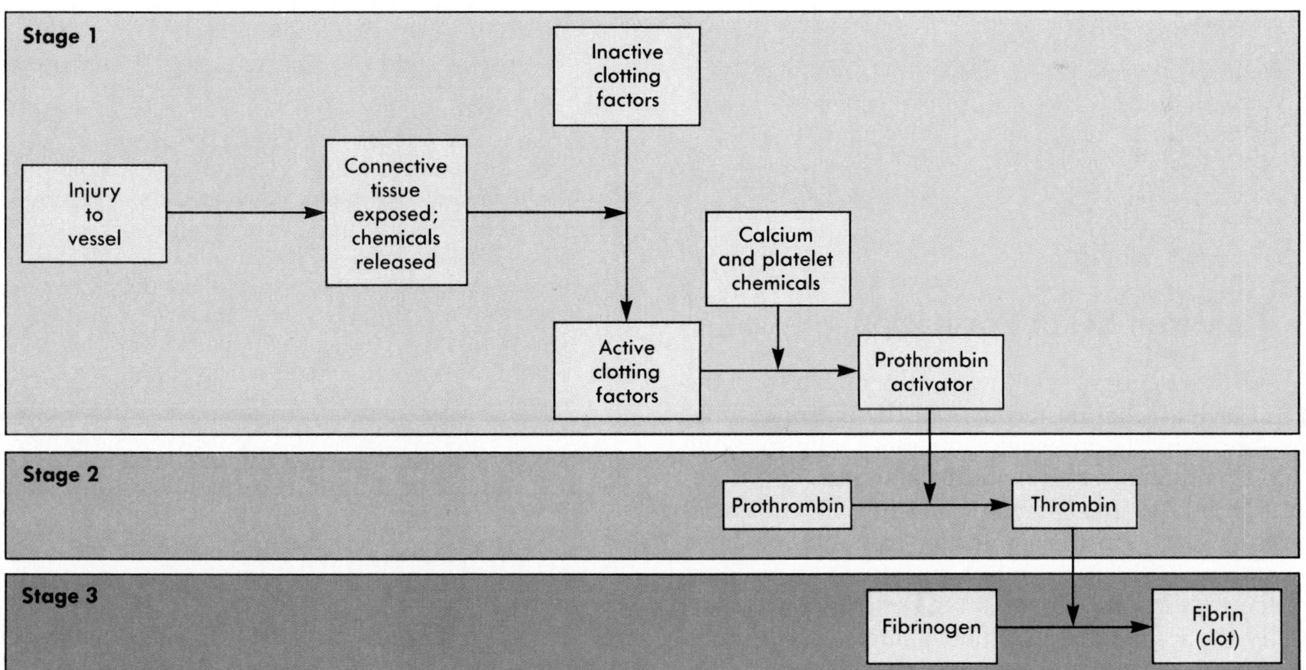

FIGURE 11-6 Clot formation.
In stage 1 inactive clotting factors are activated by exposure to connective tissue and by chemicals released from tissues. Through a series of reactions, the activated clotting factors form prothrombin activator. In stage 2 prothrombin is converted to thrombin by prothrombin activator. In stage 3 fibrinogen is converted to fibrin (the clot) by thrombin.

1. The chemical reactions can be started in two ways. Just as with platelets, the contact of inactive clotting factors with exposed connective tissue can result in their activation. Chemicals released from injured tissues can also cause activation of clotting factors. After the initial clotting factors are activated, they in turn activate other clotting factors. A series of reactions results in which each clotting factor activates the next clotting factor in the series until the clotting factor **prothrombin** (pro-throm'bin) **activator** is formed.
2. Prothrombin activator acts on an inactive clotting factor called **prothrombin.** Prothrombin is converted to its active form called **thrombin** (throm'bin).
3. Thrombin converts the inactive clotting factor **fibrinogen** (fi-brin'o-jen) into its active form, **fibrin.** The fibrin threads form a network, which traps blood cells and platelets and forms the clot.

Most of the clotting factors are manufactured in the liver, and many of them require vitamin K for their synthesis. In addition, many of the chemical reactions of clotting require calcium ions and chemicals released from platelets. Low levels of vitamin K or calcium, low numbers of platelets, or liver dysfunction can seriously impair the blood clotting process.

> Humans rely on two sources of vitamin K. About half comes from the diet and half from bacteria within the large intestine. Antibiotics sometimes kill these bacteria, reducing vitamin K levels, and resulting in bleeding problems. Vitamin K supplements may be necessary for patients on prolonged antibiotic therapy.

Control of Clot Formation

Without control, clotting would spread from the point of its initiation throughout the entire circulatory system. To prevent unwanted clotting, the blood contains several **anticoagulants** (an'tĭ-ko-ag'u-lantz), which prevent clotting factors from forming clots. **Antithrombin,** for example, destroys thrombin. Without thrombin, fibrinogen is not converted to fibrin and there is no clot formation. Normally there are enough anticoagulants in the blood to prevent clot formation. At an injury site, however, the stimulation for activating clotting factors is very strong. So many clotting factors are activated that the anticoagulants no longer can prevent a clot from forming. Away from the injury site, however, there are enough anticoagulants to prevent clot formation.

Clot Retraction and Dissolution

After a clot has formed, it begins to condense into a denser, compact structure by a process known as **clot retraction.** Serum, which is plasma without its clotting factors, is squeezed out of the clot during clot retraction. Consolidation of the clot pulls the edges of the damaged vessel together, helping to stop the flow of blood, reducing the probability of infection, and enhancing healing. The damaged vessel is repaired by the movement of fibroblasts into the damaged area and the formation of new connective tissue. In addition, epithelial cells around the wound divide and fill in the torn area (see Chapter 4).

The clot is dissolved by a process called **fibrinolysis** (fi'brin-ol'ĭ-sis). An inactive plasma protein called **plasminogen** (plaz'min'o-jen) is converted to its active form, which is called **plasmin** (plaz'min). Thrombin and other clotting factors activated during clot formation, or tissue plasminogen activator (t-PA) released from surrounding tissues, stimulate the conversion of plasminogen to plasmin. Over a period of a few days the plasmin slowly breaks down the fibrin (Figure 11-7).

A clot that forms within a blood vessel is called a **thrombus** (throm'bus). An **embolus** (em'bo-lus) is a detached clot or substance that floats through the circulatory system and becomes lodged in a blood vessel. If a thrombus or embolus blocks blood flow to essential organs, death can result.

> A heart attack can result from blockage of blood vessels that supply blood to the heart. One treatment for a heart attack caused by a thrombus is to inject into the blood chemicals that activate plasmin. Streptokinase, a bacterial enzyme, and tissue plasminogen activator produced through genetic engineering have been successfully used to dissolve thrombi. Another therapy is aspirin, which inhibits prostaglandin synthesis by platelets. This prevents platelet activation and reduces the likelihood of additional thrombus formation, because chemicals necessary for clotting are not released from the platelets.

BLOOD GROUPING

If large quantities of blood are lost during surgery or in an accident, the blood volume must be increased or the patient can go into shock and die. A **transfusion** is the transfer of blood or other solutions into the blood of the patient. In many cases the return of blood volume to normal levels is all that is necessary. This can be accomplished by the transfusion of plasma or prepared solutions that

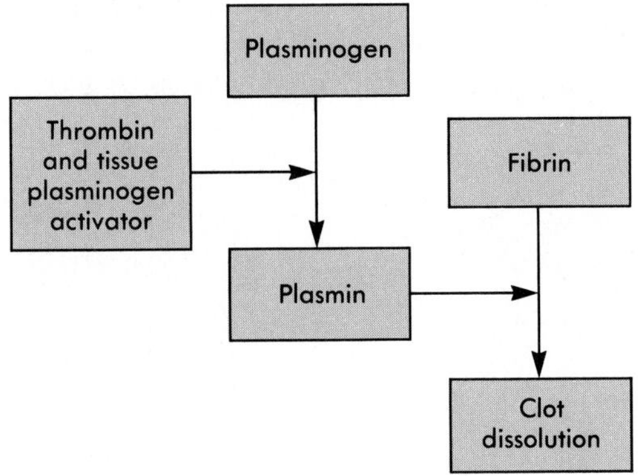

FIGURE 11-7 ■ Clot dissolution.
Thrombin and tissue plasminogen activator (t-PA) convert plasminogen into plasmin, which breaks down the fibrin in a blood clot.

have the proper amounts of solutes. When large quantities of blood are lost, however, erythrocytes must also be replaced so that the oxygen-carrying capacity of the blood is restored.

Early attempts to transfuse blood from one person to another were unsuccessful. They often resulted in transfusion reactions, which included clotting within blood vessels, kidney damage, and death. It is now known that transfusion reactions are caused by interactions between antigens and antibodies (see Chapter 14). In brief, the surface of erythrocytes has molecules called **antigens** (an'tĭ-jenz), and in the plasma there are molecules called **antibodies.** An antibody is very specific, meaning it can combine only with a certain antigen. When the antibodies in the plasma bind to the antigens on the surface of the erythrocytes, they form molecular bridges that connect the erythrocytes together. As a result, **agglutination** (ă-glu'tĭ-na'shun) or clumping of the cells occurs. The combination of the antibodies with the antigens also can initiate reactions that cause **hemolysis** (he-mol'ĭ-sis) or rupture of the erythrocytes.

The antigens on the surface of erythrocytes have been categorized into **blood groups.** Although many blood groups are recognized, the ABO and Rh blood groups are among the most important.

ABO Blood Group

In humans, blood is categorized by the ABO blood group system. The ABO antigens appear on the surface of the erythrocytes. Type A blood has type A antigens, type B blood has type B antigens, type AB blood has both types of antigens, and type O blood has neither A nor B antigens (Figure 11-8). In addition, plasma from type A blood contains antibodies against type B antigens, and plasma from type B blood contains antibodies against type A antigens. Type O blood has both A and B antibodies, and type AB blood has neither.

The reason for the presence of A and B antibodies in blood is not clearly understood. Antibodies normally do not develop against an antigen unless the body is exposed to the antigen. This means, for example, that a person with type A blood should not have type B antibodies unless he or she has received a transfusion of type B blood, which contains type B antigens. Because people with type A blood do have type B antibodies, however, even though they never have received a transfusion of type B blood, another explanation is needed. One possibility is that type A or B antigens on bacteria or food in the digestive tract stimulate the formation of antibodies against antigens that are different from one's own antigens. Thus, a person with type A blood would produce type B antibodies against the B antigens on the bacteria or food. In support of this hypothesis is the observation that A and B antibodies are not found in the blood until two months after birth.

A **donor** is a person who gives blood, and a **recipient** is a person who receives blood. Usually a donor can give blood to a recipient if they both have the same blood type. For example, a person with type A blood could donate to another person with type A blood. There would be no ABO transfusion reaction because the recipient has no antibodies against the type A antigen. On the other hand, if type A blood were donated to a person with type B blood there would be a transfusion reaction. This would occur because the person with type B blood has antibodies against the type A antigen and agglutination would result. (Figure 11-9).

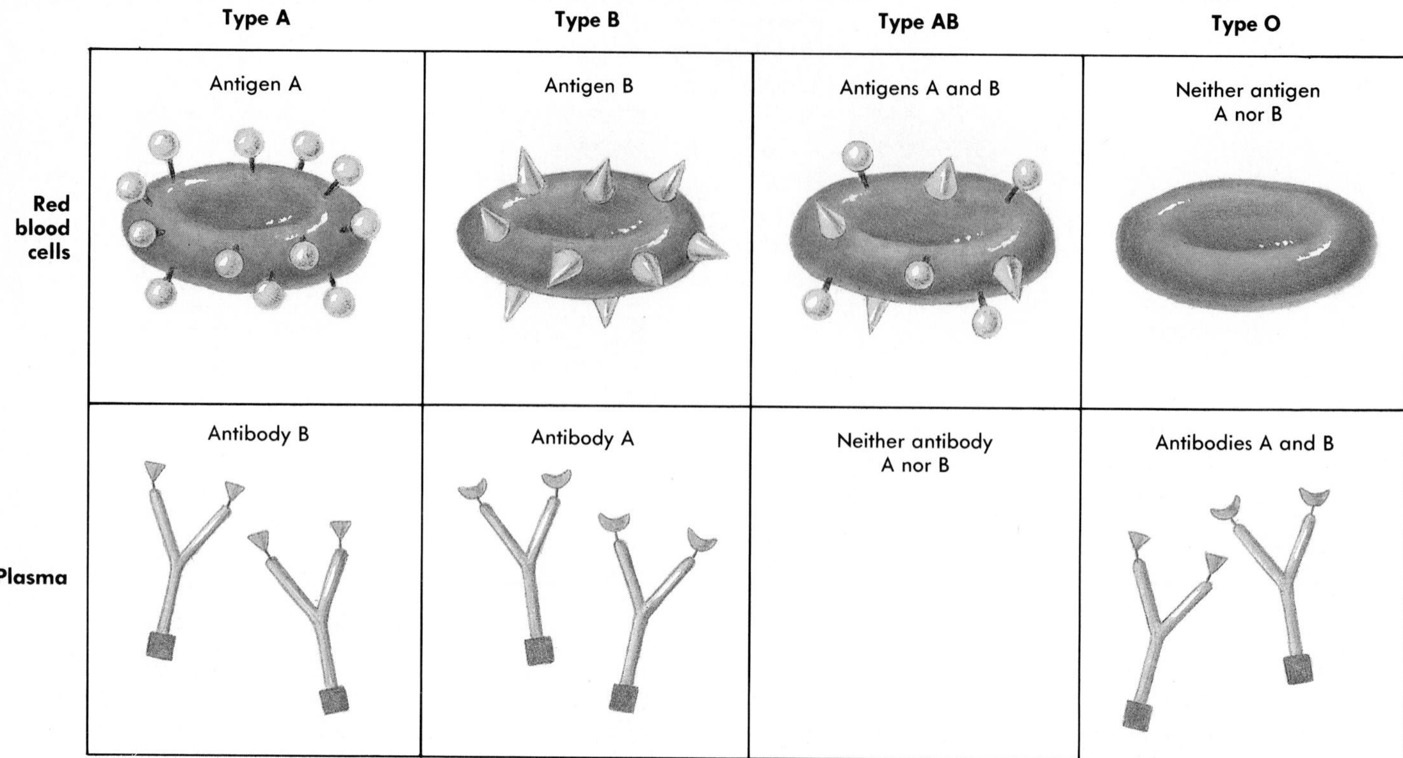

FIGURE 11-8 ABO blood groups.
The antigens found on the surface of the erythrocytes of each blood type, and the antibodies found in each blood type are shown.

Historically, people with type O blood have been called universal donors because they usually can give blood to the other ABO blood types without causing an ABO transfusion reaction. Their erythrocytes have no ABO surface antigens and therefore do not react with the recipient's A or B antibodies. For example, if type O blood is given to a person with type A blood, the type O erythrocytes do not react with the type B antibodies in the recipient's blood. In a similar fashion, if type O blood is given to a person with type B blood there would be no reaction with the recipient's type A antibodies.

It should be noted that the term universal donor is misleading. Transfusion of type O blood can produce a transfusion reaction for two reasons. First, there are other blood groups that can cause a transfusion reaction. To reduce the likelihood of a transfusion reaction, all the blood groups must be correctly matched. Second, antibodies in the blood of the donor can react with antigens in the blood of the recipient. For example, type O blood has type A and B antibodies. If type O blood is transfused into a person with type A blood, the A antibodies (in the type O blood) react against the A antigens (in the type A blood). Usually such reactions are not serious because the antibodies in the donor's blood are diluted in the blood of the recipient, and few reactions take place. Type O blood is given to a person with another blood type only in life-or-death emergency conditions, because type O blood sometimes causes transfusion reactions in these situations.

2 Historically people with type AB blood were called universal recipients. What is the rationale for this term? Explain why the term is misleading. ?

The Rh Blood Group

Another important blood group is the Rh blood group, so named because it was first studied in the rhesus monkey. People are Rh-positive if they have certain Rh antigens on the surface of their erythrocytes, and they are Rh-negative if they do not have these Rh antigens. Antibodies against the Rh antigen do not develop unless an Rh-negative person is exposed to Rh-positive blood. This can occur through a transfusion or by transfer of blood between a mother and her fetus across the placenta.

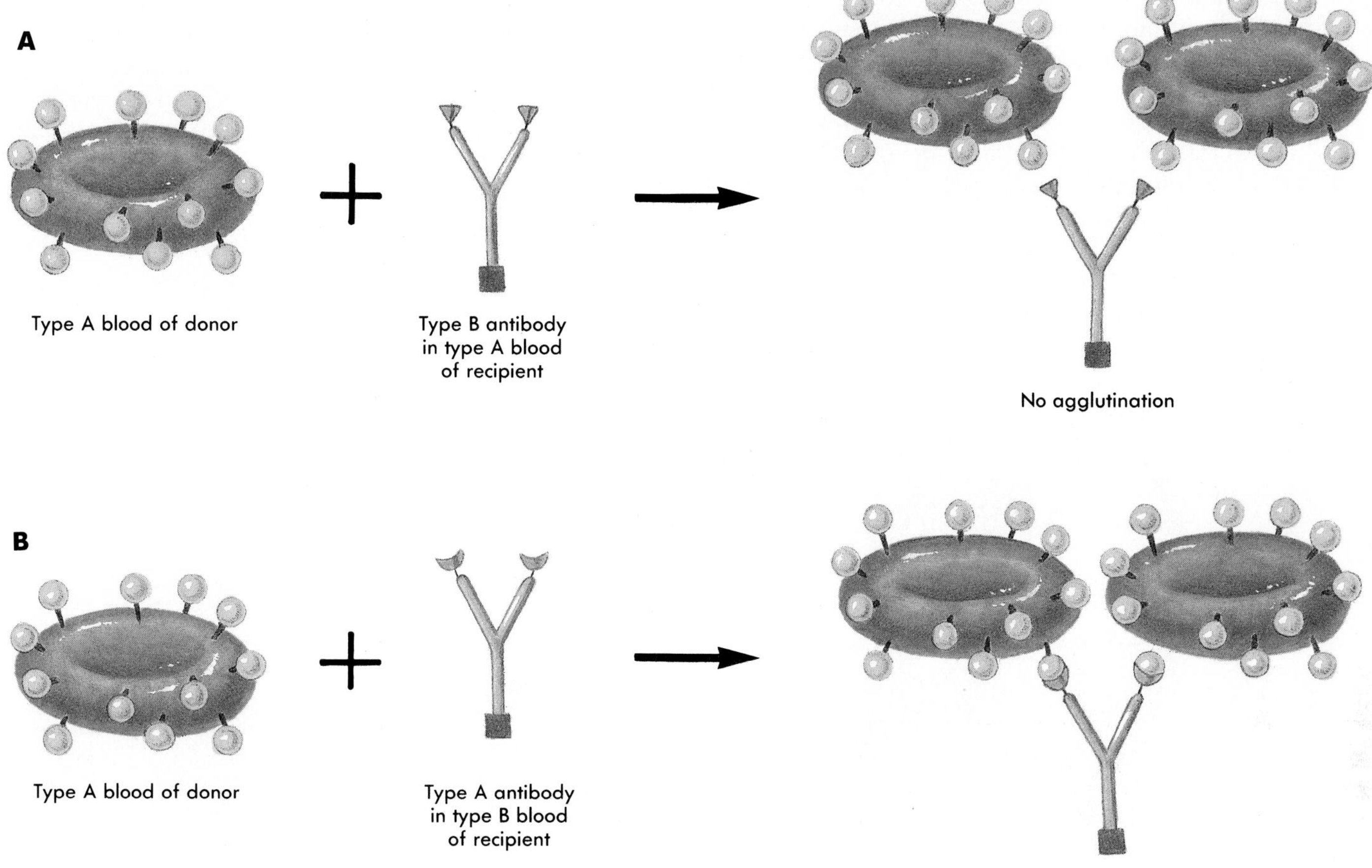

FIGURE 11-9 ■ Agglutination reaction.
A Type A blood donated to a type A recipient does not cause an agglutination reaction because the type B antibodies in the recipient do not combine with the type A antigens in the donated blood.
B Type A blood donated to a type B recipient causes an agglutination reaction because the type A antibodies in the recipient combine with the type A antigens in the donated blood.

When an Rh-negative person receives a transfusion of Rh-positive blood, the recipient becomes sensitized to the Rh antigen and produces Rh antibodies. If the Rh-negative person is unfortunate enough to receive a second transfusion of Rh-positive blood after becoming sensitized, a transfusion reaction results.

Rh incompatibility can pose a major problem in some pregnancies when the mother is Rh-negative, and the fetus is Rh-positive (Figure 11-10). If fetal blood leaks through the placenta and mixes with the mother's blood, the mother becomes sensitized to the Rh antigen. The mother produces Rh antibodies that cross the placenta and cause agglutination and hemolysis of fetal erythrocytes. This disorder is called **erythroblastosis fetalis** (ĕ-rith′ro-blas-to′sis fe-ta′lis), and it can be fatal to the fetus. In the first pregnancy, however, there is often no problem. The leakage of fetal blood is usually the result of a tear in the placenta that takes place either late in the pregnancy or during delivery. Thus there is not enough time for the mother to produce enough Rh antibodies to harm the fetus. In later pregnancies, however, there can be a problem because the mother has been sensitized to the Rh antigen. Consequently, if the fetus is Rh-positive and there is any leakage of fetal blood into the mother's blood, she rapidly produces large amounts of Rh antibodies and erythroblastosis fetalis develops.

Prevention of erythroblastosis fetalis is often possible if the Rh-negative woman is given an injection of a specific type of antibody preparation called anti-Rh_o(D) immune globulin (RhoGAM) immediately after each delivery or abortion. The injection contains antibodies against Rh antigens. The injected antibodies bind to the Rh antigens of any fetal erythrocytes that may have entered the mother's blood. This treatment inactivates the fetal Rh antigens and prevents sensitization of the mother.

If erythroblastosis fetalis develops, treatment consists of slowly removing the blood of the fetus or newborn and replacing it with Rh-negative blood. Exposure of the newborn to fluorescent light is also used because it helps break down the large amounts of bilirubin formed as a result of erythrocyte destruction. High levels of bilirubin are toxic to the nervous system and can cause destruction of brain tissue.

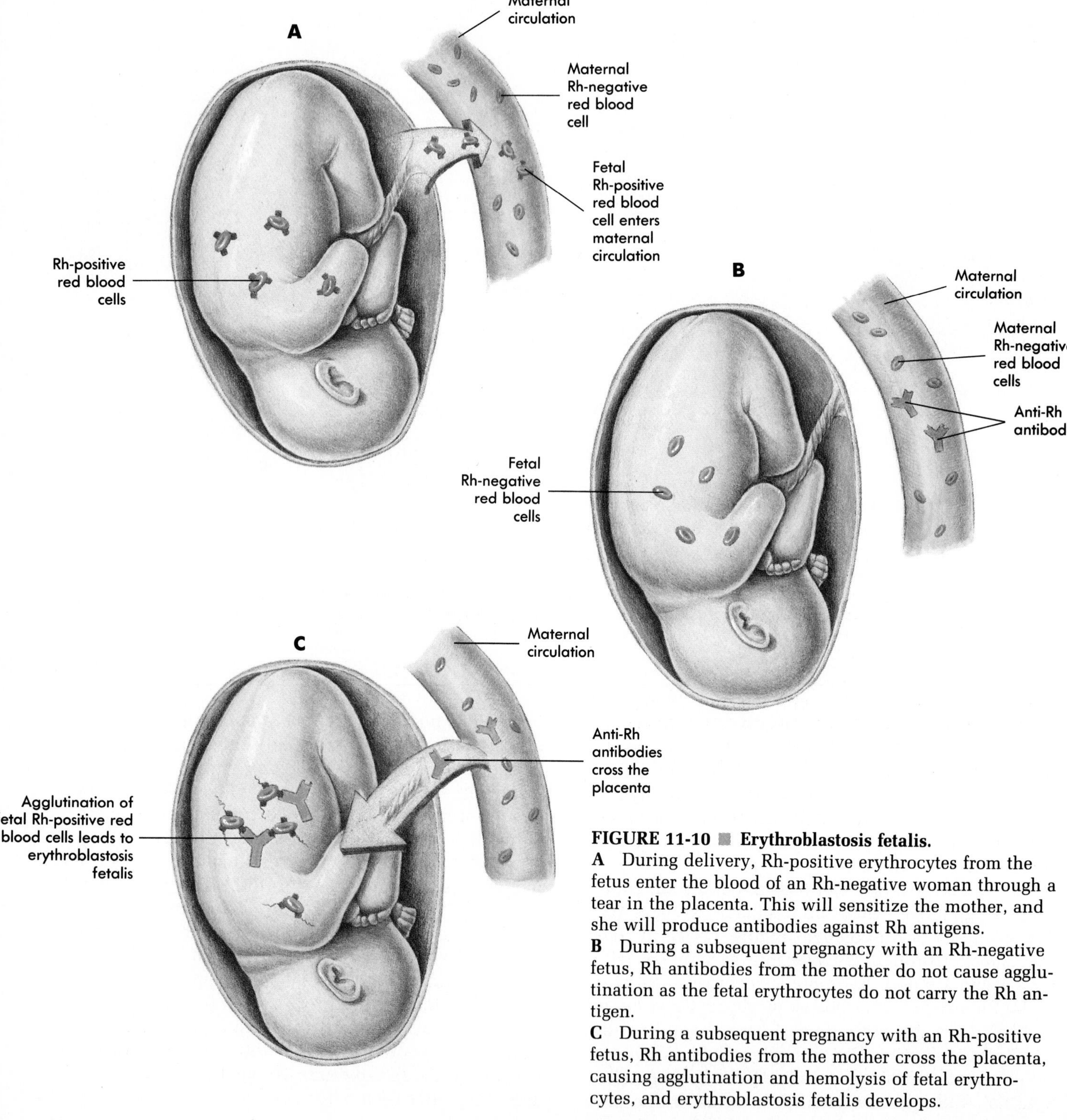

FIGURE 11-10 ■ Erythroblastosis fetalis.
A During delivery, Rh-positive erythrocytes from the fetus enter the blood of an Rh-negative woman through a tear in the placenta. This will sensitize the mother, and she will produce antibodies against Rh antigens.
B During a subsequent pregnancy with an Rh-negative fetus, Rh antibodies from the mother do not cause agglutination as the fetal erythrocytes do not carry the Rh antigen.
C During a subsequent pregnancy with an Rh-positive fetus, Rh antibodies from the mother cross the placenta, causing agglutination and hemolysis of fetal erythrocytes, and erythroblastosis fetalis develops.

DIAGNOSTIC BLOOD TESTS
Type and Cross Match

In order to prevent transfusion reactions the blood is typed, and a cross match is made. Blood typing determines the ABO and Rh blood groups of the blood sample. Typically, the cells are separated from the serum. The cells are tested with known antibodies to determine the type of antigen on the cell surface. For example, if a patient's blood cells agglutinate when mixed with type A antibodies, but do not agglutinate when mixed with type B antibodies, it is concluded that the cells have type A antigen. In a similar fashion, the serum is mixed with known cell types (antigens) to determine the type of antibodies in the serum.

Normally, donor blood must match the ABO and Rh type of the recipient. Because other blood groups can cause a transfusion reaction, however, a cross match is performed. In a cross match, the donor's blood cells are mixed with the recipient's serum, and the donor's serum is mixed with the recipient's cells. The donor's blood is considered safe for transfusion only if there is no agglutination in either match.

Complete Blood Count

The **complete blood count (CBC)** is an analysis of the blood that provides much information. It consists of a red blood cell count, hemoglobin and hematocrit measurements, and a white blood cell count.

Red blood cell count

Blood cells counts are usually done electronically with a machine, but they can be done manually with a microscope. A normal **red blood cell (RBC) count** for a male is 4.2 to 5.8 million erythrocytes per cubic millimeter of blood, and for a female it is 3.6 to 5.2 million per cubic millimeter of blood. **Polycythemia** (pol'ĭ-si-the'me-ah) is an overabundance of erythrocytes. It can result from a decreased oxygen supply, which stimulates erythropoietin production, or from red bone marrow tumors. Polycythemia makes it harder for the blood to flow through blood vessels and increases the work load of the heart. It can reduce blood flow through tissues, and if severe, can result in plugging of small blood vessels (capillaries).

Hemoglobin measurement

The **hemoglobin** measurement determines the amount of hemoglobin in a given volume of blood, usually expressed as grams of hemoglobin per 100 milliliters of blood. The normal hemoglobin for a male is 14 to 18 grams per 100 milliliters of blood, and for a female it is 12 to 16 grams per 100 milliliters of blood. Abnormally low hemoglobin is an indication of **anemia,** which is a reduced number of erythrocytes, or a reduced amount of hemoglobin in each erythrocyte.

Hematocrit measurement

The percentage of total blood volume composed of erythrocytes is the **hematocrit** (hem'ă-to-krit). One way to determine hematocrit is to place blood in a tube and spin the tube in a centrifuge. The formed elements are heavier than the plasma and are forced to one end of the tube (Figure 11-11). The erythrocytes account for 44% to 54% of the total blood volume in males and 38% to 48% in females.

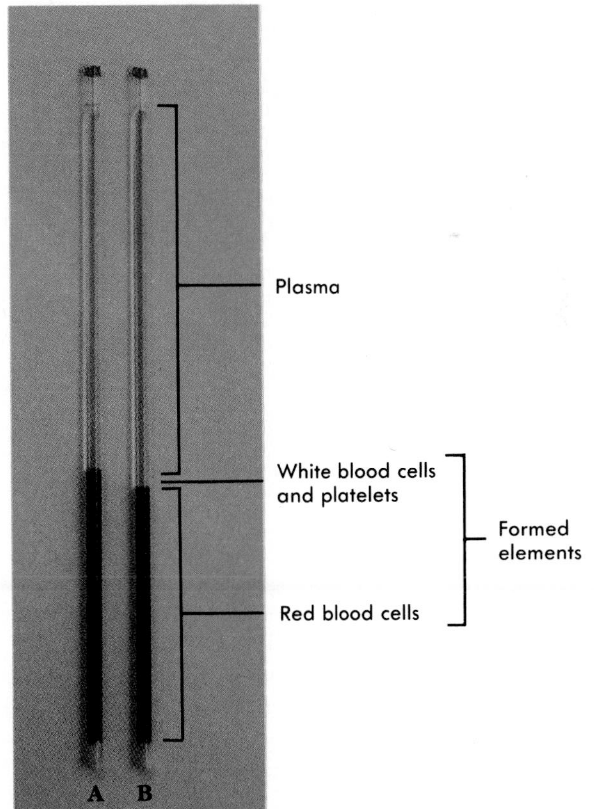

FIGURE 11-11 ■ Hematocrit.
Normal hematocrits of male **(A)** and female **(B)**. Blood is separated into plasma and formed elements. The relatively small amount of white blood cells (leukocytes) and platelets rests upon the red blood cells (erythrocytes).

Some Disorders of the Blood

ANEMIA

Anemia (a-ne'me-ah) is a deficiency of hemoglobin in the blood. It can result from a decrease in the number of erythrocytes, a decrease in the amount of hemoglobin in each erythrocyte, or both. The decreased hemoglobin reduces the ability of the blood to transport oxygen. Anemic patients suffer from a lack of energy and feel excessively tired and listless. They may appear pale and quickly become short of breath with only slight exertion.

One general cause of anemia is insufficient production of erythrocytes. **Aplastic anemia** is caused by an inability of the red bone marrow to produce erythrocytes. It is usually acquired as a result of damage to the red marrow by chemicals (for example, benzene), drugs (for example, certain antibiotics and sedatives), or radiation.

Erythrocyte production also can be lower than normal as a result of nutritional deficiencies. **Iron-deficiency anemia** results from a deficient intake or absorption of iron, or from excessive iron loss. Consequently not enough hemoglobin is produced, the number of erythrocytes decreases, and the erythrocytes that are manufactured are smaller than normal.

Another type of nutritional anemia is **pernicious** (per-nish'us) **anemia,** which is caused by inadequate vitamin B_{12}. Because vitamin B_{12} is necessary for the cell divisions that result in erythrocyte formation, a shortage of vitamin B_{12} causes reduced erythrocyte production. Although inadequate levels of vitamin B_{12} in the diet can cause pernicious anemia, the usual cause is insufficient absorption of the vitamin. Normally the stomach produces intrinsic factor, a protein that binds to vitamin B_{12}. The combined molecules pass into the lower intestine where intrinsic factor facilitates the absorption of the vitamin. Without adequate levels of intrinsic factor, insufficient vitamin B_{12} is absorbed, and pernicious anemia develops. Present evidence suggests that the inability to produce intrinsic factor is an autoimmune disease in which the body's immune system damages the cells in the stomach that produce intrinsic factor. **Folic acid deficiency** can also hinder cell divisions and cause anemia. A shortage of folic acid in the diet is the usual cause, with the disorder developing most often in the poor, in pregnant women, and in chronic alcoholics.

Another general cause of anemia is loss or destruction of erythrocytes. **Hemorrhagic anemia** results from a loss of blood such as can result from trauma, ulcers, or excessive menstrual bleeding. Chronic blood loss in which small amounts of blood are lost over a period of time can result in iron-deficiency anemia. **Hemolytic anemia** is a disorder in which erythrocytes rupture or are destroyed at an excessive rate. It can be caused by inherited defects within the erythrocytes. For example, one kind of inherited hemolytic anemia results from a defect in the cell membrane that causes erythrocytes to rupture easily. Many kinds of hemolytic anemia result from unusual damage to the erythrocytes by drugs, snake venom, artificial heart valves, autoimmune disease, or erythroblastosis fetalis.

Some anemias result from inadequate or defective hemoglobin production. **Thalassemia** (thal-ă-se'me-uh) is a hereditary disease found in people of Mediterranean, Asian, and African ancestry. It is caused by insufficient production of the globin portion of the hemoglobin molecule. The major form of the disease results in death by age 20, and the minor form is a mild anemia. **Sickle cell anemia** is a hereditary disease found mostly in blacks that results in the formation of an abnormal hemoglobin. The erythrocytes assume a rigid, sickle shape and

Because the hematocrit measurement is based on volume, it is affected by the number and size of erythrocytes. For example, a decreased hematocrit can result from a decreased number of normal-sized erythrocytes or a normal number of small-sized erythrocytes. The average size of an erythrocyte is calculated by dividing the hematocrit by the red blood cell count. A number of disorders cause erythrocytes to be smaller or larger than normal. For example, inadequate iron in the diet can impair hemoglobin production. Consequently, during their formation, erythrocytes do not fill up with hemoglobin, and they remain smaller than normal.

White blood cell count

A **white blood cell (WBC) count** measures the total number of leukocytes in the blood. There are normally 5,000 to 10,000 leukocytes per cubic millimeter of blood. **Leukopenia** (lu-ko-pe'ne-ah) is a lower than normal WBC count and often indicates depression or destruction of the red marrow (for example, as the result of radiation, drugs, tumor, or a deficiency of vitamin B_{12} or folic acid). **Leukocytosis** (lu-ko-sī-to'sis) is an abnormally high WBC count. **Leukemia,** a tumor of the red marrow, and bacterial infections often cause leukocytosis.

plug up small blood vessels. They are also more fragile than normal. In its severe form, sickle cell anemia is usually fatal before age 30, whereas in its minor, sickle cell trait, form there are usually no symptoms.

HEMOPHILIA

Hemophilia (he'mo-fil'ĭ-ah) is a genetic disorder in which clotting is abnormal or absent. It is most often found in people from northern Europe and their descendents. Hemophilia is a sex-linked trait, and it occurs almost exclusively in males. There are several types of hemophilia, each the result of a deficiency or dysfunction of a clotting factor. Treatment of hemophilia involves injection of the missing clotting factor taken from donated blood.

LEUKEMIA

Leukemia (lu-ke'me-ah) is a type of cancer in which abnormal production of one or more of the leukocyte types occurs. Because these cells are usually immature or abnormal and lack normal immunological functions, patients are very susceptible to infections. The excess production of leukocytes in the red marrow, may also interfere with erythrocyte and platelet formation and thus lead to anemia and bleeding.

INFECTIOUS DISEASES OF THE BLOOD

Normally blood is sterile. Microorganisms may be found in the blood for two reasons: transportation or multiplication. Many microorganisms gain entry through one of the body's portals and are transported by the blood to the tissues they infect. For example, the polio virus enters through the gastrointestinal portal and is carried to nervous tissue. After microorganisms are established at a site of infection, some of them can be picked up by the blood. These microorganisms can spread to other locations in the body, multiply within the blood, or be eliminated by the body's immune system.

Septicemia (sep'tĭ-se'me-ah) or blood poisoning is the multiplication of microorganisms in the blood. Often septicemia results from the introduction of microorganisms by a medical procedure such as the insertion of an intravenous tube into a blood vessel. The release of toxins by bacteria can cause **septic shock,** which is a decrease in blood pressure that can result in death.

There are a few diseases in which microorganisms actually multiply within blood cells. **Malaria** (mă-la're-ah) is caused by a protozoan that is introduced into the blood by the bite of the *Anopheles* mosquito. Part of the protozoan's development occurs inside erythrocytes. The symptoms of chills and fever are produced by toxins released when the protozoan causes the erythrocytes to rupture. **Infectious mononucleosis** (mon'o-nu'kle-o'sis) is caused by a virus that infects the salivary glands and lymphocytes. The lymphocytes are altered by the virus and the immune system attacks and destroys the lymphocytes. The immune system response is believed to produce the symptoms of fever, sore throat, and swollen lymph nodes. The **AIDS** virus also infects lymphocytes and causes immune system suppression (see Chapter 15).

The presence of microorganisms in blood is a concern when transfusions are made, because it is possible to infect the blood recipient. Blood is routinely tested in an effort to eliminate this risk, especially for AIDS and hepatitis. **Hepatitis** (hep'ă-ti'tis) is an infection of the liver caused by several different kinds of viruses. After recovering, hepatitis victims can become carriers. Although they show no signs of the disease, they release the virus into their blood or bile. To prevent infection of others, anyone who has had hepatitis is asked not to donate blood products.

White Blood Cell Differential Count

A **white blood cell differential count** determines the percentage of each of the five kinds of leukocytes in the white blood cell count. Normally neutrophils account for 60% to 70%, lymphocytes 20% to 30%, monocytes 2% to 8%, eosinophils 1% to 4%, and basophils 0.5% to 1%. Much insight about a patient's condition can be obtained from a white blood cell differential count. For example, in bacterial infections, the neutrophil count is often greatly increased, whereas in allergic reactions the eosinophil and basophil counts are elevated.

Clotting

Two measurements that test the ability of the blood to clot are the platelet count and the prothrombin time.

Platelet count

A normal **platelet count** is 150,000 to 400,000 platelets per cubic millimeter of blood. **Thrombocytopenia** (throm-bo-si-to-pe'ne-ah) is a condition in which the platelet count is greatly reduced, resulting in chronic bleeding through small vessels and capillaries. It can be caused by decreased platelet production as a result of hereditary disorders, lack of vitamin B_{12} (pernicious anemia), drug therapy, or radiation therapy.

Prothrombin time measurement

Prothrombin time is a measure of how long it takes for the blood to start clotting, which is normally 9 to 12 seconds. Because many clotting factors have to be activated to form prothrombin, a deficiency of any one of them can cause an abnormal prothrombin time. Vitamin K deficiency, certain liver diseases, and drug therapy can cause an increased prothrombin time.

Blood Chemistry

The composition of materials dissolved or suspended in the plasma can be used to assess the functioning of many of the body's systems. For example, high blood glucose levels can indicate that the pancreas is not producing enough insulin, high blood urea nitrogen (BUN) is a sign of reduced kidney function, increased bilirubin can indicate liver dysfunction, and high cholesterol levels can indicate an increased risk of developing cardiovascular disease. A number of blood chemistry tests are routinely done when a blood sample is taken, and additional tests are available.

> **3 When a patient complains of acute pain in the abdomen, the physician suspects appendicitis, which is a bacterial infection of the appendix. What blood test should be done to confirm the diagnosis?** ?

SUMMARY

Blood transports gases, nutrients, waste products, and hormones.

Blood protects against disease and is involved in temperature, fluid, and electrolyte regulation.

PLASMA

Plasma is 92% water and 8% suspended or dissolved substances.

Plasma maintains osmotic pressure, is involved in immunity, prevents blood loss, and transports molecules.

FORMED ELEMENTS

The formed elements are cells (erythrocytes and leukocytes) and cell fragments (platelets).

Production of Formed Elements

Formed elements arise (hematopoiesis) in red bone marrow from stem cells.

Erythrocytes

Erythrocytes are disk-shaped cells containing hemoglobin, which transports oxygen and carbon dioxide. Erythrocytes also contain carbonic anhydrase, which is involved with carbon dioxide transport.

In response to low blood oxygen, the kidneys produce renal erythropoietic factor, which activates erythropoietin. Erythropoietin stimulates erythrocyte production in red bone marrow.

Worn-out erythrocytes are phagocytized by macrophages in the spleen or liver. Hemoglobin is broken down, iron and amino acids are reused, and heme becomes bilirubin that is secreted in bile.

Leukocytes

Leukocytes protect the body against microorganisms and remove dead cells and debris.

Granulocytes contain cytoplasmic granules, and there are three types: Neutrophils are small phagocytic cells, eosinophilis reduce inflammation, and basophils promote inflammation.

Agranulocytes have very small granules, and there are two types: Lymphocytes are involved in antibody production and other immune system responses; monocytes become macrophages that ingest microorganisms and cellular debris.

Platelets

Platelets are cell fragments involved with preventing blood loss.

PREVENTING BLOOD LOSS

Blood Vessel Constriction

Blood vessels constrict in response to injury, resulting in decreased blood flow.

Platelet Plugs

Minor damage to blood vessels is repaired by platelet plugs.

Blood Clotting

Blood clotting or coagulation is formation of a clot (a network of protein fibers called fibrin).

There are three steps in the clotting process.

- Activation of clotting factors by connective tissue and chemicals, resulting in the formation of prothrombin activator.
- Conversion of prothrombin to thrombin by prothrombin activator.
- Conversion of fibrinogen to fibrin by thrombin.

Control of Clot Formation

Clot formation is prevented by anticoagulants in the blood such as antithrombin.

Clot Retraction and Dissolution

Clot retraction condenses the clot, pulling the edges of damaged tissue closer together.

Fibrinolysis (clot breakdown) is accomplished by plasmin.

BLOOD GROUPS

Blood groups are determined by antigens on the surface of erythrocytes.

Antibodies can bind to erythrocyte antigens, resulting in agglutination or hemolysis of erythrocytes.

ABO Blood Group

Type A blood has A antigens, type B blood has B antigens, type AB blood has A and B antigens, and type O blood does not have A or B antigens.

Type A blood has B antibodies, type B blood has A antibodies, type AB blood does not have A or B antibodies, and type O blood has A and B antibodies.

Mismatching the ABO blood group can result in transfusion reactions.

Rh Blood Group

Rh-positive blood has the Rh antigen, whereas Rh-negative blood does not.

Antibodies against the Rh antigen are produced by an Rh-negative person when the person is exposed to Rh-positive blood.

The Rh blood group is responsible for erythroblastosis fetalis.

DIAGNOSTIC BLOOD TESTS

Type and Cross Match

Blood typing determines the ABO and Rh blood groups of a blood sample. A cross match tests for agglutination reactions between donor and recipient blood.

Complete Blood Count

The complete blood count consists of the following: red blood cell count, hemoglobin measurement (grams of hemoglobin per 100 milliliters of blood), hematocrit measurement (percent volume of erythrocytes), and white blood cell count.

White Blood Cell Differential Count

The white blood cell differential count determines the percentage of each type of leukocyte.

Clotting

Platelet count and prothrombin time measure the ability of the blood to clot.

Blood Chemistry

The composition of materials dissolved or suspended in plasma (for example, glucose, urea nitrogen, bilirubin, and cholesterol) can be used to assess the functioning and status of the body's systems.

CONTENT REVIEW

1. Describe the functions of blood.
2. Define plasma. List the functions of plasma.
3. Define the formed elements and name the different types of formed elements. Explain how the formed elements arise through hematopoiesis.
4. Describe the two basic parts of a hemoglobin molecule. Which part is associated with iron? What gases are transported by each part?
5. What is the role of carbonic anhydrase in gas transport?
6. Why are vitamin B_{12} and folic acid important in erythrocyte production?
7. Explain how low blood oxygen levels result in increased erythrocyte production.
8. Where are erythrocytes broken down? What happens to the breakdown products?
9. Give two functions of leukocytes.
10. Name the five types of leukocytes and state a function for each type.
11. What are platelets, and how are they formed?
12. Describe the role of blood vessel constriction and platelet plugs in preventing bleeding.
13. What are clotting factors? Describe the three steps of their activation that result in the formation of a clot.
14. Explain the function of anticoagulants in the blood, and give an example of an anticoagulant.
15. What is clot retraction, and what does it accomplish?
16. Define fibrinolysis and name the chemicals responsible for this process.
17. What are blood groups, and how do they cause transfusion reactions? List the four ABO blood types. Why is type O blood considered a universal donor and type AB blood a universal recipient?
18. What is meant by the term Rh-positive? How can Rh incompatibility affect a pregnancy?
19. Define a thrombus and an embolus and explain why they are dangerous.
20. For each of the following tests, define the test and give an example of a disorder that would cause an abnormal test result.
    A. Type and cross match
    B. Red blood cell count
    C. Hemoglobin measurement
    D. Hematocrit measurement
    E. White blood cell count
    F. White blood cell differential count
    G. Platelet count
    H. Prothrombin time
    I. Blood chemistry tests

CONCEPT REVIEW

1. Red Packer, a physical education major, wanted to improve his performance in an upcoming marathon race. About 6 weeks before the race, 1 quart of blood was removed from his body, and the formed elements were separated from the plasma. The formed elements were frozen, and the plasma was reinfused into his body. Just before the race, the formed elements were thawed and injected into his body. Explain why this procedure, called blood doping, would help Red's performance. Can you suggest any possible bad effects?
2. Chemicals such as benzene can destroy red bone marrow, causing aplastic anemia. What symptoms would you expect to develop as a result of the lack of (1) red blood cells, (2) platelets, and (3) leukocytes?
3. E.Z. Goen habitually used barbiturates to depress feelings of anxiety. Because barbiturates suppress the respiratory centers in the brain, they cause hypoventilation (that is, slower than normal rate of breathing). What happens to the erythrocyte count of a habitual user of barbiturates? Explain.
4. What blood problems would you expect to observe in a patient after total gastrectomy (removal of the stomach)?
5. According to the old saying, "Good food makes good blood." Name three substances in the diet that are essential for "good blood." What blood disorders develop if these substances are absent from the diet?
6. Why do anemic patients often have clay-colored feces?

ANSWERS TO ? PREDICT QUESTIONS

1 *p. 277* Carbon monoxide binds to the iron of hemoglobin and prevents the transport of oxygen. The decreased oxygen stimulates the release of renal erythropoietic factor, which causes the activation of erythropoietin. The erythropoietin increases erythrocyte production in red bone marrow, and the number of erythrocytes in the blood increases.

2 *p. 282* People with type AB blood were called universal recipients because they could receive type A, B, AB, or O blood with little likelihood of a transfusion reaction. Type AB blood does not have antibodies against type A or B antigens. Therefore transfusion of these antigens in type A, B, or AB blood does not cause a transfusion reaction in a person with type AB blood. The term is misleading, however, for two reasons. First, other blood groups can cause a transfusion reaction. Second, antibodies in the donor's blood can cause a transfusion reaction. For example, type O blood contains A and B antibodies that can react against the A and B antigens in type AB blood.

3 *p. 287* An increase in the white blood cell count often indicates a bacterial infection. A white blood cell differential count with an abnormally high neutrophil percentage would confirm the diagnosis.

12

THE HEART

After reading this chapter you should be able to:

1 Describe the size, shape, and location of the heart.
2 Name the valves of the heart, state their location and function, and explain how heart sounds are produced.
3 Describe the flow of blood through the heart and name each of the chambers and structures through which the blood passes.
4 List the components of the heart wall and describe the structure and function of each.
5 Explain the structure and function of the conduction system of the heart.
6 Define each wave of the electrocardiogram and relate each of them to contractions of the heart.
7 Give the location and function of the coronary arteries.
8 Describe intrinsic and extrinsic regulation of the heart.
9 Give the conditions for which the major heart medications and treatments are administered.

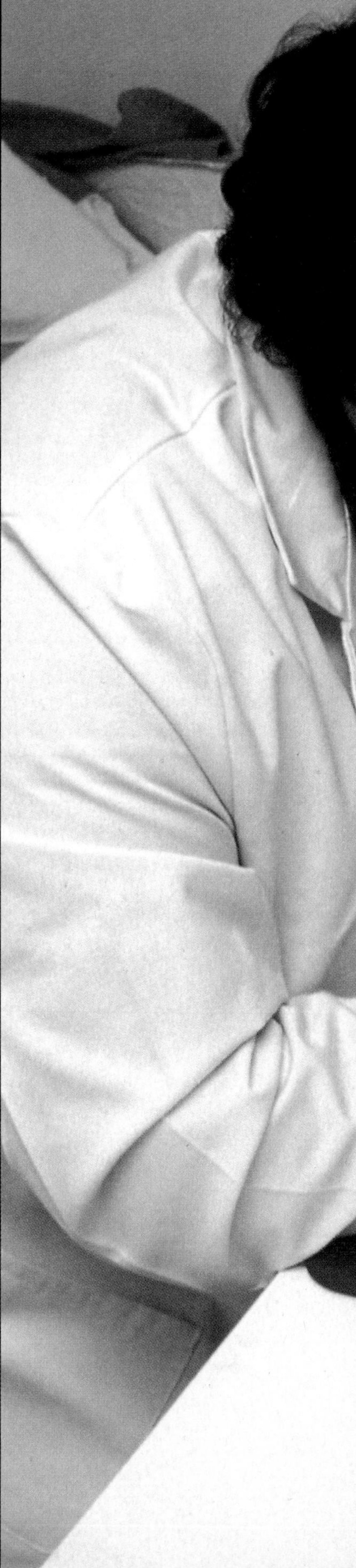

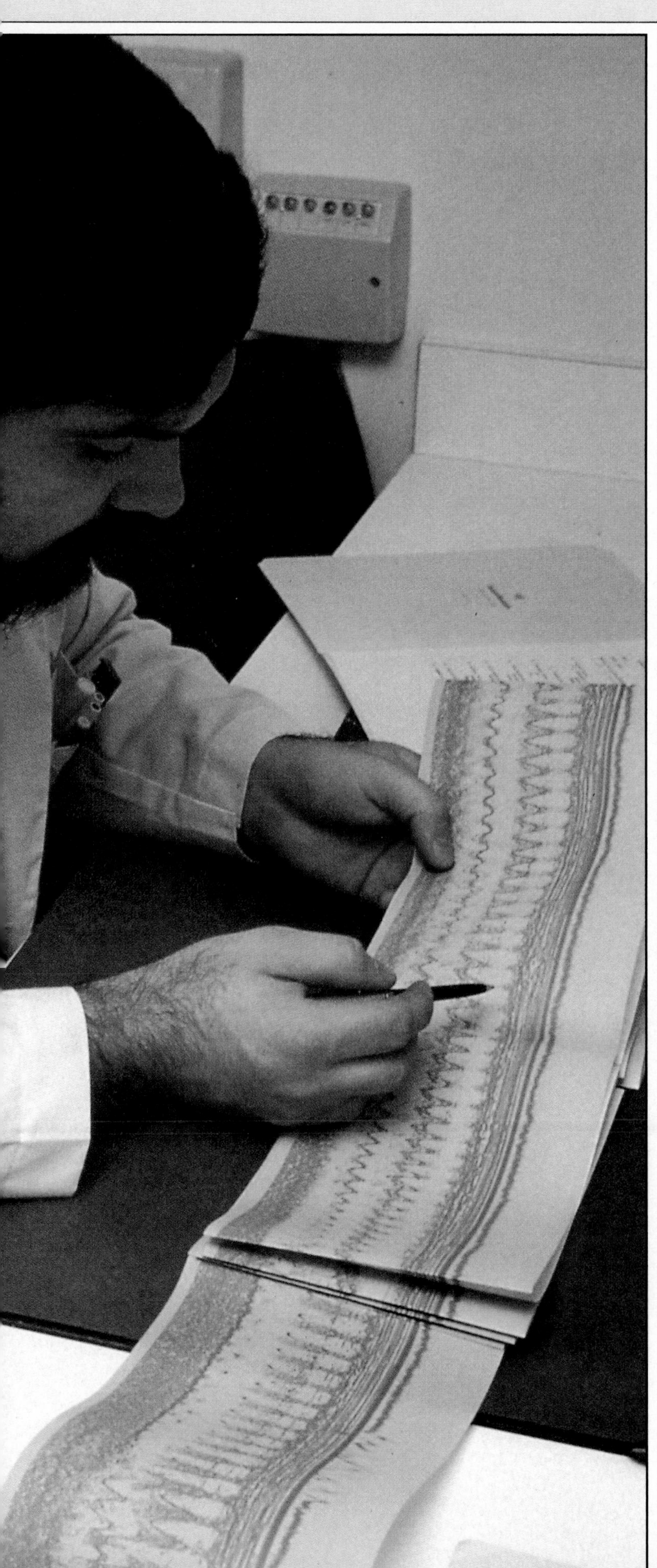

atrium, pl. atria (a'tre-ah) [L., entrance chamber] Chambers of the heart that collect blood during ventricular contraction and pump blood into the ventricles to complete ventricular filling at the end of ventricular relaxation.

atrioventricular (AV) node (a'tre-o-ven'trik'u-lar) Small collection of specialized cardiac muscle fibers located in the lower portion of the right atrium and which gives rise to the atrioventricular bundle.

baroreceptor (băr'o-re-sep'tor) Sensory nerve endings in the walls of the aorta and internal carotid arteries; sensitive to stretching of the wall caused by increased blood pressure.

bicuspid (mitral) valve Valve consisting of two cusps of tissue; located between the left atrium and left ventricle of the heart.

cardiac cycle Complete round of cardiac systole and diastole.

cardiac output (Minute volume) Volume of blood pumped by either ventricle of the heart per minute.

coronary artery (kor'o-năr-e) Artery that arises from the base of the aorta and carries blood to the muscle of the heart.

diastole (di-as'to-le) [Gr. *diastole*, dilation] Relaxation of the heart chambers during which they fill with blood; usually refers to ventricular relaxation.

electrocardiogram Graphic record of the heart's electrical currents obtained with an electronic recording instrument.

semilunar valve One of two valves in the heart composed of three semilunar-shaped cusps that prevent flow of blood back into the ventricles following ejection; located at the beginning of the aorta and pulmonary trunk.

sinoatrial (SA) node (si'no-a'tre-al) Collection of specialized cardiac muscle fibers that acts as the "pacemaker" of the cardiac conduction system.

systole (sis'to-le) [Gr. *systole*, a contracting] Contraction of the heart chambers during which blood leaves the chambers; usually refers to ventricular contraction.

tricuspid valve Valve consisting of three cusps of tissue; located between the right atrium and right ventricle of the heart.

ventricle (ven'trĭ-kul) [L. *venter*, belly] One of two chambers of the heart that pumps blood into arteries.

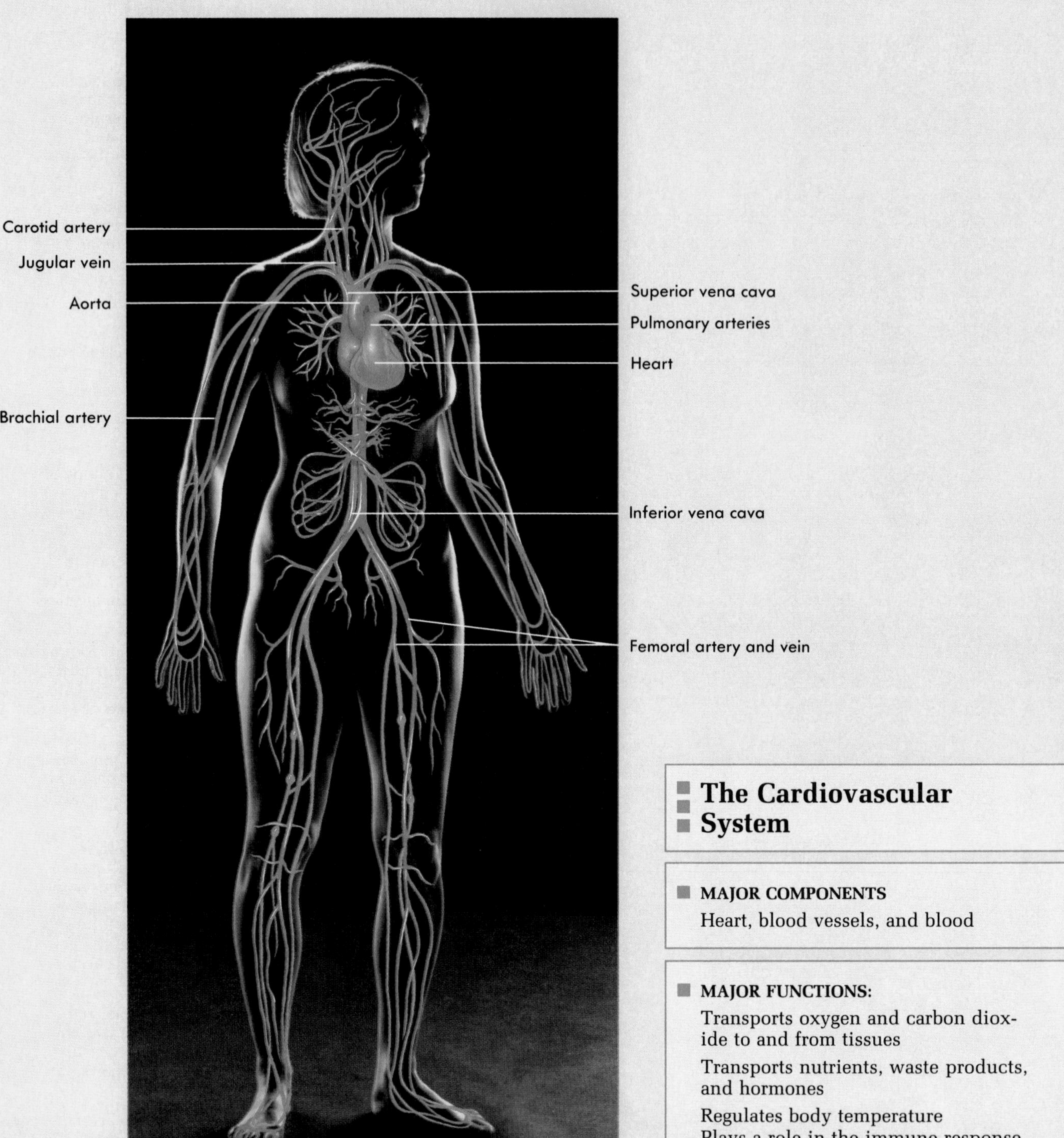

FIGURE 12-1 ■ Overview of the circulatory system.
The circulatory system consists of the pulmonary and systemic circulations. The right side of the heart pumps blood to the lungs through the pulmonary circulation (pulmonary arteries and veins). The left side of the heart receives blood from the lungs and pumps it to the body through the systemic circulation (aorta, systemic arteries, and veins). After circulating through the body, blood is returned to the right side of the heart.

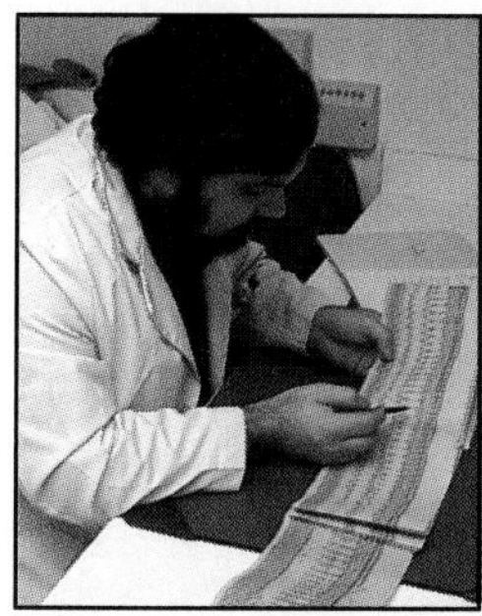

The heart contracts forcefully to pump blood through the blood vessels of the body (Figure 12-1). The heart is actually two pumps in one. One pump propels blood to the lungs (the pulmonary circulation), and the other propels blood to all other tissues of the body (the systemic circulation). The heart of a healthy adult pumps about 5 liters of blood per minute. For most people the heart continues to pump at about that rate for more than 75 years. During short periods of vigorous exercise the amount of blood pumped per minute increases several fold. If the heart loses its pumping ability for even a few minutes, however, the life of the individual is in danger.

SIZE, FORM, AND LOCATION OF THE HEART

The adult heart has the shape of a blunt cone and is about the size of a closed fist. It is located in the thoracic cavity between the lungs surrounded by a space within the mediastinum called the **pericardial** (pĕr′ĭ-kar′de-al) **cavity** (Figure 12-2; see Chapter 1). The blunt, rounded point of the cone is the **apex** (a′peks; tip), and the larger, flat portion at the opposite end of the cone is the **base.** The apex is the most inferior part of the heart, it is directed to the left, and it can be found deep to the fifth intercostal (between the ribs) space. The base is directed superiorly and slightly posteriorly. The most superior portion of the base can be found deep to the second intercostal space.

It is important to know the location of the heart. Placing a stethoscope to hear the heart sounds, placing electrodes on the chest to record an electrocardiogram (ECG), and performing cardiopulmonary resuscitation (CPR) depend on a knowledge of the heart's position.

ANATOMY OF THE HEART

The heart is a muscular pump consisting of four chambers: two **atria** (a′tre-ah; entrance chambers) and two **ventricles** (ven′trĭ-kulz) (Figure 12-3).

Pericardium

The **pericardium** (pĕr′ĭ-kar′de-um), or pericardial sac, is a double-layered closed sac that surrounds the heart and anchors it within the mediastinum. (Figure 12-4). The pericardium consists of a tough, fibrous connective tissue outer layer called the **fibrous pericardium** and an inner layer of flat epithelial cells called the **serous pericardium.** The pericardium surrounds and anchors the heart within the mediastinum. The serous pericardium

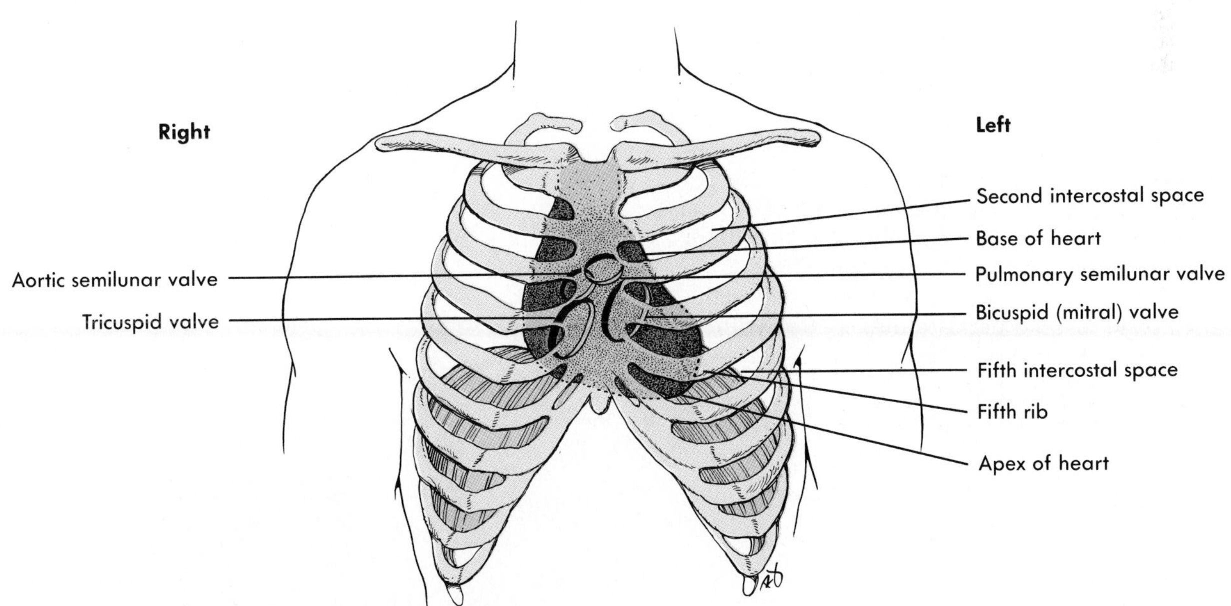

FIGURE 12-2 ■ **Location of the heart in the thorax.**
The heart is deep to and slightly left of the sternum. The apex of the heart is in the fifth intercostal space about 9 cm from the midline. The base is in the second intercostal space. The valves of the heart are located near the base of the heart.

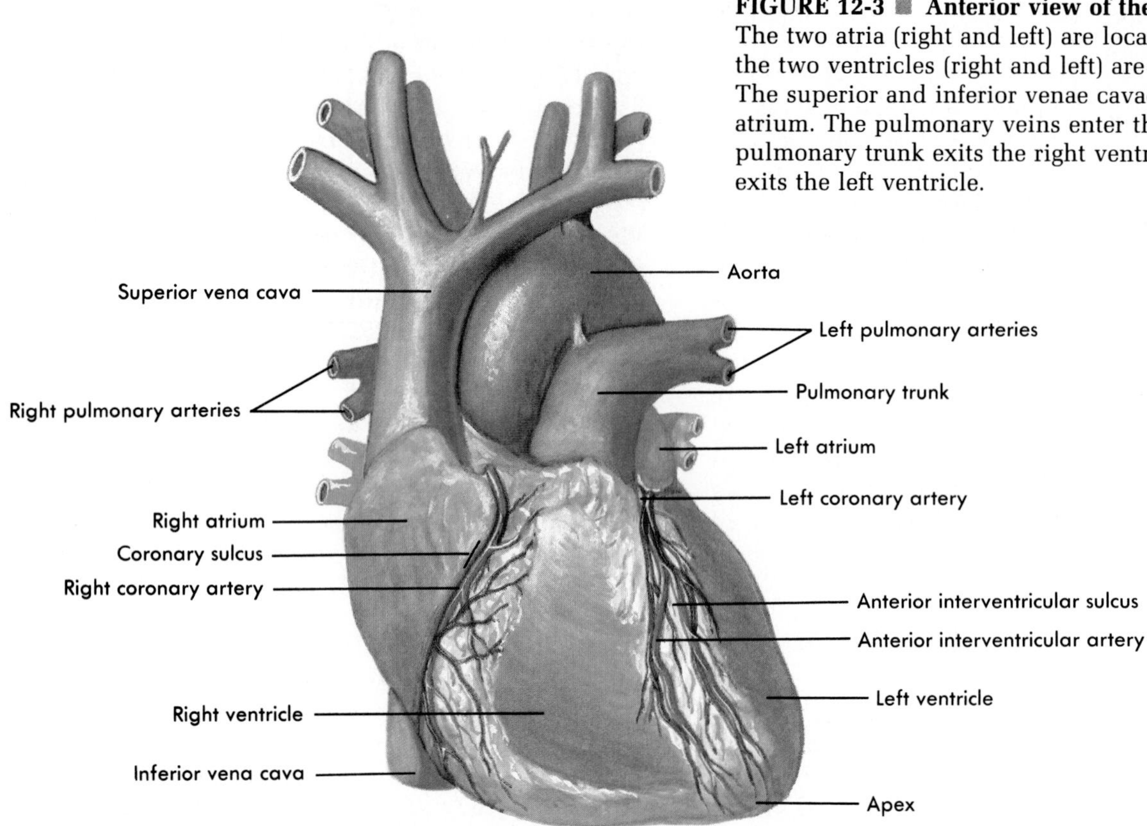

FIGURE 12-3 ■ Anterior view of the heart.
The two atria (right and left) are located superiorly and the two ventricles (right and left) are located inferiorly. The superior and inferior venae cavae enter the right atrium. The pulmonary veins enter the left atrium. The pulmonary trunk exits the right ventricle, and the aorta exits the left ventricle.

FIGURE 12-4 ■ Heart in the pericardium.
The heart is located in the pericardial sac, which consists of the outer fibrous pericardium and an inner serous pericardium. The serous pericardium has two parts: the parietal pericardium lines the fibrous pericardium and, the visceral pericardium (epicardium) lines the surface of the heart. The pericardial cavity, between the parietal and visceral pericardium, is filled with a small amount of pericardial fluid.

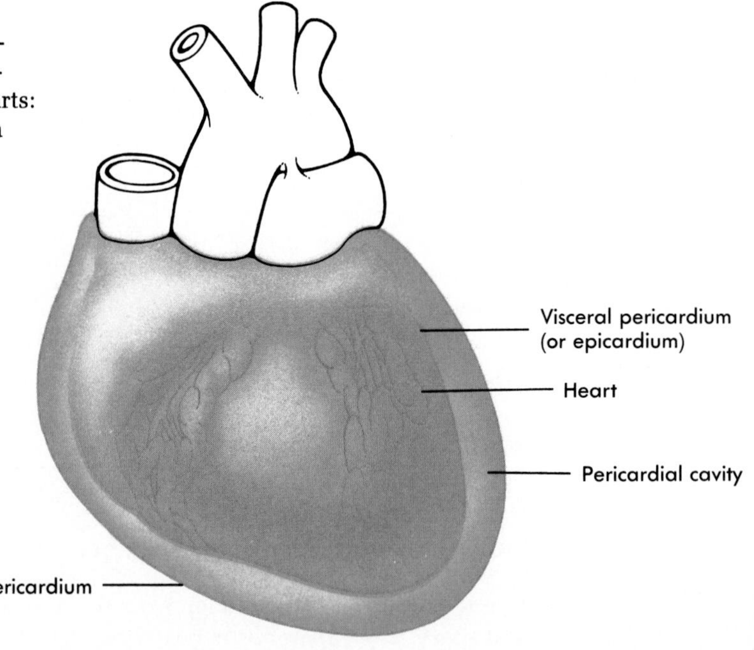

lining the fibrous pericardium is called the **parietal pericardium.** The portion of the serous pericardium covering the heart surface is called the **visceral pericardium,** or **epicardium** (ep'ĭ-kar'de-um). The parietal pericardium and visceral pericardium are continuous with each other where the great vessels enter or leave the heart. The pericardial cavity, located between the visceral and parietal pericardia, is filled with a thin layer of **pericardial fluid** that helps reduce friction as the heart moves within the pericardial sac.

External Anatomy

The thin-walled atria are located at the base of the heart, and the thick-walled ventricles extend from the base of the heart to the apex.

Six large veins carry blood to the heart: the **superior** and **inferior venae cavae** carry blood from the body to the right atrium; four **pulmonary veins** carry blood from the lungs to the left atrium. Two arteries, the **pulmonary trunk** and the **aorta,** exit the heart. The pulmonary trunk, arising from the right ventricle, splits into the right and left pulmonary arteries, which carry blood to the lungs. The aorta carries blood from the left ventricle to the body (see Figure 12-3).

A large **coronary** (kor'o-năr-e; circling like a crown) **sulcus** (sul'kus; ditch) runs around the heart, separating the atria from the ventricles (see Figure 12-3). In addition, two sulci extend inferiorly from the coronary sulcus, indicating the division between the right and left ventricles. The **anterior interventricular sulcus** is on the anterior surface of the heart, and the **posterior interventricular sulcus** is on the posterior surface of the heart.

The **coronary arteries** and their branches, which supply blood to the tissue of the heart, lie primarily within the coronary sulcus and interventricular sulci on the heart's surface. The right and left coronary arteries exit the aorta near the point where the aorta leaves the heart, and they lie within the coronary sulcus.

The major veins draining the tissue of the heart converge toward the posterior portion of the coronary sulcus and empty into a cavity called the **coronary sinus.** The coronary sinus, in turn, empties into the right atrium.

Heart Chambers

Right and left atria

The atria of the heart receive blood from veins. The atria function primarily as reservoirs where blood returning from veins collects before it enters the ventricles. Earlike projections called **auricles** extend from the margins of the atria. Contraction of the atria forces blood into the ventricles to complete ventricular filling. The right atrium has three major openings where veins enter the heart from various parts of the body: the superior vena cava, the inferior vena cava, and the coronary sinus. The left atrium has four openings that receive the four pulmonary veins. The two atria are separated from each other by the **interatrial septum.**

Right and left ventricles

The ventricles of the heart are the major pumping chambers of the heart. They eject blood into the arteries and force blood to flow through the circulatory system. The atria open into the ventricles. Each ventricle has one large outflow route located superiorly near the midline of the heart. The right ventricle opens into the pulmonary trunk, and the left ventricle opens into the aorta. The two ventricles are separated from each other by the muscular **interventricular septum** (Figure 12-5, *A*).

Heart Valves

Valves are located between the right atrium and right ventricle, and between the left atrium and left ventricle. These valves allow blood to flow from the atria into the ventricles but prevent blood from flowing back into the atria. Blood flowing from the atria into the ventricles pushes the valves open into the ventricles, but when the ventricles contract, blood pushes the valves back toward the atria. The atrioventricular openings close as the valve cusps meet. The valve between the right atrium and the right ventricle has three cusps and is called the **tricuspid valve.** The valve between the left atrium and left ventricle has two cusps and is called the **bicuspid,** or **mitral** (resembling a bishop's miter, a two-pointed hat), **valve** (see Figure 12-5, *A* and *B*).

Each ventricle contains cone-shaped muscular pillars called **papillary** (pap'ĭ-lĕr'e; nipple- or pimple-shaped) **muscles.** These muscles are attached by thin, strong connective tissue strings called **chordae tendineae** (kor'de ten'dĭ-ne; heart strings) to the cusps of the valves. When the ventricles contract the papillary muscles contract and prevent the valves from reopening into the atria by pulling on the chordae tendineae attached to the valve cusps.

The aorta and pulmonary trunk possess **aortic** and **pulmonary semilunar** (half moon–shaped) **valves** (see Figure 12-5, *B* and *C*). Each valve consists of three pocketlike semilunar cusps. Blood flowing out of the ventricles pushes against each valve, forcing it open. When blood flows back from the aorta or pulmonary trunk toward the ventricle,

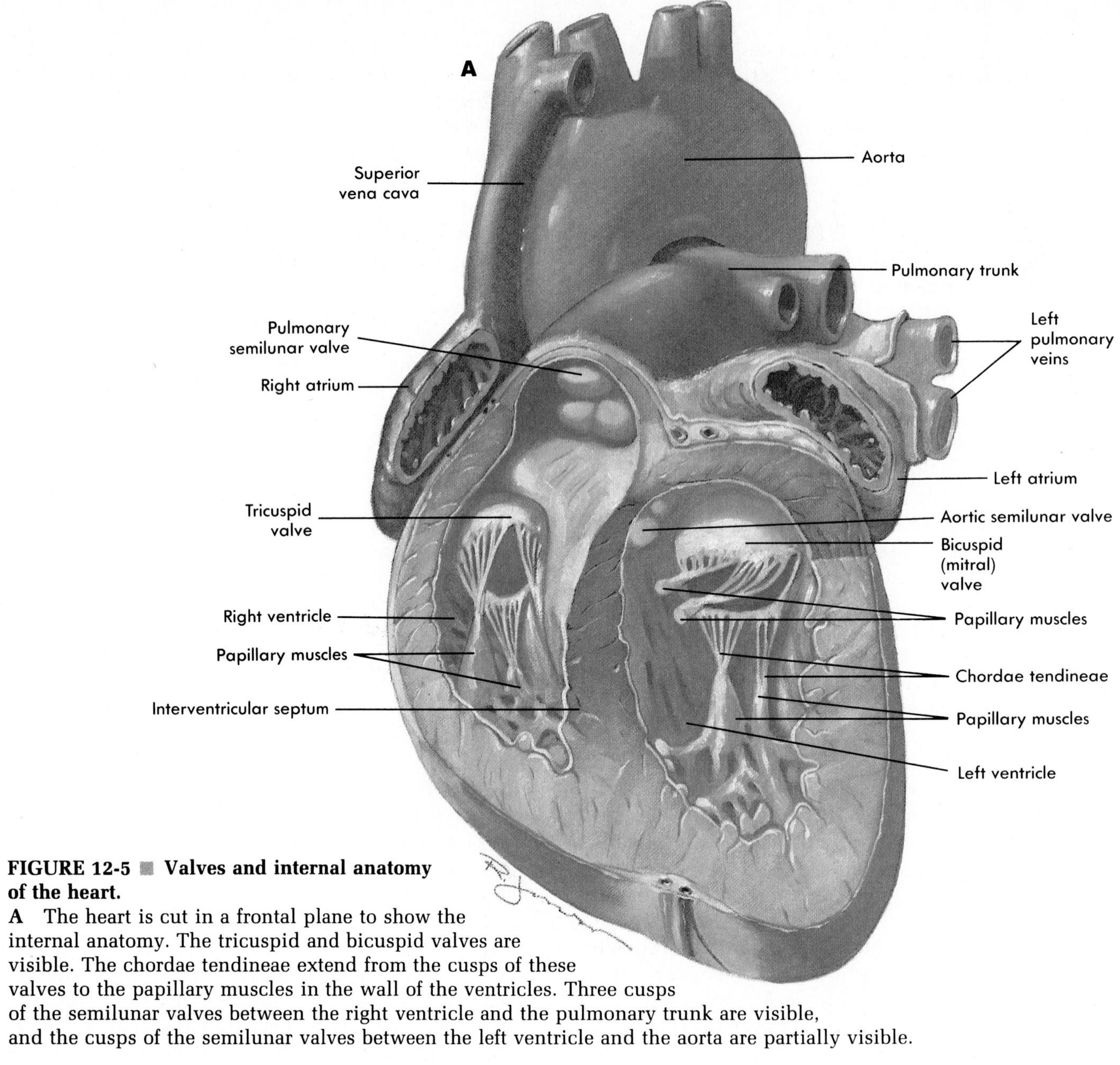

FIGURE 12-5 ■ Valves and internal anatomy of the heart.
A The heart is cut in a frontal plane to show the internal anatomy. The tricuspid and bicuspid valves are visible. The chordae tendineae extend from the cusps of these valves to the papillary muscles in the wall of the ventricles. Three cusps of the semilunar valves between the right ventricle and the pulmonary trunk are visible, and the cusps of the semilunar valves between the left ventricle and the aorta are partially visible.

it enters the pockets of the cusps, causing them to meet in the center of the aorta or pulmonary trunk, thus closing the vessels and keeping blood from flowing back into the ventricles.

Route of Blood Flow through the Heart

■ Blood flow through the heart is depicted in Figure 12-6, *A*. Even though it is more convenient to discuss blood flow through the heart one side at a time, it is important to understand that both atria contract at the same time, and both ventricles contract at the same time. This concept is particularly important when the electrical activity, pressure changes, and heart sounds are considered.

Blood enters the right atrium from the systemic circulation, which returns blood from all the tissues of the body. Blood flows from an area of higher pressure in the systemic circulation to the right atrium, which has a lower pressure. Most of the blood in the right atrium then passes into the right ventricle as the ventricle relaxes following the previous contraction. The right atrium then contracts, and the blood remaining in the atrium is pushed into the ventricle to complete right ventricular filling (Figure 12-6, *B*).

Contraction of the right ventricle pushes blood

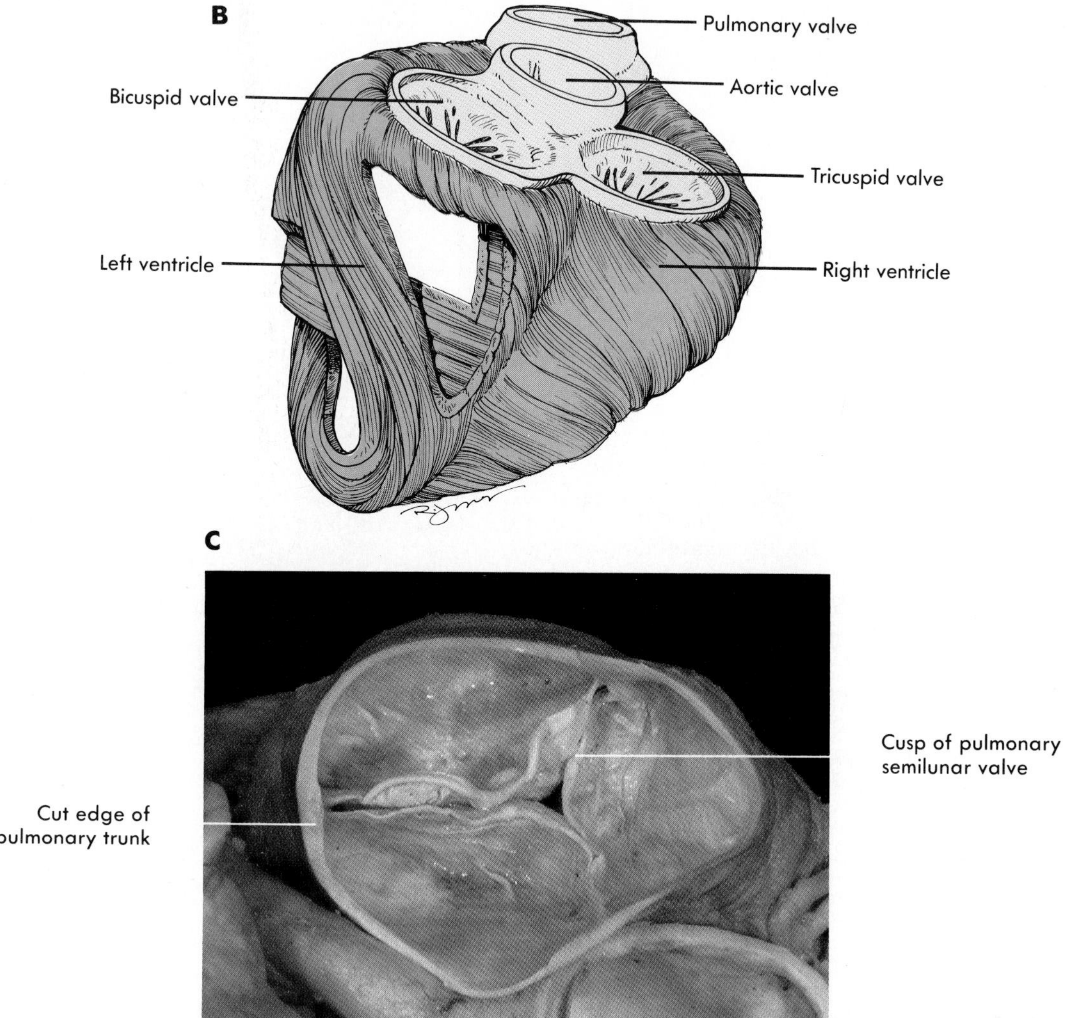

FIGURE 12-5, cont'd ■ Valves and internal anatomy of the heart.
B Posterior view of the heart with atria removed showing and relative positions of the bicuspid valve, aortic valve and pulmonary valve.
C Photograph of a pulmonary semilunar valve.

against the tricuspid valve, forcing it closed, and against the pulmonary semilunar valve, forcing it open, thus allowing blood to enter the pulmonary trunk.

The pulmonary trunk branches to form the pulmonary arteries, which carry blood to the lungs, where carbon dioxide is released and oxygen is picked up. Blood returning from the lungs enters the left atrium through the four pulmonary veins. The blood passing from the left atrium to the left ventricle opens the bicuspid valve, and contraction of the left atrium completes left-ventricular filling.

Contraction of the left ventricle pushes blood against the bicuspid valve, closing it, and against the aortic semilunar valve, opening it and allowing blood to enter the aorta. Blood flowing through the aorta is distributed to all parts of the body except the respiratory vessels in the lungs.

Heart Skeleton

A plate of fibrous connective tissue, sometimes called the **skeleton of the heart,** forms fibrous rings around the atrioventricular and semilunar valves and provides a solid support for them. This connective tissue plate also serves as electrical insulation between the atria and the ventricles and provides a rigid site of attachment for the cardiac muscle.

FIGURE 12-6 ■ Blood flow through the heart.
A Blood flows to the right atrium from systemic vessels (inferior and superior venae cavae); from the right atrium past the tricuspid valve to the right ventricle; from the right ventricle past the pulmonary semilunar valve to the pulmonary trunk; from the pulmonary trunk to the lungs where carbon dioxide is released and oxygen is picked up; from the lungs to the left atrium through four pulmonary veins; from the left atrium past the bicuspid valve to the left ventricle; from the left ventricle past the aortic semilunar valve to the aorta; and from the aorta to the remaining systemic vessels.
B During ventricular relaxation, the tricuspid and bicuspid valves are open and blood flows from the atria into the ventricles. Atrial contraction forces blood from the atria and completes ventricular filling. While the ventricles are relaxed, the semilunar valves (aortic and pulmonary) are closed.
C As the ventricles contract, the tricuspid and bicuspid valves close. After the pressure in the ventricles has increased until it exceeds the pressure in the pulmonary trunk and the aorta, the semilunar valves open, and blood is forced into the pulmonary trunk and aorta. When the ventricles begin to relax, the semilunar valves close, preventing blood from reentering the ventricles from the pulmonary trunk and aorta.

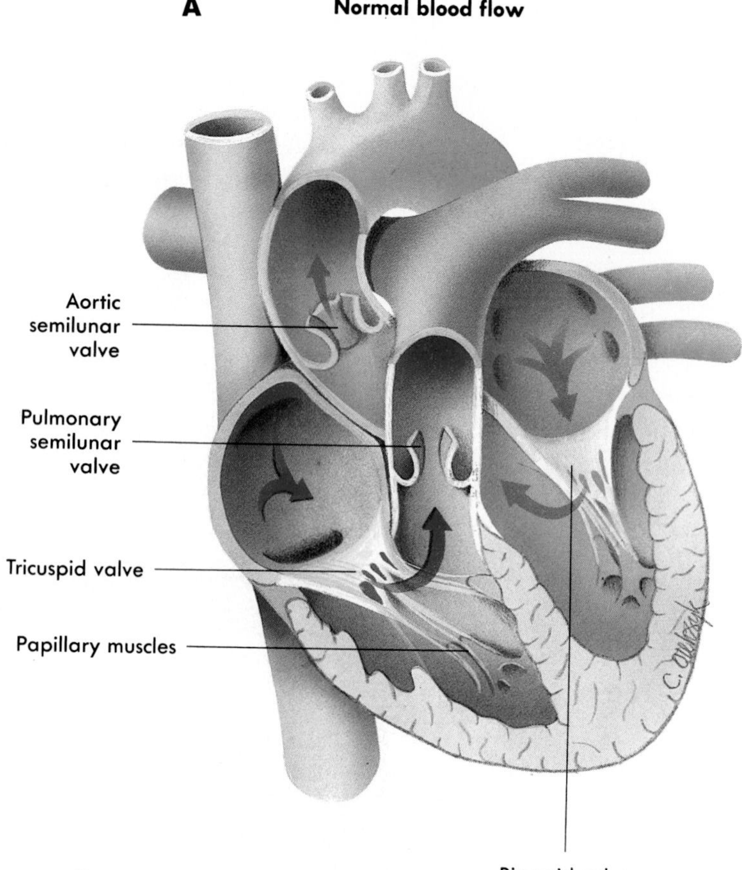

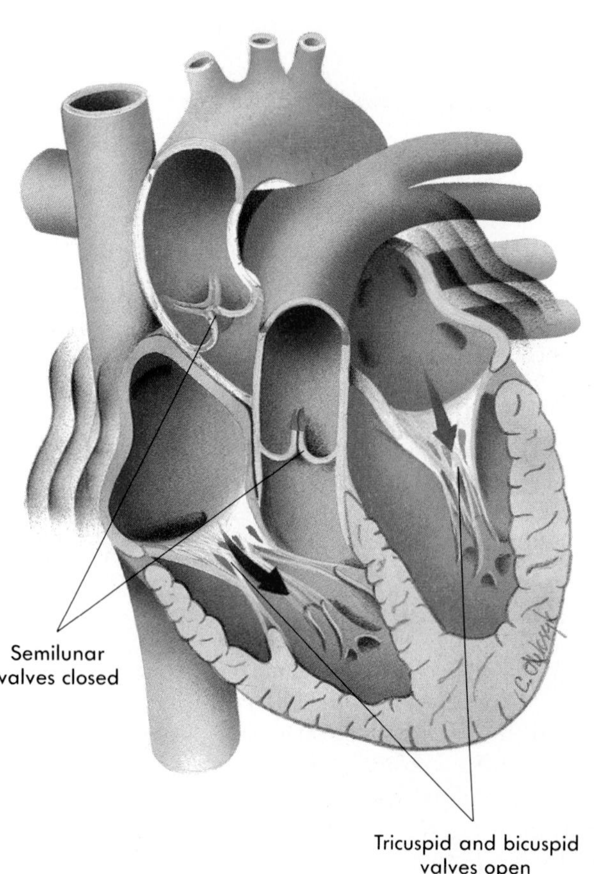

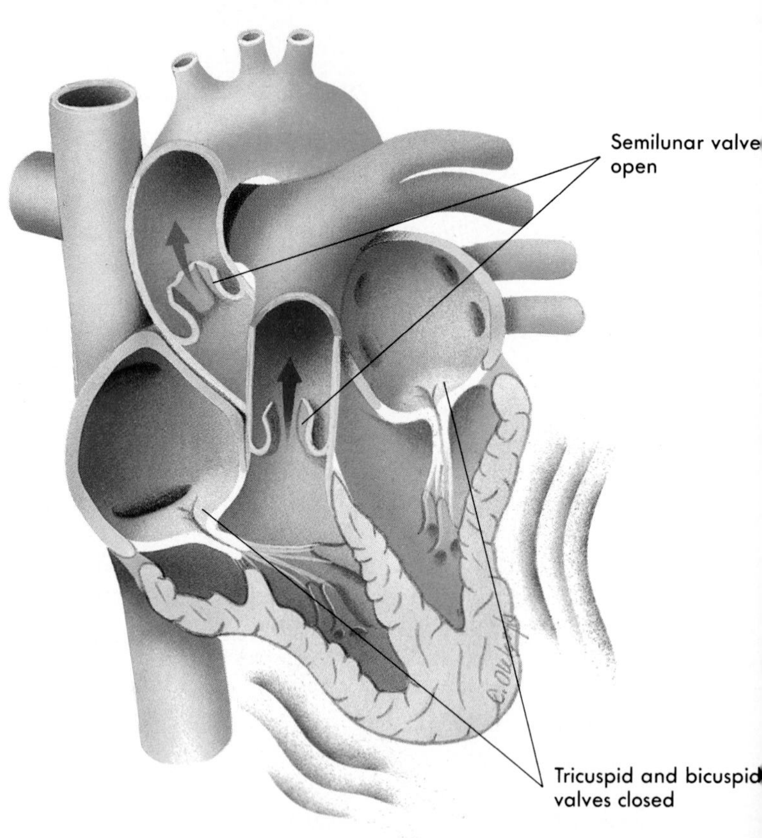

Heart Wall

The heart wall is composed of three layers of tissue: the epicardium, myocardium, and endocardium (Figure 12-7). The **epicardium** (ep′i-kar′de-um), also called the visceral pericardium, is a thin serous membrane forming the smooth outer surface of the heart. It consists of simple squamous epithelium overlying a layer of connective tissue. The thick middle layer of the heart, the **myocardium** (mi′o-kar′de-um), is composed of cardiac muscle cells and is responsible for the ability of the heart to contract. The smooth inner surface of the heart chambers is the **endocardium** (en′do-kar′de-um), which consists of simple squamous epithelium over a layer of connective tissue. The endocardium allows blood to move easily through the heart. The heart valves are formed by a fold of the endocardium, making a double layer of endocardium with connective tissue between the layers.

The interior walls of the ventricles are modified by ridges and columns called **trabeculae** (trah-bek′u-le; beams) **carneae** (kar′ne-a; meat). Smaller muscular ridges are found in portions of the atria.

Cardiac Muscle

Cardiac muscle cells are elongated, branching cells that contain one or occasionally two centrally located nuclei (Figure 12-8). Cardiac muscle cells also contain contractile proteins organized so that the cardiac muscle fibers appear striated. The striations are less regularly arranged and less numerous than in skeletal muscle.

Adenosine triphosphate (ATP) provides the energy for cardiac muscle contraction, and as in other

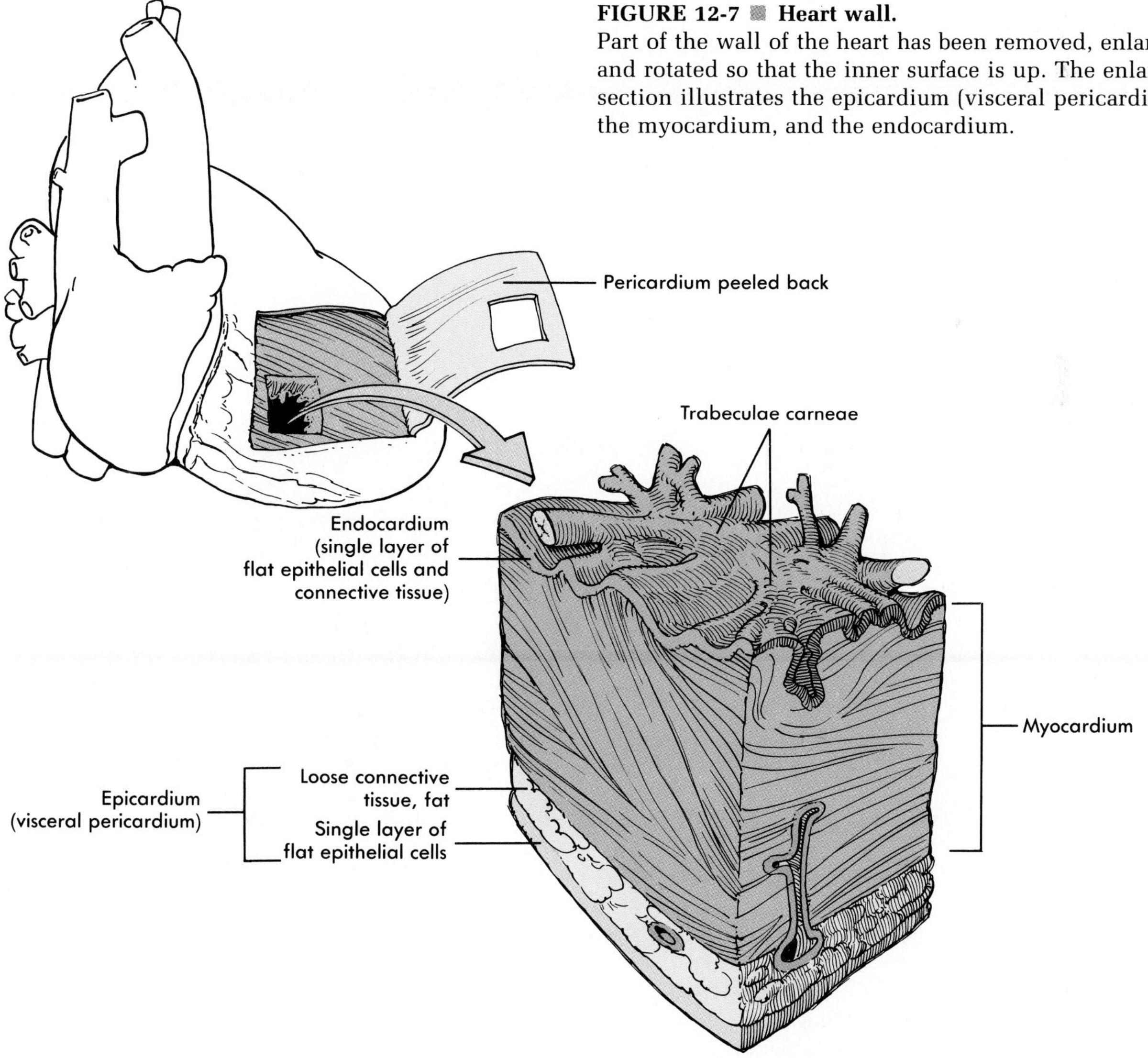

FIGURE 12-7 Heart wall.
Part of the wall of the heart has been removed, enlarged, and rotated so that the inner surface is up. The enlarged section illustrates the epicardium (visceral pericardium), the myocardium, and the endocardium.

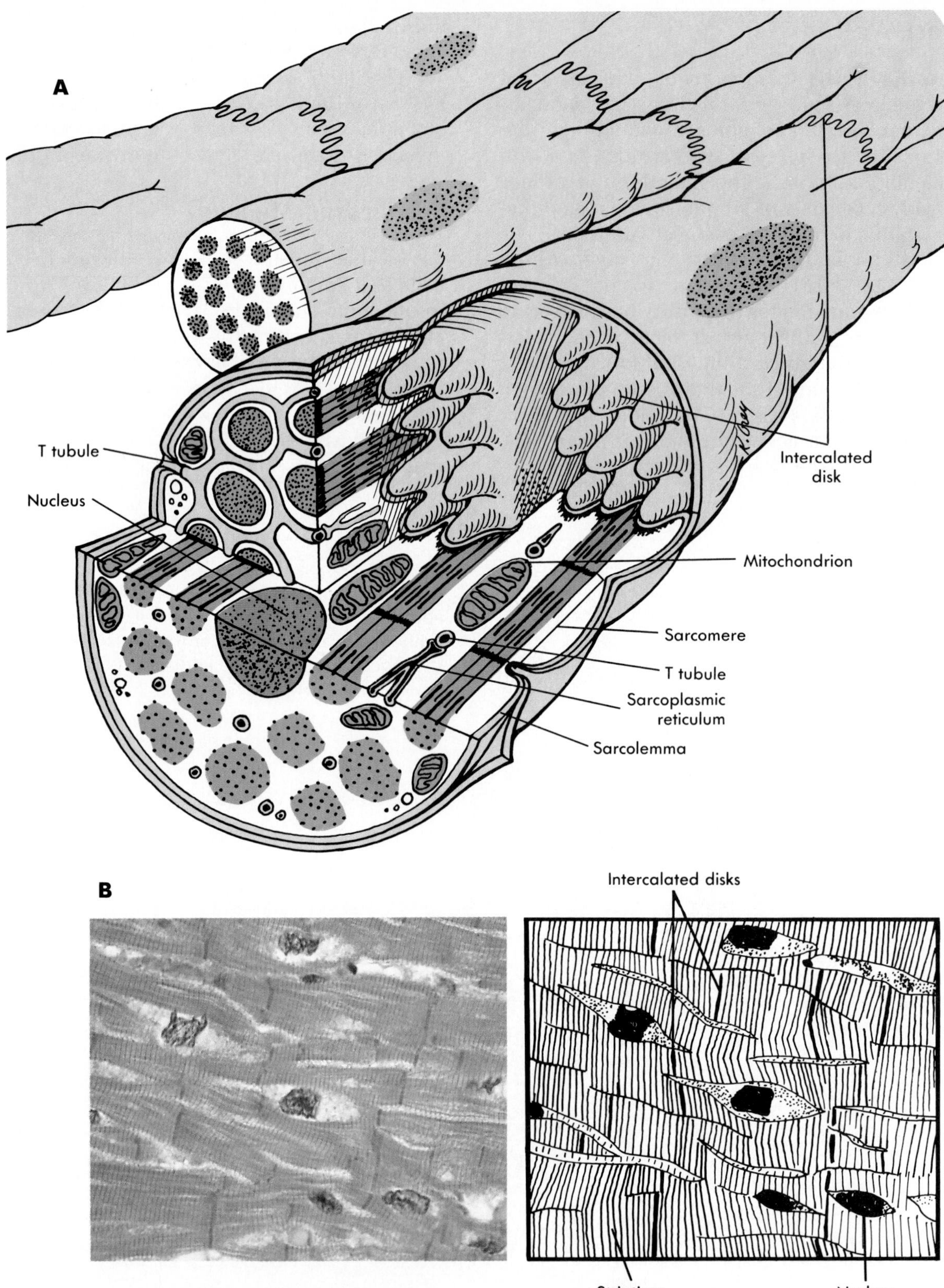

FIGURE 12-8 ■ Cardiac muscle cells.
A The structure and arrangement of individual muscle filaments make the muscle appear striated. The intercalated disks allow action potentials to pass from one cardiac muscle cell to the next.
B A light micrograph of cardiac muscle tissue. Within cardiac muscle cells, the nucleus is centrally located, and the sarcomeres, sarcoplasmic reticulum, and T tubules are not as numerous as they are in skeletal muscle fibers.

tissues, ATP production depends on oxygen availability. Cardiac muscle cells are rich in mitochondria, which produce ATP at a rate rapid enough to sustain the normal energy requirements of cardiac muscle. An extensive capillary network provides an adequate oxygen supply to the cardiac muscle cells. Unlike skeletal muscle, cardiac muscle cannot develop a significant oxygen debt. Development of an oxygen debt could result in muscular fatigue and cessation of cardiac muscle contraction.

Cardiac muscle cells are organized in spiral bundles or sheets. The cells are bound end-to-end and laterally to adjacent cells by specialized cell-to-cell contacts called **intercalated disks** (see Figure 12-8). The membranes of the disks are highly folded and the adjacent cells fit together, greatly increasing contact between them. Specialized cell membrane structures in the intercalated disks reduces electrical resistance between the cells, allowing action potentials to pass from one cell to adjacent cells. The cardiac muscle cells behave as a single electrical unit, and the highly coordinated contractions of the heart depend on this characteristic.

CONDUCTION SYSTEM OF THE HEART

The cycle of contraction and relaxation of the heart is called the **cardiac cycle**. The cardiac cycle is regulated by specialized cardiac muscle cells in the wall of the heart that form the **conduction system of the heart** (Figure 12-9). The **sinoatrial (SA)** (si′no-a′tre-al) **node,** which functions as the pacemaker of the heart, is located in the upper wall of the right atrium and initiates the contraction of the heart. Action potentials originate in the SA node and spread over the right and left atria causing them to contract.

A second area of the heart called the **atrioventricular (AV)** (a′tre-o-ven′trik′u-lar) **node** is located in the lower portion of the right atrium. When action potentials reach the AV node, they spread slowly through the AV node and into a bundle of specialized cardiac muscle called the **bundle of His** or **atrioventricular bundle**. The slow rate of action potential conduction in the AV node allows the

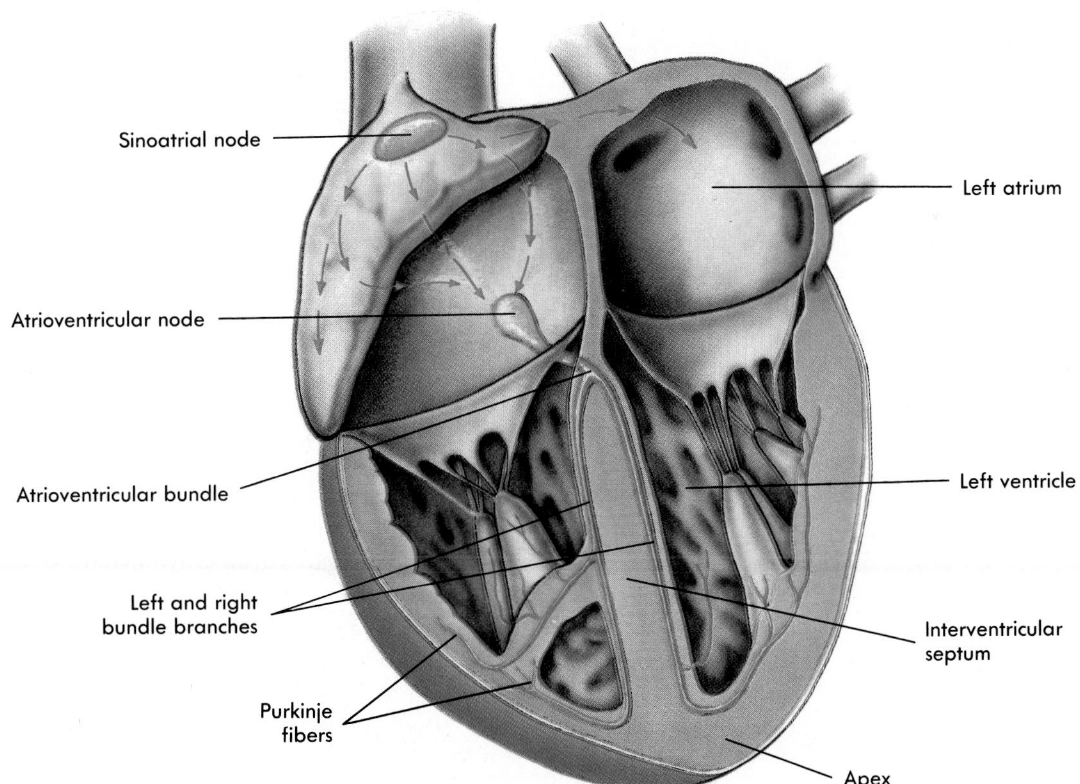

FIGURE 12-9 ■ Conducting system of the heart.
Impulses (*arrows*) travel across the wall of the right atrium from the SA node to the AV node. The atrioventricular bundle extends from the AV node to the interventricular septum, where it divides into right and left bundle branches. The bundle branches descend to the apex of each ventricle and then branch repeatedly to form Purkinje fibers, which are distributed throughout the ventricular walls.

atria to complete their contraction before action potentials are delivered to the ventricles.

After the action potentials pass through the AV node, they are transmitted rapidly through the atrioventricular bundle, which projects through the connective tissue separating the atria from the ventricles, to two branches of conducting tissue called the **left and right bundle branches**. The bundle branches are made up of specialized cardiac muscle fibers that conduct action potentials more rapidly than other cardiac muscle fibers. At the tips of the left and right bundle branches, the conducting tissue branches further to many small bundles of **Purkinje fibers**. These Purkinje fibers pass around the apex of the ventricle and extend to the cardiac muscle of the ventricle walls. The atrioventricular bundle, the bundle branches, and the Purkinje fibers rapidly deliver action potentials to all the cardiac muscle of the ventricles. The coordinated contraction of the ventricles depends on the rapid conduction of action potentials by the conduction system.

1 If blood supply were reduced in a small area of the heart through which the left bundle branch passed, predict the effect on ventricular contractions.

Following their contraction, the ventricles begin to relax. After the ventricles have completely relaxed another impulse originates in the SA node to begin the next cardiac cycle.

The SA node is the pacemaker of the heart because action potentials originate spontaneously in the SA node faster than in other areas of the heart. All cardiac muscle cells, however, are capable of producing action potentials spontaneously. For example, if the SA node is unable to function, another area of the heart such as the AV node becomes the pacemaker. The resulting heart rate would be much slower than normal. When action potentials originate in an area of the heart other than the SA node, the result is an **ectopic** (ek-top′ik) **beat**.

Cardiac muscle can also act as if there are thousands of pacemakers, each making a very small portion of the heart contract rapidly and independently of all other areas. This condition is called *fibrillation*, and it reduces the heart's output to only a few milliliters per minute when it occurs in the ventricles. Death of the individual results in a few minutes unless fibrillation of the ventricles is stopped. To stop the process of fibrillation, *defibrillation* is employed, in which a strong electrical shock is applied to the chest region. The electrical shock causes simultaneous depolarization of all cardiac muscle fibers. During depolarization the normal difference in the electrical charge across the cell membrane is reduced. Following depolarization, the SA node recovers and produces action potentials before any other area of the heart. Consequently, the normal pattern of action potential generation and the normal rhythm of contraction can be reestablished.

ELECTROCARDIOGRAM

Action potentials conducted through the heart during the cardiac cycle produce electric currents that can be measured at the surface of the body. Electrodes placed on the surface of the body and attached to a recording device can detect the small electric changes resulting from the action potentials in cardiac muscle. The record of these electrical events is an **electrocardiogram** (**ECG** or **EKG**) (Figure 12-10).

The ECG is not a direct measurement of cardiac muscle contractions and cannot be used to measure blood pressure. Each deflection in the ECG record results from an electrical event within the heart and correlates with a subsequent mechanical event. Analysis of an ECG can be used to identify abnormal heart rates or rhythms, abnormal conduction pathways, hypertrophy or atrophy of portions of the heart, and the approximate location of damaged cardiac muscle. Thus, the ECG is an extremely important diagnostic tool that is painless and easy to record, and does not require surgical procedures.

The normal ECG consists of a P wave, a QRS complex, and a T wave. The P wave results from depolarization of the atrial myocardium and precedes the onset of atrial contraction. The QRS complex consists of three individual waves: the Q, R, and S waves. The QRS complex results from depolarization of the ventricles and precedes ventricular contraction. The T wave represents repolarization of the ventricles and precedes ventricular relaxation. A wave representing repolarization of the atria cannot be seen because it occurs during the QRS complex.

The time between the beginning of the P wave and the beginning of the QRS complex is the P-Q interval, commonly called the P-R interval because the Q wave is very small. During the P-Q interval the atria contract and begin to relax. The ventricles begin to depolarize at the end of the P-Q interval.

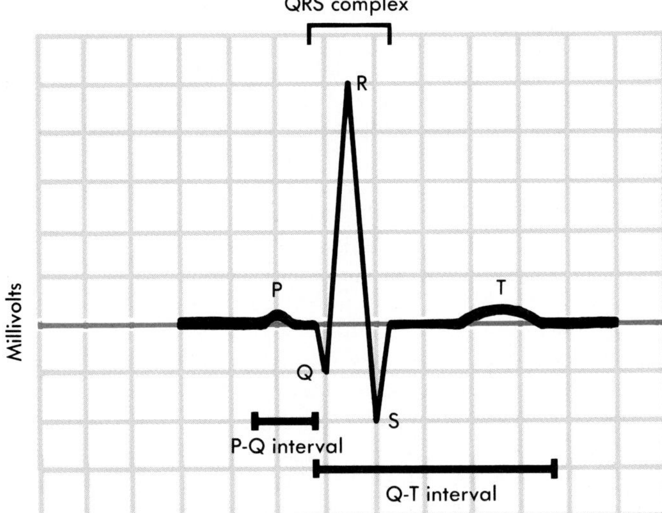

FIGURE 12-10 ■ Electrocardiogram. The major waves and intervals are labeled.

The Q-T interval extends from the beginning of the QRS complex to the end of the T wave and represents the length of time required for ventricular depolarization and repolarization. Table 12-1 describes several conditions associated with abnormal heart rhythms.

> **2 Predict the effect of the following conditions on the electrocardiogram:**
> **(A) Damage to one of the bundle branches, but not to the other.**
> **(B) Many ectopic action potentials arising in the atria.**

FUNCTIONAL CHARACTERISTICS OF THE HEART CHAMBERS

The pumping action of the heart depends on the alternating contraction of the two atria and the two ventricles. Atrial **systole** (sis′to-le; a contracting) refers to the contraction of the two atria. Ventricular systole refers to contraction of the two ventricles. Atrial **diastole** (di-as′to-le; dilation) refers to relaxation of the two atria, and ventricular diastole refers to relaxation of the two ventricles. When the terms systole and diastole are used alone they refer to ventricular contraction and relaxation because the ventricles contain more cardiac muscle and produce the pressure that forces the blood to circulate throughout the vessels of the body.

During ventricular systole, blood collects in the right and left atria. At the beginning of ventricular diastole, blood flows directly from the atria into the relaxed ventricles. When the atria contract, which occurs after the ventricles have been relaxed for a short time, both ventricles are already filled to about 70% of their volume. Atrial contraction forces additional blood to flow into the ventricles to complete their filling. After atrial systole is complete, ventricular systole begins. Almost immediately, the tricuspid and bicuspid valves close, and the pressure in the ventricles increases until it exceeds the pressure in the pulmonary trunk and aorta.

When the pressure in the ventricles exceeds the pressure in the pulmonary trunk and aorta, the pulmonary semilunar and the aortic semilunar valves are forced open, and blood is ejected into the pulmonary trunk and aorta.

During ventricular diastole, the pressure in the ventricles decreases and the pulmonary and aortic semilunar valves close so blood cannot flow back into the ventricles. The blood in the pulmonary trunk and aorta cannot flow back into the heart and must flow toward the lungs and the systemic circulation. The pressure continues to decline in the ventricles until finally the tricuspid and bicuspid valves open once again, and blood enters the ventricles from the atria.

> **3 Predict the effect of a leaky (incompetent) pulmonary semilunar valve on the volume of blood in the right ventricle just prior to ventricular contraction. Predict the effect of a severely narrowed opening through the aortic semilunar valves on the amount of work the heart must do in order to pump the normal volume of blood into the aorta during each beat of the heart.**

TABLE 12-1

Major Cardiac Arrhythmias

CONDITION	SYMPTOMS	POSSIBLE CAUSES
ABNORMAL HEART RHYTHMS		
Tachycardia	Heart rate in excess of 100 beats/min	Elevated body temperature, excessive sympathetic stimulation, toxic conditions
Bradycardia	Heart rate less than 60 beats/min	Elevated stroke volume in athletes, excessive vagus nerve stimulation, nonfunctional SA node (carotid sinus syndrome)
Sinus arrhythmia	Heart rate varies as much as 5% during respiratory cycle and up to 30% during deep respiration	Cause not always known; occasionally caused by ischemia, inflammation, or cardiac failure
Paroxysmal atrial tachycardia	Sudden increase in heart rate to 95-150 beats/min for a few seconds or even for several hours; P waves precede every QRS complex; P wave inverted and superimposed on T wave	Excessive sympathetic stimulation, abnormally elevated permeability of cardiac muscle to calcium ions
Atrial flutter	As many as 300 P waves/min and 125 QRS complexes/min; resulting in 2 or 3 P waves (atrial contractions) for every QRS complex (ventricular contraction)	Ectopic beats in the atria
Atrial fibrillation	No P waves, normal QRS and T waves, irregular timing, ventricles are constantly stimulated by atria, reduced ventricle filling; increased chance of fibrillation	Ectopic beats in the atria
Ventricular tachycardia	Frequently causes fibrillation	Often associated with damage to AV node or ventricular muscle
Ventricular fibrillation	No QRS complexes or T waves, no rhythmic contraction of myocardium, many patches of asynchronously contracting ventricular muscle	Ectopic beats in the ventricles
HEART BLOCKS		
SA node block	No P waves, low heart rate resulting from AV node acting as pacemaker, normal QRS complexes and T waves	Ischemia, tissue damage resulting from infarction; cause sometimes not known
AV node blocks		
First degree	P-R interval greater than 0.2 sec	Inflammation of AV bundle
Second degree	P-R interval 0.25 to 0.45 sec; some P waves trigger QRS complexes and others do not; examples of 2:1, 3:1, and 3:2 P wave/QRS complex ratios	Excessive vagus nerve stimulation, AV node damage
Complete heart block	P wave dissociated from QRS complex, atrial rhythm about 100 beats/min, ventricular rhythm less than 40 beats/min	Ischemia of AV node or compression of AV bundle
PREMATURE CONTRACTIONS		
Premature atrial contractions	Occasional shortened intervals between one contraction and the succeeding contraction; frequently occurs in healthy people	Excessive smoking, lack of sleep, too much coffee, alcoholism
Premature ventricular contractions (PVCs)	Prolonged QRS complex, exaggerated voltage because only one ventricle may depolarize, possible inverted T wave, increased probability of fibrillation	Ectopic beat in ventricles, lack of sleep, too much coffee, irritability; occasionally occurs with coronary thrombosis

HEART SOUNDS

A **stethoscope** (steth′o-skōp) is used to listen to heart sounds. There are two main heart sounds. The **first heart sound** can be represented by the syllable *lubb* and the **second heart sound** can be represented by *dupp*. The first heart sound has a lower pitch than the second heart sound. The first heart sound occurs at the beginning of ventricular systole and results from closure of the tricuspid and bicuspid valves. The second heart sound occurs at the beginning of ventricular diastole and results from closure of the semilunar valves. The valves usually do not make sounds when they open.

Ventricular systole occurs between the first and second heart sounds. Ventricular diastole occurs between the second heart sound and the first heart sound of the next beat. Because ventricular diastole lasts longer than ventricular systole, there is less time between the first and second heart sound than between the second heart sound and the first heart sound of the next beat.

> **4 Predict the rate of blood flow out of the ventricles between the first and second heart sounds of the same beat. Predict the rate of blood flow out of the ventricles between the second heart sound of one beat and the first heart sound of the next beat.**

Abnormal heart sounds called **murmurs** are usually a result of faulty valves. For example, if a valve fails to close tightly and blood leaks through the valve when it is closed, a murmur is heard. A murmur caused by a leaky or **incompetent valve** makes a swishing sound immediately after closure of the valve. For example, an incompetent bicuspid valve would result in a swishing sound immediately after the first heart sound.

When a valve is **stenosed**, that is, the opening of the valve is narrowed, a swishing sound precedes closure of the stenosed valve. For example, when the bicuspid valve is stenosed, a swishing sound precedes the first heart sound.

> **5 If normal heart sounds are represented by *lubb-dupp*, *lubb-dupp*, what would a heart sound represented by lubb-dupp*shhh*; lubb-dupp*shhh* represent? What would lubb-*shhh*dupp represent?**

BLOOD SUPPLY TO THE HEART

Coronary Arteries

Cardiac muscle in the wall of the heart is thick and metabolically very active. The coronary arteries supply blood to the wall of the heart (Figure 12-11, *A*). Two coronary arteries originate from the base of the aorta just above the aortic semilunar valves. The **left coronary artery** originates on the left side of the aorta. Its branches supply much of the anterior wall of the heart and most of the left ventricle. The **right coronary artery** originates on the right side and supplies most of the wall of the right ventricle.

Blood flow through the coronary arteries is greatest during ventricular diastole, when the cardiac muscle is relaxing, and the blood pressure in the aorta is still high enough to force blood to flow through the coronary vessels. Blood flow is reduced during ventricular contraction because the cardiac muscle compresses the coronary vessels.

When a blood clot or **thrombus** suddenly blocks a coronary blood vessel, a heart attack or **coronary thrombosis** occurs. The area that has been cut off from its blood supply suffers from a lack of oxygen and nutrients, and will die if the blood supply is not quickly reestablished. The region of dead heart tissue is called an **infarct** (in′farkt). If the infarct is large enough, the heart may be unable to pump enough blood to keep the person alive.

In some cases it is possible to treat heart attacks with an enzyme, such as streptokinase, that breaks down blood clots. The enzyme is injected into the circulatory system of a heart attack patient, where it reduces or removes the blockage in the coronary artery. If the clot is broken down quickly, the blood supply to cardiac muscle is reestablished and the heart may suffer little permanent damage.

Coronary arteries also may become blocked more gradually by **atherosclerotic** (ath′er-o-skle-rah′tik) **lesions.** Atherosclerotic lesions are thickenings in the walls of arteries that may contain deposits high in cholesterol and other lipids. The ability of cardiac muscle to function is reduced when it is deprived of an adequate blood supply. The person suffers from fatigue and often pain in the area of the chest and usually in the left arm with the slightest exertion. The pain is called **angina pectoris**.

Coronary Veins

The coronary veins drain blood from the cardiac muscle. They are nearly parallel to the coronary arteries and most carry blood from cardiac muscle to the coronary sinus. Blood flows from the coronary sinus into the right atrium (Figure 12-11, *B*).

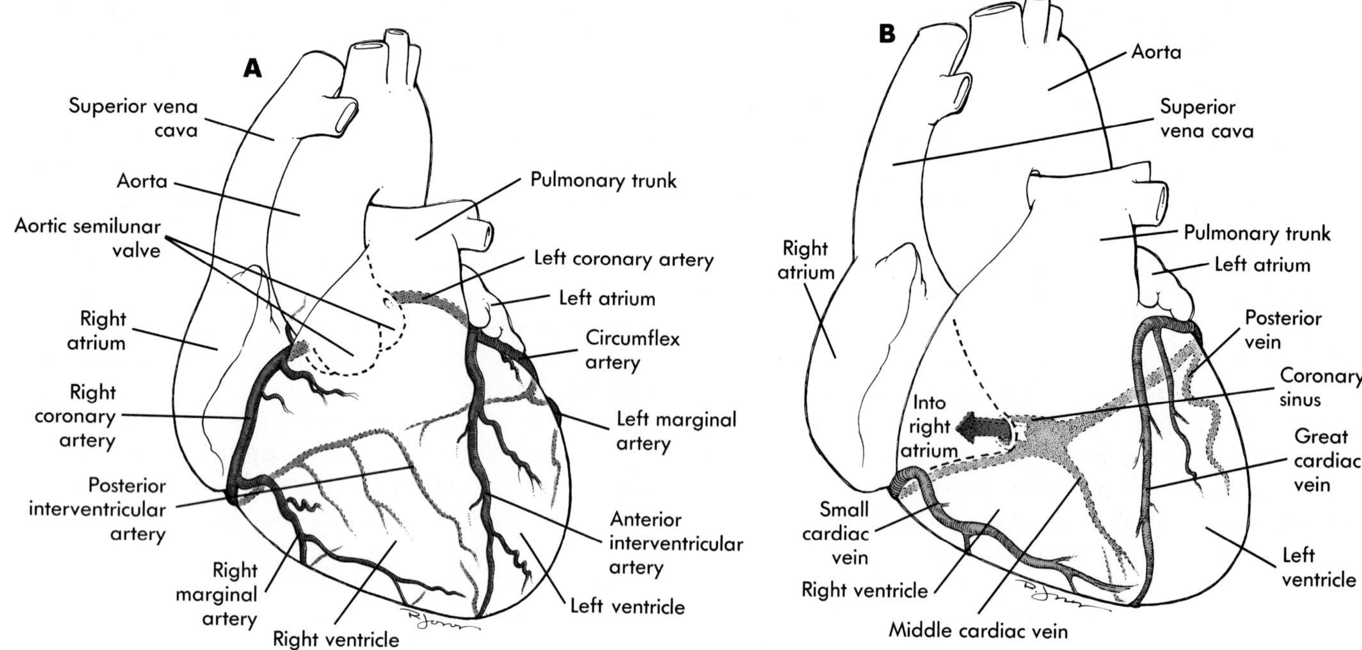

FIGURE 12-11 ■ Coronary vessels.
A Coronary arteries supply blood to the wall of the heart.
B Coronary veins carry blood from the wall of the heart back to the right atrium.
The vessels of the anterior surface are seen directly and have been given a darker color, whereas the vessels of the posterior surface are seen through the heart and have been given a lighter color.

REGULATION OF HEART FUNCTION

■ **Cardiac output** is the volume of blood pumped by either ventricle of the heart each minute. The cardiac output can be calculated by multiplying the **stroke volume** (the volume of blood pumped per ventricle each time the heart contracts), by the **heart rate** (the number of times the heart contracts each minute).

Cardiac Output	=	**Stroke Volume**	×	**Heart Rate**
ml/min		ml/beat		beats/min

There are a number of control mechanisms that modify heart rate and stroke volume.

Intrinsic Regulation of the Heart

■ **Intrinsic regulation** of the heart refers to these mechanisms contained within the heart itself and not due to either hormonal or nervous influences. **Venous return** is the amount of blood that returns to the heart. If venous return increases, the heart fills to a greater volume, which stretches the cardiac muscle fibers. In response to stretch, cardiac muscle fibers contract with a greater force. The greater force of contraction causes an increased volume of blood to be ejected from the heart, resulting in an increased stroke volume. In addition to increasing the stroke volume, stretch of cardiac muscle will cause a slight increase in the heart rate. Therefore, as venous return increases, cardiac output increases. Conversely, if venous return decreases, the cardiac output decreases. This relationship is called **Starling's law of the heart.**

■ Heart failure results from a progressive weakening of the heart muscle. In a heart failure, the heart is not capable of pumping all the blood that is returned to it. Consequently, blood backs up in the veins. For example, heart failure that affects the right ventricle, right heart failure, causes blood to back up in the veins returning blood from systemic vessels to the heart. Filling of the veins with blood causes edema, especially in the legs and feet. Edema results from the accumulation of fluid in tissues outside of blood vessels. Heart failure that affects the left ventricle, left heart failure, causes blood to back up in the veins returning blood from the lungs to the heart. Filling of these veins causes edema in the lungs. In those who are suffering from heart failure, further stretching of the cardiac muscle fibers does not increase the stroke volume of the heart. ■

Because venous return is influenced by many conditions, Starling's law of the heart has a major influence on cardiac output. For example, muscular activity during exercise causes increased venous return, and cardiac output increases. This is beneficial because an increased cardiac output is needed during exercise to supply oxygen to exercising skeletal muscles.

Extrinsic Regulation of the Heart

Extrinsic regulation of the heart refers to both nervous and hormonal regulation of the heart. Nervous influences are carried through the autonomic nervous system. Both sympathetic and parasympathetic nerve fibers innervate the SA node of the heart causing the heart rate to increase or decrease. The sympathetic nerve fibers cause the heart rate and the stroke volume to increase, whereas the parasympathetic nerve fibers cause the heart rate to decrease.

The **baroreceptor** (băr′o-re-sep′tor) **reflex** plays an important role in regulating the function of the heart. Baroreceptors are stretch receptors that monitor blood pressure in the aorta and in the wall of the internal carotid arteries, which carry blood to the brain (Figure 12-12). As blood pressure increases, the walls of these arteries are stretched, increasing the frequency of action potentials from the stretch receptors. When the blood pressure declines, there is a decrease in the frequency of action potential impulses from the stretch receptors. The action potentials are transmitted along nerve fibers from the stretch receptors to the medulla of the brain.

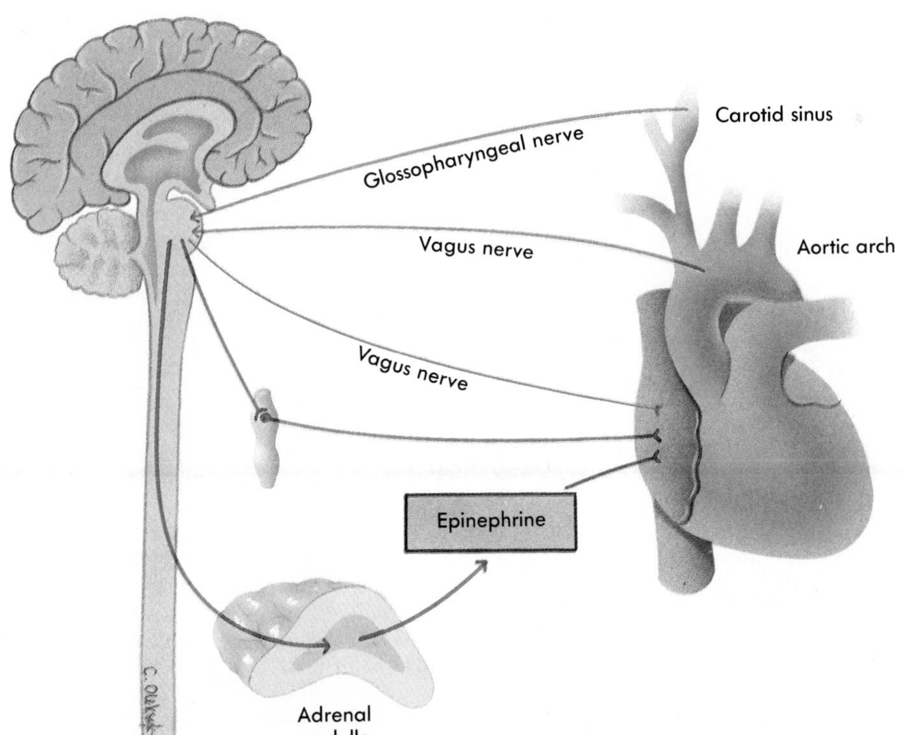

FIGURE 12-12 ■ Extrinsic regulation of the heart.
Sympathetic (*green*) and parasympathetic (*red*) nerves exit the spinal cord or medulla and extend to the heart to regulate its function. Hormonal influences such as epinephrine from the adrenal gland also help regulate the heart's action. Action potentials from baroreceptors in the aortic arch and carotid arteries are sent along nerve fibers to the medulla oblongata of the brain.

Conditions and Diseases Affecting the Heart

HEART DISEASES

Endocarditis is inflammation of the lining of the heart cavities. It affects the valves more frequently than other areas of the heart and may lead to deposition of scar tissue that causes valves to become stenosed or incompetent.

Myocarditis is inflammation of the heart muscle and can lead to heart failure.

Pericarditis is inflammation of the serous membrane of the pericardium. Pericarditis can result from bacterial or viral infections and can be extremely painful.

Rheumatic heart disease may result from a **streptoccocal infection** in young people. Toxin produced by the bacteria may cause an immune reaction called rheumatic fever about 2 to 4 weeks after the infection. The immune reaction may cause inflammation of the endocardium, called rheumatic endocarditis. The inflamed valves, especially the bicuspid valve, may become stenosed or incompetent. The effective treatment of streptococcal infections with antibiotics has reduced the frequency of rheumatic heart disease.

Coronary heart disease reduces the amount of blood that the coronary arteries are able to deliver to the myocardium. The reduction in blood flow damages the myocardium. The degree of damage depends on the size of the arteries involved, whether occlusion is partial or complete, and whether occlusion is gradual or sudden. As the walls of the arteries thicken and harden with age, the volume of blood they supply to the heart muscle declines, and the ability of the heart to pump blood decreases. Inadequate blood flow to the heart muscle may result in **angina pectoris,** which is a poorly localized sensation of pain in the region of the chest, left arm, and left shoulder.

Degenerative changes in the artery wall also may cause the inside surface of the artery to become roughened. The chance of platelet aggregation increases at the rough surface. Platelet aggregation increases the chance of coronary thrombosis. The outcome of coronary thrombosis depends on the extent of the damage and whether other blood vessels can supply enough blood to maintain the heart's function. Death may occur swiftly if the infarct is large; if the infarct is small, the heart may continue to function. In some cases the infarct weakens the wall of the heart and the wall ruptures, but in most cases scar tissue replaces damaged cardiac muscle in the area of the infarct.

People who survive infarctions often can lead a fairly normal life if they take precautions. Most cases call for moderate exercise, adequate rest, a disciplined diet, and reduced stress.

Congenital heart disease is the result of abnormal development of the heart. The following conditions are common congenital defects:

1. **Septal defect** is a hole in one of the septums between the left and right sides of the heart. The hole may be in the interatrial or interventricular septum. These defects allow blood to flow from one side of the heart to the other and as a consequence greatly reduce the pumping effectiveness of the heart.
2. **Patient ductus arteriosus** results when a blood vessel called the **ductus arteriosus,** which is present in the fetus, fails to close after birth. The ductus arteriosus extends between the pulmonary trunk and the aorta. It allows blood to pass from the pulmonary trunk to the aorta, thus bypassing the lungs. This is normal before birth because the lungs are not functioning. If the ductus arteriosus fails to close after birth, blood flows in the opposite direction, from the aorta to the pulmonary trunk. As a consequence blood flows through the lungs under high pressure and damages them. In addition, the amount of work required of the left ventricle to maintain an adequate systemic blood pressure is increased.
3. **Aortic** and **pulmonary stenosis** is a narrowed vessel or valve. In these cases the workload of the heart is increased because the ventricles must contract with a much greater force to pump blood from the ventricles. **Stenosis of the bicuspid valve** prevents the flow of blood into the left ventricle, causing blood to back up in the left atrium and in the lungs, resulting in congestion of the lungs. **Stenosis of the tricuspid valve** causes blood to back up in the right atrium and systemic veins, causing swelling in the periphery.
4. **Cyanosis** is a symptom of inadequate heart function in babies suffering from congenital heart disease. The blueness of the skin is caused by low oxygen levels in the blood in peripheral blood vessels. The term "blue baby" is sometimes used to refer to infants with cyanosis.

Heart failure is the result of progressive weakening of the heart muscle and the failure of the heart to pump blood effectively. Hypertension (high blood pressure) may cause an enlargement of the heart and may finally result in heart failure. Malnutrition, chronic infections, toxins, severe anemias, or hyperthyroidism may cause degeneration of the heart muscle. Heredity may be responsible for increased susceptibility to heart failure.

Heart function in the elderly becomes less efficient. Although the age at which the heart becomes less efficient varies considerably and depends on many factors, by the age of 70 there is often a decrease in cardiac output of about one third. Because of the decrease in reserve strength of the heart, many elderly people are often limited in their ability to respond to emergencies, infections, blood loss, or stress.

PREVENTION OF HEART DISEASE

Proper nutrition is important in reducing the risk of heart disease. A recommended diet is low in fats, especially saturated fats and cholesterol, and low in refined sugar. Diets should be high in fiber, whole grains, fruits, and vegetables. Total food intake should be limited to avoid obesity, and sodium chloride intake reduced.

Tobacco and excessive use of alcohol should be avoided. Smoking increases the risk of heart disease by at least tenfold, and excessive use of alcohol also substantially increases the risk of heart disease.

Chronic stress, frequent emotional upsets, and a lack of physical exercise may increase the risk of cardiovascular disease. Remedies include relaxation techniques and aerobic exercise programs involving gradual increases in duration and difficulty in activities such as swimming, jogging, or aerobic dancing.

Hypertension (abnormally high systemic blood pressure) affects about one fifth of the population. Regular blood pressure measurements are important, because hypertension does not produce obvious symptoms. If hypertension cannot be controlled by diet and exercise, it is important to treat the condition with prescribed drugs. The cause of hypertension in the majority of cases is unknown.

Some data suggest that taking a small dose of aspirin regularly will reduce the chance of a heart attack in males. Aspirin inhibits the synthesis of prostaglandins in platelets, which, in turn, can help avoid clot formation.

HEART MEDICATIONS AND TREATMENTS

Digitalis: Slows and strengthens contractions of the heart muscle.

Nitroglycerin: Dilates coronary blood vessels and increases the blood flow to cardiac muscle.

Beta-adrenergic blocking agents: Reduce the rate and strength of cardiac muscle contractions thus reducing the heart's oxygen demand. Beta-adrenergic blocking agents are often used to treat people who suffer from rapid heart rates and certain types of arrhythmias.

Calcium channel blockers: Reduce the rate at which calcium ions diffuse into cardiac muscle cells. Because the action potentials that produce cardiac muscle contractions depend in part on the flow of calcium ions into the cardiac muscle cells, the calcium channel blockers can be used to control the force of heart contractions, and reduce arrhythmia, tachycardia, and hypertension.

Anticoagulants: Prevent clot formation in persons with damage to heart valves or blood vessels or in persons who have had a myocardial infarction. Aspirin functions as a weak anticoagulant. One aspirin each day may benefit those who are likely to experience a coronary thrombosis.

Artificial pacemaker: An instrument placed beneath the skin that has an electrode that extends to the heart. An artificial pacemaker provides an electrical stimulus to the heart at a set frequency. Artificial pacemakers are used in patients in whom the natural pacemaker of the heart does not produce a heart rate high enough to sustain normal physical activity.

Heart lung machine: A machine that serves as a temporary substitute for the patient's heart and lungs. It pumps blood throughout the body and oxygenates and removes carbon dioxide from the blood. It has made possible many surgeries on the heart and lungs.

Coronary bypass surgery: A surgical procedure that relieves the effects of obstructions in the coronary arteries. The technique involves taking healthy segments of blood vessels from other parts of the patient's body and using them to bypass obstructions in the coronary arteries. The technique is common for those who suffer from severe occlusion of the coronary arteries.

Angioplasty: The process whereby a small balloon is threaded into the aorta and allowed to enter the coronary arteries. After the balloon has entered coronary arteries that are partially occluded, the balloon is inflated to open the occluded blood vessel. This technique is useful in improving the function of cardiac muscle in patients suffering from an inadequate blood flow to cardiac muscle through the coronary arteries.

Heart valve replacement or repair: A surgical procedure performed on those who have diseased valves that are so deformed and scarred from conditions such as endocarditis that the valves are severely incompetent or stenosed. Substitute valves made of synthetic materials such as plastic or Dacron are effective; valves transplanted from pigs are also used.

Heart transplants: Heart transplants are possible when the immune characteristics of a donor and the recipient are closely matched. The heart of a recently deceased donor is transplanted to the recipient, and the diseased heart of the recipient is removed. People who have received heart transplants must remain on drugs that suppress their immune responses for the rest of their lives. Unless they do so, their immune system will reject the transplanted heart.

Artificial hearts: Artificial hearts have been used on an experimental basis to extend the lives of individuals until an acceptable transplant can be found, or to replace the heart permanently. The technology currently available for artificial hearts has not yet reached the point at which a high quality of life can be achieved.

Within the medulla of the brain there is a **cardioregulatory center**, which receives and integrates action potentials from the baroreceptors. As blood pressure declines in the aorta, the lower frequency of action potentials sent to the medulla triggers a response in the cardioregulatory center. The cardioregulatory center responds by increasing sympathetic stimulation of the heart and decreasing parasympathetic stimulation. Consequently, the heart rate and stroke volume increase. These responses cause the blood pressure to increase toward its normal value.

When the blood pressure increases, the baroreceptors are stimulated. An increased frequency of action potentials, sent along the nerve fibers to the medulla of the brain, prompts the cardioregulatory center to increase parasympathetic stimulation of the heart and decrease sympathetic stimulation. As a result, the heart rate and stroke volume decrease, causing blood pressure to decline.

6 In response to a severe hemorrhage, blood pressure is low, the heart rate is increased dramatically, and the stroke volume is low. If low blood pressure activates a reflex that increases sympathetic stimulation of the heart, why is the stroke volume low? ?

Emotions integrated in the cerebrum of the brain can influence the heart. Excitement, anxiety, or anger can cause increased sympathetic stimulation of the heart and an increased cardiac output. Depression, on the other hand, may increase parasympathetic stimulation of the heart, causing a slight reduction in cardiac output.

Epinephrine released from the adrenal medulla in response to emotional excitement or stress also influences the heart's function (see Figure 12-12). Epinephrine binds to receptor molecules on cardiac muscle and causes an increased heart rate and stroke volume. Thus, an increased heart rate and blood pressure in response to emotional excitement are the results of both increased sympathetic stimulation of the heart and epinephrine secreted from the adrenal medulla.

The medulla oblongata of the brain also contains chemoreceptors that are sensitive to changes in pH and carbon dioxide levels. A decrease in pH and an increase in carbon dioxide result in sympathetic stimulation of the heart. Decreased blood oxygen levels also increase sympathetic stimulation of the heart, but the sensory receptors for this response are located in specialized structures found near the carotid arteries and aortic arch (see Chapter 13).

Potassium, calcium, and sodium ions, which influence other electrically excitable tissues, also affect cardiac muscle function. For example, excess potassium ions decrease the heart rate and stroke volume, and low blood calcium levels may increase the heart rate.

Body temperature affects metabolism in the heart much as it affects other tissues. Elevated body temperature increases the heart rate, and reduced body temperature slows the heart rate. For example, during fever the heart rate is usually elevated, and during heart surgery the body temperature is sometimes intentionally reduced to slow the heart rate and metabolism in the body.

SUMMARY

The heart pumps blood through the blood vessels of the body.

SIZE, FORM, AND LOCATION OF THE HEART

The heart is about the size of a fist and is located in the pericardial cavity.

The apex is inferior to the base of the heart.

ANATOMY OF THE HEART

Pericardium

The pericardial sac consists of a fibrous pericardium and is lined by the parietal pericardium.

The outer surface of the heart is lined by the visceral pericardium (epicardium).

Between the visceral and parietal pericardium is the pericardial cavity, which is filled with pericardial fluid.

External Anatomy

The inferior and superior vena cavae and the coronary sinus enter the right atrium. The four pulmonary veins enter the left atrium.

The pulmonary trunk exits the right ventricle, and the aorta exits the left ventricle.

Atria are separated externally from the ventricles by the coronary sulcus. The right and left ventricles are separated externally by the interventricular grooves.

Coronary arteries branch off the aorta to supply the heart. Blood returns from heart tissue to the right atrium through the coronary sinus.

Heart Chambers

The atria are separated internally from each other by the interatrial septum, and the ventricles are separated internally by the interventricular septum.

Heart Valves

The heart valves ensure one-way flow of blood.

The tricuspid valve (three cusps) separates the right atrium and right ventricle, and the bicuspid valve (two cusps) separates the left atrium and left ventricle.

The papillary muscles attach by the chordae tendineae to the cusps of the tricuspid and bicuspid valves.

The aorta and pulmonary trunk are separated from the ventricles by the semilunar valves.

Route of Blood Flow through the Heart

The left and right sides of the heart can be considered as separate pumps.

Blood flows from the systemic vessels to the right atrium, and from the right atrium to the right ventricle. From the right ventricle blood flows to the pulmonary trunk, and from the pulmonary trunk to the lungs. From the lungs blood flows through the pulmonary veins to the left atrium, and from the left atrium blood flows to the left ventricle. From the left ventricle blood flows into the aorta and then through the systemic vessels.

Heart Skeleton

A plate of fibrous connective tissue separates the atria from the ventricles, acts as an electrical barrier between the atria and ventricles, and supports the valves of the heart.

Heart Wall

The heart wall consists of the outer epicardium, the middle myocardium, and the inner endocardium.

Cardiac Muscle

Cardiac muscle is striated and depends on ATP for energy. It depends on aerobic metabolism.

Cardiac muscle cells are joined by intercalated discs that allow action potentials to be propagated throughout the heart.

CONDUCTION SYSTEM OF THE HEART

The conduction system of the heart is made up of specialized cardiac muscle cells.

The SA node located in the upper wall of the right atrium is the normal pacemaker of the heart.

The AV node and atrioventricular bundle conduct action potentials to the ventricles.

The right and left bundle branches conduct action potentials from the atrioventricular bundle through Purkinje fibers to the ventricular muscle.

An ectopic beat results from an action potential that originates in an area of the heart other than the SA node.

THE ELECTROCARDIOGRAM

The ECG is a record of electrical events within the heart.

The ECG can be used to detect abnormal heart rates or rhythms, conduction pathways, hypertrophy or atrophy of the heart, and the approximate location of damaged cardiac muscle.

The normal ECG consists of a P wave (atrial depolarization), the QRS complex (ventricular depolarization), and the T wave (ventricular repolarization).

Atrial contraction occurs during the P-Q interval, and the ventricles contract and relax during the Q-T interval.

FUNCTIONAL CHARACTERISTICS OF THE HEART CHAMBERS

Atrial systole is contraction of the atria, and ventricular systole is contraction of the ventricles. Atrial diastole is relaxation of the atria, and ventricular diastole is relaxation of the ventricles.

During atrial systole, filling of the right ventricle is completed.

Ventricular systole.

- The tricuspid valve closes, and blood forces open the pulmonary semilunar valve; blood flows into the pulmonary trunk.
- The bicuspid valve closes, and blood forces open the aortic semilunar valve; blood flows into the aorta.

HEART SOUNDS

The first heart sound results from closure of the tricuspid and bicuspid valves. The second heart sound results from closure of the aortic and pulmonary semilunar valves.

Abnormal heart sounds are called murmurs. They can result from leaky (incompetent) valves or narrowed (stenosed) valves.

BLOOD SUPPLY TO THE HEART

The left and right coronary arteries originate from the base of the aorta and supply blood to the wall of the heart.

Blood flow through the coronary arteries is greatest during ventricular diastole.

Coronary thrombosis results from a clot that blocks a coronary artery and produces an area that has no blood supply (an infarct).

Atherosclerotic lesions may narrow the coronary arteries and restrict blood supply to cardiac muscle.

Coronary veins are nearly parallel to the coronary arteries and carry blood from cardiac muscle to the right atrium.

REGULATION OF HEART FUNCTION

Cardiac output (volume of blood pumped per ventricle per minute) is equal to the heart rate (beats per minute) times the stroke volume (volume of blood ejected per beat).

Intrinsic Regulation of the Heart

Intrinsic regulation refers to regulation that is contained in the heart.

As venous return to the heart increases the heart wall is stretched, which increases stroke volume (Starling's law of the heart).

Extrinsic Regulation of the Heart

Extrinsic regulation refers to nervous and hormonal mechanisms.

Sympathetic stimulation increases stroke volume and heart rate; parasympathetic stimulation decreases heart rate.

The baroreceptor reflex detects changes in blood pressure and causes a decrease in heart rate and stroke volume in response to a sudden increase in blood pressure or an increase in heart rate and stroke volume in response to a sudden decrease in blood pressure.

Emotions influence heart function by increasing sympathetic stimulation of the heart in response to excitement, anxiety, or anger and by increasing parasympathetic stimulation in response to depression.

Alterations in body fluid levels of carbon dioxide, oxygen, pH, and ion concentrations, and body temperature influence heart function.

CONTENT REVIEW

1. Describe the size and location of the heart including its base and apex.
2. Describe the structure and function of the pericardium.
3. What chambers make up the left side and the right side of the heart? What are their functions?
4. Describe the structure and location of the tricuspid, bicuspid, and semilunar valves. What is the function of these values?
5. What are the functions of the atria and ventricles?
6. Starting in the right atrium, describe the flow of blood through the heart.
7. Describe the three layers of the heart and state their functions.
8. Describe the forces that cause blood to flow through the right and left side of the heart during atrial diastole, atrial systole, ventricular diastole, and ventricular systole.
9. What is the function of the conduction system of the heart? Starting with the SA node, describe the route taken by an action potential as it goes through the conduction system of the heart.
10. Explain the electrical events that generate each portion of the electrocardiogram. How do they relate to contraction events?
11. What contraction events occur during the P-Q interval and the Q-T interval of the electrocardiogram?
12. Define systole and diastole.
13. What events cause the first and second heart sounds?
14. Define murmur. Describe how either an incompetent or a stenosed valve can cause a murmur.
15. Describe the vessels that supply blood to the cardiac muscle.
16. Define coronary thrombosis and infarct. How do atherosclerotic lesions affect the heart?
17. Define cardiac output, stroke volume, and heart rate.
18. What is Starling's law of the heart? What effect does an increase or a decrease in venous return have on cardiac output?
19. Describe the effect of parasympathetic and sympathetic stimulation on heart rate and stroke volume.
20. How does the nervous system detect and respond to the following: (1) a decrease in blood pressure and (2) an increase in blood pressure?
21. What effects do the following have on cardiac output: decrease in blood

pH, increase in blood carbon dioxide, and decrease in blood oxygen levels?

22. What is the effect of epinephrine on the heart rate and stroke volume?
23. Explain how emotions affect heart function.
24. How do changes in body temperature influence the heart rate?

CONCEPT REVIEW

1. A friend tells you that her son had an ECG, and it revealed that he had a slight heart murmur. Should you be convinced that he has a heart murmur? Explain.
2. Predict the effect on Starling's law of the heart if the parasympathetic (vagus) nerves to the heart were cut.
3. An experiment was performed on a dog in which the mean arterial blood pressure was monitored before and after the common carotid arteries were clamped (at time *A*). The results are graphed below:

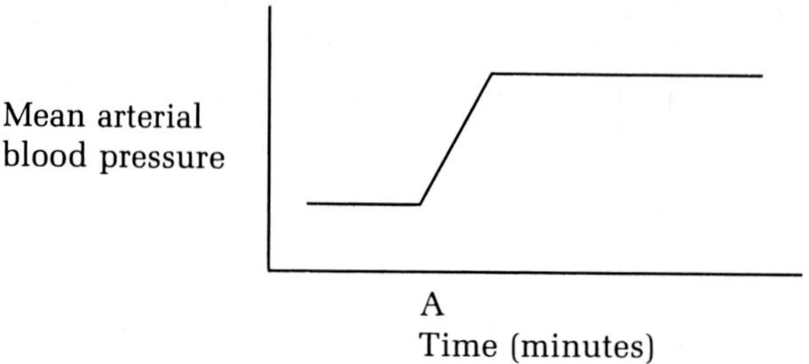

   Explain the change in mean arterial blood pressure (Hint: Baroreceptors are located in the internal carotid arteries, which are superior to the site of clamping of the common carotid arteries).
4. What would happen to cardiac output following the ingestion of a large amount of fluid?
5. A patient exhibited the following symptoms: chest pain, rapid pulse, a greater frequency of P waves than QRS complexes, and disappearance of these symptoms following the administration of a drug that blocks the effect of epinephrine on the heart (a beta adrenergic blocking agent). Is the patient suffering from a myocardial infarct, heart block, or atrial tachycardia? Explain.
6. Explain why the walls of the ventricles are thicker than the walls of the atria.

ANSWERS TO  PREDICT QUESTIONS

1 p. 304 If the normal blood supply were reduced in a small area of the heart through which the left bundle branch passed, conduction of action potentials through that side of the heart would be reduced or blocked. As a consequence, the left side of the heart would not contract at its normal rate. The right side of the heart would function more normally. The reduced rate of contraction of the left ventricle would reduce the pumping effectiveness of the left ventricle.

2 p. 305 (A) Damage to one of the bundle branches would slow the rate of action potential conduction through the side of the heart in which the damage was located. As a result the QRS complex would take longer than normal because impulses would be conducted through one of the ventricles at the normal rate and through the other ventricle at a slower rate. The QRS complex represents the time during which electrical impulses are conducted through the ventricles.

(B) If many ectopic action potentials were arising in the atria, this would increase the heart rate. Each ectopic action potential would initiate a new cardiac cycle. It is possible that some ectopic action potentials arising in the atria would occur while the ventricle was depolarized. If that were the case, those action potentials would not initiate ventricular contractions. Therefore, there may be more P waves than QRS complexes in the electrocardiogram. If ectopic action potentials do not occur in a regular fashion, they may cause the heart to beat at an irregular rate or arrhythmically.

3 p. 305 A leaky pulmonary semilunar valve results in an increased right ventricular volume just prior to ventricular contraction. During ventricular relaxation, the pulmonary semilunar valve closes in a normal person, and blood flows out of the right ventricle through the pulmonary trunk. When the pulmonary semilunar valve is incompetent, some blood leaks back into the right ventricle from the pulmonary trunk during ventricular relaxation. When this blood is added to the blood that normally enters the right ventricle from the right atrium, there is a greater than normal volume of blood in the right ventricle just prior to ventricular contraction.

A severely narrowed opening through the aortic semilunar valve increases the amount of work the heart must do in order to pump the normal volume of blood into the aorta. A greater pressure is required in the ventricle in order to force the same amount of blood through the narrowed opening during ventricular contraction.

4 p. 307 Most of ventricular contraction occurs between the first and second heart sounds of the same beat. Therefore, between the first and second heart sounds, blood is ejected from the ventricles into the pulmonary trunk and the aorta. Between the second heart sound of one beat and the first heart sound of the next beat, the ventricles are relaxing. No blood passes from the ventricles into the aorta or pulmonary trunk during that time period.

5 p. 307 The *shhh* sound made after a heart sound is created by the backward flow of blood after closure of a leaky or incompetent valve. A swishing sound immediately after the second heart sound (lubb-dupp*shhh*) represents a leaky aortic semilunar or pulmonary semilunar valve. The *shhh* sound before a heart sound is created by blood being forced through a narrowed or stenosed valve just before the valve closes. The lubb-*shhh*dupp suggests that there is a swishing sound immediately before the second heart sound, so a stenosed aortic or pulmonary semilunar valve is indicated.

6 p. 312 In response to severe hemorrhage, blood pressure decreases, which is detected by baroreceptors. A reduced frequency of action potentials is sent from the baroreceptors to the medulla oblongata. This causes the cardioregulatory center to increase sympathetic stimulation of the heart and increase the heart rate. Sympathetic stimulation of the heart also increases stroke volume as long as the volume of blood returned to the heart is adequate. Following hemorrhage, however, the blood volume in the body is reduced, and the venous return to the heart from the body is reduced. As a consequence, the volume of blood in the heart is lower than normal. Because of Starling's law, the stroke volume is reduced. Because the heart rate is increased at the same time the volume of blood returning to the heart is decreased, the stroke volume can be quite low. Thus a rapid heart rate and a reduced stroke volume is characteristic of severe hemorrhage. Increasing blood volume, such as occurs during a transfusion, increases the venous return to the heart and increases the stroke volume.

13

BLOOD VESSELS AND CIRCULATION

After reading this chapter you should be able to:

1 Describe the structure and function of arteries, capillaries, and veins.
2 Describe the changes that occur in arteries as they age.
3 Describe the pulmonary portion of the circulatory system.
4 List the major arteries that supply each of the major body areas and describe their functions.
5 List the major veins that carry blood from each of the major body areas and describe their functions.
6 Describe how blood pressure can be measured.
7 Explain how blood pressure and resistance to flow change as blood flows through the blood vessels.
8 Describe the exchange of material across the capillary.
9 Explain how local control mechanisms and nervous control regulate blood flow.
10 Describe the short-term and long-term mechanisms that regulate arterial pressure.

artery Blood vessel that carries blood away from the heart.

blood pressure [L. *pressus*, to press] The force blood exerts against the blood vessel walls; expressed relative to atmospheric pressure and reported in the form of mm Hg pressure.

capillary Minute blood vessel consisting of only simple squamous epithelium and a basement membrane; major site for the exchange of substances between the blood and tissues.

chemoreceptor reflex Process in which chemoreceptors detect changes in oxygen levels, carbon dioxide levels, and pH in the blood and produce changes in heart rate, force of heart contraction, and blood vessel diameter that return these values toward their normal levels.

hepatic portal circulation Blood flow through the veins that begin as capillary beds in the small intestine, spleen, and stomach, and that carry blood to the liver where they end as a capillary bed.

mean arterial blood pressure The average arterial blood pressure. It is slightly less than the average of the systolic and diastolic blood pressures, because diastole lasts longer than systole.

peripheral circulation Blood flow through all blood vessels that carry blood away from the heart (arteries), the capillaries, and all vessels that carry blood back to the heart (veins); consists of the systemic circulation and the pulmonary circulation; includes all blood flow except that through the heart tissue itself.

pulmonary circulation Blood flow through the system of blood vessels that carry blood from the right ventricle of the heart to the lungs and back from the lungs to the left atrium.

pulse pressure The difference between systolic and diastolic pressure.

systemic circulation Blood flow through the system of blood vessels that carry blood from the left ventricle of the heart to the tissues of the body and back from the body to the right atrium.

vein Blood vessel that carries blood toward the heart.

Blood vessels outside of the heart can be placed in two classes: (1) the **systemic vessels,** which transport blood through all parts of the body from the left ventricle and back to the right atrium, and (2) the **pulmonary vessels,** which transport blood from the right ventricle through the lungs and back to the left atrium. Blood flowing through systemic blood vessels supplies oxygen and nutrients to all tissues of the body and carries away carbon dioxide and waste. Blood flowing through pulmonary blood vessels takes up oxygen through the lungs and releases carbon dioxide. The pulmonary circulation and the systemic circulation together comprise the **peripheral circulation.** The blood vessels and the heart are regulated so that the blood supply is sufficient to meet the metabolic needs of all tissues.

GENERAL FEATURES OF BLOOD VESSEL STRUCTURE

Arteries are blood vessels that carry blood away from the heart. Blood is pumped from the ventricles of the heart into large elastic arteries, which branch repeatedly to form progressively smaller arteries. As they become smaller, the arteries undergo a gradual transition from having walls containing more elastic tissue than smooth muscle to having walls with more smooth muscle than elastic tissue. Although the arteries form a continuum from the largest to the smallest branches, they are normally classified as (1) elastic arteries, (2) muscular arteries, and (3) arterioles.

Blood flows from the arterioles into **capillaries,** where exchange occurs between the blood and tissue fluid. Capillaries have thinner walls, blood flows through them more slowly, and there are far more of them than any other blood vessel type.

From the capillaries, blood flows into veins. **Veins** are blood vessels that carry blood toward the heart. When compared to arteries, the walls of the veins are thinner and contain less elastic tissue and fewer smooth muscle cells. Going from capillaries toward the heart, small veins come together to form larger diameter veins which are fewer in number. The veins increase in diameter and decrease in number as they project toward the heart, and their walls increase in thickness. Veins are classified as (1) venules, (2) small veins, (3) medium-sized veins, and (4) large veins.

Except for the capillaries and the venules, blood vessel walls consist of three relatively distinct layers. The relative thickness and composition of each layer varies with the diameter of the blood vessel and its type. From the inside to the outer wall of the

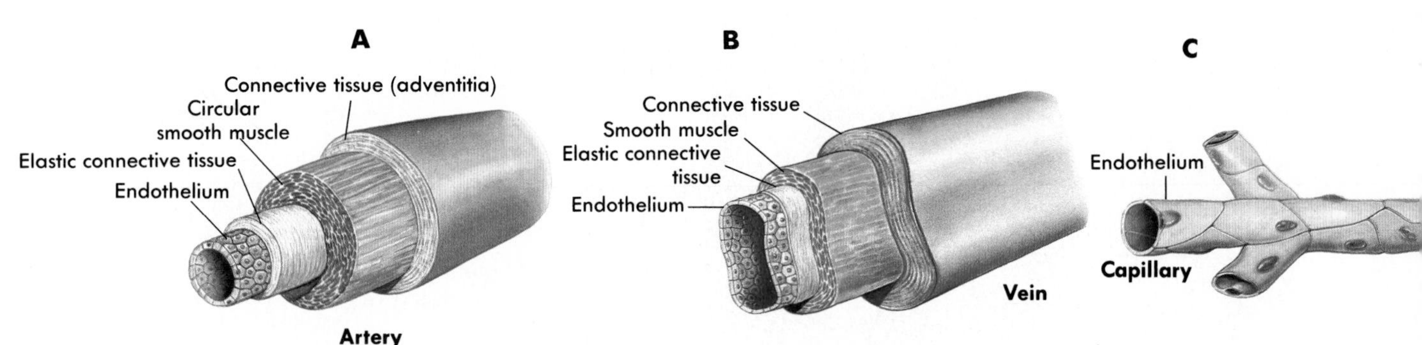

FIGURE 13-1 Blood vessel structure.
The tunica intima of arteries and veins includes the endothelium, its basement membrane, and a layer of elastic connective tissue. The tunica media includes a smooth muscle layer and variable amounts of elastic and collagen fibers. The tunica adventitia is the connective tissue layers surrounding the vessel.
A The wall of an artery. The tunica media with its smooth muscle and varying amounts of connective tissue makes up the dominant layer of the wall of arteries.
B The wall of a vein. The walls of veins are thinner than walls of arteries, and the tunica media of a vein contains less smooth muscle than an artery of equal diameter. The dominant layer in the wall of a vein is the tunica adventitia.
C The wall of a capillary. Capillaries have a smaller diameter than arteries and veins, and their walls consist of a thin layer of endothelium surrounded by a delicate layer of connective tissue. The connective tissue is not shown in the illustration.

blood vessels, the layers, or **tunics,** are (1) the tunica intima, (2) the tunica media, and (3) the tunica adventitia, or tunica externa (Figure 13-1).

The **tunica intima** consists of an **endothelium** composed of simple squamous epithelial cells and a small amount of connective tissue. The **tunica media,** or middle layer, consists of smooth muscle cells arranged circularly around the blood vessel. It also contains variable amounts of elastic and collagen fibers, depending on the size and type of the vessel. The **tunica adventitia** is composed of connective tissue, which varies from dense connective tissue adjacent to the tunica media to loose connective tissue toward the outer portion of the blood vessel wall.

Arteries

Elastic arteries are the largest-diameter arteries and have the thickest walls. A greater proportion of their walls is elastic tissue and a smaller proportion is smooth muscle compared to other arteries. Elastic arteries are stretched when the ventricles of the heart pump blood into them. The elastic recoil of the elastic arteries prevents blood pressure from falling rapidly and maintains blood flow while the ventricles are relaxed.

The muscular arteries include medium-sized and small-diameter arteries. The walls of medium-sized arteries are relatively thick when compared to their diameter. Most of the thickness of the wall results from smooth muscle cells of the tunica media. Medium-sized arteries are frequently called **distributing arteries** because the smooth muscle tissue enables these vessels to control blood flow to different regions of the body by either constricting (vasoconstriction) or dilating (vasodilation) (Figure 13-2).

Medium-sized arteries supply blood to small arteries. Small arteries have about the same structure as the medium-sized arteries except small arteries have a smaller diameter. The smallest of the small arteries have only three or four layers of smooth muscle in their walls.

Arterioles transport blood from small arteries to capillaries and are the smallest arteries in which the three tunics can be identified. The tunica media consists of only one or two layers of circular smooth muscle cells. Small arteries and the arterioles are adapted for vasodilation and vasoconstriction.

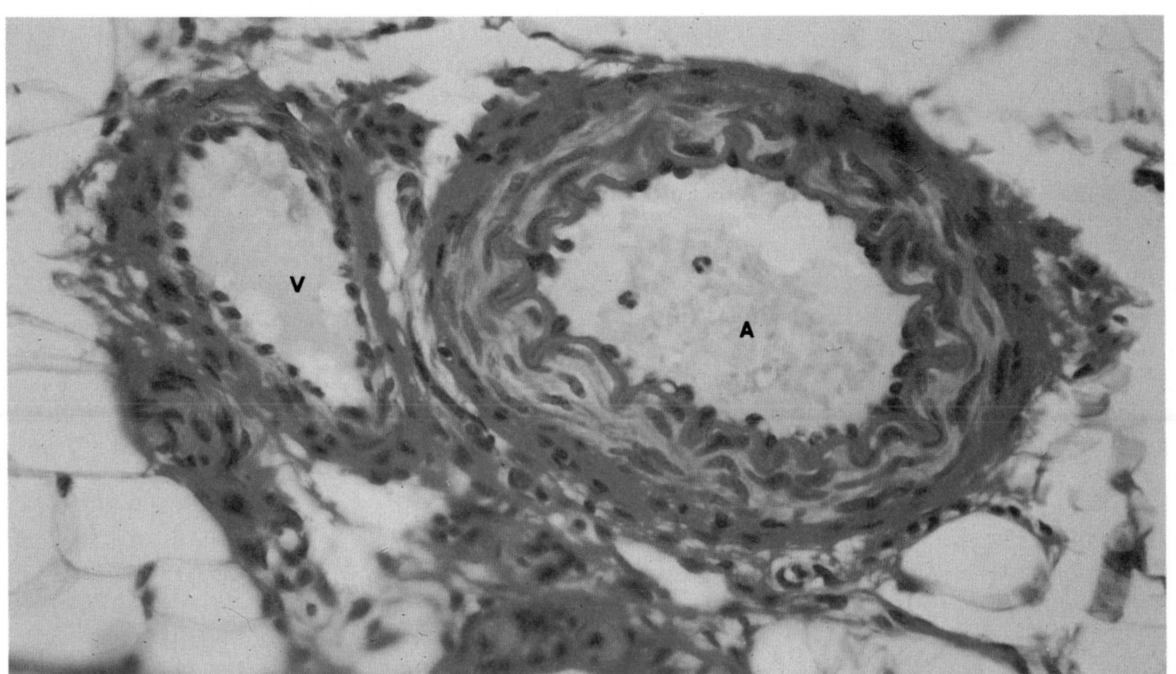

FIGURE 13-2 ■ Photomicrograph of an artery and a vein.
The typical structure of a medium-sized to small artery and vein. The predominant layer in the wall of the artery is the tunica media with its circular layers of smooth muscle. In the vein the wall is thinner, the tunica media is thinner, and the dominant layer in its wall is the tunica adventitia.

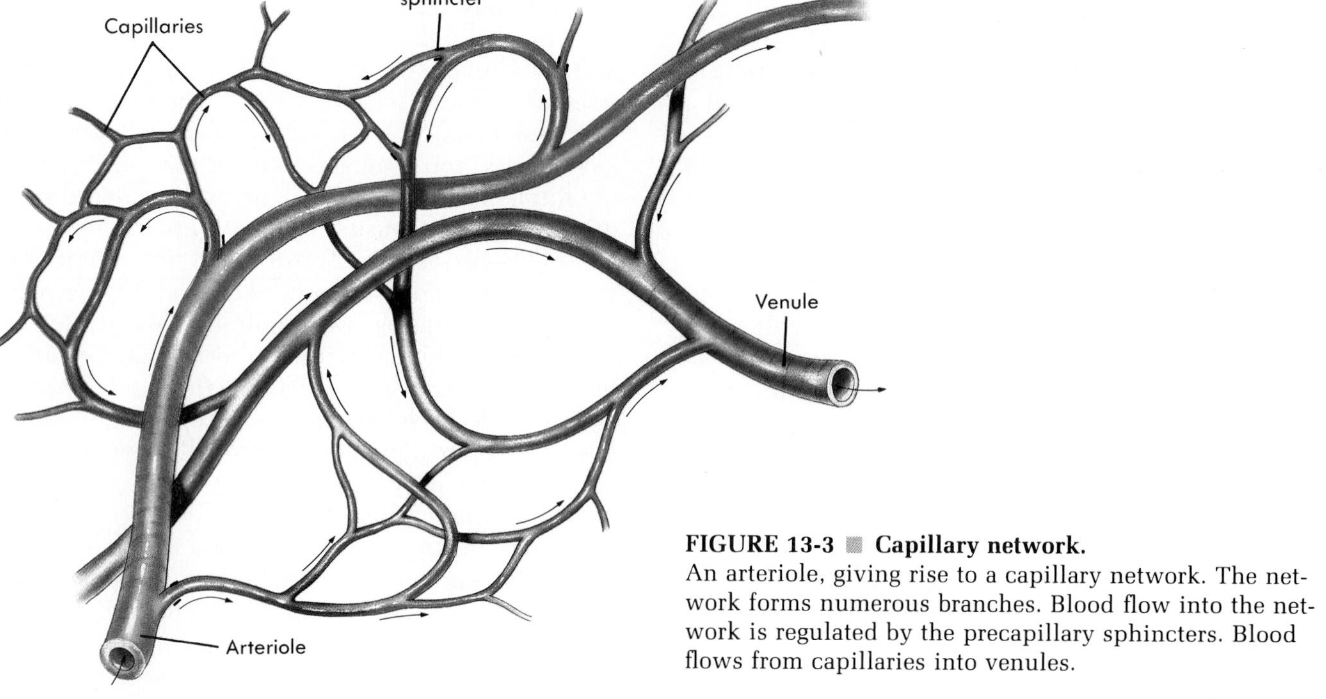

FIGURE 13-3 ■ Capillary network.
An arteriole, giving rise to a capillary network. The network forms numerous branches. Blood flow into the network is regulated by the precapillary sphincters. Blood flows from capillaries into venules.

Capillaries

The capillary wall consists of endothelium and is surrounded by a delicate layer of loose connective tissue. Capillaries are 0.5 to 1 mm long, and they branch without a change in their diameters.

Blood flows from arterioles into capillaries, which branch to form networks (Figure 13-3). Red blood cells flow through most capillaries in single file and are frequently folded as they pass through the smaller diameter capillaries. Blood flow through capillaries is regulated by smooth muscle cells called **precapillary sphincters** located at the origin of the branches. As blood flows through capillaries, blood gives up oxygen and nutrients to the tissue spaces and takes up carbon dioxide and other byproducts of metabolism. Capillary networks are more numerous and more extensive in highly metabolic tissues such as the lungs, liver, kidneys, skeletal muscle, and cardiac muscle.

Veins

Blood flows from capillaries into venules and from venules into small veins. Venules are tubes with a diameter slightly larger than capillaries and are composed of endothelium resting on a delicate connective tissue membrane. The structure of venules, except for their diameter, is very similar to capillaries. Small veins are slightly larger in diameter than venules, and their walls contain a continuous layer of smooth muscle cells.

Medium-sized veins collect blood from small veins and deliver it to large veins. Three thin but distinctive layers make up the wall of the medium-sized and large veins. The tunica intima is composed of endothelium and underlying connective tissue. The tunica media contains some circular smooth muscle and sparsely scattered elastic fibers. The predominant layer is the outer tunica adventitia, which consists primarily of dense collagen fibers (see Figure 13-2).

Veins having diameters greater than 2 mm contain valves. The valves allow blood to flow toward the heart but not in the opposite direction. Each valve consist of folds in the tunica intima that form two flaps, which are shaped like and function like the semilunar valves of the heart. There are many valves in the medium-sized veins. There are more valves in veins of the legs than in veins of the arms. This prevents the flow of blood toward the feet in response to the pull of gravity.

Varicose veins result when the veins in the legs become so dilated that the cusps of the valves no longer overlap to prevent the backflow of blood. As a consequence, venous pressure is greater than normal in the veins of the legs and can result in edema. Blood flow in the veins can become so stagnant that the blood clots. As a consequence, *phlebitis* (flĕ-bi'tis; inflammation of the veins) may result, and, if the condition becomes severe enough, it may lead to *gangrene* (necrosis of tissue resulting from a lack of blood flow; and infection of the tissue with anaerobic bacteria). The development of varicose veins is encouraged by conditions that increase the pressure in veins, causing them to stretch. An example is pregnancy, in which compression of the veins by the enlarged uterus results in increased venous pressure in the veins that drain the legs.

Aging of the Arteries

The walls of all arteries undergo changes as they age, although some arteries change more rapidly than other arteries, and some individuals are more susceptible to change than others. The most significant effects of aging occur in the large elastic arteries such as the aorta, large arteries carrying blood to the brain, and coronary arteries.

Degenerative changes in arteries that make them less elastic are referred to as **arteriosclerosis** (ar-tēr'ĭ-o-sklĕ-ro'sis; hardening of the arteries). These changes occur in nearly every individual, and they become more severe with advancing age. A related term, **atherosclerosis,** refers to the deposition of material in the walls of arteries to form plaques. The material is a fat-like substance containing cholesterol. The fatty material later may be replaced with dense connective tissue and calcium deposits.

The development of atherosclerosis is influenced by several factors. Lack of exercise, smoking, obesity, and a diet high in cholesterol and fats appear to increase the severity and the rate at which atherosclerosis develops. Severe atherosclerosis also appears to be more prevalent in some families than in others, which suggests a genetic influence.

Atherosclerosis greatly increases resistance to blood flow because the deposits reduce the inside diameter of the arteries. The increased resistance hampers normal circulation to tissues and greatly increases the work performed by the heart. The rough atherosclerotic plaques attract platelets, which adhere to them and increase the chance of thrombus formation.

PULMONARY CIRCULATION

Pulmonary circulation is the flow of blood through the system of blood vessels that carry blood from the right ventricle of the heart to the lungs and back to the left atrium. Blood from the right ventricle is pumped into the **pulmonary** (pul'mo-nĕr-e; relating to the lungs) **trunk** (Figure 13-4). This short vessel branches into the **right and left pulmonary arteries,** which extend to the right and left lungs, respectively. Poorly oxygenated blood is carried to the pulmonary capillaries, where oxygen is taken up by the blood and carbon dioxide is released. Four **pulmonary veins** exit the lung, two from each lung, and carry the oxygenated blood to the left atrium.

SYSTEMIC CIRCULATION: ARTERIES

Systemic circulation is the flow of blood through the system of blood vessels that carry blood from the left ventricle of the heart to the tissues of the body and back to the right atrium. Oxygenated blood from the pulmonary veins passes from the left atrium into the left ventricle and from the left ventricle into the aorta. Blood is distributed from the aorta to all portions of the body (see Figure 13-4).

Aorta

All arteries of the systemic circulation branch directly or indirectly from the **aorta** (a-or'tah). The aorta is usually considered in three portions: the ascending aorta, the aortic arch, and the descending aorta. The descending aorta is further divided into a thoracic aorta and an abdominal aorta (Figure 13-5).

The portion of the aorta that passes superiorly from the left ventricle is called the **ascending aorta.** The right and left coronary arteries arise from the base of the ascending aorta and supply blood to the cardiac muscle.

The aorta arches posteriorly and to the left as the **aortic arch.** Three major arteries, which carry blood to the head and upper limbs, originate from the aortic arch. They are the brachiocephalic artery, the left common carotid artery, and the left subclavian artery.

The **descending aorta** is the longest portion of the aorta. It extends through the thorax and abdomen to the upper margin of the pelvis. The portion of the descending aorta that extends through the

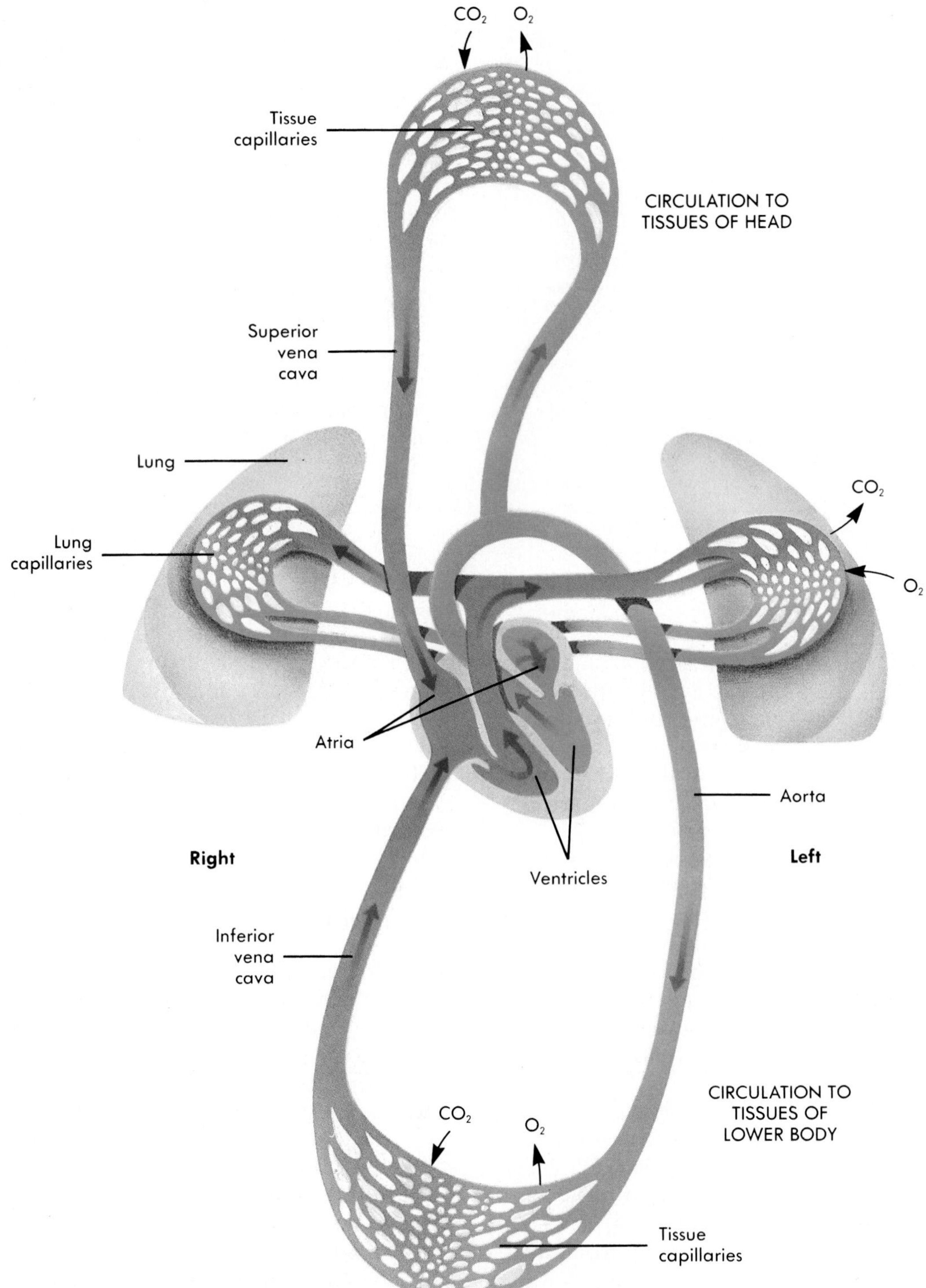

FIGURE 13-4 ■ Blood flow through the circulatory system.
Blood is returned from the body through the superior vena cava and the inferior vena cava to the right atrium. After passing from the right atrium to the right ventricle, blood is pumped into the pulmonary trunk. The pulmonary trunk divides into the right and left pulmonary arteries, which carry the oxygen-poor blood to the lungs. In the lung capillaries, carbon dioxide is given off and oxygen is picked up by the blood. Blood, now rich in oxygen, flows through two pulmonary veins from each lung to the left atrium. Blood then passes from the left atrium to the left ventricle. The left ventricle then pumps the blood into the aorta, which distributes the blood through its branches to all of the body.

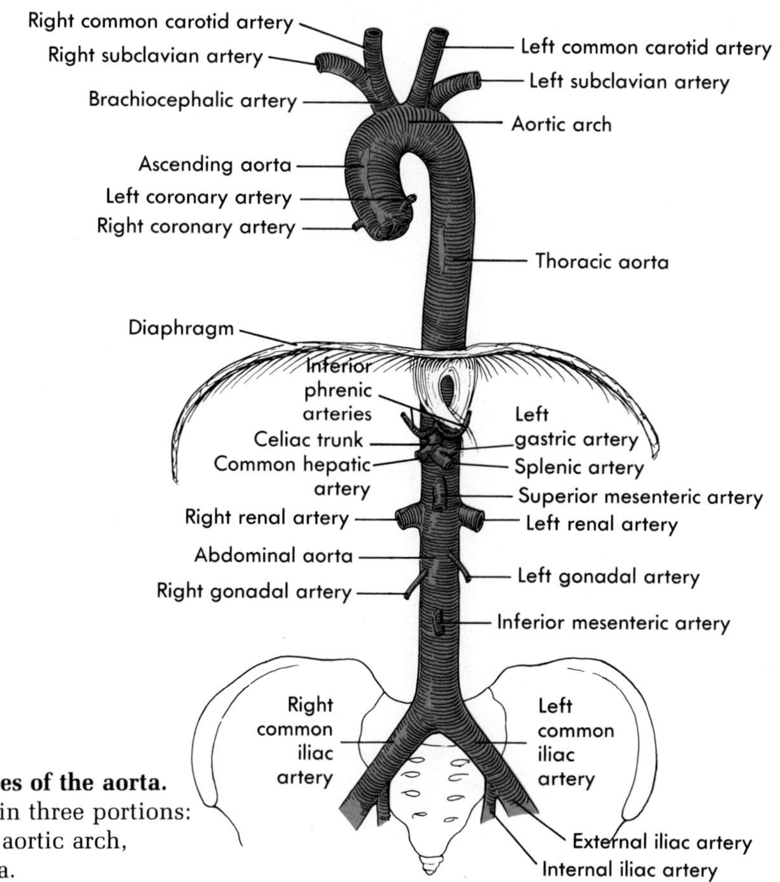

FIGURE 13-5 ■ Branches of the aorta. The aorta is considered in three portions: the ascending aorta, the aortic arch, and the descending aorta.

thorax to the diaphragm is called the thoracic aorta. The portion of the descending aorta that extends from the diaphragm to the point where it divides into the two **common iliac** (il'e-ak; relating to the flank area) **arteries** is called the abdominal aorta.

Arteries of the Head and Neck

■ The first vessel to branch from the aortic arch is the **brachiocephalic** (bra'ke-o-sĕ-fal'ik; vessel to the arm and head) **artery.** It is a short artery, and it branches at the level of the clavicle to form the **right common carotid** (kă-rot'id) **artery,** which transports blood to the right side of the head and neck, and the **right subclavian artery,** which transports blood to the right upper limb (see Figures 13-5 and 13-6).

There is no brachiocephalic artery on the left side of the body. Instead, the left common carotid and left subclavian artery branch directly off the aortic arch. They are the second and third branches of the aortic arch. The **left common carotid artery** transports blood to the left side of the head and neck, and the **left subclavian artery** transports blood to the left upper limb (see Figure 13-5).

The common carotid arteries extend superiorly along each side of the neck to the inferior angle of the mandible, where they branch into **internal** and **external carotid** arteries (see Figure 13-6). The base of the internal carotid arteries is slightly dilated to form the **carotid sinuses,** which contain structures important in monitoring blood pressure (baroreceptors). The external carotid arteries have several branches that supply the structures of the face, nose, and mouth. The internal carotid arteries supply most of the blood to the brain.

Some of the blood to the brain is supplied by the vertebral arteries, which branch from the subclavian arteries (see Figure 13-6) and pass to the head through the transverse foramina of the cervical vertebrae. The vertebral arteries then pass into the cranial vault through the foramen magnum. Branches of the vertebral arteries supply blood to the vertebrae in the neck and to muscles and ligaments associated with them.

Within the cranial vault, the vertebral arteries unite to form a single **basilar artery** located along the posterior surface of the brainstem. The basilar artery gives off branches that supply blood to the pons, cerebellum, and midbrain. It also forms right

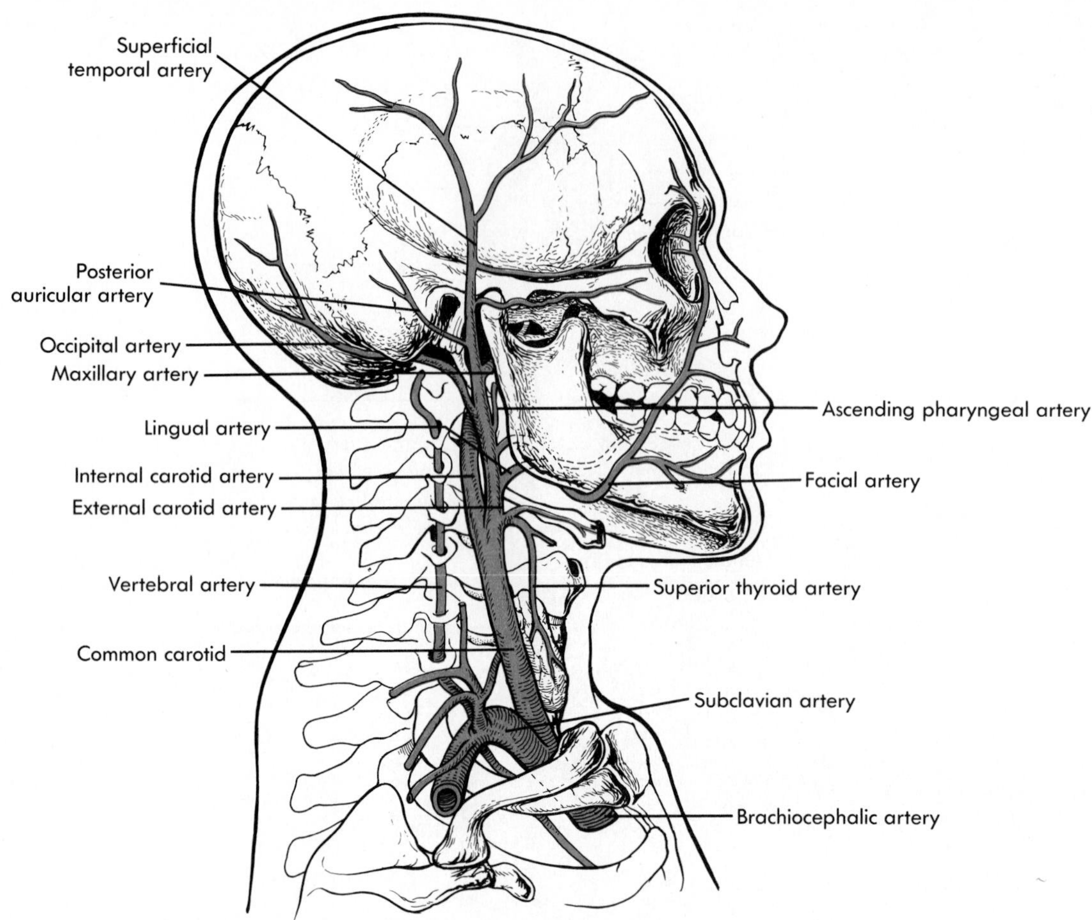

FIGURE 13-6 ■ Arteries of the head and neck.
The brachiocephalic artery, the right common carotid artery, the right subclavian artery, and their branches. The major arteries to the head are the common carotid and vertebral arteries.

and left branches that join with the right and left internal carotid arteries. These blood vessels form a system of vessels called the **Circle of Willis** at the base of the brain. The vessels that supply blood to most of the brain branch off of the Circle of Willis. The majority of the blood supply to the brain is through the internal carotid arteries. The vertebral arteries cannot supply enough blood to the brain to maintain life if the carotid arteries are blocked.

> **1 The term carotid means to put to sleep, implying that if the carotid arteries are blocked for several seconds, the patient can lose consciousness. Interruption of the blood supply for a few minutes can result in permanent brain damage. What is the physiological significance of atherosclerosis in the carotid arteries?** ?

Arteries of the Upper Limbs

The arteries of the upper limbs are named differently as they pass into different body regions even though no major branching occurs. The **subclavian** (sub-kla′ve-an) **artery,** located below the clavicle, becomes the **axillary** (ak′sĭ-lăr′e) **artery,** located in the axilla. The **brachial** (bra′ke-al) **artery,** located in the arm, is a continuation of the axillary artery (Figure 13-7). The brachial artery branches at the elbow to form the **ulnar artery** and the **radial artery,** which supply blood to the forearm and hand. The radial artery is the artery most commonly used for taking a pulse. The pulse can be detected conveniently on the thumb side of the anterior surface of the wrist.

FIGURE 13-7 The major arteries.
The major arteries that carry blood from the left ventricle of the heart to the tissues of the body.

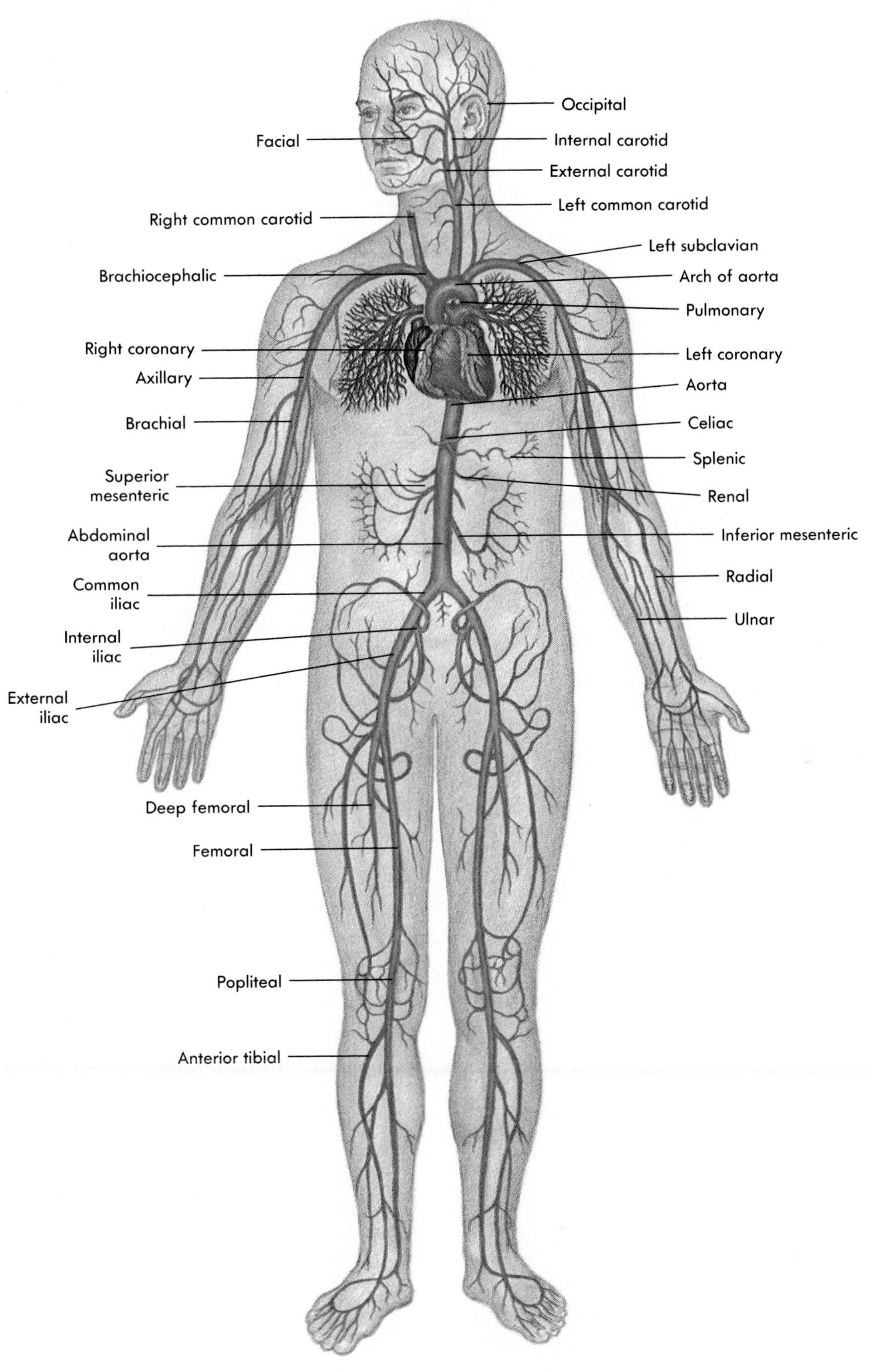

TABLE 13-1

Major Branches of the Aorta (see Figure 13-5)

ARTERIES	TISSUES SUPPLIED
ASCENDING AORTA	
Coronary arteries	Heart
AORTIC ARCH	
Brachiocephalic	Right arm, right side of head, and right side of neck
Left common carotid	Left side of head and left side of neck
Left subclavian	Left arm
DESCENDING AORTA	
THORACIC AORTA	
Visceral branches	
Bronchial	Lung tissue
Esophageal	Esophagus
Parietal branches	
Intercostal	Thoracic wall
Superior phrenic	Diaphragm
ABDOMINAL AORTA	
Visceral branches (unpaired)	
Celiac trunk	Stomach, esophagus, duodenum, liver, spleen, and pancreas
Superior mesenteric	Pancreas, small intestine, and first part of colon
Inferior mesenteric	Last part of colon and the rectum
Visceral branches (paired)	
Suprarenal	Adrenal gland
Renal	Kidney
Gonadal	
Testicular (male)	Testis and ureter
Ovarian (female)	Ovary, ureter, and uterine tube
Parietal branches	
Inferior phrenic	Diaphragm
Lumbar	Lumbar vertebrae and back muscles
Median sacral	Inferior vertebrae
Common iliac	
External iliac	Lower limb
Internal iliac	Lower back, hip, pelvis, urinary bladder, vagina, uterus, rectum, and external genitalia

The Thoracic Aorta and Its Branches

The branches of the thoracic aorta can be divided into two groups: The **visceral arteries** supply the thoracic organs, and the **parietal arteries** supply the thoracic wall (Table 13-1; see Figures 13-5 and 13-7). The visceral branches of the thoracic aorta supply the esophagus, trachea, and part of the lung. The major parietal arteries are the posterior intercostal arteries, which arise from the thoracic aorta and course through the intercostal spaces. They supply intercostal muscles, the vertebrae, the spinal cord, and deep muscles of the back (Figure 13-8). The superior phrenic artery supplies the diaphragm.

The internal thoracic artery, a branch of the subclavian artery, descends along the internal thoracic wall and gives rise to branches, called the anterior intercostal arteries, that extend from the internal thoracic artery between the ribs and supply the anterior chest wall.

The Abdominal Aorta and Its Branches

The branches of the abdominal aorta, like those of the thoracic aorta, can be divided into **visceral** and **parietal** groups. The visceral arteries are divided into paired and unpaired branches. There are three major unpaired branches: the **celiac** (se′le-ak; belly), **superior mesenteric** (mes′en-tĕr′ik; relating to the mesenteries), and **inferior mesenteric arteries.** The celiac artery supplies blood to the stomach, pancreas, spleen, upper duodenum, and liver. The superior mesenteric artery supplies blood to the small intestine and upper portion of the colon, and the inferior mesenteric artery supplies blood to the remainder of the colon. (See Figure 13-5.)

The paired visceral branches of the abdominal aorta supply the kidneys (the **renal arteries**), adrenal glands (the **suprarenal arteries**), and gonads **(testicular** or **ovarian arteries).** The parietal arteries of the abdominal aorta supply the diaphragm and abdominal wall.

The parietal arteries include the **inferior phrenic artery,** which supplies the diaphragm; the **lumbar arteries,** which supply the lumbar vertebrae and back muscles; and the **median sacral arteries,** which supply the inferior vertebrae.

Arteries of the Pelvis

The abdominal aorta divides at the level of the fifth lumbar vertebra into two **common iliac arteries.** They, in turn, divide to form the **external iliac**

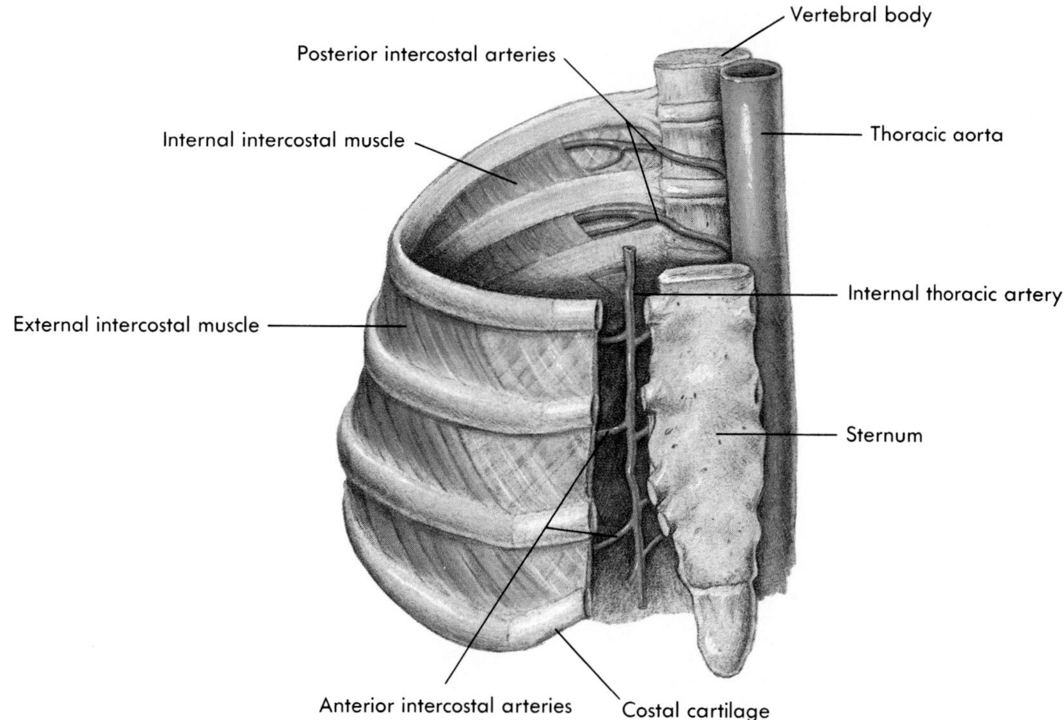

FIGURE 13-8 ■ Arteries of the thoracic wall.
The posterior intercostal arteries and the anterior intercostal arteries supply the wall of the thorax. The posterior intercostal arteries branch off from the aorta, and the anterior intercostal arteries branch off of the internal thoracic artery which, in turn, branches off of the subclavian artery.

arteries, which enter the lower limbs, and the **internal iliac arteries,** which supply the pelvic area (see Figure 13-7). Visceral branches of the internal iliac supply organs such as the urinary bladder, rectum, uterus, and vagina; parietal branches supply blood to the walls and floor of the pelvis, the lumbar, gluteal, and proximal thigh muscles, and the external genitalia.

Arteries of the Lower Limbs

■ Like the arteries of the upper limbs, arteries of the lower limbs are named differently as they pass into different body regions even though there are no major branches. The **external iliac arteries** become the **femoral** (fem'o-ral; relating to the thigh) **arteries** in the thigh, which become the **popliteal** (pop'lĭ-te-al; ham; the hamstring area posterior to the knee) **arteries** in the popliteal space (the posterior region of the knee). The popliteal arteries branch to give off the **anterior tibial arteries,** slightly inferior to the knee, and the **posterior tibial arteries.** The anterior and posterior tibial arteries give rise to arteries that supply blood to the feet (see Figure 13-7).

SYSTEMIC CIRCULATION: VEINS

■ The **superior vena cava** (ve'nah ka'vah; venus cave) returns blood from the head, neck, thorax, and upper limbs to the right atrium of the heart; and the **inferior vena cava** returns blood from the abdomen, pelvis, and lower limbs to the right atrium.

Veins of the Head and Neck

■ The two pairs of major veins that drain blood from the head and neck are the **external** and **internal jugular** (jug'u-lar; neck) **veins** (Figure 13-9). The external jugular veins are the more superficial of the two sets, and they drain blood from the posterior head and neck. The external jugular veins empty primarily into the subclavian veins. A small branch of the external jugular also delivers blood to the internal jugular vein. The internal jugular veins are much larger and deeper. They drain blood from the cranial vault and the anterior head, face, and neck. The internal jugular veins join the **subclavian veins** on each side of the body to form the **brachiocephalic veins.** The brachiocephalic veins empty into the superior vena cava.

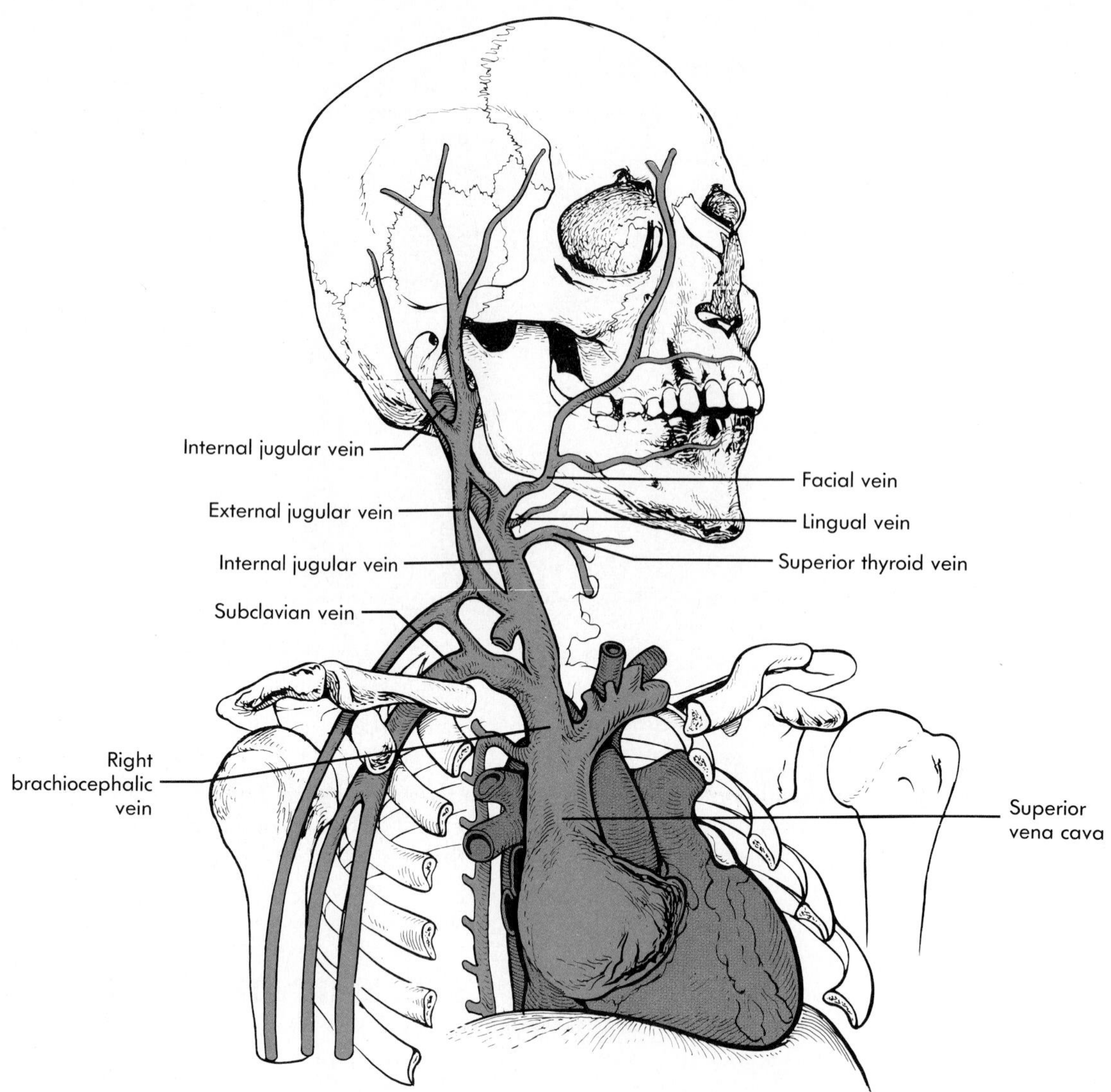

FIGURE 13-9 ■ The veins of the head and neck.
The external and internal jugular veins drain blood from the head and neck. The internal jugular veins join the subclavian veins on each side of the body to form the brachiocephalic veins. The external jugular veins drain mainly into the subclavian veins, although a branch of the external jugular veins may carry some blood directly to the internal jugular veins.

Veins of the Upper Limbs

■ The veins of the upper limbs can be divided into deep and superficial groups (Figure 13-10). The deep veins, which drain the deep structures of the upper limb, follow the same course as the arteries and are named for the arteries they accompany. The only noteworthy deep veins are the **brachial veins,** which accompany the brachial artery and empty into the axillary vein. The superficial veins drain the superficial structures of the upper limb and then empty into the deep veins. The **cephalic** (sĕ'-fal'ik; toward the head) **vein,** which empties into the subclavian vein, and the **basilic** (bah-sil'ik; toward the base of the arm) **vein,** which empties into the **axillary vein,** are the major superficial veins. Many of their tributaries in the forearm and hand can be seen through the skin. The **median cubital** (ku'bĭ-tal; elbow) **vein** usually connects the cephalic vein or its tributaries with the basilic vein. Although this vein varies in size among people, it is usually quite prominent on the anterior surface of the upper limb at the level of the elbow (cubital fossa) and is often used as a site for drawing blood.

FIGURE 13-10 ■ The major veins.
The major veins that carry blood from the tissue of the body and return it to the right atrium.

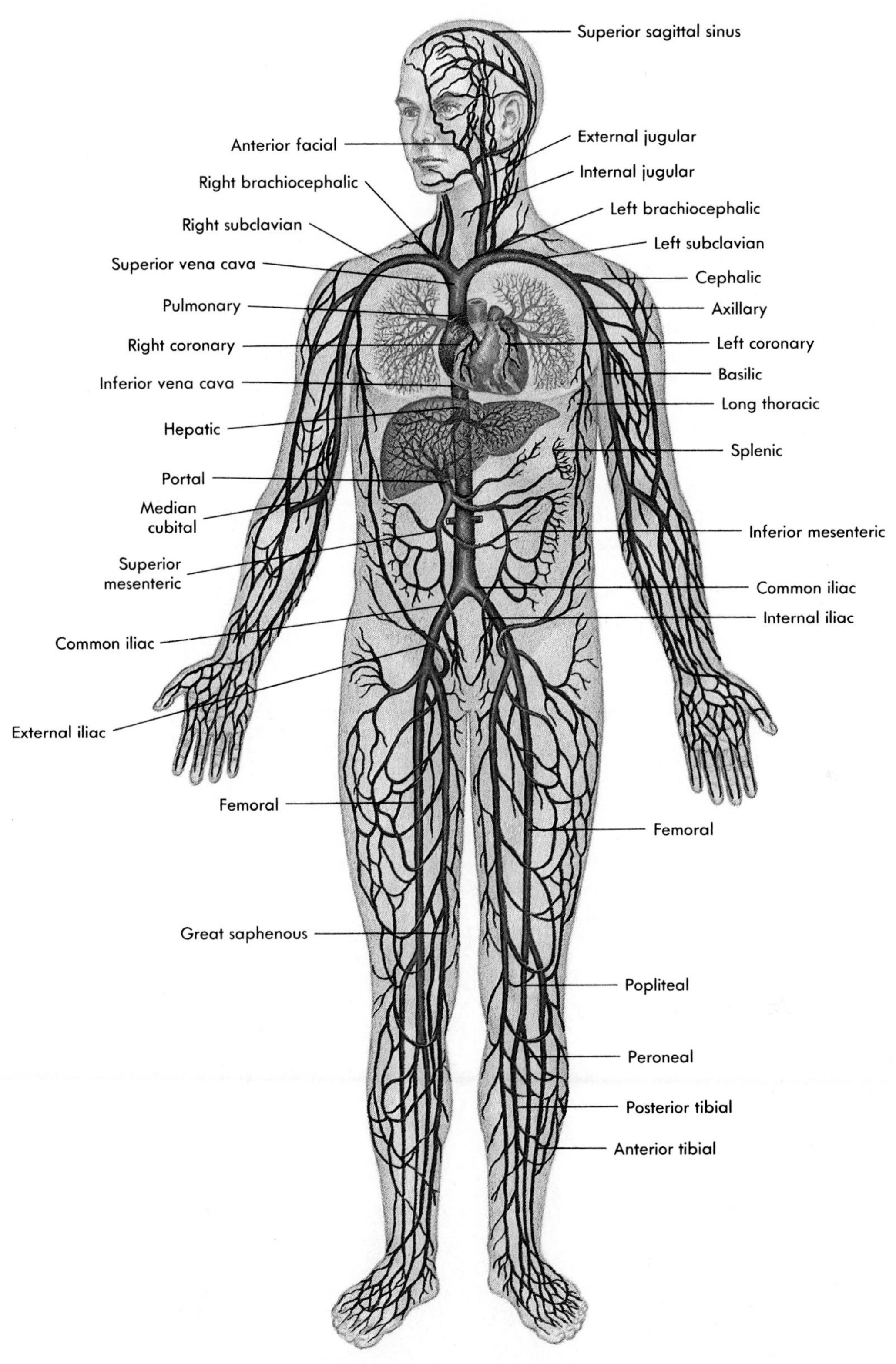

Veins of the Thorax

Three major veins return blood from the thorax to the superior vena cava: the **right** and **left brachiocephalic veins** and the **azygos** (az'ĭ-gus; unpaired) **vein.** The thoracic drainage to the brachiocephalic veins is through the anterior thoracic wall by way of the **internal thoracic veins.** Blood from the posterior thoracic wall is collected by **posterior intercostal veins** that drain into the azygos vein on the right and the **hemiazygos vein** or **accessory hemiazygos vein** on the left. The hemiazygos and accessory hemiazygos veins empty into the azygos, which drains into the superior vena cava (Figure 13-11).

Veins of the Abdomen and Pelvis

Blood from the posterior abdominal wall drains into the azygos vein. Blood from the rest of the abdomen and from the pelvis and lower limbs returns to the heart through the inferior vena cava. The gonads (testes or ovaries), kidneys, and adrenal glands are the only abdominal organs outside the pelvis that drain directly into the inferior vena cava. The **internal iliac veins** drain the pelvis and join the **external iliac veins** from the lower limbs to form the **common iliac veins.** The common iliac veins combine to form the inferior vena cava (Table 13-2; see Figure 13-10).

TABLE 13-2

Veins Draining the Abdomen and Pelvis (see Figure 13-11)

VEINS	TISSUES DRAINED
INFERIOR VENA CAVA	
Hepatic veins	Liver (see hepatic portal system)
Common iliac	
External iliac	Lower limb
Internal iliac	Pelvis and its viscera
Renal	Kidney
Suprarenal	Adrenal gland
Gonadal	
Testicular (male)	Testis
Ovarian (female)	Ovary
Phrenic	Diaphragm

Blood from the capillaries within most of the abdominal viscera such as the stomach, intestines, and spleen drains through a specialized system of

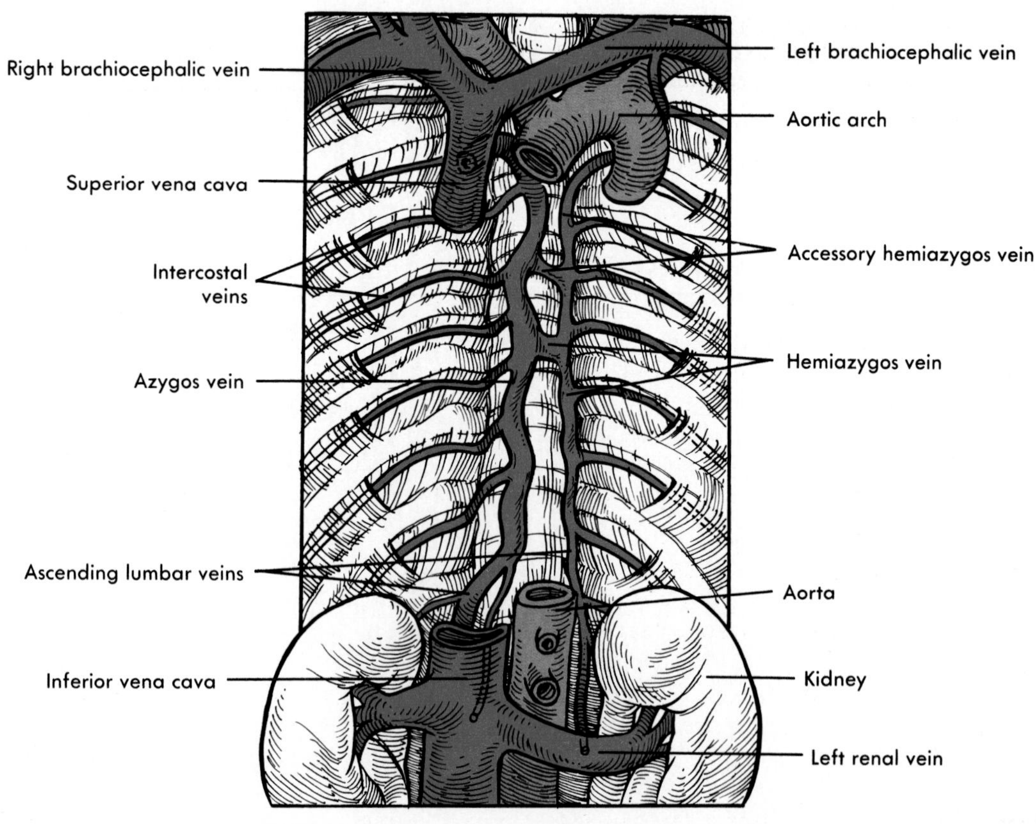

FIGURE 13-11 Veins of the thoracic wall.
The azygos and hemiazygos veins and their tributaries.

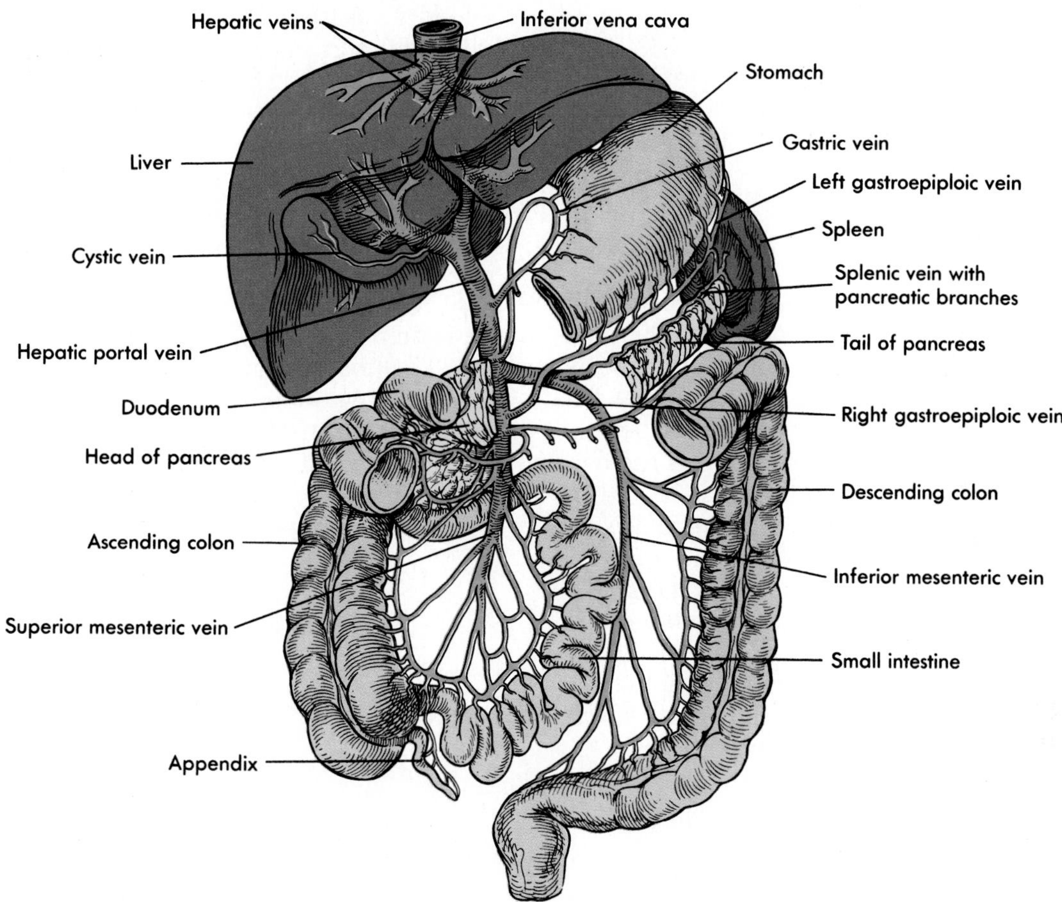

FIGURE 13-12 ■ Veins of the hepatic portal system.
The hepatic portal system begins as capillary beds in the stomach, pancreas, small intestine, and large intestine. It carries blood to a series of capillaries in the liver. Blood from the liver flows through the hepatic vein, which then enters the inferior vena cava.

blood vessels to the liver. The liver is a major processing center for substances absorbed by the intestinal tract. A **portal** (pōr′tal) **system** is a vascular system that begins and ends with capillary beds and has no pumping mechanism such as the heart between them. The **hepatic** (hĕ-pat′ik; relating to the liver) **portal system** (Figure 13-12 and Table 13-3) begins with capillaries in the viscera and ends with capillaries in the liver. The major tributaries of the hepatic portal system are the splenic vein and the superior mesenteric vein. The inferior mesenteric vein empties into the splenic vein. The splenic vein carries blood from the spleen and pancreas. The superior and inferior veins carry blood from the intestines.

Blood from the liver is collected into **hepatic veins,** which join the inferior vena cava. Blood entering the liver through the hepatic portal vein is

TABLE 13-3

Hepatic Portal System (see Figure 13-12)

VEINS	TISSUES DRAINED
HEPATIC PORTAL	
Superior mesenteric	Small intestine and most of the colon
Splenic	Spleen
Inferior mesenteric	Descending colon and rectum
Pancreatic	Pancreas
Gastroepiploic	Stomach
Gastric	Stomach
Cystic	Gallbladder

rich with nutrients collected from the intestines, but it also may contain a number of toxic substances harmful to the tissues of the body. Within the liver, nutrients are taken up and stored or modified so they can be used by other cells of the body, and toxic substances are removed from the blood.

Veins of the Lower Limbs

The veins of the lower limbs, like those of the upper limbs, consist of deep and superficial groups. The deep veins follow the same path as the arteries and are named for the arteries they accompany. The anterior tibial vein drains the deep anterior leg and dorsal foot, the posterior tibial vein drains the deep posterior leg and plantar region of the foot, and the deep peroneal vein drains the deep lateral leg and foot. The superficial veins consist of the great and small **saphenous** (să-fe′nus; visible) **veins.** The **great saphenous vein** originates over the dorsal and medial side of the foot and ascends along the medial side of the leg and thigh to empty into the femoral vein. The **small saphenous vein** begins over the lateral side of the foot and empties into the popliteal vein (see Figure 13-10), which, in turn, empties into the femoral vein. The femoral vein empties into the external iliac vein.

It is usually the great saphenous vein that is surgically removed and used in coronary bypass surgery. Portions of the saphenous vein are grafted to coronary arteries to create a route of blood flow that bypasses blocked portions of the coronary arteries. The circulation that is interrupted by the removal of the saphenous vein flows through other veins of the leg.

THE PHYSIOLOGY OF CIRCULATION

The function of the entire circulatory system is to maintain adequate blood flow to all tissues. An adequate blood flow is required to provide nutrients and oxygen to tissues and to remove waste products of metabolism from the tissues. Blood flows through the circulatory system primarily as a result of the pressure produced by contractions of the ventricles of the heart.

Blood Pressure

Blood pressure is a measure of the force blood exerts against the blood vessel walls. In arteries, blood pressure valves exhibit a cycle dependent on the rhythmic contraction of the heart. During ventricular systole, the heart pushes blood into the arteries, and the pressure reaches a maximum called the **systolic pressure.** When the ventricles relax, blood pressure in the arteries falls to a minimum value called the **diastolic pressure.** The standard measure for blood pressure is in terms of millimeters of mercury (mm Hg). If the blood pressure is 100 mm Hg, the pressure is great enough to lift a column of mercury 100 mm.

The **auscultatory** (aws-kul′tah-to′re) method of determining blood pressure is used under most clinical conditions (Figure 13-13). A blood pressure cuff connected to a **sphygmomanometer** (sfig′mo-man-ŏ′mĕ-ter) is placed around the patient's arm and a stethoscope is placed over the brachial artery.

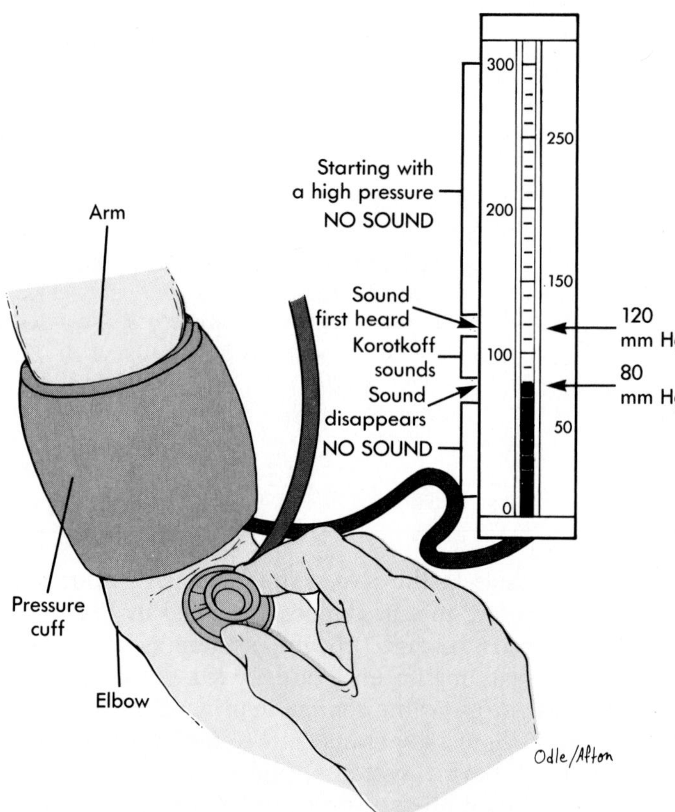

FIGURE 13-13 ■ Blood pressure measurement. Blood pressure measurement using a sphygmomanometer. The blood pressure cuff is inflated to a high pressure, and the pressure is then decreased slowly. The pressure at which turbulent blood flow is first heard is the systolic blood pressure. The pressure at which sounds disappear is the diastolic blood pressure.

The blood pressure cuff is then inflated until the brachial artery is completely blocked. Because no blood flows through the constricted area, no sounds can be heard at this point. The pressure in the cuff is then gradually lowered. As soon as the pressure in the cuff declines below the systolic pressure, blood flows through the constricted area each time the left ventricle contracts. The blood flow is turbulent, and this turbulence produces vibrations in the blood and surrounding tissues that can be heard through the stethoscope. These sounds are called **Korotkoff sounds,** and the pressure at which the first Korotkoff sound is heard represents the systolic pressure.

As the pressure in the blood pressure cuff is lowered still more, the Korotkoff sounds change tone and loudness. When the pressure has dropped until the sound disappears completely, continuous blood flow is reestablished. The pressure at which the Korotkoff sounds disappear is the diastolic pressure.

The systolic pressure represents the maximum pressure produced in the large arteries. It is also a good measure of the maximum pressure within the left ventricle. The diastolic pressure is close to the lowest pressure within large arteries. During relaxation of the left ventricle, the aortic semilunar valves close, trapping the blood that was ejected during ventricular contraction in the aorta. During ventricular relaxation, the pressure in the ventricles falls to 0 mm Hg. The blood trapped in the elastic arteries is compressed by the recoil of the elastic arteries, however, and the pressure falls slowly, reaching a minimum value of about 80 mm Hg, the diastolic pressure.

Pressure and Resistance

The values for systolic and diastolic pressure vary among healthy people, making the range of normal values quite broad. In addition, the values for blood pressure in a normal person are affected by factors such as physical activity and emotions. A standard blood pressure for a resting young adult male is 120/80 (120 mm Hg for the systolic pressure and 80 mm Hg for the diastolic pressure). Generally, if the blood pressure exceeds 150/90, it is considered too high.

As blood flows from arteries through the capillaries and the veins, the pressure falls progressively to about 0 mm Hg, or even slightly lower by the time blood is returned to the right atrium. In addition, the pressure is damped, in that the difference between the systolic and diastolic pressures is reduced in small diameter vessels. By the time blood reaches the capillaries, there is no variation in blood pressure, and only a steady pressure of about 30 mm Hg remains.

The greater the resistance in a blood vessel, the more rapidly the pressure decreases as blood flows through it. The most rapid decline in blood pressure occurs in the arterioles and capillaries because their small diameters make the resistance to blood flow high. Blood pressure declines slowly as blood flows from large to medium-sized arteries, because their diameters are larger, and the resistance to blood flow is not great.

Resistance to blood flow in veins is low because of their large diameters. Also, the valves that prevent backflow of blood in the veins, and skeletal muscle movements that periodically compress veins, force blood to flow toward the heart. Consequently, blood flows through veins even though the pressure in the veins is low.

The muscular arteries, arterioles, and precapillary sphincters are capable of constricting and dilating. If constriction occurs, resistance to blood flow increases, and the volume of blood flow through the vessels declines. Because they are able to constrict and dilate, muscular arteries help control the amount of blood flowing to each region of the body. Arterioles and precapillary sphincters, in contrast, regulate blood flow through specific tissues.

Pulse Pressure

The difference between the systolic and diastolic pressure is called the **pulse pressure.** If a person has a systolic pressure of 120 mm Hg and a diastolic pressure of 80 mm Hg, pulse pressure is 40 mm Hg. When the stroke volume increases, the systolic pressure increases more than the diastolic pressure, causing the pulse pressure to increase. During periods of exercise, the stroke volume and the pulse pressure are increased substantially.

In those who suffer from arteriosclerosis, the arteries are less elastic than normal. In these people, arterial pressure increases rapidly and falls rapidly, resulting in a large pulse pressure. The same amount of blood ejected into a less elastic artery will result in a higher systolic pressure compared to a more elastic artery. In people who suffer from arteriosclerosis, the pulse pressure is greater than normal even though the same amount of blood is ejected into the aorta as in a normal person. Because the left ventricle must produce greater pressure to eject the same amount of blood into a less elastic artery, arteriosclerosis increases the amount of work performed by the heart. In severe cases the increased workload on the heart leads to heart failure.

Ejection of blood from the left ventricle into the aorta produces a pressure wave, or **pulse,** which travels rapidly along the arteries. A pulse can be felt at locations where large arteries are close to the surface of the body. It is helpful to know the major locations where the pulse can be detected because monitoring the pulse is important clinically. The heart rate, rhythmicity, and other characteristics can be determined by feeling the pulse (Figure 13-14). For example, weak pulses usually indicate a decreased stroke volume or increased constriction of the arteries.

2 A weak pulse occurs in response to each premature beat of the heart. Explain the cause of the weak pulse under this condition.

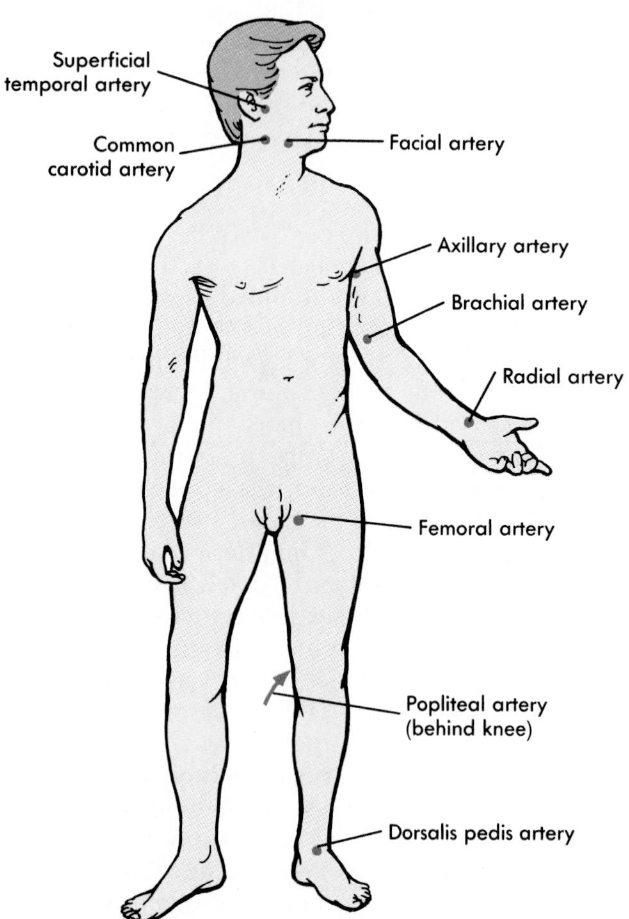

FIGURE 13-14 ■ Major points where the pulse can be monitored.
Each pulse point is named after the artery on which the pulse can be detected.

Capillary Exchange

There are about 10 billion capillaries in the body. Nutrients diffuse across the capillary walls into the interstitial spaces, and waste products diffuse in the opposite direction. In addition, a small amount of fluid is forced out of the capillaries at their arteriolar ends. Most of that fluid, but not all, reenters the capillaries at their venous ends.

Two major forces are responsible for the movement of fluid through the capillary wall (Figure 13-15). Blood pressure forces fluid out of the capillaries, and osmosis moves fluid into the capillaries. Osmosis attracts fluid into the capillaries because blood has a greater osmotic pressure than interstitial fluid. The greater the concentration of molecules dissolved in a fluid, the greater is the fluid's osmotic pressure (see Chapter 3). The greater osmotic pressure of blood is caused by the large concentration of blood proteins (see Chapter 11) that are unable to cross the capillary wall. The concentration of proteins in the tissue spaces is much lower than in blood. The capillary wall acts as a selectively permeable membrane, which prevents proteins from moving from the capillaries into the interstitial spaces but allows fluid to move across the wall of the capillaries.

At the arteriole end of capillaries, the movement of fluid out of the capillaries resulting from blood pressure is greater than the movement of fluid into capillaries due to osmosis. Consequently, there is a net movement of fluid out of the capillaries into the tissues.

At the venous end of the capillaries, blood pressure is lower than at the arteriolar end. Consequently, the movement of fluid out of the capillaries resulting from blood pressure is less than the movement of fluid into the capillaries resulting from osmosis, and there is a net movement of fluid from the tissue spaces into the capillaries.

Nine tenths of the fluid that leaves the capillaries at their arteriolar ends reenters the capillaries at their venous ends. The remaining one tenth of the fluid enters the lymphatic capillaries and is eventually returned to the general circulation.

Edema, or swelling, results from a disruption in the normal inwardly and outwardly directed pressures across the capillary walls. For example, inflammation results in an increase in the permeability of capillaries. Proteins leak out of the capillaries into the interstitial spaces. The proteins increase the osmotic pressure in the interstitial spaces, and fluid passes into capillaries by osmosis at a slower rate. Because less fluid moves by osmosis from the interstitial spaces into the capillaries, fluid accumulates in the interstitial spaces, resulting in edema.

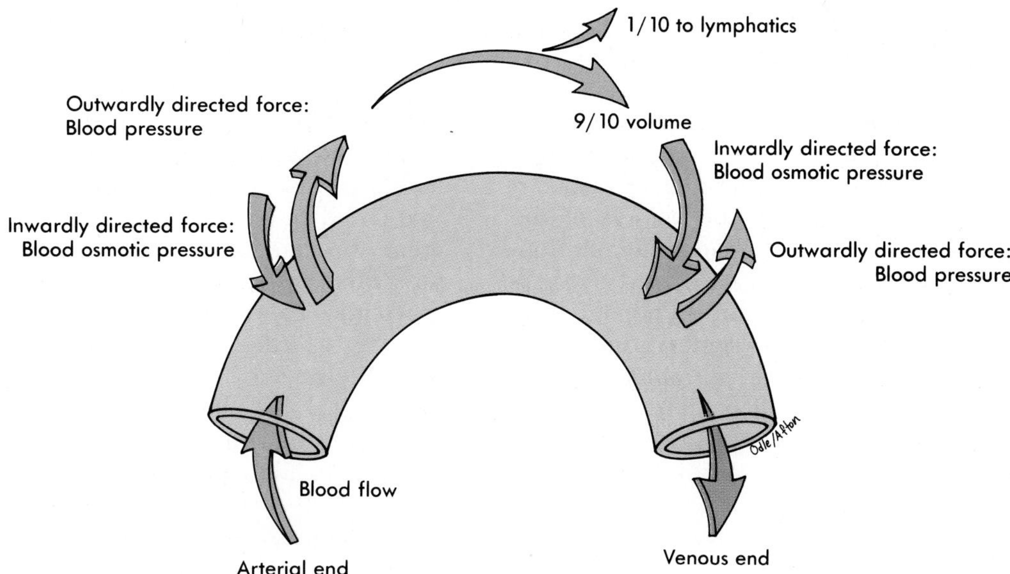

FIGURE 13-15 Capillary exchange.
At the arteriolar end of the capillary, the blood pressure that forces fluid out of the capillary is greater than the osmotic pressure that attracts fluid into the capillary. Thus, fluid is forced out of the capillaries at their arteriolar ends. At the venous ends of the capillaries, the blood pressure has decreased so that the blood pressure that forces fluid out of the capillary is less than the osmotic pressure that attracts fluid into the capillary. Thus, most of the fluid (about nine tenths) that was forced out of the capillary at its arteriolar end returns to the capillary at its venous end. The remainder of the fluid (about one tenth) enters the lymphatic vessels.

3 Explain edema, or swelling, (1) in response to a decrease in plasma protein concentration, or (2) as a result of increased blood pressure within the capillaries. ?

LOCAL CONTROL OF BLOOD VESSELS

Local control of blood flow is achieved by periodic contraction and relaxation of the precapillary sphincters. Because of the contraction and relaxation of the precapillary sphincters, blood flow through the capillaries is cyclic. The precapillary sphincters are controlled by the metabolic needs of the tissues. Blood flow increases when oxygen levels decrease or, to a lesser degree, when glucose, amino acids, fatty acids, and other nutrients decrease. Blood flow also increases when byproducts of metabolism build up in tissue spaces. An increase in carbon dioxide or a decrease in pH causes the precapillary sphincters to relax. For example, during exercise, blood flow through capillaries in exercising muscle increases dramatically because of the relaxation of the precapillary sphincters.

4 When blood flow to a tissue has been blocked for a short time, the blood flow through that tissue increases to as much as five times its normal value after the removal of the blockage. Create a reasonable explanation for that phenomenon based on what you know about the local control of blood flow. ?

In addition to the control of blood flow through existing capillaries, if the metabolic activity of a tissue increases often, additional capillaries gradually grow into the area. The additional capillaries allow local blood flow to be increased to a level that matches the metabolic demand of the tissue. For example, the density of capillaries in the well-trained skeletal muscles of athletes is greater than the density of capillaries in poorly trained skeletal muscles.

NERVOUS CONTROL OF BLOOD VESSELS

Nervous control of blood vessels is carried out primarily through the sympathetic division of the autonomic nervous system. Sympathetic vasoconstrictor fibers innervate most blood vessels of the body except the capillaries and precapillary sphincters, which have no nerve supply (Figure 13-16).

An area of the lower pons and upper medulla oblongata, called the **vasomotor center,** continually transmits a low frequency of nerve impulses to the sympathetic vasoconstrictor fibers. As a consequence, the peripheral blood vessels are continually in a partially constricted state, a condition called **vasomotor tone.** An increase in vasomotor tone causes blood pressure to increase, and a decrease in vasomotor tone causes blood pressure to decrease. Nervous control of blood vessel diameter is one way blood pressure is regulated. In addition, nervous control of blood vessels causes blood to be shunted from one large area of the body to another. For example, nervous control of blood vessels during exercise increases vasomotor tone in the viscera and skin and reduces vasomotor tone in exercising skeletal muscles. As a result, blood flow to the viscera and skin decreases, and blood flow to skeletal muscle increases.

REGULATION OF ARTERIAL PRESSURE

An adequate blood pressure is required to maintain blood flow through the blood vessels of the body, and several regulatory mechanisms assure that an adequate blood pressure is maintained. The **mean arterial blood pressure** is slightly less than the average of the systolic and diastolic pressures in the aorta. The mean arterial pressure is about 70 mm Hg at birth, is maintained at about 100 mm Hg from adolescence to middle age, and may reach 110 mm Hg in a healthy older person.

The mean arterial pressure is influenced by resistance to blood flow and by the cardiac output (CO), which is equal to the heart rate (HR) times the stroke volume (HR × SV). The resistance to blood flow in all the blood vessels is called peripheral resistance (PR). The mean arterial pressure (MAP) in the body is equal to the cardiac output times the peripheral resistance:

$$MAP = CO \times PR$$

or

$$MAP = HR \times SV \times PR$$

Blood pressure is controlled by changes in the heart rate, stroke volume, and peripheral resistance.

When blood pressure suddenly drops because of hemorrhage or some other cause, control systems attempt to reestablish blood pressure at a value consistent with life and to increase the blood volume to its normal value.

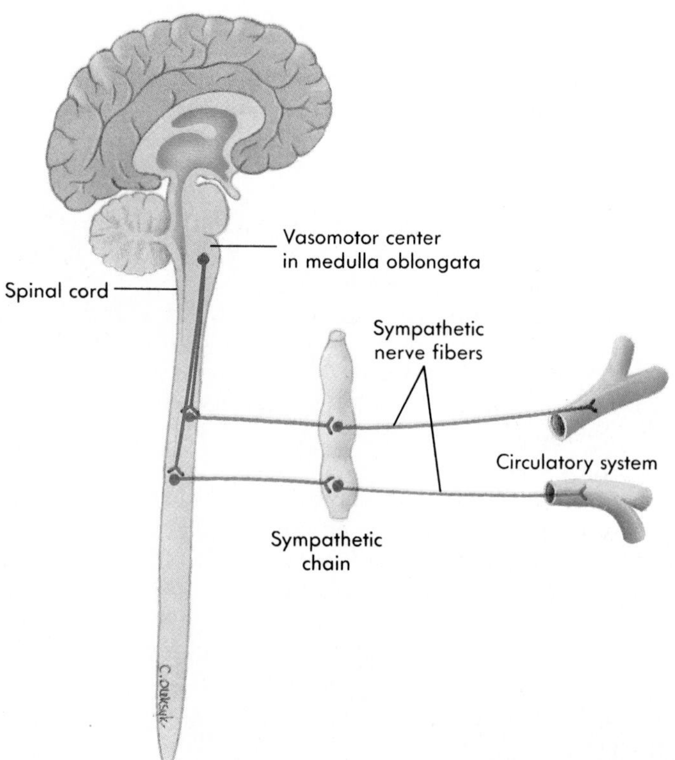

FIGURE 13-16 Nervous regulation of blood vessels. Most arteries and veins are innervated by sympathetic nerve fibers. A relatively constant frequency of action potentials is delivered to these blood vessels by the sympathetic nerve fibers, which keeps the blood vessels, especially the arteries, partially constricted; this phenomenon is called vasomotor tone. An increase in vasomotor tone causes constriction of blood vessels, which increases blood pressure. A decrease in vasomotor tone causes the blood vessels to dilate, which decreases blood pressure.

Baroreceptor Reflexes

Baroreceptors are pressure receptors that respond to stretch in arteries caused by increased pressure. They are scattered along the walls of most of the large arteries of the neck and thorax. There are many in the carotid sinus at the base of the internal carotid artery and in the walls of the aortic arch (Figure 13-17, *A*). Action potentials are transmitted from the baroreceptors to the medulla oblongata.

A sudden increase in blood pressure stretches the vessel walls and results in an increased action potential frequency in the baroreceptors. The increased action potential frequency delivered to the medulla oblongata causes responses that lower the blood pressure. One major response is a decrease in vasomotor tone, which results in vasodilation of blood vessels and a decrease in peripheral resistance. Another response is an increase in the parasympathetic stimulation of the heart, which decreases the heart rate. The decreased peripheral resistance and the decreased heart rate lower the blood pressure toward its normal value.

A sudden decrease in blood pressure results in a decreased action potential frequency in the baroreceptors. The decreased action potential frequency delivered to the medulla oblongata produces responses in the vasomotor center and cardioregulatory center, which function to raise blood pressure. A result is an increase in vasomotor tone, which causes vasoconstriction of blood vessels and an increase in peripheral resistance. In addition, there is an increase in sympathetic stimulation of the heart, which increases the heart rate. The increased peripheral resistance and increased heart rate raise the blood pressure toward its normal value (see Figure 13-17, *A*).

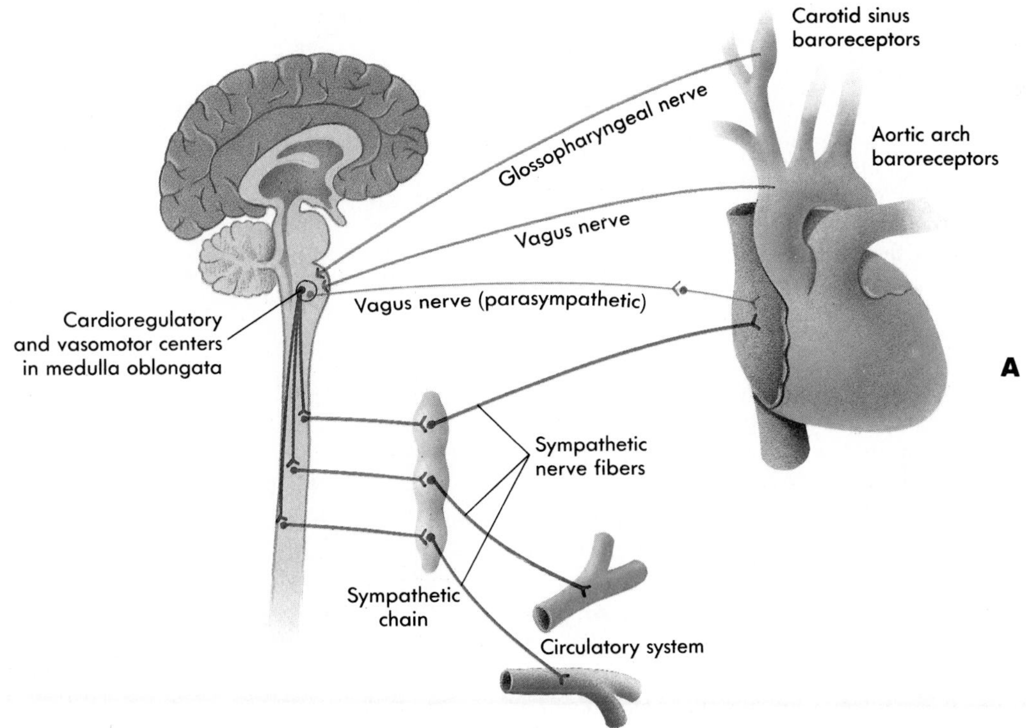

FIGURE 13-17 Nervous reflex mechanisms regulating blood pressure.
A Baroreceptor reflexes. Baroreceptors located in the carotid sinuses and aortic arch detect changes in blood pressure. Impulses from the baroreceptors are conducted to the cardioregulatory and vasomotor centers of the medulla oblongata through cranial nerves (from the carotid sinuses through the glossopharyngeal nerves and from the aortic arch through the vagus nerves). The heart rate then can be decreased by way of the parasympathetic system, or the heart rate and stroke volume can be increased by way of the sympathetic system. The sympathetic system also can constrict or dilate blood vessels. In response to a sudden decrease in blood pressure, sympathetic stimulation of the heart increases, and vasomotor tone increases. Both of these responses function to increase blood pressure back to its normal value. In response to a sudden increase in blood pressure, parasympathetic innervation of the heart increases and vasomotor tone decreases. These responses function to decrease blood pressure back to its normal value.

Continued.

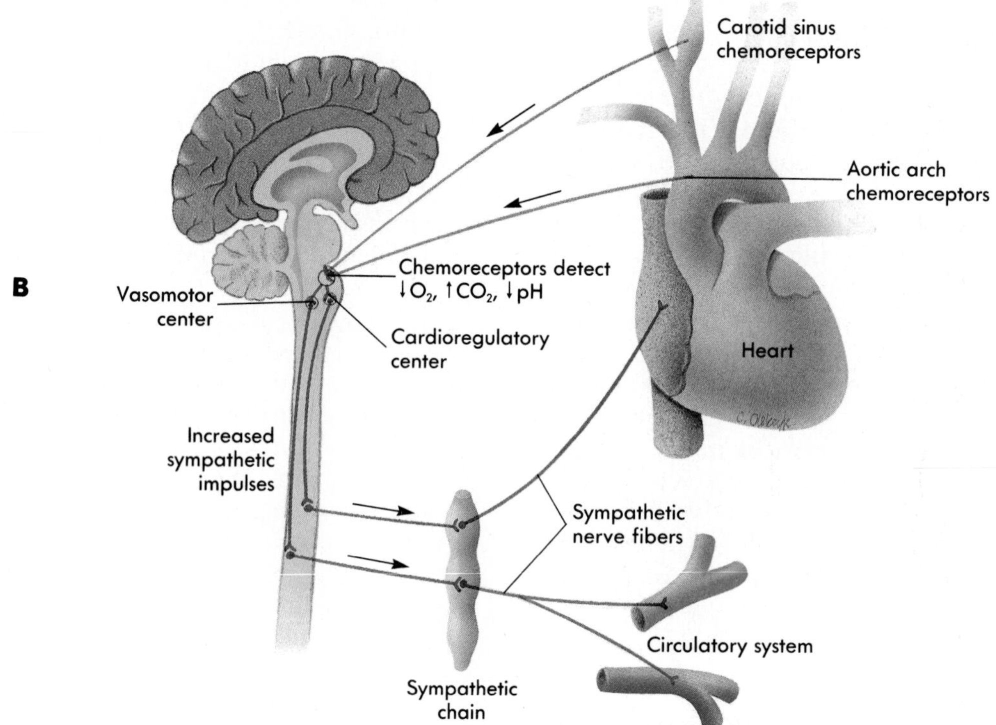

FIGURE 13-17, cont'd ■ Nervous reflex mechanisms regulating blood pressure. **B** Chemoreceptor reflexes. Chemoreceptors located in the carotid bodies, aortic bodies, and medulla oblongata detect changes in blood oxygen, carbon dioxide, and pH. Impulses are conducted to the medulla oblongata from the carotid bodies and aortic bodies by cranial nerves. Decreased oxygen, increased carbon dioxide, and decreased pH of the blood result in increased sympathetic stimulation of the heart and an increase in vasomotor tone.

These **baroreceptor reflexes** are important in regulating blood pressure on a moment-to-moment basis. For example, when a person rises rapidly from a sitting or lying position to a standing position, blood pressure in the neck and thoracic regions drops dramatically as a result of the pull of gravity on the blood. This reduction may be so great that blood flow to the brain is reduced enough to cause dizziness or loss of consciousness. The falling blood pressure activates the baroreceptor reflexes, however, which reestablish normal blood pressure within a few seconds. In a healthy person, a temporary sensation of dizziness is all that may be experienced.

5 Explain how the baroreceptor reflex would respond if a person did a headstand.

Chemoreceptor Reflexes

Carotid bodies are small structures that lie near the carotid sinuses, and **aortic bodies** are structures near the aortic arch. These structures contain sensory receptors that respond to changes in oxygen concentration, carbon dioxide concentration, and pH in blood. Because they are sensitive to chemical changes in the blood they are called **chemoreceptors.** They send action potentials along nerve fibers to the medulla oblongata (Figure 13-17, *B*). There are also chemoreceptors in the medulla oblongata.

When oxygen levels decrease, carbon dioxide levels increase, or pH decreases, the chemoreceptors respond with an increased frequency of action potentials, called the **chemoreceptor reflex.** In response, there is an increase in vasomotor tone and an increase in sympathetic stimulation to the heart, resulting in an increased blood pressure. The increased blood pressure causes a greater rate of blood flow to the lungs, which helps increase blood oxygen levels and reduce blood carbon dioxide levels (see Figure 13-17, *B*). The chemoreceptors function under emergency conditions and usually do not play an important role in the regulation of the cardiovascular system. They respond strongly only when the oxygen levels in the blood fall to very low levels, or carbon dioxide levels become substantially elevated.

Circulatory Shock

Circulatory shock is defined as an inadequate blood flow throughout the body. As a consequence, tissues suffer damage resulting from a lack of oxygen. Severe shock may damage vital body tissues and lead to death.

There are several causes of shock, but hemorrhagic shock, also known as hypovolemic shock, can be used to illustrate the general characteristics of shock. If shock is not severe, blood pressure may decrease only a moderate amount. Under these conditions the mechanisms that normally regulate blood pressure function to reestablish normal blood pressure and blood flow. The baroreceptor reflexes initiate strong sympathetic responses that result in intense vasoconstriction and increased heart rate. As a result of the reduced blood flow through the kidneys, increased amounts of renin are released. The elevated renin release results in a greater rate of angiotensin formation, causing vasoconstriction and increased aldosterone release from the adrenal cortex. The aldosterone, in turn, promotes water and salt retention by the kidneys. In response to reduced blood pressure, ADH is released from the posterior pituitary gland, and ADH also enhances the retention of water by the kidneys. An intense sensation of thirst leads to increased water intake, which helps to restore the normal blood volume.

In mild cases of shock the baroreceptor reflexes may be adequate to compensate for blood loss until the blood volume is restored, but in more severe cases all mechanisms are required to sustain life.

In more severe cases of shock, the regulatory mechanisms are not adequate to compensate for the effects of shock. As a consequence, a positive-feedback cycle begins to develop in which the blood pressure regulatory mechanisms lose their ability to compensate, and shock worsens. As shock becomes worse, the effectiveness of the regulatory mechanisms deteriorates. The positive-feedback cycle proceeds until death occurs, or until treatment is applied that terminates the cycle.

Several types of shock are classified by the cause of the condition:

1. Hemorrhagic shock, caused by internal or external bleeding.
2. Plasma loss shock, caused by reduced blood volume resulting from a loss of plasma into the interstitial spaces. Examples are loss of plasma from severely burned areas, dehydration, and severe diarrhea or vomiting.
3. Neurogenic shock, which results in vasodilation in response to emotional upset or anesthesia.
4. Anaphylactic shock, caused by an allergic response that results in the release of inflammatory substances that cause vasodilation and an increase in capillary permeability.
5. Septic shock or "blood poisoning," which results from severe infections that cause the release of toxic substances into the circulatory system, depressing the activity of the heart and leading to vasodilation and increased capillary permeability.

Hormonal Mechanisms

In addition to the rapidly acting nervous mechanisms that regulate arterial pressure, there are important hormonal mechanisms that help control blood pressure.

Adrenal medullary mechanism

Stimuli that result in increased sympathetic stimulation of the heart and the blood vessels also cause increased stimulation of the adrenal medulla. The adrenal medulla responds by releasing epinephrine into the blood. Epinephrine increases heart rate and stroke volume and causes vasoconstriction, especially of blood vessels to the skin and viscera. Epinephrine also causes vasodilation of blood vessels in skeletal muscle and cardiac muscle. Epinephrine, therefore, increases the supply of blood flowing to the skeletal muscle and cardiac muscle, and this prepares one for physical activity (Figure 13-18, *A*).

Renin-angiotensin-aldosterone mechanism

In response to reduced blood pressure, the kidneys release an enzyme called **renin** into the circulatory system (Figure 13-18, *B*). Renin acts on the blood protein angiotensinogen to produce **angiotensin.** Other enzymes in the circulatory system act on the angiotensin to convert it to its most active form. Active angiotensin is a potent vasoconstrictor substance. Thus, in response to a lack of oxygen or reduced blood pressure, the release of renin by the kidney acts to increase the blood pressure toward its normal value.

Angiotensin also acts on the adrenal cortex to increase the secretion of **aldosterone.** Aldosterone acts on the kidneys and causes them to conserve sodium ions and water. As a result, the volume of water lost from the blood into the urine is reduced. A further decrease in blood pressure is resisted because of the secretion of aldosterone.

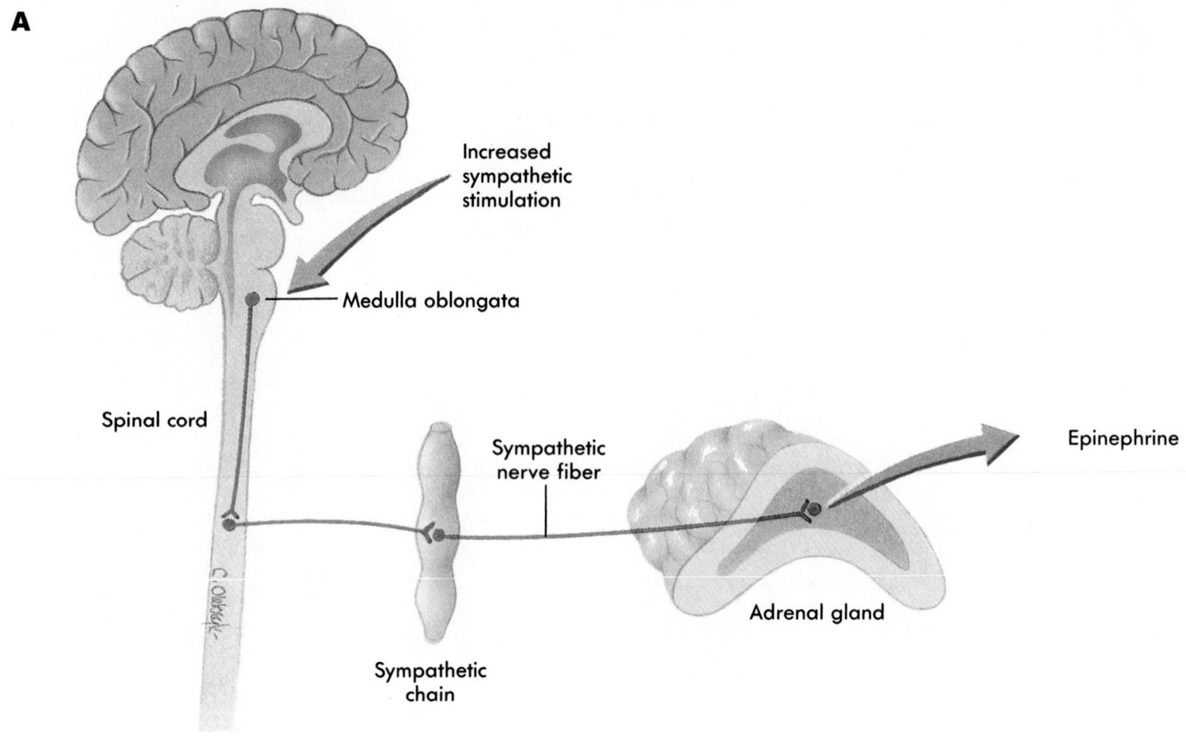

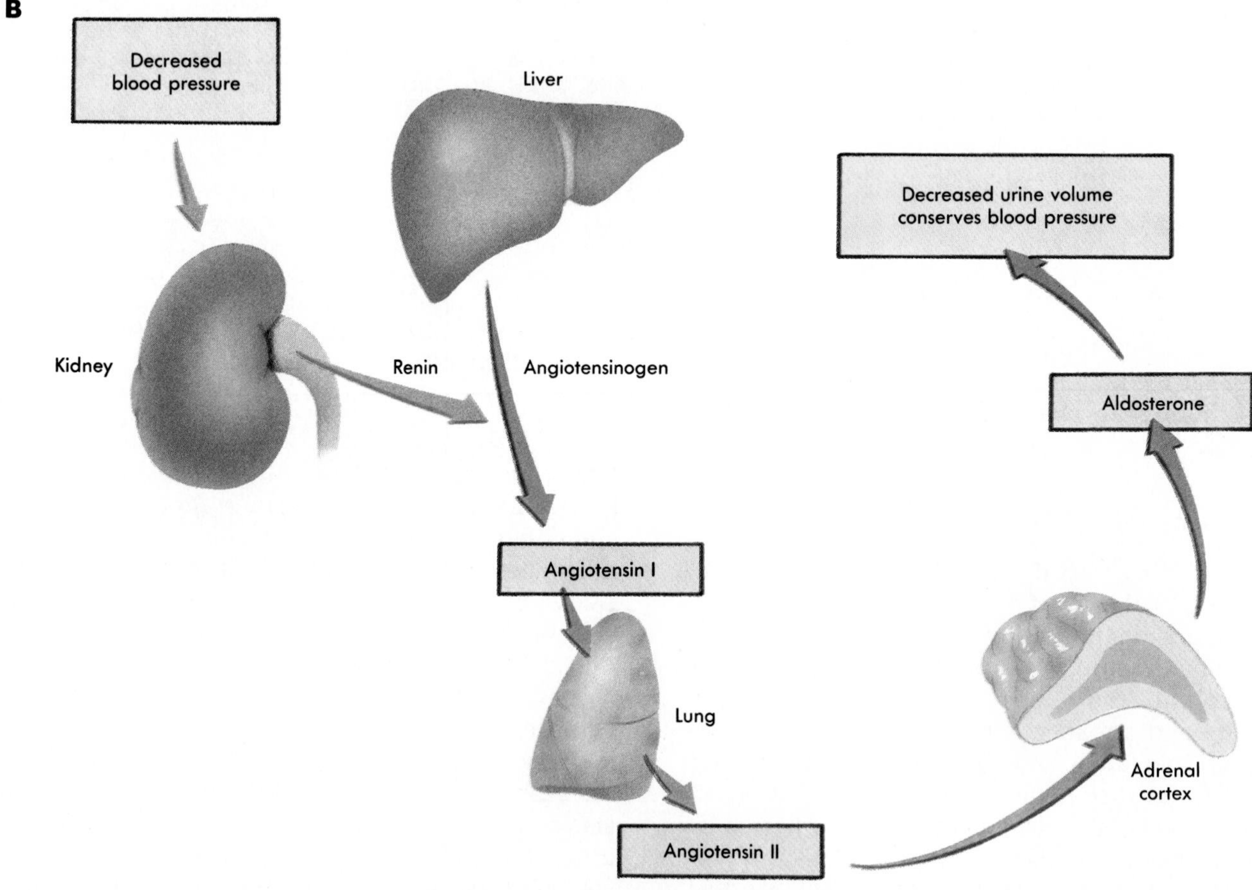

FIGURE 13-18 ■ For legend see opposite page.

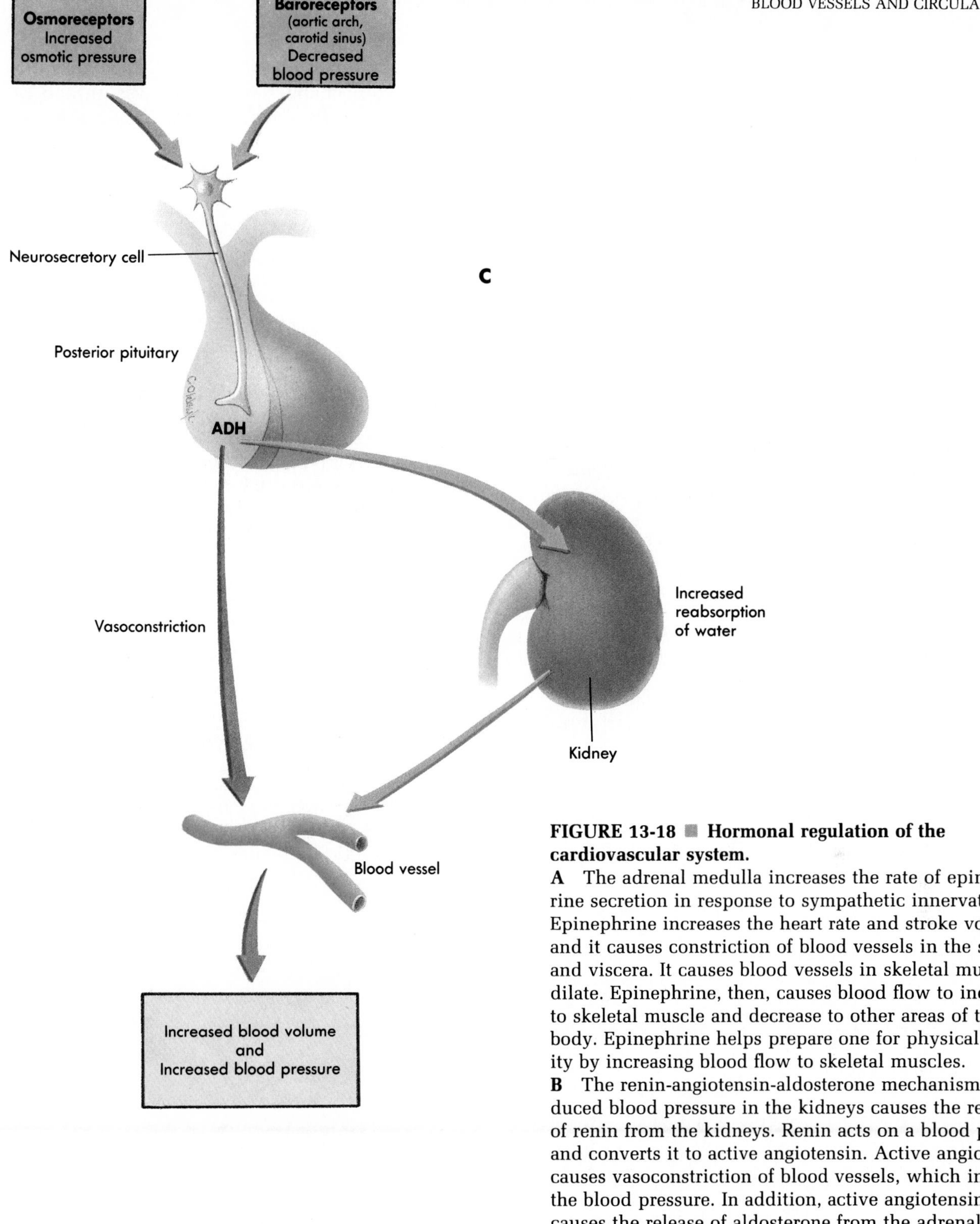

FIGURE 13-18 ■ Hormonal regulation of the cardiovascular system.

A The adrenal medulla increases the rate of epinephrine secretion in response to sympathetic innervation. Epinephrine increases the heart rate and stroke volume, and it causes constriction of blood vessels in the skin and viscera. It causes blood vessels in skeletal muscle to dilate. Epinephrine, then, causes blood flow to increase to skeletal muscle and decrease to other areas of the body. Epinephrine helps prepare one for physical activity by increasing blood flow to skeletal muscles.

B The renin-angiotensin-aldosterone mechanism. A reduced blood pressure in the kidneys causes the release of renin from the kidneys. Renin acts on a blood protein and converts it to active angiotensin. Active angiotensin causes vasoconstriction of blood vessels, which increases the blood pressure. In addition, active angiotensin causes the release of aldosterone from the adrenal cortex. Aldosterone causes the kidneys to conserve water and salt, which reduces urine volume and helps maintain blood pressure by preventing the loss of water from the blood through the kidneys as urine.

C The ADH mechanism. Either reduced blood pressure or increased blood osmotic pressure influences the hypothalamus, which increases the secretion rate of ADH from the posterior pituitary. ADH acts on the kidneys and causes them to reduce urine volume. In addition, in higher concentrations ADH does cause vasoconstriction.

Vasopressin mechanism

When blood pressure drops or the concentration of solutes in the plasma increases, neurons in the hypothalamus respond by causing the release of **antidiuretic hormone (ADH),** also called vaso pressin. ADH acts on the kidneys and causes a greater reabsorption of fluid by the kidney, thereby decreasing urine volume. This increases fluid overall, that is, less fluid is lost from the body, thus maintaining blood volume and blood pressure. (Figure 13-18, C).

Atrial natriuretic mechanism

A substance called **atrial natriuretic factor** is released primarily from specialized cells of the right atrium in response to elevated blood pressure. Atrial natriuretic factor acts on the kidneys and causes them to increase urine volume. Loss of water in the urine causes blood volume to decrease, thus decreasing the blood pressure.

Long-Term and Short-Term Regulation of Blood Pressure

The baroreceptor mechanisms are most important in controlling blood pressure on a short-term basis. They are sensitive to sudden changes in blood pressure, and they respond quickly. Hormonal mechanisms, however, such as the renin-angiotensin-aldosterone system and atrial natriuretic factor are more important in the maintenance of blood pressure on a long-term basis. They are influenced by small changes in blood pressure and respond by gradually bringing the blood pressure back to its normal range. (Figure 13-18, C).

Hypertension, or high blood pressure, affects about 20% of all people at some time in their lives. Generally, a person is considered hypertensive if the systolic blood pressure is greater than 150 mm Hg, and the diastolic blood pressure is greater than 90 mm Hg. Chronic hypertension has an adverse effect on the heart and the blood vessels. Hypertension requires the heart to perform a greater-than-normal amount of work. The extra work leads to hypertrophy of the cardiac muscle, especially in the left ventricle, and can lead to heart failure. Hypertension also increases the rate at which arteriosclerosis develops. Arteriosclerosis, in turn, increases the chance that blood clots will form and that blood vessels will rupture. Common conditions associated with hypertension are cerebral hemorrhage, coronary infarction, hemorrhage of renal blood vessels, and poor vision resulting from burst blood vessels in the retina.

Treatments that dilate blood vessels, increase the rate of urine production, or decrease cardiac output are normally used for hypertension. Low-salt diets also are normally recommended to reduce the amount of sodium chloride and water absorbed from the intestine into the bloodstream.

SUMMARY

The peripheral circulatory system can be divided into the systemic and the pulmonary vessels.

The peripheral circulatory system and the heart are regulated to maintain sufficient blood flow to tissues.

GENERAL FEATURES OF BLOOD VESSEL STRUCTURE

Blood is pumped from the heart through elastic arteries, muscular arteries, and arterioles to the capillaries.

Blood returns to the heart from the capillaries through venules, small veins, and large veins.

Except for capillaries and venules, blood vessels have three layers:

- The tunica intima consists of endothelium, basement membrane, and connective tissue.
- The tunica media, the middle layer, contains circular smooth muscle and elastic fibers.
- The outer tunica adventitia is connective tissue.

Arteries

Large elastic arteries have many elastic fibers but little smooth muscle in their walls.

Muscular arteries have much smooth muscle and some elastic fibers.

Arterioles are the smallest arteries and have smooth muscle cells and a few elastic fibers.

Capillaries

Capillaries consist of only endothelium and are surrounded by a basement membrane and loose connective tissue.

Nutrient and waste product exchange is the principal function of capillaries.

Blood is supplied to capillaries by arterioles. Precapillary sphincters regulate blood flow through capillary networks.

Veins

Venules are endothelium surrounded by a basement membrane.

Small veins are venules covered with a layer of smooth muscle.

Medium-sized and large veins contain less smooth muscle and elastic fibers than arteries of the same size.

Valves prevent the backflow of blood in the veins.

Aging of the Arteries

Arteriosclerosis results from a loss of elasticity primarily in the aorta, large arteries, and coronary arteries.

Atherosclerosis results from the deposition of plaques rich in cholesterol in the wall of blood vessels, reducing the diameter of the vessels.

PULMONARY CIRCULATION

The pulmonary circulation moves blood to and from the lungs. Pulmonary arteries carry oxygen-poor blood from the heart to the lungs, and pulmonary veins carry oxygen-rich blood from the lungs to the left atrium of the heart.

SYSTEMIC CIRCULATION: ARTERIES

Aorta

The aorta leaves the left ventricle to form the ascending aorta, aortic arch, and descending aorta, which consists of the thoracic and abdominal aorta.

Arteries of the Head and Neck	The brachiocephalic, left common carotid, and left subclavian arteries branch from the aortic arch to supply the head and the upper limbs. The common carotid arteries and the vertebral arteries supply the head. The common carotid arteries divide to form the external carotids (which supply the face and mouth) and the internal carotids (which supply the brain).
Arteries of the Upper Limbs	The subclavian artery continues as the axillary artery and then as the brachial artery, which branches to form the radial and ulnar arteries.
The Thoracic Aorta and Its Branches	The thoracic aorta has visceral branches, which supply the abdominal organs, and parietal branches, which supply the thoracic wall.
The Abdominal Aorta and Its Branches	The abdominal aorta has visceral branches, which supply the abdominal organs, and parietal branches, which supply the abdominal wall.
Arteries of the Pelvis	Branches of the internal iliac arteries supply the pelvis.
Arteries of the Lower Limbs	The common iliac arteries give rise to the external iliac arteries, and the external iliac artery continues as the femoral artery and then as the popliteal artery in the leg. The popliteal artery divides to form the anterior and posterior tibial arteries.
SYSTEMIC CIRCULATION: VEINS	The superior vena cava drains the head, neck, thorax, and upper limbs. The inferior vena cava drains the abdomen, pelvis, and lower limbs.
Veins of the Head and Neck	The internal jugular veins drain the brain, anterior head, and anterior neck. The external jugular veins drain the posterior head and posterior neck.
Veins of the Upper Limbs	The deep veins are the brachial, axillary, and subclavian; the superficial veins are the basilic, cephalic, and median cubital.
Veins of the Thorax	The left and right brachiocephalic veins and the azygos veins return blood to the superior vena cava.
Veins of the Abdomen and Pelvis	Abdominal veins join the azygos vein. Vessels from the kidneys, adrenal glands, and gonads directly enter the inferior vena cava. Vessels from the stomach, intestines, spleen, and pancreas connect with the hepatic portal vein. The hepatic portal vein tranports blood to the liver for processing. Hepatic veins from the liver join the inferior vena cava.
Veins of the Lower Limbs	The deep veins are the peroneal, anterior tibial, and posterior tibial. The superficial veins are the small and great saphenous veins.
THE PHYSIOLOGY OF CIRCULATION	
Blood Pressure	Blood pressure is a measure of the force exerted by blood against the blood vessel wall. Blood pressure moves blood through vessels. Blood pressure can be measured by listening for Korotkoff sounds produced as blood flows through arteries partially constricted by a blood pressure cuff.
Pressure and Resistance	Blood pressure fluctuates between 120 mm Hg (systolic) and 80 mm Hg (diastolic) in the aorta and drops to 0 mm Hg in the right atrium. The greatest drop occurs in the arterioles and capillaries. If constriction of blood vessels occurs, resistance to blood flow increases, and blood flow decreases.

Pulse Pressure

Pulse pressure is the difference between systolic and diastolic pressure. Pulse pressure increases when stroke volume increases.

A pulse can be detected when large arteries are near the surface of the body.

Capillary Exchange

Most exchange across the wall of the capillary is by diffusion.

Blood pressure, capillary permeability, and osmosis affect movement of fluid across the wall of the capillaries. There is a net movement of fluid from the blood into the tissues. The fluid gained by the tissues is removed by the lymphatic system.

LOCAL CONTROL OF BLOOD VESSELS

Blood flow through a tissue is usually proportional to the metabolic needs of the tissue and is controlled by the precapillary sphincters.

Nervous Regulation of Blood Vessels

The vasomotor center (sympathetic nervous system) controls blood vessel diameter. Other brain areas can excite or inhibit the vasomotor center.

Vasomotor tone is a state of partial contraction of blood vessels.

The nervous system is responsible for routing the flow of blood, except in the capillaries and precapillary sphincters, and is responsible for maintaining blood pressure.

REGULATION OF ARTERIAL PRESSURE

Mean arterial blood pressure is proportional to cardiac output times the peripheral resistance.

Baroreceptor Reflexes

Baroreceptors are sensory receptors that are sensitive to stretch.
- Baroreceptors are located in the carotid sinuses and the aortic arch.
- The baroreceptor reflex changes peripheral resistance, heart rate, and stroke volume in response to changes in blood pressure.

Chemoreceptors are sensory receptors sensitive to oxygen, carbon dioxide, and pH levels in the blood.
- Chemoreceptors are located in the carotid bodies and the aortic bodies.
- The chemoreceptor reflex increases peripheral resistance, and increases sympathetic stimulation of the heart, usually in response to low oxygen levels.

Hormonal Mechanisms

Epinephrine released from the adrenal medulla as a result of sympathetic stimulation increases heart rate, stroke volume, and vasoconstriction.

Renin is released by the kidneys in response to low blood pressure. Renin promotes the production of angiotensin which, when activated, causes vasoconstriction and an increase in aldosterone secretion. Aldosterone reduces urine output.

Antidiuretic hormone (ADH) or vasopressin released from the posterior pituitary causes vasoconstriction and reduces urine output.

Atrial natriuretic factor is released from the heart when atrial blood pressure increases. It stimulates an increase in urine production, causing a decrease in blood volume and blood pressure.

Long-Term and Short-Term Regulation of Blood Pressure

In response to an increase in blood volume, the kidneys produce more urine and decrease blood volume. The renin-angiotensin-aldosterone mechanism and atrial natriuretic factor play a major role in controlling urine volume.

CONTENT REVIEW

1. Name, in order, all the types of blood vessels, starting at the heart, going to the tissues, and returning to the heart.
2. Name the three layers of a blood vessel. What kinds of tissue are in each layer?
3. Relate the structure of the different types of arteries to their functions.
4. Describe the structure of capillaries and explain their major function.
5. Describe the structure of a capillary network. Where is the smooth muscle that regulates blood flow through the capillary network?
6. Describe the structure of veins.
7. What is the function of valves in blood vessels and in which blood vessels are valves found?
8. Describe the changes that occur in arteries as they age.
9. List the different parts of the aorta. Name the major arteries that branch from the aorta to supply the heart, the head and upper limbs, and the lower limbs.
10. Name the arteries that supply the head, upper limbs, thorax, abdomen, and lower limbs. Describe the specific areas each artery supplies.
11. Name the major vessels that return blood to the heart. What area of the body does each drain?
12. Name the major veins that return blood to the superior vena cava.
13. List the veins that drain blood from the thorax, abdomen, and pelvis. What specific area of the body does each drain? Describe the hepatic portal system.
14. List the major veins that drain the upper and lower limbs.
15. Define blood pressure and describe how it is normally measured.
16. Describe the changes in blood pressure starting in the aorta, moving through the vascular system, and returning to the right atrium.
17. Define pulse pressure and explain what information can be determined from monitoring the pulse.
18. Explain how blood pressure and osmosis affect the movement of fluid between capillaries and tissues. What happens to excess fluid that enters the tissues?
19. Explain what is meant by the local control of blood flow through tissues and describe what carries out local control.
20. Describe nervous control of blood vessels. Define vasomotor tone.
21. Define mean arterial pressure. How is it related to heart rate, stroke volume, and peripheral resistance?
22. Where are baroreceptors located? Describe the baroreceptor reflex when blood pressure increases and when it decreases.
23. Where are the chemoreceptors located? Describe what happens when oxygen levels in the blood decrease.
24. For each of the following—epinephrine, renin, angiotensin, aldosterone, ADH, and atrial natriuretic factor—state where each is produced, what stimulus causes an increased production, and what effects it has on the circulatory system.

CONCEPT REVIEW

1. For each of the following destinations, name all the arteries that a red blood cell would encounter if it started its journey in the left ventricle.
   A. The brain
   B. External portion of the skull
   C. The left hand
   D. Anterior portion of the leg

2. For each of the following starting places, name all the veins that a red blood cell would encounter on its way back to the right atrium.
   A. The brain
   B. External portion of the skull
   C. Hand
   D. Anterior portion of leg
   E. Kidney
   F. Small intestine
3. Describe the responses to a rapid decrease in blood pressure as occurs in hemorrhagic shock.
4. Hugo Faster ran a race. During the race his stroke volume and heart rate increased. Vasoconstriction occurred in his viscera and his blood pressure increased, but not dramatically. Explain these changes in his circulatory system.

ANSWERS TO  PREDICT QUESTIONS

1 *p. 326* Arteriosclerosis slowly reduces blood flow through the carotid arteries and therefore the amount of blood that flows to the brain. In advanced stages of arteriosclerosis the resistance to blood flow increases so much that the blood flow to the brain through the carotid arteries is reduced significantly. Results are confusion, loss of memory, and a reduced ability to perform other normal brain functions.

2 *p. 336* Premature beats of the heart result in contraction of the heart muscle before the heart has had time to fill to its normal capacity. Consequently, there is a reduced stroke volume, which results in a weak pulse associated with the premature contraction.

3 *p. 337* Decreased plasma protein concentration reduces the osmotic pressure of the blood. Edema results because the pressure that forces fluid out of the capillary (the blood pressure) becomes much greater than the force that moves blood into the capillary (the osmotic pressure of blood). Consequently, fluid accumulates in the interstitial spaces, resulting in edema. Fluid normally leaves the arterial end of the capillary because the capillary blood pressure is greater than the osmotic pressure. Fluid normally reenters the venous end of the capillary because the osmotic pressure is greater than the blood pressure at the venous end of the capillary. When the osmotic pressure is reduced, more fluid leaves the capillary at the arterial end, and less fluid reenters the capillary at the venous end.

Increased blood pressure within the capillaries has a similar effect. The amount of fluid that is forced from the arteriolar end of the capillaries is increased because the blood pressure has increased without an increase in osmotic pressure to counteract it. The elevated blood pressure at the venous end of the capillary will reduce the amount of fluid reentering the capillary. Even though blood pressure declines between the arterial and venous ends of the capillary, it still would be higher than normal, and it still would resist the movement of fluid back into the capillary.

4 *p. 337* When a blood vessel is blocked, oxygen and nutrients are depleted, and waste products accumulate in tissue supplied by the blocked blood vessel. The reduced supply of oxygen and nutrients and the accumulated waste products all cause relaxation of the precapillary sphincters and greatly increase blood flow through the area after the block has been removed.

5 *p. 340* During a headstand, gravity acting on the blood would cause the blood pressure to increase in the areas of the aortic arch and carotid sinus baroreceptors. The increased pressure would activate the baroreceptor reflexes, resulting in an increased parasympathetic stimulation and a decreased sympathetic stimulation of the heart. The heart would respond with a decreased heart rate. Because standing on one's head also causes blood from the periphery to run downhill to the heart, the venous return would increase, causing the stroke volume to increase due to Starling's law of the heart. Some peripheral vasodilation also might occur. The increased pressure detected by the baroreceptors would reduce sympathetic stimulation of blood vessels (reduce vasomotor tone), allowing them to dilate in an attempt to reduce the blood pressure.

14

THE LYMPHATIC SYSTEM AND IMMUNITY

After reading this chapter you should be able to:

1 Describe the functions of the lymphatic system.
2 Explain how lymph is formed and transported.
3 Describe the structure and function of tonsils, lymph nodes, the spleen, and the thymus.
4 Define nonspecific resistance and describe the cells and chemicals involved.
5 List the events that occur during an inflammatory response and explain their significance.
6 Define the term antigen.
7 Describe the origin, development, activation, and regulation of lymphocytes.
8 Define antibody-mediated immunity and cell-mediated immunity and name the cells responsible for each.
9 Diagram the structure of an antibody and describe the effects produced by antibodies.
10 Discuss the primary and secondary response to an antigen. Explain the basis for long-lasting immunity.
11 Describe the functions of T cells and null cells.
12 Explain the four ways that specific immunity can be acquired.

antibody Protein found in the plasma that is responsible for humoral (antibody-mediated) immunity; binds specifically to antigen.

antibody-mediated immunity Immunity due to B cells and the production of antibodies.

antigen (an′tĭ-jen) Any substance that induces a state of sensitivity and/or resistance to microorganisms or toxic substances after a latent period; substance that stimulates the specific immune system.

cell-mediated immunity Immunity due to the actions of T cells.

complement Group of serum proteins that stimulates phagocytosis, inflammation, and lysis of cells.

interferon (in′ter-fēr′on) A protein that inhibits viral replication.

lymph (limf) [L. *lympha*, clear spring water] Clear or yellowish fluid derived from interstitial fluid and found in lymph vessels.

lymph node Encapsulated mass of lymphatic tissue found along lymph vessels; functions to filter lymph and produce lymphocytes.

lymphokine (lim′fo-kīn) Chemicals produced by T cells that activate macrophages and other immune cells; promotes phagocytosis and inflammation.

memory cell Lymphocytes derived from B cells or T cells that have been exposed to an antigen; when exposed to the same antigen a second time, memory cells rapidly respond to provide immunity.

mononuclear phagocytic system Phagocytic cells with a single nucleus, derived from monocytes; the cells either enter a tissue by chemotaxis in response to infection or tissue damage, or are positioned to intercept microorganisms entering tissues.

nonspecific resistance Immune system response that is the same upon each exposure to an antigen; there is no ability to remember a previous exposure to a specific antigen.

specific immunity Immune system response in which there is an ability to recognize, remember, and destroy a specific antigen.

vaccine Preparation of killed microorganisms, altered microorganisms, or derivatives of microorganisms intended to produce immunity. Usually administered by injection, but sometimes ingestion is preferred.

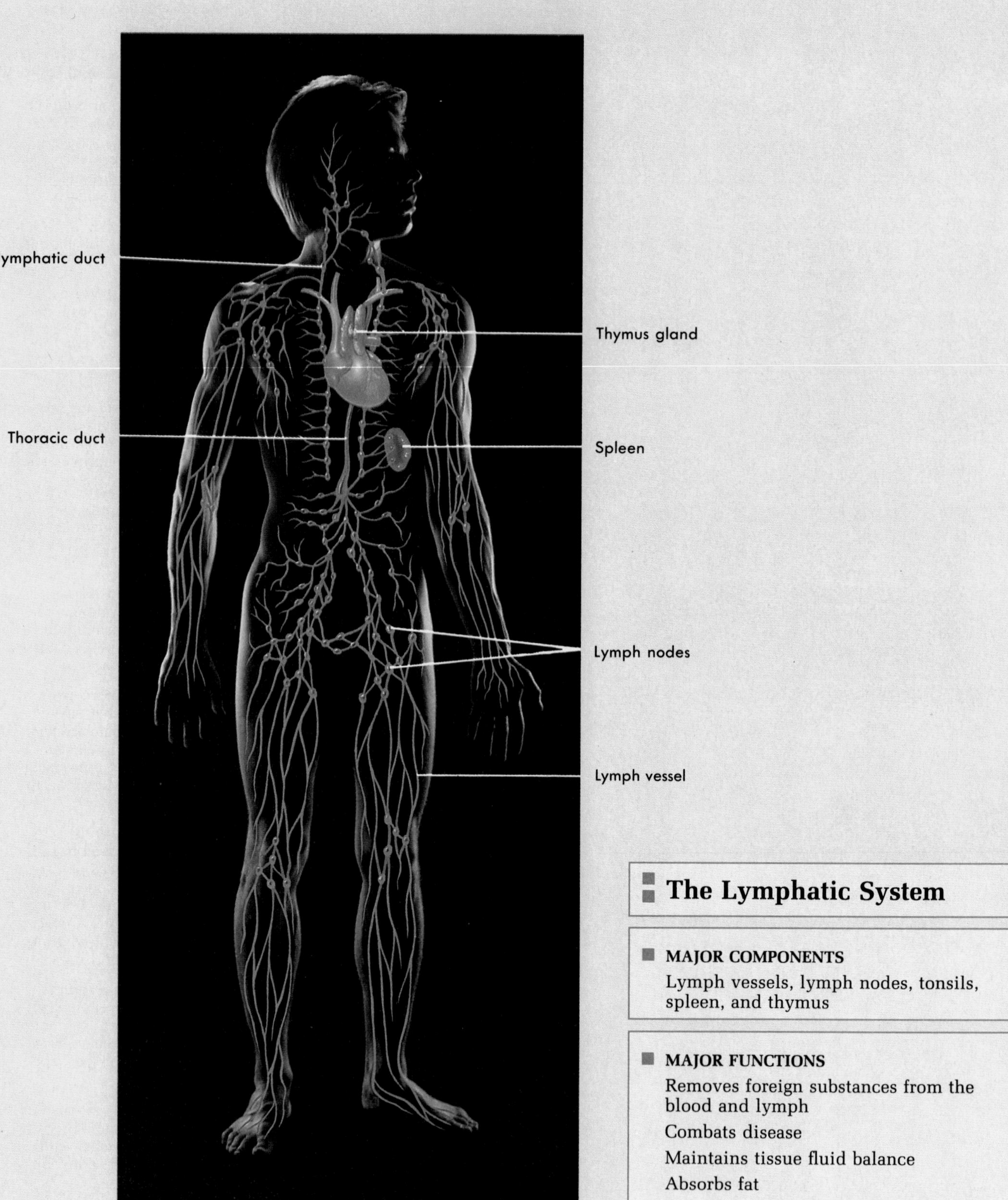

The Lymphatic System

MAJOR COMPONENTS

Lymph vessels, lymph nodes, tonsils, spleen, and thymus

MAJOR FUNCTIONS

Removes foreign substances from the blood and lymph

Combats disease

Maintains tissue fluid balance

Absorbs fat

FIGURE 14-1 ■ Lymphatic system.
The lymph vessels and major lymphatic organs are shown.

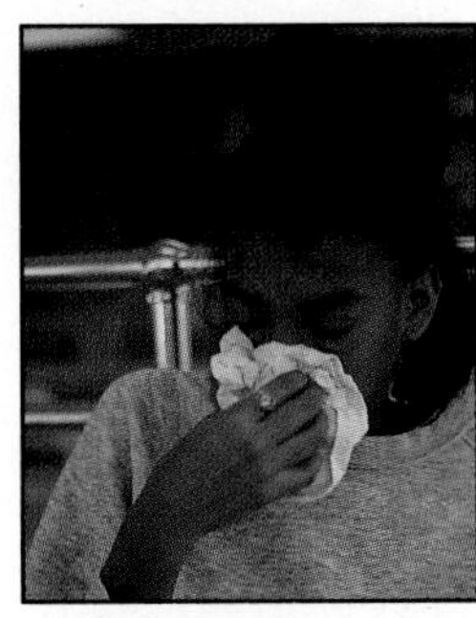

The lymphatic system includes lymph, lymphocytes, lymph vessels, lymph nodes, tonsils, the spleen, and the thymus gland (Figure 14-1). The lymphatic system performs three basic functions. First, it helps maintain fluid balance in the tissues. Approximately 30 liters of fluid pass from the blood capillaries into the interstitial spaces each day, whereas only 27 liters pass from the interstitial spaces back into the blood capillaries. If the extra 3 liters of interstitial fluid were to remain in the interstitial spaces, edema would result, causing tissue damage and eventual death. These 3 liters of fluid enter the lymphatic capillaries, where the fluid is called **lymph** (limf; clear spring water), and it passes through the lymph vessels to return to the blood. Lymph is similar in composition to plasma (see Chapter 11). In addition to water, lymph contains solutes derived from two sources. Substances in plasma, such as ions, nutrients, gases, and some proteins, pass from blood capillaries into the interstitial spaces and become part of the lymph. Substances derived from cells within the tissues such as hormones, enzymes, and waste products are also found in the lymph.

Second, the lymphatic system absorbs fats and other substances from the digestive tract (see Chapter 16). Special lymph vessels called **lacteals** (lak′te-als) are located in the lining of the small intestine. Fats enter the lacteals and pass through the lymph vessels to the venous circulation. The lymph passing through these lymph vessels has a milky appearance because of its fat content, and it is called **chyle** (kīl).

Third, the lymphatic system is part of the body's defense system. Lymphatic organs such as lymph nodes filter lymph, and the spleen filters blood, removing microorganisms and other foreign substances. In addition, lymphatic organs contain lymphocytes and other cells that are capable of destroying microorganisms and foreign substances.

LYMPHATIC SYSTEM

Lymph Vessels

The **lymphatic** (lim-fat′ik) **system,** unlike the circulatory system, only carries fluid *away* from the tissues. The lymphatic system begins in tissues as **lymph capillaries,** which are tiny, close-ended vessels consisting of simple squamous epithelium. Lymph capillaries are in almost all tissues of the body except the central nervous system, bone marrow, and tissues without blood vessels (for example, epidermis and cartilage). A superficial group of lymph capillaries drains the dermis and hypodermis, and a deep group drains muscle, viscera, and other deep structures. Fluid tends to move out of the blood capillaries into tissue spaces and then out of the tissue spaces into lymph capillaries (Figure 14-2, *A*).

The lymph capillaries join to form larger **lymph vessels** that resemble small veins. Small lymph vessels have a beaded appearance due to one-way valves that are similar to the valves of veins. When a lymph vessel is compressed, backward movement of lymph is prevented by the valves. Consequently, lymph moves forward through the lymph vessels. Three factors assist in the transport of lymph through the lymph vessels: (1) contraction of surrounding skeletal muscle during activity, (2) contraction of smooth muscle in the lymph vessel wall, and (3) pressure changes in the thorax during respiration.

The lymph vessels converge and eventually empty into the blood at two locations in the body. Lymph vessels from the upper right limb and the right half of the head, neck, and chest form the **right lymphatic duct,** which empties into the right subclavian vein. Lymph vessels from the rest of the body enter the **thoracic duct,** which empties into the left subclavian vein (Figure 14-2, *B*).

Lymphatic Organs

Lymphatic organs contain **lymphatic tissue,** which consists of many lymphocytes and other cells. The lymphocytes originate from red bone marrow (see Chapter 11) and are carried by the blood to lymph organs. When the body is exposed to microorganisms or foreign substances the lymphocytes divide and increase in number. The lymphocytes are part of the immune system response that causes the destruction of microorganisms and foreign substances. In addition to cells, lymphatic tissue has very fine collagen fibers. These fibers form an interlaced network that holds the lymphocytes and other cells in place. When lymph or blood filters through lymphatic organs, the fiber network also traps microorganisms and other items in the fluid.

Tonsils

There are three groups of **tonsils** (Figure 14-3). The **palatine tonsils** usually are referred to as "the tonsils," and they are located on each side of the posterior opening of the oral cavity. The **pharyngeal tonsils** are located near the internal opening of

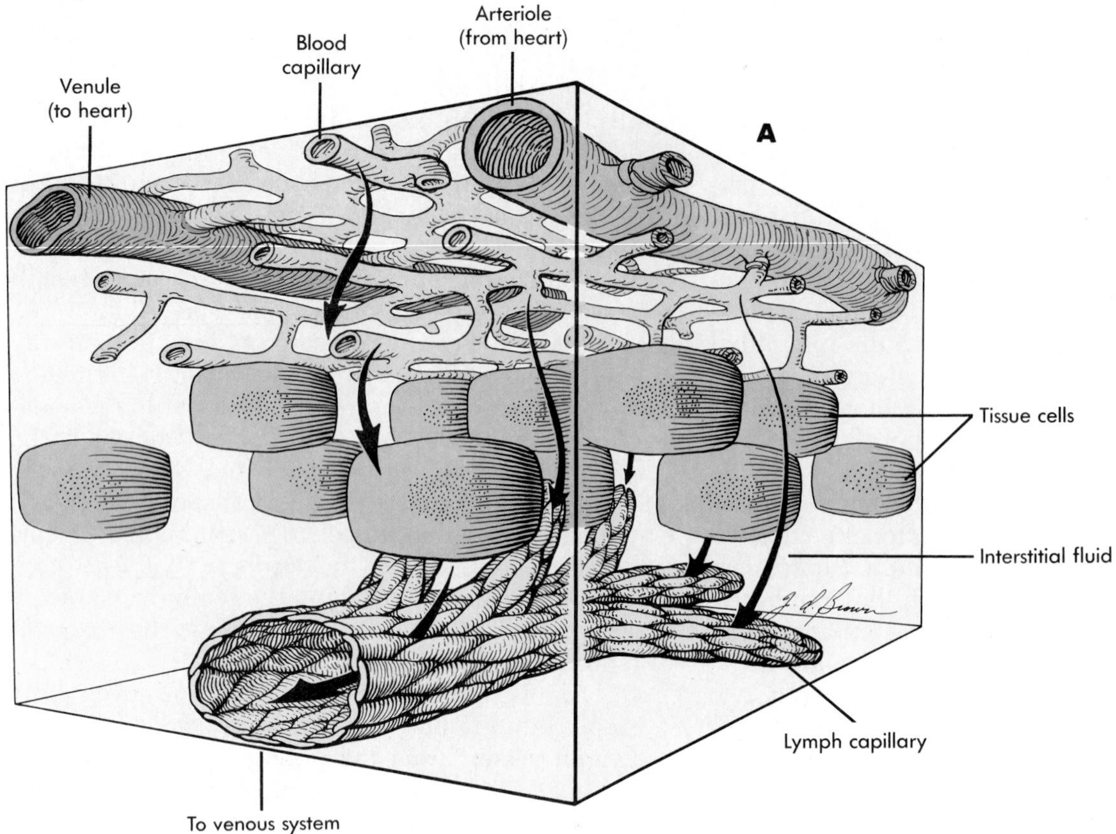

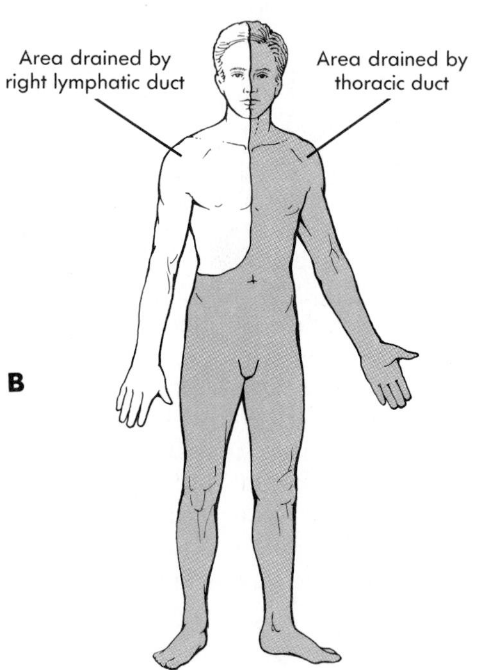

FIGURE 14-2 ■ Lymph formation and drainage.
A Movement of fluid from blood capillaries into tissues and from tissues into lymph capillaries to form lymph.
B Overall lymph drainage. Lymph from the colored area drains through the thoracic duct. Lymph from the white area drains through the right lymphatic duct.

the nasal cavity. Enlarged pharyngeal tonsils are called **adenoids** (ad'noydz), and they can interfere with normal breathing. The **lingual tonsils** are on the posterior surface of the tongue.

The tonsils form a protective ring of lymphatic tissue around the openings between the nasal and oral cavities and the pharynx. They provide protection against pathogens and other potentially harmful material entering into the nose and mouth. Sometimes the tonsils or adenoids become chronically infected and must be removed. In adults the tonsils decrease in size and may eventually disappear.

Lymph nodes

■ **Lymph nodes** are small, round structures distributed along the various lymph vessels (see Figure 14-1). Most lymph passes through at least one lymph node before entering the blood. Although lymph nodes are found throughout the body, there are three superficial aggregations of lymph nodes on each side of the body: the inguinal nodes in the groin, the axillary nodes in the axillary (armpit) region, and the cervical nodes of the neck.

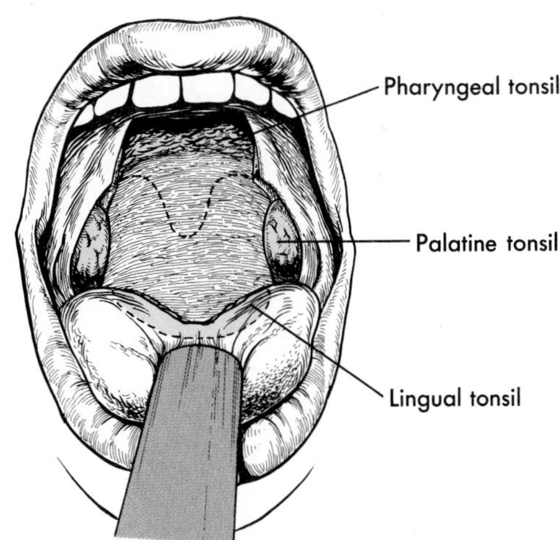

FIGURE 14-3 ■ Location of the tonsils.
Anterior view of the oral cavity with part of the palate removed *(dotted line)* to show the pharyngeal tonsils.

Lymph nodes are divided into compartments that contain lymphatic tissue separated by lymph sinuses (Figure 14-4). The lymphatic tissue consists of lymphocytes and other cells that may form dense aggregations of tissue called **lymph nodules.** A **lymph sinus** is a space between lymphatic tissue that contains macrophages on a network of fibers. Lymph enters the lymph node through afferent vessels, passes through the lymphatic tissue and sinuses, and exits through efferent vessels. As lymph moves through the lymph nodes two functions are performed. One function is activation of the immune system. Microorganisms or other foreign substances in the lymph can stimulate lymphocytes in the lymphatic tissue to start dividing. These areas of rapidly dividing lymphocytes are called **germinal centers.** The newly produced lymphocytes are released into the lymph and eventually reach the blood where they circulate and enter other lymphatic tissues. Another function of the lymph nodes is the removal of microorganisms and foreign substances from the lymph by macrophages.

> **1 Cancer cells can spread from a tumor site to other areas of the body through the lymphatic system. At first, however, as the cancer cells pass through the lymphatic system they are trapped in the lymph nodes, which filter the lymph. During radical cancer surgery, malignant (cancerous) lymph nodes are removed, and their vessels are cut and tied off to prevent the spread of the cancer. Predict the consequences of tying off the lymph vessels.** ?

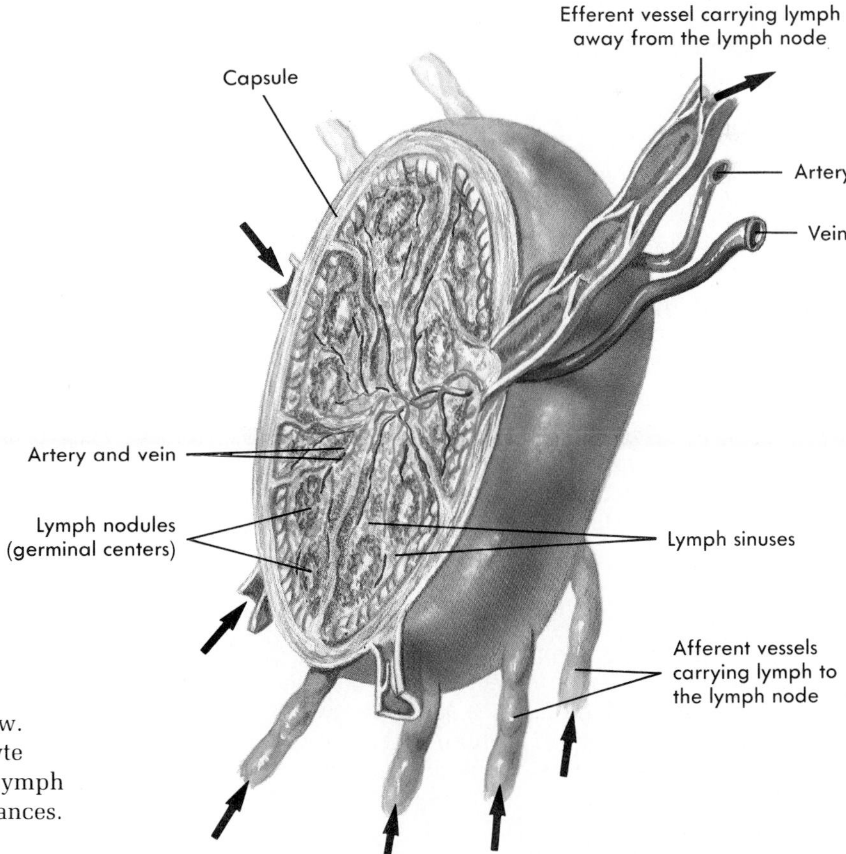

FIGURE 14-4 ■ Lymph node.
Arrows indicate the direction of lymph flow. The germinal centers are sites of lymphocyte production. As lymph moves through the lymph sinuses macrophages remove foreign substances.

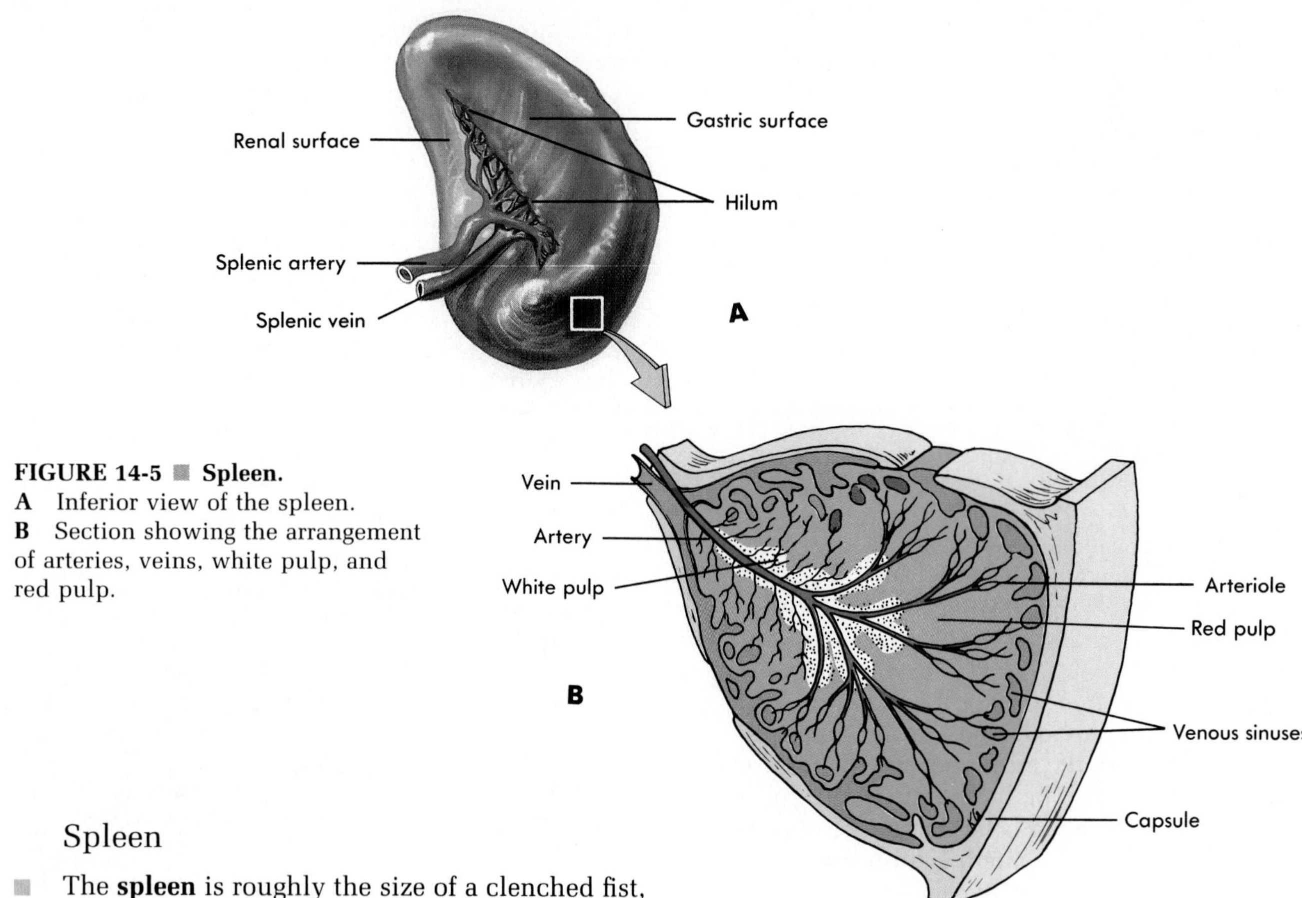

FIGURE 14-5 ■ Spleen.
A Inferior view of the spleen.
B Section showing the arrangement of arteries, veins, white pulp, and red pulp.

Spleen

■ The **spleen** is roughly the size of a clenched fist, and it is located in the left, superior corner of the abdominal cavity (Figure 14-5; see Figure 14-1). The spleen filters blood instead of lymph, and contains two specialized types of lymphoid tissue. **White pulp** surrounds the arteries within the spleen, and **red pulp** is associated with the veins.

The spleen detects and responds to foreign substances in the blood, destroys worn out erythrocytes, and acts as a blood reservoir. Lymphocytes in the white pulp can be stimulated in the same manner as in lymph nodes. Before blood leaves the spleen via veins, it passes through the red pulp. Macrophages in the red pulp remove foreign substances and worn out red blood cells through phagocytosis (see Chapter 11). In emergency situations such as hemorrhage, smooth muscle in splenic blood vessels and in the splenic capsule can contract. The result is the movement of a small amount of blood into the general circulation.

■ Although the spleen is protected by the ribs, it is often ruptured in traumatic abdominal injuries. Injury to the spleen can cause severe bleeding, shock, and possible death. A splenectomy, removal of the spleen, is performed to stop the bleeding. The spleen's functions are compensated for by other lymphatic tissue and the liver. ■

Thymus

■ The **thymus** (thi'mus) is a bilobed gland roughly triangular in shape (see Figure 14-1). It is located in the superior mediastinum, the partition dividing the thoracic cavity into left and right parts. The size of the thymus differs markedly depending on the age of the individual. In a newborn the thymus may extend halfway down the length of the thorax. The thymus continues to grow until puberty, although not as rapidly as other structures in the body. After puberty, the thymus decreases in size, and in older adults the thymus may be so small that it is difficult to find during dissection.

Large numbers of lymphocytes are produced in the thymus, but for unknown reasons most degenerate. The thymus functions as a site for the processing and maturation of lymphocytes. While in the thymus, lymphocytes do not respond to foreign substances. However, after thymic lymphocytes have matured, they enter the blood and travel to other lymphatic tissues, where they help to protect against microorganisms and other foreign substances.

Disorders of the Lymphatic System

Because the lymphatic system is involved with the production of lymphocytes that fight infectious diseases, and because the lymphatic system filters blood and lymph to remove microorganisms, it is not surprising that many infectious diseases produce symptoms associated with the lymphatic system. **Lymphadenitis** (lim-fad-ĕ-ni'tis) is an inflammation of the lymph nodes, causing them to become enlarged and tender. It is an indication that microorganisms are being trapped and destroyed within the lymph nodes. **Lymphangitis** (lim-fan-ji'tis) is an inflammation of the lymph vessels. This often results in visible red streaks in the skin that extend away from the site of infection. If the microorganisms pass through the lymph vessels and nodes to reach the blood, **septicemia,** or blood poisoning can result (see Chapter 11).

Bubonic plague and elephantiasis are diseases of the lymphatic system. In the sixth, fourteenth, and nineteenth centuries the **bubonic plague** killed large numbers of people. Fortunately, there are relatively few cases today. Bubonic plague is caused by bacteria that are transferred to humans from rats by the bite of the rat flea. The bacteria localize in the lymph nodes causing the lymph nodes to enlarge. The term bubonic is derived from a Greek word referring to the groin, since the disease often causes the inguinal lymph nodes of the groin to swell. Without treatment, septicemia followed by death rapidly occurs in 70% to 90% of those infected. **Elephantiasis** (el-ĕ-fan-ti'ă-sis) is caused by long, slender roundworms. The adult worms lodge in the lymph vessels and can cause such a blockage of lymph flow that a limb can become permanently swollen and enlarged. The resemblance of the affected limb to that of an elephant's leg is the basis for the name of the disease. The offspring of the adult worms pass through the lymphatic system into the blood. They can be transferred from an infected person to other humans by mosquitoes.

A **lymphoma** (lim-fo'mah) is a neoplasm (tumor) of lymphatic tissue that is almost always malignant. Lymphomas are usually divided into two groups: Hodgkin's disease and all other lymphomas, which are called non–Hodgkin's lymphomas. Hodgkin's disease is characterized by enlarged lymph nodes, fever, night sweats, and weight loss. Enlargement of the lymph nodes may compress surrounding structures and produce complications. The immune system is depressed, and the patient has an increased susceptibility to infections. Fortunately, treatment with drugs and radiation is effective for many people who suffer from lymphoma.

IMMUNITY

Immunity is the ability to resist damage from foreign substances such as microorganisms and harmful chemicals (for example, toxins released by microorganisms). Immunity is categorized as **nonspecific resistance** or **specific immunity.** The distinction between nonspecific resistance and specific immunity involves the concepts of specificity and memory. In nonspecific resistance each time the body is exposed to a substance, the response is the same. For example, each time a bacterial cell is introduced into the body, it is phagocytized with the same speed and efficiency. In specific immunity the response during the second exposure is faster and stronger than the response to the first exposure. For example, following initial exposure to a virus, the body may take many days to destroy the virus. During this time the virus damages tissues, producing the symptoms of disease. After the second exposure to the same virus, however, the response is very rapid and effective. The virus is destroyed before any symptoms develop, thus the person is **immune.** This immune system response is possible because of specificity and memory, the ability of the system to recognize and remember a particular substance.

NONSPECIFIC RESISTANCE

Nonspecific resistance includes mechanical mechanisms, chemicals, cells, and the inflammatory response.

Mechanical Mechanisms

Mechanical mechanisms prevent the entry of microorganisms into the body, or remove microorganisms from the body. The skin and mucous membranes form barriers that prevent the entry of microorganisms and chemicals into the body. Tears, saliva, and urine act to wash away microorganisms. Microorganisms cannot cause a disease if they cannot get into the body.

Chemicals

A variety of **chemicals** are involved in nonspecific resistance. Some chemicals that are found on the surface of cells kill microorganisms or prevent their entry into the cells. Lysozyme in tears and saliva, sebum on the skin, and mucus on the mucous membranes are examples of such chemicals. Many other chemicals (for example, histamine, complement, prostaglandins, and leukotrienes) promote inflammation by causing vasodilation, increasing vascular permeability, and stimulating phagocytosis. In addition, interferon protects cells against viral infections.

Complement

Complement is a group of at least 11 proteins found in plasma. The operation of complement proteins is very similar to that of clotting proteins (see Chapter 11). Normally, complement proteins circulate in the blood in an inactive form. Certain complement proteins can be activated by combining with foreign substances (for example, parts of a bacterial cell) or by combining with antibodies (see the discussion of the specific immune system). Once activation begins, a series of reactions results, in which each complement protein activates the next complement protein. The activated complement promotes inflammation and phagocytosis, and can directly lyse (rupture) bacterial cells.

Interferon

Interferon (in′ter-fēr′on) is a protein that protects the body against viral infections. When a virus infects a cell, the cell produces viral nucleic acids and proteins, which are assembled into new viruses. The new viruses are released from the infected cell to infect other cells. Since infected cells usually stop their normal functions or die during viral replication, viral infections are clearly harmful to the body. Fortunately, viruses often stimulate infected cells to produce interferon. Interferon does not protect the cell that produces it. Instead, interferon binds to the surface of neighboring cells where it stimulates them to produce antiviral proteins. These antiviral proteins stop viral reproduction in the neighboring cells by preventing the production of new viral nucleic acids and proteins.

Cells

Leukocytes and the cells derived from leukocytes (see Chapter 11) are the most important cellular components of the immune system. Leukocytes are produced in red bone marrow and lymphatic tissue and are released into the blood. Chemicals released from microorganisms or damaged tissues attract the leukocytes, and they leave the blood and enter affected tissues. Important chemicals known to attract leukocytes include complement, leukotrienes, kinins, and histamine. The ability to detect these chemicals and move toward them is called **chemotaxis.**

Phagocytic cells

Phagocytosis is the ingestion and destruction of particles by cells called **phagocytes** (see Chapter 3). The particles can be microorganisms or their parts, foreign substances, or dead cells from the individual's body. The most important phagocytic cells are neutrophils and macrophages, although other leukocytes have limited phagocytic ability.

Neutrophils are small phagocytic cells that are usually the first cells to enter infected tissues from the blood. However, neutrophils often die after phagocytizing a single microorganism. Pus is primarily an accumulation of dead neutrophils at a site of infection.

Macrophages are monocytes that leave the blood, enter tissues, and enlarge about fivefold. Monocytes and macrophages have a single, unlobed nucleus and are grouped together in the **mononuclear phagocytic system** (formerly referred to as the reticuloendothelial system). Sometimes macrophages are given specific names such as dust cells in the lungs, Kupffer cells in the liver, and microglia in the central nervous system. Macrophages can ingest more and larger items than neutrophils. Macrophages usually appear in infected tissues after neutrophils and are responsible for most of the phagocytic activity in the late stages of an infection, including the cleanup of dead neutrophils and other cellular debris.

In addition to leaving the blood in response to an infection, macrophages are also found in uninfected tissues. If microorganisms enter the uninfected tissue, the macrophages may phagocytize the microorganisms before they can replicate or cause damage. For example, macrophages are found beneath the surfaces of the body, such as the skin and mucous membranes, and around blood and lymph vessels. They also protect lymph in lymph nodes and blood in the spleen and liver.

Cells of inflammation

Leukocytes release chemicals that promote inflammation. In addition, there are cells found in connective tissue, called **mast cells,** that release inflammatory chemicals. Like macrophages, mast cells are located where microorganisms are likely to enter the body. On the other hand, eosinophils release chemicals that break down inflammatory chemicals. At the same time that inflammation is initiated by other leukocytes, eosinophils act to reduce and contain the inflammatory response. Inflammation is beneficial in the fight against microorganisms, but too much inflammation can be harmful, resulting in the unnecessary destruction of healthy tissues as well as the destruction of the microorganisms.

Inflammatory Response

The **inflammatory response** is a complex sequence of events involving many of the chemicals and cells previously discussed. Although they may vary in some details depending on the events producing them, most inflammatory responses are very similar. A bacterial infection is used here to illustrate an inflammatory response (Figure 14-6). The bacteria, or damage to tissues, cause the release or activation of chemical mediators such as histamine, prostaglandins, leukotrienes, complement, kinins, and others. The chemicals produce several effects: (1) vasodilation, which increases blood flow and brings phagocytes and other leukocytes to the area, (2) chemotactic attraction of phagocytes, which leave the blood and enter the tissue, and (3) increased vascular permeability, allowing fibrin and complement to enter the tissue from the blood. Fibrin prevents the spread of infection by walling off the infected area. Complement further enhances the inflammatory response and attracts additional phagocytes. This process of releasing chemical mediators and attracting phagocytes and other leukocytes continues until the bacteria are destroyed. Phagocytes (mainly macrophages) remove microorganisms and dead tissue, and the damaged tissues are repaired.

Inflammation can be localized or systemic. **Local inflammation** is an inflammatory response confined to a specific area of the body. Symptoms of local inflammation include redness, heat, swelling, pain, and loss of function. Redness, heat, and swelling result from increased blood flow and increased vascular permeability. Pain is caused by swelling and chemicals acting on nerve receptors. Loss of function is due to tissue destruction, swelling, and pain.

Systemic inflammation is an inflammatory response that occurs in many parts of the body. In ad-

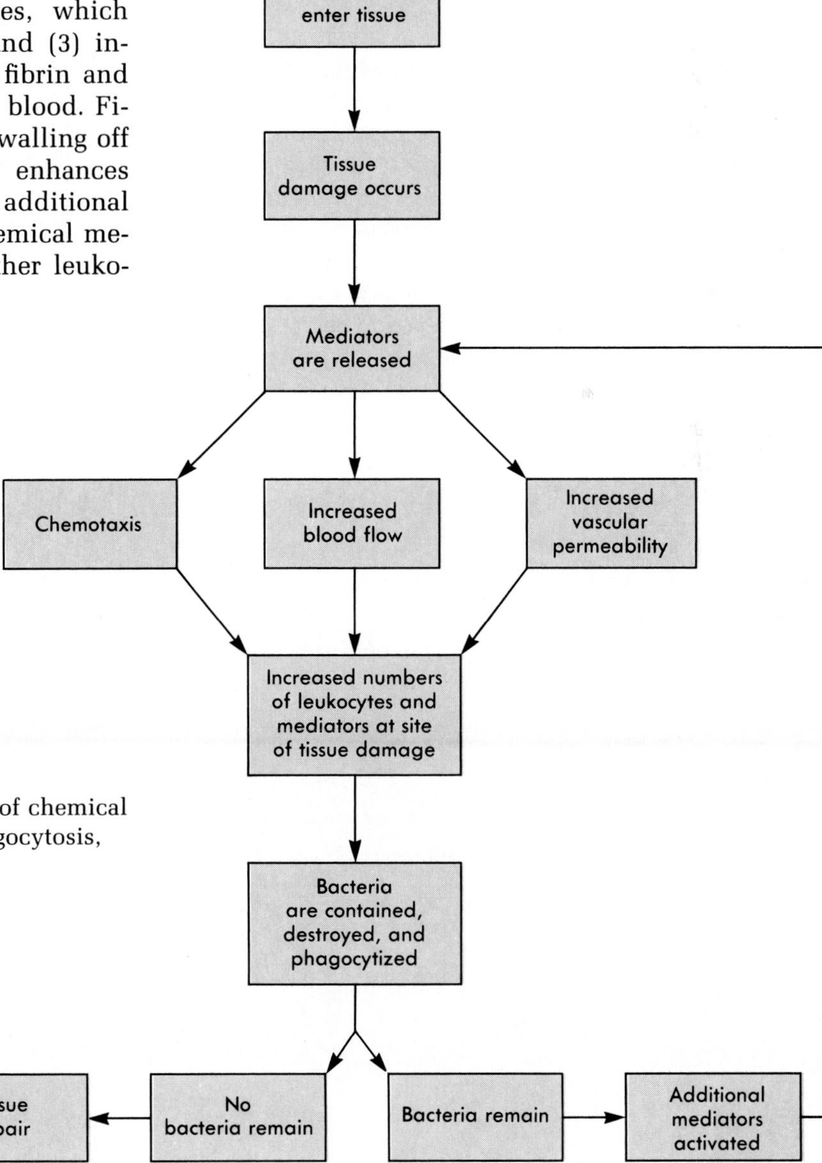

FIGURE 14-6 Inflammatory response.
Bacteria cause tissue damage and the release of chemical mediators that initiate inflammation and phagocytosis, resulting in the destruction of the bacteria.

dition to the local symptoms at the sites of inflammation, three additional features can be present. First, red bone marrow produces and releases large numbers of neutrophils that promote phagocytosis. Second, **pyrogens** (pi'ro-jenz), chemicals released by microorganisms, neutrophils, and other cells, stimulate fever production. Pyrogens affect the body temperature regulating mechanism in the hypothalamus of the brain. As a consequence heat is produced and conserved, and body temperature increases. Fever promotes the activities of the immune system, such as phagocytosis, and inhibits the growth of some microorganisms. Third, in severe cases of systemic inflammation, vascular permeability can increase so much that large amounts of fluid are lost from the blood into the tissues. The decreased blood volume can cause shock and death.

SPECIFIC IMMUNITY

Specific immunity involves the ability to recognize, respond to, and remember a particular substance. Substances that stimulate specific immunity responses are called **antigens** (an'tĭ-jenz). Antigens can be divided into two groups: foreign antigens and self antigens. **Foreign antigens** are introduced from outside the body. Components of bacteria, viruses, and other microorganisms are examples of foreign antigens that cause disease. Pollen, animal hairs, foods, and drugs are also foreign antigens and can cause an overreaction of the immune system, producing allergies. Transplanted tissues and organs that contain foreign antigens result in the rejection of the transplant. **Self antigens** are molecules produced by the body that stimulate an immune system response. The response to self antigens can be beneficial. For example, the recognition of tumor antigens can result in destruction of the tumor. The response to self antigens can also be harmful. **Autoimmune disease** results when self antigens stimulate unwanted destruction of normal tissue. An example is rheumatoid arthritis, which results in the destruction of tissue within joints.

The specific immune system response to antigens was historically divided into two parts: **humoral immunity** and **cell-mediated immunity.** Early investigators of the immune system found that when plasma from an immune animal was injected into the blood of a nonimmune animal, the nonimmune animal became immune. Since this process involved body fluids (humors), it was called humoral immunity. It was also discovered that blood cells alone could be responsible for immunity, and this process was called cell-mediated immunity.

It is now known that both types of immunity are due to the activities of lymphocytes. There are two types of lymphocytes called B cells and T cells. **B cells** produce proteins called antibodies, which are found in the plasma. Since antibodies are responsible, humoral immunity is now called **anti-**

TABLE 14-1

Comparison of Nonspecific Resistance, Antibody-Mediated Immunity, and Cell-Mediated Immunity

PRIMARY CELLS	ORIGIN OF CELLS	SITE OF MATURATION
NONSPECIFIC RESISTANCE		
Neutrophils, eosinophils, basophils, mast cells, monocytes, and macrophages	Red bone marrow	Red bone marrow (neutrophils, eosinophils, basophils, and monocytes) and tissues (mast cells and macrophages)
ANTIBODY-MEDIATED IMMUNITY		
B cells	Red bone marrow	Red bone marrow
CELL-MEDIATED IMMUNITY		
T cells	Red bone marrow	Thymus gland

body-mediated immunity. T cells are responsible for cell-mediated immunity. In addition, some T cells are involved with regulating both antibody-mediated and cell-mediated immunity.

Although the immune system is divided into different parts, it should be noted that this is an artificial division used to emphasize particular aspects of immunity. Actually, immune system responses often involve more than one part of the immune system. For example, it is possible for both the antibody-mediated and cell-mediated immune system to recognize and respond to an antigen. Once recognition has occurred, many of the events that lead to destruction of the antigen are nonspecific resistance activities such as inflammation and phagocytosis. It is the coordinated activities of all parts of the immune system that keep the body free of disease. Some of the main features of the subdivisions of the immune system are summarized in Table 14-1.

Origin and Development of Lymphocytes

All blood cells, including lymphocytes, are derived from stem cells in red bone marrow (Figure 14-7). Some stem cells give rise to pre-T cells that migrate through the blood to the thymus gland where they divide and are processed into T cells. Other stem cells produce pre-B cells, which are processed in the red bone marrow into B cells.

B cells are released from red bone marrow and T cells are released from the thymus. Both types of cells move through the blood to lymph tissue. These lymphocytes live for a few months to many years and continually circulate between the blood and the lymph tissues. Normally there are about five T cells for every B cell in the blood. When stimulated by an antigen, B cells and T cells divide, producing cells that are responsible for the destruction of the antigen (see the following discussion).

Evidence suggests that small groups of identical B cells or T cells, called **clones,** are formed during embryotic development. Although each clone can respond only to a particular antigen, there is such a large number of clones that the immune system can react to most molecules. Among the molecules to which the clones can respond, however, are self antigens. Because this response could destroy self cells, clones acting against self antigens are eliminated or suppressed. Most of this process occurs during development, but it does continue after birth and throughout life.

Activation and Regulation of Lymphocytes

To have a specific immune system response, lymphocytes must be activated by an antigen. Lymphocytes have proteins, called **antigen-binding receptors,** on their surfaces. Each receptor binds with only a specific type of antigen. Each clone consists

LOCATION OF MATURE CELLS	PRIMARY SECRETORY PRODUCT	PRIMARY ACTIONS	ALLERGIC REACTIONS
Blood, connective tissue, and lymphatic tissue	Histamine, complement, prostaglandins leukotrienes, kinins, and interferon	Inflammatory response and phagocytosis	None
Blood and lymphatic tissue	Antibodies	Protection against extracellular antigens (bacteria, toxins, and viruses outside of cells)	Immediate hypersensitivity
Blood and lymphatic tissue	Lymphokines	Protection against intracellular antigens (viruses and intracellular bacteria) and tumors responsible for graft rejection	Delayed hypersensitivity

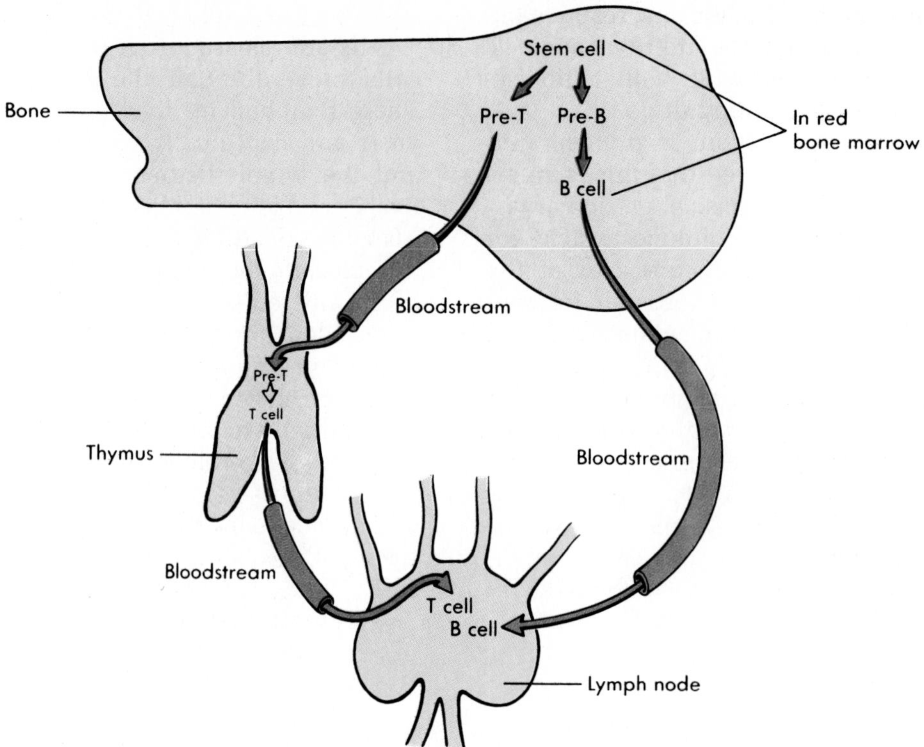

FIGURE 14-7 Origin and processing of B cells and T cells.
Both B cells and T cells originate in red bone marrow. B cells are processed in the red marrow, whereas T cells are processed in the thymus. Both cell types circulate to other lymphatic tissues.

of lymphocytes that have identical binding receptors on their surfaces. When antigens combine with the antigen-binding receptors, the lymphocytes in a clone are activated and the specific immune system response begins. The immune system can respond to virtually any antigen because the body contains many different clones.

The ability of B cells or T cells to respond to an antigen is usually regulated by other cells. For example, helper T cells are a special kind of T cell responsible for regulating the immune system. Helper T cells release a protein hormone called interleukin 2, which stimulates lymphocytes to divide. This increases the number of cells in a clone, which increases the effectiveness of the immune system response.

Current research is using genetically engineered interleukin 2 as a means of stimulating the immune system. Interleukin 2 may promote the destruction of cancer cells or boost the effectiveness of vaccinations. Conversely, decreasing the production or activity of interleukin 2 can suppress the immune system. For example, cyclosporine, a drug used to prevent the rejection of transplanted organs, inhibits the production of interleukin 2.

Antibody-Mediated Immunity

Exposure of the body to an antigen can result in activation of B cells and the production of antibodies. The antibodies bind to the antigens, and through several different mechanisms the antigens can be destroyed. Since antibodies are in body fluids, antibody-mediated immunity is effective against extracellular antigens such as bacteria, viruses (when they are outside cells), and toxins. Antibody-mediated immunity is also involved with allergic reactions.

Antibodies

Antibodies are proteins produced in response to an antigen. They are Y-shaped molecules (Fig ure 14-8). The end of each "arm" of the antibody is the **variable region,** which is the part of the antibody that combines with the antigen. The variable region of a particular antibody can only join with a particular antigen. This is similar to the lock and key model of enzymes (see Chapter 2). The rest of the antibody is the **constant region,** which has several functions. For example, the constant region can activate complement, or it can attach the antibody to cells such as macrophages, basophils, and mast cells.

Large amounts of antibodies are in plasma, although plasma contains other proteins. The plasma proteins can be separated into albumin and alpha, beta, and gamma globulin portions. Antibodies are called **gamma globulins** because they are found mostly in the gamma globulin portion of plasma. Antibodies are also called **immunoglobulins** (Ig) because they are globulin proteins involved in immunity. There are five general classes of immunoglobulins, which are denoted IgG, IgM, IgA, IgE, and IgD (Table 14-2).

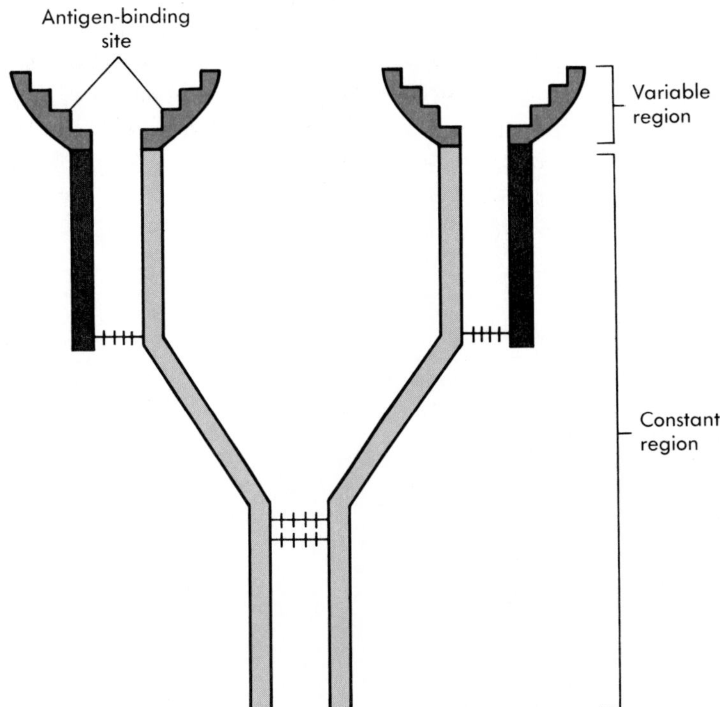

FIGURE 14-8 Structure of an antibody.
The Y-shaped antibody has two variable regions in its "arms," which function as antigen-binding sites. The constant region can activate complement or bind to other immune system cells such as macrophages, basophils, or mast cells.

TABLE 14-2

Classes of Antibodies and Their Functions

ANTIBODY	TOTAL SERUM ANTIBODY (%)	FUNCTION
IgG	80-85	Inactivates antigen, binds antigens together, facilitates phagocytosis, and activates complement; can cross the placenta and provide immune protection to the fetus and newborn; responsible for Rh reactions such as erythroblastosis fetalis
IgM	5-10	Binds antigens together, activates complement, and acts as an antigen-binding receptor on the surface of B cells; responsible for transfusion reactions in the ABO blood system; often the first antibody produced in response to an antigen
IgA	15	Inactivates antigen; secreted into saliva, tears, and onto mucous membranes, provides protection on body surfaces; found in colostrum and breast milk, provides immune protection to the newborn
IgE	0.002	Binds to mast cells and basophils and stimulates the inflammatory response
IgD	0.2	Functions as antigen-binding receptor on B cells

Effects of antibodies

■ Antibodies can affect antigens either directly or indirectly (Figure 14-9). Direct effects occur when a single antibody binds to an antigen and inactivates the antigen, or when many antigens are bound together and inactivated by many antibodies. The ability of antibodies to join antigens together is the basis for many clinical tests (for example, blood typing) because when enough antigens are bound together, they form visible clumps.

Most of the effectiveness of antibodies is due to indirect effects. After an antibody has attached to an antigen by its variable region, the constant region can activate other mechanisms that destroy the antigen. For example, macrophages can attach to the constant region of the antibody and phagocytize the antibody and antigen. When an antigen combines with the antibody, the constant region triggers a release of inflammatory chemicals from mast cells and basophils. Finally, the constant region of antibodies can activate complement, which stimulates inflammation, attracts leukocytes (chemotaxis), and lyses bacteria.

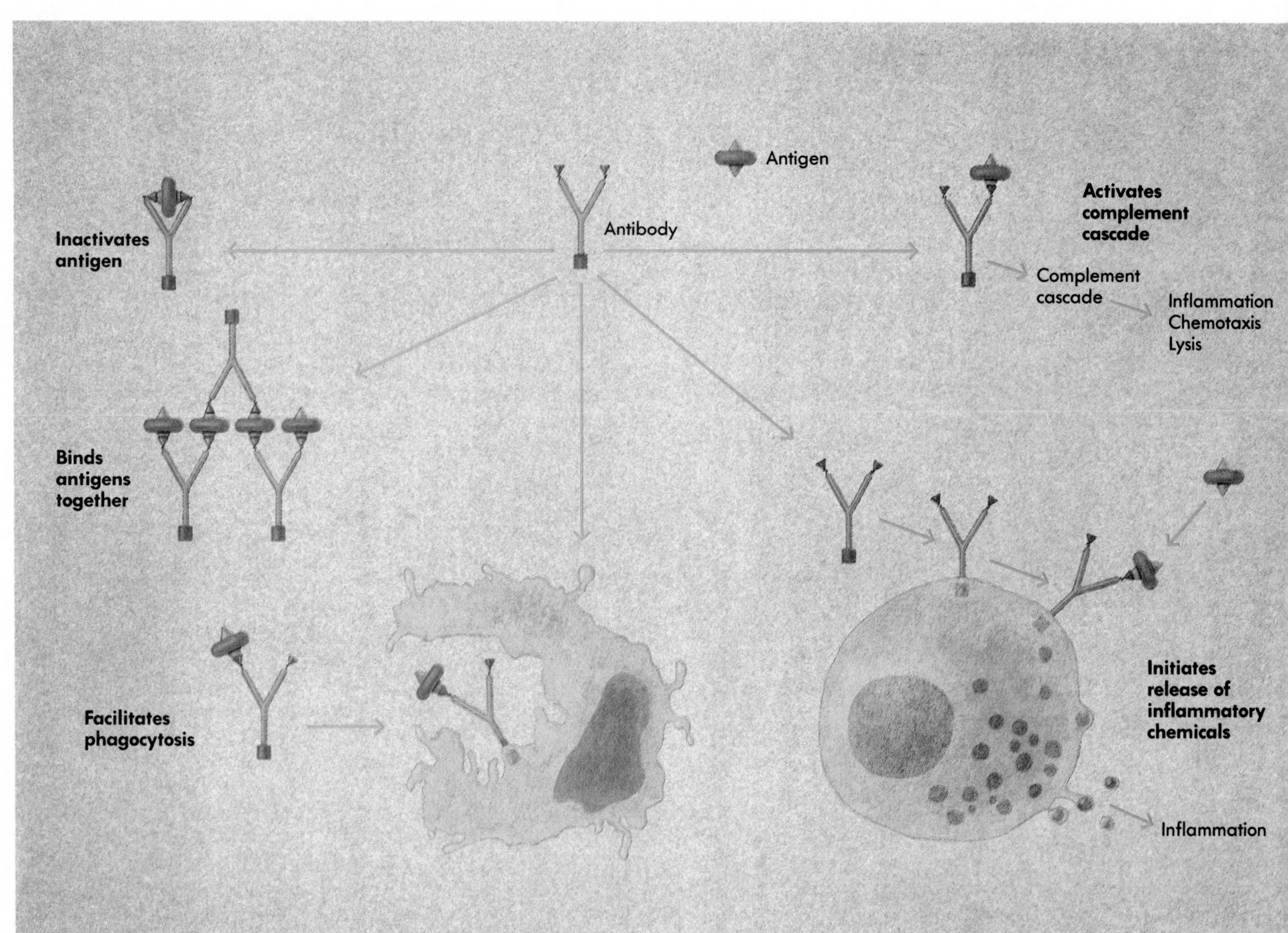

FIGURE 14-9 ■ Effects of antibodies.
Antibodies directly affect antigens by inactivating the antigens or binding the antigens together. Antibodies indirectly affect antigens by activating other mechanisms through the constant region of the antibody. Indirect mechanisms include increased phagocytosis resulting from antibody attachment to macrophages, increased inflammation due to the release of inflammatory chemicals from mast cells or basophils, and activation of complement.

Monoclonal antibodies are a pure antibody preparation that is specific for only one antigen. The antigen, injected into a laboratory animal, activates a B cell clone against the antigen. The B cells are removed from the animal and fused with tumor cells. The resulting hybridoma cells have two ideal characteristics: they divide to form large numbers of cells that produce only one kind of antibody.

Monoclonal antibodies have many applications. They are used for determining pregnancy and for diagnosing diseases such as gonorrhea, syphilis, hepatitis, rabies, and cancer. These tests are specific and rapid because the monoclonal antibodies bind only to the antigen being tested. Monoclonal antibodies may also be used to treat cancer. Anti-cancer drugs are attached to monoclonal antibodies that bind to cancer cells. The drug then kills the cancer cell. This approach has the advantage of selectively destroying cancer cells while sparing normal, healthy cells.

Antibody production

The production of antibodies after the first exposure to an antigen is different from that following a second or subsequent exposure (Figure 14-10). The **primary response** results from the first exposure of a B cell to an antigen. When the antigen binds to the antigen-binding receptor on the B cell, the B cell undergoes several divisions to form plasma cells and B memory cells. **Plasma cells** produce antibodies. The primary response normally takes 3 to 14 days to produce enough antibodies to be effective against the antigen. In the meantime, the individual usually develops disease symptoms because the antigen has had time to cause tissue damage.

B memory cells are responsible for the **secondary,** or **memory, response,** which occurs when the immune system is exposed to an antigen against which it has already produced a primary response. When exposed to the antigen, the B memory cells

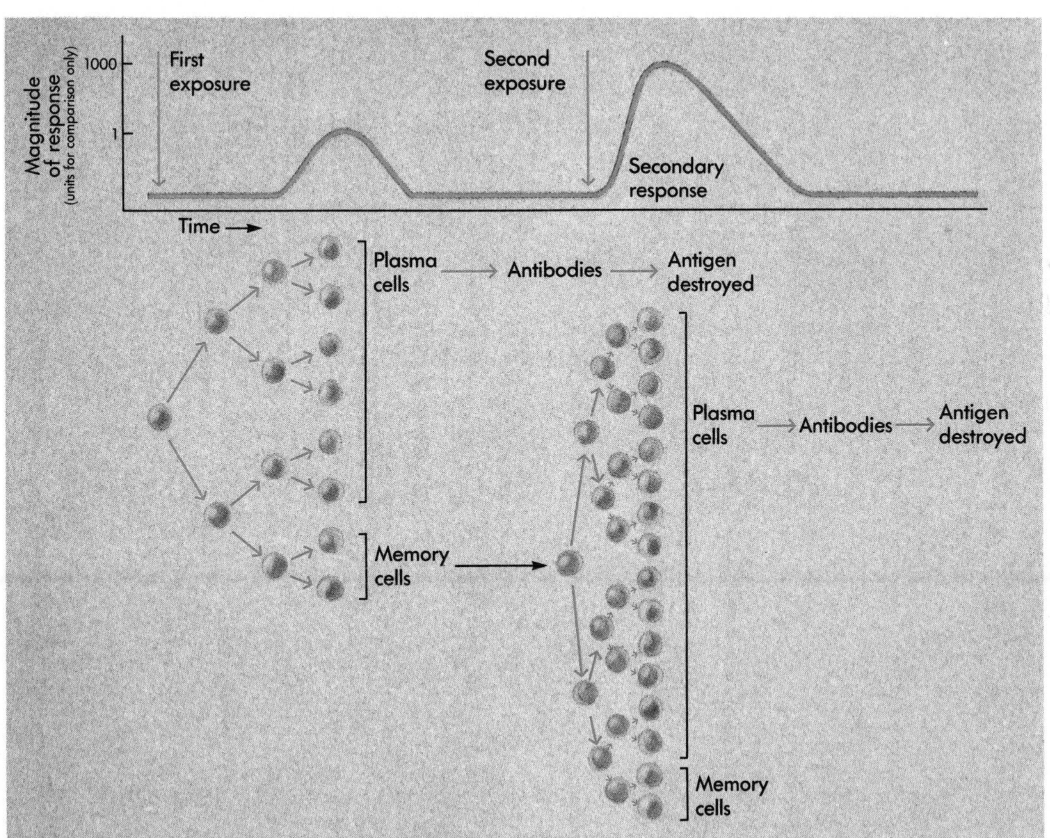

FIGURE 14-10 Antibody production.
The primary response occurs when a B cell is first activated by an antigen. The B cell proliferates to form plasma cells and B memory cells. The plasma cells produce antibodies. The secondary response occurs when another exposure to the same antigen causes the B memory cells to rapidly form plasma cells and additional B memory cells. The secondary response is faster and produces more antibodies than the primary response.

rapidly divide to form plasma cells, which rapidly produce antibodies. The secondary response provides better protection than the primary response for two reasons. First, the time required to start producing antibodies is less (hours to a few days), and second, more antibodies are produced. As a consequence, the antigen is quickly destroyed, no disease symptoms develop, and the person is immune.

The memory response also includes the formation of new memory cells, which provide protection against additional exposures to a specific antigen. Memory cells are the basis of specific immunity. After destruction of the antigen, plasma cells die, the antibodies they released are degraded, and antibody levels decline to the point where they can no longer provide adequate protection. However, memory cells persist for many years and probably for life in some cases. If memory-cell production is not stimulated, or if the memory cells produced are short-lived, it is possible to have repeated infections of the same disease. For example, the same cold virus can cause the common cold more than once in the same person.

2 One theory for long-lasting immunity assumes that humans are continually exposed to disease-causing agents. Explain how this exposure could produce lifelong immunity. ?

Cell-Mediated Immunity

Cell-mediated immunity is a function of T cells and is most effective against microorganisms that live inside the cells of the body. Viruses and some bacteria are examples of intracellular microorganisms. Cell-mediated immunity is also involved with allergic reactions and control of tumors. Like B cells, T cells have antigen-binding receptors on their surfaces. T cell receptors are especially adept at recognizing antigens on the surfaces of other cells.

After an antigen activates a T cell, the T cell undergoes a series of divisions to produce cytotoxic T cells and memory cells (Figure 14-11). **Cytotoxic T cells** are responsible for the cell-mediated immu-

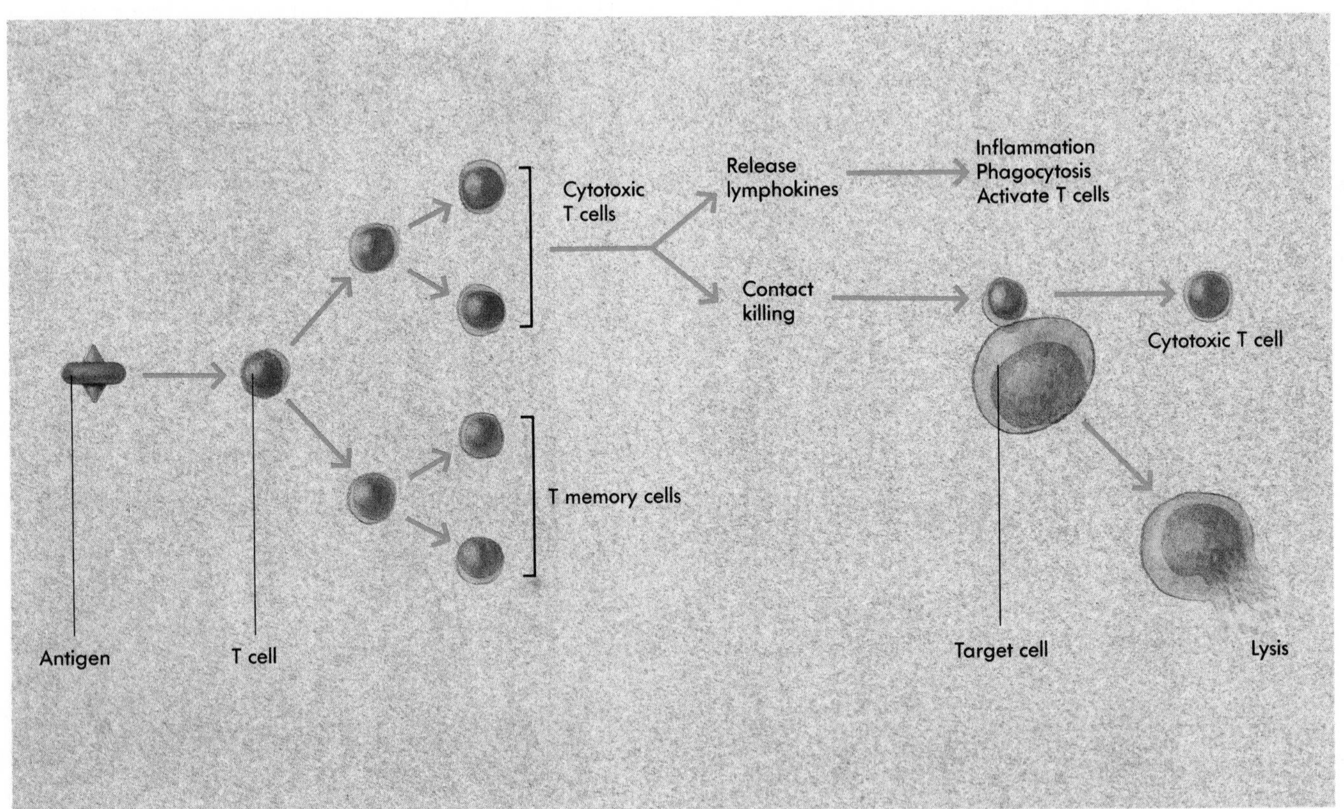

FIGURE 14-11 ■ Stimulation of and effects of T cells.
When activated, T cells form cytotoxic T cells and memory cells. The cytotoxic T cells cause lysis of target cells or release lymphokines that promote the destruction of the antigen. The memory cells are responsible for the secondary response.

nity response. The **T memory cells** provide a secondary response and long-lasting immunity in the same fashion as B memory cells.

Cytotoxic T cells have two main effects. First, they produce glycoproteins, called **lymphokines** (lim'fo-kīnz), which activate additional components of the immune system. For example, some lymphokines attract nonspecific immune system cells, especially macrophages. These cells are then responsible for phagocytosis of the antigen and the production of an inflammatory response. Lymphokines also activate additional T cells, converting them into cytotoxic T cells, which increase the effectiveness of the cell-mediated response. Second, cytotoxic T cells can come into contact with other cells and cause them to lyse. Virus-infected cells have viral antigens, tumor cells have tumor antigens, and tissue transplants have foreign antigens that can stimulate cytotoxic T cell activity. The cytotoxic T cell binds to the antigens on these cells and causes the cells to lyse.

Null Cells

Using special techniques, it is possible to identify B cells and T cells by the unique molecules each has on the surface of its cell membrane. Careful investigation of cell surface molecules has revealed a third type of lymphocyte, which is called a **null cell** because it does not have the typical B cell or T cell surface molecules. Null cells are probably produced in the red bone marrow, and they compose about 5% of the circulating lymphocytes. There are several different types of null cells, which lyse tumor cells and virus-infected cells.

ACQUIRED IMMUNITY

There are four ways to acquire specific immunity: active natural, active artificial, passive natural, and passive artificial. "Natural" and "artificial" refer to the method of exposure. Natural exposure implies that contact with the antigen occurred as part of everyday living and was not deliberate. Artificial exposure is a deliberate introduction of an antigen or antibody into the body.

"Active" and "passive" indicate whose immune system is responding to the antigen. When an individual is exposed to an antigen (either naturally or artificially), there can be a specific immune system response, which is called active immunity because the individual's own immune system is the cause of the immunity. Passive immunity occurs when another person or animal develops immunity and the immunity is transferred to a nonimmune individual.

Active Natural Immunity

Active natural immunity results from natural exposure to an antigen such as a disease-causing microorganism that causes an individual's immune system to respond against the antigen. Since the individual is not immune during the first exposure, he usually develops the symptoms of the disease.

Active Artificial Immunity

In **active artificial immunity,** an antigen is deliberately introduced into an individual to stimulate his immune system. This process is called **vaccination,** and the introduced antigen is a **vaccine.** Injection of the vaccine is the usual mode of administration. Examples of injected vaccinations are the DTP injection against diphtheria, tetanus, and pertussis (whooping cough); and the MMR injection against mumps, measles, and rubella (German measles). Sometimes the vaccine is ingested, as in the oral poliomyelitis vaccine (OPV).

The vaccine usually consists of some part of a microorganism, a dead microorganism, or a live, altered microorganism. The antigen has been changed so that it will stimulate the immune system but will not cause the symptoms of disease. Since active artificial immunity produces long-lasting immunity without disease symptoms, it is the preferred method of acquiring specific immunity.

3 In some cases, a booster shot is used as part of a vaccination procedure. A booster shot is another dose of the original vaccine given some time after the original dose was administered. Why are booster shots given? ?

Passive Natural Immunity

Passive natural immunity results from the transfer of antibodies from a mother to her child across the placenta before birth. During her life, the mother has been exposed to many antigens, either naturally or artificially, and she has antibodies against many of these antigens. These antibodies protect the mother and the developing fetus against disease. Some of the antibodies (IgG) can cross the placenta and enter the fetal blood. Following birth, the antibodies provide protection for the first few months of the baby's life. Eventually the antibodies are broken down, and the baby must rely on its own immune system. If the mother nurses her baby, antibodies (IgA) in the mother's milk may also provide some protection for the baby.

Passive Artificial Immunity

Achieving **passive artificial immunity** begins with vaccinating an animal such as a horse. After the animal's immune system responds to the antigen, antibodies are removed from the animal and are injected into the individual requiring immunity. Alternatively, a human who has developed immunity through natural exposure or vaccination is used as a source of antibodies. Passive artificial immunity provides immediate protection because the antibodies either directly or indirectly destroy the antigen. Therefore passive artificial immunity is the

Immune System Problems of Clinical Significance

ALLERGY

An **allergy**, or **hypersensitivity reaction,** is a harmful response to an antigen that does not stimulate the specific immune system of most people. Immune and allergy reactions involve the same mechanisms, and the differences between them are not clear. Both require exposure to an antigen and stimulation of antibody-mediated or cell-mediated immunity. If immunity to the antigen is established, later exposure to the antigen results in an immune system response that eliminates the antigen, and no symptoms appear. In allergy reactions the antigen is called the **allergen,** and later exposure to the allergen stimulates much the same processes that occur during the normal immune system response. However, the processes that eliminate the allergen also produce undesirable side effects such as a strong inflammatory reaction. This immune system response can be more harmful than beneficial and can produce many unpleasant symptoms.

Immediate hypersensitivities produce symptoms within a few minutes of exposure to the allergen and are caused by antibodies. The reaction takes place rapidly because the antibodies are already present due to prior exposure to the allergen. For example, in persons with hay fever, the allergens, usually plant pollens, are inhaled and absorbed through the respiratory mucous membrane. The resulting localized inflammatory response produces swelling and excess mucous production. In asthma due to an allergic reaction, the allergen combines with antibodies on mast cells or basophils. As a result, these cells release inflammatory chemicals (for example, leukotrienes and histamine) in the lungs. The chemicals cause constriction of smooth muscle in the walls of the tubes that transport air throughout the lungs. Consequently, less air flows into and out of the lungs, and the patient has difficulty breathing. Hives (urticaria) is a skin rash or localized swelling that is usually caused by an ingested allergen. Anaphylaxis is a systemic allergic reaction, often resulting from drugs (for example, penicillin) or insect stings. The chemicals released from mast cells and basophils cause systemic vasodilation, increased vascular permeability, a drop in blood pressure, and possibly death. Transfusion reactions and erythroblastosis fetalis (see Chapter 11) are also examples of immediate hypersensitivity reactions.

Delayed hypersensitivities take hours to days to develop and are caused by T cells. It takes some time for this reaction to develop because it takes time for the T cells to move by chemotaxis to the allergen. It also takes time for the T cells to release lymphokines that attract other immune system cells involved with producing inflammation. The most common type of delayed hypersensitivity reactions result from contact of the allergen with the skin or mucous membranes. For example, poison ivy, poison oak, soaps, cosmetics, and drugs can cause a delayed hypersensitivity reaction. The allergen is absorbed by epithelial cells, which are then destroyed by T cells, causing inflammation and tissue destruction. Although itching can be intense, scratching is harmful because it damages tissues and causes additional inflammation.

AUTOIMMUNE DISEASE

In **autoimmune disease** the immune system incorrectly treats self antigens as foreign antigens. Autoimmune disease operates through the same mechanisms as hypersensitivity reactions except that the reaction is stimulated by self antigens. Examples of autoimmune diseases include thrombocytopenia, lupus erythematosus, rheumatoid arthritis, rheumatic fever, and myasthenia gravis.

IMMUNODEFICIENCY

Immunodeficiency is a failure of some part of the immune system to function properly. It can be congenital (present at birth) or acquired. Congenital immunodefi-

preferred treatment when there may not be enough time for the individual to develop his own active immunity. However, the technique provides only temporary immunity since the antibodies are used or eliminated by the recipient.

Antiserum is the general term used for antibodies that provide passive artificial immunity, because the antibodies are found in serum (plasma minus the clotting factors). Antisera are available against microorganisms that cause disease (for example, rabies, hepatitis, measles), bacterial toxins (for example, tetanus, diphtheria, botulism), and venoms (for example from poisonous snakes and black widow spiders).

ciencies usually involve failure to form adequate numbers of B cells, T cells, or both. **Severe combined immunodeficiency (SCID),** in which both B cells and T cells fail to form, is probably the best known. Unless the person suffering from SCID is kept in a sterile environment or is provided with a compatible bone marrow transplant, death from infection results.

Acquired immunodeficiency can result from many different causes. For example, inadequate protein in the diet inhibits protein synthesis and therefore antibody levels decrease. The immune system can be depressed as a result of stress, drugs (for example, to prevent graft rejection), or illness. Diseases such as leukemia cause an overproduction of lymphocytes that do not function properly.

Acquired immune deficiency syndrome (AIDS) is caused by the **human immunodeficiency virus (HIV).** HIV infects primarily helper T cells, eventually resulting in the death of the helper T cells. Without adequate numbers of helper T cells, the immune system is depressed, and there is an inability to deal with intracellular microorganisms and cancer. Patients infected with HIV usually die of secondary infections (caused by microorganisms other than HIV) or Kaposi's sarcoma (cancerous growths in the skin and lymph nodes).

HIV is transmitted primarily by intimate sexual contact, by contaminated blood (for example, shared drug needles, or transfusions), or from a mother to her fetus or nursing infant. There are usually no symptoms at first. It can be 6 weeks to 1 year before the antibody test for HIV is positive. The HIV antibody test detects HIV antibodies in the blood, but does not directly detect HIV. Commonly, the first sign of AIDS is swollen lymph nodes caused by overstimulation of B cells and antibody production. As the disease progresses, helper T cell numbers decrease, and susceptibility to infections increases. At present there is no cure and all AIDS patients eventually die as a result of the disease. Present therapies interfere with HIV replication or attempt to prevent or treat secondary infections. For example, zidovidine, also known as azidothymidine (AZT), prevents the virus from replicating, and pentamidine is used to treat pneumonia caused by an intracellular protozoan. Several strategies are currently being explored to produce an AIDS vaccine.

TUMOR CONTROL

Tumor cells have tumor antigens that distinguish them from normal cells. According to the concept of **immune surveillance,** the immune system detects tumor cells and destroys them before a tumor can form. Failure of the immune system to destroy tumors as they form can result in cancer.

TRANSPLANTATION

The surface of cells in the human body contains antigens called **human lymphocyte antigens (HLAs).** The immune system can distinguish between self and foreign cells because self cells have self HLAs, whereas foreign cells have foreign HLAs. Rejection of a graft is caused by a normal immune system response to foreign HLAs.

Graft rejection can occur in two different directions. In **host vs. graft rejection,** the recipient's immune system recognizes the donor's tissue as foreign and rejects the transplant. In a **graft vs. host rejection,** the donor tissue (for example, bone marrow) recognizes the recipient's tissue as foreign, and the transplant rejects the recipient, causing destruction of the recipient's tissue and death.

To reduce graft rejection, a tissue match is performed. Only tissue with HLAs similar to the recipient's have a chance of being accepted. An exact match is possible only for a graft from one part to another part of the same person, or between identical twins. For all other graft situations, immunosuppressive drugs must be administered throughout the patient's life to prevent graft rejection.

SUMMARY

The lymphatic system consists of lymph, lymph vessels, lymphocytes, lymph nodes, tonsils, the spleen, and the thymus gland.

The lymphatic system maintains fluid balance in tissues, absorbs fats from the small intestine, and defends against foreign substances.

LYMPHATIC SYSTEM

Lymph Vessels

Lymph vessels carry lymph *away* from tissues. Valves in the vessels ensure the one-way flow of lymph.

Skeletal muscle contraction, contraction of lymph vessel smooth muscle, and thoracic pressure changes move the lymph through the vessels.

The thoracic duct and right lymphatic duct empty lymph into the blood.

Lymphatic Organs

Lymphatic tissue produces lymphocytes when exposed to foreign substances, and it filters lymph and blood.

The tonsils protect the openings between the nasal and oral cavities and the pharynx.

Lymph nodes, located along lymph vessels, filter lymph.

The white pulp of the spleen responds to foreign substances in the blood, whereas the red pulp phagocytizes foreign substances and worn out erythrocytes. The spleen also functions as a reservoir for blood.

The thymus processes lymphocytes that move to other lymphatic tissue to respond to foreign substances.

IMMUNITY

Immunity is the ability to resist the harmful effects of microorganisms and other foreign substances.

NONSPECIFIC RESISTANCE

Mechanical Mechanisms

The skin and mucous membranes are barriers that prevent the entry of microorganisms into the body.

Tears, saliva, and urine act to wash away microorganisms.

Chemicals

Chemicals kill microorganisms, promote phagocytosis, and increase inflammation.

Lysozyme in tears and complement in plasma are examples of chemicals involved in nonspecific resistance.

Interferon prevents the replication of viruses.

Cells

Chemotaxis is the ability of cells to move toward microorganisms or sites of tissue damage.

Neutrophils are the first phagocytic cells to respond to microorganisms.

Macrophages are large phagocytic cells that are active in the latter part of an infection. Macrophages are also positioned at sites of potential entry of microorganisms into tissues.

Basophils and mast cells promote inflammation, whereas eosinophils inhibit inflammation.

Inflammatory Response

Chemical mediators cause vasodilation and increase vascular permeability, allowing the entry of chemicals into damaged tissues. Chemicals also attract phagocytes.

The amount of chemical mediators and phagocytes increases until the cause of the inflammation is destroyed. Then the tissues undergo repair.

Local inflammation produces the symptoms of redness, heat, swelling, pain, and loss of function. Symptoms of systemic inflammation include an increase in neutrophil numbers, fever, and shock.

SPECIFIC IMMUNITY

Antigens are molecules that stimulate specific immunity.

B cells are responsible for humoral, or antibody-mediated, immunity. T cells are involved with cell-mediated immunity.

Origin and Development of Lymphocytes

B cells and T cells originate in red bone marrow. T cells are processed in the thymus and B cells are processed in red bone marrow.

B cells and T cells move to lymphatic tissue from their processing sites. They continually circulate from one lymphatic tissue to another.

Activation and Regulation of Lymphocytes

B cells and T cells are activated when an antigen combines with an antigen-binding receptor on the surface of the cells. Clones are lymphocytes with the same antigen-binding receptor.

Activation of B cells or T cells usually requires the presence of helper T cells.

Antibody-Mediated Immunity

Antibodies are proteins. The variable region combines with antigens and is responsible for antibody specificity. The constant region activates complement or attaches the antibody to cells. The five classes of antibodies are IgG, IgM, IgA, IgE, and IgD.

Antibodies directly inactivate antigens or cause them to clump together. Antibodies indirectly destroy antigens by promoting phagocytosis and inflammation.

The primary response results from the first exposure to an antigen. B cells form plasma cells, which produce antibodies, and memory cells.

The secondary (memory) response results from exposure to an antigen after a primary response. Memory cells quickly form plasma cells and memory cells.

Cell-Mediated Immunity

Exposure to an antigen activates cytotoxic T cells and produces memory cells.

Cytotoxic T cells lyse virus-infected cells, tumor cells, and tissue transplants. Cytotoxic T cells produce lymphokines, which promote inflammation and phagocytosis.

Null Cells

Null cells are lymphocytes that are not T cells or B cells. Null cells lyse virus-infected cells and tumor cells.

ACQUIRED IMMUNITY

Active natural immunity results from everyday exposure to an antigen against which the person's own immune system mounts a response.

Active artificial immunity results from deliberate exposure to an antigen (vaccine) to which the person's own immune system responds.

Passive natural immunity is the transfer of antibodies from a mother to her fetus or baby.

Passive artificial immunity is the transfer of antibodies from an animal or another person to a person requiring immunity.

CONTENT REVIEW

1. List the parts of the lymphatic system, and describe the three main functions of the lymphatic system.
2. What is the function of valves in lymph vessels? What causes lymph to move through lymph vessels?
3. Which parts of the body are drained by the right lymphatic duct and which by the thoracic duct?
4. Describe the cells and fibers of lymphatic tissue and explain the functions of lymphatic tissue.
5. Name the three groups of tonsils. What is their function?
6. Where are lymph nodes found? What is the function of the germinal centers within lymph nodes?
7. Where is the spleen located? What is the function of white pulp and red pulp within the spleen? What other function does the spleen perform?
8. Where is the thymus gland located and what function does it perform?
9. What is the difference between nonspecific resistance and specific immunity?
10. How do mechanical mechanisms and chemicals provide protection against microorganisms? Describe the effects of complement and interferon.
11. Describe functions of the two major phagocytic cell types of the body. What is the mononuclear phagocytic system?
12. Name the cells involved in promoting and inhibiting inflammation.
13. Describe the effects that take place during an inflammatory response. What are the symptoms of local and systemic inflammation?
14. Define antigen. What is the difference between a self antigen and a foreign antigen?
15. Which cells are responsible for antibody-mediated and for cell-mediated immunity?
16. Describe the origin and development of B cells and T cells.
17. Define antigen-binding receptor and give its function. What is the role of helper T cells in immunity?
18. What are the functions of the variable and constant regions of an antibody?
19. Describe the direct and indirect ways that antibodies function to destroy antigens.
20. What are the functions of plasma cells and B memory cells?
21. Define the primary and memory (secondary) response. How do they differ from each other in regard to speed of response and amount of antibody produced?
22. What are the functions of cytotoxic T cells and T memory cells?
23. Define null cells and explain their functions.
24. Define active natural, active artificial, passive natural, and passive artificial immunity. Give an example of each.

CONCEPT REVIEW

1. A patient is suffering from edema in the lower-right limb. Explain why massage would help to remove the excess fluid.
2. If the thymus of an experimental animal is removed immediately following birth, the animal exhibits the following characteristics: (1) it is more susceptible to infections, (2) it has decreased numbers of lymphocytes, and (3) its ability to reject grafts is greatly decreased. Explain these observations.
3. Adjuvants are substances that slow, but do not stop, the release of an antigen from an injection site into the blood. Suppose injection A of a given amount of antigen was given without an adjuvant and injection B of the same amount of antigen was given with an adjuvant that caused the release of antigen over a period of 2 to 3 weeks. Would injection A or B result in the greater amount of antibody production? Explain.

4. Durability is a measure of how long immunity lasts. Compare the durability of active immunity and passive immunity. Explain the difference between the two types of immunity. In what situations would one type be preferred over the other type?
5. Tetanus is caused by bacteria that enter the body through wounds in the skin. The bacteria produce an exotoxin that causes spastic muscle contractions. Death often results from failure of the respiration muscles. A patient comes to the emergency room after stepping on a nail. If the patient has been vaccinated against tetanus, the patient is given a tetanus booster shot, which consists of the exotoxin altered so that it is harmless. If the patient has never been vaccinated against tetanus, the patient is given an antiserum shot against tetanus. Explain the rationale for this treatment strategy. Sometimes both a booster and an antiserum shot are given, but at different locations of the body. Explain why this is done, and why the shots are given in different locations of the body.
6. A patient had many allergic reactions (see the box on p. 368). As part of the treatment scheme, it was decided to try to identify the allergens that stimulated the allergic reaction. A series of solutions, each containing an allergen that commonly causes a reaction, was composed. Each solution was then injected into the skin at different locations on the patient's back. The following results were obtained: (1) at one location, within a few minutes the injection site became red and swollen, (2) at another injection site, swelling and redness did not appear until 2 days later, and (3) no redness or swelling developed at the other sites. Explain what happened for each observation and what caused the redness and swelling.

ANSWERS TO  PREDICT QUESTIONS

1 *p. 355* Cutting and tying off the lymph vessels would prevent the movement of fluid from the affected tissue. The result would be edema.

2 *p. 366* The first exposure to the disease-causing agent (antigen) would evoke a primary immune system response. Gradually, however, the antibodies would degrade, and memory cells would die. If, before all the memory cells were eliminated, a second exposure to the antigen occurred, a secondary immune system response would result. The memory cells produced could provide immunity until the next exposure to the antigen.

3 *p. 367* The booster shot stimulates a memory (secondary) response, resulting in the formation of large amounts of antibodies and memory cells. Consequently there is better, longer-lasting immunity.

15

THE RESPIRATORY SYSTEM

After reading this chapter you should be able to:

1 Describe the anatomy of the respiratory passages beginning at the nose and ending with the alveoli.

2 Describe the lungs, the membranes that cover the lungs, and the cavities in which they lie.

3 Explain how contraction of the muscles of respiration causes air to flow into and out of the lungs during quiet respiration and during labored respiration.

4 List the pulmonary volumes and capacities and define each of them.

5 Name the components of the respiratory membrane and explain the factors that affect gas movement through it.

6 Describe how oxygen and carbon dioxide are transported in the blood.

7 Describe the partial pressure gradients for oxygen and carbon dioxide.

8 Name the neural mechanisms that control respiration and describe how they work.

9 Explain how alterations in blood carbon dioxide levels, blood pH, and blood oxygen levels affect respiration.

10 List the stimuli in addition to blood gases and pH that influence the respiratory movements and explain how they affect these respiratory movements.

alveolus,
pl. alveoli
(al've-o'lus)
: [L., cavity] The terminal saclike endings of the respiratory system, where gas exchange occurs.

bronchiole
: One of the finer subdivisions of the bronchial tubes, less than 1 mm in diameter, and having no cartilage in its wall, but relatively more smooth muscle and elastic fibers.

bronchus
(brong'kus)
: [Gr. *bronchos*, windpipe] Any one of the air ducts conducting air from the trachea to the bronchioles. A bronchus has cartilage rings or plates in its wall, and it varies in diameter from about 1 cm in the primary bronchi to about 1 mm in the smallest (tertiary) bronchus.

larynx
(lăr'ingks)
: Organ of voice production located between the pharynx and the trachea; it consists of a framework of cartilages and elastic membranes housing the vocal folds (true vocal cords) and the muscles that control the position and tension of these elements.

nasal cavity
: Cavity divided by the nasal septum, and extending from the external nares anteriorly to the nasopharynx posteriorly, and bounded inferiorly by the hard palate.

pharynx
(făr'ingks)
: [Gr. *pharynx*, throat, the joint openings of the digestive tract and the windpipe] Upper expanded portion of the digestive tube between the esophagus below and the oral and nasal cavities above and in front.

pleural cavity
(ploor'al)
: Space between the parietal and visceral layers of the pleura, normally filled with pleural fluid.

respiratory center
: Nerve cells in the medulla oblongata and pons of the brain that control inspiration and expiration.

respiratory membrane
: Membrane in the lungs across which gas exchange occurs with blood; consists of a thin layer of fluid, the alveolar epithelium, a basement membrane of the alveolar epithelium, interstitial space, the basement membrane of the capillary endothelium, and the capillary endothelium.

trachea
(tra'ke-ah)
: [Gr. *tracheia arteria*, rough artery] Air tube extending from the larynx into the thorax, where it divides to form the two primary bronchi; has 16 to 20 C-shaped rings of cartilage in its walls.

All living cells of the body require oxygen and produce carbon dioxide. The respiratory system and the cardiovascular system take oxygen from the air and transport it to the cells of the body, transport carbon dioxide from cells and release it into the air, and help regulate the pH of the body fluids. The term **respiration** refers to the following processes: (1) ventilation, the movement of air into and out of the lungs, (2) gas exchange between the air in the lungs and the blood, (3) transport of oxygen and carbon dioxide in the blood, (4) gas exchange between the blood and the tissues, and (5) cellular metabolism.

ANATOMY OF THE RESPIRATORY SYSTEM

The respiratory system consists of the nose, nasal cavity, pharynx, larynx, trachea, bronchi, and lungs (Figure 15-1). Although air frequently passes through the oral cavity, it is considered to be part of the digestive system instead of the respiratory system. Respiratory movements are accomplished by the diaphragm and the muscles of the thoracic and abdominal walls.

Nose and Nasal Cavity

The term **nose** usually refers to the visible structure that forms a prominent feature of the face and also can refer to the internal nasal cavity. The bridge of the nose consists of the nasal bones and extensions of the frontal and maxillary bones, but most of the nose is composed of cartilage. The rigid bone and cartilage are covered by connective tissue and skin. (Figure 15-2).

The **nasal cavity** extends from the external openings in the nose to the pharynx, and it is divided by the **nasal septum** into right and left sides. The external openings to the nasal cavity are the **external nares** (nă′rēz), or **nostrils,** and the internal openings from the nasal cavity into the pharynx are the **internal nares.**

The **hard palate** forms the floor of the nasal cavity (see Chapter 6). The lateral walls on each side of the nasal cavity are modified by the presence of three prominent bony ridges called **conchae** (kon′ke; resembling a conch shell).

Paranasal sinuses are air-filled spaces within the maxillary, frontal, ethmoid, and sphenoid bones of the skull. Each paranasal sinus is named after the bone in which it is located. The paranasal sinuses open into the nasal cavity and are lined with a mucous membrane. The paranasal sinuses function to reduce the weight of the skull, to produce mucus, and to influence the quality of the voice by acting as resonating chambers.

Mucus produced by the epithelium of the paranasal sinuses drains into the nasal cavity. When the mucous membranes become swollen because of nasal infections, sinus infections, or allergies, these passages can become blocked. The mucus then accumulates within the sinuses, and the increasing pressure can produce a painful sinus headache.

The **nasolacrimal ducts,** which carry tears from the eyes, also open into the nasal cavity. Sensory receptors for the sense of smell are found in the superior part of the nasal cavity (see Chapter 9).

Air enters the nasal cavity through the external nares. Just inside the external nares are hairs that trap some of the large particles of dust suspended in the air. Mucus produced by the lining of the internal nasal cavity also traps debris in the air, and the cilia on the surface of the mucous membrane sweep the mucus posteriorly to the pharynx, where it is swallowed. The air is humidified and warmed by superficial capillary networks underlying the mucous epithelium within the nasal cavity before it passes into the pharynx. This prevents damage to the more delicate linings in the rest of the respiratory passages.

> **1 Explain what happens to your throat when you sleep with your mouth open, especially when your nasal passages are plugged as a result of having a cold.** ?

Pharynx

The **pharynx** (făr′ingks; throat) is the common passageway of both the digestive and respiratory systems. It receives air from the nasal cavity and air, food, and water from the mouth. Inferiorly, the pharynx leads to the opening of the respiratory system (opening into the larynx) and the digestive system (the esophagus). The pharynx can be divided into three regions, the nasopharynx, the oropharynx, and the laryngopharynx (see Figure 15-2, *A*).

The **nasopharynx** (na′zo-făr′ingks) is the superior portion of the pharynx and extends from the internal nasal cavity to the level of the **uvula** (u′vu-lah; a grape), a soft process that extends from the posterior edge of the soft palate. The **soft palate** forms the floor of the nasopharynx. The nasopharynx is lined with a mucous membrane similar to

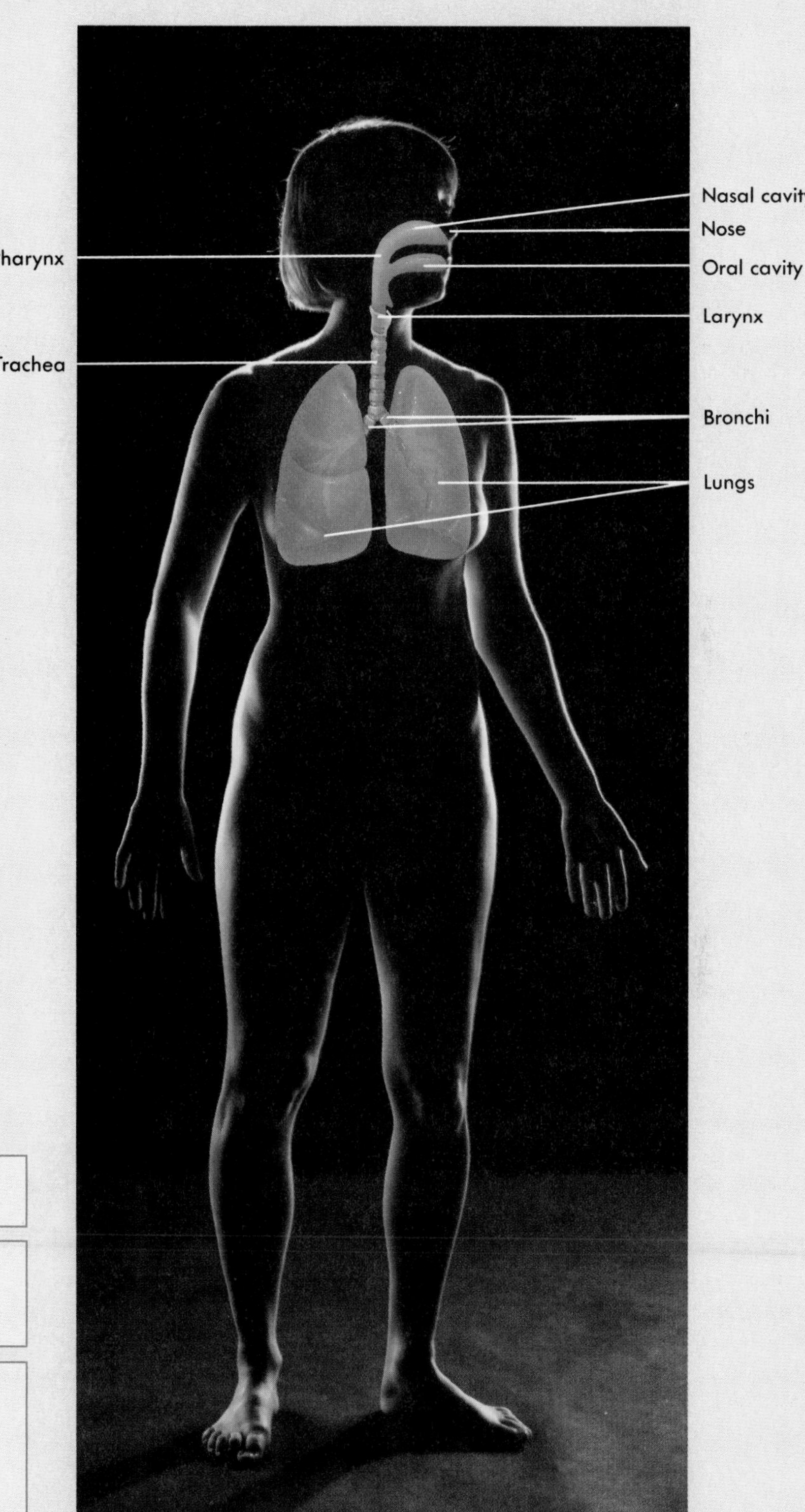

The Respiratory System

MAJOR COMPONENTS

Lungs and the respiratory passages

MAJOR FUNCTIONS

Transports air to the lungs

Exhanges gases (oxygen and carbon dioxide) between the air and blood

Regulates blood pH

FIGURE 15-1 ■ Organs of the respiratory system.
The major elements of the respiratory system are the nose, nasal cavity, pharynx, larynx, trachea, bronchi, and lungs. Air can also enter the respiratory system through the oral cavity.

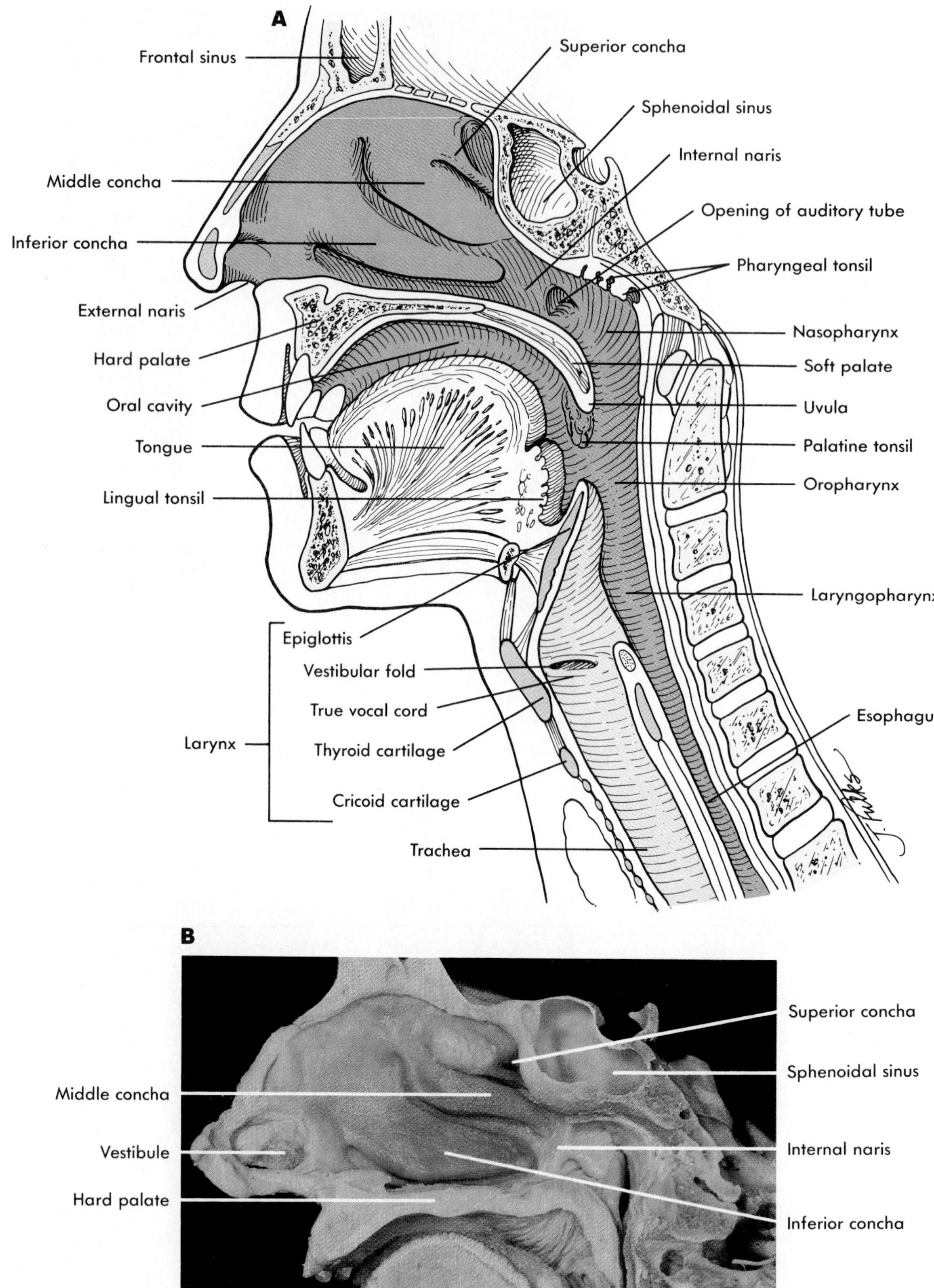

FIGURE 15-2 ■ Nasal cavity and pharynx.
A Sagittal section through the nasal cavity and pharynx viewed from the medial side.
B Photograph of a sagittal section of the nasal cavity.

that of the nasal cavity. The auditory tubes open into the nasopharynx, and the posterior portion of the nasopharynx contains the pharyngeal tonsils, which aid in defending the body against infection (see Chapter 14). The soft palate and uvula are elevated during swallowing, and this movement results in the closure of the nasopharynx, which prevents food from passing from the oral cavity into the nasopharynx.

The **oropharynx** (o'ro-făr'ingks) extends from the uvula to the epiglottis. The oral cavity opens into the oropharynx. Thus food, drink, and air all pass through the oropharynx. The oropharynx is lined with stratified squamous epithelium, which protects against abrasion. Two sets of tonsils, the palatine tonsils and the lingual tonsils, are located near the opening between the mouth and the oropharynx. The lingual tonsils are located on the surface of the posterior portion of the tongue, and the palatine tonsils are located on the lateral walls near the border of the oral cavity and the oropharynx (see Figure 14-3).

The **laryngopharynx** (lă-ring'go-făr'ingks) extends from the epiglottis to the lower margin of the larynx. The laryngopharynx, like the oropharynx, is lined with stratified squamous epithelium.

Larynx

The **larynx** (lăr'ingks) consists of an outer casing of nine cartilages that are connected to each other by muscles and ligaments (Figure 15-3). Six of the nine cartilages form three pairs of cartilages, and three cartilages are unpaired. The largest and most superior of the cartilages is the unpaired **thyroid cartilage** (thyroid means shield and refers to the shape of the cartilage), or Adam's apple (Figure 15-3, *A*).

The most inferior cartilage of the larynx is the unpaired **cricoid** (kri'koyd; ring-shaped) **cartilage,** which forms the base of the larynx, on which the other cartilages rest.

The third unpaired cartilage is the **epiglottis** (ep'ĭ-glot'is; on the glottis). It differs from the other cartilages in that it consists of elastic cartilage rather than hyaline cartilage. Its inferior margin is attached to the thyroid cartilage anteriorly, and the superior part of the epiglottis projects as a free flap toward the tongue. During swallowing, the epiglottis covers the opening of the larynx and prevents materials from entering it.

The six paired cartilages are stacked in two pillars, each consisting of three cartilages, between the

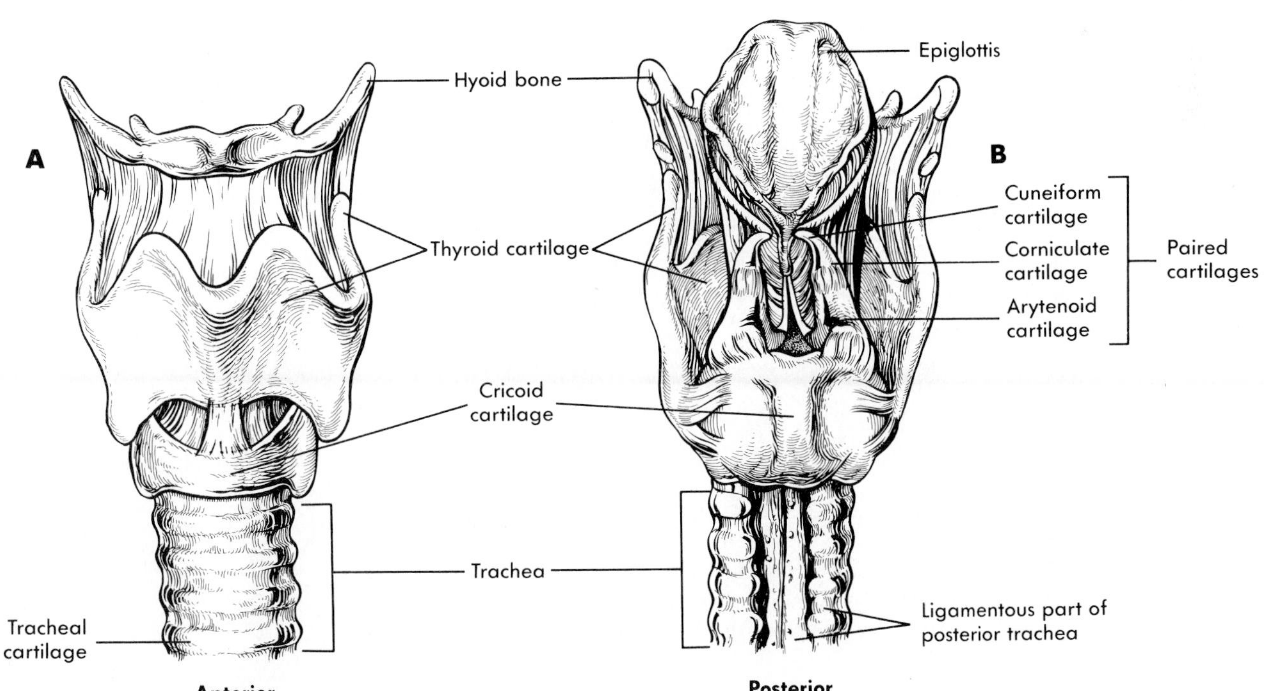

FIGURE 15-3 **Anatomy of the larynx.**
A Anterior view.
B Posterior view.

cricoid and thyroid cartilages on the posterior portion of the larynx (Figure 15-3, *B*). The top cartilage on each side is the **cuneiform** (ku′ne-ĭ-form; wedge-shaped) **cartilage,** the middle cartilage is the **corniculate** (kōr-nik′u-lat; horn-shaped) **cartilage,** and the bottom cartilage is the **arytenoid** (ăr-ĭ-te′noyd; ladle-shaped) **cartilage.** The columns of unpaired cartilages form an attachment site for the vocal folds.

Two pairs of ligaments extend from the posterior surface of the thyroid cartilage to the paired cartilages. The superior pair forms the **vestibular folds,** or **false vocal cords,** and the inferior pair composes the **vocal folds,** or **true vocal cords.** When the vestibular folds come together, they prevent air from leaving the lungs (such as when a person holds his breath) and, along with the epiglottis, they prevent food and liquids from entering the larynx (Figure 15-4).

The true vocal cords are involved in voice production. Air moving past the true vocal cords causes them to vibrate, producing sound. Muscles control the length and the tension of the true vocal cords. The force of air moving past the true vocal cords controls the loudness, and the tension of the true vocal cords controls the pitch of the voice (see Figure 15-4). An inflammation of the mucous epithelium of the true vocal cords is called **laryngitis.** Swelling of the true vocal cords during laryngitis inhibits voice production.

Trachea

The **trachea** (tra′ke-ah), or windpipe, is a membranous tube that consists of connective tissue and smooth muscle reinforced with 15 to 20 C-shaped pieces of cartilage (Figure 15-5, *A*). The adult trachea is about 1.4 to 1.6 cm in diameter. It begins immediately inferior to the cricoid cartilage of the larynx, projects through the mediastinum, and divides into the right and left primary bronchi at the level of the fifth thoracic vertebra.

The C-shaped cartilages form the anterior and lateral sides of the trachea, and they protect the trachea and maintain an open passageway for air. The posterior wall of the trachea has no cartilage and consists of a ligamentous membrane and smooth muscle. The smooth muscle can alter the diameter

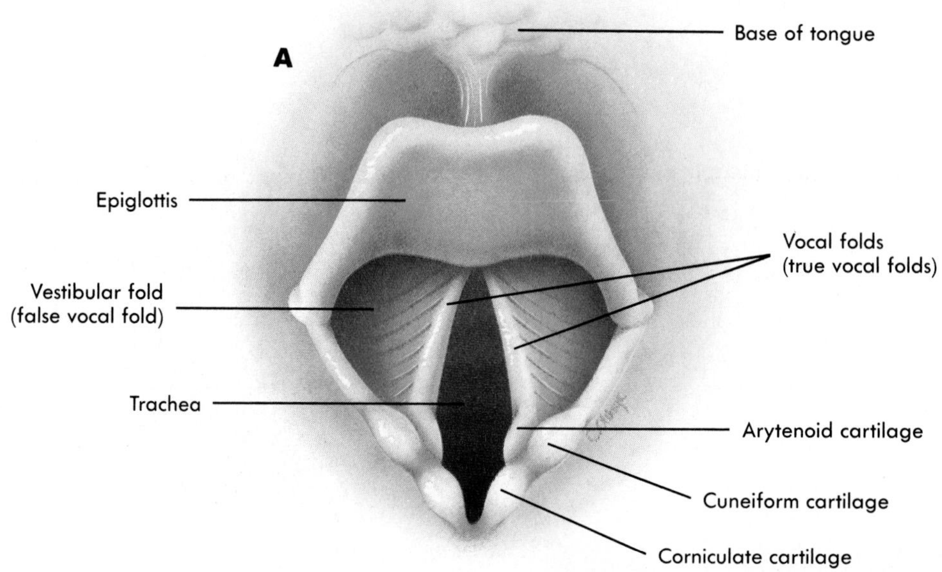

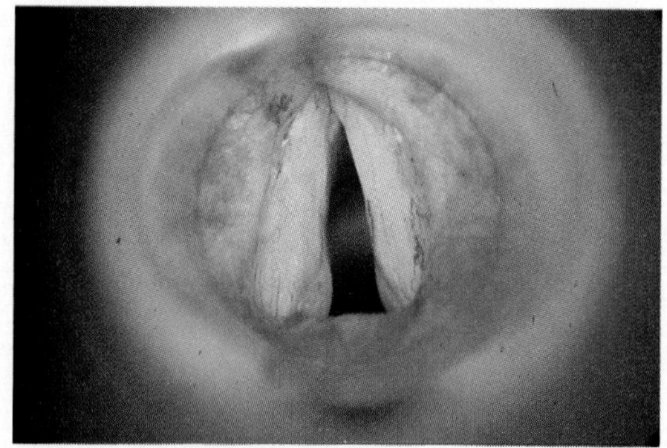

FIGURE 15-4 ■ The vocal cords.
A The vocal cords viewed from above, showing their relationship to the paired cartilages of the larynx and the epiglottis.
B Endoscopic view of the vocal cords.

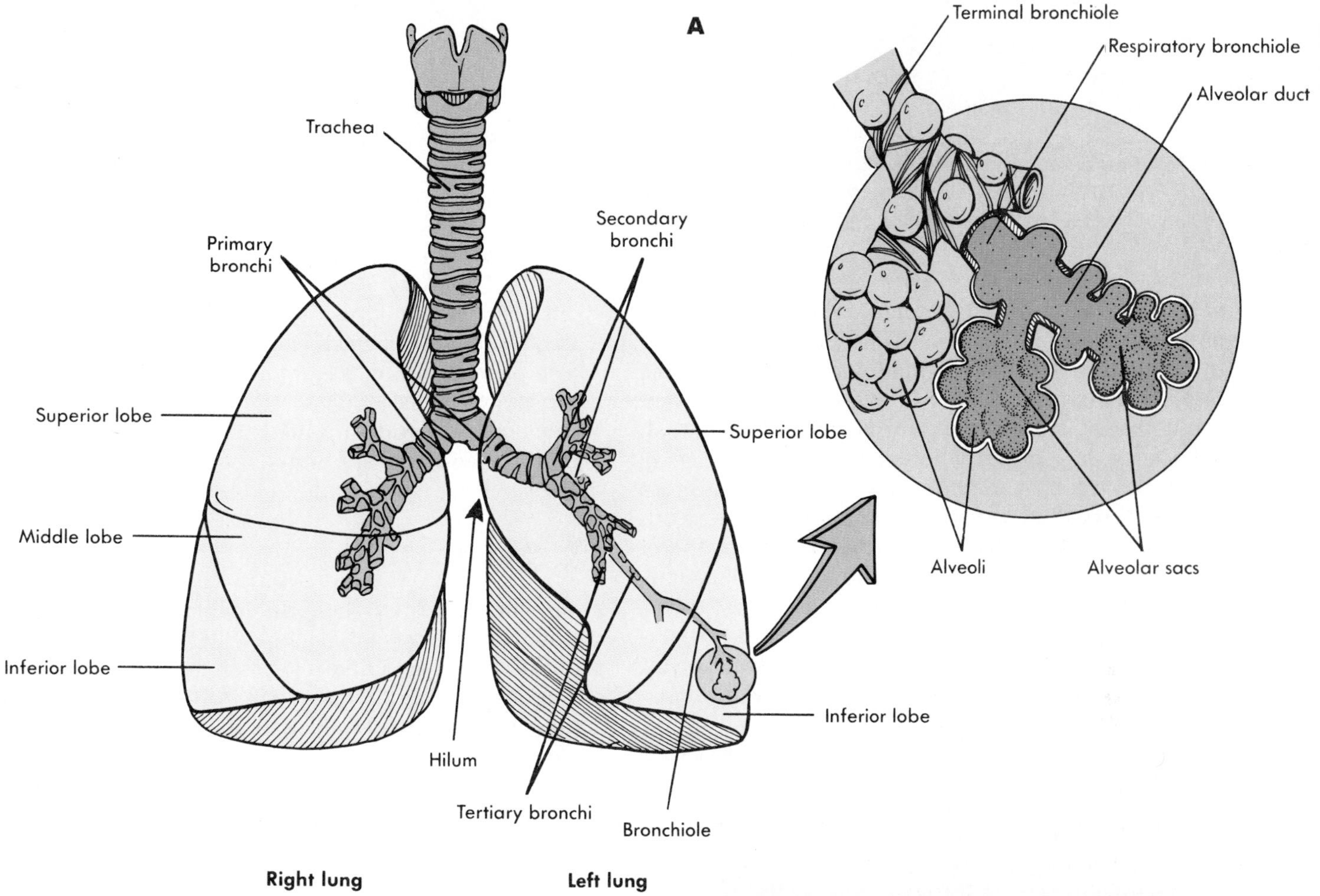

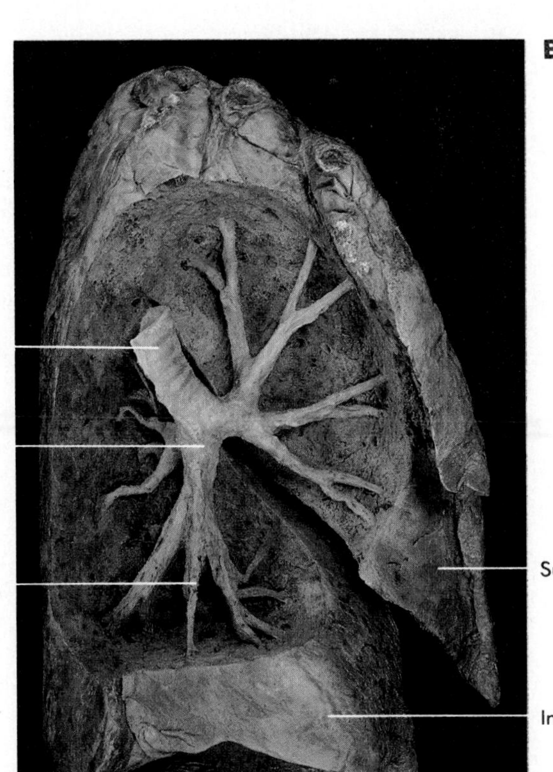

FIGURE 15-5 ▪ Anatomy of the trachea and lungs.
A Drawing of the trachea and lungs. Inset shows enlargement of a terminal bronchiole and its associated alveoli.
B Branching of the bronchi in the left lung.

of the trachea (see Figure 15-3, *B*). The esophagus lies immediately posterior to the cartilage-free posterior wall of the trachea.

> **2 Explain what happens to the shape of the trachea when a person swallows a large mouthful of food.** ?

The trachea is lined with pseudostratified columnar epithelium that contains numerous cilia and goblet cells. The cilia propel mucus produced by the goblet cells and foreign particles toward the larynx, where they enter the esophagus and are swallowed. Constant irritation to the trachea, such as occurs in people who smoke cigarettes, may cause the tracheal epithelium to change to a more resistant stratified squamous epithelium. The stratified squamous epithelium has no cilia and, therefore, lacks the ability to clear the airway of mucus and debris. The accumulations of mucus provide a place for microorganisms to grow, resulting in respiratory infections. Irritation and inflammation of the respiratory passages stimulate the cough reflex.

> In cases of extreme emergency when the upper air passageway is blocked by a foreign object to the extent that the victim cannot breathe, quick reaction is required to save the person's life. The *Heimlich maneuver* is designed to force such an object out of the air passage by the sudden application of pressure to the abdomen, forcing air up the trachea to dislodge the obstruction. The person who performs the maneuver stands behind the victim with his arms under the victim's arms and his hands over the victim's abdomen between the navel and the rib cage. With one hand formed into a fist, the other hand suddenly pulls the fist toward the abdomen with an accompanying upward motion. The pressure pushes up on the diaphram and, therefore, increases air pressure in the lungs. This maneuver, if done properly, will cause air to flow from the lungs with sufficient force to dislodge most foreign objects.
>
> In rare cases when the obstruction cannot be removed using the Heimlich maneuver, it may be necessary to form an artificial opening in the victim's air passageway to save his life. The preferred point of entry in an emergency is through the membrane between the cricoid and thyroid cartilages. This procedure is referred to as a *cricothyrotomy*. A *tracheostomy* is an incision into the trachea. A tube is inserted into the opening to facilitate the passage of air. Because the trachea has several structures overlying its anterior surface (such as arteries, nerves, and part of the thyroid gland), only trained individuals should perform these procedures.

Bronchi

The trachea divides into the left and right **primary bronchi** (brong'ki; windpipe). Because of the location of the heart in the thoracic cavity, the left primary bronchus is more horizontal than the right primary bronchus. The right primary bronchus is also shorter and wider (see Figure 15-5, *A*). Because the right primary bronchus is more vertical than the left primary bronchus, foreign objects that enter the trachea usually lodge in the right primary bronchus. The primary bronchi extend from the trachea to the lungs. Like the trachea, the primary bronchi are lined with pseudostratified ciliated columnar epithelium and are supported by C-shaped cartilage rings.

Lungs

The lungs are the principal organs of respiration. Each lung is cone-shaped, with its base resting on the diaphragm and its apex extending superiorly to a point about 2.5 cm above the clavicle. The right lung has three lobes called the superior, middle, and inferior lobes. The left lung has two lobes called the superior and inferior lobes. The lobes are separated by deep, prominent fissures on the surface of the lung (Figure 15-5, *B*). Each lobe is divided into lobules that are separated from each other by connective tissue septa, but these separations are not visible as surface fissures. Because major blood vessels and bronchi do not cross the septa, individual diseased lobules can be surgically removed, leaving the rest of the lung relatively intact. There are 9 lobules in the left lung and 10 lobules in the right lung.

The point of entry for bronchi, vessels, and nerves in each lung is called the **hilum** (hi'lum), or root, of the lung. After entering the lung, the primary bronchi branch several times to form the **bronchial tree** (see Figure 15-5, *A* and *B*). The primary bronchi first divide into secondary bronchi as they enter their respective lungs. The **secondary bronchi,** two in the left lung and three in the right lung, conduct air to each lobe. The secondary bronchi, in turn, give rise to many **tertiary bronchi,** which extend to the lobules of the lung. The bronchial tree continues to branch many times, finally giving rise to **bronchioles.** The bronchioles also subdivide numerous times to give rise to **terminal bronchioles,** which then subdivide into **respiratory bronchioles.** Each respiratory bronchiole subdivides to form **alveolar ducts** that end as clusters of air sacs called **alveoli** (al've-o'li; hollow sacs) (see Figure 15-5, *A*).

In contrast to the bronchi, bronchioles have no cartilage in their walls. They are small tubes lined with ciliated columnar epithelium that undergoes a transition to ciliated simple cuboidal epithelium as the tubes branch and then to simple squamous epithelium in the smallest branches. There is abundant smooth muscle in the walls of the bronchioles. Because of the smooth muscle, bronchioles can constrict if the smooth muscle contracts forcefully, which happens during an asthma attack. When bronchioles constrict, the resistance to air flow through the respiratory passages increases dramatically, and breathing becomes very difficult.

The walls of the alveolar ducts and the alveoli consist mainly of thin, simple squamous epithelium. It is across the epithelial lining of the alveolar ducts and the alveoli that gas exchange with blood occurs, although some exchange does occur across the wall of the respiratory bronchioles. Rapid diffusion of oxygen and carbon dioxide is possible because of the thin epithelial layer that separates the air in the alveolar ducts and alveoli from blood in the alveolar capillaries.

Pleural Cavities

■ The lungs are contained within the **thoracic cavity.** In addition, each lung is surrounded by a separate **pleural** (ploor'al; relating to the ribs) **cavity** and is attached only along its medial border at the hilum. Each pleural cavity is lined with a serous membrane called the **pleura.** The pleura consists of a parietal and visceral portion. At the hilum, the **parietal pleura,** which lines the walls of the pleural cavity, is continuous with the **visceral pleura,** which covers the surface of the lung (Figure 15-6).

The pleural cavity, between the parietal and visceral pleurae, is filled with a small volume of **pleural fluid** produced by the pleural membranes. The pleural fluid performs two functions: it acts as a lubricant, allowing the pleural membranes (the visceral and parietal pleurae) to slide past each other as the lungs and the thorax change shape during respiration, and it helps hold the pleural membranes together. The pleural fluid acts like a thin film of water between two sheets of glass (the visceral and parietal pleurae); the glass sheets can slide over each other easily, but it is difficult to sep-

3 Pleurisy is an inflammation of the pleural membranes. Explain why this condition is so painful, especially when a person takes deep breaths. ?

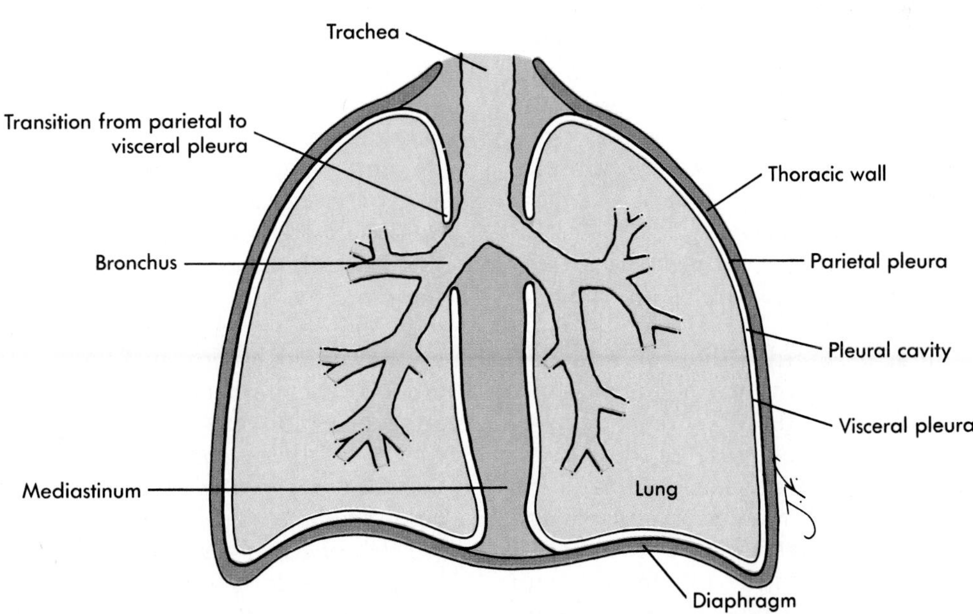

FIGURE 15-6 ■ The pleural cavities.
Lungs surrounded by pleural cavities. The parietal pleura lines the wall of each pleural cavity, and the visceral pleura covers the surface of the lungs. The pleural cavity between the parietal and visceral pleurae is small and is filled with pleural fluid.

VENTILATION AND LUNG VOLUMES

Ventilation, or breathing, is the process of moving air into and out of the lungs. There are two phases of ventilation: **inspiration,** or **inhalation,** is the movement of air into the lungs; **expiration,** or **exhalation,** is the movement of air out of the lungs.

Flow of air through the respiratory passages occurs because of pressure differences. The rate of air flow in a tube is determined by the pressure difference between one end of the tube and the other end. If the pressure is higher at one end of the tube than at the other end, air will flow from the area of high pressure toward the area of low pressure. The greater the pressure difference, the greater the rate of air flow.

During ventilation the pressure in the alveoli changes so that there is a pressure difference between the atmosphere and the alveoli, resulting in air flow. During inspiration, the pressure in the alveoli is lower than atmospheric pressure. Consequently air flows from outside the body through the respiratory passages to the alveoli. During expiration, the pressure in the alveoli is greater than atmospheric pressure. Consequently air flows from the alveoli through the respiratory passages to the outside of the body.

Mechanics of Breathing

Changes in pressure within the alveoli result from changes in the volume of the thoracic cavity. As the volume of a closed container increases, the pressure within the container decreases, and as the volume of a closed container decreases, the pressure with the container increases. It is the muscles of respiration that cause the volume of the thoracic cavity to increase and decrease.

The **diaphragm** (di'ă-fram; partition) is a large dome of skeletal muscle that separates the thoracic cavity from the abdominal cavity. When the muscles of the diaphragm contract, the dome is flattened, thus increasing the volume of the thoracic cavity (Figure 15-7). When the diaphragm relaxes, the volume of the thoracic cavity is decreased. Other muscles of respiration alter the volume of the thoracic cavity by lifting the anterior ends of the ribs and sternum, which increases the diameter of the thoracic cavity, or depressing the anterior end of the ribs and sternum, which decreases the diameter of the thoracic cavity. For example, contraction of the external intercostal muscles lifts the ribs, and contraction of the internal intercostal muscles lowers the ribs (see Figure 15-7).

Contraction of muscles of inspiration increases the volume of the thoracic cavity. The increased volume causes the lungs to expand because the visceral pleura adheres to the parietal pleura. As the lungs expand, the alveolar volume increases, resulting in a decrease in pressure in the alveoli below atmospheric pressure. Because of this pressure difference, air flows into the lungs. At the end of inspiration, the thorax and alveoli stop expanding. Air flows into the alveoli until the pressure within the alveoli becomes equal to the atmospheric pressure. During expiration, the thoracic volume decreases, producing a decrease in alveolar volume and an increase in alveolar pressure above atmospheric pressure. Consequently air flows out of the lungs. At the end of expiration, the decrease in thoracic volume stops, and the alveolar volume no longer decreases. Air flows out of the alveoli until alveolar pressure becomes equal to the atmospheric pressure.

During quiet breathing the major muscle of inspiration is the diaphragm, and the passive recoil of the lung and wall of the thorax cause expiration. It is during labored breathing that all of the muscles of expiration contract and decrease the volume of the thorax to a greater extent. During labored breathing the muscles of expiration are important in forcefully decreasing the volume of the thoracic cavity.

Factors Preventing Collapse of the Lung

The lungs tend to collapse for two reasons (1) elastic recoil caused by the elastic fibers in the connective tissue of the lungs and (2) surface tension of the film of fluid that lines the alveoli. Molecules of water attract each other, tending to form a droplet. Because water molecules of the thin layer of fluid that lines the alveoli are attracted to the surface of the alveoli, formation of a droplet causes the alveoli to collapse, producing fluid-filled alveoli with smaller volumes than air-filled alveoli.

Two factors keep the lungs from collapsing (1) surfactant and (2) the tendency of the visceral pleura to adhere to the parietal pleura. **Surfactant** (sur-fak'tant) is a mixture of the lipoprotein molecules produced by the secretory cells of the alveolar epithelium. The surfactant molecules form a single layer on the surface of the thin layer of fluid lining the alveoli to reduce surface tension.

The fluid in the pleural cavity causes the visceral pleurae to adhere to the parietal pleurae much like two pieces of glass held together by a thin film of water. Thus, when the volume of the thorax

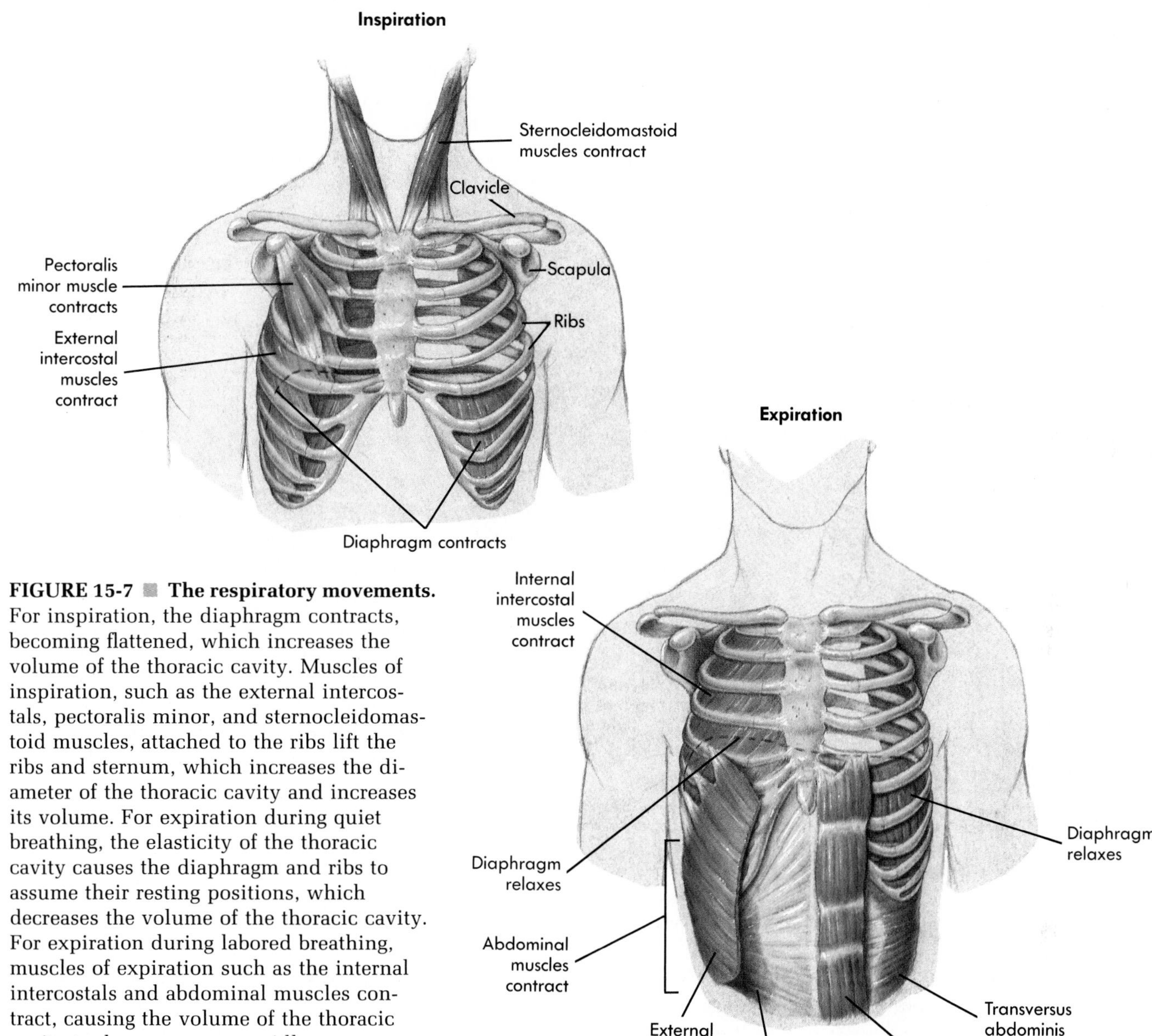

FIGURE 15-7 ■ The respiratory movements. For inspiration, the diaphragm contracts, becoming flattened, which increases the volume of the thoracic cavity. Muscles of inspiration, such as the external intercostals, pectoralis minor, and sternocleidomastoid muscles, attached to the ribs lift the ribs and sternum, which increases the diameter of the thoracic cavity and increases its volume. For expiration during quiet breathing, the elasticity of the thoracic cavity causes the diaphragm and ribs to assume their resting positions, which decreases the volume of the thoracic cavity. For expiration during labored breathing, muscles of expiration such as the internal intercostals and abdominal muscles contract, causing the volume of the thoracic cavity to decrease more rapidly.

changes, the volume of the lungs changes also. Because of the low pressure surrounding the lungs, they remain inflated and so the space between the visceral and parietal pleurae contains only a thin film of fluid.

A **pneumothorax** is the introduction of air into the pleural cavity. Air can enter by an external route when a sharp object, such as a bullet or broken rib, penetrates the thoracic wall, or air can enter the pleural cavity by an internal route if alveoli at the lung surface rupture, such as may occur in a patient with emphysema. The introduction of air into the pleural space equalizes the air pressure between the pleural space and the atmosphere. When that occurs the elastic fibers in the lung are able to create enough force to cause the lung to collapse, leaving an air-filled pleural cavity. To reinflate a lung that has collapsed, the hole must be closed. Then, if a tube is placed into the pleural cavity and suction applied to reestablish a pressure less than atomospheric pressure in the pleural cavity, the lung will reinflate.

Surfactant is a mixture of lipoprotein molecules secreted by cells of the alveolar epithelium. The surfactant molecules form a layer over the thin layer of fluid within the alveoli and reduce surface tension and the tendency of the lungs to collapse. Surfactant is not produced in adequate quantities until about the seventh month of gestation. Thereafter, the amount produced increases as the fetus matures. In premature infants, *hyaline membrane disease*, or *respiratory distress syndrome*, caused by too little surfactant, is common, especially for infants delivered before the seventh month of pregnancy. If too little surfactant has been produced by the time of birth, the lungs tend to collapse, and a great deal of energy must be exerted by the muscles of respiration to keep the lungs inflated and, even then, inadequate ventilation occurs. Without specialized treatment, such as forcing enough air into the lungs to inflate them, most babies with this condition die soon after birth as a result of inadequate ventilation of the lungs and fatigue of the respiratory muscles.

Pulmonary Volumes and Capacities

Spirometry (spĭ-rom′ĕ-tre) is the process of measuring volumes of air that move into and out of the respiratory system, and the **spirometer** is the device that is used to measure these pulmonary volumes (Figure 15-8, *A*). Measurements of the respiratory volumes can provide information about the health of the lungs. The four **pulmonary volumes** and their normal values (Figure 15-8, *B*) for a young adult male are as follows:

1. **Tidal volume:** the volume of air inspired or expired during quiet breathing (about 500 ml)
2. **Inspiratory reserve volume:** the amount of air that can be inspired forcefully after inspiration of the normal tidal volume (about 3000 ml)
3. **Expiratory reserve volume:** the amount of air that can be expired forcefully after expiration of the normal tidal volume (about 1100 ml)
4. **Residual volume:** the volume of air still remaining in the respiratory passages and lungs after a maximum expiration (about 1200 ml)

A **pulmonary capacity** is the sum of two or more pulmonary volumes (see Figure 15-8, *B*). For example:

1. **Vital capacity:** the sum of the inspiratory reserve volume, the tidal volume, and the expiratory reserve volume; it is the maximum volume of air that a person can expel from his respiratory tract after a maximum inspiration (about 4600 ml)
2. **Inspiratory capacity:** the tidal volume plus the inspiratory reserve volume; the amount of air that a person can inspire maximally after a normal expiration (about 3500 ml)
3. **Functional residual capacity:** the expiratory reserve volume plus the residual volume; the amount of air remaining in the lungs at the end of a normal expiration (about 2300 ml)
4. **Total lung capacity:** the sum of the inspiratory and expiratory reserve volumes plus the tidal volume and the residual volume (about 5800 ml)

4 If a resting person had a tidal volume of 500 ml and a respiratory rate of 12 respirations per minute, and an exercising person had a tidal volume of 4000 ml and a respiratory rate of 24 respirations per minute, what would be the difference in the total amount of air respired per minute between them? ?

The tidal volume may increase when a person is more active. If the tidal volume increases, the inspiratory reserve volume and the expiratory reserve volume decrease. Because the maximum volume of the respiratory system does not change from moment to moment, any increase or decrease in tidal volume affects the inspiratory and expiratory reserve volumes.

Factors such as sex, age, body size, and physical conditioning influence the respiratory volumes and capacities. For example, the vital capacity of adult females is usually 20% to 25% less than that of adult males. The vital capacity reaches its maximum amount in the young adult, and it gradually decreases in the elderly. Tall people usually have a greater vital capacity than short people, and thin people have a greater vital capacity than obese people. Well-trained athletes can have a vital capacity 30% to 40% above average; that is, 6 to 7 liters instead of the average 4.6 liters in males. In people with paralysis of their respiratory muscles resulting from spinal cord injury or diseases, such as poliomyelitis or muscular dystrophy, the vital capacity may be reduced so much (to less than 500 to 1000 ml) that the individual cannot survive. Factors that reduce the elasticity of the lungs and thorax also reduce the vital capacity.

The **forced expiratory vital capacity** is the rate at which lung volume changes during direct measurement of the vital capacity. It is a simple and clinically important pulmonary test. The individual inspires maximally and then exhales maximally

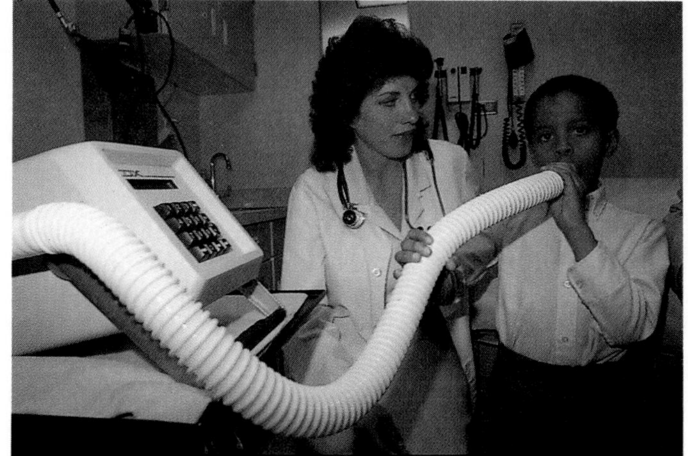

FIGURE 15-8 **Lung volumes and capacities.**
A Photograph of a spirometer used to measure lung volumes and capacities.
B The values for the lung volumes and capacities. The values shown in the diagram are values during quiet breathing.

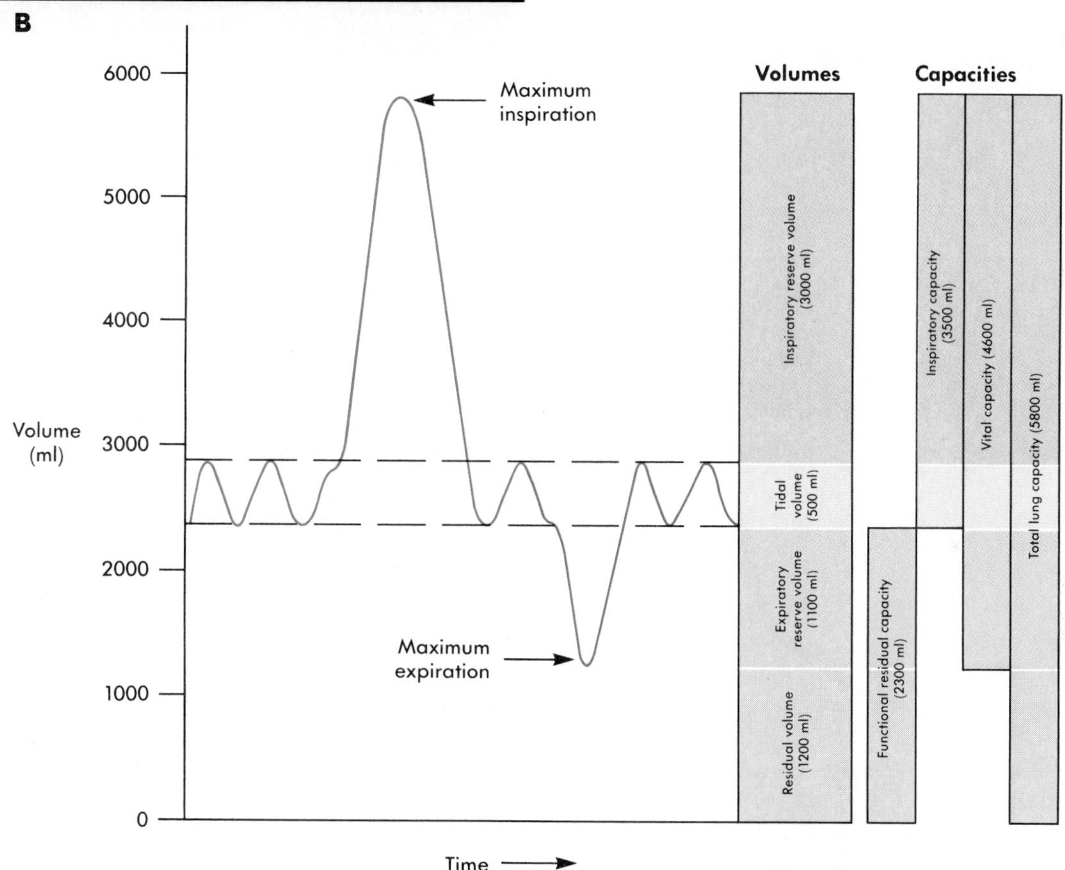

and as rapidly as possible into a spirometer. The spirometer records the volume of air expired per second. This test can be used to help identify conditions in which the vital capacity may not be affected, but in which the expiratory flow rate is reduced. Abnormalities that reduce the expiratory flow rate include conditions that reduce the ability of the lungs to deflate. For example, in people who have fibrosis (connective tissue replacement of lung tissue) or silicosis (connective tissue encapsulation of silicon particles in lung tissue) of the lungs, much of the lung tissue has been replaced with connective tissue with reduced elasticity. A greater than normal pressure is required to force air out of these lungs after they have become inflated. In addition, in people who suffer from asthma, contraction of the smooth muscle in the bronchioles increases the resistance to air flow. The increased resistance to air flow slows the rate at which air can be forced out of the lungs.

GAS EXCHANGE

Ventilation supplies atmospheric air to the alveoli. The next step in the process of respiration is the diffusion of gases between the alveoli and the blood in the pulmonary capillaries.

The **respiratory membranes** are all of those areas where gas exchange between air and blood occurs. Although the major area of gas exchange is in the alveoli, some gas exchange also occurs in the respiratory bronchioles and alveolar ducts. There are about 300 million alveoli in each lung. Surrounding each alveolus is a network of capillaries arranged so air in the alveolus is separated by a thin respiratory membrane from the blood contained within the capillaries. The total surface area of the respiratory membrane is about 70 m^2 (about the area of one half of a tennis court) in the normal adult.

The respiratory membrane (Figure 15-9) consists of (1) a thin layer of fluid containing surfactant that lines the alveolus, (2) the alveolar epithelium comprised of simple squamous epithelium, (3) the basement membrane of the alveolar epithelium, (4) a thin interstitial space, (5) the basement membrane of the capillary endothelium, and (6) the capillary endothelium comprised of simple squamous epithelium.

The exchange of gases across the respiratory membrane is influenced by the thickness of the membrane, the total surface area of the membrane, the concentration gradients for gases across the membrane, and the rate at which each gas diffuses through the fluids. Increasing the thickness of the respiratory membrane decreases the rate of diffu-

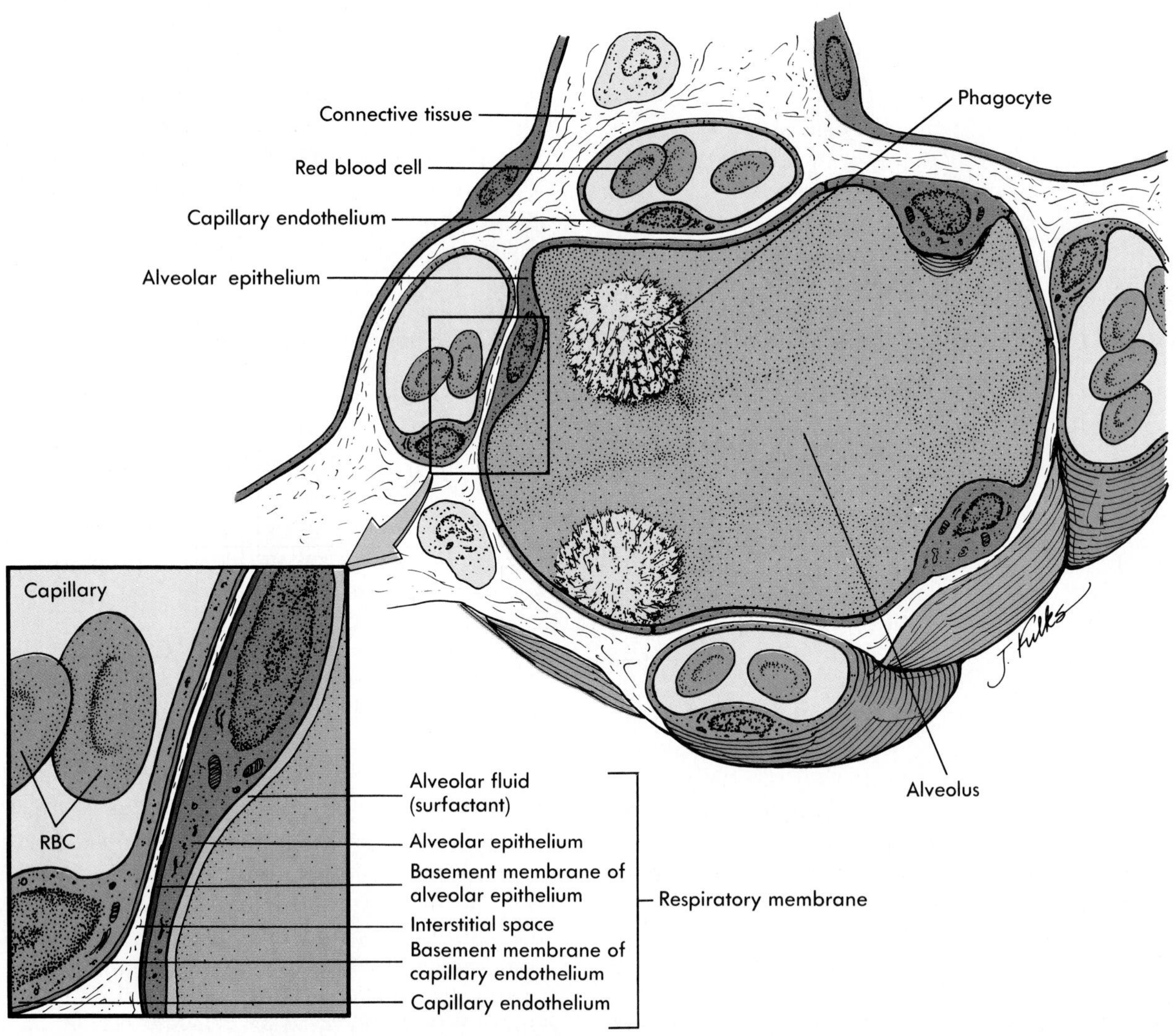

FIGURE 15-9 ■ The respiratory membrane.
The structure of an alveolus. The inset shows a magnification of the interface between the alveolus and a capillary wall.

TABLE 15-1

Partial Pressures of Gases at Sea Level

	DRY AIR		HUMIDIFIED AIR		ALVEOLAR AIR		EXPIRED AIR	
GASES	mm HG	%	mm HG	%	mm HG	%	mm HG	%
Nitrogen	600.2	78.98	563.4	74.09	569.0	74.9	566.0	74.5
Oxygen	159.5	20.98	149.3	19.67	104.0	13.6	120.0	15.7
Carbon dioxide	0.3	0.04	0.3	0.04	40.0	5.3	27.0	3.6
Water vapor	0.0	0.0	47.0	6.20	47.0	6.2	47.0	6.2

sion across the membrane. The thickness of the respiratory membrane increases during certain respiratory diseases. For example, in patients with pulmonary edema, fluid accumulates in the alveoli, and gases must diffuse through a thicker-than-normal layer of fluid. If the thickness of the respiratory membrane is doubled or tripled, the rate of gas exchange is markedly decreased.

The surface area of the respiratory membrane is decreased by several respiratory diseases, including emphysema and lung cancer. Even small decreases in this surface area adversely affect the respiratory exchange of gases during strenuous exercise. When the surface area of the respiratory membrane is decreased to one third or one fourth of normal, the exchange of gases is significantly restricted even under resting conditions.

The pressure exerted by a specific gas in a mixture of gases such as air is usually reported as that gas's **partial pressure.** For example, if the total pressure of all gases in a mixture of gases is 760 mm Hg (atmospheric pressure at sea level), and 21% of the mixture is made up of oxygen, then the partial pressure for oxygen is (21% × 760 mm Hg) 160 mm Hg. On the other hand, if the composition of air is 0.04% carbon dioxide at sea level, the partial pressure for carbon dioxide is (.04% × 760 mm Hg) 0.3 mm Hg (Table 15-1).

When air comes into contact with a liquid such as water, gases such as carbon dioxide and oxygen in the air dissolve in the liquid. The gases dissolve in the liquid until the partial pressure of each gas in the liquid is equal to the partial pressure of each gas in the air. In a liquid, gases diffuse from areas of high partial pressure toward areas of low partial pressure until the partial pressures of each gas are equal throughout the liquid. Thus it is the difference in partial pressures that determines the direction a gas will diffuse.

The difference in the partial pressure of a gas across the respiratory membrane influences the rate of gas exchange. The cells of the body use oxygen and produce carbon dioxide. Therefore, the partial pressure of oxygen in the alveoli is greater than the partial pressure of oxygen in the alveolar capillaries. Thus oxygen diffuses from the alveoli into the alveolar capillaries. The partial pressure of carbon dioxide, in contrast, is greater in the alveolar capillaries than in the alveoli. Thus carbon dioxide diffuses from the alveolar capillaries into the alveoli.

Because of the mixing of air in the respiratory airways during breathing, the partial pressure of oxygen is greater in the air than in the alveoli, and the partial pressure of carbon dioxide is greater in the alveoli than in air. Therefore, an increase in volume of atmospheric air entering and leaving the alveoli causes the higher partial pressure of oxygen to increase and the partial pressure of carbon dioxide to decrease in the alveoli. Increasing the rate of ventilation keeps the partial pressure of oxygen higher in the alveoli than during slow breathing. The rate of oxygen diffusion into the alveolar capillaries increases because the partial pressure of oxygen in the alveoli is increased, and the difference in partial pressure between the alveoli and the alveolar capillaries is also increased.

Increasing the rate of ventilation also keeps the partial pressure of carbon dioxide lower in the alveoli than during slow breathing. Because the partial pressure of carbon dioxide in the alveoli is lower, the difference in partial pressure between the alveoli and the alveolar capillary is increased. Thus carbon dioxide diffuses from the alveolar capillaries into the alveoli at a greater rate.

Inadequate ventilation causes a smaller difference in the partial pressure of oxygen and carbon dioxide across the respiratory membrane. The lower rate of ventilation results in a decrease in the difference of the partial pressures for oxygen and carbon dioxide across the respiratory membrane. Therefore the rate of oxygen and carbon dioxide diffusion across the membrane decreases. Carbon dioxide accumulates in the blood, and oxygen levels in the blood decline.

Figure 15-10 illustrates the differences in partial pressures for oxygen and carbon dioxide across the

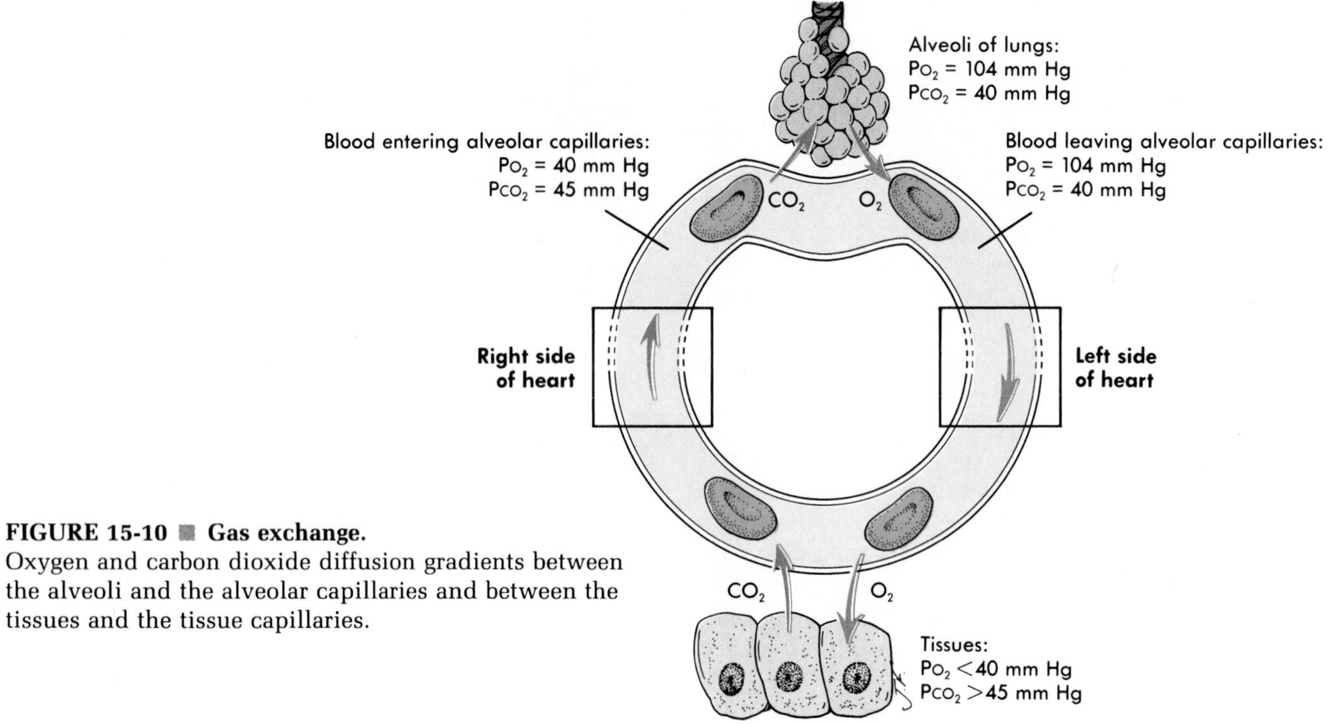

FIGURE 15-10 ■ Gas exchange.
Oxygen and carbon dioxide diffusion gradients between the alveoli and the alveolar capillaries and between the tissues and the tissue capillaries.

respiratory membrane. The partial pressures of oxygen and carbon dioxide in the blood are maintained because of the diffusion of oxygen from the alveoli into the blood and the diffusion of carbon dioxide from the blood into the alveoli. The partial pressures of oxygen and carbon dioxide in the alveoli are maintained because of the movement of air into and out of the alveoli. By the time blood flows through the first third of the pulmonary capillary, an equilibrium is achieved, and the partial pressures of oxygen and carbon dioxide in the alveoli and the lung capillaries are equal. Therefore there is an efficient exchange of these gases between blood and the air in the alveoli.

Blood flows from the lungs through the left side of the heart to the tissue capillaries. Figure 15-10 also illustrates the partial pressure differences for oxygen and carbon dioxide across the wall of the tissue capillaries. Oxygen diffuses because of its high partial pressure in the tissue capillaries to the interstitial fluid and from the interstitial fluid into the cells of the body, where the partial pressure of oxygen is lower. Within the cells, oxygen is used in aerobic metabolism. Because oxygen is continuously used by cells, there is a constant difference in partial pressure from the tissue capillaries to the cells. Because carbon dioxide is continuously produced by cells, there is a constant diffusion gradient for carbon dioxide from the cells to the tissue capillaries. Therefore carbon dioxide diffuses from cells into the interstitial fluid and from interstitial fluid into tissue capillaries.

OXYGEN AND CARBON DIOXIDE TRANSPORT IN THE BLOOD

After oxygen diffuses across the respiratory membrane into the blood, about 97% of the oxygen combines reversibly with the iron-containing heme groups of hemoglobin (see Chapter 11). About 3% of the oxygen remains dissolved in the plasma. Hemoglobin with oxygen bound to its heme groups is called **oxyhemoglobin.** The oxygen bound to hemoglobin does not affect the partial pressure of oxygen in blood. Because oxygen binds to hemoglobin, the amount of oxygen contained within a given volume of blood is greatly increased without an increase in the partial pressure of oxygen in the blood. Hemoglobin carries oxygen from the alveolar capillaries through the blood vessels to the tissue capillaries. The combination of oxygen with hemoglobin is reversible, and in the tissue spaces where the partial pressure of oxygen is low, oxygen diffuses away from hemoglobin molecules and into the tissue spaces. Oxygen then diffuses into cells and is used by the cells in aerobic metabolism.

The amount of oxygen released from oxyhemoglobin is influenced by several factors. More oxygen is released from hemoglobin if the partial pressure of oxygen is low, the partial pressure of carbon dioxide is high, the pH is low, and the temperature is high. Increased muscular activity results in a decreased partial pressure of oxygen, an increased partial pressure of carbon dioxide, a reduced pH, and an increased temperature. Consequently more oxygen is released from hemoglobin in skeletal muscles during periods of physical exercise.

Carbon dioxide diffuses from cells, where it is produced, into the tissue capillaries. After it enters the blood, carbon dioxide is transported in three principal ways: About 8% is transported as carbon dioxide dissolved in the plasma, 20% is transported in combination with blood proteins, primarily hemoglobin, and 72% is transported in the form of bicarbonate ions.

Carbon dioxide reacts with water to form carbonic acid, which then dissociates to form hydrogen ions and bicarbonate ions.

$$CO_2 + H_2O \rightleftarrows H_2CO_3 \rightleftarrows H^+ + HCO_3^-$$

An enzyme called **carbonic anhydrase** inside red blood cells increases the rate at which carbon dioxide reacts with water to form hydrogen ions and bicarbonate ions in the tissue capillaries (Figure 15-11). In capillaries of the lungs, the process is reversed so that the bicarbonate and hydrogen ions combine to produce carbonic acid, which then dissociates to form carbon dioxide and water. The carbon dioxide diffuses into the alveoli and is expired.

Carbon dioxide has an important effect on the pH of blood. As carbon dioxide levels increase, the blood pH decreases (becomes more acidic) because carbon dioxide reacts with water to form carbonic acid. Hydrogen ions that result from the formation of carbonic acid are responsible for the decrease in pH. Conversely, as blood levels of carbon dioxide decline, the blood pH increases (becomes less acidic or more basic).

5 What effect would a rapid rate of respiration have on blood pH? What effect would holding one's breath have on blood pH? Explain. ?

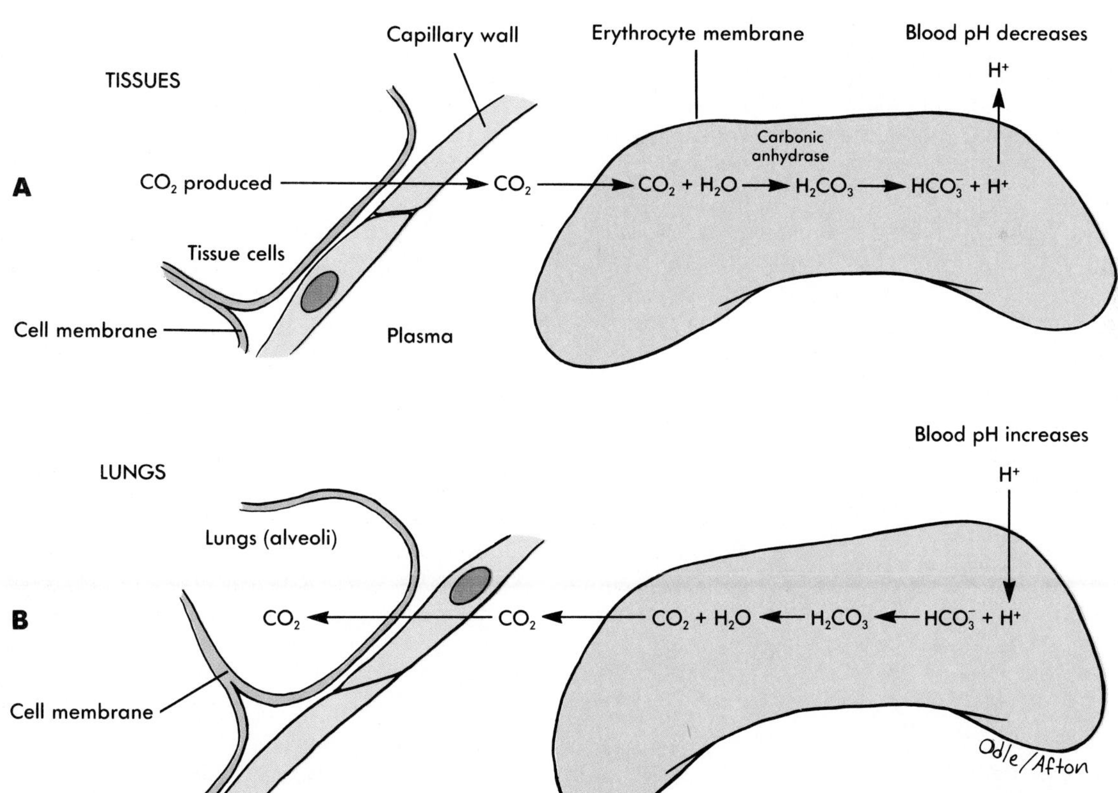

FIGURE 15-11 Carbon dioxide transport as bicarbonate.
A In the tissues carbon dioxide enters the red blood cells and combines with water to form carbonic acid. The enzyme carbonic anhydrase catalyzes the reaction. Carbonic acid then dissociates to form hydrogen ions and bicarbonate ions.
B In the lung capillaries, hydrogen ions and bicarbonate ions combine to form carbonic acid. Carbonic acid then dissociates to form water and carbon dioxide. The carbon dioxide diffuses into the alveoli of the lungs.

Abnormalities that Reduce Alveolar Ventilation and the Oxygen-Carrying Capacity of Blood

PARALYSIS OF THE RESPIRATORY MUSCLES

Paralysis of the respiratory muscles can result from **poliomyelitis,** a viral infection that damages neurons of the respiratory center or motor neurons that stimulate the muscles of respiration. Another cause is transection of the spinal cord in the cervical or thoracic regions, interrupting nerve impulses that normally pass to the muscles of respiration. Transection of the spinal cord in the upper cervical region eliminates innervation of all respiratory muscles, whereas transection of the spinal cord in the lower cervical region or upper thorax region leaves the innervation of the diaphragm through the phrenic nerves intact, but results in paralysis of other respiratory muscles. Transection of the spinal cord often results from traumatic accidents such as occurs when one dives into water that is too shallow. Anesthetics or central nervous system depressants can temporarily depress the function of the respiratory center. If they are taken or administered in large enough doses they can cause death by stopping respiratory movements.

INCREASED RESISTANCE IN THE RESPIRATORY PASSAGES

Asthma results in the release of inflammatory chemicals into the circulatory system that result in contraction of the smooth muscle cells in the bronchioles. Contraction of these smooth muscle cells causes the bronchioles to constrict, making it difficult to move air in and out of the lungs. **Emphysema** results in increased airway resistance because the bronchioles are obstructed as a result of inflammation and because damaged bronchioles collapse during expiration, trapping air within the alveoli. **Cancer** may also block respiratory passages as lung tissue is replaced by the tumor.

DECREASED ELASTICITY OF THE LUNGS AND THE THORACIC WALL

Elasticity is the ability of the lungs and thorax to expand and then recoil to their original shape. Loss of elasticity makes it more difficult for the lungs to increase and decrease in volume, resulting in difficulty in breathing. Replacement of lung tissue with fibrous connective tissue or encapsulation of materials or organisms in the lung make it less elastic. For example, **silicosis** (sil'ĭ-ko-sis) and **asbestosis** (as'bes-tō'sis) in which silicon or asbestos particles are encapsulated by connective tissue elements reduce the lungs' elasticity. **Tuberculosis** and **pneumonia** are infections that result in pulmonary inflammation and edema, and both of these conditions reduce lung elasticity. Cancer replaces lung tissue with a very inelastic tumor, resulting in reduced lung elasticity.

Decreased elasticity of the thoracic cavity may be caused by severe **arthritis,** or by conditions resulting in severe curvature of the spine, such as **scoliosis** (sko'le-o'sis), and **kyphosis** (ki-fo'sis). These conditions reduce the ability of the thoracic wall to increase its volume when the muscles of inspiration contract, thereby increasing the muscular effort required for inspiration.

DECREASED SURFACE AREA FOR GAS EXCHANGE

A decreased surface area for gas exchange results from the surgical removal of lung tissue, the destruction of lung tissue by cancer, the degeneration of the alveolar walls by emphysema, or the replacement of lung tissue by connective tissue caused by tuberculosis. More acute conditions (for example, pneumonia, pulmonary edema as a result of failure of the left ventricle, and collapse of the lung) that cause the alveoli to fill with fluid also reduce the surface area for gas exchange.

INCREASED THICKNESS OF THE RESPIRATORY MEMBRANE

Pulmonary edema resulting from failure of the left side of the heart is the most common cause of an increase in the thickness of the respiratory membrane. The increased venous pressure in the pulmonary capillaries results in the accumulation of fluid in the alveoli. As a consequence, the efficiency of gas diffusion across the respiratory membrane is decreased. Conditions (for example, tuberculosis, pneumonia, and silicosis) that result in inflammation of the lung tissues also cause pulmonary edema.

VENTILATION AND THE PERFUSION OF LUNG TISSUE WITH BLOOD

In some alveoli there is too little ventilation to allow blood that flows through the alveolar capillaries to become saturated with oxygen. Diseases that reduce ventilation include asthma, which causes excess resistance of pulmonary airways, and emphysema, which damages lung tissues. In other alveoli the ventilation may be adequate, but blood flow through the alveolar capillaries may be inadequate. Disorders that reduce perfusion of lung tissue with blood include thrombosis of pulmonary arteries and reduced cardiac output that results from heart attack or shock.

REDUCED CAPACITY OF THE BLOOD TO TRANSPORT OXYGEN

Anemias that result in a reduction of the total amount of hemoglobin available to transport oxygen also reduce the capacity of blood to transport oxygen. **Carbon monoxide** binds more strongly to the hemoglobin molecule than oxygen does. It occupies binding sites, making them unavailable for oxygen transport. Thus carbon monoxide poisoning decreases the ability of hemoglobin to transport oxygen, even though it does not affect the total hemoglobin concentration in the blood.

CONTROL OF RESPIRATION

The normal rate of respiration in adults is between 12 and 20 respirations per minute. In children the rates are higher and may vary from 20 to 40 per minute.

A collection of neurons in the medulla oblongata and in the lower portion of the pons together make up the **respiratory center,** which controls the rate and depth of respiration (Figure 15-12). If the respiratory center is damaged, respiratory movements stop. Action potentials pass from the respiratory center to the diaphragm along nerve fibers of the phrenic nerve. The phrenic nerve arises from the cervical region of the spinal cord and passes to the diaphragm. Action potentials pass from the respiratory center to the other muscles of respiration along nerve fibers of the spinal nerves.

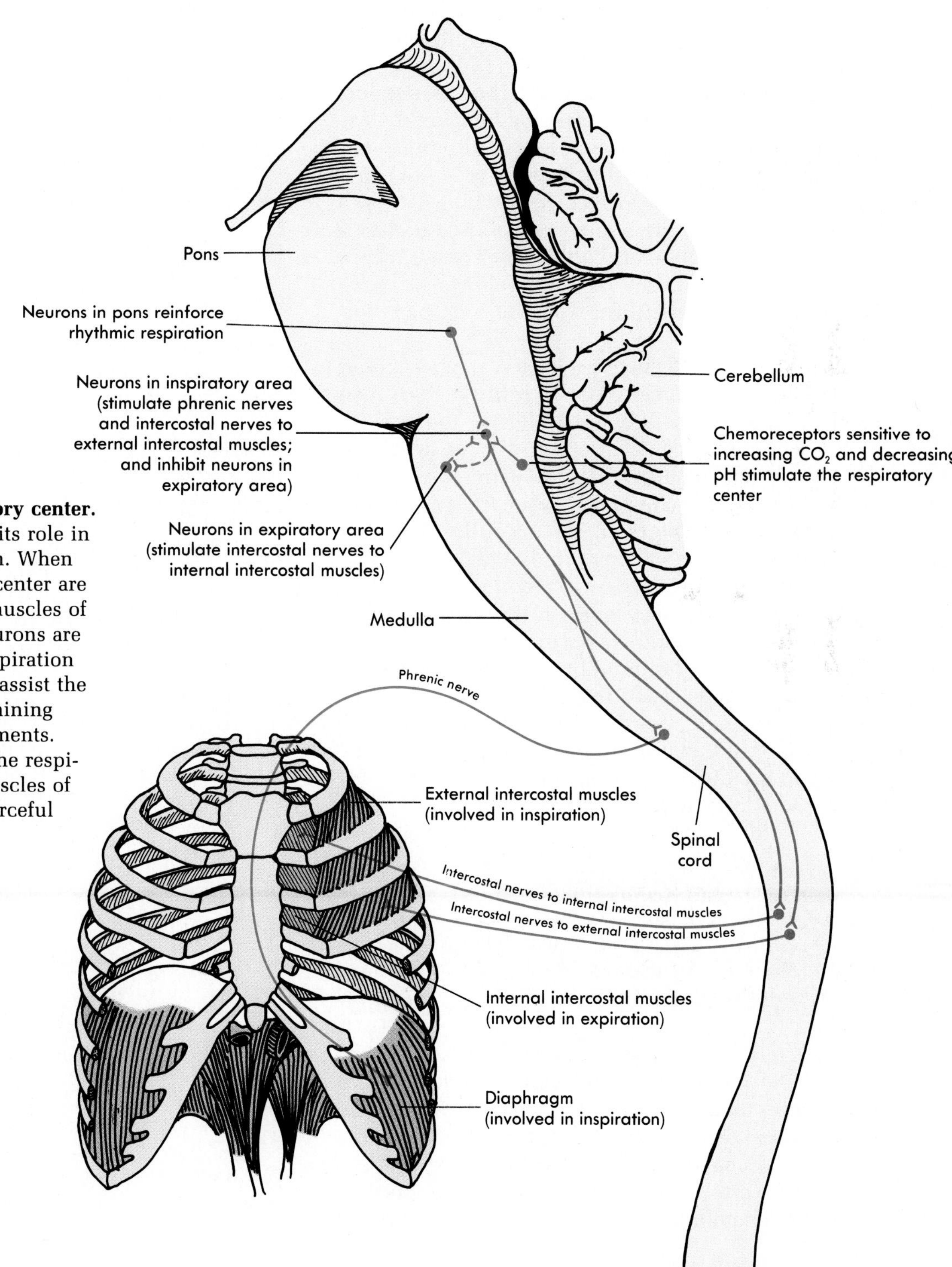

FIGURE 15-12 Respiratory center. The respiratory center and its role in the regulation of respiration. When neurons in the respiratory center are active, they stimulate the muscles of inspiration; when these neurons are inactive, the muscles of inspiration relax. Neurons in the pons assist the respiratory center in maintaining rhythmic respiratory movements. During labored breathing, the respiratory center stimulates muscles of expiration to cause more forceful expirations.

In a healthy person, neurons in the medulla oblongata produce action potentials for a short time and then fatigue; this process results in inspiration followed by expiration. During the period of action potential production, the muscles of inspiration are stimulated, and inspiration results. During the period of fatigue, the muscles of inspiration are no longer stimulated, and expiration results. The neurons in the lower portion of the pons also produce action potentials rhythmically. The rhythmic pattern of action potentials from the pons reinforces the rhythmic breathing produced by neurons of the medulla oblongata.

The **Hering-Breuer reflex** functions to support rhythmic respiratory movements by limiting the extent of inspiration. As muscles of inspiration contract, the lungs fill with air. Sensory receptors that respond to stretch are located in the lungs. As the lungs fill with air, the stretch receptors are stimulated. Action potentials from the lung stretch receptors are then sent to the medulla oblongata, where they inhibit the respiratory center neurons and cause expiration.

Stimuli that influence respiration, such as blood levels of carbon dioxide, blood pH, blood oxygen levels, and emotions, do so by altering the activity of the respiratory center. Carbon dioxide levels in the blood and blood pH play an extremely important role in influencing activity of the respiratory center. **Chemoreceptors** in the medulla oblongata are sensitive to small changes in blood carbon dioxide levels and pH. Increasing carbon dioxide levels in the blood and a decreasing pH strongly stimulate the respiratory center, resulting in an increased rate and depth of respiration. In contrast, decreasing carbon dioxide levels in the blood and an increasing pH result in a slower rate and depth of respiration. Respiration may even stop for a period of time until carbon dioxide levels increase and pH decreases to a level that stimulates respiration.

6 Explain why a person who breathes rapidly and deeply (hyperventilates) for several seconds experiences a short period of time in which respiration does not occur (apnea) before normal breathing resumes. ?

Chemoreceptors in the carotid and aortic bodies also provide input to the respiratory center. These chemoreceptors are sensitive to changes in oxygen, carbon dioxide, and pH. However, they are more sensitive to changes in blood levels of oxygen and less sensitive to changes in carbon dioxide and pH than the chemoreceptors in the medulla. When blood levels of oxygen decline to low levels, the chemoreceptors of the carotid and aortic bodies are strongly stimulated. They send action potentials to the respiratory center and produce an increase in the rate and depth of respiration. Changes in blood levels of oxygen do not normally act as a stimulus to the respiratory center. Blood oxygen levels become important under conditions when they decline, but carbon dioxide levels are not increased. Examples are emergency conditions such as asphyxiation, exposure to high altitudes, or diseases such as emphysema.

■ Carbon dioxide is much more important than oxygen as a regulator of normal respiration, but under certain circumstances, blood levels of oxygen do play an important stimulatory role. During conditions of shock in which the blood pressure is very low, the blood level of oxygen may decrease to levels low enough to strongly stimulate the carotid and aortic body sensory receptors.

Air is composed of 21% oxygen at low and high altitudes. At sea level the partial pressure of oxygen is about 160 mm Hg (760 mm Hg [the atmospheric pressure] × 21% = 160 mm Hg). At a higher altitude where the barometric pressure is low, the partial pressure of oxygen is also decreased. For example, if the barometric pressure is 500 mm Hg, the partial pressure of oxygen would be 105 mm Hg (500 mm Hg × 21% = 105 mm Hg). Because the partial pressure of oxygen is lower at high altitudes, the blood levels of oxygen may decline enough to stimulate the carotid and aortic bodies. Oxygen then becomes an important stimulus for an elevated rate and depth of respiration. At high altitudes, the ability of the respiratory system to eliminate carbon dioxide is not adversely affected by the low barometric pressure. Thus the blood carbon dioxide levels become lower than normal because of the increased rate and depth of respiration stimulated by the low blood levels of oxygen. The decreased blood levels of carbon dioxide cause the blood pH to rise to abnormally high levels.

A similar situation may exist in people who have emphysema. Carbon dioxide diffuses across the respiratory membrane more readily than oxygen. The decrease in surface area of the respiratory membrane, caused by the disease, results in low blood levels of oxygen without elevated blood levels of carbon dioxide. The elevated rate and depth of respiration is the result, primarily, of the stimulatory effect of low blood levels of oxygen on the carotid and aortic bodies. More severe emphysema, in which the surface area of the respiratory membrane is reduced to a minimum, can also result in elevated blood levels of carbon dioxide. ■

Infectious Diseases of the Respiratory System

Respiratory system infections are the most common types of infections. Most are relatively mild, but some are among the most damaging types of infection. Many of the respiratory infections are spread by direct contact with droplets produced by sneezing and coughing and by contact of contaminated objects with hands and fingers, which then come into contact with the mouth and nasal passages.

The major respiratory diseases are bacterial and viral, although some are fungal or protozoan infections. Respiratory infections affect the upper respiratory system (nasal passages, pharynx, and auditory tubes [Eustachian tubes]) or the lower respiratory system (larynx, trachea, bronchi, and alveoli).

INFECTIONS OF THE UPPER RESPIRATORY SYSTEM

Strep throat is caused by a streptoccocal bacteria and is characterized by inflammation of the pharynx and fever. Frequently, inflammation of the tonsils and middle ear are involved. Without a throat analysis, the infection cannot be distinguished from viral causes of pharyngeal inflammation. Current techniques allow rapid diagnosis within minutes to hours, and antibiotics are effective in treating strep throat. **Scarlet fever** occurs in response to certain streptococcal bacteria. This infection is characterized by fever and a pinkish-red skin rash produced by a circulating toxin released by the bacteria.

Diptheria was once a major cause of death among children. It is caused by a bacterium. A grayish membrane forms in the throat and can block the respiratory passages totally. A vaccine against diptheria is part of the normal immunization program for children in the United States.

Middle ear infection (otitis media) can result from a number of bacterial organisms that find access to the middle ear. There is a build-up of pus or serous fluid that causes pressure within the middle ear and results in an earache. Antibiotics effective against a broad range of bacteria are used to treat middle ear infections.

The **common cold** is the result of a viral infection. Symptoms include sneezing, excessive nasal secretions, and congestion. The infection easily can spread to sinus cavities, lower respiratory passages, and the middle ear. Laryngitis and otitis media are common complications. The common cold usually runs its course to recovery in about one week.

DISEASES OF THE LOWER RESPIRATORY SYSTEM

Many of the same infections that mainly affect the upper respiratory system can cause laryngitis and bronchitis (inflammation of the bronchi).

Whooping cough (pertussis) is a bacterial infection. The infection causes a loss of cilia of the respiratory epithelium. Mucus accumulates and the infected person attempts to cough up the mucous accumulations. The coughing can be severe. A vaccine for whooping cough is part of the normal vaccination procedure for children in the United States.

Tuberculosis is caused by a tuberculosis bacterium. In the lung, the tuberculosis bacteria form lesions called tubercles. The small lumps contain degenerating macrophages and tuberculosis bacteria. An immune reaction is directed against the tubercles, which causes the formation of larger lesions and inflammation. The tubercles can rupture, releasing additional bacteria, which infect other parts of the lung or body.

Pneumonia refers to many infections of the lung. Most pneumonias are bacterial, but some are viral. Symptoms include fever, difficulty in breathing, and chest pain. Inflammation of the lungs results in pulmonary edema and poor inflation of the lungs with air. A protozoal infection that results in **Pneumocystosis pneumonia** is rare except in persons who have a compromised immune system. This type of pneumonia has become one of the infections commonly suffered by persons who have AIDS.

Flu (influenza) is a viral infection of the respiratory system and does not affect the digestive system as is commonly assumed. Flu is characterized by chills, fever, headache, and muscular aches in addition to respiratory symptoms. There are several strains of flu viruses. Flu vaccines are of limited use because the time required to develop an effective vaccine usually is too long for it to be effective during the major epidemic. The mortality rate is about 1%, and most of those deaths are among the very old and very young. During a flu epidemic the infection rate is so rapid and the disease is so widespread, the total number of deaths is substantial even though the percentage of deaths is relatively low.

A number of fungal diseases affect the respiratory system. The fungal spores usually enter the respiratory system through dust particles. Spores in soil and feces of certain animals make the rate of infection higher in farm workers and gardeners in certain areas of the country (usually the south and southwest). The infections usually result in minor respiratory infections, but in some cases they can become generalized.

There is some conscious control over respiration. It is possible to breathe voluntarily or to stop respiratory movements voluntarily. Some people can hold their breath until they lose consciousness as a result of the lack of oxygen in the brain. (Some children have used this strategy to encourage parents to give them what they want.) As soon as conscious control of respiration is lost, however, the automatic control of respiration resumes, and the person starts to breathe again.

Emotions can also have an important influence on respiratory movements. Some people respond to stressful situations with uncontrolled and exaggerated respiratory movements (hyperventilation). In response to hyperventilation, blood carbon dioxide levels become very low, and blood pH increases. These changes can cause the person to faint. The elevated pH causes blood vessels in the periphery to dilate and blood pools in the abdomen and legs instead of returning to the heart. As a consequence, blood pressure falls dramatically, the brain suffers from a lack of blood flow, and the brain malfunctions as a result of the lack of oxygen, causing the person to faint. Depression, in contrast to stress, can cause a slight reduction in the respiratory rate.

During exercise, respiration rate and depth greatly increase. Only during very heavy exercise, however, do blood carbon dioxide, oxygen, and pH levels change very much from their normal values. The respiratory rate increases are influenced by stimuli from proprioceptors in the joints (proprioceptors detect joint movements). Action potentials are transmitted from the proprioceptors to the medulla oblongata, where they cause an increase in the rate and depth of respiration. Movements of the limbs have a strong stimulatory influence on the respiratory center. Touch, thermal, and pain receptors in the skin also stimulate the respiratory center, which explains the gasp in response to being splashed with cold water or being pinched. Increased activity in the motor cortex also results in stimulation of the respiratory center. It is the increased activity in the motor cortex that causes the respiratory rate to increase in anticipation of physical exercise.

The brain "learns" after a period of training to match the rate of respiration with the intensity of the exercise. Well-trained athletes unconsciously match their respiratory movements more efficiently with their level of physical activity than do untrained individuals. Thus centers of the brain involved in learning have an indirect influence on the respiratory center, but the exact mechanism for this kind of regulation is not clear.

SUMMARY

Respiration includes the movement of air into and out of the lungs, the exchange of gases between the air and the blood, the transport of gases in the blood, and the exchange of gases between the blood and tissues.

ANATOMY OF THE RESPIRATORY SYSTEM

Nose and Nasal Cavity

The bridge of the nose is bone, and most of the external nose is cartilage.

The nasal cavity.

- The external nares open to the outside, and the internal nares lead to the pharynx.
- The nasal cavity is divided by the nasal septum into right and left portions.
- The paranasal sinuses and the nasolacrimal duct open into the nasal cavity.
- Hairs just inside the external nares trap debris.
- The nasal cavity is lined with pseudostratified ciliated epithelium that traps debris and moves it to the pharynx.
- The superior part of the nasal cavity contains the sensory cells for the sense of smell.

Pharynx

The nasopharynx joins the nasal cavity through the internal nares and contains the opening to the auditory tube and the pharyngeal tonsils.

The oropharynx joins the oral cavity and contains the palatine and lingual tonsils.

The laryngopharynx opens into the larynx and the esophagus.

Larynx

Cartilage.

- There are three unpaired cartilages. The thyroid cartilage and cricoid cartilage form most of the larynx. The epiglottis covers the opening of the larynx during swallowing.
- There are six paired cartilages.

Sound.

- Sounds are produced as the vocal cords vibrate when air passes through the larynx.
- The cords produce sounds of different pitch when their length is varied.

Trachea

The trachea connects the larynx to the primary bronchi.

Bronchi

The primary bronchi extend from the trachea to each lung.

Lungs

There are two lungs.

The airway passages of the lungs branch and decrease in size.

- The primary bronchi form the secondary bronchi, which go to each lobe of the lungs.
- The secondary bronchi form the tertiary bronchi, which go to each lobule of the lungs.
- The tertiary bronchi branch many times to form the bronchioles.
- The bronchioles branch to form the terminal bronchioles, which become the respiratory bronchioles, from which the alveoli branch.

Important features of the tube system.

The epithelium from the trachea to the terminal bronchioles is ciliated to facilitate removal of debris.

Cartilage helps to hold the tube system open (from the trachea to the bronchioles).

Smooth muscle controls the diameter of the tubes (especially the bronchioles).

The alveoli are sacs formed by simple squamous epithelium, and they facilitate diffusion of gases.

Pleural Cavities

The pleural membranes surround the lungs and provide protection against friction.

VENTILATION AND LUNG VOLUMES

Ventilation is the movement of air into and out of the lungs.

Air moves from an area of higher pressure to an area of lower pressure.

Pressure in the lungs decreases as the volume of the lungs increases, and pressure increases as lung volume decreases.

Mechanics of Breathing

Inspiration occurs when the diaphragm contracts and the external intercostal muscles lift the ribcage, thus increasing the volume of the thoracic cavity.

Expiration occurs when the diaphragm relaxes and the internal intercostal muscles depress the ribcage, thus decreasing the volume of the thoracic cavity.

Factors Preventing Collapse of the Lung

Lungs tend to collapse because of the elastic recoil of the connective tissue and surface tension of the fluid lining the alveoli.

The lungs normally do not collapse because surfactant reduces the surface tension of the fluid lining the alveoli, and the visceral pleura tends to adhere to the parietal pleura.

Pulmonary Volumes and Capacities

There are four pulmonary volumes: tidal volume, inspiratory reserve, expiratory reserve, and residual volume.

Pulmonary capacities are the sum of two or more pulmonary volumes and include inspiratory capacity, functional residual capacity, vital capacity, and total lung capacity.

The forced expiratory vital capacity measures the rate at which air can be expelled from the lungs.

GAS EXCHANGE

The respiratory membranes are thin and have a large surface area that facilitates gas exchange.

The components of the respiratory membrane include a film of water, the walls of the alveolus and the capillary, and an interstitial space.

The rate of diffusion depends on the thickness of the respiratory membrane, the surface area of the membrane, and the partial pressure of the gases in the alveoli and the blood.

OXYGEN AND CARBON DIOXIDE TRANSPORT IN THE BLOOD

Oxygen diffuses from a high partial pressure in the alveoli to a lower partial pressure in the alveolar capillary. Blood is saturated with oxygen when it leaves the capillary.

Oxygen diffuses from a higher partial pressure in the tissue capillaries to a lower partial pressure in the tissue spaces.

Oxygen is transported by hemoglobin and is dissolved in plasma.

Carbon dioxide is transported as bicarbonate ions, in combination with blood proteins, and in solution in plasma.

In tissue capillaries, carbon dioxide combines with water inside the red blood cells to form carbonic acid that dissociates to form bicarbonate ions and hydrogen ions.

In lung capillaries, bicarbonate ions and hydrogen ions move into red blood cells. Bicarbonate ions combine with hydrogen ions to form carbonic acid. The carbonic acid dissociates to form carbon dioxide that diffuses out of red blood cells.

CONTROL OF RESPIRATION

The respiratory center in the medulla oblongata and pons stimulates the muscles of inspiration to contract. When stimulation of the muscles of inspiration stops, expiration occurs passively.

During labored breathing the muscles of inspiration are stimulated to a greater degree, and the muscles of expiration are also stimulated to increase expiration.

The Hering-Breuer reflex inhibits the inspiratory center when the lungs are stretched during inspiration.

Carbon dioxide is the major chemical regulator of respiration. An increase in carbon dioxide or a decrease in pH of the blood can directly stimulate chemoreceptors in the medulla oblongata, causing a greater rate and depth of respiration.

Low blood levels of oxygen can stimulate chemoreceptors in the carotid and aortic bodies, which then stimulate the respiratory center.

It is possible to consciously control ventilation.

Input from higher brain centers and from proprioceptors stimulates the respiratory center during exercise.

CONTENT REVIEW

1. What are the functions of the respiratory system?
2. Define respiration.
3. Describe the structure of the nasal cavity.
4. Name the three parts of the pharynx. With what structures does each part communicate?
5. Name and describe the three unpaired cartilages of the larynx.
6. How do the vocal cords produce sounds of different loudness and pitch?
7. What is the function of the C-shaped cartilages in the trachea? What happens to the amount of cartilage in the tube system of the respiratory system as the tubes become smaller? Explain why breathing becomes more difficult during an asthma attack.
8. What is the function of the ciliated epithelium in the trachea, primary bronchi, and lungs?
9. Distinguish between the lungs, a lobe of the lung, and a lobule.
10. Describe the pleura of the lungs. What is their function?
11. How does movement of the ribs and diaphragm affect thoracic volume?
12. Describe the pressure changes that cause air to move into and out of the lungs. What causes these pressure changes?
13. Why do the lungs tend to collapse? What two factors keep the lungs from collapsing?
14. Define tidal volume, inspiratory reserve, expiratory reserve, and residual volume. Define inspiratory capacity, functional residual capacity, vital capacity, and total lung capacity.
15. List the components of the respiratory membrane. Describe the factors that affect the diffusion of gases across the respiratory membrane. Give some examples of diseases that decrease diffusion by altering these factors.
16. List the ways that oxygen and carbon dioxide are transported in the blood.
17. What is the effect of going from a low altitude to a high altitude on the partial pressure of oxygen in the air?
18. How can changes in respiration affect blood pH?
19. How do the respiratory center and the Hering-Breuer reflex control respiration?
20. Describe the effect of blood carbon dioxide, blood oxygen, and blood pH on respiratory movements. Explain the role of chemoreceptors in regulating respiration.
21. During exercise, how is respiration regulated?

CONCEPT REVIEW

1. A patient has pneumonia, and fluids accumulate within the alveoli. Explain why this results in an increased rate of respiration that can be returned to normal with oxygen therapy.
2. A patient has severe emphysema that has extensively damaged the alveoli and reduced the surface area of the respiratory membrane. Although the patient is receiving oxygen therapy, he still has a tremendous urge to take a breath, that is, he does not feel as if he is getting enough air. Why does this occur?
3. Patients with diabetes mellitus who are not being treated with insulin therapy rapidly metabolize lipids, and there is an accumulation of acidic byproducts of lipid metabolism in the circulatory system. What effect would this have on respiration? Why is the change in respiration beneficial?
4. Ima Anxious is hysterical and is hyperventilating. The doctor makes her breathe into a paper bag. Because you are an especially astute student you say to the doctor, "When Ima was hyperventilating, she was reducing blood carbon dioxide levels, and when she breathed into the paper bag, carbon dioxide was trapped in the bag, and she was rebreathing it, thus causing blood carbon dioxide levels to increase. The urge to breathe should have increased after breathing into the paper bag instead of causing her to breathe more slowly." How do you think the doctor would respond? (Hint: Recall that the effect of decreased blood carbon dioxide on the vasomotor center will result in vasodilation and a sudden decrease in blood pressure).
5. Hardy Climber got up in the morning and climbed a very high mountain. When he arrived at the top of the mountain, he noticed he was breathing more rapidly than usual, even when he was resting. Later, he explained to his friend, who happened to be a physiologist, that the air was so thin at that altitude that he was unable to get enough oxygen. Do you think his physiologist friend agreed? Explain.
6. The blood pH of a runner was monitored during a race. It was noticed that shortly after the beginning of the race her blood pH increased for a short time. Propose an explanation that would account for the increased pH values following the start of the race.

ANSWERS TO ? PREDICT QUESTIONS

1 *p. 376* When you sleep with your mouth open, air does not pass through the nasal passages. This is especially true when nasal passages are plugged as a result of having a cold. As a consequence air is not humidified and warmed. The dry air dries the throat and the trachea, thus irritating them.

2 *p. 382* When a large mouthful of food is swallowed, the esophagus is enlarged in the area through which the food passes. The bulge in the esophagus applies pressure on the trachea, which is immediately anterior to the esophagus. Because the C-shaped cartilages of the trachea have their open portion facing the esophagus, the trachea collapses momentarily as the food passes. Thus the passage of food through the esophagus is not hampered by the trachea.

3 *p. 383* During respiratory movements, the parietal and visceral pleura slide over the surface of each other. Normally the pleural fluid in the pleural cavities lubricates the surfaces of these membranes. When the pleural membranes are inflamed, they rub against each other and create an intense pain. The pain is exacerbated when a person takes a deep breath because the movement of the membranes is greater than during normal breaths.

4 *p. 386* If a resting person had a tidal volume of 500 ml and a respiratory rate of 12 respirations per minute, the total amount of air respired per minute would be 6000 ml. If an exercising person had a tidal volume of 4000 ml and a respiratory rate of 24 respirations per minute, the total amount of air respired per minute would be 96,000 ml. The difference between the two would be 90,000 ml, which would mean that the exercising person respired 90,000 ml more air per minute than the person at rest.

5 *p. 391* A rapid rate of respiration would increase the blood pH, because carbon dioxide would be eliminated from the blood more rapidly during rapid respiration. As carbon dioxide is lost, hydrogen ions and bicarbonate ions would combine to form carbonic acid, which would, in turn, dissociate to form carbon dioxide and water. The decrease in hydrogen ions would cause an increase in blood pH. Holding one's breath would result in a decrease in pH, because carbon dioxide would accumulate in the blood. The carbon dioxide would combine with water to form carbonic acid, which would dissociate to form hydrogen ions and bicarbonate ions. The increase in hydrogen ions would cause a decrease in blood pH.

6 *p. 394* When a person breathes rapidly and deeply for several seconds, the carbon dioxide levels decrease and blood pH increases. Carbon dioxide is an important regulator of respiratory movements. A decrease in blood carbon dioxide and an increase in blood pH result in a reduced stimulus to the respiratory center. As a consequence, respiratory movements stop until blood carbon dioxide levels build up again in the body fluid. This normally takes only a short time.

16

THE DIGESTIVE SYSTEM

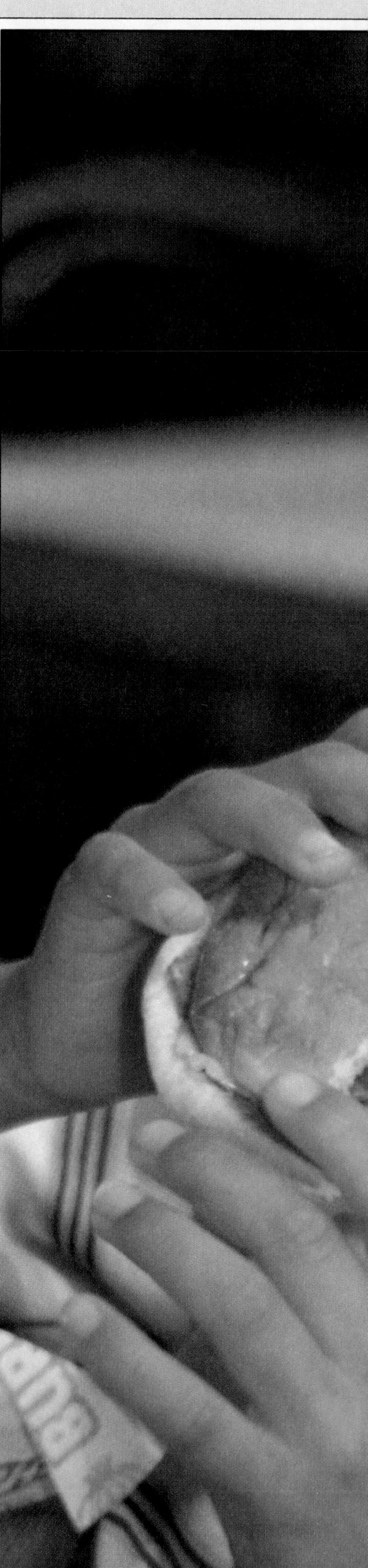

After reading this chapter you should be able to:

1 List the organs that make up the digestive tract and describe the structure of each.
2 Name the teeth and describe the structure of an individual tooth.
3 Describe the major salivary glands. Compare their structures and functions.
4 Outline the stomach's anatomical and physiological characteristics that are most important to its function.
5 List the anatomical and histological characteristics of the small intestine that account for its large surface area.
6 Describe the peritoneum and the mesenteries.
7 Explain the functions of the structures in the oral cavity. Describe mastication (chewing) and deglutition (swallowing).
8 List the stomach secretions, describe their functions, and explain how they are regulated.
9 Describe gastric movements and stomach emptying, and their regulation.
10 Describe the secretions and movements that occur in the small and large intestines.
11 List the major functions of the pancreas and liver, and explain how they are regulated.

bile Fluid secreted from the liver, stored in the gallbladder, and released into the duodenum; consists of bile salts, bile pigments, bicarbonate ions, fats, hormones, and more.

chyme (kīm) [Gr. *chymos*, juice] Semifluid mass of partly digested food passed from the stomach into the duodenum.

colon (ko'lon) Division of the large intestine that extends from the cecum to the rectum.

defecation [L. *defaeco*, to purify] Discharge of feces from the rectum.

deglutition (dĕ'glu-tish'un) [L. *de-* + *glutio*, to swallow] The act of swallowing.

duodenum (du-odĕ'num) [L. *duodeni*, twelve] First division of the small intestine; connects to the stomach.

esophagus (e-sof'ă-gus) [Gr. *oisophagos*, gullet] The part of the digestive tract between the pharynx and stomach.

intramural plexus (in'trah-mu'ral) [L., within the wall] A nerve plexus within the walls of the gastrointestinal tract.

lacteal (lak'tē-al) Lymphatic vessel in the wall of the small intestine that carries lymph from the intestine and absorbs fat.

mastication (mas'tĭ-ka'shun) [L. *mastico*, to chew] The process of chewing.

mucosa (mu-ko'sah) Mucous membrane consisting of the epithelium and connective tissue. In the digestive tract there is also a layer of smooth muscle.

parietal peritoneum (pĕr'it-o-ne'um) [L., wall] That portion of the serous membranes of the abdominal cavity lining the inner surface of the body wall.

peristalsis (pĕr'ĭ-stal'sis) A wave of contraction and relaxation moving along a tube; propels food along the digestive tube.

pharynx (făr'ingks) [Gr., throat] The part of the digestive and respiratory tubes superior to the larynx and esophagus and inferior and posterior to the oral and nasal cavities.

visceral peritoneum (pĕr'it-o-ne'um) [L., organ] That part of the serous membrane in the abdominal cavity covering the surface of some abdominal organs.

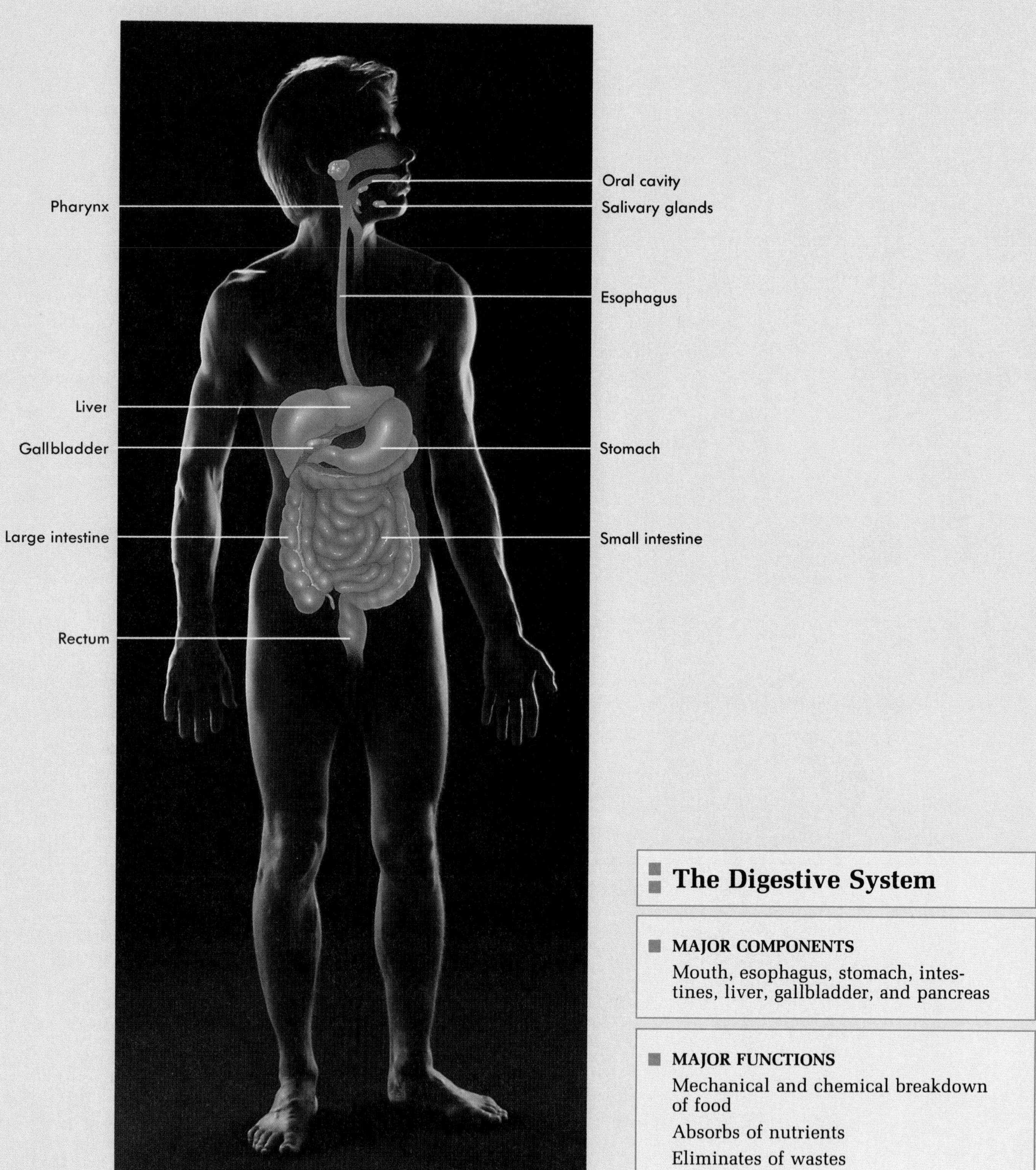

FIGURE 16-1 ■ The digestive system.
The digestive system consists of a tube extending from mouth to anus, as well as the accessory organs that empty their secretions into the tube.

The function of the **digestive system** (Figure 16-1) is to take in food, break it down into smaller units, and absorb those units so that the body can use them. This process provides the body with water, electrolytes, and other nutrients. The digestive system consists of the digestive tract, a tube extending from the mouth to the anus, plus the associated organs, which secrete fluids into the digestive tract. The digestive tract is also often referred to as the gastrointestinal (GI) tract. The inside of the digestive tract is continuous with the outside environment where it opens at the mouth and anus. In this chapter, the structure and function of the digestive organs are described. The breakdown and absorption of food molecules are described in Chapter 17.

ANATOMY AND HISTOLOGY OF THE DIGESTIVE SYSTEM

The **digestive tract** consists of the mouth, esophagus, stomach, small intestine, large intestine, rectum, and anus. Various portions of the digestive tube are specialized for different functions, but nearly all portions of the digestive tube consist of four layers or tunics (Figure 16-2). These will be described in order from the inside of the tube. The innermost tunic, the **mucosa** (mu-ko'sah), consists of **mucous epithelium,** a loose, irregular connective tissue called the **lamina propria**, and a thin smooth muscle layer, the **muscularis mucosa.** The epithelium is thickened in the mouth and esophagus to resist abrasion, and is thin (simple) in the intestine for absorption and secretion.

The **submucosa** lies just outside the mucosa. It is a thick layer of loose connective tissue containing

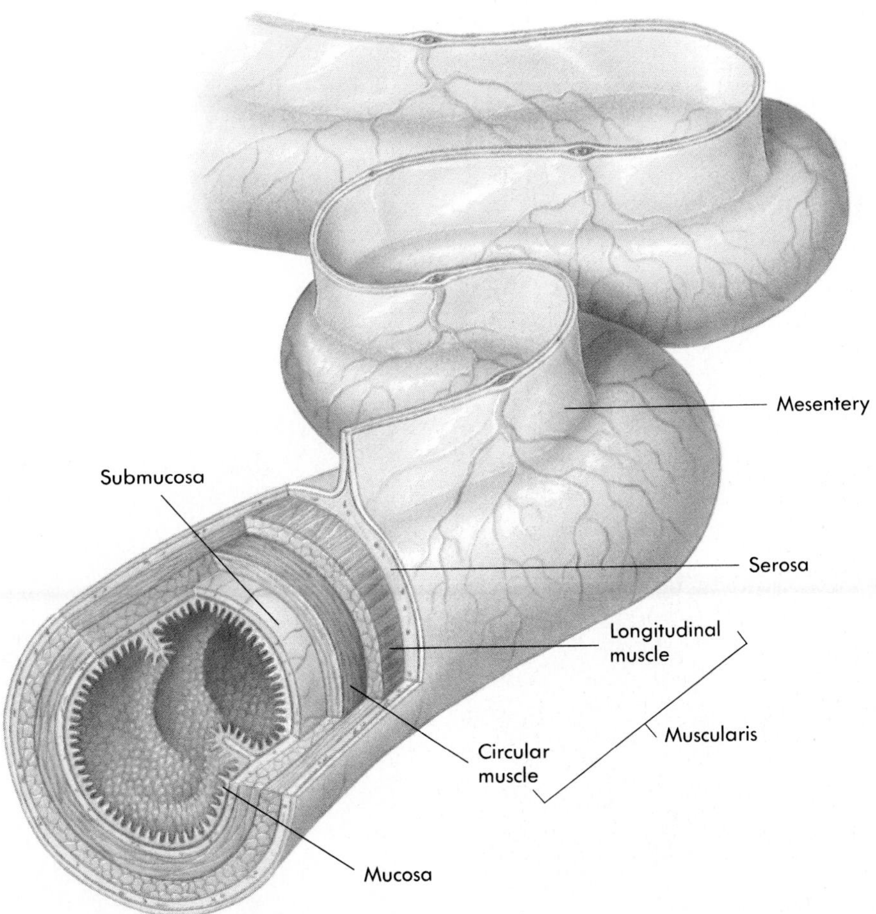

FIGURE 16-2 ■ Digestive tract histology.
The four tunics are the mucosa, submucosa, muscularis, and serosa or adventitia.

nerves, blood vessels, and small glands. Nerves of the submucosa form a **plexus** (network) consisting of parasympathetic nerve fibers and cell bodies.

The next tunic is the **muscularis,** which, in most parts of the digestive tube, consists of an inner layer of circular smooth muscle and an outer layer of longitudinal smooth muscle. Another nerve plexus, also made up of parasympathetic nerve fibers and cell bodies, lies between the two muscle layers. Together, the nerve plexuses of the submucosa and muscularis compose the **intramural** (in'trah-mu'ral; within the walls) **plexus.** This plexus is extremely important in the control of movement and secretion within the gastrointestinal tract.

The fourth, outermost layer of the digestive tract is either an epithelial layer called the **serosa** or a connective tissue layer called the **adventitia** (ad'ven-tish'yah; foreign; coming from outside).

Accessory glands (see Figure 16-2), such as the liver and pancreas, are associated with certain parts of the intestinal tract, especially the small intestine. Those glands will be described with the part of the tract to which they are attached.

Oral Cavity

■ The **oral cavity** (Figure 16-3), or mouth, is the first portion of the digestive tract. It is bounded by the lips and cheeks, and contains the teeth and tongue. The **lips** are muscular folds, covered internally by mucosa and externally by stratified squamous epithelium. The epithelial covering of the lips is relatively thin and the color from the underlying blood vessels can be seen through the transparent epithelium, giving the lips a reddish-pink appearance. The cheeks form the lateral walls of the oral cavity. The substance of the cheeks is contributed by the **buccinator muscle,** which flattens the cheek against the teeth. The lips and cheeks are important in the process of **mastication** (mas'tĭ-ka'shun; chewing food) and **speech.** They help manipulate the food within the mouth and hold the food in place while the teeth crush or tear it. They also help form words during the speech process.

The **tongue** is a large, muscular organ that occupies most of the oral cavity. The tongue's major attachment is through its posterior portion. The anterior portion of the tongue is relatively free and is attached to the floor of the mouth by a thin fold of tissue called the **frenulum** (fren'u-lum) (see Figure 16-3, *B*). (The muscles associated with the tongue are described in Chapter 7.)

The tongue moves food in the mouth and, in cooperation with the lips and cheeks, holds the food in place during mastication. It also plays a major role in the process of swallowing. The tongue is a major sensory organ for taste, and it is one of the major organs of speech.

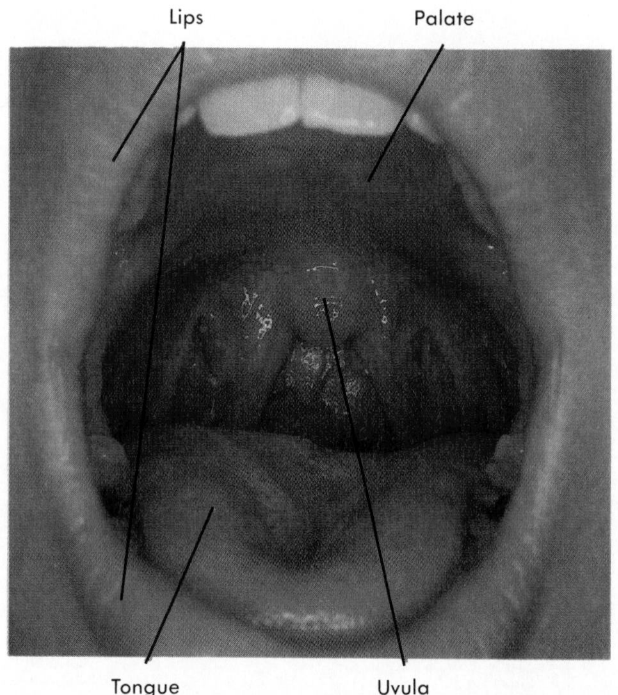

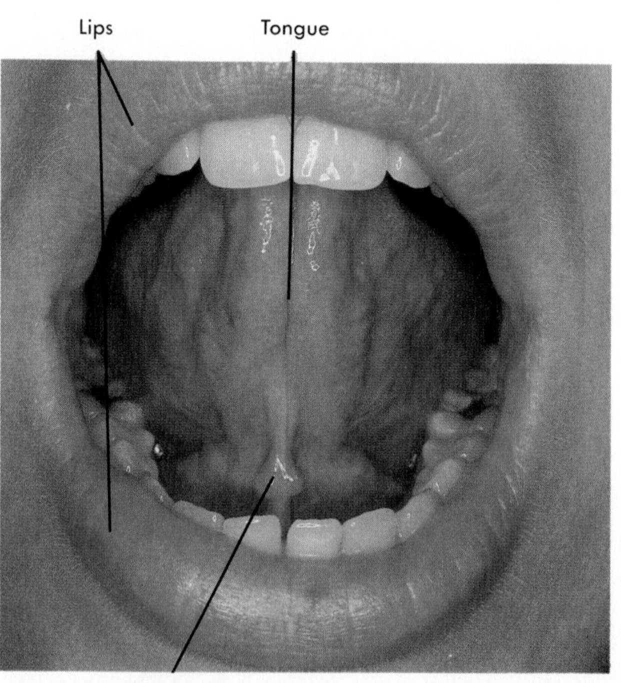

FIGURE 16-3 ■ The oral cavity, with the tongue depressed (A) and the tongue elevated (B).

Teeth

■ There are 32 **teeth** in the normal adult mouth, located in the maxillary and mandibular dental arches. The teeth in the right and left halves of each dental arch are roughly mirror images of each other. As a result, the teeth can be divided into four quadrants: right upper, left upper, right lower, and left lower quadrants. Each quadrant contains one central and one lateral **incisor,** one **canine,** first and second **premolars,** and first, second, and third **molars.** The third molars are referred to as **wisdom teeth** because they usually appear when the person is in his late teens or early twenties, when a person is thought to have acquired some degree of wisdom.

The teeth of the adult mouth are **permanent,** or **secondary, teeth** (Figure 16-4, *A*). Most of them are replacements of the 20 **primary,** or **deciduous, teeth** (dĕ-sid'u-us; those that fall out; also called milk teeth) (Figure 16-4, *B*) that are lost during childhood.

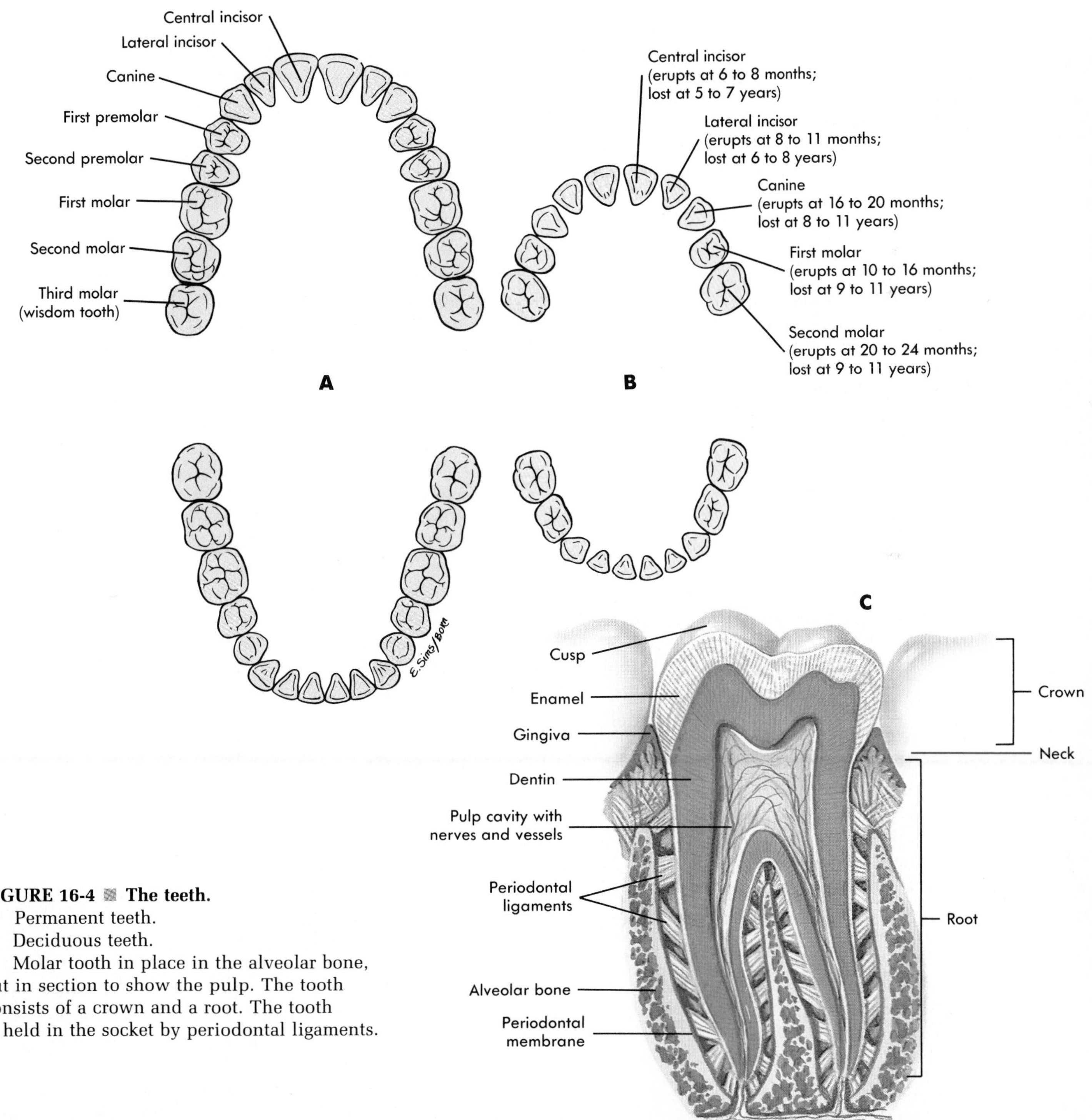

FIGURE 16-4 ■ The teeth.
A Permanent teeth.
B Deciduous teeth.
C Molar tooth in place in the alveolar bone, cut in section to show the pulp. The tooth consists of a crown and a root. The tooth is held in the socket by periodontal ligaments.

Each tooth (Figure 16-4, C) consists of a **crown** with one or more cusps (points), a **neck,** and a **root.** The center of the tooth is a **pulp cavity** that is filled with blood vessels, nerves, and connective tissue called **pulp.** The pulp cavity is surrounded by a living, cellular, calcified tissue called **dentin.** The dentin of the tooth crown is covered by an extremely hard, acellular substance called **enamel,** which protects the tooth against abrasion and acids produced by bacteria in the mouth.

The teeth are rooted within **alveoli** (al-ve'o-li; sockets) along the alveolar ridges of the mandible and maxillae. The alveolar ridges are covered by dense, fibrous connective tissue, and moist stratified squamous epithelium, referred to as the **gingiva** (jin'jĭ-vah; gums). The teeth are held in place by **periodontal** (pĕr'e-o-don'tal; around the teeth) **ligaments,** and the alveolar walls are lined with a **periodontal membrane.**

Formation of dental caries or tooth decay is the result of breakdown of enamel by acids that are produced by bacteria on the tooth surface. Enamel is nonliving and cannot repair itself. Consequently a dental filling is necessary to prevent further damage. Periodontal disease is an inflammation and degeneration of the periodontal structures, gingiva, and alveolar bone. This disease is the most common cause of tooth loss in adults.

Palate and tonsils

The palate, or roof of the oral cavity, consists of two portions, an anterior bony portion, the **hard palate,** and a posterior portion, the **soft palate,** which consists of skeletal muscle and connective tissue. The **uvula** (u'vu-lah; a grape) projects from the posterior edge of the soft palate. The palate is important in preventing food from passing into the nasal cavity during the swallowing process.

The **tonsils,** also called the palatine tonsils, are in the lateral posterior walls of the oral cavity. Other tonsilar tissue is located in the nasopharynx and on the posterior surface of the tongue. The functions of the tonsils are described in Chapter 14.

Salivary glands

There are three pairs of **salivary glands:** the parotid, submandibular, and sublingual glands (Figure 16-5). They produce **saliva,** which is a mixture of serous (watery) and mucous fluids, containing digestive enzymes. Saliva helps keep the oral cavity moist and begins the process of digestion. All the large salivary glands are alveolar glands (branching glands with clusters of alveoli that resemble grapes; see Chapter 4).

The largest of the salivary glands, the **parotid** (pă-rot'id; beside the ear) **glands,** are serous glands located just anterior to each ear. A parotid duct exits each gland, crosses the masseter muscle, and enters the oral cavity adjacent to the second upper molar.

A viral infection resulting in parotiditis (inflammation of the parotid gland) is called *mumps.*

The **submandibular** (below the mandible) **glands** are mixed glands, producing more serous than mucous secretions. Each gland can be felt as a soft lump along the inferior border of the mandible. The submandibular ducts exit the gland and open into the oral cavity on each side of the frenulum of the tongue. In certain people, if the mouth is opened and the tip of the tongue is elevated, saliva may squirt out of the mouth from the ducts of these glands.

The **sublingual** (below the tongue) **glands,** the smallest of the three paired salivary glands, are mixed glands producing primarily mucous secretions. They lie immediately below the mucous membrane in the floor of the oral cavity. Each sublingual gland has 10 to 12 small ducts opening onto the floor of the oral cavity.

Pharynx

The **pharynx** (făr'ingks),or throat, which connects the mouth with the esophagus, consists of three parts: the nasopharynx, oropharynx, and laryngopharynx (see Chapter 15). Normally, only the oropharynx and laryngopharynx transmit food. The posterior walls of the oropharynx and laryngopharynx are formed by **pharyngeal constrictor muscles.**

Esophagus

The **esophagus** (e-sof'ă-gus) is a muscular tube that extends from the pharynx to the stomach. It is about 25 cm long and lies anterior to the vertebrae and posterior to the trachea within the mediastinum. It passes through the diaphragm and ends at the stomach. The esophagus transports food from the pharynx to the stomach. Upper and lower

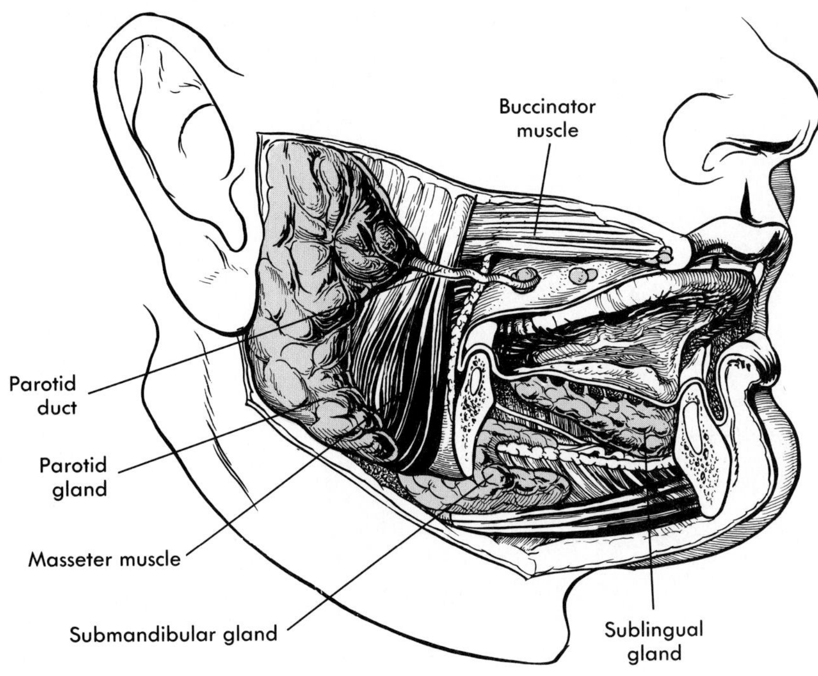

FIGURE 16-5 ■ Salivary glands.
The large salivary glands are the parotid glands, the submandibular glands, and the sublingual glands.

esophageal sphincters regulate the movement of food into and out of the esophagus. Numerous mucous glands produce a thick, lubricating mucus that coats the inner surface of the esophagus.

Stomach

■ The **stomach** (Figure 16-6) is an enlarged segment of the digestive tract in the left superior portion of the abdomen. The opening from the esophagus into the stomach is called the **cardiac opening** because it is near the heart. The region of the stomach around the cardiac opening is called the **cardiac region.** The most superior portion of the stomach is the **fundus** (fun′dus; the bottom of a round-bottomed leather bottle). It is located to the left and superior to the cardiac region of the stomach. The largest portion of the stomach is the **body,** which turns to the right, forming a **greater curvature** and a **lesser curvature.** The opening from the stomach into the small intestine is the **pyloric** (pi-lor′ik; gatekeeper) **opening,** which is surrounded by a relatively thick ring of smooth muscle called the **pyloric sphincter.** The region of the stomach near the pyloric opening is the **pyloric region.**

The muscular tunic of the stomach is different from other regions of the GI tract in that it consists of three layers: an outer longitudinal layer, a middle circular layer, and an inner oblique layer. The submucosa and mucosa of the stomach are thrown into large folds called **rugae** (ru′ge; wrinkles) (see Figure 16-6, *A*) when the stomach is empty. These folds allow the mucosa and submucosa to stretch, and the folds disappear as the stomach is filled.

The stomach is lined with simple columnar epithelium. The mucosal surface forms numerous tubelike **gastric pits** (see Figure 16-6, *B*), which are the openings for the **gastric glands.** The epithelial cells of the stomach can be divided into five groups. The first group, consisting of **surface mucous cells,** which produce mucus that coats and protects the stomach lining, is on the inner surface of the stomach and lines the gastric pits. The remaining four cell types are in the gastric glands. They are **mucous neck cells,** which produce mucus; **parietal cells,** which produce hydrochloric acid and intrinsic factor; **endocrine cells,** which produce regulatory hormones; and **chief cells,** which produce pepsinogen, a precursor of the protein-digesting enzyme pepsin.

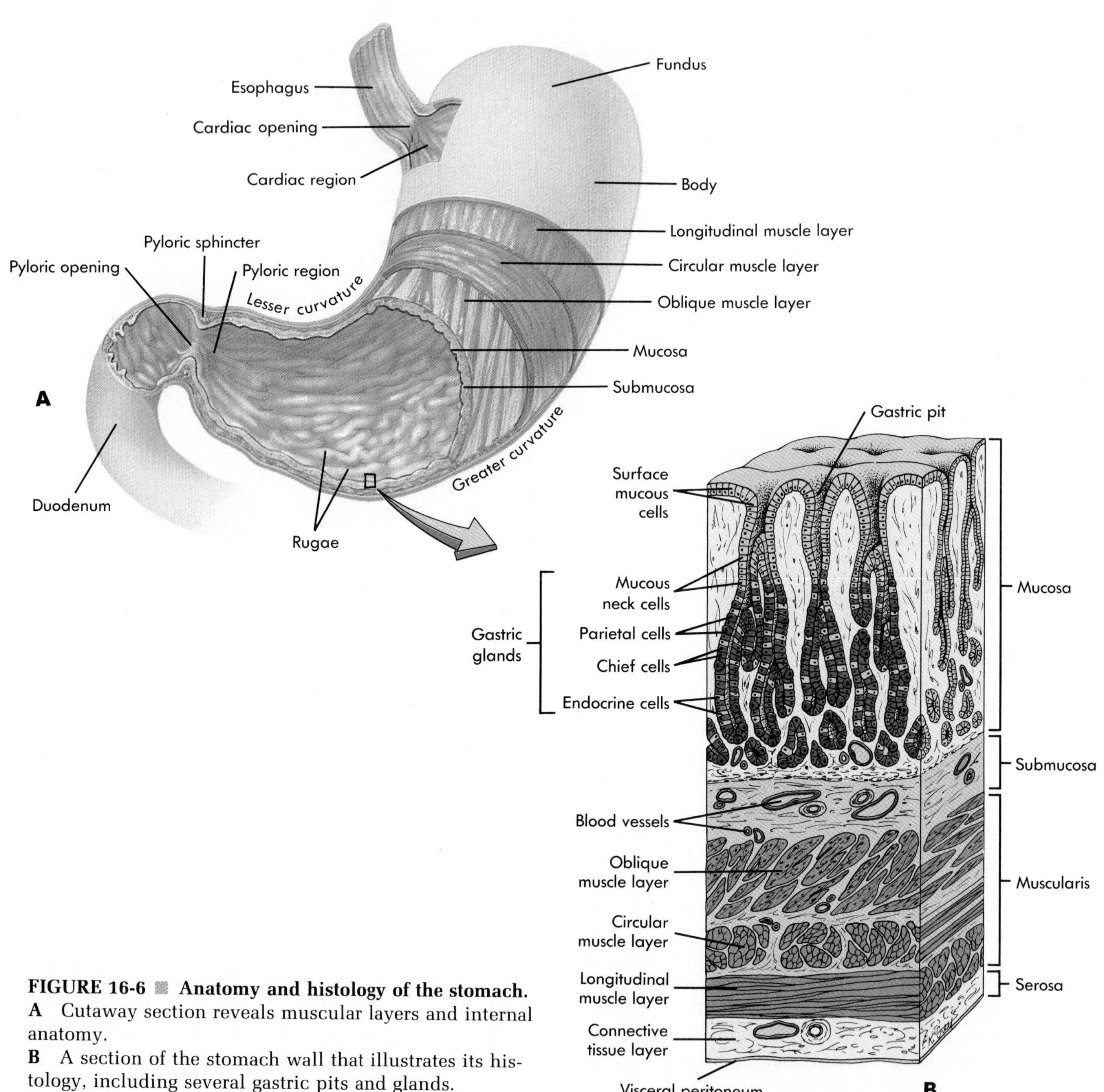

FIGURE 16-6 ■ Anatomy and histology of the stomach.
A Cutaway section reveals muscular layers and internal anatomy.
B A section of the stomach wall that illustrates its histology, including several gastric pits and glands.

Small Intestine

The **small intestine** is about 6 meters long and consists of three portions: the **duodenum** (du-od'ĕ-num, or du-o-de'num), **jejunum** (jĕ-ju'num), and **ileum** (il'e-um) (Figure 16-7). The duodenum is about 25 cm long (the term duodenum means twelve, suggesting that it is 12 inches long). The jejunum is about 2.5 meters long and makes up about two fifths of the total length of the small intestine. The ileum is about 3.5 meters long and makes up about three fifths of the small intestine.

The duodenum nearly completes a 180-degree arc as it curves within the abdominal cavity. The head of the pancreas lies within this arc. The **common bile duct,** from the liver, and the **pancreatic duct,** from the pancreas, join each other and empty into the duodenum.

The surface of the duodenum has several modifications that increase surface area about 600-fold (Figure 16-8). The increased surface area allows more efficient digestion and absorption of food. The mucosa and submucosa form a series of **circular folds** that run perpendicular to the long axis of the digestive tract (Figure 16-8, *A*). Tiny fingerlike projections of the mucosa form numerous **villi** (vil'e; shaggy hair), which are 0.5 to 1.5 mm long (Figure 16-8, *B*). Each villus is covered by simple columnar epithelium and contains a blood capillary network and a lymph capillary called a **lacteal** (lak'te-al) (Figure 16-8, *C*). The blood capillary network and the lacteal, located inside the villi, are very important in transporting absorbed nutrients (see Chapter 17). Most of the cells composing the surface of the villi have numerous cytoplasmic extensions, called **microvilli** (Figure 16-8, *D* and *E*), which further increase the surface area.

The submucosa of the duodenum contains many mucous glands, called **duodenal glands** (Brunner's glands). Other cells of the duodenum produce regulatory hormones.

The jejunum and ileum are similar in structure to the duodenum except that there is a gradual decrease in the diameter of the small intestine, in the

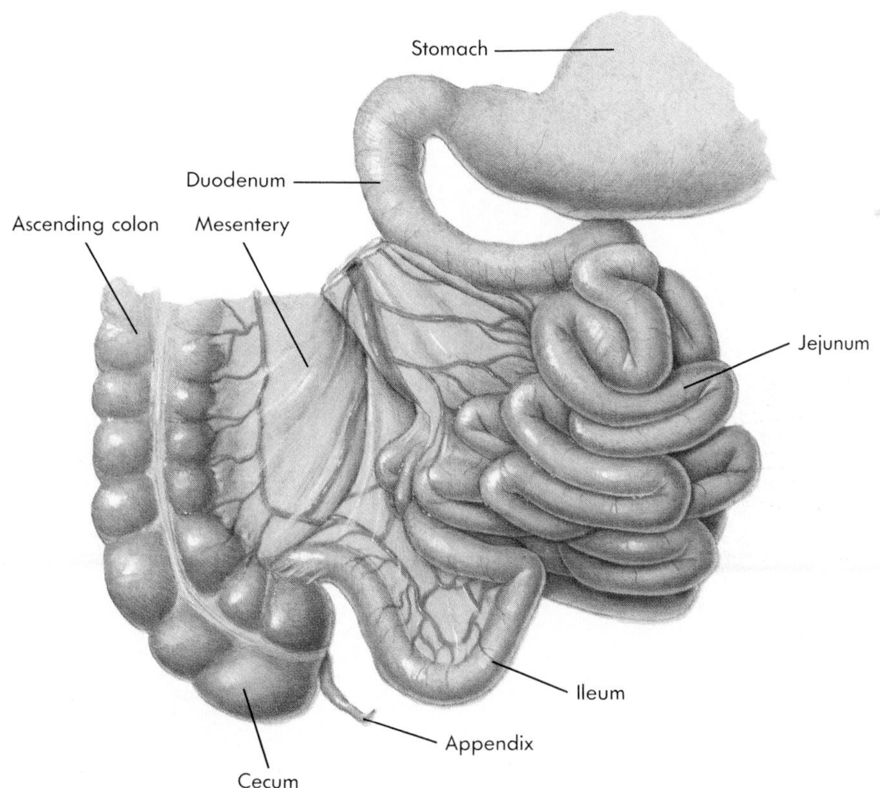

FIGURE 16-7 ■ The small intestine.
The duodenum is attached to the stomach and is continuous with the jejunum. The jejunum is continuous with the ileum, which empties into the cecum.

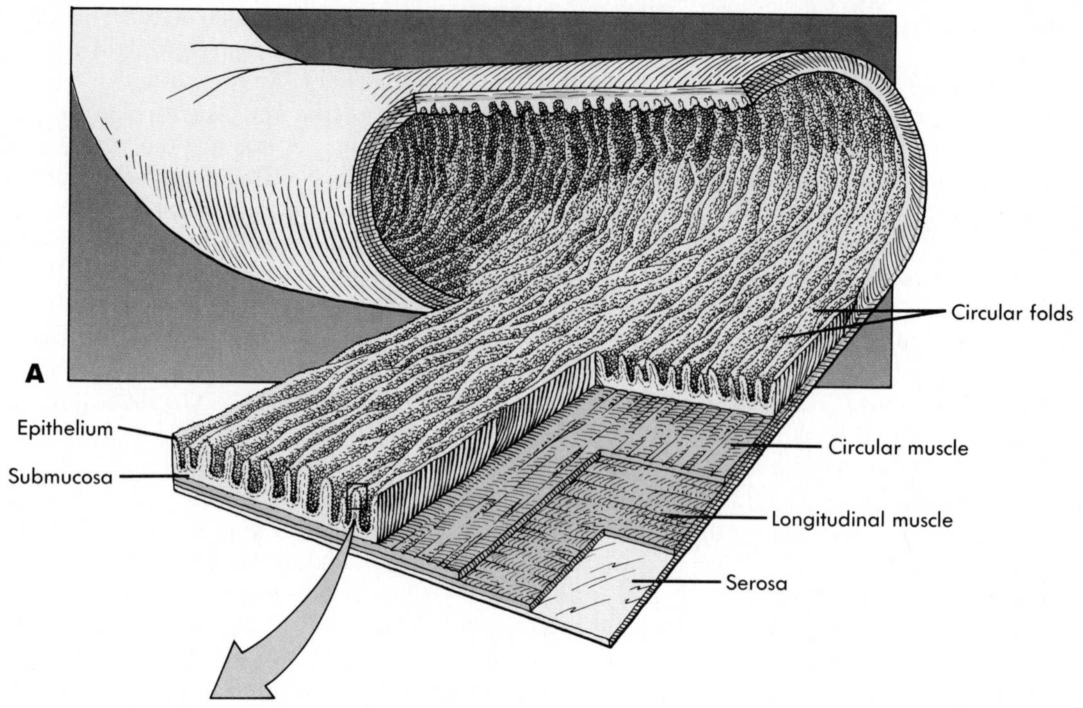

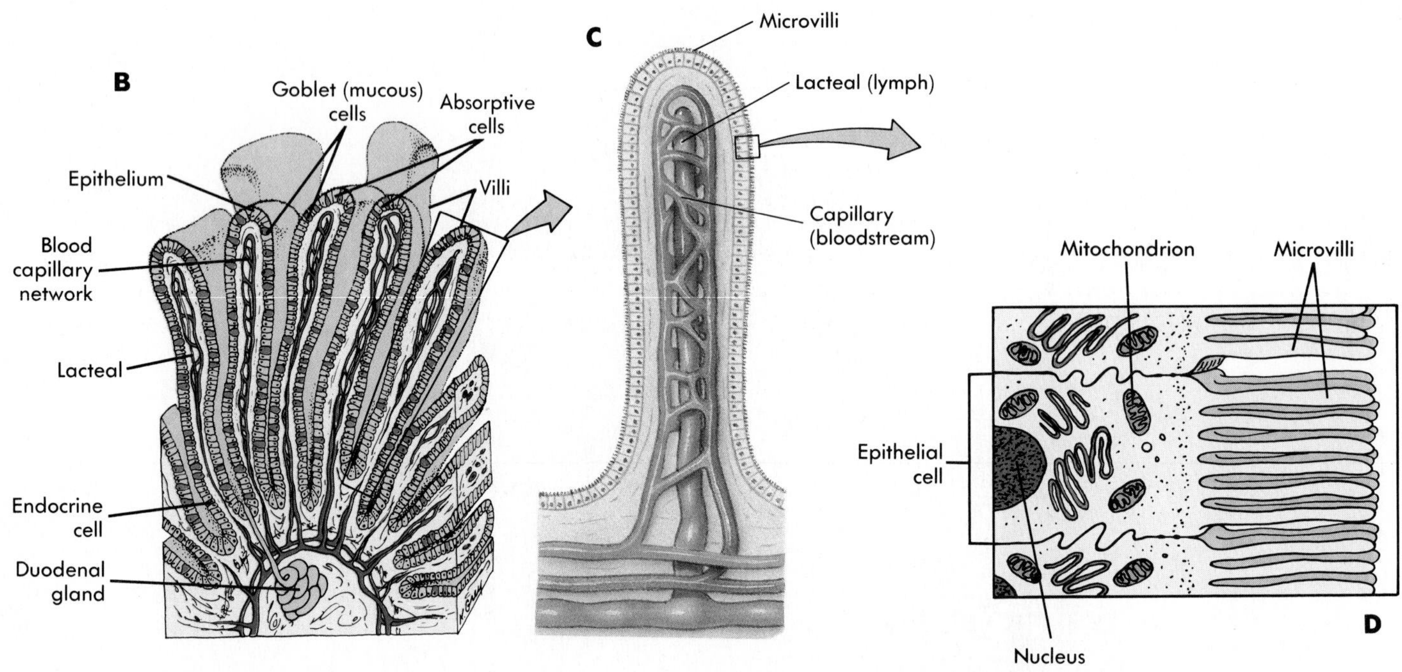

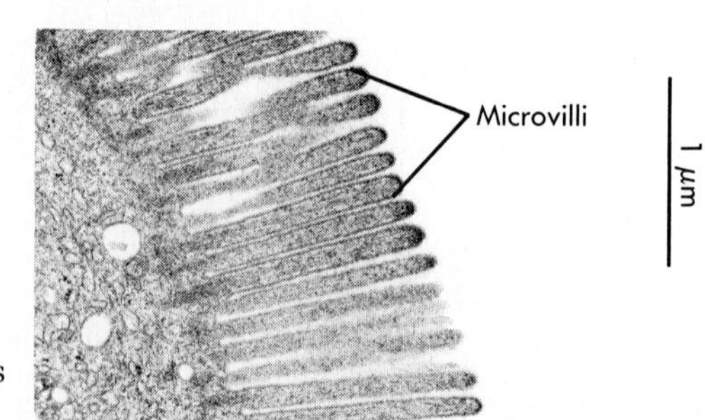

FIGURE 16-8 ■ Interior view and histology of the duodenum.
A The wall of the duodenum has been opened to reveal the circular folds.
B The villi.
C A single villus showing the lacteal and capillary.
D The microvilli.
E Electron micrograph of the microvilli. The extensive surface area allows more efficient absorption of nutrients.

thickness of the intestinal wall, in the number of circular folds, and in the number of villi as one progresses through the small intestine. Lymph nodules are common along the entire length of the GI tract. Clusters of lymph nodules, called **Peyer's patches,** are numerous in the ileum. These lymphatic tissues in the intestine help protect the intestinal tract from harmful microorganisms.

The junction between the ileum and the large intestine is the **ileocecal junction.** It has a ring of smooth muscle, the **ileocecal sphincter,** and a one-way **ileocecal valve** (see Figure 16-12).

Liver

The **liver** (Figure 16-9 and see Figure 16-1) weighs about 1.36 kilograms (3 pounds) and is located in the upper-right quadrant of the abdomen, tucked against the inferior surface of the diaphragm. The liver consists of two major **lobes, left** and **right,** and two minor lobes, **caudate** and **quadrate.** The right and left lobes are separated by a connective tissue septum, the **falciform ligament,** which is continuous with the round ligament. The round ligament is the remnant of the umbilical vein (see Chapter 20) and extends from the liver to the inner surface of the umbilicus.

The **hepatic** (hĕ-pat'ik; associated with the liver) **portal vein,** the **hepatic artery,** and a small hepatic nerve plexus enter the liver at a region on its inferior surface called the **porta** (gate). Lymphatic vessels and two **hepatic ducts** exit the liver at the porta. **Hepatic veins** drain blood from the liver, but they are not located in the porta. Several hepatic veins empty into the inferior vena cava. (See Figure 16-9; the hepatic veins cannot be seen in the figure because they lie deep to the inferior vena cava.) The hepatic ducts transport bile out of the liver. The right and left hepatic ducts unite to form a single **common hepatic duct.** The common hepatic duct is joined by the **cystic duct** from the gallbladder to form the **common bile duct,** which joins the pancreatic duct to empty into the duodenum (Figure 16-10). The **gallbladder** is a small sac on the inferior surface of the liver that stores bile (see Figure 16-9).

Connective tissue septa divide the liver into **lobules** with **portal triads** at the corners, so named because three vessels—the hepatic portal vein, hepatic artery, and hepatic duct—are commonly located in the triads (Figure 16-11). There is a **central vein** in the center of each lobule. The central veins unite to form the hepatic veins.

Hepatic cords, formed by platelike groups of cells, called **hepatocytes,** are located between the central vein and the septa of each lobule. A cleftlike lumen, the **bile canaliculus** (kan'ă-lik'u-lus; little canal), is between the cells of each cord. Bile, produced by the hepatocytes, flows through the bile canaliculi toward the triad and exits the liver through the hepatic ducts. The hepatic cords are separated from each other by blood channels called **hepatic sinusoids.** The sinusoid epithelium contains phagocytic cells that help remove foreign particles from the blood. Blood from the hepatic portal vein and the hepatic artery flows into the sinusoids and becomes mixed. The mixed blood then flows into the central vein, where it exits the lobule and then exits the liver through the hepatic veins. Therefore blood flows from the triad toward the center of each lobule, whereas bile flows away from the center of the lobule toward the triad.

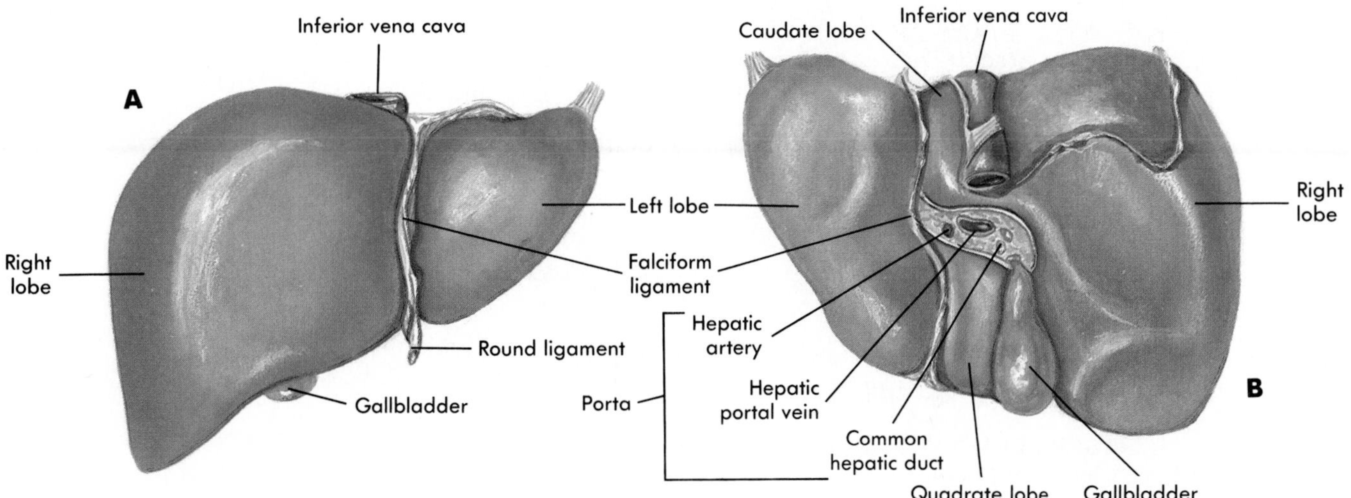

FIGURE 16-9 ■ **The liver—anterior view (A) and inferior view (B), showing the porta.**

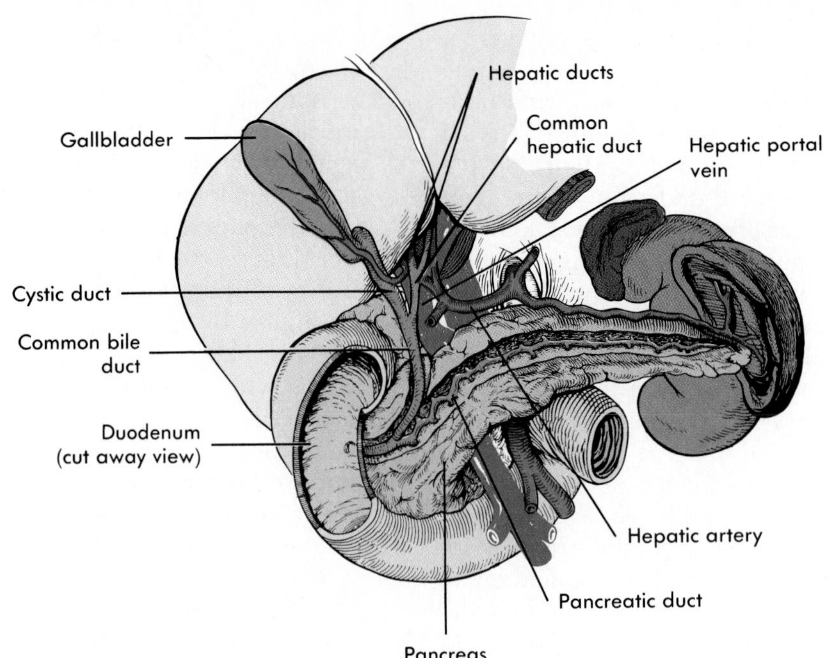

FIGURE 16-10 ■ The liver, gallbladder, pancreas, and duct system *(shown in green).* The hepatic ducts join to form the common hepatic duct, which joins the cystic duct to form the common bile duct. The common bile duct joins the pancreatic duct and enters the duodenum.

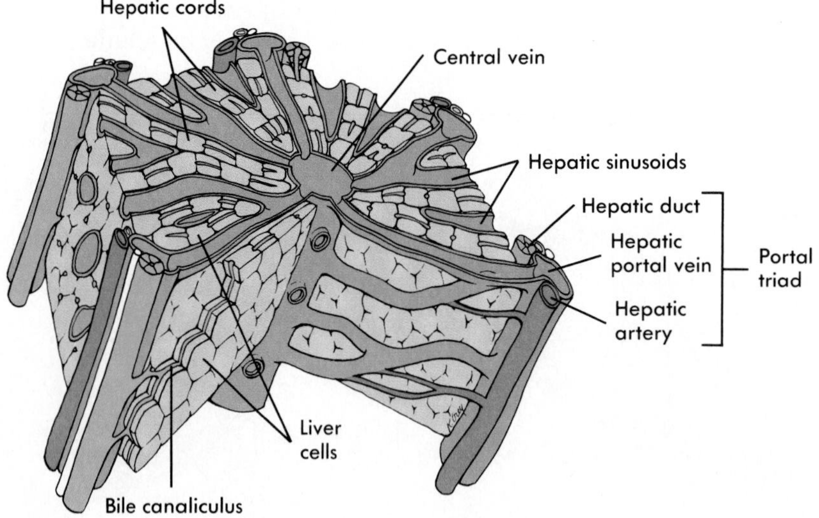

FIGURE 16-11 ■ Liver histology.
A lobule of the liver with triads at each corner and a central vein in the middle.

Pancreas

The **pancreas** (see Figure 16-10) is a complex organ composed of both endocrine and exocrine tissues that perform several functions. The endocrine portion of the pancreas consists of **pancreatic islets** (islets of Langerhans). The islet cells produce insulin and glucagon, which are very important in controlling blood levels of nutrients such as glucose and amino acids (see Chapter 10).

The exocrine portion of the pancreas consists of **acini** (as′ĭ-ne; grapes) (see discussion of glands in Chapter 4), which produce digestive enzymes. Clusters of acini are connected by small ducts, which join to form larger ducts, and the larger ducts join to form the **pancreatic duct.** The pancreatic duct joins the common bile duct and empties into the duodenum.

Large Intestine

The **large intestine** (Figure 16-12 and see Figure 16-1) consists of the cecum, colon, rectum, and anal canal.

Cecum

The **cecum** (se′kum; blind) (see Figure 16-12) is the proximal end of the large intestine and is where the large and small intestines meet. The cecum is a blind sac that extends inferiorly about 6 cm past the ileocecal junction. Attached to the cecum is a small blind tube about 9 cm long called the **appendix.**

Colon

The **colon** (ko′lon) (see Figure 16-12) is about 1.5 to 1.8 meters long and consists of four portions: the ascending colon, transverse colon, descending colon, and sigmoid colon. The mucosal lining of the colon contains numerous straight tubular glands called **crypts,** which contain many mucus-producing goblet cells.

The circular muscle layer of the colon is complete, and the longitudinal muscle layer is incomplete. The longitudinal layer does not completely envelope the intestinal wall but forms three bands that run the length of the colon, called the **teniae**

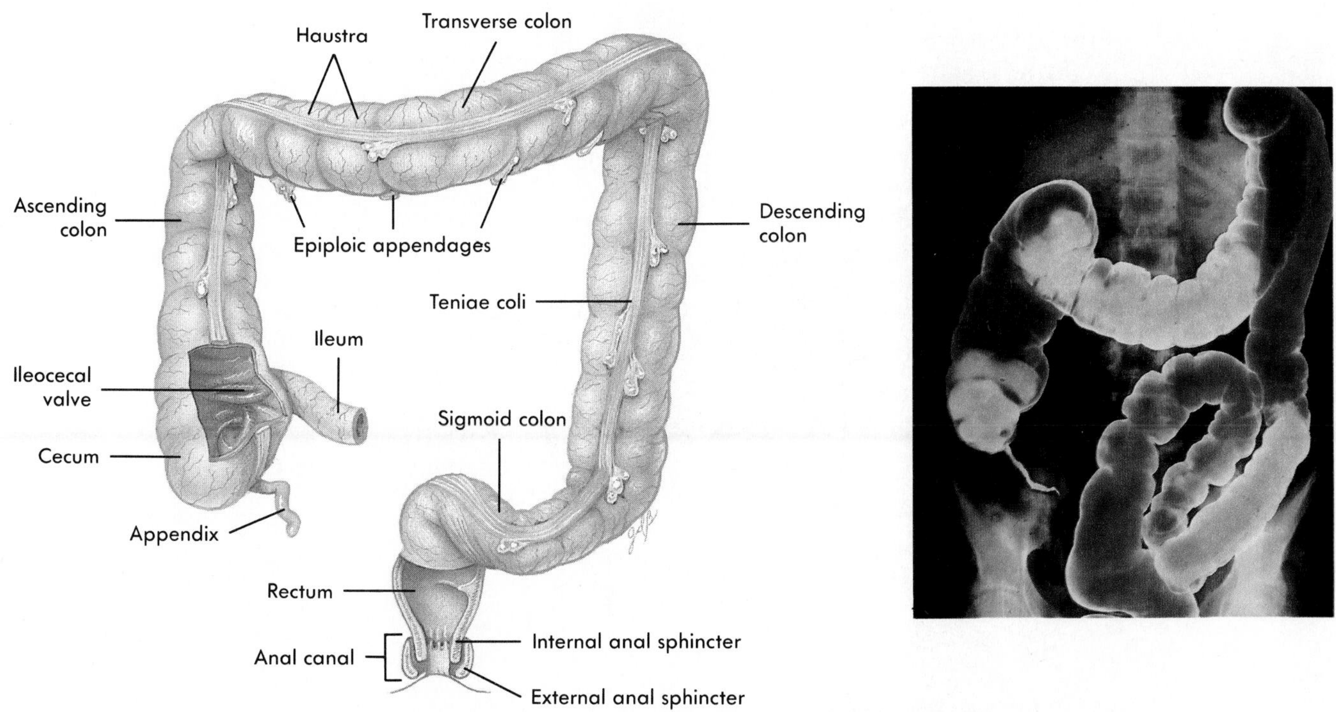

FIGURE 16-12 ■ The large intestine (cecum, colon, and rectum) and anal canal. The teniae coli and epiploic appendages can be seen along the length of the colon. The photograph shows the large intestine in place in the body.

coli (te′ne-ē ko′le; a band or tape along the colon). Contractions of the teniae coli cause pouches called **haustra** (haw′strah; to draw up) to form along the length of the colon, giving it a puckered appearance. Small, fat-filled connective tissue pouches called **epiploic** (ep′ĭ-plo′ik; related to the omentum) **appendages** are attached to the outer surface of the colon along its length.

Rectum

■ The **rectum** is a straight, muscular tube that begins at the termination of the sigmoid colon and ends at the anal canal (see Figure 16-12). The muscular tunic is relatively thick in the rectum compared to the rest of the GI tract.

Anal canal

■ The last 2 to 3 cm of the digestive tract is the **anal canal.** It begins at the inferior end of the rectum and ends at the anus (external GI tract opening). The smooth muscle layer of the anal canal is even thicker than that of the rectum and forms the **internal anal sphincter** at the superior end of the anal canal. The **external anal sphincter** at the inferior end of the anal canal is formed by skeletal muscle.

Peritoneum

■ The body walls and organs of the abdominal cavity are lined with **serous membranes** (Figure 16-13). The serous membrane that covers the organs

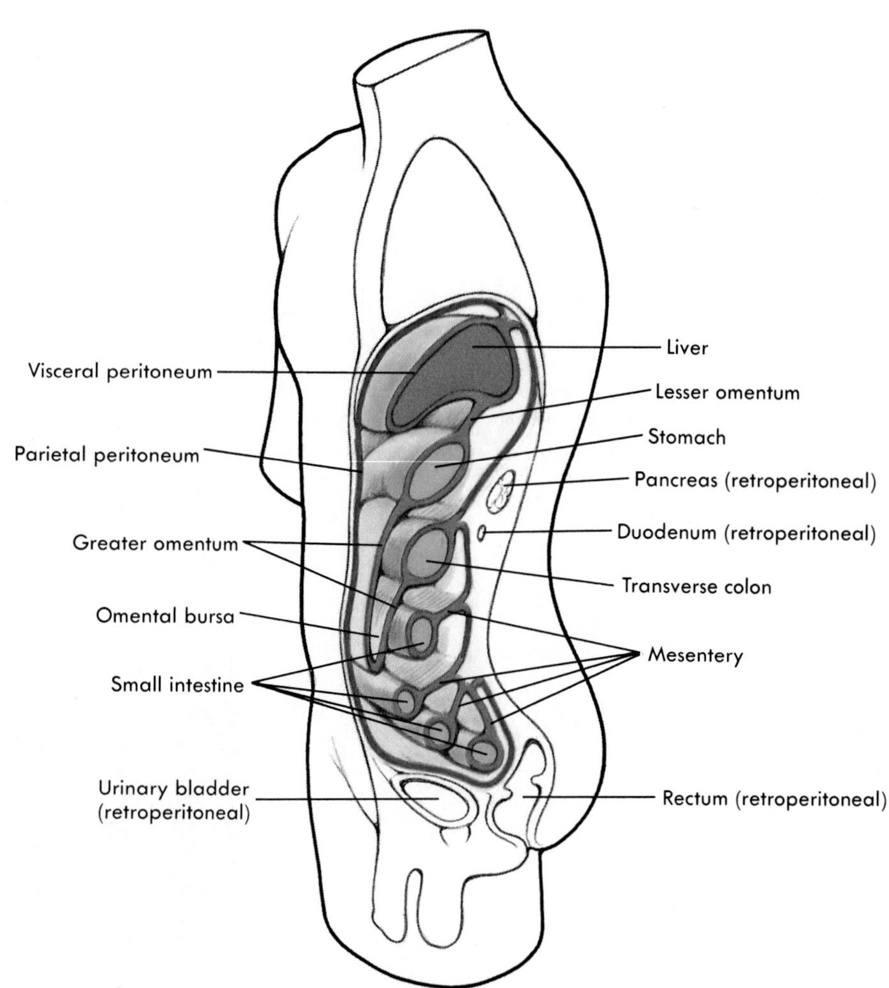

FIGURE 16-13 ■ Peritoneum and mesenteries.
The parietal peritoneum lines the abdominal cavity, and the visceral peritoneum covers abdominal organs. Retroperitoneal organs are deep to the parietal peritoneum. The mesenteries are membranes that connect abdominal organs to each other and to the body wall.

is the **visceral peritoneum** (pěr′it-o-ne′um; to stretch over), and the serous membrane that covers the interior of the body wall is the **parietal peritoneum.**

Many of the organs of the abdominal cavity are held in place by connective tissue sheets called **mesenteries** (mes′en-těr′ēz; middle intestine). The mesenteries consist of two layers of serous membranes with a thin layer of loose connective tissue in between. Specific mesenteries are given names. The mesentery connecting the lesser curvature of the stomach to the liver and diaphragm is the **lesser omentum** (o-men′tum; membrane of the bowels), and the mesentery connecting the greater curvature of the stomach to the transverse colon and posterior body wall is the **greater omentum.** The greater omentum is unusual in that it is a long, double fold of mesentery that extends inferiorly from the stomach to create a cavity or pocket called the **omentaı bursa** (bur′sah; pocket). Fat accumulates in the greater omentum, giving it the appearance of a fat-filled apron that covers the anterior surface of the abdominal viscera.

1 If you placed a pin completely through both folds of the greater omentum, through how many layers of simple squamous epithelium would the pin pass? ?

Other abdominal organs lie against the abdominal wall, have no mesenteries, and are described as **retroperitoneal** (rě′tro-pěr′ĭ-to-ne′al; behind the peritoneum). The retroperitoneal organs include the duodenum, pancreas, ascending colon, descending colon, rectum, kidneys, adrenal glands, and urinary bladder.

MOVEMENTS AND SECRETIONS IN THE DIGESTIVE SYSTEM

As food moves through the digestive tract, secretions are added to liquify and digest the food and to provide lubrication. Each segment of the digestive tract is specialized to assist in moving its contents from the oral end to the anal end. Portions of the digestive system are also specialized to absorb molecules from the digestive tract into the circulation. The processes of secretion, movement, and absorption are regulated by elaborate neural and hormonal mechanisms.

Oral Cavity, Pharynx, and Esophagus

Secretions of the oral cavity

Saliva is secreted at the rate of approximately 1 to 1.5 liters per day. The serous portion of saliva contains a digestive enzyme called **salivary amylase** (am′ĭ-lās; starch-splitting enzyme) (Table 16-1), which breaks the covalent bonds between glucose molecules in starch and other polysaccharides to produce the disaccharides maltose and isomaltose. Maltose and isomaltose have a sweet taste.

Only about 5% of the total carbohydrates are digested in the mouth. Most starches are contained in cellulose-covered globules and are inaccessible to salivary amylase. Humans lack the necessary enzymes to digest cellulose. Cooking and thorough chewing of food destroy the cellulose covering and increase the efficiency of the digestive process.

Cellulose is a polysaccharide that is abundant in vegetable matter. Even though humans cannot digest cellulose, it is important to normal digestive function. Cellulose provides bulk, or fiber, in the diet. The presence of this bulk facilitates movement of chyme (partially digested food) in the small intestine and feces in the large intestine by providing mass against which the muscular wall of the digestive tract can push. In the 1950s some nutritionists dreamed that all the nutrients we need could be eventually reduced into a single tablet and that we no longer would have to eat food. It is now known that the indigestible, bulk components of food are as important to normal digestion as the digestible components.

Saliva prevents bacterial infection in the mouth by washing the oral cavity, and it contains lysozyme (an enzyme that lyses, "breaks open," some bacterial cells), which has a weak antibacterial action. A lack of salivary gland secretion increases the chance of ulceration and infection of the oral mucosa and caries formation in the teeth.

The mucous secretions of the submandibular and sublingual glands contain a large amount of **mucin** (mu′sin), a proteoglycan that gives a lubricating quality to the secretions of the salivary glands. Salivary-gland secretion is regulated by the autonomic nervous systems, with parasympathetic innervation being the most important. Brainstem salivary nuclei stimulate saliva secretions via parasympathetic fibers of the facial and glossopharyngeal cranial nerves (VII and IX). Salivary secretions increase in response to a variety of stimuli, such as

TABLE 16-1

Functions of Digestive Secretions

FLUID OR ENZYME	SOURCE	FUNCTION
MOUTH		
Saliva	Salivary glands	Moistens and lubricates food
Salivary amylase	Salivary glands	Digests starch
STOMACH		
Hydrochloric acid	Gastric glands	Kills bacteria, activates pepsin
Pepsinogen	Gastric glands	Active form, pepsin, digests protein
Mucus	Mucous cells	Protects stomach lining
Intrinsic factor	Gastric glands	Binds to vitamin B_{12}, aiding in its absorption
Gastrin	Gastric glands	Regulates stomach secretions
SMALL INTESTINE AND ASSOCIATED GLANDS		
Bile salts	Liver	Emulsify fats
Bicarbonate ions	Pancreas	Neutralize stomach acid
Trypsin	Pancreas	Digests protein
Pancreatic amylase	Pancreas	Digests starch
Lipase	Pancreas	Digests lipid
Nucleases	Pancreas	Digest nucleic acid
Mucus	Duodenal glands and goblet cells	Protects duodenum from stomach acid and digestive enzymes
Secretin	Duodenum	Inhibits gastric secretions
Cholecystokinin	Duodenum	Inhibits gastric motility, stimulates gallbladder and pancreas secretions
Gastric inhibitory polypeptide	Duodenum	Inhibits gastric motility and secretion, stimulates gallbladder contraction
Peptidases	Small intestine	Digest polypeptide
Amylase	Small intestine	Digests starch
Lipase	Small intestine	Digests lipid
Sucrase	Small intestine	Digests sucrose
Lactase	Small intestine	Digests lactose
Maltase	Small intestine	Digests maltose

tactile stimulation in the oral cavity and certain tastes, especially sour. Higher brain centers also affect the activity of the salivary glands. Odors that trigger thoughts of food or the sensation of hunger can increase salivary secretions.

Mastication

Food taken into the mouth is chewed, or **masticated** (mas'tĭ-ka'ted) by the teeth. The anterior teeth, that is, the incisors and the canines, primarily cut and tear food, whereas the premolars and molars primarily crush and grind food. Mastication breaks large food particles into small particles, which have a much larger total surface area than would a few large particles. Because digestive enzymes act on food molecules only at the surface of the particles, mastication increases the efficiency of digestion.

Deglutition

Deglutition (dĕ'glu-tish'un), or swallowing, can be divided into three separate phases: the voluntary phase, the pharyngeal phase, and the esophageal phase (Figure 16-14). During the **voluntary phase,** a **bolus,** or mass of food, is formed in the mouth. The bolus is pushed by the tongue against the hard pal-

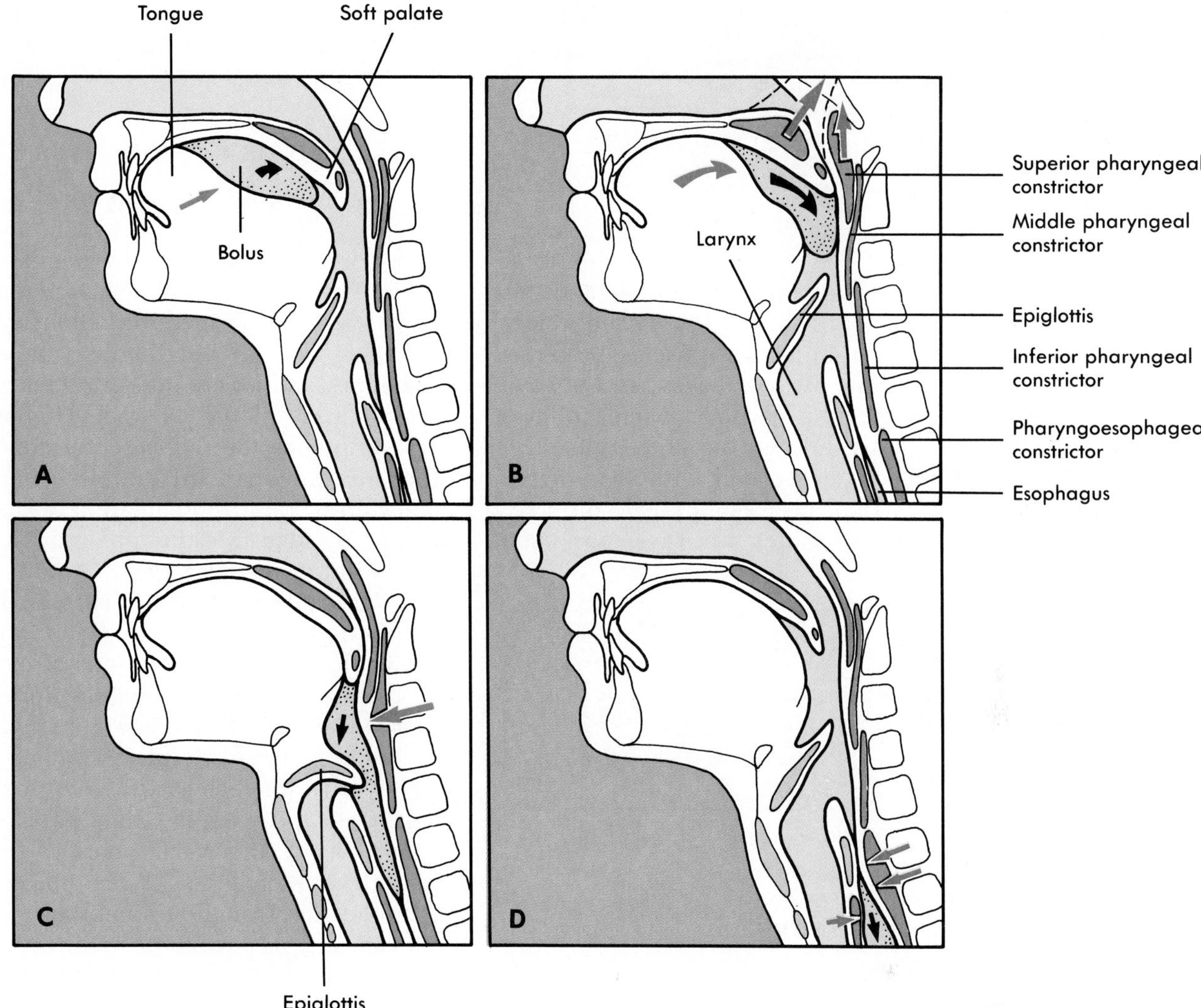

FIGURE 16-14 ■ Three phases of swallowing (deglutition).
A During the voluntary phase, a bolus of food *(yellow)* is pushed by the tongue against the hard palate and posteriorly toward the pharynx (*black arrow:* movement of the bolus).

During the pharyngeal phase,
B The soft palate and pharynx are elevated *(red arrows).*
C Contraction of the pharyngeal constrictors (*red arrow:* direction of force) forces the bolus through the pharynx and into the esophagus. As this occurs, the epiglottis is bent down over the opening of the larynx largely by the force of the bolus pressing against it.
D During the esophageal phase, the bolus is moved by successive constrictions of the esophagus toward the stomach.

ate, forcing the bolus toward the posterior portion of the mouth and into the oropharynx.

The **pharyngeal phase** of swallowing is a reflex that is initiated by stimulation of receptors in the oropharynx. This phase of swallowing begins with the elevation of the soft palate, which closes the passage between the nasopharynx and oropharynx. The pharynx elevates to receive the bolus of food from the mouth. The three **pharyngeal constrictor muscles** (superior, middle, and inferior) then contract in succession, forcing the food through the pharynx. At the same time, the upper esophageal sphincter relaxes, and food is pushed into the esophagus. As food passes through the pharynx, the **epiglottis** (ep′ĭ-glot′is; upon the glottis) is tipped posteriorly so that the opening into the larynx is covered, preventing food from passing into the larynx.

2 Why is it important to close off the nasopharynx during swallowing? What may happen if a person has an explosive burst of laughter while trying to swallow a liquid? What happens if you try to swallow and speak at the same time? ?

The **esophageal phase** of swallowing is responsible for moving food from the pharynx to the stomach. Muscular contractions of the esophagus occur in **peristaltic** (pĕr′ĭ-stal′tik) **waves.** A wave of relaxation of the circular esophageal muscles precedes the bolus of food down the esophagus, and a wave of strong contraction of the circular muscles follows and propels the bolus through the esophagus. The peristaltic contractions associated with swallowing cause relaxation of the lower esophageal sphincter in the esophagus as the peristaltic waves approach the stomach.

■ Gravity assists the movement of material through the esophagus, especially when liquids are swallowed. However, the peristaltic contractions that move material through the esophagus are sufficiently forceful to allow a person to swallow even while standing on his head. ■

Stomach

Secretions of the stomach

The stomach functions primarily as a storage and mixing chamber for ingested food. As food enters the stomach, it is mixed with stomach secretions to become a semifluid mixture called **chyme** (kīm; juice). Although some digestion and a small amount of absorption occur in the stomach, they are not its principal functions.

Stomach secretions from the gastric glands include mucus, hydrochloric acid, pepsinogin, and regulatory molecules such as intrinsic factor and gastrin (see Table 16-1). A thick layer of **mucus** lubricates and protects the epithelial cells of the stomach wall from the damaging effect of the acidic chyme and pepsin. Irritation of the stomach mucosa stimulates the secretion of a greater volume of mucus. **Hydrochloric acid** produces a low pH in the stomach, which acts as an antimicrobial agent that kills microorganisms. **Pepsinogen,** a protein secreted by chief cells, is converted by hydrochloric acid to the active enzyme pepsin. **Pepsin** is an enzyme that breaks proteins into smaller peptide chains. Pepsin exhibits optimum enzymatic activity at a pH of 3 or below, the pH produced in the stomach by the presence of hydrochloric acid. **Intrinsic factor** binds with vitamin B_{12} and makes it more readily absorbed in the ileum. Vitamin B_{12} is important in DNA synthesis. **Gastrin** (gas′trin) is a hormone, secreted by the stomach into the bloodstream, that helps regulate stomach secretions.

Regulation of stomach secretions

Approximately 2 to 3 liters of gastric secretions (gastric juice) are produced each day. Both nervous and hormonal mechanisms regulate gastric secretions. The neural mechanisms involve reflexes integrated within the medulla oblongata, and local reflexes, which do not involve the central nervous system but rather are integrated within the intramural plexuses of the GI tract. Higher brain centers also influence the reflexes. Hormones produced by the stomach (gastrin) and intestine also regulate stomach secretions. Regulation of stomach secretion can be divided into three phases: the cephalic, gastric, and intestinal phases.

The **cephalic phase** (Figure 16-15, *A*) is anticipatory and prepares the stomach to receive food. In the cephalic phase, sensations of taste and smell of food, stimulation of tactile receptors during the process of chewing and swallowing, and pleasant thoughts of food stimulate centers within the medulla that influence gastric secretions. Impulses are sent from the medulla along parasympathetic neurons within the vagus nerves to the stomach. Within the stomach wall, the preganglionic neurons stimulate postganglionic neurons in the intramural plexus. The postganglionic neurons stimulate secretory activity in the cells of the stomach mucosa. Gastrin is secreted by the stomach as a result of the parasympathetic stimulation. The gastrin enters the circulation and is carried back to the stomach, where it stimulates additional gastric secretion.

During the **gastric phase** (Figure 16-15, *B*), food is present in the stomach and is being mixed with gastric secretions. The gastric phase is responsible for the greatest volume of gastric secretion. It is initiated by the presence of food in the stomach. Distention of the stomach results in the stimulation of mechanoreceptors. Impulses generated by these receptors initiate reflexes integrated in the medulla oblongata of the central nervous system and local reflexes, integrated in the intramural plexuses, resulting in secretion of hydrochloric acid and pepsin by the gastric glands. Pepsin breaks down proteins into peptides. The peptides stimulate the secretion of gastrin, which, in turn, stimulates additional hydrochloric acid secretion.

The **intestinal phase** (Figure 16-15, *C*) regulates the entry of chyme into the small intestine. This phase of gastric secretion is controlled by the entrance of acidic chyme into the duodenum. The presence of chyme in the duodenum initiates both neural and hormonal mechanisms, which first stim-

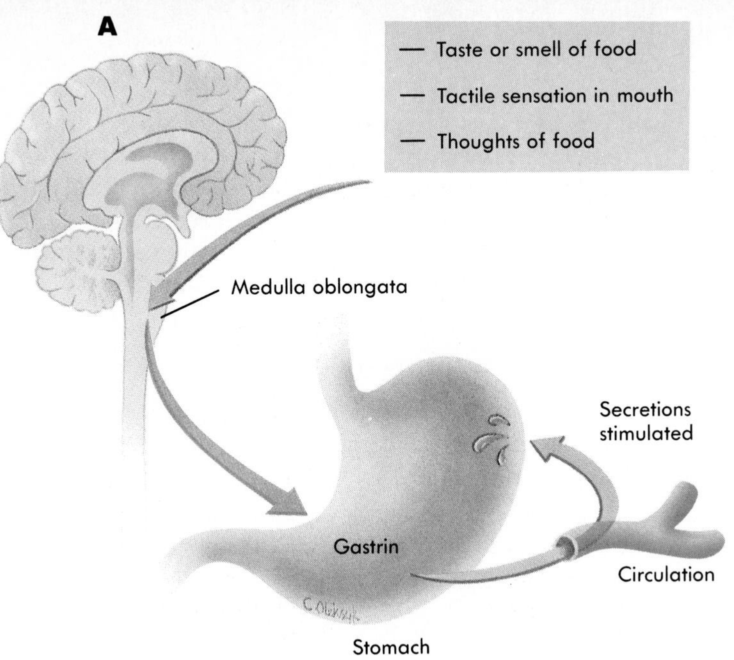

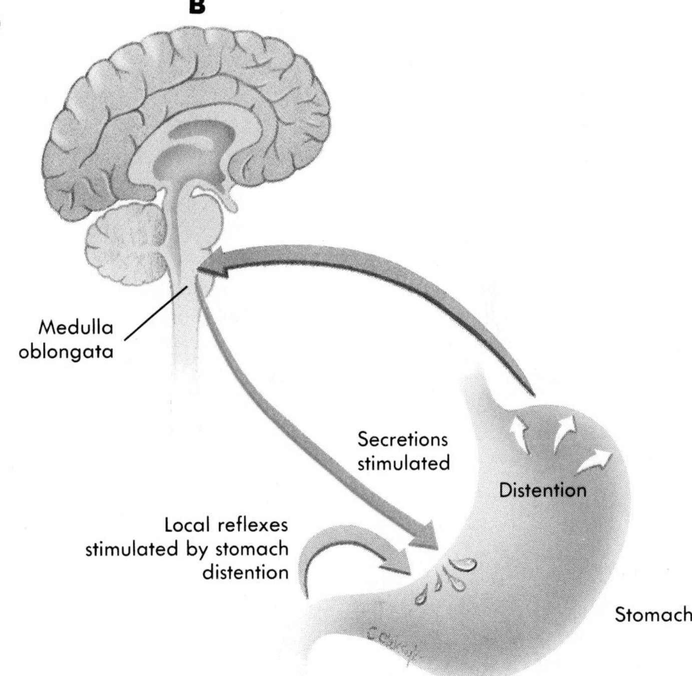

FIGURE 16-15 ■ The three phases of gastric secretion.
A Cephalic phase. Taste, smell, and tactile sensations stimulate the medulla oblongata. Parasympathetic fibers of the vagus nerve directly stimulate secretion. Gastrin, produced in the gastric glands, is carried by the circulation back to the stomach where it also stimulates secretion.
B Gastric phase. Distention of the stomach stimulates local reflexes and vagus nerve impulses to the medulla. Efferent parasympathetic vagal impulses and local reflexes stimulate secretion.
C Intestinal phase. The presence in the duodenum of chyme with a pH greater than 3 stimulates gastric secretions through vagus nerve pathways and though secretion of gastrin. Chyme with a pH less than 2 inhibits gastric secretions.

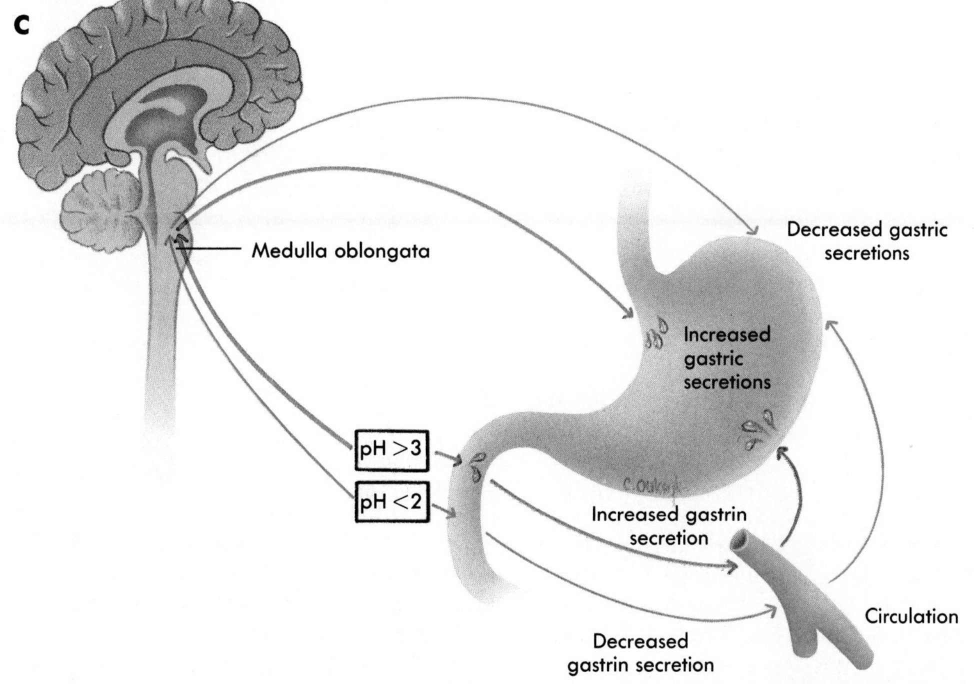

ulate and then inhibit gastric secretions. When the pH of the chyme is 3 or above, the stimulatory response of the intestinal phase is greatest. Gastrin, released from the duodenum rather than from the stomach, is carried in the blood to the stomach, where it stimulates gastric secretions. Reflexes mediated through the medulla oblongata also increase gastric secretions during the intestinal phase.

When the pH is 2 or below, the inhibitory influence of the intestinal phase is greatest. The hormone **secretin** (se-kre′tin), which inhibits gastric secretions, is released from the duodenum. Fatty acids and certain other lipids in the duodenum initiate the release of two hormones, **cholecystokinin** (ko-le-sis-to-kīn′in) and **gastric inhibitory polypeptide,** that also inhibit gastric secretions. Acidic chyme in the duodenum also initiates medullary and local reflexes that inhibit gastric secretion.

Movement in the stomach

■ Two types of stomach movement occur: mixing waves and peristaltic waves (Figure 16-16). Both types of movements result from peristaltic contrac-

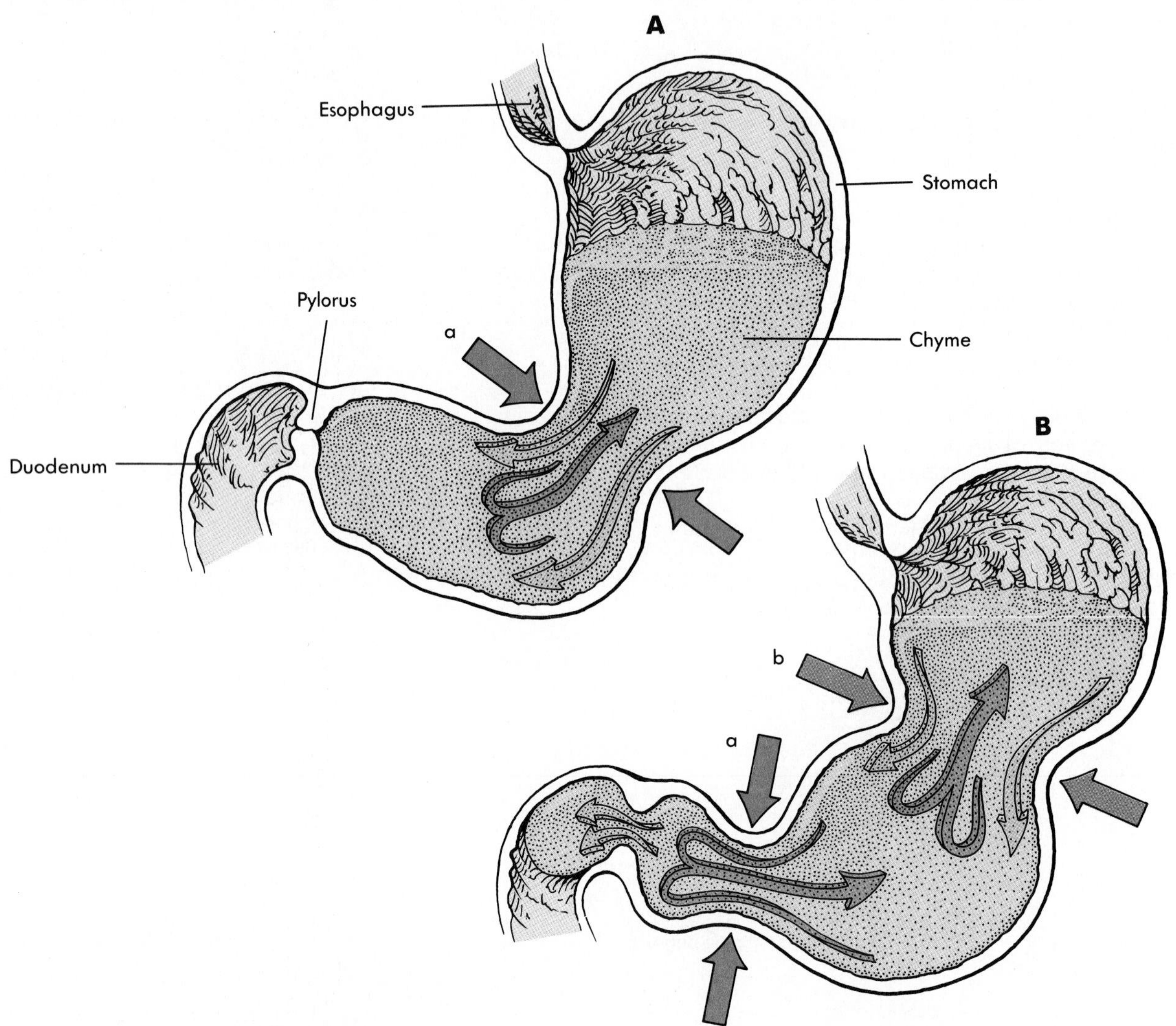

FIGURE 16-16 ■ **Movement in the stomach.**
A Mixing waves that are initiated in the body of the stomach *(a, red arrows)* progress toward the pylorus. The more fluid portion of the chyme is pushed toward the pylorus *(blue arrows)*, whereas the more solid center of the chyme squeezes past the peristaltic constrictions back toward the body *(brown arrow)*.
B More powerful peristaltic waves *(b, purple arrows)* move in the same direction and in the same way as described in **A.** Some of the moist fluid chyme is squeezed through the pylorus into the duodenum *(small blue arrows)*, whereas most of the chyme is forced back toward the body for further mixing *(brown arrow)*.

tions that occur about every 20 seconds. The contractions proceed from the body of the stomach toward the pyloric sphincter. **Mixing waves** are relatively weak and cause ingested food to be thoroughly mixed with stomach secretions to form chyme. **Peristaltic waves** are significantly more powerful than the normal mixing movements and force the chyme toward the pyloric sphincter. The pyloric sphincter usually remains partially closed because of mild tonic contraction. Each peristaltic contraction is sufficiently strong to force a small amount of chyme through the pyloric opening and into the duodenum. This movement of chyme through the partially closed opening by the force of peristaltic contractions is called the **pyloric pump.**

■ Hunger contractions are peristaltic contractions that approach tetany for periods of about 2 to 3 minutes. The contractions are increased by a decrease in blood glucose levels and are sufficiently strong to create an uncomfortable sensation often called a "hunger pang" or "hunger pain." Hunger pangs usually begin 12 to 24 hours after the previous meal. If nothing is ingested, they reach their maximum intensity within 3 or 4 days and then become progressively weaker. ■

If the stomach empties too fast, the efficiency of digestion and absorption is reduced, and the acidic chyme may damage the intestinal wall; if the rate of emptying is too slow, the highly acidic contents of the stomach may damage the stomach wall and reduce the rate at which nutrients are digested and absorbed. Stomach emptying is regulated to prevent these two extremes. Many of the hormonal and neural mechanisms that regulate stomach secretions are also involved with controlling stomach motility.

Small Intestine

Secretions of the small intestine

The mucosa of the small intestine produces secretions that contain primarily mucus, electrolytes, and water. Intestinal secretions lubricate and protect the intestinal wall from the acidic chyme and the action of digestive enzymes. They also keep the chyme in the small intestine in a liquid form to facilitate the digestive process. Most of the secretions entering the small intestine are produced by the intestinal mucosa, but the secretions of the liver and the pancreas also enter the small intestine and play an important role in the process of digestion.

The epithelial cells in the walls of the small intestine have enzymes bound to them that play a sig-

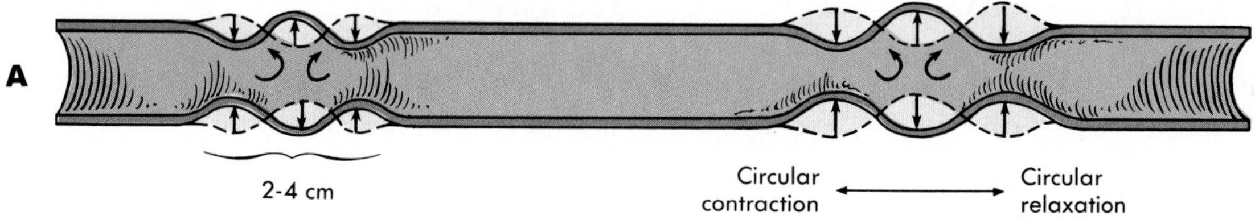

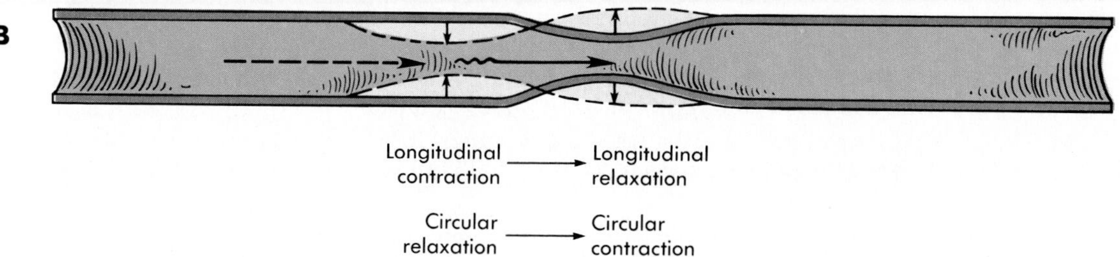

FIGURE 16-17 ■ Movement in the small intestine.
The short arrows directed inward depict circular contractions; those directed outward depict circular relaxation. The short curved arrows in **A** depict mixing movements or segmental contractions. The long arrows in **B** depict longitudinal movement or peristaltic contractions.

nificant role in the final steps of digestion. Some of these enzymes (peptidases) break down peptides into single amino acids. Peptides result from the previous digestion of proteins. Disaccharides, such as maltose and isomaltose, result from the digestion of starch. Disaccharides, such as sucrose or lactose, are ingested as disaccharides. Disaccharides are broken down into monosaccharides. These digestion products can be absorbed by the intestinal epithelium (see Table 16-1).

Mucus is produced by **duodenal glands** and by **goblet cells,** which are dispersed throughout the epithelial lining of the entire small intestine and within **intestinal glands.** Hormones released from the intestinal mucosa stimulate hepatic (from the liver) and pancreatic secretions. Secretion by duodenal glands is stimulated by the vagus nerve, secretin release, and chemical or tactile irritation of the duodenal mucosa.

Movement in the small intestine

Mixing and propulsion of chyme are the primary mechanical events that occur in the small intestine. **Segmental contractions** are propagated for only short distances, and function to mix intestinal contents (Figure 16-17, *A*). **Peristaltic contractions** proceed along the length of the intestine for variable distances, and cause the chyme to move along the small intestine (Figure 16-17, *B*).

The ileocecal sphincter at the juncture of the ileum and the large intestine remains mildly contracted most of the time, but peristaltic contractions reaching it from the small intestine cause it to relax and allow movement of chyme from the small intestine into the cecum. The ileocecal valve is a one-way valve that allows chyme to move from the ileum into the large intestine, but does not allow movement from the large intestine back into the ileum.

Liver

The liver performs important digestive and excretory functions, stores and processes nutrients, synthesizes new molecules, and detoxifies harmful chemicals (Table 16-2).

The liver secretes about 600 to 1000 ml of **bile** each day. Although bile contains no digestive enzymes, it plays a role in digestion by diluting and neutralizing stomach acid and by increasing the efficiency of fat digestion and absorption. Bile salts **emulsify** fats, reducing surface tension on fat globules and breaking the globules into smaller droplets that are more easily digested and absorbed, much like the action of detergents in dishwater (see Tables 16-1 and 16-2). Digestive enzymes cannot act efficiently on large fat globules. Emulsification breaks fat into small droplets that can be attacked by digestive enzymes. Bile also contains excretory products such as bile pigments (like bilirubin, which results from the breakdown of hemoglobin), cholesterol, and fats.

Bile secretion by the liver is stimulated by secretin, released from the duodenum (Figure 16-18). Cholecystokinin stimulates the gallbladder to con-

TABLE 16-2

Functions of the Liver

FUNCTION	EXPLANATION
Digestion	Bile neutralizes stomach acid and emulsifies fats, which facilitates digestion
Excretion	Bile contains excretory products, such as cholesterol, fats, or bile pigments like bilirubin resulting from hemoglobin breakdown
Nutrient storage	Liver cells remove sugar from the blood and store it in the form of glycogen; also store fat, vitamins (A, B_{12}, D, E, and K), copper, and iron
Nutrient conversion	Liver can convert some nutrients into others, for example, amino acids can be converted to lipids or glucose; fats can be converted to phospholipids; vitamin D is converted to its active form
Detoxification of harmful chemicals	Liver cells remove ammonia from the circulation and convert it to urea, which is eliminated in the urine; other substances are detoxified and secreted in the bile
Synthesis of new molecules	Synthesizes blood proteins such as albumin, fibrinogen, globulin, and clotting factors

tract and release bile into the duodenum. Parasympathetic stimulation through the vagus nerve also stimulates bile secretion and release. Bile salts also increase bile secretion through a positive-feedback system. Most bile salts are reabsorbed in the ileum, and the blood carries them back to the liver, where they are once again secreted into the bile. The loss of bile salts in the feces is reduced by this recycling process.

The liver can remove sugar from the blood and store it in the form of glycogen (see Table 16-2). It can also store fat, vitamins, copper, and iron. This storage function is usually short term.

Another function that the liver performs is the interconversion of nutrients (see Table 16-2). Ingested foods are not always in the proportion that are needed by the tissues. If this is the case, the liver can convert some nutrients into others. If, for

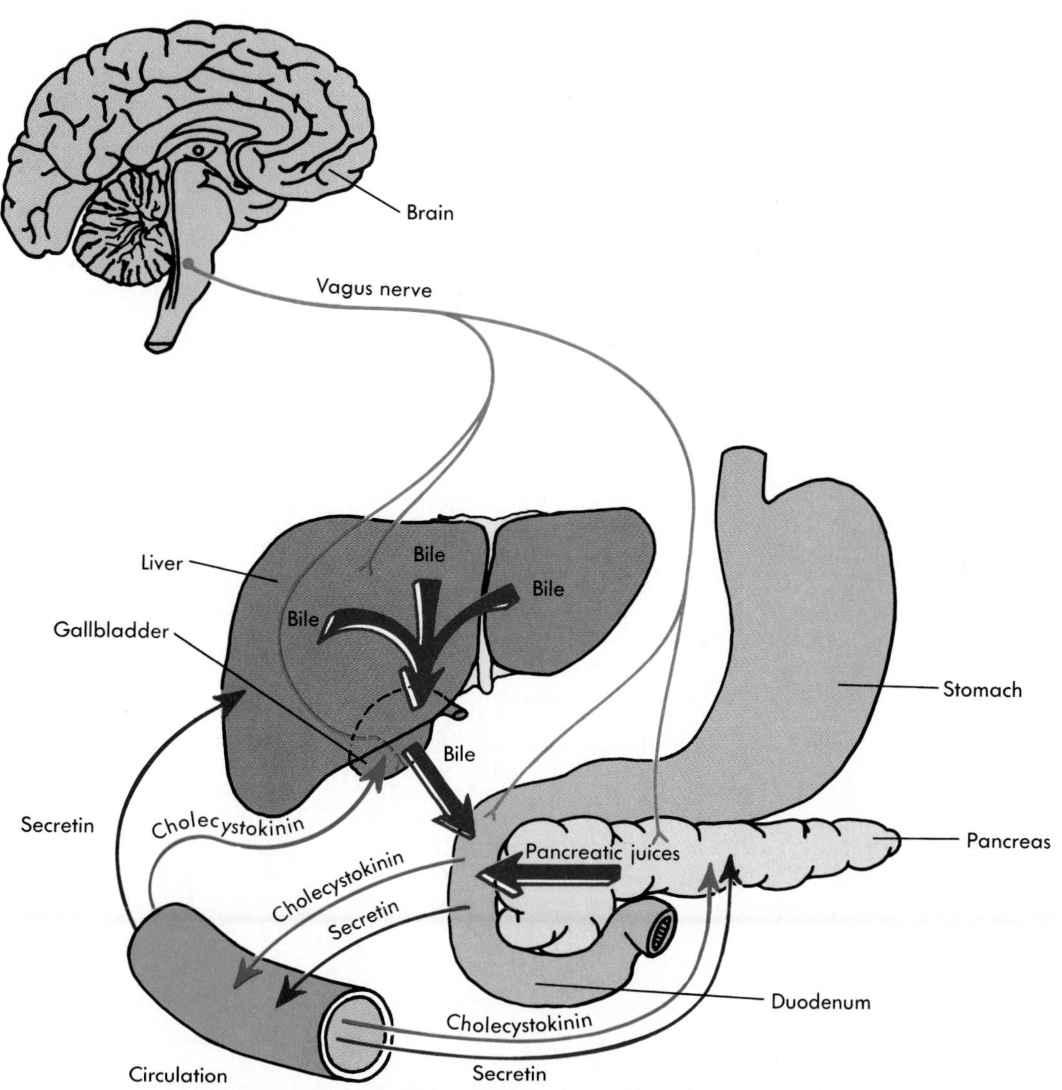

FIGURE 16-18 ■ Regulation of intestinal, hepatic, and pancreatic secretions. Cholecystokinin stimulates the pancreas to secrete an enzyme-rich solution and stimulates the gallbladder to contract, releasing large amounts of stored bile into the intestine. Secretin stimulates the pancreas to secrete a watery solution and stimulates the liver to produce bile. Parasympathetic stimulation from the medulla oblongata via the vagus nerve also causes pancreatic, hepatic, and intestinal secretions.

Disorders of the Digestive Tract

ORAL CAVITY

Inflammation of the parotid gland is called parotiditis. The most common type of parotiditis, caused by a viral infection, is **mumps.**

Dental caries (tooth decay) involve bacteria that adhere to the surface of the teeth. The bacteria convert sucrose and other carbohydrates into acid, which dissolves the tooth enamel and dentin. If the caries are not treated, bacteria can enter the pulp and kill the tooth. A root canal may be performed to remove the dead pulp.

STOMACH

Vomiting can result from irritation (for example, overdistention or overexcitation) anywhere along the GI tract. Impulses travel through visceral afferent nerves to the vomiting center in the medulla oblongata. After the vomiting center is stimulated and the reflex is initiated, the following events occur:

1. A deep breath is taken.
2. The hyoid bone and larynx are elevated, opening the upper esophageal sphincter.
3. The opening of the larynx is closed.
4. The soft palate is elevated, closing the posterior nares.
5. The diaphragm and abdominal muscles are forcefully contracted, strongly compressing the stomach and increasing the intragastric pressure.
6. The lower esophageal sphincter is relaxed, and the gastric contents are forcefully expelled.

ULCERS

Peptic ulcer is a condition in which the stomach acids digest the mucosal lining of the GI tract itself. The most common site of a peptic ulcer is near the pylorus, usually on the duodenal side (that is, a **duodenal ulcer**). Ulcers occur less frequently along the lesser curvature of the stomach or at the point where the esophagus enters the stomach (a **gastric ulcer**). The most common cause of a duodenal ulcer is the oversecretion of gastric juice relative to the degree of mucous and alkaline protection of the small intestine.

People experiencing severe anxiety for a long period of time are the most prone to develop duodenal ulcers. They often have a high rate of gastric secretion (as much as 15 times the normal amount) between meals. Cerebral impulses enhance vagal nerve activity, which stimulates excessive gastric acid secretion. This secretion results in highly acidic chyme entering the duodenum. The duodenum is usually protected by sodium bicarbonate (secreted mainly by the pancreas), which neutralizes the chyme. When large amounts of acid enter the duodenum, however, the sodium bicarbonate is not adequate to neutralize the acid.

Duodenal gland secretion is inhibited by sympathetic nerve stimulation, thus reducing the duodenal wall's coating of mucus, which protects it against acid and gastric enzymes. Therefore, if a person is highly stressed, elevated sympathetic activity may inhibit duodenal gland secretion and increase his susceptibility to a duodenal ulcer.

In people with gastric ulcers, there are often normal or even low levels of hydrochloric acid secretion. In these cases, however, the stomach has a reduced resistance to its own acid. Such inhibited resistance can result from excessive ingestion of alcohol or aspirin. Reflux of duodenal contents into the pylorus also can cause peptic ulcers. In this case, bile, which is present in the reflux, has a detergent effect that removes some of the mucus and reduces gastric mucosal resistance to acid.

LIVER

Hepatitis is an inflammation of the liver that may result from alcohol consumption or viral infection. If not corrected, liver cells can die and be replaced by scar tissue, resulting in loss of liver function. Death can result from liver failure.

Viral hepatitis is the second most frequently reported infectious disease in the United States. Infectious hepatitis (hepatitis A) is usually transmitted by poor sanitation practices or from mollusks, such as oysters, living in contaminated waters. Serum hepatitis (hepatitis B) is usually transmitted through blood or other body fluids, such as when blood is transferred through a contaminated hypodermic needle. Symptoms include nausea, diarrhea, loss of appetite, abdominal pain, fever, chills, and malaise. Jaundice is seen in about two thirds of the cases, with yellowing of the skin and sclera of the eyes resulting from the accumulation of bile pigments in those tissues.

Cholesterol, secreted by the liver into the bile, may precipitate in the gallbladder to produce **gallstones.** Occasionally a gallstone may pass out of the gallbladder and enter the cystic duct, blocking release of the bile. Such a condition interferes with normal digestion and often the gallbladder must be removed surgically.

SMALL INTESTINE

Malabsorption syndrome (sprue) is a spectrum of disorders of the small intestine that result in abnormal nutrient absorption. In some people, one type of malabsorption results from the toxic effects of gluten present in certain types of grains. The reaction to gluten may destroy newly formed epithelial cells, causing the villi to become blunted and the intestinal surface area to decrease. As a result, the intestinal epithelium is less capable of absorbing nutrients. Another type of malabsorption (called tropical) is apparently caused by bacteria, although no specific bacterium has been identified.

Appendicitis is an inflammation of the appendix and usually occurs because of obstruction of the appendix. Secretions from the appendix cannot pass the obstruction and accumulate, causing enlargement and pain. Bacteria in the area cause the appendix to become infected. If the appendix bursts, the infection can spread throughout the peritoneal cavity with life-threatening results. The right inferior quadrant of the abdomen becomes very tender in people with acute appendicitis as a result of pain referred from the inflamed appendix to the body surface.

Staphylococcal food poisoning occurs when toxin from the bacteria *Staphylococcus aureus* is ingested. The bacteria usually come from the hands of a person preparing the food. If food is cooked in large volumes at low temperatures (below 60° C [140° F]) or is allowed to sit for an extended period of time, the bacteria can reproduce and form toxins. Reheating can eliminate the bacteria but not the toxins. Staphylococcal food poisoning is characterized by nausea, vomiting, and diarrhea from 1 to 6 hours after the contaminated food is ingested.

Salmonellosis is a disease caused by *Salmonella* bacteria. They are ingested with contaminated food (usually meat, poultry, or milk) and grow in the digestive tract. The disease symptoms may not be seen for up to 36 hours after the contaminated food has been consumed. Symptoms include nausea, fever, abdominal pain, and diarrhea. The bacteria are generally destroyed by normal cooking with internal temperatures greater than 68° C (155° F).

Typhoid fever is caused by a particularly virulent strain of *Salmonella* bacteria. The bacteria can cross the intestinal wall and invade other tissues. The incubation period is normally about 2 weeks. Symptoms include severe fever and headaches, as well as diarrhea. Poor sanitation practices are the main source of contamination, and typhoid fever is still a leading cause of death in many underdeveloped countries.

Cholera is caused by a bacterium that infects the small intestine. The bacteria produce a toxin that stimulates the secretion of chlorides, bicarbonates, and water from the intestinal tract. The loss of fluid and electrolytes (as much as 12 to 20 liters of fluid loss per day) causes shock, collapse, and even death. Cholera was common in the United States and Europe in the 1800s but is not very common in western countries today. Cholera is still a major problem in Asia, however, particularly India.

Giardiasis is a disease caused by a protozoan that invades the intestine. Symptoms include nausea, abdominal cramps, weakness, weight loss, and malaise, and may last for several weeks. The disease is carried by humans and wild animals, especially beaver, and commonly affects persons who drink unfiltered water from wilderness streams.

Intestinal parasites are not uncommon in humans, especially under conditions of poor sanitation. **Tapeworms** can infect the digestive tract by way of undercooked beef, pork, or fish. The tapeworms attach to the intestinal wall by suckers and may live in the intestine for 25 years, reaching lengths of 6 meters. There are few symptoms beyond a vague abdominal discomfort. **Pinworms** are common in humans. The tiny worm lives in the digestive tract but migrates out of the anus to lay its eggs. This causes a local itching, and the eggs can be spread by contaminated fingers to numerous surfaces. Eggs resist dehydration and can be picked up from contaminated surfaces by other people. It is common for entire households to be contaminated if one child contracts the disease.

Hookworms attach to the intestinal wall and feed on the blood and tissue of the host, rather than on partially digested food as other parasites do. Infection can cause anemia and lethargy. Because hookworms are spread through fecal contamination of the soil and bare skin contact with contaminated soil, improved sanitation and the practice of wearing shoes has greatly decreased the incidence of hookworm infection.

Ascariasis is caused by a roundworm and is fairly common in the United States. Ingested eggs hatch in the upper intestine into wormlike larvae that pass into the bloodstream and then into the lungs, where they may cause pulmonary symptoms. Extremely large numbers may cause pneumonia. The larvae enter the throat and are swallowed, whereby they return to the intestinal tract. Adults in the intestinal tract cause few symptoms. The adult worms, however, measuring up to 30 cm, migrate. In some cases, they may emerge from the anus, or they may cut their way through the intestinal wall and infect the abdominal cavity. They may exit the body through the umbilicus of a child or through the nostrils of a sleeping person.

LARGE INTESTINE

Constipation is the slow movement of feces through the large intestine. The feces often become dry and hard because of the increased fluid absorption during the extended time they are retained in the large intestine. Constipation often results from irregular defecation patterns that develop after a prolonged time of inhibiting normal defecation reflexes. Spasms of the sigmoid colon resulting from irritation also can result in slow feces movement and constipation.

When the large intestine is irritated and inflamed, such as in patients with **enteritis** (bacterial infection of the bowel), the intestinal mucosa secretes large amounts of water and electrolytes in addition to mucus. This condition is called **diarrhea,** and although it increases fluid and electrolyte loss, it also moves the infected feces out of the intestine more rapidly and speeds recovery from the disease.

Dysentery is a severe form of diarrhea in which blood or mucus is present in the feces. Dysentery can be caused by bacteria, protozoa, or amoebae.

Hemorrhoids are enlarged or inflamed veins located in the wall of the anal canal.

example, a person is on a fad diet that is very high in protein, an oversupply of protein and an undersupply of lipids and carbohydrates are delivered to the liver. The liver breaks down the amino acids and cycles many of them through metabolic pathways so they can be used to produce ATP, lipids, and glucose.

The liver also transforms substances that cannot be used by cells into more readily usable substances. Ingested fats, for example, are combined with choline and phosphorus in the liver to produce phospholipids, which are essential components of cell membranes.

Many ingested substances are harmful to the cells of the body. In addition, the body itself produces many byproducts of metabolism that, if accumulated, are toxic. The liver is one line of defense against many of those harmful substances. It detoxifies many substances by altering their structure, making their excretion easier (see Table 16-2). For example, ammonia, a byproduct of amino acid metabolism, is toxic and is not readily removed from the circulation by the kidneys. The liver removes ammonia from the circulation and converts it to urea, which is then secreted into the circulation and eliminated by the kidneys in the urine. Other substances are removed from the circulation and are excreted by the liver into the bile.

The liver can also produce its own unique new compounds (see Table 16-2). Many of the blood proteins (for example, albumins, fibrinogen, globulins, and clotting factors) are produced by the liver and are released into the circulation.

Pancreas

The exocrine secretions of the pancreas include bicarbonate ions, which neutralize the acidic chyme that enters the small intestine from the stomach. The increased pH resulting from the secretion of bicarbonate ions stops pepsin digestion but provides the proper environment for the function of pancreatic enzymes. Pancreatic enzymes, also present in the exocrine secretions, are important for the digestion of all major classes of food (see Table 16-1). Without the enzymes produced by the pancreas, lipids, proteins, and carbohydrates are not adequately digested.

The major proteolytic enzyme is **trypsin,** which continues the protein digestion that started in the stomach. **Pancreatic amylase** (am′ĭ-lās) continues the polysaccharide digestion that was initiated in the oral cavity. The pancreatic enzymes also include a group of lipid-digesting enzymes called pancreatic **lipases** (li′pās-ez). **Nucleases,** enzymes that reduce DNA and RNA to their component nucleotides, are also among the pancreatic enzymes.

The exocrine secretory activity of the pancreas is controlled by both hormonal and neural mechanisms (see Figure 16-18). Secretin initiates the release of a watery pancreatic solution that contains a large amount of bicarbonate ions. The primary stimulus for secretin release is the presence of acidic chyme in the duodenum. Cholecystokinin stimulates the pancreas to release an enzyme-rich solution. The primary stimulus for cholecystokinin release is the presence of fatty acids and amino acids in the duodenum.

3 Explain why secretin production in response to acidic chyme and the stimulation of bicarbonate ion secretion by secretin constitute a negative feedback mechanism. ?

The presence of fatty acids and amino acids in the intestine, or parasympathetic stimulation through the vagus nerves, stimulates the secretion of pancreatic juices rich in pancreatic enzymes. Sympathetic impulses inhibit secretion.

Large Intestine

Normally 18 to 24 hours are required for material to pass through the large intestine in contrast to the 3 to 5 hours required for movement of chyme through the small intestine. While in the colon, chyme is converted to **feces.** Absorption of water and salts, the secretion of mucus, and extensive action of microorganisms are involved in the formation of feces, which the colon stores until they are eliminated by the process of **defecation.**

Numerous microorganisms inhabit the colon. They reproduce rapidly and ultimately compose about 30% of the dry weight of the feces. Some bacteria in the intestine synthesize vitamin K, which is passively absorbed in the colon, and break down a small amount of cellulose to glucose.

At widely spaced intervals (8 to 12 hours apart), large portions of the colon undergo several strong peristaltic contractions called **mass movements,** which propel its contents considerable distances toward the anus. These mass movements are very common following some meals, especially breakfast.

Distention of the rectum wall by feces acts as a stimulus that initiates the **defecation reflex.** Local reflexes cause weak contractions of the rectum and relaxation of the internal anal sphincter. Parasympathetic reflexes, integrated in the sacral portion of the spinal cord, cause stronger contractions of the rectum and are normally responsible for most of the defecation reflex. The external anal sphincter is composed of skeletal muscle and is under conscious cerebral control. If this sphincter is relaxed voluntarily, feces are expelled. Defecation is usually accompanied by voluntary movements that support the expulsion of feces. These voluntary movements include a large inspiration of air followed by closure of the larynx and forceful contraction of the abdominal muscles. As a consequence, the pressure in the abdominal cavity increases and forces the contents of the colon through the anal canal and out of the anus.

4 Explain how an enema stimulates defecation. ?

SUMMARY

The digestive system provides the body with water, electrolytes, and other nutrients.

ANATOMY AND HISTOLOGY OF THE DIGESTIVE SYSTEM

The gastrointestinal tract is composed of four tunics: mucosa, submucosa, muscularis, and serosa or adventitia.

Oral Cavity

The lips and cheeks are involved in facial expression, mastication, and speech.

The tongue is involved in speech, taste, mastication, and swallowing.

There are 32 permanent teeth, including incisors, canines, premolars, and molars. Each tooth consists of a crown, neck, and root.

The roof of the oral cavity is divided into the hard and soft palates.

Salivary glands produce serous and mucous secretions. The three pairs of large salivary glands are the parotid, submandibular, and sublingual glands.

Pharynx

The pharynx consists of the nasopharynx, oropharynx, and laryngopharynx.

Esophagus

The esophagus connects the pharynx to the stomach. The upper and lower esophageal sphincters regulate movement.

Stomach

There are cardiac (to the esophagus) and pyloric (to the duodenum) openings in the stomach.

The wall of the stomach consists of three muscle layers: longitudinal, circular, and oblique.

Gastric glands produce mucus, hydrochloric acid, pepsin, and gastrin.

Small Intestine

The small intestine is divided into the duodenum, jejunum, and ileum.

Circular folds, villi, and microvilli greatly increase the surface area of the intestinal lining.

Duodenal glands produce mucus.

Liver

The liver is the largest internal organ of the body.
Bile leaves the liver through the hepatic duct system.
The cystic duct from the gallbladder joins the hepatic duct from the liver to form the common bile duct.
The common bile duct joins the pancreatic duct where the common bile duct empties into the duodenum.

Pancreas

The pancreas is an endocrine and exocrine gland. Its endocrine function is to control blood nutrient levels. Its exocrine function is to produce bicarbonate ions and digestive enzymes.

Large Intestine

The cecum forms a blind sac at the junction of the small and large intestines. The appendix is a blind sac off the cecum.
The colon consists of ascending, transverse, descending, and sigmoid portions.
Smooth muscles of the large intestine wall are arranged into bands called teniae coli that contract to produce pouches called haustra.
The large intestine contains mucus-producing crypts.
The rectum is a straight tube that ends at the anus.

Anal Canal

The anal canal is surrounded by an internal anal sphincter (smooth muscle) and an external anal sphincter (skeletal muscle).

Peritoneum

The peritoneum is a serous membrane that lines the abdominal cavity and organs.
Mesenteries are peritoneum that extend from the body wall to many of the abdominal organs.
Retroperitoneal organs are located behind the peritoneum.

MOVEMENTS AND SECRETIONS IN THE DIGESTIVE SYSTEM

The digestive system is regulated by neural and hormonal mechanisms. Intramural plexuses are responsible for local reflexes.

Oral Cavity, Pharynx, and Esophagus

Amylase in saliva starts starch digestion. Mucin provides lubrication.
Mastication is accomplished by the teeth, which cut, tear, and crush the food.
During the voluntary phase of deglutition, a bolus of food is moved by the tongue from the oral cavity to the pharynx.
During the pharyngeal phase of deglutition, the soft palate closes the nasopharynx, and the epiglottis closes the opening into the larynx. Pharyngeal muscles move the bolus to the esophagus.
During the esophageal phase of deglutition, a wave of constriction (peristalsis) moves the food down the esophagus to the stomach.

Stomach

Secretions of the stomach.
- Mucus protects the stomach lining.
- Hydrochloric acid kills microorganisms and activates pepsin.
- Pepsin digests proteins.
- Intrinsic factor aids in vitamin B_{12} absorption.
- Gastrin helps regulate stomach secretions and movements.

Regulation of stomach secretions.
- During the cephalic phase, the stomach secretions are initiated by the sight, smell, taste, or thought of food.
- During the gastric phase, distention of the stomach also promotes secretion.
- During the intestinal phase, acidic chyme in the duodenum stimulates neuronal reflexes and the secretion of hormones that induce and then inhibit gastric secretions.

Movement in the stomach.
- Mixing waves mix the stomach contents with the stomach secretions to form chyme.
- Peristaltic waves move the chyme into the duodenum.

Small Intestine

Secretions of the small intestine.
- Mucus protects against digestive enzymes and stomach acids.
- Chemical or tactile irritation, vagal stimulation, and secretion stimulate intestinal secretion.

Movement in the small intestine.
- Segmental contractions occur over short distances and mix the intestinal contents.
- Peristaltic contractions occur the length of the intestine and propel chyme through the intestine.

Liver

The liver produces bile, which contains bile salts that emulsify fats.

The liver stores and processes nutrients, produces new molecules, and detoxifies molecules.

The liver produces blood proteins.

Pancreas

The pancreas produces bicarbonate ions and digestive enzymes.

Acidic chyme stimulates the release of a watery bicarbonate solution that neutralizes acidic chyme. Fatty acids and amino acids in the duodenum stimulate the release of pancreatic enzymes.

Large Intestine

It takes much longer for material to move through the large intestine than the small intestine.

In the colon, chyme is converted to feces.

Mass movements occur three to four times a day.

Defecation is the elimination of feces. Reflex activity moves feces through the internal anal sphincter. Voluntary activity regulates movement through the external anal sphincter.

CONTENT REVIEW

1. What are the functions of the digestive system?
2. What are the major layers or tunics of the digestive tract?
3. List the functions of the lips, cheeks, and tongue.
4. What are the deciduous and permanent teeth? Name the different kinds of teeth.
5. Describe the parts of a tooth. What are dentin, enamel, and pulp?
6. What are the hard and soft palates? What is the function of the palate?
7. Name and give the location of the three largest salivary glands.
8. Distinguish the three portions of the pharynx.
9. Where is the esophagus located?
10. Describe the parts of the stomach. How are the stomach muscles different from those in the esophagus?
11. What are gastric pits and gastric glands? Name the secretions they produce.
12. Name and describe the three parts of the small intestine.
13. What are circular folds, villi, and microvilli in the small intestine? What are their functions?
14. Describe the anatomy and location of the liver and pancreas. Describe their duct system.
15. Describe the parts of the large intestine. What are teniae coli, haustra, and crypts?
16. What are the peritoneum, mesenteries, and retroperitoneal organs?
17. What are the functions of saliva?
18. Describe the three phases of swallowing.
19. List the stomach secretions and give their functions.
20. Describe the three phases of stomach secretion.
21. What are the two kinds of stomach movements? What do they accomplish?

22. List the secretions of the small intestine and give their functions.
23. Describe the kinds of movements in the small intestine and explain what they accomplish.
24. Describe the functions of the liver.
25. Name the exocrine secretions of the pancreas. What are their functions?
26. How is chyme converted to feces?
27. Describe the defecation reflex.

CONCEPT REVIEW

1. While anesthetized, patients sometimes vomit. Given that the anesthetic eliminates the swallowing reflex, explain why vomiting when anesthetized can be dangerous.
2. Achlorhydria is a condition in which the stomach stops producing hydrochloric acid and other secretions. What effect would achlorhydria have on the digestive process?
3. Victor Worrystudent developed a duodenal ulcer during final examination week. Describe the possible reasons. Explain what habits could have caused the ulcer and what you would recommend to remedy the problem.
4. Many people have a bowel movement shortly after a meal, especially breakfast. Why does this occur?

ANSWERS TO  PREDICT QUESTIONS

1 *p. 417* Four. Each portion of the mesentery has two layers, with a layer of connective tissue in between. The mesentery is folded back on itself to form the greater omentum, with the omental bursa in between the folds.

2 *p. 420* It is important to close off the nasopharynx during swallowing so that food, and especially liquid, doesn't pass into the nasal cavity. If a person has an explosive burst of laughter while trying to swallow a liquid, the liquid may be explosively expelled from the mouth and even from the nose. Speaking requires that the epiglottis be elevated so that air can pass out of the larynx. If you do this while you are swallowing, the food, and especially liquid, may pass into the larynx, causing you to choke.

3 *p. 428* Secretin production in response to acidic chyme stimulates bicarbonate ion secretion, which neutralizes the acidic chyme. Thus, secretin prevents the acid levels in the chyme from becoming too great. This type of regulation is negative feedback.

4 *p. 429* Introducing fluid into the rectum by way of an enema causes distention of the rectum. Distention stimulates the defecation reflex.

17

NUTRITION AND METABOLISM

After reading this chapter you should be able to:

1 Define metabolism, anabolism, catabolism, and nutrition.
2 List the common vitamins and minerals, and give a function for each.
3 Define digestion, absorption, and transport.
4 Describe the digestion of and list the breakdown products of carbohydrates, lipids, and proteins.
5 Describe the movement of water in relation to intestinal osmotic pressure.
6 Define cellular metabolism.
7 Describe the basic steps in glycolysis, and name the principal products.
8 Describe the citric acid cycle and the electron-transport chain, and explain how ATP is produced in these processes.
9 Explain how 2 ATP molecules are produced in anaerobic respiration and 36 ATP molecules are produced in aerobic respiration from 1 molecule of glucose.
10 Describe the basic steps involved in using lipids and amino acids as an energy source.
11 Define metabolic rate.
12 Describe heat production and regulation in the body.

aerobic respiration Breakdown of glucose in the presence of oxygen to produce carbon dioxide, water, and 36 ATP molecules.

anaerobic respiration Breakdown of glucose in the absence of oxygen to produce lactic acid and two ATP molecules.

carbohydrate Organic molecule made up of one or more monosaccharides chemically bound together; sugars and starches.

citric acid cycle Series of chemical reactions in which citric acid (six-carbon molecule) is converted into a four-carbon molecule, carbon dioxide is formed, and energy is released as ATP.

electron-transport chain Series of energy transfer molecules in the inner mitochondrial membrane; they receive energy and use it in the formation of ATP and water.

glycolysis (gli-kol'ĭ-sis) [Gr. *glykys*, sweet + *lysis*, a loosening] Anaerobic process during which one glucose molecule is converted to two pyruvic acid molecules; a net of two ATP molecules is produced during glycolysis.

lipid [Gr. *lipos*, fat] Substance composed principally of carbon, oxygen, and hydrogen; generally soluble in nonpolar solvents; fats and cholesterol.

metabolic rate The total amount of energy produced and used by the body per unit of time.

metabolism (mĕ-tab'o-lizm) [Gr. *metabole*, change] Sum of the chemical changes that occur in tissues, consisting of the breakdown of food to produce energy (catabolism) and the buildup of molecules (anabolism, which requires energy).

mineral Inorganic nutrient necessary for normal metabolic functions.

nutrition Process by which nutrients are obtained and used in the body.

protein [Gr. *proteios*, primary] Large molecule consisting of long sequences of amino acids (polypeptides) linked by peptide bonds.

vitamin (vi'tah-min) [L. *vita*, life + *amine*, from ammonia] One of a group of organic substances, present in minute amounts in natural foods, that are essential to normal metabolism.

Energy is the ability to do work. All body functions require energy, and that energy is obtained from food. For cells to have the necessary energy to conduct their normal functions, food must be taken into the body, broken down, and absorbed through the digestive tract. The intake and initial breakdown of food is described in Chapter 16. The more complete breakdown of molecules, their absorption in the gastrointestinal tract, and their metabolism in cells are the topics of this chapter.

Metabolism (mĕ-tab′o-lizm; change) is the total of all the chemical reactions that occur in the body. It consists of **anabolism** (ah-nab′o-lizm), the energy-requiring process by which small molecules are joined to form larger molecules, and **catabolism** (kah-tab′o-lizm), the energy-releasing process by which large molecules are broken down into smaller molecules. Anabolism occurs in all cells of the body as they move, contract, divide to form new cells, produce molecules that perform specific functions within the cell, and produce molecules that are exported, such as hormones, neurotransmitters, or extracellular matrix molecules. Catabolism begins during the process of digestion and is concluded within individual cells, where the energy released by the breaking of chemical bonds is used to produce ATP and heat. The energy derived from catabolism is used to drive anabolic reactions.

Nutrition is the process by which food is taken into and used by the body. The process includes digestion, absorption, transport, and metabolism. During the process of nutrition, tissues are built up and energy is released. Nutrition can also be defined as the evaluation of food and drink requirements for normal body function.

NUTRIENTS

Nutrients are the chemicals taken into the body that provide energy and building blocks for new molecules. Some substances in food are not nutrients but provide bulk (fiber) in the diet. Nutrients can be divided into six major classes: carbohydrates, lipids, proteins, vitamins, minerals, and water. Carbohydrates, proteins, and lipids are the major organic nutrients and are broken down by enzymes into their individual subunits during digestion. Subsequently, many of these subunits are broken down further to supply energy. Others are used as building blocks for other molecules. Vitamins, minerals, and water are taken into the body without being digested. They are essential participants in the chemical reactions necessary to maintain life. Some nutrients are required in fairly substantial quantities, and others are required in minute amounts (trace elements).

A balanced diet consists of enough nutrients in the correct proportions to support normal body functions. A balanced intake of carbohydrates, lipids, proteins, vitamins, minerals, and water is necessary to sustain those normal functions.

Vitamins

Vitamins (vi′tah-minz; life-giving chemicals) are organic molecules that exist in minute quantities in food and are essential to normal metabolism (Table 17-1). **Essential vitamins** cannot be produced by the body and must be obtained through the diet. Because no single food item or nutrient class provides all the essential vitamins, it is necessary to maintain a balanced diet by eating a variety of foods. The absence of an essential vitamin in the diet can result in a specific deficiency disease. A few vitamins (for example, vitamin K) are produced by intestinal bacteria, and a few can be formed by the body from substances called **provitamins.** A provitamin is a portion of a vitamin that can be assembled or modified by the body into a functional vitamin. Carotene is an example of a provitamin that can be modified by the body to form vitamin A.

Vitamins are not broken down by catabolism but are used by the body in their original or slightly modified forms. After the chemical structure of a vitamin is destroyed, its function is usually lost. The chemical structure of many vitamins is destroyed by heat, such as when food is overcooked. Vitamins function as coenzymes (chemicals necessary for normal enzyme function), parts of coenzymes, or parts of enzymes in various metabolic reactions. Many vitamins are critical to the production of energy, whereas others (for example, folic acid and cyanocobalamin) are involved in nucleic acid synthesis. Retinol, thiamine, pyridoxine, cyanocobalamin, ascorbic acid (vitamin C), and vitamins D and E are necessary for growth. Vitamin K is necessary for the synthesis of proteins involved in blood clotting.

> **1 Predict what would happen if vitamins were broken down during the process of digestion rather than being absorbed intact into the circulation.** ?

TABLE 17-1

The Principal Vitamins

VITAMIN	SOURCE	FUNCTION	SYMPTOMS OF DEFICIENCY	MINIMUM DAILY REQUIREMENTS (mg)
WATER-SOLUBLE VITAMINS				
B_1 (thiamine)	Yeast, grains, and milk	Carbohydrate and protein metabolism; growth	Beriberi—muscle weakness, neuritis, and paralysis	1.6
B_2 (riboflavin)	Green vegetables, liver, wheat germ, milk, and eggs	Energy transport and citric acid cycle	Eye disorders and skin cracking at corners of mouth	1.8
Pantothenic acid (part of B_2 complex)	Green vegetables, liver, yeast, grains, and intestinal bacteria	Part of coenzyme A; glucose production	Neuromuscular dysfunction and fatigue	Unknown
B_3 (niacin)	Fish, liver, red meat, yeast, grains, peas, beans, and nuts	Energy transport, glycolysis, and citric acid cycle	Pellagra—diarrhea, dermatitis, and mental disturbance	20
B_6 (pyridoxine)	Fish, liver, yeast, tomatoes, and intestinal bacteria	Amino acid metabolism	Dermatitis, retarded growth, and nausea	Unknown
Folic acid	Green vegetables, liver, and intestinal bacteria	Nucleic acid synthesis and hematopoiesis	Anemia involving enlarged red blood cells	0.25
B_{12} (cobalamin)	Liver, red meat, milk, and eggs	Nucleic acid synthesis and hematopoiesis	Pernicious anemia and nervous system disorders	1.2
C (ascorbic acid)	Citrus fruit, tomatoes, and green vegetables	Collagen synthesis and protein metabolism	Scurvy—defective bone growth and poor wound healing	80
H (biotin)	Liver, yeast, eggs, and intestinal bacteria	Fatty acid and protein synthesis; movement of pyruvic acid into citric acid cycle	Mental and muscle dysfunction; fatigue and nausea	Unknown
FAT-SOLUBLE VITAMINS				
A (retinol)	From carotene (a provitamin) in vegetables	Vision, skin, bones, and teeth	Night blindness, retarded growth, and skin disorders	1.7
D (cholecalciferol)	Fish liver oil, enriched milk; provitamin D converted by sunlight to vitamin D	Calcium and phosphorus absorption; bone and teeth formation	Rickets—poorly developed, weak bones; bone resorption	11 (for children, or women during pregnancy)
E (alphatocopherol)	Wheat germ, cottonseed, palm, and rice oils; grain, liver, and lettuce	Prevents catabolism of certain fatty acids	Hemolysis of red blood cells and nerve destruction	Unknown
K (phylloquinone)	Alfalfa, liver, spinach, vegetable oils, cabbage, and intestinal bacteria	Synthesis of several clotting factors	Excessive bleeding resulting from retarded blood clotting	Unknown (70-140 μg recommended)

There are two major classes of vitamins—**fat soluble** and **water soluble.** Fat-soluble vitamins such as vitamins A, D, E, and K are absorbed from the intestine along with lipids. Some of them can be stored in the body for a long time. Because they can be stored, it is possible to accumulate an overdose of these vitamins in the body (hypervitaminosis) to the point of toxicity. Water-soluble vitamins such as the B complex vitamins and vitamin C are absorbed with water from the intestinal tract and remain in the body only a short time before being excreted.

Minerals

A number of inorganic nutrients, **minerals,** are necessary for normal metabolic functions. Minerals are taken into the body by themselves or in combination with organic molecules. They compose about 4% to 5% of the total body weight and are involved in a number of important functions, such as adding mechanical strength to bones, combining with organic molecules, or acting as coenzymes, buffers, or regulators of osmotic pressure. Table 17-2 lists some minerals and their functions.

TABLE 17-2

Important Minerals

MINERAL	FUNCTION	SYMPTOMS OF DEFICIENCY	MINIMUM DAILY REQUIREMENTS
Calcium (Ca)	Bone and teeth formation, blood clotting, and muscle and nerve function	Spontaneous nerve discharge and tetany	0.8-1.2 g
Chlorine (Cl)	Blood acid-base balance; HCl production in stomach	Acid-base imbalance	1.7-5.1 g
Cobalt (Co)	Part of vitamin B_{12}; erythrocyte production	Anemia	Unknown
Copper (Cu)	Hemoglobin production and electron-transport chain	Anemia and loss of energy	2.0-3.0 mg
Fluorine (F)	Extra strength in teeth and prevention of tooth decay	No real pathology	1.5-4.0 mg
Iodine (I)	Thyroid hormone production and maintenance of normal metabolic rate	Decrease in normal metabolism	150 mg
Iron (Fe)	Component of hemoglobin; ATP production in electron-transport system	Anemia, decreased oxygen transport, and energy loss	10-18 mg
Magnesium (Mg)	Coenzyme constituent; bone formation, and muscle and nerve function	Increased nervous system irritability, vasodilation, and arrhythmias	300-350 mg
Manganese (Mn)	Hemoglobin synthesis, growth, and activation of several enzymes	Tremors and convulsions	2.5-5.0 mg
Phosphorus (P)	Bone and teeth formation, ATP production, and part of nucleic acids	Loss of energy and cellular function	800-1200 mg
Potassium (K)	Muscle and nerve function	Muscle weakness and abnormal electrocardiogram	1.8-5.6 g
Sodium (Na)	Osmotic pressure regulation, and nerve and muscle function	Nausea, vomiting, exhaustion, and dizziness	1.1-3.3 g
Sulfur (S)	Component of proteins, vitamins, and hormones	Unknown	Unknown
Zinc (Zn)	Part of several enzymes; carbon dioxide transport; necessary for protein metabolism	Deficient carbon dioxide transport and deficient protein metabolism	15 mg

DIGESTION, ABSORPTION, AND TRANSPORT

Digestion is the chemical breakdown of organic molecules into their component parts: carbohydrates into monosaccharides, proteins into amino acids, and fats into fatty acids and glycerol (Figure 17-1). Digestive enzymes break the chemical bonds between the component parts.

Absorption begins in the stomach, where some very small molecules (for example, alcohol and aspirin) can pass through the stomach epithelium into the circulation. Most absorption occurs in the duodenum and jejunum, although some absorption occurs in the ileum. Some molecules can diffuse through the intestinal wall. Others must be transported across the intestinal wall. **Transport** requires a carrier molecule. If the transport is active, energy is required to move the transported molecule across the intestinal wall.

Carbohydrates

Ingested **carbohydrates** consist primarily of starches, glycogen, sucrose (table sugar), and small amounts of lactose (milk sugar) and fructose (fruit sugar). **Polysaccharides** are large carbohydrates consisting of many sugars linked by chemical bonds. Polysaccharides are broken down by enzymes. Salivary **amylase** (am'ĭ-lās) begins the digestion of carbohydrates in the mouth. The carbohydrates then pass to the stomach, where almost no carbohydrate digestion occurs. In the duodenum, pancreatic amylase continues the digestion of carbohydrates, and absorption begins. The amylases break down polysaccharides to **disaccharides** (two sugars chemically linked; see Chapter 16). A series of **disaccharidases** that are bound to the microvilli of the intestinal epithelium break down the disaccharides to monosaccharides.

The **monosaccharides** (single sugars) are taken up by the intestinal epithelial cells and are carried by the hepatic portal system to the liver, where different types of monosaccharides are converted to **glucose.** Glucose is carried by the circulation to the cells that require energy. Glucose enters the cells through facilitated diffusion. The rate of glucose transport into most types of cells is greatly influenced by **insulin** and may increase tenfold in the presence of insulin.

> In people with diabetes mellitus, insulin is lacking, and sufficient glucose is not transported into many cells of the body. As a result, the cells do not have enough energy for normal function, blood glucose levels become elevated, and large amounts of glucose are released into the urine.

Lipids

Lipids are molecules that are insoluble or only slightly soluble in water (see Chapter 2). They include triglycerides, phospholipids, steroids, and fat-soluble vitamins. **Triglycerides** (tri-glis'er-īdz), the most common type of lipid, consist of three fatty acids bound to glycerol. The first step in lipid digestion is **emulsification** (e-mul'sĭ-fĭ-ka'shun),

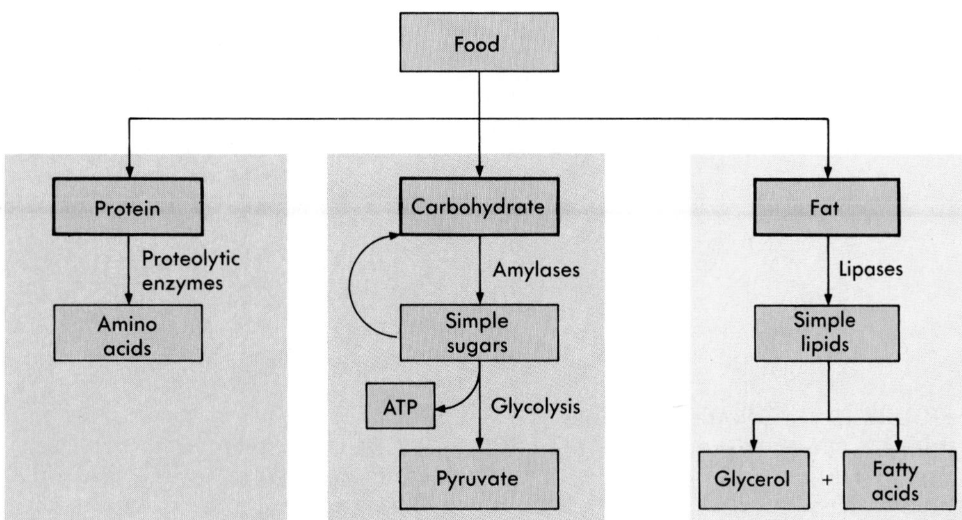

FIGURE 17-1 **Digestion of food molecules.**
Food consists primarily of protein, carbohydrate, and fat. Proteins are broken into amino acids, carbohydrates into monosaccharides, and fats into fatty acids and glycerol.

which is the transformation of large lipid droplets into much smaller droplets. The enzymes that digest lipids are soluble in water and can digest the lipids only by acting at the surface of the droplets. The emulsification process increases the surface area of the lipid exposed to the digestive enzymes by decreasing the droplet size. Emulsification is accomplished by **bile salts** secreted by the liver.

Lipase (li′pās) secreted by the pancreas digests, along with intestinal lipase, lipid molecules. The primary products of this digestive process are free fatty acids and glycerol. Cholesterol and phospholipids also are products of lipid digestion.

■ *Cystic fibrosis* is a hereditary disorder that results, in addition to other symptoms, in blockage of the pancreatic ducts so that the pancreatic digestive enzymes are prevented from reaching the duodenum. Thus fats, which can be digested only by these enzymes, are not digested. As a result, fats and fat-soluble vitamins are not absorbed, and the patient suffers from vitamin A, D, E, and K deficiencies. These deficiencies result in conditions such as night blindness, skin disorders, rickets, and excessive bleeding. Therapy consists of administering the missing vitamins to the patient and reducing dietary fat intake. ■

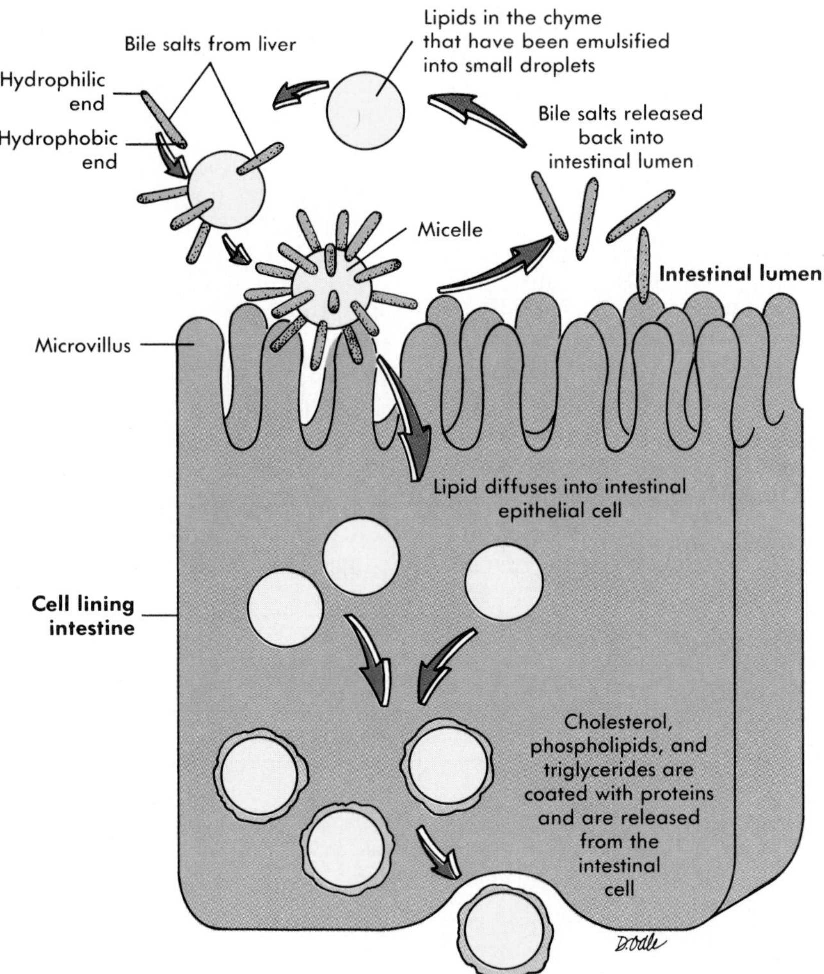

FIGURE 17-2 ■ Micelle formation.
Bile salts *(green rods)* produced in the liver, attach to and surround lipid droplets *(yellow balls)*. The hydrophobic ends of the bile salts are directed inward toward the lipid core, and the hydrophilic ends are directed outward toward the water environment of the digestive tract. When a micelle contacts the microvilli of an epithelial cell *(purple)*, the lipids diffuse into the cell, and the bile salts are released. Inside the cell, the lipids are coated with proteins and are released into the circulation.

In the intestine, bile salts aggregate around small droplets of digested lipids to form **micelles** (mi-sēlz'; a small morsel) (Figure 17-2). The hydrophobic (water-fearing) ends of the bile salts are directed toward the lipid particles, and the hydrophilic (water-loving) ends are directed outward toward the water environment. When a micelle comes into contact with the epithelial cells of the small intestine, the micelle's contents pass, by means of simple diffusion, through the lipid cell membrane of the epithelial cells.

The digested lipid droplets are packaged inside a protein coat within the epithelial cells of the intestinal villi. The packaged lipids leave the epithelial cells and enter the lacteals that run in the center of the villi (see Chapter 16). The lacteals are part of the lymphatic system, and lipid-rich lymph is called **chyle** (kīl; milky lymph). The lymphatic system carries the lipids to the bloodstream. Packaged lipids in the blood may be transported to adipose tissue, where they are stored until an energy source is needed elsewhere in the body. Lipids are also transported by the hepatic portal system to the liver, where they are stored, converted into other molecules, or used as energy.

■ *Cholesterol* levels in the blood are of great concern to many older adults. Cholesterol levels of less than 200 mg/dl (milligrams per deciliter) are considered low, which is good. Cholesterol levels of 200-239 mg/dl are considered borderline high if there are no other risk factors, such as coronary artery disease, cigarette smoking, hypertension, diabetes mellitus, or greater than 30% overweight. Cholesterol levels of 200-239 mg/dl are considered high if one or more of these other risk factors are present. Cholesterol levels of greater than 239 mg/dl are considered high in anyone. People with high blood cholesterol levels run a much greater risk of heart disease and stroke than people with low cholesterol levels. People with high levels should reduce intake of foods rich in cholesterol and other fats.

Fats are not soluble in water so they are transported in the blood as lipid-protein complexes, or *lipoproteins*. Low-density lipoproteins (LDLs) carry cholesterol to the tissues for use by the cells. When LDLs are in excess, cholesterol is deposited on arterial walls. High-density lipoproteins (HDLs), on the other hand, transport cholesterol from the tissues to the liver where cholesterol is removed from the bloodstream, and broken down or excreted in bile. A high HDL/LDL ratio in the bloodstream is related to a lower risk of heart disease. Aerobic exercise is one way to elevate the level of HDL. ■

Proteins

■ **Pepsin** is a **proteolytic** enzyme secreted by the stomach. It breaks down **proteins,** producing smaller polypeptide chains. Only about 10% to 20% of the total ingested protein is digested by pepsin. After the remaining proteins and polypeptide chains leave the stomach and enter the small intestine, the proteolytic enzyme, **trypsin,** produced by the pancreas continues the digestive process. Trypsin produces short amino acid chains, called peptides, that are broken down further into **amino acids** by **peptidases.** Peptidases are digestive enzymes bound to the microvilli of the small intestine, which act on smaller peptides to release amino acids.

Absorption of individual amino acids occurs through the intestinal epithelial cells by active transport. The amino acids then leave the epithelial cells and enter the hepatic portal system, which transports them to the liver. The amino acids may be modified in the liver, or they may be released into the bloodstream and distributed throughout the body.

Amino acids are actively transported into the various cells of the body. This transport is stimulated by growth hormone and insulin. Most amino acids are used as building blocks to form new proteins, but some amino acids may be used for energy. The body cannot store amino acids, so they are partially broken down to their component molecules. Those molecules are used to synthesize glycogen or fat, which can be stored.

Water and Minerals

■ Water can move in either direction across the wall of the small intestine. The direction of its movement is determined by osmotic gradients across the epithelium. When the chyme is dilute, water moves out of the intestine into the blood. If the chyme is concentrated and contains little water, water moves into the lumen of the small intestine. Nearly 90% of the water that enters the small intestine either by way of stomach or intestinal secretions is reabsorbed as chyme moves through the small and large intestines.

The minerals sodium, potassium, calcium, magnesium, and phosphate are actively transported from the small intestine. Vitamin D is required for the transport of calcium. Negatively charged chloride ions move passively through the wall of the duodenum and jejunum with the positively charged sodium ions, but chloride ions are actively transported from the ilium.

CELLULAR METABOLISM

Metabolism can be divided into the chemical changes that occur during digestion and the chemical processes that occur after the products of digestion are taken up by cells. The chemical processes that occur within cells are often referred to as **cellular metabolism.** The digestive products of carbohydrates, proteins, and lipids taken into cells are further catabolized. The released energy is used to combine **adenosine diphosphate (ADP)** and an inorganic phosphate group (P_i) to form **adenosine triphosphate** or **ATP** (Figure 17-3).

ATP is often called the energy currency of the cell. When the terminal phosphate is cleaved from ATP, and the ATP is converted back to ADP, the released energy can be used to drive chemical reactions such as those involved in active transport, muscle contraction, and synthesis of molecules.

When food is digested, energy is released by **oxidation** from the chemical bonds within the nutrient molecules. The energy that is released is used to synthesize ATP. In some of the metabolic steps, ATP can be produced directly from the released energy, but in most cases the energy must be transported by carrier molecules (such as **nicotinamide adenine dinucleotide; NADH**) to the electron-transport chain (described later in this chapter), where the energy is used to produce ATP.

Monosaccharides

Monosaccharides are the breakdown products of carbohydrate digestion. Glucose is the most important of the monosaccharides as far as cellular metabolism is concerned. Glucose is transported in the circulation to all tissues of the body where it is used to produce energy. If there is excess glucose in the blood following a meal, it can be used to form **glycogen,** or it can be partially broken down and the components used to form fat. Glycogen is a short-term energy storage molecule, and fat is a long-term energy storage molecule. Most of the body's glycogen is in skeletal muscle and in the liver.

Glycolysis

Glycolysis (gli-kol'ĭ-sis) (Figure 17-4) is a series of chemical reactions that occurs in the cytoplasm of most cells and results in the breakdown of glucose (which has six carbon atoms) to two **pyruvic acid** molecules. Two ATP molecules are required to start the process, and four ATP molecules are produced during the process, for a net gain of two ATP molecules. In the first step of glycolysis the glucose molecule is modified by the addition of two phosphate groups from two ATP molecules. This modified glucose molecule is then cleaved into two three-carbon molecules, each with a phosphate

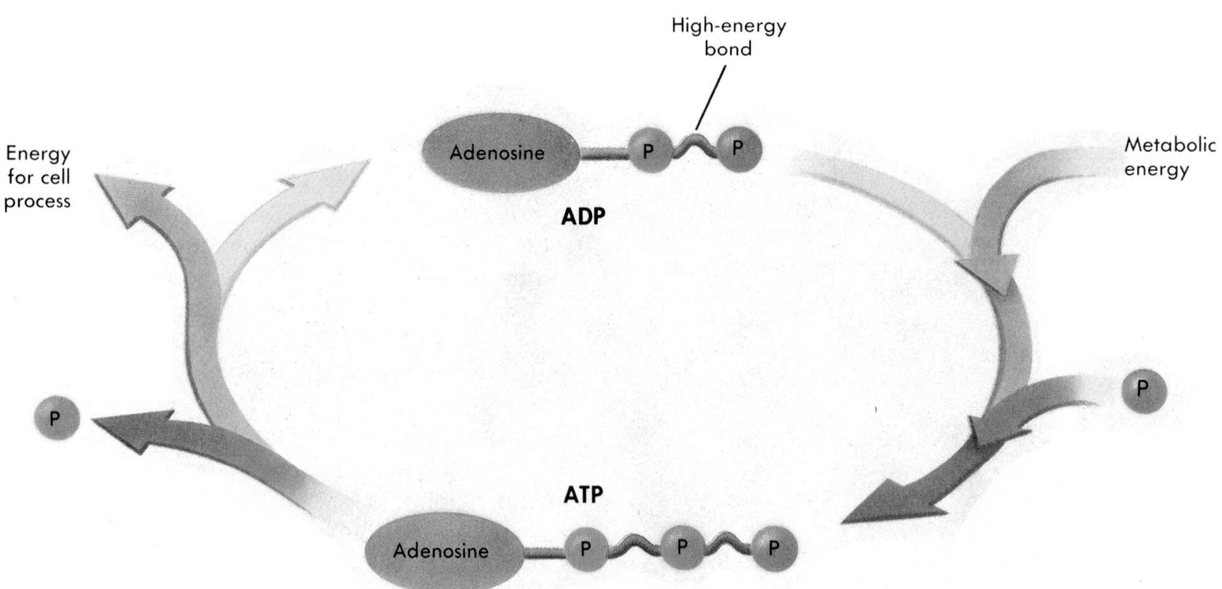

FIGURE 17-3 The interconversion of ADP and ATP.
Energy from metabolism and phosphate (P) are required to form ATP from ADP. Energy and a phosphate are given off when ATP is converted back to ADP. The wavy bars represent high-energy bonds.

group. Each three-carbon molecule is then oxidized, and the released energy can be used in other metabolic pathways.

If the cell has adequate amounts of oxygen, the pyruvic acid and energy produced in glycolysis will be used in aerobic respiration to produce additional ATP. In the absence of oxygen they will be processed by anaerobic respiration.

Anaerobic respiration

Anaerobic respiration (Figure 17-5) is the breakdown of glucose in the absence of oxygen to produce two molecules of **lactic acid** and two molecules of ATP. The ATP thus produced is a source of energy during activities such as intense exercise when insufficient oxygen is delivered to tissues. The first phase of anaerobic respiration is glycoly-

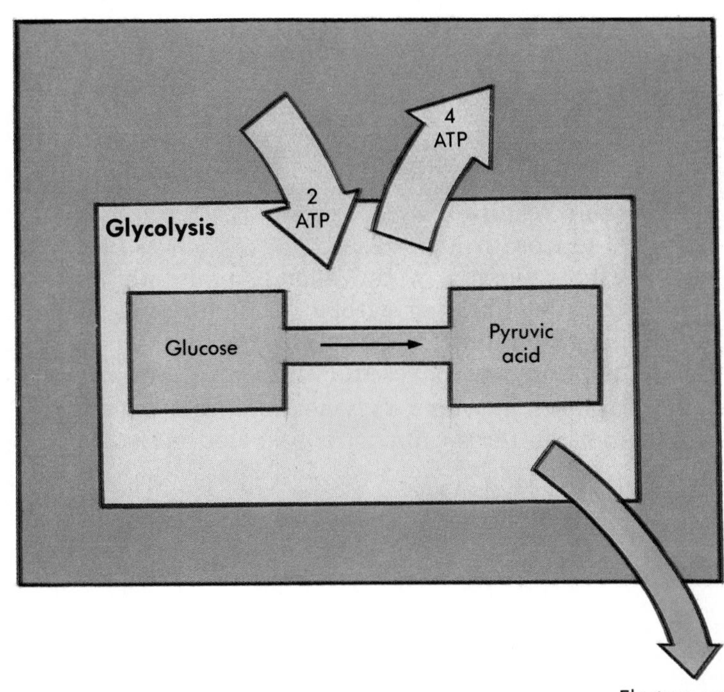

FIGURE 17-4 Glycolysis.
Through the process of glycolysis, one glucose molecule is converted to two molecules of pyruvic acid. Two molecules of ATP are required to start the process, and four molecules of ATP are produced, for a net gain of two ATP molecules. Electrons are also released in the process and are stored by electron carrier molecules.

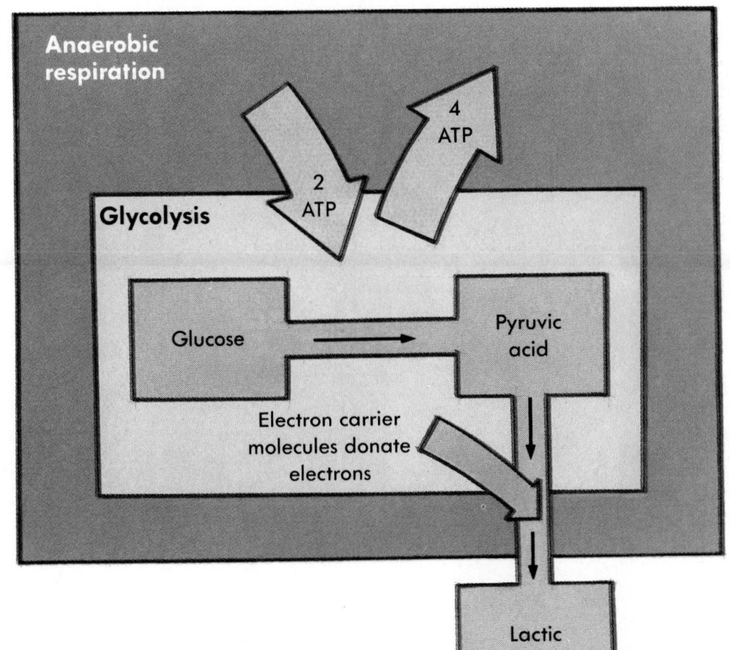

FIGURE 17-5 Anaerobic respiration.
In the absence of oxygen, the pyruvic acid produced in glycolysis is converted to lactic acid. This conversion requires energy, which can be obtained from the electrons on carrier molecules generated in glycolysis.

sis. The second phase is the conversion of pyruvic acid to lactic acid, a reaction that requires the input of energy.

Lactic acid is released from the cells that produce it and is transported by the blood to the liver. When oxygen becomes available, the lactic acid in the liver can be converted through a series of chemical reactions into glucose. The glucose then can be released from the liver and transported in the blood to cells that use glucose as an energy source. Some of the reactions that convert lactic acid into glucose require the input of ATP (energy) produced by aerobic respiration. The oxygen necessary for the synthesis of glucose from lactic acid is called the **oxygen debt** (see Chapter 7).

Aerobic respiration

■ **Aerobic respiration** (Figure 17-6) is the breakdown of glucose in the presence of oxygen to produce carbon dioxide, water, and 36 molecules of ATP. The first phase of aerobic respiration, as in anaerobic respiration, is glycolysis. In the second phase, pyruvic acid moves from the cytoplasm into a mitochondrion, where enzymes remove a carbon atom from the three-carbon pyruvic acid molecule to form carbon dioxide and a two-carbon acetyl group. Energy is released in the reaction, which may be used later to produce ATP. The acetyl group combines with coenzyme A (CoA), derived from part of the vitamin B_2 complex, to form **acetyl CoA.** In the third phase, acetyl CoA combines with a four-carbon molecule to form a six-carbon citric acid molecule, which enters the citric acid cycle.

The **citric acid cycle** (also called the Krebs cycle) (Figure 17-7) is a series of reactions in which citric acid is converted back into a four-carbon molecule. The four-carbon molecule then can combine with another acetyl CoA to reinitiate the cycle. During the conversion, two carbon atoms are lost as carbon dioxide, and energy is released. The energy can be transported to the electron-transport chain and used to produce ATP. The carbon dioxide is transported to the lungs where it is expired.

The **electron-transport chain** (Figure 17-8) is a series (chain) of energy-relay molecules attached to the inner mitochondrial membrane, in which electrons are transferred from one molecule to the next. Energy, released during glycolysis, during the formation of the two-carbon acetyl group, and during the citric acid cycle, and transported by carrier molecules to the electron-transport chain is used to

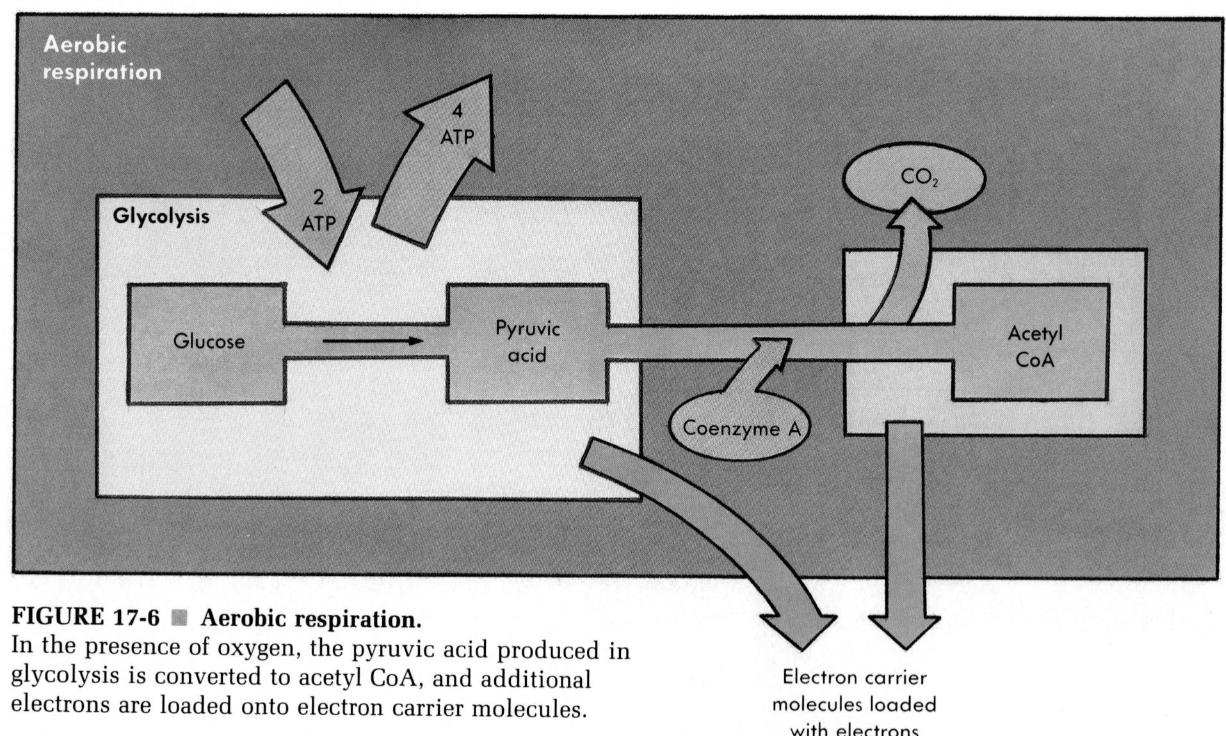

FIGURE 17-6 ■ **Aerobic respiration.**
In the presence of oxygen, the pyruvic acid produced in glycolysis is converted to acetyl CoA, and additional electrons are loaded onto electron carrier molecules.

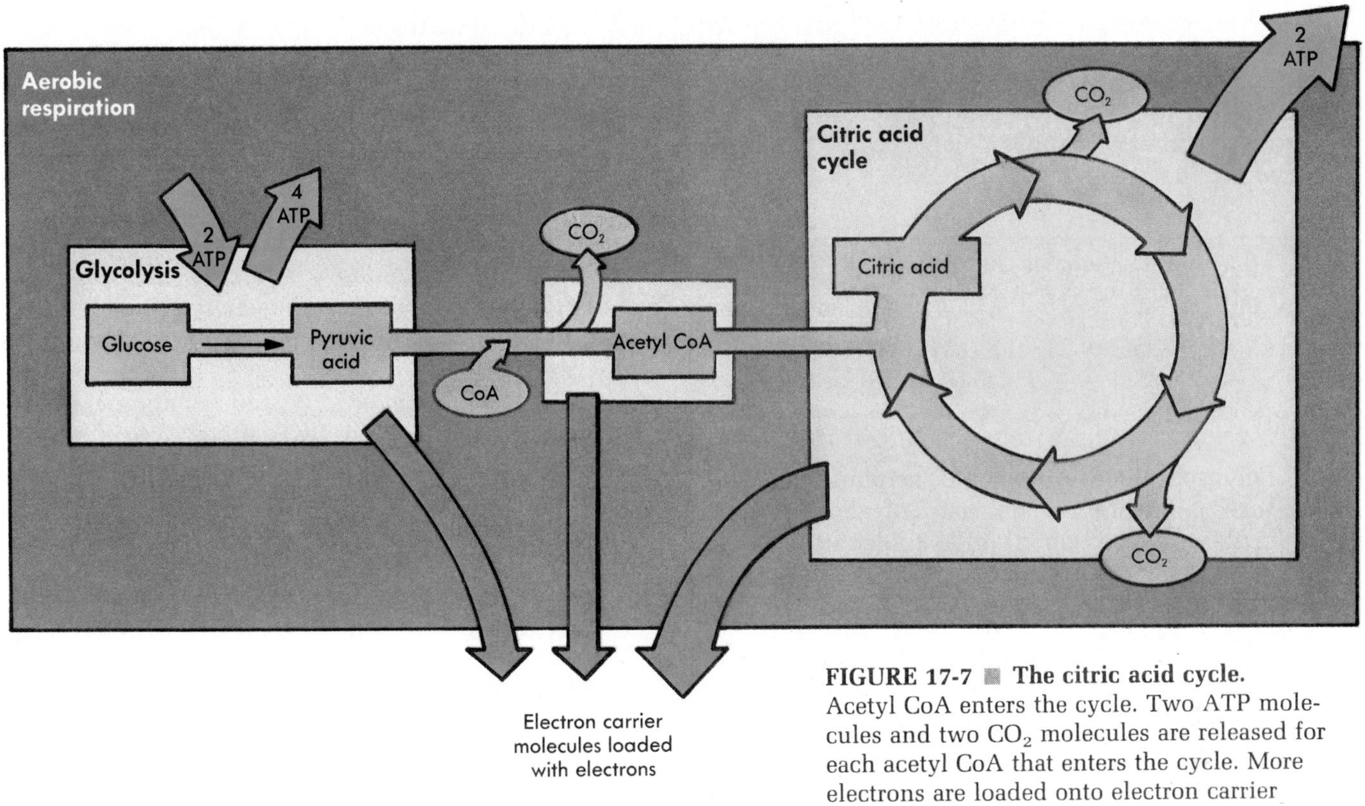

Aerobic respiration

Glycolysis

2 ATP

4 ATP

Glucose

Pyruvic acid

CoA

CO_2

Acetyl CoA

Citric acid cycle

Citric acid

CO_2

2 ATP

CO_2

Electron carrier molecules loaded with electrons

FIGURE 17-7 ■ The citric acid cycle. Acetyl CoA enters the cycle. Two ATP molecules and two CO_2 molecules are released for each acetyl CoA that enters the cycle. More electrons are loaded onto electron carrier molecules.

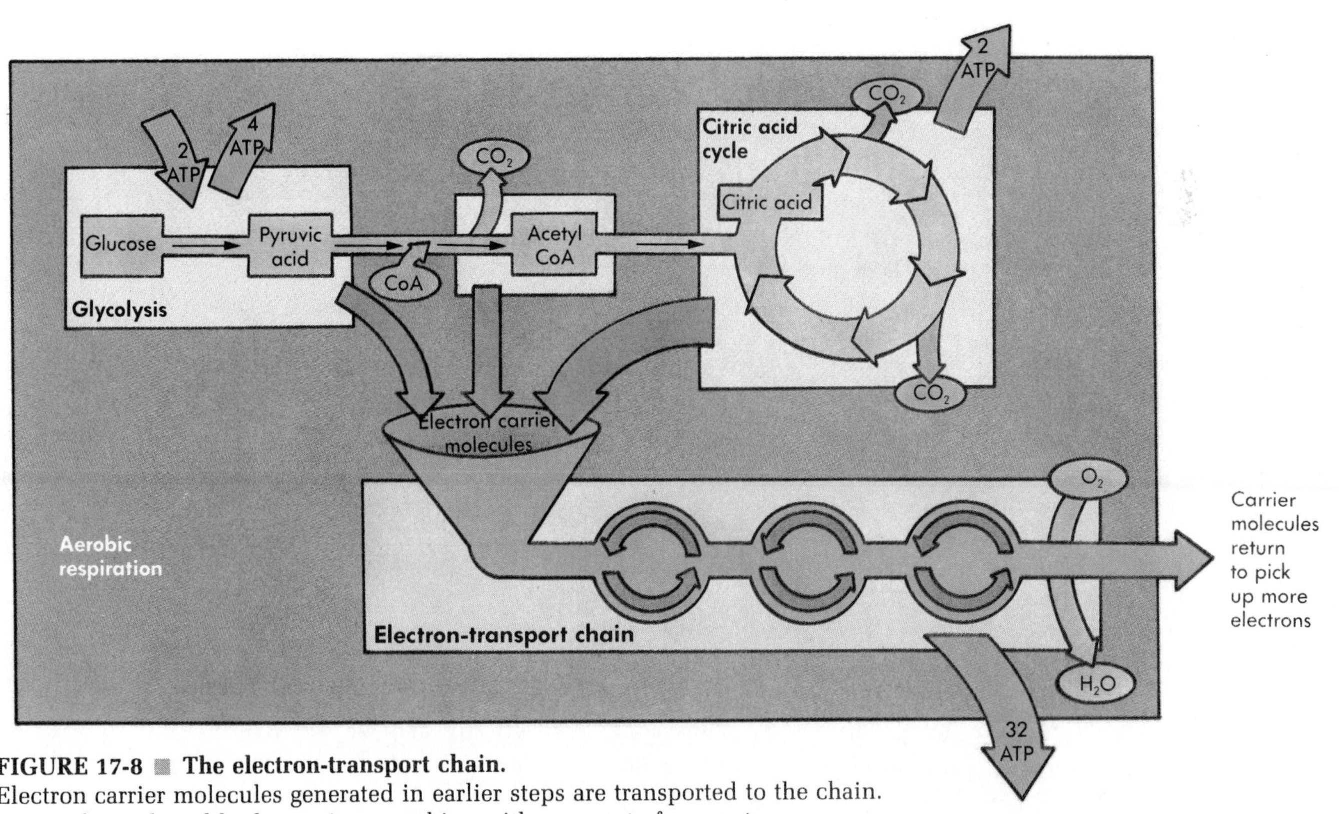

FIGURE 17-8 ■ The electron-transport chain.
Electron carrier molecules generated in earlier steps are transported to the chain. ATP is formed, and hydrogen ions combine with oxygen to form water.

convert ADP to ATP. In the last step of the electron-transport chain, hydrogen ions combine with oxygen to form water. Without **oxygen** to accept the hydrogen ions, the electron-transport chain (and thus aerobic respiration) stops. That is why it is necessary for us to breathe.

> **2 Many poisons function by blocking certain steps in the metabolic pathways. For example, cyanide blocks the last step in the electron-transport chain. Explain why this blockage would cause death.** ?

For each glucose molecule, aerobic respiration produces a theoretical net gain of about 36 ATP molecules. Six carbon dioxide molecules and six molecules of water are also produced in aerobic respiration.

> Even though the theoretical optimum yield for aerobic respiration is 36 molecules of ATP, the body is not a perfect machine. The actual yield has been estimated to be somewhat lower than 36 molecules of ATP per molecule of glucose.

Fatty Acids

Lipids are the body's main energy storage molecules. In a normal person, lipids are responsible for about 99% of the body's energy storage, and glycogen accounts for about 1%.

Between meals, when lipids are broken down in adipose tissue, some of the fatty acids produced are released into the blood where they are called **free fatty acids.** Other tissues, especially skeletal muscle and the liver, use the free fatty acids as a source of energy.

The metabolism of **fatty acids** occurs by a series of reactions in which two carbon atoms are removed from the end of a fatty acid chain to form acetyl CoA (Figure 17-9). The process continues, and two carbon atoms are removed at a time until the entire fatty acid chain is converted into acetyl CoA. Acetyl CoA can enter the citric acid cycle and be used to generate ATP. In the liver, two acetyl CoA molecules can also combine to form **ketone bodies.** The ketone bodies are released into the blood and travel to other tissues, especially skeletal muscle. In these tissues, the ketone bodies are converted back to acetyl CoA, which enters the citric acid cycle to produce ATP.

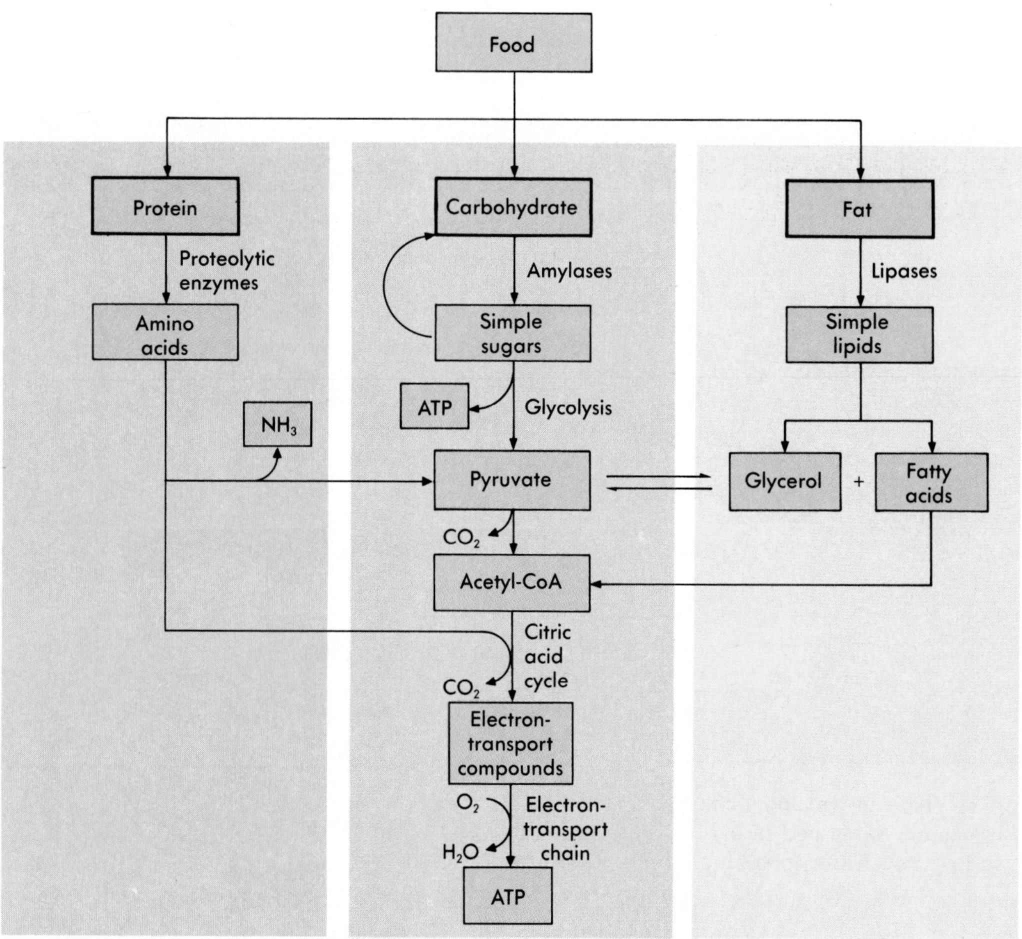

FIGURE 17-9 The overall pathways for the metabolism of food. Carbohydrates, fats, and proteins enter metabolic pathways to produce energy in the cell.

Starvation and Obesity

Starvation is the inadequate intake of nutrients or the inability to metabolize or absorb nutrients. Starvation can result from a number of causes, such as prolonged fasting, anorexia, deprivation, or disease. No matter what the cause, starvation takes about the same course and consists of three phases. The events of the first two phases occur even during relatively short periods of fasting or dieting. The third phase occurs in prolonged starvation and ends in death.

During the first phase of starvation, blood glucose levels are maintained through the production of glucose from glycogen, proteins, and fats. At first glycogen is broken down into glucose. Enough glycogen is stored in the liver to last only a few hours, however. Thereafter, blood glucose levels are maintained by the breakdown of proteins and fats. Fats are decomposed into fatty acids and glycerol. Fatty acids can be used as a source of energy, especially by skeletal muscle, thus decreasing the use of glucose by tissues other than the brain. The brain cannot use fatty acids as an energy source, so the conservation of glucose is critical to normal brain function. Glycerol can be used to make a small amount of glucose, but most of the glucose is formed from the amino acids of proteins. In addition, some amino acids can be used directly for energy.

In the second stage, which can last for several weeks, fats are the primary energy source. The liver metabolizes fatty acids into ketone bodies that can be used as a source of energy. After about a week of fasting, the brain begins to use ketone bodies as well as glucose for energy. This usage decreases the demand for glucose, and the rate of protein breakdown diminishes but does not stop. In addition, there is a selective use of proteins; those proteins not essential for survival are used first.

The third stage of starvation begins when the fat reserves are nearly depleted, and there is a switch to proteins as the major energy source. Muscles, the largest source of protein in the body, are rapidly depleted. At the end of this stage, proteins essential for cellular functions are broken down, and cell function degenerates. Death can occur very rapidly.

Obesity, the storage of excess fat, results from the ingestion of more food than is necessary for the body's energy needs. In most cases of obesity no specific reason can be detected for the condition, but hypothalamic defects such as tumors can result in obesity. If the balance between the hunger and satiety centers in the hypothalamus is improper, a person may have a constant or prolonged urge to eat (bulimia). Emotional stress also may cause some people to overeat.

In **hypertrophic obesity** (also called adult-onset obesity) people who were thin or of average weight and quite active when young become less active as they become older. They gain weight at age 20 to 40, and, although they no longer use as many calories, they still consume the same amount of food as when they were younger. The excess calories (see discussion of metabolic rate) are used to synthesize fat. In this type of obesity, the amount of fat in each adipocyte (fat cell) increases, but the total number of adipocytes does not increase.

In **hyperplastic obesity** (occurring early in life) the total number of adipocytes increases. People with hyperplastic obesity are obese as children and become more obese with age. This type of obesity is a major health problem in school-aged children.

Amino Acids

Once absorbed into the body, **amino acids,** the products of protein digestion, are quickly taken up by cells, especially in the liver. Amino acids are used primarily to synthesize needed proteins and only secondarily as a source of energy. If amino acids are used as a source of energy, an amine group (NH_2) is removed from the amino acid, leaving ammonia and a keto acid. In the process, energy is released, which can enter the electron-transport chain to produce ATP (see Figure 17-9). Ammonia is toxic to cells, and it is converted by the liver into urea, which is carried by the blood to the kidneys, where it is eliminated. The keto acid can enter the citric acid cycle or can be converted into pyruvic acid, acetyl CoA, or glucose. Although proteins can be used as an energy source, they are not considered storage molecules because the breakdown of proteins normally involves the loss of necessary tissue.

METABOLIC RATE

The energy available in foods and released through metabolism is expressed as a measure of heat called **calories** (cal). A calorie is the amount of heat (energy) required to raise the temperature of 1000 grams of water from 14° C to 15° C. Calories are used to express the large amounts of energy involved in metabolism, and the calories in a serving of food are commonly listed on food packages. For example, one slice of white bread contains about 75 cal, one cup of whole milk contains 150 cal, a banana contains 100 cal, a hot dog contains 170 cal (not counting the bun and dressings), a McDonald's Big Mac has 563 cal, and a soft drink adds another 145 cal. For each gram of carbohydrate or protein metabolized by the body, about 4 cal of energy is released. Fats contain more energy per unit of weight than carbohydrates and proteins, and yield about 9 cal per gram. A typical diet in the United States consists of 50% to 60% carbohydrates, 35% to 45% fats, and 10% to 15% protein.

The **metabolic rate** is the total amount of energy produced and used by the body per unit of time. Metabolic rate is usually estimated by measuring the amount of oxygen used per minute. One liter of oxygen consumed by the body is assumed to produce 4.825 cal of energy.

Metabolic energy can be used in three ways: for basal metabolism, for muscle contraction, and for the assimilation of food (production of digestive enzymes, active transport, and so forth). The **basal metabolic rate** (BMR) is the metabolic rate calculated in expended calories per square meter of body surface area per hour, and is measured when a person is awake but restful and has not eaten for 12 hours. A typical BMR for a 70-kilogram (154-pound) man would be 38 cal/m^2/hr.

BMR is the energy needed to keep the resting body functional. Active transport mechanisms, muscle tone, maintenance of body temperature, beating of the heart, and other activities are supported by basal metabolism. A number of factors can affect the BMR. Males have a greater BMR than females, younger people have a higher BMR than older people, and fever can increase BMR. Greatly reduced caloric input (for example, during dieting or fasting) depresses BMR.

The daily input of energy should equal the energy demand of metabolism; otherwise, a person will gain or lose weight. For a 23-year-old, 70-kilogram (154-pound) man to maintain his weight, the input should be 2700 cal per day; for a 58-kilogram (128-pound) woman of the same age, 2000 cal per day are necessary. For every 3500 cal above the energy requirement (not necessarily in one day), a pound of body fat can be gained, whereas for every 3500 cal below the requirement (usually over several days), a pound of fat can be lost. Clearly, adjusting caloric input is an important way to control body weight.

The other way to control weight is through energy expenditure. Physical activity through skeletal muscle movement greatly increases metabolic rates. In the average person, basal metabolism accounts for about 60% of energy expenditure, muscular activity 30%, and assimilation of food about 10%. Of these amounts, energy loss through muscular activity is the only component that a person reasonably can control. A comparison of the number of calories gained from food and the number of calories burned during exercise reveals why losing weight can be such a difficult task. For example, it takes 20 minutes of brisk walking to burn off the calories in one slice of bread (about 75 cal).

HEAT PRODUCTION AND REGULATION

Free energy is the total amount of energy that can be liberated by the complete catabolism of food. Only a portion (about 43%) of the total energy released by catabolism is used to accomplish biological work such as anabolism, muscular contraction, and other cellular activities. The remaining energy is lost as heat.

> **3 Explain why we become warm during exercise and why we shiver when it is cold.** ?

Humans are **homeotherms** (ho'me-o-thermz; uniform warming), or **warm-blooded** animals, and we can maintain a constant body temperature even though the environmental temperature varies. Maintenance of a constant body temperature is very important to homeostasis. Most enzymes are very temperature sensitive and function only within narrow temperature ranges. Environmental temperatures are too low for normal enzyme function, and the heat produced by metabolism and muscle contraction helps maintain the body temperature at a steady, elevated level that is high enough for normal enzyme function. Excessively high temperatures can alter enzyme structure, resulting in the loss of the enzyme's function.

Normal body temperature is a range like any other homeostatically controlled condition in the body. The average normal temperature is usually considered to be 37° C (98.6° F) when it is measured orally, and 37.6° C (99.7° F) when it is measured rectally. Rectal temperature comes closer to the true core body temperature, but an oral temperature is more easily obtained in older children and adults, and therefore is the preferred measure.

Body temperature is maintained by balancing heat input with heat loss (Figure 17-10). Heat may be exchanged with the environment in a number of ways. **Radiation** is the loss of heat as infrared energy, a type of electromagnetic radiation. For example, the coals in a fire give off radiant heat that can be felt some distance away from the fire. **Conduction** is the exchange of heat between objects that are in direct contact with each other (for example, the bottom of the feet and the floor). **Convection** is a transfer of heat between the body and the air. A cool breeze results in movement of air over the body and loss of heat from the body. **Evaporation** is the loss of water from the body; the water carries heat away with it.

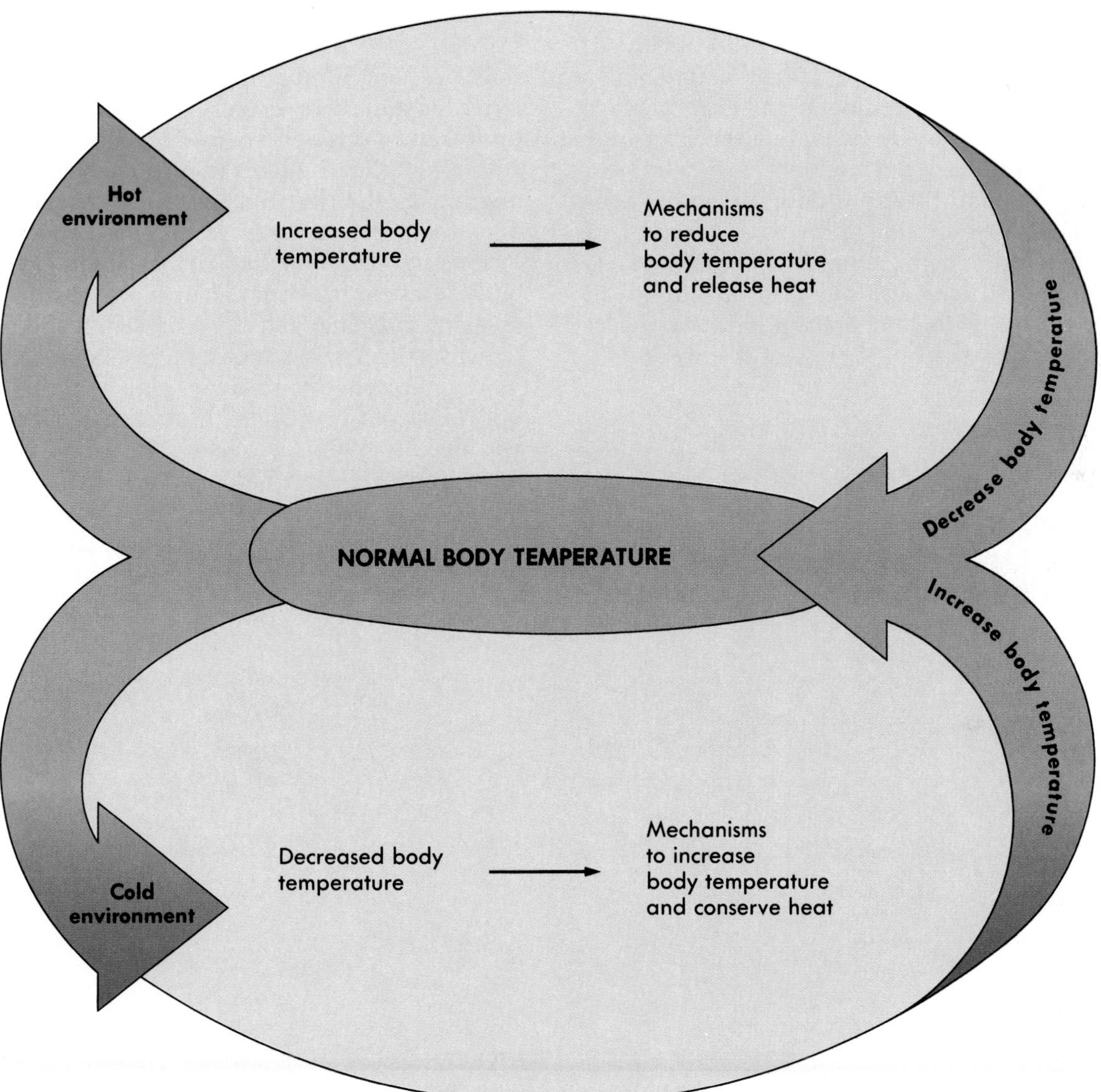

FIGURE 17-10 ■ Temperature regulation.
When environmental factors *(arrows on left side of figure)* move the body temperature out of its normal range, regulatory mechanisms *(arrows on right side of figure)* tend to bring the temperature back to its normal range.

The amount of heat exchanged between the environment and the body is determined by the difference in temperature between the body and the environment. The greater the temperature difference, the greater is the rate of heat exchange. Control of the temperature difference can be used to regulate body temperature. If environmental temperature is very cold (for example, on a winter day), there is a large temperature difference between the body and the environment, and there is a large loss of heat. The loss of heat can be decreased by behaviorally selecting a warmer environment (for example, inside a heated house) or by insulating the exchange surface (for example, putting on extra clothes). Physiologically, temperature difference can be controlled through vasodilation and vasoconstriction of blood vessels in the skin. When these blood vessels vasodilate, they bring warm blood to the surface of the body, raising skin temperature, whereas vasoconstriction decreases blood flow and lowers skin temperature.

4 Explain why vasoconstriction of skin blood vessels on a cold winter day is beneficial. ?

When environmental temperature is greater than body temperature, vasodilation brings blood to the skin, causing an increase in skin temperature that decreases the gain of heat from the environment. At the same time, evaporation carries away excess heat to prevent heat gain and overheating.

Body temperature regulation is an example of a negative-feedback system (see Figure 17-10). Maintenance of a specific body temperature is accomplished by a "set point" located in the hypothalamus. A small area in the anterior part of the hypothalamus can detect slight increases in body temperature through changes in blood temperature. As a result, mechanisms that cause heat loss (for example, vasodilation and sweating) are activated, and body temperature decreases. A small area in the posterior hypothalamus can detect slight decreases in body temperature and can initiate heat gain by increasing muscular activity (shivering) and by initiating vasoconstriction of blood vessels in the skin.

Under some conditions, the hypothalamus set point is actually changed. For example, during a fever the set point is raised. Heat-conserving and heat-producing mechanisms are stimulated, and body temperature increases. In recovery from a fever, the set point is reduced to normal. Heat-loss mechanisms are initiated, and body temperature decreases.

SUMMARY

Metabolism consists of anabolism and catabolism. Anabolism is the building up of molecules and requires energy. Catabolism is the breaking down of molecules and gives off energy.

Nutrition is the taking in and the use of food.

NUTRIENTS

Nutrients are the chemicals used by the body and consist of carbohydrates, lipids, proteins, vitamins, minerals, and water.

Vitamins

Most vitamins are not produced by the body and must be obtained in the diet. Some vitamins can be formed from provitamins.

Vitamins are important in energy production, nucleic acid synthesis, growth, and blood clotting.

Vitamins are classified as either fat soluble or water soluble.

Minerals

Minerals are necessary for normal metabolism and add mechanical strength to bones.

DIGESTION, ABSORPTION, AND TRANSPORT

Digestion is the chemical breakdown of organic molecules into their component parts. After the molecules are digested, they diffuse through the intestinal wall. Some molecules must be transported across the intestinal wall.

Carbohydrates

Polysaccharides are split into disaccharides by salivary and pancreatic amylases.

Disaccharides are broken down to monosaccharides by disaccharidases on the surface of the intestinal epithelium.

Monosaccharides are absorbed in the blood and are carried to the liver.

Glucose is carried in the blood and enters cells by facilitated diffusion. Insulin increases the rate of glucose transport.

Lipids

Bile salts emulsify lipids.

Pancreatic lipase breaks down lipids. The breakdown products aggregate with bile salts to form micelles.

Micelles come into contact with the intestinal epithelium, and their contents diffuse into the cells, where they are packaged and released into the lymph and blood.

Lipids are stored in adipose tissue and in the liver, which release the lipids into the blood when energy sources are needed elsewhere in the body.

Proteins

Proteins are split into small polypeptides by enzymes secreted by the stomach (pepsin), the pancreas, and the intestine.

Peptidases on the surface of intestinal epithelial cells complete the digestive process.

Amino acids are absorbed into intestinal epithelial cells.

Amino acids are actively transported into cells under the influence of growth hormone and insulin.

Amino acids are used to build new proteins or are used as a source of energy.

Water and Minerals

Water can move either direction across the intestinal wall, depending on osmotic conditions.

Most minerals are actively transported across the intestinal wall.

CELLULAR METABOLISM The energy in carbohydrates, lipids, and proteins is used to produce ATP.

Monosaccharides

Glycolysis is the breakdown of glucose to two pyruvic acid molecules. Two ATP molecules are also produced.

Anaerobic respiration is the breakdown of glucose in the absence of oxygen to two lactic acid molecules and two ATP molecules.

Lactic acid can be converted to glucose using aerobically produced ATP; the necessary oxygen is the oxygen debt.

Aerobic respiration is the breakdown of glucose in the presence of oxygen to produce carbon dioxide, water, and 36 molecules of ATP.
- The first phase is glycolysis.
- The second phase is the conversion of pyruvic acid to acetyl CoA.
- The third phase is the citric acid cycle.

During aerobic respiration energy is produced, which can enter the electron-transport chain and can be used in the synthesis of ATP.

Fatty Acids

Lipids are broken down in adipose tissue and released as free fatty acids into the blood.

Free fatty acids are taken up by cells and broken down into acetyl CoA, which can enter the citric acid cycle.

Amino Acids

Amino acids are used to synthesize proteins.

Amino acids can be used for energy, and ammonia is produced as a byproduct. Ammonia is converted to urea and is excreted by the kidneys.

METABOLIC RATE

A calorie (cal) is the energy required to raise the temperature of 1000 grams of water from 14° C to 15° C. A calorie is the unit of measurement used to express the energy content of food.

Metabolic rate is the total energy expenditure per unit of time. Metabolic energy is used for basal metabolism, muscular activity, and the assimilation of food.

HEAT PRODUCTION AND REGULATION

Body temperature is a balance between heat gain and heat loss.
- Heat is produced through metabolism.
- Heat is exchanged through radiation, conduction, convection, and evaporation.

The greater the temperature difference, the greater is the rate of heat exchange.

Body temperature is regulated by a "set point" in the hypothalamus.

CONTENT REVIEW

1. Define metabolism, anabolism, and catabolism.
2. Define a nutrient and list the six major classes of nutrients.
3. What are vitamins and provitamins? Name the water-soluble vitamins and the fat-soluble vitamins. List some of the functions of vitamins.
4. List some of the minerals and give their functions.
5. Describe carbohydrate digestion.
6. Describe the role of bile salts in lipid digestion, absorption, and transport in the blood.
7. Describe protein digestion. What enzymes are responsible?
8. Describe glycolysis. Although four ATP molecules are produced in glycolysis, explain why there is a net gain of only two molecules of ATP.
9. What determines whether the pyruvic acid produced in glycolysis becomes lactic acid or acetyl CoA?
10. Describe the two phases of anaerobic respiration. How many ATP molecules are produced? What happens to the lactic acid produced when oxygen becomes available?

11. Define aerobic respiration and list the products it produces.
12. Why is the citric acid cycle a cycle?
13. What is the function of the electron-transport chain?
14. What is meant by metabolic rate? Describe its three component parts. Define basal metabolic rate.
15. Describe how heat is produced by and lost from the body. How is body temperature regulated?

CONCEPT REVIEW

1. Why does a vegetarian have to be more careful about his or her diet than a person who includes meat in the diet?
2. Why can some people lose weight on a 1200 cal per day diet, and other people cannot?
3. Lotta Bulk, a muscle builder, wanted to increase her muscle mass. Knowing that proteins are the main components of muscle, she consumed large amounts of protein daily (high protein diet), along with small amounts of lipid and carbohydrate. Explain why this strategy will or will not work.

ANSWERS TO  PREDICT QUESTIONS

1 *p. 436* If vitamins were broken down during the process of digestion, their structures would be destroyed, and as a result, their ability to function would be lost.

2 *p. 446* If electrons in the electron-transport chain cannot be donated to oxygen atoms (the last step), the entire electron-transport chain stops, no ATP can be produced aerobically, and the patient dies because too little energy is available for the body to maintain vital functions. Anaerobic metabolism can provide energy for only very short periods of time and cannot sustain life very long.

3 *p. 448* When muscles contract, they must use ATP as the energy source for the contractions. As more ATP is produced, heat is also produced. During exercise, the large amounts of heat can raise body temperature, and we feel warm. Shivering consists of small, rapid muscle contractions that produce heat in an effort to prevent a decrease in body temperature in the cold. When the body temperature declines below normal, shivering is initiated involuntarily.

4 *p. 450* Vasoconstriction reduces blood flow to the skin, which cools as a result. The benefit is that less heat is lost through the skin to the environment, and the internal body temperature is maintained. As the difference in temperature between the skin and the environment decreases, less heat is lost.

18

URINARY SYSTEM AND FLUID BALANCE

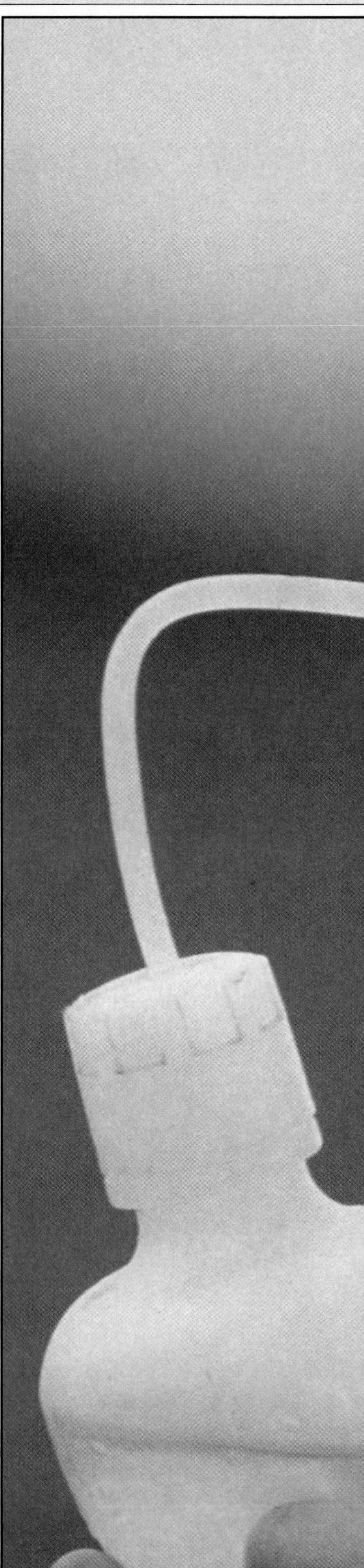

After reading this chapter you should be able to:

1 List the structures that make up the urinary system and describe the overall functions it performs.
2 Describe the location and anatomy of the kidneys.
3 Describe the structure of the nephron and the location of the parts of the nephron in the kidney.
4 List the components of the filtration barrier and describe the composition of the filtrate.
5 Describe the ureters, urinary bladder, and urethra.
6 Identify the principal factors that influence filtration pressure and explain how they affect the rate of filtrate formation.
7 Give the function of the proximal convoluted tubule, descending limb and ascending limb of Henle's loop, distal convoluted tubule, and collecting duct. Discuss how the movement of substances across the walls of these structures influences the composition of the filtrate.
8 Explain how antidiuretic hormone, aldosterone, and atrial natriuretic factor influence the volume and concentration of urine.
9 Describe the micturition reflex.
10 Describe the mechanisms by which sodium ions, potassium ions, and calcium ions are regulated in the extracellular fluid.
11 Illustrate how the mechanisms that regulate the body fluid pH function by explaining how they respond to a decreasing and an increasing pH in the body fluids.

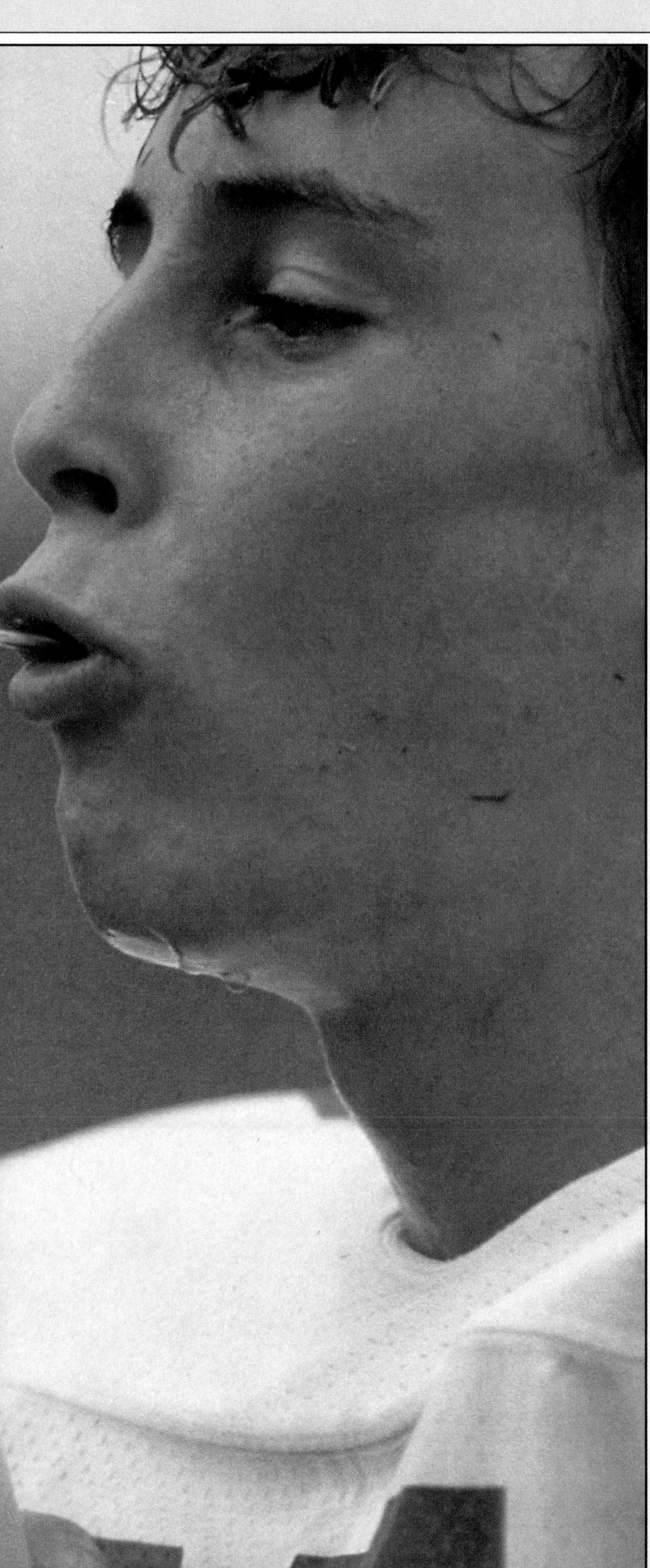

acidosis Condition characterized by a lower-than-normal blood pH (pH of 7.35 or lower).

aldosterone (al-dos′ter-ōn) Steroid hormone produced by the adrenal cortex that facilitates exchange of potassium for sodium in the distal convoluted tubule and collecting duct, causing sodium reabsorption, and potassium and hydrogen ion secretion.

alkalosis Condition characterized by a higher-than-normal blood pH (pH of 7.45 or above).

antidiuretic hormone (ADH) Hormone secreted from the posterior pituitary that acts on the kidney to reduce the output of urine; also called vasopressin.

atrial natriuretic factor Hormone released from cells in the atrial wall of the heart in response to increased blood pressure; acts to lower blood pressure by increasing the rate of urine production.

calcitonin (cal′sĭ-to′nin) Hormone released from cells of the thyroid gland that acts on tissues, especially bone, to cause a decrease in blood levels of calcium ions.

filtration Movement of a liquid under pressure through a filter, which prevents some or all of the substances in the liquid from passing.

filtration membrane Membrane formed by the glomerular capillary endothelium, the basement membrane, and the podocytes of Bowman's capsule.

micturition reflex Contraction of the urinary bladder stimulated by stretching of the urinary bladder wall; results in emptying of the urinary bladder.

nephron (nef′ron) [Gr. *nephros*, kidney] Functional unit of the kidney, consisting of the Bowman's capsule, the proximal convoluted tubule, the loop of Henle, and the distal convoluted tubule.

parathyroid hormone Hormone produced by the parathyroid gland; increases bone breakdown and blood calcium levels.

renin Enzyme secreted by the kidney that converts angiotensinogen to angiotensin I.

tubular reabsorption Movement of materials, by means of diffusion or active transport, from the filtrate into the blood.

tubular secretion Movement of materials, by means of active transport, from the blood into the filtrate of a nephron.

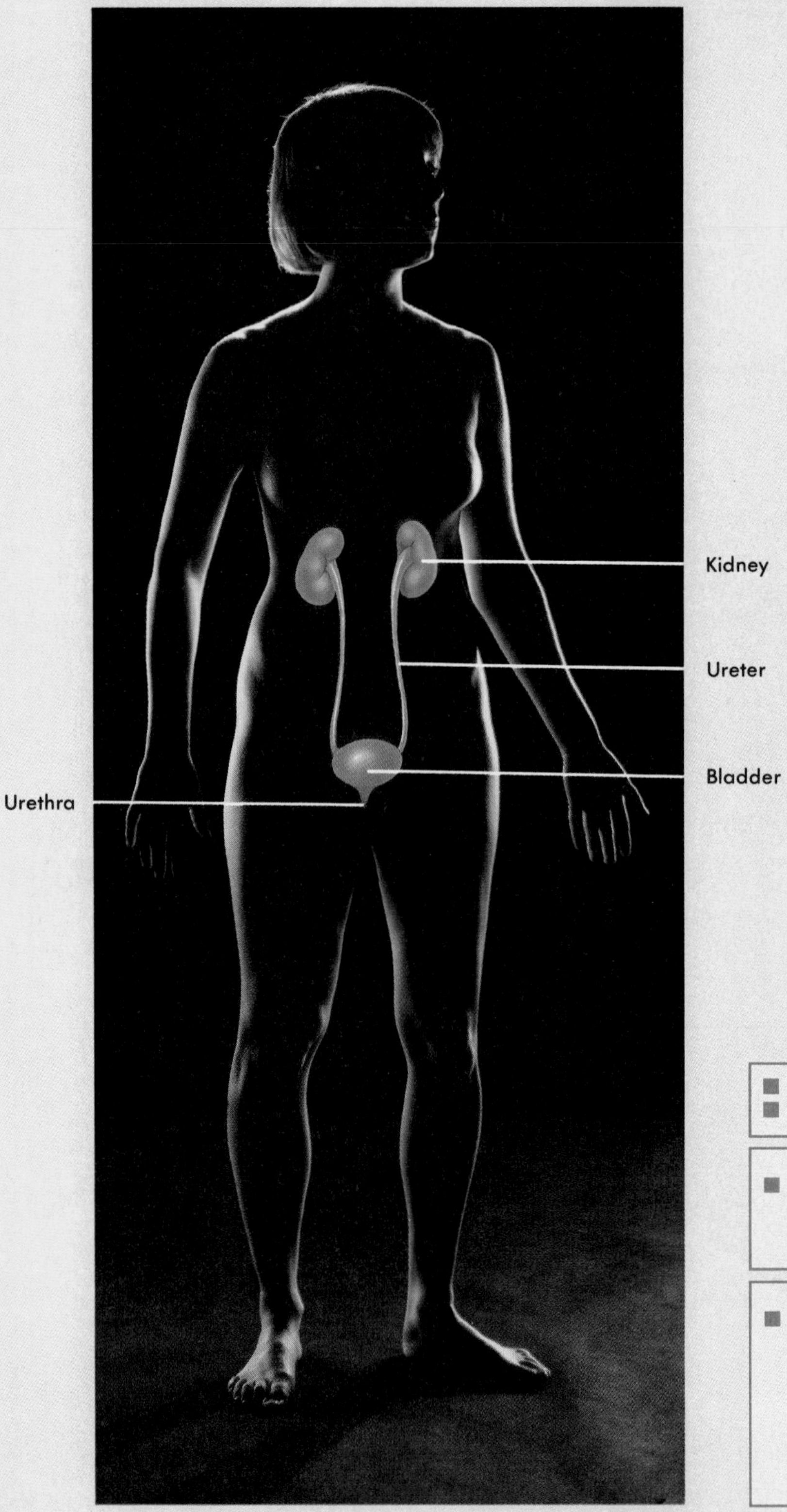

Kidney

Ureter

Bladder

Urethra

The Urinary System

MAJOR COMPONENTS

Kidneys, ureters, urinary bladder, and urethra

MAJOR FUNCTIONS

Removes waste products from the circulatory system

Regulates blood pH, ion balance, and fluid balance

Assists in regulating blood pressure

FIGURE 18-1 ■ The urinary system.
The urinary system consists of two kidneys, two ureters, a urinary bladder, and a urethra.

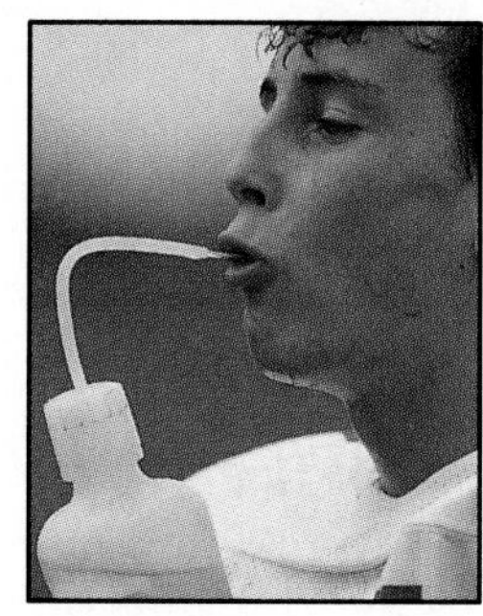

The urinary system consists of the kidneys, ureters, urinary bladder, and urethra (Figure 18-1). The kidneys remove waste products, many of which are toxic, from the blood and play a major role in controlling the volume, red blood cell concentration, ion concentration, and pH of the blood. Although the kidneys are the major organs of excretion, the skin, liver, lungs, and intestines also eliminate wastes. If the kidneys fail to function, however, the other structures cannot compensate adequately.

URINARY SYSTEM

Kidneys

The kidneys are bean-shaped organs, each about the size of a tightly clenched fist. They lie on the posterior abdominal wall behind the peritoneum to either side of the vertebral column (Figure 18-2). A connective tissue **renal capsule** surrounds each kidney. Around the renal capsule is a thick layer of fat called the **renal fat pad,** which protects the kidney from mechanical shock.

On the medial side of each kidney is the **hilum** (hi'lum; a small amount), where the renal artery and nerves enter and the renal vein and ureter exit the kidney. The hilum opens into a cavity called

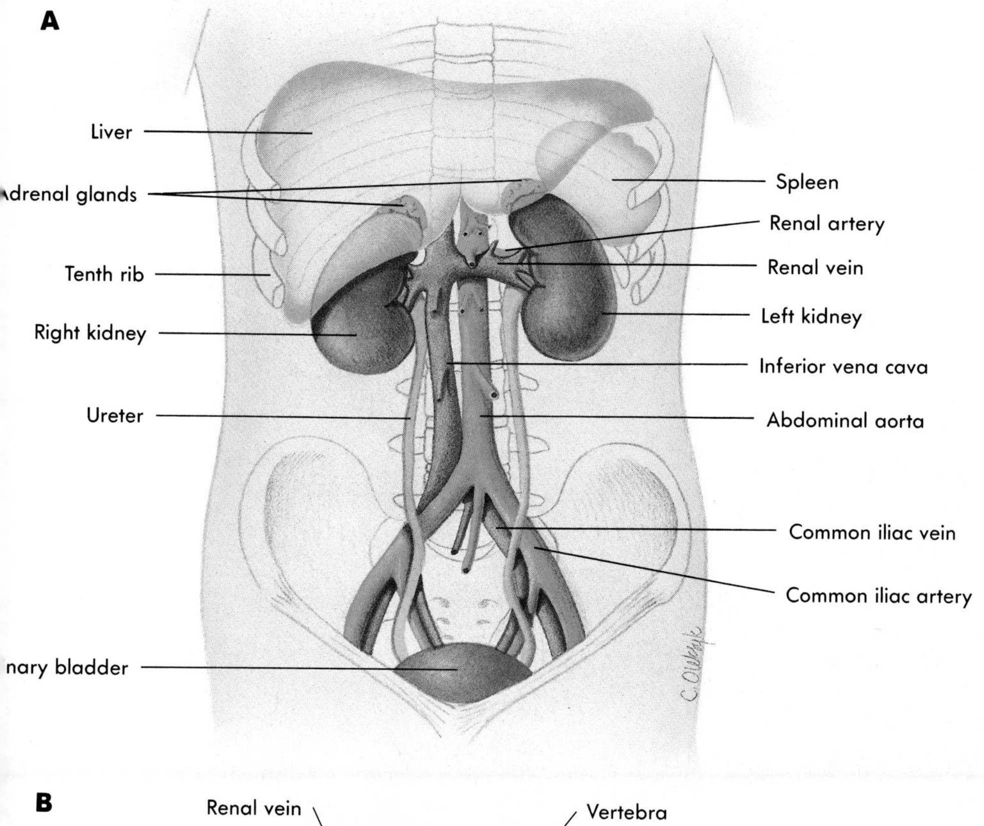

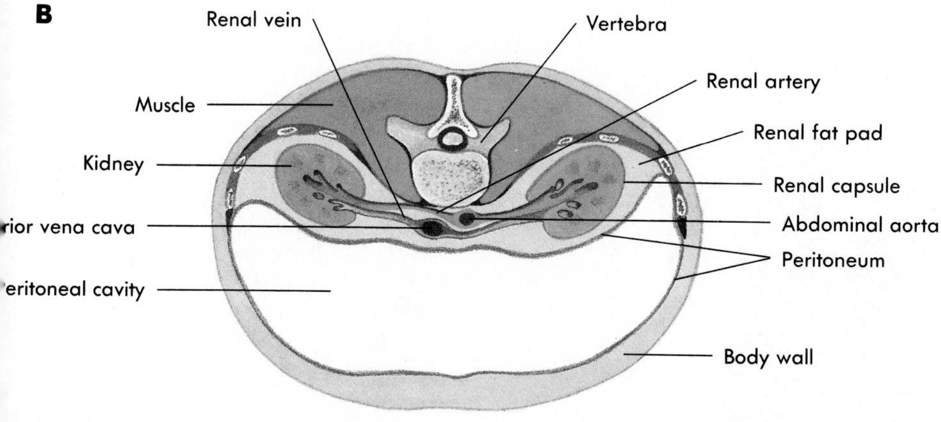

FIGURE 18-2 ■ Anatomy of the kidney.
A The kidneys are located in the abdominal cavity, with the right kidney just below the liver and the left kidney below the spleen. The ureters extend from the kidneys to the urinary bladder within the pelvic cavity. An adrenal gland is located at the superior pole of each kidney.
B The kidneys are located behind the peritoneum. Surrounding each kidney is the renal fat pad. The renal arteries extend from the abdominal aorta to each kidney, and the renal veins extend from the kidneys to the inferior vena cava.

the **renal sinus,** which is filled with fat and connective tissue (Figure 18-3). Within the renal sinus the urinary channel is enlarged to form the **renal pelvis** (basin). Several funnel-shaped structures called **calyces** (kal'ĭ-sēz; flower petals; singular **calyx**) extend from the renal pelvis to the kidney tissue. At the hilum the renal pelvis narrows to form the **ureter** (ur-re'ter).

The kidney is divided into an outer **cortex** and an inner **medulla.** The medulla consists of a number of cone-shaped **renal pyramids.** The base of each renal pyramid extends into the cortex, and the apex of each pyramid projects into the medulla, where its tip is surrounded by a calyx.

The functional unit of the kidney is the **nephron** (nef'ron) (Figure 18-4), which consists of an enlarged terminal end called a **renal corpuscle,** a **proximal convoluted tubule,** a **loop of Henle,** and a **distal convoluted tubule.** The distal convoluted tubule empties into a **collecting duct,** which carries

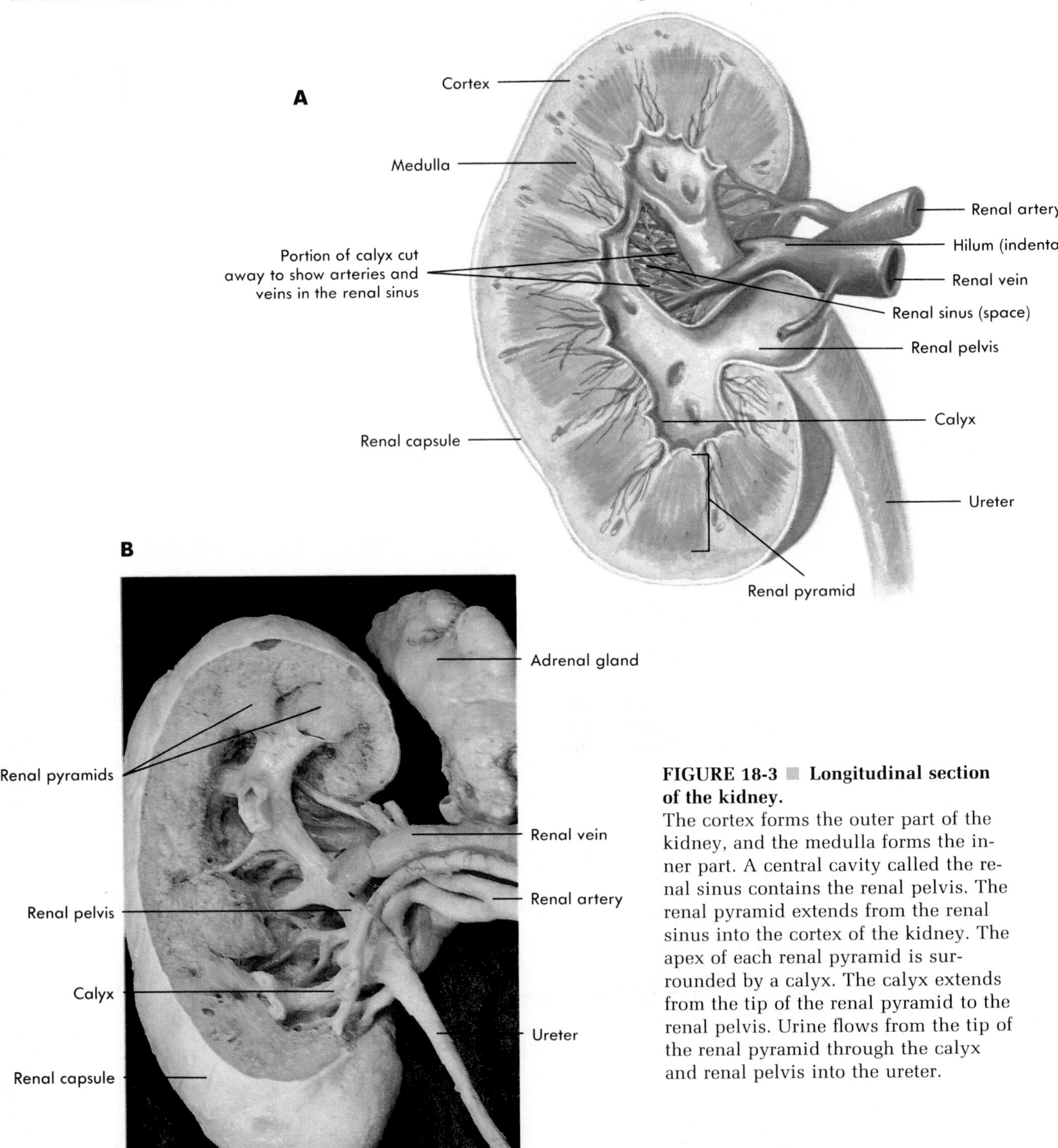

FIGURE 18-3 Longitudinal section of the kidney.
The cortex forms the outer part of the kidney, and the medulla forms the inner part. A central cavity called the renal sinus contains the renal pelvis. The renal pyramid extends from the renal sinus into the cortex of the kidney. The apex of each renal pyramid is surrounded by a calyx. The calyx extends from the tip of the renal pyramid to the renal pelvis. Urine flows from the tip of the renal pyramid through the calyx and renal pelvis into the ureter.

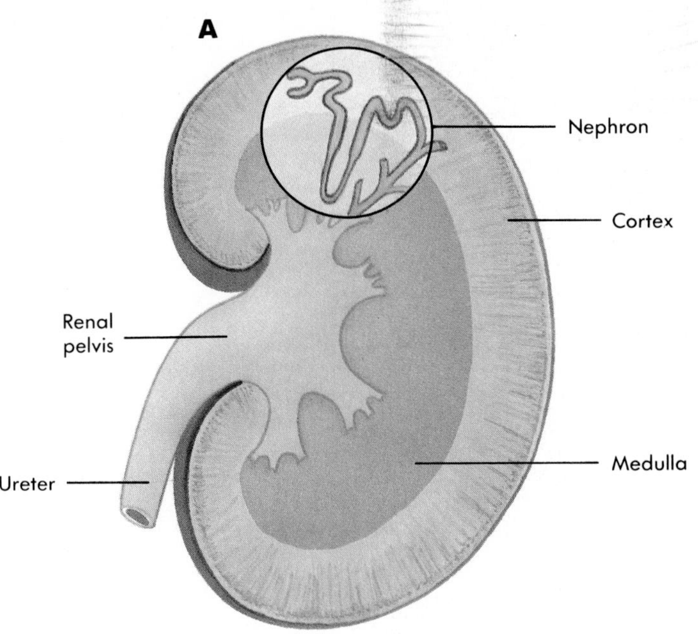

FIGURE 18-4 ■ The nephron.
A A longitudinal section of a kidney showing the location of a nephron in the kidney.
B Each nephron consists of a renal corpuscle, a proximal convoluted tubule, a loop of Henle, and a distal convoluted tubule. The distal convoluted tubules join the collecting duct that extends to the tip of the renal pyramid. The renal corpuscle, the proximal convoluted tubule, and the distal convoluted tubule are in the cortex of the kidney. The loops of Henle and the collecting ducts extend into the medulla of the kidney. An afferent arteriole carries blood to the glomerulus in the renal corpuscle. An efferent arteriole carries blood from the glomerulus and gives rise to peritubular capillaries, which surround the nephron.

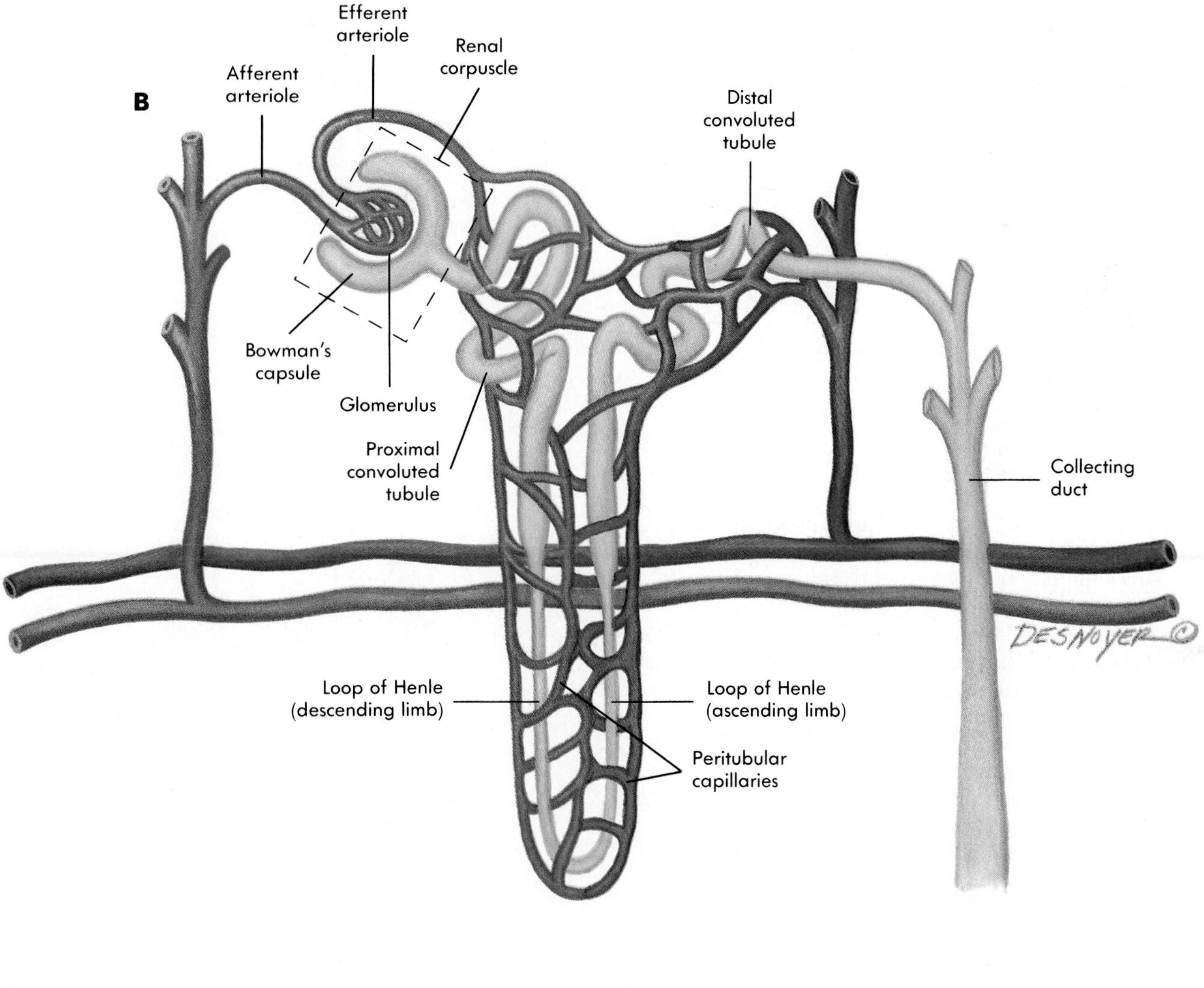

B
Efferent arteriole
Renal corpuscle
Afferent arteriole
Distal convoluted tubule
Bowman's capsule
Glomerulus
Proximal convoluted tubule
Collecting duct
DESNOYER ©
Loop of Henle (descending limb)
Loop of Henle (ascending limb)
Peritubular capillaries

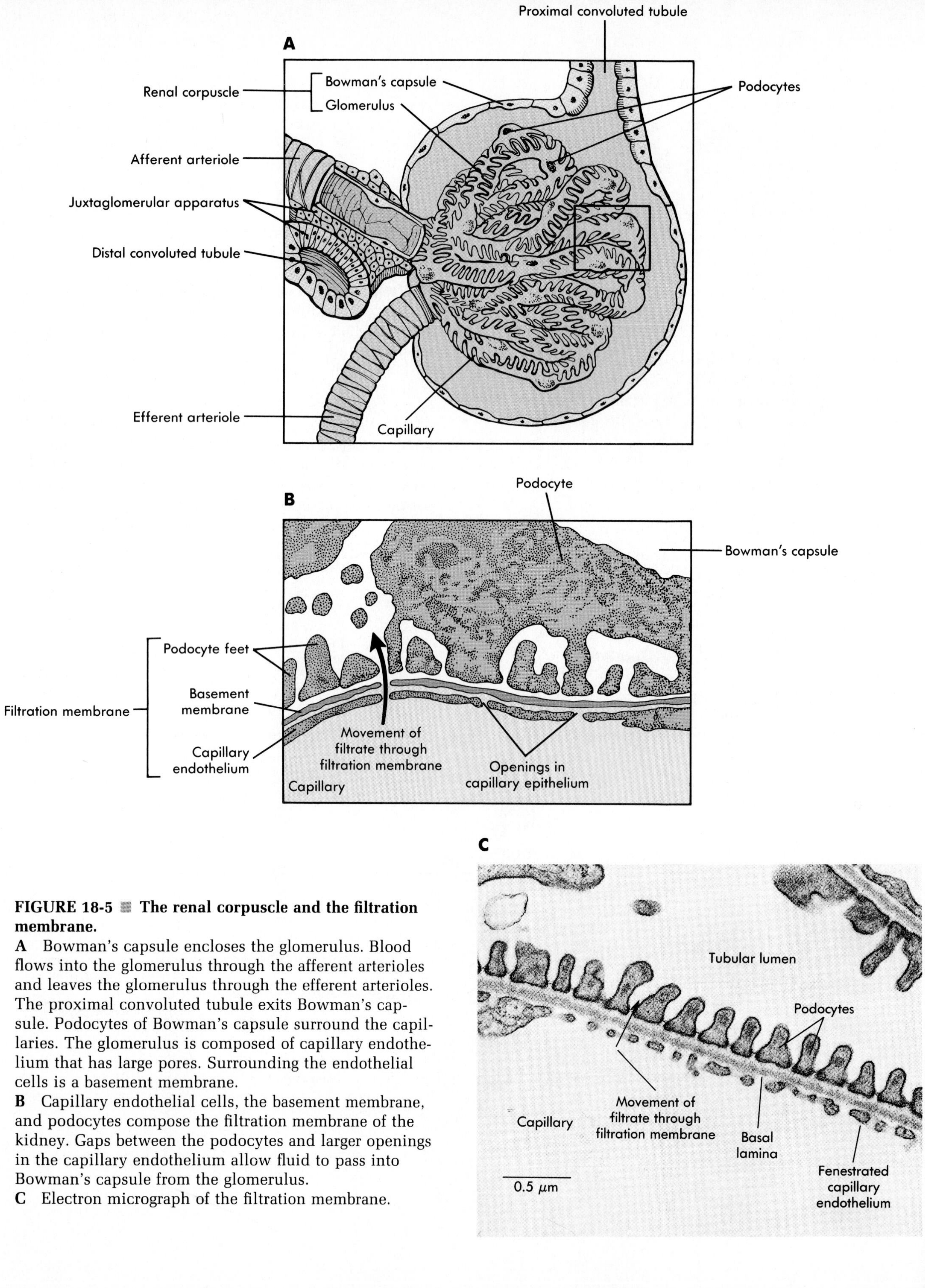

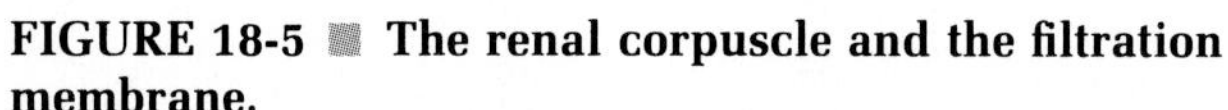

FIGURE 18-5 ■ The renal corpuscle and the filtration membrane.

A Bowman's capsule encloses the glomerulus. Blood flows into the glomerulus through the afferent arterioles and leaves the glomerulus through the efferent arterioles. The proximal convoluted tubule exits Bowman's capsule. Podocytes of Bowman's capsule surround the capillaries. The glomerulus is composed of capillary endothelium that has large pores. Surrounding the endothelial cells is a basement membrane.

B Capillary endothelial cells, the basement membrane, and podocytes compose the filtration membrane of the kidney. Gaps between the podocytes and larger openings in the capillary endothelium allow fluid to pass into Bowman's capsule from the glomerulus.

C Electron micrograph of the filtration membrane.

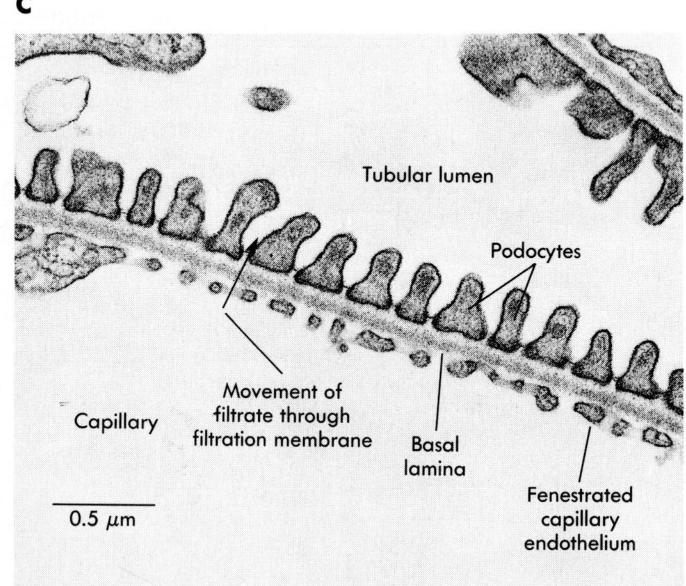

urine from the cortex of the kidney to a calyx. The renal corpuscle and both convoluted tubules are in the renal cortex. The collecting tubule and loop of Henle enter the medulla. About one third of the 1.3 million nephrons in each kidney must be functional to ensure survival.

The renal corpuscle of the nephron consists of **Bowman's capsule** and the **glomerulus** (glo-měr'u-lus) (Figures 18-4 and 18-5). The wall of Bowman's capsule is indented to form a double-walled chamber. The indentation is occupied by a tuft of capillaries called the glomerulus, which resembles a ball of yarn. The cavity of Bowman's capsule opens into a proximal convoluted tubule, which carries fluid away from the capsule. The inner layer of Bowman's capsule surrounds the glomerulus and consists of specialized cells called **podocytes** (pod'o-sītz).

The glomerular capillaries have pores in their walls, and the podocytes have gaps between them. The walls of the glomerular capillaries, the podocytes, and the basement membrane between them form a **filtration membrane.** In the first step of urine formation, fluid called **filtrate** is filtered from the glomerular capillaries into Bowman's capsule through the filtration membrane.

After entering Bowman's capsule, filtrate flows into the proximal convoluted tubule (see Figure 18-4). From there, filtrate flows into the loop of Henle, which is a continuation of the proximal convoluted tubule. Each loop has a **descending limb,** which extends toward the renal sinus and an **ascending limb,** which extends back toward the cortex.

The ascending limb of Henle's loop gives rise to the distal convoluted tubule, which joins a collecting duct. The collecting duct extends from the cortex through the medulla and empties its contents into a calyx. A large portion of the descending limb has very thin walls, as it is made up of simple squamous epithelium. The remainder of the nephron and collecting duct is made up of simple cuboidal epithelium.

Arteries and Veins

The **renal arteries** branch off the abdominal aorta and enter the renal sinus of each kidney (Figure 18-6). The renal arteries give rise to branches to form several **interlobar arteries,** which pass between the renal pyramids. The interlobar arteries, in turn, give rise to arteries that turn and project between the cortex and medulla, the **arcuate arteries. Interlobular arteries** branch off the arcuate arteries to project into the cortex. The **afferent arterioles** arise from branches of the interlobular arteries and extend to the glomerular capillaries. **Efferent arterioles** extend from the glomerular capillaries to the **peritubular capillaries,** which surround the proximal and distal convoluted tubules and the loop of Henle (see Figure 18-4). Blood from the peritubular capillaries enters the renal veins in the renal sinus. The veins of the kidney run parallel to the arteries and have similar names (see Figure 18-6).

A specialized structure called the **juxtaglomerular apparatus** is formed where the distal convoluted tubule projects between the afferent arteriole and the efferent arteriole of a nephron. The specialized walls of the afferent arteriole and distal convoluted tubule come into contact to form the juxtaglomerular apparatus (see Figure 18-5).

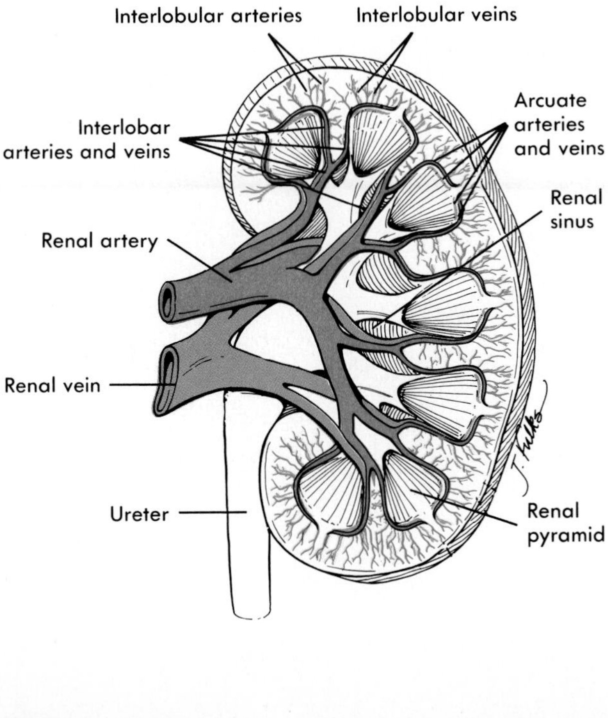

FIGURE 18-6 ■ Blood flow through the kidney. Renal arteries project to the renal sinus. Branches extend through the renal sinus to give rise to the interlobar arteries. At the boundary between the medulla and the cortex, the arteries turn to become the arcuate arteries, which are parallel with the surface of the kidney. The interlobular arteries branch off the arcuate arteries and extend into the cortex. Branches of the interlobular arteries give rise to the afferent arterioles. Interlobular veins drain blood from peritubular capillaries. The interlobular veins, arcuate veins, interlobar veins, and renal veins are parallel to the arteries with the same names.

Ureters, Urinary Bladder, and Urethra

The **ureters** are small tubes that carry urine from the renal pelvis to the posterior inferior portion of the urinary bladder (Figure 18-7). The **urinary bladder** is a hollow muscular container that lies in the pelvic cavity just posterior to the pubic symphysis. The urinary bladder functions to store urine, and the size of the urinary bladder depends on the quantity of urine present. The urinary bladder can hold from a few milliliters to a maximum of about 1000 milliliters of urine. The urinary bladder is normally emptied, however, when it reaches a volume of a few hundred milliliters. The **urethra** is a tube that exits the urinary bladder inferiorly and anteriorly near the entrance of the two ureters. The urethra carries urine from the urinary bladder to the outside of the body.

The ureters and the urinary bladder are lined with transitional epithelium. The transitional epithelium is specialized to stretch as the volume of the urinary bladder increases. The walls of the ureter and urinary bladder are composed of layers of smooth muscle. Regular waves of smooth muscle contractions produce the force that causes urine to flow from the kidneys through the ureters to the urinary bladder. Contractions of smooth muscle in the urinary bladder wall also force urine to flow from the urinary bladder through the urethra to exit the body.

At the junction of the urinary bladder and the urethra, smooth muscle of the bladder wall forms the **internal urinary sphincter.** The **external urinary sphincter** is skeletal muscle that surrounds the urethra as the urethra extends through the pelvic floor. The sphincters regulate the flow of urine through the urethra.

In the male the urethra extends to the end of the penis, where it opens to the outside. The female urethra is much shorter than the male urethra and opens into the vestibule anterior to the vaginal opening.

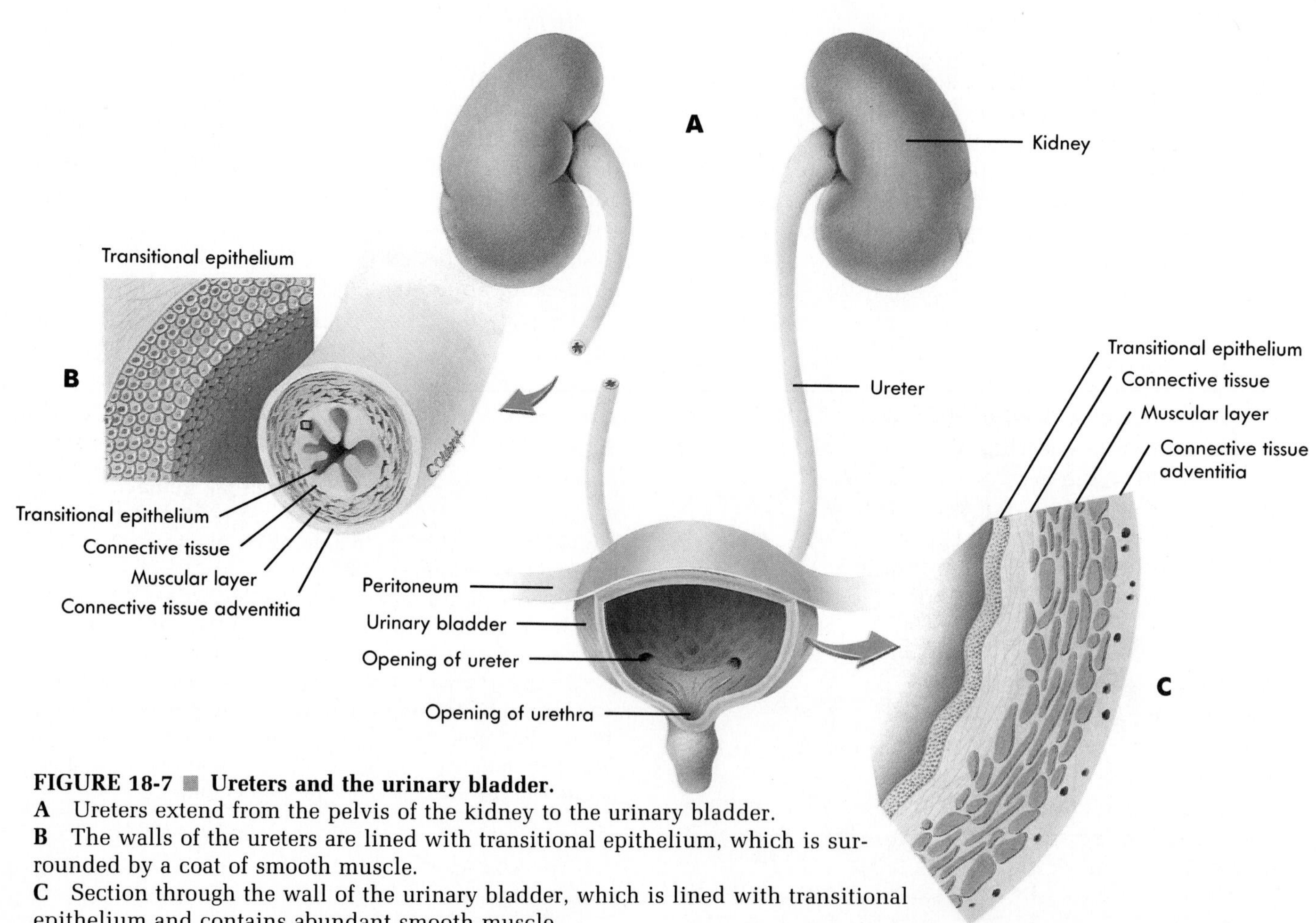

FIGURE 18-7 Ureters and the urinary bladder.
A Ureters extend from the pelvis of the kidney to the urinary bladder.
B The walls of the ureters are lined with transitional epithelium, which is surrounded by a coat of smooth muscle.
C Section through the wall of the urinary bladder, which is lined with transitional epithelium and contains abundant smooth muscle.

1 Urinary bladder infection (cystitis) often occurs when bacteria from outside the body enter the bladder. Who is more prone to urinary bladder infection, males or females? Explain.

URINE PRODUCTION

Urine is mostly water and contains organic waste products such as urea, uric acid, and creatinine; and excess ions such as sodium, potassium, chloride, bicarbonate, and hydrogen. The three processes critical to the formation of urine are filtration, reabsorption, and secretion.

Filtration is the movement of plasma across the filtration membrane of the renal corpuscle. Blood cells and large molecules such as proteins cannot cross the filtration membrane. The portion of the plasma entering the nephron becomes the filtrate. **Tubular reabsorption** is the movement of substances from the filtrate back into the blood of the peritubular capillaries. In general, the useful substances that enter the filtrate are reabsorbed, and metabolic waste products remain in the filtrate and are eliminated. For example, when proteins are metabolized, ammonia is a byproduct. The ammonia is converted into urea by the liver. The urea forms part of the filtrate and is eliminated in the urine. **Tubular secretion** is the transport of substances, usually waste products, into the filtrate. Therefore, urine produced by the nephrons consists of the substances that are filtered and secreted into the nephron minus those substances that are reabsorbed.

Filtration

An average of 21% of the cardiac output flows through the kidneys, and about 19% of the plasma that flows through a glomerulus passes through the filtration membrane into Bowman's capsule to become filtrate. In all of the nephrons of both kidneys, about 180 liters of filtrate are produced each day.

The filtration membrane functions as a filtration barrier that prevents the entry of blood cells and proteins into Bowman's capsule. Water and solutes of small molecular diameter readily pass from the glomerular capillaries through the filtration barrier into Bowman's capsule. Albumin, a blood protein with a diameter slightly less than the pores of the filtration barrier, enters the filtrate in very small amounts. Consequently the filtrate contains no cells and only a small amount of protein.

The formation of filtrate depends on a pressure difference called the **filtration pressure,** which forces fluid from the glomerular capillaries through the filtration membrane into Bowman's capsule. In general, when the filtration pressure increases, the volume of the filtrate increases, and the urine volume increases. When the filtration pressure decreases, the filtrate volume decreases, and the urine volume decreases.

The filtration pressure is influenced by factors such as the blood pressure in the glomerular capillary, the blood protein concentration, and the solutes that enter Bowman's capsule. If the blood pressure in the glomerular capillaries is high, the filtration pressure is high, and if the blood pressure is low, the filtration pressure is low. Sudden decreases in blood pressure during cardiovascular shock reduce the filtration pressure to nearly zero. In addition, strong sympathetic stimulation of blood vessels during periods of excitement or vigorous physical activity result in vasoconstriction of renal blood vessels and reduced filtration pressure.

During cardiovascular shock, renal blood vessels constrict and blood flow to the kidneys is decreased to a very low rate. One of the dangers of shock is that the renal blood flow may be so low that the kidneys suffer from a lack of oxygen. If the oxygen level remains too low for a long enough time, permanent kidney damage or complete failure of the kidneys will result. One important reason for treating shock quickly is to avoid damage to the kidneys.

Reabsorption

As the filtrate flows through the proximal convoluted tubule, loop of Henle, distal convoluted tubule, and collecting ducts, many of the substances in the filtrate are reabsorbed. Inorganic salts, organic molecules, and about 99% of the filtrate volume leave the nephron and enter the interstitial fluid. These substances enter the peritubular capillaries and flow through the renal veins to enter the general circulation. Only about 1% of the filtrate becomes urine, which contains a high concentration of waste products not reabsorbed from the filtrate.

The proximal convoluted tubule is a primary site for the reabsorption of solutes and water. Substances actively transported from the proximal convoluted tubule to the interstitial spaces include proteins, amino acids, glucose, and fructose molecules, as well as sodium, potassium, calcium, bicarbonate, and chloride ions. The proximal convoluted tubule is permeable to water. As solute molecules are actively transported from the proximal convoluted tubule to the interstitial spaces, water moves by osmosis in the same direction. Consequently 65% of the filtrate volume is reabsorbed in the proximal convoluted tubule (Figure 18-8).

The descending limb of the loop of Henle functions to concentrate the filtrate by removing water and adding solutes. The descending limb is permeable to water and moderately permeable to solutes. As the descending limb extends into the medulla of the kidney, the interstitial fluid surrounding it becomes highly concentrated. Consequently as the filtrate passes through the descending limb, water moves out of the nephron by osmosis, and solutes move into the nephron. By the time the filtrate has passed through the descending limb, another 15% of the filtrate volume has been reabsorbed, and the filtrate is as concentrated as the interstitial fluid.

The ascending limb of the loop of Henle functions to dilute the filtrate by removing solutes. The ascending limb is not permeable to water or most solutes. Chloride ions (by active transport) and sodium ions (by diffusion), however, pass from the nephron into the interstitial spaces. As a result, solutes, but little water, are removed from the filtrate.

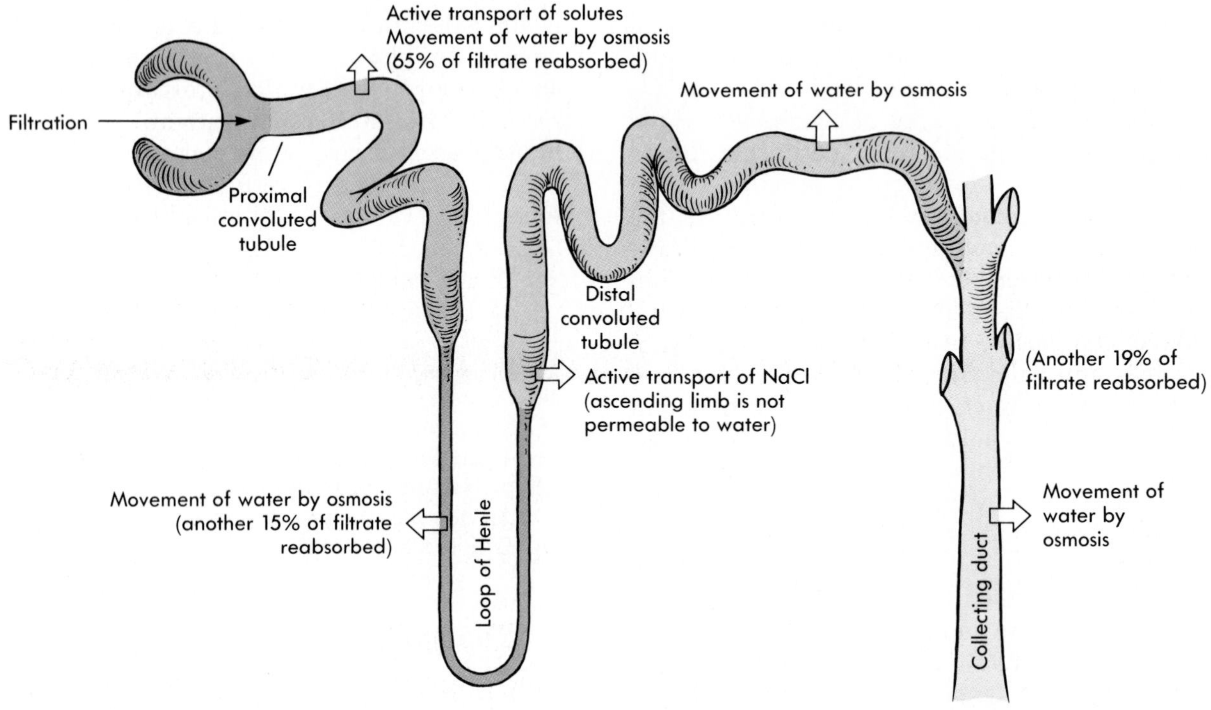

Filtrate produced	100%
Filtrate reabsorbed	
Proximal convuluted tubule	65%
Descending loop of Henle	15%
Distal convoluted tubule and collecting duct	19%
Filtrate remaining as urine	1%

FIGURE 18-8 ■ Water reabsorption from the filtrate. About 180 liters of filtrate are produced each day in all of the nephrons, but only about 1 to 2 liters of urine are produced. In the proximal convoluted tubule, solutes such as glucose, amino acids, and sodium chloride are actively transported from the nephron. About 65% of the filtrate volume follows as a result of osmosis. The descending limb of Henle's loop extends into the medulla of the kidney, where the extracellular fluid is very concentrated. Because the descending limb is permeable to water, water moves out of the nephron by osmosis, which decreases the filtrate volume by another 15%. The ascending limb is not permeable to water. The wall of the ascending limb actively transports chloride from the nephron and sodium ions follow. By the time the filtrate flows to the distal convoluted tubule, it is very dilute. Water leaves the distal convoluted tubule and collecting duct by osmosis, reducing the filtrate volume by another 19%. Only 1% of the original filtrate volume remains as urine.

The highly concentrated filtrate that enters the ascending limb of Henle's loop is converted to a dilute solution by the time it reaches the distal convoluted tubule. As the filtrate enters the distal convoluted tubule, it is more dilute than the interstitial fluid, and only about 20% of the original volume remains.

The distal convoluted tubule and collecting duct function to remove water and solutes. Solutes, such as sodium ions and chloride ions, are actively removed, and 19% of the original filtrate volume is reabsorbed by osmosis, leaving about 1% of the original filtrate as urine (see Figure 18-8).

2 People who suffer from untreated diabetes mellitus may experience very high blood levels of glucose (blood sugar). The glucose can easily cross the filtration membrane into Bowman's capsule. Although some glucose can be reabsorbed across the wall of the nephron, if the concentration of glucose in the nephron gets too high, not all of the glucose can be reabsorbed. How will the volume of urine produced by a person with untreated diabetes mellitus differ from that of a normal person?

In summary, most of the useful solutes that pass through the filtration membrane into Bowman's capsule are reabsorbed in the proximal convoluted tubule. Filtrate volume is reduced by 65% in the proximal convoluted tubule and 15% in the descending limb of the loop of Henle. In the ascending limb of the loop of Henle, sodium chloride, but little water, is removed from the filtrate. Consequently the filtrate becomes dilute. In the distal convoluted tubule and the collecting duct, additional sodium chloride is removed, water moves out by osmosis, and the filtrate volume is reduced by another 19%, leaving 1% of the original filtrate volume as urine.

Secretion

Some substances, including byproducts of metabolism that become toxic in high concentrations and drugs or molecules not normally produced by the body, are secreted into the nephron. As with tubular reabsorption, tubular secretion can be either active or passive. For example, ammonia diffuses into the lumen of the nephron, whereas hydrogen ions, potassium ions, creatinine, histamine, and penicillin are actively transported into the nephron.

Hydrogen ions are actively secreted in the proximal portion of the nephron. The epithelial cells secrete large quantities of hydrogen ions across the wall of the nephron into the filtrate. The secretion of hydrogen ions plays an important role in the regulation of the body fluid pH.

In the proximal portion of the nephron, potassium ions are actively reabsorbed. In the distal convoluted tubule and collecting duct, potassium ions are secreted. As sodium ions are actively reabsorbed from the nephron, a negative charge develops within the nephron. The negative charge attracts potassium ions, which move across the wall of the nephron into the filtrate.

REGULATION OF URINE CONCENTRATION AND VOLUME

The volume and composition of urine changes, depending on conditions in the body. Urine production plays an important role in maintaining blood concentration and blood volume. If blood concentration increases above normal levels, the kidneys produce a smaller-than-normal amount of concentrated urine. This eliminates solutes and conserves water, both of which help to lower blood concentration back to normal. On the other hand, if blood concentration decreases, the kidneys produce a large volume of dilute urine. As a result, water is lost, solutes are conserved, and blood concentration increases.

Urine production also maintains blood volume and, therefore, blood pressure. An increase in blood volume can increase blood pressure, and a decrease in blood volume can decrease blood pressure. When blood volume increases above normal, the kidneys produce a large amount of dilute urine. The loss of water in the urine lowers blood volume. Conversely, if blood volume decreases below normal, the kidneys produce smaller-than-normal amounts of concentrated urine to conserve water and maintain blood volume.

Hormonal Mechanisms

Antidiuretic hormone

Antidiuretic hormone (ADH), secreted by the posterior pituitary gland, passes through the circulatory system to the kidneys. ADH regulates the amount of water reabsorbed by the distal convoluted tubules and collecting ducts. When ADH levels increase, the permeability to water of the distal convoluted tubules and collecting ducts increases, and more water is reabsorbed from the filtrate. Con-

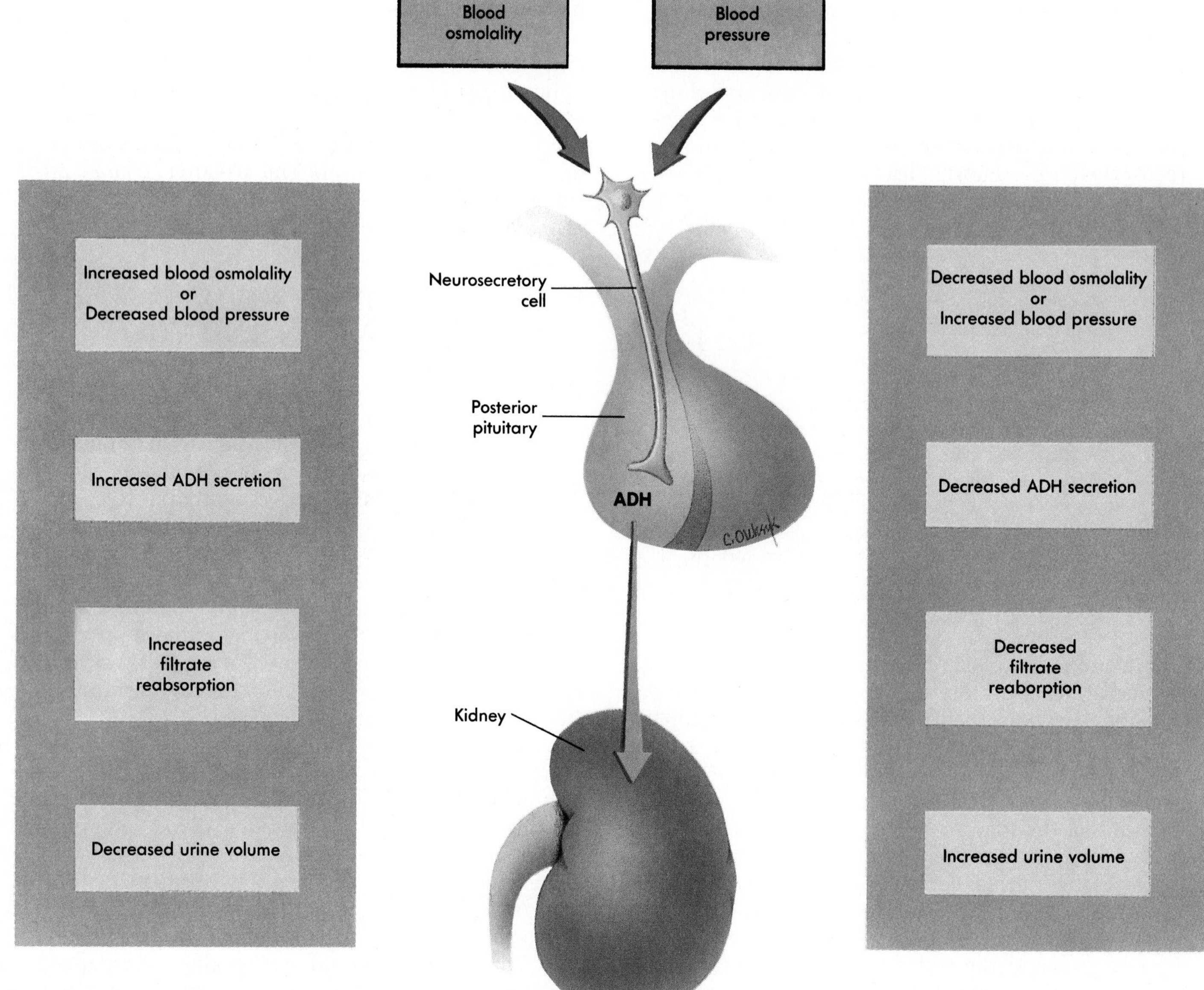

FIGURE 18-9 ■ Control of ADH secretion and its effect on the nephron. Increased blood osmolality or decreased blood pressure results in increased ADH secretion from the posterior pituitary. The increased ADH acts on the distal convoluted tubule and collecting duct, causing an increased reabsorption of water from the nephron and the production of a smaller-than-normal volume of concentrated urine. Decreased blood osmolality or increased blood pressure results in decreased ADH secretion from the posterior pituitary. The decreased ADH causes a decreased reabsorption of water from the nephron and the production of a large volume of dilute urine.

Diabetes insipidis is a pathological condition in which the posterior pituitary fails to secrete ADH, or the kidney tubules do not respond to the presence of ADH. In the absence of a normal response to ADH, much of the filtrate entering the proximal and distal convoluted tubules becomes urine. People with this condition may produce as much as 20 to 30 liters of urine each day. Because they lose so much water, they are continually in danger of severe dehydration. Even though they produce dilute urine, producing such a large volume of urine results in the loss of large amounts of sodium ions, calcium ions, and other ions. Resulting ionic imbalances cause the nervous system and cardiac muscle to function abnormally.

sequently an increase in ADH results in the production of a more concentrated urine. On the other hand, when ADH levels decrease, the distal convoluted tubules and collecting ducts become less permeable to water. As a result, less water is reabsorbed, and a dilute urine is produced (Figure 18-9).

The release of ADH from the posterior pituitary is regulated by the hypothalamus. Cells of the hypothalamus are sensitive to changes in the osmolality of the interstitial fluid. If the osmolality of the blood and interstitial fluid increases, action potentials are sent along the axons of the ADH-secreting neurons to the posterior pituitary, causing ADH to be released from the ends of the axons (see Chapter 10). A reduced osmolality of the interstitial fluid within the hypothalamus causes inhibition of ADH secretion.

Pressure receptors that monitor blood pressure also influence ADH secretion. Increased blood pressure causes a decrease in ADH secretion, and decreased blood pressure increases ADH secretion (see Chapter 13).

Aldosterone

Aldosterone (al-dos′ter-ōn), secreted by the adrenal gland, regulates the rate of active transport in the distal portion of the nephron and collecting ducts. When aldosterone is present, sodium ions and chloride ions are actively absorbed from the lumen of the nephron. In the absence of aldosterone, large amounts of sodium ions and chloride ions remain in the nephron and become part of the urine. Unless active transport of sodium ions and chloride ions occurs, the urine volume increases, and the urine contains large concentrations of sodium and chloride.

Renin and angiotensin

Renin and angiotensin help regulate aldosterone secretion. Renin is secreted by cells of the juxtaglomerular apparatus in the kidney. When blood pressure decreases, when the concentration of sodium ions in the blood gets too low, or when concentrations of potassium ions in the blood get too high, renin secretion increases. Renin is an enzyme that acts on a protein produced by the liver. Amino acids are removed from the protein leaving angiotensin I. Angiotensin I is rapidly converted to a smaller peptide called angiotensin II. Angiotensin II acts on the adrenal gland causing it to secrete aldosterone.

When blood pressure suddenly decreases, renin is released from the kidney. The resultant increase in aldosterone causes sodium ions to be reabsorbed from the nephron. Along with the sodium, more water is also reabsorbed. Thus, the volume of fluid lost as urine declines. The conservation of water helps prevent a further decline in blood pressure.

3 Drugs that increase the urine volume are called diuretics. Some diuretics inhibit the active transport of sodium ions and chloride ions in the nephron. Explain how these diuretic drugs could cause an increase in urine volume. ?

Alcohol and caffeine are examples of diuretics. Alcohol inhibits the secretion of ADH from the posterior pituitary. Consequently the consumption of alcoholic beverages results in the production of a large volume of dilute urine. The volume of urine lost easily can exceed the volume of water consumed with the alcohol. Consequently several hours after drinking alcoholic beverages (such as the morning after an intense celebration) one may experience some dehydration and intense thirst. Caffeine also is a diuretic. Caffeine and related substances act on the kidney by increasing blood flow to the kidney and by increasing the loss of sodium and chloride in the urine. Both the increased blood flow to the kidney and the increased loss of sodium and chloride in the urine increases urine volume.

Atrial natriuretic factor

Atrial natriuretic factor is secreted from cells in the right atrium of the heart when blood pressure in the right atrium increases. Atrial natriuretic factor reduces the ability of the kidneys to concentrate

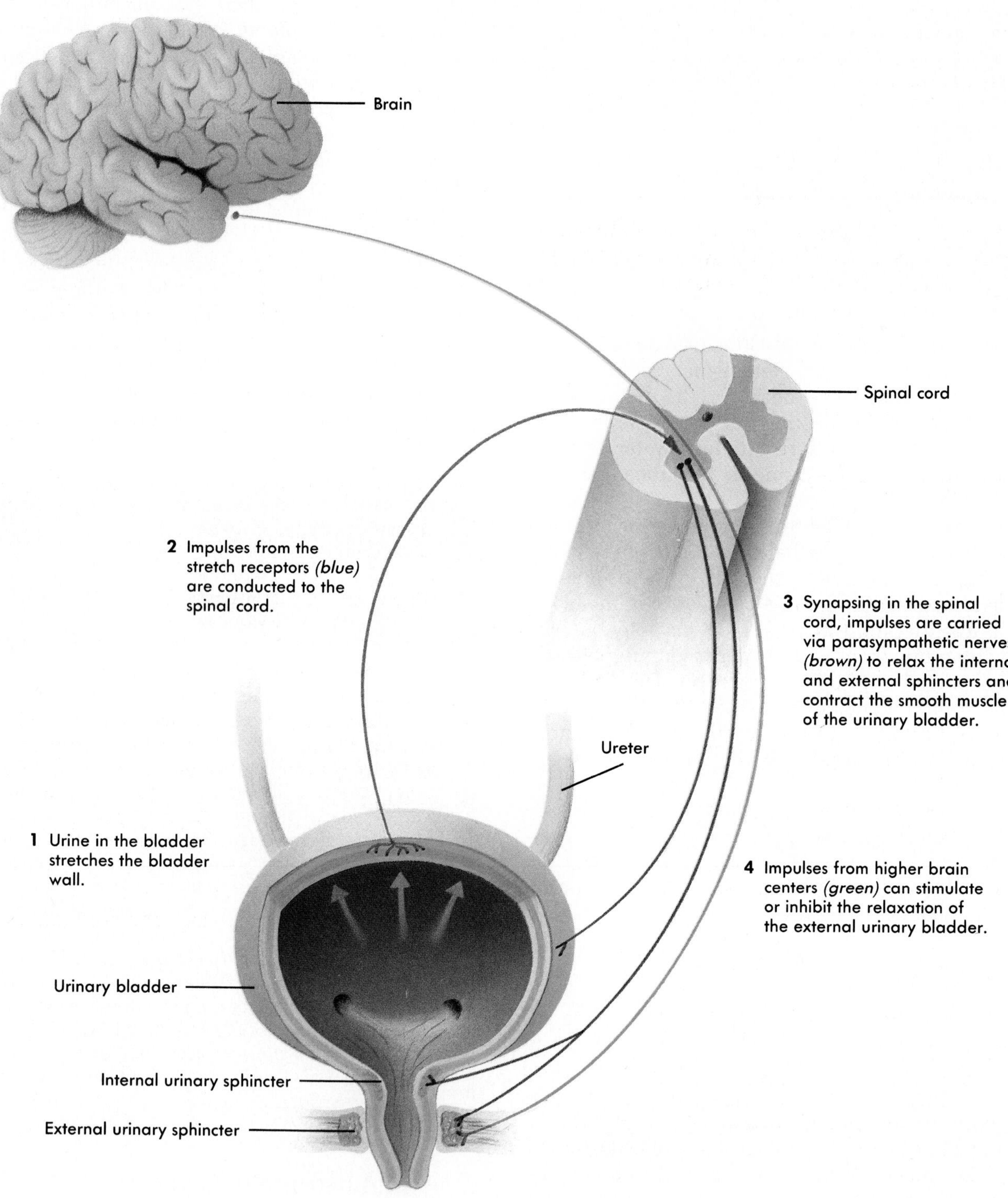
Brain
Spinal cord
2 Impulses from the stretch receptors (blue) are conducted to the spinal cord.
3 Synapsing in the spinal cord, impulses are carried via parasympathetic nerves (brown) to relax the internal and external sphincters and contract the smooth muscle of the urinary bladder.
Ureter
1 Urine in the bladder stretches the bladder wall.
4 Impulses from higher brain centers (green) can stimulate or inhibit the relaxation of the external urinary bladder.
Urinary bladder
Internal urinary sphincter
External urinary sphincter

urine, which leads to the production of a large volume of urine. The resulting decrease in blood volume causes a decrease in blood pressure.

4 Predict the effect of high blood pressure on atrial natriuretic factor secretion and on the rate of urine production.

Effect of Sympathetic Innervation on Kidney Function

Sympathetic neurons that have norepinephrine as their neurotransmitter substance innervate the blood vessels of the kidney. Sympathetic stimulation constricts the arteries, causing a decrease in renal blood flow and filtrate formation. Intense sympathetic stimulation, for example, during shock or intense exercise, causes the rate of filtrate formation to decrease to only a few milliliters per minute.

URINE MOVEMENT

The **micturition reflex** is initiated by stretching of the bladder wall. As the bladder fills with urine, pressure increases, and stretch receptors in the wall of the bladder are stimulated. Action potentials are conducted from the bladder to the spinal cord through the pelvic nerves. Integration of the reflex occurs in the spinal cord, and action potentials are conducted along parasympathetic nerve fibers to the urinary bladder. Parasympathetic action potentials cause the urinary bladder to contract and the internal and external urinary sphincters to relax (Figure 18-10).

The micturition reflex is an automatic reflex, but it can be inhibited or stimulated by higher centers in the brain. The higher brain centers prevent micturition by sending impulses through the spinal cord to decrease the intensity of urinary bladder contractions and to stimulate efferent neurons that keep the external urinary sphincter tonically contracted. The ability to voluntarily inhibit micturition develops at the age of 2 to 3 years.

When the desire to urinate exists, the higher brain centers send impulses to the spinal cord to facilitate the micturition reflex and inhibit the external urinary sphincter. The desire to urinate is initiated because stretch of the urinary bladder stimulates ascending fibers in the spinal cord. Irritation of the urinary bladder or the urethra by bacterial infections or other conditions may also initiate the urge to urinate, even though the bladder may be nearly empty.

BODY FLUID COMPARTMENTS

For a 70-kilogram adult, approximately 40 liters, or 57% of the total body weight, consists of water. The water and its electrolytes are distributed in two major compartments. Water and electrolytes move between these compartments, but their movement is regulated.

The **intracellular fluid compartment** includes all the water and electrolytes inside the cells of the body. The cell membranes of the individual cells enclose the intracellular compartments, which actually consist of trillions of small compartments. However, the composition of intracellular fluid is similar and the regulation of water and electrolyte movement across cell membranes is similar. Ap-

FIGURE 18-10 The micturition reflex.
An increased volume of urine in the bladder stretches the bladder wall, sending impulses *(blue)* along the spinal nerves to the spinal cord. The impulses result in increased parasympathetic impulses *(brown)*, which return to the bladder causing the bladder wall to contract and the urinary sphincters to relax. Impulses are also sent to the brain from the spinal cord stimulating the desire to urinate. The relaxation of the external urinary sphincter can be inhibited by impulses *(green)* from the brain. When convenient, the brain sends impulses to the external urinary sphincter *(green)* causing it to relax. The bladder wall contracts and urine is expelled.

proximately 25 of the 40 liters of body fluid, or 63% of the total body water, is contained within cell membranes.

The **extracellular fluid compartment** includes all the fluid outside the cells. It constitutes approximately 37% of the total body water. The extracellular fluid compartment includes the water found in: the tissue spaces (interstitial fluid), the plasma within blood vessels, and fluid in the lymph vessels. A small portion of the extracellular fluid volume is separated by epithelial membranes into subcompartments. These special subcompartments contain fluid with a different composition from the remainder of the extracellular fluid. Included among the subcompartments are the aqueous and vitreous humor of the eye, cerebrospinal fluid, synovial fluid in joint cavities, serous fluid in the body cavities, and fluid secreted by glands.

Composition of the Fluid in the Body Fluid Compartments

Intracellular fluid has a similar composition from cell to cell. The intracellular fluid contains relatively high concentration of electrolytes such as potassium, phosphate, magnesium, and sulfate ions compared to the extracellular fluid. It has a lower concentration of sodium, chloride, and bicarbonate ions than the extracellular fluid. The concentration of protein in the intracellular fluid is also greater than the concentration of proteins in the extracellular fluid.

Like intracellular fluid, the extracellular fluid has a fairly consistent composition, also. The concentrations of sodium, chloride, and bicarbonate ions are higher, and the concentrations of potassium, calcium, magnesium, phosphate, and sulfate ions and protein are lower in the extracellular fluid when compared to the intracellular fluid.

Exchange between Body Fluid Compartments

The cell membranes that separate the body fluid compartments are selectively permeable. Water can pass through them, but they are much less permeable to electrolytes dissolved in the water. Water moves between the compartments continually, but the movement is regulated mainly by pressure differences and osmotic differences between the compartments.

For example, water moves across the wall of the capillary at the capillary's arteriolar end because the blood pressure is great enough to force fluid through the wall of the capillary into the interstitial space. At the venous end of the capillary, the blood pressure is much lower. Fluid returns to the capillary at its venous end because the osmotic pressure is higher inside the capillary than outside the capillary.

The major influence controlling the movement of water between the intracellular and extracellular spaces is osmotic pressure. For example, if the extracellular concentration of electrolytes increases, water moves because of osmosis from the cells across the cell membrane into the extracellular spaces. Because so much of the total body water is contained within cells, the intracellular fluid can help maintain the extracellular fluid volume if it is depleted. When a person becomes dehydrated, the concentration of electrolytes in the extracellular fluid increases. As a consequence, fluid moves from the intracellular fluid into the extracellular fluid, thus, maintaining the extracellular fluid volume. Because blood is an important component of the extracellular fluid volume, this process helps maintain blood volume. Movement of water from the intracellular fluid compartment to the extracellular fluid compartment can help prolong the time a person can survive conditions such as dehydration or cardiovascular shock.

If the concentration of electrolytes in the extracellular fluid decreases, water moves, by osmosis, from the extracellular fluid into the cells. The water movement causes the cells to swell. Movement of water between the intracellular and extracellular fluid compartments occurs continuously. Under most conditions, the movement is maintained within limits that are consistent with survival of the individual.

REGULATION OF EXTRACELLULAR FLUID COMPOSITION

Homeostasis requires that the intake of substances such as water and electrolytes equal their elimination. Ingestion of water and electrolytes adds them to the body, whereas they are excreted from the body by organs such as the kidneys and liver or, to a lesser degree, the skin. Greater quantities of water and electrolytes are lost from the body in the form of perspiration on warm days than cool days, and varying amounts of water and electrolytes can be lost in the form of feces. Over a long period, the total amount of water and electrolytes in the body does not change unless the individual is growing, gaining weight, or losing weight. The regulation of water and electrolytes involves the coordinated participation of several organ systems, but the most important organ in regulating the loss of water and electrolytes from the body is the kidney.

Thirst

Water intake is controlled by neurons in the hypothalamus collectively called the thirst center. When the osmolality of blood increases, the thirst center responds by initiating the sensation of thirst. When water or some other dilute solution is consumed, the osmolality of the blood decreases, and the sensation of thirst also decreases. When the blood pressure decreases, such as during shock, the thirst center is activated and the sensation of thirst is triggered. Consumption of water increases the blood volume and allows the blood pressure to increase toward its normal value. Other stimuli besides changes in osmolality and blood pressure can temporarily trigger the sensation of thirst. For example, if the mucosa of the mouth becomes dry, the thirst center is activated. Thirst is one of the important means of regulating the extracellular fluid volume and concentration.

Ions

Transport mechanisms, the permeability of the cell membrane, and the concentration of water and solutes in the extracellular fluid influence the concentration of water and solutes inside cells. The kidney and other organ systems function to regulate the composition of the extracellular fluid. If the water content or concentration of ions in the extracellular fluid deviate from their normal range, cells cannot control the movement of substances across their cell membranes or the composition of intracellular fluid. The consequence is abnormal cell function and death. Therefore keeping the extracellular fluid composition within a normal range of values is required to sustain life.

The concentrations of positively charged ions such as sodium, potassium, and calcium ions in the body fluids are particularly important. A normal range of concentrations is necessary for the conduction of action potentials, contraction of muscles, and maintenance of normal cell membrane permeability. Important control mechanisms regulate the concentration of these ions in the body. Negatively charged ions such as chloride are secondarily regulated by the mechanisms that control the positively charged ions. The negatively charged ions are attracted to positively charged ions. Consequently when the positively charged ions are transported, the negatively charged ions move with them.

Sodium ions

Sodium ions are one of the dominant extracellular ions. About 90% to 95% of the osmotic pressure of the extracellular fluid results from sodium ions and the negative ions associated with sodium ions.

The recommended intake of sodium is 2.4 grams per day, which is still more than is needed. Most people in the United States consume 20 to 30 times the amount of sodium that is needed. The kidneys provide the major route by which the excess sodium ions are excreted.

The sodium ions enter the nephron as part of the filtrate. If little sodium is reabsorbed from the nephron, a large amount is lost in the urine. On the other hand, if sodium is intensively reabsorbed by the nephron, little sodium is lost in the urine.

When **aldosterone** is present, the reabsorption of sodium ions from the distal convoluted tubules and collecting duct is very efficient. When aldosterone is absent, reabsorption of sodium in the nephron is greatly reduced, and the amount of sodium lost in the urine increases.

Sodium ions are also excreted from the body in **sweat.** Normally, only a small quantity of sodium is lost each day in the form of sweat, but the amount increases during conditions of heavy exercise in a warm environment.

The primary mechanisms that regulate the sodium ion concentration in the extracellular fluid are changes in blood pressure and changes in the osmolality of extracellular fluid (Figure 18-11).

An increase in extracellular sodium results in increased water reabsorption in the kidneys. Consequently blood volume and blood pressure increase. Elevated blood pressure results in increased water excretion because it stimulates baroreceptors, which send signals to the hypothalamus of the brain to cause a reduction in ADH secretion. At the same time, elevated blood pressure inhibits renin secretion from the kidney. A reduced rate of renin secretion leads to a reduced rate of angiotensin II formation. The lower angiotensin II concentration leads to a reduction in the rate of aldosterone secretion. A decline in aldosterone increases the rate of sodium loss in the urine and increases the volume of urine produced. Atrial natriuretic factor is also secreted in response to elevated blood pressure within the atria. Atrial natriuretic factor increases

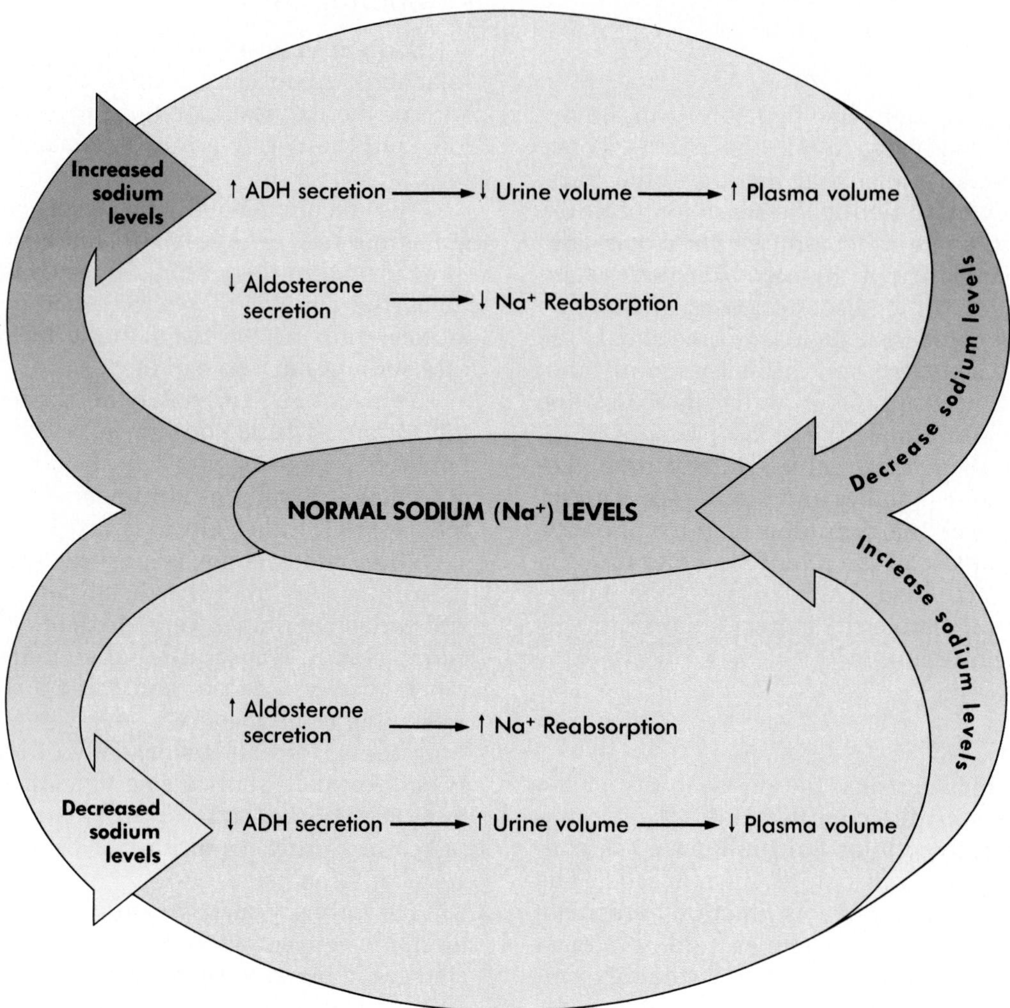

FIGURE 18-11 ■ Regulation of sodium levels in the blood and other extracellular fluid. As sodium levels in the extracellular fluid increase, the increased concentration affects the hypothalamus and results in increased ADH secretion from the posterior pituitary. The ADH acts on the kidney to increase water reabsorption from the nephron. Increased sodium levels also result in reduced aldosterone secretion. When aldosterone secretion declines, the rate at which sodium is reabsorbed from the nephrons also declines, resulting in the loss of sodium in the urine. In response to elevated sodium levels, a smaller-than-normal volume of urine containing a large concentration of sodium is produced.

When sodium levels decline in the blood, the concentration of the blood fluid is reduced. In response to the reduced concentration, ADH secretion is reduced, and aldosterone secretion increases. As a consequence the kidneys produce a large volume of dilute urine, which causes the concentration of the blood to increase. Aldosterone increases the rate of sodium ion reabsorption in the kidney, which causes the concentration of the blood to increase also.

urine production, and it increases sodium loss in the urine. As a result of the increased sodium loss, extracellular sodium levels decrease. The water loss that accompanies the sodium loss causes a decrease in blood volume and blood pressure.

Potassium ions

■ Electrically excitable tissues such as muscle and nerve are highly sensitive to slight changes in the extracellular potassium concentration. The extracellular concentration of **potassium ions** must be maintained within a narrow range for these tissues to function normally.

Aldosterone plays a major role in regulating the concentration of potassium ions in extracellular fluid. Circulatory system shock resulting from plasma loss, dehydration, and tissue damage (for example, in burn patients) causes extracellular potassium ions to become more concentrated than normal. In response, aldosterone secretion increases and causes potassium secretion to increase (Figure 18-12).

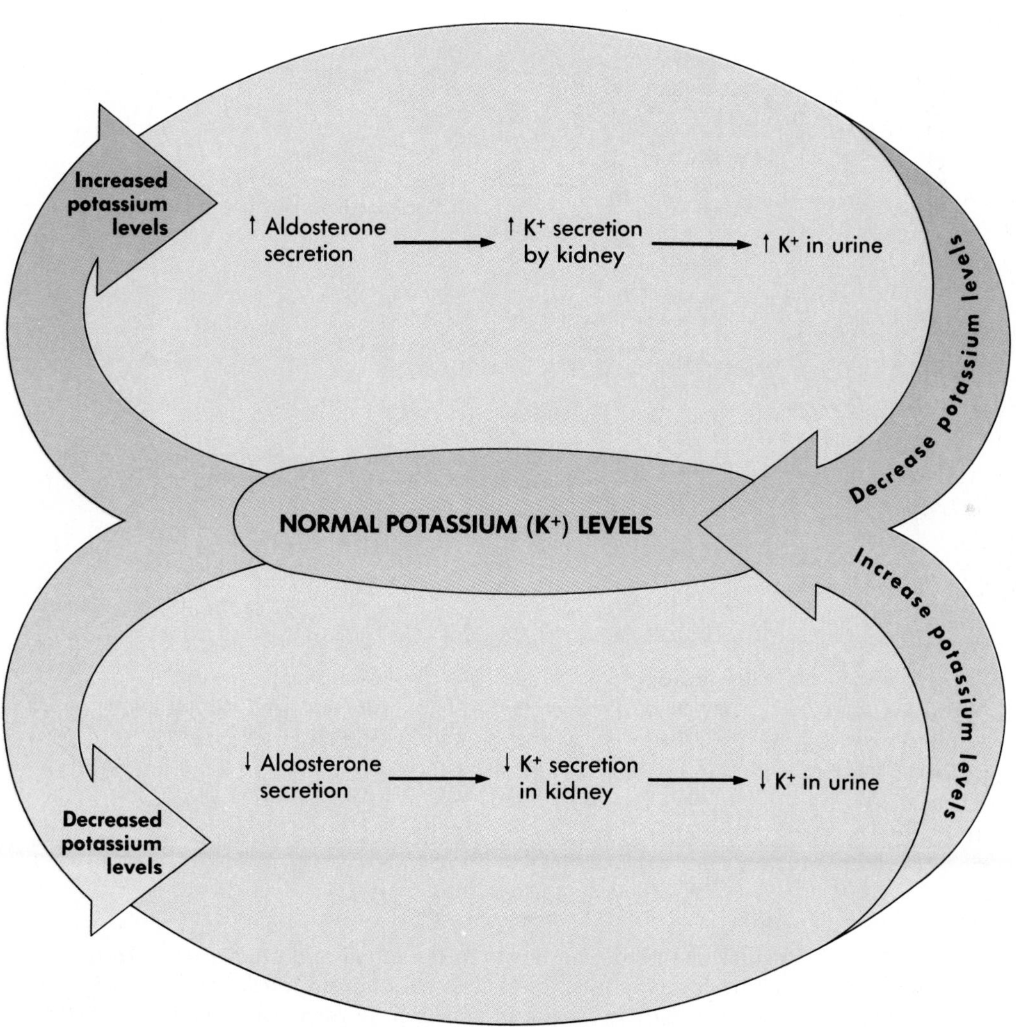

FIGURE 18-12 ■ Regulation of potassium levels in the blood and other extracellular fluid. Increased extracellular potassium stimulates aldosterone secretion. Aldosterone acts on the kidneys, resulting in increased potassium secretion into the nephron, which lowers the extracellular potassium concentration.

Decreased extracellular potassium inhibits aldosterone secretion, resulting in a reduced rate of potassium secretion into the nephron. The result is an increase in extracellular potassium.

Calcium ions

The extracellular concentration of **calcium ions,** like that of potassium ions, is maintained within a narrow range. Increases and decreases in the extracellular concentration of calcium ions have dramatic effects on the electrical properties of excitable tissues.

Parathyroid hormone, secreted by the parathyroid glands, increases extracellular calcium levels. The rate of parathyroid hormone secretion is regulated by the extracellular calcium ion level (Figure 18-13). Elevated calcium ion levels inhibit, and reduced levels stimulate, the secretion of parathyroid hormone. Parathyroid hormone causes osteoclasts to degrade bone and to release calcium and phosphate ions into the body fluids. It also increases the rate of calcium ion reabsorption from nephrons in the kidneys and increases the rate of calcium absorption from the intestine.

Vitamin D increases calcium levels in the blood

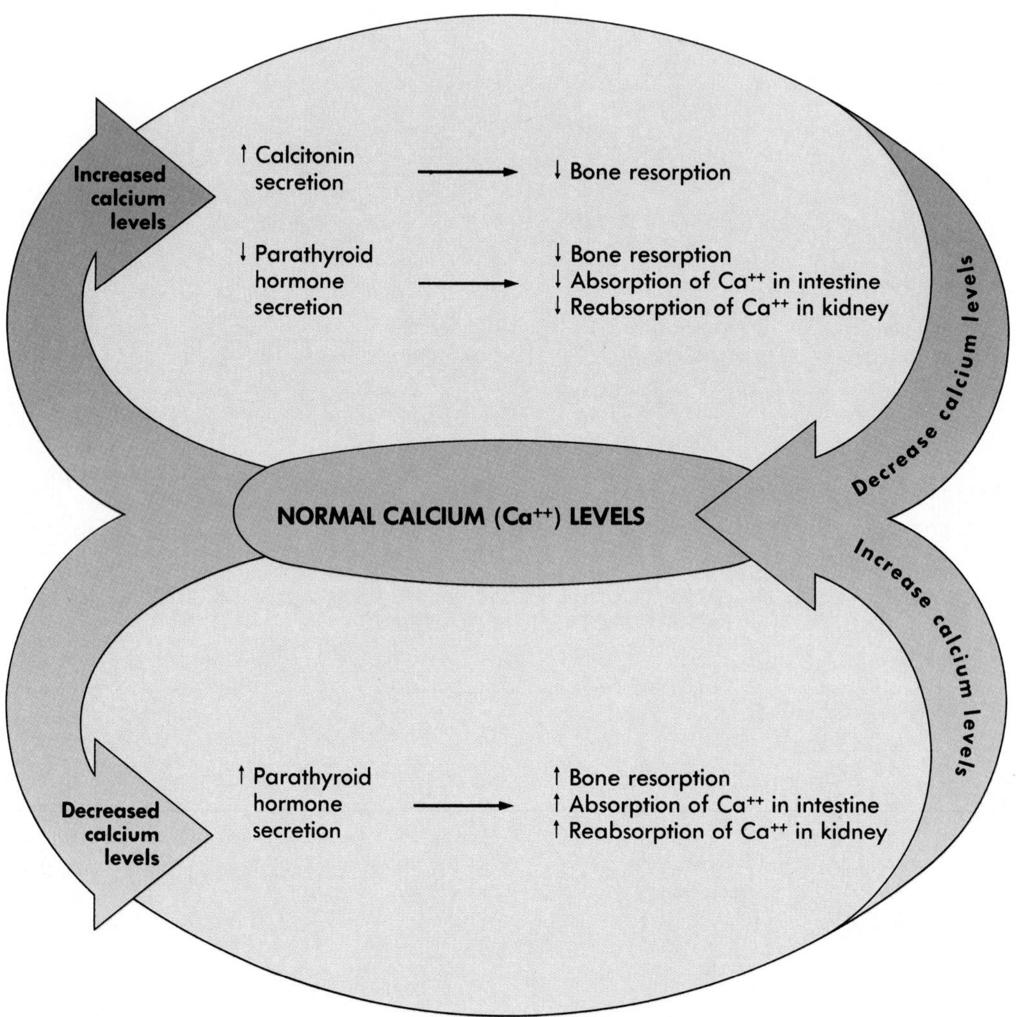

FIGURE 18-13 Regulation of calcium levels in the blood and other extracellular fluid. Increased plasma calcium levels reduce the rate of parathyroid hormone secretion, resulting in reduced extracellular calcium levels. In addition, in response to increased extracellular calcium levels, calcitonin is secreted by the thyroid gland. The increased calcitonin secretion also affects blood calcium levels through its action on bone. Calcitonin acts on bone to slow the rate of bone resorption, which also causes extracellular calcium levels to decrease.

Decreased plasma calcium levels stimulate parathyroid hormone secretion. Parathyroid hormone increases calcium absorption in the intestinal tract, resorption of calcium ions from bone, and reabsorption of calcium ions from filtrate in the kidney. Parathyroid hormone also increases the rate of active vitamin D synthesis in the kidney, which also stimulates calcium absorption in the gastrointestinal tract.

by increasing the rate of calcium absorption by the intestine and calcium reabsorption by the kidneys. Some vitamin D is consumed in food, and the rest is produced by the body (see Chapter 17). Parathyroid hormone affects the intestinal uptake of calcium ions because parathyroid hormone increases the rate of vitamin D production in the body.

Calcitonin (cal′sĭ-to′nin) is secreted by the thyroid gland. Calcitonin reduces blood levels of calcium when they are too high. Elevated blood levels of calcium cause the thyroid gland to secrete calcitonin, and low blood levels of calcium inhibit calcitonin secretion. Calcitonin increases the rate at which osteoblasts deposit calcium in bone, and reduces the rate of calcium removal from bone (see Figure 18-13).

Phosphate and sulfate ions

Some ions, such as phosphate ions and sulfate ions, are reabsorbed by active transport in the kidneys. The rate of reabsorption is slow, so that if the concentration of these ions in the filtrate exceeds the ability of the nephron to reabsorb them, the excess is excreted into the urine. As long as the concentration of these ions is low, nearly all of them are reabsorbed by active transport. This mechanism plays a major role in regulating the concentration of phosphate ions and sulfate ions in the body fluid.

REGULATION OF ACID-BASE BALANCE

The concentration of hydrogen ions in the body fluids is reported as the pH of the body fluids. The pH of body fluids is controlled by buffers in the body fluids, the respiratory system, and the kidneys. The pH of the body fluids is maintained between 7.35 and 7.45. Any deviation from that range is life threatening. Consequently the mechanisms that regulate body pH are critical for survival.

Buffers

Buffers are chemicals that resist a change in the pH of a solution when either acids or bases are added to the solution. The buffers found in the body fluids contain salts of either weak acids or weak bases that combine with hydrogen ions when excess hydrogen ions are added to a solution, or release hydrogen ions when bases are added to a solution. Because of these characteristics, buffers tend to keep the hydrogen ion concentration and thus the pH within a narrow range of values. The three principal classes of buffers in the body fluids are the proteins, the phosphate buffer system, and the bicarbonate buffer system.

Proteins and phosphate ions in the body fluids combine with a large number of hydrogen ions. When hydrogen ion levels increase, proteins and phosphate ions prevent a decrease in pH by combining with the hydrogen ions. Conversely, when hydrogen ion levels decrease, proteins and phosphate ions release hydrogen ions, preventing an increase in pH.

The following reaction illustrates how phosphate buffers work:

$$\underset{\text{monohydrogen phosphate ion}}{HPO_4^{-2}} + \underset{\text{hydrogen ion}}{H^+} \rightleftarrows \underset{\text{dihydrogen phosphate ion}}{H_2PO_4^-}$$

Monohydrogen phosphate (HPO_4^{-2}) combines with hydrogen ions to form dihydrogen phosphate ($H_2PO_4^{-1}$) when excess hydrogen ions are present. When hydrogen ion concentration declines, some of the hydrogens separate from the dihydrogen phosphate ions.

Proteins are able to function as buffers because amino acids in protein chains have side chains that function as weak acids and weak bases. There are many side chains that contain carboxyl groups (—COOH) or amino groups (—NH2). Both of these groups are able to function as buffers because of the following reactions:

$$\underset{\text{carboxyl group (ionized)}}{—COO^-} + \underset{\text{hydrogen ion}}{H^+} \rightleftarrows \underset{\text{carboxyl group}}{—COOH}$$

$$\underset{\text{amino group}}{—NH_2} + \underset{\text{hydrogen ion}}{H^+} \rightleftarrows \underset{\text{ammonium group}}{—NH_3^+}$$

The bicarbonate buffer system is unable to combine with as many hydrogen ions as can proteins and phosphate buffers, but it is critical because it can be regulated by the respiratory and urinary systems. Carbon dioxide (CO_2) combines with water (H_2O) to form carbonic acid (H_2CO_3), which, in turn, forms hydrogen ions (H^+) and bicarbonate ions (HCO_3^-) as follows:

$$\underset{\text{water}}{H_2O} + \underset{\text{carbon dioxide}}{CO_2} \rightleftarrows \underset{\text{carbonic acid}}{H_2CO_3} \rightleftarrows \underset{\text{hydrogen}}{H^+} + \underset{\text{bicarbonate}}{HCO_3^-}$$

The reaction between carbon dioxide and water is catalyzed by an enzyme, **carbonic anhydrase,** that is found in red blood cells (see Chapter 15). The enzyme accelerates the rate at which the reaction proceeds in either direction.

The higher the concentration of carbon dioxide, the greater is the amount of carbonic acid that is formed, and the greater the number of hydrogen ions and bicarbonate ions formed. This results in a

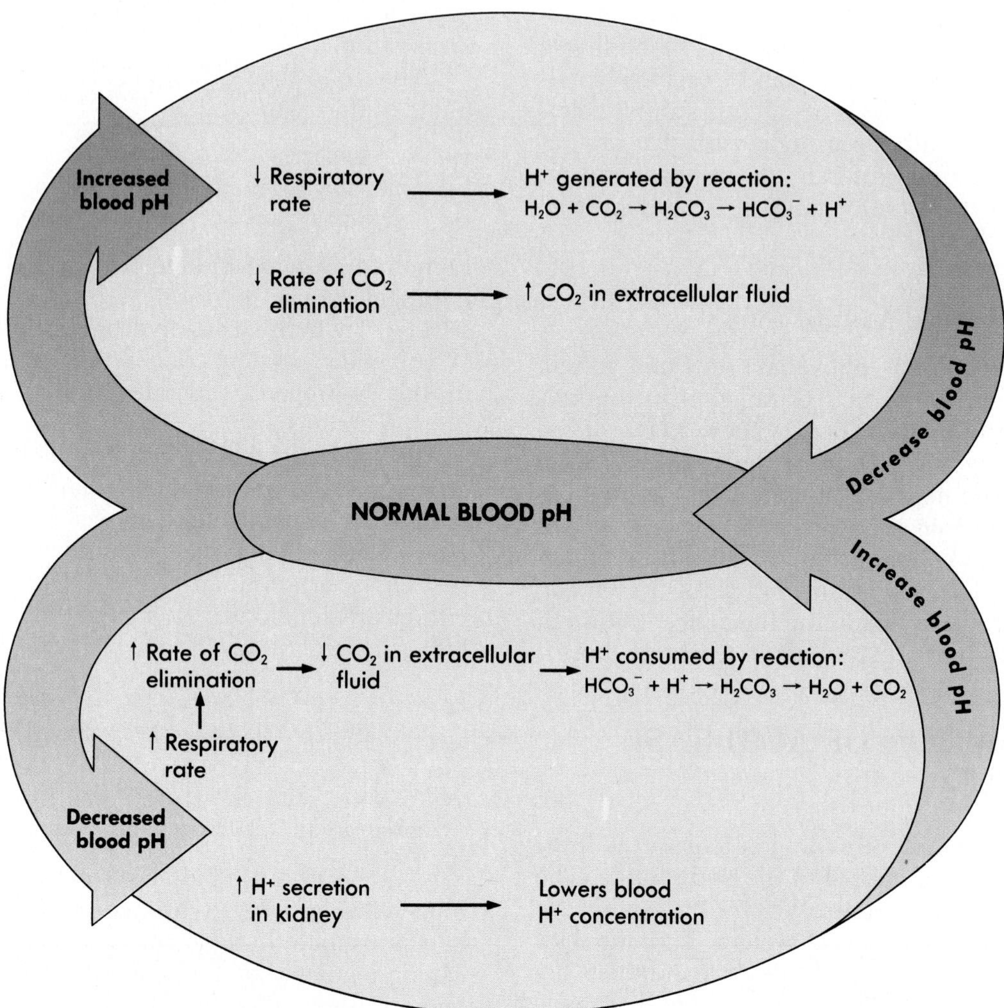

FIGURE 18-14 ■ Control of extracellular pH.
When pH increases, the rate of respiration slows, resulting in the accumulation of carbon dioxide in the circulatory system. The carbon dioxide reacts with water to produce carbonic acid, which dissociates to form hydrogen ions and bicarbonate ions, resulting in a decrease in pH. In addition, the rate of hydrogen ion secretion by the kidney declines.

When pH decreases, the rate of respiration increases, resulting in the elimination of carbon dioxide from the circulatory system. Because carbon dioxide is eliminated, the pH of the extracellular fluid increases. In addition, as the pH of the extracellular fluid declines, the rate at which hydrogen ions are secreted into the nephron increases, which also causes the pH to increase.

decreased pH. On the other hand, the reaction is reversible. If carbon dioxide levels decline, the equilibrium shifts in the opposite direction; that is, hydrogen and bicarbonate ions combine to form carbonic acid, which then dissociates to form carbon dioxide and water. Consequently the pH increases.

Respiratory System

The **respiratory system** responds rapidly to a change in pH and functions to bring the pH of body fluids back toward its normal range. Increasing carbon dioxide levels and decreasing body fluid pH stimulate neurons in the respiratory center of the brain and cause the rate and depth of ventilation to increase. As a result of the increased rate and depth of ventilation, carbon dioxide is eliminated from the body through the lungs at a greater rate, and the concentration of carbon dioxide in the body fluids decreases. As carbon dioxide levels decline, the concentration of hydrogen ions also declines. Therefore the pH is increased back to its normal range (Figure 18-14).

5 Under stressful conditions some people hyperventilate. Predict the effect of the rapid rate of ventilation on the pH of body fluids. In addition, explain why a person who is hyperventilating may benefit from breathing into a paper bag. ?

If carbon dioxide levels become too low or the pH of the body fluids is elevated, the rate and depth of respiration decline. As a consequence the rate at which carbon dioxide is eliminated from the body is reduced. Carbon dioxide then accumulates in the body fluids because it is continually produced as a by product of metabolism. As carbon dioxide accumulates in the body fluids, so do hydrogen ions, resulting in a decreased pH.

Kidneys

The nephrons of the kidneys secrete hydrogen ions into the urine and therefore can regulate pH of the body fluids directly. The kidney is a powerful regulator of pH, but it responds more slowly than the respiratory system. Cells in the walls of the tubules of the nephron are primarily responsible for the secretion of hydrogen ions. As the pH of the body fluids decreases below normal, the rate at which the kidneys secrete hydrogen ions increases. At the same time, reabsorption of bicarbonate ions increases. The increased rate of hydrogen ion secretion and the increased rate of bicarbonate ion reabsorption both function to cause the pH of the blood to increase toward its normal value. On the other hand, as the pH of the body fluids increases above normal, the rate of hydrogen ion secretion by the kidneys declines, and the amount of bicarbonate ions lost in the urine increases. Consequently the pH of the blood decreases toward its normal value (see Figure 18-14).

Acidosis and Alkalosis

Failure of the buffer systems, the respiratory system, or the urinary system to maintain normal pH levels can result in acidosis or alkalosis.

Acidosis

Acidosis occurs when the pH of the blood falls below 7.35. The central nervous system malfunctions, and the individual becomes disoriented and, as the condition worsens, possibly comatose. Acidosis is separated into two categories. **Respiratory acidosis** results when the respiratory system is unable to eliminate adequate carbon dioxide. Carbon dioxide accumulates in the circulatory system, causing the pH of the body fluids to decline. **Metabolic acidosis** results from the excessive production of acidic substances because of increased metabolism, or a decreased ability of the kidneys to eliminate hydrogen ions in the urine.

Alkalosis

Alkalosis occurs when the pH of the blood increases above 7.45. A major effect of alkalosis is hyperexcitability of the nervous system. Peripheral nerves are affected first, resulting in spontaneous nervous stimulation of muscles. Spasms and tetanic contractions result, as may extreme nervousness or convulsions. Tetany of respiratory muscles may cause death. **Respiratory alkalosis** results from hyperventilation, such as may occur in response to stress. **Metabolic alkalosis** usually results from the rapid elimination of hydrogen ions from the body such as during severe vomiting, or when excess aldosterone is secreted by the adrenal glands.

Disorders of the Urinary System

Glomerulonephritis is the inflammation of the filtration membrane within the renal corpuscle. The permeability of the filtration membrane increases, and plasma proteins and white blood cells enter the filtrate. The plasma proteins cause the urine volume to increase because they increase the osmotic concentration of the filtrate. Acute glomerular nephritis often occurs 1 to 3 weeks after a severe bacterial infection such as streptococcal sore throat or scarlet fever. Antigen-antibody complexes associated with the disease become deposited in the filtration membrane and cause inflammation. Acute glomerular nephritis normally subsides after several days.

Chronic glomerulonephritis is long term and usually progressive. The filtration membrane thickens and eventually is replaced by connective tissue. In its early stages chronic glomerular nephritis resembles the acute form. In the advanced stages, many of the renal corpuscles are replaced by connective tissue, and the kidney eventually becomes nonfunctional.

Cystitis is inflammation of the urinary bladder, usually resulting from bacterial infection. The inflammation causes diffuse pain in the lower back and a burning sensation during urination. The irritation of the urinary bladder results in a frequent urge to urinate.

Kidney stones are precipitates of substances such as uric acid and usually form in the renal pelvis. They can cause irritation and increase the chance of kidney infection, and small parts can break off and pass through the ureter. Passage of a kidney stone through the ureter is usually very painful. Ultrasound techniques have been developed to pulverize kidney stones without the use of surgery.

Renal failure may result from any condition that interferes with kidney function. **Acute renal failure** occurs when damage to the kidney is rapid and extensive. It leads to the accumulation of urea and other metabolites in the blood and to acidosis. If renal failure is complete, death may occur in 1 to 2 weeks. Acute renal failure may result from acute glomerular nephritis, or it may be caused by damage to or blockage of the renal tubules. Lack of blood supply or exposure to certain toxic substances can cause damage to the epithelial cells of the nephron and lead to acute renal failure.

Chronic renal failure is the result of permanent damage to so many nephrons that the remaining nephrons are inadequate for normal kidney function. Chronic renal failure may result from chronic glomerular nephritis, trauma to the kidneys, tumors, urinary tract obstruction by kidney stones, or severe lack of blood supply resulting from arteriosclerosis. Chronic renal failure leads to the inability to eliminate toxic metabolic byproducts. Water retention and edema result from the accumulation of solutes in the body fluids. Potassium levels become elevated and acidosis develops. The toxic effects of accumulated metabolic waste products are mental confusion, coma, and finally death when chronic renal failure is severe.

Dialysis is used when a person is suffering from severe acute or chronic kidney failure. The procedure substitutes for the excretory functions of the kidney. Renal dialysis is based on blood flow through tubes composed of a selectively permeable membrane. Blood is usually taken from an artery, passed through the tubes of the dialysis machine, and then returned to a vein. On the outside of the dialysis tubes is a fluid that contains the same concentration of solutes as the plasma except for the metabolic waste products. As a consequence, the metabolic wastes diffuse from the blood to the dialysis fluid. The dialysis membrane has pores that are too small to allow the plasma proteins to pass through them, and because the dialysis fluid contains the same beneficial solutes as the plasma, the net movement of these substances is zero.

Kidney transplants are performed on people who suffer from severe renal failure. A transplantation requires a donor kidney from an individual who has an immune system similar to that of the recipient. Usually the donor has suffered an accidental death and had granted permission to have his or her kidneys used for transplantation. In most cases, the transplanted kidney functions well. The recipient does have to take medication to prevent his or her immune system from rejecting the transplanted kidney. The major cause of kidney transplant failure is rejection by the immune system of the recipient.

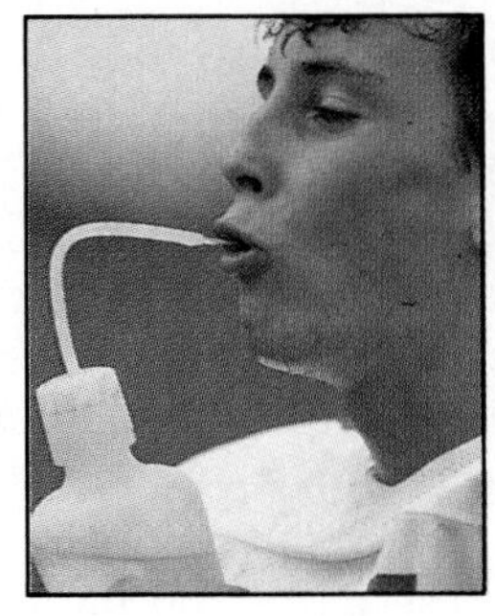

SUMMARY

The urinary system consists of the kidneys, ureters, urinary bladder, and urethra.

The urinary system eliminates wastes, controls blood volume, regulates blood ion concentration and pH, and regulates red blood cell production.

URINARY SYSTEM

Kidneys

Each kidney is behind the peritoneum, and surrounded by a renal capsule and a renal fat pad.

The ureter expands to form the renal pelvis within the renal sinus, and the renal pelvis has extensions called calyces.

The kidney is divided into an outer cortex and an inner medulla.

- Each renal pyramid in the medulla has a base that extends into the cortex.
- The apex of each renal pyramid projects to a calyx.

The functional unit of the kidney is the nephron. The parts of a nephron are the renal corpuscle, the proximal convoluted tubule, the loop of Henle, and the distal convoluted tubule.

The filtration membrane is formed by the glomerular capillaries, basement membrane, and the podocytes of Bowman's capsule.

Arteries and Veins

Arteries branch as follows: renal arteries give rise to branches that lead to afferent arterioles.

Afferent arterioles supply the glomeruli.

Efferent arterioles carry blood from glomeruli to peritubular capillaies.

Blood from the peritubular capillaries flows to the renal vein.

Ureters, Urinary Bladder, and Urethra

Ureters carry urine from the renal pelvis to the urinary bladder. The urethra carries urine from the urinary bladder to the outside of the body.

The ureters and urinary bladder are lined with transitional epithelium and have smooth muscle in their walls.

The internal and external urinary sphincter muscles regulate the flow of urine through the urethra.

URINE PRODUCTION

Urine is produced by the processes of filtration, reabsorption, and secretion.

Filtration

The renal filtrate passes from the glomerulus into Bowman's capsule and contains no blood cells and few blood proteins.

Filtration pressure is responsible for filtrate formation.

Reabsorption

About 99% of the filtrate volume is reabsorbed; 1% becomes urine.

Proteins, amino acids, glucose, fructose, sodium, potassium, calcium, bicarbonate, and chlorides are among the substances reabsorbed.

About 80% of the volume is reabsorbed in the proximal convoluted tubule and descending limb of the loop of Henle. About 19% is reabsorbed in the distal convoluted tubule and collecting duct.

Secretion

Some byproducts of metabolism and some drugs are actively secreted into the nephron.

REGULATION OF URINE CONCENTRATION AND VOLUME

Hormonal Mechanisms

ADH is secreted from the posterior pituitary when osmolality of blood increases or when blood pressure decreases. ADH increases the permeability to water of the distal convoluted tubule and collecting duct. It increases water reabsorption by the kidney.

Aldosterone increases the rate of sodium chloride reabsorption, potassium secretion, and hydrogen ion secretion.

Atrial natriuretic factor, secreted from the right atrium in response to increases in blood pressure, acts on the kidney to increase sodium and water loss in the urine.

Effect of Sympathetic Innervation on Kidney Function

Increased sympathetic activity decreases blood flow to the kidney, decreases filtration pressure, and decreases filtrate and urine formation.

URINE MOVEMENT

Increased volume in the urinary bladder stretches its wall and activates the micturition reflex.

Parasympathetic impulses cause contraction of the urinary bladder and relaxation of the internal and external urinary sphincters.

Higher brain centers control the micturition reflex. Stretch of the urinary bladder stimulates ascending neurons that carry impulses to the brain to create the desire to urinate.

BODY FLUID COMPARTMENTS

Water and its electrolytes are distributed in two major compartments.

Approximately 63% of the total body water is found within cells.

Approximately 37% of the total body water is found outside cells, mainly in tissue spaces, plasma of blood, and lymph.

Composition of Fluid in the Body Fluid Compartments

Intracellular fluid contains more potassium, phosphate, magnesium, and sulfate ions and more protein than extracellular fluid.

Extracellular fluid contains more sodium, chloride, and bicarbonate ions than intracellular fluid.

Exchange between Body Fluid Compartments

Water moves between compartments continually in response to pressure differences and osmotic differences between the compartments.

REGULATION OF EXTRACELLULAR FLUID COMPOSITION

The total amount of water and electrolytes in the body does not change unless the person is growing, gaining weight, or losing weight.

Thirst

The sensation of thirst increases if blood pressure decreases or the extracellular fluid becomes more concentrated.

IONS

Sodium ions

Sodium ions are one of the dominant extracellular ions.

Aldosterone increases sodium reabsorption from the filtrate.

Atrial natriuretic factor increases sodium loss in the urine.

Potassium ions

Aldosterone increases potassium secretion in the urine.

Increased blood levels of potassium stimulate, and decreased blood levels of potassium inhibit, aldosterone secretion.

Calcium ions

Parathyroid hormone secreted from the parathyroid glands increases extracellular calcium levels by causing bone resorption and increased calcium transport in the kidney and gut.

Parathyroid hormone increases vitamin D synthesis.

Calcitonin, secreted by the thyroid gland, inhibits bone resorption and functions to lower blood calcium levels.

Phosphate and sulfate ions

When phosphate and sulfate levels in the filtrate are low, nearly all phosphate and sulfate ions are reabsorbed. When levels are high, excess is lost in the urine.

REGULATION OF ACID-BASE BALANCE

Buffers There are three principal classes of buffers in the circulatory system that resist changes in the pH: protein, bicarbonate, and phosphate buffers.

Respiratory System The respiratory system functions to regulate pH. It responds rapidly. An increased respiratory rate raises the pH because the rate of carbon dioxide elimination is increased, and a reduced respiratory rate reduces the pH because the rate of carbon dioxide elimination is reduced.

Kidneys The kidneys excrete hydrogen ions in response to a decreasing blood pH, and they reabsorb hydrogen ions in response to an increasing blood pH.

Acidosis and Alkalosis Acidosis is a decrease in pH below normal, and alkalosis is an increase in pH above normal.

Acidosis Acidosis occurs when the pH of the blood falls below 7.35. The two major types are respiratory acidosis and metabolic acidosis.

Alkalosis Alkalosis occurs when the pH of the blood increases above 7.45. The two major types are respiratory alkalosis and metabolic alkalosis.

CONTENT REVIEW

1. Name the structures that make up the urinary system. List the functions of the urinary system.
2. What structures surround the kidney?
3. Describe the relationships of the renal pyramids, calyces, renal pelvis, and ureter.
4. What is the functional unit of the kidney? Name its parts.
5. Describe the blood supply of the kidney.
6. What are the functions of the ureters, urinary bladder, and urethra? Describe their structure.
7. Name the three general processes that are involved in the production of urine.
8. Describe the filtration barrier. What substances do not pass through it?
9. How do changes in blood pressure in the glomerulus affect the volume of filtrate produced?
10. What substances are reabsorbed in the nephron? What happens to most of the filtrate volume that enters the nephron?
11. In what parts of the nephron are large volumes of filtrate reabsorbed? In what part of the nephron is no filtrate reabsorbed?
12. In general, what substances are secreted into the nephron?
13. What effect does ADH have on urine volume? Name the factors that cause an increase in ADH secretion.
14. Where is aldosterone produced and what effect does it have on urine volume? What factors stimulate aldosterone secretion?
15. Where is atrial natriuretic factor produced, and what effect does it have on urine production?
16. What effect does sympathetic stimulation have on the kidneys?
17. Describe the micturition reflex. How is voluntary control of micturition accomplished?
18. What stimuli result in an increased sensation of thirst?
19. Describe how sodium levels are regulated in the body fluids.
20. Describe how potassium levels are regulated in the body fluids.

21. Describe how calcium levels are regulated in the body fluids.
22. Explain how the respiratory system and the kidneys respond to changes in the pH of body fluids.
23. Define respiratory acidosis and metabolic acidosis, and respiratory alkalosis and metabolic alkalosis.

CONCEPT REVIEW

1. Mucho McPhee decided to do an experiment after reading the urinary system chapter in his favorite anatomy and physiology textbook. He drank 2 liters of water in 15 minutes and then monitored the rate of urine production and urine concentration over the next two hours. What did he observe?
2. A man ate a full bag of salty (sodium chloride) potato chips but drank no liquids. What effect did this have on urine concentration and the rate of urine production? Explain the mechanisms involved.
3. During severe exertion in a hot environment, a person can lose up to 4 liters of sweat per hour (sweat is less concentrated than extracellular fluid in the body). What effect would this loss have on urine concentration and rate of production? Explain the mechanisms involved.
4. Which of the following symptoms are consistent with reduced secretion of aldosterone: excessive urine production, low blood pressure, high plasma potassium levels, and high plasma sodium levels. Explain.
5. Propose as many ways as you can to decrease glomerular filtration rate.
6. Swifty Trotts has an enteropathogenic *E. coli* infection that produces severe diarrhea. Diarrhea causes the production of a large volume of mucus that contains high concentrations of bicarbonate. What would this diarrhea do to his blood pH, urine pH, and respiration rate?
7. When Spanky and his mother went to a grocery store, Spanky eyed some candy he decided he wanted. His mother refused to buy it, so Spanky became angry. He held his breath for two minutes. What effect did this have on his body fluid pH? After the two minutes, what mechanisms were most important in reestablishing the normal body fluid pH?

ANSWERS TO  PREDICT QUESTIONS

1 *p. 463* Because the urethra of females is much shorter than the urethra of males, the female urinary bladder is more accessible to bacteria from the exterior. This accessibility is one of the reasons that urinary bladder infection is more common in women than in men.

2 *p. 465* If large amounts of glucose enter the nephron and are not reabsorbed, the glucose causes the concentration of solutes in the filtrate to increase. As a result the urine volume will increase because of the osmotic effect of the increased concentration of the filtrate. The glucose molecules attract water and, because they are trapped in the nephron, the amount of water that remains in the nephron is increased. A large volume of urine that contains glucose is a symptom of diabetes mellitus.

3 *p. 467* Without the normal active transport of sodium ions and chloride ions, their concentration within the nephron remains elevated. The normal movement of water out of the nephron cannot occur because of the osmotic effects of the sodium and chloride ions trapped in the nephron. The result is an increased urine volume.

4 *p. 469* Elevated blood pressure stretches the walls of the atria of the heart. In response, the atria release atrial natriuretic factor. The atrial natriuretic factor affects the kidneys and causes increased sodium excretion and increased urine volume.

5 *p. 477* Hyperventilation results in a greater-than-normal rate of carbon dioxide loss from the circulatory system. Because carbon dioxide is lost from the circulatory system, hydrogen ion concentration decreases, and the pH of body fluids increases. A person benefits from breathing into a paper bag, because this causes the person to rebreathe air that has a higher concentration of carbon dioxide. The result is an increase in carbon dioxide in the body. Consequently the hydrogen ion concentration increases, and pH decreases toward normal levels.

19

THE REPRODUCTIVE SYSTEM

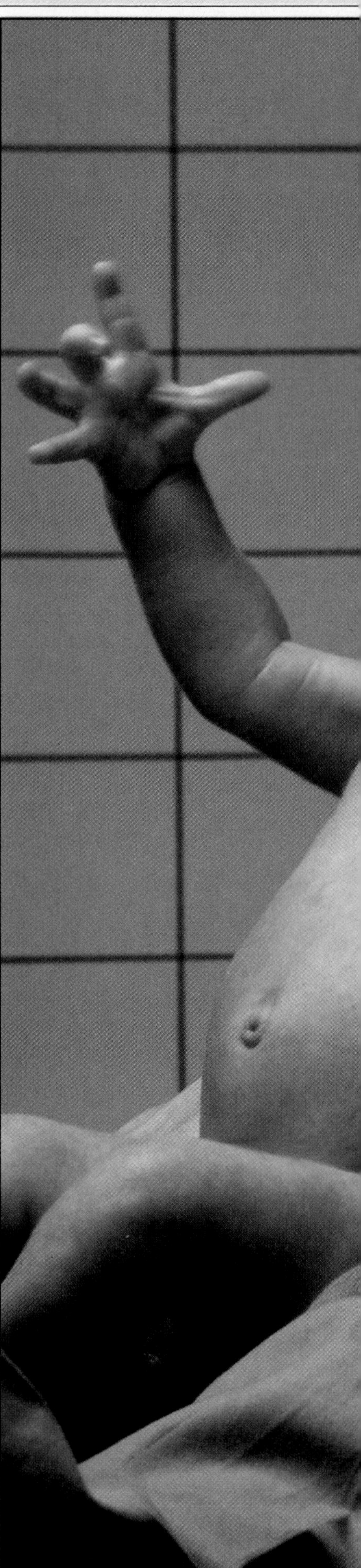

After reading this chapter you should be able to:

1 Describe the scrotum and explain the role of the dartos and cremaster muscles in temperature regulation of the testes.
2 Describe the structure of the testes, name the specialized cells of the testes, and give their functions.
3 Describe the process of spermatogenesis.
4 Describe the route sperm cells follow from the site of their production to the outside of the body.
5 Describe the structure of the penis.
6 Name the male reproductive glands, state where they empty into the duct system, and describe their secretions.
7 List the hormones that influence the male reproductive system and describe their functions.
8 Explain the events that occur during the male sexual act.
9 Name the organs of the female reproductive system and describe their structure.
10 Describe the anatomy and histology of the ovaries.
11 Discuss the development of the follicle and the oocyte, and describe the processes of ovulation and fertilization.
12 Describe the changes that occur in the ovary and uterus during the menstrual cycle.
13 List the hormones of the female reproductive system and explain how their secretion is regulated.
14 Define menopause and describe the changes that occur as a result.

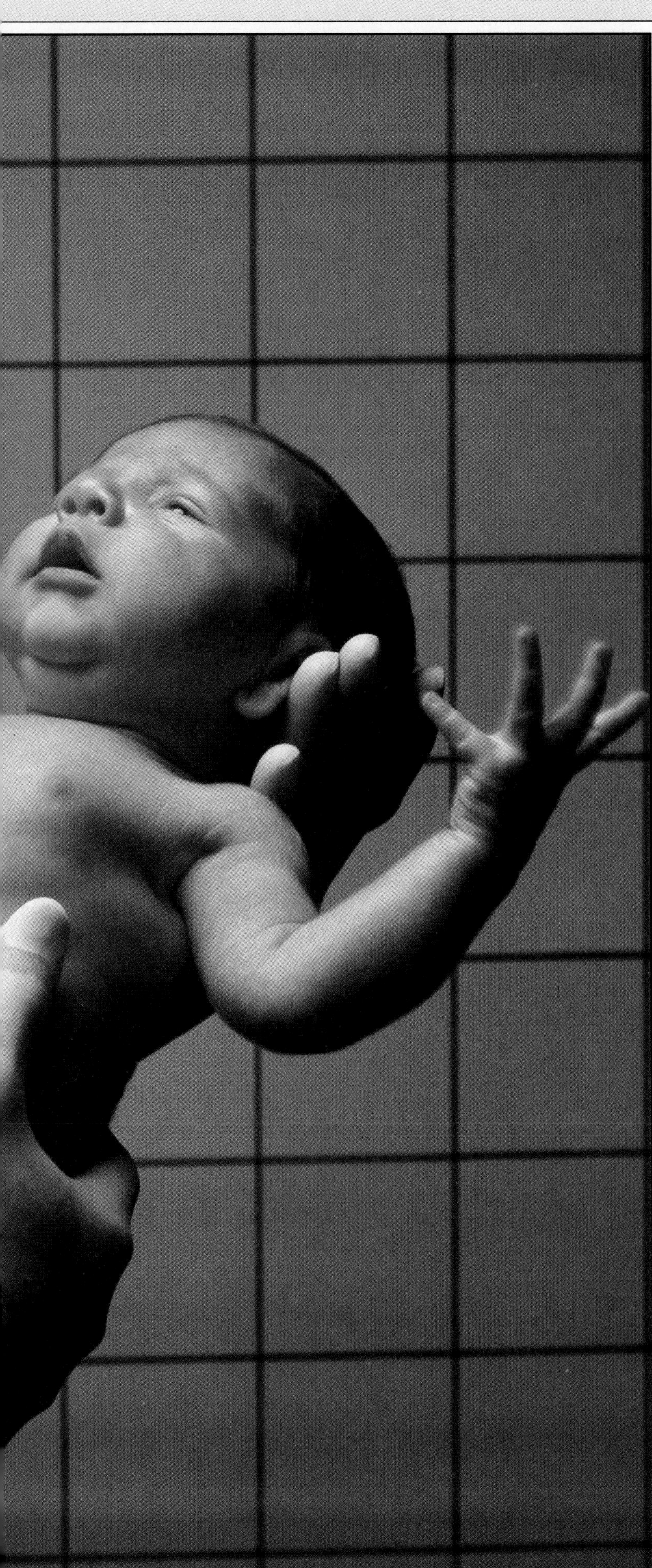

meiosis (mi-o'sis) Type of cell division in the testes and ovaries that produces sex cells, each having half the number of chromosomes as the parent cells.

menopause [Gr. *mensis*, month + *pausis*, cessation] Permanent cessation of the menstrual cycle.

menses [L. *mensis*, month] Loss of blood and tissue as the endometrium of the uterus sloughs away at the end of the menstrual cycle; occurring at approximately 28-day intervals in the nonpregnant female of reproductive age.

menstrual cycle Series of changes that occur in sexually mature, nonpregnant females resulting in menses; specifically includes the cyclical changes that occur in the uterus and ovary.

ovary One of two female reproductive glands located in the pelvic cavity; produces the ovum, estrogen, and progesterone.

ovulation Release of an oocyte from the mature follicle.

puberty [L. *pubertas*, grown up] Series of events that transform a child into a sexually mature adult; involves an increase in the secretion of all reproductive hormones.

semen [L., seed] Penile ejaculate; thick, yellowish-white, viscous fluid containing sperm cells and secretions of the testes, seminal vesicles, prostate gland, and bulbourethral glands.

spermatogenesis Formation and development of sperm cells.

spermatozoon, pl. spermatozoa [Gr., *sperma*, seed + *zoon*, animal] Sperm cell. Male gamete or sex cell, composed of a head, midpiece, and tail. Spermatozoa contain the genetic information transmitted by the male.

testis, pl. testes One of two male reproductive glands located in the scrotum; produces testosterone and sperm cells.

testosterone Hormone secreted primarily by the testes; aids in spermatogenesis, controls maintenance and development of male reproductive organs and secondary sexual characteristics, and influences sexual behavior.

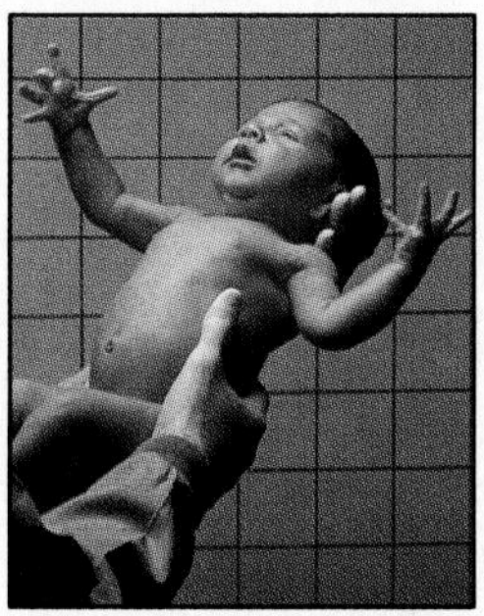

The major function of the male and female reproductive systems is to produce offspring. Although a functional reproductive system is not required for the survival of the individual, it is essential for the survival of the species. Some individuals must produce offspring to keep the species from becoming extinct.

Reproductive organs in males and females produce sex cells. The reproductive organs sustain the sex cells, transport them to the site where fertilization may occur, and, in the female, nurture the developing offspring both before and, for a time, after birth.

Reproductive organs produce hormones that play important roles in the development and maintenance of the reproductive system. These hormones help determine sexual characteristics, influence sexual behavior, and play a major role in regulating the physiology of the reproductive system (Figure 19-1).

The formation of sex cells in males and females occurs by a special type of cell division called **meiosis** (mi-o'sis). For both males and females meiosis begins in cells that contain 23 pairs of chromosomes (46 chromosomes). Before the beginning of cell division the genetic material is duplicated. During meiosis two cell divisions occur. In the male, for each cell that begins the process four sex cells, called **spermatozoa** or **sperm cells,** are produced. Each sex cell contains 23 chromosomes with one chromosome coming from each of the original 23 pairs of chromosomes. In the female, during each of the meiotic divisions, distribution of the cytoplasm among the sex cells is unequal. Most of the cytoplasm remains with one of the resulting cells, forming the female sex cell, or **oocyte**, with 23 chromosomes. The cells receiving little cytoplasm are called **polar bodies,** and they are not functional as sex cells.

Fertilization results when a male sex cell and a female sex cell unite. In the process, the 23 chromosomes from each of the sex cells combine to form 23 pairs of chromosomes. All of the cells of the body contain 23 pairs of chromosomes except for the sex cells.

MALE REPRODUCTIVE SYSTEM

The male reproductive system consists of the primary reproductive organs, the testes, and the secondary reproductive organs, which include the scrotum, epididymis, ductus deferens, seminal vesicles, urethra, prostate gland, bulbourethral glands, and penis (see Figure 19-1, *A*). The sperm cells, are very heat sensitive and must develop at a temperature slightly less than normal body temperature. The testes, in which the sperm cells develop, are located outside the body cavity in the scrotum, where the temperature is lower. Sperm cells are transported from the testes to the epididymis, which lies on the external surface of each testis, and then through the ductus deferens into the pelvic cavity. The ductus deferens joins the ducts of the seminal vesicles to form the ejaculatory duct, which projects through the prostate gland, and empties into the urethra. The urethra exits from the pelvis and passes through the penis to the outside of the body.

Scrotum

The scrotum contains the testes and is divided into two internal compartments by a connective tissue septum. Externally the scrotum consists of skin. Beneath the skin is a layer of loose connective tissue and a layer of smooth muscle called the **dartos** (dar'tōs; to skin) **muscle.**

In cold temperatures the dartos muscle contracts, causing the skin of the scrotum to become firm and wrinkled and reducing the scrotum's overall size. At the same time extensions of the abdominal muscles called **cremaster muscles,** which extend into the scrotum, contract (see Figure 19-4). Consequently the testes are pulled nearer to the body and their temperature is raised. During warm weather or exercise, the dartos and cremaster muscles relax, the skin of the scrotum becomes loose and thin, and the testes descend away from the body, which lowers their temperature. The response of the dartos and cremaster muscles is important in the regulation of temperature in the testes. If the testes becomes too warm or too cold, normal spermatogenesis does not occur.

Testes

The **testes** are oval organs, each about 4 to 5 cm long, within the scrotum (Figure 19-2). The outer portion of each testis consists of a thick, white connective tissue capsule. Extensions of the capsule project into the interior of the testis and divide each testis into about 250 cone-shaped **lobules** (Figure 19-3, *A*). The lobules contain **seminiferous** (sem'ĭ-nif'er-us; seed carriers) **tubules,** in which sperm cells develop. Delicate connective tissue surrounding the tubules contains clusters of endocrine cells called **interstitial cells** or **cells of Leydig,** which secrete testosterone.

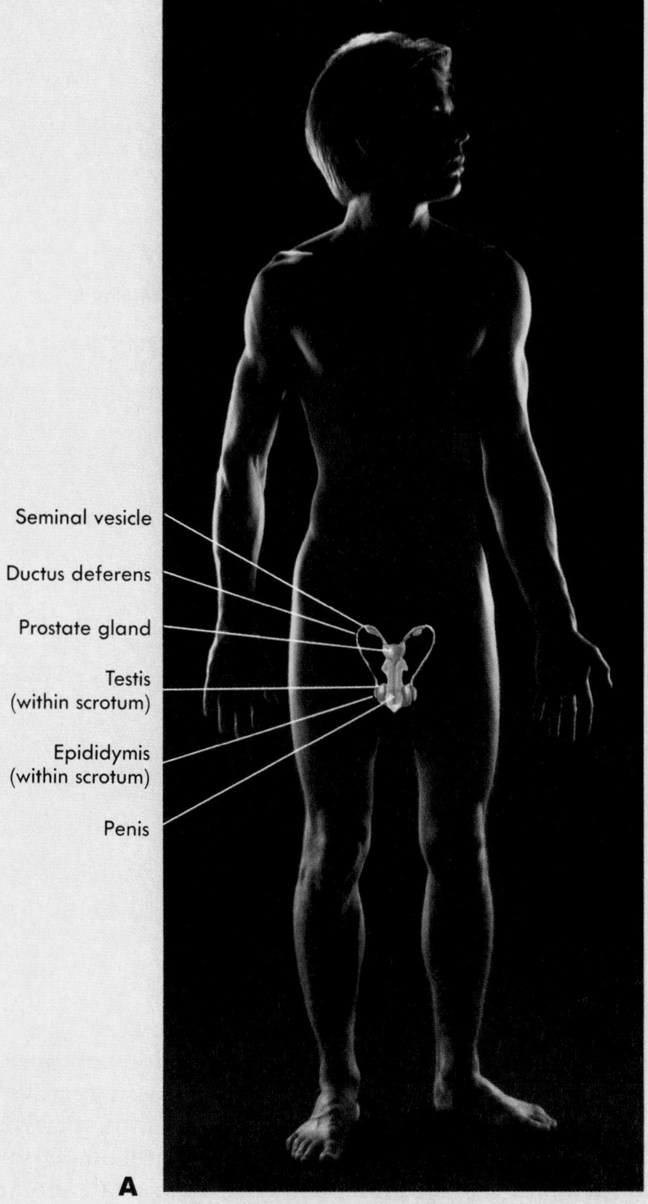

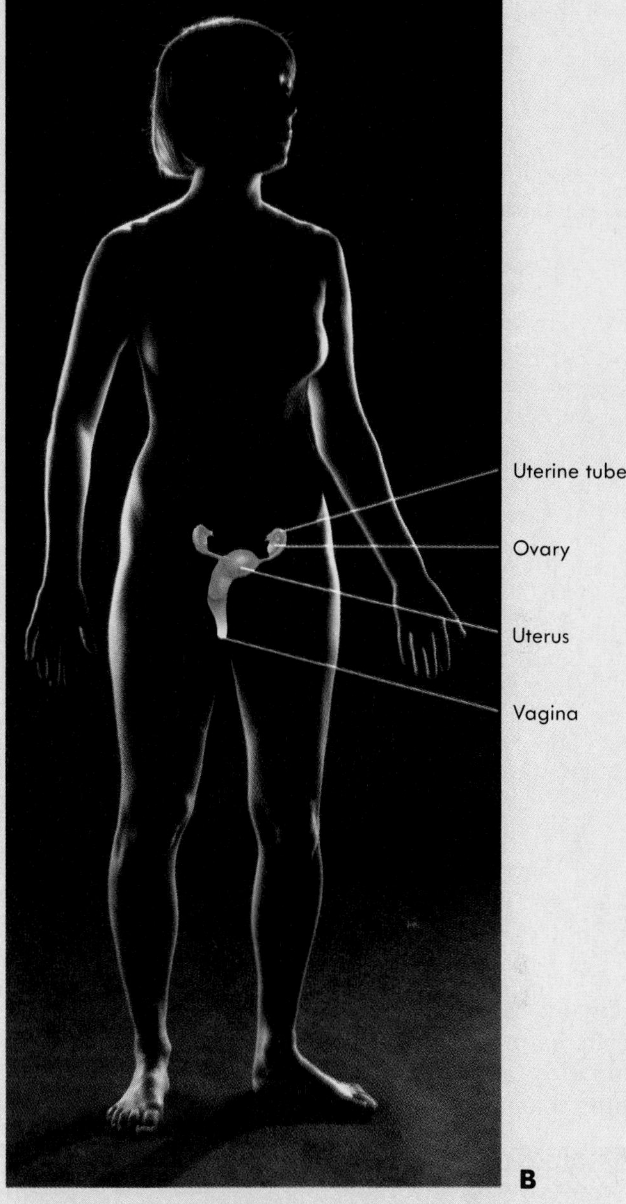

FIGURE 19-1 ■ Organs of the reproductive systems.
A The male reproductive system.
B The female reproductive system.

The Reproductive System

■ **MAJOR COMPONENTS**

MALE

Penis, testes, epididymis, ductus deferens, seminal vesicle, and prostate gland

FEMALE

Vagina, ovary, uterine tube, and uterus

■ **MAJOR FUNCTIONS**

Performs the processes of reproduction
Controls sexual functions and behavior

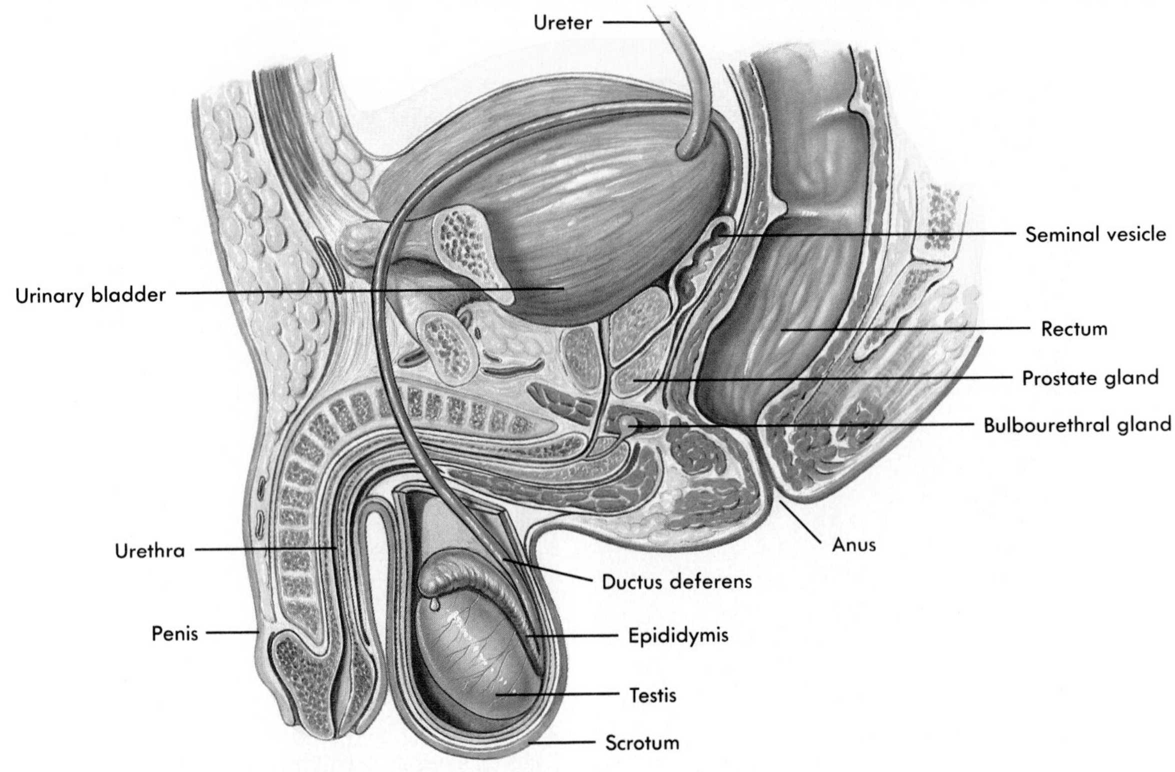

FIGURE 19-2 ■ Sagittal view of the male pelvis.
Sagittal section of the male pelvis showing the male reproductive structures.

The seminiferous tubules empty into a tubular network called the **rete** (re′te; net) **testis.** The rete testis empties into 15 to 20 tubules called **efferent ductules,** and the efferent ductules exit the testis into the epididymis (see Figure 19-3, *A*).

> ■ The testes develop in the abdominopelvic cavity. They move from the abdominopelvic cavity through the *inguinal canal* to the scrotum. The descent of the testes occurs either before birth or, sometimes, shortly after birth. Failure of the testes to descend into the scrotal sac causes sterility because of the inhibiting effect of normal body temperature on spermatogenesis. After the testes descend, the inguinal canal narrows permanently, but it represents a weak spot in the abdominal wall. If the inguinal canal enlarges or ruptures, this can cause an inguinal hernia through which a loop of intestine can protrude. This herniation can be quite painful and even very dangerous, especially if the inguinal canal compresses the intestine and cuts off its blood supply. Fortunately, inguinal hernias can be repaired surgically. ■

Spermatogenesis

■ **Spermatogenesis** is the formation of sperm cells. Before puberty, the testes remain relatively simple and unchanged from the time of their initial development. The interstitial cells are not prominent and the seminiferous tubules are small and not yet functional. At the time of puberty, the interstitial cells increase in number and size. The seminiferous tubules enlarge, and spermatogenesis begins.

The seminiferous tubules contain **germ cells** and **Sertoli** (ser-to′le; named for an Italian histologist) **cells** (see Figure 19-3, *B*). Sertoli cells are large cells that extend from the periphery to the lumen of the seminiferous tubule. They nourish the germ cells and probably produce a number of hormones.

Germ cells are scattered between the Sertoli cells. The most peripheral cells are **spermatogonia** (sper′mă-to-go′ne-ah), which divide through mitosis. Some daughter cells produced from these mitotic divisions remain as spermatogonia and continue to divide by mitosis. Other daughter cells form **primary spermatocytes**, which divide by meiosis.

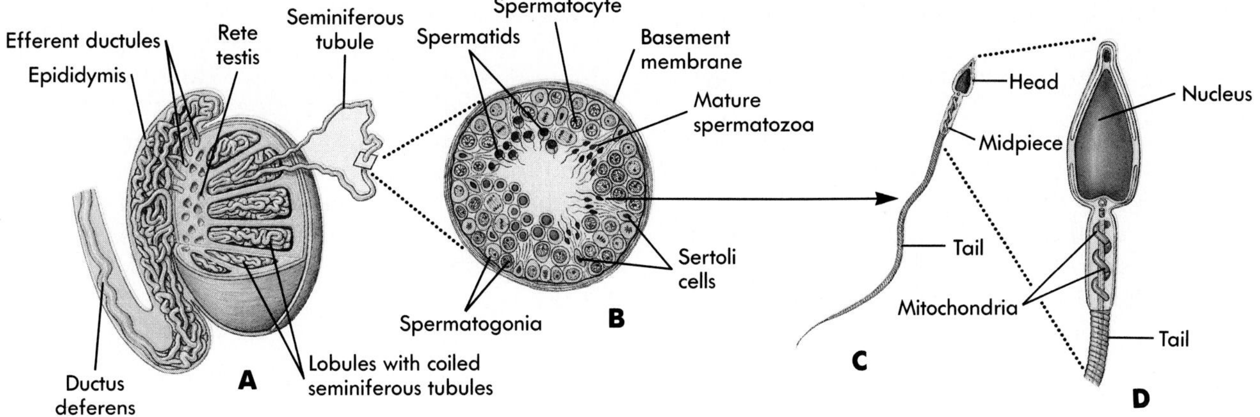

FIGURE 19-3 ■ **Structure of the testis.**
A Gross anatomy of the testis with a section cut away to reveal internal structures.
B Cross section of a seminiferous tubule. Spermatogonia are near the periphery, and mature sperm cells are near the lumen of the seminiferous tubule.
C The head, midpiece, and tail of a sperm cell.
D The head of a sperm cell contains the nucleus.

Each primary spermatocyte (with 23 pairs of chromosomes) passes through the first meiotic division to produce two **secondary spermatocytes.** Each secondary spermatocyte undergoes a second meiotic division to produce two smaller sex cells called **spermatids,** each having 23 chromosomes with one chromosome coming from each of the 23 pairs of chromosomes in the primary spermatocyte. Each spermatid develops a head, midpiece, and flagellum (tail) and becomes a sperm cell or spermatozoon (see Figure 19-3, *B-D*). Thus, each sperm cell contains half the number of chromosomes found in other cells of the body.

Ducts

After their production in the seminiferous tubules, sperm cells leave the testes and pass through a series of ducts to reach the exterior of the body.

Epididymis

The **efferent ductules** lead from the testis to a coiled series of tubules that form a comma-shaped structure on the posterior side of the testis called the **epididymis** (ep-ĭ-did′ĭ-mis) (see Figures 19-2 and 19-3, *A*). Maturation of the sperm cells continues within the epididymis. Sperm cells taken directly from the testes are not capable of fertilizing ova, but after spending several days in the epididymis, the sperm cells mature and develop the capacity to function as sex cells.

Ductus deferens

The **ductus deferens** or **vas deferens** extends from the epididymis through the abdominal wall by way of the inguinal canal. The ductus deferens then crosses the lateral wall of the pelvic cavity, travels over the ureter, and loops over the posterior surface of the urinary bladder to approach the prostate gland (see Figures 19-2 and 19-4). Just before reaching the prostate gland, the ductus deferens increases in diameter to become the **ampulla of the ductus deferens** (see Figure 19-4). The wall of the ductus deferens contains smooth muscle. Peristaltic contractions of these smooth muscles propel the sperm from the epididymis through the ductus deferens.

As the ductus deferens emerges from the epididymis and ascends along the posterior side of the testis, it becomes associated with the blood vessels and nerves that supply the testis. These structures and their coverings constitute the **spermatic cord** (see Figure 19-4). The spermatic cord consists of the ductus deferens, testicular artery and veins, lymph vessels, testicular nerve, cremaster muscle, and a connective tissue sheath.

Ejaculatory duct

Adjacent to the ampulla of each ductus deferens is a sac-shaped gland called the **seminal vesicle.** A short duct from the seminal vesicle joins the distal ductus deferens at the ampulla. The duct formed by the joining of the ductus deferens and the seminal vesicle duct is called the **ejaculatory duct.** Each ejaculatory duct extends through the prostate gland and ends by opening into the urethra (see Figure 19-4).

Urethra

The male **urethra** (u-re'thrah) extends from the urinary bladder to the distal end of the penis (see Figures 19-2 and 19-4). The urethra is a passageway for both urine and male reproductive fluids. However, urine and male reproductive fluids do not exit the urethra at the same time. While male reproductive fluids are passing through the urethra, a reflex causes the urinary sphincter muscles to contract tightly to keep urine from passing from the urinary bladder through the urethra. Many minute mucus-secreting glands, located in the epithelial lining of the urethra, empty their secretions into the urethra.

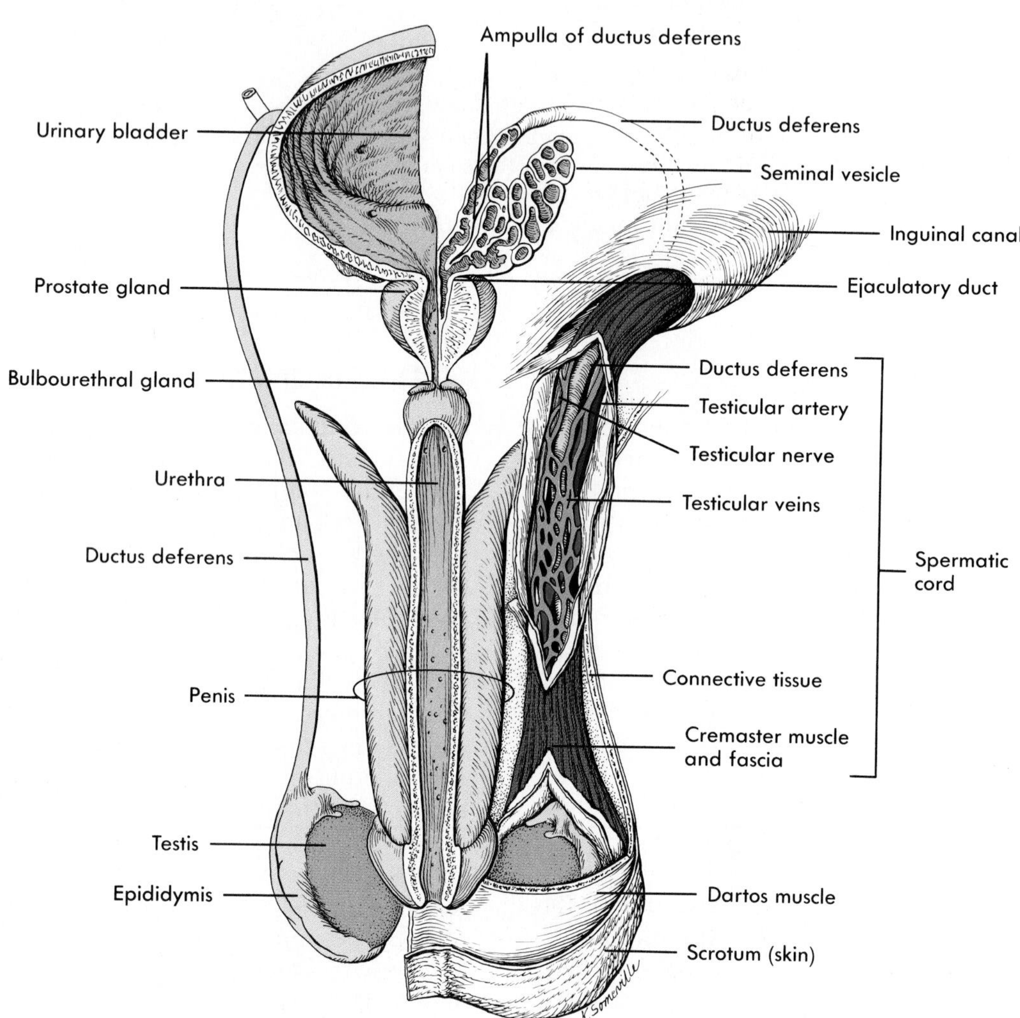

FIGURE 19-4 Frontal view of the male reproductive organs.
Frontal view of the testes, epididymis, ductus deferens, penis, and glands of the male reproductive system. The testis is viewed in the scrotal sac with the dartos and cremaster muscles on one side. The ductus deferens extends from the epididymis in the scrotal sac, and passes through the inguinal canal to the prostate gland. Note that the ductus deferens joins the artery, vein, cremaster muscles, and nerves that supply the testes to form the spermatic cord.

Penis

The penis contains three columns of erectile tissue (Figure 19-5), and engorgement of this erectile tissue with blood causes the penis to enlarge and become firm, a process called **erection.** The penis is the male organ of copulation and functions in the transfer of sperm cells from the male to the female. Two of the erectile columns form the dorsal portion and the sides of the penis and are called the **corpora cavernosa.** The third and smaller erectile column occupies the ventral portion of the penis and is called the **corpus spongiosum**, which expands over the distal end of the penis to form a cap, the **glans penis.** The urethra passes through the corpus spongiosum, penetrates the glans penis, and opens as the **external urethral orifice.**

The shaft of the penis is covered by skin that is loosely attached to the connective tissue of the penis. The skin is firmly attached at the base of the glans penis, and a thinner layer of skin tightly covers the glans penis. The skin of the penis, especially the glans penis, is well supplied with sensory receptors. A loose fold of skin called the **prepuce** (pre′pus), or **foreskin,** covers the glans penis. **Circumcision** is the surgical removal of the prepuce.

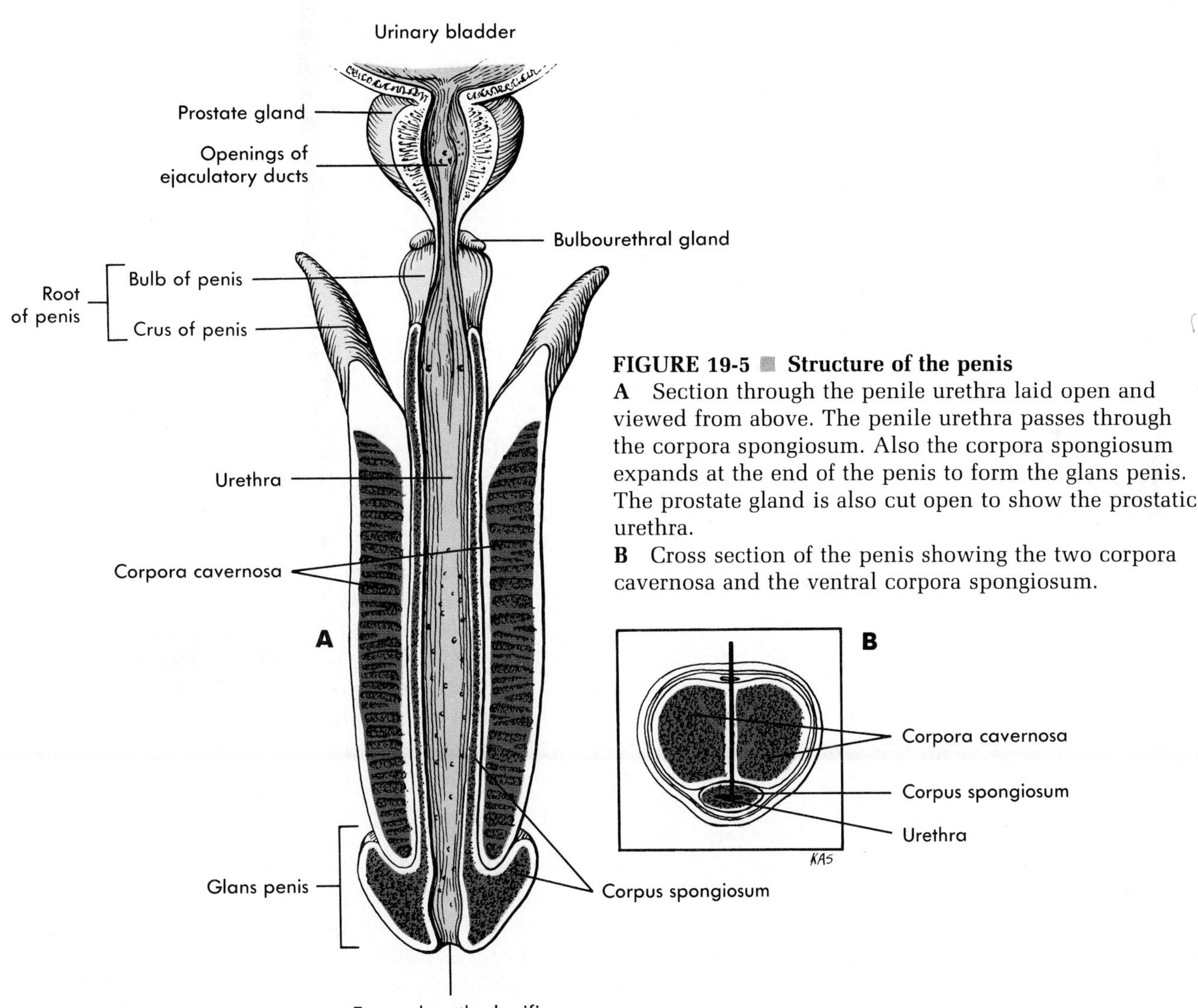

FIGURE 19-5 Structure of the penis
A Section through the penile urethra laid open and viewed from above. The penile urethra passes through the corpora spongiosum. Also the corpora spongiosum expands at the end of the penis to form the glans penis. The prostate gland is also cut open to show the prostatic urethra.
B Cross section of the penis showing the two corpora cavernosa and the ventral corpora spongiosum.

Glands

The **seminal vesicles** are glands consisting of many saclike structures located next to the ampulla of the ductus deferens (see Figures 19-2 and 19-4). Each seminal vesicle is about 5 cm long and tapers into a short duct that joins the ductus deferens to form an ejaculatory duct.

The **prostate** (pros′tāt) **gland** consists of both glandular and muscular tissue and is about the size and shape of a walnut (see Figures 19-2, 19-4, and 19-5, *A*). The prostate surrounds the urethra and the two ejaculatory ducts. The gland consists of a capsule and numerous partitions. The cells lining the partitions secrete prostatic fluid. There are 10 to 20 short ducts that carry secretions of the prostate gland to the urethra.

1 The prostate gland can enlarge for several reasons, including infections and tumors. Noncancerous enlargement of the prostate occurs in many elderly men. Cancer of the prostate is the second most common cause of male death from cancer in the United States (less than lung cancer and more than colon cancer). The detection of enlargement or changes in the prostate is important. Suggest a way that the prostate gland can be examined by palpation for any abnormal changes (see Figure 19-2). ?

The **bulbourethral glands** are a pair of small glands located near the base of the penis (see Figures 19-2 and 19-5). In young adults they are each about the size of a pea, but they decrease in size with age. A single duct from each gland enters the urethra.

Secretions

Semen is a mixture of sperm cells and secretions from the male reproductive glands. The seminal vesicles produce about 60% of the fluid, the prostate gland contributes approximately 30%, the testes contribute 5%, and the bulbourethral glands contribute 5%. **Emission** is the discharge of semen into the urethra. **Ejaculation** is the forceful expulsion of semen from the urethra caused by the contraction of smooth muscle in the wall of the urethra and skeletal muscles surrounding the base of the penis.

The bulbourethral glands and the mucous glands of the urethra produce a mucous secretion up to several minutes before ejaculation. This mucus lubricates the urethra, neutralizes the contents of the normally acidic urethra, provides a small amount of lubrication during intercourse, and helps reduce the acidity in the vagina.

Testicular secretions include sperm cells and a small amount of fluid. The thick, mucoid secretion of the seminal vesicles has a relatively low pH and contains nutrients that nourish the sperm cells. The seminal vesicle secretions also contain proteins that weakly coagulate after ejaculation, and prostaglandins, which may cause contractions of the female reproductive tract that help transport sperm cells through the female reproductive tract.

The thin, milky secretions of the prostate have a rather high pH and neutralize the acidic urethra and the acidic secretions of the testes, the seminal vesicles, and the vagina. The increased pH is important for normal sperm cell function.

Before ejaculation, the ductus deferens begins to contract rhythmically, propelling sperm cells and testicular fluid from the epididymis through the ductus deferens. Contractions of the ductus deferens, seminal vesicles, and ejaculatory ducts cause the sperm cells, testicular secretions, and seminal fluid to move into the urethra, where they mix with prostatic secretions released as a result of contraction of the prostate.

PHYSIOLOGY OF MALE REPRODUCTION

The male reproductive system depends on both hormonal and neural mechanisms to function normally. Hormones are primarily responsible for the development of reproductive structures and maintenance of their functional capacities, for the development of secondary sexual characteristics, for the control of spermatogenesis, and they influence sexual behavior. Neural mechanisms are primarily involved in controlling the sexual act and in the expression of sexual behavior.

Regulation of Sex Hormone Secretion

Hormonal mechanisms that influence the male reproductive system involve the hypothalamus of the brain, the pituitary gland, and the testes (Figure 19-6). **Gonadotropin-releasing hormone (GnRH)** is released from neurons in the hypothalamus and passes to the anterior pituitary gland. In response to GnRH, cells in the anterior pituitary gland secrete two hormones, **luteinizing** (lu′te-ĭ-nīz-ing) **hormone (LH)** and **follicle-stimulating hormone (FSH),** into

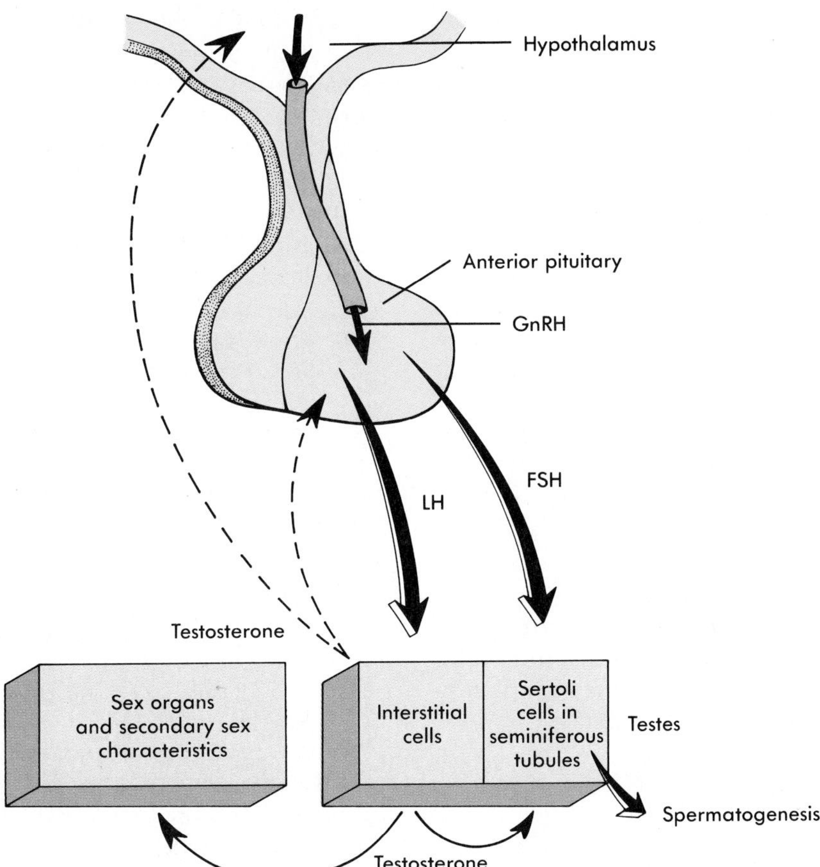

FIGURE 19-6 ■ Regulation of reproductive hormone secretion in males. Gonadotropin-releasing hormone *(GnRH)* from the hypothalamus stimulates the secretion of luteinizing hormone *(LH)* and follicle-stimulating hormone *(FSH)* from the anterior pituitary. LH stimulates the interstitial cells to secrete testosterone. FSH binds to the seminiferous tubules and stimulates spermatogenesis. Testosterone has a negative-feedback effect on the hypothalamus and pituitary gland to reduce LH and FSH secretion. Testosterone has a stimulatory effect on the sex organs and secondary sex characteristics, as well as on the Sertoli cells.

the blood. LH and FSH are named for their functions in females, but they are also important in males.

LH binds to the interstitial cells of Leydig in the testes and causes them to secrete testosterone. This hormone was once referred to as interstitial cell–stimulating hormone (ICSH) because it stimulated interstitial cells of the testes to secrete testosterone, but it was later discovered to be identical to LH found in the female. Consequently it is now simply called LH. FSH binds primarily to Sertoli cells in the seminiferous tubules and promotes spermatogenesis.

■ For GnRH to stimulate LH and FSH release, the pituitary gland must be exposed to a series of brief increases and decreases in GnRH. GnRH maintained at a high level in the circulatory system for days or weeks causes the anterior pituitary cells to become insensitive to GnRH. GnRH can be produced synthetically and is useful in treating some people who are infertile. The synthetic GnRH must be administered in small amounts in frequent pulses or surges. GnRH can also inhibit reproduction, since chronic (long-term) administration of GnRH can sufficiently reduce LH and FSH levels to prevent sperm production in males or ovulation in females. ■

Puberty

Puberty is the sequence of events by which a child is transformed into a young adult. The reproductive system matures and assumes its adult functions, and the structural differences between adult males and females become more apparent. The first signs of puberty may appear as early as age 8 in girls, and the process is largely completed by age 16; in boys, puberty commonly begins at ages 10 to 12 and is largely completed by age 18. Before puberty small amounts of testosterone secreted by the testes and the adrenal gland inhibit GnRH, LH, and FSH secretion. At puberty, the hypothalamus and anterior pituitary gland become less sensitive to the inhibitory effect of testosterone, and the rate of GnRH, LH, and FSH secretion increases. Elevated FSH levels promote spermatogenesis, and elevated LH levels cause the interstitial cells to secrete larger amounts of testosterone. Testosterone still has a negative-feedback effect on the hypothalamus and anterior pituitary gland, but it does not completely suppress GnRH, LH, and FSH secretion.

Effects of Testosterone

Testosterone is the major male hormone secreted by the testes. During puberty, testosterone encourages the development of male secondary sexual characteristics; it causes the enlargement and differentiation of the male genitals and reproductive duct system and is necessary for spermatogenesis. The **secondary sexual characteristics** are those structural and behavioral changes, other than reproductive organs, that develop at puberty and that distinguish males from females (Table 19-1). After puberty, testosterone maintains the adult structure of the male genitals, reproductive ducts, and secondary sex characteristics.

Some athletes, especially those who depend on muscle strength, ingest synthetic androgens (hormones that have testosterone-like effects), commonly called anabolic steroids or simply *steroids,* in an attempt to increase muscle mass. Many of the synthetic androgens are structurally different from testosterone. Their effect on muscle is greater than their effect on the reproductive organs. However, they are often taken in large amounts, and they do have some ability to influence the reproductive system. Large doses of synthetic androgens have a negative-feedback effect on the hypothalamus and pituitary. Consequently, GnRH, LH, and FSH levels are reduced. As a result, the testes may atrophy, and sterility may develop. Other side effects of large doses of synthetic androgens include kidney and liver damage, heart attack, and stroke. Taking synthetic androgens is highly discouraged by the medical profession and is a violation of the rules for most athletic organizations.

TABLE 19-1

Effects of Testosterone on Target Tissues

TARGET TISSUE	RESPONSE
Penis and scrotum	Enlargement and differentiation
Hair follicles	Hair growth and coarser hair: pubic area, legs, chest, axillary region, the face, and occasionally the back; male pattern baldness on the head if the person has the appropriate genetic makeup
Skin	Coarser texture of skin; increased rate of secretion of sebaceous glands, frequently resulting in acne at the time of puberty; increased secretion of sweat glands in axillary regions
Larynx	Enlargement of larynx and deeper masculine voice
Most tissues	Increased rate of metabolism
Red blood cells	Increased rate of red blood cell production; red blood cell count increased by about 20% as a result of increased erythropoietin secretion
Kidney	Retention of sodium and water to a small degree, resulting in increased extracellular fluid volume
Skeletal muscle	Skeletal muscle mass increases at puberty; the average is greater in men than in women
Bone	Rapid bone growth resulting in increased rate of growth and in early cessation of growth; males who mature sexually at a later age do not exhibit a rapid period of growth, but they grow for a longer period of time and may become taller than men who mature earlier

Male Sexual Behavior and the Male Sexual Act

Testosterone is required for normal sexual behavior. Testosterone enters certain cells within the brain, especially within the hypothalamus, and influences their functions. The blood levels of testosterone remain relatively constant throughout the lifetime of a male from puberty until about 40 years of age. Thereafter, the levels slowly decline to approximately 20% of this value by 80 years of age, causing a slow decrease in sex drive and fertility.

> **2 Predict the effect on secondary sexual characteristics, external genitalia, and sexual behavior, if the testes failed to produce normal amounts of testosterone at puberty.** ?

The male sexual act is a complex series of reflexes that result in erection of the penis, secretion of mucus into the urethra, emission, and ejaculation. Sensations that are normally interpreted as pleasurable occur during the male sexual act and result in a climax sensation, **orgasm** or **male climax,** associated with ejaculation. After ejaculation, a phase called resolution occurs in which the penis becomes flaccid, an overall feeling of satisfaction exists, and the male is unable to achieve erection and a second ejaculation.

Afferent impulses and integration

Sensory action potentials from the genitals are carried to the sacral region of the spinal cord, where reflexes that result in the male sexual act are integrated. Action potentials also travel from the spinal cord to the cerebrum to produce conscious sexual sensations.

Rhythmic massage of the penis, especially the glans, and surrounding tissues (the scrotum, anal, and pubic region) provide important sources of sensory action potentials. Engorgement of the prostate gland and seminal vesicles with secretions, or irritation of the urethra, urinary bladder, ductus deferens, and testes can also cause sexual sensations.

Psychic stimuli such as sight, sound, odor, or thoughts have a major effect on male sexual reflexes. Ejaculation while sleeping (nocturnal emission) is a relatively common event in young males and is thought to be triggered by psychic stimuli associated with dreaming. The inability to concentrate on sexual sensations may result in **impotence** or the inability to accomplish the sexual act. Impotence can also be caused by physical factors such as inability of the erectile tissue to fill with blood.

Erection, emission, and ejaculation

Erection is the first major component of the male sexual act. Parasympathetic action potentials from the sacral region of the spinal cord cause the arteries that supply blood to the erectile tissues to dilate. Blood then fills small venous sinuses called sinusoids in the erectile tissue and compresses the veins, which reduces blood flow from the penis. The increased blood pressure in the sinusoids causes the erectile tissue to become inflated and rigid. Parasympathetic action potentials also cause the mucous glands within the penile urethra and the bulbourethral glands to secrete mucus.

Emission is controlled by sympathetic action potentials that originate in the lumbar region of the spinal cord. Action potentials cause contractions of the reproductive ducts and stimulate the seminal vesicles and prostate gland to release secretions. Consequently semen accumulates in the urethra.

Action potentials are finally sent to the skeletal muscles that surround the base of the penis. Rhythmic contractions are produced that force the semen out of the urethra, and ejaculation results. In addition, there is an increase in muscle tension throughout the body.

The normal volume of semen is 2 to 3 ml. The normal sperm cell count is about 100 million sperm cells per milliliter of semen. If the sperm cell count falls below about 20 million sperm cells per milliliter of semen, sterility usually results even though the millions of sperm cells that are present appear to be normal.

FEMALE REPRODUCTIVE SYSTEM

The female reproductive organs consist of the ovaries, uterine tubes (or fallopian tubes), uterus, vagina, external genitalia, and mammary glands (see Figure 19-1, *B*). The internal reproductive organs of the female (Figures 19-7 and 19-8) are located within the pelvis between the urinary bladder and the rectum. The **uterus** and the **vagina** are in the midline with an **ovary** to each side of the uterus. The internal reproductive organs are held in place within the pelvis by a group of ligaments. The most conspicuous is the **broad ligament,** which spreads out on both sides of the uterus and to which the ovaries and uterine tubes attach.

Ovaries

The ovaries are small organs attached to ligaments that suspend them in the pelvic cavity. The ovaries are attached to the posterior surface of the

FIGURE 19-7 Sagittal view of the female pelvis.
Anatomy of the female reproductive organs.

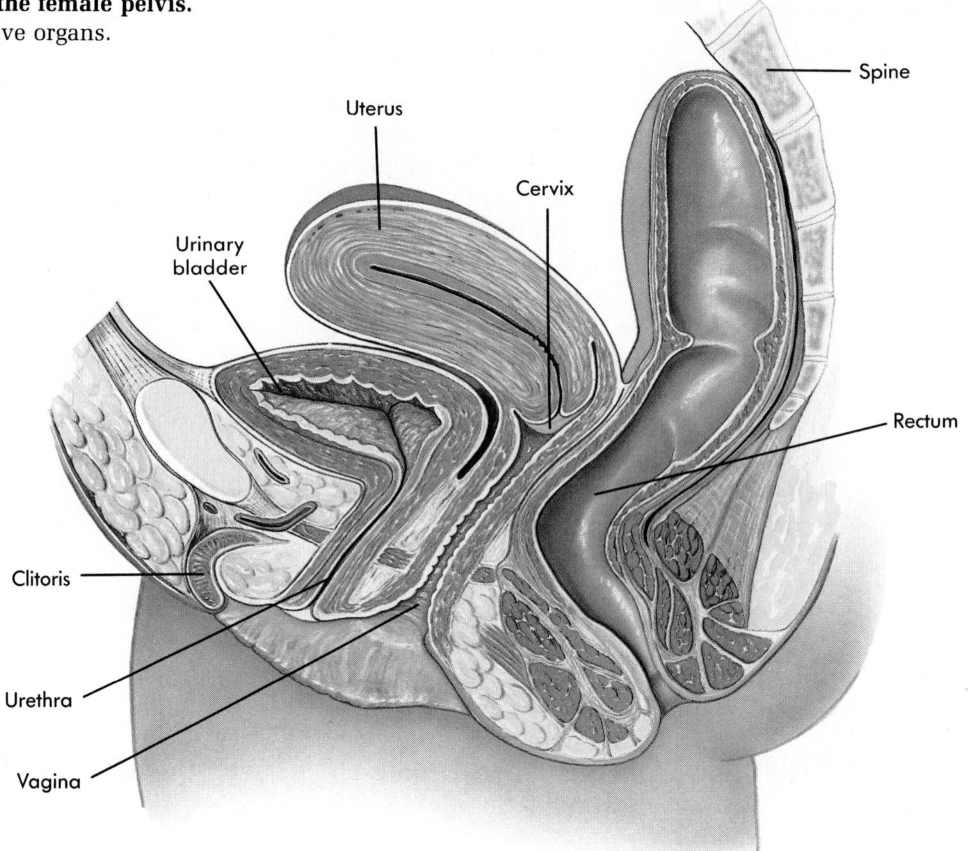

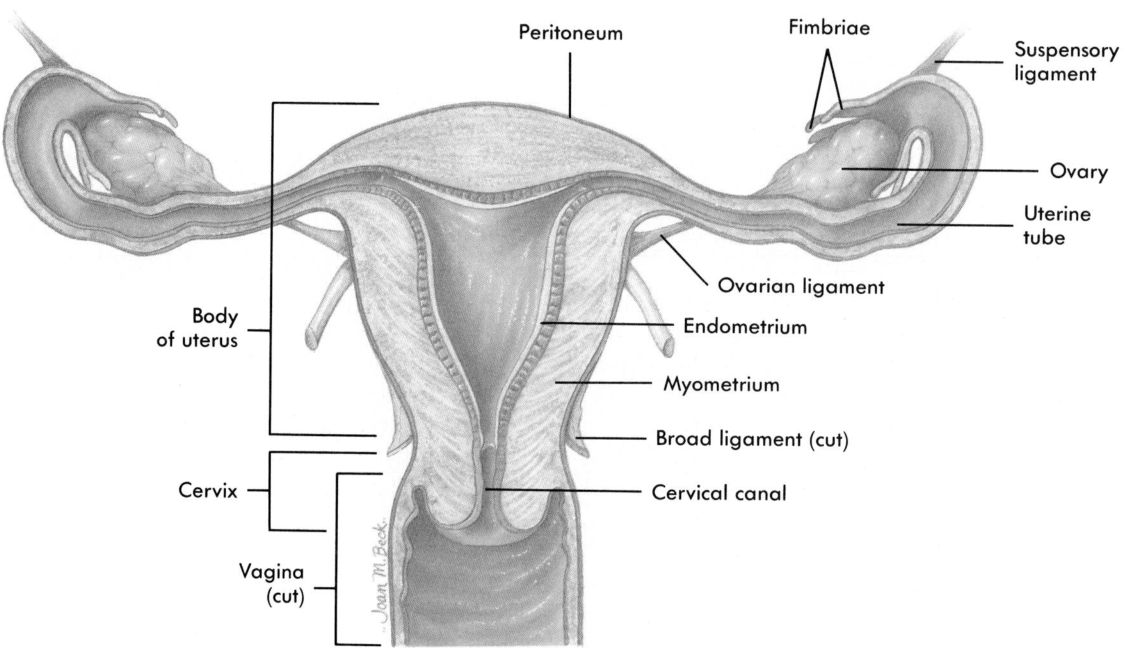

FIGURE 19-8 Frontal view of female reproductive organs.
Uterus, vagina, uterine tubes, ovaries, and supporting ligaments. The uterus, uterine tubes, and vagina are cut in section to show the internal anatomy.

broad ligament by a fold in the peritoneum called the **mesovarium** (mes′o-va′rĭ-um; mesentery of the ovary). Two other ligaments are associated with the ovary. The **suspensory ligament** extends from the ovary to the lateral body wall, and the **ovarian ligament** attaches the ovary to the superior margin of the uterus (see Figure 19-8). The ovarian arteries, veins, and nerves traverse the suspensory ligament and enter the ovary through the mesovarium. A layer of visceral peritoneum covers the surface of the ovary. The outer portion of the ovary is made up of dense connective tissue and contains **ovarian follicles.** Each of the ovarian follicles contains an **oocyte,** the female germ cell. Looser connective tissue makes up the inner portion of the ovary, where blood vessels, lymph vessels, and nerves enter the ovary.

Follicle and oocyte development

By the time a female is born she may have about 2 million follicles in her ovaries. Oocytes have begun the process of meiosis, but the process is stopped early in the first meiotic division and does not proceed until ovulation of each of the follicles occurs. At this time the cell is called a **primary oocyte.** The primary oocyte is surrounded by a layer of cells called the **granulosa cells,** and the entire structure is called a **primary follicle** (Figures 19-9 and 19-10). From birth to puberty, the number of primary follicles declines to about 400,000; of these only about 450 continue to develop and ovulate. Those follicles that do not ovulate degenerate.

Beginning during puberty, approximately every 28 days, hormonal changes stimulate some of the follicles to continue to develop (see Figure 19-10).

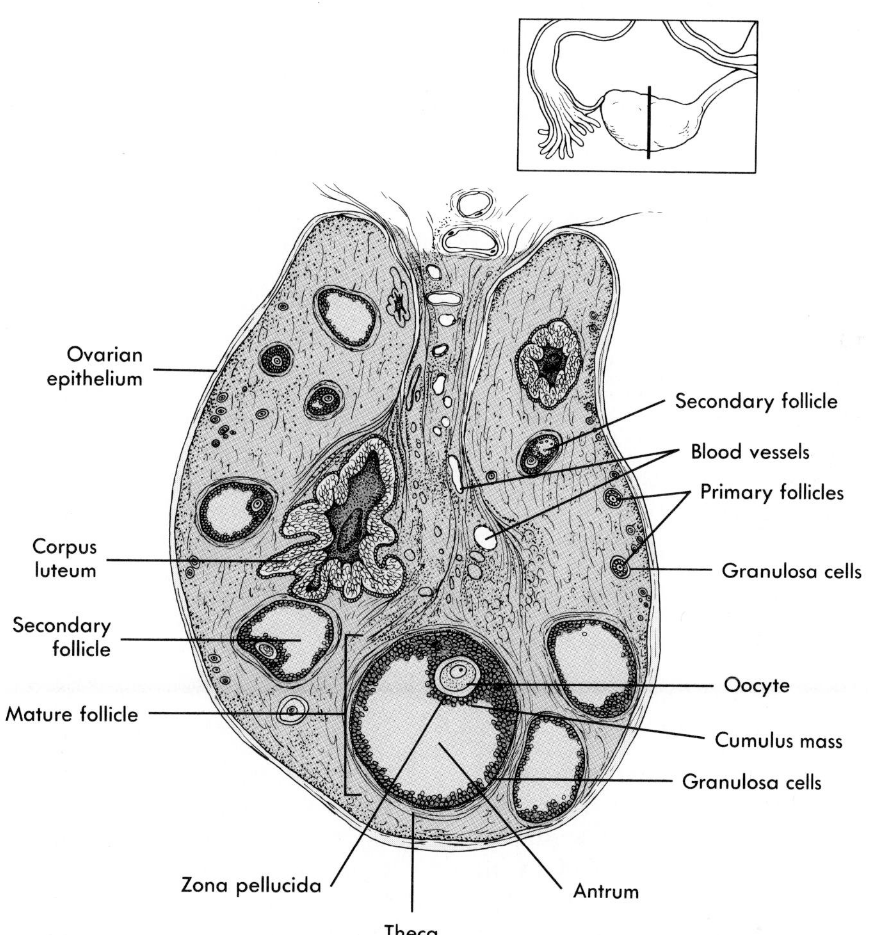

FIGURE 19-9 Structure of the ovary.
The ovary is sectioned to illustrate its internal structure (inset shows plane of section). Ovarian follicles from each major stage of development are present, and a corpus luteum is present.

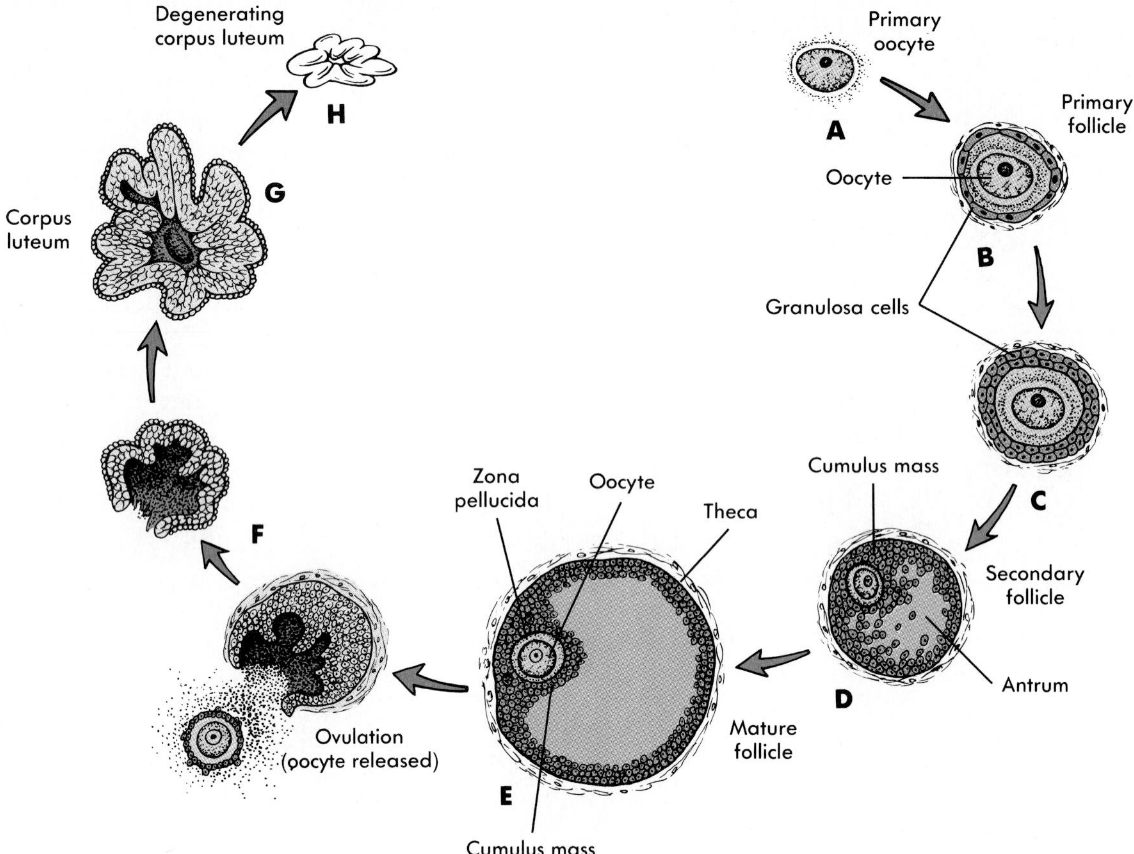

FIGURE 19-10 ■ Maturation of the follicle, oocyte, and corpus luteum.
A Primary oocytes begin to mature one or two menstrual cycles before they are ovulated. Several follicles begin to mature at the same time, but only one reaches the final stage of development and undergoes ovulation.
B The primary follicle enlarges.
C Granulosa cells form more than one layer.
D An antrum begins to form and to fill with fluid to form a secondary follicle.
E When a follicle becomes mature, it enlarges to its maximum size, and a large antrum is present.
F During ovulation the ovum is released from the follicle along with some surrounding granulosa cells.
G Subsequently, the granulosa cells divide rapidly and enlarge to form the corpus luteum.
H The corpus luteum degenerates if fertilization does not occur.

The granulosa cells that surround the oocyte multiply and form an increasing number of layers. The center of the follicle becomes a chamber, the **antrum,** which is filled with fluid produced by the granulosa cells. After formation of the antrum has started, the follicle is called a **secondary follicle.** The oocyte is pushed off to one side of the follicle and lies in a mass of follicular cells called the **cumulus mass.**

The secondary follicle continues to enlarge, the antrum fills with additional fluid, and the follicle forms a lump on the surface of the ovary. The follicle is then called a **mature follicle.** As the follicle enlarges, tissue surrounding the follicle forms a layer called the **theca.** The oocyte enlarges because of an accumulation of yolk, and a clear layer called the **zona pellucida** forms around the oocyte.

Follicles are stimulated to develop by FSH secreted by the anterior pituitary gland. FSH causes several follicles to begin developing during each menstrual cycle, but normally only one ovulates. The remainder of the mature follicles degenerate.

Also, it takes more than one menstrual cycle for a follicle to undergo the transition from a primary follicle to a mature follicle. Thus a follicle ovulated during one menstrual cycle started developing in response to FSH possibly two menstrual cycles earlier.

The developing follicles secrete small amounts of a hormone called **estrogen.** Estrogen plays an important role in coordinating the menstrual cycle and preparing the uterus to receive the fertilized ovum.

Ovulation

As the follicle matures, it can be seen on the surface of the ovary as a blister. The mature follicle expands and ruptures, a process called **ovulation,** forcing a small amount of blood, follicular fluid, and the oocyte out of the ruptured follicle and into the peritoneal cavity. Ovulation occurs in response to LH secreted by the anterior pituitary gland.

Near the time of ovulation, the first meiotic division is completed and the second meiotic division begins. However the process of meiosis is stopped during the second meiotic division, and proceeds to completion only if fertilization occurs. If fertilization occurs, the oocyte completes the second meiotic division.

After ovulation, the ruptured follicle becomes transformed into a glandular structure called the **corpus luteum** (lu′te-um; yellow) (see Figures 19-9 and 19-10). Cells of the corpus luteum begin to secrete hormones—progesterone and smaller amounts of estrogen.

If pregnancy occurs, the corpus luteum enlarges and becomes the **corpus luteum of pregnancy.** Maintenance of pregnancy depends on progesterone secreted by the corpus luteum of pregnancy for about the first trimester (first third or 12 weeks) of pregnancy. After the first trimester of pregnancy, progesterone is produced by the placenta, and the corpus luteum is no longer essential for the maintenance of pregnancy. If pregnancy does not occur, the corpus luteum lasts for about 10 to 12 days and then begins to degenerate.

Uterine Tubes

There are two **uterine tubes,** also called **fallopian tubes** or **oviducts.** One uterine tube is associated with each ovary. The uterine tubes extend from the area of the ovaries to the uterus, and they open directly into the peritoneal cavity near the ovary to receive the oocyte. The opening of the uterine tube is surrounded by long, thin processes called **fimbriae** (fim′brĭ-ah; fringe) (see Figure 19-8).

The fimbriae nearly surround the surface of the ovary. As a result, as soon as the oocyte is ovulated, it comes into contact with the surface of the fimbriae. Cilia on the fimbriae surface sweep the oocyte into the uterine tube. Fertilization usually occurs in the portion of the uterine tube near the ovary.

Uterus

The **uterus** is the size of a medium-sized pear (see Figures 19-7 and 19-8). It is slightly flattened anteroposteriorly and is oriented in the pelvic cavity with the larger, rounded portion called the **body of the uterus** directed superiorly, and the narrower portion, the **cervix** (ser′viks; neck), directed inferiorly. Internally, the **uterine cavity** in the uterine body continues as the **cervical canal,** which opens into the vagina.

The uterine wall is composed of three layers: a serous layer, a muscular layer, and the endometrium (see Figure 19-8). The outer layer, or **serous layer,** of the uterus is the peritoneum. The middle layer is the **muscular layer** or **myometrium,** which consists of smooth muscle, is quite thick, and accounts for the bulk of the uterine wall. The innermost layer of the uterus is the **endometrium.** The endometrium consists of simple, columnar epithelial cells and a connective tissue layer. Simple tubular glands are formed by folds of the epithelium. The superficial layers of the endometrium are sloughed off during menstruation.

A Pap smear is a diagnostic test used to determine if a woman is suffering from cancer of the uterine cervix. A smear of epithelial cells is taken from the area of the cervix by inserting a swab through the vagina. The smear is placed on a glass slide and stained. The cells are then examined microscopically to determine whether some of them show signs of being cancerous. Cells that are cancerous appear to be more immature than the characteristic epithelial cells of the cervix. The more embryonic the cells appear the more severe the cancer.

Vagina

The **vagina** is the female organ of copulation and functions to receive the penis during intercourse. It also allows menstrual flow and childbirth. The vagina extends from the uterus to the outside of the body (see Figures 19-7 and 19-8). The superior portion of the vagina is attached to the sides of the cervix so that a portion of the cervix extends into the vagina.

The wall of the vagina consists of an outer muscular layer and an inner mucous membrane. The muscular layer is smooth muscle that allows the vagina to increase in size. Thus the vagina can accommodate the penis during intercourse and can stretch greatly during childbirth. The mucous membrane is moist, stratified squamous epithelium that forms a protective surface layer. Most of the lubricating secretions produced by the female during intercourse are produced by the vaginal epithelium.

In young females the vaginal opening is covered by a thin mucous membrane called the **hymen.** The hymen may completely close the vaginal orifice in which case it must be removed to allow menstrual flow. More commonly, the hymen may be perforated by one or several holes. The openings in the hymen are usually greatly enlarged during the first sexual intercourse. The hymen may also be perforated or torn at some earlier time in a young woman's life (for example, during strenuous exercise). Therefore, the condition of the hymen is not a reliable indicator of virginity.

External Genitalia

The external female genitalia, also called the **vulva** or **pudendum,** consist of the vestibule and its surrounding structures (Figure 19-11). The **vestibule** is the space into which the vagina and urethra open. The urethra opens just anterior to the vagina. It is bordered by a pair of thin, longitudinal skin folds called the **labia** (la'be-ah; lips) **minora.** A small erectile structure called the **clitoris** (klit'o-ris) is located in the anterior margin of the vestibule. Anteriorly, the two labia minora unite over the clitoris to form a fold of skin called the **prepuce.**

The clitoris consists of a shaft and a distal glans. It is well supplied with sensory receptors and functions to initiate and intensify levels of sexual tension. Like the penis, the clitoris is made up of erectile tissue. Additional erectile tissue is located on either side of the vaginal opening.

On each side of the vestibule, between the vaginal opening and the labia minora, are openings of the **vestibular glands.** They produce a lubricating

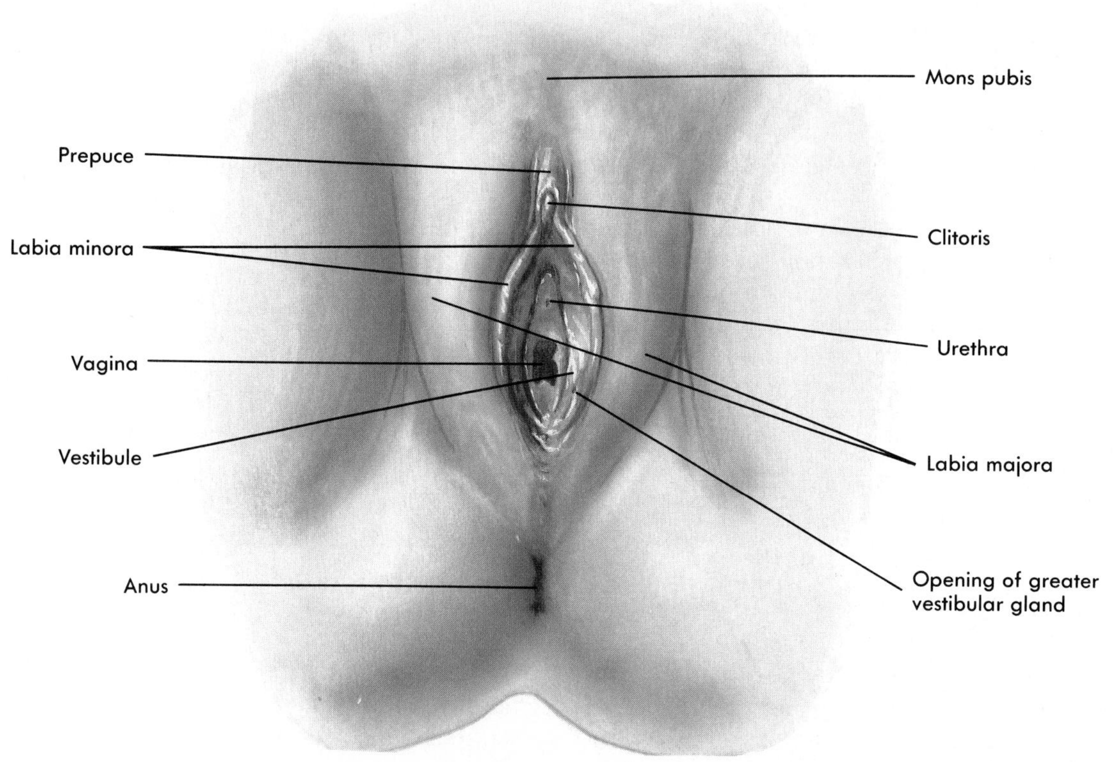

FIGURE 19-11 ■ Female external genitalia.

fluid that helps to maintain the moistness of the vestibule.

Lateral to the labia minora are two prominent, rounded folds of skin called the **labia majora.** The two labia majora unite anteriorly in an elevation over the pubic symphysis called the **mons** (mound) **pubis.** The lateral surfaces of the labia majora, as well as the surface of the mons pubis are covered with coarse hair. The medial surfaces of the labia majora are covered with numerous sebaceous and sweat glands. The space between the labia majora is called the **pudendal cleft.** Most of the time, the labia majora are in contact with each other across the midline, closing the pudendal cleft and concealing the deeper structures within the vestibule.

The region between the vagina and the anus is the **clinical perineum.** The skin and muscle of this region may tear during childbirth. To prevent such tearing, an incision called an **episiotomy** is sometimes made in the clinical perineum.

Mammary Glands

The **mammary glands** are the organs of milk production and are located in the **breasts** or **mammae** (Figure 19-12). The mammary glands are modified sweat glands. Externally, the breast of both males and females has a raised **nipple** surrounded by a circular, pigmented **areola** (a-re′o-lah).

In prepubescent children, the general structure of the breasts is similar, and both males and females possess a rudimentary glandular system. The female breasts begin to enlarge during puberty under the influence of estrogens and progesterone. On rare occasions, the breasts of a male become enlarged; a condition called **gynecomastia.**

Each adult female mammary gland usually consists of 15 to 20 glandular **lobes** covered by a considerable amount of adipose tissue. It is primarily this superficial fat that gives the breast its form. Each lobe possesses a single duct that opens independently on the surface of the nipple. The duct of each lobe subdivides into smaller ducts, forming a **lobule.** Within a lobule, the ducts branch and become even smaller. In the milk-producing breast, the ends of these small ducts expand to form secretory sacs called **alveoli.**

The nipples are very sensitive to tactile stimulation and contain smooth muscle that can contract, causing the nipple to become erect in response to stimulation. These smooth-muscle fibers respond similarly to general sexual arousal.

> Cancer of the breast is a serious, often fatal disease in women. The use of mammography and regular self-examination of the breast can lead to early detection of breast cancer and effective treatment. Mammography utilizes low-intensity x-rays to detect tumors in the soft tissue of the breast. With modern techniques, tumors can often be edentified before they can be detected by palpation. Once a tumor is identified, a biopsy normally is performed to determine whether the tumor is benign or malignant. Most tumors of the mammary glands are benign. Those that are malignant do have the potential to spread to other areas of the body and ultimately lead to death.

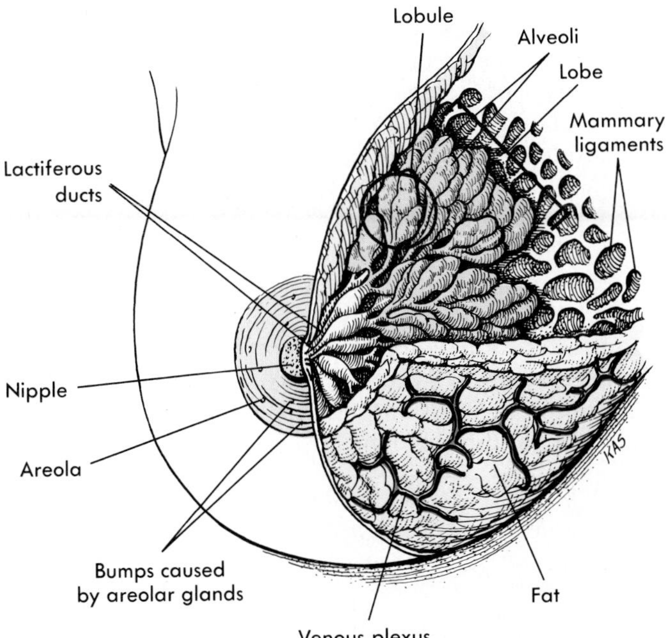

FIGURE 19-12 Anatomy of the breast. The section illustrates the duct system, secretory units, and adipose tissue of the breast.

PHYSIOLOGY OF FEMALE REPRODUCTION

As in the male, female reproduction is under hormonal and neural regulation.

Puberty

Puberty in females is marked by the first episode of menstrual bleeding, which is called **menarche.** During puberty, the vagina, uterus, uterine tubes, and external genitalia begin to enlarge. Fat is deposited in the breasts and around the hips, causing them to enlarge and assume an adult form. In addition, pubic and axillary hair grows. Development of a sexual drive is also associated with puberty.

The changes associated with puberty are primarily the result of the elevated rate of estrogen and progesterone secretion by the ovaries. Before puberty, estrogen and progesterone are secreted in very small amounts. At puberty the cyclic adult pattern of hormone secretion is established.

Before puberty, the rate of GnRH secretion from the hypothalamus and the rate of LH and FSH secretion from the anterior pituitary are very low. Estrogen and progesterone from the ovaries have a strong negative-feedback effect on the hypothalamus and pituitary. After the onset of puberty the hypothalamus and anterior pituitary secrete larger amounts of GnRH, LH, and FSH. Estrogen and progesterone have less of a negative-feedback effect on the hypothalamus and pituitary and in low concentrations they have a positive-feedback effect. The normal cyclic pattern of reproductive hormone secretion that occurs during the menstrual cycle becomes established. The initial change that results in puberty appears to be maturation of the hypothalamus of the brain.

Menstrual Cycle

The term **menstrual cycle** refers to the series of changes that occur in sexually mature, nonpregnant females that culminate in menses. *Menses* (derived from a Latin word meaning month) is a period of mild hemorrhage during which the endometrium is sloughed and expelled from the uterus. Typically, the menstrual cycle is about 28 days long, although it may be as short as 18 days or as long as 40 days (Figure 19-13 and Table 19-2).

The first day of menstrual bleeding (menses) is considered to be day 1, and menses typically lasts 4 or 5 days. Ovulation occurs on about day 14 of the menstrual cycle, although the timing of ovulation varies from individual to individual and may vary within an individual from one menstrual cycle to the next.

The time between the ending of menses and ovulation is called the **proliferative phase** (proliferation of the uterine mucosa). During the proliferative phase, follicles in the ovary mature and, as they do so, they secrete small amounts of estrogen. Estrogen acts on the uterus and causes the cells of the endometrium to divide rapidly. The endometrium thickens and tubular glands form.

The small amounts of estrogen secreted by the developing follicles stimulates GnRH secretion from the hypothalamus, and GnRH triggers LH and FSH secretion from the anterior pituitary gland. LH and FSH stimulate a greater rate of estrogen secretion from the developing follicles. This positive-feedback loop produces a series of larger and larger surges of LH and FSH secretion. Ovulation occurs in response to the large increases in LH levels that normally occur on about day 14 of the menstrual cycle.

The large increase in LH secretion that occurs on about day 14 of the menstrual cycle causes a mature follicle to undergo ovulation. Following ovulation the corpus luteum begins to secrete progesterone and smaller amounts of estrogen. Progesterone acts on the uterus causing the cells of the endometrium to become larger and to secrete a small amount of fluid. The progesterone and estrogen, produced by the corpus luteum, inhibits LH and FSH secretion. Thus, LH and FSH levels decline to low levels after ovulation.

The time between ovulation and the next menses is called the **secretory phase** of the menstrual cycle because of the small amount of fluid secreted by the cells of the endometrium. During the secretory phase the lining of the uterus reaches its greatest degree of development.

By 7 or 8 days following ovulation (day 21 or 22 of the menstrual cycle), the endometrium is prepared to receive the developing blastocyst. The developing blastocyst becomes implanted in the endometrium where development proceeds. If the oocyte is not fertilized, the endometrium sloughs away as a result of declining blood progesterone levels. Unless the oocyte is fertilized, the corpus luteum begins to produce less progesterone by day 24 or 25 of the menstrual cycle. By day 28 the declining progesterone causes the endometrium to slough away to begin menses and the next menstrual cycle.

3 Predict the effect of administering a relatively large amount of estrogen and progesterone just before the increase in LH that precedes ovulation.

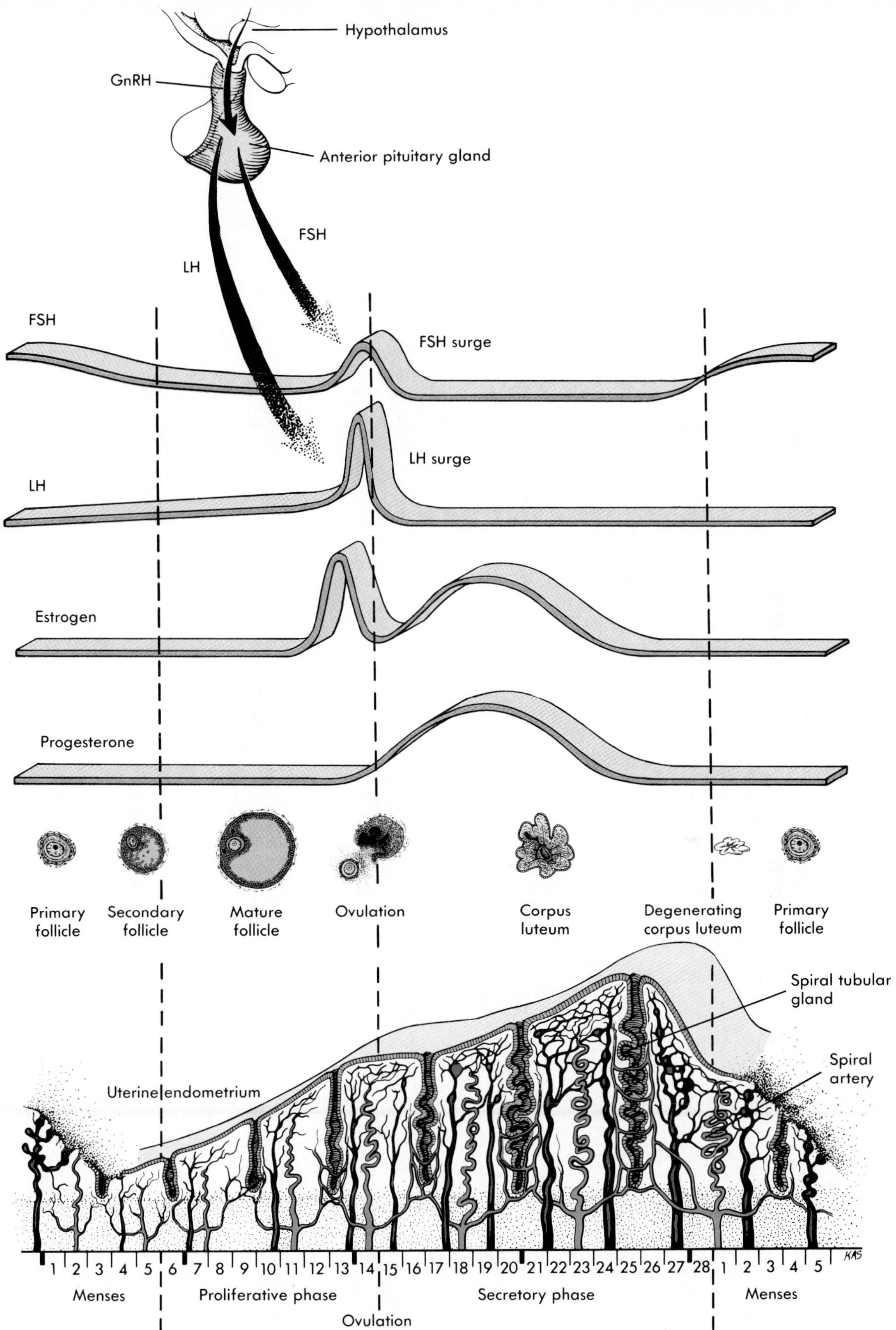

FIGURE 19-13 ■ The menstrual cycle.
Hormonal changes that occur during the menstrual cycle, changes that occur in the ovary, and changes that occur in the endometrium of the uterus arranged relative to each other.

TABLE 19-2

Events during the Menstrual Cycle

MENSES (day 1 to day 4 or 5 of the menstrual cycle)	
Pituitary gland	The rate of LH and FSH secretion is low.
Ovary	The rate of estrogen and progesterone secretion is low after degeneration of the corpus luteum produced during the previous menstrual cycle.
Uterus	In response to declining progesterone levels, the endometrial lining of the uterus sloughs off, resulting in menses followed by repair of the endometrium.
PROLIFERATIVE PHASE (from day 4 or 5 until ovulation on about day 14)	
Pituitary gland	The rate of FSH and LH secretion is low during most of the proliferative phase; FSH and LH secretion increase near the end of the proliferative phase in response to increasing estrogen secretion from the ovaries.
Ovary	Developing follicles secrete small amounts of estrogen, especially near the end of the proliferative phase; increasing FSH and LH cause additional estrogen secretion from the ovaries near the end of the proliferative phase.
Uterus	Estrogen causes endometrial cells of the uterus to divide. The endometrium of the uterus thickens and tube-like glands form. Estrogen causes the cells of the uterus to be more sensitive to progesterone by increasing the number of progesterone receptors in uterine tissues.
OVULATION (about day 14)	
Pituitary gland	The rate of FSH and LH secretion increases rapidly just before ovulation in response to increasing estrogen levels. Increasing FSH and LH levels stimulate estrogen secretion resulting in a positive-feedback cycle.
Ovary	LH causes final maturation of a mature follicle and initiates the process of ovulation. FSH acts on immature follicles and causes several of them to begin to enlarge.
Uterus	The endometrium continues to divide in response to estrogen.
SECRETORY PHASE (from about day 14 to day 28)	
Pituitary gland	Estrogen and progesterone reach levels high enough to inhibit LH and FSH secretion from the pituitary gland.
Ovary	After ovulation the follicle is converted to the corpus luteum; the corpus luteum secretes large amounts of progesterone and smaller amounts of estrogen from shortly after ovulation until about day 24 or 25. If fertilization does not occur, the corpus luteum degenerates after about day 25 and the rate of progesterone secretion rapidly declines to low levels.
Uterus	In response to progesterone the endometrial cells enlarge, the endometrial layer thickens, and the glands of the endometrium reach their greatest degree of development; the endometrial cells secrete a small amount of fluid. After progesterone levels decline, the endometrium begins to degenerate.
MENSES (day 1 to day 4 or 5 of the next menstrual cycle)	
Pituitary gland	The rate of LH and FSH secretion remains low.
Ovary	The rate of estrogen and progesterone secretion is low.
Uterus	In response to declining progesterone levels, the endometrial lining of the uterus sloughs off, resulting in menses followed by repair of the endometrium.

Menstrual cramps are the result of strong myometrial contractions that occur before and during menstruation. The cramps may result from excessive production of inflammatory substances such as prostaglandins. The production of inflammatory substances is a consequence of the degeneration of the endometrium as it is sloughed. In some women, menstrual cramps are extremely uncomfortable. Many women can alleviate painful menstrual cramps by taking aspirin-like drugs, which inhibit the process of inflammation. Aspirin-like drugs inhibit prostaglandin production. Prostaglandins are inflammatory substances. Aspirin-like drugs, however, are not effective in treating painful menstruation in all women, especially when the cause of pain is secondary to a disease process.

Some women suffer from more or less severe changes in mood that can result in aggression or other socially undesirable behaviors just before menses. This condition is called the *premenstrual syndrome (PMS)*. The fluctuations in estrogen and progesterone associated with the menstrual cycle may trigger these mood changes. The precise mechanisms responsible for the condition are unknown.

Menopause

When a woman is 40 to 50 years old, the menstrual cycles become less regular, and ovulation does not consistently occur during each cycle. Eventually the cycles stop completely. The cessation of menstrual cycles is called **menopause,** and the whole time period from the onset of irregular cycles to their complete cessation is called the **female climacteric.**

The major cause of menopause is age-related changes in the ovary. The number of follicles remaining in the ovaries of menopausal women is small. In addition, the follicles that remain become less sensitive to stimulation by LH and FSH. As the ovaries become less responsive to stimulation by FSH and LH, fewer mature follicles and corpora lutea are produced. Gradual changes occur in women in response to the reduced amount of estrogen and progesterone produced by the ovaries (Table 19-3).

TABLE 19-3

Possible Changes in Post-menopausal Women Caused by Decreased Ovarian Hormone Secretion

	CHANGES
Menstrual cycle	5-7 years before menopause the cycle becomes irregular; the number of cycles in which ovulation does not occur and corpora lutea do not develop increases.
Uterus	Irregular menstruation gradually is followed by no menstruation; the endometrium finally atrophies, and the uterus becomes smaller.
Vagina and external genitalia	Epithelial lining becomes thinner; external genitalia become thinner and less elastic; labia majora becomes smaller; pubic hair decreases; reduced secretion leads to dryness; the vagina is more easily inflamed and infected.
Skin	Epidermis becomes thinner; melanin synthesis increases and skin becomes darker.
Cardiovascular system	Hypertension and atherosclerosis occur more frequently.
Vasomotor instability	Hot flashes and increased sweating are correlated with vasodilation of cutaneous blood vessels; hot flashes are related to decreased estrogen levels.
Libido	Temporary changes, usually a decrease, in libido are associated with the onset of menopause.
Fertility	Fertility begins to decline about 10 years before the onset of menopause; by age 50 almost all oocytes and follicles have degenerated.
Pituitary function	The low levels of estrogen and progesterone produced by the ovary cause the pituitary gland to secrete larger than normal amounts of LH and FSH. The increased levels of these hormones have little effect on the postmenopausal ovary.

Reproductive Disorders

INFECTIOUS DISEASES
SEXUALLY TRANSMITTED DISEASES

Sexually transmitted diseases (STDs) are a class of infectious diseases spread by intimate sexual contact between individuals. These diseases include the major venereal diseases such as nongonococcal urethritis, trichomoniasis, gonorrhea, genital herpes, syphilis, and acquired immunodeficiency syndrome.

Nongonococcal urethritis refers to any inflammation of the urethra that is not gonorrhea. Factors such as trauma, or passage of a nonsterile catheter through the urethra can cause this condition, but many cases are acquired through sexual contact. In most cases a bacterium such as *Chlamydia trachomatis* is responsible, but other bacteria may be involved. Antibiotics are usually effective in treating the condition.

Trichomonas is a protozoan commonly found in the vagina of females and the urethra of males. If the normal acidity of the vagina is disturbed *trichomonas* can grow rapidly. The condition is more common in females than in males. The rapid growth of these organisms results in inflammation and a greenish-yellow discharge characterized by a foul odor.

Gonorrhea is caused by *Neisseria gonorrhoeae*. The organisms attach to the epithelial cells of the vagina or the male urethra. The invasion of bacteria establishes an inflammatory response in which pus is formed. Males become aware of a gonorrheal infection by painful urination and the discharge of pus-containing material from the urethra. Symptoms appear within a few days to a week. Recovery may eventually occur without complication, but when complications do occur, they can be serious. The urethra may become partially blocked or sterility may result from blockage of reproductive ducts with scar tissue. In some cases other organ systems such as the heart, meninges of the brain, or joints may become infected. In females, the early stages of infection may not be noticeable, but the infection can lead to pelvic inflammatory disease. Gonorrheal eye infections may occur in newborn children of women with gonorrheal infections. Antibiotics are usually effective in treating gonorrheal infections, and the immune system often successfully combats gonorrheal infections in untreated individuals.

Genital herpes is a viral infection by Herpes simplex type 2. Lesions appear after an incubation period of about 1 week and cause a burning sensation. After this, blisterlike areas of inflammation appear. In males and females, urination can be painful, and walking or sitting may be unpleasant depending on the location of the lesions. Usually the vesicles heal in about 2 weeks. The lesions may reoccur. The viruses exist in latent condition in the infected tissues, and may initiate periods of inflammation in response to factors such as menstruation, emotional stress, or illness. If active lesions are present in the mother's vaginal or external genitalia at the time of delivery, a caesarean delivery is performed, to prevent newborns from becoming infected with the herpes virus.

Syphilis is caused by the bacterium *Treponema pallidum*, which can be spread by sexual contact of all kinds. Syphilis exhibits an incubation period from 2 weeks to several months. The disease progresses through several recognized stages. In the primary stage, the initial symptom is a small, hard-based chancre, or sore which usually appears at the site of infection. Several weeks after the primary stage, the disease enters the secondary stage, characterized

During the climacteric, some women experience "hot flashes," irritability, fatigue, anxiety, and occasionally, severe emotional disturbances. Many of these symptoms can be treated successfully by administering small amounts of estrogen and then gradually decreasing the dosage, or by providing psychological counseling. A potential side effect of estrogen therapy is a slightly increased possibility of the development of breast and uterine cancer in some women (recent reports suggest that the low dosage currently used in estrogen therapy may not increase the probability of breast cancer).

Female Sexual Behavior and the Female Sex Act

Sexual drive in females, like sexual drive in males, is dependent on hormones. Testosterone-like hormones and possibly estrogens affect brain cells (especially in the area of the hypothalamus) and influence sexual behavior. Testosterone-like hormones are produced primarily in the adrenal gland. Psychic factors also play a role in sexual behavior. The female neural pathways, both afferent and efferent, involved in controlling sexual responses are similar to those found in the male.

mainly by skin rashes and mild fever. The symptoms of secondary syphilis usually subside after a few weeks, and the disease enters a latent period in which no symptoms are present. In less than half the cases, a tertiary stage develops after many years. In the tertiary stage, many lesions develop that can cause extensive tissue damage. Syphilis can be passed on to newborns if the mother is infected. Damage to mental development and other neurological symptoms are among the more serious consequences. Females who have syphilis in the latent phase are most likely to have babies who are infected. Antibiotics are used to treat syphilis, although some strains are very resistant to certain antibiotics.

Acquired immunodeficiency syndrome (AIDS) is caused by infection with the human immunodeficiency virus (HIV) that ultimately results in destruction of the immune system. HIV appears to infect cells of the immune system. After the initial infection there may be a latent period lasting up to several years. Eventually, the virus becomes activated and destroys cell types that are essential to the maintenance of the immune system. The victims do not die directly from HIV. The destruction of the immune system by AIDS makes the immune system incapable of protecting the individual from other infections. It is the effect of these opportunistic infections that ultimately causes the death of the AIDS victim. The current treatments that show some promise appear to prolong the life of the AIDS victim, but there is no cure for AIDS at this time. Essentially everyone who has AIDS will die from the condition. People who are infected with HIV produce an antibody to HIV. The current tests for AIDS depend on the detection of this antibody in the blood. A person may live for several years with the presence of the antibody in their blood being the only indication of infection. It is thought that essentially all of the people who have developed the HIV antibody in response to infection of HIV will eventually develop AIDS and die from the condition. Preventative measures against HIV infection are the only effective treatment for avoiding AIDS at this time. The most common mechanisms of transmission of the virus are through sexual contact with a person infected with HIV and through sharing needles with an infected person during the administration of illicit drugs. Screening techniques now implemented make the transmission of HIV through blood transfusions very rare. Some documented cases of transmission of HIV through accidental needle sticks in hospitals and other health care facilities exist, but the frequency is rare. There is no evidence that casual contact with a person who has AIDS or who is infected with HIV will result in transmission of the disease. Transmission appears to require exposure to body fluids of an infected person in a way that allows HIV into the interior of another person. Normal casual contact, including touching an HIV-infected person, does not increase the risk of infection.

OTHER INFECTIOUS DISEASES

Pelvic inflammatory disease (PID) is a bacterial infection of the pelvic organs. It usually involves the uterus, uterine tubes, or ovaries. A vaginal or uterine infection may spread throughout the pelvis. PID is commonly caused by gonorrhea, but other bacteria can be involved. Early symptoms of PID include increased vaginal discharge and pelvic pain. Early treatment with antibiotics can stop the spread of PID, but lack of treatment results in a life-threatening infection.

During sexual excitement, erectile tissue within the clitoris and around the vaginal opening become engorged with blood. The mucous glands within the vestibule, especially the vestibular glands, secrete small amounts of mucus. Large amounts of mucuslike fluid are also extruded into the vagina through its wall. These secretions provide lubrication to allow easy entry and easy movement of the penis in the vagina during intercourse. The tactile stimulation of the female's genitals that occurs during sexual intercourse, as well as psychological stimuli, normally triggers an **orgasm,** or the **female climax.** The vaginal, uterine, and surrounding muscles contract rhythmically and there is an increase in muscle tension throughout much of the body. After the sexual act there is a period of **resolution,** which is characterized by an overall sense of satisfaction and relaxation. However, unlike most males, females are often receptive to further immediate stimulation and can often experience successive orgasms. Although orgasm is a pleasurable component of sexual intercourse, it is not necessary for females to experience an orgasm for fertilization to occur.

Control of Pregnancy

Many methods are used to prevent or terminate pregnancy (Figure 19-A), including methods that prevent fertilization (contraception), prevent implantation of the developing embryo (IUDs), or remove the implanted embryo or fetus (abortion). Many of these techniques are quite effective when done properly and used consistently. For example abstinence, when practiced consistently, is a sure way to prevent pregnancy, but it is not an effective method when used only occasionally.

BEHAVIORAL METHODS

Coitus interruptus is removal of the penis from the vagina just before ejaculation. This is a very unreliable method of preventing pregnancy, since it requires perfect awareness and willingness to withdraw the penis at the correct time. It also ignores the fact that some sperm cells are found in pre-ejaculatory emissions.

The rhythm method requires abstaining from sexual intercourse near the time of ovulation. A major factor in the success of this method is the ability to predict accurately the time of ovulation. Although the rhythm method provides some protection against becoming pregnant, it has a relatively high rate of failure that is a result of both the inability to predict the time of ovulation and the failure to abstain during the period of fertility.

BARRIER METHODS

A condom is a sheath of animal membrane, rubber, or plastic that is placed over the erect penis. The condom is a barrier device, since the semen is collected within the condom instead of within the vagina. Condoms also provide protection against sexually transmitted diseases.

Methods to prevent sperm cells from reaching the oocyte once they are in the vagina include use of a diaphragm, spermicidal agents, and a vaginal sponge. A diaphragm is a flexible plastic or rubber dome that is placed over the cervix within the vagina, where it prevents passage of sperm cells from the vagina through the cervical canal of the uterus. The most commonly used spermicidal agents are foams or creams that are inserted within the vagina before sexual intercourse. They kill the sperm cells. A sponge, either natural or synthetic, is permeated with spermicidal agents and is placed over the cervix where it acts as a barrier and kills the sperm cells. When used in combination with another technique, condoms, foams, creams, and sponges are much more effective than when they are used alone. After intercourse, spermicidal douches, which remove and kill sperm cells, are sometimes used. Spermicidal douches used alone and after intercourse are not very effective.

CHEMICAL METHODS

Synthetic estrogen and progesterone in oral contraceptives (birth control pills) effectively suppress fertility in females. These substances may have more than one action, but they reduce LH and FSH release from the anterior pituitary. The estrogen and progesterone are present in high enough concentrations to have a negative-feedback effect on the pituitary, which prevents the large increase in LH and FSH secretion that triggers ovulation at the end of the secretory phase of the menstrual cycle. Consequently, oral contraceptives prevent ovulation in many cases. Over the years the dose of estrogen and progesterone in birth control pills has been reduced. The current lower dose birth control pills have fewer side effects than earlier dosages. There is an increased risk of heart attack or stroke in women using oral contraceptives and who smoke or those who have a history of hypertension or coagulation disorders. For most women, the pill is effective and has a minimum frequency of complications until at least age 35.

Lactation prevents the menstrual cycle for a few months after childbirth. Action potentials sent to the hypothalamus that cause the release of oxytocin and prolactin also inhibit GnRH release and, therefore, reduce LH and FSH release, which prevents ovulation. Despite continual lactation, the ovarian and uterine cycles eventually resume. Since ovulation normally precedes menstruation, relying on lactation to prevent pregnancy is not consistently effective.

SURGICAL METHODS

Vasectomy is a common method used to render males permanently incapable of fertilization without affecting the performance of the sexual act. Vasectomy is a surgical procedure used to cut and tie the ductus deferens within the scrotal sac, preventing sperm cells from becoming part of the ejaculate. Since such a small volume of ejaculate comes from the testis and epididymis, vasectomy has little effect on the volume of the ejaculated semen. The sperm cells are reabsorbed in the epididymis.

A common method of permanent birth control in females is tubal ligation, a procedure in which the uterine tubes are tied and cut or clamped through an incision made through the wall of the abdomen. This procedure closes off the pathway between the sperm cells and the oocyte. Laparoscopy, in which a special instrument is inserted into the abdomen through a small incision, is commonly used so that only small openings are required to perform the operation.

In some cases, pregnancies are terminated by surgical procedures called abortions. The most common method for performing abortions is the insertion of an instrument through the cervix into the uterus. The instrument scrapes the endometrial surface, and at the same time, a strong suction is applied. The endometrium and the embedded embryo are disrupted and sucked out of the uterus. This technique is normally used only in pregnancies that have progressed less than 3 months.

PREVENTION OF IMPLANTATION

Intrauterine devices (IUDs) are inserted into the uterus through the cervix, and they prevent normal implantation of the developing embryo within the endometrium. Some early IUD designs produced serious side effects such as perforation of the uterus and, as a result, many IUDs have been removed from the market. Data indicate, however, that IUDs are effective in preventing pregnancy.

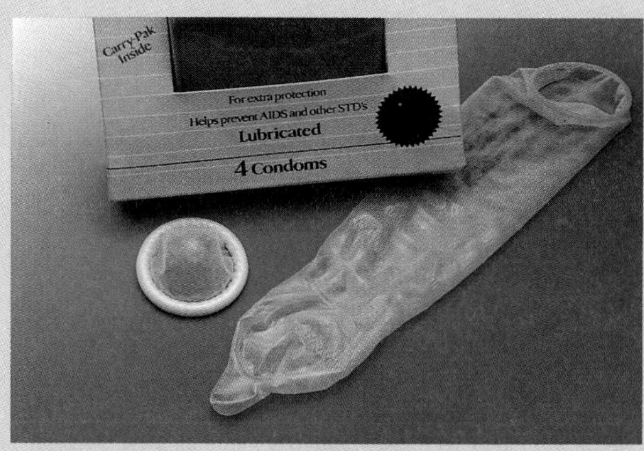

FIGURE 19-A ■ Contraceptive devices and techniques.
A Condom.
B Diaphragm with spermicidal jelly.
C Vaginal sponge.
D Spermicidal foam.
E Oral contraceptives.
F Vasectomy.
G Tubal ligation.

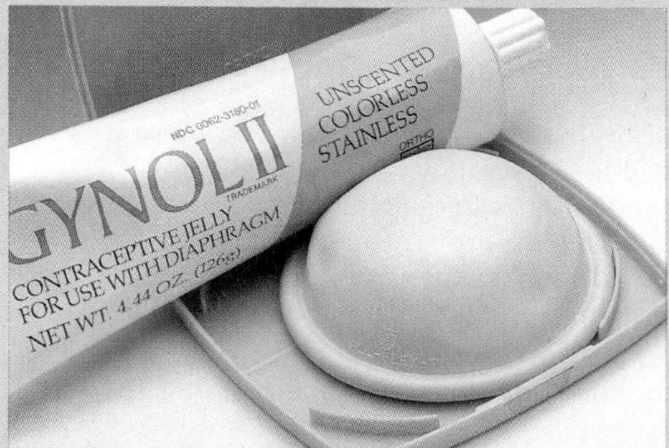

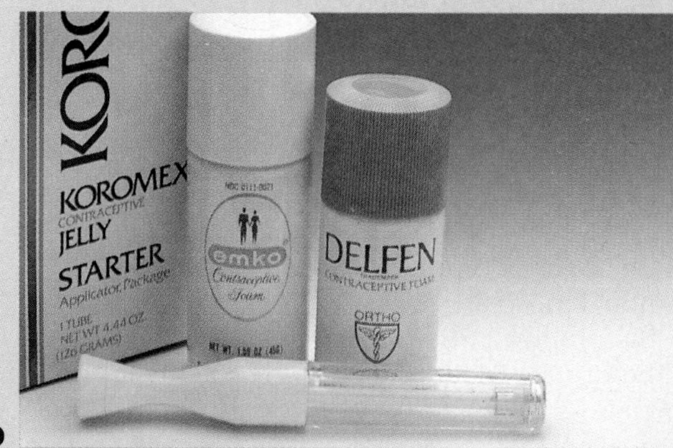

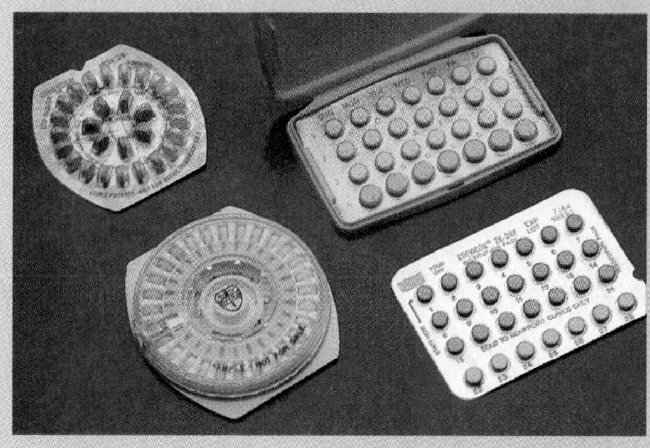

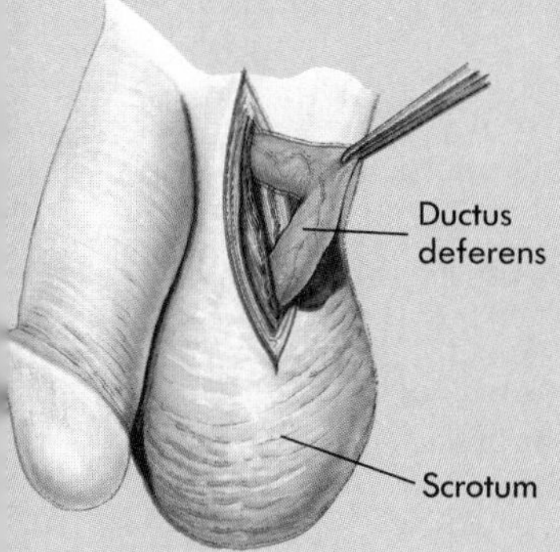

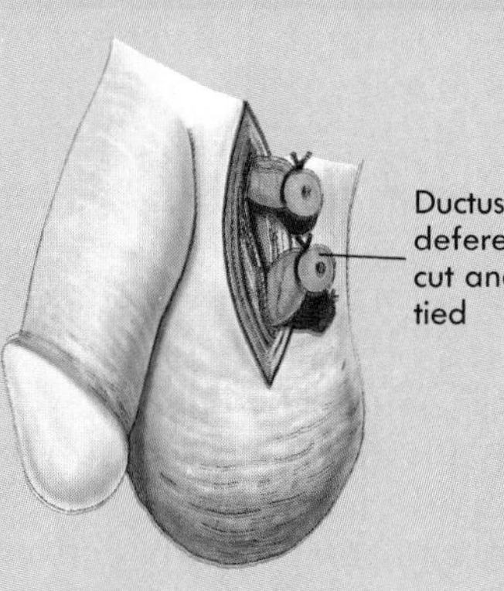

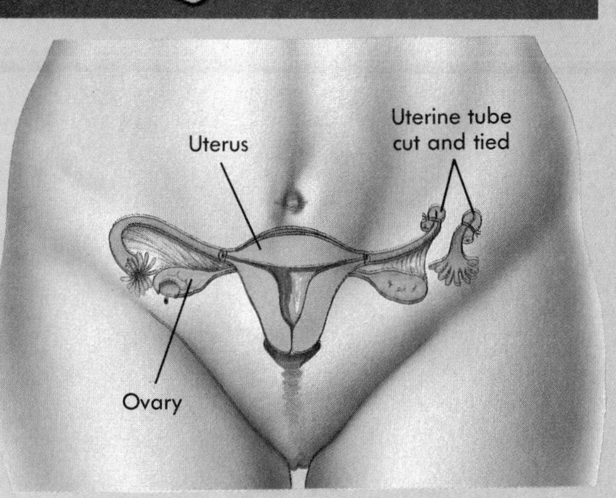

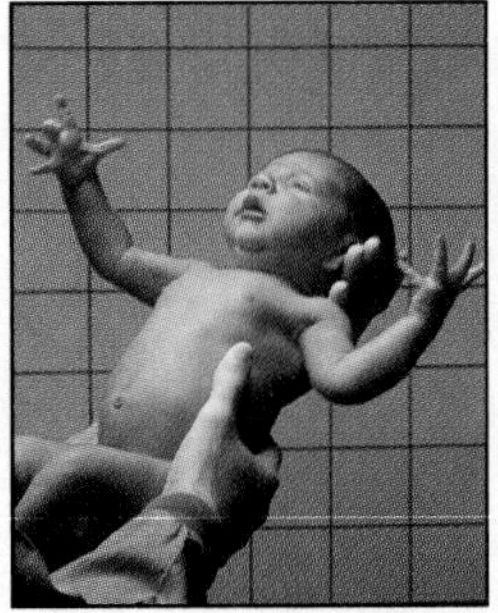

SUMMARY

The male reproductive system produces sperm cells and transfers them to the female. The female reproductive system produces the oocyte and nurtures the developing child.

MALE REPRODUCTIVE SYSTEM

Scrotum

The scrotum is a two-chambered sac that contains the testes.
The dartos and cremaster muscles help to regulate testes temperature.

Testes

The testes are divided into lobules containing the seminiferous tubules and interstitial cells of Leydig.
The seminiferous tubules lead to the rete testis. The rete testis opens into the efferent ductules that extend to the epididymis.
During development the testes pass from the abdominal cavity through the inguinal canal to the scrotum.
Sperm cells are produced in the seminiferous tubules.
- Germ cells divide (mitosis) to form primary spermatocytes.
- Primary spermatocytes divide by meiosis to produce sperm cells.
- Sertoli cells nourish the spermatozoa and produce small amounts of hormones.

Spermatogenesis

Spermatogenesis begins in the seminiferous tubules at the time of puberty.
Spermatogonia give rise to spermatocytes, which divide by means of meiosis to form spermatids.
Spermatids develop a head, midpiece, and flagellum to become a sperm cell.

Ducts

The epididymis is a coiled tube system, located on the testis, that is the site of sperm maturation.
The ductus deferens passes from the epididymis into the abdominal cavity.
The ejaculatory duct is formed by the joining of the ductus deferens and the duct from the seminal vesicle.
The ejaculatory ducts join the urethra in the prostate gland.
The urethra extends from the urinary bladder through the penis to the outside of the body.

Penis

The penis consists of erectile tissue.
- The two corpora cavernosa form the dorsum and the sides.
- The corpus spongiosum forms the ventral portion and the glans penis.

The prepuce covers the glans penis.

Glands

The seminal vesicles empty into the ejaculatory duct.
The prostate gland consists of glandular and muscular tissue and empties into the urethra.
The bulbourethral glands are compound mucous glands that empty into the urethra.

Secretions

Semen is a mixture of gland secretions and spermatozoa.
The bulbourethral glands and the urethral mucous glands produce mucus that neutralizes the acidic pH of the urethra.
The testicular secretions contain sperm cells.
The seminal vesicle fluid contains nutrients and proteins that coagulate.
The prostate fluid contains nutrients, and it neutralizes the pH of the vagina.

PHYSIOLOGY OF MALE REPRODUCTION

Regulation of Sex Hormone Secretion

Gonadotropin-releasing hormone (GnRH) is produced in the hypothalamus and is released in surges.

GnRH stimulates release of luteinizing hormone (LH) and follicle-stimulating hormone (FSH) from the anterior pituitary.

- LH stimulates the interstitial cells of Leydig to produce testosterone.
- FSH stimulates spermatogenesis.

Puberty

Before puberty small amounts of testosterone inhibit GnRH release.

During puberty testosterone does not completely suppress GnRH release, resulting in increased production of FSH, LH, and testosterone.

Effects of Testosterone

Testosterone causes enlargement of the genitals and is necessary for spermatogenesis.

Testosterone is responsible for the development of secondary sex characteristics.

Male Sexual Behavior and the Male Sex Act

Testosterone is required for normal sex drive.

Stimulation of the sexual act can be tactile or psychic.

Afferent impulses pass to the sacral region of the spinal cord.

Efferent stimulation causes erection, mucus production, emission, and ejaculation.

FEMALE REPRODUCTIVE SYSTEM

Ovaries

The ovaries are covered by the peritoneum.

Follicles consist of an oocyte surrounded by granulosa cells and a thecal layer.

As follicles mature in response to FSH they enlarge, an antrum forms, and the oocyte increases in size. Granulosa cells form a cumulus layer that surrounds the oocyte, and a zona pellucida (a thin noncellular layer) forms around the oocyte.

About the time of ovulation the first meiotic division is completed.

Ovulation is the release of the oocyte from the ovary.

- In response to LH the follicle completes its growth and ruptures, and the oocyte is released from the ovary.
- The second meiotic division is completed when the oocyte unites with a sperm to form a zygote.

Fate of the follicle.

- The ovulated follicle becomes the corpus luteum.
- If fertilization occurs, the corpus luteum persists. If there is no fertilization, it degenerates.

Uterine Tubes

The uterine tubes transport the oocyte or zygote from the ovary to the uterus.

The ovarian end of the uterine tube is surrounded by fimbriae.

Cilia on the fimbriae move the oocyte into the uterine tube.

Fertilization usually occurs in the upper portion of the uterine tube.

Uterus

The uterus is a pear-shaped organ. The uterine cavity and the cervical canal are the spaces formed by the uterus.

The wall of the uterus consists of the serous layer, myometrium (smooth muscle), and endometrium.

Vagina

The vagina connects the uterus (cervix) to the vestibule.

The vagina consists of a layer of smooth muscle and an inner lining of moist, stratified squamous epithelium.

The hymen covers the vestibular opening of the vagina.

External Genitalia

The vestibule is a space into which the vagina and the urethra open.
The clitoris is composed of erectile tissue and contains many sensory organs important in detecting sexual stimuli.
The labia minora are folds that cover the vestibule and form the prepuce.
The vestibular glands produce a mucous fluid, and lubricating fluid is produced by the wall of the vagina.
The labia majora cover the labia minora, and the pudendal cleft is a space between the labia majora.
The mons pubis is an elevated area superior to the labia majora.

Mammary Glands

The mammary glands are modified sweat glands that consist of glandular lobes and adipose tissue.
The lobes connect to the nipple through ducts. The nipple is surrounded by the areola.

PHYSIOLOGY OF FEMALE REPRODUCTION

Puberty

Puberty begins with the first menstrual bleeding (menarche).
Puberty begins when GnRH, LH, and FSH levels increase.

Menstrual Cycle

The cyclical changes in the uterus are controlled by estrogen and progesterone produced by the ovary.
Cyclic changes in the uterus.
- Menses (day 1 to days 4 or 5). Menses is composed of sloughed cells, secretions, and blood.
- Proliferative phase (day 5 to day 14). Epithelial cells multiply and form glands.
- Secretory phase (day 15 to day 28). The endometrium becomes thicker, and endometrial glands secrete. The uterus is prepared for implantation of the fertilized ovum by day 21.
- Estrogen stimulates proliferation of the endometrium, and progesterone causes hypertrophy of the endometrium. Decreased progesterone causes menses.

FSH initiates the development of the follicles.
LH stimulates ovulation and formation of the corpus luteum.
A positive-feedback mechanism causes FSH and LH levels to increase near the time of ovulation.
- Estrogen produced by the theca cells of the follicle stimulates GnRH, FSH, and LH secretion.
- FSH and LH stimulate more estrogen secretion, and so on.

Menopause

The cessation of the menstrual cycle is called menopause.

Female Sexual Behavior and the Female Sex Act

Female sex drive is partially influenced by testosterone (produced by the adrenal gland) and estrogens produced by the ovary.
Autonomic nerves cause erectile tissue to become engorged with blood, the vestibular glands to secrete mucus, and the vagina to produce a lubricating fluid.

CONTENT REVIEW

1. What is the scrotum? Explain the function of the dartos and cremaster muscles.
2. Where, specifically, are spermatozoa produced in the testes? Describe the process of spermatogenesis.
3. Name the ducts the spermatozoa traverse to go from their site of production to the outside of the body.
4. Where do sperm develop their ability to fertilize?
5. Describe the erectile tissue of the penis.

6. State where the seminal vesicles, prostate gland, and bulbourethral glands empty into the male reproductive duct system.
7. Define emission and ejaculation.
8. Define semen. What structures give rise to secretions that make up the semen?
9. Where are GnRH, FSH, LH, and testosterone produced?
10. Describe the effects of testosterone during puberty and on the adult male.
11. Describe the male sexual act.
12. Describe the process of ovulation, and formation of the corpus luteum.
13. What is the corpus luteum? What happens to the corpus luteum if fertilization occurs? If fertilization does not occur?
14. Describe the normal pathway followed by the oocyte after ovulation. Where does fertilization usually take place?
15. Describe the relationship between the uterus, vagina, vestibule, and external genitalia.
16. Describe the labia minora, the prepuce, the labia majora, the pudendal cleft, and the mons pubis.
17. What are the effects of estrogen and progesterone on the uterus?
18. Describe the hormonal changes that result in ovulation. Explain the sequence of events during each phase of the menstrual cycle.
19. Define menopause, female climacteric, and the postmenopausal period. What causes these changes?

CONCEPT REVIEW

1. If an adult male were castrated by having his testes removed, what would happen to the levels of GnRH, FSH, LH, and testosterone in his blood?
2. If the ovaries were removed from a woman, what would happen to the levels of GnRH, FSH, LH, estrogen, and progesterone in her blood?
3. Birth control pills for women contain estrogen and progesterone compounds. Explain how these hormones can prevent pregnancy.
4. During the secretory phase of the menstrual cycle, you would normally expect
   A. the highest levels of progesterone that occur during the menstrual cycle.
   B. a follicle present in the ovary that is ready to undergo ovulation.
   C. the endometrium reaches its greatest degree of development.
   D. A and B
   E. A and C
5. During menopause which reproductive hormones are reduced in the blood and which are increased?

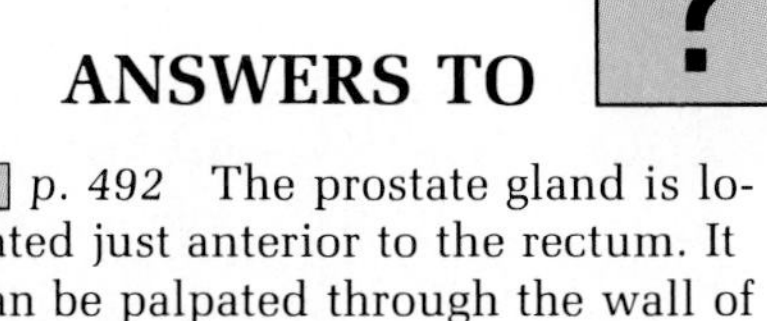

ANSWERS TO PREDICT QUESTIONS

1 *p. 492* The prostate gland is located just anterior to the rectum. It can be palpated through the wall of the rectum. A physician can insert fingers into the rectum and palpate the prostate through the wall of the rectum. The procedure does not require surgical procedures and involves relatively minor discomfort.

2 *p. 495* Since secondary sexual characteristics, external genitalia, and sexual behavior develop in response to testosterone, if the testes failed to produce normal amounts of testosterone at puberty, they would not develop normally. Secondary sexual characteristics and external genitalia would remain juvenile and normal adult sexual behavior would not develop.

3 *p. 502* Administration of a large amount of estrogen and progesterone just before the preovulatory LH surge would inhibit the release of GnRH, LH, and FSH. Consequently ovulation could not occur. However, small amounts of estrogen, given without progesterone, when administered before the preovulatory LH surge could stimulate GnRH, LH, and FSH secretion.

20

DEVELOPMENT, HEREDITY, AND AGING

After reading this chapter you should be able to:

1 List the prenatal periods and state the major events associated with each.
2 Describe the events of fertilization.
3 Describe the blastocyst and the process of implantation and placental formation.
4 List the three germ layers, describe their formation, and list the adult derivatives of each layer.
5 Describe the formation of the neural tube and neural crest.
6 Describe the formation of the gastrointestinal tract, the limbs, and the face.
7 Explain how the single heart tube is divided into four chambers.
8 Explain the events that occur during parturition and lactation.
9 Discuss the circulatory, digestive, and other changes that occur in the newborn at the time of birth.
10 List the stages of life and describe the major events that are associated with each stage. Explain aging.
11 Define genetics and explain how chromosomes are related to genetics.
12 Describe the three major patterns of inheritance.

blastocyst (blas′to-sist) [Gr. *blastos*, germ + *kystis*, bladder] Early stage of mammalian embryo development consisting of a hollow ball of cells with an inner cell mass and an outer trophoblast layer.

ectoderm (ek′to-derm) Outermost of the three germ layers of the embryo.

embryo In prenatal development, the developing human between approximately 14 and 60 days of development.

endoderm (en′do-derm) Innermost of the three germ layers of the embryo.

fertilization Union of the sperm and oocyte to form a zygote.

fetus In prenatal development, the developing human between approximately 60 days and birth.

genotype Genetic makeup of the individual.

heterozygous Having two different genes for a given trait.

homozygous Having two identical genes for a given trait.

lactation [L. *lactatio*, suckle] Period following childbirth during which milk is formed in the breasts.

mesoderm (mez′o-derm) Middle of the three germ layers of the embryo.

neural tube Tube formed from the neuroectoderm in the embryo by closure of the neural groove; develops into the brain and spinal cord.

parturition [L., *parturio*, to be in labor] Childbirth; the delivery of an infant at the end of pregnancy.

phenotype [Gr., *phaino*, to display + *typos*, model] Characteristic observed in the individual due to expression of the genotype.

primitive streak A shallow groove in the ectodermal surface of the embryonic disc; cells migrating through the streak become mesoderm.

zygote (zi′gōt) [Gr. *zygotos*, yoked] The single-celled, diploid cell product of fertilization, resulting from union of the sperm and oocyte.

The lifespan of a person is usually considered the time from birth to death. However, the 9 months before birth comprise a critical part of an individual's existence, and the events that occur during that period have profound effects on the rest of the person's life. Most people develop normally and are born without defects. However, approximately 10 out of every 100 people are born with some type of birth defect, and 3 out of every 100 people are born with a birth defect so severe that it requires medical attention during the first year of life. Later in life many more people discover unknown congenital (present at birth) problems such as the tendency to develop asthma, certain brain disorders, or cancer.

PRENATAL DEVELOPMENT

The **prenatal period,** the period from conception to birth, can be divided into three portions: (1) the germinal period—approximately the first 2 weeks of development during which the primitive germ layers are formed; (2) the embryonic period—from about the second to the eighth week of development during which the major organ systems come into existence; and (3) the fetal period—the last 7 months of the prenatal period during which the organ systems grow and become more mature.

The medical community in general uses the **last menstrual period (LMP)** to calculate the **clinical age** of the unborn child. Therefore, a given embryo or fetus is considered to be a certain number of days post-LMP. Most embryologists, on the other hand, use **developmental age,** which begins with fertilization, to describe the timing of developmental events. Since fertilization is assumed to occur approximately 14 days after LMP, it is assumed that developmental age is 14 days less than clinical age. The times presented in this chapter are based on developmental age.

Fertilization

After sperm cells are ejaculated into the vagina, they are transported through the cervix, the body of the uterus, and the uterine tubes where fertilization occurs. The swimming ability of the sperm and muscular contractions of the uterus and uterine tubes are responsible for the movement of sperm cells through the female reproductive tract. Oxytocin released by the female pituitary and prostaglandins within the semen both stimulate contractions in the uterus and uterine tubes.

While passing through the uterus and the uterine tubes, the sperm cells undergo **capacitation,** a process that enables the sperm to release enzymes that allow penetration through the cumulus cells and the oocyte cell membrane.

The oocyte is capable of being fertilized for up to 24 hours after ovulation, and some sperm remain viable in the female reproductive tract for up to 72 hours, although most of them degenerate by 24 hours. Therefore, for fertilization to result, sexual intercourse must occur approximately between 3 days before and 1 day following ovulation.

Thousands of **sperm** reach the **oocyte,** but normally a change occurs in the oocyte cell membrane that prevents more than one sperm from entering the oocyte. The oocyte undergoes the second mitotic division only after a sperm cell enters it. The nucleus that results from the second meiotic division moves to the center of the cell, where it meets the enlarged sperm head. Both of these nuclei are **haploid** (each having one chromosome from each chromosome pair), and their fusion, which completes the process of fertilization, restores the **diploid** number of chromosomes. **Fertilization** is defined as the union of the sperm and oocyte, with subsequent union of the genetic material (nuclei). The product of fertilization (Figure 20-1, *A*) is the **zygote** (zi'gōt).

> The zygote may, in rare cases, split, resulting in "identical" or monozygotic twins. Identical twins, therefore, have identical genetic information in their cells. Occasionally a woman may ovulate two or more oocytes at the same time. Fertilization of multiple oocytes by different sperm results in "fraternal" or dizygotic twins. Multiple ovulation can occur naturally or can be stimulated by injection of drugs that stimulate gonadotropin release. These drugs are sometimes used to treat certain forms of infertility.

Early Cell Division

About 18 to 36 hours after fertilization, the zygote divides to form two cells. Those two cells divide to form four cells, which divide to form eight, and so on (Figure 20-1, *B-D*). Even though the number of cells increases, the total mass remains about the same as the zygote. The cells of this dividing embryonic mass have the ability to develop into a wide range of tissues. As a result, the total number of embryonic cells can be decreased, increased, or reorganized during this period without affecting the normal development of the embryo.

Embryo Transfer

In a small number of women normal pregnancy is not possible because of some anatomical or physiological condition. In 87% of these cases the uterine tubes are incapable of transporting the zygote to the uterus or of allowing sperm to reach the oocyte. *In vitro* fertilization and embryo transfer have made pregnancy possible in hundreds of such women since 1978.

The woman is first induced to superovulate (an LH-like substance is injected, causing more than one oocyte to be ovulated at one time). Before the follicles rupture, the oocytes are surgically removed from the ovary. The oocytes are incubated in a dish (thus the term *in vitro*, in glass) and maintained at body temperature for 6 hours; then sperm are added to the dish.

After 24 to 48 hours when the zygotes have divided to form two- to eight-cell embryonic masses, several of the embryonic masses are transferred to the uterus. Implantation and subsequent development then proceed in the uterus as they would for natural implantation. However, the woman is usually required to lie perfectly still for several hours after the embryonic masses have been introduced into the uterus to prevent possible expulsion before implantation can occur. It is not fully understood why such expulsion does not occur in natural fertilization and implantation.

The success rate of embryo transfer varies from clinic to clinic but is increasing steadily (the success rate at the best U.S. clinic in 1987 to 1988 was 27%). Multiple births have occurred frequently following embryo transfer because of the practice of introducing more than one embryonic mass into the uterus in an attempt to increase the success rate as much as possible.

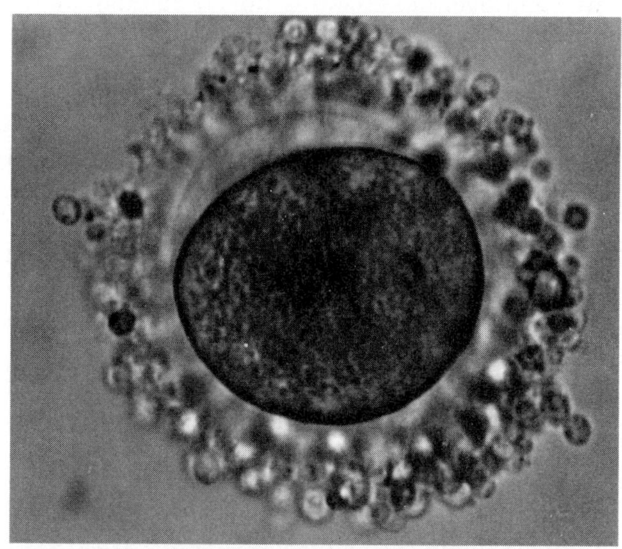
A

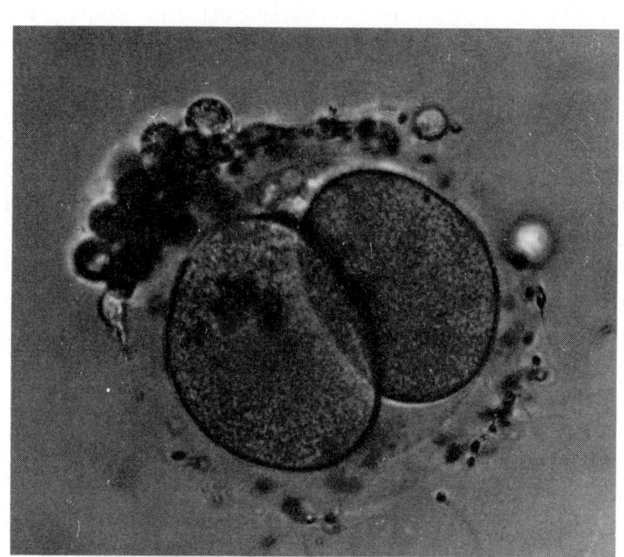
B

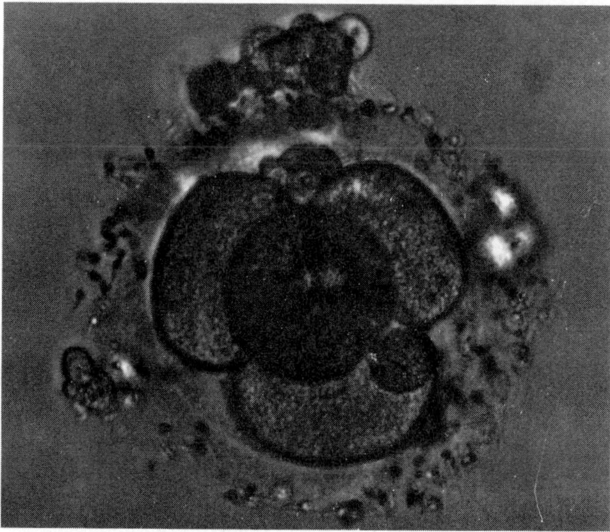
C

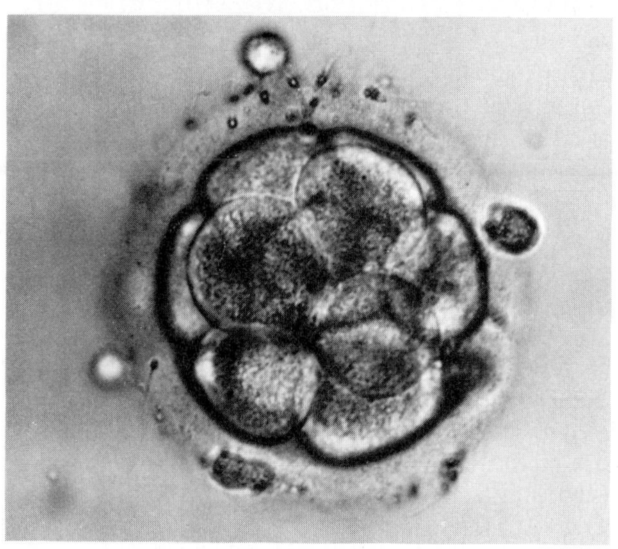
D

FIGURE 20-1 ■ Early stages of human development.
A Zygote (120 mm in diameter).
B to **D** During the early cell divisions, the zygote divides into more and more cells, but the total mass remains relatively constant.

Blastocyst

About 3 or 4 days after fertilization the embryonic mass consists of about 32 cells; and, when a cavity begins to appear in the midst of the cellular mass, the mass is called a **blastocyst** (blas'to-sist) (Figure 20-2). The fluid-filled cavity is called the **blastocele** (blas'to-sēl). A single layer of cells, the **trophoblast** (tro'fo-blast; feeding layer), surrounds most of the blastocele, but at one end of the blastocyst the cells are several layers thick. The thickened area is the **inner cell mass** and is the tissue from which the embryo will develop. The trophoblast forms the placenta and the membranes surrounding the embryo.

Implantation of the Blastocyst and Development of the Placenta

All the events of the early germinal phase, including the first cell division through formation of the blastocele, occur as the embryonic mass moves from the site of fertilization in the uterine tube to the site of implantation in the uterus. By 7 or 8 days after ovulation (day 21 or 22 of the menstrual cycle), the endometrium of the uterus is prepared for implantation. About 7 days after fertilization, the blastocyst attaches itself to the uterine wall and begins the process of **implantation.** The trophoblast cells of the blastocyst digest the uterine tissues as the blastocyst burrows into the uterine wall.

As the blastocyst invades the uterine wall, trophoblast cells, now called the **chorion,** form the embryonic portion of the **placenta.** Fingerlike projections, called **chorionic villi,** protrude into cavities containing pools of maternal blood, called **lacunae** (lă-ku'ne), formed within the maternal endometrium. In the mature placenta, the embryonic blood supply is separated from the maternal blood supply by the embryonic capillary wall, a basement membrane, and a thin layer of chorion (Figure 20-3, *A*). As a result, the embryonic and maternal blood does not mix, but nutrients and waste products can pass between the two circulations.

The developing human between 14 and 60 days of development is called an **embryo.** Initially the embryo is attached to the placenta by a connecting stalk (Figure 20-3, *B*). As the embryo matures, the connecting stalk elongates and becomes known as the **umbilical cord.** Within the umbilical cord, two umbilical arteries, which originate in the embryo's iliac arteries, carry blood from the embryo to the mother and one umbilical vein carries blood back to the embryo's liver.

The chorion secretes **human chorionic gonadotropin (HCG),** which is transported in blood to the maternal ovary and causes the corpus luteum to remain functional. The secretion of HCG increases rapidly and reaches a peak about 8 or 9 weeks after fertilization. Subsequently, HCG levels decline to a lower level and are maintained at a low level throughout the remainder of the pregnancy (Figure 20-4). Most pregnancy tests are designed to detect HCG in either urine or blood.

The estrogen and progesterone secreted by the corpus luteum are essential for the maintenance of the endometrium for the first 3 months of pregnancy. After the placenta forms, it also begins to secrete estrogens and progesterone. By the third month of pregnancy the placenta has become an endocrine gland that secretes sufficient quantities of estrogen and progesterone to maintain pregnancy. Estrogen and progesterone levels increase in the mother's blood throughout pregnancy.

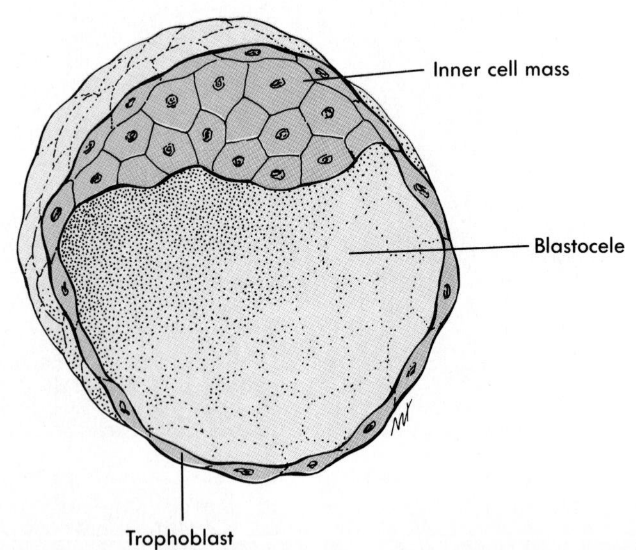

FIGURE 20-2 Blastocyst.
The orange cells of the inner cell mass will become the embryo. All the green cells will become placenta and membranes (amnion and yolk sac).

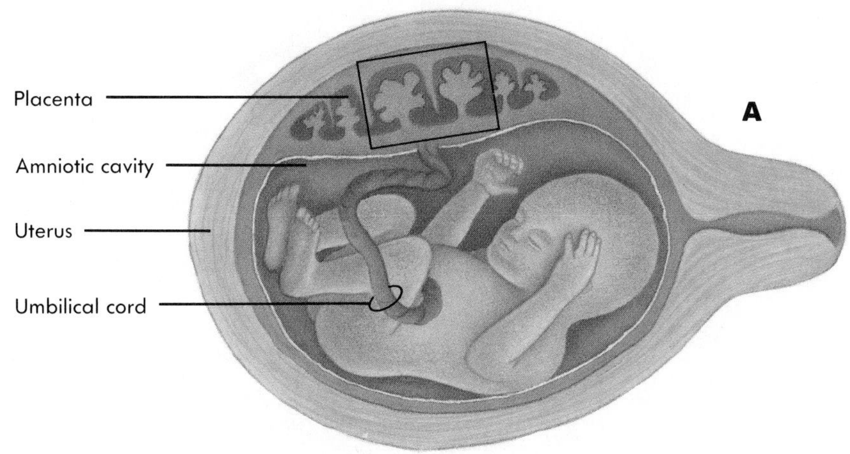

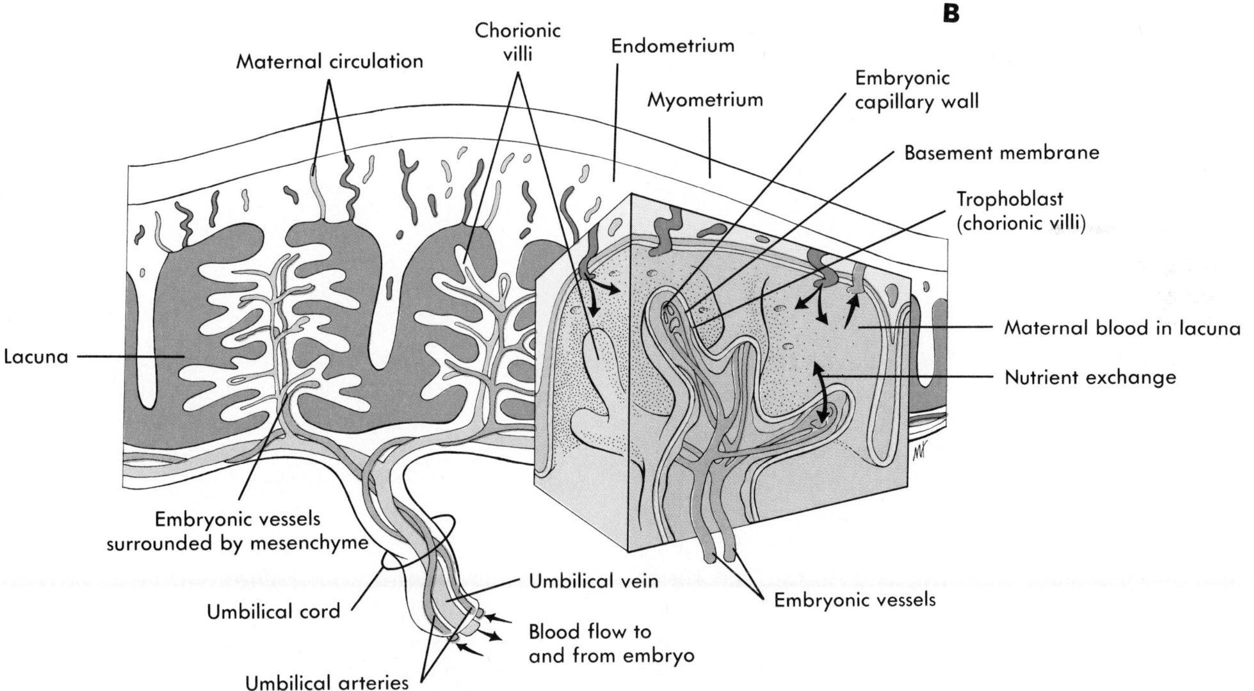

FIGURE 20-3 ■ The interface between maternal and fetal circulation.
A Location of the placenta and umbilical cord.
B As maternal blood vessels are encountered by the trophoblast (chorionic villi), lacunae (cavities) are formed and are filled with maternal blood. In the mature placenta, the embryonic mesoderm and blood vessels form chorionic villi, and nutrients are exchanged between embryonic and maternal blood. At no time, under normal conditions, do the maternal and fetal blood mix during development.

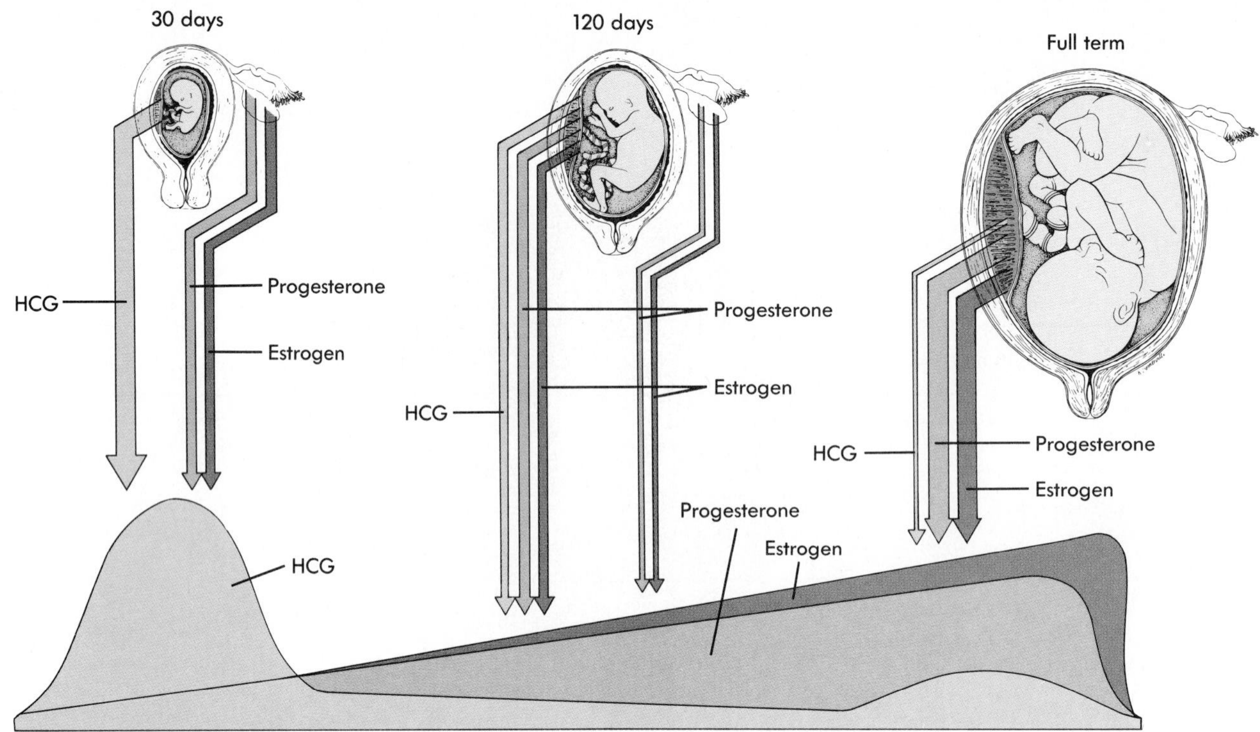

FIGURE 20-4 Hormone levels (indicated by the size of the arrows) released from the ovary and placenta during pregnancy.
The levels of HCG, which is always produced in the placenta, decline during pregnancy. There is a shift in progesterone and estrogen synthesis from the ovary early in pregnancy to the placenta late in pregnancy.

Formation of the Germ Layers

After implantation a new cavity called the **amniotic cavity** forms inside the inner cell mass and causes the part of the inner cell mass nearest the blastocele to separate as a flat disk of tissue called the **embryonic disk** (Figure 20-5). The amniotic cavity will enlarge as a fluid-filled amniotic sac into which the embryo will grow. The amniotic fluid within the cavity forms a protective cushion around the embryo. The embryonic disk is composed of two layers of cells—an **ectoderm** (ek'to-derm; outside layer) adjacent to the amniotic cavity and an **endoderm** (en'do-derm; inside layer) on the side of the disk opposite the **amniotic cavity.** A third cavity, the **yolk sac,** forms inside the blastocele from the endoderm.

At about 14 days after fertilization, the embryonic disk has become a slightly elongated oval structure. Proliferating cells of the ectoderm migrate toward the center of the disk, forming a thickened line called the **primitive streak.** The embryo will form around this primitive streak. Some ectoderm cells migrate through the primitive streak and emerge between the ectoderm and endoderm as a new germ layer, the **mesoderm** (mez'o-derm; middle layer) (Figure 20-6). The embryo is now three-layered, having ectoderm, mesoderm, and endoderm; all tissues of the adult can be traced to these three germ layers (Table 20-1). A cordlike structure called the **notochord** extends from the cephalic end of the primitive streak.

1 Predict the results of two primitive streaks forming in one embryonic disk. What if the two primitive streaks are touching each other? ?

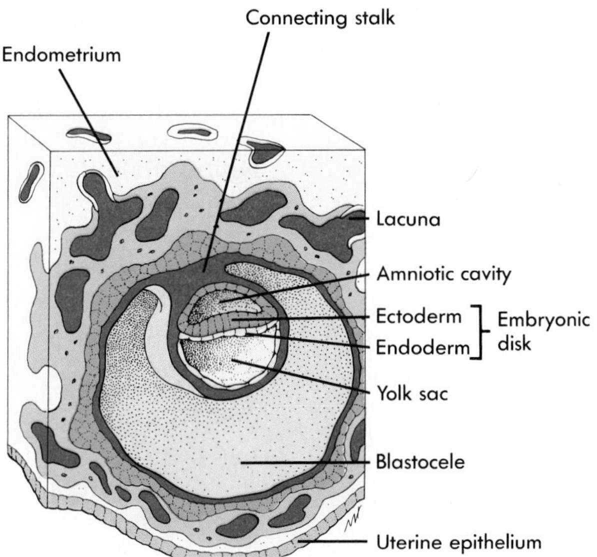

FIGURE 20-5 ■ Embryonic disc consisting of ectoderm and endoderm, with amniotic cavity and yolk sac.
The connecting stalk, which attaches the embryo to the uterus, will become part of the umbilical cord.

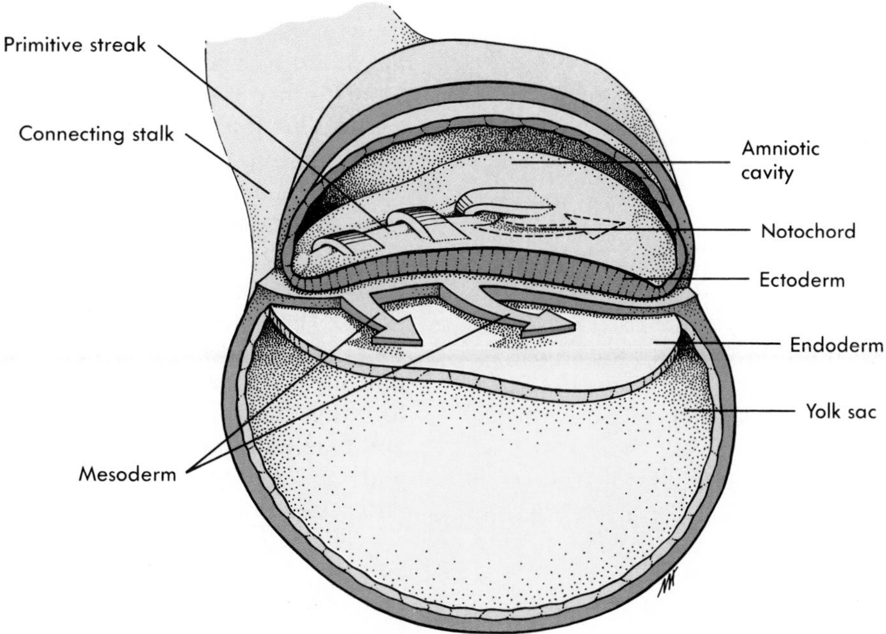

FIGURE 20-6 ■ Embryonic disk with a primitive streak.
The head of the embryo will develop over the notochord.

TABLE 20-1

Germ Layer Derivatives

ENDODERM
- Lining of gastrointestinal tract
- Lining of lungs
- Lining of hepatic, pancreatic, and other exocrine ducts
- Kidney ducts and bladder
- Adenohypophysis
- Thymus gland
- Thyroid gland
- Parathyroid gland
- Tonsils

ECTODERM
- Epidermis of skin
- Tooth enamel
- Lens and cornea of eye
- Outer ear
- Nasal cavity
- Neuroectoderm
  - Brain and spinal cord
  - Somatomotor neurons
  - Preganglionic autonomic neurons
- Neural crest cells
  - Melanocytes
  - Sensory neurons
  - Postganglionic autonomic neurons
  - Adrenal medulla
  - Facial bones
  - Teeth: dentin and pulp
  - Skeletal muscles in head

MESODERM
- Dermis of skin
- Circulatory system
- Parenchyma (substance) of glands
- Kidneys
- Gonads
- Muscle
- Bones (except facial)

Neural Tube and Neural Crest Formation

At about 18 days after fertilization, the ectoderm overlying the notochord thickens to form the **neural plate** (Figure 20-7). The lateral edges of the plate begin to rise like two ocean waves coming together. These edges are called the **neural crests,** and a **neural groove** lies between them. The neural crests begin to meet in the midline and fuse into a **neural tube,** which is completely closed by 26 days. The cells of the neural tube are called **neuroectoderm** (see Table 20-1). Neuroectoderm becomes the brain, the spinal cord, and parts of the peripheral nervous system.

As the neural crests come together and fuse, a population of cells breaks away from the neuroectoderm all along the margins of the crests. Most of these **neural crest cells** become part of the peripheral nervous system or become melanocytes of the skin. In the head, neural crest cells perform additional functions; they contribute to the skull, the dentin of teeth, blood vessels, and general connective tissue.

Formation of the General Body Structure

Arms and legs first appear at about 28 days after fertilization as **limb buds** (Figure 20-8) and quickly begin to elongate. At about 35 days, expansions, called hand and foot plates, form at the ends of the limb buds. Radial zones of cell death within the hand and foot plates sculpture the fingers and toes.

The face develops by fusion of five masses of tissue. One mass forms the forehead, nose, and center of the upper jaw and lip. The nose begins as two structures, one on each side of this forehead mass. Two masses form the lateral portions of the upper jaw and lip, and two other masses form the lower jaw and lip (Figure 20-9, *A* and *B*).

As the brain enlarges and the face matures, the two portions of the nose approach each other in the midline and fuse (Figure 20-9, *C-E*). The two masses forming the upper jaw expand toward the midline and fuse with part of the nose to form the upper jaw and lip. A **cleft lip** results from failure of these structures to fuse.

The roof of the mouth, or palate, begins to form as vertical shelves of tissue that grow on the inside of the maxillary masses. These shelves swing to a horizontal position and begin to fuse with each other at about 56 days of development. If the palate does not fuse, a midline cleft in the roof of the mouth called a **cleft palate** results.

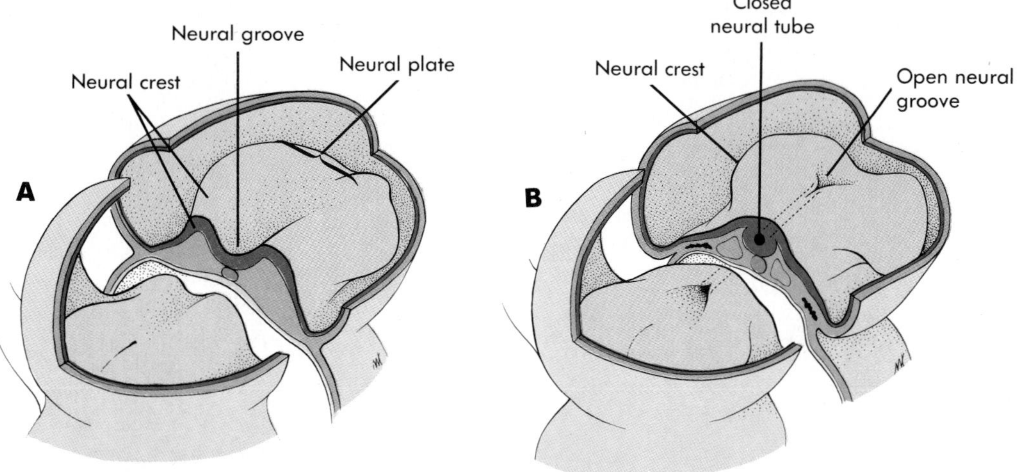

FIGURE 20-7 ■ Formation of the neural tube.
A Neural plate with the neural crests located laterally and the neural groove in the center.
B The neural crests have come together in the center of the embryo to form the neural tube; neural grooves can be seen at each end.

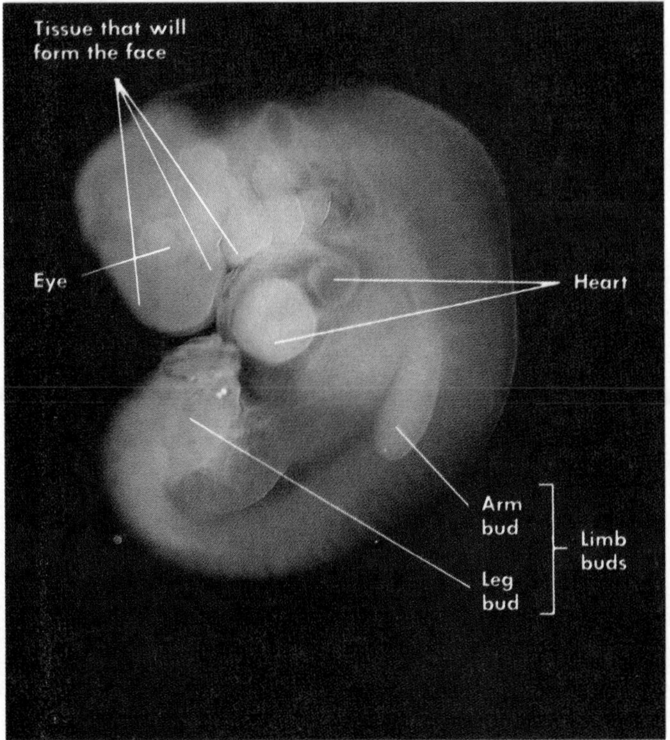

FIGURE 20-8 ■ Human embryo at 35 days after fertilization.

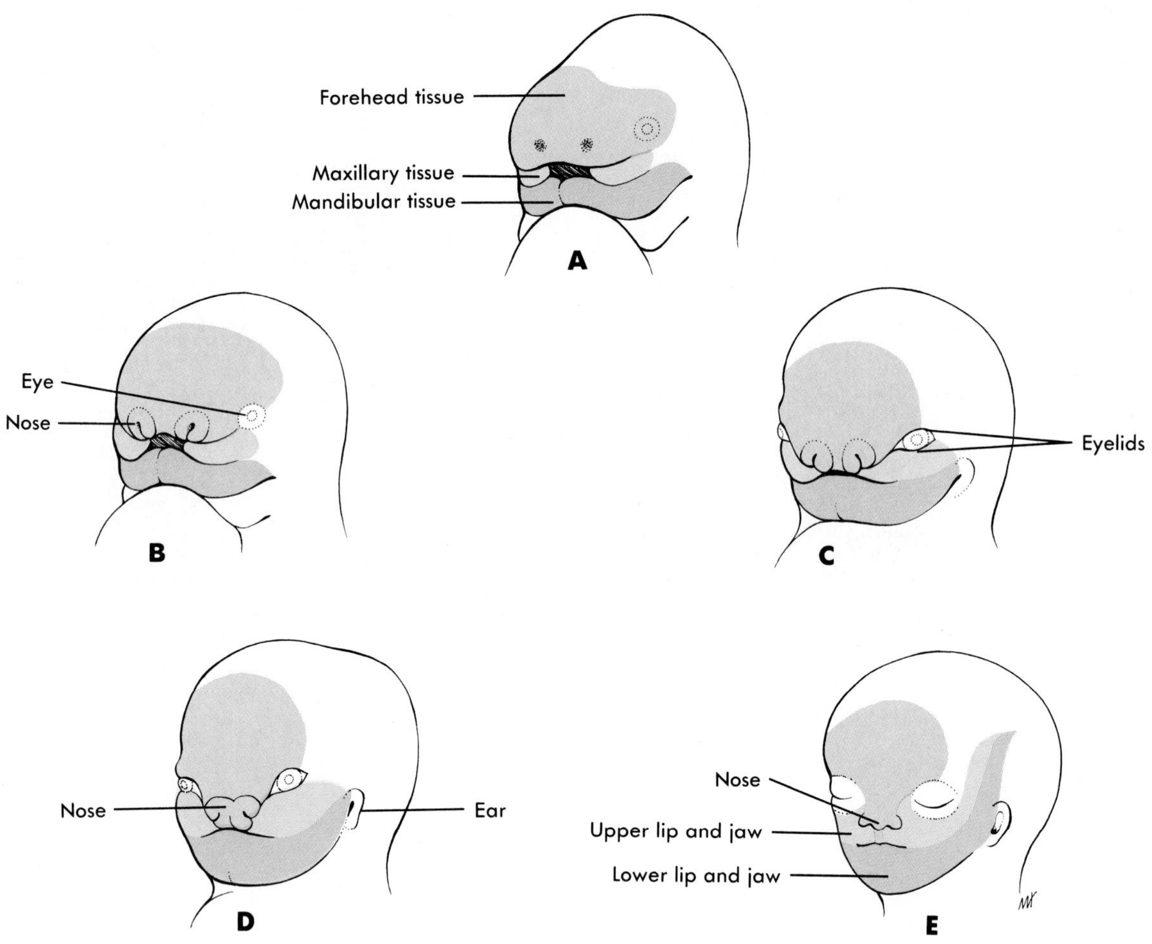

FIGURE 20-9 Development of the face.
Ages indicate developmental days; colors show the contributions of each process to the adult face.
A 28 days. The face develops from five tissue masses: forehead (1, *blue*), maxilla (2, *tan*), and mandible (2, *brown*).
B 33 days. Nose appears on the forehead mass.
C 40 days. Maxillary masses enlarge and move toward the midline. The two halves of the nose also move toward the midline and fuse with the maxillary masses to form the upper jaw and lip.
D 48 days. Continued growth brings structures more toward the midline.
E 14-week fetus.

Development of the Organ Systems

The major organ systems appear and begin to develop during the embryonic period. Therefore the period between 14 and 60 days is called the period of **organogenesis.** The individual organ systems will not be described in the text but are listed in Table 20-2 (pp. 526-529). Only general comments about a few select systems will be presented in the text.

At the same time the neural tube is forming (18 to 26 days), the embryo itself is folding to form a tube along the upper portion of the yolk sac (Figure 20-10). This tube, formed from the upper portion of the yolk sac, is the beginning of the gastrointestinal (GI) tract.

A considerable number of outpocketings appear at about 28 days after fertilization along the entire length of the GI tract (Figure 20-11). A surprisingly large number of important internal organs develops from those outpocketings, including the auditory tube, tonsils, thymus gland, anterior pituitary gland, thyroid gland, parathyroid glands, lungs, liver, pancreas, and urinary bladder.

Text continued on p. 528.

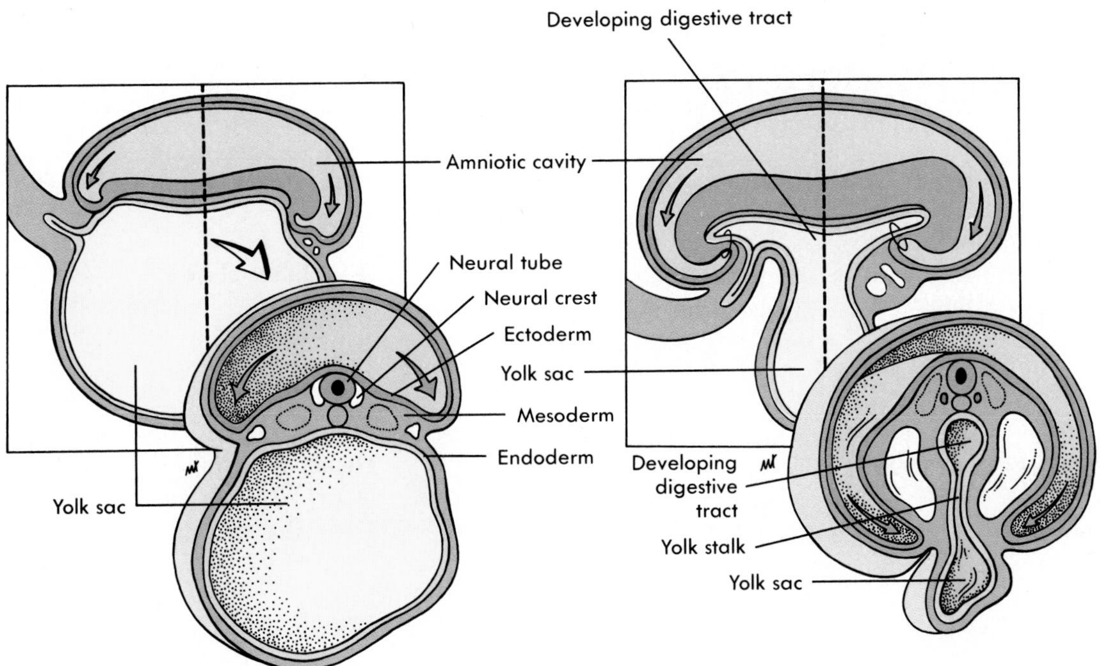

FIGURE 20-10 ■ Development of the digestive tract *(yellow)* as the body folds into a tube *(arrows)*.

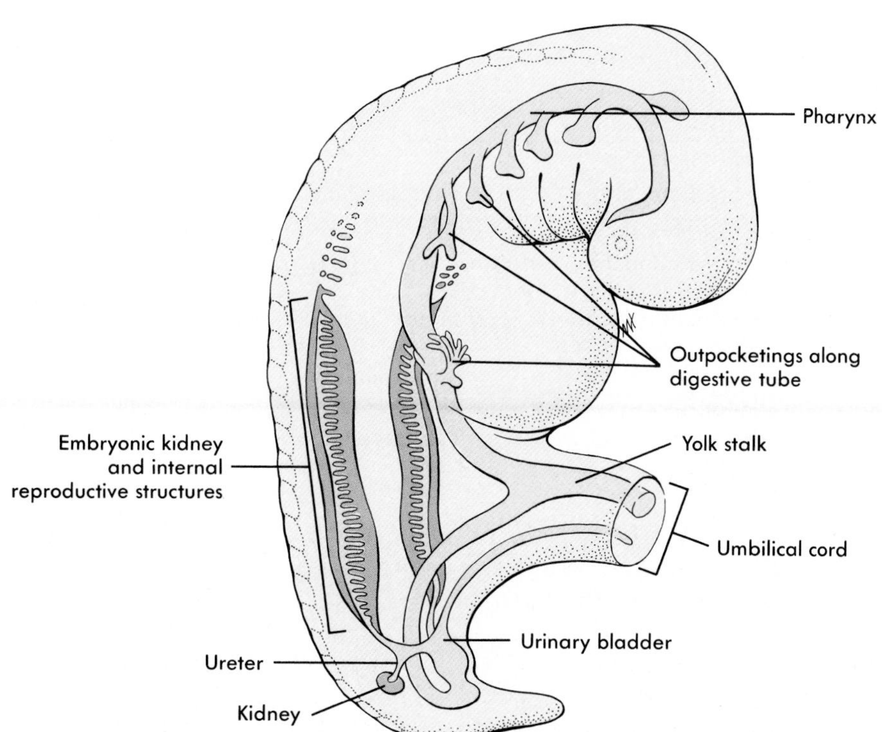

FIGURE 20-11 ■ Outpocketings of the digestive tract *(yellow)* and formation of the embryonic *(tan)* and adult kidneys *(purple)*.

TABLE 20-2

Development of the Organ Systems

	AGE (DAYS SINCE FERTILIZATION)					
	1-5	6-10	11-15	16-20	21-25	26-30
GENERAL FEATURES	Fertilization Blastocyst	Blastocyst implants	Primitive streak Three germ layers	Neural plate	Neural tube closed	Limb buds and other "buds" appear
INTEGUMENTARY SYSTEM			Ectoderm Mesoderm		Melanocytes from neural crest	
SKELETAL SYSTEM			Mesoderm		Neural crest	Limb buds
MUSCULAR SYSTEM			Mesoderm	Somites (body segments) begin to form		Somites all formed
NERVOUS SYSTEM			Ectoderm	Neural plate	Neural tube complete Neural crest Eyes and ears begin to form	Lens appears
ENDOCRINE SYSTEM			Ectoderm Mesoderm Endoderm	Thyroid gland begins to develop		Parathyroid glands and pancreas appear
CARDIOVASCULAR SYSTEM			Mesoderm	Blood islands form Two-tubed heart	Single-tubed heart, begins to beat	Interatrial septum forms
LYMPHATIC SYSTEM			Mesoderm			Thymus appears

AGE (DAYS SINCE FERTILIZATION)					
31-35	36-40	41-45	46-50	51-55	56-60
Hand and foot plates on limbs	Fingers and toes appear Lips formed Embryo 15 mm	External ear forming Embryo 20 mm	Embryo 25 mm	Limbs elongate to adult relation Embryo 35 mm	Face is distinctly human in appearance
Sensory receptors appear in skin		Collagen fibers clearly present in skin		Extensive sensory nerve endings in skin	
Mesoderm condensation in areas of future bone	Cartilage in site of future humerus	Cartilage in site of future ulna and radius	Cartilage in site of future hand and fingers		Ossification begins in clavicle and then in other bones
Muscle precursor cells enter limb buds			Functional muscle		Nearly all muscles appear in adult form
Nerve processes enter limb buds		External ear forming Olfactory nerve begins to form		Semicircular canals in inner ear complete	Eyelids form Cochlea in inner ear complete
Pituitary appears as evaginations from brain and mouth	Gonads begin to form Adrenal glands form		Pineal gland appears	Thyroid gland in adult position	Pituitary loses its connection to mouth
Interventricular septum begins to form		Interventricular septum complete	Interatrial septum complete but has opening until birth		
Large lymph vessels form in neck	Spleen appears			Adult lymph pattern formed	

Continued.

TABLE 20-2—cont'd

Development of the Organ Systems

	AGE (DAYS SINCE FERTILIZATION)					
	1-5	6-10	11-15	16-20	21-25	26-30
RESPIRATORY SYSTEM			Mesoderm Endoderm		Diaphragm begins to form	Trachea Lung buds
DIGESTIVE SYSTEM			Endoderm		Foregut and hindgut form	Liver and pancreas appear as buds
URINARY SYSTEM			Mesoderm Endoderm		Embryonic kidney appears	Embryonic kidney elongates
REPRODUCTIVE SYSTEM			Mesoderm Endoderm	Primordial germ cells on yolk sac	Male reproductive ducts appear External genital structures begin to form.	

The heart develops from two tubes, which fuse about 21 days after fertilization into a single, midline heart tube. At about this time, the primitive heart begins to beat. Blood vessels form from "blood islands" on the surface of the yolk sac and inside the embryo. These islands expand and fuse to form the circulatory system.

The major chambers of the heart, the atrium and ventricle, expand rapidly. The single ventricle is subdivided into two chambers by the development of an **interventricular septum** (Figure 20-12). If the interventricular septum does not grow enough to completely separate the ventricles, a ventricular septal defect (VSD) results.

An **interatrial septum** forms to separate the two atria. An opening in the interatrial septum, called the **foramen ovale** (o-val'e), connects the two atria and allows blood to flow from the right to the left atrium in the embryo and fetus. Because of the presence of the foramen ovale, most of the blood in the fetus passes from the right atrium to the left atrium and bypasses the right ventricle and the lungs. The foramen ovale normally closes off at the time of birth, and blood then circulates through the right ventricle and the lungs. If this does not occur, an interatrial septal defect (or atrial septal defect [ASD]) occurs. Either an interatrial septal defect or a ventricular septal defect usually result in a heart murmur.

The kidneys develop from mesoderm located along the lateral wall of the body cavity (see Figure 20-11). The embryonic kidney is much more extensive than the adult kidney, extending the entire length of the body cavity, and is closely associated with internal reproductive organs such as the gonads (ovaries or testes) and reproductive ducts (uterine tubes or vas deferens). Most of the embryonic kidney degenerates with only a very small caudal portion forming the adult kidney.

AGE (DAYS SINCE FERTILIZATION)					
31-35	36-40	41-45	46-50	51-55	56-60
Secondary bronchi to lobes form	Tertiary bronchi to lobules form		Tracheal cartilage begins to form		
Mouth opens to outside		Palate begins to form Tooth buds begin to form			Palate begins to fuse (fusion complete by 90 days); anus opens
Adult kidney begins to develop				Embryonic kidney degenerates	
	Gonads begin to form	Primordial germ cells enter gonads	Female reproductive ducts appear		Uterus forming External genitalia begins to differentiate in male and female

Growth of the Fetus

The embryo becomes a fetus about 60 days after fertilization (Figure 20-13). The beginning of the fetal period is marked by the beginning of bone ossification. In the embryo, most of the organ systems are developing, whereas in the fetus the organs are present. During the fetal period, most of the organ systems enlarge and mature. The fetus grows from about 3 cm and 2.5 grams (0.09 ounce) at 60 days to 50 cm and 3500 grams (7 pounds 11 ounces) at the end of pregnancy. The growth during the fetal period represents more than a 15-fold increase in length and a 1400-fold increase in weight.

Fine, soft hair called **lanugo** covers the fetus, and a waxy coat of sloughed epithelial cells called **vernix caseosa** protects the fetus from the somewhat toxic amniotic fluid. The amniotic fluid contains toxic waste products from the fetus.

Subcutaneous fat that accumulates in the fetus provides a nutrient reserve, helps insulate, and aids the newborn infant in sucking by strengthening and supporting the cheeks so that a small vacuum can be developed in the oral cavity.

Peak body growth occurs late in gestation, but as placental size reaches a maximum, the oxygen and nutrient supply to the fetus becomes limited. Growth of the placenta essentially stops at about 35 weeks, restricting further fetal growth.

At approximately 38 weeks of development the fetus has progressed to the point that it can survive outside the mother. The average weight at this point is 3250 grams (7 pounds 2 ounces) for a female fetus and 3300 grams (7 pounds 4 ounces) for a male fetus.

2 How many days (developmental time) does it take an infant to develop from fertilization to parturition?

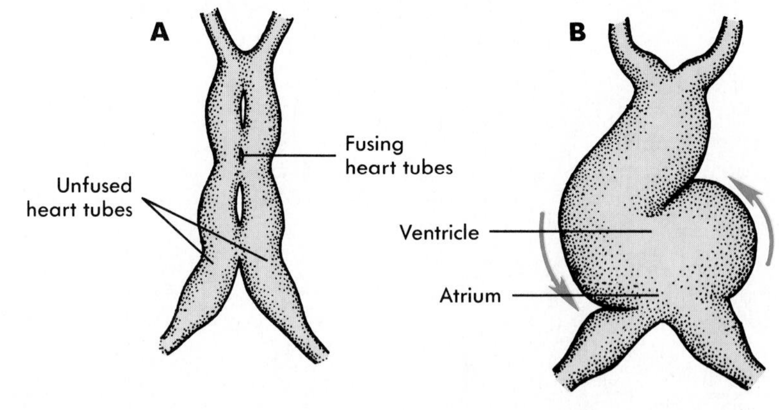

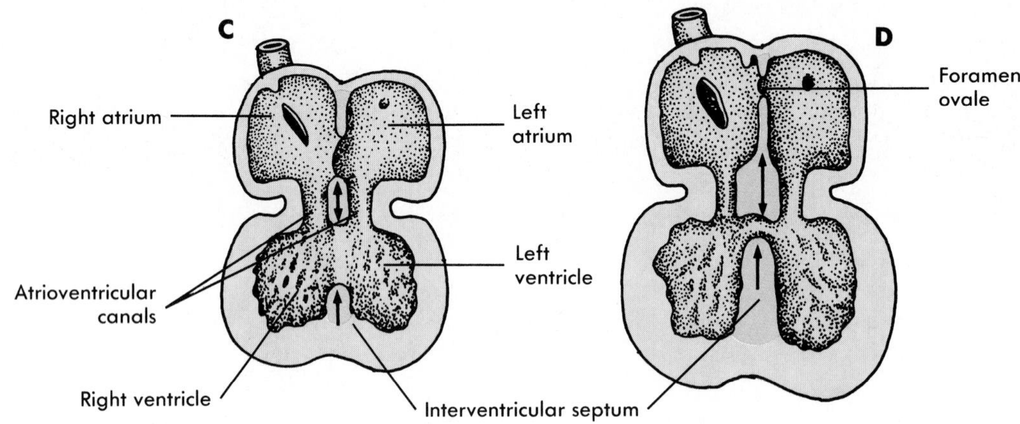

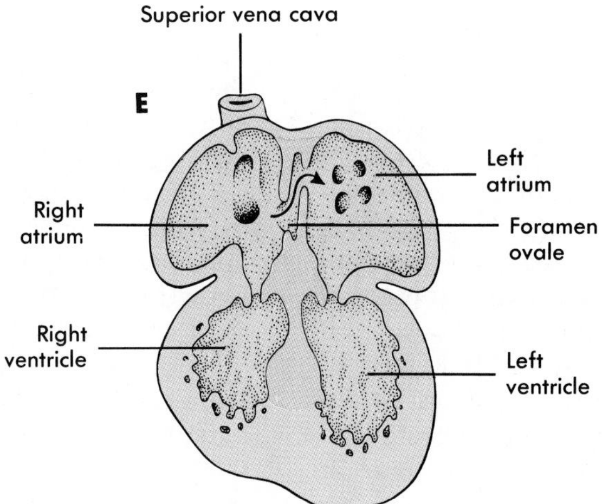

FIGURE 20-12 ■ Formation of the heart.
A Two heart tubes fuse to form a single heart tube.
B The heart tube bends *(arrows)* pulling the atrium superior to the ventricle.
C to **E** Septa form inside the heart to separate the right and left chambers.

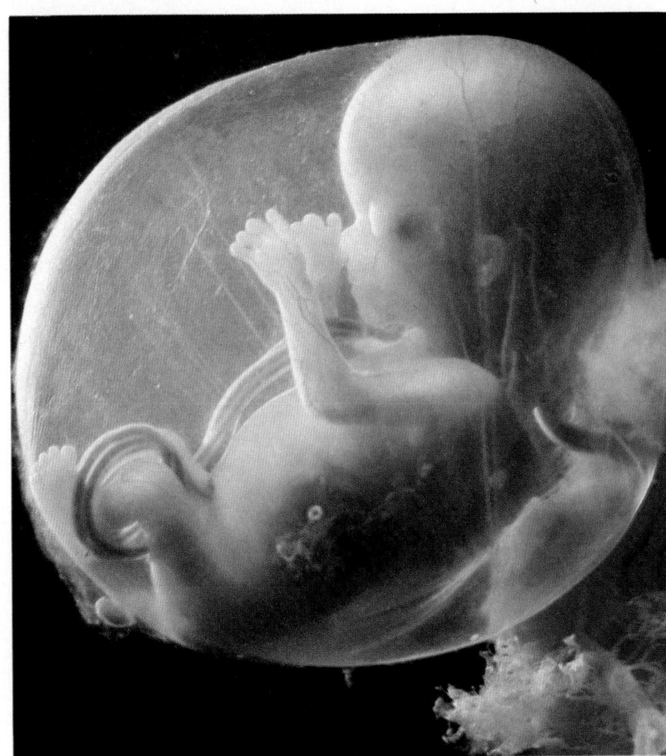

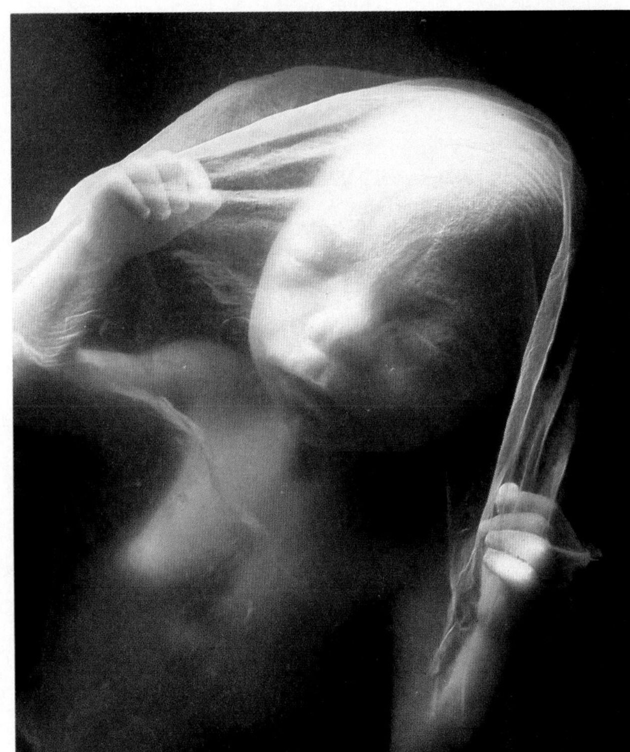

FIGURE 20-13 Fetus.
A At 3 months of development.
B At 4 months of development.

PARTURITION

Physicians usually calculate the gestation period (length of pregnancy) as 280 days (40 weeks or 10 lunar months) from the last menstrual period (LMP) to the date of delivery of the fetus. **Parturition** refers to the process by which the baby is born (Figure 20-14). Near the end of pregnancy, the uterus becomes progressively more irritable and usually exhibits occasional contractions that become stronger and more frequent until parturition is initiated. The cervix gradually dilates. Finally, strong uterine contractions complete cervical dilation and ultimately expel the fetus from the uterus through the vagina. Before expulsion of the fetus from the uterus, the membranous sac surrounding the fetus (the amniotic sac) ruptures, and amniotic fluid flows through the vagina to the exterior.

Labor is the period during which uterine contractions that result in expulsion of the fetus occur. Although labor may differ greatly from woman to woman, it can usually be divided into three stages. The **first stage** begins with the onset of regular uterine contractions and extends until the cervix dilates to a diameter about the size of the fetus' head (10 cm). This stage takes approximately 24 hours, but it may be as short as a few minutes in some women who have had more than one child. The **second stage** of labor lasts from the time of maximum cervical dilation until the time that the baby exits the vagina. This stage may last from 1 minute to up to 1 hour. During this stage, contraction of the abdominal muscles assists the uterine contractions. The **third stage** of labor involves the expulsion of the placenta from the uterus. Contractions of the uterus cause the placenta to tear away from the wall of the uterus. Some bleeding occurs because of the intimate contact between the placenta and the uterus. However, bleeding is normally restricted because uterine smooth muscle contractions compress the blood vessels.

Blood levels of estrogen and progesterone fall dramatically after parturition. Once the placenta has been dislodged from the uterus, the source of these hormones is gone. During the 4 or 5 weeks following parturition, the uterus becomes much smaller, but it remains somewhat larger than it was before pregnancy. A vaginal discharge composed of

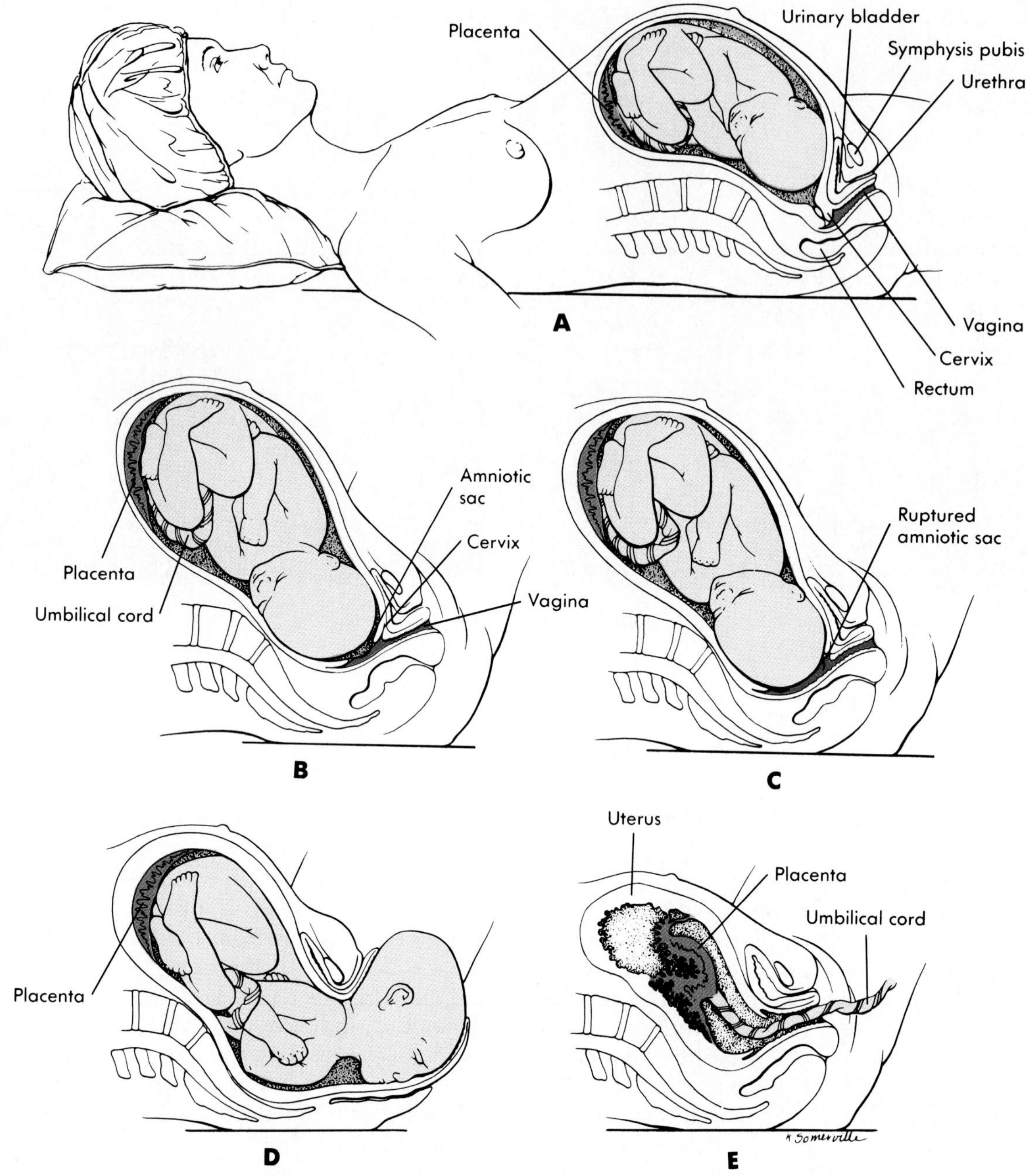

FIGURE 20-14 ■ Parturition.
A The relation of the fetus to the mother.
B The fetus moves into the opening of the birth canal and the cervix begins to dilate.
C Complete dilation of the cervix.
D The fetus is expelled from the uterus.
E The placenta is expelled.

small amounts of blood and degenerating endometrium may persist for several days after parturition.

The precise signal that triggers parturition is not known, but many factors that support parturition have been identified (Figure 20-15). Before parturition, the progesterone concentration in the mother's blood has reached its highest level. Progesterone has an inhibitory effect on uterine smooth muscle cells. However, estrogen levels are rapidly increasing in the maternal circulation, and estrogens have an excitatory influence on uterine smooth muscle. The inhibitory influence of progesterone on smooth muscle may be overcome by the stimulatory effect of estrogens near the end of pregnancy.

The adrenal gland of the fetus is greatly enlarged before parturition. Secretions of the fetal adrenal gland, initiated by stress on the fetus and stimulated by ACTH from the fetal pituitary may also stimulate parturition. The adrenal steroids reduce progesterone secretion and increase estrogen secretion and prostaglandin production by the placenta. Prostaglandins strongly stimulate uterine contractions.

During parturition, oxytocin is released from the mother's posterior pituitary. Stretching of the cervix produces action potentials that are sent to the hypothalamus and cause the release of oxytocin. Oxytocin also stimulates uterine contractions.

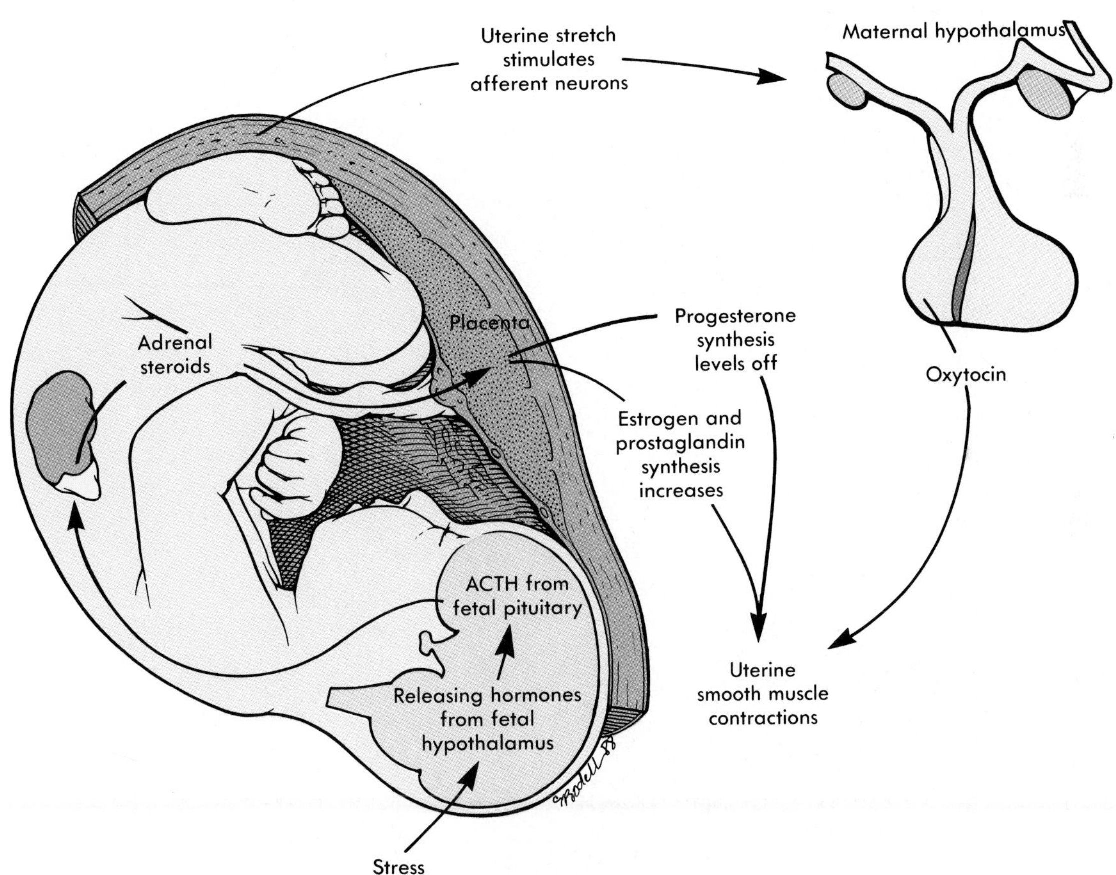

FIGURE 20-15 ■ Factors that influence the process of parturition.
As the fetus increases in size, becoming more and more confined within the uterus, it becomes stressed. This stress stimulates releasing hormones in the hypothalamus, which stimulate the fetal pituitary. The fetal pituitary secretes adrenocorticotropic hormone (ACTH) in greater amounts near parturition. ACTH causes the fetal adrenal gland to secrete greater quantities of adrenal cortical steroids, which travel in the umbilical blood to the placenta. There the adrenal steroids cause progesterone synthesis to level off and estrogen and prostaglandin synthesis to increase, making the uterus more irritable. Stretch of the uterus causes neural impulses to the brain through ascending pathways and stimulates the secretion of oxytocin. Oxytocin also causes the uterine smooth muscle to contract. Although the precise control of parturition in humans is not known, these changes appear to play a role.

THE NEWBORN

The newborn infant, or **neonate,** experiences several dramatic changes at the time of birth. The major and earliest changes deal with the separation of the infant from the maternal circulation and transfer from a fluid to a gaseous environment. The large, forced gasps of air that occur when the infant cries at the time of delivery help to inflate the lungs.

Circulatory Changes

The initial inflation of the lungs causes important changes in the circulatory system (Figure 20-16). Expansion of the lungs reduces the resistance to blood flow through the lungs, resulting in increased blood flow from the right ventricle of the heart through the pulmonary arteries. Consequently, an increased amount of blood flows from the right atrium to the right ventricle and into the

FIGURE 20-16 Circulatory changes at the time of birth. **A** Before birth the umbilical vein and arteries attach to the placenta, blood passes through the foramen ovale in the heart, and the ductus arteriosus shunts blood from the pulmonary trunk to the aorta.

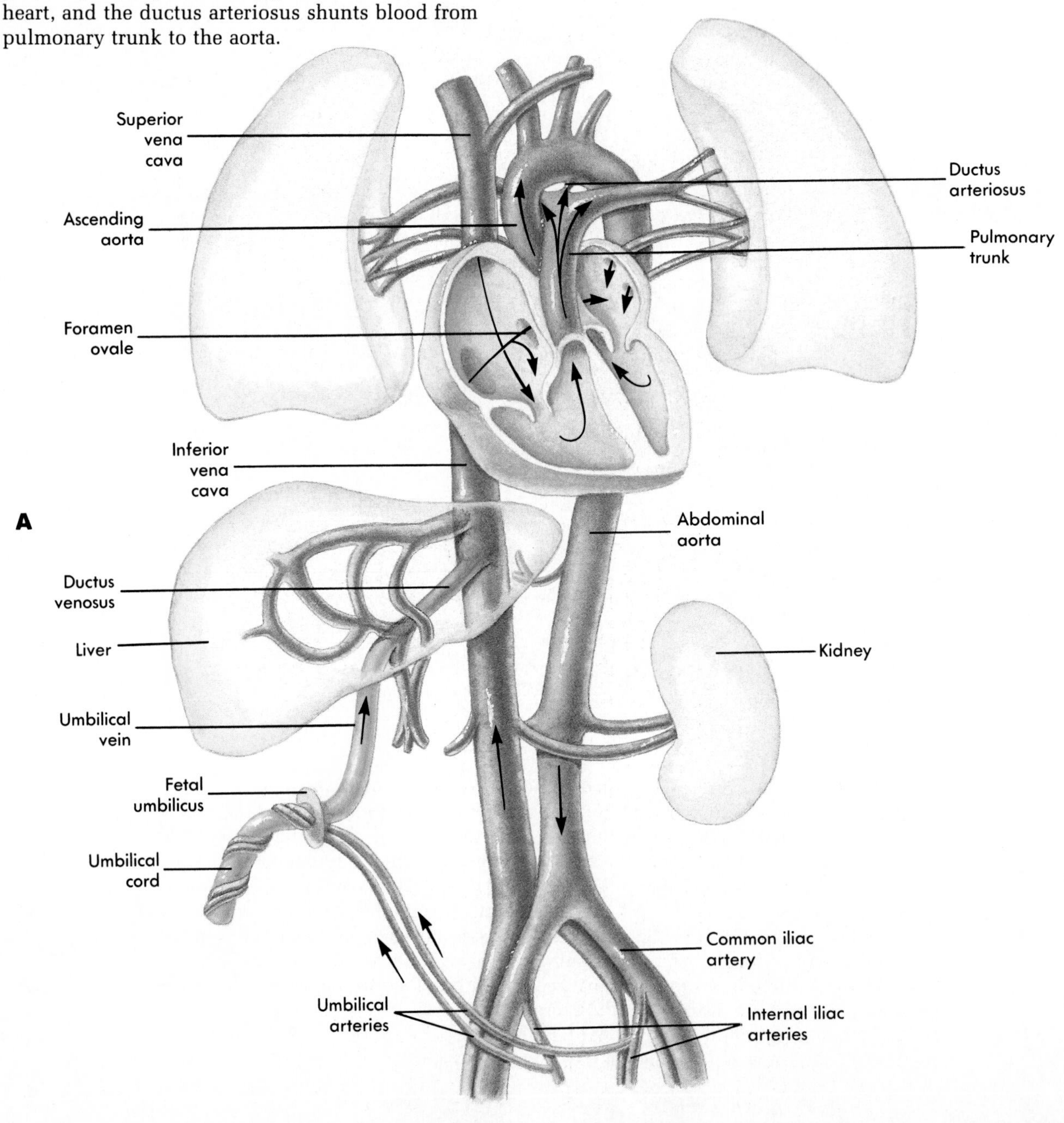

pulmonary arteries, and less blood flows from the right atrium through the foramen ovale to the left atrium. The increasing volume of blood returning from the lungs through the pulmonary veins to the left atrium increases the pressure in the left atrium. The increased left atrial pressure forces blood against the interatrial septum, closing a flap of tissue that develops in that region over the foramen ovale. This action functionally completes the separation of the heart into two pumps—the right side of the heart and the left side of the heart. A short artery, the **ductus arteriosus,** connects the pulmonary trunk to the aorta. Before birth, the ductus arteriosis carries blood from the pulmonary trunk to the aorta, bypassing the fetal lungs. At birth, this artery closes off, forcing blood to flow through the lungs.

The fetal blood supply passes to the placenta through **umbilical arteries,** which originate in the iliac arteries, and returns through an **umbilical vein.** The umbilical vein passes through the liver, but bypasses the sinusoids of the liver by way of the

FIGURE 20-16, cont'd ■ Circulatory changes at the time of birth.
B After birth the placenta is gone, the umbilical vein and arteries degenerate, the foramen ovale closes, and the ductus arteriosus closes.

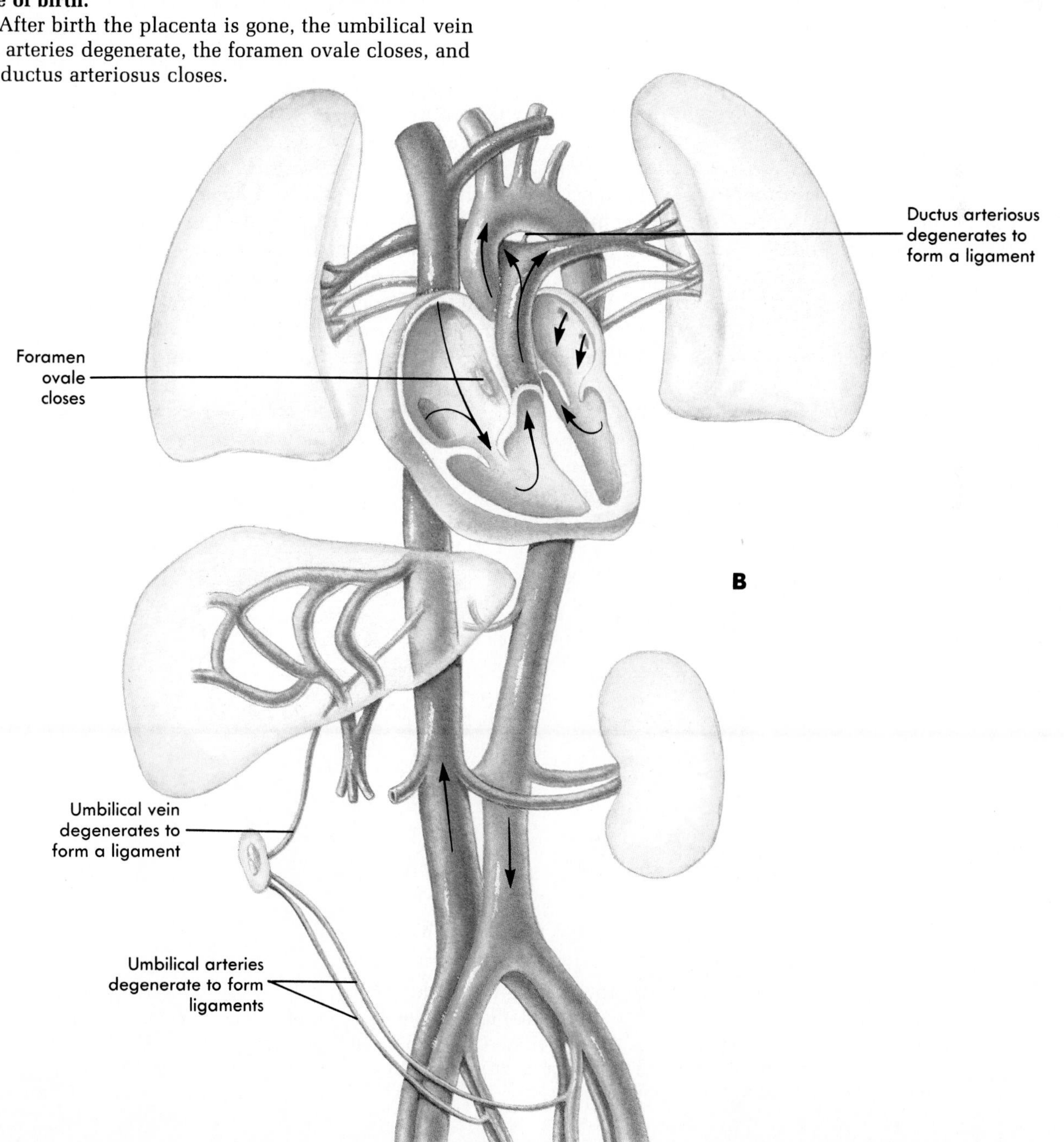

ductus venosus, and joins the inferior vena cava. When the umbilical cord is tied and cut, no more blood flows through the umbilical vein and arteries, and they degenerate. The remnant of the umbilical vein becomes the round ligament of the liver.

Digestive Changes

The fetus swallows amniotic fluid from time to time late in gestation. Shortly after birth this swallowed fluid plus cells sloughed from the mucosal lining, mucus produced by intestinal mucous glands, and bile from the liver pass from the GI tract as a greenish anal discharge called **meconium** (me-ko'ni-um).

When a child is born, it is suddenly separated from its source of nutrients provided by the maternal circulation. Because of this separation and the shock of birth and new life, the neonate usually loses 5% to 10% of its total body weight during the first few days of life. Although the digestive system of the fetus becomes somewhat functional late in development, it is still very immature in comparison to that of the adult, and only a limited number of food types can be digested.

The newborn digestive system is capable of digesting lactose (milk sugar) from the time of birth. The pancreatic secretions are sufficiently mature for a milk diet, but the digestive system only gradually develops the ability to digest more solid foods over the first year or two. Therefore new foods should be introduced gradually during the first 2 years. It is also advised that only one new food at a time be introduced into the infant's diet so that, if an allergic reaction occurs, the cause is more easily determined.

Amylase secretion by the salivary glands and the pancreas remains low until after the first year. Lactase activity in the small intestine is high at birth but declines during infancy, although the levels still exceed those in adults. In many adults, lactase activity is lost, and an intolerance to milk can develop.

LACTATION

Lactation is the production of milk by the breasts (Figure 20-17). It normally occurs in women following parturition and may continue for 2 or 3 years, provided suckling occurs often and regularly.

During pregnancy, the high concentration and continuous presence of estrogens and progesterone cause development of the duct system and the secretory units within the breast. Other hormones, including prolactin, produced by the placenta help support the development of the breasts. Additional adipose tissue also is deposited, so the size of the breasts increases substantially throughout pregnancy. Estrogens and progesterones prevent the secretory portion of the breast from producing milk.

After parturition, in the absence of estrogens and progesterones, prolactin produced by the anterior pituitary stimulates milk production. During suckling, sensory action potentials are sent from the nipple to the brain and result in the release of pro-

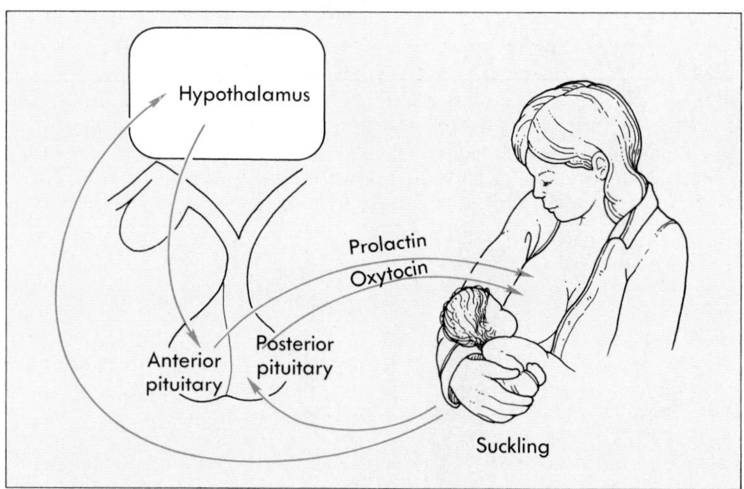

FIGURE 20-17 Hormonal activities during lactation.
Suckling stimulates afferent nerves to the mother's brain. The brain signals the posterior pituitary to release oxytocin and the posterior pituitary to release prolactin. The prolactin stimulates milk production. The oxytocin stimulates alveolar cells to contract, causing milk to flow from the breasts.

Disorders of Pregnancy and Birth

ECTOPIC PREGNANCY

The term **ectopic** means out of place, and an ectopic pregnancy is one that occurs outside the uterus. The most common site of an ectopic pregnancy is the uterine tube (tubal pregnancy). The blastocyst may not have reached the uterine cavity by the time it is ready to implant and may implant into the wall of the uterine tube. The uterine tube cannot expand enough to accommodate the growing fetus and if the fetus is not removed, the tube will eventually rupture. The ruptured uterine tube causes life-threatening internal bleeding.

MISCARRIAGE

It is estimated that as many as 50% of all zygotes are lost before delivery. Most are lost before implantation. Approximately 15% of all pregnancies end in **miscarriage,** or **spontaneous abortion,** which results from the death or extreme premature delivery of the fetus (before about 24 weeks) before it can be viable outside the uterus. After 24 weeks, (but before 37 weeks) the infant is referred to as **premature.**

Although there is a higher incidence of birth defects among aborted fetuses, the vast majority of the fetuses appear to be normal. There are many factors that may cause a miscarriage, many of which do not directly involve the fetus, and many of which are unknown. One common cause of miscarriage is improper implantation of the blastocyst in the uterus. In most cases, the blastocyst implants in the upper part of the uterus, but occasionally a blastocyst may implant near the opening into the cervical canal, a condition called **placenta previa.** As the fetus grows and the uterus stretches, the previa placenta may tear away from the uterine wall, a condition called **placental abruption.** When this occurs, the fetus often dies. The associated hemorrhaging may be life threatening to the mother as well.

PREGNANCY-INDUCED HYPERTENSION

One reason the mother's weight is carefully monitored during pregnancy is that a sudden weight gain associated with edema may be a sign of **pregnancy-induced hypertension** (toxemia of pregnancy). The cause of the disorder is unknown, but, it can result in convulsions, kidney failure, and death of both the mother and the fetus.

TERATOGENS

Teratogens are drugs that can cross the placenta and cause birth defects in the developing embryo. The most famous teratogen is thalidomide, an over-the-counter drug that was given to thousands of pregnant women in the early 1960s. The drug inhibited normal limb development, and resulted in several thousand children born (mostly in Germany and England) with severely reduced, or even absent arms and/or legs.

Fetal alcohol syndrome (FAS) is a major concern today. This syndrome, which consists of brain dysfunction, growth retardation, and facial peculiarities, is seen in children of women who consumed substantial amounts of alcohol during the pregnancy. It has been estimated that FAS may occur as often as 1 in 350 births, and may account for as much as 33% of all mental retardation. **Fetal alcohol effect** includes brain dysfunction without the facial characteristics. The effects may be three times as common as FAS.

Cocaine addiction in the newborn can occur in infants whose mothers are cocaine users. A fetus may also suffer strokelike symptoms if the mother ingests cocaine during the latter part of pregnancy.

INFECTIONS

Infections may occur in the mother or infant, or both. Maternal death associated with childbirth is often the result of infection. As many as 12% of all women giving birth have suffered such deaths. Cleanliness associated with childbirth procedures has reduced the infection rate to about 6% in the United States, and antibiotics have greatly reduced the number of fatal infections.

If a pregnant woman contracts **German measles,** or **rubella,** during pregnancy, the fetus may be severely affected. These effects may occur even if the mother suffers only a mild case of measles. Defects in the newborn may include visual and hearing defects, as well as mental retardation.

Neonatal gonorrheal ophthalmia is a severe form of conjunctivitis that is contracted by an infant as it passes through the birth canal of a mother with gonorrhea. This infection carries a high risk of blindness. The treatment of newborn eyes with silver nitrate or antibiotics is effective in preventing the disease.

Chlamydial conjunctivitis is also contracted as an infant passes through the birth canal, if the mother has a chlamydial infection. This infection is not affected by silver nitrate eye drops, so in many places, newborns are treated with antibiotics against both chlamydia and gonorrhea.

If a woman has genital herpes and has open lesions in the birth canal near the time of parturition, the baby may be removed by cesarean section to prevent infection of the baby by the herpes virus.

Human immunodeficiency virus (HIV), the virus that causes **acquired immunodeficiency syndrome (AIDS),** can cross the placenta and infect the fetus in utero, can infect the infant during parturition, and/or can infect the infant through breast milk. Approximately 30% to 50% of the infants born to HIV-infected mothers will be infected. About 20% of those infected die of AIDS within the first 18 postnatal months.

lactin from the anterior pituitary (see Figure 20-17). For the first few days following parturition, the mammary glands secrete **colostrum,** which contains little fat and less lactose than milk. Eventually, more nutritious milk is produced. Colostrum and milk provide nutrition and antibodies that help protect the nursing baby from infections.

Repeated stimulation of prolactin release makes nursing possible for several years. However, if nursing is stopped, within a few days, the ability of the breast to respond to prolactin is lost and milk production ceases.

At the time of nursing, milk contained in the alveoli and ducts of the breast is forced out of the breast by contractions of the walls of these structures. Mechanical stimulation of the breasts produces nerve impulses that cause the release of oxytocin from the posterior pituitary (see Figure 20-17). Oxytocin stimulates cells surrounding the alveoli to contract; milk then flows from the breasts, a process that is called **milk "letdown."** Higher brain centers can cause the release of oxytocin in response to such things as hearing an infant cry or thinking about breastfeeding.

3 While nursing her baby, a woman noticed that she developed uterine "cramps." Explain what was happening. ?

THE FIRST YEAR FOLLOWING BIRTH

A great number of changes occur in the infant from the time of birth until 1 year of age. The time when these changes occur may vary considerably from child to child, and the dates given are only rough estimates. The brain is still developing at this time, and much of what the infant can accomplish depends on the amount of brain development achieved. It is estimated that the total adult number of neurons is present in the central nervous system at birth, but subsequent growth and maturation of the brain involves the addition of new neuroglial cells and myelin sheaths and the addition of new connections between neurons, which may continue throughout life.

By 6 weeks the baby is usually able to hold up his head when placed in a prone position and begins to smile in response to people or objects. At 3 months of age, the infant's limbs are exercised aimlessly. However, the arms and hands are in enough control that voluntary thumb-sucking can occur. The infant can follow a moving person with his eyes. At 4 months the infant begins to do push-ups (that is, raises himself by his arms). The infant can begin to grasp things placed in his hands, coo and gurgle, roll from back to side, listen quietly when hearing a person's voice or music, hold his head erect, and play with his hands. At 5 months the infant can usually laugh out loud, reach for objects, turn his head to follow an object, lift his head and shoulders, sit with support, and roll over. At 8 months the infant can recognize familiar people, sit up without support, and reach for specific objects. At 12 months the infant may pull himself to a standing position and may be able to walk without support. The child can pick up objects in his hands and examine them carefully. A 12-month-old child can understand much of what is said and may say several words.

LIFE STAGES

The stages of life from fertilization to death are as follows: (1) the **germinal period**—fertilization to 14 days; (2) the **embryo**—14 to 60 days after fertilization; (3) the **fetus**—60 days after fertilization to birth; (4) **neonate**—birth to 1 month after birth; (5) **infant**—1 month to 1 or 2 years (the end of infancy is sometimes set at the time that the child begins to walk); (6) **child**—1 or 2 years to puberty; (7) **adolescent**—teenage years, puberty (age 11 to 14) to 20 years; (8) **adult**—age 20 to death. Adulthood is sometimes divided into three periods: young adult, age 20 to 40; middle age, age 40 to 65; and older adult or senior citizen, age 65 to death. Much of this designation is associated more with social norms than with physiology.

During childhood the individual grows in size and develops considerably. Many of the emotional characteristics that a person possesses throughout his or her life are formed during early childhood.

Major physical and physiological changes occur during adolescence, and many of these changes also affect the emotions and behavior of the individual. Other emotional changes occur as the adolescent attempts to fit into an adult world. **Puberty** is the time when maturation of reproductive cells begins and when gonadal hormones are first secreted. These hormones stimulate the development and maturation of secondary sex characteristics, such as enlargement of the female breasts and growth of body hair in both sexes. Puberty usually occurs in females at about 8 to 16 years, which is somewhat earlier than in males (about 10 to 18 years). The onset of puberty is usually accompanied by a growth spurt, followed by a period of slower growth. Full adult stature is usually achieved before age 17 or 18 in females and before age 19 or 20 in males.

AGING

Development of a new and usually unique human being begins at fertilization, as does the process of aging. Cell proliferation occurs at an extremely rapid rate during early development and then begins to slow as various cells become committed to specific functions within the body.

Many cells of the body continue to proliferate throughout life, replacing dead or damaged tissue, but other cells, such as the neurons in the brain, cease to proliferate once they have reached a certain number. Damage or death of these cells is irreversible. After the number of neurons reaches a peak (at approximately the time of birth), their numbers begin to decline. Neuronal loss is most rapid early in life and decreases to a slower, steady rate.

The physical elasticity (that is, the state of being soft and pliable) of young embryonic tissues is due to relatively small amounts of collagen. Furthermore, the collagen that is present is not highly cross-linked; thus the tissues are very flexible and elastic. However, many of the collagen fibers produced during development are permanent components of the individual; and, as the individual ages, more and more cross-links form between the collagen molecules, rendering the tissues more rigid and less flexible.

The tissues with the highest collagen content and the greatest dependency on collagen for their function are the most severely affected by collagen cross-linking and tissue rigidity associated with aging. The lens of the eye is one of the first structures to exhibit pathological changes as a result of this increased rigidity. Vision of close objects becomes more difficult with advancing age until most middle-aged people require reading glasses. Loss of plasticity also affects other tissues, including the joints, kidneys, lungs, and heart, and greatly reduces the functional ability of these organs.

Like nervous tissue, muscle does not normally proliferate after birth. As a result, the total number of skeletal and cardiac muscle fibers declines with age. The strength of skeletal muscle reaches a peak between 20 to 35 years of life and declines steadily thereafter. Furthermore, like the collagen of connective tissue, the macromolecules of muscle undergo biochemical changes during aging, rendering the muscle tissue less functional. However, a good exercise program can slow or even reverse this process.

The decline in muscular function also contributes to the decline in cardiac function with advancing age. The heart loses elastic recoil ability and muscular contractility. As a result, total cardiac output declines and less oxygen and nutrients reach cells such as neurons of the brain and cartilage cells of the joints, contributing to the decline in these tissues. Reduced cardiac function also may result in decreased blood flow to the kidneys, contributing to decreases in the kidney's filtration ability. Degeneration of the connective tissues as a result of collagen cross-linking and other factors also decreases the filtration efficiency of the glomerular basement membrane.

Atherosclerosis (ath'er-o-sklĕ-ro'sis) is the deposit of lipids in the tunica intima of large and medium-sized arteries (Figure 20-18). These deposits then become fibrotic and calcified, resulting in **arte-**

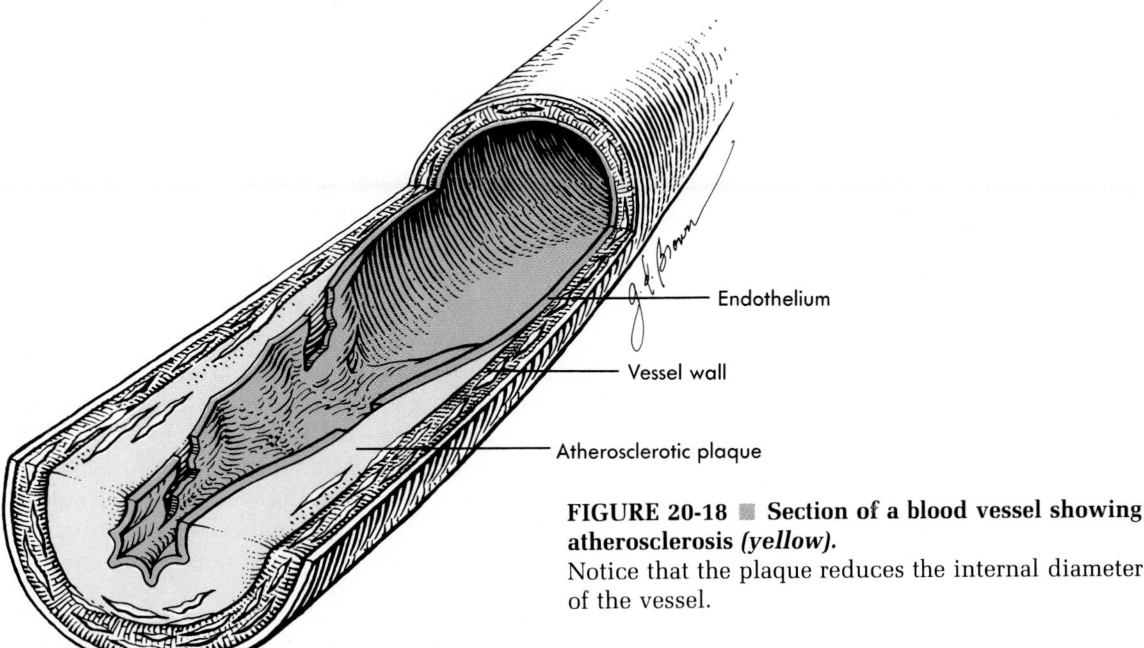

FIGURE 20-18 Section of a blood vessel showing atherosclerosis *(yellow)*.
Notice that the plaque reduces the internal diameter of the vessel.

riosclerosis (ar-ter′ĭ-o-sklĕ-ro′sis; hardening of the arteries). Arteriosclerosis interferes with normal blood flow and may result in a **thrombosis,** which is a clot or plaque formed inside a vessel. An **embolus** is a piece of the clot that has broken loose and floats through the circulation. An embolus can lodge in smaller arteries to cause myocardial infarctions or strokes. Although atherosclerosis occurs to some extent in all middle-aged and elderly people and may occur even in certain young people, some people appear more at risk because of high blood cholesterol levels. This condition seems to have a hereditary component, and blood tests are available to screen people for high blood cholesterol levels.

Many other organs such as the liver, pancreas, stomach, and colon undergo degenerative changes with age. The ingestion of harmful agents may accelerate such changes. Examples include the degenerative changes induced in the lungs (aside from lung cancer) by cigarette smoke and sclerotic changes in the liver as a result of alcohol consumption.

In addition to the previously described changes associated with aging, cellular wear and tear, or cytologic aging, is another factor that contributes to aging. Progressive damage from many sources such as radiation and toxic substances may result in irreversible cellular insults and may be one of the major factors leading to aging. It has been speculated that ingestion of moderate amounts of vitamins E and C in combination may help slow this portion of aging by stimulating cell repair. Vitamin C also stimulates collagen production and may slow the loss of tissue elasticity associated with aging collagen.

Immune system changes may also be a major contributing factor to aging. The aging immune system loses its ability to respond to outside antigens but begins to be more sensitive to the body's own antigens. These autoimmune changes add to the degeneration of the tissues already described and may be responsible for such things as arthritic joint disorders, chronic glomerular nephritis, and hyperthyroidism. In addition, T lymphocytes tend to lose their functional capacity with aging and cannot destroy abnormal cells as efficiently. This change may be one reason that certain types of cancer occur more readily in older people.

One of the greatest disadvantages of aging is the increasing lack of ability to adjust to stress. Elderly people have a far more precarious homeostatic balance than younger people, and eventually some stress is encountered that is so great that the body's ability to recover is surpassed and death results.

DEATH

Death is usually not attributed to old age. Some other problem such as heart failure, renal failure, or stroke is usually listed as the cause of death.

Death was once defined as the loss of heartbeat and respiration. However, in recent years more precise definitions of death have been developed, since both the heart and the lungs can be kept working artificially and the heart can even be replaced by an artificial device. Modern definitions of death are based on the permanent cessation of life functions and the cessation of integrated tissue and organ function. The most widely accepted indication of death in humans is whole brain death, which is manifested clinically by the absence of response to stimulation, the absence of natural respiration and heart function, and an isoelectric ("flat") electroencephalogram for at least 30 minutes (that is, in the absence of known central nervous system poisoning or hypothermia).

GENETICS

Genetics is the study of heredity, that is, those characteristics inherited by children from their parents. The functional unit of heredity is the **gene,** and DNA is the molecule responsible for heredity. Each gene consists of a certain portion of a DNA molecule but not necessarily a continuous stretch of DNA.

There are two major types of genes: structural and regulatory. Structural genes are those DNA sequences that code for specific amino acid sequences in proteins such as enzymes, hormones, or structural proteins such as collagen (see Chapter 3). Regulatory genes are segments of DNA involved in controlling which structural genes are expressed. The function of regulatory genes is not understood as well as that of structural genes.

Human genetics is the study of inherited human traits. A major objective of geneticists is to prevent genetically caused birth defects by identifying people who are carriers of potentially harmful genes. Identification of carriers can lead to prevention of genetic birth defects if those carriers accept genetic counseling. Genetic counseling includes talking to parents of children with genetic disorders about the possible treatment of the disorder and the possible outcome; predicting the possible results of matings involving carriers of harmful genes; and providing information concerning available options. Genetic counselors often diagram a **pedigree**

(family tree) of the family to aid in making predictions.

One common error concerning birth defects (made surprisingly often even among medical personnel) is the belief that all or most congenital disorders are genetic. Congenital means "present at birth" and does not imply a genetic component. Only about 15% of all congenital disorders have a known genetic cause, and about 70% of all birth defects are of unknown cause.

Chromosomes

The DNA of each cell is packed into chromosomes within the nucleus. Each normal human somatic cell (that is, all cells except certain reproductive cells) contains 23 pairs of chromosomes or 46 chromosomes. Two of the 46 chromosomes are called **sex chromosomes,** and the remaining 44 chromosomes are called **autosomes.** A normal female has two X sex chromosomes (XX) in each somatic cell, whereas a normal male has one X and one Y sex chromosome (XY) in each somatic cell. The presence of a Y chromosome makes a person male. The absence of a Y chromosome makes a person female. For convenience, the autosomes are numbered in pairs from 1 through 22.

> Some genetic disorders involve abnormal numbers of chromosomes. For example, Down syndrome involves an extra chromosome 21. The most prominent characteristics of a person with Down syndrome are "mongoloid" facial features, a short, stocky build, and mental retardation. Turner syndrome involves a missing X or Y chromosome (XO). Because the person with Turner syndrome has no Y chromosome, even though she has only one X chromosome, she is a female. Turner syndrome includes dwarfism, webbing of the neck, and underdevelopment of secondary sex characteristics.

Each chromosome contains thousands of genes, and each gene occupies a specific position on the chromosome. Both chromosomes of a given pair contain similar genes so that each gene is paired. If the two genes for a trait are identical, the person is **homozygous** for the trait. If the two genes for a trait are different, the person is **heterozygous** for the trait.

During reproduction, the female contributes (through the oocyte) one set of 23 chromosomes to the new zygote, and the male contributes (through the sperm) one set of 23 chromosomes. Therefore, for any given pair of genes controlling a given trait, the female contributes one gene, and the male contributes one gene. An exception occurs in males with the X and Y chromosomes—there are no genes on the Y chromosome for most of the genes on the X chromosome.

Patterns of Inheritance

For many genetic traits, only one functioning copy of a given gene is necessary for the expression of a given trait. In such cases, a given characteristic in one parent may also exist in the child without regard to the contrasting characteristic of the other parent. A trait that is expressed when only one gene is required is said to be **dominant** over the opposite, **recessive** trait. By convention, dominant traits are indicated by upper case letters and recessive traits are indicated by lower case letters. Certain types of polydactyly (extra fingers and/or toes) are examples of dominant traits. Dominant polydactyly probably involves a regulatory gene, and the actual gene products are not known.

If a person with dominant polydactyly mates with a person without this trait, the probability is that either half or all of their children will also have polydactyly. If dominant polydactyly is designated *D* and the normal recessive condition is designated *d*, then the parent with polydactyly can be either homozygous (two copies of the same gene, *DD*) for the trait or heterozygous (different genes, *Dd*) for the trait. The normal parent will be homozygous recessive, *dd*. Polydactyly or normal is the **phenotype** (the trait that can be seen), and *DD*, *Dd*, or *dd* are the possible **genotypes** (the actual genetic condition). If one parent is homozygous for polydactyly, *DD*, all the children will have polydactyly, the same phenotype, and the same genotype, *Dd*, since *D* came from the parent with polydactyly and *d* came from the normal parent. However, if the parent with polydactyly is heterozygous, *Dd*, approximately half the reproductive cells from that person will carry the *D* gene and half will carry the *d* gene. As a result, the probability is that half the children will have polydactyly (*Dd*), and half will be normal (*dd*), since the normal parent contributed *d* in both cases (Figure 20-19, *A*).

> **4 Will the normal offspring *(dd)* from a parent with polydactyly *(Dd)* have an increased risk of having a child with polydactyly?**
>
> 

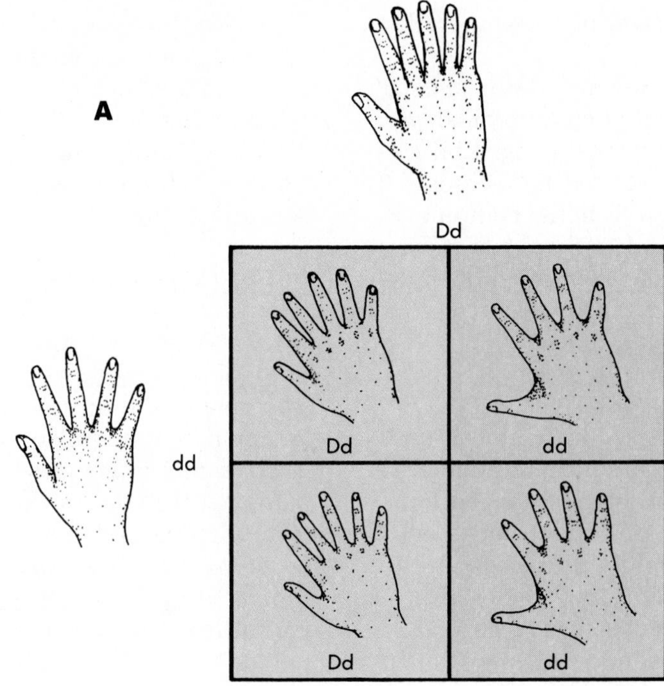

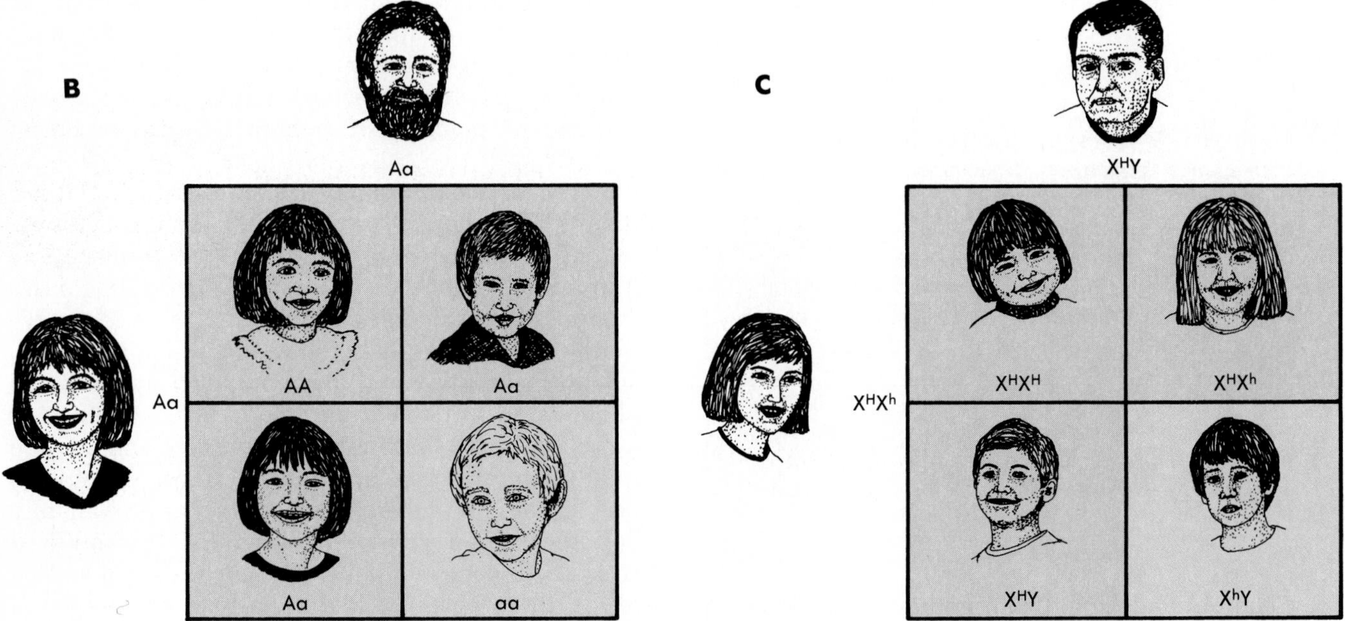

FIGURE 20-19 ■ Examples of inheritance patterns.
A A dominant trait (polydactyly where *D* represents the dominant condition with extra fingers or toes and *d* represents the recessive normal condition). The figure represents a mating between a normal person and a person with dominant polydactyly.
B A recessive trait (albinism where *A* represents the normal, pigmented condition and *a* represents the recessive unpigmented condition). The figure represents a mating between two normal carriers.
C An X-linked trait (hemophilia, where X^H represents the normal X chromosome condition with all clotting factors, and X^h represents the X chromosome lacking a gene for one clotting factor). The figure represents a mating between a normal male and a normal carrier female.

Albinism (absence of pigment) is a trait that is recessive to the normal, dominant condition. Albinism involves a "structural" gene. Melanin, a pigment responsible for skin, hair, and eye color, is made in the body under the influence of enzymes. Each enzyme is a protein coded for in the DNA (a gene). In albinism, one of those enzymes is abnormal and cannot function in the production of melanin. If both copies of a given enzyme are abnormal, no melanin is produced, and albinism results.

A person with albinism has only one possible genotype *(aa)*, whereas a normal person can be either homozygous *(AA)* or heterozygous *(Aa)*. If an albino person *(aa)* mates with a homozygous normal person *(AA)*, all the children will be phenotypically normal with a heterozygous genotype *(Aa)*. If an albino person *(aa)* mates with a heterozygous normal person *(Aa)*, the probability is that approximately half the children will be albino *(aa)*, and half will be normal heterozygous carriers *(Aa)*. A **carrier** is a person with an abnormal gene, but with a normal phenotype. Albinism can result from matings of two phenotypically normal carriers, with genotypes *Aa*. Approximately half the children will be carriers, like the parents *(Aa)*, one fourth will be homozygous normal *(AA)* and will not be carriers of the trait, and one fourth will be albino *(aa)* (Figure 20-19, *B*).

Simple dominant and recessive traits are located on the autosomes. Genes on the sex chromosomes are called **X-linked** (sex-linked) or **Y-linked.** There are very few Y-linked traits. They are passed only from male to male and never appear in the female. In the case of an X-linked trait, females with two genes function as though the gene were recessive or dominant, except for those traits that make the possessor male. However, males with only one X chromosome, have only one copy of the gene. Therefore, a heterozygous carrier female will have sons of whom approximately half do not have the trait and half do, because the male receives only one X chromosome with no comparable gene on the Y chromosome. An example of an X-linked trait is hemophilia (lacking one of the clotting factors), where ^{H} represents the normal condition, and ^{h} represents the recessive hemophilia (Figure 20-19, *C*). A heterozygous woman (X^HX^h) will have either normal sons (X^HY) or sons with hemophilia (X^hY).

> **5 Describe the inheritance pattern for a girl with Turner syndrome (XO; that is, with only one X chromosome and no other sex chromosome) whose mother is a carrier for hemophilia.** ?

SUMMARY

Prenatal development is an important part of an individual's life.

About 10 of every 100 people are born with some type of birth defect.

PRENATAL DEVELOPMENT

- Prenatal development is divided into the germinal, embryonic, and fetal periods.
- Developmental age is 14 days less than clinical age.

Fertilization

- Fertilization, the union of the oocyte and sperm, results in a zygote.

Early Cell Division

- The zygote undergoes divisions until it becomes a mass of cells.

Blastocyst

- The mass of cells develops a cavity and is known as the blastocyst.
- The blastocyst consists of a trophoblast and an inner cell mass.

Implantation of the Blastocyst and Development of the Placenta

- The blastocyst implants into the uterus about 7 days after fertilization.
- The placenta is derived from the trophoblast of the blastocyst.

Formation of the Germ Layers

- The embryo forms around the primitive streak, which forms about 14 days after fertilization.
- All tissues of the body are derived from three primary germ layers: ectoderm, mesoderm, and endoderm.

Neural Tube and Neural Crest Formation	The nervous system develops from a neural tube that forms in the ectodermal surface of the embryo and from neural crest cells derived from the developing neural tube.
Formation of the General Body Structure	The limbs develop as outgrowths called limb buds. The face develops by the fusion of five tissue masses.
Development of the Organ Systems	The gastrointestinal (GI) tract develops as the developing embryo closes off part of the yolk sac. The heart develops as two tubes fuse into a single tube that develops septa to form four chambers. The kidneys and reproductive system are closely related in their development.
Growth of the Fetus	The fetus increases 15-fold in length and 1400-fold in weight.
PARTURITION	Uterine contractions force the infant out of the uterus during labor. Increased estrogens, decreasing progesterone levels, and secretions from the fetal adrenal gland initiate parturition. Stretching of the uterus stimulates oxytocin secretion, which stimulates uterine contractions.
THE NEWBORN	Several dramatic changes occur shortly after birth.
Circulatory Changes	Inflation of the lungs at birth results in closure of the foramen ovale and the ductus arteriosis. When the umbilical cord is cut, blood no longer flows through the umbilical vessels.
Digestive Changes	The newborn digestive system only gradually develops the ability to digest a variety of foods.
LACTATION	Estrogens and progesteron help stimulate the growth of the breasts during pregnancy. Suckling stimulates prolactin and oxytocin synthesis. Prolactin stimulates milk production, and oxytocin stimulates milk "letdown."
THE FIRST YEAR FOLLOWING BIRTH	Many important changes occur during the first year after birth. Many of those changes are linked to continued development of the brain.
LIFE STAGES	The eight stages of life are: germinal period (fertilization to 14 days), embryo (14 to 60 days after fertilization), fetus (60 days after fertilization to birth), neonate (birth to 1 month), infant (1 month to 1 or 2 years), child (1 or 2 years to puberty), adolescent (puberty [age 11-14] to 20 years), adult (20 years to death).
AGING	Aging occurs as irreplaceable cells wear out and the tissue becomes more brittle and less able to repair damage. Atherosclerosis is the deposit of lipids in the arteries. Arteriosclerosis is hardening of the arteries.
DEATH	Death is defined as the absence of brain response to stimulation, the absence of natural respiration and heart function, and a flat electroencephalogram for 30 minutes.
GENETICS	Genetics is the study of heredity. Human genetics concentrates on genetic birth defects and includes genetic counseling.
Chromosomes	Humans have 46 chromosomes in 23 pairs; 22 pairs of autosomes and 1 pair of sex chromosomes. The chromosomes contain genes. Genes are paired.
Patterns of Inheritance	Most genetic traits involve dominant and recessive genes. X-linked genes affect males far more commonly than females.

CONTENT REVIEW

1. Define clinical age and developmental age, and distinguish the two.
2. What are the events during the first week after fertilization? Define zygote and blastocyst.
3. How does the placenta develop?
4. Describe the formation of the germ layers and the role of the primitive streak.
5. How are the neural tube and neural crest cells formed? What do they become?
6. Describe the formation of the limbs and face.
7. Describe the formation of the digestive tract.
8. How does the single heart tube become four-chambered?
9. What major events distinguish embryonic and fetal development?
10. Describe the hormonal changes that take place before and during parturition.
11. What changes take place in the newborn's circulatory system and digestive system shortly after birth?
12. What hormones are involved in preparing the breasts for lactation? What hormones are involved in milk production and milk "letdown"?
13. Describe the changes in motor and language skills that take place during the first year of life.
14. List the different life stages, starting with the germinal stage and ending with the adult.
15. How does the loss of cells that are not replaced affect the aging process? Give examples.
16. How does the loss of tissue elasticity affect the aging process? Give examples.
17. How does aging affect the immune system?
18. Define death.
19. How are genes related to chromosomes?
20. Define homozygous, heterozygous, dominant, recessive, and X-linked in genetics.

CONCEPT REVIEW

1. A woman is told by her physician that her pregnancy has progressed 44 days since her last menstrual period (LMP). How many days has the embryo been developing, and what developmental events are occurring?
2. A high fever can prevent neural tube closure. If a woman has a high fever about 35 to 45 days post-LMP, what kinds of birth defects may be seen in the developing embryo?
3. A drug that would stop the production of milk in the breast after a few days probably has which effect?
   A. Inhibits prolactin secretion.
   B. Inhibits oxytocin secretion.
   C. Increases estrogen secretion.
   D. Increases progesterone secretion.
   E. Increases prolactin secretion.
4. Dimpled cheeks are inherited as a dominant trait. Is it possible for two parents, each of whom have dimpled cheeks, to have a child without dimpled cheeks? Explain.

ANSWERS TO  PREDICT QUESTIONS

1 *p. 520* Two primitive streaks forming in one embryonic disk could result in twins. If the two streaks are touching, the twins will be conjoined (Siamese), or attached to each other. This attachment may be fairly simple, and the twins may be separated fairly easily by surgery, or the attachment may be extensive involving internal organs and may not be corrected easily.

2 *p. 529* 266 days (280 minus 14).

3 *p. 530* Suckling causes a reflex release of oxytocin from the mother's posterior pituitary. Oxytocin causes expulsion of milk from the breast, but it also causes contraction of the uterus. Contraction of the uterus is responsible for the sensation of cramps.

4 *p. 541* The normal *dd* offspring from a parent with polydactyly *(Dd)* will not have any greater risk of having a child with polydactyly than any other normal person, since that person does not possess the abnormal *D* gene to pass on.

5 *p. 543* A girl with Turner syndrome (XO; that is, with only one X chromosome and no other sex chromosome) whose mother is a carrier for hemophilia would run the same risk of developing hemophilia as a male, since she has only one X chromosome. If her mother is a carrier for hemophilia, the daughter would be either X^HO (normal), or X^hO (hemophiliac).

Appendix A
TABLE OF MEASUREMENTS

UNIT	METRIC EQUIVALENT	SYMBOL	U.S. EQUIVALENT
MEASURES OF LENGTH			
1 kilometer	= 1000 meters	km	0.62137 mile
1 meter	= 10 decimeters or 100 centimeters	m	39.37 inches
1 decimeter	= 10 centimeters	dm	3.937 inches
1 centimeter	= 10 millimeters	cm	0.3937 inch
1 millimeter	= 1000 micrometers	mm	
1 micrometer	= 1/1000 millimeter or 1000 nanometers	µm	
1 nanometer	= 10 angstroms or 1000 picometers	nm	No U.S. equivalent
1 angstrom	= 1/10,000,000 millimeter	Å	
1 picometer	= 1/1,000,000,000 millimeter	pm	
MEASURES OF VOLUME			
1 cubic meter	= 1000 cubic decimeters	m^3	1.308 cubic yards
1 cubic decimeter	= 1000 cubic centimeters	dm^3	0.03531 cubic foot
1 cubic centimeter	= 1000 cubic millimeters or 1 milliliter	cm^3 (cc)	0.06102 cubic inch
MEASURES OF CAPACITY			
1 kiloliter	= 1000 liters	kl	264.18 gallons
1 liter	= 10 deciliters	l	1.0567 quarts
1 deciliter	= 100 milliliters	dl	0.4227 cup
1 milliliter	= volume of 1 gram of water at standard temperature and pressure	ml	0.3381 ounces
MEASURES OF MASS			
1 kilogram	= 1000 grams	kg	2.2046 pounds
1 gram	= 100 centigrams or 1000 milligrams	g	0.0353 ounces
1 centigram	= 10 milligrams	cg	0.1543 grain
1 milligram	= 1/1000 gram	mg	

Note that a micrometer was formerly called a micron (µ), and a nanometer was formerly called a millimicron (mµ).

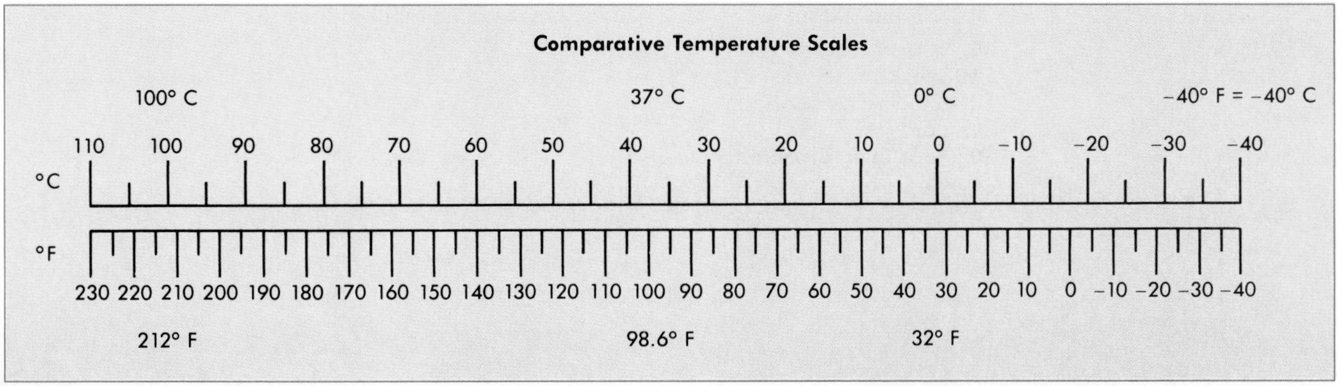

Appendix B SOME REFERENCE LABORATORY VALUES

TABLE B-1

Blood, Plasma, or Serum Values

TEST	NORMAL VALUES	CLINICAL SIGNIFICANCE
Acetoacetate plus acetone	0.32-2 mg/100 ml	Values increase in diabetic acidosis, fasting, high fat diet, and toxemia of pregnancy
Ammonia	80-110 μg/100 ml	Values decrease with proteinuria and as a result of severe burns and increase in multiple myeloma
Amylase	4-25 U/ml	Values increase in acute pancreatitis, intestinal obstruction, and mumps; values decrease in cirrhosis of the liver, toxemia of pregnancy, and chronic pancreatitis
Barbiturate	0 Coma level: phenobarbital, approximately 10 mg/100 ml; most other drugs, 1-3 mg/100 ml	
Bilirubin	0.4 mg/100 ml	Values increase in conditions causing red blood cell destruction or biliary obstruction or liver inflammation
Blood volume	8.5%-9% of body weight in kilograms	
Calcium	8.5-10.5 mg/ml	Values increase in hyperparathyroidism, vitamin D hypervitaminosis; values decrease in hypoparathyroidism, malnutrition, and severe diarrhea
Carbon dioxide content	24-30 mEq/L 20-26 mEq/L in infants (as HCO_3^-)	Values increase in respiratory diseases, vomiting, and intestinal obstruction; they decrease in acidosis, nephritis, and diarrhea
Carbon monoxide	Symptoms with over 20% saturation	
Chloride	100-106 mEq/L	Values increase in Cushing's syndrome, nephritis, and hyperventilation; they decrease in diabetic acidosis, Addison's disease, and diarrhea and after severe burns
Creatine phosphokinase (CPK)	Female 5-35 mU/ml Male 5-55 mU/ml	Values increase in myocardial infarction and skeletal muscle diseases such as muscular dystrophy
Creatinine	0.6-1.5 mg/100 ml	Values increase in certain kidney diseases
Ethanol	0.3%-0.4%, marked intoxication 0.4%-0.5%, alcoholic stupor 0.5% or over, alcoholic coma	

TABLE B-1—cont'd

Blood, Plasma, or Serum Values

TEST	NORMAL VALUES	CLINICAL SIGNIFICANCE
Glucose	Fasting 70-110 mg/100 ml	Values increase in diabetes mellitus, liver diseases, nephritis, hyperthyroidism, and pregnancy; they decrease in hyperinsulinism, hypothyroidism, and Addison's disease
Iron	50-150 μg/100 ml	Values increase in various anemias and liver disease; they decrease in iron deficiency anemia
Lactic acid	0.6-1.8 mEq/L	Values increase with muscular activity and in congestive heart failure, severe hemorrhage, shock, and anaerobic exercise
Lactic dehydrogenase	60-120 U/ml	Values increase in pernicious anemia, myocardial infarction, liver diseases, acute leukemia, and widespread carcinoma
Lipids	Cholesterol 120-220 mg/100 ml Cholesterol esters 60%-75% of cholesterol Phospholipids 9-16 mg/100 ml as lipid phosphorus Total fatty acids 190-420 mg/100 ml Total lipids 450-1000 mg/100 ml Triglycerides 40-150 mg/100 ml	
Lithium	Toxic levels 2 mEq/L	
Osmolality	285-295 mOsm/kg water	
Oxygen saturation (arterial) see Po_2 values	96%-100%	
Pco_2	35-43 mm Hg	Values decrease in acidosis, nephritis, and diarrhea; they increase in respiratory diseases, intestinal obstruction, and vomiting
pH	7.35-7.45	Values decrease as a result of hypoventilation, severe diarrhea, Addison's disease, and diabetic acidosis; values increase due to hyperventilation, Cushing's syndrome, and vomiting
Po_2	75-100 mm Hg (breathing room air)	Values increase in polycythemia and decrease in anemia and obstructive pulmonary diseases
Phosphatase (acid)	Male: total 0.13-0.63 Sigma U/ml Female: total 0.01-0.56 Sigma U/ml	Values increase in cancer of the prostate gland, hyperparathyroidism, some liver diseases, myocardial infarction, and pulmonary embolism
Phosphatase (alkaline)	13-39 IU/L (infants and adolescents up to 104 IU/L)	Values increase in hyperparathyroidism, some liver diseases, and pregnancy
Phosphorus (inorganic)	3-4.5 mg/100 ml (infants in first year up to 6 mg/100 ml)	Values increase in hypoparathyroidism, acromegaly, vitamin D hypervitaminosis, and kidney diseases; they decrease in hyperparathyroidism
Potassium	3.5-5 mEq/100 ml	
Protein	Total 6-8.4 g/100 ml Albumin 3.5-5 g/100 ml Globulin 2.3-3.5 g/100 ml	Total protein values increase in severe dehydration and shock; they decrease in severe malnutrition and hemorrhage

Continued.

TABLE B-1—cont'd

Blood, Plasma, or Serum Values

TEST	NORMAL VALUES	CLINICAL SIGNIFICANCE
Salicylate	20-25 mg/100 ml	
Therapeutic	Over 30 mg/100 ml	
Toxic	Over 20 mg/100 ml after age 60	
Sodium	135-145 mEq/L	Values increase in nephritis and severe dehydration; they decrease in Addison's disease, myxedema, kidney disease, and diarrhea
Sulfonamide Therapeutic	5-15 mg/100 ml	
Urea nitrogen	8-25 mg/100 ml	Values increase in response to increased dietary protein intake; values decrease in impaired renal function
Uric acid	3-7 mg/100 ml	Values increase in gout and toxemia of pregnancy and as a result of tissue damage

TABLE B-2

Blood Count Values

TEST	NORMAL VALUES	CLINICAL SIGNIFICANCE
Clotting (coagulation) time	5-10 minutes	Values increase in afibrinogenemia and hyperheparinemia, severe liver damage
Fetal hemoglobin	Newborns: 60%-90% Before age 2: 0%-4% Adults: 0%-2%	Values increase in thalassemia, sickle-celled anemia, and leakage of fetal blood into maternal bloodstream during pregnancy
Hemoglobin	Male: 14-16.5 g/100 ml Female: 12-15 g/100 ml Newborn: 14-20 g/100 ml	Values decrease in anemia, hyperthyroidism, cirrhosis of the liver, and severe hemorrhage; values increase in polycythemia, congestive heart failure, obstructive pulmonary disease, high altitudes
Hematocrit	Male: 40%-54% Female: 38%-47%	Values increase in polycythemia, severe dehydration, and shock; values decrease in anemia, leukemia, cirrhosis, and hyperthyroidism
Ketone bodies	0.3-2 mg/100 ml Toxic level: 20 mg/100 ml	Values increase in ketoacidosis, fever, anorexia, fasting, starvation, high fat diet
Platelet count	250,000-400,000/mm^3	Values decrease in anemias and allergic conditions and during cancer chemotherapy; values increase in cancer, trauma, heart disease, and cirrhosis
Prothrombin time	11-15 seconds	Values increase in prothrombin and vitamin deficiency, liver disease, and hypervitaminosis A
Red blood cell count	Males: 5.4 million/mm^3 Females: 4.8 million/mm^3	Values decrease in systemic lupus erythematosus, anemias, and Addison's disease; values increase in polycythemia and dehydration and following hemorrhage
Reticulocyte count	0.5%-1.5%	Values decrease in iron-deficiency and pernicious anemia and radiation therapy; values increase in hemolytic anemia, leukemia, and metastatic carcinoma
White blood cell count, differential	Neutrophils 60%-70% Eosinophils 2%-4% Basophils 0.5%-1% Lymphocytes 20%-25% Monocytes 3%-8%	Neutrophils increase in acute infections; eosinophils and basophils increase in allergic reactions; monocytes increase in chronic infections; lymphocytes increase during antigen-antibody reactions
White blood cell count, total	5000-9000/mm^3	Values decrease in diabetes mellitus, anemias, and following cancer chemotherapy; values increase in acute infections, trauma, some malignant diseases, and some cardiovascular diseases

TABLE B-3

Urine Values

TEST	NORMAL VALUES	CLINICAL SIGNIFICANCE
Acetone and acetoacetate	0	Values increase in diabetic acidosis and during fasting
Albumin	0 to trace	Values increase in glomerular nephritis and hypertension
Ammonia	20-70 mEq/L	Values increase in diabetes mellitus and liver disease
Bacterial count	Uncer 10,000/ml	Values increase in urinary tract infection
Bile and bilirubin	0	Values increase in biliary tract obstruction
Calcium	Under 250 mg/24 h	Values increase in hyperparathyroidism and decrease in hypoparathyroidism
Chloride	110-254 mEq/24 h	Values decrease in pyloric obstruction, diarrhea; values increase in Addison's disease and dehydration
Potassium	25-100 mEq/L	Values decrease in diarrhea, malabsorption syndrome, and adrenal cortical insufficiency; values increase in chronic renal failure, dehydration, and Cushing's syndrome
Sodium	75-200 mg/24 h	Values decrease in diarrhea, acute renal failure, and Cushing's syndrome; values increase in dehydration, starvation, and diabetic acidosis
Creatinine clearance	100-140 ml/minute	Values increase in renal diseases
Creatinine	1-2 g/24 h	Values increase in infections and decrease in muscular atrophy, anemia, and certain kidney diseases
Glucose	0	Values increase in diabetes mellitus and certain pituitary gland disorders
Urea clearance	Over 40 ml of blood cleared of urea per minute	Values increase in certain kidney diseases
Urea	25-35 g/24 h	Values decrease in complete biliary obstruction and severe diarrhea; values increase in liver diseases and hemolytic anemia
Uric acid	0.6-1 g/24 h	Values increase in gout and decrease in certain kidney diseases
Casts		
Epithelial	Occasional	Increase in nephrosis and heavy metal poisoning
Granular	Occasional	Increase in nephritis and pyelonephritis
Hyaline	Occasional	Increase in glomerular membrane damage and fever
Red blood cell	Occasional	Values increase in pyelonephritis; blood cells appear in urine in response to kidney stones and cystitis
White blood cell	Occasional	Values increase in kidney infections
Color	Amber, straw, transparent yellow	Varies with hydration, diet, and disease states
Odor	Aromatic	Becomes acetone-like in diabetic ketosis
Osmolality	500-800 mOsm/kg water	Values decrease in aldosteronism and diabetes insipidus; values increase in high protein diets, heart failure, and dehydration
pH	4.6-8	Values decrease in acidosis, emphysema, starvation, and dehydration; values increase in urinary tract infections and severe alkalosis

TABLE B-4

Hormone Levels

TEST	NORMAL VALUES
Steroid hormones	
Aldosterone	Excretion: 5-19 μg/24 h
Fasting at rest, 210 mEq sodium diet	Supine: 48± 29 pg/ml Upright: 65 ± 23 pg/ml
Fasting at rest, 10 mEq sodium diet	Supine: 175 ± 75 pg/ml Upright: 532 ± 228 pg/ml
Cortisol	
Fasting	8 a.m.: 5-25 μg/100 ml
At rest	8 p.m.: Below 10 μg/100 ml
Testosterone	Adult male: 300-1100 ng/100 ml Adolescent male: over 100 ng/100 ml Female: 25-90 ng/100 ml
Peptide hormones	
Adrenocorticotropin (ACTH)	15-170 pg/ml
Calcitonin	Undetectable in normals
Growth hormone (GH)	
Fasting, at rest	Below 5 ng/ml
After exercise	Children: over 10 ng/ml Male: below 5 ng/ml Female: up to 30 ng/ml
Insulin	
Fasting	6-26 μU/ml
During hypoglycemia	Below 20 μU/ml
After glucose	Up to 150 μU/ml
Luteinizing hormone (LH)	Male: 6-18 mU/ml Preovulatory or postovulatory female: 5-22 mU/ml Midcycle peak 30-250 mU/ml
Parathyroid hormone	Less than 10 microl equiv/L
Prolactin	2-15 ng/ml
Renin activity	
Normal diet	
Supine	1.1 ± 0.8 ng/ml/h
Upright	1.9 ± 1.7 ng/ml/h
Low-sodium diet	
Supine	2.7 ± 1.8 ng/ml/h
Upright	6.6 ± 2.5 ng/ml/h
Thyroid-stimulating hormone (TSH)	0.5-3.5 μU/ml
Thyroxine-binding globulin	15.25 μg T_4/100 ml
Total thyroxine	4-12 μg/100 ml

GLOSSARY

Many of the words in this glossary and throughout the text are followed by a simplified phonetic spelling showing pronunciation. The pronunciation key reflects standard clinical usage as presented in *Stedman's Medical Dictionary*, which has long been a leading reference volume in the health sciences.

As a general rule, vowels will be unmarked. If an unmarked vowel is at the end of a syllable, the sound is long (like the "a" in mate). If an unmarked vowel is followed by a consonant, the sound is short (like the "a" in mat). Accent marks follow stressed syllables.

Page numbers indicate where entries may be found in the text.

a at the end of a syllable, long, as in day (da), before a consonant (except r), short, as in hat (hat).

ă as in hat (flă-jel·ah).

ā as in mate (māt).

ah as in father (fah·ther).

ar as in far (far).

ăr as in fair (făr).

aw as in fall (fawl).

e at the end of a syllable, long, as in bee (be); before a consonant (except r), short, as in met.

ĕ as in met; (ap·ĕ-tīt).

ē as in meet (mēt).

er as in term.

ĕr as in merry (mĕr·e).

i at the end of a syllable, long, as in pie (pi); before a consonant (except r), short, as in pit.

ĭ as in pit; (kar·tĭ-lij)

ī as in pine (pīn).

ir as in firm.

ĭr as in mirror (mĭr·or).

īr as in fire (fīr).

o at the end of a syllable, long, as in no; before a consonant, short, as in not.

ŏ as in occult (ŏ-kult).

ō as in note (nōt).

ŏŏ as in food.

ōr as in for.

ōr as in fore, four (fōr).

ow as in cow.

oy as in boy; as the oi in mastoid (mas·toyd).

u at the end of a syllable, as in u·nit; before a consonant, as in bud.

ŭ as in bud.

ū as in tune (tūn).

ur as in fur.

ūr as in pūre.

abdominal cavity space bounded by the diaphragm, the abdominal wall, and the pelvis. (p. 14)
abdominopelvic cavity the abdominal and pelvic cavities considered together. (p. 15)
abduction [L., uncertain origin] movement away from the midline. (p. 142)
absorption the movement of digested molecules across the intestinal wall and into the blood stream. (p. 439)
accommodation increase in the thickness and convexity of the lens in order to focus an object on the retina as the object moves closer to the eye; decreasing sensitivity of a nerve cell to a stimulus of constant strength. (p. 234)
acetabulum (a'sŭ-tab'u-lum) [L., shallow vinegar vessel or cup] cup-shaped depression on the lateral surface of the coxa, where the head of the femur articulates. (p. 133)
acetyl CoA acetyl coenzyme A; formed by the combination of the two-carbon acetyl group with coenzyme A; the molecule that combines with a four-carbon molecule to enter the citric acid cycle. (p. 444)
acetylcholine (as'e-til-ko'lēn) neurotransmitter substance released from motor neurons that innervates skeletal muscle, all autonomic preganglionic neurons, all postganglionic parasympathetic neurons, some postganglionic sympathetic neurons, and some central nervous system neurons. (p. 150)
Achilles tendon common tendon of the calf muscles that attaches to the heel (calcaneus). (p. 174)
acid any substance that is a proton donor; any substance that releases hydrogen ions. (p. 32)
acidic solution solution with more hydrogen ions than hydroxide ions; has a pH of less than 7. (p. 32)
acidosis condition characterized by a lower-than-normal blood pH (pH of 7.35 or lower). (p. 477)
acinus (as'ĭ-nus) [L., berry, grape] grape-shaped secretory portion of a gland. (p. 25)
acromegaly (ak'ro-meg'al-e) [Gr. *acro* + *megas*, large] disorder marked by progressive enlargement of the bones of the head, face, hands, feet, and thorax as a result of excessive secretion of growth hormone by the anterior pituitary. (p. 255)
action potential the all-or-none change in membrane potential in an excitable tissue that is propagated as an electrical signal. (p. 148)
activation energy energy that must be added to atoms or molecules to start a chemical reaction. (p. 31)
active transport carrier-mediated process that requires ATP and can move substances against a concentration gradient. (p. 50)
active vitamin D fat-soluble vitamin produced from precursor molecules in skin exposed to ultraviolet light; increases calcium and phosphate uptake from the intestines. (p. 101)
adduction [L. *adductus*, to bring toward] movement toward the midline. (p. 142)
adductor [L. *adductus*, to bring toward] a muscle causing movement toward the midline. (p. 174)
adenosine triphosphate adenosine, an organic base, with three phosphate groups attached to it; energy stored in adenosine triphosphate (ATP) is used in nearly all the energy-requiring reactions in the body. (p. 442)
adipose (ad'ĭ-pōs) [L. *adeps*, fat] fat. (p. 80)
adrenal cortex the outer portion of the adrenal gland, which secretes the following steroid hormones: glucocorticoids, mainly cortisol; mineralocorticoids, mainly aldosterone; and androgens. (p. 258)
adrenal gland (ă-dre'nal) [L. *ad*, to + *ren*, kidney] one of two endocrine glands located on the superior pole of each kidney; secretes the hormones epinephrine, norepinephrine, aldosterone, cortisol, and androgens. (p. 258)
adrenal medulla the inner portion of the adrenal gland, which secretes mainly epinephrine but also small amounts of norepinephrine. (p. 258)
adrenalin synonym for epinephrine. (p. 258)
adrenocorticotropic hormone (ă-dre'no-kor'tĭ-ko-tro'pik) (ACTH) hormone of the anterior pituitary that stimulates the adrenal cortex to secrete cortisol. (p. 255)
adventitia (ad'ven-tish'yah) [L. *adventicius*, coming from abroad, foreign] outermost covering of an organ that is continuous with the surrounding connective tissue. (p. 406)
aerobic (a-ro'bik) **respiration** breakdown of glucose in the presence of oxygen to produce carbon dioxide, water, and 36 ATP molecules; includes glycolysis, the citric acid cycle, and the electron-transport chain. (p. 154)
afferent arteriole a small artery in the renal cortex that supplies blood to the glomerulus. (p. 461)
afferent fiber nerve fiber going from the peripheral to the central nervous system; sensory fiber. (p. 183)
agglutination (ă-glu'tĭ-na'shun) [L. *ad*, to + *gluten*, glue] the process by which cells stick together to form clumps. (p. 281)
aldosterone (al-dos'ter-ōn) steroid hormone produced by the adrenal cortex that facilitates potassium exchange for sodium in the distal convoluted tubule and collecting duct, causing sodium reabsorption and potassium and hydrogen ion secretion. (p. 261)

alkaline solution see basic solution. (p. 32)

alkalosis condition characterized by a higher-than-normal blood pH (pH of 7.45 or above). (p. 477)

alveolar duct part of the respiratory passages beyond a respiratory bronchiole; from it arise alveolar sacs and alveoli. (p. 382)

alveolar sac two or more alveoli that share a common opening. (p. 382)

alveolus pl. alveoli (al-ve'o-lus) [L., cavity] cavity; examples include the sockets into which the teeth fit and the terminal ends of the respiratory system. (p. 75, 408)

amino acid class of organic acids containing an amine group (NH_2) that makes up the building blocks of proteins. (p. 36)

amniotic cavity fluid-filled cavity surrounding and protecting the developing embryo. (p. 520)

amylase (am'ĭ-lās) one of a group of starch-splitting enzymes that cleave starch, glycogen, and related polysaccharides. (p. 439)

anabolism (ah-nab'o-lizm) [Gr. *anabole*, a raising up] all the synthesis reactions that occur within the body; requires energy. (p. 436)

anaerobic (an'ă-ro'bik) **respiration** breakdown of glucose in the absence of oxygen to produce lactic acid and two ATP molecules; consists of glycolysis and the reduction of pyruvic acid to lactic acid. (p. 154)

anatomical position position in which a person is standing erect with the feet facing forward, arms hanging to the sides, and the palms of the hands facing forward with the thumbs to the outside. (p. 11)

anatomy [Gr. *ana*, up + *tome*, a cutting] scientific discipline that investigates the structure of the body. (p. 2)

anemia (ă-ne'me-ah) [Gr. *an*, without + *haima*, blood] any condition that results in less than normal hemoglobin in the blood or a lower than normal number of erythrocytes. (p. 285)

angioplasty a technique used to dilate the coronary arteries by threading a small balloon-like device into a partially blocked coronary artery and then inflating the balloon to enlarge the diameter of the vessel. (p. 311)

angiotensin angiotensin I is a peptide derived when renin acts on angiotensinogen; angiotensin II is formed from angiotensin I; angiotensin II is a potent vasoconstrictor, and it stimulates the secretion of aldosterone from the adrenal cortex. (p. 262)

antagonist a muscle that works in opposition to another muscle. (p. 157)

anterior [L., to go before] that which goes first; in humans, toward the belly or front. (p. 11)

anterior horn the part of the spinal cord gray matter containing motor neurons; also called ventral horn. (p. 200)

anterior pituitary portion of the pituitary derived from the oral epithelium. (p. 254)

antibody protein found in the plasma that is responsible for antibody-mediated (humoral) immunity; binds specifically to antigen. (p. 281)

antibody-mediated immunity immunity resulting from B cells and the production of antibodies. (p. 360)

anticoagulant (an'tĭ-ko-ag'u-lant) chemical that prevents coagulation or blood clotting; an example is antithrombin. (p. 280)

antidiuretic (an'tĭ-di-u-rĕ-tik) **hormone** (ADH) hormone secreted from the posterior pituitary that acts on the kidney to reduce the output of urine; also called vasopressin. (p. 256)

antigen (an'tĭ-jen) any substance that induces a state of sensitivity and/or resistance to microorganisms or toxic substances after a latent period; substance that stimulates the specific immune system; self antigens are produced by the body, and foreign antigens are introduced into the body. (p. 281)

antigen-binding receptor molecule on the surface of lymphocytes that specifically binds antigens. (p. 361)

aorta (a-or'tah) [Gr. *aorte*, from *aeiro*, to lift up] large elastic artery that is the main trunk of the systemic arterial system that carries blood from the left ventricle of the heart and passes through the thorax and abdomen. (p. 297)

aortic semilunar valve the semilunar valve consisting of three cusps of tissue located at the base of the aorta where it arises from the left ventricle; the cusps overlap during ventricular diastole to prevent leakage of blood from the aorta into the left ventricle. (p. 297)

apex (a'-peks) [L., tip] extremity of a conical or pyramidal structure; the apex of the heart is the rounded tip directed anteriorly and slightly inferiorly. (p. 295)

apocrine (ăp'o-krin) [Gr. *apo*, away from + *krino*, to separate] gland whose cells contribute cytoplasm to its secretion; sweat glands that produce organic secretions traditionally are called apocrine. These sweat glands now are known, however, actually to be merocrine glands; see merocrine and holocrine. (p. 99)

appendicular (ap'pen-dik'u-lar) [L. *appendo*, to hang something on] relating to an appendage, as the limbs and their associated girdles. (p. 12)

appendix [L. *appendo*, to hang something on] a smaller structure usually attached by one end to a larger structure; a small blind extension of the colon attached to the cecum. (p. 415)

appositional growth [L. *ap* + *pono*, to put or place] to place one layer of bone, cartilage, or other connective tissue against an existing layer; increases the width or diameter of bones. (p. 114)

aqueous humor watery, clear fluid that fills the anterior compartment of the eye. (p. 231)

arachnoid layer (ar-ak'noyd) [Gr. *arachne*, spider-like, cobweb] thin, cobweb-like meningeal layer surrounding the brain and spinal cord; the middle of three layers. (p. 203)

areolar (ah-re'o-lar) relating to connective tissue with small spaces within it; loose connective tissue. (p. 80)

arrector pili (ah-rek'tor pĭ le) [L., that which raises hair] smooth muscle attached to the hair follicle and dermis that raises the hair when it contracts. (p. 98)

arteriosclerosis (ar-tēr'ĭ-o-sklĕ-ro'sis) [L. *arterio-* + Gr. *sklerosis*, hardness] hardness of the arteries. (p. 323)

arteriosclerotic lesion (ar-tēr'ĭ-o-skle-ro'tik) [L. *arterio* + Gr. *sklerosis*, hardness] a lesion or growth in arteries that narrows the lumen, or passage, and makes the walls of the arteries less elastic; commonly called hardening of the arteries. (p. 323)

artery blood vessel that carries blood away from the heart. (p. 320)

articulation the place where two bones come together; a joint. (p. 136)

artificial heart a mechanical pump used to replace a diseased heart. (p. 311)

artificial pacemaker an electronic device implanted beneath the skin with an electrode that extends to the heart; provides periodic electrical stimuli to the heart and substitutes for a faulty SA node. (p. 310)

astrocyte (as'tro-sīt) [Gr. *astron*, star + *kytos*, a cell] star-shaped neuroglia cell involved in forming the blood-brain barrier. (p. 187)

atherosclerosis (ath'er-o-sklĕ-ro'sis) lipid deposits (plaques) in the tunica intima of large and medium-sized arteries. (p. 323)

atom [Gr. *atomos*, indivisible, uncut] smallest particle into which an element can be divided using conventional methods; composed of neutrons, protons, and electrons. (p. 24)

atomic number the number of protons in an element. (p. 24)

atrial natriuretic factor hormone released from cells in the atrial wall of the heart when atrial blood pressure is increased; acts to lower blood pressure by increasing the rate of urine production. (p. 344)

atrioventricular (AV) node (a'tre-o-ven'trik'u-lar) small collection of specialized cardiac muscle fibers located in the lower portion of the right atrium; gives rise to the atrioventricular bundle of the conduction system of the heart, which projects through the connective tissue separating the atria from the ventricles. (p. 303)

atrioventricular bundle bundle of modified cardiac muscle fibers that projects from the AV node through the interventricular septum; conducts action potentials from the AV node rapidly through the interventricular septum; also called the bundle of His. (p. 303)

atrium, pl. atria (a'tre-ah) [L., entrance chamber] one of the two chambers of the heart that collect blood during ventricular contraction and pump blood into the ventricles to complete ventricular filling at the end of ventricular relaxation; the right atrium receives blood from the inferior and superior vena cavae and from the coronary sinus, and delivers blood to the right ventricle; the left atrium receives blood from the pulmonary veins and delivers blood to the left ventricle. (p. 295)

auditory relating to the hearing. (p. 235)

auditory ossicles bones of the middle ear; the malleus, incus, and stapes. (p. 236)

auditory tube an air-filled passageway between the middle ear and pharynx. (p. 235)

autoimmune disease disorder resulting from a specific immune system reaction against self antigens. (p. 360)

autonomic nervous system (ANS) that part of the peripheral nervous system composed of efferent fibers that reach from the central nervous system to smooth muscle, cardiac muscle, and glands. (p. 212)

autosome [Gr. *auto-*, self + *soma*, body] any chromosome other than a sex chromosome; normally occur in pairs in somatic cells and singly in gametes. (p. 57)

axial (ak'se-al) [L. *axle*, axis] head, neck, and trunk as distinguished from the extremities. (p. 12)

axon [Gr., axis] main process of a neuron; usually conducts action potentials away from the neuron cell body. (p. 82)

baroreceptor (pressoreceptor) sensory nerve endings in the walls of the atria of the heart, aortic arch, and carotid sinuses; sensitive to stretching of the wall caused by increased blood pressure. (p. 339)

baroreceptor reflex process in which baroreceptors detect changes in blood pressure and produce changes in heart rate, force of heart contraction, and blood vessel diameter that return blood pressure toward normal levels. (p. 309)

basal ganglia nuclei at the base of the cerebrum, diencephalon, and midbrain involved in controlling motor functions. (p. 199)

base any substance that is a proton acceptor; any substance that binds to hydrogen ions. (p. 32) Lower part or bottom of a structure; the base of the heart is the flat portion directed posteriorly and superiorly; veins and arteries project into and out of the base, respectively. (p. 295)

basement membrane the structure that attaches most epithelia (exceptions include lymph vessels and the liver sinusoids) to underlying tissue; consists of carbohydrates and proteins secreted by the epithelia and the underlying connective tissue. (p. 68)

basic solution solution with less hydrogen ions than hydroxide ions; has a pH greater than 7. (p. 32)

basilar membrane one of two membranes forming the cochlear duct; supports the organ of Corti. (p. 236)

basophil (ba'so-fil) [Gr. *basis*, base + *phileo*, to love] leukocyte with granules that stain purple with basic dyes; promotes inflammation and prevents clot formation. (p. 278)

belly the largest portion of a muscle, between the origin and insertion. (p. 157)

beta-adrenergic blocking agent drug that binds to and prevents adrenergic receptors from responding to adrenergic compounds that normally bind to beta-adrenergic receptors and cause them to function; beta-adrenergic blocking agents are used to treat certain arrhythmias in the heart and to treat tachycardia (rapid heart rate). (p. 310)

biceps brachii muscle in the anterior arm with two heads or origins on the scapula and an insertion onto the radius; flexes and supinates the forearm. (p. 169)

bicuspid (mitral) valve valve closing the opening between the left atrium and left ventricle of the heart; has two cusps. (p. 297)

bile fluid secreted from the liver, stored in the gallbladder, and released into the duodenum; consists of bile salts, bile pigments, bicarbonate ions, fats, hormones, and more. (p. 424)

bile salt organic salt secreted by the liver that functions to emulsify lipids. (p. 424)

blastocele (blas'to-sēl) [Gr. *blastos*, germ + *koilos*, hollow] cavity in the blastocyst. (p. 518)

blastocyst (blas'to-sist) [Gr. *blastos*, germ + *kystis*, bladder] stage of mammalian embryo development consisting of a hollow ball of cells with an inner cell mass and an outer trophoblast layer. (p. 518)

blood group a category of erythrocytes based on the type of antigen on the surface of the erythrocyte; for example, the ABO blood group is involved with transfusion reactions. (p. 281)

blood pressure [L. *pressus*, to press] the force blood exerts against the blood vessel walls; expressed relative to atmospheric pressure and reported in the form of mm Hg pressure. (p. 334)

blood-brain barrier a cellular and matrix barrier made up primarily of blood vessel endothelium, with some help from the surrounding astrocytes; it allows some (usually small) substances to pass from the circulation into the brain, but does not allow other (larger) substances to pass. (p. 187)

Bowman's capsule the enlarged terminal end of the nephron; Bowman's capsule and the glomerulus make up the renal corpuscle. (p. 461)

brachialis muscle of the anterior arm that originates on the humerus and inserts onto the ulna; flexes the forearm. (p. 169)

brachial plexus [L. *brachium*, arm] the nerve plexus to the upper limb. (p. 211)

brainstem portion of the brain consisting of the midbrain, pons, and medulla oblongata. (p. 192)

bronchiole one of the finer subdivisions of the bronchial tubes, less than 1 mm in diameter, and having no cartilage in its wall, but relatively more smooth muscle and elastic fibers than do larger bronchial tubes. (p. 382)

bronchus (brong'kus) [Gr. *bronchos*, windpipe] any one of the air ducts conducting air from the trachea to the bronchioles. (p. 382)

buccinator (buk'sĭ-na'tor) muscle making up the lateral sides of the oral cavity; flattens the cheeks. (p. 161)

buffer a chemical that resists changes in pH when either an acid or base is added to a solution containing the buffer. (p. 33)

bundle of His see atrioventricular bundle. (p. 303)

burn a lesion caused by heat, acid, or other agents; a partial-thickness burn of the skin damages only the epidermis (first-degree burn) or the epidermis and part of the dermis (second-degree burn); a full-thickness (third-degree) burn destroys the epidermis and the dermis and sometimes the underlying tissue as well. (p. 101)

bursa (bur'sah) [L., purse] closed sac or pocket containing synovial fluid, usually found in areas where friction occurs. (p. 137)

calcaneus (kal-ka'ne-us) [L., the heel] the largest tarsal bone, forming the heel. (p. 134)

calcitonin (kal-sĭ-to'nin) hormone released from cells of the thyroid gland that acts on tissues, especially bone, to cause a decrease in blood levels of calcium ions. (p. 475)

calcium channel blocker a class of drugs that specifically block channels in cell membranes through which calcium passes; calcium channel blockers are used to treat some kinds of cardiac arrhythmias. (p. 310)

callus [L., hard skin] thickening of the stratum corneum of skin in response to friction. The hard bonelike substance that develops at the site of a broken bone. (pp. 95, 114)

calorie [L. *calor*, heat] unit of heat or energy content; the quantity of energy required to raise the temperature of 1 gram of water 1° C. (p. 447)

calyx, pl. calyces (kal'iks; flower petal) the small containers into which urine flows as it leaves the collecting ducts at the tip of the renal pyramids; the calyces come together to form the renal pelvis. (p. 458)

canaliculus (kan-ă-lik'u-lus) tiny canal in bone between osteocytes containing osteocyte cell processes. (p. 110)

cancellous bone (kan'sĕ-lus) [L., grating or lattice] bone with lattice-like appearance; spongy bone. (p. 110)

capillary minute blood vessel consisting only of simple squamous epithelium and a basement membrane; major site for the exchange of substances between the blood and tissues. (p. 320)

carbohydrate organic molecule made up of one or more monosaccharides chemically bound together; sugars and starches. (p. 35)

carcinoma (kar-sĭ-no'mah) [Gr. *karkinoma*, cancer + *oma*, tumor] a malignant tumor derived from epithelial tissue. (p. 103)

cardiac cycle one complete sequence of cardiac systole and diastole. (p. 303)

cardiac output volume of blood pumped by either ventricle of the heart per minute; about 5 liters/min for the heart of a healthy adult at rest. (p. 308)

cardioregulatory center specialized area within the medulla oblongata of the brain that receives sensory input and functions to control parasympathetic and sympathetic innervation to the heart. (p. 312)

carotene (kar'o-tēn) type of yellow or orange plant pigment that is ingested by humans and used as a source of vitamin A. (p. 96)

carotid bodies small organ near the carotid sinuses that detects changes in blood oxygen, carbon dioxide, and pH. (p. 340)

carotid sinus enlargement of the internal carotid artery near the point where the internal carotid artery branches from the common carotid artery; contains baroreceptors. (p. 325)

carpal (kar'pul) [Gr. *karpos*, wrist] associated with the wrist; bones of the wrist. (p. 131)

cartilage (kar'tĭ-lij) [L., cartilage, gristle] firm, smooth, resilient, nonvascular connective tissue. (p. 80)

catabolism (kah-tab'o-lizm) [Gr. *katabole*, a casting down] all the decomposition reactions that occur in the body; releases energy. (p. 436)

catalyst a substance that increases the rate of a chemical reaction; in the process the catalyst is not permanently changed or used up. (p. 31)

cecum (se'kum) [L. *caecus*, blind] a blind sac forming the beginning of the large intestine. (p. 415)

cell [L. *cella*, chamber] basic living unit of all plants and animals. (p. 8)

cell-mediated immunity immunity resulting from the actions of T cells. (p. 360)

central nervous system (CNS) the brain and spinal cord. (p. 183)

cerebellum (ser'ĕ-bel'um) [L., little brain] a part of the brain attached to the brainstem at the pons and important in maintaining muscle tone, balance, and coordination of movements. (p. 199)

cerebral aqueduct a small connecting tube through the midbrain between the third and fourth ventricles. (p. 205)

cerebrospinal fluid (ser-e-bro-spi'nal) fluid filling the ventricles and surrounding the brain and spinal cord. (p. 205)

cerebrum [L., brain] the largest part of the brain, consisting of two hemispheres and including the cortex, internal white matter, and basal ganglia. (p. 195)

cerumen (se-ru'men) a specific type of sebum produced in the external auditory meatus; earwax. (p. 235)

cervical neck. (p. 123)

cervical plexus the nerve plexus of the neck. (p. 210)

chemical formula symbols of atoms used to indicate the composition of a molecule. (p. 24)

chemical reaction process by which atoms or molecules interact to form or break chemical bonds. (p. 27)

chemistry [Gr. *chemeia*, alchemy] science dealing with the atomic composition of substances and the reactions they undergo. (p. 24)

chemoreceptor reflex process in which chemoreceptors detect changes in oxygen levels, carbon dioxide levels, and pH in the blood and produce changes in heart rate, force of heart contraction, and blood vessel diameter that return these values toward their normal levels. (p. 340)

chordae tendineae (kor'de ten'dĭ-ne) [L., cord] tendinous strands running from the papillary muscles to the free margin of the cusps that make up the tricuspid and bicuspid valves; prevent the cusps of these valves from extending up into the atria during ventricular contraction. (p. 297)

choroid (ko'royd) portion of the vascular tunic associated with the sclera of the eye, functions to prevent scattering of light (p. 228)

choroid plexus (ko'royd) [Gr. *chorioeides,* membrane-like] specialized group of ependymal cells in the ventricles; secretes cerebrospinal fluid. (p. 205)

chromatid (kro'mah-tid) one of a pair of duplicated chromosomes joined by the centromere, which separates from its partner during cell division. (p. 59)

chromosome (kro'mo-sōm) [Gr. *chroma,* color + Gr. *soma,* body] one of the bodies (normally 46 in humans) in the cell nucleus that carry the cell's genetic information. (p. 51)

chyle (kīl) [Gr., chylos, juice] milky colored lymph with a high fat content. (p. 353)

chyme (kīm) [Gr. *chymos,* juice] semifluid mass of partly digested food passed from the stomach into the duodenum. (p. 420)

ciliary body (sil'e-ăr-e) structure continuous with the choroid layer at its anterior margin that contains smooth muscle cells and is attached to the lens by suspensory ligaments; regulates thickness of the lens and produces aqueous humor (p. 228)

citric acid cycle series of chemical reactions in which citric acid (six-carbon molecule) is converted into a four-carbon molecule, carbon dioxide is formed, and energy is released; the released energy is used to form ATP; the four-carbon molecule can combine with acetyl CoA (two carbon) to form citric acid and start the cycle again. (p. 444)

clavicle (klav'ĭ-kl) [L., a small key] the bone between the sternum and shoulder; the collar bone. (p. 127)

clot retraction condensation of the clot into a denser, compact structure. (p. 280)

clotting factor one of many proteins found in the blood in an inactive state; activated in a series of chemical reactions that result in the formation of a blood clot. (p. 279)

cochlea (ko'kle-ah) the portion of the inner ear involved in hearing; shaped like a snail shell. (p. 236)

codon sequence of three nucleotides in mRNA that codes for a specific amino acid in a protein. (p. 57)

collagen (kol'lă-jen) [Gr *koila,* glue + *gen,* producing] ropelike protein of the extracellular matrix. (p. 80)

collecting duct straight tubule that extends from the cortex of the kidney to the tip of the renal pyramid; filtrate from the distal convoluted tubules enters the collecting duct and is carried to the calyces. (p. 458)

colliculus (kol-lik'u-lus) [L. *collis,* hill] one of four small mounds on the dorsal side of the midbrain; the superior two are involved in visual reflexes, and the inferior two are involved in hearing. (p. 194)

colon (ko'lon) division of the large intestine that extends from the cecum to the rectum. (p. 415)

common bile duct duct formed by the union of the common hepatic and cystic ducts; it joins the pancreatic duct and empties into the duodenum. (p. 411)

common hepatic duct duct formed by union of the right and left hepatic ducts; it joins the cystic duct to form the common bile duct. (p. 413)

compact bone bone that is more dense and has fewer spaces than cancellous bone. (p. 110)

complement group of serum proteins that stimulates phagocytosis, inflammation, and lysis of cells. (p. 358)

compound a molecule containing two or more different kinds of atoms. (p. 24)

concha (kon'kah) [L., shell] structure resembling a shell in shape; the three bony ridges on the lateral wall of the nasal cavity. (p. 376)

condyle (kon'dīl) [Gr. *kondyles,* knuckle] rounded articulating surface of a joint. (p. 118)

cone photoreceptor cell in the retina of the eye with cone-shaped photoreceptive process; important in color vision and visual acuity. (p. 229)

conjunctiva (kon-junk-ti'vah) [L. *conjungo,* to bind together] mucous membrane covering the anterior surface of the eye and the inner lining of the eyelids. (p. 227)

connective tissue one of the four major tissue types; consists of cells surrounded by large amounts of extracellular material; functions to hold other tissues together and provides a supporting framework for the body. (p. 75)

constant region part of an antibody that does not combine with an antigen and is the same in different antibodies; responsible for activation of complement and binding the antibody to cells such as macrophages, basophils, and mast cells. (p. 363)

corn [L. *cornu,* horn] thickening of the stratum corneum of the skin over a bony projection in response to friction or pressure. (p. 95)

cornea (kor'ne-ah) transparent, anterior portion of the fibrous tunic of the eye through which light enters the eye. (p. 228)

coronal (ko-ro'nal) [Gr. *korone,* crown] plane separating the body into anterior and posterior portions; frontal section. (p. 11)

coronary artery (kor'o-năr-e) an artery that carries blood to the muscles of the heart; the left and right coronary arteries arise from the aorta. (p. 297)

coronary bypass surgery in which a vein from some other part of the body is grafted to a coronary artery in such a way as to bypass a blocked coronary artery. (p. 310)

coronary vein vein that carries blood from the heart muscle primarily to the right atrium. (p. 307)

corpus callosum (kor'pus kah-lo'sum) [L., body; callous] a large, thick nerve fiber tract connecting the two cerebral hemispheres. (p. 198)

corpus luteum (lu'te-um) yellow endocrine body formed in the ovary in the site of a ruptured follicle immediately after ovulation; secretes progesterone and estrogen. (p. 499)

cortex [L., bark] the outer portion of an organ such as the brain. (p. 187)

cortisol (kor'tĭ-sol) steroid hormone released by the adrenal cortex; increases blood glucose and inhibits inflammation; it is a glucocorticoid. (p. 255)

covalent bond chemical bond that is formed when two atoms share a pair of electrons. (p. 26)

coxa (kok'sah) [L., hip] the bone of the hip. (p. 131)

cranial nerves peripheral nerves originating in the brain. (p. 208)

cranial vault eight skull bones that surround and protect the brain; brain case. (p. 118)

cremaster muscle extension of abdominal muscles; in the male it raises the testicles. (p. 486)

cricoid cartilage (kri'koyd) most inferior laryngeal cartilage. (p. 379)

crown that part of the tooth that is formed of and covered by enamel. (p. 408)

crypt [Gr. *kryptos*, hidden] a pitlike depression. (p. 415)

cupula (ku'pu-lah) [L. *cupa*, a tub] gelatinous mass that overlies the hair cells of the cristae ampullares of the semicircular canals; responds to fluid movement. (p. 241)

cyanosis (si-ă-no'sis) [Gr., dark blue color] blue coloration of the skin and mucous membranes caused by insufficient oxygenation of blood. (p. 96)

cystic duct duct from the gallbladder; it joins the common hepatic duct to form the common bile duct. (p. 413)

cytoplasm (si'to-plazm) cellular material surrounding the nucleus. (p. 44)

deciduous teeth (dĕ-sid'u-us) [L. *deciduus*, falling off] the primary teeth that fall out to be replaced by the permanent teeth. (p. 407)

decomposition reaction the breakdown of a larger molecule into smaller molecules, ions, or atoms. (p. 29)

deep [O.E. *deop*, deep] away from the surface, internal. (p. 11)

defecation [L. *defaeco*, to purify] discharge of feces from the rectum. (p. 428)

deglutition (dĕ'glu-tish'un) [L. *de-* + *glutio*, to swallow] the act of swallowing. (p. 418)

deltoid triangular muscle over the shoulder; inserts onto the humerus; abducts the arm. (p. 169)

denaturation the change in shape of a protein caused by breaking hydrogen bonds; agents that cause denaturation include heat and changes in pH. (p. 37)

dendrite (den'drīt) [G. *dendrite*, tree] short, tree-like cell process of a neuron; receives stimuli. (p. 82)

dentin bonelike material forming the mass of the tooth. (p. 408)

deoxyribonucleic acid (de-ox'sĭ-ri'bo-nu-kle'ik) type of nucleic acid containing the sugar deoxyribose; the genetic material of cells; DNA. (p. 39)

dermis (der'mis) [Gr. *derma*, skin] dense connective tissue that forms the deep layer of the skin; responsible for the structural strength of the skin. (p. 93)

desmosome (dez'mo-sōm) a point of adhesion between two cells. (p. 75)

diaphragm (di'ă-fram) muscular separation between the thoracic and abdominal cavities; its contraction results in inspiration. (p. 163)

diaphysis (di-af'ĭ-sis) [Gr., growing between] shaft of a long bone. (p. 110)

diastole (di-as'to-le) [Gr. *diastole*, dilation] relaxation of the heart chambers during which they fill with blood; usually refers to ventricular relaxation. (p. 305)

diencephalon [G. *dia*, through + *enkephalos*, brain] part of the brain inferior to and nearly surrounded by the cerebrum, and connecting posteriorly and inferiorly to the brainstem. (p. 194)

diffusion [L. *diffundo*, to pour in different directions] tendency for solute molecules to move from an area of high concentration to an area of low concentration in solution; the product of the constant random motion of all atoms, molecules, or ions in a solution. (p. 46)

digestion the breakdown of carbohydrates, lipids, proteins, and other large molecules to their component parts. (p. 439)

digestive tract the tract from the mouth to the anus, including the stomach and intestines, where food is taken in, broken down, and absorbed. (p. 405)

digitalis (dij'ĭ-tal'is) [L., relating to finger-like flowers] a steroid used in the treatment of heart diseases such as heart failure; increases the force of contraction of the heart; extracted from the foxglove plant. (p. 311)

diploid the condition in which there are two copies of each autosome and two sex chromosomes (46 total chromosomes in humans). (p. 57)

disaccharide two monosaccharides chemically bound together; glucose and fructose chemically join to form sucrose. (p. 35)

dissociate [L. *dis-* + *socio*, to disjoin, separate] the separation of positive and negative ions when they dissolve in water and are surrounded by water molecules. (p. 27)

distal [L. *di-* + *sto*, to be distant] farther from the point of attachment to the body than another structure. (p. 11)

distal convoluted tubule convoluted tubule of the nephron that extends from the ascending limb of the loop of Henle and ends in a collecting duct. (p. 458)

DNA see deoxyribonucleic acid. (p. 39)

dominant [L. *dominus*, a master] in genetics, a gene that is expressed phenotypically to the exclusion of a contrasting recessive trait. (p. 541)

dorsal [L. *dorsum*, back] back surface of the body; in humans, synonymous with posterior. (p. 11)

dorsal root sensory (afferent) root of a spinal nerve. (p. 200)

ductus deferens (vas deferens) duct of the testicle, running from the epididymis to the ejaculatory duct; the vas deferens. (p. 489)

duodenum (du-od'ĕ-num or du-o-de'num) [L. *duodeni*, twelve] first division of the small intestine; connects to the stomach. (p. 411)

dura mater (du'rah ma'ter) [L., tough mother] tough, fibrous membrane forming the outermost meningeal covering of the brain and spinal cord. (p. 202)

eardrum see tympanic membrane. (p. 235)

eccrine (ek'rin) [Gr. *ek*, out + *krino*, to separate] exocrine; refers to water-producing sweat glands; see merocrine. (p. 98)

ectoderm (ek'to-derm) outermost of the three germ layers of the embryo. (p. 520)

ectopic beat (ek-top'ik) a heart beat that originates from an area of the heart other than the SA node. (p. 304)

edema (e-de'mah) [Gr. *oidema*, a swelling] excessive accumulation of fluid, usually causing swelling. (p. 83)

efferent arteriole vessel that carries blood from the glomerulus to the peritubular capillaries. (p. 461)

efferent fiber nerve fiber going from the central nervous system toward the peripheral nervous system; motor fiber. (p. 183)

ejaculation reflexive expulsion of semen from the penis. (p. 492)

ejaculatory duct duct formed by the union of the ductus deferens and the excretory duct of the seminal vesicle, which opens into the urethra. (p. 492)

electrocardiogram graphic record of the heart's electrical currents obtained with an electronic recording instrument; an ECG. (p. 304)

electrolyte [Gr. *electro-* + *lytos*, soluble] positive and negative ions that conduct electricity in solution. (p. 26)

electron negatively charged particle found in the orbitals of atoms. (p. 24)

electron-transport chain series of energy transfer molecules in the inner mitochondrial membrane; they receive energy and use it in the formation of ATP and water. (p. 444)

element [L. *elementum*, a rudiment] substance composed of only one kind of atoms. (p. 24)

embolus (em'bo-lus) [Gr. *embolos*, a plug] a detached clot or other foreign body that occludes a blood vessel. (p. 280)

embryo in prenatal development, the developing human between 14 and 60 days of development. (p. 518)

emission [L. *emissio*, to send out] discharge; formation and accumulation of semen prior to ejaculation. (p. 492)

emulsification (e-mul'sĭ-fĭ-ka'shun) the dispersal of one liquid, or very small globules of the liquid, within another liquid. (p. 424)

emulsify to form an emulsion, which is one liquid dispersed in another liquid. (p. 424)

enamel hard substance covering the exposed portion of the tooth. (p. 408)

endergonic reaction (en-der-gon'ik) [L. *endo-*, inside + Gr. *ergon*, work] a chemical reaction resulting in the absorption of energy. (p. 30)

endocardium (en'do-kar'dĭ-um) innermost layer of the heart, including endothelium and connective tissue. (p. 301)

endochondral ossification bone formation within cartilage. (p. 114)

endocrine (en'do-krin) [Gr. *endon*, inside + *krino*, to separate] ductless gland that secretes internally, usually into the circulatory system. (p. 75)

endocytosis (en'do-si'to-sis) bulk uptake of material through the cell membrane by taking it into a vesicle. (p. 50)

endoderm (en'do-derm) innermost of the three germ layers of the embryo. (p. 520)

endolymph [Gr. *endo-*, inside, + *lympha*, clear fluid] the fluid inside the membranous labyrinth of the inner ear. (p. 236)

endometrium mucous membrane comprising the inner layer of the uterine wall; consists of a simple columnar epithelium and a lamina propria that contains simple tubular uterine glands. (p. 499)

endoplasmic reticulum (en′do-plaz′mik re-tik′u-lum) membranous network inside the cytoplasm; rough endoplasmic reticulum has ribosomes attached to the surface; smooth endoplasmic reticulum does not have ribosomes attached. (p. 53)

endosteum (en-dos′te-um) [Gr. *endo*, within + *osteon*, bone] membranous lining of the medullary cavity and the cavities of spongy bone. (p. 110)

enzyme (en′zīm) [Gr. *en*, in + *zyme*, leaven] a protein molecule that increases the rate of a chemical reaction without being permanently altered; an organic catalyst. (p. 31)

eosinophil (e-o-sin′o-fil) [Gr. *eos*, dawn + *phileo*, to love] leukocyte with granules that stain red with acidic dyes; inhibits inflammation. (p. 278)

ependymal (ep-en′dĭ-mal) the neuroglial cell layer lining the ventricles of the brain. (p. 187)

epicardium (ep′ĭ-kar′dĭ-um) [Gr. *epi-* + *kardia*, heart] serous membrane covering the surface of the heart; also called the visceral pericardium. (p. 297)

epicondyle [Gr. *epi-*, on + *kondyles*, knuckle] projection upon (usually to the side of) a condyle. (p. 127)

epidermis (ep′ĭ-der′mis) [Gr. *epi*, upon + *derma*, skin] outer portion of the skin formed of epithelial tissue that rests on the dermis; resists abrasion and forms a permeability barrier. (p. 93)

epididymis (ep-ĭ-did′ĭ-mis) [Gr. *epi*, on + *didymos*, twin] elongated structure connected to the posterior surface of the testis, site of storage and maturation of the spermatozoa. (p. 489)

epiglottis (ep′ĭ-glot′is) [Gr. *epi*, on + *glottis*, the mouth of the windpipe] plate of elastic cartilage, covered with mucous membrane, that serves as a valve over the opening of the larynx during swallowing to prevent materials from entering the larynx. (p. 379)

epinephrine hormone similar in structure to the neurotransmitter norepinephrine; major hormone released from the adrenal medulla; increases cardiac output and blood glucose levels. (p. 258)

epiphyseal line dense plate of bone in a bone that is no longer growing, indicating the former site of the epiphyseal plate. (p. 110)

epiphyseal plate site at which bone growth in length occurs; located between the epiphysis and diaphysis of a long bone; area of cartilage where cartilage growth is followed by endochondral ossification; also called the growth plate. (p. 110)

epiphysis (e-pif′ĭ-sis) [Gr. *epi*, on + *physis*, growth] the end of a bone; separated from the remainder of the bone by the epiphyseal plate or epiphyseal line. (p. 110)

epiploic appendage (ep′ĭ-plo′ik) one of a number of little fat-filled processes of peritoneum projecting from the serous coat of the large intestine. (p. 416)

epithelial tissue one of the four major tissue types consisting of cells with a basement membrane (exceptions are lymph vessels and liver sinusoids), little extracellular material, and no blood vessels; covers the surfaces of the body and forms glands. (p. 68)

erector spinae common name of the muscle group of the back; holds the back erect. (p. 163)

erythroblastosis fetalis (ĕ-rith′ro-blast-to′sis feta′lis) [erythroblast + Gr. *-osis*, condition] destruction of erythrocytes in the fetus or newborn caused by antibodies produced in the Rh-negative mother acting on the Rh-positive blood of the fetus or newborn. (p. 283)

erythrocyte (ĕ-rith′ro-sīt) [Gr. *erythro*, red + *kytos*, cell] red blood cell; biconcave disk that contains hemoglobin, which transports oxygen and carbon dioxide; erythrocyte does not have a nucleus. (p. 273)

erythropoietin (ĕ-rith′ro-poy′ĕ-tin) protein hormone that stimulates erythrocyte formation in red bone marrow. (p. 276)

esophagus (e-sof′ă-gus) [Gr. *oisophagos*, gullet] the part of the digestive tract between the pharynx and stomach. (p. 408)

estrogen steroid hormone secreted primarily by the ovaries; involved in maintenance and development of female reproductive organs, secondary sexual characteristics, and the menstrual cycle. (p. 265)

Eustachian tube see auditory tube. (p. 235)

exchange reaction a combination of a decomposition reaction, in which molecules are broken down, and a synthesis reaction, in which the products of the decomposition reaction are combined to form new molecules. (p. 29)

exergonic reaction (ek′ser-gon′ik) [L. *exo-*, outside + Gr. *ergon*, work] a chemical reaction resulting in the release of energy. (p. 29)

exocrine (ek′so-krin) [Gr. *exo-*, outside + *krino*, to separate] gland that secretes to a surface or outward through a duct. (p. 75)

exocytosis (eks-o-si-to′sis) elimination of material from a cell through the formation of vesicles. (p. 50)

extension [L. *extensio*] to stretch out; usually to straighten out a joint. (p. 142)

extracellular matrix nonliving chemical substances located between connective tissue cells; consists of protein fibers, ground substance, and fluid. (p. 75)

extrinsic muscles muscles located outside of the structure upon which they act. (p. 161)

extrinsic regulation regulation of the heart that involves mechanisms outside the heart, including nervous and hormonal regulation. (p. 309)

facet (fas'et) [Fr., little face] a small, smooth articular surface. (p. 124)

facilitated diffusion carrier-mediated process that does not require ATP and moves substances into or out of cells from a high to a low concentration. (p. 49)

fascia (fash'e-ah) [L., band or fillet] loose areolar connective tissue found beneath the skin (hypodermis), or dense connective tissue that encloses and separates muscles. (p. 148)

fasciculus (fă-sik'u-lus) [L. *fascis*, bundle] band or bundle of nerve or muscle fibers bound together by connective tissue. (p. 148)

fat greasy, soft-solid lipid found in animal tissues and many plants; composed of glycerol and fatty acids. (p. 36)

fatty acid straight chain of carbon atoms with a carboxyl group (-COOH) attached at one end; a building block of fats. (p. 36)

feces matter discharged from the digestive tract during defecation, consisting of the undigested residue of food, epithelial cells, intestinal mucus, bacteria, and waste material. (p. 428)

fertilization union of the sperm and oocyte to form a zygote. (p. 516)

fetus in prenatal development, the developing human between 60 days and birth. (p. 538)

fibrin (fi'brin) [L. *fibra*, fiber] a threadlike protein fiber derived from fibrinogen by the action of thrombin; forms a clot, that is, a network of fibers that traps blood cells, platelets, and fluid, which stops bleeding. (p. 279)

fibrinolysis (fi'brĭn-ol'ĭ-sis) [L. *fibra*, fiber + Gr. *lysis*, dissolution] the breakdown of a clot by plasmin. (p. 280)

filtration movement, resulting from a pressure difference, of a liquid through a filter, which prevents some or all of the substances in the liquid from passing. (p. 48)

filtration membrane membrane formed by the glomerular capillary endothelium, the basement membrane, and the podocytes of Bowman's capsule. (p. 461)

first heart sound the heart sound that results from the simultaneous closure of the tricuspid and bicuspid valves. (p. 307)

flagella (flă-jel'ah) [L., whip] whiplike locomotor organelles similar to cilia except longer and there is usually only one per cell. (p. 55)

flexion [L. *flectus*] to bend. (p. 142)

focal point the point at which light rays cross after passing through a concave lens. (p. 231)

follicle-stimulating hormone (FSH) hormone of the anterior pituitary that, in the female, stimulates the follicles of the ovary, assists in maturation of the follicle, and causes secretion of estrogen from the follicle; in the male, stimulates the epithelium of the seminiferous tubules and is partially responsible for inducing spermatogenesis. (p. 256)

fontanel (fon'tă-nel) [Fr., fountain] one of several membranous gaps between bones of the skull. (p. 136)

foramen (fo-ra'men) a hole; referring to a hole or opening in a bone. (p. 116)

foramen ovale (o-val'e) in the fetal heart, the oval opening in the interatrial septum with a valve that allows blood to flow from the right to left atrium but not in the opposite direction; becomes the fossa ovalis after birth. (p. 528)

formed element one of the cells (for example, erythrocytes or leukocytes) or cell fragments (for example, platelets) in blood. (p. 272)

fovea centralis (fo've-ah) depression in the center of the macula of the eye, which has the greatest visual acuity and where there are only cones. (p. 230)

free energy total amount of energy that can be liberated by the complete catabolism of food. (p. 448)

frenulum (fren'u-lum) [L. *frenum*, bridle] fold extending from the floor of the mouth to the middle of the under surface of the tongue. (p. 406)

frontal see coronal. (p. 11)

full-thickness burn burn that destroys the epidermis and the dermis and sometimes the underlying tissue as well; sometimes called a third-degree burn. (p. 102)

fundus (fun'dus) [L., bottom] the bottom, or area farthest from the opening, of a hollow organ such as the stomach or uterus. (p. 409)

ganglion (gan'gle-on) [Gr., knot] a group of nerve cell bodies in the peripheral nervous system. (p. 187)

gastric gland a gland within the stomach. (p. 409)

gastrin (gas'trin) hormone secreted in the mucosa of the stomach and duodenum that stimulates secretion of hydrochloric acid by the gastric glands. (p. 420)

genetics the branch of science that deals with heredity. (p. 540)

genotype genetic makeup of an individual. (p. 541)

giantism abnormal growth in young people due to hypersecretion of growth hormone by the pituitary gland. (p. 255)

gingiva (jin'jĭ-vah) dense fibrous tissue, covered by mucous membrane, that covers the alveolar processes of the upper and lower jaws and surrounds the necks of the teeth. (p. 408)

girdle a bony ring or belt that attaches a limb to the body. (p. 12)

gland a single cell or a multicellular structure that secretes substances into the blood, a cavity, or onto a surface. (p. 75)

glia see neuroglia. (p. 187)

glomerulus (glo-mĕr'u-lus) [L. *glomus*, ball of yarn] mass of capillary loops at the beginning of each nephron, nearly surrounded by Bowman's capsule. (p. 461)

glucagon (glu'kă-gon) hormone secreted from the pancreatic islets of the pancreas that acts primarily on the liver to release glucose into the circulatory system. (p. 263)

glucocorticoid hormones from the adrenal cortex capable of increasing the rate at which lipids are broken down to fatty acids and proteins are broken down to amino acids; elevates blood glucose levels, and acts as an antiinflammatory substance. (p. 261)

glycerol (glis'er-ol) a three-carbon molecule with a hydroxyl group attached to each carbon; a building block of fats. (p. 36)

glycolysis (gli-kol'ĭ-sis) [Gr. *glykys*, sweet + *lysis*, a loosening] anaerobic process during which one glucose molecule is converted to two pyruvic acid molecules; a net of two ATP molecules is produced during glycolysis. (p. 442)

glycoprotein an organic molecule composed of a protein and a carbohydrate.

goblet cell epithelial cell that has its apical end distended with mucin. (p. 75)

goiter (goy'ter) [L. *guttur*, throat] an enlargement of the thyroid gland, not due to a neoplasm, usually caused by a lack of iodine in the diet. (p. 256)

Golgi apparatus (gōl'je) named for Camillo Golgi, Italian histologist and Nobel laureate, 1843-1926; specialized endoplasmic reticulum that concentrates and packages materials for secretion from the cell. (p. 53)

gonadotropin (gon'ă-do-tro'pin) hormone capable of promoting gonadal growth and function; two major gonadotropins are luteinizing hormone (LH) and follicle-stimulating hormone (FSH). (p. 255)

gonadotropin-releasing hormone (GnRH) hypothalamic hormone that stimulates the secretion of LH and FSH from the anterior pituitary. (p. 492)

granulation tissue vascular connective tissue formed in wounds. (p. 84)

growth hormone (GH) protein hormone of the anterior pituitary; it promotes body growth, fat mobilization, and inhibition of glucose utilization. (p. 255)

gyrus (ji'rus) [L. *gyros*, circle] rounded elevation or fold on the surface of the brain. (p. 195)

hair a threadlike outgrowth of the skin consisting of columns of dead keratinized epithelial cells. (p. 96)

hair cells cells of the inner ear containing hair-like processes (microvilli) that respond to bending of the hairs by depolarizing. (p. 236)

hamstring the posterior thigh. (p. 173)

haploid the condition in which a cell has one copy of each autosome and one sex chromosome (23 total chromosomes in humans); characteristic of gametes. (p. 516)

haustra (haw'strah) sacs of the colon, formed by the teniae coli, which are slightly shorter than the gut, so that the gut is thrown into pouches. (p. 416)

haversian canal (hă-ver'shan) named for 17th century English anatomist, Clopton Harvers; canal containing blood vessels, nerves, and loose connective tissue and running parallel to the long axis of the bone. (p. 110)

haversian system a single haversian canal, with its contents, and the associated lamellae and osteocytes surrounding it. (p. 110)

heart lung machine a machine that pumps blood and carries out the process of gas exchange; it substitutes for the heart and lungs during heart surgery. (p. 310)

heart rate the number of complete cardiac cycles (heart beats) per minute. (p. 308)

heart transplant the process of taking a healthy heart from a recently deceased donor and transplanting it into a recipient who has a diseased heart. (p. 311)

hematocrit (hem'ă-to-krit) [Gr. *hemato*, blood + *krino*, to separate] the percentage of total blood volume composed of erythrocytes. (p. 285)

hematopoiesis (hem'ă-to-poy-e'sis) [Gr. *haima*, blood + *poiesis*, a making] production of blood cells. (p. 273)

hemoglobin (he'mo-glo'bin) red protein of erythrocytes consisting of four globin proteins with an iron-containing red pigment heme bound to each globin protein; transports oxygen and carbon dioxide. (p. 275)

hemolysis (he-mol'ĭ-sis) [Gr. *hemo*, blood + *lysis*, destruction] the rupture of erythrocytes. (p. 281)

hemorrhage (hem'ŏ-rij) [Gr. *haima*, blood + *rhegnymi*, to burst forth] rupture or leaking of blood from vessels.

hepatic portal circulation blood flow through the veins that begin as capillary beds in the small intestine, spleen, and stomach and carry blood to the liver, where they end as a capillary bed. (p. 333)

hepatic portal vein (hĕ-pat′ik) [Gr. *hepar*, liver + L. *porta*, gate] the vein that carries blood from the intestines, stomach, spleen, and pancreas to the liver. (p. 413)

Hering-Breuer reflex process in which action potentials from stretch receptors in the lungs arrest inspiration, and expiration then occurs. (p. 394)

heterozygous having two different genes for a given trait. (p. 541)

hilum (hi′lum) [L., a small amount or trifle] Part of an organ where the nerves and vessels enter and leave. (p. 382)

histology (his-tol′o-je) [Gr. *histo*, web (tissue) + *logos*, study] the science that deals with the structure of cells, tissues, and organs in relation to their function. (p. 68)

homeostasis (ho′me-o-sta′sis) [Gr. *homoio*, like + *stasis*, a standing] state or maintenance of equilibrium in the body with respect to functions and the composition of fluids and tissues. (p. 8)

homeotherm (ho′me-o-therm) [Gr. *homoiois*, like + *thermos*, warm] (warm-blooded animals) any animal, including mammals and birds, that tends to maintain a constant body temperature. (p. 448)

homozygous having two identical genes for a given trait. (p. 541)

hormone [Gr. *hormon*, to set into motion] substance secreted by endocrine tissues into the blood that acts on a target tissue to produce a specific response. (p. 249)

humerus [L., shoulder] the bone of the arm. (p. 127)

humoral immunity see antibody-mediated immunity. (p. 360)

hydrogen bond the weak attraction between the oppositely charged ends of two polar covalent molecules; the weak attraction between the end of a polar covalent molecule and an ion. (p. 27)

hyoid (hi′oyd) [Gr., shaped like the letter upsilon, υ] the U-shaped bone in the throat. (p. 118)

hypertonic (hi′per-ton′ik) [Gr. *hyper*, above + *tonos*, tension] solution that causes cells to shrink. (p. 48)

hypodermis (hi′po-der′mis) [Gr. *hypo*, under + *dermis*, skin] loose connective tissue under the dermis that attaches the skin to muscle and bone. (p. 93)

hypophysis (hi-pof′ĭ-sis) [Gr., an undergrowth] endocrine gland attached to the hypothalamus by the infundibulum; the pituitary gland. (p. 254)

hypothalamic-pituitary portal system (hi′po-thal′ă-mik–pit-u′ĭ-tĕr-e) series of blood vessels that carry blood from the area of the hypothalamus to the anterior pituitary; originates from capillary beds in the hypothalamus and terminates as a capillary bed in the anterior pituitary. (p. 255)

hypothalamus (hi′po-thal′ă-mus) [Gr. *hypo*, under, below + *thalamus*, bedroom] important autonomic and endocrine control center of the brain located beneath the thalamus of the brain. (p. 195)

hypotonic (hi′po-ton′ik) [Gr. *hypo* + *tonos*, tension] solution that causes cells to swell. (p. 48)

ileocecal junction the junction of the ileum of the small intestine and the cecum of the large intestine. (p. 413)

implantation attachment of the blastocyst to the endometrium of the uterus; occurring 6 or 7 days after fertilization of the ovum. (p. 518)

incompetent valve a leaky valve; usually refers to a leaky valve in the heart that allows blood to flow through it when it is closed. (p. 307)

incus (ing′kus) [L., anvil] the middle bone of the middle ear; the anvil. (p. 236)

infarct (in′farkt) area of necrosis resulting from a sudden insufficiency of arterial blood supply. (p. 307)

inferior [L., lower] down, or lower, with reference to the anatomical position. (p. 11)

inferior vena cava (ve′nah ca′vah) receives blood from the lower limbs and the greater part of the pelvic and abdominal organs and empties into the right atrium of the heart. (p. 332)

inflammatory response complex sequence of events involving chemicals and immune system cells that results in the isolation and destruction of foreign substances such as bacteria; symptoms include redness, heat, swelling, pain, and disturbance of function. (p. 359)

infundibulum (in-fun-dib′u-lum) [L., funnel] funnel-shaped structure or passage, for example, the infundibulum that attaches the pituitary to the hypothalamus; funnel-like expansion of the uterine tube near the ovary. (p. 195)

inner cell mass group of cells at one end of the blastocyst from which the embryo develops. (p. 518)

inorganic molecules that do not contain carbon atoms; originally defined as molecules that came from nonliving sources; the original definition is no longer valid, because carbon dioxide produced by living organisms is considered an inorganic molecule. (p. 34)

insertion the more movable attachment point of a muscle. (p. 157)

insulin (in′su-lin) protein hormone secreted from the pancreas that increases the uptake of glucose and amino acids by most tissues. (p. 262)

interatrial septum the cardiac muscle partition separating the right and left atria. (p. 297)

intercostal muscles muscles located between ribs. (p. 163)

interferon (in′ter-fēr′on) a protein released by virally infected cells that binds to other cells and stimulates them to produce antiviral proteins that inhibit viral replication. (p. 358)

interstitial cells (cells of Leydig) cells between the seminiferous tubules of the testes; secrete testosterone. (p. 486)

interventricular septum the cardiac muscle partition separating the right and left ventricles. (p. 528)

intestinal glands tubular glands in the mucous membrane of the small intestine. (p. 424)

intramembranous ossification bone formation within connective tissue membranes. (p. 114)

intramural plexus (in′trah-mu′ral) [L., within the wall] a nerve plexus within the walls of the gastrointestinal tract. (p. 406)

intrinsic factor factor secreted by the gastric glands and required for adequate absorption of vitamin B_{12}. (p. 420)

intrinsic muscles muscles located within the structure upon which they act. (p. 161)

ion (i′on) atom or group of atoms carrying a charge of electricity because of a loss or gain of one or more electrons. (p. 26)

ionic bond chemical bond that is formed when one atom loses an electron and another atom accepts that electron. (p. 24)

iris specialized part of the vascular tunic of the eye; the "colored" part of the eye that can be seen through the cornea; consists of smooth muscles that regulate the amount of light entering the eye. (p. 228)

isometric contraction muscle contraction where the length of the muscle does not change but the amount of tension increases. (p. 155)

isotonic (i′so-ton′ik) [Gr. *iso*, equal + *tonos*, tension] solution that causes cells to neither shrink nor swell. (p. 48)

isotonic contraction muscle contraction where the amount of tension is constant and the muscle shortens. (p. 155)

isotopes elements that have the same number of protons and electrons, but have a different number of neutrons. (p. 28)

jaundice (jawn′dĭs) [Fr. *jaune*, yellow] yellowish staining of the skin, sclerae, and deeper tissues and excretions with bile pigments. (p. 101)

keratinization (kĕr′ah-tin-ĭ-za′shun) production of keratin and changes in the structure and shape of epithelial cells as they move to the skin surface. (p. 94)

Korotkoff sound sound heard over an artery when blood pressure is determined by the auscultatory method. (p. 335)

labia majora one of two rounded folds of skin surrounding the labia minora and vestibule. (p. 501)

labia minora one of two narrow longitudinal folds of mucous membrane enclosed by the labia majora; anteriorly they unite to form the prepuce. (p. 500)

labyrinth (lab′ĭ-rinth) a series of membranous and bony tunnels in the temporal bone; part of the inner ear involved in hearing and balance. (p. 236)

lacrimal (lak′rĭ-mal) [L., a tear] relating to tears or tear production. (p. 119)

lactation [L. *lactatio*, suckle] period following childbirth during which milk is formed in the breasts. (p. 536)

lacteal (lak′te-al) lymphatic vessel in the wall of the small intestine that absorbs fats and carries chyle from the intestine. (p. 353)

lactic acid three-carbon molecule derived from pyruvic acid as a product of anaerobic respiration. (p. 443)

lamella (lă-mel′ah) [L. *lamina*, plate, leaf] a thin sheet or layer of bone. (p. 110)

lamina propria (lam′ĭ-nah pro′pre-ah) layer of connective tissue underlying the epithelium of a mucous membrane. (p. 405)

laryngitis inflammation of the mucous membrane of the larynx. (p. 380)

laryngopharynx (lă-ring′go-făr′ingks) part of the pharynx lying below the tip of the epiglottis extending to the level of the cricoid cartilage of the larynx. (p. 379)

larynx (lăr′ingks) organ of voice production located between the pharynx and the trachea; it consists of a framework of cartilages and elastic membranes housing the vocal folds (true vocal cords) and the muscles that control the position and tension of these elements. (p. 379)

lateral [L. *latus*, side] away from the middle or midline of the body. (p. 11)

lateral horn the small, lateral extension of spinal cord gray matter; located only in spinal cord regions T1-L2 and S2-S4; containing preganglionic autonomic nerve cell bodies. (p. 200)

lens the biconvex structure in the anterior part of the eye capable of being flattened or thickened to adjust the focus of light entering the eye. (p. 228)

leukemia (lu-ke′me-ah) [Gr. *leukos*, white + *haima*, blood] a tumor of the red bone marrow that results in the production of large numbers of abnormal leukocytes; often accompanied by decreased production of erythrocytes and platelets. (p. 286)

leukocyte (lu′ko-sīt) [Gr. *leukos*, white + *kytos*, cell] white blood cell; round, nucleated cell involved in immunity; the five types of leukocytes are neutrophils, eosinophils, basophils, lymphocytes, and monocytes. (p. 273)

leukocytosis (lu-ko-sī-to'sis) [leukocyte + Gr. *-osis*, a condition] a higher-than-normal number of white blood cells. (p. 273)

leukopenia (lu-ko-pe'ne-ah) [leukocyte + Gr. *penia*, poverty] a lower-than-normal number of white blood cells. (p. 286)

ligament a tough connective tissue band usually connecting bone to bone.

limbic system (lim'bik) [L. *limbus*, a border] a primitive part of the brain involved in visceral and emotional response to odor. (p. 80)

linea alba (lin'e-ah al'bah) white line in the center of the abdomen where muscles of the abdominal wall insert. (p. 166)

lipase (li'pās) an enzyme that breaks down lipids. (p. 428)

lipid [Gr. *lipos*, fat] substance composed principally of carbon, oxygen, and hydrogen; generally soluble in nonpolar solvents; fats and cholesterol. (p. 36)

local inflammation inflammation confined to a specific area of the body; symptoms include redness, heat, swelling, pain, and loss of function. (p. 359)

longitudinal section a cut made through the long axis of an organ. (p. 12)

loop of Henle U-shaped part of the nephron extending from the proximal to the distal convoluted tubule and consisting of descending and ascending limbs; many of the loops of Henle extend into the renal pyramids. (p. 458)

lower motor neuron a motor neuron located in the brainstem or spinal cord, as opposed to the cerebral cortex. (p. 202)

lumbosacral plexus the nerve plexus that innervates the lower limbs. (p. 211)

lunula (lu'nu-lah) [L. *luna*, moon] white, crescent-shaped portion of the nail matrix visible through the proximal end of the nail. (p. 99)

luteinizing hormone (LH) (lu'te-ĭ-nīz-ing) hormone of the anterior pituitary that, in the female, initiates final maturation of the follicles, their rupture to release the ovum, the conversion of the ruptured follicle into the corpus luteum, and the secretion of progesterone; in the male, stimulates the secretion of testosterone in the testes, and is sometimes referred to as interstitial cell-stimulating hormone (ICSH). (p. 255)

lymph (limf) [L. *lympha*, clear spring water] clear or yellowish fluid derived from interstitial fluid and found in lymph vessels. (p. 353)

lymph node encapsulated mass of lymph tissue found along lymph vessels; functions to filter lymph and produce lymphocytes. (p. 354)

lymphocyte (lim'fo-sīt) nongranulocytic leukocyte involved in the immune system; there are several types of lymphocytes with diverse functions, including antibody production, allergic reactions, graft rejections, tumor control, and regulation of the immune system. (p. 278)

lymphokine (lim'fo-kīn) a class of chemicals produced by T cells that activate macrophages and other immune cells; promote phagocytosis and inflammation. (p. 367)

lysosome (li'so-sōm) [Gr. *lysis*, a loosening + *soma*, body] membrane-bound vesicle containing intracellular digestive enzymes. (p. 54)

macrophage (mak'ro-fāj) [Gr. *makros*, large + *phagein*, to eat] any large mononuclear, phagocytic cell. (p. 80)

macula (mak'u-lah) one of the sensory structures in the vestibule, consisting of hair cells and a gelatinous mass embedded with otoliths; responds to gravity. (p. 240)

macula lutea (mak'u-lah lu'te-ah) [L., a yellow spot] small yellow spot in the posterior retina of the eye where the cones are concentrated; has no red tint because it is devoid of blood vessels. (p. 230)

malleus (mal'e-us) [L., hammer] the most lateral of the middle ear bones, attached to the tympanic membrane; the hammer. (p. 236)

mammary gland breast; the organ of milk secretion, located in the breast or mamma; one of two hemispheric projections of variable size situated in the subcutaneous layer over the pectoralis major muscle on each side of the chest. (p. 501)

mastication (mas'tĭ-ka-shun) [L. *mastico*, to chew] process of chewing. (p. 161)

matrix the substance between the cells of a tissue. (p. 110)

matter anything that occupies space. (p. 24)

mean arterial blood pressure the average of the arterial blood pressure; it is slightly less than the average of the systolic and diastolic blood pressure, because diastole lasts longer than systole. (p. 338)

meatus (me-a'tus) [L., to go, pass] passageway or tunnel. (p. 118)

medial [L. *medialis*, middle] toward the middle or midline of the body. (p. 11)

mediastinum (me'de-as-ti'num) [L., middle septum] the middle wall of the thorax consisting of the trachea, esophagus, thymus, heart, and other structures. (p. 15)

mediators of inflammation chemicals released or activated by injured tissues and adjacent blood vessels; produce vasodilation, increase vascular permeability, and attract blood cells; include histamine, kinins, prostaglandins, and leukotrienes. (p. 83)

medulla (mĕ-dul'ah) [L. *medius,* middle, marrow] the center or core of an organ. (p. 95)

medulla oblongata (ob'long-gah'tah) inferior portion of the brainstem that connects the spinal cord with the brain; contains nuclei of cranial nerves plus autonomic control centers for heart rate, respiration, etc. (p. 192)

medullary cavity large, marrow-filled cavity in the diaphysis of a long bone. (p. 110)

meiosis (mi-o'sis) type of cell division in the testes and ovaries that produces sex cells, each having half the number of chromosomes as the parent cell. (p. 486)

melanin (mel'ah-nin) [Gr. *melas,* black] brown to black pigment responsible for skin and hair color. (p. 95)

melanocyte (mel'ă-no-sīt) [Gr. *melas,* black + *kytos,* cell] cells found mainly in the stratum basale of skin that produce the brown or black pigment melanin. (p. 95)

melanocyte-stimulating hormone (MSH) peptide hormone secreted by the anterior pituitary; increases melanin production by melanocytes, making the skin darker in color. (p. 256)

melanoma (mel'ă-no'mah) [Gr. *melas,* black + *oma,* tumor] a malignant tumor derived from melanocytes. (p. 103)

melatonin hormone secreted by the pineal body; may inhibit gonadotropin-releasing hormone secretion from the hypothalamus. (p. 265)

membranous labyrinth the membrane-bound set of tunnels and chambers of the inner ear. (p. 236)

memory cell lymphocytes derived from B cells, or T cells that have been exposed to an antigen; when exposed to the same antigen a second time, memory cells rapidly respond to provide immunity. (p. 365)

memory response immune response that occurs when the immune system is exposed to an antigen against which it has already had a primary response; results in the production of large amounts of antibodies and memory cells; also called a secondary response. (p. 365)

meninges (mĕ-nin'jēz) [Gr., membrane] a series of three connective tissue membranes; the dura mater, arachnoid, and pia mater; surround and protect the brain and spinal cord. (p. 202)

menopause [Gr. *mensis,* month + *pausis,* cessation] permanent cessation of the menstrual cycle. (p. 505)

menses [L. *mensis,* month] loss of blood and tissue as the endometrium of the uterus sloughs away at the end of the menstrual cycle; occurring at about 28-day intervals in the nonpregnant female of reproductive age. (p. 502)

menstrual cycle series of changes that occur in sexually mature, nonpregnant females and result in menses; specifically includes the cyclical changes that occur in the uterus and ovary. (p. 502)

merocrine (mĕr'o-krin) [Gr. *meros,* part + *krino,* to separate] gland that secretes products with no loss of cellular material; an example is water-producing sweat glands; see apocrine and holocrine. (p. 98)

mesentery (mes'en-tĕr'e) [Gr. *mesos,* middle + *enteron,* intestine] double layer of peritoneum extending from the abdominal wall to the abdominopelvic organs; conveys blood vessels and nerves to abdominopelvic organs; holds and supports abdominopelvic organs. (p. 16)

mesoderm (mez'o-derm) middle of the three germ layers of the embryo. (p. 520)

metabolic rate the total amount of energy produced and used by the body per unit of time. (p. 448)

metabolism (mĕ-tab'o-lizm) [Gr. *metabole,* change] sum of the chemical changes that occur in tissues, consisting of the breakdown of food to produce energy (catabolism) and the buildup of molecules (anabolism; which requires energy). (p. 436)

micelle (mi-sēl') [L. *micella,* small morsel] droplet of digested lipid surrounded by bile salts in the small intestine. (p. 441)

microglia (mi-krog'le-ah) [Gr. *micro,* small + *glia,* glue] small neuroglial cells that become phagocytic and mobile in response to inflammation; considered to be macrophages of the central nervous system. (p. 187)

microtubule hollow tube composed of tubulin; microtubules help provide support to the cytoplasm of the cell and are components of certain cell organelles such as cilia and flagella. (p. 55)

microvillus (mi'kro-vil'us) one of the minute projections of the cell membrane that greatly increase the surface area of the cell membrane. (p. 56)

micturition reflex contraction of the urinary bladder stimulated by stretching of the urinary bladder wall; results in emptying of the urinary bladder. (p. 469)

midbrain the superior end of the brainstem; located between the pons and diencephalon; contains fibers crossing from the brain to the spinal cord and vice versa, as well as nuclei and visual reflex centers. (p. 194)

midsagittal (mid'saj'ĭ-tal) plane running vertically through the body and dividing it into equal right and left parts. (p. 11)

mineral inorganic nutrient necessary for normal metabolic functions. (p. 438)

mineralocorticoids steroid hormones released from the adrenal cortex; act on the kidney to increase the rate of sodium reabsorption from the nephron and potassium and hydrogen ion secretion into the nephron of the kidney; an example is aldosterone. (p. 261)

mitochondria (mi′to-kon′dre-ah) [Gr. *mitos*, thread + *chandros*, granule] small, spherical, rod-shaped or thin filamentous structures in the cytoplasm that are sites of ATP production. (p. 55)

mitosis (mi-to′sis) [Gr., thread] division of the nucleus; process of cell division that results in two daughter cells with exactly the same number and type of chromosomes as the parent cell. (p. 57)

mitral valve see bicuspid valve. (p. 297)

molecule two or more atoms of the same or different type joined by a chemical bond. (p. 24)

monocyte (mon′o-sīt) a type of leukocyte that transforms to become a macrophage. (p. 278)

mononuclear phagocytic system phagocytic cells with a single nucleus, derived from monocytes; the cells either enter a tissue by chemotaxis in response to infection or tissue damage, or are positioned to intercept microorganisms entering tissues. (p. 358)

monosaccharide the basic building block from which more complex carbohydrates are constructed; for example, glucose and fructose. (p. 35)

mons pubis [L., mountain] prominence formed by a pad of fatty tissue over the symphysis pubis in the female. (p. 501)

motor unit a single motor neuron and all the skeletal muscle fibers it innervates. (p. 150)

mucin (mu′sin) secretion containing mucopolysaccharides (proteoglycans), produced by mucous gland cells. (p. 417)

mucosa (mu-ko′sah) mucous membrane consisting of the epithelium and connective tissue; in the digestive tract there is also a layer of smooth muscle. (p. 405)

mucous membrane thin sheet consisting of epithelium and connective tissue that lines cavities opening to the outside of the body; many contain mucous glands, which secrete mucous. (p. 82)

mucus (mu′kus) viscous secretion produced by and covering mucous membranes; lubricates and protects the mucous membrane, and traps foreign substances. (p. 69)

murmur an abnormal sound produced within the heart. (p. 307)

muscle fiber muscle cell. (p. 148)

muscle tissue one of the four major tissue types; consists of cells with the ability to contract; includes skeletal, cardiac, and smooth muscle. (p. 81)

muscle twitch contraction of a whole muscle in response to a stimulus that causes an action potential in one or more muscle fibers. (p. 153)

muscularis the outermost smooth muscle coat of a hollow organ. (p. 406)

muscularis mucosa the inner, thin layer of smooth muscle found in most parts of the digestive tube outside the lamina propria. (p. 405)

myelinated [Gr. *myelos*, medulla, marrow] nerve fibers having a myelin sheath; a myelin sheath is a lipoprotein envelope made by wrappings of the cell membrane of a Schwann cell or oligodendrocyte around an axon. (p. 187)

myocardium (mi′o-kar′dĭ-um) [myo- + Gr. *kordin*, heart] middle layer of the heart, consisting of cardiac muscle. (p. 301)

myofibril a fine longitudinal fibril of skeletal muscle consisting of sarcomeres, composed of thick (myosin) and thin (actin) myofilaments, placed end to end. (p. 148)

myofilament an ultramicroscopic protein thread helping to form myofibrils in skeletal muscle; thin myofilaments are composed of actin, and thick myofilaments are composed of myosin. (p. 148)

myometrium muscular wall of the uterus, composed of smooth muscle. (p. 499)

nail a thin horny plate at the ends of the fingers and toes, consisting of several layers of dead epithelial cells containing a hard keratin. (p. 99)

nasal cavity cavity divided by the nasal septum, and extending from the external nares anteriorly to the nasopharynx posteriorly; bounded inferiorly by the hard palate. (p. 119)

nasolacrimal duct duct that leads from the lacrimal sac to the nasal cavity. (p. 227)

nasopharynx (na′zo-făr′ingks) part of the pharynx that lies above the soft palate; anteriorly it opens into the nasal cavity. (p. 376)

negative feedback mechanism by which any deviation from an ideal normal value is resisted; returns a parameter to its normal range and thereby maintains homeostasis. (p. 8)

neonate [Gr. *neos*, new + L. *natalis*, relating to birth] newborn, from birth to one month. (p. 534)

nephron (nef′ron) [Gr. *nephros*, kidney] functional unit of the kidney, consisting of the renal corpuscle, the proximal convoluted tubule, the loop of Henle, and the distal convoluted tubule. (p. 458)

nerve cell a cell capable of receiving a stimulus and propagating an action potential; a neuron. (p. 183)

nerve tract bundle of axons, their sheaths, and accompanying connective tissues located in the central nervous system. (p. 187)

nervous tissue one of the four major tissue types; consists of neurons, which have the ability to conduct action potentials, and neuroglia, which are support cells. (p. 82)

neural crest cell cell derived during embryonic development from the crests of the neural folds; give rise to facial structures, pigment cells, and peripheral nerve ganglia. (p. 522)

neural tube tube formed from the neuroectoderm in the embryo by closure of the neural groove; develops into the brain and spinal cord. (p. 522)

neuroectoderm that part of the ectoderm that forms the neural tube and neural crest. (p. 522)

neuroglia (nu-rog'le-ah) [G. *neuro,* nerve + *glia,* glue] cells of the nervous system other than neurons; play a support role in the nervous system; include astrocytes, ependymal cells, microglia, oligodendrocytes, and Schwann cells; also called glia. (p. 187)

neuromuscular junction the synaptic junction between a nerve axon and a muscle fiber. (p. 150)

neuron [Gr., nerve] a nerve cell. (p. 82)

neurotransmitter [Gr. *neuro,* nerve + L. *tramitto,* to send across] a chemical that is released by a presynaptic cell into the synaptic cleft and that acts upon the postsynaptic cell to cause a response. (p. 189)

neutral solution solution with equal numbers of hydrogen and hydroxide ions; has a pH of 7. (p. 32)

neutron [L. *neuter,* neither] electrically neutral particle found in the nucleus of atoms. (p. 24)

neutrophil (nu'tro-fil) [L. *neuter,* neither + Gr. *phileo,* to love] leukocyte with granules that stain with neither basic nor acidic dyes; phagocytic white blood cell. (p. 83)

nicotinamide adenine dinucleotide (NADH) a base-containing organic molecule capable of accepting hydrogen atoms and of transferring energy from glycolysis to the electron transport chain. (p. 442)

nitroglycerin glyceryl trinitrate used as a vasodilator, especially in angina pectoris. (p. 311)

nonspecific resistance immune system response that is the same upon each exposure to an antigen; there is no ability to remember a previous exposure to a specific antigen. (p. 357)

norepinephrine (nor'ep-ĭ-nef'rin) neurotransmitter substance released from most of the postganglionic neurons of the sympathetic division; hormone released from the adrenal cortex that increases cardiac output and blood glucose levels. (p. 190)

notochord [Gr. *notor,* back + *chords,* cord] small rod of tissue lying ventral to the neural tube; characteristic of all vertebrates; in humans it becomes the nucleus pulposus of the intervertebral disks. (p. 520)

nuclease an enzyme that breaks down nucleic acids. (p. 428)

nucleic acid molecule consisting of many nucleotides chemically bound together; deoxyribonucleic acid and ribonucleic acid. (p. 39)

nucleotide basic building block of nucleic acids consisting of sugar (either ribose or deoxyribose), one of several types of organic bases, and a phosphate group. (p. 39)

nucleus [L., inside of a thing] cell organelle containing most of the genetic material of the cell; center of an atom consisting of protons and neutrons (p. 24); collection of nerve cell bodies in the central nervous system. (p. 187)

null cell a type of lymphocyte that acts against tumor- and virus-infected cells; a lymphocyte that lacks the surface molecules of B cells and T cells. (p. 367)

nutrient [L. *nutriens,* to nourish] chemical taken into the body that is used to produce energy, provide building blocks for new molecules, or function in other chemical reactions. (p. 436)

nutrition process by which nutrients are obtained and used in the body. (p. 436)

oblique section a cut made at other than a right angle to the long axis of an organ. (p. 12)

obturator [L., to occlude or stop up] any occluding structure or a foramen so occluded, as with the obturator foramen of the hip. (p. 132)

occipital (ok-sĭ'pĭ-tal) the back of the head. (p. 118)

olecranon (o-lek'ră-non) the point of the elbow. (p. 130)

olfactory relating to smell. (p. 224)

oligodendrocyte (o-lig'o-den'dro-sīt) neuroglial cells with multiple cell processes that form myelin sheaths around axons in the central nervous system. (p. 187)

omental bursa (bur'sah) the pocket-like sac inside the fold of the greater omentum. (p. 417)

omentum (o-men'tum) a fold of peritoneum extending from the stomach to another organ. (p. 417)

oocyte [Gr. *oon,* egg + *kytos,* cell] the female reproductive cell. (p. 497)

optic relating to vision. (p. 226)

optic disc the region in the posterior wall of the eye where the optic nerve exits the eye; the blind spot. (p. 230)

orbit seven skull bones that surround and protect the eye; eye socket. (p. 119)

organ [Gr. *organon,* tool] part of the body composed of two or more tissue types and performing one or more specific functions. (p. 8)

organ of Corti specialized region of the cochlear duct consisting of hair cells; produces action potentials in response to sound waves. (p. 236)

organ system group of organs classified as a unit because of a common function or set of functions. (p. 8)

organelle specialized part of a cell performing one or more specific functions. (p. 3)

organic molecules that contain a carbon atom (carbon dioxide is an exception); originally defined as molecules extracted from living organisms; the original definition became obsolete when it became possible to manufacture these molecules in the laboratory. (p. 34)

organism any living thing considered as a whole, whether composed of one cell or many. (p. 8)

organogenesis the formation of organs during embryonic development. (p. 524)

orgasm [Gr. *orgao*, to swell, be excited] climax of the sexual act, associated with a pleasurable sensation. (p. 495)

origin the less movable attachment point of a muscle. (p. 157)

oropharynx (o'ro-făr'ingks) portion of the pharynx that lies posterior to the mouth; it is continuous above with the nasopharynx and below with the laryngopharynx. (p. 379)

osmosis (os-mo'sis) [Gr. *osmos*, thrusting or an impulsion] diffusion of solvent (water) through a selectively permeable membrane from a less concentrated solution to a more concentrated solution. (p. 47)

osmotic pressure force required to prevent the movement of water across a selectively permeable membrane. (p. 47)

ossification (os'ĭ-fĭ-ka'shun) [L. *os*, bone + *facio*, to make] bone formation. (p. 114)

osteoblast (os'te-o-blast) a cell that makes bone. (p. 114)

osteoclast (os'te-o-klast) a cell that digests and removes bone. (p. 114)

osteocyte (os'te-o-sīt) [Gr. *osteon*, bone + *kytos*, cell] mature bone cell surrounded by bone matrix. (p. 80)

otolith [Gr. *ous*, ear + *lithos*, stone] small protein and calcium carbonate weights in the maculae of the vestibule. (p. 240)

ovary one of two female reproductive glands located in the pelvic cavity; produces the oocyte, estrogen, and progesterone. (p. 495)

ovulation release of an oocyte from the mature follicle. (p. 499)

oxygen debt the oxygen required to produce enough ATP to break down lactic acid produced during anaerobic respiration and to replenish creatine phosphate stores. (p. 154)

oxytocin (ok-sĭ-to'sin) peptide hormone secreted by the posterior pituitary that increases uterine contraction and stimulates milk ejection from the mammary glands. (p. 256)

pancreatic duct the duct of the pancreas; it joins the common bile duct to empty into the duodenum. (p. 411)

pancreatic islet cellular mass in the tissue of the pancreas; composed of different cell types that comprise the endocrine portion of the pancreas and are the source of insulin and glucagon. (p. 262)

papilla (pă-pil'ah) [L., nipple] a small nipplelike process; projection of the dermis, containing blood vessels and nerves, into the epidermis; projections on the surface of the tongue. (p. 93, 225)

papillary muscle (pap'ĭ-lĕr'e) a raised area of cardiac muscle in the ventricle to which the chordae tendineae attach. (p. 297)

paranasal sinus air-filled cavity within certain skull bones that connects to the nasal cavity; the four sets of paranasal sinuses are the frontal, maxillary, sphenoid, and ethmoid. (p. 120)

parasympathetic subdivision of the autonomic nervous system with preganglionic neurons in the brainstem and sacral portion of the spinal cord; involved in involuntary (vegetative) functions such as digestion, defecation, and urination. (p. 214)

parathyroid gland (păr-ă-thi'royd) one of four glandular masses imbedded in the posterior surface of the thyroid gland; secretes parathyroid hormone. (p. 258)

parathyroid hormone peptide hormone produced by the parathyroid gland that increases bone breakdown and blood calcium levels. (p. 258)

parietal (pă-ri'ĕ-tal) [L. *paries*, wall] relating to the wall of any cavity; parietal serous membranes are in contact with the walls of cavities. (p. 16)

parietal peritoneum (pĕr'it-o-ne'um) [L., wall] that portion of the serous membranes of the abdominal cavity lining the inner surface of the body wall. (p. 417)

parotid gland (pă-rot'id) the largest of the salivary glands; located anterior and inferior to the ear. (p. 408)

partial pressure pressure exerted by a single gas in a mixture of gases. (p. 389)

partial-thickness burn burn that damages only the epidermis (first-degree burn) or the epidermis and part of the dermis (second-degree burn). (p. 101)

parturition [L. *parturio*, to be in labor] childbirth; the delivery of an infant at the end of pregnancy. (p. 531)

patella [L. *patina*, shallow disk] kneecap. (p. 133)

pelvic cavity space completely surrounded by the pelvic bones. (p. 14)

pepsin [Gr. *pepsis*, digestion] principal digestive enzyme produced by the stomach; digests proteins into smaller peptide chains. (p. 420)

peptidase an enzyme capable of breaking peptide chains into smaller chains and amino acids. (p. 441)

pericardial cavity (pĕr-ĭ-kar'de-al) space between the visceral and parietal pericardium, filled with pericardial fluid; a cavity that surrounds the heart. (p. 16)

pericardial fluid the serous fluid found within the pericardial cavity. (p. 297)

pericardium (pĕr-ĭ-kar-dĭ-um) [Gr. *pericardion*, the membrane around the heart] the membrane consisting of the epicardium and parietal pericardium (of the serous layers) and the outer fibrous pericardium that forms the pericardial sac. (p. 295)

perilymph [Gr. *peri-*, around, + *lympha*, clear fluid] fluid contained between the bony labyrinth and the membranous labyrinth of the inner ear. (p. 236)

perineum (per'ĭ-ne'um) area inferior to the pelvic diaphragm between the thighs; extends from the coccyx to the pubis. (p. 166)

periodontal (pĕr'e-o-don'tal) [Gr. *peri*, around + *odous*, tooth] referring to structures surrounding the tooth, primarily in the alveolus. (p. 408)

periosteum (per'e-os'te-um) [Gr. *peri*, around + *osteon*, bone] thick, double-layered connective tissue sheath covering the entire surface of a bone except the articular surface, which is covered with cartilage. (p. 110)

peripheral circulation blood flow through all blood vessels that carry blood away from the heart (arteries), the capillaries, and all vessels that carry blood back to the heart (veins); consists of the pulmonary circulation and the systemic circulation; includes all blood flow except that through the heart tissue itself. (p. 320)

peripheral nervous system the part of the nervous system not surrounded by the skull or vertebral column; consisting of nerves and ganglia. (p. 183)

peristalsis (pĕr'ĭ-stal'sis) a wave of relaxation followed by a wave of contraction moving along a tube; propels food along the digestive tube. (p. 420)

peritoneal cavity (pĕr'ĭ-to-ne'al) space between the visceral and parietal peritoneum, filled with peritoneal fluid; cavity that surrounds many abdominopelvic organs. (p. 16)

peritubular capillary the capillary network located in the cortex of the kidney; associated with the distal and proximal convoluted tubules. (p. 461)

Peyer's patches lymph nodes found in the distal half of the small intestine and in the appendix. (p. 413)

pH scale a measure of the hydrogen ion concentration of a solution; the scale extends from 0 to 14 with a pH of 7 being neutral, a pH of less than seven acidic, and a pH of greater than 7 basic. (p. 32)

phagocytosis (fag'o-si-to'sis) [Gr. *phagein*, to eat + *kytos*, cell + *osis* condition] process of ingestion and digestion by cells of substances such as other cells, bacteria, cell debris, and foreign particles. (p. 50)

pharynx (făr'ingks) [Gr. *pharynx*, throat, the joint openings of the digestive tract and the windpipe] upper expanded portion of the digestive tube between the esophagus below and the oral and nasal cavities above and in front. (p. 376)

phenotype [Gr. *phaino*, to display + *typos*, model] characteristic observed in the individual resulting from expression of the genotype. (p. 541)

phospholipid lipid with phosphorus resulting in a molecule with a polar and a nonpolar end; main component of cell membranes. (p. 45)

physiology [Gr. *physis*, nature + *logos*, study] scientific discipline that deals with the processes or functions of living things. (p. 2)

pia mater (pe'ah) [L., affectionate mother] the innermost meningeal layer; tightly attached to the brain and spinal cord. (p. 203)

pineal body (pi'ne-al) [L. *pineus*, pine cone-shaped] a small endocrine gland attached to the dorsal surface of the diencephalon; may influence the onset of puberty and may play a role in some long-term cycles. (p. 194)

pinocytosis (pin'o-si-to'sis) [Gr. *pineo*, to drink + *kytos*, cell + *osis*, condition] cell drinking; uptake of liquid by a cell. (p. 50)

pituitary (pit-u'ĭ-tĕr-e) endocrine gland attached to the hypothalamus by the infundibulum; secretes hormones that influence the function of several other glands and tissues. (p. 254)

pituitary dwarf an individual of short stature, of relatively normal proportion, as the result of insufficient growth hormone secreted from the anterior pituitary. (p. 255)

placenta structure derived from embryonic and maternal tissues by which the embryo and fetus are attached to the uterus. (p. 518)

plasma fluid portion of blood; blood minus the formed elements. (p. 44)

plasma membrane (plaz'mah) cell membrane; outermost component of the cell, surrounding and binding the rest of the cell contents. (p. 45)

platelet plug accumulation of platelets that stick to connective tissue and to each other and prevent blood loss from damaged blood vessels. (p. 278)

pleural cavity (ploo'ral) space between the visceral and parietal pleura, filled with pleural fluid; a cavity that surrounds the lungs. (p. 16)

plexus (plek'sus) [L., a braid] an intertwining of nerves or blood vessels. (p. 210)

podocyte (pod'o-sīt) [Fr. *pous, podos,* foot + *kytos,* a hollow (cell)] epithelial cell of Bowman's capsule attached to the outer surface of the glomerular capillary basement membrane; forms part of the filtration membrane. (p. 461)

polar covalent bond chemical bond in which electrons are shared unequally between two atoms. (p. 26)

polycythemia (pol'ĭ-si-the'me-ah) [Gr. *polys,* many + *kytos,* cell] increase in erythrocyte numbers above the normal value. (p. 285)

polysaccharide many monosaccharides chemically bound together, such as glycogen and starch. (p. 35)

pons [L., bridge] the part of the brainstem between the medulla oblongata and midbrain; contains relays between the cerebrum and cerebellum, as well as ascending and descending tracts. (p. 192)

portal system system of vessels in which blood, after passing through one capillary bed, is conveyed through a second capillary network. (p. 333)

positive feedback mechanism by which any deviation from an ideal normal value is made greater. (p. 9)

posterior [L. *posterus,* following] that which follows; in humans, toward the back. (p. 11)

posterior horn the posterior extension of spinal cord gray matter; contains nerve cell bodies that receive input from primary sensory neurons and relay that input to the brain; also called the dorsal horn. (p. 200)

posterior pituitary the posterior portion of the pituitary gland, which consists of processes of nerve cells that have their cell bodies located in the hypothalamus; secretes oxytocin and antidiuretic hormone. (p. 254)

postganglionic autonomic neurons whose cell bodies are located outside the central nervous system and that receive synaptic stimulation from preganglionic autonomic neurons. (p. 212)

preganglionic autonomic neurons whose cell bodies are located in the central nervous system and that synapse with postganglionic neurons. (p. 212)

premenstrual syndrome (PMS) in some women of reproductive age, the regular monthly experience of physiological and emotional distress, usually during the few days preceding menses, typically involving fatigue, edema, irritability, tension, anxiety, and depression. (p. 505)

prenatal period [L. *prae,* before + *natalis,* relating to birth] the period before birth. (p. 516)

prepuce (pre'pus) in the male, a free fold of skin that almost completely covers the glans penis; the foreskin; in the female, a fold of skin that covers the clitoris. (p. 491)

primary response immune response that occurs as a result of the first exposure to an antigen; results in the production of antibodies and memory cells. (p. 365)

primary union tissue repair in which the edges of a wound are close together, and tissue repair is rapid with a minimum of scar tissue formation. (p. 84)

prime mover muscle that plays the principal role in accomplishing a movement. (p. 157)

primitive streak a shallow groove in the ectodermal surface of the embryonic disk; cells migrating through the streak become mesoderm. (p. 520)

process projection on a bone. (p. 118)

product substance produced in a chemical reaction. (p. 27)

progesterone steroid hormone secreted by the ovaries; necessary for uterine and mammary gland development and function. (p. 265)

prolactin (pro-lak'tin) hormone of the anterior pituitary that stimulates the secretion of milk. (p. 256)

pronation (pro-na'shun) [L. *pronare,* to bend forward] rotation of the forearm so that the anterior surface is down (prone). (p. 142)

pronator a muscle that pronates, that is, turns the hand palm down or posterior. (p. 169)

proprioception (pro'pre-o-sep'shun) [L. *proprius,* one's own + *capio,* to take] information about the position of the body and its various parts. (p. 199)

prostaglandins (pros'tă-glan'dinz) class of physiologically active substances present in many tissues; effects include vasodilation, stimulation and contraction of uterine smooth muscle, and promotion of inflammation and pain. (p. 265)

prostate gland (pros'tāt) [Gr. *prostates,* one standing before] gland that surrounds the beginning of the urethra in the male. The secretion of the gland is a milky fluid that is discharged into the urethra as part of the semen. (p. 492)

protein [Gr. *proteios,* primary] large molecule consisting of long sequences of amino acids (polypeptides) linked by peptide bonds. (p. 36)

proteoglycan (pro'te-o-gli'kan) macromolecule consisting of numerous polysaccharides attached to a common protein core. (p. 110)

proteolytic an enzyme capable of digesting proteins or polypeptides. (p. 441)

proton [Gr. *protos,* first] positively charged particle found in the nucleus of atoms. (p. 24)

provitamin substance that can be converted into a vitamin. (p. 436)

proximal [L. *proximus*, nearest] closer to the point of attachment to the body than another structure. (p. 11)

proximal convoluted tubule convoluted portion of the nephron that extends from Bowman's capsule to the descending limb of Henle's loop. (p. 458)

pterygoid (těr'ĭ-goyd) [Gr. *pteryx*, wing] wing-shaped structure; two of the muscles of mastication, attached to wing-shaped bony projections. (p. 161)

puberty [L. *pubertas*, grown up] series of events that transform a child into a sexually mature adult; involves an increase in the secretion of all reproductive hormones. (p. 494)

pudendal cleft cleft between the labia majora. (p. 501)

pulmonary capacity the sum of two or more pulmonary volumes. (p. 386)

pulmonary circulation blood flow through the system of blood vessels that carry blood from the right ventricle of the heart to the lungs and back from the lungs to the left atrium. (p. 320)

pulmonary semilunar valve the semilunar valve found at the base of the pulmonary trunk where it exits from the right ventricle. (p. 297)

pulmonary trunk (pul'mo-něr-e) large elastic artery that carries blood from the right ventricle of the heart to the right and left pulmonary arteries. (p. 297)

pulmonary volumes lung volumes, measured by spirometry; deviations from normal values can be used to diagnose certain lung diseases; the pulmonary volumes are the tidal volume, inspiratory reserve volume, expiratory reserve volume, and residual volume. (p. 386)

pulp [L. *pulpa*, flesh] the soft tissue inside a tooth, consisting of connective tissue, blood vessels, nerves, and lymphatics. (p. 408)

pulse pressure the difference between systolic and diastolic pressure. (p. 335)

pupil [L. *pupa*, a doll, because you can see a little reflection or doll in the pupil of another person's eye] opening in the iris of the eye through which light enters. (p. 228)

pyloric sphincter [Gr., gatekeeper] a thickened ring of smooth muscle at the distal end of the stomach. (p. 409)

pyrogen (pi'ro-jen) chemical released by microorganisms, neutrophils, monocytes, and other cells that stimulates fever production by acting on the hypothalamus. (p. 360)

pyruvic acid three-carbon end product of glycolysis; two pyruvic acid molecules are produced from each glucose molecule. (p. 442)

reactant substance taking part in a chemical reaction. (p. 27)

receptor a molecule that is located in the membrane or cytoplasm of cells of a target tissue to which a hormone binds; each receptor binds to a specific type of hormone, neurotransmitter, or other substance. (p. 249)

recessive in genetics, a gene that may not be expressed phenotypically because of the expression of a contrasting dominant gene. (p. 541)

rectum [L. *rectus*, straight] the last, straight part of the large intestine; between the colon and the anal canal. (p. 416)

rectus straight. (p. 157)

regeneration tissue repair in which the damaged cells are replaced by cells of the same type as those damaged. (p. 84)

releasing hormone a hormone released from neurons in the hypothalamus of the brain that travels through the hypothalamic-pituitary portal system to the anterior pituitary, where it causes the release of a specific anterior pituitary hormone. (p. 255)

renal capsule the connective tissue capsule that surrounds each kidney. (p. 457)

renal corpuscle the structure composed of a Bowman's capsule and its glomerulus. (p. 458)

renal erythropoietic factor an enzyme produced by the kidney in response to decreased blood oxygen levels; converts inactive erythropoietin into active erythropoietin. (p. 276)

renal fat pad the layer of adipose tissue that surrounds each kidney. (p. 457)

renal pyramid cone-shaped structure that extends from the renal sinus, where the apex is located, into the cortex of the kidney, where the base is located. (p. 458)

renal sinus the cavity central to the medulla of the kidney that is filled with adipose tissue and contains the renal pelvis. (p. 458)

renin (re'nin) enzyme secreted by the kidney that converts a plasma protein (angiotensinogen) to angiotensin I, which is subsequently converted to angiotensin II. (p. 262)

replacement tissue repair in which the damaged cells are replaced by cells of a type different from those damaged. (p. 84)

respiratory center nerve cells in the medulla oblongata and pons of the brain that control inspiration and expiration. (p. 393)

respiratory membrane membrane in the lungs across which gas exchange occurs with blood; consists of a thin layer of fluid, the alveolar epithelium, a basement membrane of the alveolar epithelium, interstitial space, the basement membrane of the capillary endothelium, and the capillary endothelium. (p. 388)

resting membrane potential the charge difference across the membrane of a resting cell (that is, a cell that has not been stimulated to produce an action potential). (p. 148)

reticular formation (rĕ-tik'u-lar) [L. *rete,* net] a loose network of nerve cell bodies scattered throughout the brainstem; involved in the regulation of cycles such as the sleep-wake cycle. (p. 194)

retina [L. *rete,* a net] the inner, light-sensitive tunic of the eye; nervous tunic. (p. 229)

retinaculum (ret'ĭ-nak'u-lum) [L., band, halter, to hold back] dense, regular connective tissue sheath holding down the tendons at the wrist, ankle, or other sites. (p. 170)

retroperitoneal (rĕ'tro-pĕr'ĭ-to-ne'al) located behind the parietal peritoneum; includes the kidneys, adrenal glands, pancreas, portions of the intestines, and urinary bladder. (p. 16)

reversible reaction chemical reaction in which the reaction can proceed from reactants to products, or from products to reactants; the amount of reactants relative to products is constant at equilibrium. (p. 31)

ribonucleic acid (ri'bo-nu-kle'ik) type of nucleic acid containing the sugar ribose; involved in protein synthesis; RNA. (p. 39)

ribosomal RNA (rRNA) RNA that is associated with certain proteins to form ribosomes. (p. 57)

ribosome small, spherical, cytoplasmic organelle where protein synthesis occurs. (p. 53)

right lymphatic duct lymphatic duct that empties into the right subclavian vein; drains the right side of the head and neck, the right upper thorax, and the right upper limb. (p. 353)

RNA see ribonucleic acid. (p. 39)

rods photoreceptor cells in the retina of the eye with rod-shaped photoreceptive processes; very light sensitive cells that are important in dim light. (p. 229)

rotator cuff four deep muscles that attach the humerus to the scapula. (p. 169)

rugae (ru'ge) ridges or folds in the mucous membrane of the stomach. (p. 409)

sagittal (saj'ĭ-tal) [L. *sagitta,* the flight of an arrow] plane running vertically through the body and dividing it into right and left parts. (p. 11)

saliva a fluid containing enzymes and mucus, produced by the salivary glands and released into the oral cavity. (p. 408)

salivary gland gland opening into the mouth and producing saliva. (p. 408)

salt molecule consisting of a positively charged ion other than hydrogen, and negatively charged ion other than hydroxide. (p. 33)

sarcomere (sar'ko-mēr) [Gr. *sarco,* flesh means muscle + *meros,* part] the part of a myofibril formed of actin and myosin myofilaments, extending from Z line to Z line; the structural and functional unit of a muscle. (p. 148)

scapula (skap'u-lah) the shoulder blade. (p. 127)

Schwann cell neuroglial cell forming myelin sheaths around axons in the peripheral nervous system. (p. 187)

sclera (skler'ah) [L. *skleros,* hard] the dense, white, opaque posterior portion of the fibrous tunic of the eye; white of the eye. (p. 228)

scrotum musculocutaneous sac containing the testes. (p. 486)

sebaceous gland (sĕ-ba'shus) [L. *sebum,* tallow] gland of the skin that produces sebum; usually associated with a hair follicle. (p. 98)

sebum (se'bum) [L., tallow] oily, white, fatty substance produced by the sebaceous glands; oils hair and the surface of the skin. (p. 98)

secondary response see memory response. (p. 365)

secondary union tissue repair in which the edges of a wound are far apart, and tissue repair takes a relatively long time, often with substantial scar tissue formation. (p. 84)

sella turcica (sel'ah tur'sĭ-kah) [L., saddle + Turkish] the saddle-shaped depression in the inner surface of the skull where the pituitary is located. (p. 122)

semen [L., seed] penile ejaculate; thick, yellowish-white, viscous fluid containing spermatozoa and secretions of the testes, seminal vesicles, prostate gland, and bulbourethral glands. (p. 492)

semicircular canal one of three canals in each temporal bone; involved in the detection of motion. (p. 240)

semilunar valve one of two valves in the heart composed of three semilunar-shaped cusps that prevent flow of blood into the ventricles following ejection; located at the beginning of the aorta and pulmonary trunk. (p. 297)

seminal vesicle one of two glandular structures that empty into the ejaculatory ducts; its secretion is one of the components of semen. (p. 489)

seminiferous tubule (sem'ĭ-nif'er-us) tubule in the testis in which spermatozoa develop. (p. 486)

serosa the smooth, outermost covering of an organ where it faces a cavity and is not surrounded by connective tissue. (p. 406)

serous membrane (sēr'us) thin sheet consisting of epithelium and connective tissue that lines cavities not opening to the outside of the body; does not contain glands but does secrete serous fluid. (p. 16)

Sertoli cell (ser-to'le) cell in the wall of the seminiferous tubules to which spermatogonia and spermatids are attached. (p. 488)

serum fluid portion of blood after the removal of fibrin and formed elements. (p. 272)

sex chromosome a chromosome other than an autosome; responsible for sex determination. (p. 57)

sinoatrial (SA) node (si'no-a'tre-al) mass of specialized cardiac muscle fibers located in the right atrium near the opening of the superior vena cava that acts as the "pacemaker" of the cardiac conduction system. (p. 303)

sliding filament mechanism mechanism by which actin and myosin myofilaments slide over one another during muscle contraction. (p. 152)

solute [L. *solutus,* dissolved] dissolved substance in a solution. (p. 46)

solution homogeneous mixture formed when a solute dissolves in a solvent (liquid). (p. 46)

solvent [L. *solvens,* to dissolve] liquid that holds another substance in solution. (p. 46)

somatomotor (so-mă'to-mo'tor) [Gr. *soma,* body] motor (efferent) neurons of the peripheral nervous system that innervate skeletal muscle. (p. 212)

somesthetic (so'mes-thet'ik) [Gr. *soma,* body + *aisthesis,* sensation] consciously perceived body sensations. (p. 196)

somesthetic cortex (so'mes-thet'ik) [Gr. *soma,* body + *aisthesis,* sensation] that portion of the cerebral cortex involved with the conscious perception and localization of general body sensations. (p. 196)

specific immunity immune system response in which there is an ability to recognize, remember, and destroy a specific antigen. (p. 357)

sperm male reproductive cell. (p. 489)

spermatocyte cell arising from a spermatogonium and destined to give rise to spermatozoa. (p. 488)

spermatogenesis formation and development of spermatozoa. (p. 488)

spermatozoon, pl. spermatazoa [Gr. *sperma,* seed + *zoon,* animal] male gamete or sex cell, composed of a head, midpiece, and tail; spermatozoa contain the genetic information transmitted by the male. (p. 489)

spinal nerve peripheral nerve exiting from the spinal cord. (p. 200)

spirometer [L. *spiro,* to breath + Gr. *metron,* measure] gasometer used for measuring the volume of respiratory gases; usually consisting of a counterbalanced cylindrical bell sealed by dipping into a circular trough of water. (p. 386)

spirometry (spĭ-rom'ĕ-tre) process of making pulmonary measurements with a spirometer. (p. 386)

spleen large lymphatic organ in the upper part of the abdominal cavity on the left side, between the stomach and diaphragm, composed of white and red pulp; responds to foreign substances in the blood, destroys worn out erythrocytes, and is a reservoir for blood. (p. 356)

squamous (skwa'mus) [L. *squama,* a scale] scale-like, flat. (p. 68)

stapes (sta'pēz) [L., stirrup] the third of the three middle ear bones; attached to the membrane of the oval window; the stirrup. (p. 236)

Starling's law of the heart force of contraction of cardiac muscle is a function of the length of its muscle fibers at the end of diastole; the greater the degree of filling of the heart (the greater the venous return), the greater the force of contraction of the cardiac muscle. (p. 308)

stenosed valve a valve that has its opening narrowed or partially closed. (p. 307)

sternum [L. *sternon,* chest] breastbone. (p. 126)

steroids large family of lipids, including some hormones, vitamins, and cholesterol. (p. 250)

stethoscope (steth'o-skōp) [Gr. *stetho-* chest + *skopeo,* to view] an instrument originally devised for aid in hearing the respiratory and cardiac sounds in the chest and now used in hearing other sounds in the body as well. (p. 307)

strabismus (stră-biz'mus) [Gr. *strabismos,* a squinting] lack of parallelism of the visual axes of the eyes. (p. 233)

stratum [L., bed cover, layer] layer of tissue. (p. 94)

stroke volume the volume of blood ejected from either the right or left ventricle during each heart beat. (p. 308)

styloid (sti'loyd) [G. *stylos,* pillar] a slender, pencil-shaped process. (p. 122)

subarachnoid space the fluid-filled space below the arachnoid layer covering the brain and spinal cord; contains cerebrospinal fluid. (p. 203)

subcutaneous (sub'ku-ta'ne-us) [L. *sub,* under + *cutis,* skin] under the skin; same tissue as the hypodermis. (p. 93)

sublingual glands the paired salivary glands located below the tongue. (p. 408)

submandibular glands the paired salivary glands located below the mandible. (p. 408)

submucosa the layer of connective tissue deep to the mucous membrane. (p. 405)

sulcus (sul'kus) [L., ditch] a groove on the surface of the brain between gyri. (p. 195)

superficial [L. *superficialis,* surface] toward or on the surface. (p. 11)

superior [L., higher] up, or higher, with reference to the anatomical position. (p. 11)

supination (su'pin-a'shun) [L. *supino*, to place something on its back] rotation of the forearm so that the anterior surface is up (supine). (p. 142)

suture (su'chur) [L. *surtura*, a seam] fibrous joint between flat bones of the skull. (p. 136)

sweat gland usually means a secretory organ that produces a watery secretion called sweat that is released onto the surface of the skin; some sweat glands, however, produce an organic secretion. (p. 98)

sympathetic subdivision of the autonomic nervous system with preganglionic nerve cell bodies located in the thoracic and lumbar regions of the spinal cord; generally involved in preparing the body for immediate physical activity. (p. 214)

synapse [Gr. *syn*, together + *haptein*, to clasp] junction between a nerve cell and some other cell; in a chemical synapse, chemicals are released from the nerve cell as a result of an action potential in the nerve cell, the chemicals cross the cleft between the cells, and they cause some response in the postsynaptic cell. (p. 189)

synergist (sin'er-jist) a muscle that works with another muscle to cause a movement. (p. 157)

synovial fluid (sĭ-no-ve-al) [Gr. *syn*, coming together + *ovia*, resembling egg albumin] a somewhat viscous substance serving as a lubricant in movable joints, tendon sheaths, and bursae. (p. 137)

synthesis reaction the combination of atoms, ions, or molecules to form a new and larger molecule. (p. 27)

systemic circulation blood flow through the system of blood vessels that carry blood from the left ventricle of the heart to the tissues of the body and back from the body to the right atrium. (p. 320)

systemic inflammation inflammation that occurs in many areas of the body; in addition to the symptoms of local inflammation, can include increased neutrophil numbers in the blood, fever, and shock. (p. 359)

systole (sis'to-le) [Gr. *systole*, a contracting] contraction of the heart chambers during which blood leaves the chambers; usually refers to ventricular contraction. (p. 305)

target tissue tissue upon which a hormone acts. (p. 249)

tarsal (tar'sal) [Gr. *tarsos*, sole of foot] bones of the instep of the foot. (p. 134)

taste bud sensory structure, mostly on the tongue, that functions as a taste receptor. (p. 225)

tectorial membrane (tek-tor'e-al) membrane attached to the spiral lamina and extending over the hair cells; hairs of the hair cells have their tips embedded in the membrane. (p. 236)

temporal [L. *tempus*, time] indicating the temple; the temple of the head is so named because it is there that the hair first begins turning white, indicating the passage of time. (p. 118)

tendinous inscriptions (in-scrip'shunz) bands of connective tissue crossing the rectus abdominus muscle and attaching it to adjacent connective tissue. (p. 166)

tendon a tough connective tissue band connecting a muscle to bone. (p. 157)

teniae coli (te'ne-ĕ ko'le) [Gr. *tainia*, band, tapeworm + *coli*, colon] the incomplete, longitudinal smooth muscle layer of the colon. (p. 415)

tensor fascia latae muscle that tenses the fibrous band on the outside of the thigh and abducts the thigh. (p. 173)

testis, pl. testes one of two male reproductive glands located in the scrotum; produces testosterone and spermatozoa. (p. 486)

testosterone steroid hormone secreted primarily by the testes; aids in spermatogenesis, controls maintenance and development of male reproductive organs and secondary sexual characteristics, and influences sexual behavior. (p. 494)

tetany a condition in muscle contraction where there is no relaxation between muscle twitches. (p. 153)

thalamus [Gr., a bedroom] a large mass of gray matter making up the bulk of the diencephalon; involved in the relay of sensory input to the cerebral cortex. (p. 194)

thoracic cavity space bounded by the neck, the thoracic wall, and the diaphragm. (p. 14)

thoracic duct largest lymph vessel in the body; drains the left side of the head and neck, the left upper thorax, the left upper limb, and the inferior half of the body into the left subclavian vein. (p. 353)

thrombocyte (throm'bo-sīt) [Gr. *thrombos*, clot + *kytos*, cell] a cell fragment involved in platelet plug and clot formation; also called a platelet. (p. 273)

thrombus (throm'bus) [Gr. *thrombos*, a clot] a clot formed within the cardiovascular system. (p. 280)

thymosin (thi'mo-sin) a hormone secreted from the thymus gland that helps activate the immune system. (p. 265)

thymus (thi'mus) [Gr. *thymos*, sweetbread] bilobed lymphatic organ located in the inferior neck and superior mediastinum; involved with the maturation of T cells. (p. 356)

thyroid cartilage largest laryngeal cartilage; forms the laryngeal prominence or Adam's apple. (p. 379)

thyroid gland (thi′royd) [Gr. *thyreoeides*, shield] endocrine gland located inferior to the larynx and consisting of two lobes connected by a narrow band; secretes the thyroid hormones. (p. 256)

thyroid hormones the hormones secreted by the thyroid gland; especially those such as thyroxine that contain iodine and function to regulate metabolism and maturation of tissues. (p. 256)

thyroid-stimulating hormone (TSH) hormone released from the hypothalamus that stimulates thyroid hormone secretion from the thyroid gland. (p. 255)

thyroxine (thi-rok′sin) one of the thyroid hormones that contain iodine atoms. (p. 256)

tissue [L. *texo*, to weave] a collection of cells with similar structure and function and the substances between the cells. (p. 8)

tissue repair substitution of viable cells for damaged or dead cells by regeneration or replacement. (p. 84)

tonsil any collection of lymphoid tissue; usually refers to large collections of lymphoid tissue beneath mucous membranes of the oral cavity and pharynx; lingual, pharyngeal, and palatine tonsils. (p. 353)

trabecula (tră-bek′u-lah) [L. *trabs*, beam] a beam or plate of cancellous bone or other tissue. (p. 113)

trachea (tra′ke-ah) [Gr. *tracheia arteria*, rough artery] air tube extending from the larynx into the thorax, where it divides to form bronchi; has 16 to 20 C-shaped rings of cartilage in its walls. (p. 380)

transfer RNA (tRNA) RNA that attaches to individual amino acids and transports them to the ribosomes where they are connected to form a protein polypeptide chain. (p. 57)

transverse plane separating the body into superior and inferior parts: a cross section. (p. 11)

transverse section a cut made at right angles to the long axis of an organ. (p. 12)

trapezius back muscle, shaped like a trapezium (a four-sided geometric figure where no two sides are parallel), that rotates the scapula. (p. 167)

triceps brachii a three-headed muscle in the posterior arm that extends the forearm. (p. 169)

tricuspid valve valve closing the opening between the right atrium and right ventricle of the heart. (p. 297)

trochanter (tro′kan-ter) [Gr., a runner] one of the large tubercles of the proximal femur. (p. 133)

trophoblast (tro′fo-blast) [Gr. *trophe*, nourishment + *blastos*, germ] the outer portion of the blastocyst; invades the uterus and becomes the embryonic portion of the placenta. (p. 518)

trypsin an enzyme released from the pancreas that digests proteins. (p. 428)

tubercle lump on a bone. (p. 118)

tuberosity lump on a bone, usually larger than a tubercle. (p. 118)

tubular reabsorption movement of materials, by means of diffusion or active transport, from the filtrate within a nephron to the blood. (p. 463)

tubular secretion movement of materials, by means of active transport, from the blood into the filtrate of a nephron. (p. 463)

tunic [L., coat] one of the three enveloping layers of the wall of the eye; the three tunics are the fibrous, vascular, and nervous tunics. (p. 228)

tunica adventitia outermost fibrous coat of a vessel or an organ that is derived from the surrounding connective tissue. (p. 321)

tunica intima innermost layer of a blood or lymphatic vessel; consists of endothelium and a small amount of connective tissue. (p. 321)

tunica media middle, usually muscular, coat of an artery or other tubular structure. (p. 321)

tympanic membrane (tim-pan′ik) cellular membrane that covers the inner opening of the external auditory meatus and separates the middle and external ears; vibrates in response to sound waves; the eardrum. (p. 235)

umbilical vein vein in the umbilical cord of the fetus by which the fetus receives nourishment from the maternal system; becomes the round ligament of the liver in the adult. (p. 535)

upper motor neuron a motor neuron located in the cerebral cortex and synapsing with a lower motor neuron in the brainstem or spinal cord. (p. 202)

ureter (ur-re′ter) [Gr. *oureter*, urinary canal] tube conducting urine from the kidney to the urinary bladder. (p. 458)

uterus hollow muscular organ in which the fertilized ovum develops into a fetus. (p. 495)

uvula (u′vu-lah) [L. *uva*, grape] small grape-like appendage at the posterior margin of the soft palate. (p. 376)

vaccine preparation of killed microorganisms, altered microorganisms, or derivatives of microorganisms intended to produce immunity; usually administered by injection, but sometimes ingestion is preferred. (p. 367)

vagina [L., sheath] genital canal in the female, extending from the uterus to the vulva. (p. 495)

variable region part of an antibody that combines with an antigen; responsible for the specificity of the antibody. (p. 363)

vasomotor tone the relatively constant frequency of sympathetic impulses that keep blood vessels partially constricted in the periphery. (p. 338)

vein blood vessel that carries blood toward the heart. (p. 320)

ventral [L. *ventr,* belly] in humans, synonymous with anterior. (p. 11)

ventral root motor (efferent) root of a spinal nerve. (p. 200)

ventricle (ven'trĭ-kul) [L. *venter,* belly] a cavity; in the brain, one of four cavities filled with cerebrospinal fluid; one of two chambers of the heart that pump blood into arteries; there is a left and a right ventricle. (p. 295)

villus (vil'us) [L., shaggy hair] projection of the mucous membrane in the small intestine that increases surface area. (p. 411)

visceral (vis'er-al) [L. *viscus,* the soft parts, internal organs] relating to the internal organs. (p. 16)

visceral peritoneum (pĕr'it-o-ne'um) [L., organ] that part of the serous membrane in the abdominal cavity covering the surface of some abdominal organs. (p. 417)

vitamin (vi'tah-min) [L. *vita,* life + *amine,* from ammonia] one of a group of organic substances, present in minute amounts in natural foods, that are essential to normal metabolism; insufficient amounts in the diet may cause deficiency diseases. (p. 436)

vitamin D fat-soluble vitamin produced from precursor molecule in skin exposed to ultraviolet light; increases calcium and phosphate uptake in the intestine. (p. 101)

X-linked a trait caused by a gene on the X chromosome. (p. 543)

yolk sac highly vascular endodermal layer surrounding the yolk of an embryo. (p. 520)

zygomatic (zi-go-mat'ik) [Gr. *zygon,* yoke] the cheek bone; a bony arch created by the junction of the zygomatic and temporal bones. (p. 118)

zygomaticus (zi'go-mat'ĭ-kus) a muscle originating on the zygomatic bone and inserting onto the corner of the mouth, involved in smiling. (p. 161)

zygote (zi'gōt) [Gr. *zygotos,* yoked] the single-celled, diploid cell product of fertilization, resulting from union of the sperm and oocyte. (p. 516)

CREDITS

CHAPTER 1

1-1 Thomas D. Sims; **1-2, 1-3** Trent Stephens; **1-4, 1-5*A*, 1-6** Terry Cockerham, Synapse Media Production; **1-5*B*, 1-7** through **1-10** Michael P. Schenk; **1-A** through **1-K** Cynthia Turner Alexander/Terry Cockerham, Synapse Media Production.

CHAPTER 2

2-1, 2-2*B*, 2-5, 2-6, 2-7, 2-9, 2-14*B,D,E* Ronald J. Ervin; **2-4, 2-13 *C,D*** William Ober; **2-14** Trent Stephens; **2-10** Brenda Eisenberg, PhD.

CHAPTER 3

3-1, 3-3, 3-4, 3-5, 3-7, 3-9 through **3-12*A*, 3-21** Ronald J. Ervin; **3-2, 3-6*B*, 3-8, 3-13** William Ober; **3-6*A*** Barbara Stackhouse; **3-8** Birgit Satir; **3-12*B*** Charles Flickinger; **3-14, 3-15** Andrew Grivas.

CHAPTER 4

4-1, 4-6 Christine Oleksyk; **4-2*A*, *C*, *D*, 4-5*A*, *E*** Trent Stephens; **4-2*B*, 4-5*B*, *C*** Ed Reschke; **4-2*F*, 4-5*D*, *F*** Carolina Biological Supply Co.; **4-2 *(locations)*, 4-5 *(locations)*** Eileen Draper; **4-3, 4-4, 4-8, 4-9** Michael P. Schenk.

CHAPTER 5

5-1, 5-8 Ronald J. Ervin; **5-2** Ed Reschke; **5-4, 5-5, 5-6** Joan M. Beck; **5-7** William Ober; **5-9 *A-C*** Thomas P. Habif.

CHAPTER 6

6-1*A*, 6-4, 6-5, 6-6*A*, 6-8, 6-9, 6-10, 6-12 through **6-15, 6-17, 6-18, 6-20** through **6-24, 6-26, 6-27, 6-28, 6-*A*, *c*, *d*, *e*** David J. Mascaro & Associates; **6-1*B*, *C*, 6-2, 6-3** John V. Hagen; **6-6*C*, 6-7** Karen Waldo; **6-6*B*, 6-16, 6-19, 6-25, 6-31** Terry Cockerham, Synapse Media Production; **6-29** Rusty Jones; **6-30** Scott Bodell; **6-A** Ewing Galloway; **6-B, *a*, *b*** Christine Oleksyk; **6-C** Paul R. Manske.

CHAPTER 7

7-1*A*, 7-8 through **7-16, 7-18, 7-19** John V. Hagen; **7-1*B*, *C*, *D*, 7-4** Joan M. Beck; **7-2** Richard Rodewald; **7-3, 7-5** Joan M. Beck/Andrew Grivas; **7-6** Trent Stephens; **7-17, 7-20** Terry Cockerham, Synapse Media Production.

CHAPTER 8

8-1 Cynthia Turner Alexander/Terry Cockerham, Synapse Media Production; **8-3** through **8-6, 8-14, 8-15** Scott Bodell; **8-7, 8-8** Trent Stephens; **8-9*A*** Joan M. Beck; **8-9*B*, 8-10, 8-11, 8-12,** Scott Bodell; **8-13, 8-16** Courtesy Branislav Vidić; **8-17** William Ober; **8-18*A*** Photo Researchers; **8-18*B*** William E. Rosenfield, MD; **8-19, 8-20, 8-22, 8-26, 8-27** Scott Bodell; **8-23** Christine Oleksyk; **8-24, 8-25, 8-28, 8-29** Michael P. Schenk; **8-31** John Martini.

CHAPTER 9

9-1 Janis K. Atlee/Michael P. Schenk; **9-2, 9-16*C*, *D*, 9-17*C*, *D*, *E*, 9-A** G. David Brown; **9-3*A*, 9-4*A*, 9-5*B*, 9-7, 9-8*B*, 9-16*A*, 9-17*A*** Marsha J. Dohrmann; **9-3*B*** Christine Oleksyk; **9-4*B*, *C*, 9-16*B*, 9-17*B*,** Kathy Mitchell Grey; **9-5*A*** Terry Cockerham, Synapse Media Production; **9-6** John V. Hagen; **9-8*A*** Christine Oleksyk; **9-B** S. Isihara, Tests for Colour-Blindness, Tokyo, Japan, 1973, Kanehara Shuppan Co., Ltd, provided by Washington University Department of Ophthamology.

CHAPTER 10

10-1 Terry Cockerham, Synapse Media Production; **10-2, 10-3, 10-4** Andrew Grivas; **10-5** Barbara Stackhouse; **10-6*A*, *B*** Nadine Sokol; **10-6*C*, 10-8** Trent Stephens; **10-10** Courtesy Branislav Vidić.

CHAPTER 11

11-1 Kathryn A. Born; **11-2** David Phillips/Visuals Unlimited; **11-3, 11-4** Christine Oleksyk; **11-5** Carolina Biological Supply Co.; **11-8, 11-9, 11-10** Molly Babich/John Daugherty; **11-11** Trent Stephens; **Table 11-2** William Ober.

CHAPTER 12

12-1 Cynthia Turner Alexander/Terry Cockerham, Synapse Media Production; **12-2, 12-5*A*,*B*, 12-7, 12-11** Rusty Jones; **12-3** David Mascaro & Associates; **12-4** Michael P. Schenk **12-5*C*** Courtesy Branislav Vidić; **12-6, 12-12** Christine Oleksyk; **12-8*A*** Kathy Mitchell Grey; **12-8*B*, 12-10** Trent Stephens; **12-9** Ronald J. Ervin.

CHAPTER 13

13-1, 13-3 William Ober; **13-2** Ed Reschke; **13-4** Christine Oleksyk; **13-5, 13-8** John Daugherty; **13-6, 13-9** David J. Mascaro & Associates; **13-7, 13-10** George J. Wassilchenko; **13-11, 13-12** Karen Waldo; **13-13, 13-15** Joan M. Beck/Donna Odle; **13-14** G. David Brown; **13-16 13-17, 13-18** Christine Oleksyk.

CHAPTER 14

14-1 Cynthia Turner Alexander/Terry Cockerham, Synapse Media Production; **14-2** G. David Brown; **14-3** David J. Mascaro & Associates; **14-4** Glanze, Walter D., editor *Mosby's Medical & Nursing Dictionary*, ed 2, St Louis, 1986, The CV Mosby Co.; **14-5** Kathy Mitchell Grey; **14-7, 14-8** Michael P. Schenk; **14-9, 14-10, 14-11** John Daugherty.

CHAPTER 15

15-1 Cynthia Turner Alexander/Terry Cockerham, Synapse Media Produuction; **15-2*A*, 15-5*A*, 15-6, 15-9** Jody L. Fulks; **15-2*B*, 15-5*B*** Courtesy Branislav Vidić; **15-3** David J. Mascaro & Associates; **15-4*A*** Christine Oleksyk; **15-4*B*, 15-8*A*** Custom Medical Stock Photo; **15-7** John Daugherty; **15-10** Joan M. Beck/Donna Odle; **15-11** Joan M. Beck; **15-12** Trent Stephens.

CHAPTER 16

16-1 Cynthia Turner Alexander/Terry Cockerham, Synapse Media Production; **16-2, 16-8*C*** William Ober; **16-3, 16-18** Trent Stephens; **16-4*A*, *B*** Kathryn A. Born; **16-4*C*, 16-5, 16-9, 16-10** David J. Mascaro & Associates; **16-6*A*, 16-12, 16-14, 16-16*A*, *B*** G. David Brown; **16-6*B*, 16-8*A*, *B*, *D*, 16-11** Kathy Mitchell Grey; **16-7** John Daugherty; **16-8*E*** Joan M. Beck/Donna Odle; **16-12 *(inset)*** Science Photo Library/Photo Researchers; **16-13** Michael P. Schenk; **16-15** Christine Oleksyk; **16-17** Barbara Stackhouse.

CHAPTER 17

17-2 Joan M. Beck/Donna Odle; **17-3** through **17-8** Christine Oleksyk.

CHAPTER 18

18-1 Cynthia Turner Alexander/Terry Cockerham, Synapse Media Production; **18-2, 18-7, 18-9, 18-10** Christine Oleksyk; **18-3 *(top)*** David J. Mascaro & Associates; **18-3 *(bottom)*** Courtesy Branislav Vidić; **18-4, 18-8** Barbara Stackhouse; **18-5*A*, *B*, 18-6** Jody L. Fulks; **18-5*C*** Andrew P. Evan.

CHAPTER 19

19-1 Cynthia Turner Alexander/Terry Cockerham, Synapse Media Production; **19-2, 19-7, 19-*A*, *B*, *C*** Ronald J. Ervin; **19-3** William Ober; **19-4, 19-5, 19-12, 19-13** Kevin A. Somerville; **19-6** Trent Stephens; **19-8, 19-9** Kevin A. Somerville/Kathy Mitchell Grey; **19-10** Kevin A. Somerville/Kathy Mitchell Grey/Scott Bodell; **19-11** David J. Mascaro & Associates; **19-A, *a*, *b*, *c*, *d*, *e*** Joel Gordon.

CHAPTER 20

20-1 Lucinda L. Veeck; **20-2, 20-3, 20-5, 20-7, 20-9, 20-10, 20-11, 20-12*E*** Maria Hartsock; **20-3*A*** Barbara Stackhouse; **20-4, 20-14** Kevin A. Sommerville; **20-6** Michael P. Schenk; **20-8, 20-13** Lennart Nilsson; **20-12*A*, *B*, *C*, *D*** David J. Mascaro & Associates; **20-15** Scott Bodell; **20-16** Molly Babich/John Daugherty; **20-18** Ronald J Ervin; **20-19** Eileen Draper.

OPENERS

1 Michel Tcherevkoff, The Image Bank; **2** Comstock, Inc; **3** Comstock, Inc; **4** Comstock, Inc; **5** Linda H. Hopon, Visuals Unlimited; **6** Bonnie Mamen, Comstock, Inc; **7** Larry Dale Gordon, The Image Bank; **8** Mike & Carol Werner, Comstock, Inc; **9** Jack Spratt, The Image Works; **10** Bob Daemmrich, The Image Works; **11** Comstock, Inc; **12** Comstock, Inc; **13** Bob Daemmrich, The Image Works; **14** Willie Hill, Jr, The Image Works; **15** Alan Carey, The Image Works; **16** Arthur R. Hill, Visuals Unlimited; **17** Dion Ogust, The Image Works; **18** Bob Daemmich, The Image Works; **19** Comstock, Inc.; **20** Comstock, Inc.

INDEX

P

Q

R

S

T

U

Prefixes, Suffixes, And Combining Forms

Term	Meaning	Example
intra-	Within	Intraocular (within the eye)
-ism	Condition, state of	Dimorphism (condition of two forms)
iso-	Equal	Isotonic (same tension)
-itis	Inflammation	Gastritis (inflammation of the stomach)
-ity	Expressing condition	Acidity (condition of acid)
kerato-	Cornea or horny tissue	Keratinization (formation of a hard tissue)
-kin-	Move	Kinesiology (study of movement)
leuko-	White	Leukocyte (white blood cell)
-liga-	Bind	Ligament (structure that binds bone to bone)
lip-	Fat	Lipolysis (breakdown of fats)
-logy	Study	Histology (study of tissue)
-lysis	Breaking up, dissolving	Glycolysis (breakdown of sugar)
macro-	Large	Macrophage (large phagocytic cell)
mal-	Bad	Malnutrition (bad nutrition)
malaco-	Soft	Osteomalacia (soft bone)
mast-	Breast	Mastectomy (excision of the breast)
mega-	Great	Megacolon (large colon)
melano-	Black	Melanocyte (black pigment producing skin cell)
meso-	Middle, mid	Mesoderm (middle skin)
meta-	Beyond, after, change	Metastasis (beyond original position)
micro-	Small	Microorganism (small organism)
mito-	Thread, filament	Mitosis (referring to threadlike chromosomes during cell division)
mono-	One, single	Monosaccharide (one sugar)
-morph-	Form	Morphology (study of form)
multi-	Many, much	Multinucleated (two or more nuclei)
myelo-	Marrow, spinal cord	Myeloid (derived from bone marrow)
myo-	Muscle	Myocardium (heart muscle)
narco-	Numbness	Narcotic (drug producing stupor or weakness)
neo-	New	Neonatal (first four weeks of life)
nephro-	Kidney	Nephrectomy (removal of a kidney)
neuro-	Nerve	Neuritis (inflammation of a nerve)
oculo-	Eye	Oculomotor (movement of the eye)
odonto-	Tooth or teeth	Odontomy (cutting a tooth)
-oid	Expressing resemblance	Epidermoid (resembling epidermis)
oligo-	Few, scanty, little	Oliguria (little urine)
-oma	Tumor	Carcinoma (cancerous tumor)
-op-	See	Myopia (nearsighted)
ophthalm-	Eye	Ophthalmology (study of the eye)
ortho-	Straight, normal	Orthodontics (discipline dealing with the straightening of teeth)
-ory	Referring to	Olfactory (relating to the sense of smell)
-ose	Full of	Adipose (full of fat)
-osis	A condition of	Osteoporosis (porous condition of bone)
osteo-	Bone	Osteocyte (bone cell)
oto-	Ear	Otolith (ear stone)
-ous	Expressing material	Serous (composed of serum)
para-	Beside, beyond, near to	Paranasal (near the nose)